CW01220227

Discrete-Time Signals and Systems

Drawing on author's 30+ years of teaching experience, "Discrete-Time Signals and Systems: A MATLAB Integrated Approach" represents a novel and comprehensive approach to understanding signals and systems theory. Many textbooks use MATLAB as a computational tool, but Alkin's text employs MATLAB both computationally and pedagogically to provide interactive, visual reinforcement of fundamental concepts important in the study of discrete-time signals and systems.

In addition to 204 traditional end-of-chapter problems and 160 solved examples, the book includes hands-on MATLAB modules consisting of:

- 108 MATLAB-based homework problems and projects (coordinated with the traditional end-of-chapter problems)
- 44 live scripts and GUI-based interactive apps that animate key figures and bring core concepts to life
- Downloadable MATLAB code for most of the solved examples
- 92 fully detailed MATLAB exercises that involve step by step development of code to simulate the relevant signal and/or system being discussed, including some case studies on topics such as real-time audio processing, synthesizers, electrocardiograms, sunspot numbers, etc.
- The ebook+ version includes clickable links that allow running MATLAB code associated with solved examples and exercises in a browser, using the online version of MATLAB. It also includes audio and video files for some of the examples.

Each module or application is linked to a specific segment of the text to ensure seamless integration between learning and doing. The aim is to not simply give the student just another toolbox of MATLAB functions, but to use the development of MATLAB code as part of the learning process, or as a litmus test of students' understanding of the key concepts. All relevant MATLAB code is freely available from the publisher. In addition, a solutions manual, figures, presentation slides and other ancillary materials are available for instructors with qualifying course adoption.

Discrete-Time Signals and Systems

A MATLAB-Integrated Approach

Second Edition

Oktay Alkin

CRC Press is an imprint of the
Taylor & Francis Group, an **informa** business

Second edition published 2025
by CRC Press
2385 NW Executive Center Drive, Suite 320, Boca Raton FL 33431

and by CRC Press
4 Park Square, Milton Park, Abingdon, Oxon, OX14 4RN

CRC Press is an imprint of Taylor & Francis Group, LLC

ISBN: 978-1-032-94390-9 (hbk)
ISBN: 978-1-032-94391-6 (pbk)
ISBN: 978-1-003-57046-2 (ebk)
ISBN: 978-1-003-57049-3 (eBook+)

DOI: 10.1201/9781003570462

Typeset in Utopia-Regular font
by KnowledgeWorks Global Ltd.

Publisher's note: This book has been prepared from camera-ready copy provided by the authors.

Access the Instructor and Student Resources: Will be provided later

To Nathan, Ender, Lunara, Esmeray, and Zeki

CONTENTS

Preface **xvii**

1 Discrete-Time Signal Representation and Modeling **1**

Chapter Objectives 1

1.1 Introduction 1

1.1.1 Discrete-time vs. digital signals 4

1.1.2 Examples of digital signals 5

1.1.3 Why study discrete-time signals and systems instead of digital? 6

1.2 Basic Signal Operations 7

1.2.1 Arithmetic operations 7

1.2.2 Time shifting 11

1.2.3 Time scaling 13

1.2.4 Time reversal 16

1.3 Basic Building Blocks for Discrete-Time Signals 17

1.3.1 Unit step function 19

1.3.2 Unit ramp function 22

1.3.3 Sinusoidal signals 24

1.4 Impulse Decomposition for Discrete-Time Signals 28

1.5 Signal Classifications 29

1.5.1 Real vs. complex signals 29

1.5.2 Periodic vs. non-periodic signals 30

1.5.3 Periodicity of discrete-time sinusoidal signals 30

1.5.4 Deterministic vs. random signals 34

1.6 Energy and Power Definitions 34

1.6.1 Energy of a signal . . . 34
1.6.2 Time averaging operator . . . 36
1.6.3 Average power of a signal . . . 36
1.6.4 Energy signals vs. power signals . . . 38
1.7 Symmetry Properties . . . 39
1.7.1 Even and odd symmetry . . . 39
1.7.2 Decomposition into even and odd components . . . 39
1.7.3 Symmetry properties for complex signals . . . 41
1.7.4 Decomposition of complex signals . . . 42
Summary of Key Points . . . 42
Further Reading . . . 45
MATLAB Exercises with Solutions . . . 46
Computing and graphing discrete-time signals . . . 46
Graphing a signal from its analytical description . . . 48
Functions for basic building blocks . . . 49
Testing the functions for basic building blocks . . . 50
Periodic extension of a discrete-time signal . . . 52
Working with signals in tabular form . . . 53
Working with signals in tabular form – revisited . . . 55
Case study – Sunspot numbers; working with large data sets . . . 56
Working with audio files . . . 58
Case study – Synthesizing music with sinusoidal signals . . . 59
Objects for working with audio signals . . . 61
Case study – Audio synthesizer object . . . 64
Problems . . . 66
MATLAB Problems . . . 68
MATLAB Projects . . . 70

2 Analyzing Discrete-Time Systems in the Time Domain 73
Chapter Objectives . . . 73
2.1 Introduction . . . 73
2.2 Linearity and Time Invariance . . . 74
2.2.1 Linearity in discrete-time systems . . . 75
2.2.2 Time invariance in discrete-time systems . . . 77
2.2.3 DTLTI systems . . . 79
2.3 Difference Equations for Discrete-Time Systems . . . 79
2.4 Constant-Coefficient Linear Difference Equations . . . 89
2.5 Solving Difference Equations . . . 94
2.5.1 Finding the natural response of a discrete-time system . . . 95
2.5.2 Finding the forced response of a discrete-time system . . . 105

2.6 Block Diagram Representation of Discrete-Time Systems . . . 110
2.7 Impulse Response and Convolution . . . 115
2.7.1 Finding impulse response of a DTLTI system . . . 116
2.7.2 Convolution operation for DTLTI systems . . . 118
2.8 Causality in Discrete-Time Systems . . . 129
2.9 Stability in Discrete-Time Systems . . . 130
Summary of Key Points . . . 133
Further Reading . . . 135
MATLAB Exercises with Solutions . . . 135
Writing functions for moving average filters . . . 135
Testing moving average filtering functions . . . 137
Moving average filter functions revisited . . . 138
Writing and testing a function for exponential smoother . . . 140
Iteratively solving a difference equation . . . 141
Implementing a discrete-time system from its block diagram . . . 143
Implementation of a discrete-time system revisited . . . 144
Discrete-time convolution . . . 145
Developing our own convolution function . . . 147
Implementing a moving average filter through convolution . . . 148
Sunspot numbers – Using convolution and downsampling . . . 149
Frame-based processing of audio signals . . . 151
Case study – Echoes and reverberation, part 1 . . . 153
Problems . . . 156
MATLAB Problems . . . 164
MATLAB Projects . . . 166

3 Fourier Analysis for Discrete-Time Signals and Systems 170
Chapter Objectives . . . 170
3.1 Introduction . . . 171
3.2 Analysis of Periodic Discrete-Time Signals . . . 171
3.2.1 Discrete-time Fourier series (DTFS) . . . 171
3.2.2 Finding DTFS coefficients . . . 176
3.2.3 Properties of the DTFS . . . 180
3.3 Analysis of Non-Periodic Discrete-Time Signals . . . 190
3.3.1 Discrete-time Fourier transform (DTFT) . . . 190
3.3.2 Developing further insight . . . 193
3.3.3 Existence of the DTFT . . . 197
3.3.4 DTFT of some signals . . . 197
3.3.5 Properties of the DTFT . . . 205
3.3.6 Applying DTFT to periodic signals . . . 221

3.4 Energy and Power in the Frequency Domain . . . 227
3.4.1 Parseval's theorem . . . 227
3.4.2 Energy and power spectral density . . . 228
3.4.3 Energy or power in a frequency range . . . 234
3.4.4 Autocorrelation . . . 234
3.4.5 Properties of the autocorrelation function . . . 238
3.5 System Function Concept . . . 239
3.6 DTLTI Systems with Periodic Input Signals . . . 243
3.6.1 Response of a DTLTI system to complex exponential signal . . . 243
3.6.2 Response of a DTLTI system to sinusoidal signal . . . 244
3.6.3 Response of a DTLTI system to periodic input signal . . . 247
3.7 DTLTI Systems with Non-Periodic Input Signals . . . 249
Summary of Key Points . . . 250
Further Readings . . . 255
MATLAB Exercises with Solutions . . . 256
Writing functions for DTFS analysis and synthesis . . . 256
Testing DTFS functions . . . 257
Improving performance of DTFS functions . . . 258
A function to implement periodic convolution . . . 259
Steady-state response of DTLTI system to sinusoidal input . . . 260
Case study – Echoes and reverberation, part 2 . . . 261
Problems . . . 263
MATLAB Problems . . . 272
MATLAB Project . . . 273

4 Sampling and Reconstruction **275**
Chapter Objectives . . . 275
4.1 Introduction . . . 275
4.2 Sampling of a Continuous-Time Signal . . . 277
4.2.1 Nyquist sampling criterion . . . 284
4.2.2 DTFT of sampled signal . . . 286
4.2.3 Sampling of sinusoidal signals . . . 288
4.2.4 Practical issues in sampling . . . 294
4.3 Reconstruction of a Signal from Its Sampled Version . . . 299
4.4 Resampling Discrete-Time Signals . . . 306
4.4.1 Reducing the sampling rate by an integer factor . . . 307
4.4.2 Increasing the sampling rate by an integer factor . . . 310
Summary of Key Points . . . 314
Further Reading . . . 316
MATLAB Exercises with Solutions . . . 317

Spectral relations in impulse sampling . . . 317
DTFT of discrete-time signal obtained through sampling . . . 319
Sampling a sinusoidal signal . . . 320
Natural sampling . . . 321
Zero-order hold sampling . . . 322
Graphing signals for natural and zero-order hold sampling . . . 323
Reconstruction of right-sided exponential . . . 325
Frequency spectrum of reconstructed signal . . . 328
Resampling discrete-time signals . . . 330
Problems . . . 332
MATLAB Problems . . . 336
MATLAB Projects . . . 337

5 The z-Transform 339

Chapter Objectives . . . 339
5.1 Introduction . . . 339
5.2 Characteristics of the Region of Convergence . . . 362
5.3 Properties of the z-Transform . . . 368
5.3.1 Linearity . . . 368
5.3.2 Time shifting . . . 372
5.3.3 Time reversal . . . 374
5.3.4 Multiplication by an exponential signal . . . 375
5.3.5 Differentiation in the z-domain . . . 377
5.3.6 Convolution property . . . 380
5.3.7 Initial value . . . 384
5.3.8 Correlation property . . . 385
5.3.9 Summation property . . . 387
5.4 Inverse z-Transform . . . 389
5.4.1 Inversion integral . . . 390
5.4.2 Partial fraction expansion . . . 391
5.4.3 Long division . . . 398
5.5 Using the z-Transform with DTLTI Systems . . . 404
5.5.1 Relating the system function to the difference equation . . . 405
5.5.2 Response of a DTLTI system to complex exponential signal . . . 408
5.5.3 Response of a DTLTI system to exponentially damped sinusoid . . . 410
5.5.4 Graphical interpretation of the pole-zero plot . . . 412
5.5.5 System function and causality . . . 420
5.5.6 System function and stability . . . 420
5.5.7 Comb filters . . . 424
5.5.8 All-pass filters . . . 426

5.5.9 Inverse systems . . . 431
5.6 Implementation Structures for DTLTI Systems . . . 434
5.6.1 Direct-form implementations . . . 434
5.6.2 Cascade and parallel forms . . . 439
5.7 Unilateral z-Transform . . . 443
Summary of Key Points . . . 448
Further Reading . . . 451
MATLAB Exercises with Solutions . . . 452
Three-dimensional plot of z-transform . . . 452
Computing the DTFT from the z-transform . . . 454
Graphing poles and zeros . . . 455
Using convolution function for polynomial multiplication . . . 456
Partial fraction expansion with MATLAB . . . 457
Developing a function for long division . . . 457
Frequency response of a system from pole-zero layout . . . 459
Frequency response from pole-zero layout revisited . . . 460
Preliminary calculations for a cascade-form block diagram . . . 462
Cascade-form block diagram revisited . . . 463
Preliminary calculations for a parallel-form block diagram . . . 464
Implementing a system using second-order sections . . . 465
Solving a difference equation through z-transform . . . 467
Case Study – Echoes and reverberation, part 3 . . . 468
Problems . . . 472
MATLAB Problems . . . 482
MATLAB Projects . . . 484

6 State-Space Analysis of Discrete-Time Systems **487**
Chapter Objectives . . . 487
6.1 Introduction . . . 488
6.2 State-Space Modeling of Discrete-Time Systems . . . 489
6.3 State-Space Models for DTLTI Systems . . . 491
6.3.1 Obtaining state-space model from a difference equation . . . 492
6.3.2 Obtaining state-space model from a system function . . . 494
6.3.3 Alternative state-space models . . . 499
6.3.4 DTLTI systems with multiple inputs and/or outputs . . . 500
6.4 Solution of State-Space Model . . . 501
6.5 Obtaining System Function from State-Space Model . . . 503
6.6 Discretization of Continuous-Time State-Space Model . . . 503
Summary of Key Points . . . 507
Further Reading . . . 508

MATLAB Exercises with Solutions 508
Obtaining system function from discrete-time state-space model 508
Discretization of state-space model 509
Discretization using Euler method 511
Problems 512
MATLAB Problems 514

7 Discrete Fourier Transform 516
Chapter Objectives 516
7.1 Introduction 517
7.2 Discrete Fourier Transform (DFT) 517
7.2.1 Relationship of the DFT to the DTFT 520
7.2.2 Zero padding 524
7.2.3 Detection of sinusoidal signals with the DFT 526
7.2.4 Frequency spacing vs. frequency resolution 531
7.2.5 Using the DFT to approximate the EFS coefficients 533
7.2.6 Using the DFT to approximate the continuous Fourier transform 535
7.2.7 Matrix Formulation of the DFT 537
7.3 Properties of the DFT 539
7.3.1 Linearity 542
7.3.2 Time shifting 542
7.3.3 Time reversal 543
7.3.4 Conjugation property 544
7.3.5 Symmetry of the DFT 545
7.3.6 Frequency shifting 547
7.3.7 Circular convolution 548
7.4 Special Uses of the DFT 555
7.4.1 Goertzel algorithm 555
7.4.2 Chirp transform algorithm (CTA) 558
7.4.3 Short time Fourier transform (STFT) 559
7.5 Improving Computational Efficiency of the DFT 562
7.5.1 Length-4 DFT 564
7.5.2 Length-8 DFT 567
7.5.3 Radix-2 Decimation-In-Time Fast Fourier Transform 568
7.5.4 Generalizing Decimation-In-Time Algorithms 570
7.5.5 Radix-2 Decimation-In-Frequency Fast Fourier Transform 571
7.6 Using FFT to Implement Convolution in Real Time 572
Summary of Key Points 575
Further Reading 578
MATLAB Exercises with Solutions 579

Writing and testing a function for DFT . . . 579
Using the DFT to compute DTFS coefficients . . . 581
Exploring the relationship between the DFT and the DTFT . . . 582
Using the DFT to approximate the DTFT . . . 584
Writing functions for circular time shifting and time reversal . . . 584
Circular conjugate symmetric and antisymmetric components . . . 585
Using the symmetry properties of the DFT . . . 586
Circular and linear convolution using the DFT . . . 586
Writing a convolution function using the DFT . . . 587
Exponential Fourier series approximation using the DFT . . . 588
Testing the EFS approximation function . . . 589
Fourier transform approximation using the DFT . . . 590
Writing a function for DFT using matrix formulation . . . 592
Implementing Goertzel algorithm . . . 593
Implementing chirp transform algorithm (CTA) . . . 594
Short time Fourier transform . . . 595
Writing a function for radix-2 decimation in time FFT . . . 596
Writing a function for radix-2 decimation-in-frequency FFT . . . 598
Convolution in real-time processing using overlap-add method . . . 599
Problems . . . 600
MATLAB Problems . . . 606
MATLAB Projects . . . 610

8 Analysis and Design of Discrete-Time Filters **614**
Chapter Objectives . . . 614
8.1 Introduction . . . 615
8.2 Discrete-Time Processing of Continuous-Time Signals . . . 616
8.3 Pole-Zero Placement Design of Filters . . . 617
8.3.1 Resonant bandpass filters . . . 617
8.3.2 Notch filters . . . 622
8.4 IIR Filters . . . 626
8.4.1 IIR filter specifications . . . 628
8.4.2 Review of analog filter design formulas . . . 629
8.4.3 Analog to discrete-time transformation methods . . . 635
8.4.4 Impulse invariance . . . 636
8.4.5 Transformations based on rectangular approximation to integrals . . . 641
8.4.6 Bilinear transformation . . . 646
8.4.7 Obtaining analog prototype specifications . . . 650
8.5 FIR Filters . . . 652
8.5.1 Linear phase in FIR filters . . . 653

8.5.2 Design of FIR filters . . . 662
8.5.3 Fourier series design using Window functions . . . 666
8.5.4 Frequency sampling design . . . 680
8.5.5 Least-squares design . . . 683
8.5.6 Parks-McClellan technique for FIR filter design . . . 685
Summary of Key Points . . . 686
Further Reading . . . 688
MATLAB Exercises with Solutions . . . 689
Discrete-time processing of analog audio signals . . . 689
Writing functions for resonant bandpass filters . . . 690
Resonators in real-time processing . . . 692
Writing functions for notch filters . . . 694
Case Study – ECG signal with sinusoidal interference . . . 696
Impulse invariant design . . . 698
IIR filter design using bilinear transformation . . . 699
IIR filter design using bilinear transformation – revisited . . . 700
A complete IIR filter design example . . . 701
Second-order sections for IIR design and implementation . . . 703
FIR filter design using Fourier series method . . . 706
FIR filter design using Fourier series method – revisited . . . 707
FIR filter design by frequency sampling . . . 709
FIR filter design by least-squares method . . . 710
FIR filter design using Parks-McClellan algorithm . . . 712
Case Study – Plucked-string filter . . . 713
Problems . . . 720
MATLAB Problems . . . 725
MATLAB Projects . . . 733

A Complex Numbers and Euler's Formula **735**
A.1 Introduction . . . 735
A.2 Arithmetic with Complex Numbers . . . 737
A.2.1 Addition and subtraction . . . 737
A.2.2 Multiplication and division . . . 739
A.3 Euler's Formula . . . 739

B Mathematical Relations **740**
B.1 Trigonometric Identities . . . 740
B.2 Indefinite Integrals . . . 741
B.3 Laplace Transform Pairs . . . 742
B.4 z-Transform Pairs . . . 743

C Closed Forms for Sums of Geometric Series **744**
C.1 Infinite-Length Geometric Series 744
C.2 Finite-Length Geometric Series 745
C.3 Finite-Length Geometric Series (Alternative Form) 745

D Orthogonality of Basis Functions **746**
D.1 Orthogonality for Trigonometric Fourier Series 746
D.2 Orthogonality for Exponential Fourier Series 748
D.3 Orthogonality for Discrete-Time Fourier Series 748

E Partial Fraction Expansion **750**
E.1 Partial Fractions for Continuous-Time Signals and Systems 750
E.2 Partial Fraction Expansion for Discrete-Time Signals and Systems 756

F Review of Matrix Algebra **757**

G Answers/Partial Solutions to Selected Problems **761**
G.1 Chapter 1 Problems 761
G.2 Chapter 2 Problems 763
G.3 Chapter 3 Problems 765
G.4 Chapter 4 Problems 768
G.5 Chapter 5 Problems 770
G.6 Chapter 6 Problems 775
G.7 Chapter 7 Problems 776
G.8 Chapter 8 Problems 777

Index **781**

PREFACE

The theory of discrete-time signals and systems finds wide ranging applications in electrical engineering, computer engineering, mechanical engineering, aerospace engineering, and bioengineering. It encompasses analysis, design, and implementation of systems as well as problems involving signal-system interaction. A solid background in discrete-time signals and systems is essential for the student to be able to venture into fields such as signal processing, image processing, multimedia, medical imaging and instrumentation, communications, control systems, robotics, power systems, and so on.

This textbook was written with the goal of providing a modern treatment of discrete-time signals and systems at the undergraduate level. It can be used as the main textbook for a one-semester course. It can also be used for self study. Writing style is student-friendly, starting with the basics of each topic and advancing gradually. There is not a lot of hand-waving; not many paragraphs start with "It can be shown that ..." without actually showing how or why something is the way it is. On the other hand, proofs and derivations that may safely be skipped at a first reading are clearly indicated through the use of color-coded frames. Also, important concepts and conclusions are highlighted to stand out. No prior signals and systems knowledge is assumed. The level of presentation is appropriate for a second or third year engineering student with differential calculus background. Some knowledge of introductory circuit theory is also helpful.

There are many excellent textbooks available for use in undergraduate level courses in this area of study. Some have matured over long periods of time, decades rather than just years, and have gained great popularity and traction among instructors as well as students. Consequently, the writing of a new textbook in this area is a difficult task as one must consider the questions of what new ideas the book would employ, and what value it would add to the teaching of the subject. This textbook resulted from the author's efforts over the past three decades in trying to find ways to incorporate software into the teaching of the material not only as a computational tool but also as a pedagogical one. It utilizes MATLAB software due to its popularity in the engineering community and its availability for a variety of operating systems. Basic familiarity with MATLAB software is assumed. For readers who have had no prior exposure to MATLAB, it would be helpful to spend a couple of hours going through some of the beginner-level tutorials freely available on the website of MathWorks, the publisher of MATLAB.

Each chapter contains a number of fully solved pencil-and-paper type examples as well as a set of traditional end-of-chapter problems. In addition, software use is integrated into the material at several levels:

1. Interactive MATLAB apps and live scripts: Interactive apps are graphical user interface-based MATLAB programs that allow key concepts of signals and systems to be visualized. We are all visual learners; we tend to remember an interesting scene in a movie much better than something we hear on the radio or something we read in a textbook. Taking this one step further, if we were also able to control the scene in a movie through our actions, we would have an even stronger grasp of the cause-effect relationships between our actions and the results they create. This is perhaps the main reason why children can become very proficient in video games. A large number of interactive apps are available for download with this textbook. Some allow a static figure in the text to come alive on a computer where the student can change key parameters and observe the results of such changes to understand cause-effect relationships. Some take a solved example in the text and expand it in an open-ended manner to allow the student to set up "what if" scenarios. Some of the interactive apps provide animations for concepts that are difficult to teach. Examples of this group include convolution, Fourier series representation of signals, three-dimensional visualization of z-transform, and relationships between DTFT, DTFS and DFT. Live scripts serve a similar purpose. Each one is like a MATLAB-based notebook that develops a key concept section by section, displaying results of calculations numerically and graphically. Modifying one or more parameters results in all calculations and graphics being updated as appropriate.
2. MATLAB code for solved examples: Most of the solved examples in each chapter have associated MATLAB listings available for download. These listings include detailed comments and are useful in a variety of ways: They help students check their work against a computer solution. They reinforce good coding practices. They also allow experimentation by changing parameter values and running the code again.
3. MATLAB exercises: In addition to the code listings associated with solved examples, there is a section at the end of each chapter which contains stand-alone MATLAB exercises that take the student through exploration and/or simulation of a concept by developing the necessary code step by step. Intermediate steps are explained in detail, and good coding practices are enforced throughout. Exercises are designed to help students become more proficient with MATLAB while working on problems in the context of signals and systems. Furthermore, MATLAB exercises are synchronized with the coverage of the material. At specific points in the text the reader is referred to MATLAB exercises relevant to the topic being discussed. The goal is to not just provide cookbook style quick solutions to problems, but rather to develop additional insight and deeper understanding of the material through the use of software. Some of the MATLAB exercises designated as *case studies* are somewhat longer and more complex than others, and highlight practical applications of the concepts covered in that particular chapter.

   Overall, MATLAB exercises were designed to reinforce students' understanding of the theory of signals and systems, and to develop software skills that would be useful in more advanced courses and research projects. It is the author's hope that they can also be motivating and fun.
4. MATLAB based problems: In addition to traditional end-of-chapter problems, each chapter contains a section with problems that require MATLAB solutions.
5. MATLAB projects: Most chapters contain a section with project ideas that are MATLAB-based. These can be used by instructors as the basis of computer assignments.

6. MATLAB files downloaded from the support website of the textbook integrate with the help browser of MATLAB software upon installation. This allows the student to have the textbook and a computer running MATLAB side by side while studying. Additionally, it gives the instructor the freedom to display MATLAB exercises on a projector while lecturing, without the need to type and run any code, if preferred.
7. For the ebook version: At the end of each solved example or Matlab exercise, there are clickable links for the script files used. Clicking on a link runs the online version of MATLAB in the default browser for the system, and loads the selected script ready to run. It should be noted that a user account needs to be created with MathWorks in order to use the online version of MATLAB. At the time of writing this text, mobile phones and tablets were unable to run MATLAB in a browser. A desktop or a laptop computer is needed for that. Also, some scripts that produce real-time audio do not work in a browser due to limitations of browsers to interact with the audio hardware of the system.

While software is an integral part of the textbook, and one that is meant to distinguish it from other works in the same field, it "never gets in the way". If desired, one can ignore all MATLAB related content and use this textbook as a traditional one with which to teach or learn the theory of signals and systems. Obviously, this would not be the recommended use. The main point is that the coverage of theory is not cluttered with code segments. Instead, all MATLAB exercises, problems and projects are presented in their own sections at the end of each chapter, with appropriate references provided within the narrative. Apart from the software use, the textbook includes plenty of solved pencil-and-paper type examples and traditional end-of-chapter problems.

Organization of the Material

In Chapter 1 we deal with mathematical modeling of discrete-time signals. Basic building blocks for signals are presented as well as mathematical operations applied to signals. Classification methods for signals are discussed; definitions of signal energy and signal power are given. The idea of impulse decomposition of a discrete-time signal is presented in preparation for the discussion of convolution operation in later chapters. Even and odd symmetry properties for real-valued signals are introduced, and the idea of decomposing a signal into its even and odd components is discussed. Notions of even and odd symmetry are generalized to complex-valued signals in the form of conjugate symmetry and conjugate antisymmetry.

Time-domain analysis methods for discrete-time systems are covered in Chapter 2. After introducing linearity and time-invariance from a discrete-time system perspective, the chapter proceeds with analysis of systems by means of linear constant-coefficient difference equations. Solution methods for difference equations are summarized. Representation of a difference equation by a block diagram is discussed. Impulse response and convolution concepts are developed for discrete-time systems and their relationship to the difference equation of the system is shown. Stability and causality concepts are detailed for discrete-time systems.

Chapter 3 is on Fourier analysis of discrete-time signals and systems. Analysis of periodic discrete-time signals through the use of discrete-time Fourier series (DTFS) is presented. Afterward the discrete-time Fourier transform (DTFT) is developed by generalizing the discrete-time Fourier series. The relationship between the DTFS coefficients of a periodic signal and the DTFT of a single isolated period of it is emphasized to highlight the link between the indices of the DTFS coefficients and the angular frequencies of the DTFT spectrum. Energy and power spectra concepts for deterministic discrete-time signals are discussed. DTFT-based system function is introduced. Response of linear and time invariant systems to both periodic and non-periodic input signals is studied.

Chapter 4 is dedicated to the topic of sampling. First the concept of impulse sampling is introduced. The relationship between the Fourier transforms of the original signal and its impulse-sampled version is explored and the Nyquist sampling criterion is derived. DTFT of the discrete-time signal obtained through sampling is related to the Fourier transform of the original continuous-time signal. The aliasing phenomenon is explained through special emphasis given to the sampling of sinusoidal signals. Practical forms of sampling such as natural sampling and zero-order hold sampling are discussed. Reconstruction of a continuous signal from its discrete-time version by means of zero- and first-order hold as well as bandlimited interpolation is studied. Resampling of discrete-time signals and the concepts of decimation and interpolation are introduced.

In Chapter 5, the z-transform is introduced as a more general version of the discrete-time Fourier transform (DTFT) covered in Chapter 3. The relationship between the z-transform and the DTFT is illustrated. Convergence characteristics of the z-transform are explained and the importance of the region of convergence concept is highlighted. Fundamental properties of the z-transform are detailed along with solved examples of their use. The problem of finding the inverse z-transform using the inversion integral, partial fraction expansion, and long division is discussed. Examples are provided with detailed solutions in using the partial fraction expansion and long division to compute the inverse transform for all possible types of the region of convergence. Application of the z-transform to the analysis of linear time-invariant systems is treated in detail. Connections between the z-domain system function, the difference equation and the impulse response are explored to provide insight. Graphical interpretation of pole-zero diagrams for the purpose of determining the frequency response of DTLTI systems is illustrated. Causality and stability concepts are discussed from the perspective of the z-domain system function. Characteristics of comb filters, all-pass systems, minimum-phase systems, and inverse systems are outlined. Direct-form, cascade and parallel implementation structures for DTLTI systems are derived from the z-domain system function. The unilateral version of the z transform is introduced as a practical tool for use in solving constant-coefficient linear difference equations with specified initial conditions.

Chapter 6 deals with state-space analysis for discrete-time systems. The concept of state variables is developed by starting with a difference equation and expressing it in terms of first-order difference equations. Methods of obtaining state space models from a system function, or from a block diagram are also detailed. Various solution methods for the state equations are developed. State transition matrix and its properties are discussed. Methods for obtaining alternative state-space models through the use of similarity transformations are detailed.

The discrete Fourier transform (DFT) is the subject of Chapter 7. It is introduced as a generalization of the discrete-time Fourier series (DTFS) discussed in Chapter 3. The relationship of DFT to the discrete-time Fourier transform (DTFT) is explored. Zero padding the input signal and its effects on the transform are studied. Concepts of DFT frequency spacing and frequency resolution are discussed in the context of detection of sinusoidal signals from finite-length observations. The use of the DFT for approximating exponential Fourier series for periodic continuous-time signals and for approximating the Fourier transform for continuous-time finite-length signals are explored. Special uses of the DFT through Goertzel algorithm, chirp transform algorithm, and short-time Fourier transform are introduced. The subject of improving the computational efficiency of the DFT is covered, leading to the fast Fourier transform (FFT) through the use of decimation-in-time and decimation-in-frequency techniques. Finally, applying the FFT for efficient implementation of systems in real time is discussed.

Chapter 8 focuses on analysis and design of discrete-time filters. The idea of processing continuous-time signals using discrete-time filters is introduced. Pole-zero placement design methods are discussed for two types of filters, namely resonant bandpass filters and notch filters. Afterward, the distinction between infinite impulse response (IIR) and finite impulse response (FIR) filters is given. Design of IIR filters by transforming analog prototypes to discrete-time filters is discussed. A brief review of analog prototype filter design methods is provided. Various transformation methods are discussed including the impulse invariance method and the bilinear transformation. The coverage of FIR filter design begins with a discussion of the significance of having a linear phase characteristic and the types of symmetries in the impulse response that lead to linear phase. Various design methods for FIR filters are covered including Fourier series design using window functions, frequency sampling design method, least squares design method, and the Parks-McClellan technique.

A number of appendices are provided for complex numbers and Euler's formula, various mathematical relations, proofs of orthogonality for the basis functions used in Fourier analysis, partial fraction expansion, a brief review of matrix algebra and solutions to selected end-of-chapter problems.

Supplementary Materials

The following supplementary materials are available to the instructors who adopt the textbook for classroom use:

- A solutions manual in pdf format that contains solutions to the problems at the end of each chapter, including the MATLAB problems,
- Presentation slides in pdf format created using LaTeX and Beamer,
- Image files in pdf format for the figures in the book.

The following supplementary materials are available to all users of the textbook:

- A downloadable archive containing MATLAB code files for GUI-based interactive apps, exercises, and examples in the textbook as well as the files needed for integrating them into MATLAB help browser.

CHAPTER 1

Discrete-Time Signal Representation and Modeling

Chapter Objectives

- Understand the concept of a *discrete-time signal* and how to work with mathematical models of discrete-time signals.
- Discuss fundamental signal types and signal operations used in the study of signals and systems. Experiment with methods of simulating discrete-time signals with MATLAB.
- Learn basic building blocks for discrete-time signals and ways to express other signals in terms of these building blocks.
- Learn various ways of classifying discrete-time signals and discuss symmetry properties.
- Explore characteristics of discrete-time sinusoidal signals.
- Understand the decomposition of signals using unit impulse functions.
- Learn energy and power definitions applicable to discrete-time signals.

1.1 Introduction

In contrast with continuous-time signals, discrete-time signals are not defined at all time instants. Instead, they are defined only at time instants that are integer multiples of a fixed time increment T_s, that is, at instants $t = nT_s$. Consequently, the mathematical model for a discrete-time signal is a function $x[n]$ in which the independent variable n is an integer and is referred to as the *sample index*. Consider, for example, the voltage signal $x_a(t)$ that plays your favorite song when applied to a loudspeaker. Suppose we measure the voltage $x_a(t)$ every T_s seconds,

DOI: 10.1201/9781003570462-1

and write down the measurements. The result is an indexed sequence of numbers each of which is referred to as a *sample* of the signal. In our example, samples of the signal $x[n]$ would be

$$x[0] = x_a(0), \quad x[1] = x_a(T_s), \quad x[2] = x_a(2T_s), \quad \ldots, \quad x[k] = x_a(kT_s)$$

The act of measuring the continuous-time signal at periodic intervals is called *sampling* and will be studied in detail later in Chapter 4. The time increment T_s between successive samples is called the *sampling period* or the *sampling interval* and its reciprocal $f_s = 1/T_s$ is called the *sampling frequency* or the *sampling rate.* Discrete-time signals are often illustrated graphically using *stem* plots. An example is shown in Fig. 1.1.

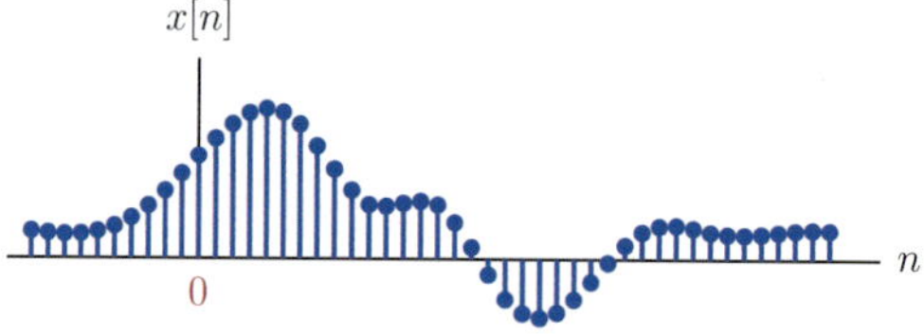

Figure 1.1 – A discrete-time signal.

It's also possible to visualize the measurements that make up a discrete-time signal as a table or a spreadsheet where each measurement has a value as well as a unique integer index associated with it. A compact way of tabulating a signal is by listing the significant signal samples between a pair of braces, and separating them with commas:

$$x[n] = \{\ 3.7,\ \underset{\uparrow}{1.3},\ -1.5,\ 3.4,\ 5.9\ \}$$

The up-arrow indicates the position of the sample index $n = 0$, so we have $x[-1] = 3.7$, $x[0] = 1.3$, $x[1] = -1.5$, $x[2] = 3.4$, and $x[3] = 5.9$. If the significant range of signal samples to be tabulated does not include $n = 0$, then we specify which index the up-arrow indicates. For example

$$x[n] = \{\ 1.1,\ \underset{\substack{\uparrow \\ n=5}}{2.5},\ 3.7,\ 3.2,\ 2.6\ \}$$

indicates that $x[4] = 1.1$, $x[5] = 2.5$, $x[6] = 3.7$, $x[7] = 3.2$, and $x[8] = 2.6$. Sometimes discrete-time signals are modeled using mathematical functions. For example

$$x[n] = 3\sin[0.2n]$$

is a discrete-time sinusoidal signal any sample of which can be computed using the expression provided. Yet another method of expressing a discrete-time signal is by using multiple mathematical expressions each applicable in a particular range of the index variable. An example of that is given below.

$$x[n] = \begin{cases} 3\sin[0.2n], & 0 \le n \le 19 \\ 0, & \text{otherwise} \end{cases}$$

Examples of discrete-time signals can also be found outside engineering disciplines. Financial markets, for example, rely on certain economic indicators for investment decisions. One such widely used indicator is the Dow Jones Industrial Average (DJIA) which is computed as a weighted average of the stock prices of 30 large publicly owned companies. Day-to-day variations of the DJIA can be used by investors in assessing the health of the economy and in making buying or selling decisions. Daily closing values of the Dow Jones Industrial Average for the first three months of 2014 are shown in Fig. 1.2. In this case the sampling interval T_s corresponds to a day, and values of the sample index n correspond to subsequent days of trading.

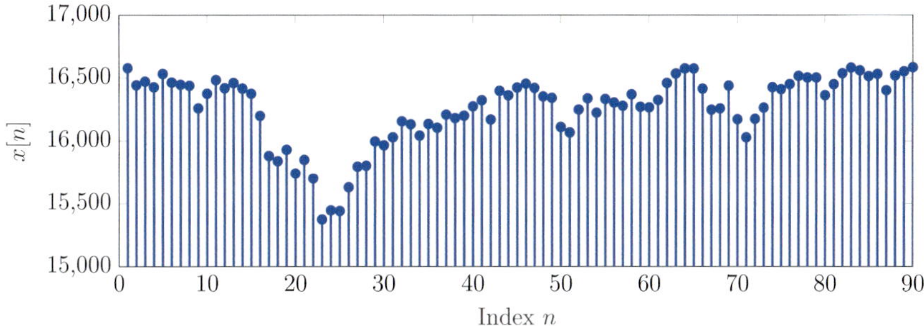

Figure 1.2 – Dow Jones Industrial Average for the first three months of 2014.

Representing market data as a discrete-time signal allows us to apply numerical analysis techniques to the signal in an attempt to understand the dynamics of market behavior.

Another interesting example of a discrete-time signal is the *sunspot numbers.* Astronomers have been observing and counting the dark spots on the visible face of the sun since the early 17th century. These spots are of varying quantity and size, and may occur as individual spots or in clusters. In 1848, Rudolf Wolf, a Swiss astronomer and mathematician, devised a formula to standardize the data collected by different observatories. He did this by taking into account the numbers of individual spots and clusters as well as the variations in the equipment used for collecting the data. The result is known as the *Wolf number* or the *Zurich relative number.* Recorded data in various formats is available dating back to the year 1700. Sunspot number data is seen as an indicator of solar activity, and has significant impact on satellite and radio communications. It also has some degree of impact on the climate. Numbers seem to be cyclic with a period of about 11 years. Fig. 1.3 shows the monthly international sunspot numbers from 1975 to 2010.

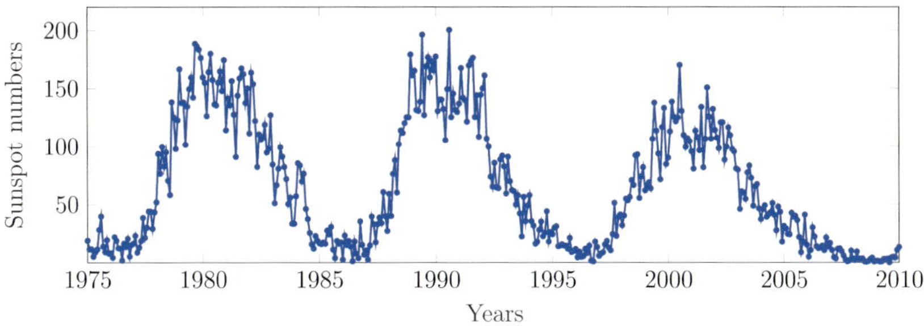

Figure 1.3 – Monthly international sunspot numbers recorded between the beginning of 1975 and the end of 2010. Data for the plot was downloaded from U.S. National Oceanic and Atmospheric Administration website at `https://www.ngdc.noaa.gov`.

For consistency in handling discrete-time signals we will assume that any discrete-time signal $x[n]$ described in one of the ways detailed above has an infinite number of samples for the index range $-\infty < n < \infty$. Sample amplitudes not resolved by the signal description will be taken as zero, unless there is obvious justification against doing that as in signals of Figs. 1.2 and 1.3 above.

Software resources: See MATLAB Exercises 1.1 and 1.2.

1.1.1 Discrete-time vs. digital signals

In a discrete-time signal the time variable is discrete, yet the amplitude of each sample is continuous, meaning it can take on any value. Even if the range of amplitude values may be limited, any amplitude value within the prescribed range is typically allowed. Discrete-time signals are ideal for mathematical representation and manipulation.

If, in addition to limiting the time variable to integer multiples of an increment, we also limit the amplitude values to a discrete set, then the resulting signal is called a *digital* signal. In the simplest case there are only two possible values for the amplitude of each sample, typically indicated by "0" and "1". The corresponding signal is called a *binary* signal. Each sample of a binary signal is called a *bit* which stands for *binary digit.* Alternatively each sample of a digital signal could take on a value from a set of M allowed values, and the resulting digital signal is called an *M-ary signal.*

A discrete-time signal can be converted to a digital signal through the use of the *quantization* operation. Suppose the line-out voltage $x_a(t)$ of an audio amplifier is limited to ± 2 V peak to peak, and is sampled at periodic intervals of T_s to obtain the discrete-time signal $x[n] = x_a(nT_s)$. Even though the range of amplitudes is limited, it is clear that there is an infinite number of possibilities for the value of each sample of $x[n]$. Now further suppose that we divide the amplitude range into eight equal intervals, and associate each sample with the interval it lands in. This is illustrated in Fig. 1.4.

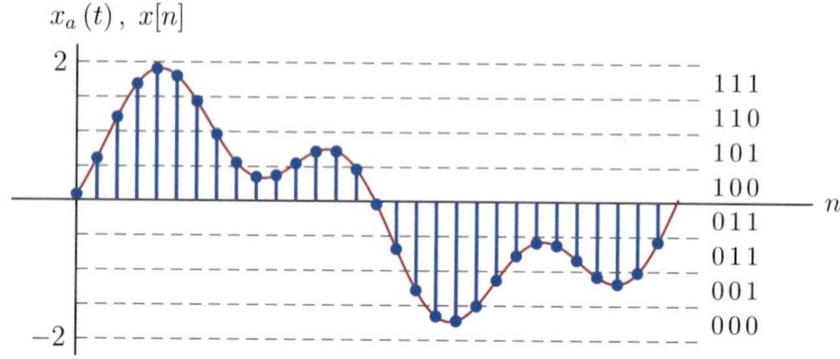

Figure 1.4 – Quantizing a discrete-time signal.

For this particular example, three bits would be sufficient to represent the eight quantization intervals we have, and one possible scheme of assigning a 3-bit word to each quantization interval is shown in Fig. 1.4. This is referred to as *encoding.* Each sample of the signal can thus be represented, written to computer storage media, and numerically manipulated through its assigned 3-bit word. The first few samples of the signal in Fig. 1.4 could be stored in binary form as

100 101 110 111 111 111 110 101 101 100 100 ...

Once a signal is stored in quantized form, the only information we have about each sample of that signal is the quantization interval in which the sample falls. We no longer know where exactly the tip of the sample is within the interval. Two samples that fall in the same interval are indistinguishable from each other even if they originally had different amplitude values. Now suppose that, at some point, we need to convert the digital signal back to an analog form. We can approximate each sample by setting its amplitude to the midpoint of its associated interval. Let the resulting approximate discrete-time signal be $\hat{x}[n]$. Connecting sample amplitudes of $\hat{x}[n]$ leads to a continuous-time signal $\hat{x}_a(t)$ as an approximation to the original signal $x_a(t)$. This process is referred to as *reconstruction*, and is illustrated in Fig. 1.5.

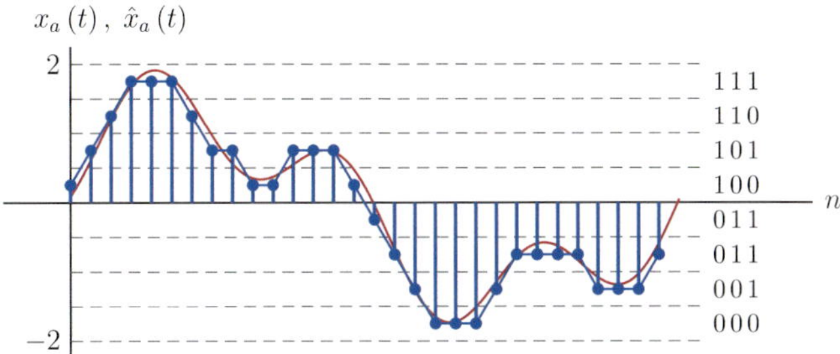

Figure 1.5 – Obtaining an approximation to the analog signal from quantized samples.

The difference between $\hat{x}[n]$ and $x[n]$ is the *quantization error.*

$$\epsilon[n] = \hat{x}[n] - x[n]$$

If Δ is the width of each quantization level, then sample amplitudes of the quantization error must be in the range

$$-\frac{\Delta}{2} < \epsilon[n] < \frac{\Delta}{2}$$

A reduction in the quantization error can be achieved by making Δ smaller which, in turn, requires increasing the number of quantization levels and, consequently, the number of bits assigned to each level. If 4 bits are used for representing each sample, we can have 16 quantization levels. If we allocate 8 bits for each sample, we get 256 quantization levels.

1.1.2 Examples of digital signals

We encounter digital signals in almost every part of our lives. Our computers and mobile phones work with digital signals. Audio recordings that we buy either in compact disc format or through streaming services are provided as digital signals, typically obtained by sampling analog audio with a sampling interval of $T_s = 22.68$ μs corresponding to a sampling rate of 44,100 samples per second, and then coding each sample into 16, 24, or 32 bits. Some audio formats include compression to facilitate efficient transmission of the signals. Digital signals are also used in systems that process speech waveforms and respond to voice commands as well as in systems that synthesize natural sounding speech from written text.

Another field that benefits from the use of digital signals is medical science. One particular example is the electrocardiogram (ECG) which is essentially a recording of the heartbeats of a patient that is sampled and digitized. Modern portable ECG systems consist of a series of sensors interfaced to a laptop computer with the appropriate software. The ECG signals acquired can be viewed by a physician or cleaned up, analyzed, and classified by software. An example is shown in Fig. 1.6.

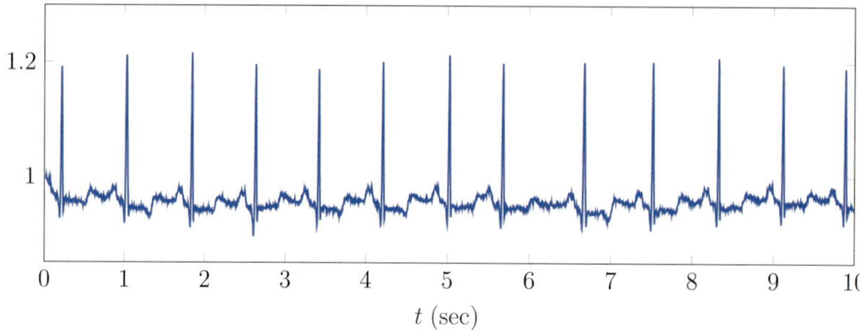

Figure 1.6 – ECG signal sampled at a rate of 360 samples/sec and digitized. Data for the signal was obtained from MIT-BIH arrhythmia database posted to `www.physionet.org`, used under Open Data Commons Attribution License v1.0.

For some signals the independent variable is not time, but rather displacement from a set point that is designated as *the origin*. The most common example of this is a photograph. Consider a gray-scale photograph printed from a film negative obtained using an old film camera. Each point on the photograph has a shade of gray ranging from pure black to pure white. If we can represent shades of gray numerically, say with values ranging from 0.0 to 1.0, then the photograph can be mathematically modeled as a two-dimensional continuous-time signal. The amplitude of the signal at each point is a function of two independent variables, one representing the horizontal distance from the origin and the other representing the vertical distance from the same point.

If a digital camera is to be used to capture the image, the resulting signal is slightly different. The image sensor of a digital camera is made up of a finite number of photo-sensitive cells arranged in a rectangular grid pattern. Each cell measures the intensity of light it receives and produces a voltage that is proportional to it. Thus, the signal that represents the image consists of light intensity values at a discrete set of horizontal and vertical distances from the origin.

If color images are desired, the image sensor becomes more complex since both intensity and color information must be captured at each point on the grid. The number of cells on the image sensor and, consequently, the number of data points per unit area of the resulting image determine the level of detail in the photo. Color accuracy of each point in the photo depends on the size and the sensitivity of each cell in the image sensor.

1.1.3 Why study discrete-time signals and systems instead of digital?

If a lot of signals that we encounter in our daily lives are digital, the question may be raised as to why we study discrete-time signals and systems rather than digital ones. The reason for that has to do with our desire to establish a unified theoretical framework for the study of signals and systems. In Section 1.1.1 we have briefly discussed how a discrete-time signal can be converted to a digital signal by first quantizing each sample into a preset number of quantization levels and then subsequently encoding these quantized samples into binary form. Clearly, these operations are dependent on the specific hardware being used. Quantization may be achieved in a variety of ways. Quantization levels may be spaced linearly or in a logarithmic fashion. In the particular example illustrated in Figs. 1.4 and 1.5 we have used 8 quantization levels with equal spacing, and assigned a 3-bit word to each level. We could have used a logarithmic formula with narrow quantization levels in the middle that get wider gradually toward both ends. The encoding assignments shown in Fig. 1.5 could have been done differently. More than 8 quantization levels could have been used. There are 8-, 16-, 32-, and 64-bit systems and many others in use today.

Some microprocessors and microcontrollers utilize integer arithmetic; others use fixed-point or floating-point numbers. Different manufacturers represent numerical values and operations in different ways.

In this text, our primary goal is to understand mathematical modeling of signals and to explore and develop techniques for manipulating those signals using numerical analysis methods. Toward that goal, we use numbers in decimal format in the development of the theory of signals and systems. This allows us to study signals and systems in a manner that is not dependent on any particular hardware or any manufacturer preferences. When we need to implement one of the algorithms discussed in this text on a particular hardware platform, we need to adapt it to the specific requirements of that platform. In most cases software tools such as assemblers, interpreters, or compilers are available for the target platform, so that we don't need to do very low level programming.

1.2 Basic Signal Operations

In this section we will discuss fundamental signal operations for discrete-time signals. Arithmetic operations such as addition and multiplication for discrete-time signals bear strong resemblance to their continuous-time counterparts. In addition, discrete-time versions of time shifting, time scaling, and time reversal operations will also be discussed. Technically the use of the word *time* is somewhat inaccurate for these operations since the independent variable for discrete-time signals is the sample index n which may or may not correspond to time. We will, however, use the established terms of time shifting, time scaling, and time reversal while keeping in mind the distinction between t and n. More advanced signal operations such as convolution will be covered in later chapters.

1.2.1 Arithmetic operations

Consider a discrete-time signal $x[n]$. A constant offset value can be added to this signal to obtain

$$g[n] = x[n] + A$$

The offset A is added to each sample of the signal $x[n]$ to yield the corresponding sample of $g[n]$. This is illustrated in Fig. 1.7 for both positive and negative values of the offset A. Multiplication of the signal $x[n]$ with gain factor B is expressed in the form

$$g[n] = B\,x[n]$$

Each sample of the signal $g[n]$ is equal to the product of the corresponding sample of $x[n]$ and the constant gain factor B. This is illustrated in Fig. 1.8 for both positive and negative values of the gain factor B. Often, an offset and a gain factor are used in combination. This will be illustrated in Example 1.2.

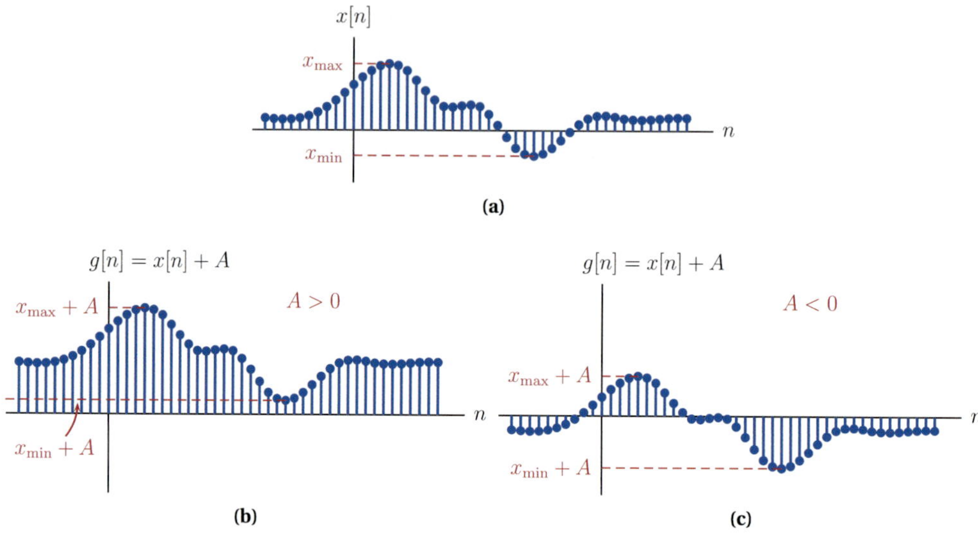

Figure 1.7 – Adding an offset A to signal $x[n]$: **(a)** Original signal $x[n]$, **(b)** $g[n] = x[n] + A$ with $A > 0$, and **(c)** $g[n] = x[n] + A$ with $A < 0$.

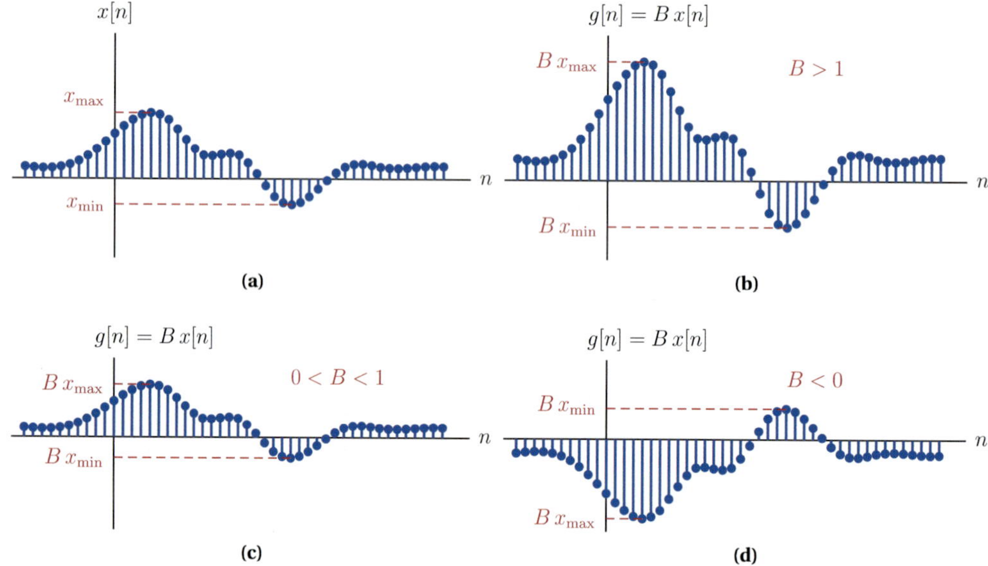

Figure 1.8 – Multiplying signal $x[n]$ with a constant gain factor B: **(a)** Original signal $x[n]$, **(b)** $g[n] = B\,x[n]$ with $B > 1$, **(c)** $g[n] = B\,x[n]$ with $0 < B < 1$, and **(d)** $g[n] = B\,x[n]$ with $B < 0$.

Addition of two discrete-time signals is expressed in the form

$$g[n] = x_1[n] + x_2[n]$$

and is accomplished by adding the amplitudes of the corresponding samples of the two signals as illustrated in Fig. 1.9.

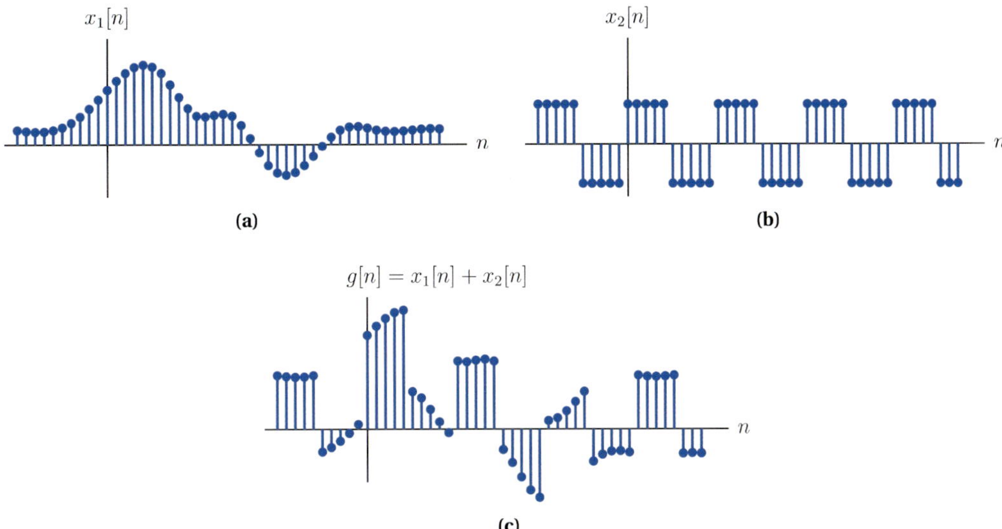

Figure 1.9 – Adding discrete-time signals.

Example 1.1: Adding discrete-time signals

Two discrete-time signals $x_1[n]$ and $x_2[n]$ are given below. Determine their sum, that is, $g[n] = x_1[n] + x_2[n]$.

$$x_1[n] = \{\, 1.5,\ 1.9,\ \underset{\substack{\uparrow \\ n=0}}{3.2},\ -1.1,\ 4.8,\ 0.4 \,\}$$

$$x_2[n] = \{\, 2.8,\ 4.1,\ \underset{\substack{\uparrow \\ n=3}}{1.7},\ -3.3,\ -1.8,\ 2.1 \,\}$$

Solution: Arrows point to different indices. The first step is to line up the two signals properly in terms of their index values.

$$\begin{aligned} x_1[n] &= \{\, 1.5,\ 1.9,\ \underset{\substack{\uparrow \\ n=0}}{3.2},\ -1.1,\ 4.8,\ 0.4 \qquad\qquad\qquad \} \\ x_2[n] &= \{ \qquad\qquad\qquad 2.8,\ 4.1,\ \underset{\substack{\uparrow \\ n=3}}{1.7},\ -3.3,\ -1.8,\ 2.1 \,\} \end{aligned}$$

Now overlapping signal samples can be added to obtain $g[n]$ as

$$g[n] = \{\, 1.5,\ 1.9,\ \underset{\substack{\uparrow \\ n=0}}{3.2},\ 1.7,\ 8.9,\ 2.1,\ -3.3,\ -1.8,\ 2.1 \,\}$$

Software resource: `exdt_1_1.m`

Example 1.2: Adding discrete-time signals with scaling and offset

For the discrete-time signals $x_1[n]$ and $x_2[n]$ given in Example 1.1 determine the signal $g[n] = 3x_1[n] - 2x_2[n] + 5$.

Solution: The first step is to obtain the scaled signals $3x_1[n]$ and $-2x_2[n]$.

$$3x_1[n] = \{ 4.5, 5.7, \underset{\substack{\uparrow \\ n=0}}{9.6}, -3.3, 14.4, 1.2 \}$$

$$-2x_2[n] = \{ -5.6, -8.2, \underset{\substack{\uparrow \\ n=3}}{-3.4}, 6.6, 3.6, -4.2 \}$$

The next step is to line up the signals:

$$\begin{array}{rcl} 3x_1[n] &=& \{ \; 4.5, \; 5.7, \; \underset{\substack{\uparrow \\ n=0}}{9.6}, \; -3.3, \; 14.4, \; 1.2 \qquad\qquad\qquad \} \\ -2x_2[n] &=& \{ \qquad\qquad\qquad -5.6, \; -8.2, \; \underset{\substack{\uparrow \\ n=3}}{-3.4}, \; 6.6, \; 3.6, \; -4.2 \} \\ +5\,(\text{const}) &=& \{ \ldots, \; 5.0, \; 5.0, \; 5.0, \; 5.0, \; 5.0, \; 5.0, \; 5.0, \; 5.0, \; 5.0, \ldots \} \end{array}$$

Note that we have added the constant offset term as a signal every sample of which has an amplitude of 5. Now the overlapping signal samples can be added up to obtain $g_1[n]$ as

$$g_1[n] = \{ 9.5, 10.7, \underset{\substack{\uparrow \\ n=0}}{14.6}, -3.9, 11.2, 2.8, 11.6, 8.6, 0.8 \}$$

The intermediate signal $g_1[n]$ represents the result we seek only in the interval $-2 \le n \le 6$. All samples of $g[n]$ outside this interval have an amplitude of 5 due to the constant offset term. Therefore the signal $g[n]$ is

$$g[n] = \begin{cases} g_1[n], & n = -2, \ldots, 6 \\ 5, & \text{otherwise} \end{cases}$$

Software resource: exdt_1_2.m

Finally, two discrete-time signals can be multiplied in a similar manner. The product of two signals $x_1[n]$ and $x_2[n]$ is expressed as

$$g[n] = x_1[n]\, x_2[n]$$

Each sample of the result $g[n]$ is obtained by multiplying the corresponding samples of $x_1[n]$ and $x_2[n]$ as shown in Fig. 1.10.

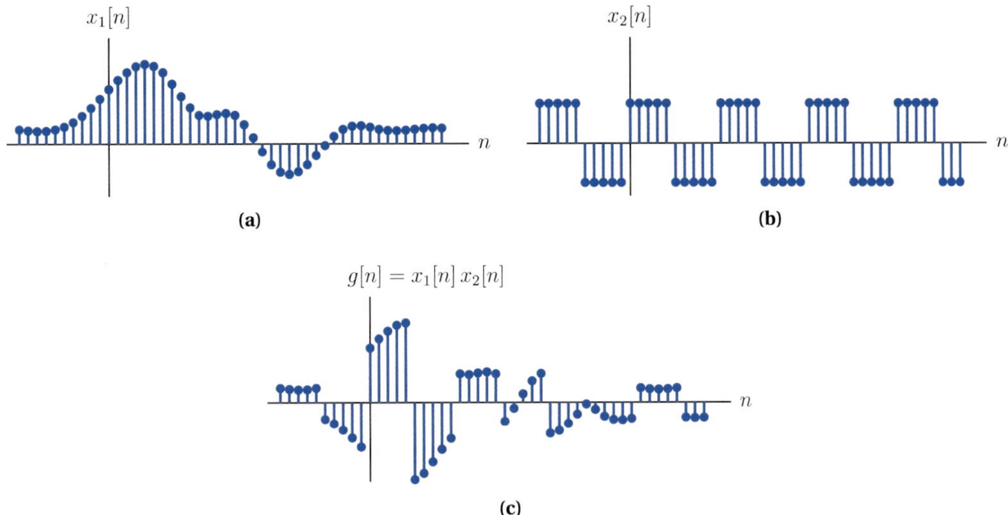

Figure 1.10 – Multiplying discrete-time signals.

Example 1.3: Multiplying discrete-time signals

For the two discrete-time signals $x_1[n]$ and $x_2[n]$ described in Example 1.1, determine the product signal $g[n] = x_1[n]\,x_2[n]$.

Solution: The first step is to line up the two signals:

$$x_1[n] = \{\ 1.5,\ 1.9,\ \underset{\substack{\uparrow\\ n=0}}{3.2},\ -1.1,\ 4.8,\ 0.4\ \}$$

$$x_2[n] = \{\ 2.8,\ 4.1,\ \underset{\substack{\uparrow\\ n=3}}{1.7},\ -3.3,\ -1.8,\ 2.1\ \}$$

The only significant overlaps occur at indices $n = 1,2,3$ with all other index positions exhibiting at least one factor with zero amplitude. Thus the product signal is

$$g[n] = \{\ \underset{\substack{\uparrow\\ n=1}}{-3.08},\ 19.68,\ 0.68\ \}$$

Software resource: exdt_1_3.m

1.2.2 Time shifting

For discrete-time signals time shifting operations must utilize integer shift parameters. A time shifted version $g[n]$ of the signal $x[n]$ is obtained as

$$g[n] = x[n-k] \tag{1.1}$$

where k is any positive or negative integer. This relationship is illustrated in Fig. 1.11.

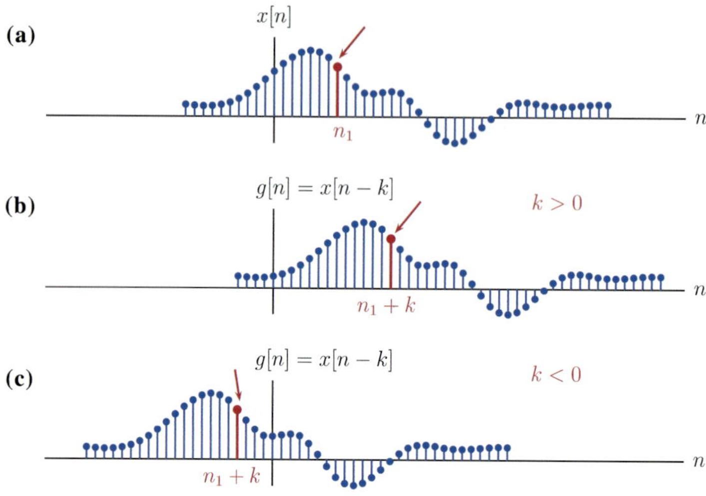

Figure 1.11 – Time shifting a discrete-time signal.

In part (a) of Fig. 1.11 the sample of $x[n]$ at index n_1 is marked with a thicker and red-colored stem. Let that sample correspond to a special event in the signal $x[n]$. Substituting $n = n_1 + k$ in Eqn. (1.1) we have

$$g[n_1 + k] = x[n_1]$$

The event that takes place in $x[n]$ at index $n = n_1$ takes place in $g[n]$ at index $n = n_1 + k$. If k is positive, this corresponds to a delay by k samples. Conversely, a negative k implies an advance.

Example 1.4: Time shifting a signal

Determine the signal $g[n] = x[n] + 0.5\,x[n-4]$ for a signal $x[n]$ defined in tabular form as follows:

$$x[n] = \left\{ \; 0.2,\, 0.4,\, 0.6,\, \underset{\underset{n=0}{\uparrow}}{0.8},\, 0.8,\, 0.8,\, 0.8,\, 0.4 \; \right\}$$

Solution: Based on Fig. 1.11, to apply the effect of a 4-sample shift to the signal $x[n]$, we could simply pick one particular sample of $x[n]$ and update its index from n to $n+4$. For example, we could pick the index $n = 0$ and make it the index $n = 4$ as follows:

$$x[n-4] = \left\{ \; 0.2,\, 0.4,\, 0.6,\, \underset{\underset{n=4}{\uparrow}}{0.8},\, 0.8,\, 0.8,\, 0.8,\, 0.4 \; \right\}$$

If we adjust the up arrow so that it points to $n = 0$ instead of $n = 4$, we obtain

$$x[n-4] = \left\{ \; \underset{\underset{n=0}{\uparrow}}{0.0},\, 0.2,\, 0.4,\, 0.6,\, 0.8,\, 0.8,\, 0.8,\, 0.8,\, 0.4 \; \right\}$$

Now the two components of $g[n]$ can be lined up with consistent index values.

$$\begin{aligned} x[n] &= \left\{ \; 0.2,\, 0.4,\, 0.6,\, \underset{\underset{n=0}{\uparrow}}{0.8},\, 0.8,\, 0.8,\, 0.8,\, 0.4, \; \right\} \\ 0.5\,x[n-4] &= \left\{ \; \underset{\underset{n=0}{\uparrow}}{0.0},\, 0.1,\, 0.2,\, 0.3,\, 0.4,\, 0.4,\, 0.4,\, 0.4,\, 0.2 \; \right\} \end{aligned}$$

We are ready to add overlapping signal samples to obtain $g[n]$ as

$$g[n] = \{\ 0.2,\ 0.4,\ 0.6,\ \underset{\substack{\uparrow \\ n=0}}{0.8},\ 0.9,\ 1.0,\ 1.1,\ 0.8,\ 0.4,\ 0.4,\ 0.4,\ 0.2\ \}$$

Software resource: exdt_1_4.m

1.2.3 Time scaling

For discrete-time signals we will consider time scaling in the following two forms:

$$g[n] = x[kn]\,,\quad k\text{: a positive integer} \tag{1.2}$$

and

$$g[n] = x[n/k]\,,\quad k\text{: a positive integer} \tag{1.3}$$

Let us first consider the form of time scaling in Eqn. (1.2). This is referred to as *downsampling*. The relationship between $x[n]$ and $g[n] = x[kn]$ is illustrated in Fig. 1.12 for $k = 2$ and $k = 3$.

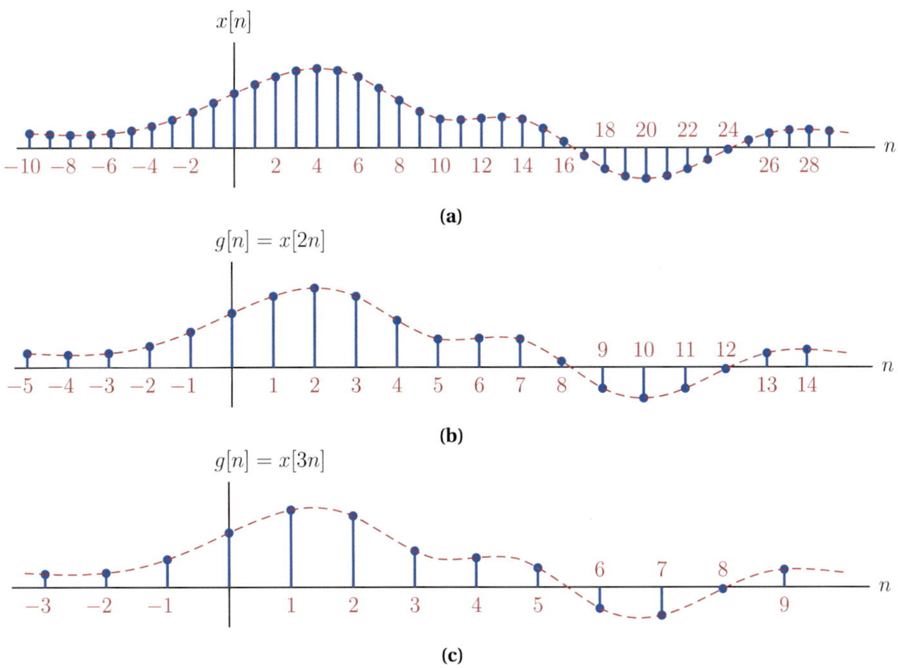

Figure 1.12 – Time scaling a signal $x[n]$ to obtain $g[n] = x[kn]$: **(a)** Original signal $x[n]$, **(b)** $g[n] = x[2n]$, and **(c)** $g[n] = x[3n]$.

It will be interesting to write this relationship for several values of the index n. For $k = 2$ we have

$$\ldots\quad g[-1] = x[-2]\,,\quad g[0] = x[0]\,,\quad g[1] = x[2]\,,\quad g[2] = x[4]\,,\quad \ldots$$

which suggests that $g[n]$ retains every other sample of $x[n]$, and discards the samples between them. This relationship is further illustrated in Fig. 1.13.

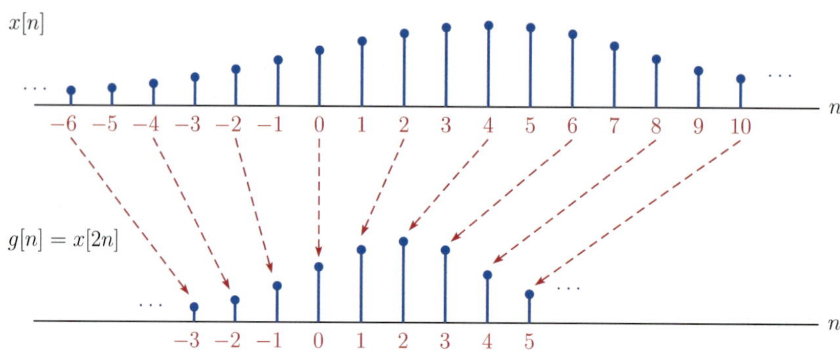

Figure 1.13 – Illustration of time scaling (downsampling) a discrete-time signal by a factor of 2.

For $k = 3$, samples of $g[n]$ are

$$\ldots \quad g[-1] = x[-3]\,, \quad g[0] = x[0]\,, \quad g[1] = x[3]\,, \quad g[2] = x[6]\,, \quad \ldots$$

In this case every third sample of $x[n]$ is retained, and the samples between them are discarded, as shown in Fig. 1.14.

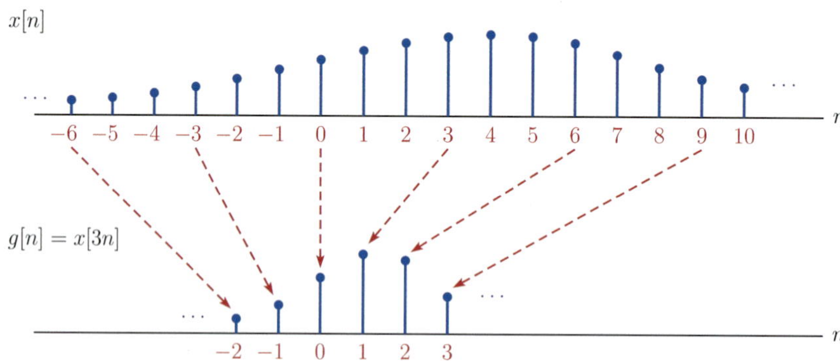

Figure 1.14 – Illustration of time scaling (downsampling) a discrete-time signal by a factor of 3.

This raises an interesting question: Does the act of discarding samples lead to a loss of information, or were those samples redundant in the first place? The answer depends on the characteristics of the signal $x[n]$, and will be explored further when we discuss *downsampling* and *decimation* in Chapter 4. For some signals, discarding samples may indeed lead to a loss of information, and for others it may just be an act of discarding redundant samples.

Example 1.5: Downsampling a signal

Consider the signal $x[n]$ given below:

$$x[n] = \{\, 0.8,\ 1.0,\ 1.3,\ 0.9,\ 0.7,\ \underset{\substack{\uparrow \\ n=0}}{1.2},\ 1.4,\ 1.1,\ 0.8,\ 0.6,\ 0.5,\ -0.2 \,\}$$

The signal $g[n]$ is obtained by downsampling $x[n]$ by a factor of 2, that is, $g[n] = x[2n]$. Determine $g[n]$.

Solution: Significant samples of $x[n]$ are in the index range $-5 \le n \le 6$. Application of the downsampling relationship in this range of indices yields

$$g[-2] = x[-4], \quad g[-1] = x[-2], \quad g[0] = x[0],$$
$$g[1] = x[2], \quad g[2] = x[4], \quad g[3] = x[6]$$

Thus, the signal $g[n]$ is

$$g[n] = \{\, 1.0,\ 0.9,\ \underset{\substack{\uparrow \\ n=0}}{1.2},\ 1.1,\ 0.6,\ -0.2 \,\}$$

This is equivalent to dropping out every other sample of $x[n]$ as shown below:

$$g[n] = \{\, \cancel{0.8,}\ 1.0,\ \cancel{1.3,}\ 0.9,\ \cancel{0.7,}\ \underset{\substack{\uparrow \\ n=0}}{1.2},\ \cancel{1.4,}\ 1.1,\ \cancel{0.8,}\ 0.6,\ \cancel{0.5,}\ -0.2 \,\}$$

Software resource: exdt_1_5.m

An alternative form of time scaling for a discrete-time signal was given by Eqn. (1.3). Consider for example, the signal $g[n]$ defined based on Eqn. (1.3) with $k = 2$:

$$g[n] = x[n/2] \tag{1.4}$$

Since the index of the signal on the right side of the equal sign is $n/2$, the relationship between $g[n]$ and $x[n]$ is defined only for values of n that make $n/2$ an integer. We can write

$$\ldots \quad g[-2] = x[-1]\,, \quad g[0] = x[0]\,, \quad g[2] = x[1]\,, \quad g[4] = x[2]\,, \quad \ldots$$

Eqn. 1.4 does not provide a complete definition of the signal $g[n]$ since its sample amplitudes cannot be determined for odd values of the index n. In order to have a completely defined signal, we will expand Eqn. 1.4 as follows:

$$g[n] = \begin{cases} x[n/2]\,, & \text{if } n/2 \text{ is integer} \\ 0\,, & \text{otherwise} \end{cases} \tag{1.5}$$

Fig. 1.15 illustrates the process of obtaining $g[n]$ from $x[n]$ based on Eqn. (1.5). The relationship between $g[n]$ and $x[n]$ is known as *upsampling*. The signal $g[n]$ is said to be an upsampled version of the signal $x[n]$. Upsampling will be discussed in more detail in Chapter 4 in the context of *interpolation*.

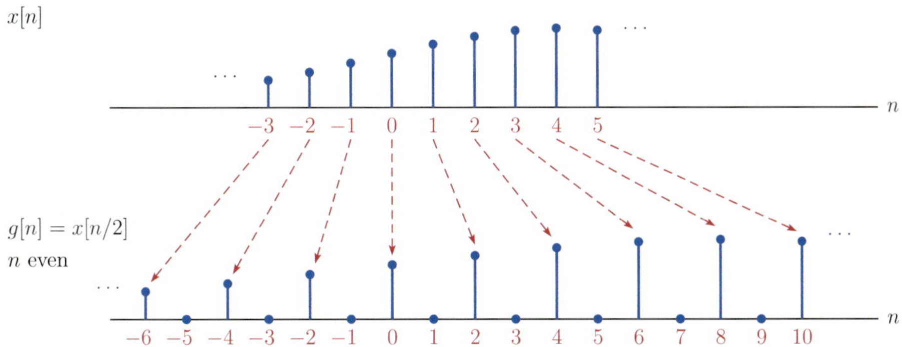

Figure 1.15 – Illustration of time scaling (upsampling) a discrete-time signal by a factor of 2.

Example 1.6: Upsampling a signal

For the signal $x[n]$ given in tabular form as

$$x[n] = \{\ 2.3,\ -0.7,\ -1.6,\ \underset{\substack{\uparrow \\ n=0}}{1.3},\ 1.7,\ 2.1,\ 2.9\ \}$$

determine the upsampled signal

$$g[n] = \begin{cases} x[n/3]\,, & \text{if } n/3 \text{ is integer} \\ 0\,, & \text{otherwise} \end{cases}$$

Solution: For the specified samples of the signal $x[n]$ we have the following relationships:

$$\begin{array}{llll} g[-9] = x[-3], & g[-6] = x[-2], & g[-3] = x[1], & g[0] = x[0] \\ g[9] = x[3], & g[6] = x[2], & g[3] = x[1] & \end{array}$$

The signal $g[n]$ is

$$g[n] = \{\ 2.3,\ 0,\ 0,\ -0.7,\ 0,\ 0,\ -1.6,\ 0,\ 0,\ \underset{\substack{\uparrow \\ n=0}}{1.3},\ 0,\ 0,\ 1.7,\ 0,\ 0,\ 2.1,\ 0,\ 0,\ 2.9\ \}$$

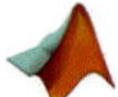

Software resource: exdt_1_6.m

1.2.4 Time reversal

A time reversed version of the signal $x[n]$ is

$$g[n] = x[-n] \tag{1.6}$$

An event that takes place at index value $n = n_1$ in the signal $x[n]$ takes place at index value $n = -n_1$ in the signal $g[n]$. Graphically this corresponds to folding or flipping the signal $x[n]$ around $n = 0$ axis as illustrated in Fig. 1.16.

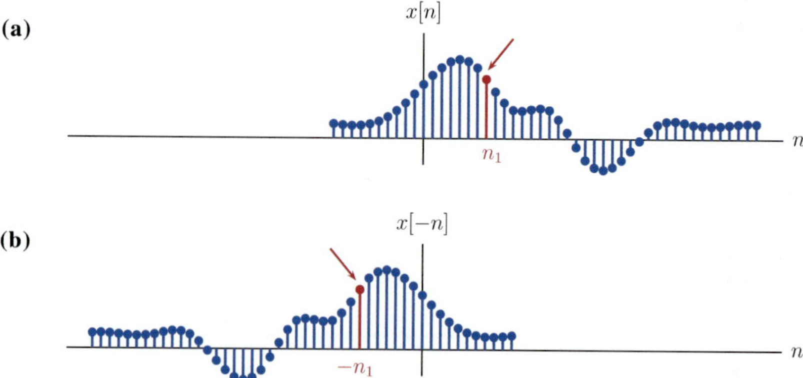

Figure 1.16 – Time reversal of a discrete-time signal.

Example 1.7: Time reversal of a signal

For the signal $x[n]$ given below, determine the signal $g[n] = x[n] + x[-n]$.

$$x[n] = \{ 0.1, 0.2, 0.3, 0.4, 0.5, 0.6, 0.7, \underset{\substack{\uparrow \\ n=0}}{0.8}, 0.9, 1.0 \}$$

Solution: The time reversed version of $x[n]$ is

$$x[-n] = \{ 1.0, 0.9, \underset{\substack{\uparrow \\ n=0}}{0.8}, 0.7, 0.6, 0.5, 0.4, 0.3, 0.2, 0.1 \}$$

The first step is to line up $x[n]$ and $x[-n]$ so that corresponding samples can be added together.

$$\begin{aligned} x[n] &= \{ 0.1, 0.2, 0.3, 0.4, 0.5, 0.6, 0.7, \underset{\substack{\uparrow \\ n=0}}{0.8}, 0.9, 1.0 \qquad\qquad\qquad\qquad\qquad \} \\ x[-n] &= \{ \qquad\qquad\qquad\qquad\qquad 1.0, 0.9, \underset{\substack{\uparrow \\ n=0}}{0.8}, 0.7, 0.6, 0.5, 0.4, 0.3, 0.2, 0.1 \} \end{aligned}$$

The signal $g[n]$ is found to be

$$g[n] = \{ 0.1, 0.2, 0.3, 0.4, 0.5, 1.6, 1.6, \underset{\substack{\uparrow \\ n=0}}{1.6}, 1.6, 1.6, 0.5, 0.4, 0.3, 0.2, 0.1 \}$$

Software resource: `exdt_1_7.m`

1.3 Basic Building Blocks for Discrete-Time Signals

In this section we will look at basic discrete-time signal building blocks that are used in constructing mathematical models for discrete-time signals with higher complexity. We will see that many of the continuous-time signal building blocks have discrete-time counterparts that are defined similarly, and that have similar properties. There are also some fundamental differences between continuous-time and discrete-time versions of the basic signals, and these will be indicated throughout our discussion.

Unit impulse function

It will be apparent in the rest of this chapter as well as in later chapters of this text that the unit impulse function plays an important role in mathematical modeling and analysis of signals and linear systems. The discrete-time unit impulse function is defined as follows:

Unit impulse function:

$$\delta[n] = \begin{cases} 1, & n = 0 \\ 0, & n \neq 0 \end{cases} \tag{1.7}$$

It is shown graphically in Fig. 1.17.

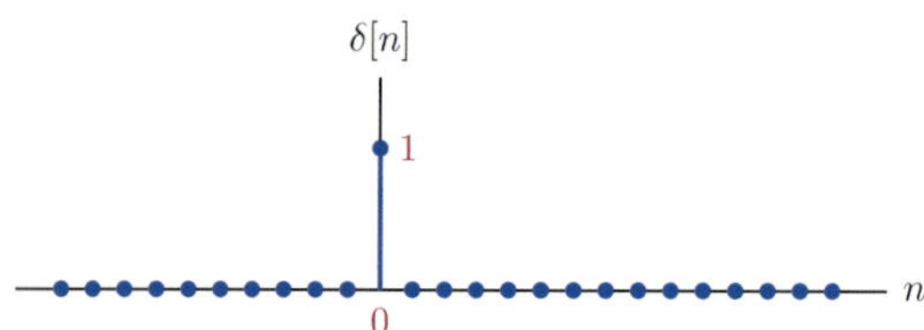

Figure 1.17 – Discrete-time unit impulse signal.

As seen from the definition in Eqn. (1.7) and Fig. 1.17, the discrete-time unit impulse function does not have the complications associated with its continuous-time counterpart. The signal $\delta[n]$ is unambiguously defined for all integer values of the sample index n.

Shifted and scaled versions of the discrete-time unit impulse function are used often in problems involving signal-system interaction. A unit impulse function that is scaled by a and time shifted by n_1 samples is described below and is is shown in Fig. 1.18.

Scaled and shifted unit impulse function:

$$a\delta[n-n_1] = \begin{cases} a, & n = n_1 \\ 0, & n \neq n_1 \end{cases} \tag{1.8}$$

The fundamental properties of the continuous-time unit impulse function can be readily adapted to its discrete-time counterpart. The *sampling property* of the discrete-time unit impulse function is expressed as

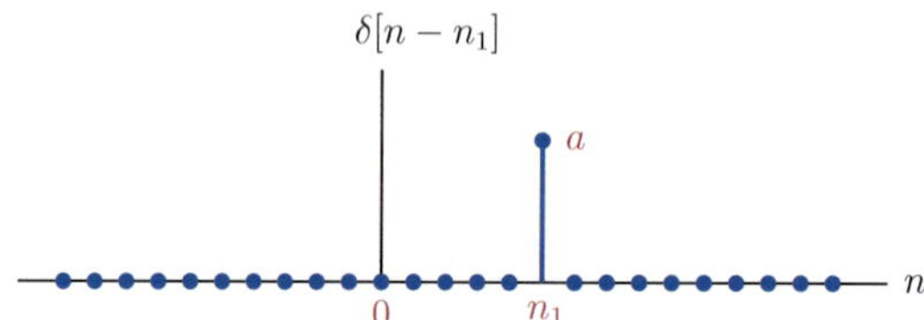

Figure 1.18 – Scaled and time shifted discrete-time unit impulse signal.

$$x[n]\,\delta[n-n_1] = x[n_1]\,\delta[n-n_1] \tag{1.9}$$

It is important to interpret Eqn. (1.9) correctly: $x[n]$ and $\delta[n-n_1]$ are both infinitely long discrete-time signals, and can be graphed in terms of the sample index n as shown in Fig. 1.19.

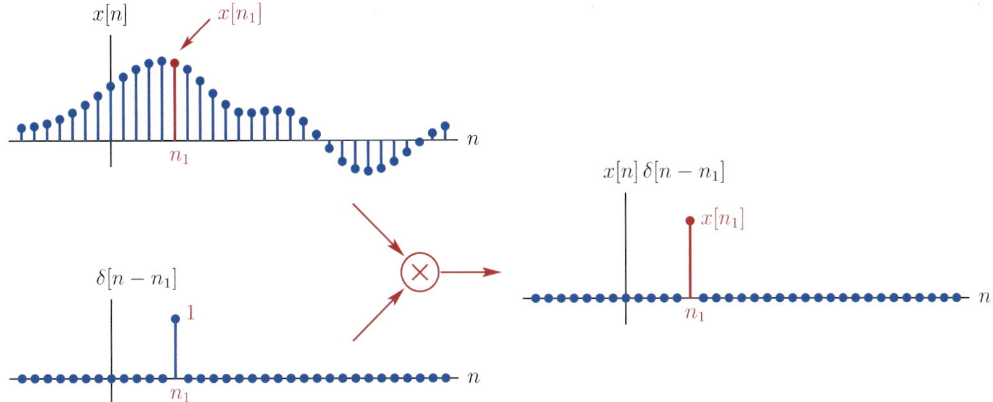

Figure 1.19 – Illustration of the sampling property of the discrete-time unit impulse signal.

The claim in Eqn. (1.9) is easy to justify: If the two signals $x[n]$ and $\delta[n-n_1]$ are multiplied on a sample-by-sample basis, the product signal is equal to zero for all but one value of the sample index n, and the only non-zero amplitude occurs for $n = n_1$. Mathematically we have

$$x[n]\,\delta[n-n_1] = \begin{cases} x[n_1]\,, & n = n_1 \\ 0\,, & n \neq n_1 \end{cases} \tag{1.10}$$

which is equivalent to the right side of Eqn. (1.9). The *sifting property* for the discrete-time unit impulse function is expressed as

$$\sum_{n=-\infty}^{\infty} x[n]\,\delta[n-n_1] = x[n_1] \tag{1.11}$$

which easily follows from Eqn. (1.9). Substituting Eqn. (1.9) into Eqn. (1.11) we obtain

$$\begin{aligned} \sum_{n=-\infty}^{\infty} x[n]\,\delta[n-n_1] &= \sum_{n=-\infty}^{\infty} x[n_1]\,\delta[n-n_1] \\ &= x[n_1] \sum_{n=-\infty}^{\infty} \delta[n-n_1] \\ &= x[n_1] \end{aligned} \tag{1.12}$$

where we have relied on the sum of all samples of the impulse signal being equal to unity. The result of the summation in Eqn. (1.11) is a scalar, the value of which equals sample n_1 of the signal $x[n]$.

1.3.1 Unit step function

The unit step function is useful in situations where we need to model a signal that is turned on or off at a specific sample index. The discrete-time unit step function is defined in a way similar to its continuous-time version:

Unit step function:

$$u[n] = \begin{cases} 1, & n \geq 0 \\ 0, & n < 0 \end{cases} \tag{1.13}$$

The function $u[n]$ is shown in Fig. 1.20.

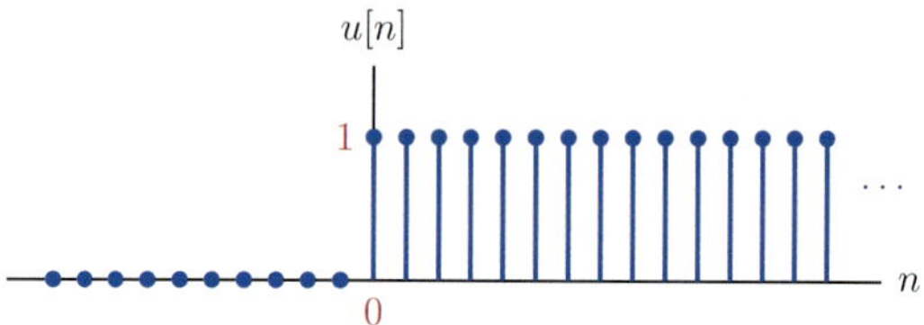

Figure 1.20 – Discrete-time unit step signal.

As in the case of the discrete-time unit impulse function, this function also enjoys a clean definition without any of the complications associated with its continuous-time counterpart. Eqn. (1.13) provides a complete definition of the discrete-time unit step function for all integer values of the sample index n. A time shifted version of the discrete-time unit step function can be written as

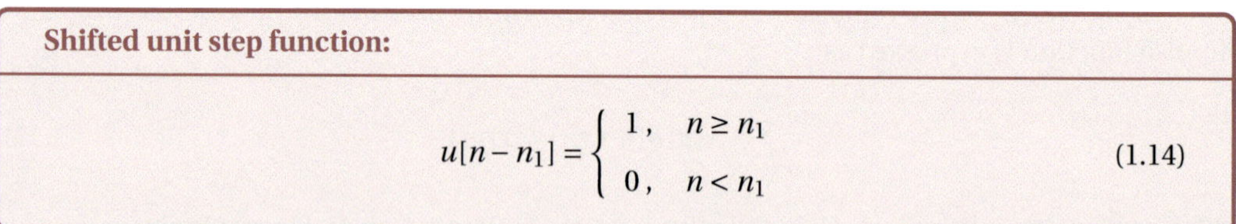

Shifted unit step function:

$$u[n-n_1] = \begin{cases} 1, & n \geq n_1 \\ 0, & n < n_1 \end{cases} \tag{1.14}$$

This is illustrated in Fig. 1.21.

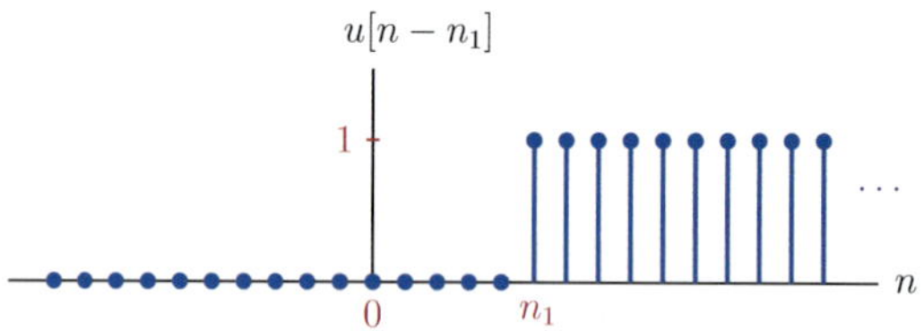

Figure 1.21 – Time shifted discrete-time unit step signal.

Recall that a continuous-time unit impulse could be obtained as the first derivative of the continuous-time unit step. An analogous relationship exists between the discrete-time counterparts of these signals. It is possible to express a discrete-time unit impulse signal as the *first difference* of the discrete-time unit step signal:

$$\delta[n] = u[n] - u[n-1] \tag{1.15}$$

This relationship is illustrated in Fig. 1.22.

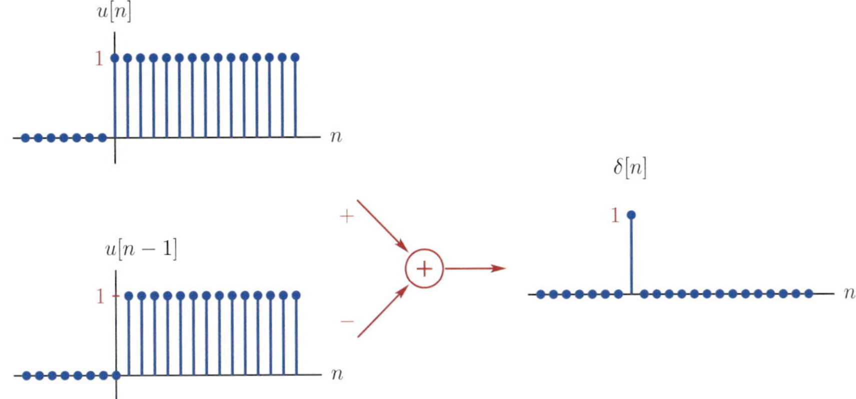

Figure 1.22 – Obtaining a discrete-time unit impulse from a discrete-time unit step through first difference.

Conversely, a unit step signal can be constructed from unit impulse signals through a *running sum* in the form

$$u[n] = \sum_{k=-\infty}^{n} \delta[k] \tag{1.16}$$

This is analogous to the running integral relationship between the continuous-time versions of these signals. In Eqn. (1.16) we are adding the samples of a unit step signal $\delta[k]$ starting from $k = -\infty$ up to and including the sample for $k = n$. If n, the upper limit of the summation, is zero or positive, the summation includes the only sample with unit amplitude and the result is equal to unity. If $n < 0$, the summation ends before we reach that sample and the result is zero. This is shown in Fig. 1.23.

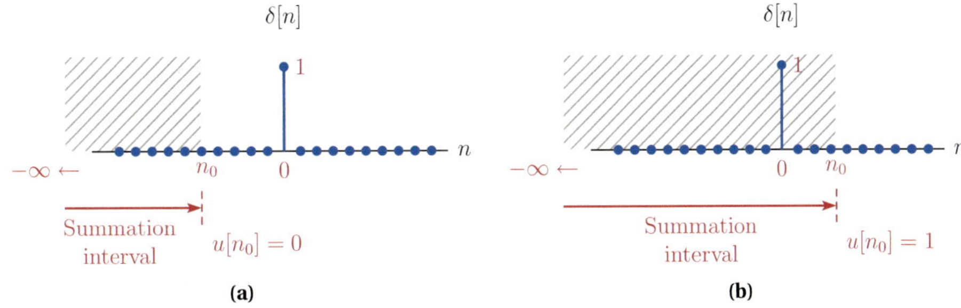

Figure 1.23 – Obtaining a discrete-time unit step from a discrete-time unit impulse through a running sum: **(a)** $n = n_0 < 0$ and **(b)** $n = n_0 > 0$.

An alternative approach for obtaining a unit step from a unit impulse is to use

$$u[n] = \sum_{k=0}^{\infty} \delta[n-k] \tag{1.17}$$

in which we add the signals $\delta[n], \delta[n-1], \delta[n-2], \ldots, \delta[n-k]$ to construct a unit step signal. This is an example of impulse decomposition that will be discussed in Section 1.4.

Example 1.8: Constructing a signal using unit step signals

Express the signal $x[n]$ shown in Fig. 1.24 using scaled and time shifted unit step signals.

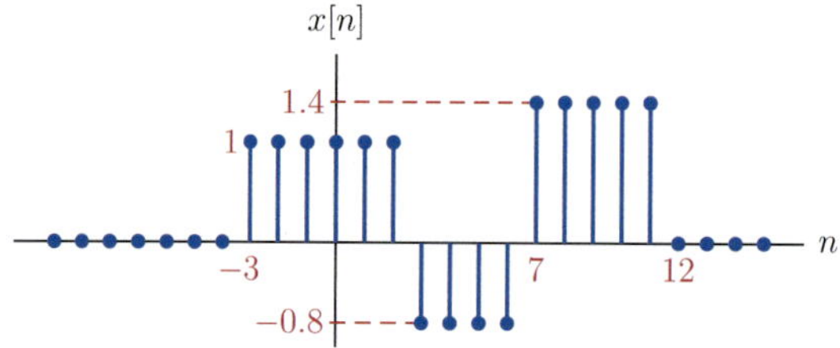

Figure 1.24 – The signal x[n] for Example 1.8.

Solution:

$$x[n] = u[n+3] - 1.8\,u[n-3] + 2.2\,u[n-7] - 1.4\,u[n-12]$$

Software resource: exdt_1_8.m

Software resources: See MATLAB Exercises 1.3 and 1.4.

1.3.2 Unit ramp function

The discrete-time unit ramp function has zero amplitude for $n < 0$ and unit slope for $n \geq 0$. Its definition is similar to that of its continuous-time counterpart:

Unit ramp function:

$$r[n] = \begin{cases} n\,, & n \geq 0 \\ 0\,, & n < 0 \end{cases} \tag{1.18}$$

Unit ramp function is illustrated in Fig. 1.25.

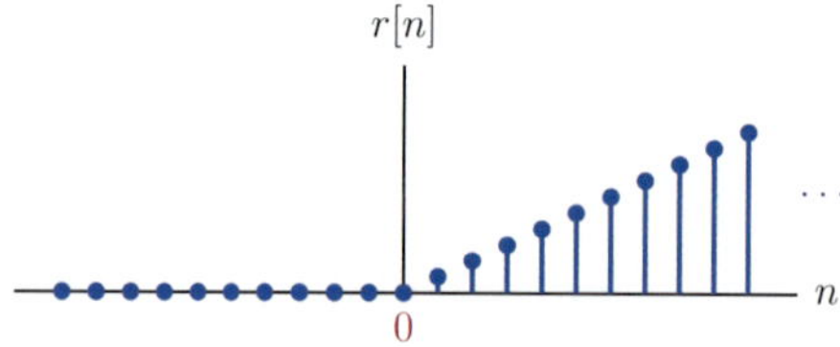

Figure 1.25 – Discrete-time unit ramp function.

The definition in Eqn. (1.18) can be written in a more compact form as the product of the linear signal $g[n] = n$ and the unit step function. We can write $r[n]$ as

$$r[n] = n\,u[n] \tag{1.19}$$

which is illustrated in Fig. 1.26. Alternatively, the discrete-time unit ramp function can be expressed as a running summation applied to the discrete-time unit step function in the form

$$r[n] = \sum_{k=-\infty}^{n-1} u[k] \tag{1.20}$$

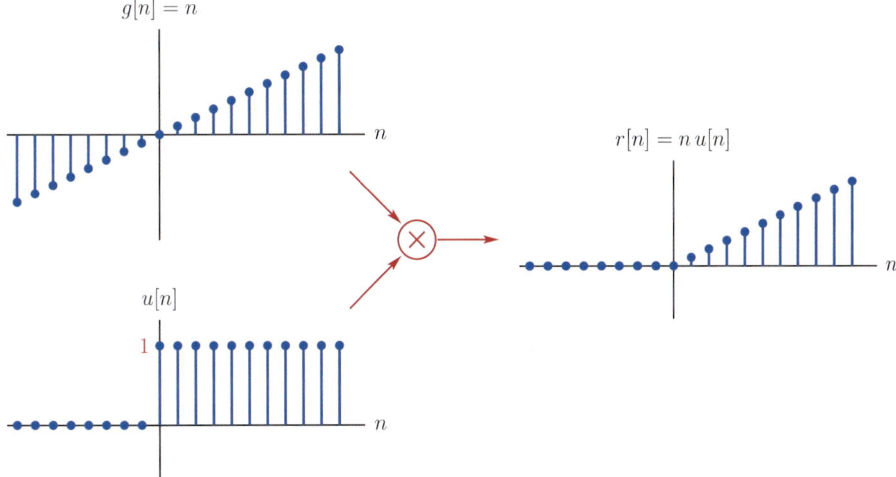

Figure 1.26 – Obtaining a discrete-time unit ramp.

By trying out the summation in Eqn. (1.20) for a few different values of the index n it is easy to see that it produces values consistent with the definition of the unit ramp function given by Eqn. (1.19). This is analogous to the relationship between the continuous-time versions of these signals where a running integral is used for obtaining a continuous-time unit ramp signal from a continuous-time unit step signal. The process of obtaining a discrete-time unit ramp signal through a running sum is illustrated in Fig. 1.27.

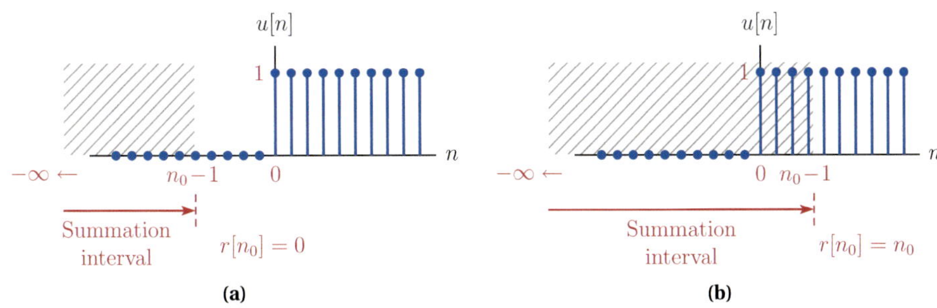

Figure 1.27 – Obtaining a discrete-time unit ramp from a discrete-time unit step through a running sum: **(a)** $n = n_0 < 0$ and **(b)** $n = n_0 > 0$.

Example 1.9: Constructing a signal using unit ramp signals

Express the signal $x[n]$ shown in Fig. 1.28 using scaled and time shifted unit ramp signals.

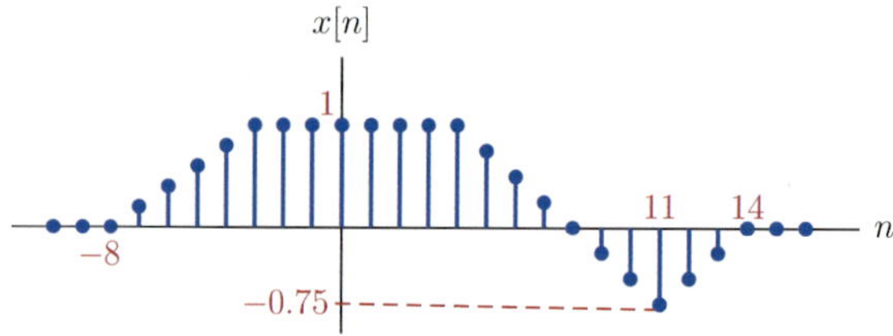

Figure 1.28 – The signal x[n] for Example 1.9.

Solution:

$$x[n] = \frac{1}{5}\,r[n+8] - \frac{1}{5}\,r[n+3] - \frac{1}{4}\,u[n-4] + \frac{1}{2}\,r[n-11] - \frac{1}{4}\,r[n-14]$$

Software resource: `exdt_1_9.m`

1.3.3 Sinusoidal signals

Sinusoidal signals play a very important role in the study of discrete-time signals and systems. This will become apparent as we look at examples of working with audio signals in this chapter, and as we study frequency domain analysis techniques in later chapters. A discrete-time sinusoidal signal is in the general form shown below.

Sinusoidal signal:

$$x[n] = A\cos(\Omega_0 n + \theta) \tag{1.21}$$

A is the *amplitude*, Ω_0 is the *angular frequency* in radians, and θ is the phase angle which is also in radians. The angular frequency Ω_0 can be expressed as

$$\Omega_0 = 2\pi F_0 \tag{1.22}$$

The parameter F_0, a dimensionless quantity, is referred to as the *normalized frequency* of the sinusoidal signal. Fig. 1.29 illustrates discrete-time sinusoidal signals for various values of Ω_0. At this point we will note a fundamental difference between a discrete-time sinusoidal signal and its continuous-time counterpart:

1. A continuous-time sinusoidal signal is defined as follows:

$$x_a(t) = A\cos(\omega_0 t + \theta) \tag{1.23}$$

Its parameter ω_0 is in rad/s since it appears next to the time variable t, and the product $\omega_0 t$ must be in radians to qualify as the argument of a trigonometric function.

2. In contrast, the parameter Ω_0 is in radians since it appears next to a dimensionless index parameter n, and the product $\Omega_0 n$ must be in radians. For this reason, Ω_0 is referred to as the angular frequency of the discrete-time sinusoidal signal.

Even though the word *frequency* is used, Ω_0 is not really a frequency in the traditional sense, but rather an angle. In support of this assertion, we will see in later parts of this text that values of Ω_0 outside the range $-\pi \le \Omega_0 < \pi$ are mathematically indistinguishable from values within that range. Similar reasoning applies to the parameter F_0: It is not really a frequency but a dimensionless quantity that can best be thought of as a *percentage*. In order to elaborate further on the meanings of the parameters Ω_0 and F_0, we will consider the case of obtaining a discrete-time sinusoidal signal from a continuous-time sinusoidal signal given by Eqn. 1.23. Let us evaluate the amplitude of $x_a(t)$ at time instants that are integer multiples of T_s, and construct a discrete-time signal with the results:

$$x[n] = x_a(nT_s) = A\cos(\omega_0 T_s n + \theta) = A\cos\left(2\pi f_0 T_s n + \theta\right) \tag{1.24}$$

This is illustrated in Fig. 1.30. Since the signal $x_a(t)$ is evaluated at intervals of T_s, the number of samples taken per unit time is

$$f_s = \frac{1}{T_s} \tag{1.25}$$

Substituting Eqn. (1.25) into Eqn. (1.24) we obtain

$$x[n] = A\cos\left(2\pi\left[\frac{f_0}{f_s}\right]n + \theta\right) = A\cos(2\pi F_0 n + \theta) \tag{1.26}$$

Based on Eqn. (1.26), the normalized frequency F_0 is simply the analog frequency expressed as a percentage of the sampling rate, i.e.,

$$F_0 = \frac{f_0}{f_s} \tag{1.27}$$

The act of constructing a discrete-time signal by evaluating a continuous-time signal at uniform intervals is called *sampling* and will be discussed in more detail in Chapter 4. The parameters f_s and T_s are the *sampling rate* and the *sampling interval*, respectively. Eqn. (1.27) suggests that the normalized frequency F_0 is essentially the frequency f_0 of the continuous-time sinusoid expressed as a percentage of the sampling rate f_s.

Software resource: See MATLAB Exercise 1.9.

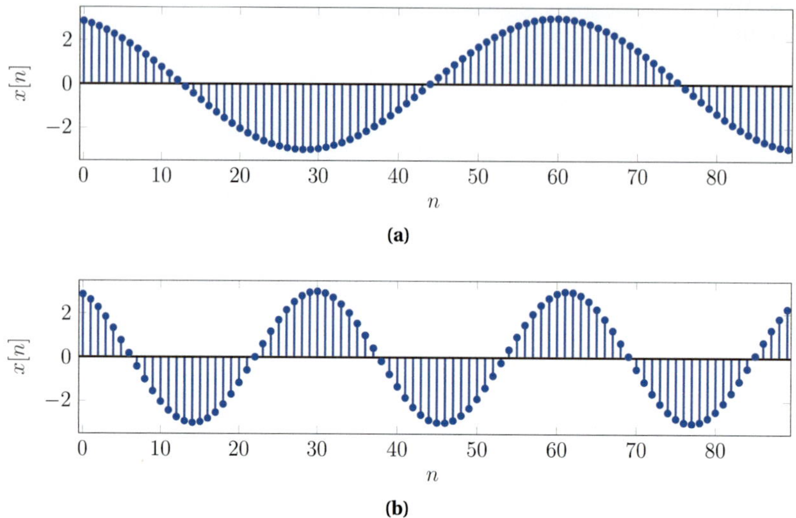

Figure 1.29 – Discrete-time sinusoidal signal $x[n] = 3\cos(\Omega_0 n + \pi/10)$ for **(a)** $\Omega_0 = 0.1$ rad and **(b)** $\Omega_0 = 0.2$ rad.

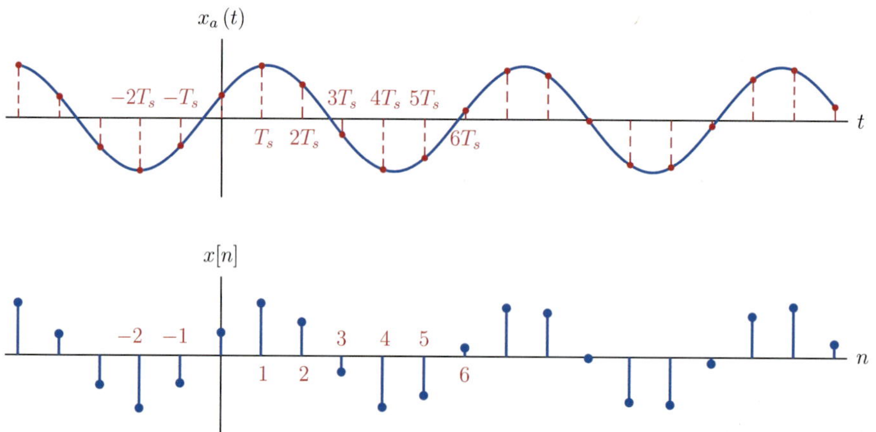

Figure 1.30 – Obtaining a discrete-time sinusoidal signal from a continuous-time sinusoidal signal.

Example 1.10: Compact form of a sinusoidal signal

Write the signal $x[n]$ given by

$$x[n] = A_1 \cos(\Omega_0 n) + A_2 \sin(\Omega_0 n)$$

in the compact form

$$x[n] = B\cos(\Omega_0 n + \theta)$$

Determine the parameters B and θ in terms of A_1 and A_2.

Solution: We will make use of a trigonometric identity found in Eqn. (B.1) of Appendix B which is repeated here for convenience:

$$\text{Eqn. (B.1):} \qquad \cos(a \pm b) = \cos(a)\cos(b) \mp \sin(a)\sin(b)$$

Applying this identity to the desired form of the signal with the substitutions $a = \Omega_0 n$ and $b = \theta$ yields

$$\begin{aligned} x[n] =& B\cos(\Omega_0 n + \theta) \\ =& B\cos(\Omega_0 n)\cos(\theta) - B\sin(\Omega_0 n)\sin(\theta) \end{aligned} \tag{1.28}$$

Matching the coefficients in Eqn. (1.28) with those of the original signal $x[n]$ we obtain

$$A_1 = B\cos(\theta) \quad \text{and} \quad A_2 = -B\sin(\theta)$$

from which we can solve for the parameters B and θ as

$$B = \sqrt{A_1^2 + A_2^2} \qquad \text{and} \qquad \theta = -\tan^{-1}\left(\frac{A_2}{A_1}\right)$$

Example 1.11: Adding two sinusoidal signals with the same angular frequency

Write the signal $x[n]$ given by

$$x[n] = A_1\cos(\Omega_0 n + \theta_1) + A_2\cos(\Omega_0 n + \theta_2)$$

in the compact form

$$x[n] = B\cos(\Omega_0 n + \theta)$$

Determine the parameters B and θ in terms of A_1, A_2, θ_1, θ_2.

Solution: Again using the trigonometric identity found in Eqn. (B.1) of Appendix B the signal $x[n]$ can be written as

$$\begin{aligned} x[n] =& A_1\cos(\Omega_0 n + \theta_1) + A_2\cos(\Omega_0 n + \theta_2) \\ =& \left[A_1\cos(\theta_1) + A_2\cos(\theta_2)\right]\cos(\Omega_0 n) - \Big[A_1\sin(\theta_1) + A_2\sin(\theta_2)\Big]\sin(\Omega_0 n) \end{aligned}$$

Let C_1 and C_2 be defined as

$$C_1 = A_1\cos(\theta_1) + A_2\cos(\theta_2) \qquad \text{and} \qquad C_2 = A_1\sin(\theta_1) + A_2\sin(\theta_2)$$

and we have

$$x[n] = C_1\cos(\Omega_0 n) - C_2\sin(\Omega_0 n)$$

which reverts to the same form we have observed in Example 1.10. Thus, the signal $x[n]$ can be written as

$$x[n] = B\cos(\Omega_0 n + \theta)$$

with

$$B = \sqrt{C_1^2 + C_2^2} \qquad \text{and} \qquad \theta = -\tan^{-1}\left(\frac{C_2}{C_1}\right)$$

Interactive App: Waveform explorer

This interactive app in `appWaveExplorerDT.m` allows experimentation with the basic signal building blocks discussed in Section 1.3. The signal $x[n]$ is constructed using up to five terms in the form

$$x[n] = x_1[n] + x_2[n] + x_3[n] + x_4[n] + x_5[n] \tag{1.29}$$

and is graphed in the lower part of the user interface. Each term $x_i[n]$, $i = 1,\ldots,5$ in Eqn. (1.29) can be a scaled, time shifted and optionally time reversed unit step or unit ramp function. If you do not need all five terms for constructing $x[n]$, unneeded terms can be kept off.

a. Using the app with two unit step functions, verify the relationship shown in Fig. 1.22.

b. Construct the signal shown in Fig. 1.24 using unit step functions with appropriate scaling and time shifting parameters. Ideally, work this out with pencil and paper first, and then use the app to verify your solution.

c. Construct the signal shown in Fig. 1.28 using unit ramp functions with appropriate scaling and time shifting parameters. First do a pencil and paper sketch of the signals involved, and then use the app to verify your solution.

d. On paper, devise a method of constructing a symmetric triangle with a unit amplitude peak at $n = 0$ and two zero amplitude corners at $n = -10$ and $n = 10$ using unit ramp signals with appropriate scaling and time shifting parameters. Afterward use the app to verify your pencil and paper solution.

Software resource: `appWaveExplorerDT.m`

1.4 Impulse Decomposition for Discrete-Time Signals

Consider an arbitrary discrete-time signal $x[n]$. Let us define a new signal $x_k[n]$ by using the k-th sample of the signal $x[n]$ in conjunction with a time shifted unit impulse function as

$$x_k[n] = x[k]\,\delta[n-k] = \begin{cases} x[k]\,, & n = k \\ 0\,, & n \neq k \end{cases} \tag{1.30}$$

The signal $x_k[n]$ is a scaled and time shifted impulse signal, the only non-trivial sample of which occurs at index $n = k$ with an amplitude of $x[k]$. If we were to repeat Eqn. (1.30) for all possible values of k, we would obtain an infinite number of signals $x_k[n]$ for $k = -\infty,\ldots,\infty$. In each of these signals there would only be one non-trivial sample the amplitude of which equals the amplitude of the corresponding sample of $x[n]$. For example, consider the signal

$$x[n] = \{\, 3.7,\, \underset{\uparrow}{1.3},\, -1.5,\, 3.4,\, 5.9 \,\}$$

Suppose that we would like to write $x[n]$ as the sum of five scaled and time shifted unit impulse signals even if that may seem like a trivial and pointless task at the moment. The signals $x_k[n]$ for this case would be as follows:

$$\begin{array}{rcl}
& \vdots & \\
x_{-1}[n] & = & \{3.7, \underset{\uparrow}{0}, 0, 0, 0\} \\
x_0[n] & = & \{0, \underset{\uparrow}{1.3}, 0, 0, 0\} \\
x_1[n] & = & \{0, \underset{\uparrow}{0}, -1.5, 0, 0\} \\
x_2[n] & = & \{0, \underset{\uparrow}{0}, 0, 3.4, 0\} \\
x_3[n] & = & \{0, \underset{\uparrow}{0}, 0, 0, 5.9\} \\
& \vdots &
\end{array}$$

The signal $x[n]$ can be reconstructed by adding these components together.

$$x[n] = \sum_{k=-\infty}^{\infty} x_k[n] = \sum_{k=-\infty}^{\infty} x[k]\,\delta[n-k] \tag{1.31}$$

Eqn. (1.31) represents an impulse decomposition of the discrete-time signal $x[n]$. We will use it in deriving the convolution relationship later in Chapter 2.

1.5 Signal Classifications

In this section we will summarize various methods and criteria for classifying the types of discrete-time signals that will be useful in our future discussions.

1.5.1 Real vs. complex signals

A discrete-time signal may be real or complex valued. A real signal is one in which the amplitude is real-valued at all time instants. In contrast, a complex signal is one in which the signal amplitude may also have an imaginary part. A complex signal may be written in *Cartesian form* using its real and imaginary parts as

$$x[n] = x_r[n] + j\,x_i[n] \tag{1.32}$$

or in *polar form* using its magnitude and phase as

$$x[n] = \big|x[n]\big|\, e^{j\measuredangle x[n]} \tag{1.33}$$

The two forms in Eqns. (1.32) and (1.33) can be related to each other through the following set of equations:

$$\big|x[n]\big| = \left[x_r^2[n] + x_i^2[n]\right]^{1/2} \tag{1.34}$$

$$\measuredangle x[n] = \tan^{-1}\left[\frac{x_i[n]}{x_r[n]}\right] \tag{1.35}$$

$$x_r[n] = \big|x[n]\big| \cos\left(\measuredangle x[n]\right) \tag{1.36}$$

$$x_i[n] = \big|x[n]\big| \sin\left(\measuredangle x[n]\right) \tag{1.37}$$

In deriving Eqns. (1.36) and (1.37) we have used Euler's formula.[1] Even though we will mostly focus on the use of real-valued signals in our discussion in the rest of this text, some of the transforms such as the DFT and the FFT will lead to complex valued sequences for which conversions in Eqns. 1.32 to 1.37 will prove useful.

1.5.2 Periodic vs. non-periodic signals

A discrete-time signal is said to be periodic if it satisfies

$$x[n] = x[n+N] \tag{1.38}$$

for all values of the integer index n and for a specific value of $N \neq 0$. The parameter N is referred to as the *period* of the signal. An example of a periodic discrete-time signal is shown in Fig. 1.31.

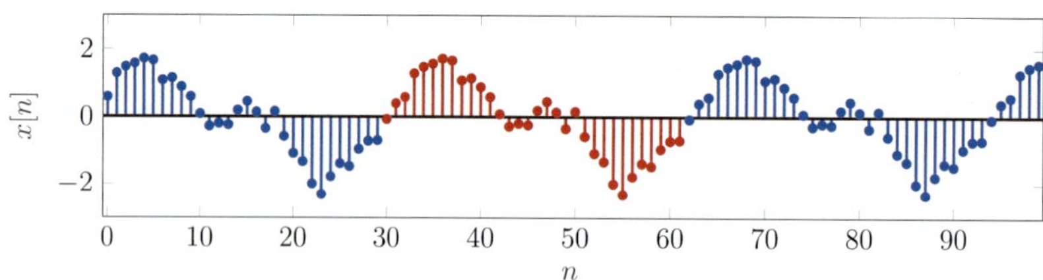

Figure 1.31 – Example of a discrete-time signal that is periodic.

A discrete-time signal that is periodic with a period of N samples is also periodic with periods of $2N, 3N, \ldots, kN$ for any positive integer k. Applying the periodicity definition of Eqn. (1.38) to $x[n+N]$ instead of $x[n]$ yields

$$x[n+N] = x[n+2N] \tag{1.39}$$

Substituting Eqn. (1.39) into Eqn. (1.38) we obtain

$$x[n] = x[n+2N] \tag{1.40}$$

and, through repeated use of this process, we can show that

$$x[n] = x[n+kN] \tag{1.41}$$

where k is any integer. Thus we conclude that a signal that is periodic with a period of N samples is also periodic with a period of kN samples where k is any integer. The smallest positive value of N that satisfies Eqn. (1.38) is called the *fundamental period.* To avoid ambiguity, when we refer to the period of a signal, fundamental period will be implied unless it is specifically stated otherwise. The *normalized fundamental frequency* of a discrete-time periodic signal is the reciprocal of its fundamental period, i.e.,

$$F_0 = \frac{1}{N} \tag{1.42}$$

1.5.3 Periodicity of discrete-time sinusoidal signals

The general form of a discrete-time sinusoidal signal $x[n]$ was given by Eqn. (1.26). For $x[n]$ to be periodic, it needs to satisfy the periodicity condition given by Eqn. (1.38). Specifically we need

$$A\cos(2\pi F_0 n + \theta) = A\cos(2\pi F_0 [n+N] + \theta) \tag{1.43}$$

[1] $e^{\pm ja} = \cos(a) \pm j\sin(a)$.

For Eqn. (1.43) to hold, the arguments of the cosine functions must differ by an integer multiple of 2π. This requirement results in

$$2\pi F_0 N = 2\pi k$$

and consequently

$$N = \frac{k}{F_0} \tag{1.44}$$

for the period N. Since we are dealing with a discrete-time signal, there is an added requirement that the period N obtained from Eqn. (1.44) must be an integer value. Thus, the discrete-time sinusoidal signal defined by Eqn. (1.26) is periodic provided that Eqn. (1.44) yields an integer value for N. The fundamental period of the sinusoidal signal is then obtained by using the smallest integer value of k, if any, that results in N being an integer.

It should be obvious from the foregoing discussion that, contrary to a continuous-time sinusoidal signal always being periodic, a discrete-time sinusoidal signal may or may not be periodic. The signal will not be periodic, for example, if the normalized frequency F_0 is an irrational number so that no value of k produces an integer N in Eqn. (1.44).

Software resource: See MATLAB Exercise 1.5.

Example 1.12: Periodicity of a discrete-time sinusoidal signal

Check the periodicity of the following discrete-time signals:

a. $x[n] = \cos(0.2n)$

b. $x[n] = \cos(0.2\pi n + \pi/5)$

c. $x[n] = \cos(0.3\pi n - \pi/10)$

Solution:

a. The angular frequency of this signal is $\Omega_0 = 0.2$ radians which corresponds to a normalized frequency of

$$F_0 = \frac{\Omega_0}{2\pi} = \frac{0.2}{2\pi} = \frac{0.1}{\pi}$$

This results in a period

$$N = \frac{k}{F_0} = 10\pi k$$

Since no value of k would produce an integer value for N, the signal is not periodic.

b. In this case the angular frequency is $\Omega_0 = 0.2\pi$ radians, and the normalized frequency is $F_0 = 0.1$. The period is

$$N = \frac{k}{F_0} = \frac{k}{0.1} = 10k$$

For $k = 1$ we have $N = 10$ samples as the fundamental period. The signal $x[n]$ is shown in Fig. 1.32.

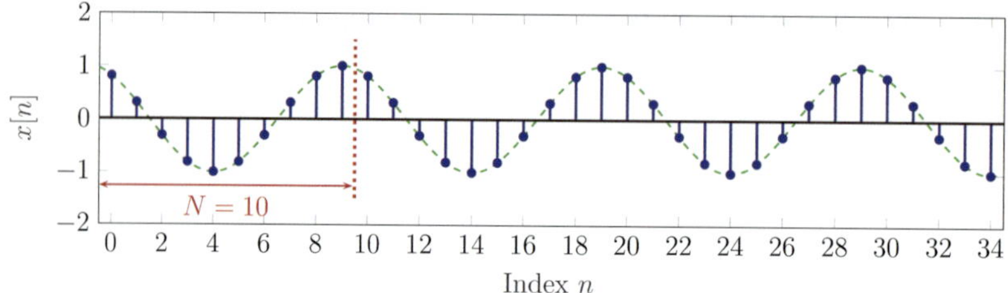

Figure 1.32 – The signal $x[n]$ for part (b) of Example 1.12.

c. For this signal the angular frequency is $\Omega_0 = 0.3\pi$ radians, and the corresponding normalized frequency is $F_0 = 0.15$. The period is

$$N = \frac{k}{F_0} = \frac{k}{0.15}$$

The smallest positive integer k that would result in an integer value for the period N is $k = 3$. Therefore, the fundamental period is $N = 3/0.15 = 20$ samples. The signal $x[n]$ is shown in Fig. 1.33.

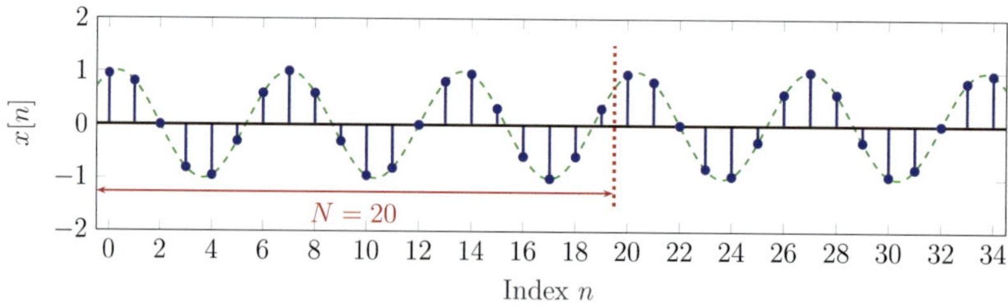

Figure 1.33 – The signal $x[n]$ for part (c) of Example 1.12.

It is interesting to observe from Fig. 1.33 that the period of $N = 20$ samples corresponds to three full cycles of the continuous-time sinusoidal signal from which $x[n]$ may have been derived (see the outline shown in the figure). This is due to the fact that $k = 3$ and, based on Eqn. (1.44), the argument of the cosine function is advanced by

$$2\pi F_0 N = 2\pi k = 6\pi \quad \text{radians}$$

after one period of 20 samples.

Software resources: exdt_1_12a.m , exdt_1_12b.m , exdt_1_12c.m

Example 1.13: Periodicity of a multi-tone discrete-time sinusoidal signal

Comment on the periodicity of the two-tone discrete-time signal

$$x[n] = 2\cos(0.4\pi n) + 1.5\sin(0.48\pi n)$$

Solution: The signal is in the form

$$x[n] = x_1[n] + x_2[n]$$

with

$$x_1[n] = 2\cos(\Omega_1 n)\,, \quad \Omega_1 = 0.4\pi \text{ rad}$$

and

$$x_2[n] = 1.5\sin(\Omega_2 n)\,, \quad \Omega_2 = 0.48\pi \text{ rad}$$

Corresponding normalized frequencies are $F_1 = 0.2$ and $F_2 = 0.24$. The period of each component can be found as follows:

$$\text{For } x_1[n]: \quad N_1 = \frac{k_1}{F_1} = \frac{k_1}{0.2}\,, \qquad k_1 = 1\,, \quad N_1 = 5$$

$$\text{For } x_2[n]: \quad N_2 = \frac{k_2}{F_2} = \frac{k_2}{0.24}\,, \qquad k_2 = 6\,, \quad N_2 = 25$$

Thus the fundamental period for $x_1[n]$ is $N_1 = 5$ samples, and the fundamental period for $x_2[n]$ is $N_2 = 25$ samples. The period of the total signal $x[n]$ is $N = 25$ samples. Within the period of $N = 25$ samples, the first component $x_1[n]$ completes 5 cycles since $N = 5N_1$. The second component completes 6 cycles since $k_2 = 6$ to get an integer value for N_2. This is illustrated in Figs. 1.34 and 1.35.

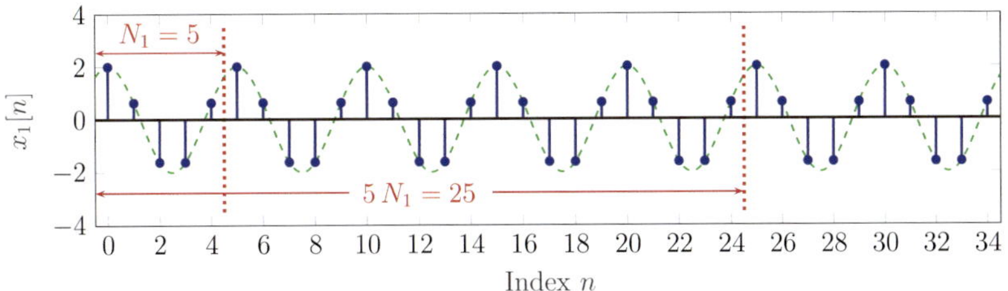

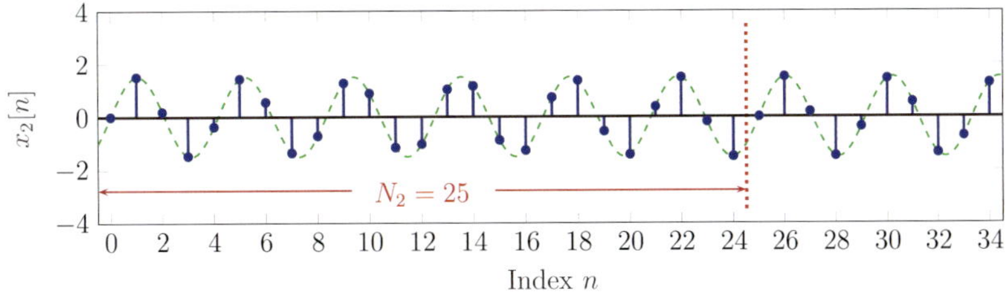

Figure 1.34 – The signals used in Example 1.13: **(a)** $x_1[n]$ and **(b)** $x_2[n]$.

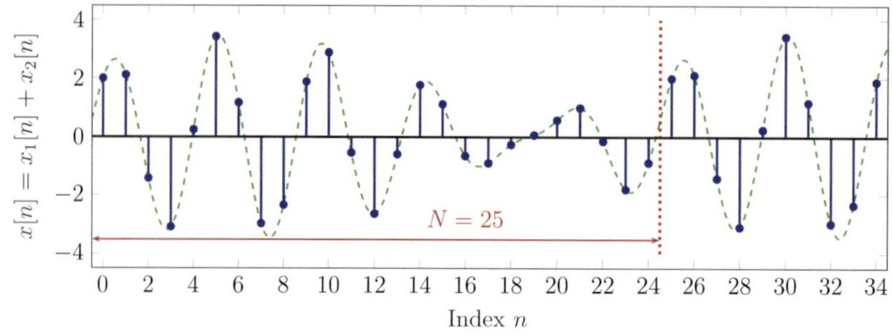

Figure 1.35 – The signal $x[n] = x_1[n] + x_2[n]$ for Example 1.13.

Software resource: `exdt_1_13.m`

1.5.4 Deterministic vs. random signals

Deterministic signals are signals that can be described completely in analytical form in the time domain. Random signals are signals that cannot be modeled analytically. They can be analyzed in terms of their statistical properties. The study of random signals is beyond the scope of this text, and the reader is referred to one of the many excellent texts available on the subject [5, 6, 8].

1.6 Energy and Power Definitions

1.6.1 Energy of a signal

The energy of a real-valued discrete-time signal is

$$E_x = \sum_{n=-\infty}^{\infty} x^2[n] \tag{1.45}$$

provided that it can be computed. If the signal under consideration is complex-valued, its energy is given by

$$E_x = \sum_{n=-\infty}^{\infty} \left|x[n]\right|^2 \tag{1.46}$$

where $\left|x[n]\right|$ represents the norm of a complex signal defined by Eqn. 1.34. Substituting this definition into Eqn. 1.46 we get

$$E_x = \sum_{n=-\infty}^{\infty} \left[x_r^2[n] + x_i^2[n]\right] \tag{1.47}$$

For a real signal $x[n]$, the imaginary part is $x_i[n] = 0$ and Eqn. 1.47 reduces to Eqn. 1.45. Therefore, we can use Eqn. 1.46 as the definition of the energy of a discrete-time signal for all cases.

Energy of a signal:

$$E_x = \sum_{n=-\infty}^{\infty} \left|x[n]\right|^2 , \quad \text{if it can be computed.}$$

Example 1.14: Energy of an exponential signal

Find the energy of the signal

$$x[n] = a^n \, u[n]$$

where the parameter a is a real-valued constant. Determine the range of values for a for which the signal energy can be computed.

Solution: Using Eqn. (1.45) the signal energy is

$$E_x = \sum_{n=-\infty}^{\infty} \left(a^n \, u[n]\right)^2 = \sum_{n=0}^{\infty} \left(a^2\right)^n$$

The term inside the summation is a geometric series which can be put in closed form (see Section C.1 of Appendix C) as

$$E_x = \frac{1}{1-a^2} \tag{1.48}$$

provided that

$$\left|a^2\right| < 1 \qquad \Longrightarrow \qquad |a| < 1$$

 **Software resource:** exdt_1_14.m

Example 1.15: Energy of an exponentially decaying sinusoidal signal

Find the energy of the signal

$$x[n] = a^n \cos\left(\Omega_0 n\right) \, u[n] \, , \qquad -1 < a < 1$$

where the parameter a is a real-valued constant.

Solution: Using Eqn. (1.45) the signal energy is

$$E_x = \sum_{n=-\infty}^{\infty} \left(a^n \cos\left(\Omega_0 n\right) \, u[n]\right)^2 = \sum_{n=0}^{\infty} \left(a^n \cos\left(\Omega_0 n\right)\right)^2 \tag{1.49}$$

The term inside the summation in Eqn. (1.49) can be written using Euler's formula as

$$\begin{aligned}\left(a^n \cos\left(\Omega_0 n\right)\right)^2 &= \left(\frac{1}{2}\, a^n \, e^{j\Omega_0 n} + \frac{1}{2}\, a^n \, e^{-j\Omega_0 n}\right)^2 \\ &= \frac{1}{2}\, a^{2n} + \frac{1}{4}\left(a^2 \, e^{j2\Omega_0}\right)^n + \frac{1}{4}\left(a^2 \, e^{-j2\Omega_0}\right)^n\end{aligned}$$

The signal energy is

$$E_x = \frac{1}{2}\sum_{n=0}^{\infty} a^{2n} + \frac{1}{4}\sum_{n=0}^{\infty}\left(a^2\,e^{j2\Omega_0}\right)^n + \frac{1}{4}\sum_{n=0}^{\infty}\left(a^2\,e^{-j2\Omega_0}\right)^n$$

$$= \frac{1/2}{1-a^2} + \frac{1/4}{1-a^2\,e^{j2\Omega_0}} + \frac{1/4}{1-a^2\,e^{-j2\Omega_0}}$$

which can be simplified to

$$E_x = \frac{1}{2\left(1-a^2\right)} + \frac{1-a^2\cos\left(2\Omega_0\right)}{2\left(1+a^4-2a^2\cos\left(2\Omega_0\right)\right)}$$

A quick way of checking the consistency of this result with that of Example 1.14 would be to set $\Omega_0 = 0$ in which case the signal $x[n]$ simplifies to the one used in the previous example. The expression for E_x should also simplify to the result found in Example 1.14.

 **Software resource:** `exdt_1_15.m`

1.6.2 Time averaging operator

In preparation for defining the average power in a signal, we will first define the time average of a signal. We will use the operator $\langle\ldots\rangle$ to indicate time average. If the signal $x[n]$ is periodic with period N, its time average can be computed as

$$\left\langle x[n]\right\rangle = \frac{1}{N}\sum_{n=0}^{N-1} x[n] \tag{1.50}$$

For a signal $x[n]$ that is non-periodic, the definition of time average in Eqn. (1.50) can be generalized with the use of the limit operator as

$$\left\langle x[n]\right\rangle = \lim_{M\to\infty}\left[\frac{1}{2M+1}\sum_{n=-M}^{M} x[n]\right] \tag{1.51}$$

One way to make sense of Eqn. (1.51) is to view the non-periodic signal as though it is periodic with an infinitely large period so that we never get to see the signal pattern repeat itself.

Time average of a signal:

For a periodic signal: $\left\langle x[n]\right\rangle = \frac{1}{N}\sum_{n=0}^{N-1} x[n]$

For a non-periodic signal: $\left\langle x[n]\right\rangle = \lim_{M\to\infty}\left[\frac{1}{2M+1}\sum_{n=-M}^{M} x[n]\right]$

1.6.3 Average power of a signal

The average power of a real-valued discrete-time signal is computed as

$$P_x = \left\langle x^2[n]\right\rangle \tag{1.52}$$

This definition works for both periodic and non-periodic signals. For a periodic signal, Eqn. (1.50) can be used with Eqn. (1.52) to yield

$$P_x = \frac{1}{N}\sum_{n=0}^{N-1} x^2[n] \tag{1.53}$$

For a non-periodic signal, Eqn. (1.51) can be substituted into Eqn. (1.52) to yield

$$P_x = \lim_{M\to\infty}\left[\frac{1}{2M+1}\sum_{n=-M}^{M} x^2[n]\right] \tag{1.54}$$

Eqns. (1.53) and (1.54) apply to signals that are real-valued. If the signal under consideration is complex, then Eqns. (1.53) and (1.54) can be generalized by using the squared norm of the signal $x[n]$, i.e.,

$$P_x = \left\langle \left|x[n]\right|^2 \right\rangle \tag{1.55}$$

Thus, the power of a periodic complex signal is

$$P_x = \frac{1}{N}\sum_{n=0}^{N-1} \left|x[n]\right|^2 \tag{1.56}$$

and the power of a non-periodic complex signal is

$$P_x = \lim_{M\to\infty}\left[\frac{1}{2M+1}\sum_{n=-M}^{M} \left|x[n]\right|^2\right] \tag{1.57}$$

Note that, for a real-valued signal, the squared norm of the signal is simply the signal squared, and we can write

$$\left|x[n]\right|^2 = x^2[n]$$

Consequently, Eqn. (1.55) can be used for both real- and complex-valued signals.

Average power of a signal:

$$P_x = \left\langle \left|x[n]\right|^2 \right\rangle$$

Example 1.16: Average power of a unit step signal

Find the average power of the signal

$$x[n] = A\,u[n]$$

where the parameter A is a real-valued constant.

Solution: The signal $x[n]$ is non-periodic, so we will use Eqn. (1.57).

$$\sum_{n=-M}^{M} \left|x[n]\right|^2 = \sum_{n=-M}^{M} \left|A\,u[n]\right|^2 = \sum_{n=-M}^{-1} 0 + \sum_{n=0}^{M} A^2 = (M+1)\,A^2$$

The average power of $x[n]$ is found as

$$P_x = \lim_{M\to\infty}\left[\frac{(M+1)\,A^2}{2M+1}\right] = \frac{A^2}{2}$$

Example 1.17: Average power of a sinusoidal signal

Find the average power of the signal

$$x[n] = A\cos(\Omega_0 n)$$

where the parameter A is a real-valued constant, and the angular frequency Ω_0 is such that the resulting sinusoidal signal is periodic (see Example 1.12).

Solution: Using Eqn. (1.53) the average power of the signal is

$$P_x = \frac{1}{N}\sum_{n=0}^{N-1}\left(A\cos(\Omega_0 n)\right)^2$$

where we have used N as the period of the sinusoidal signal. Using the trigonometric identity

$$\cos^2(a) = \frac{1}{2}\left[1+\cos(2a)\right]$$

the average power can be written as

$$\begin{aligned}
P_x &= \frac{1}{N}\sum_{n=0}^{N-1}\left(\frac{A^2}{2}+\frac{A^2}{2}\cos(2\Omega_0 n)\right) \\
&= \frac{A^2}{2}+\frac{A^2}{2N}\sum_{n=0}^{N-1}\cos(2\Omega_0 n) \\
&= \frac{A^2}{2}
\end{aligned}$$

where we have recognized that the term $\cos(2\Omega_0 n)$ has two full periods in the index range $n=0,\ldots,N-1$, and therefore sums up to zero.

1.6.4 Energy signals vs. power signals

In Examples 1.14 through 1.17 we have observed that the concept of signal energy is useful for some signals and not for others. Same can be said for average signal power. Based on our observations, we can classify signals encountered in practice into two categories as *energy signals* and *power signals*. Defining characteristics of the two types of signals are as follows:

- Energy signals are those that have finite energy and zero average power, i.e., $E_x < \infty$, and $P_x = 0$.
- Power signals are those that have finite average power and infinite energy, i.e., $E_x \to \infty$, and $P_x < \infty$.

1.7 Symmetry Properties

Some signals have certain symmetry properties that could be utilized in a variety of ways in the analysis. For example, we will see in the study of Fourier analysis that, signals with certain symmetry properties have simpler transforms. More importantly, a signal that may not have any symmetry properties can still be written as a linear combination of signals with certain symmetry properties, creating opportunities for simplified analysis and better understanding of signal and transform relationships.

1.7.1 Even and odd symmetry

A real-valued discrete-time signal is said to be *even symmetric* if it satisfies

$$x[-n] = x[n] \tag{1.58}$$

for all integer values of the sample index n. In contrast, a discrete-time signal is said to be *odd symmetric* if it satisfies

$$x[-n] = -x[n] \tag{1.59}$$

for all n. A signal $x[n]$ that is even remains unchanged when it is reversed in time. On the other hand, for a signal $x[n]$ with odd symmetry, time reversing it causes the signal to be negated. Examples of discrete-time signals with even and odd symmetry are shown in Fig. 1.36.

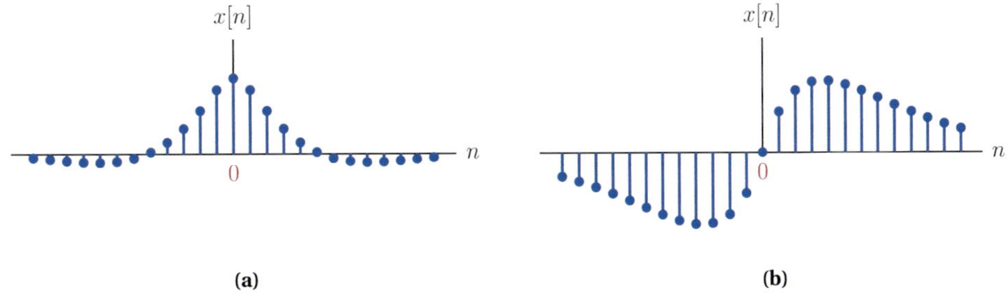

Figure 1.36 – **(a)** Discrete-time signal with even symmetry and **(b)** discrete-time signal with odd symmetry.

1.7.2 Decomposition into even and odd components

As in the case of continuous-time signals, any real-valued discrete-time signal $x[n]$ can be written as the sum of two signals, one with even symmetry and one with odd symmetry.

The signal $x[n]$ does not have to have any symmetry properties for this to work. Consider the following representation of the signal $x[n]$:

$$x[n] = x_e[n] + x_o[n] \tag{1.60}$$

The signal $x_e[n]$ is the even component defined as

$$x_e[n] = \frac{x[n] + x[-n]}{2} \tag{1.61}$$

and $x_o[n]$ is the odd component defined as

$$x_o[n] = \frac{x[n] - x[-n]}{2} \tag{1.62}$$

It can easily be verified that $x_e[-n] = x_e[n]$ and $x_o[-n] = -x_o[n]$. Furthermore, the signals $x_e[n]$ and $x_o[n]$ always add up to $x[n]$ owing to the way they are defined.

Example 1.18: Even and odd components of a discrete-time pulse

Find even and odd components of the signal

$$x[n] = \begin{cases} 1\,, & n = 0, \ldots, N-1 \\ 0\,, & \text{otherwise} \end{cases}$$

Solution: The first step is to write $x[-n]$:

$$x[-n] = \begin{cases} 1\,, & n = -N+1, \ldots, 0 \\ 0\,, & \text{otherwise} \end{cases}$$

The even component, computed using Eqn. (1.61), is

$$x_e[n] = \begin{cases} 0.5\,, & n = -N+1, \ldots, -1 \\ 1\,, & n = 0 \\ 0.5\,, & n = 1, \ldots, N-1 \\ 0\,, & \text{otherwise} \end{cases}$$

which can be put in closed form as

$$x_e[n] = 0.5\,\delta[n] + 0.5\,\big(u[n+N-1] - u[n-N]\big)$$

Similarly, the odd component is found using Eqn. (1.62) as

$$x_o[n] = \begin{cases} -0.5\,, & n = -N+1, \ldots, -1 \\ 0.5\,, & n = 1, \ldots, N-1 \\ 0\,, & \text{otherwise} \end{cases}$$

Signals involved in the computation of even and odd components of $x[n]$ are shown in Fig. 1.37 for $N = 7$.

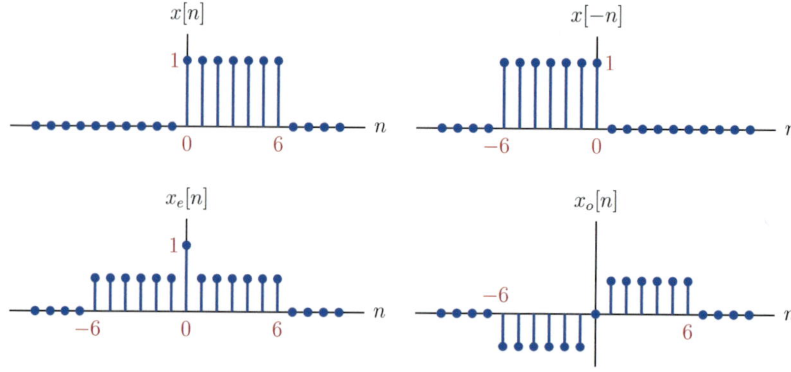

Figure 1.37 – Signals for Example 1.18.

Software resource: exdt_1_18.m

Example 1.19: Even and odd components of an exponentially damped sinusoid

Find even and odd components of the signal

$$x[n] = a^n \cos(\Omega_0 n)\, u[n]$$

where the parameter a is a real-valued constant, and $|a| < 1$.

Solution: Using Eqn. (1.61) the even component is

$$x_e[n] = \tfrac{1}{2}\, a^n \cos(\Omega_0 n)\, u[n] + \tfrac{1}{2}\, a^{-n} \cos(-\Omega_0 n)\, u[-n]$$

Taking advantage of the even symmetry of the cosine function, $x_e[n]$ can be written as

$$\begin{aligned} x_e[n] &= \tfrac{1}{2} \cos(\Omega_0 n) \left(a^n\, u[n] + a^{-n}\, u[-n]\right) \\ &= \begin{cases} 1\,, & n = 0 \\ \tfrac{1}{2} \cos(\Omega_0 n)\, a^{|n|}\,, & \text{otherwise} \end{cases} \end{aligned}$$

This result can be put in a more compact form as

$$x_e[n] = \tfrac{1}{2}\,\delta[n] + \tfrac{1}{2} \cos(\Omega_0 n)\, a^{|n|}$$

The odd component of $x[n]$ is

$$x_o[n] = \tfrac{1}{2}\, a^n \cos(\Omega_0 n)\, u[n] - \tfrac{1}{2}\, a^{-n} \cos(-\Omega_0 n)\, u[-n]$$

Again using the even symmetry of the cosine function, we have

$$\begin{aligned} x_o[n] &= \tfrac{1}{2} \cos(\Omega_0 n) \left(a^n\, u[n] - a^{-n}\, u[-n]\right) \\ &= \begin{cases} \tfrac{1}{2} \cos(\Omega_0 n)\, a^n\,, & n > 0 \\ -\tfrac{1}{2} \cos(\Omega_0 n)\, a^{-n}\,, & n < 0 \\ 0\,, & n = 0 \end{cases} \end{aligned}$$

Software resource: exdt_1_19.m

1.7.3 Symmetry properties for complex signals

Even and odd symmetry definitions given by Eqns. (1.58) and (1.59) for a real-valued signal $x[n]$ can be extended to apply to complex-valued signals as well. A complex-valued signal $x[n]$ is conjugate symmetric if it satisfies

$$x[-n] = x^*[n] \tag{1.63}$$

for all n. Similarly, $x[n]$ is conjugate antisymmetric if it satisfies

$$x[-n] = -x^*[n] \tag{1.64}$$

for all n. If $x[n]$ is conjugate symmetric, time reversing it has the same effect as conjugating it. For a conjugate antisymmetric $x[n]$, time reversal is equivalent to conjugation and negation applied simultaneously.

If the signal $x[n]$ is real-valued, then its complex conjugate is equal to itself, that is, $x^*[n] = x[n]$. In this case the definition of conjugate symmetry reduces to that of even symmetry as can be seen from Eqns. (1.63) and (1.58). Similarly, conjugate antisymmetry property reduces to odd symmetry for a real-valued $x[n]$.

1.7.4 Decomposition of complex signals

Any complex signal $x[n]$ can be expressed as the sum of two signals of which one is conjugate symmetric and the other is conjugate antisymmetric. The component $x_E[n]$ defined as

$$x_E[n] = \frac{x[n] + x^*[-n]}{2} \tag{1.65}$$

is always conjugate symmetric and the component $x_O[n]$ defined as

$$x_O[n] = \frac{x[n] - x^*[-n]}{2} \tag{1.66}$$

is always conjugate antisymmetric. The relationship

$$x[n] = x_E[n] + x_O[n] \tag{1.67}$$

holds due to the way $x_E[n]$ and $x_O[n]$ are defined.

Summary of Key Points

☞ A discrete-time signal is an indexed set of amplitude values. Most discrete-time signals are obtained by sampling an analog signal at integer multiples of a fixed time increment.

☞ In a discrete-time signal the time variable is discrete, yet the amplitude of each sample is continuous. In contrast, digital signals have sample amplitude values that come from a discrete set. A binary digital signal is one in which each sample is allowed to have one of two amplitude levels. An M-ary digital signal has sample amplitudes that come from a set of M possible values.

☞ A discrete-time signal can be converted to a digital signal through the use of quantization as illustrated in Figs. 1.4 and 1.5. In general, errors caused by the quantization operation can be reduced by increasing the number of bits allocated to each sample.

☞ Arithmetic operations such as constant offset and gain, addition and multiplication of discrete-time signals are discussed in Section 1.2.1. Constant offset and gain are applied to each sample of the signal. Addition and multiplication of two discrete-time signals is accomplished by adding or multiplying the samples with the same index.

☞ Time shifting a discrete-time signal involves replacing index n with $n-k$ as given by Eqn. (1.1) and illustrated in Fig. 1.11. Depending on the sign of parameter k, the time shift may correspond to a delay or an advance.

☞ Time scaling operations are downsampling and upsampling, given by Eqns. (1.2) and (1.5), and illustrated by Figs. 1.12 through 1.15. Downsampling a signal by a factor of M amounts to taking every M-th sample and discarding the samples in between. It may or may not result in a loss of information depending on the characteristics of the signal. Upsampling by M is equivalent to taking each sample of the signal followed by $M-1$ zero-amplitude samples.

☞ Time reversal of a signal is achieved by replacing the index n with $-n$ as detailed by Eqn. (1.6) and illustrated by Fig. 1.16.

☞ The discrete-time unit impulse function is defined by

Eqn. (1.7): $$\delta[n] = \begin{cases} 1, & n = 0 \\ 0, & n \neq 0 \end{cases}$$

☞ The *sampling property* of the discrete-time unit impulse function is expressed as

Eqn. (1.9): $$x[n]\,\delta[n-n_1] = x[n_1]\,\delta[n-n_1]$$

☞ The *sifting property* for the discrete-time unit impulse function is expressed as

Eqn. (1.11): $$\sum_{n=-\infty}^{\infty} x[n]\,\delta[n-n_1] = x[n_1]$$

☞ The discrete-time unit step function is defined as

Eqn. (1.13): $$u[n] = \begin{cases} 1, & n \geq 0 \\ 0, & n < 0 \end{cases}$$

It can be constructed from unit impulse signals through a *running sum* in the form

Eqn. (1.16): $$u[n] = \sum_{k=-\infty}^{n} \delta[k]$$

Conversely, a discrete-time unit impulse signal can be expressed as the *first difference* of the discrete-time unit step signal:

Eqn. (1.15): $$\delta[n] = u[n] - u[n-1]$$

☞ The discrete-time unit ramp function is defined as

Eqn. (1.18): $$r[n] = \begin{cases} n, & n \geq 0 \\ 0, & n < 0 \end{cases} = n\,u[n]$$

☞ A discrete-time sinusoidal signal is in the general form

Eqn. (1.21): $$x[n] = A\cos(\Omega_0 n + \theta)$$

The parameter Ω_0 is the *angular frequency* in radians and is expressed as

Eqn. (1.22): $$\Omega_0 = 2\pi F_0$$

where F_0 is a dimensionless quantity referred to as the *normalized frequency*.

☞ A discrete-time signal can be expressed as a sum of scaled and time-shifted unit impulse signals, a representation form referred to as impulse decomposition.

Eqn. (1.30): $$x_k[n] = x[k]\,\delta[n-k] = \begin{cases} x[k], & n = k \\ 0, & n \neq k \end{cases}$$

This forms the basis for the discrete-time convolution operation to be explored in Chapter 2.

☞ Complex discrete-time signals can be expressed either in Cartesian form as

Eqn. (1.32): $$x[n] = x_r[n] + j\, x_i[n]$$

or in polar form as

Eqn. (1.33): $$x[n] = \big|x[n]\big|\, e^{j\measuredangle x[n]}$$

☞ A discrete-time signal is said to be periodic if it satisfies $x[n] = x[n+N]$ for all values of n and for a specific value of $N \neq 0$. A discrete-time signal that is periodic with a period of N samples is also periodic with kN samples for any positive integer k. The smallest positive value of N that satisfies the periodicity condition is called the *fundamental period.* Its reciprocal is the fundamental frequency.

☞ A discrete-time sinusoidal signal is periodic only if an integer value for the period N can be found such that $N = k/F_0$ where F_0 is the normalized frequency of the signal and k is any positive integer.

☞ The energy of a discrete-time signal is

Eqn. (1.46): $$E_x = \sum_{n=-\infty}^{\infty} \big|x[n]\big|^2$$

provided that it can be computed. This definition works for both real- and complex-valued signals.

☞ Average power of a discrete-time signal is the time average of its squared magnitude. This definition also works for both real- and complex-valued signals. How the time average is computed depends on the type of the signal. For periodic signals, time average is computed as

Eqn. (1.50): $$\big\langle x[n]\big\rangle = \frac{1}{N}\sum_{n=0}^{N-1} x[n]$$

For non-periodic signals, time average is

Eqn. (1.51): $$\big\langle x[n]\big\rangle = \lim_{M\to\infty}\left[\frac{1}{2M+1}\sum_{n=-M}^{M} x[n]\right]$$

☞ Signals can be classified as *energy signals* and *power signals.* Energy signals are those that have finite energy and zero average power, i.e., $E_x < \infty$ and $P_x = 0$. Power signals are those that have finite average power and infinite energy, i.e., $E_x \to \infty$, and $P_x < \infty$.

☞ A real-valued signal $x[n]$ has even symmetry if $x[-n] = x[n]$ for all n.

☞ A real-valued signal $x[n]$ has odd symmetry if $x[-n] = -x[n]$ for all n.

☞ A complex-valued signal $x[n]$ is conjugate symmetric if $x[-n] = x^*[n]$ for all n.

☞ A complex-valued signal $x[n]$ is conjugate antisymmetric if $x[-n] = -x^*[n]$ for all n.

☞ Any real-valued discrete-time signal $x[n]$ can be written as the sum of two signals, one with even symmetry and one with odd symmetry. The even component of the signal is

Eqn. 1.61: $$x_e[n] = \frac{x[n] + x[-n]}{2}$$

and the odd component is

Eqn. 1.62: $$x_o[n] = \frac{x[n] - x[-n]}{2}$$

☞ Any complex signal $x[n]$ can be expressed as the sum of two signals of which one is conjugate symmetric and the other is conjugate antisymmetric. Conjugate symmetric component is

Eqn. 1.65: $$x_E[n] = \frac{x[n] + x^*[-n]}{2}$$

and the conjugate antisymmetric component is

Eqn. 1.66: $$x_O[n] = \frac{x[n] - x^*[-n]}{2}$$

Further Reading

[1] Ary L. Goldberger et al. "PhysioBank, PhysioToolkit, and PhysioNet". In: *Circulation* 101.23 (June 2000). [Online; accessed 2023-06-11].

[2] R.P. Kanwal. *Generalized Functions: Theory and Applications.* Birkhèauser, 2004.

[3] G.B. Moody and R.G. Mark. "The impact of the MIT-BIH Arrhythmia Database". In: *IEEE Engineering in Medicine and Biology Magazine* 20.3 (2001). [Online; accessed 2023-06-11], pp. 45–50.

[4] Peter V. O'Neil. *Advanced Engineering Mathematics.* Cengage Learning, 2017.

[5] A. Papoulis and S. Unnikrishna Pillai. *Probability, Random Variables, and Stochastic Processes.* 4th. McGraw-Hill, 2002.

[6] P.Z. Peebles. *Probability, Random Variables, and Random Signal Principles.* McGraw-Hill, 2001.

[7] A. Spanias, T. Painter, and V. Atti. *Audio Signal Processing and Coding.* Wiley, 2006.

[8] Henry Stark and John W. Woods. *Probability, Statistics, and Random Processes for Engineers.* 4th. Prentice Hall, 2011.

[9] Kenneth Steiglitz. *Digital Signal Processing Primer.* Courier Dover Publications, 2020.

[10] Kenneth Arthur Stroud and Dexter J. Booth. *Engineering Mathematics.* Bloomsbury Publishing, 2020.

[11] M. Tohyama. *Sound and Signals.* Springer, 2011.

[12] U. Zölzer. *Digital Audio Signal Processing.* Wiley, 2008.

MATLAB Exercises with Solutions

MATLAB Exercise 1.1: Computing and graphing discrete-time signals

In graphing continuous-time signals with MATLAB we typically use the `plot()` function which connects dots with straight lines so that the visual effect is consistent with a signal defined at all time instants. When we work with discrete-time signals, however, we will use a *stem plot* to emphasize the discrete-time nature of the signal. Built-in function `stem()` will be utilized for this purpose.

a. Compute and graph the signal

$$x_1[n] = \{\ 1.1,\ \underset{\underset{n=5}{\uparrow}}{2.5},\ 3.7,\ 3.2,\ 2.6\ \}$$

for the range of the sample index $4 \le n \le 8$.

Solution: Type the following three lines into MATLAB command window to obtain the graph shown in Fig. 1.38.

```
>> n = [4:8];
>> x1 = [1.1, 2.5, 3.7, 3.2, 2.6];
>> stem(n,x1);
```

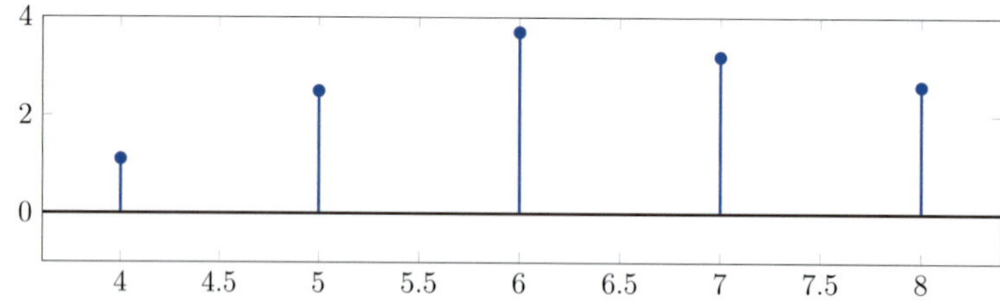

Figure 1.38 – Graph obtained in MATLAB Exercise 1.1 part (a).

b. Compute and graph the signal

$$x_2[n] = \sin(0.2n)$$

for the index range $n = 0, 1, \ldots, 99$.

Solution: This signal can be computed and graphed with the following statements:

```
>> n = [0:99];
>> x2 = sin(0.2*n);
>> stem(n,x2);
```

The result is shown in Fig. 1.39.

c. Compute and graph the signal

$$x_3[n] = \begin{cases} \sin(0.2n)\ , & n = 0, \ldots, 39 \\ 0\ , & \text{otherwise} \end{cases}$$

for the interval $n = -20, \ldots, 59$.

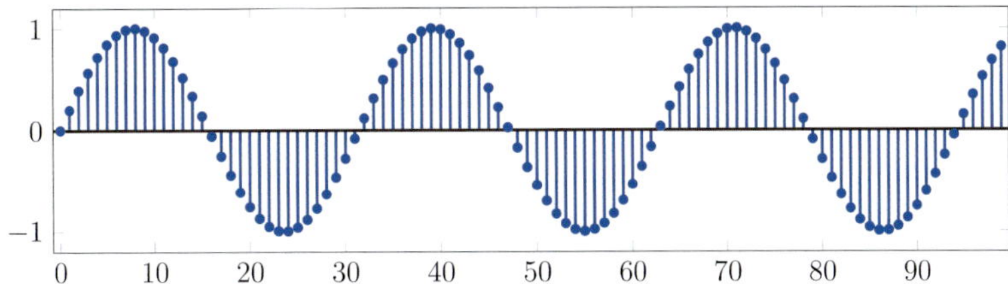

Figure 1.39 – Graph obtained in MATLAB Exercise 1.1 part (b).

Solution: The signal $x_3[n]$ to be graphed is similar to the signal of part (b), but is truncated outside the range $0 \le n \le 39$. We are asked to graph the signal for the sample index range $-20 \le n \le 59$ which will include some of the zero-amplitude samples. In order to do this efficiently, we will use the appropriate logic operators on the vector `n`. Type the following statements into the command window to compute and graph the signal $x_3[n]$.

```
>> n = [-20:59];
>> w = (n>=0)&(n<=39);
>> x3 = sin(0.2*n).*w;
>> stem(n,x3);
```

The result is shown in Fig. 1.40. Some observations are in order at this point:

- Line 2 of the listing above tests the index vector `n` to check if its elements are in the range $0 \le n \le 39$.
- The result of this test is a *logical vector* `w` that has the same number of elements as the vector `n`.
- The vector `w` has 1 (true) at index positions where the condition is satisfied, and 0 (false) at index positions where it is not.
- Element by element multiplication of the sinusoidal signal with the vector `w` achieves the truncation effect needed.

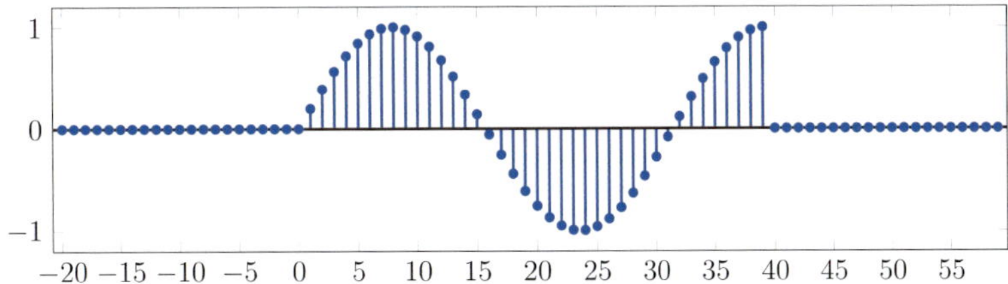

Figure 1.40 – Graph obtained in MATLAB Exercise 1.1 part (c).

Common mistake: It is important to use the *element-by-element multiplication* operator "`.*`" and not the *vector multiplication* operator "$*$" in line 3 of the listing above. The expression for computing the vector `x3` involves two vectors of the same size, namely `sin(0.2*n)` and `w`. The operator "`.*`" creates a vector of the same size by multiplying the corresponding elements of the two vectors. The operator "$*$", on the other hand, is used for computing the *scalar product* of two vectors, and would not work in this case.

Software resources: mexdt_1_1a.m , mexdt_1_1b.m , mexdt_1_1c.m

MATLAB Exercise 1.2: Graphing a signal from its analytical description

a. Compute and graph the signal

$$x[n] = \begin{cases} 0\,, & n<0 \\ 1-(0.8)^n\,, & 0 \le n \le 7 \\ 4.96\,(0.8)^n\,, & n>7 \end{cases}$$

for the range of the sample index $-10 \le n \le 49$.

Solution: Type the following lines into MATLAB command window:

```
>> n = [-10:49];
>> w1 = (n>=0 & n<=7);
>> w2 = (n>7);
>> x = (1-(0.8).^n).*w1 + (4.96*(0.8).^n).*w2;
>> stem(n,x);
```

The resulting stem plot is shown in Fig. 1.41.

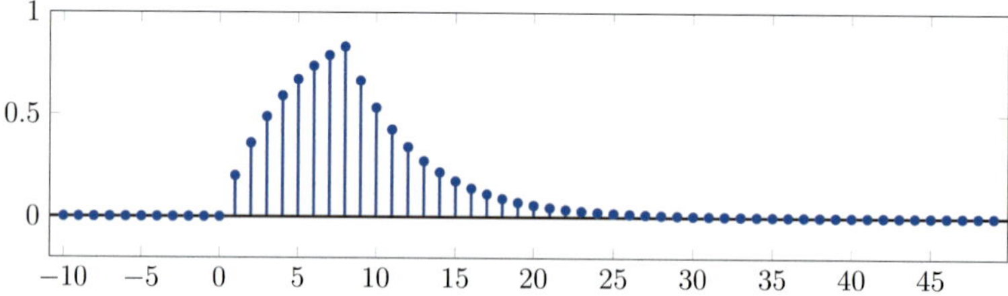

Figure 1.41 – Graph obtained in MATLAB Exercise 1.2.

Similar to what was done in part (c) of MATLAB Exercise 1.1, logical vectors w1 and and w2 are used for isolating different parts of the signal definition. It is important to remember to use the element by element operator ".^" for raising 0.8 to the n-th power when n is a vector.

b. We will repeat the exercise above, this time using an *anonymous function* to express the signal $x[n]$. The script mexdt_1_2b.m listed below accomplishes that.

```
% Script: mexdt_1_2b.m
x = @(n) (1-(0.8).^n).*(n>=0 & n<=7)+(4.96*(0.8).^n).*(n>7);
n = [-10:49];
stem(n,x(n));
```

Expressing a signal in the form of a function has the added benefit that common signal operations can be directly applied to the function. For example, try typing the following into the command line to compute and graph the signal $x[n]+0.5\,x[n-20]$:

```
>> stem(n,x(n)+0.5*x(n-20));
```

Software resources: mexdt_1_2a.m , mexdt_1_2b.m

MATLAB Exercise 1.3: Functions for basic building blocks

In this exercise we will develop simple functions for generating the basic signal building blocks such as the unit impulse, unit step, and unit ramp functions discussed in Section 1.3. A few design considerations are listed below:

- When we write our own reusable MATLAB functions we will adopt the rule of starting function names with the prefix "ss_" for "signals and systems" in order to avoid potential naming conflicts with existing MATLAB functions.
- The code for each function will be saved in a file that carries the same name with the function, and uses the extension ".m".
- In keeping with good programming practices we will try to avoid the use of looping structures if and when it is possible and practical to do so.

a. Develop a function to compute samples of the unit impulse signal. Afterward write a script that uses this function to compute and graph samples of the signal $x[n]=\delta[n]$ for $n=-5,\ldots,10$.

Solution: Our first instinct may be to write a function ss_imp() similar to the following:

```
function x = ss_imp(n)
  if (n==0)
    x = 1;    % If index is 0, set value to 1
  else
    x = 0;    % Otherwise set value to 0
  end
end
```

You may have seen similarly written functions in some textbooks. The main problem with the function ss_imp() is that it can only take a scalar value for the index n, and returns the amplitude x of the corresponding sample. If, for example, samples of a unit impulse signal are needed to be computed and graphed for $n=-5,\ldots,10$, then they must be obtained by looping through the index values as shown in the script below:

```
% Script: mexdt_1_3a.m
x = [];                   % Start with empty vector
for n=-5:10               % Loop over indices
  x = [x,ss_imp(n)];      % Append each value to the vector
end
stem([-5:10],x);
```

This is clearly inefficient and not very useful. It is also not easy to time shift or time reverse the result of the function. A better example of writing a function for the unit impulse signal is as follows:

```
function x = ss_dimpulse(n)
  x = 1*(n==0);  % Eqn. (1.7)
end
```

The advantage of this implementation is that it can take a vector of index values and return a vector of signal amplitudes. Now the problem of computing and graphing $x[n]$ can be solved with the following much simpler script.

```
% Script: mexdt_1_3b.m
n = [-5:10];           % Vector of sample index values
x = ss_dimpulse(n);  % Compute impulse signal amplitudes
stem(n,x);             % Graph the result
```

b. Use the function `ss_dimpulse()` to compute and graph the signal

$$x[n] = \delta[n] + 0.5\delta[n-3] - 0.8\delta[n-5]$$

Solution: The function `ss_dimpulse()` allows scaling and shifting operations to be coded in a natural manner as shown in the script below.

```
% Script: mexdt_1_3c.m
n = [-5:10];
x = ss_dimpulse(n)+0.5*ss_dimpulse(n-3)-0.8*ss_dimpulse(n-5);
stem(n,x);
```

c. Develop MATLAB functions to compute samples of unit step and unit ramp signals for a specified set of indices.

Solution: Functions `ss_dstep()` and `ss_dramp()` can be written using using the same approach that was employed for the function `ss_dimpulse()`.

```
function x = ss_dstep(n)
  x = 1*(n>=0);  % Eqn. (1.13)
end
```

```
function x = ss_dramp(n)
  x = n.*(n>=0);  % Eqn. (1.18)
end
```

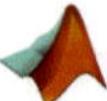

Software resources: mexdt_1_3a.m , mexdt_1_3b.m , mexdt_1_3c.m , ss_dimpulse.m , ss_dstep.m , ss_dramp.m

MATLAB Exercise 1.4: Testing the functions for basic building blocks

a. Develop and test a function `x = ss_dpulse(n,M)` that generates a unit pulse signal with length M that starts at $n = 0$.

Solution: The unit pulse signal described above can be written in terms of unit step signals as

$$p_M[n] = \begin{cases} 1, & n = 0, \ldots, M-1 \\ 0, & \text{otherwise} \end{cases}$$

$$= u[n] - u[n-M] \tag{1.68}$$

We can implement the function `ss_dpulse()` by making use of the function `ss_dstep()` that was developed in MATLAB Exercise 1.3:

```
function x = ss_dpulse(n,M)
  x = ss_dstep(n)-ss_dstep(n-M);  % Eqn. (1.68)
end
```

The script `mexdt_1_4a.m` listed below uses the new function `ss_dpulse()` to compute and display a unit pulse signal of width 10 for the index range $n = -10, \ldots, 30$.

```
% Script: mexdt_1_4a.m
n = [-10:30];
x = ss_dpulse(n,10);
stem(n,x);
```

For another example of using the function `ss_dpulse()`, let the signal $g[n]$ be defined as

$$g[n] = \begin{cases} 1, & n = 0, \ldots, 9 \\ -0.8, & n = 10, \ldots, 24 \\ 0, & \text{otherwise} \end{cases}$$

$$= p_{10}[n] - 0.8\, p_{15}[n-10]$$

The signal $g[n]$ can be computed and graphed using the script `mexdt_1_4b.m` listed below.

```
% Script: mexdt_1_4b.m
n = [-10:30];
x = ss_dpulse(n,10)-0.8*ss_dpulse(n-10,15);
stem(n,x);
```

b. Develop and test a function `x = ss_dtriangle(n,M)` to generate a symmetric triangle signal with unit amplitude peak at the origin, and two corners at $n = \pm M$.

Solution: The triangle described can be written in terms of scaled and shifted unit ramp signals as

$$q_M[n] = \begin{cases} (n+M)/M, & n = -M, \ldots, -1 \\ -(n-M)/M, & n = 0, \ldots, M \\ 0, & \text{otherwise} \end{cases}$$

$$= \frac{1}{M}\left(r[n+M] - 2r[n] + r[n-M]\right) \tag{1.69}$$

Following is an implementation of the function `ss_dtriangle()` using the function `ss_dramp()`.

```
function x = ss_dtriangle(n,M)
  x = 1/M*(ss_dramp(n+M)-2*ss_dramp(n)+ss_dramp(n-M));  % Eqn. (1.69)
end
```

The script `mexdt_1_4c.m` listed below uses the function `ss_dtriangle()` to compute and display the unit triangle with its corners at $n = \pm 10$, that is $x[n] = q_{10}[n]$.

```
% Script: mexdt_1_4c.m
n = [-20:20];
x = ss_dtriangle(n,10);
stem(n,x);
```

Software resources: mexdt_1_4a.m , mexdt_1_4b.m , mexdt_1_4c.m , ss_dpulse.m , ss_dtriangle.m

MATLAB Exercise 1.5: Periodic extension of a discrete-time signal

Sometimes a periodic discrete-time signal is specified using samples of just one period. Let $x[n]$ be a length-N signal with samples in the interval $n = 0, \ldots, N-1$. Let us define the signal $\tilde{x}[n]$ as the periodic extension of $x[n]$ so that

$$\tilde{x}[n] = \sum_{m=-\infty}^{\infty} x[n+mN]$$

In this exercise we will develop a MATLAB function to periodically extend a discrete-time signal. The function `ss_dper()` given below takes a vector `x` that holds one period (N samples) of $x[n]$. The second argument `n` is a vector that holds the indices at which the periodic extension signal should be evaluated. The samples of the periodic signal $\tilde{x}[n]$ are returned in the vector `xtilde`.

```
function xtilde = ss_dper(x,n)
  Nper = length(x);     % Period of the signal
  nmod = mod(n,Nper);   % Modulo indexing
  nn = nmod+1;          % MATLAB indices start with 1
  xtilde = x(nn);       % Return periodic signal
end
```

Consider a length-5 signal $x[n]$ given by

$$x[n] = n\,, \quad n = 0, 1, 2, 3, 4$$

The periodic extension $\tilde{x}[n]$ can be computed and graphed for $n = -15, \ldots, 15$ with the following statements:

```
>> x = [0,1,2,3,4];
>> n = [-15:15];
>> xtilde = ss_dper(x,n);
>> stem(n,xtilde);
```

What if we need to compute and graph a time reversed version $\tilde{x}[-n]$ in the same interval? That can also be accomplished easily using the following statements:

```
>> xtilde = ss_dper(x,-n);
>> stem(n,xtilde);
```

Software resources: mexdt_1_5.m , ss_dper.m

MATLAB Exercise 1.6: Working with signals in tabular form

In this exercise we will explore methods of working with signals expressed in tabular form. Consider the problem encountered in Example 1.7 earlier in this chapter. Given the signal $x[n]$, repeated here for convenience, we were asked to find the signal $g[n] = x[n] + x[-n]$.

$$x[n] = \{ \ 0.1, 0.2, 0.3, 0.4, 0.5, 0.6, 0.7, \underset{\substack{\uparrow \\ n=0}}{0.8}, 0.9, 1.0 \ \}$$

The simplest method of representing the signal $x[n]$ in MATLAB would be to enter it as a row vector by typing the following line into the command window:

```
>> xn = [0.1,0.2,0.3,0.4,0.5,0.6,0.7,0.8,0.9,1]

xn =
Columns 1 through 7
0.1000    0.2000    0.3000    0.4000    0.5000    0.6000    0.7000
Columns 8 through 10
0.8000    0.9000    1.0000
```

It is somewhat cumbersome to work with this representation since our indexing of the signal and MATLAB's indexing of it are quite different. The leftmost element of the vector xn is the sample $x[-7]$ for our signal, but is accessed as xn(1) in MATLAB notation. The eighth element of the vector xn is the sample $x[0]$ of the signal, accessed as xn(8) in MATLAB. Time-reversal operation can be implemented using the built-in function fliplr().

```
>> xmn = fliplr(xn)

xmn =
Columns 1 through 7
1.0000    0.9000    0.8000    0.7000    0.6000    0.5000    0.4000
Columns 8 through 10
0.3000    0.2000    0.1000
```

In this case the 3rd element of the vector xmn, accessed as xmn(3) in MATLAB, is the sample $x[0]$ for our signal. Because of this added complexity with differing locations of the time origin, we cannot simply add the vectors xn and xmn to find the solution. We first need to align the origins of the two vectors by adding zero-amplitude samples to the end of the vector xn and to the beginning of the vector xmn.

```
>> xn = [xn,0,0,0,0,0];
>> xmn = [0,0,0,0,0,xmn];
>> g = xn+xmn;
```

Now the 8th element of the vector g, accessed as g(8) in MATLAB, has the sample $g[0]$. It should be clear from this discussion that, while it works, the method presented is quite messy. It is also not scalable, that is, it would be impractical for use in problems involving multiple signals or datasets with large numbers of samples in each.

Our first attempt to improve on this solution will be to write a *local function* to represent the signal $x[n]$ with its natural indexing. Recall that a local function is a function that appears at the end of a MATLAB script and is only visible within that script. Consider the script `mexdt_1_6a.m` given below:

```
1  % Script: mexdt_1_6a.m
2  idx=[-10:10];   % Index range for the result
3  g = [];         % Start with empty vector
4  for n=idx                           % Loop over indices
5    value = mysig(n)+mysig(-n);       % Compute each sample
6    g = [g,value];                    % Append to the vector
7  end
8  g                                   % Display the vector g
9  stem(idx,g);                        % Graph the signal g
10 
11 function x = mysig(n)
12   xData = [0.1,0.2,0.3,0.4,0.5,0.6,0.7,0.8,0.9,1];
13   initIndex = -7;   % Initial index
14   finalIndex = 2;   % Final index
15   if (n>=initIndex & n<=finalIndex)  % Check if n is in range
16     x = xData(n-initIndex+1);        % If yes, adjust index for MATLAB
17   else
18     x = 0;                           % Otherwise return 0
19   end
20 end
```

The local function `mysig()`, given between lines 11 and 20, returns one sample of the signal $x[n]$ for one specified value of the sample index. In line 15 we check if the index is in the range $-7 \leq n \leq 2$. If it is, we add the necessary offset to it, essentially mapping the natural index range of $[-2,7]$ to MATLAB's array index range of $[1,10]$. We then return the corresponding sample amplitude. If the index is out of the range $-7 \leq n \leq 2$, we simply return an amplitude of zero. Now the signal $g[n]$ can be computed one sample at a time using function `mysig()` within the loop structure between lines 4 and 7.

This method is certainly more streamlined than the one before, however, it has a major drawback. The function `mysig()` can only return one sample at a time, necessitating the use of a loop in order to compute the entire result. For a large dataset the use of such a loop may be slow and inefficient. Since MATLAB is an interpreted language, the use of loops should be avoided whenever possible. If we could find a way to rewrite the function `mysig()` so that it takes a vector of index values and returns a vector of amplitudes, then the loop would no longer be necessary. The script `mexdt_1_6b.m` shown below accomplishes that.

```
% Script: mexdt_1_6b.m
n = [-10:10];  % Index range for the result
g = mysig(n)+mysig(-n)
stem(n,g);

function x = mysig(n)
  xData = [0.1,0.2,0.3,0.4,0.5,0.6,0.7,0.8,0.9,1];
  initIndex = -7;     % Initial index
  finalIndex = 2;     % Final index
```

```
10   numSamples = 10;     % Number of samples in xData
11   xData = [xData,0];  % Append a zero for all out of range samples
12                       % (this will be needed in line 15)
13   n1 = (n>=initIndex & n<= finalIndex);  % Indices that are in range
14   n(n1) = n(n1)-initIndex+1;  % Convert to MATLAB index
15   n(~n1) = numSamples+1;      % Fix indices that are out of range
16   x = xData(n);               % Return data
17 end
```

The revised function `mysig()` is shown between lines 6 and 17. In line 11 we append a zero to the vector `xData` for reasons that will become apparent later. Lines 12 through 14 are critical. Line 12 tests the vector of indices to see if its elements are in the range $-7 \le n \le 2$. The result is the vector `n1` which has 1 (true) in positions that pass the test, and 0 (false) in positions that don't. In line 13, the notation `n(n1)` refers to those elements of the index vector `n` that pass the test. We simply convert those index values from actual indices to MATLAB indices. In line 14, the notation `n(~n1)` refers to those elements of `n` that fail the test. These positions are mapped to the index of the zero-valued sample that was appended to `xData` on line 11. This practice is referred to as *logical indexing*. We will see more examples of it in MATLAB Exercises 1.7 and 1.8.

Software resources: `mexdt_1_6a.m` , `mexdt_1_6b.m`

MATLAB Exercise 1.7: Working with signals in tabular form – revisited

Convert the local function `mysig()` developed in MATLAB Exercise 1.6 into a general purpose function `ss_dsignal()` that can be used with any data vector with a specified initial index. Afterward use this new function to verify the solution found in Example 1.4 and the solution found in Example 1.5.

Solution: The listing for the function `ss_dsignal()` is given below:

```
function x = ss_dsignal(n,xData,initIndex)
  numSamples = max(size(xData));  % Number of samples
  finalIndex = initIndex+numSamples-1;
  xData = [xData,0];  % Append a zero for all out of range samples
                      % (this will be needed in line 8)
  n1 = (n>=initIndex & n<= finalIndex);  % Indices that are in range
  n(n1) = n(n1)-initIndex+1;  % Convert to MATLAB index
  n(~n1) = numSamples+1;      % Fix indices that are out of range
  x = xData(n);               % Return data
end
```

The problem of Example 1.4 can be solved using the script `mexdt_1_7a.m` below.

```
% Script: mexdt_1_7a.m
xData = [0.2,0.4,0.6,0.8,0.8,0.8,0.8,0.4];
n = [-10:10];  % Index range for the result
g = ss_dsignal(n,xData,-3)+0.5*ss_dsignal(n-4,xData,-3)
stem(n,g);
```

The script `mexdt_1_7b.m` listed below can be used for verifying the solution found in Example 1.5. Note that, for the downsampling operation, we have used `2*n` as the first argument passed to the function `ss_dsignal()`.

```
% Script: mexdt_1_7b.m
xData = [0.8,1.0,1.3,0.9,0.7,1.2,1.4,1.1,0.8,0.6,0.5,-0.2];
n = [-10:10];  % Index range for the result
g = ss_dsignal(2*n,xData,-5);
stem(n,g);
```

Software resources: mexdt_1_7a.m , mexdt_1_7b.m , ss_dsignal.m

MATLAB Exercise 1.8: Case study – Sunspot numbers; working with large data sets

In this exercise we will consider the problem of loading a large set of data into MATLAB, and extracting a specified range of it to analyze and graph. Refer to the discussion of sunspot numbers in Section 1.1 of this Chapter. Part of the monthly data obtained from U.S. National Oceanic and Atmospheric Administration website at `https://www.ngdc.noaa.gov` was graphed in Fig. 1.3 for the years 1975 to 2010. The data, originally in comma-separated text form, has been converted and saved into a MATLAB data file "`sundata_monthly.mat`", and is available for use in this exercise. Type the following two lines into the command window to load the data file and get information about its contents:

```
>> load 'sundata_monthly.mat';
>> whos

  Name          Size       Bytes    Class     Attributes
  sspots_mo     3197x3     76728    double
```

The data file contains one variable with the name `sspots_mo` which is a matrix with 3197 rows and 3 columns. Now type the following two lines to display the first 15 rows and all 3 columns of the data:

```
>> format shortg
>> sspots_mo(1:15,:)

  ans =
  1749           1          58
  1749           2        62.6
  1749           3          70
  1749           4        55.7
  1749           5          85
  1749           6        83.5
  1749           7        94.8
  1749           8        66.3
  1749           9        75.9
  1749          10        75.5
  1749          11       158.6
  1749          12        85.2
  1750           1        73.3
  1750           2        75.9
  1750           3        89.2
```

For the row index we have used `1:15`, indicating that we want to display rows 1 through 15. In contrast, the colon `:` for the column index refers to *all columns*. Inspection of the displayed data reveals that column 1 has the year, column 2 has the months within that year, and column 3 has the average sunspot number for each month. The last 4 rows of the variable `sspots_mo` can be displayed with the following:

```
>> sspots_mo(end-3:end,:)

  ans =
  2015          2          44.8
  2015          3          38.4
  2015          4          54.4
  2015          5          58.8
```

It appears that the available data extends from January of 1749 to May of 2015. Suppose we want to graph the data from the beginning of the year 1810 to the end of the year 1850. The script listed below accomplishes that.

```
1  % Script: mexdt_1_8a.m
2  load 'sundata_monthly.mat';        % Load the data
3  year = sspots_mo(:,1);             % Leftmost column of data matrix
4  span = year>=1810 & year<=1850;    % Span of years to be selected
5  x = sspots_mo(span,:);             % Copy specified range into variable "x"
6  years = x(:,1);                    % Vector of years
7  months = x(:,2);                   % Vector of months
8  index = years + (months-1)/12;     % Create a composite index
9  % Graph the data from 1810 to 1850
10 plot(index,x(:,3));                % Graph spot count versus index
11 axis([1810,1850,0,250]);           % Set axis limits
```

In line 3 of the script, we copy column 1 of the matrix `sspots_mo` into a new vector named `year`. In line 4, a logical vector named `span` is created by checking each element of this vector to see if it is in the desired range of 1810 to 1850. In line 5, the rows of the matrix `sspots_mo` for which the test of line 4 succeeds are copied into a new matrix named `x`. This is logical indexing as we have used earlier in MATLAB Exercises 1.6 and 1.7. Lines 6 through 8 create a composite index against which the monthly numbers can be graphed. The resulting graph is shown in Fig. 1.42.

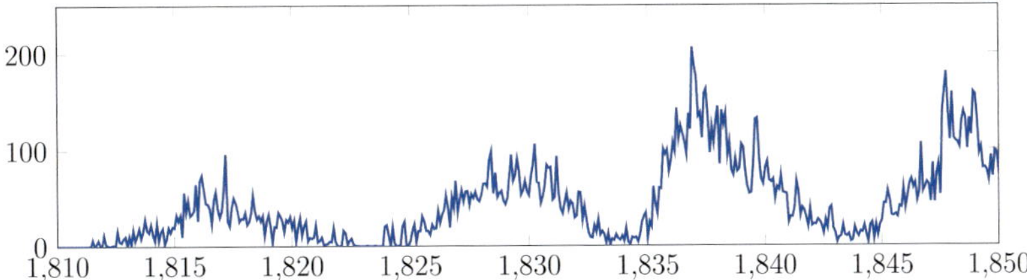

Figure 1.42 – Graph obtained in MATLAB Exercise 1.8.

It is possible to make the script shorter and eliminate the need for some of the intermediate variables, perhaps at the expense of readability. Lines 3 through 5 can be combined into one line. Similarly, lines 6 through 8 can also be combined, resulting in the compact script `mexdt_1_8b.m` listed below.

```
% Script: mexdt_1_8b.m
load 'sundata_monthly.mat';        % Load the data
x = sspots_mo(sspots_mo(:,1)>=1810 & sspots_mo(:,1)<=1850,:);
index = x(:,1)+(x(:,2)-1)/12;
% Graph the data from 1810 to 1850
plot(index,x(:,3));                % Graph spot count versus index
axis([1810,1850,0,250]);           % Set axis limits
```

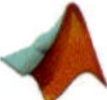

Software resources: mexdt_1_8a.m , mexdt_1_8b.m

MATLAB Exercise 1.9: Working with audio files

A song that is stored on a computer with a filename extension such as *wav, mp3,* or *flac* is essentially a discrete-time signal $x[n]$ obtained from an analog audio signal. Some formats also include compression to reduce the size of resulting file. A commonly used value for the sampling interval is $T_s = 22.68$ μs which corresponds to a sampling rate of $f_s = 44,100$ Hz, or 44,100 samples per second. This value of T_s is used for compact discs and digital streaming services. In this exercise we will work with digitized audio files. A couple of music files in *flac* format, used with permission, are provided with the accompanying software:

- Ballad_44100_Hz.flac
- Ballad_22050_Hz.flac

The numbers in file names refer to the sampling rates used. Other works of the composer, Charles Shomo, Sr., can be found at

https://pixabay.com/users/caffeine_creek_band-11181297/

Type the following line into the MATLAB command window to obtain information about one of the audio files:

```
>> audioinfo('Ballad_22050_Hz.flac')

  ans =
  struct with fields:
           Filename: 'C:\Files\Ballad_22050_Hz.flac'
  CompressionMethod: 'FLAC'
        NumChannels: 2
         SampleRate: 22050
       TotalSamples: 4199424
           Duration: 190.4501
              Title: 'Soothing Ballad (Sampled at 22050 Hz)'
            Comment: []
             Artist: 'Charles Shomo'
      BitsPerSample: 16
```

The built-in function audioinfo() returns a *structure* with fields that contain specific details about the file. Functions audioread() and sound() can be used for loading the audio data into a MATLAB variable and for playing it back respectively. One little caveat is in order: The function audioread() reads the entire file into memory by default. The result would be a matrix with 2 columns (for left and right channels) and over 4 million rows (190.4501 seconds times 22,050 samples per second). Instead, we will load only the first 10 seconds of the song into memory by specifying a range of sample indices that encompass $10 f_s$ samples:

```
>> [x,fs] = audioread('Ballad_22050_Hz.flac',[1,220500]);
>> sound(x,fs);
```

Once an audio file is read into a MATLAB variable x as shown above, it is possible to process it in some desired way and write the result into a new audio file using the built-in function `audiowrite()`. For example, the script `mexdt_1_9.m` listed below reads the file "`Ballad_22050_Hz.flac`", swaps the left and the right channels, and writes out a new file with the name "`Ballad_22050_Hz_Flipped.flac`". Try the script only if you are confident that the computer you use has sufficient memory to handle it.

```
% Script: mexdt_1_9.m
clear all;
[x,fs] = audioread('Ballad_22050_Hz.flac');
audiowrite('Ballad_22050_Hz_Flipped.flac',[x(:,2),x(:,1)],fs);
```

The newly created audio file can be played back with the following two lines:

```
>> [x,fs] = audioread('Ballad_22050_Hz_Flipped.flac',[1,220500]);
>> sound(x,fs);
```

The method of processing a signal after it is fully loaded into memory is known as *post processing*. This mode of processing may not always be convenient as it is slow, and quite demanding of CPU and memory resources. A lot of the algorithms we will study in later chapters require the signal to be partitioned into smaller, more manageable, blocks of samples referred to as *frames*. This makes *real-time processing* possible where the audio signal is read one *frame* at a time, processed, and played back before the next frame is read. We will explore this further in MATLAB Exercises 1.11 and 1.12.

Software resources: `mexdt_1_9.m` , `Ballad_22050_Hz.flac` , `Ballad_44100_Hz.flac`

MATLAB Exercise 1.10: Case study – Synthesizing music with sinusoidal signals

The subject of sampling a continuous-time sinusoidal signal to obtain a discrete-time signal was discussed in Section 1.3.3. Given the continuous-time signal

$$x_a(t) = A \sin\left(2\pi f_0 t + \theta\right)$$

its discrete-time version can be obtained by evaluating $x_a(t)$ at integer multiples of a sampling interval T_s as

$$x[n] = x_a(nT_s) = A \sin(2\pi F_0 n + \theta) \qquad (1.70)$$

where F_0 is the normalized frequency of the sinusoid, that is,

$$F_0 = \frac{f_0}{f_s} = f_0 T_s$$

Let the sampling rate be chosen as $f_s = 44,100$ Hz. In the following script, we will choose the analog frequency to be $f_0 = 261.63$ Hz, and produce 22,050 samples of $x[n]$ for 0.5 seconds of audio, and play them back using the built-in function `sound()`.

```
1 % Script: mexdt_1_10a.m
2 f0 = 261.63;                % Audio frequency
3 fs = 44100;                 % Sampling rate
4 F0 = f0/fs;                 % Normalized frequency
5 n = [0:22049];              % Sample indices (for 1/2 second of audio)
6 x = 0.6*sin(2*pi*F0*n);     % Eqn. (1.70)
7 sound(x,fs);                % Play the sound
```

Upon running the code you should hear the sinusoidal signal being played as a sound. For those familiar with keyboard instruments, the frequency we used, namely $f = 261.63$ Hz, corresponds to the *middle* C sound also known as C_4 owing to the fact that it's part of the fourth octave on most keyboard instruments. It is possible to play different notes by simply changing the frequency of the sinusoidal signal. Table 1.1 lists frequencies for musical notes C_4 through C_5, covering one full octave. For those familiar with piano keyboards, these correspond to the white keys.

Table 1.1 – Frequencies of some music notes in Hz.

C_4	D_4	E_4	F_4	G_4	A_4	B_4	C_5
261.63	293.66	329.63	349.23	392.00	440.00	493.88	523.25

If a different sampling rate is desired, it can be achieved by modifying lines 3 and 5 of the script file. For example, the script listed below generates and plays the same tone with a sampling rate of $f_s = 20,000$ Hz.

```
% Script: mexdt_1_10b.m
f0 = 261.63;                % Audio frequency
fs = 20000;                 % Sampling rate
F0 = f0/fs;                 % Normalized frequency
n = [0:9999];               % Sample indices (for 1/2 second of audio)
x = 0.6*sin(2*pi*F0*n);     % Eqn. (1.70)
sound(x,fs);                % Play the sound
```

There should be no noticeable difference in the sounds produced by scripts `mexdt_1_10a.m` and `mexdt_1_10b.m` although the latter has less than half the samples compared to the former. The fundamental difference is on the highest analog frequency that can be represented. This will be explored in Chapter 4.

The script `mexdt_1_10c.m` below synthesizes a signal with notes C_4 through C_5 played in sequence. A modulating envelope $q_a(t)$ shown in Fig. 1.43 is sampled to obtain $q[n] = q_a(nT_s)$, and applied to each tone.

```
% Script: mexdt_1_10c.m
fs = 44100;        % Sampling rate
Ts = 1/fs;         % Sampling interval
n = [0:fs/2-1];    % Sample indices (for 1/2 second of audio)
% Set corner points for the modulating envelope
tq = [0,0.05,0.15,0.4,0.45,0.5];
```

```
xq = [0,1.5,0.9,0.9,0,0];
% Obtain the modulating envelope q(n) through interpolation/sampling
q = interp1(tq,xq,n*Ts,'linear');
% Compute signals for each note
C4 = tone(261.63,fs,q,n);
D4 = tone(293.66,fs,q,n);
E4 = tone(329.63,fs,q,n);
F4 = tone(349.23,fs,q,n);
G4 = tone(392.00,fs,q,n);
A4 = tone(440.00,fs,q,n);
B4 = tone(493.88,fs,q,n);
C5 = tone(523.25,fs,q,n);
% Play the sound
sound([C4,D4,E4,F4,G4,A4,B4,C5],fs);

function x = tone(f,fs,q,n)
  F = f/fs;   % Normalized frequency
  x = 0.6*sin(2*pi*F*n).*q;
end
```

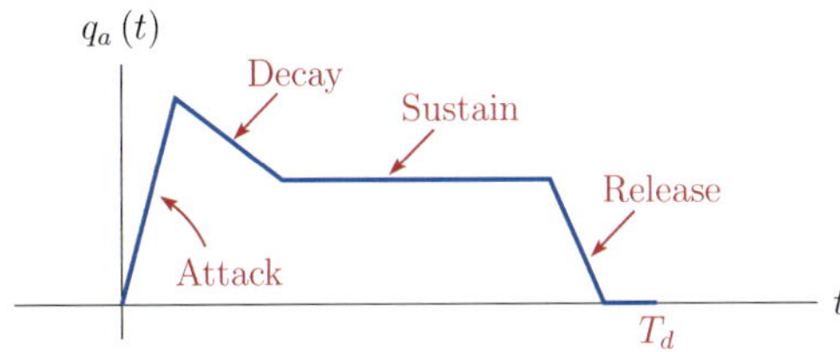

Figure 1.43 – Modulating envelope signal used with a sinusoidal tone.

Software resources: mexdt_1_10a.m , mexdt_1_10b.m , mexdt_1_10c.m

MATLAB Exercise 1.11: Objects for working with audio signals

In MATLAB Exercise 1.9 we have used functions `audioread()`, `audiowrite()` and `sound()` to load, save, and play back digital audio files in post processing mode. In this exercise we will use object-oriented features of MATLAB which will allow us to process audio signals in a way that mimics the behavior of real-time systems. Our approach will be as follows:

1. Receive a *frame* of N samples of audio data from the input stream.
2. Analyze and/or modify the current frame as needed.
3. Put the frame of audio data into the output stream to be played back or saved into a file or both.
4. Repeat steps 1 through 3 until no more input frames are left.

DSP System Toolbox of MATLAB provides the functionality to read, write and process audio files on a frame-by-frame basis. Consider a discrete-time signal $x[n]$. If frame size chosen is

N, then the first frame of the signal would include samples for $n = 0, \ldots, N-1$. The second frame would include samples for $n = N, \ldots, 2N-1$, and so on.

a. Introduction to the use of objects

An *object* is an encapsulation of data specific to that object as well as methods that act on that data. In the script `mexdt_1_11a.m` listed below, we create an audio file reader object with the name `sReader` to read data from the file `Ballad_22050_Hz.mp3`. Lines 5 through 9 demonstrate how various properties of the object are accessed.

```
1  % Script: mexdt_1_11a.m
2  % Create an "audio file reader" object
3  sReader = dsp.AudioFileReader('Ballad_22050_Hz.flac');
4  % Get and display some information about the audio file.
5  fileName = sReader.Filename               % Name for the audio file
6  sampleRate = sReader.SampleRate           % Sampling rate of the audio file
7  frameSize = sReader.SamplesPerFrame       % Number of samples in each frame
8  dataType = sReader.OutputDataType         % Type of audio data
9  readRange = sReader.ReadRange             % Range of samples to be read
10 release(sReader);   % We are finished with the input audio file
```

b. A simple audio player

The script `mexdt_1_11b.m` listed below reads the audio data in a file one frame at a time and sends it to the audio playback device.

```
1  % Script: mexdt_1_11b.m
2  % Create an "audio file reader" object
3  sReader = dsp.AudioFileReader('Ballad_22050_Hz.flac');
4  sReader.ReadRange = [1,441000];   % Read 20 seconds of music
5  % Create an "audio player" object
6  sPlayer = audioDeviceWriter('SampleRate',sReader.SampleRate);
7
8  while ~isDone(sReader)
9    x = sReader();   % Get the next frame of data from the reader
10   sPlayer(x);      % Play back the frame
11 end
12
13 release(sReader);  % We are finished with the input audio file
14 release(sPlayer);  % We are finished with the audio output device
```

We are already familiar with the reader object created in line 3. Line 4 ensures that we only receive data for the first 20 seconds of the song. It can be commented out if the entire song data is to be received. Line 6 of the code creates an audio player object that interfaces with the sound card of the computer, and initializes it to have the same sampling rate as the audio reader object. The loop structure in lines 8 through 11 takes a frame from the reader object and passes it on to the player object, repeating this process until the reader object has no frames left. Note that the variable `x` represents one frame of data. In this case it has 1024 (frame size) rows and two columns for left and right channels.

If the computer being used is a relatively old one with a slow processor, the audio playback may be choppy. On the other hand, most newer computers with higher end processors should be able to run this script with either 22.05 kHz or 44.1 kHz sampling rate.

c. Graphing the signals while audio is played

In script `mexdt_1_11c.m` below, we begin with the script `mexdt_1_11b.m` as the basis, and insert additional code to graph the left and right channel signals of each frame. It is important to prepare the graph with axis limits and labels before entering the loop. This is done in lines 9 through 17. Within the loop we simply update the `YData` property of each graph with data from the current frame.

```
1  % Script: mexdt_1_11c.m
2  % Create an "audio file reader" object
3  sReader = dsp.AudioFileReader('Ballad_22050_Hz.flac');
4  sReader.ReadRange = [1,441000];    % Read 20 seconds of music
5  % Create an "audio player" object
6  sPlayer = audioDeviceWriter('SampleRate',sReader.SampleRate);
7
8  % Prepare the plot before the loop
9  figure;
10 frameSize = sReader.SamplesPerFrame;    % The frame size put out by sReader
11 x = zeros(frameSize,2);            % Dummy matrix for the plot
12 n = [0:frameSize-1];               % Array of indices
13 p1 = plot(n,x(:,1),n,x(:,2));      % p1 is a length-2 array of handles
14 axis([0,frameSize-1,-1,1]);        % Set axis limits
15 xlabel('Sample index');
16 ylabel('Amplitude');
17 drawnow;
18
19 while ~isDone(sReader)
20   x = sReader();     % Get the next frame of data from the reader
21   sPlayer(x);        % Play back the frame
22   set(p1(1),'YData',x(:,1));  % Update left channel graph
23   set(p1(2),'YData',x(:,2));  % Update right channel graph
24   drawnow limitrate nocallbacks;
25 end
26
27 release(sReader);   % We are finished with the input audio file
28 release(sPlayer);   % We are finished with the audio output device
```

d. Displaying average power of each frame

The script `mexdt_1_11d.m` listed below is a modification of `mexdt_1_11c.m` that computes and displays the average power in left and right channel signals in the form of a live stem plot. Pay attention to how average signal power for left and right channels is computed on lines 23 and 24.

```
1  % Script: mexdt_1_11d.m
2  % Create an "audio file reader" object
3  sReader = dsp.AudioFileReader('Ballad_22050_Hz.flac');
4  sReader.ReadRange = [1,441000];
5  % Create an "audio player" object
6  sPlayer = audioDeviceWriter('SampleRate',sReader.SampleRate);
7
8  % Prepare the average power plot
9  figure;
10 hPlot = stem([0,0],'diamondr','linewidth',3);
11 axis([0.5,2.5,0,0.1]);
```

```
ylabel('Normalized average power');
xlabel('Channel');
set(gca,'XTick',[1,2])
set(gca,'XTickLabel',{'Left','Right'});
drawnow;

while ~isDone(sReader)
  x = sReader();       % Get the next frame of data from the reader
  sPlayer(x);          % Play back the frame
  xLeft = x(:,1);      % Left channel data
  xRight = x(:,2);     % Right channel data
  pLeft = xLeft'*xLeft/1024;          % Power for left channel
  pRight = xRight'*xRight/1024;       % Power for right channel
  set(hPlot,'YData',[pLeft,pRight]);  % Update plot values
  drawnow limitrate nocallbacks;
end

release(sReader);  % We are finished with the input audio file
release(sPlayer);  % We are finished with the audio output device
```

Software resources: mexdt_1_11a.m , mexdt_1_11b.m , mexdt_1_11c.m , mexdt_1_11d.m

MATLAB Exercise 1.12: Case study – Audio synthesizer object

In MATLAB Exercise 1.10 we have established a framework for experimenting with synthesizing audio through the use of sinusoidal signals at musical frequencies and modulating envelope functions. In this exercise we will expand that framework with the `audioSynthesizer` object that is provided as part of the software accompanying the textbook.

The script `mexdt_1_12a.m` listed below creates a synthesizer object `sSource` using default values of all parameters. By default, the object produces a one octave sequence from C_4 to C_5 similar to what was done in MATLAB Exercise 1.10. It uses a sampling rate of 44.1 kHz, a frame size of 1024 samples, purely sinusoidal tones at 100 beats per minute and the default modulating envelope.

```
% Script: mexdt_1_12a.m
sSource = audioSynthesizer;
sPlayer = audioDeviceWriter('SampleRate',sSource.SampleRate);

while ~isDone(sSource)  % Loop through until done:
  x = step(sSource);    %   Get the next frame of data from synthesizer
  sPlayer(x);           %   Play back the frame
end
release(sPlayer);
```

In the script `mexdt_1_12b.m` we provide a brief passage of music notation to be played, in the form of a cell array `sheetMusic` with three columns. Column 1 holds each note to be played. Column 2 holds the duration of play for each note in terms of "beats". For those familiar with music notation, a quarter-note would be one beat, a half-note would be two beats, and so on. Column 3 is optional; if it exists, it specifies the amplitude scale for each note.

```
% Script: mexdt_1_12b.m
sheetMusic = {
  "F4",  1.5, 1;
  "rest",0.5, 1;
  "A4",  1.5, 1;
  "F4",  0.5, 1;
  "C5",  1,   1;
  "Bb4", 0.5, 1;
  "A4",  0.5, 1;
  "G4",  0.5, 1;
  "F4",  0.5, 1;
  "F#4", 1,   1;
  "G4",  1,   1;
  "rest",2,   1;
  "C4",  2,   1;
  "E4",  1,   1;
  "G4",  1,   1;
  "Bb4", 2.5, 1;
  "D5",  0.5, 1;
  "C5",  0.5, 1;
  "Bb4", 0.5, 1;
  "G#4", 0.5, 1;
  "A4",  0.5, 1;
  "A4",  1,   1;
  "rest",2,   1 }

sSource = audioSynthesizer('Notes',sheetMusic,'BeatsPerMin',100);
sPlayer = audioDeviceWriter('SampleRate',sSource.SampleRate);

while ~isDone(sSource)  % Loop through until done:
  x = step(sSource);    %   Get the next frame of data from synthesizer
  sPlayer(x);           %   Play back the frame
end
release(sPlayer);
```

The script `mexdt_1_12c.m` below allows experimentation with different waveform types in tone generation. In line 3 the synthesizer object is initiated to use a square wave signal. As we will see in the next chapter, a square wave signal has a very rich set of frequencies compared to a sinusoid which contains only one frequency. During playback the difference in sound should be quite noticeable.

In line 10 the object is rewound, and the signal type is changed to a sawtooth waveform. The playback speed is changed to 125 beats per minute.

Finally, in line 18, the object is rewound again. The playback speed is changed to 150 beats per minute. Signal type is changed to custom so that a user-defined function can be utilized for generating each note. The user function `mysignal()` is called by the object `sSource` as needed, and is given amplitude and fundamental frequency information, and a time vector. It should compute and return the vector `x` which must have the same dimensions as the vector `t`.

```
% Script: mexdt_1_12c.m
sSource = audioSynthesizer('BeatsPerMin',100,'SignalType','Square');
sPlayer = audioDeviceWriter('SampleRate',sSource.SampleRate);

while ~isDone(sSource)                % Loop through until done:
  x = step(sSource);                  %   Get the next frame of data
  sPlayer(x);                         %   Play back the frame
end

rewind(sSource);                      % Rewind the source so we can play again
sSource.SignalType = 'Sawtooth';      % Change to sawtooth waveform
sSource.BeatsPerMin = 125;            % A little bit faster
while ~isDone(sSource)                % Loop through until done:
  x = step(sSource);                  %   Get the next frame of data
  sPlayer(x);                         %   Play back the frame
end

rewind(sSource);                      % Rewind the source so we can play again
sSource.SignalType = 'Custom';        % Change to custom waveform
sSource.SignalGen = @mysignal;        % Use custom function for each tone
sSource.BeatsPerMin = 150;            % A little bit faster
while ~isDone(sSource)                % Loop through until done:
  x = step(sSource);                  %   Get the next frame of data
  sPlayer(x);                         %   Play back the frame
end
release(sPlayer);

function x = mysignal(ampl,freq,t)
  % Mix sinusoid and square wave
  x = 0.45*ampl*sin(2*pi*freq*t)+0.1*ampl*square(2*pi*freq*t);
end
```

Software resources: mexdt_1_12a.m , mexdt_1_12b.m , mexdt_1_12c.m , mexdt_1_12a_Alt.m , mexdt_1_12b_Alt.m , mexdt_1_12c_Alt.m

Problems

1.1. For the signal $x[n]$ shown in Fig. P.1.1, sketch the following signals.

a. $g_1[n] = x[n-3]$

b. $g_2[n] = x[2n-3]$

c. $g_3[n] = x[-n]$

d. $g_4[n] = x[2-n]$

e. $g_5[n] = \begin{cases} x[n/2]\,, & \text{if } n/2 \text{ is integer} \\ 0\,, & \text{otherwise} \end{cases}$

f. $g_6[n] = x[n]\,\delta[n]$

g. $g_7[n] = x[n]\,\delta[n-3]$

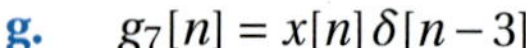

h. $g_8[n] = x[n]\,\big(u[n+2]-u[n-2]\big)$

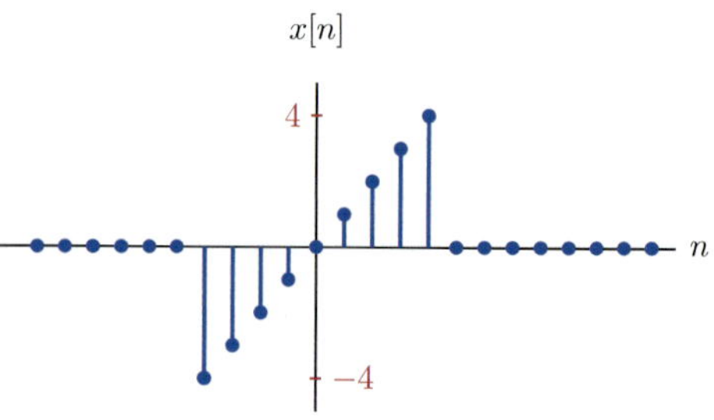

Figure P. 1.1

1.2. Carefully sketch each signal described below.

a. $x[n] = \{ \underset{\uparrow}{1}, -1, 3, -2, 5 \}$

b. $x[n] = (0.8)^n \, u[n]$

c. $x[n] = u[n] - u[n-10]$

d. $x[n] = r[n] - 2r[n-5] + r[n-10]$

1.3. Two discrete-time signals $x_1[n]$ and $x_2[n]$ are given below.

$$x_1[n] = \{ 1.2, 0.2, 0.7, 3.1, 2.7, \underset{\substack{\uparrow \\ n=0}}{2.2}, 1.7, 1.9, 2.3. 2.5, 1.9, 1.1, 0.8 \}$$

$$x_2[n] = \{ -1.2, 0.2, \underset{\substack{\uparrow \\ n=3}}{1.3}, 1.8, 0.7, 0.2, -0.4 \}$$

Determine the following signals.

a. $g_1[n] = x_1[n] + 2\,x_2[n]$

b. $g_2[n] = x_1[n]\,x_2[n-2]$

c. $g_3[n] = 2\,x_1[n-3] + x_2[-n-2]$

d. $g_4[n] = x_1[2n-1]$

e. $g_5[n] = x_1[-3n+1]$

f. $g_6[n] = \begin{cases} x_2[n/2]\,, & \text{if } n/2 \text{ is integer} \\ 0\,, & \text{otherwise} \end{cases}$

1.4. For the signal $x[n]$ shown in Fig. P.1.4, sketch the following signals.

a. $g_1[n] = x[n]\,u[n]$

b. $g_2[n] = x[n]\,\big(u[n] - u[n-5]\big)$

c. $g_3[n] = x[-n]\,\big(u[n+2] - u[n-4]\big)$

d. $g_4[n] = \delta[n] + x[n]\,\delta[n-3] + x[n]\,\delta[n-5]$

e. $g_5[n] = n^2\,x[n]\,u[n]$

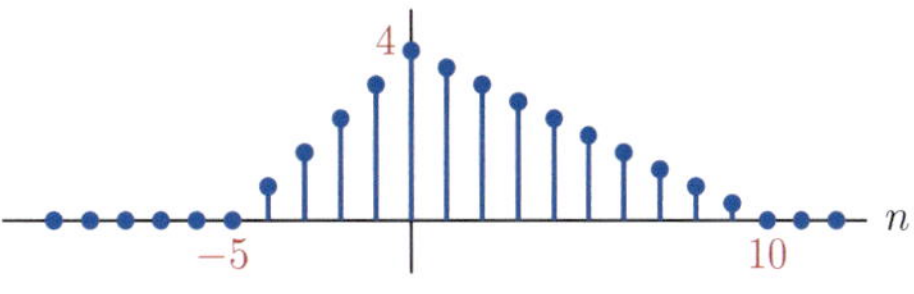

Figure P. 1.4

1.5. Consider the signals shown in Fig. 1.5. Express each signal using unit impulse, unit step, and unit ramp functions as needed.

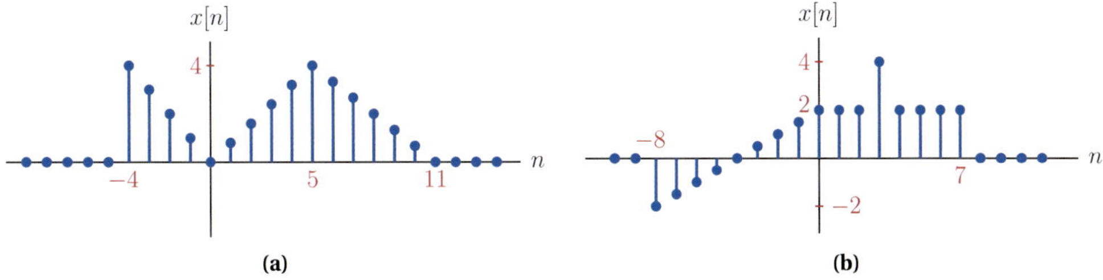

Figure P. 1.5

1.6. Consider the sinusoidal discrete-time signal

$$x[n] = 5\cos\left(\frac{3\pi}{23}n + \frac{\pi}{4}\right)$$

Is the signal periodic? If yes, determine the fundamental period.

1.7. Determine the energy of each signal described in Problem 1.2.

1.8. Consider a signal $x[n]$. Show that

a. If $x[n]$ is even, then $\sum_{n=-M}^{M} x[n] = x[0] + 2\sum_{n=1}^{M} x[n]$

b. If $x[n]$ is odd, then $\sum_{n=-M}^{M} x[n] = 0$

1.9. Let $x[n] = x_1[n]\, x_2[n]$. Show that

a. If both $x_1[n]$ and $x_2[n]$ are even, then $x[n]$ is even.

b. If both $x_1[n]$ and $x_2[n]$ are odd, then $x[n]$ is even.

c. If $x_1[n]$ is even and $x_2[n]$ is odd, then $x[n]$ is odd.

1.10. Determine and sketch even and odd components of the signal $x[n]$ shown in Fig. P.1.10.

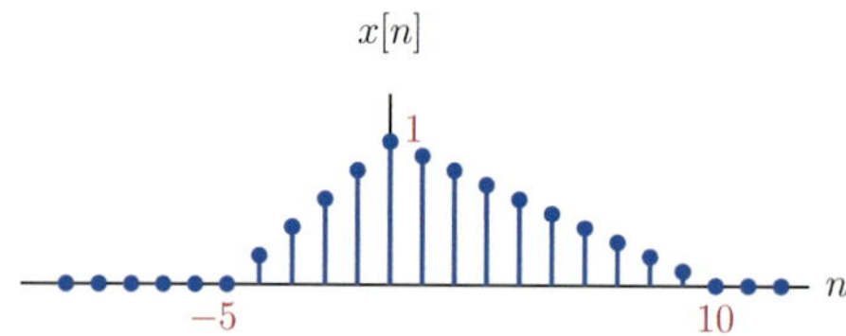

Figure P. 1.10

1.11. A complex valued signal $x[n]$ is defined as

$$x[n] = (0.9 + j\,0.2)^n\, u[n]$$

a. Express the signal $x[n]$ in Cartesian complex form. Sketch real and imaginary parts of the signal.

b. Determine conjugate symmetric and conjugate antisymmetric components of the signal.

MATLAB Problems

1.12. Consider the discrete-time signal $x[n]$ used in Problem 1.1 and graphed in Fig. P.1.1. Express this signal through an anonymous MATLAB function `x`. Graph the signal $x[n]$ for the index range $n = -15,\ldots,15$. Afterward, express each of the signals in parts (a) through (h) of Problem 1.1 in terms of the anonymous function `x`. Graph signals $g_1[n]$ through $g_8[n]$ for the index range $n = -15,\ldots,15$. Hint: For part (e), use built-in function `mod()` along with logical indexing.

1.13. Develop a script to compute and graph the eight signals $g_1[n]$ through $g_8[n]$ of Problem 1.1 on the same screen. Instead of using an anonymous function `x` as in Problem 1.12, this time use a local function `mysignal()` to represent the signal $x[n]$. Use functions `tiledlayout()` and `nexttile` to arrange the stem plots for the signals $g_1[n]$ through $g_8[n]$ as a grid with two rows and four columns. Each signal should be graphed for the index range $n = -15,\ldots,15$. Hint: For part (e), use built-in function `mod()` along with logical indexing.

1.14. For each part of Problem 1.2 develop a script to compute and graph the signal $x[n]$ for the index range $n = -5,\ldots,20$. Your script for each part should contain a local function `mysignal()` to compute sample amplitudes of the signal $x[n]$ for a specified vector of sample indices. The function `mysignal()` should make use of the functions `ss_dimpulse()`, `ss_dstep()`, `ss_dramp()` and `ss_dsignal()` as needed.

1.15.

a. Develop a function `ss_signalProb_1_3_x1()` to compute sample amplitudes of the signal $x_1[n]$ of Problem 1.3 for a specified vector of sample indices. Also write a script to test your function by computing and graphing $x_1[n]$ for $n = -10,\ldots,10$.

b. Repeat the requirements of part (a) with the signal $x_2[n]$ of Problem 1.3. The function to be developed for this part should be called `ss_signalProb_1_3_x2()`.

1.16. For each part of Problem 1.3 develop a script to compute and graph the signal $g[n]$ for the index range $n = -15,\ldots,15$. Your script for each part should contain a local function `mysignal()` to compute sample amplitudes of the signal $g[n]$ for a specified vector of sample indices. The function `mysignal()` should make use of the functions `ss_signalProb_1_3_x1()` and `ss_signalProb_1_3_x2()` developed in Problem 1.15 for the computation of the signal $g[n]$. Hint: For part (f), use built-in function `mod()` along with logical indexing.

1.17. Express the signal $x[n]$ of Problem 1.4 using an anonymous function. Afterward, compute and graph the signal $g_i[n]$ for each part of Problem 1.4 using your anonymous function along with functions `ss_dimpulse()`, `ss_dstep()` and `ss_dramp()` as needed. All graphs should be done for the index range $n = -15,\ldots,15$.

1.18. Verify your pencil and paper solutions of Problem 1.5 by developing two functions with names `ss_signalProb_1_5a()` and `ss_signalProb_1_5b()` for the signals of parts a and b, respectively. Use the functions `ss_dimpulse()`, `ss_dstep()` and `ss_dramp()` as needed. Write a script file for each part of the problem to compute and graph the signals for the index range $n = -15,\ldots,15$.

1.19. Consider the discrete-time signal $x[n]$ used in Problem 1.10 and graphed in Fig. P.1.10.

a. Express this signal through an anonymous MATLAB function that utilizes the function `ss_dramp()`, and graph the result for index range $n = -15,\ldots,15$.

b. Write a script to compute and graph even and odd components of this signal.

1.20.

a. Develop a function to clip the peaks of a signal at desired levels. Consider a signal $x[n]$ from which a new signal $y[n]$ is obtained through the following relationship:

$$y[n] = \begin{cases} X_{min}\,, & \text{if } x[n] \le X_{min} \\ X_{max}\,, & \text{if } x[n] \ge X_{max} \\ x[n]\,, & \text{otherwise} \end{cases}$$

Your function should have the syntax

```
y = ss_dclip(x,Xmin,Xmax);
```

Use logical indexing as detailed in MATLAB Exercises 1.6 through 1.8.

b. Write a script to test your function. Generate a discrete-time sinusoidal signal $x[n]$ for $n = 0,\ldots,399$ with unit amplitude and a normalized frequency of $F_0 = 0.01$. Clip the peaks of the signal with $X_{min} = -0.8$ and $X_{max} = 0.9$. Graph the resulting signal $y[n]$.

1.21. Start with the script given in part c of MATLAB Exercise 1.12. Set the signal type for the synthesizer object `sSource` to `'Custom'`, and write a custom tone generating function `mysignal()` to return a sinusoidal tone with peaks clipped at 90 percent of the maximum amplitude. Use the function `ss_dclip()` developed in Problem 1.20. Listen to the result, and compare to the sound of a purely sinusoidal tone. Repeat with peaks clipped at 80 percent and 60 percent. Comment on the results. How does clipping affect the sound?

1.22. Develop a MATLAB program to read in an audio file and create a monophonic version of it which has only one channel. If the original audio file has left and right channel signals $x_L[n]$ and $x_R[n]$, respectively, the produced audio file should have only one channel with

$$x[n] = \frac{x_L[n] + x_R[n]}{2}$$

Use one of the audio files provided with the accompanying software as input. Create and use an audio file writer object `sWriter` to facilitate the production of a new audio file containing the monophonic version of the song. Play back the monophonic version while simultaneously writing it into the output file. Use `dsp.AudioFileWriter()` system object. Name your output file by adding "`_Mono`" to the end of the original file name. For example, if the input file is "`Ballad_22050_Hz.flac`", the output file should be named "`Ballad_22050_Hz_Mono.flac`".

MATLAB Projects

1.23. Early standards for two channel (stereophonic) recordings were introduced in 1958, and several companies started selling vinyl recordings in stereo. Most of the music released prior to that was in single channel (monophonic) format, with only a few experimental releases in stereo, mostly on reel-to-reel tape. Vinyl recordings in stereo became popular rather quickly and were practically the industry standard by the end of 1959.

The quick success of stereo recordings created an interesting problem for recording companies, namely how to create stereo versions of the mono recordings they had accumulated over the years. The incentive was that they could charge higher prices for stereo recordings compared to mono recordings. An obvious solution would be to record the same performances again, this time using two microphones instead of one. This proved to be expensive, impractical, or sometimes even impossible. An alternative solution would be to take a single channel recording and electronically produce two channels from it that are slightly different from each other to create a *pseudo-stereo* effect. RCA Victor devised a system called *ESR (Electronic Stereo Reprocessing)* for this purpose. Capitol Records devised an alternative system referred to as *Duophonic Sound.* Typically the missing channel would be created using time delays or filtering to separate low and high frequencies from each other. Consumer reaction to these efforts was somewhat mixed, leading to terms such as *fake stereo* or *mock stereo.*

In this project, we will explore the problem of obtaining a synthetic stereo effect by starting with an audio file in monophonic form and creating left and right channels from it using a variety of approaches. Write a MATLAB script to read in the audio file `Ballad_20050_Hz_Mono.flac` obtained in Problem 1.22. Within the loop structure of the code, create a two-channel frame matrix, play it back, and write it out to a new file. Use a frame size of 1100 samples. Use the following methods to create the synthetic stereo effect, and comment on the results.

a. Initially, duplicate the existing channel to create a two-channel signal with both channels containing the same exact audio, i.e.,

$$x_L[n] = x_R[n] = x[n]$$

b. One method of creating separation is to time-shift the right channel signal by m samples while leaving the left channel signal unchanged, so that we have

$$x_R[n] = x[n] \qquad \text{and} \qquad x_L[n] = x[n-m]$$

Let $m = 1100$. This corresponds to a delay of about 50 ms. Since we picked the amount of delay to be the same as the frame size, implementing the delay is easy: Simply keep track of the data $x[n]$ for the previous frame, and use it as the right channel of the new frame.

$$x_L[n] = x[n] \qquad \text{and} \qquad x_R[n] = x_{PREV}[n]$$

c. Repeat part b with both the frame size and the delay amount changed to 2200 samples, corresponding to a delay of about 100 ms.

d. Let both channels use current and previous frame data as follows:

$$x_L[n] = \alpha\, x[n] + (1-\alpha)\; x_{PREV}[n]$$
$$x_R[n] = \alpha\, x[n] - (1-\alpha)\; x_{PREV}[n]$$

Experiment with various values of $0 < \alpha < 1$.

1.24. The `audioSynthesizer` object introduced in MATLAB Exercise 1.12 generates a single-channel signal. In this project we will construct a two-channel stereo signal by using two synthesizer objects playing the same notes, but with different sounds. One will be used for the left channel. The other will be delayed slightly, and used for the right channel.

Start with the script given in part c of MATLAB Exercise 1.12. Create two `audioSynthesizer` objects named `sSource1` and `sSource2`. The former should use the default signal type (sinusoid), and the latter should use a custom function `mysignal()` for tone generation. Let the function `mysignal()` return a sinusoidal signal at *twice* the frequency requested, i.e.,

```
x = ampl*sin(4*pi*freq*t);
```

This essentially amounts to playing the same notes an octave higher. Within the loop structure, use the signal of `sSource1` object for the left channel. Use the signal of `sSource2` object *from the previous frame* for the right channel. This will require that you save each frame from the object `sSource2` for use in the subsequent turn. The default frame size of 1024 samples should provide a delay of about 23 ms.

1.25. In some digital systems such as inexpensive toys, sinusoidal signals at specific frequencies may be generated through table look-up, thus eliminating the need for the system to include the capability to compute trigonometric functions. Samples corresponding to exactly one period of the sinusoidal signal are computed ahead of time and stored in memory. When the signal is needed, the contents of the look-up table are played back repeatedly. Consider a sinusoidal signal

$$x_a = \cos\left(2\pi f_1\, t\right)$$

with $f_1 = 1336$ Hz. Let $x_1[n]$ be a discrete-time sinusoidal signal obtained by evaluating $x_a(t)$ at intervals of 125 μs corresponding to a *sampling rate* of 8000 samples per second.

$$x_1[n] = x_a\left(125\times 10^{-6}\,n\right) = x_a\left(\frac{n}{8000}\right)$$

a. Determine the fundamental period N_1 for the signal $x_1[n]$.

b. Write a MATLAB script to accomplish the following:

- Create a vector `x1` that holds exactly one period of the signal.
- By repeating the vector `x1` as many times as needed, create a vector `x` that holds 8000 samples of the sinusoidal signal that corresponds to a time duration of 1 second.
- Play back the resulting sound using the `sound()` function of MATLAB. If the fundamental period was computed properly in part (a) then the repeated sections of the signal should fit together seamlessly, and you should hear a clean tone. Use the syntax

```
sound(x,8000)
```

c. Modify the script in part (b) so that the number of samples used in creating the vector `x1` is 10 samples fewer than the correct value found in part (a). Create a vector `x` by repeating this imperfect vector `x1` as many times as necessary. Listen to the audio playback of the vector `x` and comment on the difference. Also, graph the resulting signal. Using the zoom tool, zoom into the transition area between two repetitions, and observe the flaw in the transition that is due to incorrect period length.

d. Repeat parts (a) through (c) using the sinusoidal signal

$$x_b = \cos\left(2\pi f_b t\right)$$

with $f_1 = 852$ Hz, and the corresponding discrete-time signal

$$x_2[n] = x_b\left(125\times 10^{-6}\,n\right) = x_b\left(\frac{n}{8000}\right)$$

e. Consider the dual-tone signal

$$x[n] = x_1[n] + x_2[n]$$

Using the table look-up method with the vectors "`x1`" and "`x2`" obtained above, generate a vector that holds 8000 samples of the dual-tone signal. Play back the resulting vector and comment.

CHAPTER 2

ANALYZING DISCRETE-TIME SYSTEMS IN THE TIME DOMAIN

Chapter Objectives

- Develop the notion of a *discrete-time system.* Learn simplifying assumptions made in the analysis of discrete-time systems, the concepts of *linearity, time invariance,* and their significance.
- Explore the use of difference equations for representing discrete-time systems. Develop methods for solving difference equations. Use these methods to compute the output signal of a system in response to a specified input signal.
- Learn to represent a difference equation in the form of a block diagram that can be used as the basis for simulating or realizing a system.
- Discuss the significance of the *impulse response* as an alternative description form for linear and time-invariant systems.
- Learn how to compute the output signal for a linear and time-invariant system using the convolution operation, and learn the steps involved in carrying it out.
- Learn the concepts of causality and stability as they relate to physically realizable and usable systems.

2.1 Introduction

The definition of a discrete-time system is similar to that of its continuous-time counterpart. Signal-system interaction involving a single-input single-output discrete-time system is illustrated in Fig. 2.1.

DOI: 10.1201/9781003570462-2

Discrete-time system:

In general, a discrete-time system is a mathematical formula, method, or algorithm that defines a cause-effect relationship between a set of discrete-time input signals and a set of discrete-time output signals.

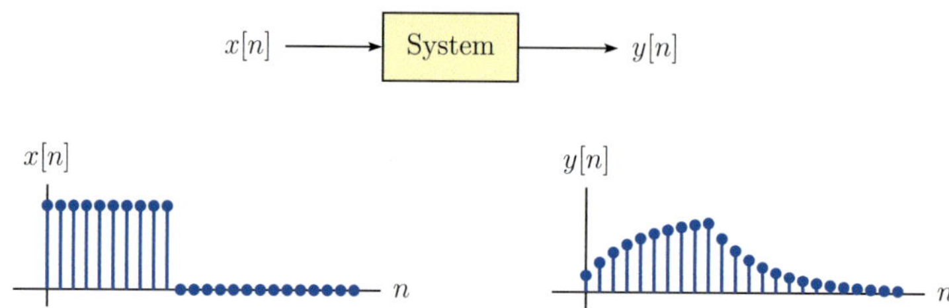

Figure 2.1 – Discrete-time signal-system interaction.

The input-output relationship of a discrete-time system may be expressed in the form

$$y[n] = \text{Sys}\{x[n]\} \tag{2.1}$$

where $\text{Sys}\{\ldots\}$ represents the transformation that defines the system in the time domain. A very simple example is a system that simply multiplies its input signal by a constant gain factor K

$$y[n] = K\,x[n]$$

or a system that delays its input signal by m samples

$$y[n] = x[n-m]$$

or one that produces an output signal proportional to the square of the input signal

$$y[n] = K\left(x[n]\right)^2$$

A system with higher complexity can be defined using a difference equation that establishes the relationship between input and output signals.

Two commonly used simplifying assumptions used in studying mathematical models of systems, namely *linearity* and *time invariance,* will be the subject of Section 2.2. Section 2.3 deals with the issue of deriving a difference equation model for a discrete-time system. Section 2.4 discusses the characteristics of constant-coefficient linear difference equations. Solution methods for constant-coefficient linear difference equations are presented in Section 2.5. Block diagrams for realizing discrete-time systems are introduced in Section 2.6. In Section 2.7 we discuss the significance of the impulse response for discrete-time systems. We also discuss its use in the context of the convolution operator for determining the output signal. Concepts of causality and stability of systems are discussed in Sections 2.8 and 2.9, respectively.

2.2 Linearity and Time Invariance

In most of this textbook we will focus our attention on a particular class of systems referred to as *linear and time-invariant* systems. Linearity and time invariance will be two important properties which, when present in a system, will allow us to analyze the system using well-established techniques of the linear system theory. In contrast, the analysis of systems that are not linear and time-invariant tends to be more difficult, and often relies on methods that are specific to the types of systems being analyzed.

2.2.1 Linearity in discrete-time systems

Linearity property will be very important as we analyze and design discrete-time systems. A discrete-time system is said to be linear if it satisfies the two conditions given below.

Conditions for linearity:

$$\text{Sys}\left\{x_1[n]+x_2[n]\right\}=\text{Sys}\left\{x_1[n]\right\}+\text{Sys}\left\{x_2[n]\right\} \tag{2.2}$$

$$\text{Sys}\left\{\alpha_1\,x_1[n]\right\}=\alpha_1\,\text{Sys}\left\{x_1[n]\right\} \tag{2.3}$$

Eqns. (2.2) and (2.3), referred to as the *additivity rule* and the *homogeneity rule* respectively, must be satisfied for any two discrete-time signals $x_1[n]$ and $x_2[n]$ as well as any arbitrary constant α_1. These two criteria can be combined into one equation known as the *superposition principle.*

Superposition principle:

$$\text{Sys}\left\{\alpha_1\,x_1[n]+\alpha_2\,x_2[n]\right\}=\alpha_1\,\text{Sys}\left\{x_1[n]\right\}+\alpha_2\,\text{Sys}\left\{x_2[n]\right\} \tag{2.4}$$

Verbally expressed, Eqn. (2.4) implies that the response of the system to a weighted sum of any two input signals is equal to the same weighted sum of the individual responses of the system to each of the two input signals. Fig. 2.2 illustrates this concept.

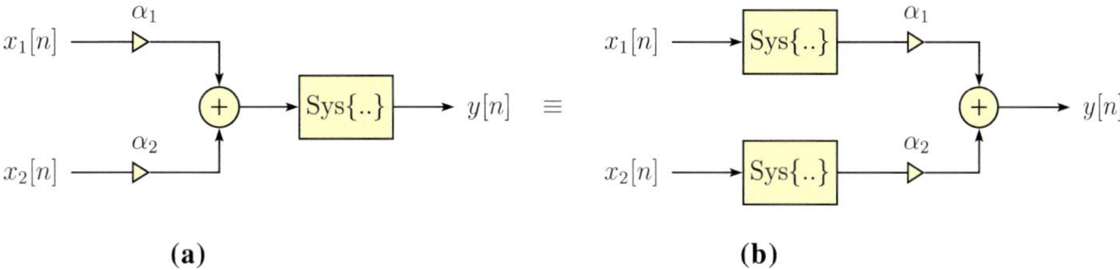

Figure 2.2 – Illustration of Eqn. (2.4). The two configurations shown are equivalent if the system under consideration is linear.

A generalization of the principle of superposition for the weighted sum of N discrete-time signals is expressed as

$$\text{Sys}\left\{\sum_{i=1}^{N}\alpha_i\,x_i[n]\right\}=\sum_{i=1}^{N}\alpha_i\,\text{Sys}\left\{x_i[n]\right\} \tag{2.5}$$

The response of a linear system to a weighted sum of N arbitrary signals is equal to the same weighted sum of individual responses of the system to each of the N signals. Let $y_i[n]$ be the response to the input term $x_i[n]$ alone, that is $y_i[n]=\text{Sys}\left\{x_i[n]\right\}$ for $i=1,\ldots,N$. Superposition principle implies that

$$y[n]=\text{Sys}\left\{\sum_{i=1}^{N}\alpha_i\,x_i[n]\right\}=\sum_{i=1}^{N}\alpha_i\,y_i[n] \tag{2.6}$$

This is illustrated in Fig. 2.3.

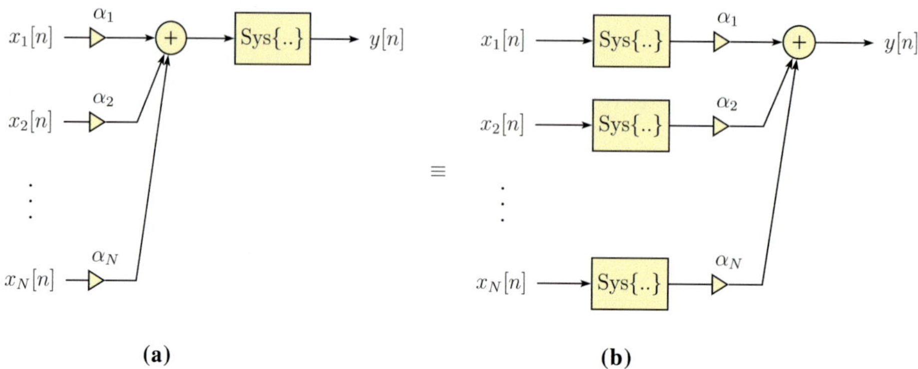

Figure 2.3 – Illustration of Eqn. (2.4). The two configurations shown are equivalent if the system under consideration is linear.

Example 2.1: Testing linearity in discrete-time systems

For each of the discrete-time systems described below, determine whether the system is linear or not:

a. $y[n] = 3\,x[n] + 2x[n-1]$

b. $y[n] = 3\,x[n] + 2\,x[n+1]\,x[n-1]$

c. $y[n] = a^{-n}\,x[n]$

Solution:

a. In order to test the linearity of the system we will think of its responses to the two discrete-time signals $x_1[n]$ and $x_2[n]$ as

$$y_1[n] = \text{Sys}\left\{x_1[n]\right\} = 3\,x_1[n] + 2\,x_1[n-1]$$

and

$$y_2[n] = \text{Sys}\left\{x_2[n]\right\} = 3\,x_2[n] + 2\,x_2[n-1]$$

The response of the system to the linear combination signal $x[n] = \alpha_1\,x_1[n] + \alpha_2\,x_2[n]$ is computed as

$$\begin{aligned} y[n] =& \text{Sys}\left\{\alpha_1\,x_1[n] + \alpha_2\,x_2[n]\right\} \\ =& 3\left(\alpha_1\,x_1[n] + \alpha_2\,x_2[n]\right) + 2\left(\alpha_1\,x_1[n-1] + \alpha_2\,x_2[n-1]\right) \\ =& \alpha_1\left(3\,x_1[n] + 2\,x_1[n-1]\right) + \alpha_2\left(3\,x_2[n] + 2\,x_2[n-1]\right) \\ =& \alpha_1\,y_1[n] + \alpha_2\,y_2[n] \end{aligned}$$

Superposition principle holds true; therefore the system in question is linear.

b. Again using the test signals $x_1[n]$ and $x_2[n]$ we have

$$y_1[n] = \text{Sys}\left\{x_1[n]\right\} = 3\,x_1[n] + 2\,x_1[n+1]\,x_1[n-1]$$

and

$$y_2[n] = \text{Sys}\left\{x_2[n]\right\} = 3\,x_2[n] + 2\,x_2[n+1]\,x_2[n-1]$$

Use of the linear combination signal $x[n] = \alpha_1\,x_1[n] + \alpha_2\,x_2[n]$ as input to the system yields the output signal

$$\begin{aligned} y[n] =& \text{Sys}\left\{\alpha_1\,x_1[n] + \alpha_2\,x_2[n]\right\} \\ =&3\left(\alpha_1\,x_1[n] + \alpha_2 x_2[n]\right) \\ &+2\left(\alpha_1\,x_1[n+1] + \alpha_2\,x_2[n+1]\right)\left(\alpha_1\,x_1[n-1] + \alpha_2\,x_2[n-1]\right) \end{aligned}$$

In this case the superposition principle does not hold true. The system in part (b) is therefore not linear.

c. The responses of the system to the two test signals are

$$y_1[n] = \text{Sys}\left\{x_1[n]\right\} = a^{-n}\,x_1[n]$$

and

$$y_2[n] = \text{Sys}\left\{x_2[n]\right\} = a^{-n}\,x_2[n]$$

and the response to the linear combination signal $x[n] = \alpha_1\,x_1[n] + \alpha_2\,x_2[n]$ is

$$\begin{aligned} y[n] =& \text{Sys}\left\{\alpha_1\,x_1[n] + \alpha_2\,x_2[n]\right\} \\ =&a^{-n}\left(\alpha_1\,x_1[n] + \alpha_2\,x_2[n]\right) \\ =&\alpha_1\,a^{-n}\,x_1[n] + \alpha_2\,a^{-n}\,x_2[n] \\ =&\alpha_1\,y_1[n] + \alpha_2\,y_2[n] \end{aligned}$$

Superposition principle holds true. The system in question is linear.

2.2.2 Time invariance in discrete-time systems

Another important concept in the analysis of systems is time invariance. A system is said to be *time-invariant* if its behavior characteristics do not change in time. Let a discrete-time system be described with the input-output relationship $y[n] = \text{Sys}\left\{x[n]\right\}$. For the system to be considered time-invariant, the only effect of time shifting the input signal should be to cause an equal amount of time shift in the output signal.

Condition for time invariance:

$$\text{Sys}\left\{x[n]\right\} = y[n] \quad \text{implies that} \quad \text{Sys}\left\{x[n-k]\right\} = y[n-k] \tag{2.7}$$

This relationship is depicted in Fig. 2.4. Alternatively, the time-invariant nature of a system can be characterized by the equivalence of the two configurations shown in Fig. 2.5. If a delayed version of the original input signal is applied to the system, a delayed version of the original output signal is obtained in return. Therefore, a delay of k samples can be placed before or after the system without impacting the overall input-output relationship of the cascade combination.

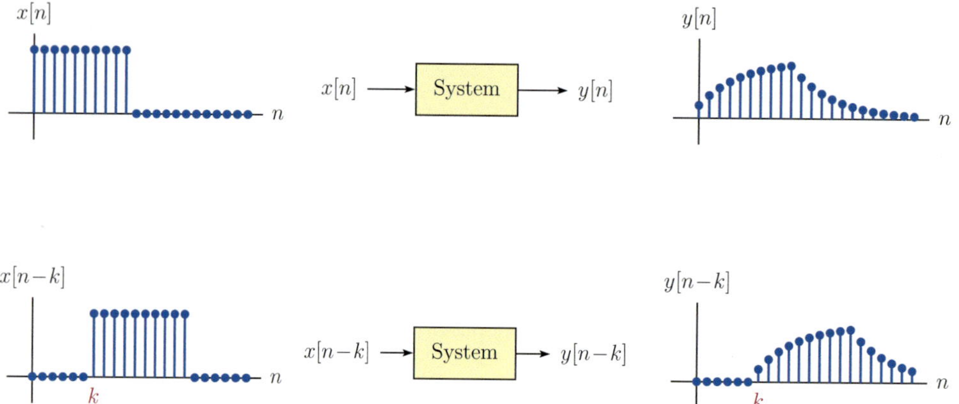

Figure 2.4 – Illustration of time invariance for a discrete-time system.

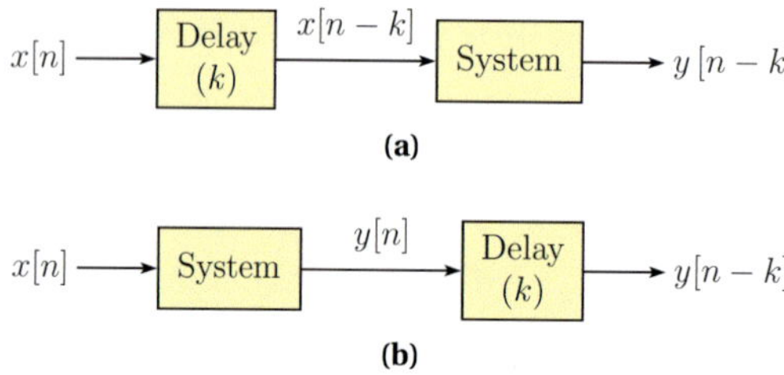

Figure 2.5 – Another interpretation of time invariance. The two configurations shown are equivalent for a time-invariant system.

Example 2.2: Testing time invariance in discrete-time systems

For each of the discrete-time systems described below, determine whether the system is time-invariant or not:

a. $y[n] = y[n-1] + 3\,x[n]$

b. $y[n] = x[n]\,y[n-1]$

c. $y[n] = n\,x[n-1]$

Solution:

a. We will test the time invariance property of the system by time shifting both the input and the output signals by the same number of samples, and see if the input-output relationship still holds. Replacing the index n by $n-k$ in the arguments of all input and output terms we obtain

$$\text{Sys}\left\{x[n-k]\right\} = y[n-k-1] + 3\,x[n-k] = y[n-k]$$

The input-output relationship holds, therefore the system is time-invariant.

b. Proceeding in a similar fashion we have

$$\text{Sys}\left\{x[n-k]\right\} = x[n-k]\,y[n-k-1] = y[n-k]$$

This system is time-invariant as well.

c. Replacing the index n by $n-k$ in the arguments of all input and output terms yields

$$\text{Sys}\left\{x[n-k]\right\} = n\,x[n-k-1] \neq y[n-k]$$

This system is clearly not time-invariant since the input-output relationship no longer holds after input and output signals are time-shifted.

Should we have included the factor n in the time shifting operation when we wrote the response of the system to a time shifted input signal? In other words, should we have written the response as

$$\text{Sys}\left\{x[n-k]\right\} \stackrel{?}{=} (n-k)\;x[n-k-1]$$

The answer is no. The factor n that multiplies the input signal is part of the system definition. It is not part of either the input signal or the output signal. Therefore we cannot include it in the process of time shifting input and output signals.

2.2.3 DTLTI systems

Discrete-time systems that are both linear and time-invariant will play an important role in the rest of this textbook. We will develop time- and frequency-domain analysis and design techniques for working with such systems. To simplify the terminology, we will use the acronym *DTLTI* to refer to *discrete-time linear and time-invariant* systems.

2.3 Difference Equations for Discrete-Time Systems

Most discrete-time systems can be modeled with difference equations involving current, past, or future samples of input and output signals. We will begin our study of discrete-time system models based on difference equations with a few examples. Each example will start with a verbal description of a system, and lead to a system model in the form of a difference equation. Some of the examples will lead to systems that will be of fundamental importance in the rest of this text. Other examples will lead to nonlinear or time-varying systems that we will not consider further. In the following sections, we will focus our attention on difference equations for DTLTI systems. We will also develop solution techniques for them.

Example 2.3: Moving average filter

A length-N *moving average filter* is a simple system that produces an output equal to the arithmetic average of the most recent N samples of the input signal. Let the discrete-time output signal be $y[n]$. If the current sample index is 100, the current output sample $y[100]$ would be equal to the arithmetic average of the current input sample $x[100]$ and $(N-1)$

previous input samples. Mathematically we have

$$y[100] = \frac{x[100] + x[99] + \ldots + x[100-(N-1)]}{N} = \frac{1}{N}\sum_{k=0}^{N-1} x[100-k]$$

The general expression for the length-N moving average filter is obtained by expressing the n-th sample of the output signal in terms of the relevant samples of the input signal as

$$y[n] = \frac{x[n] + x[n-1] + \ldots + x[n-(N-1)]}{N} = \frac{1}{N}\sum_{k=0}^{N-1} x[n-k] \tag{2.8}$$

Eqn. (2.8) is a difference equation describing the input-output relationship of the moving average filter as a discrete-time system. The operation of the length-N moving average filter is best explained using the analogy of a window, as illustrated in Fig. 2.6.

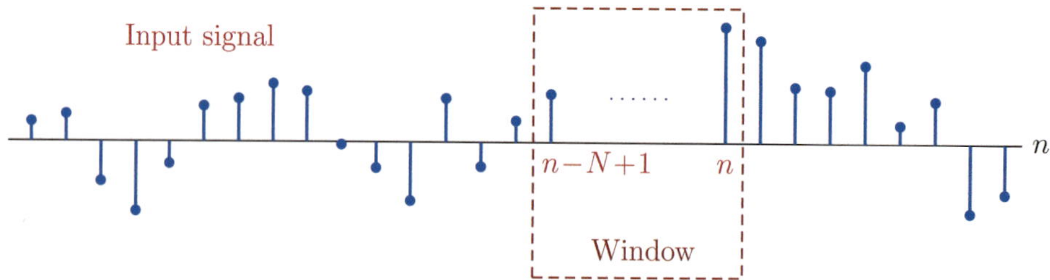

Figure 2.6 – Length-N moving average filter.

Suppose that we are observing the input signal through a window that is wide enough to hold N samples of it at any given time. Let the window be stationary, and let the input signal $x[n]$ move to the left one sample at a time, similar to a film strip. The current sample of the input signal is always the rightmost sample visible through the window. The current output sample is the arithmetic average of all input samples that are visible through the window.

Moving average filters are used in practical applications to smooth the variations in a signal. One example is in analyzing the changes in a financial index such as the Dow Jones Industrial Average. An investor might use a moving average filter on a signal that contains the values of the index for each day. A 50-day or a 100-day moving average window may be used for producing an output signal that disregards the day-to-day fluctuations of the input signal and focuses on the slowly varying trends instead. The degree of smoothing is dependent on N, the size of the window. In general, a 100-day moving average is smoother than a 50-day moving average. Let $x[n]$ be the signal that holds the daily closing values of the index for the calendar year 2014. The 50-day moving averages are computed by

$$y_1[n] = \frac{1}{50}\sum_{k=0}^{49} x[n-k]$$

and 100-day moving averages are computed by

$$y_2[n] = \frac{1}{100}\sum_{k=0}^{99} x[n-k]$$

Fig. 2.7 shows the daily values of the index for the calendar year 2014 as a discrete-time signal as well as the outputs produced by 50-day and 100-day moving average filters.

In Fig. 2.8 all three signals are graphed on the same coordinate system for comparison.

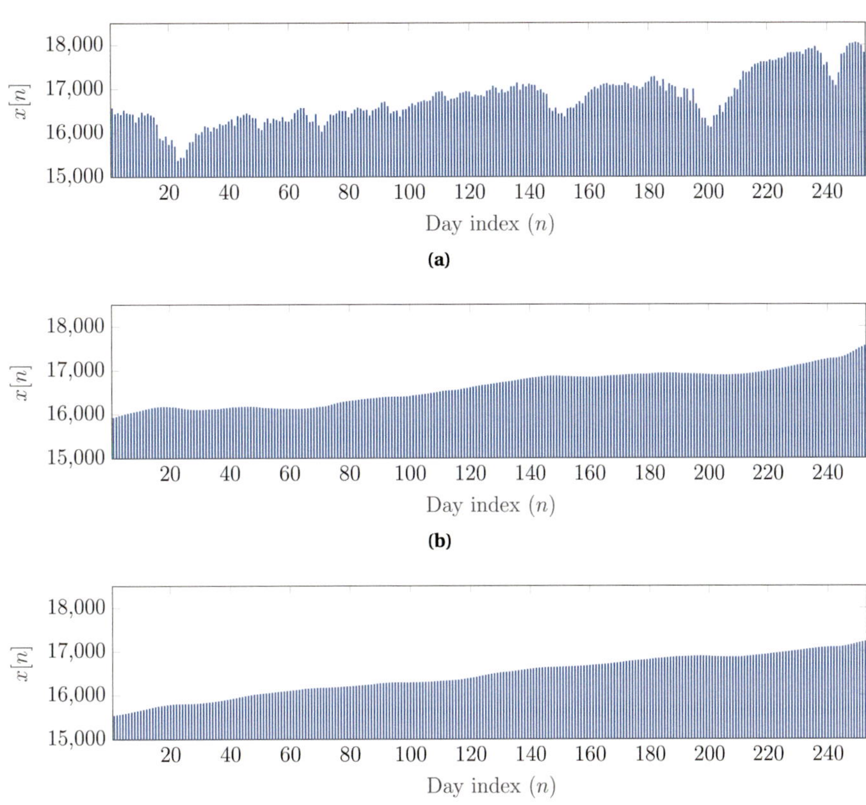

Figure 2.7 – **(a)** Signal $x[n]$ representing the Dow Jones Industrial Average daily values for the calendar year 2014, **(b)** signal $y_1[n]$ holding 50-day moving average values for the index, and **(c)** signal $y_2[n]$ holding 100-day moving average values for the index.

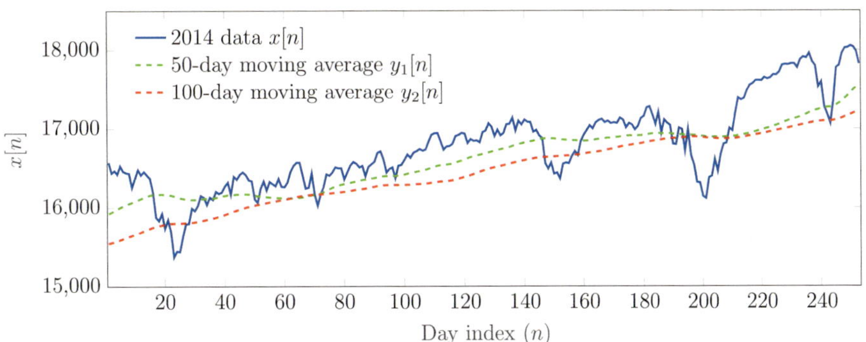

Figure 2.8 – Signals $x[n]$, $y_1[n]$ and $y_2[n]$ graphed as line graphs on the same set of axes for comparison

Note that, in the computation of the output signal samples $y_1[0],\ldots,y_1[48]$ as well as the output signal samples $y_2[0],\ldots,y_2[98]$, the averaging window had to be supplied with values from 2013 data for the index. For example, to compute the 50-day moving average on

January 1 of 2014, previous 49 samples of the data from December and part of November of 2013 had to be used. We will develop better insight for the type of smoothing achieved by moving average filters when we discuss frequency domain analysis methods for discrete-time systems.

Interactive App: Length-N moving average filter

The interactive app in `appMovingAvg1.m` illustrates the length-N moving average filter discussed in Example 2.3. Because of the large number of samples involved, the two discrete-time signals are shown through line plots as opposed to stem plots. The first plot is the Dow Jones Industrial Average data for the year 2014 which is preceded by partial data from the year 2013. First trading day of 2014 corresponds to $n = 0$. The second plot is the smoothed output signal of the moving average filter. The current sample index n and the length N of the moving average filter can each be specified through the user interface controls. The current sample index is marked on each plot with a red dot. Additionally, the window for the moving average filter is shown on the first plot, superimposed with the input data. The green horizontal line within the window as well as the green arrows indicate the average amplitude of the samples that fall within the window.

a. Increment the sample index and observe how the window slides to the right each time. Observe that the rightmost sample in the window is the current sample and the window accommodates a total of N samples.

b. For each position of the window, the current output sample is the average of all input samples that fall within the window. Observe how the position of the window relates to the value of the current output sample.

c. The length of the filter, and consequently the width of the window, relates to the degree of smoothing achieved by the moving average filter. Vary the length of the filter and observe the effect on the smoothness of the output signal.

Software resource: `appMovingAvg1.m`

Example 2.4: Length-2 moving average filter

A length-2 moving average filter produces an output by averaging the current input sample and the previous input sample. This action translates to a difference equation in the form

$$y[n] = \tfrac{1}{2}\,x[n] + \tfrac{1}{2}\,x[n-1] \tag{2.9}$$

and is illustrated in Fig. 2.9. The window through which we look at the input signal accommodates two samples at any given time, and the current input sample is close to the right edge of the window. As in the previous example, we will assume that the window is stationary, and the system structure shown is also stationary. We will imagine the input signal moving to the left one sample at a time like a film strip. The output signal in the lower part of the figure also acts like a film strip and moves to the left one sample at a time, in sync with the input signal.

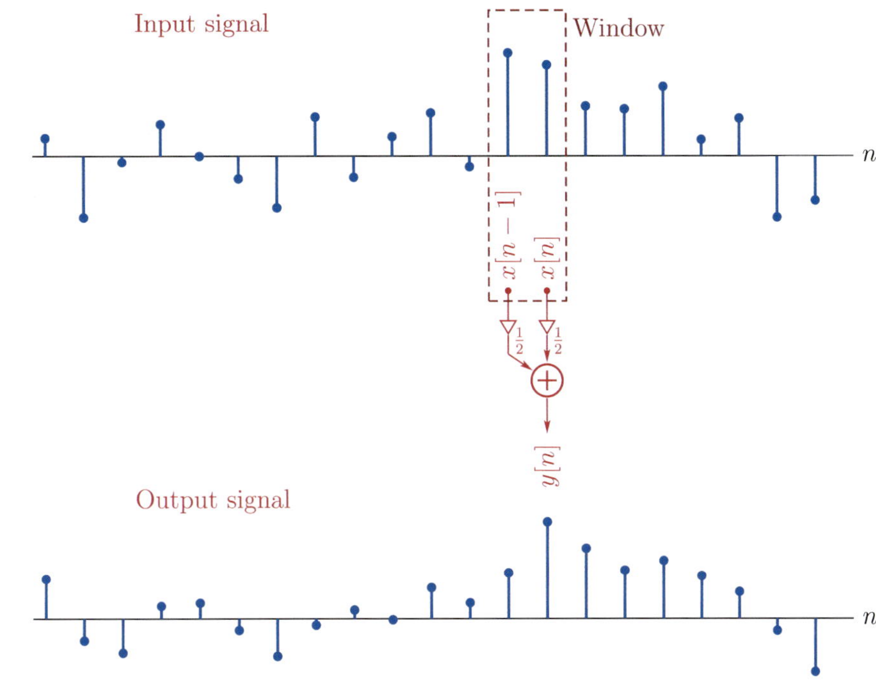

Figure 2.9 – Illustration of length-2 moving average filter.

Interactive App: Length-2 moving average filter

The interactive app in `appMovingAvg2.m` allows experimentation with the length-2 moving average filter discussed in Example 2.4. The input and output signals follow the film strip analogy and can be moved by changing the current index through the user interface. Input signal samples that fall into the range of the length-2 window are shown in red color. Several choices are available for the input signal for experimentation.

a. Set the input signal to a unit impulse, and observe the output signal of the length-2 moving average filter. What is the *impulse response* of the system?

b. Set the input signal to a unit step and observe the output signal of the system. Pay attention to the output samples as the input signal transitions from a sample amplitude of 0 to 1. The smoothing effect should be most visible during this transition.

 **Software resource:** `appMovingAvg2.m`

Example 2.5: Length-4 moving average filter

A length-4 moving average filter produces an output by averaging the current input sample and the previous three input samples. This action translates to a difference equation in the form

$$y[n] = \tfrac{1}{4}\,x[n] + \tfrac{1}{4}\,x[n-1] + \tfrac{1}{4}\,x[n-2] + \tfrac{1}{4}\,x[n-3] \tag{2.10}$$

and is illustrated in Fig. 2.10.

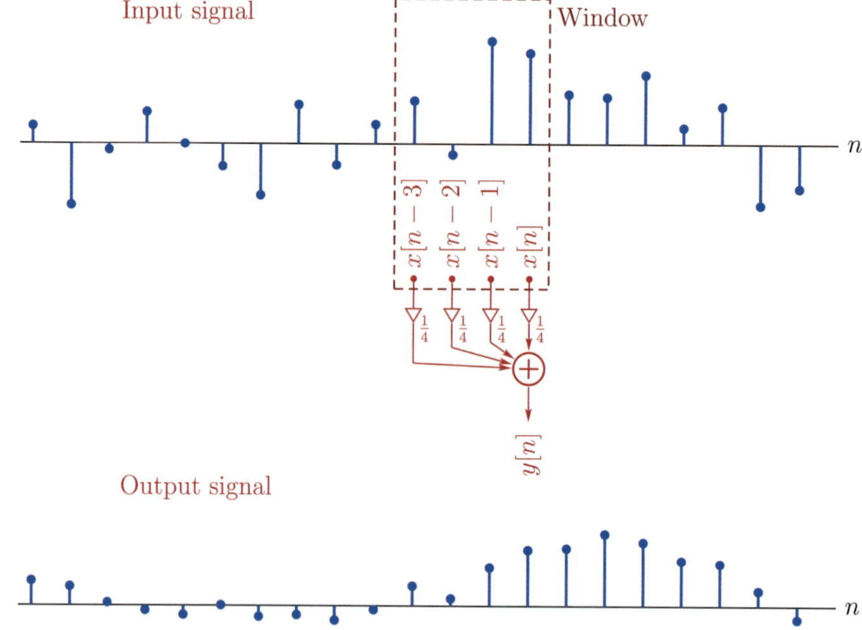

Figure 2.10 – Illustration of length-4 moving average filter.

Interactive App: Length-4 moving average filter

The interactive app in `appMovingAvg3.m` allows experimentation with the length-4 moving average filter discussed in Example 2.5. Its operation is very similar to that of the interactive app `appMovingAvg2` discussed earlier.

a. Set the input signal to a unit impulse, and observe the output signal of the length-4 moving average filter. What is the *impulse response* for this system?

b. Set the input signal to a unit step and observe the output signal of the system. Pay attention to the output samples as the input signal transitions from a sample amplitude of 0 to 1. How does the result differ from the output signal of the length-2 moving average filter in response to a unit step input signal?

Software resource: `appMovingAvg3.m`

Software resource: See MATLAB Exercises 2.1, 2.2 and 2.3.

Computer implementation of moving average filters has some challenging aspects. Some of the details that come up in practical implementation may be tricky to overcome. These will be discussed in MATLAB exercises 2.1, 2.2, 2.3, and 2.12 at the end of this chapter.

Example 2.6: Exponential smoother

Another method of smoothing a discrete-time signal is through the use of an *exponential smoother* which employs a difference equation with feedback. The current output sample is computed as a mix of the current input sample and the previous output sample using

$$y[n] = (1-\alpha)\, y[n-1] + \alpha\, x[n] \tag{2.11}$$

The parameter α is a constant in the range $0 < \alpha < 1$, and it controls the degree of smoothing. According to Eqn. (2.11), the current output sample $y[n]$ has two contributors, namely the current input sample $x[n]$ and the previous output sample $y[n-1]$. The contribution of the current input sample is proportional to α and the contribution of the previous output sample is proportional to $1-\alpha$. Smaller values of α lead to smaller contributions from each input sample, and therefore a smoother output signal. Writing the difference equation given by Eqn. (2.11) for several values of the sample index n we obtain

$$\begin{aligned} n=0: &\quad y[0] = (1-\alpha)\, y[-1] + \alpha\, x[0] \\ n=1: &\quad y[1] = (1-\alpha)\, y[0] + \alpha\, x[1] \\ n=2: &\quad y[2] = (1-\alpha)\, y[1] + \alpha\, x[2] \end{aligned}$$

and so on. Since the difference equation in Eqn. (2.11) is linear with constant coefficients, the exponential smoother would be an example of a DTLTI system provided that it is initially relaxed which, in this case, implies that $y[-1] = 0$. Figs. 2.11 and 2.12 illustrate the application of the linear exponential smoother to the 2014 Dow Jones Industrial Average data for $\alpha = 0.1$ and $\alpha = 0.3$.

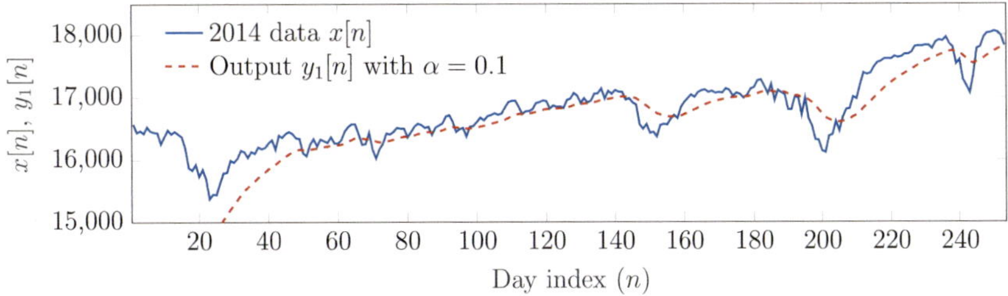

Figure 2.11 – Input signal $x[n]$ representing the Dow Jones Industrial Average daily values for the calendar year 2014 compared to the output of the linear exponential smoother with $\alpha = 0.1$.

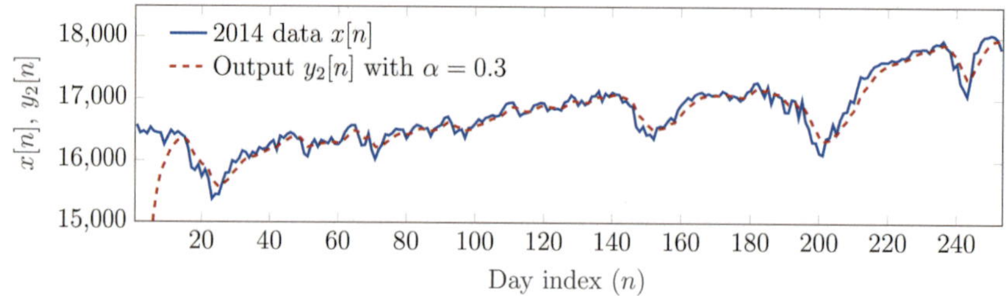

Figure 2.12 – Input signal $x[n]$ representing the Dow Jones Industrial Average daily values for the calendar year 2014 compared to the output of the linear exponential smoother with $\alpha = 0.3$.

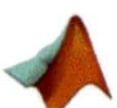

Software resource: See MATLAB Exercise 2.4.

Interactive App: Exponential smoother

The interactive app in `appExpSmooth.m` illustrates the operation of the exponential smoother discussed in Example 2.6. The input and output signals follow the film strip analogy and can be moved by changing the current index through the user interface. The smoothing parameter α can also be varied. Several choices are available for the input signal for experimentation.

a. Set $\alpha = 0.1$. Set the input signal to a unit impulse and observe the output signal of the linear exponential smoother. Increment the sample index n and observe values of output samples. What is the length of the impulse response for this system?

b. Increase the value of the parameter α and observe its effect on the response to a unit impulse.

c. Set $\alpha = 0.1$. Set the input signal to a unit step and observe the output signal. How long does it take for the output signal to approximately reach unit amplitude? Increase the value of α and observe the transition of the output signal from zero to unit amplitude.

Software resource: `appExpSmooth.m`

Example 2.7: Loan payments

A practical everyday use of a difference equation can be found in banking: borrowing money for the purchase of a house or a car. The scenario is familiar to all of us. We borrow the amount that we need to purchase that dream house or dream car, and then pay back a fixed amount for each of a number of periods, often measured in terms of months. At the end of each month the bank will compute our new balance by taking the balance of the previous month, increasing it by the monthly interest rate, and subtracting the payment we made for that month. Let $y[n]$ represent the amount we owe at the end of the n-th month, and let $x[n]$ represent the payment we make in month n. If c is the monthly interest rate expressed as a fraction (for example, 0.01 for a 1 percent monthly interest rate), then the loan balance may be modeled as the output signal of a system with the following difference equation as shown in Fig. 2.13:

$$y[n] = (1 + c)\ y[n-1] - x[n] \tag{2.12}$$

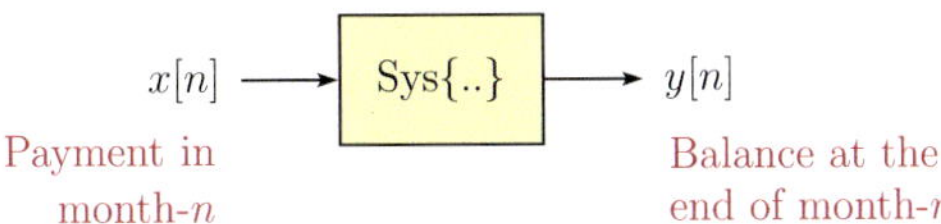

Figure 2.13 – System model for loan balance

Expressing the input-output relationship of the system through a difference equation allows us to analyze it in a number of ways. Let A represent the initial amount borrowed in month $n = 0$, and let the monthly payment be equal to B for months $n = 1, 2, \ldots$. One method of finding $y[n]$ would be to solve the difference equation in Eqn. (2.12) with the input signal

$$x[n] = B\,u[n-1]$$

and the initial condition $y[0] = A$. Alternatively we can treat the borrowed amount as a negative payment in month-0, and solve the difference equation with the input signal

$$x[n] = -A\,\delta[n] + B\,u[n-1] \tag{2.13}$$

and the initial condition $y[-1] = 0$. In later parts of this text, as we develop the tools we need for analysis of systems, we will revisit this example. After we learn the techniques for solving linear constant-coefficient difference equations, we will be able to find the output of this system in response to the input signal given by Eqn. (2.13).

Example 2.8: A nonlinear dynamics example

An interesting example of the use of difference equations is seen in chaos theory and its applications to nonlinear dynamic systems. Let the output $y[n]$ of the system represent the population of a particular kind of species in a particular environment, for example, a certain type of plant in the rain forest. The value $y[n]$ is normalized to be in the range $0 < y[n] < 1$ with the value 1 corresponding to the maximum capacity for the species which may depend on the availability of resources such as food, water, direct sunlight, etc. In the *logistic growth* model, the population growth rate is assumed to be proportional to the remaining capacity $1 - y[n]$, and population change from one generation to the next is given by

$$y[n] = r\big(1 - y[n-1]\big)\ y[n-1] \tag{2.14}$$

where r is a constant. This is an example of a nonlinear system that does not have an input signal. Instead, it produces an output signal based on its initial state.

Software resource: exdt_2_8.m

Example 2.9: Newton-Raphson method for finding a root of a function

In numerical analysis, one of the simplest methods for finding the real roots of a well-behaved function is the Newton-Raphson technique. Consider a function $u = f(w)$ shown in Fig. 2.14. Our goal is to find the value of w for which $u = f(w) = 0$. Starting with an initial guess $w = w_0$ for the solution, we draw the line that is tangent to the function at that point, as shown in Fig. 2.14. The value $w = w_1$ where the tangent line intersects the real axis is our next, improved, guess for the root. The slope of the tangent line is $f'(w_0)$ and it passes through the point $\big(w_0, f(w_0)\big)$. Therefore, the equation of the tangent line is

$$u - f(w_0) = f'(w_0)\,(w - w_0) \tag{2.15}$$

At the point where the tangent line intersects the horizontal axis we have $u = 0$ and $w = w_1$. Substituting these values into Eqn. (2.15) we have

$$-f(w_0) = f'(w_0)\,(w_1 - w_0)$$

and solving for w_1 results in

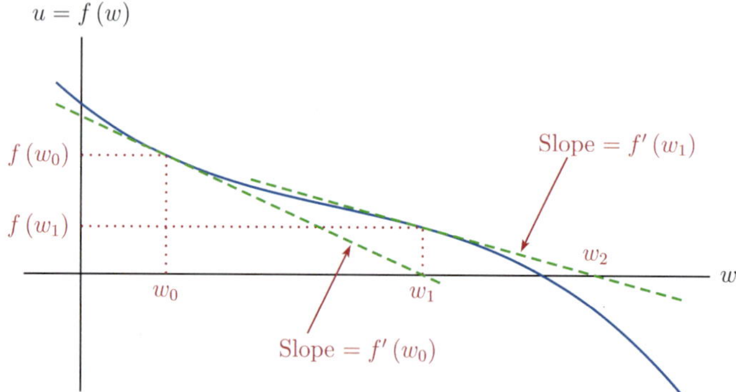

Figure 2.14 – Newton Raphson root finding technique.

$$w_1 = w_0 - \frac{f(w_0)}{f'(w_0)} \tag{2.16}$$

Thus, from an initial guess w_0 for the solution, we find a better guess w_1 through the use of Eqn. (2.16). Repeating this process, an even closer guess can be found as

$$w_2 = w_1 - \frac{f(w_1)}{f'(w_1)}$$

and so on. The technique can be modeled as a discrete-time system. Let the output signal $y[n]$ represent successive guesses for the root, that is, $y[n] = w_n$. The next successive guess can be obtained from the previous one through the difference equation

$$y[n] = y[n-1] - \frac{f\big(y[n-1]\big)}{f'\big(y[n-1]\big)} \tag{2.17}$$

As an example of converting this procedure to a discrete-time system, let the function be

$$u = f(w) = w^2 - A$$

where A is any positive real number. We can write

$$f'(w) = 2w \quad \text{and} \quad \frac{f(w)}{f'(w)} = \frac{w}{2} - \frac{A}{2w} \tag{2.18}$$

The difference equation in Eqn. (2.17) becomes

$$\begin{aligned} y[n] &= y[n-1] - \frac{y[n-1]}{2} + \frac{A}{2y[n-1]} \\ &= \frac{1}{2}\left(y[n-1] + \frac{A}{y[n-1]}\right) \end{aligned} \tag{2.19}$$

Obviously Eqn. (2.19) represents a nonlinear difference equation. Let us use this system to iteratively find the square-root of 10 which we know is $\sqrt{10} = 3.162278$. The function the root of which we are seeking is $f(w) = w^2 - 10$. Starting with an initial guess of $y[0] = 1$ we have

$$y[1] = \frac{1}{2}\left(y[0] + \frac{10}{y[0]}\right) = \frac{1}{2}\left(1 + \frac{10}{1}\right) = 5.5$$

The next iteration produces

$$y[2] = \frac{1}{2}\left(y[1] + \frac{10}{y[1]}\right) = \frac{1}{2}\left(5.5 + \frac{10}{5.5}\right) = 3.659091$$

Continuing in this fashion we obtain $y[3] = 3.196005$, $y[4] = 3.162456$, and $y[5] = 3.162278$ which is accurate as the square-root of 10 up to the sixth decimal digit.

Software resource: `exdt_2_9.m`

2.4 Constant-Coefficient Linear Difference Equations

DTLTI systems can be modeled with constant-coefficient linear difference equations. A linear difference equation is one in which current, past, and perhaps even future samples of the input and the output signals can appear as linear terms. Furthermore, a constant coefficient linear difference equation is one in which the linear terms involving input and output signals appear with coefficients that are constant, independent of time or any other variable.

The moving-average filters we have explored in Examples 2.3, 2.4, and 2.5 were described by constant-coefficient linear difference equations:

$$\begin{aligned} &\text{Length-2:} \quad y[n] = \frac{1}{2}\,x[n] + \frac{1}{2}\,x[n-1] \\ &\text{Length-4:} \quad y[n] = \frac{1}{4}\,x[n] + \frac{1}{4}\,x[n-1] + \frac{1}{4}\,x[n-2] + \frac{1}{4}\,x[n-3] \\ &\text{Length-}N\text{:} \quad y[n] = \frac{1}{N}\sum_{k=0}^{N-1} x[n-k] \end{aligned}$$

A common characteristic of these three difference equations is that past or future samples of the output signal do not appear on the right side of any of them. In each of the three difference equations the output $y[n]$ is computed as a function of current and past samples of the input signal. In contrast, reconsider the difference equation for the exponential smoother

$$y[n] = (1-\alpha)\; y[n-1] + \alpha\, x[n]$$

or the difference equation for the system that computes the current balance of a loan

$$y[n] = (1+c)\; y[n-1] - x[n]$$

Both of these systems also have constant-coefficient linear difference equations (we assume that parameters α and c are constants). What sets the last two systems apart from the three moving average filters above is that they also have *feedback* in the form of past samples of the output signal appearing on the right side of the difference equation. The value of $y[n]$ depends on the past output sample $y[n-1]$.

Examples 2.8 and 2.9, namely the logistic growth model and the Newton-Raphson algorithm for finding a square root, utilized the difference equations

$$y[n] = r\left(1 - y[n-1]\right)\, y[n-1]$$

and

$$y[n] = \frac{1}{2}\left(y[n-1] + \frac{A}{y[n-1]}\right)$$

Both of these difference equations are nonlinear since they contain nonlinear terms of $y[n-1]$. A general constant-coefficient linear difference equation representing a DTLTI system is in the form

$$\begin{aligned} a_0\, y[n] + a_1\, y[n-1] + \ldots + a_{N-1}\, y[n-N+1] + a_N\, y[n-N] = \\ b_0\, x[n] + b_1\, x[n-1] + \ldots + b_{M-1}\, x[n-M+1] + b_M\, x[n-M] \end{aligned} \tag{2.20}$$

or in closed summation form as shown below.

Constant-coefficient linear difference equation:

$$\sum_{k=0}^{N} a_k\, y[n-k] = \sum_{k=0}^{M} b_k\, x[n-k] \tag{2.21}$$

The order of the difference equation (and therefore the order of the system it represents) is the larger of N and M. For example, the length-2 moving average filter discussed in Example 2.4 is a first-order system. Similarly, the orders of the length-4 and the length-N moving average filters of Examples 2.5 and 2.3 are 3 and $(N-1)$, respectively.

A note of clarification is in order here: The general form we have used in Eqns. (2.20) and (2.21) includes current and past samples of $x[n]$ and $y[n]$ but no future samples. This is for practical purposes only. The inclusion of future samples in a difference equation for the computation of the current output would not affect the linearity and the time invariance of the system represented by that difference equation, as long as the future samples also appear as linear terms and with constant coefficients. For example, the difference equation

$$y[n] = y[n-1] + x[n+2] - 3x[n+1] + 2x[n]$$

is still a constant-coefficient linear difference equation and it may still correspond to a DTLTI system. We just have an additional challenge in computing the output signal through the use of this difference equation: We need to know future values of the input signal. For example, computation of $y[45]$ requires the knowledge of $x[46]$ and $x[47]$ in addition to other terms. We will explore this further when we discuss the causality property later in this chapter.

Example 2.10: Checking linearity and time invariance of a difference equation

Determine whether the first-order constant-coefficient linear difference equation in the form

$$a_0\, y[n] + a_1\, y[n-1] = b_0\, x[n]$$

represents a DTLTI system.

Solution: Assume that two input signals $x_1[n]$ and $x_2[n]$ produce the corresponding output signals $y_1[n]$ and $y_2[n]$, respectively. Each of the signal pairs $x_1[n] \leftrightarrow y_1[n]$ and $x_2[n] \leftrightarrow y_2[n]$ must satisfy the difference equation, so we have

$$a_0\, y_1[n] + a_1\, y_1[n-1] = b_0\, x_1[n] \tag{2.22}$$

and

$$a_0\, y_2[n] + a_1\, y_2[n-1] = b_0\, x_2[n] \tag{2.23}$$

Let a new input signal be constructed as a linear combination of $x_1[n]$ and $x_2[n]$ as

$$x_3[n] = \alpha_1\, x_1[n] + \alpha_2\, x_2[n]$$

For the system described by the difference equation to be linear, its response to the input signal $x_3[n]$ must be

$$y_3[n] = \alpha_1\, y_1[n] + \alpha_2\, y_2[n] \tag{2.24}$$

and the input output signal pair $x_3[n] \leftrightarrow y_3[n]$ must also satisfy the difference equation. Substituting $y_3[n]$ into the left side of the difference equation yields

$$a_0\, y_3[n] + a_1\, y_3[n-1] = a_0\, \big(\alpha_1\, y_1[n] + \alpha_2\, y_2[n]\big) + a_1\, \big(\alpha_1\, y_1[n-1] + \alpha_2\, y_2[n-1]\big) \tag{2.25}$$

By rearranging the terms on the right side of Eqn. (2.25) we get

$$a_0\, y_3[n] + a_1\, y_3[n-1] = \alpha_1\, \big(a_0\, y_1[n] + a_1\, y_1[n-1]\big) + \alpha_2\, \big(a_0\, y_2[n] + a_1\, y_2[n-1]\big) \tag{2.26}$$

Substituting Eqns. (2.22) and (2.23) into Eqn. (2.26) leads to

$$\begin{aligned} a_0\, y_3[n] + a_1\, y_3[n-1] =& \alpha_1\, \big(b_0\, x_1[n]\big) + \alpha_2\, \big(b_0\, x_2[n]\big) \\ =& b_0\, \big(\alpha_1\, x_1[n] + \alpha_2\, x_2[n]\big) \\ =& b_0\, x_3[n] \end{aligned}$$

proving that the input-output signal pair $x_3[n] \leftrightarrow y_3[n]$ satisfies the difference equation.

Before we can claim that the difference equation in question represents a linear system, we need to check the initial value of $y[n]$. We know from previous discussion that a first-order difference equation such as the one given in the problem statement can be solved iteratively starting at a specified value of the index $n = n_0$ provided that the value of the output sample at index $n = n_0 - 1$ is known. For example, let $n_0 = 0$, and let $y[n_0 - 1] = y[-1] = A$. Starting with the specified value of $y[-1]$ we can determine $y[0]$ as

$$y[0] = \left(-\frac{a_1}{a_0}\right) A + b_0\, x[0]$$

Having determined the value of $y[0]$ we can find $y[1]$ as

$$\begin{aligned} y[1] &= \left(-\frac{a_1}{a_0}\right) y[0] + b_0\, x[1] \\ &= \left(-\frac{a_1}{a_0}\right) \left[\left(-\frac{a_1}{a_0}\right) A + b_0\, x[0]\right] + b_0\, x[1] \end{aligned}$$

and continue in this fashion. Clearly the result obtained is dependent on the initial value $y[-1] = A$. Since the $y_1[n]$, $y_2[n]$, and $y_3[n]$ used in the development above are all solutions of the difference equation for input signals $x_1[n]$, $x_2[n]$, and $x_3[n]$, respectively, they must each satisfy the specified initial condition, that is,

$$y_1[-1] = A\,, \quad y_2[-1] = A\,, \quad y_3[-1] = A$$

In addition, the linearity condition in Eqn. (2.24) must be satisfied for all values of the index n including $n = -1$:

$$y_3[-1] = \alpha_1\, y_1[-1] + \alpha_2\, y_2[-1]$$

Thus we are compelled to conclude that the system in question is linear only if $y[-1] = 0$. In the general case, we need $y[n_0 - 1] = 0$ if the solution is to start at index $n = n_0$.

Our next task is to check the time invariance property of the system described by the difference equation. If we replace the index n with $n - m$, the difference equation becomes

$$a_0\, y[n-m] + a_1\, y[n-m-1] = b_0\, x[n-m]$$

Delaying the input signal $x[n]$ by m samples causes the output signal $y[n]$ to be delayed by the same amount. The system is time-invariant.

In Example 2.10 we have verified that the first-order constant-coefficient linear difference equation corresponds to a DTLTI system provided that its initial state is zero. We are now ready to generalize that result to the constant-coefficient linear difference equation of any order. Let the two input signals $x_1[n]$ and $x_2[n]$ produce the output signals $y_1[n]$ and $y_2[n]$, respectively. The input-output signal pairs $x_1[n] \longrightarrow y_1[n]$ and $x_2[n] \longrightarrow y_2[n]$ satisfy the difference equation, so we can write

$$\sum_{k=0}^{N} a_k\, y_1[n-k] = \sum_{k=0}^{M} b_k\, x_1[n-k] \tag{2.27}$$

and

$$\sum_{k=0}^{N} a_k\, y_2[n-k] = \sum_{k=0}^{M} b_k\, x_2[n-k] \tag{2.28}$$

To test linearity of the system we will construct a new input signal as a linear combination of $x_1[n]$ and $x_2[n]$:

$$x_3[n] = \alpha_1\, x_1[n] + \alpha_2\, x_2[n]$$

If the system described by the difference equation is linear, its response to the input signal $x_3[n]$ must be

$$y_3[n] = \alpha_1\, y_1[n] + \alpha_2\, y_2[n] \tag{2.29}$$

On the linearity of a constant-coefficient difference equation:

We will test the input-output signal pair $x_3[n] \longrightarrow y_3[n]$ through the difference equation. Substituting $y_3[n]$ into the left side of the difference equation yields

$$\sum_{k=0}^{N} a_k\, y_3[n-k] = \sum_{k=0}^{N} a_k \left(\alpha_1\, y_1[n-k] + \alpha_2\, y_2[n-k]\right) \tag{2.30}$$

Rearranging the terms on the right side of Eqn. (2.30) and separating them into two separate summations yields

$$\sum_{k=0}^{N} a_k\, y_3[n-k] = \alpha_1 \sum_{k=0}^{N} a_k\, y_1[n-k] + \alpha_2 \sum_{k=0}^{N} a_k\, y_2[n-k] \tag{2.31}$$

The two summations on the right side of Eqn. (2.31) can be substituted with their equivalents from Eqns. (2.27) and (2.28), resulting in

$$\sum_{k=0}^{N} a_k\, y_3[n-k] = \alpha_1 \sum_{k=0}^{M} b_k\, x_1[n-k] + \alpha_2 \sum_{k=0}^{M} b_k\, x_2[n-k] \tag{2.32}$$

Finally, we will combine the two summations on the right side of Eqn. (2.32) back into one summation to obtain

$$\begin{aligned}\sum_{k=0}^{N} a_k\, y_3[n-k] &= \sum_{k=0}^{M} b_k \left(\alpha_1\, x_1[n-k] + \alpha_2\, x_2[n-k]\right) \\ &= \sum_{k=0}^{M} b_k\, x_3[n-k] \end{aligned} \tag{2.33}$$

We conclude that the input-output signal pair $x_3[n] \longrightarrow y_3[n]$ also satisfies the difference equation. The restriction discussed in Example 2.10 regarding the initial conditions will be applicable here as well. If we are interested in finding a unique solution for $n \ge n_0$, then the initial values

$$y[n_0-1]\,, \quad y[n_0-2]\,,\ldots, \quad y[n_0-N]$$

are needed. The linearity condition given by Eqn. (2.29) must be satisfied for all values of n including index values $n = n_0-1, n_0-2, \ldots, n_0-N$. Consequently, the system that corresponds to the difference equation in Eqn. (2.21) is linear only if all the initial conditions are zero.

Linearity of a constant-coefficient difference equation:

The constant-coefficient difference equation

$$\sum_{k=0}^{N} a_k\, y[n-k] = \sum_{k=0}^{M} b_k\, x[n-k]$$

represents a linear system provided that all initial conditions are zero. If we are looking for a unique solution starting at index $n = n_0$, then for linearity we need

$$y[n_0-k] = 0$$

for $k = 1, \ldots, N$.

Next we need to check for time invariance. Replacing the index n with $n-m$ in Eqn. (2.21) we get

$$\sum_{k=0}^{N} a_k\, y[n-m-k] = \sum_{k=0}^{M} b_k\, x[n-m-k] \tag{2.34}$$

indicating that the input-output signal pair $x[n-m] \longrightarrow y[n-m]$ also satisfies the difference equation. Thus, the constant-coefficient linear difference equation is time-invariant.

2.5 Solving Difference Equations

The output signal of a discrete-time system in response to a specified input signal can be determined by solving the corresponding difference equation. In some of the examples of Section 2.3 we have already experimented with one method of solving a difference equation, namely the *iterative method.* Consider again the difference equation for the exponential smoother of Example 2.6. By writing the difference equation for each value of the index n we were able to obtain the output signal $y[n]$ one sample at a time. Given the initial value $y[-1]$ of the output signal, its value for $n=0$ is found by

$$y[0] = (1-\alpha)\; y[-1] + \alpha\, x[n]$$

Setting $n=1$ we obtain

$$y[1] = (1-\alpha)\; y[0] + \alpha\, x[1]$$

Repeating for $n=2$ leads to

$$y[2] = (1-\alpha)\; y[1] + \alpha\, x[2]$$

and we can continue in this fashion indefinitely. The function `ss_smooth()` developed in MATLAB Exercise 2.4 is an implementation of the iterative solution of this difference equation.

The iterative solution method is not limited to DTLTI systems; it can also be used for solving the difference equations of nonlinear and/or time-varying systems. Consider, for example, the nonlinear difference equation for the logistic growth model of Example 2.8. For a specified parameter value r and initial value $y[-1]$, the output at $n=0$ is

$$y[0] = r\left(1 - y[-1]\right)\, y[-1]$$

Next, $y[1]$ is computed from $y[0]$ as

$$y[1] = r\left(1 - y[0]\right)\, y[0]$$

and so on. Fig. 2.15 shows the first 50 samples of the solution obtained in this fashion for $r=3.1$ and $y[-1]=0.3$.

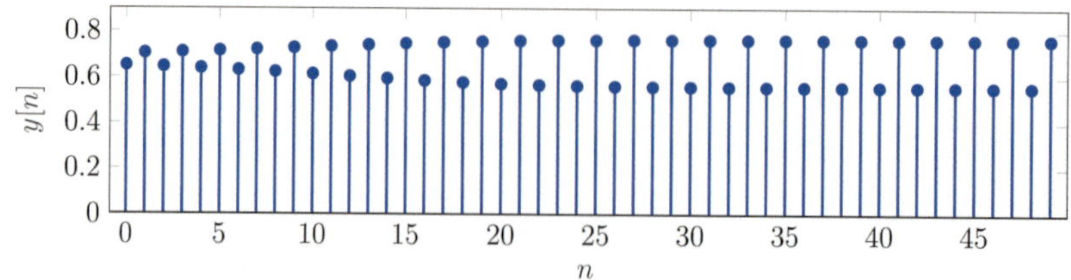

Figure 2.15 – First 50 samples of the iterative solution of the difference equation for the logistic growth model.

Iterative solution of a difference equation can be also used as the basis of implementing a discrete-time system on real-time signal processing hardware. One shortcoming of this approach, however, is the lack of a complete analytical solution. Each time we iterate through the difference equation we obtain one more sample of the output signal, but we do not get an expression or a formula for computing the output for an arbitrary value of the index n. If we need to know $y[1527]$, we must iteratively compute samples $y[0]$ through $y[1526]$ first.

Software resource: See MATLAB Exercise 2.5.

In the rest of this section we will concentrate our efforts on developing an analytical method for solving constant-coefficient linear difference equations. Analytical solution of nonlinear and/or time-varying difference equations is generally difficult or impossible, and will not be considered further in this text.

The solution method we are about to present exhibits a lot of similarities to the method employed for solving constant-coefficient ordinary differential equations for continuous-time systems. We will recognize two separate components of the output signal $y[n]$ in the form

$$y[n] = y_h[n] + y_p[n] \tag{2.35}$$

The term $y_h[n]$ is the solution of the *homogeneous difference equation* found by setting $x[n] = 0$ in Eqn. (2.21) for all values of n:

$$\sum_{k=0}^{N} a_k\, y[n-k] = 0 \tag{2.36}$$

Thus, $y_h[n]$ is the signal at the output of the system when no input signal is applied to it. As in the continuous-time case, we will refer to $y_h[n]$ as the *homogeneous solution* of the difference equation or, equivalently, as the *natural response* of the system to which it corresponds. It depends on the structure of the system which is expressed through the set of coefficients a_i for $i = 0,\ldots,N$. Furthermore, it depends on the initial state of the system that is expressed through the output samples $y[n_0-1], y[n_0-2],\ldots,y[n_0-N]$. (Recall that n_0 is the beginning index for the solution; usually we will use $n_0 = 0$.) When we discuss the stability property of DTLTI systems in Section 2.9 we will discover that, for a stable system, $y_h[n]$ approaches zero for large positive and negative values of the index n.

The second term $y_p[n]$ in Eqn. (2.35) is the part of the solution that is due to the input signal $x[n]$ applied to the system. It is referred to as the *particular solution* of the difference equation. It depends on both the input signal $x[n]$ and the internal structure of the system. It is independent of the initial state of the system. The combination of the homogeneous solution and the particular solution is referred to as the *forced solution* or the *forced response*.

2.5.1 Finding the natural response of a discrete-time system

We will begin the discussion of how to solve the homogeneous equation by revisiting the linear exponential smoother first encountered in Example 2.6. Recall that the solution of the homogeneous difference equation is the natural response of the system.

Example 2.11: Natural response of exponential smoother

Find the natural response of the exponential smoother defined in Example 2.6 if $y[-1]=2$.

Solution: The difference equation for the exponential smoother was given in Eqn. (2.6). The homogeneous difference equation is found by setting $x[n]=0$:

$$y[n]=(1-\alpha)\,y[n-1] \tag{2.37}$$

The natural response $y_h[n]$ yet to be determined must satisfy the homogeneous difference equation. We need to start with an educated guess for the type of signal $y_h[n]$ must be, and then adjust any relevant parameters. Therefore, looking at Eqn. (2.37), we ask the question: "What type of discrete-time signal remains proportional to itself when delayed by one sample?" A possible answer is a signal in the form

$$y_h[n]=c\,z^n \tag{2.38}$$

where z is a yet undetermined constant. Delaying $y[n]$ of Eqn. (2.38) by one sample we get

$$y_h[n-1]=c\,z^{n-1}=z^{-1}c\,z^n=z^{-1}y[n] \tag{2.39}$$

Substituting Eqns. (2.38) and (2.39) into the homogeneous difference equation yields

$$c\,z^n=(1-\alpha)\,z^{-1}c\,z^n$$

or, equivalently

$$c\,z^n\left[1-(1-\alpha)\,z^{-1}\right]=0$$

which requires one of the following conditions to be true for all values of n:

a. $c\,z^n=0$

b. $\left[1-(1-\alpha)\,z^{-1}\right]=0$

We cannot use the former condition since it leads to the trivial solution $y[n]=c\,z^n=0$, and is obviously not very useful. Furthermore, the initial condition $y[-1]=2$ cannot be satisfied using this solution. Therefore we must choose the latter condition and set $z=(1-\alpha)$ to obtain

$$y_h[n]=c\,(1-\alpha)^n\,,\quad \text{for } n\ge 0 \tag{2.40}$$

The constant c is determined based on the desired initial state of the system. We want $y[-1]=2$, so we impose it as a condition on the solution found in Eqn. (2.40):

$$y_h[-1]=c\,(1-\alpha)^{-1}=2$$

This yields $c=2\,(1-\alpha)$ and

$$y_h[n]=2\,(1-\alpha)\,(1-\alpha)^n=2\,(1-\alpha)^{n+1}$$

The natural response found is shown in Figs. 2.16 and 2.17 for $\alpha=0.1$ and $\alpha=0.2$.

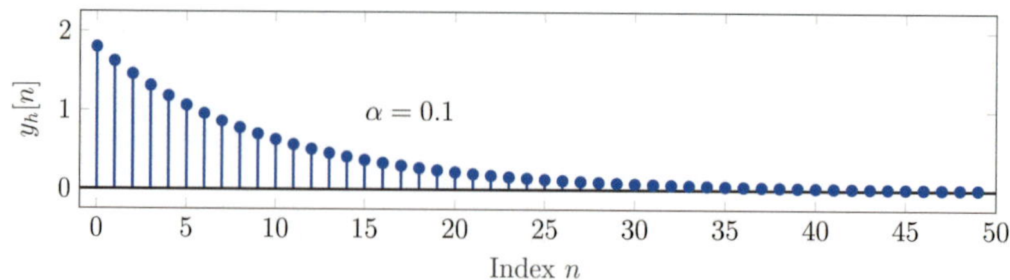

Figure 2.16 – The natural response of the linear exponential smoother for $\alpha=0.1$.

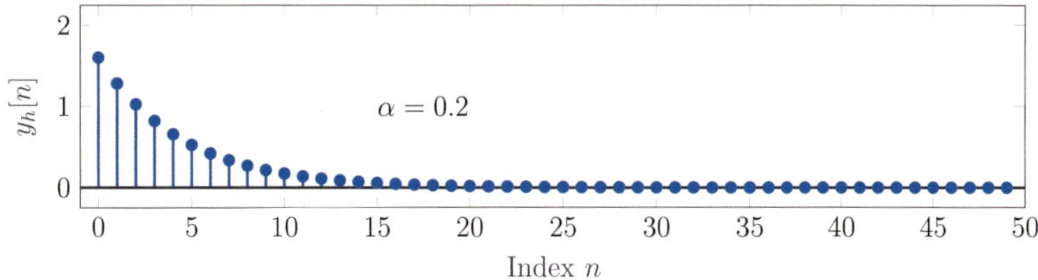

Figure 2.17 – The natural response of the linear exponential smoother for $\alpha = 0.2$.

Software resources: exdt_2_11.m , exdt_2_11_Live.mlx

Let us now consider the solution of the general homogeneous difference equation in the form

$$\sum_{k=0}^{N} a_k \, y[n-k] = 0 \tag{2.41}$$

We will start with the same initial guess that we used in Example 2.11:

$$y_h[n] = c\, z^n \tag{2.42}$$

Shifted versions of the prescribed homogeneous solution are

$$\begin{aligned} y_h[n-1] &= c\, z^{n-1} = z^{-1} c\, z^n \\ y_h[n-2] &= c\, z^{n-2} = z^{-2} c\, z^n \\ y_h[n-3] &= c\, z^{n-3} = z^{-3} c\, z^n \end{aligned}$$

which can be expressed in the general form

$$y_h[n-k] = c\, z^{n-k} = z^{-k} c\, z^n \tag{2.43}$$

Which value (or values) of z can be used in the homogeneous solution? We will find the answer by substituting Eqn. (2.43) into Eqn. (2.41):

$$\sum_{k=0}^{N} a_k \, z^{-k} c\, z^n = 0 \tag{2.44}$$

The term cz^n is independent of the summation index k, and can be factored out to yield

$$c\, z^n \sum_{k=0}^{N} a_k \, z^{-k} = 0 \tag{2.45}$$

There are two ways to satisfy Eqn. (2.45):

1. $c\, z^n = 0$

   This leads to the trivial solution $y[n] = 0$ for the homogeneous equation and is therefore not very interesting. Also, we have no means of satisfying any initial conditions with this solution other than $y[-i] = 0$ for $i = 1, \ldots, N$.

2. $\sum_{k=0}^{N} a_k z^{-k} = 0$

   This is called the *characteristic equation* of the system. Values of z that are the solutions of the characteristic equation can be used in exponential functions as solutions of the homogeneous difference equation.

The characteristic equation:

$$\sum_{k=0}^{N} a_k z^{-k} = 0 \tag{2.46}$$

The characteristic equation for a DTLTI system is found by starting with the homogeneous difference equation and replacing delayed versions of the output signal with the corresponding negative powers of the complex variable z.

To obtain the characteristic equation, substitute:

$$y[n-k] \rightarrow z^{-k} \tag{2.47}$$

The characteristic equation can be written in open form as

$$a_0 + a_1 z^{-1} + \ldots + a_{N-1} z^{-N+1} + a_N z^{-N} = 0$$

If we want to work with non-negative powers of z, we could simply multiply both sides of the characteristic equation by z^N to obtain

$$a_0 z^N + a_1 z^{N-1} + \ldots + a_{N-1} z^1 + a_N = 0 \tag{2.48}$$

The polynomial on the left side of the equal sign in Eqn. (2.48) is the *characteristic polynomial* of the DTLTI system. Let the roots of the characteristic polynomial be $z_1, z_2, \ldots, z_N$ so that Eqn. (2.48) can be written as

$$a_0\,(a - z_1)\,(a - z_2)\ldots(a - z_N) = 0 \tag{2.49}$$

Any of the roots of the characteristic polynomial can be used in a signal in the form

$$y_i[n] = c_i z_i^n\,, \quad i = 1, \ldots, N \tag{2.50}$$

which satisfies the homogeneous difference equation:

$$\sum_{k=0}^{N} a_k y_i[n-k] = 0 \quad \text{for } i = 1, \ldots, N \tag{2.51}$$

Furthermore, any linear combination of all valid terms in the form of Eqn. (2.50) satisfies the homogeneous difference equation as well, so we can write

$$y_h[n] = c_1 z_1^n + c_2 z_2^n + \ldots + c_N z_N^n = \sum_{k=1}^{N} c_k z_k^n \tag{2.52}$$

The coefficients $c_1, c_2, \ldots, c_N$ are determined from the initial conditions. The exponential terms z_i^n in the homogeneous solution given by Eqn. (2.52) are the *modes of the system.* In later parts of this text we will see that the modes of a DTLTI system correspond to the *poles of the system function* and the *eigenvalues of the state matrix.*

Example 2.12: Natural response of second-order system

A second-order system is described by the difference equation

$$y[n] - \frac{5}{6}\, y[n-1] + \frac{1}{6}\, y[n-2] = 0$$

Determine the natural response of this system for $n \ge 0$ subject to initial conditions

$$y[-1] = 19\,, \quad \text{and } y[-2] = 53$$

Solution: The characteristic equation is

$$z^2 - \frac{5}{6}\, z + \frac{1}{6} = 0$$

with roots $z_1 = 1/2$ and $z_2 = 1/3$. Therefore the homogeneous solution of the difference equation is

$$y_h[n] = c_1 \left(\frac{1}{2}\right)^n + c_2 \left(\frac{1}{3}\right)^n$$

for $n \ge 0$. The coefficients c_1 and c_2 need to be determined from the initial conditions. We have

$$y_h[-1] = c_1 \left(\frac{1}{2}\right)^{-1} + c_2 \left(\frac{1}{3}\right)^{-1} = 2\,c_1 + 3\,c_2 = 19 \tag{2.53}$$

and

$$y_h[-2] = c_1 \left(\frac{1}{2}\right)^{-2} + c_2 \left(\frac{1}{3}\right)^{-2} = 4\,c_1 + 9\,c_2 = 53 \tag{2.54}$$

Solving Eqns. (2.53) and (2.54) yields $c_1 = 2$ and $c_2 = 5$. The natural response of the system is

$$y_h[n] = 2 \left(\frac{1}{2}\right)^n u[n] + 5 \left(\frac{1}{3}\right)^n u[n]$$

 **Software resources:** `exdt_2_12.m` , `exdt_2_12_Live.mlx`

In Example 2.12 the characteristic equation obtained from the homogeneous difference equation had two distinct roots that were both real-valued, allowing the homogeneous solution to be written in the standard form of Eqn. (2.52). There are other possibilities as well. We will consider three possible scenarios:

Case 1. All roots are distinct and real-valued.
This leads to the homogeneous solution

$$y[n] = \sum_{k=1}^{N} c_k z_k^n \tag{2.55}$$

for $n \ge n_0$ as we have seen in Example 2.12. The value of the real root z_k determines the type of contribution made to the homogeneous solution by the term $c_k z_k^n$. If $|z_k| < 1$ then z_k^n decays exponentially over time. Conversely, $|z_k| > 1$ leads to a term z_k^n that grows exponentially. A negative value for z_k causes the corresponding term in the homogeneous solution to have alternating positive and negative sample amplitudes. Possible forms of the contribution z_k^n are shown in Fig. 2.18.

Case 2. Characteristic polynomial has complex-valued roots.
Since the difference equation and its characteristic polynomial have only real-valued coefficients, any complex roots of the characteristic polynomial must appear in conjugate pairs. Therefore, if

$$z_{1a} = r_1\, e^{j\Omega_1}$$

is a complex root, then its conjugate

$$z_{1b} = z_{1a}^* = r_1\, e^{-j\Omega_1}$$

must also be a root. Let the part of the homogeneous solution that is due to these two roots be

$$\begin{aligned} y_{h1}[n] =& c_{1a}\, z_{1a}^n + c_{1b}\, z_{1b}^n \\ =& c_{1a}\, r_1^n\, e^{j\Omega_1 n} + c_{1b}\, r_1^n\, e^{-j\Omega_1 n} \end{aligned} \tag{2.56}$$

The coefficients c_{1a} and c_{1b} are yet to be determined from the initial conditions. Since the coefficients of the difference equation are real-valued, the solution $y_h[n]$ must also be real. Furthermore, $y_{h1}[n]$, the part of the solution that is due to the complex conjugate pair of roots we are considering, must also be real. This implies that the coefficients c_{1a} and c_{1b} must form a complex conjugate pair. We will write the two coefficients in polar complex form as

$$c_{1a} = |c_1|\, e^{j\theta_1}\,, \quad \text{and } c_{1b} = |c_1|\, e^{-j\theta_1} \tag{2.57}$$

Substituting Eqn. (2.57) into Eqn. (2.56) we obtain

$$\begin{aligned} y_{h1}[n] =& |c_1|\, r_1^n\, e^{j(\Omega_1 n+\theta_1)} + |c_1|\, r_1^n\, e^{-j(\Omega_1 n+\theta_1)} \\ =& 2\,|c_1|\, r_1^n \cos(\Omega_1 n+\theta_1) \end{aligned} \tag{2.58}$$

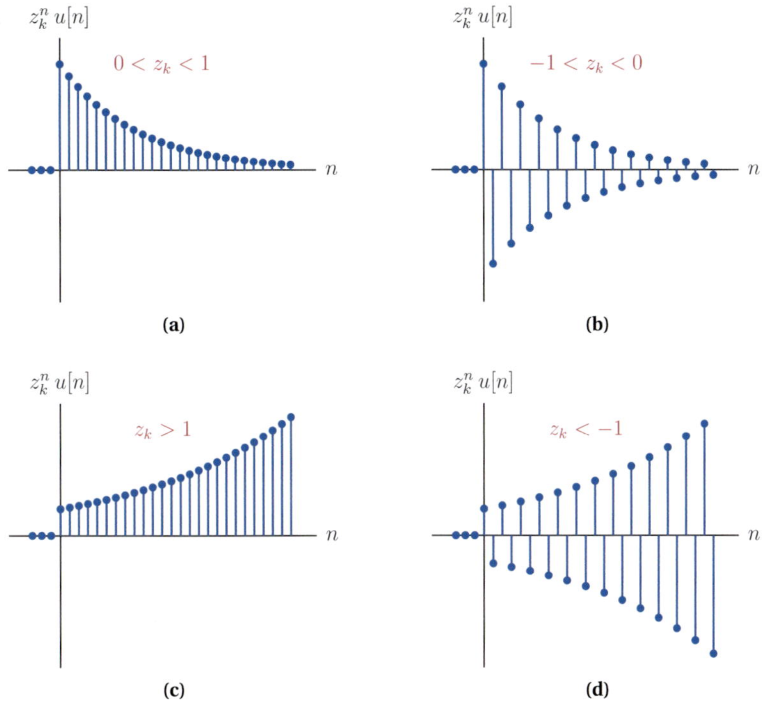

Figure 2.18 – The term $z_k^n\, u[n]$ for **(a)** $0 < z_k < 1$, **(b)** $-1 < z_k < 0$, **(c)** $z_k > 1$, and **(d)** $z_k < -1$.

The contribution of a complex conjugate pair of roots to the solution is in the form of a cosine signal multiplied by an exponential signal. The oscillation frequency of the discrete-time cosine signal is determined by Ω_1. The magnitude of the complex conjugate roots, r_1, impacts the amplitude behavior. If $r_1 < 1$, then the amplitude of the cosine signal decays exponentially over time. If $r_1 > 1$ on the other hand, the amplitude of the cosine signal grows exponentially over time. These two possibilities are illustrated in Fig. 2.19. With the use of the appropriate trigonometric identity[1], Eqn. (2.58) can also be written in the alternative form

$$\begin{aligned} y_{h1}[n] =& 2\,|c_1|\cos(\theta_1)\, r_1^n \cos(\Omega_1 n) - 2\,|c_1|\sin(\theta_1)\, r_1^n \sin(\Omega_1 n) \\ =& d_1\, r_1^n \cos(\Omega_1 n) + d_2\, r_1^n \sin(\Omega_1 n) \end{aligned} \tag{2.59}$$

Case 3. Characteristic polynomial has some multiple roots.

Consider again the factored version of the characteristic equation first given by Eqn. (2.49):

$$a_0\,(z - z_1)\,(z - z_2)\ldots(z - z_N) = 0$$

What if the first two roots are equal, that is, $z_2 = z_1$? If we were to ignore the fact that the two roots are equal, we would have a natural response in the form

[1] $\cos(a+b) = \cos(a)\cos(b) - \sin(a)\sin(b)$.

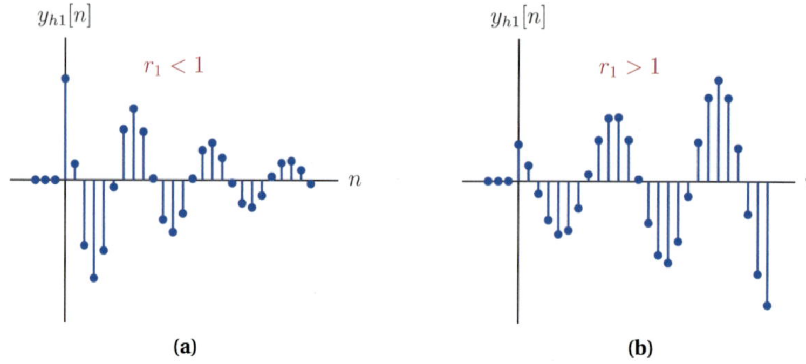

Figure 2.19 – Terms corresponding to a pair of complex conjugate roots of the characteristic equation: **(a)** $r_1 < 0$ and **(b)** $r_1 > 1$.

$$\begin{aligned} y_h[n] &= c_1\, z_1^n + c_2\, z_2^n + \text{other terms} \\ &= c_1\, z_1^n + c_2\, z_1^n + \text{other terms} \\ &= (c_1 + c_2)\, z_1^n + \text{other terms} \\ &= \tilde{c}_1\, z_1^n + \text{other terms} \end{aligned} \tag{2.60}$$

The equality of two roots leads to loss of one of the coefficients that we will need in order to satisfy the initial conditions. In order to gain back the coefficient we have lost, we need an additional term for the two roots at $z = z_1$ and it can be obtained by considering a solution in the form

$$y_h[n] = c_{11}\, z_1^n + c_{12}\, n z_1^n + \text{other terms} \tag{2.61}$$

In general, a root of multiplicity r requires r terms in the homogeneous solution. If the characteristic polynomial has a factor $(z - z_1)^r$ then resulting homogeneous solution is

$$y_h[n] = c_{11}\, z_1^n + c_{12}\, n z_1^n + \ldots + c_{1r}\, n^{r-1} z_1^n + \text{other terms} \tag{2.62}$$

Example 2.13: Natural response of second-order system revisited

Determine the natural response of each of the second-order systems described by the difference equations below:

a. $y[n] - 1.4\, y[n-1] + 0.85\, y[n-2] = 0$
with initial conditions $y[-1] = 5$ and $y[-2] = 7$.

b. $y[n] - 1.6\, y[n-1] + 0.64\, y[n-2] = 0$
with initial conditions $y[-1] = 2$ and $y[-2] = -3$.

Solution:

a. The characteristic equation is

$$z^2 - 1.4\, z + 0.85 = 0$$

which can be solved to yield

$$z_{1,2} = 0.7 \pm j0.6$$

Thus, the roots of the characteristic polynomial form a complex conjugate pair. They can be written in polar complex form as

$$z_{1,2} = 0.922\, e^{\pm j0.7086}$$

which leads us to a homogeneous solution in the form

$$y_h[n] = d_1\ (0.922)^n \cos(0.7086n) + d_2\ (0.922)^n \sin(0.7086n)$$

for $n \geq 0$. The coefficients d_1 and d_2 need to be determined from the initial conditions. Evaluating $y_h[n]$ for $n = -1$ and $n = -2$ we have

$$\begin{aligned} y_h[-1] =& d_1\ (0.922)^{-1} \cos(-0.7086) + d_2\ (0.922)^{-1} \sin(-0.7086) \\ =& 0.8235\, d_1 - 0.7058\, d_2 = 5 \end{aligned} \tag{2.63}$$

and

$$\begin{aligned} y_h[-2] =& d_1\ (0.922)^{-2} \cos(-1.4173) + d_2\ (0.922)^{-2} \sin(-1.4173) \\ =& 0.1800\, d_1 - 1.1625\, d_2 = 7 \end{aligned} \tag{2.64}$$

Solving Eqns. (2.63) and (2.64) yields $d_1 = 1.05$ and $d_2 = -5.8583$. The natural response of the system is

$$y_h[n] = 1.05\ (0.922)^n \cos(0.7086\, n)\ u[n] - 5.8583\ (0.922)^n \sin(0.7086\, n)\ u[n]$$

and is graphed in Fig. 2.20.

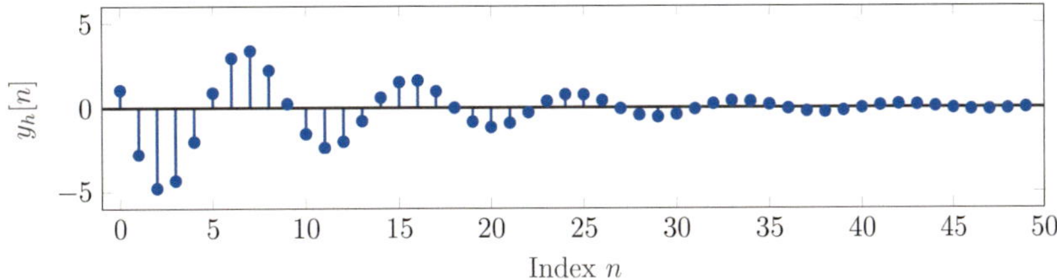

Figure 2.20 – Natural response of the system in Example 2.13 part (a).

b. For this system the characteristic equation is

$$z^2 - 1.6\, z + 0.64 = 0$$

The roots of the characteristic polynomial are $z_1 = z_2 = 0.8$. Therefore we must look for a homogeneous solution in the form of Eqn. (2.61):

$$y_h[n] = c_1\ (0.8)^n + c_2\, n\, (0.8)^n$$

for $n \geq 0$. Imposing the initial conditions at $n = -1$ and $n = -2$ we obtain

$$\begin{aligned} y_h[-1] =& c_1\ (0.8)^{-1} + c_2\ (-1)\, (0.8)^{-1} \\ =& 1.25\, c_1 - 1.25\, c_2 = 2 \end{aligned} \tag{2.65}$$

and

$$\begin{aligned} y_h[-2] =& c_1\,(0.8)^{-2} + c_2\,(-2)\,(0.8)^{-2} \\ =& 1.5625\,c_1 - 3.125\,c_2 = -3 \end{aligned} \tag{2.66}$$

Eqns. (2.65) and (2.66) can be solved to obtain the coefficient values $c_1 = 5.12$ and $c_2 = 3.52$. The natural response is

$$y_h[n] = 5.12\,(0.8)^n + 3.52\,n\,(0.8)^n$$

and is graphed in Fig. 2.21.

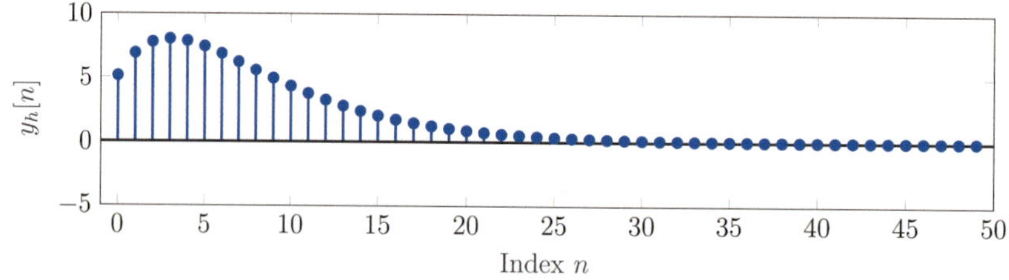

Figure 2.21 – Natural response of the system in Example 2.13 part (b).

Software resources: `exdt_2_13a.m` , `exdt_2_13b.m` , `exdt_2_13_live.mlx`

Interactive App: Natural response of second-order discrete-time system

The interactive app `appNatRespDT.m` illustrates different types of homogeneous solutions for a second-order discrete-time system based on the roots of the characteristic polynomial. Three possible scenarios were explored above, namely distinct real roots, complex conjugate roots, and identical real roots.

In the app the two roots can be specified in terms of their norms and angles using slider controls, and the corresponding natural response can be observed. If the roots are both real, then they can be controlled independently. If the roots are complex, then they move simultaneously to keep their complex conjugate relationship. The locations of the two roots z_1 and z_2 are marked on the complex plane. A circle with unit radius that is centered at the origin of the complex plane is also shown. The difference equation, the characteristic equation, and the analytical solution for the natural response are displayed and updated as the roots are moved.

a. Start with two complex conjugate roots as given in part (a) of Example 2.13. Set the roots as

$$z_{1,2} = 0.922\,e^{\pm 40.5998^\circ}$$

Set the initial values as they were set in the example, that is, $y[-1] = 5$ and $y[-2] = 7$. The natural response displayed should match the result obtained in Example 2.13.

b. Gradually increase the norm of the first root. Since the roots are complex, the second root will also change to keep the complex conjugate relationship of the roots. Observe the natural response as the norm of the complex roots become greater than unity and cross over the circle to the outside. What happens when the roots cross over?

c. Bring the norm of the roots back to $z_{1,2} = 0.922$. Gradually decrease the angle of z_1. The angle of z_2 will also change. How does this impact the shape of the natural response?

d. Set the angle of the z_1 equal to zero, so that the angle of z_2 also becomes zero, and the roots can be moved individually. Set the norms of the two roots as

$$|z_1| = 0.8 \quad \text{and} \quad |z_2| = 0.5$$

and observe the natural response.

e. Gradually increase $|z_1|$ and observe the changes in the natural response, especially as the root moves outside the circle.

Software resource: `appNatRespDT.m`

2.5.2 Finding the forced response of a discrete-time system

In the preceding section we focused our efforts on determining the homogeneous solution of the constant-coefficient linear difference equation. This is equivalent to the natural response $y_h[n]$ of the system when no external input signal exists. As stated in Eqn. (2.35), the complete solution is the sum of the homogeneous solution with the particular solution that corresponds to the input signal applied to the system. The procedure for finding a particular solution for a difference equation is similar to that employed for a differential equation. We start with an educated guess about the form of the particular solution we seek, and then adjust the values of its parameters so that the difference equation is satisfied. The form of the particular solution picked should include the input signal $x[n]$ as well as the delayed input signals $x[n-k]$ that differ in form. For example, if the input signal is $x[n] = K\cos(\Omega_0 n)$, then we assume a particular solution in the form

$$y_p[n] = k_1 \cos(\Omega_0 n) + k_2 \sin(\Omega_0 n)$$

Both cosine and sine terms are needed since $x[n-1]$ is in the form

$$\begin{aligned} x[n-1] =& K\cos(\Omega_0 [n-1]) \\ =& K\cos(\Omega_0)\cos(\Omega_0 n) + K\sin(\Omega_0)\sin(\Omega_0 n) \end{aligned}$$

Other delays of $x[n]$ do not produce any terms that differ from these. If the input signal is in the form $x[n] = n^m$ then the delays of $x[n]$ would contain the terms $n^{m-1}, n^{m-2}, \ldots, n^1, n^0$, and the particular solution is in the form

$$y_p[n] = k_m n^m + k_{m-1} n^{m-1} + \ldots + k_1 n + k_0$$

Table 2.1 lists some of the common types of input signals and the forms of particular solutions to be used for them.

Table 2.1 – Choosing a particular solution for various discrete-time input signals.

Input signal	Particular solution
K (constant)	k_1
$K\,e^{an}$	$k_1\,e^{an}$
$K\,\cos(\Omega_0 n)$	$k_1\,\cos(\Omega_0 n) + k_2\,\sin(\Omega_0 n)$
$K\,\sin(\Omega_0 n)$	$k_1\,\cos(\Omega_0 n) + k_2\,\sin(\Omega_0 n)$
$K\,n^m$	$k_m n^m + k_{m-1} n^{m-1} + \ldots + k_1 n + k_0$

The unknown coefficients k_i of the particular solution are determined from the difference equation by assuming all initial conditions are zero (recall that the particular solution does not depend on the initial conditions of the difference equation, or the initial state of the system). Initial conditions of the difference equation are imposed in the subsequent step for determining the unknown coefficients of the homogeneous solution, not the coefficients of the particular solution. The procedure for determining the complete forced solution of the difference equation is summarized below:

Finding the forced solution:

1. Write the homogeneous difference equation, and then find the characteristic equation by replacing delays of the output signal with corresponding negative powers of the complex variable z.
2. Solve for the roots of the characteristic equation and write the homogeneous solution in the form of Eqn. (2.55). If some of the roots appear as complex conjugate pairs, then use the form in Eqn. (2.59) for those roots. If there are any multiple roots, use the procedure outlined in Eqn. (2.62). Leave the homogeneous solution in parametric form with undetermined coefficients; do not attempt to compute the coefficients $c_1, c_2, \ldots$ of the homogeneous solution yet.
3. Find the form of the particular solution by either picking the appropriate form of it from Table 2.1, or by constructing it as a linear combination of the input signal and its delays. (This latter approach requires that delays of the input signal produce a finite number of distinct signal forms.)
4. Try the particular solution in the non-homogeneous difference equation and determine the coefficients $k_1, k_2, \ldots$ of the particular solution. At this point the particular solution should be uniquely determined. However, the coefficients of the homogeneous solution are still undetermined.
5. Add the homogeneous solution and the particular solution together to obtain the forced solution. Impose the necessary initial conditions and determine the coefficients $c_1, c_2, \ldots$ of the homogeneous solution.

Example 2.14: Forced response of exponential smoother for unit step input

Find the forced response of the exponential smoother of Example 2.6 when the input signal is a unit step function, and $y[-1] = 2.5$.

Solution: In Example 2.11 the homogeneous solution of the difference equation for the exponential smoother was determined to be in the form

$$y_h[n] = c\,(1-\alpha)^n$$

For a unit step input, the particular solution is in the form

$$y_p[n] = k_1$$

The particular solution must satisfy the difference equation. Substituting $y_p[n]$ into the difference equation we get

$$k_1 = (1-\alpha)\,k_1 + \alpha$$

and consequently $k_1 = 1$. The forced solution is the combination of homogeneous and particular solutions:

$$\begin{aligned} y[n] =& y_h[n] + y_p[n] \\ =& c\,(1-\alpha)^n + 1 \end{aligned}$$

The constant c needs to be adjusted to satisfy the specified initial condition $y[-1] = 2.5$.

$$y[-1] = c\,(1-\alpha)^{-1} + 1 = 2.5$$

results in $c = 1.5\,(1-\alpha)$, and the forced response of the system is

$$\begin{aligned} y_h[n] =& 1.5\,(1-\alpha)\,(1-\alpha)^n + 1 \\ =& 1.5\,(1-\alpha)^{n+1} + 1\,, \quad \text{for } n \geq 0 \end{aligned}$$

This signal is shown in Fig. 2.22 for $\alpha = 0.1$.

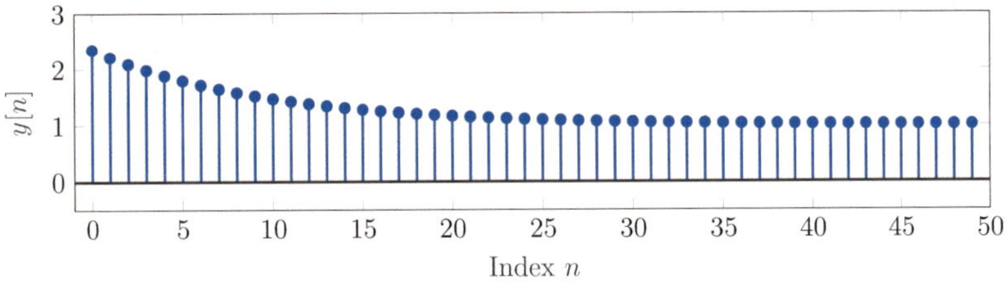

Figure 2.22 – Forced response of the system in Example 2.14.

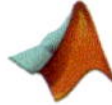

Software resource: exdt_2_14.m

Example 2.15: Forced response of exponential smoother for sinusoidal input

Find the forced response of the exponential smoother of Example 2.6 when the input signal is a sinusoidal function in the form

$$x[n] = A\cos(\Omega n)$$

Use parameter values $A = 20$, $\Omega = 0.2\pi$, and $\alpha = 0.1$. The initial value of the output signal is $y[-1] = 2.5$.

Solution: Recall that the difference equation of the exponential smoother is

$$y[n] = (1-\alpha)\; y[n-1] + \alpha\, x[n]$$

The homogeneous solution is in the form

$$y_h[n] = c\,(1-\alpha)^n$$

For a sinusoidal input signal, the form of the appropriate particular solution is obtained from Table 2.1 as

$$y_p[n] = k_1\,\cos(\Omega n) + k_2\,\sin(\Omega n) \tag{2.67}$$

The particular solution must satisfy the difference equation of the exponential smoother. Therefore we need

$$y_p[n] - (1-\alpha)\; y_p[n-1] = \alpha\, x[n] \tag{2.68}$$

The term $y_p[n-1]$ is needed in Eqn. (2.68). Time shifting both sides of Eqn. (2.67)

$$y_p[n-1] = k_1\,\cos(\Omega[n-1]) + k_2\,\sin(\Omega[n-1]) \tag{2.69}$$

Using the appropriate trigonometric identities Eqn. (2.69) can be written as

$$y_p[n-1] = [k_1 \cos(\Omega) - k_2 \sin(\Omega)]\,\cos(\Omega n) + [k_1 \sin(\Omega) + k_2 \cos(\Omega)]\,\sin(\Omega n) \tag{2.70}$$

Let us define

$$\beta = 1-\alpha$$

to simplify the notation. Substituting Eqns. (2.67) and (2.70) along with the input signal $x[n]$ into the difference equation in Eqn. (2.68) we obtain

$$\begin{aligned}
&\left[k_1 - \beta\, k_1\,\cos(\Omega) + \beta\, k_2\,\sin(\Omega)\right]\,\cos(\Omega n)\\
&\quad + \left[k_2 - \beta\, k_1\,\sin(\Omega) - \beta\, k_2\,\cos(\Omega)\right]\,\sin(\Omega n)\\
&\quad = \alpha A\,\cos(\Omega n)
\end{aligned} \tag{2.71}$$

Since Eqn. (2.71) must be satisfied for all values of the index n, coefficients of $\cos(\Omega n)$ and $\sin(\Omega n)$ on both sides of the equal sign must individually be set equal to each other. This leads to the two equations

$$k_1\,\left[1-\beta\,\cos(\Omega)\right] + k_2\,\beta\,\sin(\Omega) = A \tag{2.72}$$

and

$$-k_1\,\beta\,\sin(\Omega) + k_2\,\left[1-\beta\,\cos(\Omega)\right] = 0 \tag{2.73}$$

Eqns. (2.72) and (2.73) can be solved for the unknown coefficients k_1 and k_2 to yield

$$k_1 = \frac{\alpha A\left[1-\beta\,\cos(\Omega)\right]}{1-2\beta\,\cos(\Omega)+\beta^2} \quad \text{and} \quad k_2 = \frac{\alpha A \beta\,\sin(\Omega)}{1-2\beta\,\cos(\Omega)+\beta^2} \tag{2.74}$$

Now the forced solution of the system can be written by combining the homogeneous and particular solutions as

$$\begin{aligned}
y[n] =& y_h[n] + y_p[n]\\
=& c\,\beta^n + \frac{\alpha A\left[1-\beta\,\cos(\Omega)\right]}{1-2\beta\,\cos(\Omega)+\beta^2}\,\cos(\Omega n) + \frac{\alpha A \beta\,\sin(\Omega)}{1-2\beta\,\cos(\Omega)+\beta^2}\,\sin(\Omega n)
\end{aligned} \tag{2.75}$$

Using the specified parameter values of $A = 20$, $\Omega = 0.2\pi$, $\alpha = 0.1$, and $\beta = 0.9$, the coefficients k_1 and k_2 are evaluated to be

$$k_1 = 1.5371 \quad \text{and} \quad k_2 = 2.9907$$

and the forced response of the system is

$$y[n] = c\,(0.9)^n + 1.5371\,\cos(0.2\pi n) + 2.9907\,\sin(0.2\pi n)$$

We need to impose the initial condition $y[-1] = 2.5$ to determine the remaining unknown coefficient c. For $n = -1$ the output signal is

$$\begin{aligned} y[-1] =& c\,(0.9)^{-1} + 1.5371\,\cos(-0.2\pi) + 2.9907\,\sin(-0.2\pi) \\ =& 1.1111\,c - 0.5144 = 2.5 \end{aligned}$$

Solving for the coefficient c we obtain $c = 2.7129$. The forced response can now be written in complete form:

$$y[n] = 2.7129\,(0.9)^n + 1.5371\,\cos(0.2\pi n) + 2.9907\,\sin(0.2\pi n) \tag{2.76}$$

for $n \geq 0$. The forced response consists of two components. The first term in Eqn. (2.76) is the *transient response*

$$y_t[n] = 2.7129\,(0.9)^n \tag{2.77}$$

which is due to the initial state of the system. It disappears over time. The remaining terms in Eqn. (2.76) represent the *steady-state response* of the system:

$$y_{ss}[n] = 1.5371\,\cos(0.2\pi n) + 2.9907\,\sin(0.2\pi n) \tag{2.78}$$

Signals $y[n]$, $y_t[n]$, and $y_{ss}[n]$ are shown in Figs. 2.23 through 2.25.

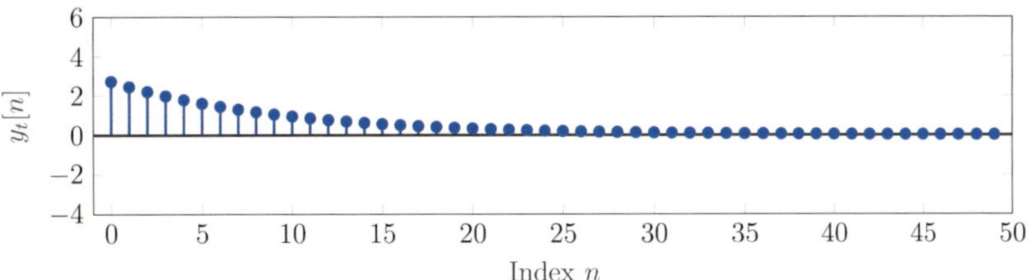

Figure 2.23 – Transient response $y_t[n]$ for Example 2.15.

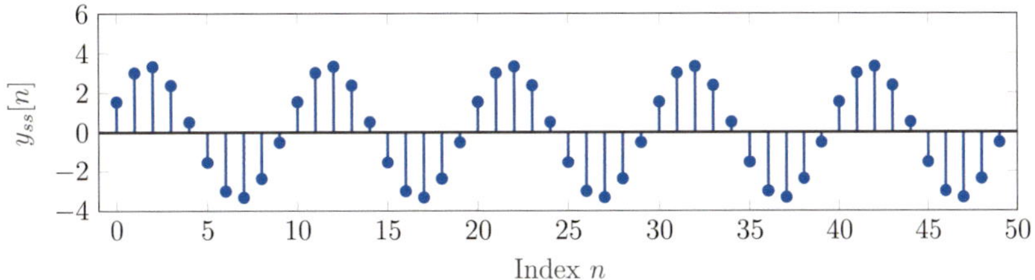

Figure 2.24 – Steady state response $y_{ss}[n]$ for Example 2.15.

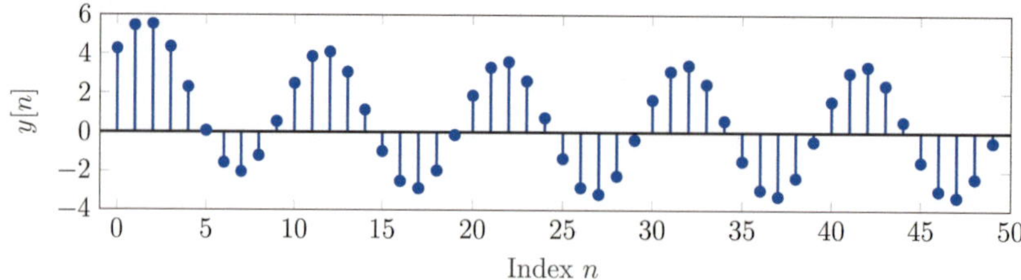

Figure 2.25 – Forced response $y[n]$ for Example 2.15.

Software resource: exdt_2_15.m

Interactive App: Forced response of exponential smoother for sinusoidal input

The interactive app appForcedRespDT.m is based on Example 2.15, and allows experimentation with parameters of the problem. The amplitude A is fixed at $A = 20$ so that the input signal is

$$x[n] = 20\cos(\Omega n)$$

The exponential smoother parameter α, the angular frequency Ω, and the initial output value $y[-1]$ can be varied using slider controls. The effect of parameter changes on the transient response $y_t[n]$, the steady-state response $y_{ss}[n]$, and the total forced response $y_t[n] + y_{ss}[n]$ can be observed.

a. Start with the settings $\alpha = 0.1$, $\Omega = 0.2\pi = 0.62832$ radians, and $y[-1] = 2.5$. Observe the peak amplitude value of the steady-state component. Confirm that it matches with what was found in Example 2.15, Fig. 2.24.

b. Now gradually increase the angular frequency Ω up to $\Omega = 0.5\pi = 1.5708$ radians, and observe the change in the peak amplitude of the steady-state component of the output. Compare with the result obtained in Eqn. (2.75).

c. Set parameter values back to $\alpha = 0.1$, $\Omega = 0.2\pi = 0.62832$ radians, and $y[-1] = 2.5$. Pay attention to the transient response, and how many samples it takes for it to become negligibly small.

d. Gradually decrease the value of α toward $\alpha = 0.05$ and observe the changes in the transient behavior. How does the value of α impact the number of samples it takes for the output signal to reach steady state?

Software resource: appForcedRespDT.m

2.6 Block Diagram Representation of Discrete-Time Systems

A discrete-time system can also be represented with a block diagram. Multiple block diagrams can be found for the same system, and these are functionally equivalent. In this section we will discuss just one particular technique for obtaining a block diagram and the discussion of other techniques will be deferred until Chapter 5.

Block diagrams are useful for discrete-time systems not only because they provide additional insight into the operation of a system, but also because they allow implementation of the system on a digital computer. We often use the block diagram as the first step in developing the computer code for implementing a discrete-time system. An example of this is given in MATLAB Exercise 2.6.

Three types of operators are utilized in the constant-coefficient linear difference equation of Eqn. (2.21): multiplication of a signal by a constant gain factor, addition of two signals, and time shift of a signal. Consequently, the fundamental building blocks for use in block diagrams of discrete-time systems are constant-gain amplifier, signal adder, and one-sample delay element as shown in Fig. 2.26.

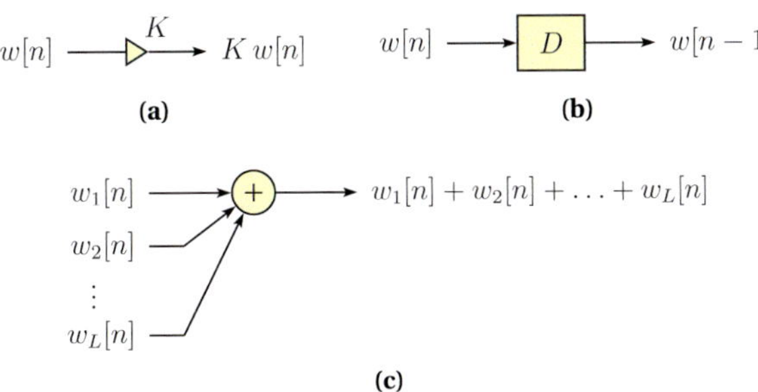

Figure 2.26 – Block diagram components for discrete-time systems: **(a)** constant-gain amplifier, **(b)** one-sample delay, and **(c)** signal adder.

We will begin the discussion with a simple third-order difference equation expressed as

$$y[n] + a_1\, y[n-1] + a_2\, y[n-2] + a_3\, y[n-3] = b_0\, x[n] + b_1\, x[n-1] + b_2\, x[n-2] \qquad (2.79)$$

This is a constant-coefficient linear difference equation in the standard form of Eqn. (2.21) with parameter values $N = 3$ and $M = 2$. Additionally, the coefficient of the term $y[n]$ is chosen to be $a_0 = 1$. This is not very restricting since, for the case of $a_0 \neq 1$, we can always divide both sides of the difference equation by a_0 to satisfy this condition. It can be shown that the following two difference equations that utilize an intermediate signal $w(t)$ are equivalent to Eqn. (2.79):

$$w[n] + a_1\, w[n-1] + a_2\, w[n-2] + a_3\, w[n-3] = x[n] \qquad (2.80)$$

$$y[n] = b_0\, w[n] + b_1\, w[n-1] + b_2\, w[n-2] \qquad (2.81)$$

Proof of Eqns. (2.80) and (2.81):

The proof is straightforward. The terms on the right side of Eqn. (2.79) can be written using Eqn. (2.80) and its time shifted versions. The following can be written:

$$b_0\, x[n] = b_0\, \big(w[n] + a_1\, w[n-1] + a_2\, w[n-2] + a_3\, w[n-3]\big) \qquad (2.82)$$

$$b_1\, x[n-1] = b_1\, \big(w[n-1] + a_1\, w[n-2] + a_2\, w[n-3] + a_3\, w[n-4]\big) \qquad (2.83)$$

$$b_2\, x[n-2] = b_2\, \big(w[n-2] + a_1\, w[n-3] + a_2\, w[n-4] + a_3\, w[n-5]\big) \qquad (2.84)$$

We are now in a position to construct the right side of Eqn. (2.79) using Eqns. (2.82) through (2.84):

$$\begin{aligned} b_0\, x[n]+b_1\, x[n-1]+b_2\, x[n-2] = \\ & b_0\left(w[n]+a_1\, w[n-1]+a_2\, w[n-2]+a_3\, w[n-3]\right) \\ & +b_1\left(w[n-1]+a_1\, w[n-2]+a_2\, w[n-3]+a_3\, w[n-4]\right) \\ & +b_2\left(w[n-2]+a_1\, w[n-3]+a_2\, w[n-4]+a_3\, w[n-5]\right) \end{aligned} \tag{2.85}$$

Rearranging the terms of Eqn. (2.85) we can write it in the form

$$\begin{aligned} b_0\, x[n]+b_1\, x[n-1]+b_2\, x[n-2] = \\ & \left(b_0\, w[n]+b_1\, w[n-1]+b_2\, w[n-2]\right) \\ & +a_1\left(b_0\, w[n-1]+b_1\, w[n-2]+b_2\, w[n-3]\right) \\ & +a_2\left(b_0\, w[n-2]+b_1\, w[n-3]+b_2\, w[n-4]\right) \\ & +a_3\left(b_0\, w[n-3]+b_1\, w[n-4]+b_2\, w[n-5]\right) \end{aligned} \tag{2.86}$$

and recognizing that the terms on the right side of Eqn. (2.86) are time shifted versions of $y[n]$ from Eqn. (2.81) we obtain

$$b_0\, x[n]+b_1\, x[n-1]+b_2\, x[n-2] = y[n]+a_1\, y[n-1]+a_2\, y[n-2]+a_3\, y[n-3] \tag{2.87}$$

which is the original difference equation given by Eqn. (2.79). Therefore, Eqns. (2.80) and (2.81) form an equivalent representation of the system described by Eqn. (2.79).

One possible block diagram implementation of the difference equation in Eqn. (2.80) is shown in Fig. 2.27. It takes the discrete-time signal $x[n]$ as input, and produces the intermediate signal $w[n]$ as output. Three delay elements are used for obtaining the samples $w[n-1]$, $w[n-2]$, and $w[n-3]$. The intermediate signal $w[n]$ is then obtained via an adder that adds scaled versions of the three past samples of $w[n]$. Keep in mind that our ultimate goal is to obtain the signal $y[n]$ which is related to the intermediate signal $w[n]$ through Eqn. (2.81). The computation of $y[n]$ requires the knowledge of $w[n]$ as well as its two past samples $w[n-1]$ and $w[n-2]$, both of which are available in the block diagram of Fig. 2.27. The complete block diagram for the system is obtained by adding the necessary connections to the block diagram in Fig. 2.27 and is shown in Fig. 2.28.

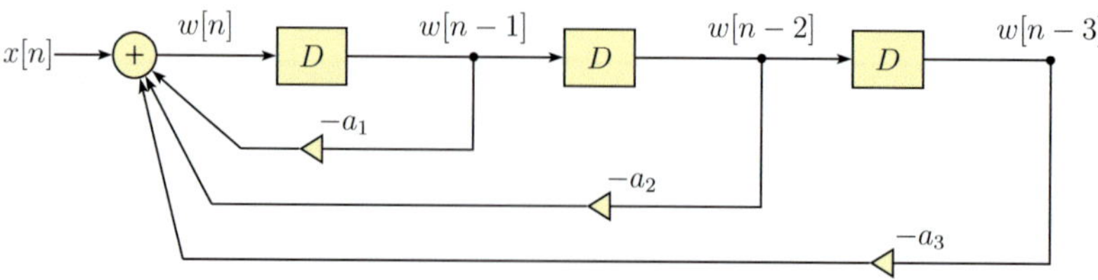

Figure 2.27 – The block diagram for Eqn. (2.81).

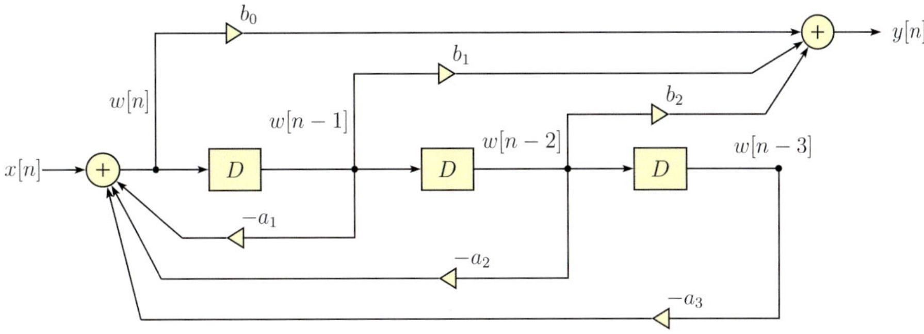

Figure 2.28 – The completed block diagram for Eqn. (2.79).

The development above was based on a third-order difference equation, however, the extension of the technique to a general constant-coefficient linear difference equation is straightforward. In Fig. 2.28 the feed-forward gains of the block diagram are the right-side coefficients $b_0, b_1, \ldots, b_M$ of the difference equation in Eqn. (2.79). Feedback gains of the block diagram are the negated left-side coefficients $-a_1, -a_2, \ldots, -a_N$ of the difference equation. Recall that we must have $a_0 = 1$ for this to work.

Imposing initial conditions

Initial conditions can easily be incorporated into the block diagram. The third-order difference equation given by Eqn. (2.79) would typically be solved subject to initial values specified for $y[-1]$, $y[-2]$, and $y[-3]$. For the block diagram we need to determine the corresponding values of $w[-1]$, $w[-2]$ and $w[-3]$ through the use of Eqns. (2.80) and (2.81). This will be illustrated in Example 2.16.

Example 2.16: Block diagram for discrete-time system

Construct a block diagram to solve the difference equation

$$y[n] - 0.7\,y[n-1] - 0.8\,y[n-2] + 0.84\,y[n-3] = 0.1\,x[n] + 0.2\,x[n-1] + 0.3\,x[n-2]$$

with the input signal $x[n] = u[n]$ and subject to initial conditions

$$y[-1] = 0.5\,, \quad y[-2] = 0.3\,, \quad y[-3] = -0.4$$

Solution: Using the intermediate variable $w[n]$ as outlined in the preceding discussion we can write the following pair of difference equations that are equivalent to the original difference equation:

$$w[n] - 0.7\,w[n-1] - 0.8\,w[n-2] + 0.84\,w[n-3] = x[n] \tag{2.88}$$

$$y[n] = 0.1\,w[n] + 0.2\,w[n-1] + 0.3\,w[n-2] \tag{2.89}$$

The block diagram can now be constructed as shown in Fig. 2.29.

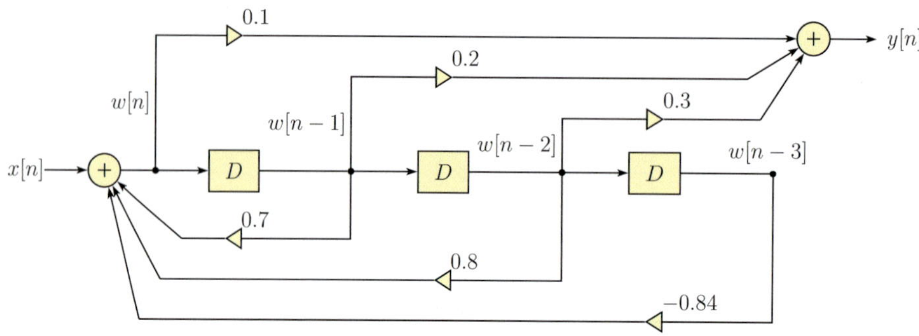

Figure 2.29 – Block diagram for Example 2.16.

Initial conditions specified in terms of the values of $y[-1]$, $y[-2]$, and $y[-3]$ need to be translated to corresponding values of $w[-1]$, $w[-2]$, and $w[-3]$. Writing Eqn. (2.89) for $n = -1, -2, -3$ yields the following three equations:

$$y[-1] = 0.1\,w[-1] + 0.2\,w[-2] + 0.3\,w[-3] = 0.5 \tag{2.90}$$

$$y[-2] = 0.1\,w[-2] + 0.2\,w[-3] + 0.3\,w[-4] = 0.3 \tag{2.91}$$

$$y[-3] = 0.1\,w[-3] + 0.2\,w[-4] + 0.3\,w[-5] = -0.4 \tag{2.92}$$

These equations need to be solved for $w[-1]$, $w[-2]$, and $w[-3]$. The two additional unknowns, namely $w[-4]$ and $w[-5]$, need to be obtained in other ways. Writing Eqn. (2.88) for $n = -1$ yields

$$w[-1] - 0.7\,w[-2] - 0.8\,w[-3] + 0.84\,w[-4] = x[-1] \tag{2.93}$$

Since $x[n] = u[n]$ we know that $x[-1] = 0$, therefore $w[-4]$ can be expressed as

$$w[-4] = -1.1905\,w[-1] + 0.8333\,w[-2] + 0.9524\,w[-3] \tag{2.94}$$

which can be substituted into Eqn. (2.91) to yield

$$y[-2] = -0.3571\,w[-1] + 0.35\,w[-2] + 0.4857\,w[-3] = 0.3 \tag{2.95}$$

Similarly, writing Eqn. (2.88) for $n = -2$ yields

$$w[-2] - 0.7\,w[-3] - 0.8\,w[-4] + 0.84\,w[-5] = x[-2] \tag{2.96}$$

We know that $x[-2] = 0$, therefore

$$\begin{aligned} w[-5] =& -1.1905\,w[-2] + 0.8333\,w[-3] + 0.9524\,w[-4] \\ =& -1.1905\,w[-2] + 0.8333\,w[-3] \\ &\quad + 0.9524\left(-1.1905\,w[-1] + 0.8333\,w[-2] + 0.9524\,w[-3]\right) \\ =& -1.1338\,w[-1] - 0.3968\,w[-2] + 1.7404\,w[-3] \end{aligned} \tag{2.97}$$

Substituting Eqns. (2.94), and (2.97) into Eqn. (2.92) we obtain

$$y[-3] = -0.5782\,w[-1] + 0.0476\,w[-2] + 0.8126\,w[-3] = -0.4 \tag{2.98}$$

Eqns. (2.90), (2.95), and (2.98) can be solved simultaneously to determine the initial conditions in terms of $w[n]$ as

$$w[-1] = 1.0682\,, \quad w[-2] = 1.7149\,, \quad w[-3] = 0.1674$$

In the block diagram of Fig. 2.29 the outputs of the three delay elements should be set equal to these values before starting the simulation.

Interactive App: Block diagram for a discrete-time system

This interactive app in `appBlockDgmDT.m` illustrates the solution of the difference equation of Example 2.16 through the use of the block diagram constructed and shown in Fig. 2.29. Two choices are given for the input signal $x[n]$: a unit step signal and a periodic sawtooth signal both of which have $x[-1] = x[-2] = 0$. Numerical values of the node variables are shown on the block diagram. For $n = 0$, initial values $w[n-1] = w[-1]$, $w[n-2] = w[-2]$ and $w[n-3] = w[-3]$ are shown on the diagram as computed in Example 2.16.

Incrementing the sample index n by clicking the button to the right of the index field causes the node values to be updated in an animated fashion, illustrating the operation of the block diagram. Additionally, input and output samples of the system are shown as stem plots with the current samples indicated in red color.

a. Ensure that the initial values $w[-1]$, $w[-2]$, and $w[-3]$ are set to the same values that were found in Example 2.16. Using pencil and paper, write an expression for $w[0]$ and another expression for the output sample $y[0]$. Verify the numerical values shown on the diagram for $w[0]$ and $y[0]$.

b. Using the button on the right side of the sample index field, change the current index to $n = 1$ while watching the numerical values of node variables shift to the right followed by updates to the values of $w[n]$ and $y[n]$, in this case $w[1]$ and $y[1]$ respectively.

c. Write expressions for $w[1]$ and $y[1]$, and verify the numerical values shown on the diagram.

Software resource: `appBlockDgmDT.m`

Software resource: See MATLAB Exercises 2.6 and 2.7.

2.7 Impulse Response and Convolution

A constant-coefficient linear difference equation is sufficient for describing a DTLTI system such that the output signal of the system can be determined in response to any arbitrary input signal. However, we will often find it convenient to use additional description forms for DTLTI systems. One of these additional description forms is the impulse response which is simply the forced response of the system under consideration when the input signal is a unit impulse. This is illustrated in Fig. 2.30.

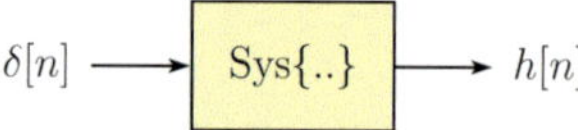

Figure 2.30 – Computation of the impulse response for a DTLTI system.

The impulse response also constitutes a complete description of a DTLTI system. The response of a DTLTI system to any arbitrary input signal $x[n]$ can be uniquely determined from the knowledge of its impulse response.

In the next section we will discuss how the impulse response of a DTLTI system can be obtained from its difference equation. The reverse is also possible, and will be discussed in later chapters.

2.7.1 Finding impulse response of a DTLTI system

Finding the impulse response of a DTLTI system amounts to finding the forced response of the system when the forcing function is a unit impulse, that is, $x[n] = \delta[n]$. In the case of a difference equation with no feedback, the impulse response is found by direct substitution of the unit impulse input signal into the difference equation. If the difference equation has feedback, then finding an appropriate form for the particular solution may be a bit more difficult. This difficulty can be overcome by finding the unit step response of the system as an intermediate step, and then determining the impulse response from the unit step response.

The problem of determining the impulse response of a DTLTI system from the governing difference equation will be explored in the next two examples.

Example 2.17: Impulse response of moving average filters

Find the impulse response of the length-2, length-4, and length-N moving average filters discussed in Examples 2.3, 2.4, and 2.5.

Solution: Let us start with the length-2 moving average filter. The governing difference equation is

$$y[n] = \tfrac{1}{2}\,x[n] + \tfrac{1}{2}\,x[n-1]$$

Let $h_2[n]$ denote the impulse response of the length-2 moving average filter (we will use the subscript to indicate the length of the window). It is easy to compute $h_2[n]$ is by setting $x[n] = \delta[n]$ in the difference equation:

$$h_2[n] = \text{Sys}\left\{\delta[n]\right\} = \tfrac{1}{2}\,\delta[n] + \tfrac{1}{2}\,\delta[n-1]$$

The result can also be expressed in tabular form as

$$h_2[n] = \{\ \underset{\uparrow}{1/2}, 1/2\ \}$$

Similarly, for a length-4 moving average filter with the difference equation

$$y[n] = \tfrac{1}{4}\,x[n] + \tfrac{1}{4}\,x[n-1] + \tfrac{1}{4}\,x[n-2] + \tfrac{1}{4}\,x[n-3]$$

the impulse response is

$$h_4[n] = \text{Sys}\left\{\delta[n]\right\} = \tfrac{1}{4}\,\delta[n] + \tfrac{1}{4}\,\delta[n-1] + \tfrac{1}{4}\,\delta[n-2] + \tfrac{1}{4}\,\delta[n-3]$$

or in tabular form

$$h_4[n] = \{ \underset{\uparrow}{1/4}, 1/4, 1/4, 1/4 \}$$

These results are easily generalized to a length-N moving average filter. The difference equation of the length-N moving average filter is

$$y[n] = \sum_{k=0}^{N-1} x[n-k] \tag{2.99}$$

Substituting $x[n] = \delta[n]$ into Eqn. (2.99) we get

$$h_N[n] = \text{Sys}\{\delta[n]\} = \sum_{k=0}^{N-1} \delta[n-k] \tag{2.100}$$

The result in Eqn. (2.100) can be written in alternative forms as well. One of those alternative forms is

$$h_N[n] = \begin{cases} \dfrac{1}{N}, & n = 0, \ldots, N-1 \\ 0, & \text{otherwise} \end{cases}$$

and another one is

$$h_N[n] = \frac{1}{N}\Big(u[n] - u[n-N]\Big)$$

Example 2.18: Impulse response of exponential smoother

Find the impulse response of the exponential smoother of Example 2.6 with $y[-1] = 0$.

Solution: With the initial condition $y[-1] = 0$, the exponential smoother described by the difference equation in Eqn. (2.11) is linear and time-invariant (refer to the discussion in Section 2.4). These two properties will allow us to use superposition in finding its impulse response. Recall that the unit impulse function can be expressed in terms of unit step functions as

$$\delta[n] = u[n] - u[n-1]$$

As a result, the impulse response of the linear exponential smoother can be found through the use of superposition in the form

$$\begin{aligned} h[n] = \text{Sys}\{\delta[n]\} &= \text{Sys}\{u[n] - u[n-1]\} \\ &= \text{Sys}\{u[n]\} - \text{Sys}\{u[n-1]\} \end{aligned}$$

We will first find the response of the system to a unit step signal. The homogeneous solution of the difference equation at hand was already found in Example 2.11 as

$$y_h[n] = c\,(1-\alpha)^n$$

For a unit step input, the particular solution is in the form

$$y_p[n] = k_1$$

Using this particular solution in the difference equation we get

$$k_1 = (1-\alpha)\,k_1 + \alpha$$

which leads to $k_1 = 1$. Combining the homogeneous and particular solutions, the forced solution is found to be in the form

$$\begin{aligned} y[n] =& y_h[n] + y_p[n] \\ =& c\,(1-\alpha)^n + 1 \end{aligned} \tag{2.101}$$

with coefficient c yet to be determined. If we now impose the initial condition $y[-1] = 0$ on the result found in Eqn. (2.101), we get

$$y[-1] = c\,(1-\alpha)^{-1} + 1 = 0$$

and consequently

$$c = -\,(1-\alpha)$$

Thus, the unit step response of the linear exponential smoother is

$$y[n] = \text{Sys}\left\{u[n]\right\} = 1 - (1-\alpha)\,(1-\alpha)^n$$

for $n \ge 0$. In compact notation, this result can be written as

$$y[n] = \left[1 - (1-\alpha)^{n+1}\right]\,u[n] \tag{2.102}$$

Since the system is time-invariant, its response to a delayed unit step input is simply a delayed version of $y[n]$ found in Eqn. (2.102):

$$\begin{aligned} \text{Sys}\left\{u[n-1]\right\} =& y[n-1] \\ =& \left[1 - (1-\alpha)^n\right]\,u[n-1] \end{aligned} \tag{2.103}$$

The impulse response of the linear exponential smoother is found using Eqns. (2.102) and (2.103) as

$$h[n] = y[n] - y[n-1] = \alpha\,(1-\alpha)^n\,u[n]$$

and is shown in Fig. 2.31 for $\alpha = 0.1$.

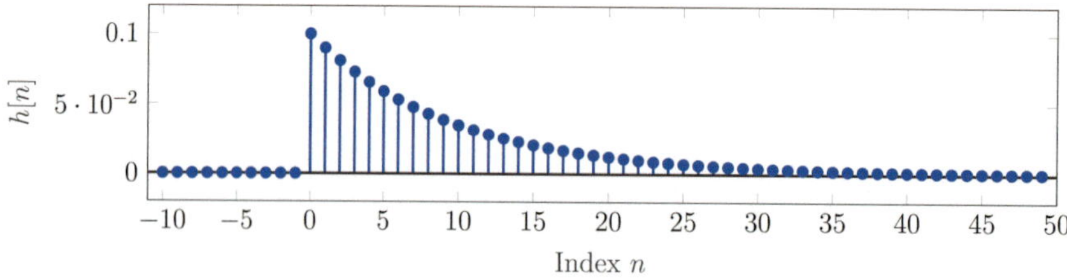

Figure 2.31 – Impulse response of the linear exponential smoother of Example 2.18 for $\alpha = 0.1$.

Software resource: exdt_2_18.m

2.7.2 Convolution operation for DTLTI systems

The development of the convolution operation for DTLTI systems will be based on the impulse decomposition of a discrete-time signal. It was established in Section 1.4 that any arbitrary discrete-time signal can be written as a sum of scaled and shifted unit impulse signals, leading to a decomposition in the form

$$\text{Eqn. (1.31):} \qquad x[n] = \sum_{k=-\infty}^{\infty} x[k]\,\delta[n-k]$$

Time shifting a unit impulse signal by k samples and multiplying it by the amplitude of the k-th sample of the signal $x[n]$ produces the signal $x_k[n] = x[k]\,\delta[n-k]$. Repeating this process for all integer values of k and adding the resulting signals leads to Eqn. (1.31).

If $x[n]$ is the input signal applied to a system, the output signal can be written as

$$y[n] = \text{Sys}\left\{x[n]\right\} = \text{Sys}\left\{\sum_{k=-\infty}^{\infty} x[k]\,\delta[n-k]\right\} \tag{2.104}$$

The input signal is the sum of shifted and scaled impulse functions. If the system under consideration is linear, then using the additivity rule given by Eqn. (2.2) we can write

$$y[n] = \sum_{k=-\infty}^{\infty} \text{Sys}\left\{x[k]\,\delta[n-k]\right\} \tag{2.105}$$

Furthermore, using the homogeneity rule given by Eqn. (2.3), Eqn. (2.105) becomes

$$y[n] = \sum_{k=-\infty}^{\infty} x[k]\,\text{Sys}\left\{\delta[n-k]\right\} \tag{2.106}$$

For the sake of discussion, let us assume that we know the system under consideration to be linear, but not necessarily time-invariant. The response of the linear system to any arbitrary input signal $x[n]$ can be computed through the use of Eqn. (2.106), provided that we already know the response of the system to unit impulse signals shifted in by all possible delay amounts. The knowledge necessary for determining the output of a linear system in response to an arbitrary signal $x[n]$ is

$$\left\{\text{Sys}\left\{\delta[n-k]\right\}\,, \quad \text{all } k\right\}$$

Eqn. (2.106) provides us with a viable, albeit impractical, method of determining system output $y[n]$ for any input signal $x[n]$. The amount of the prerequisite knowledge that we must possess about the system to be able to use Eqn. (2.106) diminishes its usefulness. Things improve, however, if the system under consideration is also time-invariant.

Let the *impulse response* of the system be defined as

$$h[n] = \text{Sys}\left\{\delta[n]\right\} \tag{2.107}$$

If, in addition to being linear, the system is also known to be time-invariant, then the response of the system to any shifted impulse signal can be derived from the knowledge of $h[n]$ through

$$\text{Sys}\left\{\delta[n-k]\right\} = h[n-k] \tag{2.108}$$

consistent with the definition of time invariance given by Eqn. (2.7). This reduces the prerequisite knowledge to just the impulse response $h[n]$, and we can compute the output signal as

$$y[n] = \sum_{k=-\infty}^{\infty} x[k]\,h[n-k] \tag{2.109}$$

Eqn. (2.109) is known as the *convolution sum* for discrete-time signals. The output signal $y[n]$ of a DTLTI system is obtained by *convolving* the input signal $x[n]$ and the impulse response $h[n]$ of the system. This relationship is expressed in compact notation as

$$y[n] = x[n] * h[n] \tag{2.110}$$

where the symbol $*$ represents the *convolution operator*. We will show later in this section that the convolution operator is commutative, that is, the relationship in Eqn. (2.109) can also be written in the alternative form

$$\begin{aligned} y[n] =& h[n] * x[n] \\ =& \sum_{k=-\infty}^{\infty} h[k]\, x[n-k] \end{aligned} \tag{2.111}$$

by swapping the roles of $h[n]$ and $x[n]$ without affecting the end result.

Discrete-time convolution summary:

$$\begin{aligned} y[n] =& x[n] * h[n] = \sum_{k=-\infty}^{\infty} x[k]\, h[n-k] \\ =& h[n] * x[n] = \sum_{k=-\infty}^{\infty} h[k]\, x[n-k] \end{aligned}$$

Example 2.19: A simple discrete-time convolution example

A discrete-time system is described through the impulse response

$$h[n] = \{ \underset{\substack{\uparrow \\ n=0}}{4}, 3, 2, 1 \}$$

Use the convolution operation to find the response of the system to the input signal

$$x[n] = \{ \underset{\substack{\uparrow \\ n=0}}{-3}, 7, 4 \}$$

Solution: Consider the convolution sum given by Eqn. (2.109). Let us express the terms inside the convolution summation, namely $x[k]$ and $h[n-k]$, as functions of k.

$$x[k] = \{ \underset{\substack{\uparrow \\ k=0}}{-3}, 7, 4 \}$$

$$h[-k] = \{ 1, 2, 3, \underset{\substack{\uparrow \\ k=0}}{4} \}$$

$$h[n-k] = \{ 1, 2, 3, \underset{\substack{\uparrow \\ k=n}}{4} \}$$

In its general form both limits of the summation in Eqn. (2.109) are infinite. On the other hand, $x[k] = 0$ for negative values of the summation index k, so setting the lower limit of the summation to $k = 0$ would have no effect on the result. Similarly, the last significant sample of $x[k]$ is at index $k=2$, so the upper limit can be changed to $k=2$ without affecting the result as well, leading to

$$y[n]=\sum_{k=0}^{2} x[k]\,h[n-k] \tag{2.112}$$

If $n<0$, we have $h[n-k]=0$ for all terms of the summation in Eqn. (2.112) and the output amplitude is zero. Therefore we will only concern ourselves with samples for which $n \ge 0$. The factor $h[n-k]$ has significant samples in the range

$$0 \le n-k \le 3$$

which can be expressed in the alternative form

$$n-3 \le k \le n$$

The upper limit of the summation in Eqn. (2.112) can be set equal to $k=n$ without affecting the result, however, if $n>2$ then we should leave it at $k=2$. Similarly, the lower limit can be set to $k=n-3$ provided that $n-3>0$, otherwise it should be left at $k=0$. So, a compact version of the convolution sum adapted to the particular signals of this example would be

$$y[n]=\sum_{k=\max(0,n-3)}^{\min(2,n)} x[k]\,h[n-k]\,, \quad \text{for } n \ge 0 \tag{2.113}$$

where the lower limit is the larger of $k=0$ and $k=n-3$, and the upper limit is the smaller of $k=2$ and $k=n$. We will use this result to compute the convolution of $x[n]$ and $h[n]$.

$$n=0 \quad \Rightarrow \quad y[0]=\sum_{k=0}^{0} x[k]\,h[0-k] = x[0]\,h[0] = -12$$

$$n=1 \quad \Rightarrow \quad y[1]=\sum_{k=0}^{1} x[k]\,h[1-k] = x[0]\,h[1]+x[1]\,h[0] = 19$$

$$n=2 \quad \Rightarrow \quad y[2]=\sum_{k=0}^{2} x[k]\,h[2-k] = x[0]\,h[2]+x[1]\,h[1]+x[2]\,h[0] = 31$$

$$n=3 \quad \Rightarrow \quad y[3]=\sum_{k=0}^{2} x[k]\,h[3-k] = x[0]\,h[3]+x[1]\,h[2]+x[2]\,h[1] = 23$$

$$n=4 \quad \Rightarrow \quad y[4]=\sum_{k=1}^{2} x[k]\,h[4-k] = x[1]\,h[3]+x[2]\,h[2] = 15$$

$$n=5 \quad \Rightarrow \quad y[5]=\sum_{k=2}^{2} x[k]\,h[5-k] = x[2]\,h[3] = 4$$

Thus, the output signal $y[n]$ is obtained as

$$y[n] = \{ \underset{\substack{\uparrow \\ n=0}}{-12}, 19, 31, 23, 15, 4 \}$$

Example 2.20: Simple discrete-time convolution example revisited

Rework the convolution problem of Example 2.19 with the following modifications applied to the two signals:

$$h[n] = \{ \underset{\substack{\uparrow \\ n=N_2}}{4}, 3, 2, 1 \}$$

and

$$x[n] = \{ \underset{\substack{\uparrow \\ n=N_1}}{-3}, 7, 4 \}$$

Assume the starting indices N_1 and N_2 are known constants.

Solution: We need to readjust limits of the summation index. Significant samples, that is, samples with non-zero amplitude, of $x[k]$ are in the index range

$$N_1 \le k \le N_1 + 2 \tag{2.114}$$

We also need to find the index range for the samples of $h[n-k]$ with non-zero amplitude. For that, we use the inequality

$$N_2 \le n - k \le N_2 + 3$$

the terms of which can be rearranged to yield

$$n - N_2 - 3 \le k \le n - N_2 \tag{2.115}$$

Using the inequalities in Eqns. (2.114) and (2.115) the convolution sum can be written as

$$y[n] = \sum_{k=K_1}^{K_2} x[k]\, h[n-k] \tag{2.116}$$

with the limits

$$K_1 = \max(N_1, n - N_2 - 3) \quad \text{and} \quad K_2 = \min(N_1 + 2, n - N_2) \tag{2.117}$$

For example, suppose we have $N_1 = 5$ and $N_2 = 7$. Using Eqn. (2.116) with the limits in Eqn. (2.117) we can write the convolution sum as

$$y[n] = \sum_{k=\max(5,n-10)}^{\min(7,n-7)} x[k]\, h[n-k]$$

For the summation to contain any significant terms, the lower limit must not be greater than the upper limit, that is, we need

$$\max(5, n-10) \le \min(7, n-7)$$

As a result, the leftmost significant sample of $y[n]$ will occur at index $n = 12$. In general, it can be shown (see Problem 2.25 at the end of this chapter) that the leftmost significant sample will be at the index $n = N_1 + N_2$. The sample $y[12]$ is computed as

$$\begin{aligned} y[12] &= \sum_{k=5}^{5} x[k]\, h[12-k] \\ &= x[5]\, h[7] = (-3)\,(4) = -12 \end{aligned}$$

Other samples of $y[n]$ can be computed following the procedure demonstrated in Example 2.19 and yield the same pattern of values with the only difference being the starting index. The complete solution is

$$y[n] = \{ \underset{\substack{\uparrow \\ n=12}}{-12}, 19, 31, 23, 15, 4 \}$$

Software resource: See MATLAB Exercise 2.8.

Example 2.21: A different perspective for the simple convolution example

Consider again the simple convolution problem we have explored in Examples 2.19 and 2.20. A discrete-time system is described through the impulse response

$$h[n] = \{ \underset{\substack{\uparrow \\ n=0}}{4}, 3, 2, 1 \}$$

We are asked to find the response of this system to the input signal

$$x[n] = \{ \underset{\substack{\uparrow \\ n=0}}{-3}, 7, 4 \}$$

The system with impulse response $h[n]$ is essentially a *finite impulse response (FIR)* type of system that we will study in more detail in later chapters of this textbook. Let us write the alternative form of the convolution equation given by Eqn. (2.111) for this particular case:

$$\begin{aligned} y[n] &= \sum_{k=0}^{3} h[k]\, x[n-k] \\ &= h[0]\, x[n] + h[1]\, x[n-1] + h[2]\, x[n-2] + h[3]\, x[n-3] \end{aligned} \tag{2.118}$$

We have set summation limits knowing that $h[n] = 0$ outside the interval $k = 0,\ldots,3$. The relationship of Eqn. (2.118) can be expressed with the block diagram of Fig. 2.32. For each sample index, the buffer holds the most recent four samples of $x[n]$. Samples of $x[n]$ enter the buffer from the top, and push the contents of the buffer down. Fig. 2.33 depicts the situation for $n = 0, 1, 2$.

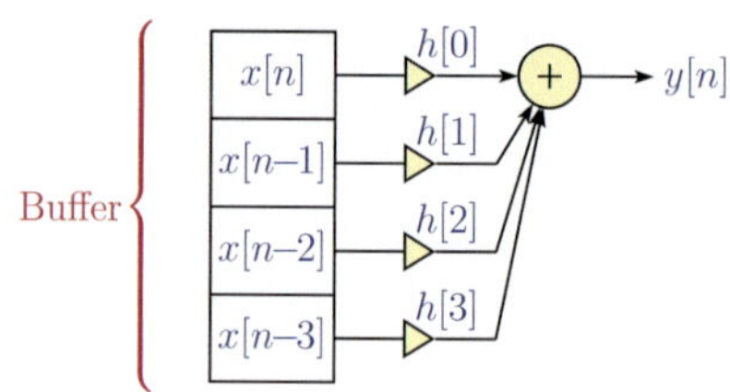

Figure 2.32 – Signals involved in the convolution sum of Example 2.22.

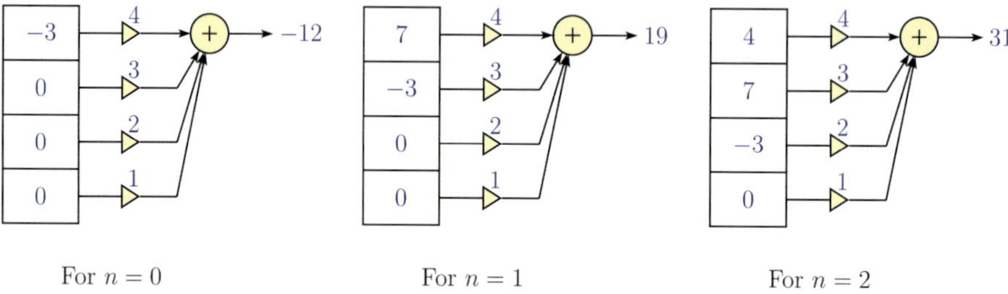

Figure 2.33 – Convolution of signals $h[n]$ and $x[n]$ for $n = 0, 1, 2$.

At this point, the signal $x[n]$ does not have any non-zero samples left. However, the buffer still has some values in it and the convolution operation will continue to produce non-zero output samples until the buffer is completely flushed out. Fig. 2.34 depicts the situation for $n = 3, 4, 5$.

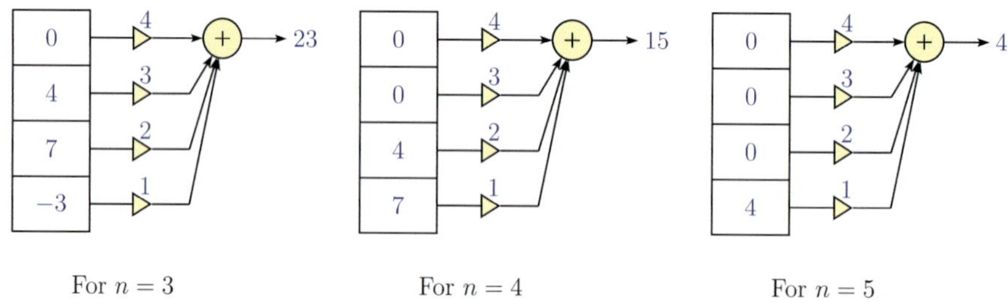

Figure 2.34 – Convolution of signals $h[n]$ and $x[n]$ for $n = 3, 4, 5$.

As expected, we get the same output signal we found before:

$$y[n] = \{ \underset{\substack{\uparrow \\ n=0}}{-12}, 19, 31, 23, 15, 4 \}$$

This interpretation of the convolution sum is useful in implementing the convolution operation in real-time processing applications. We will also use this approach in writing our own convolution function in MATLAB Exercise 2.9.

Software resource: See MATLAB Exercise 2.9.

In computing the convolution sum it is helpful to sketch the signals. The graphical steps involved in computing the convolution of two signals $x[n]$ and $h[n]$ at a specific index value n can be summarized as follows:

Steps in computing the convolution sum:

1. For one specific value of n, sketch the signal $h[n-k]$ as a function of the independent variable k. This task can be broken down into two steps as follows:

   1a. Sketch $h[-k]$ as a function of k. This step amounts to time reversal of the signal $h[k]$.

   1b. In $h[-k]$ substitute $k \rightarrow k-n$. This step yields

   $$h[-k]\Big|_{k\rightarrow k-n} = h[n-k] \tag{2.119}$$

   and amounts to time shifting $h[-k]$ by n samples.

   See Fig. 2.35 for an illustration of the steps for obtaining $h[n-k]$.
2. Sketch the signal $x[k]$ as a function of the independent variable k. This corresponds to a simple name change on the independent variable, and the graph of the signal $x[k]$ appears identical to the graph of the signal $x[n]$. (See Fig. 2.36.)
3. Multiply the two signals sketched in 1 and 2 to obtain the product $x[k]\,h[n-k]$.
4. Sum the sample amplitudes of the product $x[k]\,h[n-k]$ over the index k. The result is the amplitude of the output signal at the index n.
5. Repeat steps 1 through 4 for all values of n that are of interest.

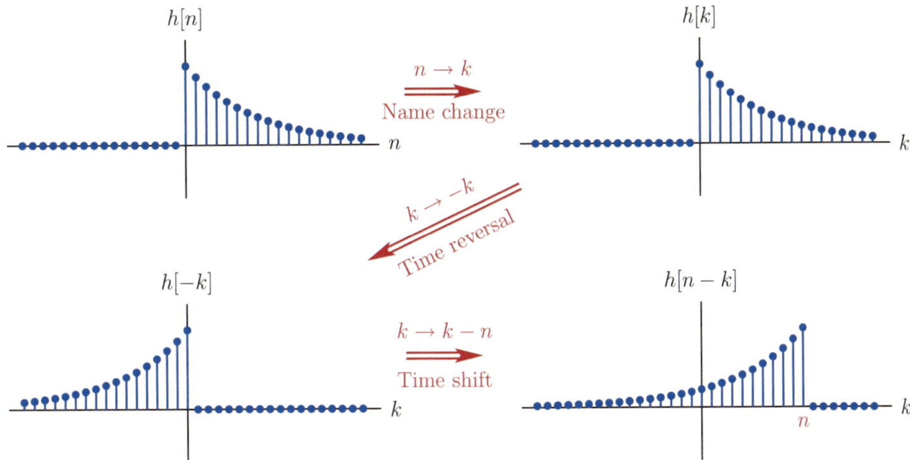

Figure 2.35 – Obtaining $h[n-k]$ for the convolution sum.

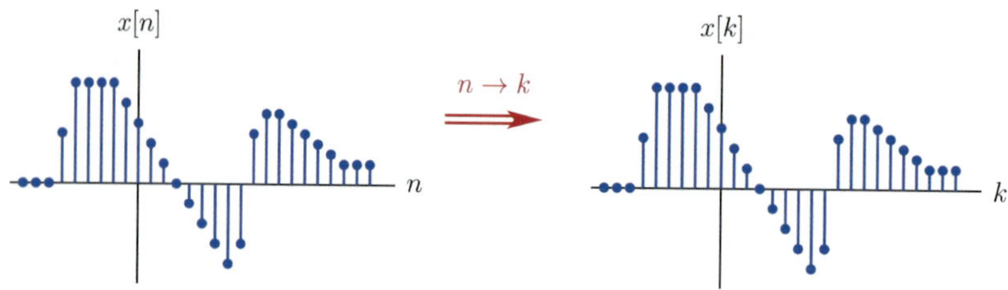

Figure 2.36 – Obtaining $x[k]$ for the convolution sum.

In the next example we will illustrate the graphical details of the convolution operation for DTLTI systems.

Example 2.22: A more involved discrete-time convolution example

A discrete-time system is described through the impulse response

$$h[n] = (0.9)^n \; u[n]$$

Use the convolution operation to find the response of a system to the input signal

$$x[n] = u[n] - u[n-7]$$

Signals $x[n]$ and $h[n]$ are shown in Fig. 2.37.

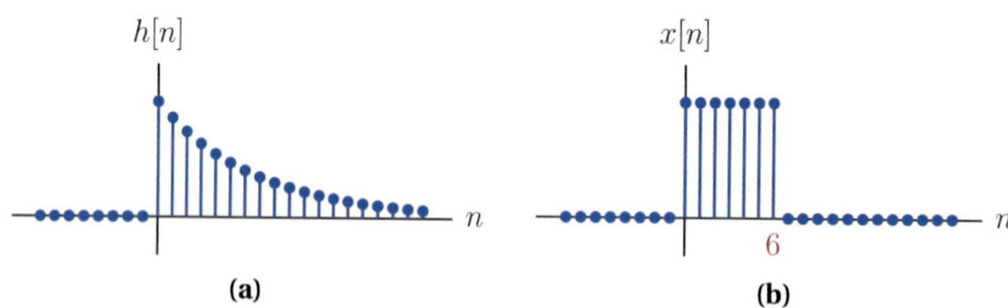

Figure 2.37 – The signals for Example 2.22: **(a)** Impulse response $h[n]$ and **(b)** input signal $x[n]$.

Solution: Again we will find it useful to sketch the functions $x[k]$, $h[n-k]$ and their product before we begin evaluating the convolution result. Such a sketch is shown in Fig. 2.38. It reveals three distinct possibilities for the overlap of $x[k]$ and $h[n-k]$. The convolution sum needs to be set up for each of the three regions of index n.

Case 1: $n < 0$

There is no overlap between the two functions in this case, therefore their product equals zero for all values of k. The output signal is

$$y[n] = 0 \quad \text{for} \quad n < 0$$

Case 2: $0 \le n \le 6$

For this case, the two functions overlap for the range of the index $0 \le k \le n$. Setting summation limits accordingly, the output signal can be written as

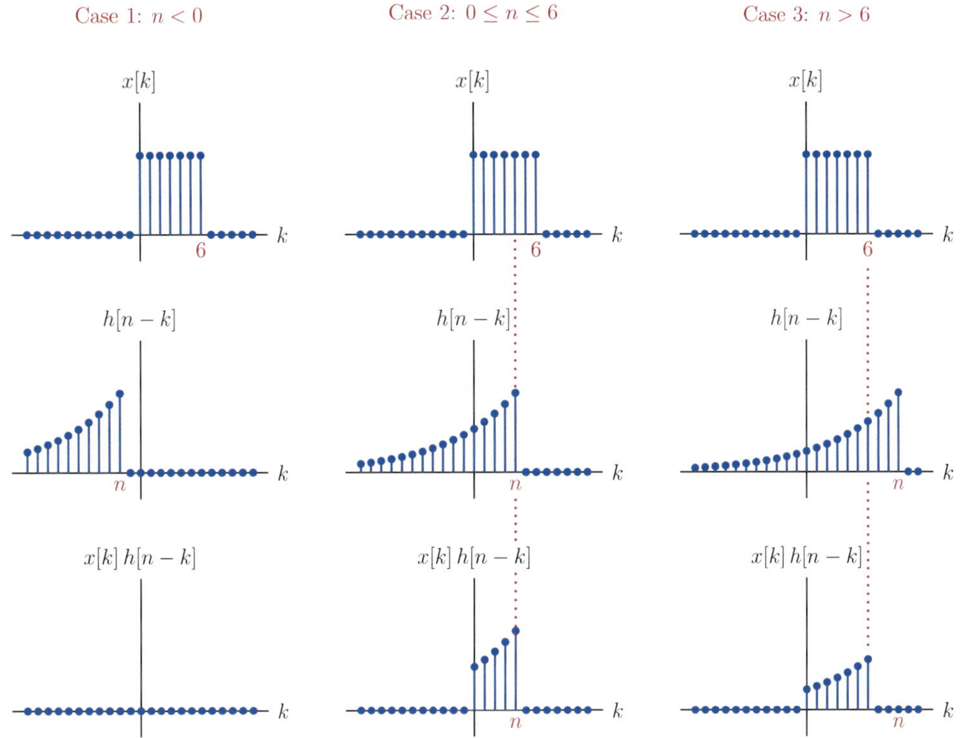

Figure 2.38 – Signals involved in the convolution sum of Example 2.22.

$$y[n] = \sum_{k=0}^{n} (1)\,(0.9)^{n-k}$$

This result can be simplified by factoring out the common term $(0.9)^n$ and using the geometric series formula (see Appendix C) to yield

$$\begin{aligned} y[n] &= (0.9)^n \sum_{k=0}^{n} (0.9)^{-k} \\ &= (0.9)^n \, \frac{1-(0.9)^{-(n+1)}}{1-(0.9)^{-1}} \\ &= -9\left[(0.9)^n - \frac{1}{0.9}\right] \quad \text{for} \quad 0 \le n \le 6 \end{aligned}$$

Case 3: $n > 6$

For $n > 6$ the overlap of the two functions will occur in the range $0 \le k \le 6$, so the summation limits need to be adjusted.

$$y[n] = \sum_{k=0}^{6} (1)\,(0.9)^{n-k}$$

Again factoring out the $(0.9)^n$ term and using the geometric series formula we get

$$\begin{aligned} y[n] &= (0.9)^n \sum_{k=0}^{6} (0.9)^{-k} \\ &= (0.9)^n \frac{1-(0.9)^{-7}}{1-(0.9)^{-1}} \\ &= 9.8168\,(0.9)^n \quad \text{for} \quad n>6 \end{aligned}$$

Thus, we have computed the output signal in each of the three distinct intervals we have identified. Putting these three partial solutions together, the complete solution for the output signal is obtained as

$$y[n] = \begin{cases} 0\,, & n<0 \\ -9\left[0.9^n - \dfrac{1}{0.9}\right], & 0 \le n \le 6 \\ 9.8168\,(0.9)^n\,, & n>6 \end{cases}$$

and is shown in Fig. 2.39.

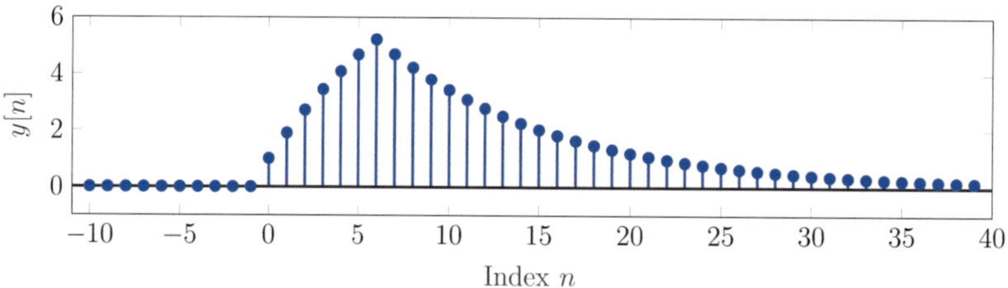

Figure 2.39 – Convolution result for Example 2.22.

Software resource: exdt_2_22.m

Interactive App: Discrete-time convolution example

The interactive app in `appConvDT.m` is based on the discrete-time convolution problem solved in Example 2.22. It facilitates visualization of the overlapping samples between the functions $x[k]$ and $h[n-k]$ as the index n is varied. The impulse response used in the app is in the general form

$$h[n] = a^n\, u[n]$$

and can be made identical to the impulse response used in Example 2.22 by setting $a = 0.9$.

a. Set the parameter value to $a = 0.9$ and the current sample index to $n = -5$. Observe the signals $x[k]$, $h[n-k]$ and the product $x[k]\,h[n-k]$.

b. Increase the sample index to $n = -4, -3, -2, -1$ and verify that the output signal amplitude is still zero.

c. Set the current sample index to $n = 3$. How many significant samples of $x[k]$ and $h[n-k]$ overlap? Verify the amplitude of the output sample $y[3]$ with pencil and paper calculations.

d. Set the current sample index to $n = 8$. How many significant samples of $x[k]$ and $h[n-k]$ overlap now?

Software resource: `appConvDT.m`

Software resource: See MATLAB Exercises 2.10 and 2.11.

2.8 Causality in Discrete-Time Systems

Causality is an important feature of physically realizable systems. A system is said to be causal if the current value of the output signal depends only on current and past values of the input signal, but not on its future values. An example of a causal discrete-time system is one defined by the difference equation

$$y[n] = y[n-1] + x[n] - 3\,x[n-1]$$

In contrast, a system defined by the difference equation

$$y[n] = y[n-1] + x[n] - 3\,x[n+1]$$

is non-causal. Causal systems can be implemented in *real-time processing* mode where a sample of the output signal is computed in response to each incoming sample of the input signal. On the order hand, implementation of non-causal systems may only be possible in *post processing* mode where the entire input signal must be observed and recorded before processing can begin.

In the case of a DTLTI system, the causality property can easily be related to the impulse response. Recall that the output signal $y[n]$ of a DTLTI system can be computed as the convolution of its input signal $x[n]$ with its impulse response $h[n]$ as

$$y[n] = h[n] * x[n] = \sum_{k=-\infty}^{\infty} h[k]\,x[n-k] \tag{2.120}$$

For the system under consideration to be causal, the computation of $y[n]$ should not require any future samples of the input signal. Thus, the term $x[n-k]$ in the summation should not contain index values in the future. This requires

$$n - k \le n \quad \Longrightarrow \quad k \ge 0 \tag{2.121}$$

The product $h[k]\,x[n-k]$ should not have any nonzero values for $k < 0$. Therefore, the impulse response $h[n]$ should be equal to zero for all negative index values.

Condition for causality of a DTLTI system:

$$h[n] = 0 \quad \text{for all} \quad n < 0 \tag{2.122}$$

For a causal DTLTI system, the convolution relationship in Eqn. (2.120) can be written in right-sided form by setting the lower limit of the summation index equal to zero:

$$y[n] = \sum_{k=0}^{\infty} h[k]\, x[n-k] \tag{2.123}$$

A non-causal DTLTI system in which the impulse response $h[n]$ has a finite number of significant samples for $n < 0$ can be converted to a causal system by adding a delay to $h[n]$ so that the new impulse response satisfies the condition in Eqn. (2.122).

Software resource: See MATLAB Exercise 2.3.

2.9 Stability in Discrete-Time Systems

A system is said to be *stable* in the *bounded-input bounded-output (BIBO)* sense if any bounded input signal is guaranteed to produce a bounded output signal.

A discrete-time input signal $x[n]$ is said to be bounded if an upper bound B_x exists such that

$$\left|x[n]\right| < B_x < \infty \tag{2.124}$$

for all values of the integer index n. A discrete-time system is stable if a finite upper bound B_y exists for the output signal in response to any input signal bounded as described in Eqn. (2.124).

Condition for stability of a discrete-time system:

$$\left|x[n]\right| < B_x < \infty \quad \text{implies that} \quad \left|y[n]\right| < B_y < \infty \tag{2.125}$$

If the system under consideration is DTLTI, it is possible to relate the stability condition given by Eqn. (2.125) to the impulse response of the system as well as its difference equation.

Derivation of the necessary condition for stability:

The output signal of a DTLTI system is found from its input signal and impulse response through the use of the convolution sum expressed as

$$y[n] = \sum_{k=-\infty}^{\infty} h[k]\, x[n-k]$$

The absolute value of the output signal is

$$|y[n]| = \left| \sum_{k=-\infty}^{\infty} h[k]\, x[n-k] \right|$$

Absolute value of a sum is less than or equal to the sum of the absolute values, so an inequality involving $y[n]$ can be written as

$$|y[n]| \le \sum_{k=-\infty}^{\infty} |h[k]\, x[n-k]| \tag{2.126}$$

The summation term in Eqn. (2.126) can be expressed as

$$|h[k]\, x[n-k]| = |h[k]|\, |x[n-k]| \tag{2.127}$$

and the inequality in Eqn. (2.126) becomes

$$|y[n]| \le \sum_{k=-\infty}^{\infty} |h[k]|\, |x[n-k]| \tag{2.128}$$

Replacing the term $|x[n-k]|$ in Eqn. (2.128) with B_x makes the term on the right side of the inequality even greater, so we have

$$|y[n]| \le \sum_{k=-\infty}^{\infty} |h[k]|\, B_x$$

or, by factoring out the common factor B_x

$$|y[n]| \le B_x \sum_{k=-\infty}^{\infty} |h[k]| < \infty \tag{2.129}$$

Condition for stability:

Eqn. (2.129) leads us to the conclusion

$$\sum_{k=-\infty}^{\infty} |h[k]| < \infty \tag{2.130}$$

For a DTLTI system to be stable, its impulse response must be *absolute summable.*

Example 2.23: Stability of a length-2 moving-average filter

Comment on the stability of the length-2 moving-average filter described by the difference equation

$$y[n] = \tfrac{1}{2}\, x[n] + \tfrac{1}{2}\, x[n-1]$$

Solution: We will approach this problem in two different ways. First, let's check directly to see if any arbitrary bounded input signal is guaranteed to produce a bounded output signal.

The absolute value of the output signal is

$$\left|y[n]\right| = \left|\tfrac{1}{2}\,x[n] + \tfrac{1}{2}\,x[n-1]\right| \tag{2.131}$$

Since absolute value of a sum is less than or equal to sum of absolute values, we can derive the following inequality from Eqn. (2.131):

$$\left|y[n]\right| \le \tfrac{1}{2}\left|x[n]\right| + \tfrac{1}{2}\left|x[n-1]\right| \tag{2.132}$$

Furthermore, since we assume $\left|x[n]\right| < B_x$ for all n, replacing $\left|x[n]\right|$ and $\left|x[n-1]\right|$ terms on the right side of the inequality in Eqn. (2.132) does not affect the validity of the inequality, so we have

$$\left|y[n]\right| \le \tfrac{1}{2}\,B_x + \tfrac{1}{2}\,B_x = B_x$$

proving that the output signal $y[n]$ is bounded as long as the input signal $x[n]$ is. An alternative way to attack the problem would be to check the impulse response for the absolute summability condition in Eqn. (2.130). The impulse response of the length-2 moving-average filter is

$$h[n] = \tfrac{1}{2}\,\delta[n] + \tfrac{1}{2}\,\delta[n-1]$$

which is clearly summable in an absolute sense.

Example 2.24: Stability of the loan balance system

Comment on the stability of the loan balance system discussed in Example 2.7.

Solution: Recall that the governing difference equation is

$$y[n] = (1+c)\;y[n-1] - x[n]$$

where c is a positive constant interest rate. This system can be analyzed in terms of its stability using the same approach that we have used in the previous example. However, we will take a more practical approach and analyze the system from a layperson's perspective. What if we borrowed the money and never made a payment? The loan balance would keep growing each period. If the initial value of the output signal is $y[0] = A$ and if $x[n] = 0$ for $n \ge 1$, then we would have

$$\begin{aligned} y[1] &= (1+c)\;A \\ y[2] &= (1+c)^2\;A \\ &\;\vdots \\ y[n] &= (1+c)^n\;A \end{aligned}$$

which indicates that the system is unstable since we were able to find at least one bounded input signal that leads to an unbounded output signal.

It is also worth noting that the characteristic equation is

$$z - (1+c) = 0$$

The only solution of the characteristic equation, and the only mode of the system, is at $z_1 = 1+c$. The significance of this will become apparent when we study the z-transform in Chapter 5.

The stability of a DTLTI system can also be associated with the modes of the difference equation that governs its behavior. We have seen in Section 2.5.1 that, for a causal DTLTI system, real and complex roots z_k of the characteristic polynomial for which $|z_k| \geq 1$ are associated with unstable behavior. Thus, for a causal DTLTI system to be stable, the magnitudes of all roots of the characteristic polynomial must be less than unity. If a circle is drawn on the complex plane with its center at the origin and its radius equal to unity, all roots of the characteristic polynomial must lie inside the circle for the corresponding causal DTLTI system to be stable.

A side note:

In describing the associations between the roots of the characteristic polynomial and stability, we referred to a *causal* DTLTI system. If we were to consider an *anti-causal* system, one the impulse response of which proceeds in the negative direction toward $n \to -\infty$, then the associations described above would have to be reversed. In that case roots outside the unit circle of the complex plane would lead to stable behavior.

Why is stability important for a DTLTI system?

An unstable DTLTI system is one that is capable of producing an unbounded output signal in response to at least some bounded input signals. What is the practical significance of this? Since we are dealing with numerical algorithms implemented via software on a computer, an unstable algorithm may lead to an output signal containing numbers that keep growing until they reach magnitudes that can no longer be handled by the number representation conventions of the processor or the operating system. When this occurs, the software that implements the system ceases to function in the proper way.

Summary of Key Points

☞ A discrete-time system is a mathematical formula, method or algorithm that defines a cause-effect relationship between a set of discrete-time input signals and a set of discrete-time output signals.

☞ A system is said to be linear if it satisfies both the additivity rule and the homogeneity rule. Additivity rule requires that the response of the system to the sum of any two input signals be the same as the sum of its individual responses to each of the input signals

Eqn. 2.2: $$\text{Sys}\{x_1[n] + x_2[n]\} = \text{Sys}\{x_1[n]\} + \text{Sys}\{x_2[n]\}$$

Homogeneity rule requires that the only effect of multiplying the input signal by a constant scale factor should be multiplying the output signal by the same scale factor.

Eqn. 2.3: $$\text{Sys}\{\alpha_1\, x_1[n]\} = \alpha_1\, \text{Sys}\{x_1[n]\}$$

☞ In checking the linearity of a system, homogeneity and additivity rules can be combined into the superposition principle.

Eqn. 2.4: $$\text{Sys}\{\alpha_1\, x_1[n] + \alpha_2\, x_2[n]\} = \alpha_1\, \text{Sys}\{x_1[n]\} + \alpha_2\, \text{Sys}\{x_2[n]\}$$

☞ A system is said to be time-invariant if the only effect of time shifting the input signal by a constant amount is a time shift of the output signal by the same amount.

Eqn. 2.7: $$\text{Sys}\{x[n]\} = y[n] \quad \text{implies that} \quad \text{Sys}\{x[n-k]\} = y[n-k]$$

☞ The acronym DTLTI is used for systems that are discrete-time, linear, and time-invariant.

☞ A discrete-time system can be represented by a difference equation that expresses the current output sample of the system in terms of any or all of the following:

- Current, past, and future samples of the input signal.
- Past and future samples of the output signal.

☞ A length-N moving average filter is an example of a DTLTI system in which the current sample of the output signal is the arithmetic average of the current sample and the past $N-1$ samples of the input signal. The difference equation for the length-N moving average filter is

Eqn. 2.8:
$$y[n] = \frac{x[n] + x[n-1] + \ldots + x[n-(N-1)]}{N} = \frac{1}{N} \sum_{k=0}^{N-1} x[n-k]$$

Moving average filters are typically used for smoothing a signal.

☞ Another simple example of a DTLTI system is an exponential smoother in which the current sample of the output signal is a linear combination of the current sample of the input signal and the previous sample of the output signal. The difference equation for the exponential smoother is given in Eqn. (2.11).

Eqn. 2.11:
$$y[n] = (1-\alpha)\, y[n-1] + \alpha\, x[n]$$

☞ A constant-coefficient linear difference equation is in the form

Eqn. 2.21:
$$\sum_{k=0}^{N} a_k\, y[n-k] = \sum_{k=0}^{M} b_k\, x[n-k]$$

It represents a discrete-time system that is always time-invariant, and may or may not be linear. For such a system to be linear, the initial conditions must all be equal to zero.

☞ Natural response of a system represented by a difference equation is the solution that is due to the initial conditions alone, without any contribution from any external input signals. To find the natural solution, we solve the homogeneous difference equation obtained by setting all input terms equal to zero.

☞ Forced response of a system represented by a difference equation is the part of the solution that is due to the input signal and is independent of the initial conditions.

☞ A discrete-time system can be represented by a block diagram. In most cases, the block diagram consists of constant gain amplifiers, delay elements, and signal adders. Block diagrams are useful for translating the behavior of a system into code that can be used on a digital computer for the implementation of the system.

☞ The impulse response of a system is the output signal of the system when the input signal is a unit impulse. For a DTLTI system, the impulse response provides a complete description of the system.

☞ The output signal of a DTLTI system in response to an arbitrary input signal can be determined as the convolution of the input signal and the impulse response. Convolution operation is given by Eqn. (2.109).

Eqn. 2.109:
$$y[n] = \sum_{k=-\infty}^{\infty} x[k]\, h[n-k]$$

☞ A system is said to be stable in bounded-input bounded-output (BIBO) sense if any bounded input signal is guaranteed to produce a bounded output signal. For a DTLTI system BIBO stability requires that the impulse response be absolute summable.

☞ A causal system is one in which the output signal depends only on current and past samples of the input signal but not its future samples. For a DTLTI system causality requires that the impulse response be equal to zero for all negative values of the sample index. Causal systems are suitable for real-time processing.

Further Reading

[1] S.C. Chapra. *Applied Numerical Methods with MATLAB for Engineers and Scientists.* McGraw-Hill, 2017.

[2] S.C. Chapra and R.P. Canale. *Numerical Methods for Engineers.* McGraw-Hill, 2020.

[3] S. Elaydi. *An Introduction to Difference Equations.* Springer Undergraduate Texts in Mathematics and Technology. Springer, 2005.

[4] W.G. Kelley and A.C. Peterson. *Difference Equations: An Introduction With Applications.* Harcourt/Academic Press, 2001.

[5] V. Lakshmikantham. *Theory of Difference Equations Numerical Methods and Applications.* Marcel Dekker, 2002.

[6] Ronald E. Mickens. *Difference Equations: Theory and Applications.* CRC Press, 2022.

[7] Ken C. Pohlmann. *Principles of Digital Audio.* McGraw-Hill Professional, 2000.

[8] A.N. Sharkovsky, Y.L. Maistrenko, and E.Y. Romanenko. *Difference Equations and Their Applications.* Vol. 250. Springer Science & Business Media, 2012.

[9] Kenneth Steiglitz. *Digital Signal Processing Primer.* Courier Dover Publications, 2020.

[10] U. Zölzer. *Digital Audio Signal Processing.* Wiley, 2008.

MATLAB Exercises with Solutions

MATLAB Exercise 2.1: Writing functions for moving average filters

In this exercise we will develop MATLAB functions for implementing length-2 and length-4 moving average filters that were discussed in Examples 2.4 and 2.5, respectively. Our purpose is not the development of fastest and most efficient functions. MATLAB already includes some powerful functions for implementing a wide variety of discrete-time systems, and those could certainly be used for the simulation of moving average filters as well. Two of these built-in functions, namely `conv()` and `filter()`, will be utilized in later exercises. Our purpose in developing our own functions for moving average filters is to provide further insight into their operation, and to build a foundation that can be used for implementing these filters on any hardware platform using any programming language. Therefore, we will sacrifice execution speed and efficiency in favor of a better understanding of coding a difference equation for a moving average filter.

Before developing any MATLAB function, it would be a good idea to consider how that function would eventually be used. Our development of the MATLAB code in this exercise will parallel the idea of *real-time processing* where the input and the output signals exist in the form of *data streams* of unspecified length. Algorithmically, we will execute the following steps:

1. Pick the *current input sample* from the input stream.
2. Process the *current input sample* through the moving average filter to compute the *current output sample.*
3. Put the *current output sample* into the output stream.
4. Repeat steps 1 through 3 until no more input samples are left.

Let us begin by developing a function named `ss_movavg2()` to implement step 2 above for a length-2 moving average filter. The function will be used with the syntax

```
y = ss_movavg2(x)
```

where `x` represents the *current input sample* and `y` is the *current output sample.* Once developed, the function can be placed into the loop described above to simulate a length-2 moving average filter with any input data stream.

The difference equation of a length-2 moving average filter was given in Eqn. (2.9) which is repeated here for convenience:

$$\text{Eqn. (2.9):} \qquad y[n] = \tfrac{1}{2}\,x[n] + \tfrac{1}{2}\,x[n-1]$$

A local variable named `xnm1` will be used for keeping track of the previous sample $x[n-1]$. The following two lines compute the output sample and then update the previous input sample for future use.

```
y = (x+xnm1)/2;
xnm1 = x;
```

Herein lies our first challenge: In MATLAB, local variables created within functions are discarded when the function returns. The next time the same function is called, its local variable `xnm1` would not have the value previously placed in it. The solution is to declare the variable `xnm1` to be a *persistent* variable, that is, one that retains its value between function calls. Following is a listing of the completed function `ss_movavg2()`:

```
1 function y = ss_movavg2(x)
2   persistent xnm1;      % Persistent variable to hold x[n-1]
3   if isempty(xnm1)      % If the function is called for the first time
4     xnm1 = 0;           %   Initialize xnm1
5   end;
6   y = (x+xnm1)/2;       % Eqn. (2.9)
7   xnm1 = x;             % Bookkeeping to prepare for next call
8 end
```

When the function `ss_movavg2()` is called for the first time, the persistent variable `xnm1` is created as an empty matrix. Lines 3 through 5 of the code above check for this condition and set `xnm1` equal to zero if it happens to be empty.

The function `ss_movavg4()` for a length-4 moving average filter can be developed in a similar manner. Consider Eqn. (2.10). In addition to $x[n-1]$ we also need to keep track of two input samples prior to that, namely $x[n-2]$ and $x[n-3]$. Therefore we will define and initialize two additional persistent variables with the names xnm2 and xnm3. The code for the function ss_movavg4() is listed below.

```
1  function y = ss_movavg4(x)
2    persistent xnm1 xnm2 xnm3;  % Variables to hold x[n-1], x[n-2], x[n-3]
3    if isempty(xnm1)     % If the function is called for the first time
4      xnm1 = 0;          %   Initialize xnm1
5      xnm2 = 0;          %   Initialize xnm2
6      xnm3 = 0;          %   Initialize xnm3
7    end;
8    y = (x+xnm1+xnm2+xnm3)/4;    % Eqn. (2.10)
9    % Bookkeeping to prepare for next call
10   xnm3 = xnm2;
11   xnm2 = xnm1;
12   xnm1 = x;
13 end
```

In lines 10 through 12 of the code we perform bookkeeping, and prepare the persistent variables for the next call to the function. The value $x[n]$ we have now will be the "previous sample" $x[n-1]$ the next time this function is called. Similar logic applies to the other two variables as well. Therefore the values of x, xnm1, and xnm2 need to be moved into xnm1, xnm2, and xnm3 respectively. In doing this, the order of assignments is critical. Had we used the order

```
xnm1 = x;
xnm2 = xnm1;
xnm3 = xnm2;
```

all three variables would have ended up (incorrectly) with the same value.

A final detail regarding persistent variables is that sometimes we may need to clear them from memory. Suppose we want to use the function ss_movavg2() with another data stream, and need to clear any persistent variables left over from a previous simulation. The following command accomplishes that:

```
>> clear ss_movavg2
```

Software resources: ss_movavg2.m , ss_movavg4.m

MATLAB Exercise 2.2: Testing moving average filtering functions

In this exercise we will develop the code for testing the moving average filtering functions developed in Exercise 2.1. Daily values of the Dow Jones Industrial Average data for the years 2013 and 2014 are available in MATLAB data file "djia_data.mat" in variables x2013 and x2014. The following script computes the length-4 moving average for the 2014 data, and graphs both the input and the output signals.

```
% Script: mexdt_2_2.m
clear ss_movavg4;              % Clear persistent variables
load 'djia_data.mat';          % Load the input data stream
xstream = [x2013(251:253),x2014];  % Input stream with 3 extra samples
```

```
5  ystream = [];                 % Create an empty output stream
6  nsamp = length(xstream);      % Number of samples in the input stream
7  for n=1:nsamp
8    x = xstream(n);             % "x" is the current input sample
9    y = ss_movavg4(x);          % "y" is the current output sample
10   ystream = [ystream,y];      % Append "y" to the output stream
11 end;
12 % Graph input and output signals
13 ystream = ystream(4:nsamp);   % Drop the first 3 samples
14 plot([0:252],x2014,[0:252],ystream);
15 axis([0,252,15000,18500]);
```

Note that the function `ss_movavg4()` begins with its persistent variables `xnm1`, `xnm2`, and `xnm3` each set equal to 0. As a result, the first three samples of the output will be inaccurate. This problem can be alleviated by prepending the last three samples of the year 2013 data to the year 2014 data before we begin processing. That is done on line 4 of the script. Line 13 of the script drops the first 3 samples of the output stream.

The MATLAB script above can be easily modified to simulate a length-2 moving average filter by substituting function `ss_movavg2()` in place of function `ss_movavg4()`.

Software resources: `mexdt_2_2.m` , `djia_data.mat`

MATLAB Exercise 2.3: Moving average filter functions revisited

In this exercise we will develop a function to implement a moving average filter of any length. Afterward we will use the function to apply a length-50 moving average filter to the 2014 Dow Jones data.

In previous exercises involving length-2 and length-4 moving average filters we have used persistent scalar variables to keep track of past samples of $x[n]$. For a length-50 filter this would be impractical as it would require 49 persistent variables to hold $x[n-1], x[n-2], \ldots, x[n-49]$. Instead, we will use a *buffer* vector to hold the contents of the "window" discussed in Example 2.3. Recall that, in Fig. 2.6 we have used the analogy of a window that displays the most recent N samples of the input signal for a moving average filter of length N. The rightmost sample visible through the window would be $x[n]$ and the leftmost sample visible would be $x[n-N+1]$. In a sense, each time a new sample comes in, it enters from the right side of the window. At the same time, the oldest sample leaves the window from the left side. From a visual perspective we will find it more convenient to think in terms of a vertical *buffer*, that is, a *column vector*. Samples enter from the top and fall off the bottom. See the listing for the function `ss_movavg()` below:

```
function y = ss_movavg(x,N)
  persistent buffer;        % Vector to hold the 'buffer'
  if isempty(buffer)        % If the function is called for the first time
    buffer = zeros(N,1);    %   Initialize buffer
  end;
  buffer = [x;buffer(1:end-1)];  % Update buffer with new incoming sample
  y = sum(buffer)/N;        % Eqn. (2.8)
end
```

In line 6, the new input sample $x[n]$ takes the top position in the buffer, and the oldest sample $x[n-50]$ which is no longer needed is pushed out. Afterward, in line 7, the output sample $y[n]$ is computed as the arithmetic average of all samples in the buffer. The listing `mexdt_2_3a.m` below uses the function `ss_movavg()` on 2014 Dow Jones data. The loop between lines 7 and 11 processes the input signal `xstream` one sample at a time to produce the output signal `ystream`. Compare the graph generated by the script to that in Fig. 2.7.

```
1  % Script: mexdt_2_3a.m
2  clear ss_movavg;              % Clear persistent variables
3  load 'djia_data.mat';         % Load the input data stream
4  xstream = [x2013(205:253),x2014];  % Input stream with 49 extra samples
5  ystream = [];                 % Create an empty output stream
6  nsamp = length(xstream);      % Number of samples in the input stream
7  for n=1:nsamp                 % Repeat for the specified range:
8    x = xstream(n);             %   "x" is the current input sample
9    y = ss_movavg(x,50);        %   "y" is the current output sample
10   ystream = [ystream,y];      %   Append "y" to the output stream
11 end;
12 % Graph input and output signals
13 ystream = ystream(50:nsamp);    % Drop the first 49 samples
14 plot([0:252],x2014,[0:252],ystream);
15 axis([0,252,15000,18500]);
```

Concerns about function `ss_movavg()`:

The major concern about the function `ss_movavg()` involves the persistent variables. The use of persistent variables, while useful for implementing moving average filters, is not always ideal. They could lead to programming errors that may be hard to diagnose. We may forget to clear them before starting a new filter. Also, it would be impossible to use the function `ss_movavg()` within a complex system that employs more than one moving average filter.

There is a better way to implement a moving average filter: What if we eliminate the need for persistent variables by passing the buffer vector to the function as one of its arguments? The function then updates the buffer vector as needed, and returns it back to the calling script. If multiple moving average filters are needed in a complex system, each could have its own buffer vector independent of the others. Such an alternative implementation, `ss_movavgb()` , is given below:

```
function [y,buffer] = ss_movavgb(x,N,buffer)
  buffer = [x;buffer(1:end-1)];    % Update buffer with new incoming sample
  y = sum(buffer)/N;               % Eqn. (2.8)
end
```

The script `mexdt_2_3b.m` uses the function `ss_movavgb()` to compute the output signal. The buffer vector is created and initialized by the calling script. As an added bonus, we can place the last 50 samples of 2013 data into the buffer (see line 6 of the code) before we start the processing loop.

The concept of a buffer is quite handy. We will also utilize it in implementing other types of digital filters in later parts of this book.

```
1  % Script: mexdt_2_3b.m
2  load 'djia_data.mat';         % Load the input data stream
3  xstream = x2014;              % Input stream
```

```
ystream = [];                   % Create an empty output stream
nsamp = length(xstream);        % Number of samples in the input stream.
buffer = x2013(204:253)';       % Set up length-50 buffer with 2013 data
for n=1:nsamp                   % Repeat for the specified range:
  x = xstream(n);               %   "x" is the current input sample
  [y,buffer] = ss_movavgb(x,50,buffer); %   "y" is the current output sample
  ystream = [ystream,y];        %   Append "y" to the output stream
end;
% Graph input and output signals
plot([0:252],xstream,[0:252],ystream);
axis([0,252,15000,18500]);
```

Software resources: mexdt_2_3a.m , mexdt_2_3b.m , ss_movavg.m , ss_movavgb.m

MATLAB Exercise 2.4: Writing and testing a function for exponential smoother

Consider the exponential smoother introduced in Example 2.6 with the difference equation given by Eqn. (2.11) repeated here for convenience:

$$\text{Eqn. (2.11):} \qquad y[n] = (1-\alpha)\; y[n-1] + \alpha\, x[n]$$

We will implement the exponential smoother in MATLAB and test it with the Dow Jones Industrial Average data using the same algorithmic approach as in earlier exercises. The code for the function `ss_smooth()` is listed below:

```
function y = ss_smooth(x,alpha)
  persistent ynm1;    % Persistent variable to hold y[n-1]
  if isempty(ynm1)    % If the function is called for the first time
    ynm1 = 0;         %   Initialize ynm1
  end;
  y = (1-alpha)*ynm1+alpha*x;  % Eqn. (2.9)
  ynm1 = y;                    % Bookkeeping to prepare for next call
end
```

The following script computes the output of the exponential smoother for the 2014 Dow Jones data as input, and graphs both the input and the output signals.

```
% Script: mexdt_2_4a.m
clear ss_smooth;                % Clear persistent variables
load 'djia_data.mat';           % Load the input data stream
xstream = x2014;                % Input stream
ystream = [];                   % Create an empty output stream
nsamp = length(xstream);        % Number of samples in the input stream
alpha = 0.1;                    % Parameter for exponential smoother
for n=1:nsamp                   % Repeat for the specified range:
  x = xstream(n);               %   "x" is the current input sample
  y = ss_smooth(x,alpha);       %   "y" is the current output sample
  ystream = [ystream,y];        %   Append "y" to the output stream
end;
% Graph input and output signals
plot([0:nsamp-1],xstream,[0:nsamp-1],ystream);
axis([0,252,15000,18500]);
```

The concerns that were raised in MATLAB Exercise 2.3 regarding the use of persistent variables apply here as well. An alternative exponential smoother function that uses a buffer is listed below. In this case the buffer is simply a scalar to hold the value of ynm1.

```
function [y,ynm1] = ss_smoothb(x,alpha,ynm1)
  y = (1-alpha)*ynm1+alpha*x;  % Eqn. (2.9)
  ynm1 = y;                    % Bookkeeping to prepare for next call
end
```

A slightly modified version of the test script for this case is given below.

```
% Script: mexdt_2_4b.m
load 'djia_data.mat';       % Load the input data stream
xstream = x2014;            % Input stream
ystream = [];               % Create an empty output stream
nsamp = length(xstream);    % Number of samples in the input stream
alpha = 0.1;                % Parameter for exponential smoother
ynm1 = 0;                   % Initialize the buffer variable
for n=1:nsamp                             % Repeat for the specified range:
  x = xstream(n);                         %    "x" is the current input
  [y,ynm1] = ss_smoothb(x,alpha,ynm1);    %    "y" is the current output
  ystream = [ystream,y];                  %    Append "y" to the output stream
end;
% Graph input and output signals
plot([0:nsamp-1],xstream,[0:nsamp-1],ystream);
axis([0,252,15000,18500]);
```

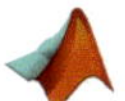

Software resources: mexdt_2_4a.m , mexdt_2_4b.m , ss_smooth.m , ss_smoothb.m

MATLAB Exercise 2.5: Iteratively solving a difference equation

It was shown in Section 2.5 that one way to solve a difference equation is to iterate through it one sample index at a time. In this exercise we will explore this concept in MATLAB, using the difference equation for the loan balance problem introduced in Example 2.7. While the specific problem at hand is quite simplistic, it will help us highlight some of the challenges encountered in adapting a signal-system interaction problem for MATLAB coding, in particular with the vector indexing scheme of MATLAB.

The system under consideration accepts monthly payment data $x[n]$ as input and produces an output signal $y[n]$ that represents the balance at the end of each month. The difference equation for the system was given by Eqn. (2.12) and will be repeated here:

$$\text{Eqn. (2.12):} \qquad y[n] = (1+c)\, y[n-1] + x[n]$$

Let the amount borrowed be \$10,000 which we will place into $y[0]$ as the initial value of the output signal. Let us also assume that \$100 will be paid each month starting with month 1. Thus, the input signal would be

$$x[n] = 100\, u[n-1]$$

Using a monthly interest rate of 0.5 percent, meaning $c = 0.5/100 = 0.005$, we will compute $y[n]$ for $n = 1, \ldots, 18$. Following statements create two vectors xvec and yvec:

```
>> xvec = [0,100*ones(1,18)];
>> yvec = [10000,zeros(1,18)];
```

The length-19 vector xvec contains payment data. Its first element is zero, and all other elements are equal to 100. The vector yvec holds samples of the output signal. Its first element is initialized to 10,000, the loan amount, and other elements are set equal to zero as place holders to be modified later.

One of the difficulties encountered in representing discrete-time signals with vectors is that MATLAB vectors do not have an element with index 0; the first element of a vector always has an index of 1. When we translate the difference equation of the system into code, we need to remember that

xvec(1) corresponds to $x[0]$,
xvec(2) corresponds to $x[1]$,

$\vdots$

yvec(1) corresponds to $y[0]$,
yvec(2) corresponds to $y[1]$,

$\vdots$

and so on. Fortunately, this is not a difficult problem to overcome. The method we will use in this exercise is not the most elegant, but it is one that will keep us continuously aware of the indexing differences between MATLAB vectors and our signals. The script listed below iterates through the difference equation. Notice how the MATLAB variable offset is used in line 8 to deal with the indexing problem.

```
1  % Script: mexdt_2_5a.m
2  xvec = [0,100*ones(1,18)];    % Vector to hold input signal
3  yvec = [10000,zeros(1,18)];   % Vector to hold output signal
4  c = 0.005;                    % Interest rate
5  offset = 1;                   % Offset to fix index issues
6  % Start the loop to compute the output signal
7  for n=1:18
8    yvec(offset+n)=(1+c)*yvec(offset+n-1)-xvec(offset+n);  % Eqn. (2.12)
9  end;
10 % Display the output signal
11 tmp = [[0:18]',yvec'];
12 disp(tmp);
```

The last two lines

```
tmp = [[0:18]',yvec'];
disp(tmp);
```

create a 19 by 2 matrix tmp in which the first column contains the indices from 0 to 18, and the second column contains the corresponding sample amplitudes of the output signal. A better looking tabulated display of the results can be produced using the function fprintf(). The modified script mexdt_2_5b.m tabulates the results a bit more cleanly, and also graphs the output signal $y[n]$.

```
% Script: mexdt_2_5b.m
xvec = [0,100*ones(1,18)];    % Vector to hold input signal
yvec = [10000,zeros(1,18)];   % Vector to hold output signal
c = 0.005;                    % Interest rate.
```

```
offset = 1;                    % Offset to fix index issues.
fprintf(1,'Index    Input  Output\n'); % Print header.
% Start the loop to compute and print the output signal.
for n=1:18
  yvec(offset+n)=(1+c)*yvec(offset+n-1)-xvec(offset+n); % Eqn. (2.12)
  fprintf(1,'%5d  %5.2f  %5.2f\n',n,xvec(offset+n),yvec(offset+n));
end;
% Graph the output signal.
stem([0:18],yvec);
```

Software resources: mexdt_2_5a.m , mexdt_2_5b.m

MATLAB Exercise 2.6: Implementing a discrete-time system from its block diagram

In this exercise we will implement the discrete-time system for which a block diagram was obtained in Example 2.16, Fig. 2.29. A MATLAB function named `ss_system1()` will be developed with the syntax

```
y = ss_system1(x)
```

where x is the current input sample and y is the current output sample. Let the outputs $w[n-1]$, $w[n-2]$, and $w[n-3]$ of the three delay elements from left to right be represented by MATLAB variables wnm1, wnm2 and wnm3, respectively. These are the persistent variables. A listing of the completed function `ss_system1()` is given below:

```
function y = ss_system1(x)
  persistent wnm1 wnm2 wnm3;
  if isempty(wnm1)
    wnm1 = 1.0682;  % Initial value w[-1]
    wnm2 = 1.7149;  % Initial value w[-2]
    wnm3 = 0.1674;  % Initial value w[-3]
  end;
  wn = x+0.7*wnm1+0.8*wnm2-0.84*wnm3;  % Eqn. (2.88)
  y = 0.1*wn+0.2*wnm1+0.3*wnm2;        % Eqn. (2.89)
  % Bookkeeping to prepare for next call
  wnm3 = wnm2;
  wnm2 = wnm1;
  wnm1 = wn;
end
```

The script listed below can be used for testing the function `ss_system1()` with a unit step input signal and graphing the resulting output signal. When the function is first called, the initial values obtained in Example 2.16 are placed into the persistent variables.

```
% Script: mexdt_2_6.m
clear ss_system1;            % Clear persistent variables
xstream = ones(1,50);        % Input data stream
ystream = [];                % Create an empty output stream.
nsamp = length(xstream);     % Number of samples in the input stream
for n=1:nsamp                % Repeat for the specified range:
  x = xstream(n);            %   "x" is the current input sample
```

```
  y = ss_system1(x);        %   "y" is the current output sample
  ystream = [ystream,y];    %   Append "y" to the output stream
end;
% Graph the output signal
stem([0:nsamp-1],ystream);
title('The output signal');
xlabel('Index n');
ylabel('y[n]');
```

The graph produced by this script is shown in Fig. 2.40. Compare it to the result displayed in the interactive app `appBlockDgmDT.m`.

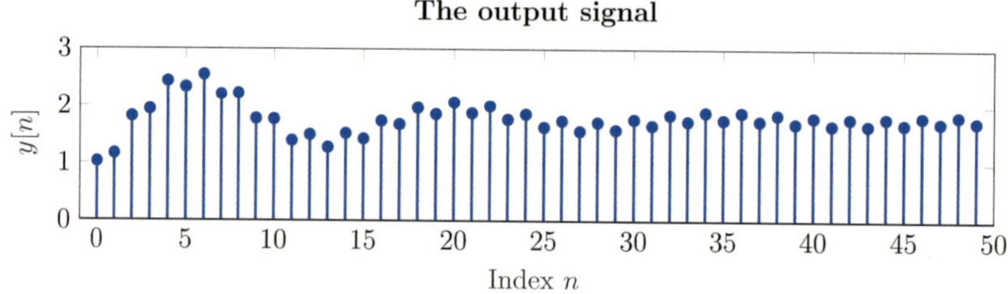

Figure 2.40 – Unit step response of the system in MATLAB Exercise 2.6.

Software resources: `mexdt_2_6.m` , `ss_system1.m`

MATLAB Exercise 2.7: Implementation of a discrete-time system revisited

Consider again the system that was implemented in MATLAB Exercise 2.6. In this exercise we will implement the same system using the built-in function `filter()`. Recall that the system in question was introduced in Example 2.16. Its difference equation is

$$y[n]-0.7\,y[n-1]-0.8\,y[n-2]+0.84\,y[n-3]=0.1\,x[n]+0.2\,x[n-1]+0.3\,x[n-2]$$

Our goal is to determine the unit step response of the system subject to the initial conditions

$$y[-1]=0.5\,,\quad y[-2]=0.3\,,\quad y[-3]=-0.4$$

The function `filter()` solves a difference equation in the general form

$$\begin{aligned}a_0\,y[n]+a_1\,y[n-1]+\ldots+a_{N-1}\,y[n-N+1]+a_N\,y[n-N]=\\ b_0\,x[n]+b_1\,x[n-1]+\ldots+b_{M-1}\,x[n-M+1]+b_M\,x[n-M]\end{aligned}$$

using a *transposed direct-form II* implementation which is different than the block diagram we have used in MATLAB Exercise 2.6. We should, however, be able to find the exact same solution as in Example 2.16 and MATLAB Exercise 2.6, independent of the block diagram implementation form. The syntax of the function `filter()` is

```
y = filter(b,a,x,zi)
```

Meanings of the parameters in the function call are explained below:

b : Vector of feed-forward coefficients $b_0, b_1, \ldots, b_M$

a : Vector of feedback coefficients $a_0, a_1, \ldots, a_N$

x : Vector containing samples of the input signal

y : Vector containing samples of the output signal

zi : Vector containing initial values of the delay elements for the *transposed direct-form II* block diagram

In both Example 2.16 and MATLAB Exercise 2.6 we have converted the initial output values $y[-1]$, $y[-2]$, and $y[-3]$ to values $w[-1]$, $w[-2]$, and $w[-3]$ for the block diagram in Fig. 2.29. This step will need to be revised when using the function `filter()`. The vector `zi` is obtained from the initial values of input and output samples through a call to the function `filtic()`. The script `mexdt_2_7.m` computes and graphs the output signal. It can easily be verified that the output signal obtained is identical to that of MATLAB Exercise 2.6.

```
% Script: mexdt_2_7.m
a = [1,-0.7,-0.8,0.84];       % Feedback coefficients
b = [0.1,0.2,0.3];            % Feed-forward coefficients
y_init = [0.5,0.3,-0.4];      % y[-1], y[-2], and y[-3]
x_init = [0,0,0];             % x[-1], x[-2], and x[-3]
xstream = ones(1,50);         % Unit step input signal
zi = filtic(b,a,y_init,x_init);     % Convert initial conditions
ystream = filter(b,a,xstream,zi);   % Compute the output signal
% Graph the output signal.
stem([0:49],ystream);
title('The output signal');
xlabel('Index n');
ylabel('y[n]');
```

Software resources: mexdt_2_7.m

MATLAB Exercise 2.8: Discrete-time convolution

In Example 2.19 we have discussed the convolution operation applied to two finite-length signals, namely

$$h[n] = \{ \underset{\substack{\uparrow \\ n=0}}{4}, 3, 2, 1 \} \qquad \text{and} \qquad x[n] = \{ \underset{\substack{\uparrow \\ n=0}}{-3}, 7, 4 \}$$

We will use the built-in function `conv()` for convolving signals $h[n]$ and $x[n]$. The following set of statements create two vectors `h` and `x` corresponding to the signals we are considering, and then compute their convolution:

```
>> h = [4,3,2,1];
>> x = [-3,7,4];
>> y = conv(h,x)

y =
   -12    19    31    23    15     4
```

The convolution operator is commutative, that is, the roles of $h[n]$ and $x[n]$ can be reversed with no effect on the result. Therefore, the statement

```
>> y = conv(x,h)
```

produces exactly the same output. The vector y produced by the function conv() starts with index 1, a characteristic of all MATLAB vectors. Consequently, an attempt to graph it with a statement like

```
>> stem(y);
```

results in the incorrectly indexed stem graph shown in Fig. 2.41.

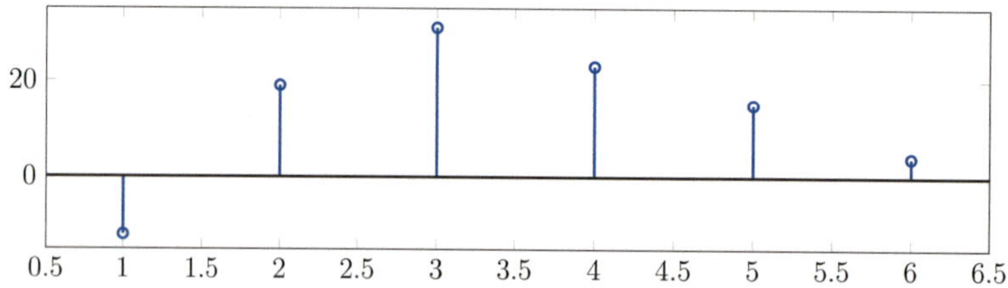

Figure 2.41 – The graph of the convolution result with incorrect sample indices.

A correctly indexed stem graph can be obtained by the statements

```
>> n = [0:5];
>> stem(n,y);
```

and is shown in Fig. 2.42.

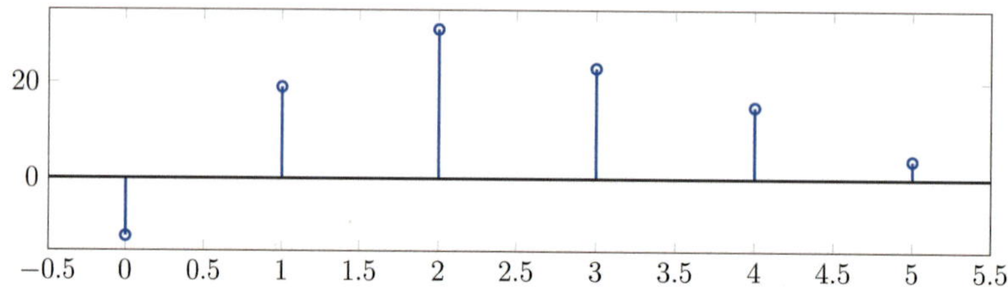

Figure 2.42 – The graph with corrected sample indices.

What if the starting indices of the two signals $h[n]$ and $x[n]$ are not necessarily at $n = 0$, but are specified as arbitrary values N_h and N_x respectively? (See Example 2.20.) We could simply adjust the vector n to accommodate the change. Instead, we will use this as an opportunity to develop a *wrapper function* that utilizes the built-in conv() function, but also automates the process of generating the appropriate index vector n to go with it. The listing for the function ss_conv() is given below:

```
function [y,n] = ss_conv(h,x,Nh,Nx)
  y = conv(h,x);          % Compute the convolution
  Ny = length(y);         % Number of samples in y[n]
  nFirst = Nh+Nx;         % Correct index for the first sample in y[n]
  nLast = nFirst+Ny-1;    % Correct index for the last sample in y[n]
  n = [nFirst:nLast];     % Vector of corrected indices
end
```

Parameters Nh and Nx are the correct indices of first samples of the vectors h and x respectively. Example usage of the function ss_conv() is given below:

```
>> [y,n] = ss_conv(h,x,5,7)
>> stem(n,y);
```

Software resources: mexdt_2_8.m , ss_conv.m

MATLAB Exercise 2.9: Developing our own convolution function

Recall the discussion about block diagram interpretation of convolving two very short signals in Example 2.21. We will use the ideas presented in that example for writing a convolution function named ss_convolve(). MATLAB has a built-in function conv() that is faster and more efficient, and it should preferred over our function ss_convolve(). On the other hand, writing our own version of a function is instructive. It provides insight into how the convolution operation works. The code listing for the function ss_convolve() is given below.

```
1  function y = ss_convolve(h,x)
2    Nh = length(h);                      % Length of vector 'h'
3    Nx = length(x);                      % Length of vector 'x'
4    buffer = zeros(Nh,1);                % Create all zero buffer
5    y = [];                              % Empty vector for 'y'
6    for n = 1:Nx
7      buffer = [x(n);buffer(1:end-1)];   % Update buffer with next sample
8      out = h*buffer;                    % Compute dot product
9      y = [y,out];                       % Append to vector 'y'
10   end
11   % Flush out the buffer
12   for n = 1:Nh-1
13     buffer = [0;buffer(1:end-1)];      % Insert zero into buffer
14     out = h*buffer;                    % Compute dot product
15     y = [y,out];                       % Append to vector 'y'
16   end
17 end
```

The arguments h and x should be row vectors. The output y returned by the function is also a row vector. The loop between lines 6 and 10 corresponds to Fig. 2.33 with new samples of vector x entering the buffer. The dot product in line 8 implements the summation in Eqn. (2.118). Finally, the loop between lines 12 and 16 corresponds to Fig. 2.34. There are no more non-zero samples of vector x coming in, and the buffer is flushed out. The script mexdt_2_9.m can be used for testing the function ss_convolve().

```
% Script: mexdt_2_9.m
h = [4,3,2,1];
x = [-3,7,4];
y = ss_convolve(h,x);
```

Software resources: mexdt_2_9.m , ss_convolve.m

MATLAB Exercise 2.10: Implementing a moving average filter through convolution

Consider again the problem of applying a length-50 moving-average filter to the Dow Jones data for the calendar year 2014. In MATLAB Exercise 2.3 we developed two custom functions, `ss_movavg()` and `ss_movavgb()` for this purpose. It is also possible to use the built-in function `conv()` to achieve the same result. The impulse response of the length-50 moving-average filter is

$$h[n] = \frac{1}{50}\left(u[n] - u[n-50]\right) = \begin{cases} 1/50\,, & 0 \le n \le 49 \\ 0\,, & \text{otherwise} \end{cases}$$

which can be represented with a MATLAB vector `h` obtained by statement

```
>> h = 1/50*ones(1,50);
```

The following set of statements statements load the MATLAB data file "`djia_data.mat`" into memory, and list the contents of MATLAB workspace so far.

```
>> load 'djia_data.mat';
>> whos

  Name        Size               Bytes  Class     Attributes
  h           1x50                 400  double
  x2013       1x253               2024  double
  x2014       1x253               2024  double
```

We are assuming that we have started a fresh session for this exercise; otherwise there may be additional variables listed above. The data for the year 2014 is in vector `x2014` and has 253 samples. Thus, we have an input signal $x[n]$ for $n = 0,\ldots,252$. Its convolution with the impulse response in vector `h` can be obtained with the statements

```
>> y2014 = conv(h,x2014);
```

There is one practical problem with this result. Recall the discussion in Example 2.3 about observing the input signal through a window that holds as many samples as the length of the filter, and averaging the window contents to obtain the output signal. For the first 49 samples of the output signal, the window is only partially full. (Refer to Fig. 2.6.) In the computation of $y[0]$, the only meaningful sample in the window is $x[0]$, and all samples to the left of it are zero-valued. The next output sample $y[1]$ will be based on averaging two meaningful samples $x[1]$ and $x[0]$ along with 48 zero-amplitude samples, and so on. As a result, the first meaningful sample of the output signal (meaningful in the sense of a moving average) will occur at index $n = 49$ when, for the first time, the window is full.

To circumvent this problem and obtain meaningful moving-average values for all 2014 data, we will prepend the signal $x[n]$ with an additional set of 49 samples borrowed from the last part of the data for calendar year 2013. The following MATLAB statement accomplishes that:

```
>> x = [x2013(205:253),x2014];
```

The resulting vector `x` represents the signal $x[n]$ that starts with index $n = -49$. We can now convolve this signal with the impulse response of the length-50 moving-average filter to obtain the output signal $y[n]$ which will also begin at index $n = -49$. The statement

```
>> y = conv(h,x);
```

produces a vector y with 351 elements. Its first 49 elements correspond to signal samples $y[-49]$ through $y[-1]$, and represent averages computed with incomplete data due to empty slots on the left side of the window. In addition, the last 49 elements of the vector y corresponding to signal samples $y[253]$ through $y[301]$ also represent averages computed with incomplete data, in this case with empty slots appearing on the right side of the window. We will obtain the smoothed data for the index range $n = 0,\ldots,252$ by discarding the first and the last 49 samples of the vector y.

```
>> y = y(50:302);
>> plot([0:252],x2014,'-',[0:252],y,'--');
```

Software resources: mexdt_2_10.m

MATLAB Exercise 2.11: Sunspot numbers – Using convolution and downsampling

In this exercise we will start with the monthly average sunspot numbers listed in MATLAB data file sundata_monthly.mat, and determine yearly averages. Type the following three lines into the command window to load the data file and list the first 15 rows of the matrix sspots_mo:

```
>> load 'sundata_monthly.mat';
>> format shortg
>> sspots_mo(1:15,:)

  ans =
  1749           1         58
  1749           2       62.6
  1749           3         70
  1749           4       55.7
  1749           5         85
  1749           6       83.5
  1749           7       94.8
  1749           8       66.3
  1749           9       75.9
  1749          10       75.5
  1749          11      158.6
  1749          12       85.2
  1750           1       73.3
  1750           2       75.9
  1750           3       89.2
```

If we were to use a length-12 moving average filter on the sunspot numbers, its output in response to the sample at year 1749 month 12 would be the arithmetic average of all monthly data for the year 1749. Similarly, the output in response to the sample at year 1750 month 12 would be the average of all monthly data for the year 1750. This leads to a simple idea for computing the yearly averages: We could convolve the monthly sunspot data with the impulse response of a length-12 moving average filter and then downsample the result, retaining only the 12th month of each year. The script mexdt_2_11.m listed below accomplishes that.

```
1 % Script: mexdt_2_11.m
2 % Load the data file
3 load('sundata_monthly.mat');
4 % Set up hn, the impulse response of length-12 moving average filter
5 hn = ones(12,1)/12;
6 y = conv(hn,sspots_mo(:,3));        % Convolve hn with 3rd column data
7 y = y(1:end-11);                     % Drop the last 11 samples
8 sspots_mo(:,3) = y;                  % Replace 3rd column data
9 x = downsample(sspots_mo,12,11);    % Downsample to keep mo 12 of each yr
10 % Graph yearly averages
11 plot(x(:,1),x(:,3));
12 axis([1749,2014,0,200]);
13 xlabel('Year');
14 ylabel('Sunspot numbers');
```

Note that the convolution result computed in line 6 is has 11 extra samples at the end. These are discarded in line 7. In line 9 of the code we downsample the entire matrix `sspots_mo` keeping every 12th row. The third argument to the function `downsample()` is the *offset* that specifies the number of rows skipped before the first retained row. Without the offset parameter, we would end up keeping month 1 data of each year. At this point it's illustrative to display the first 15 rows of the resulting matrix:

```
>> x(1:15,:)

  ans =
  1749        12       80.925
  1750        12       83.392
  1751        12       47.658
  1752        12         47.8
  1753        12       30.675
  1754        12       12.217
  1755        12        9.567
  1756        12       10.192
  1757        12       32.425
  1758        12         47.6
  1759        12       53.967
  1760        12       62.858
  1761        12        85.85
  1762        12        61.15
  1763        12       45.117
```

The graph produced by the script `mexdt_2_11.m` is shown in Fig. 2.43.

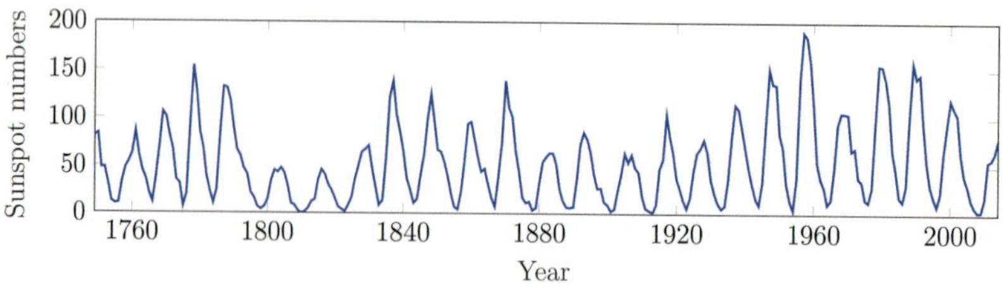

Figure 2.43 – Graph obtained in MATLAB Exercise 2.11.

Software resources: mexdt_2_11.m

MATLAB Exercise 2.12: Frame-based processing of audio signals

The use of objects for frame-by-frame processing of audio signals was introduced in MATLAB Exercise 1.11. The script used in part b of that exercise provided the boilerplate code for reading a frame of data from the reader object, sending the frame to the player object, and repeating this sequence of actions until no more frames are available. In this exercise we will insert an additional step into the code, and modify the frame of data coming from the reader object before it is sent to the player object.

In the early parts of this chapter we have explored two basic systems, namely a moving average filter and an exponential smoother. MATLAB functions `ss_movavgb()` and `ss_smoothb()` were developed for implementing these systems. These two functions are not suitable for frame-by-frame processing since they were designed for use on a sample-by-sample basis. We will need to develop new versions of them for this exercise.

a. Moving average filter

First step is to write a function `ss_movavgbf()` that takes a frame of data containing one or more channels of the audio signal, apply a moving average filter of specified length to it, and produce an output frame with the results. The code for the function is given below.

```
1 function [y,buffer] = ss_movavgbf(x,N,buffer)
2   y = zeros(size(x));
3   frameSize = size(x,1);                    % Get the frame size
4   for i = 1:frameSize                       % Repeat for the specified range
5     buffer = [x(i,:);buffer(1:N-1,:)]; %   Update buffer
6     y(i,:) = sum(buffer)/N;                 %   Eqn. (2.8)
7   end;
8 end
```

The matrix `buffer` has N rows for a length-N moving average filter. It has as many columns as the number of channels in the audio signal. In line 5 we update `buffer` by dropping its last row which is no longer needed and adding the current row of the frame data in `x` on top. This ensures that `buffer` has the set of samples we need to average in order to compute the output signal in line 6. The script `mexdt_2_12a.m` listed below uses the function `ss_movavgbf()` to apply a length-8 moving average signal to the audio file `Ballad_22050_Hz.flac`. Pay attention to the loss of high frequency detail in the audio signal. Larger values of N result in heavier smoothing.

```
% Script: mexdt_2_12a.m
% Create an "audio file reader" object
sReader = dsp.AudioFileReader('Ballad_22050_Hz.flac','ReadRange',[1,441000]);
% Create an "audio player" object
sPlayer = audioDeviceWriter('SampleRate',sReader.SampleRate);
% Create a "time scope" object to display signals
sScope = timescope('YLimits',[-1,1]);

N = 8;
```

```
numChannels = info(sReader).NumChannels;  % Number of channels
buffer = zeros(N,numChannels);            % Initial buffer matrix
while ~isDone(sReader)                    % Repeat until no more frames left:
  x = sReader();                          %   Get the next frame of data
  [y,buffer] = ss_movavgbf(x,N,buffer); %   Compute moving averages
  sScope(y);                              %   Display frame data
  sPlayer(y);                             %   Play back the frame
end

release(sReader);  % We are finished with the input audio file
release(sPlayer);  % We are finished with the audio output device
release(sScope);   % We are finished with the time scope
```

b. Exponential smoother

A modified version of the function `ss_smoothb()` is given below as `ss_smoothbf()`. It accepts a frame of audio data with one or more channels, applies exponential smoothing to it and returns a compatible output frame. The vector `ynm1` represents $y[n-1]$ for each channel, and therefore has as many elements as the number of channels.

```
function [y,ynm1] = ss_smoothbf(x,alpha,ynm1)
  y = zeros(size(x));
  frameSize = size(x,1);     % Get the frame size
  for i = 1:frameSize        % Repeat for the specified range
    y(i,:) = (1-alpha)*ynm1+alpha*x(i,:);  %   Eqn. (2.9)
    ynm1 = y(i,:);           %   Bookkeeping to prepare for next call
  end
end
```

The script `mexdt_2_12b.m` listed below uses the function `ss_smoothbf()` to apply exponential smoothing to the audio file `Ballad_22050_Hz.flac`. Pay attention to the loss of high frequency detail in the audio signal. Smaller values of α result in heavier smoothing.

```
% Script: mexdt_2_12b.m
% Create an "audio file reader" object
sReader = dsp.AudioFileReader('Ballad_22050_Hz.flac','ReadRange',[1,441000]);
% Create an "audio player" object
sPlayer = audioDeviceWriter('SampleRate',sReader.SampleRate);
% Create a "time scope" object to display signals
sScope = timescope('YLimits',[-1,1]);

alpha = 0.2;
ynm1 = zeros(1,info(sReader).NumChannels);
while ~isDone(sReader)                    % Repeat until no more frames left
  x = sReader();                          %   Get the next frame of data
  [y,ynm1] = ss_smoothbf(x,alpha,ynm1); %   Compute filter outputs
  sScope(y);                              %   Display frame data
  sPlayer(y);                             %   Play back the frame
end

release(sReader);  % We are finished with the input audio file
release(sPlayer);  % We are finished with the audio output device
release(sScope);   % We are finished with the time scope
```

Software resources: mexdt_2_12a.m , mexdt_2_12b.m , ss_movavgbf.m , ss_smoothbf.m

MATLAB Exercise 2.13: Case study – Echoes and reverberation, part 1

Our perception of characteristics of sound depends greatly on the environment we are in. Imagine a musical instrument such as a guitar being played at a specific distance from us. The sound we perceive when the instrument is played in open air would be quite different than when it is played in a small room, or a large concert hall, or a narrow canyon. In spaces with reflective surfaces, multiple echoes (reflections of sound from various surfaces) reach our ears in addition to the direct sound. An overly simplified illustration of this is shown in Fig. 2.44. Typical sound perceived in an enclosed space would have a large number of echoes each arriving with a different time delay and a different amount of attenuation. The collective effect of all these echoes is referred to as *reverberation*. Each concert hall, arena, or church has its own reverberation characteristics that are dependent on the size and type of reflective surfaces, and even the way people and instruments are positioned within the space.

The music produced in an anechoic recording studio (a small room where walls are covered with sound dampening material to eliminate or greatly reduce echoes) sounds a bit flat and somewhat boring. In the mixing stage, audio engineers often add reverberation effects to the recording to make the end result sound more lively. In this exercise, we will explore the basics of enhancing audio signals with echoes and reverberation. We will first consider a single echo added to the direct sound. This can easily be modeled with the difference equation

$$y[n] = x[n] + r\,x[n-L] \tag{2.133}$$

where $x[n]$ is the original sound. It is combined with an echo that has a delay of L samples, and an gain factor of r. We typically have $r < 1$ since each reflection causes the sound to be attenuated. The difference equation in Eqn. (2.133) represents a DTLTI system. A block diagram implementation of this system is shown in Fig. 2.45. The block with the notation D_L represents a delay of L samples.

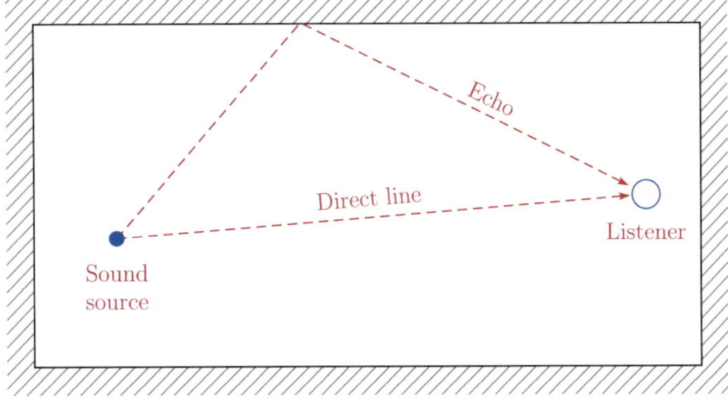

Figure 2.44 – A simplified depiction of sound propagation in an enclosed space.

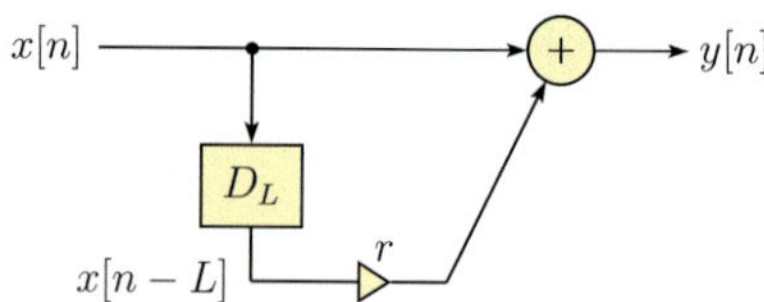

Figure 2.45 – A block diagram for implementing the system with the difference equation given by Eqn. (2.133).

This is one form of *feed-forward comb filter* that will be studied in more detail in Chapter 5. The function `ss_echo()` listed below can be used for iteratively solving the difference equation of Eqn. (2.133) one sample at a time.

```
1 function [y,buffer] = ss_echo(x,r,buffer)
2   y = x+r*buffer(end);                % Eqn. (2.133)
3   buffer = [x;buffer(1:end-1)];  % Update buffer with new incoming sample
4 end
```

The first two input arguments for `ss_echo()`, namely `x` and `r`, are scalars for the current input sample and the gain factor, respectively. The third argument `buffer` is a length-L column vector that holds the past L samples of the input signal. The sample $x[n-L]$ is the last sample of this vector, accessed in MATLAB with the statement `buffer(end)`. Thus, line 2 is a direct implementation of the difference equation. Once $y[n]$ is computed, the vector `buffer` is prepended with $x[n]$, and its last sample is discarded.

In the script `mexdt_2_13a.m` listed below, we first load the music signal from the file "`AG_Duet_22050_Hz.flac`" into the vector `x`. We extract the left channel of audio, and then use the function `ss_echo()` to add a single echo to it with parameters $L = 1024$ and $r = 0.8$. Finally we play back the resulting audio signal. The sampling rate of the audio signal is $f_s = 22{,}050$ Hz. Consequently, a time delay of $L = 1024$ samples corresponds to $t_d = 46.4$ ms for the corresponding analog signal. The audio used was downloaded from `www.freesound.org`, was converted to FLAC format, and is used under Creative Commons Attribution License 4.0 available at `https://creativecommons.org/licenses/by/4.0/`.

```
% Script: mexdt_2_13a.m
[x,fs] = audioread('AG_Duet_22050_Hz.flac');
xLeft = x(:,1);                    % Left channel
numSamples = size(xLeft,1);        % Number of samples
yLeft = zeros(size(xLeft));        % Create output vector
buffer = zeros(1024,1);            % L = 1024
for i=1:numSamples                 % Loop through audio samples
  [yLeft(i),buffer] = ss_echo(xLeft(i),0.8,buffer);
end
sound(yLeft,fs);                   % Play back the output vector
```

We have discussed in previous MATLAB exercises that, for efficiency purposes, real-time processing of audio signals is often done on a frame-by-frame basis rather than one sample at a time. In order to facilitate the use of audio processing objects introduced in MATLAB Exercises 1.11 and 2.12, we need to develop a version of the function `ss_echo()` that can process one frame of data at a time. The listing for the function `ss_echof()` is given below.

```
function [y,buffer] = ss_echof(x,r,buffer)
  y = zeros(size(x));                      % Placeholder for output frame
```

```
  frameSize = size(x,1);
  for i=1:frameSize
    y(i,:) = x(i,:)+r*buffer(end,:);       % Eqn. (2.133)
    buffer = [x(i,:);buffer(1:end-1,:)];   % Update buffer
  end
end
```

In this case, x is a matrix with as many rows as the frame size, and as many columns as the number of audio channels. The third argument, buffer is now a matrix with L rows, and as many columns as the number of audio channels. The script mexdt_2_13b.m listed below uses the new function ss_echof() to add an echo to the audio signal in real-time.

```
% Script: mexdt_2_13b.m
% Create an "audio file reader" object
sReader = dsp.AudioFileReader('AG_Duet_22050_Hz.flac','ReadRange',[1,661500]);
% Create an "audio player" object
sPlayer = audioDeviceWriter('SampleRate',sReader.SampleRate);

L = 1024;                                 % Corresponds to 46 ms delay
r = 0.8;                                  % Gain factor
numChannels = info(sReader).NumChannels;  % Number of channels
buffer = zeros(L,numChannels);            % Initial buffer matrix
while ~isDone(sReader)                    % Repeat until no more frames left
  x = sReader();                          %   Get the next frame of data
  [y,buffer] = ss_echof(x,r,buffer);      %   Compute output frame with echo
  sPlayer(x);                             %   Play back the frame
end

release(sReader);  % We are finished with the input audio file
release(sPlayer);  % We are finished with the audio output device
```

It is possible to approximate the reverberation effect with multiple echoes. As an example, consider the difference equation

$$y[n] = x[n] + r\,x[n-L] + r^2\,x[n-2L] + r^3\,x[n-3L] \tag{2.134}$$

The system represented by this difference equation can be implemented with the block diagram shown in Fig. 2.46.

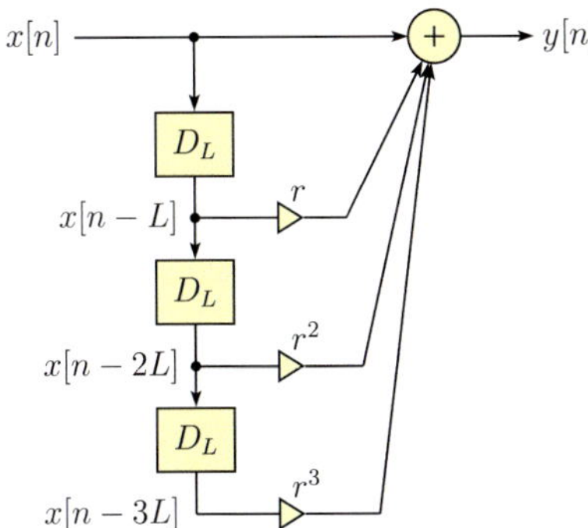

Figure 2.46 – A block diagram for implementing the system with the difference equation given by Eqn. (2.134).

A modified version of the function `ss_echof()` is given below with the name `ss_echo3f()` to implement this system. In the case, the matrix `buffer` has $3L$ rows holding the past $3L$ samples of the input signal.

```
function [y,buffer] = ss_echo3f(x,L,r,buffer)
  y = zeros(size(x));  % Placeholder for output frame
  frameSize = size(x,1);
  % Coefficients for the difference equation:
  a = r;
  b = a*r;  % Compute r^2 before the loop
  c = b*r;  % Compute r^3 before the loop
  for i=1:frameSize
    % Eqn. (2.134):
    y(i,:) = x(i,:)+a*buffer(L,:)+b*buffer(2*L,:)+c*buffer(3*L,:);
    buffer = [x(i,:);buffer(1:end-1,:)];  % Update buffer
  end
end
```

The script `mexdt_2_13c.m` listed below utilizes the function `ss_echo3f()` to add three echoes to the audio signal in real-time.

```
% Script: mexdt_2_13c.m
% Create an "audio file reader" object
sReader = dsp.AudioFileReader('AG_Duet_22050_Hz.flac','ReadRange',[1,661500]);
% Create an "audio player" object
sPlayer = audioDeviceWriter('SampleRate',sReader.SampleRate);

L = 1024;                                % Corresponds to 46 ms delay
r = 0.8;                                 % Attenuation factor
numChannels = info(sReader).NumChannels; % Number of channels
buffer = zeros(3*L,numChannels);         % Initial buffer matrix
while ~isDone(sReader)                   % Repeat until no more frames left
  x = sReader();                         %   Get the next frame of data
  [y,buffer] = ss_echo3f(x,L,r,buffer);  %   Compute output frame with echo
  sPlayer(y);                            %   Play back the frame
end

release(sReader);  % We are finished with the input audio file
release(sPlayer);  % We are finished with the audio output device
```

Software resources: `mexdt_2_13a.m` , `mexdt_2_13b.m` , `mexdt_2_13c.m` , `ss_echo.m` , `ss_echof.m` , `ss_echo3f.m`

Problems

2.1. A number of discrete-time systems are specified below in terms of their input-output relationships. For each case determine if the system is linear and/or time-invariant.

a. $y[n] = x[n]\,u[n]$

b. $y[n] = 3\,x[n] + 5$

c. $y[n] = 3\,x[n] + 5\,u[n]$

d. $y[n] = n\,x[n]$

e. $y[n] = \cos(0.2\pi\,n)\,x[n]$

f. $y[n] = x[n] + 3\,x[n-1]$

g. $y[n] = x[n] + 3\,x[n-1]\,x[n-2]$

2.2. A number of discrete-time systems are described below. For each case determine if the system is linear and/or time-invariant.

a. $y[n] = \sum\limits_{k=-\infty}^{n} x[k]$

b. $y[n] = \sum\limits_{k=0}^{n} x[k]$

c. $y[n] = \sum\limits_{k=n-2}^{n} x[k]$

d. $y[n] = \sum\limits_{k=n-2}^{n+2} x[k]$

2.3. Consider the cascade combination of two systems shown in Fig. P.2.3(a).

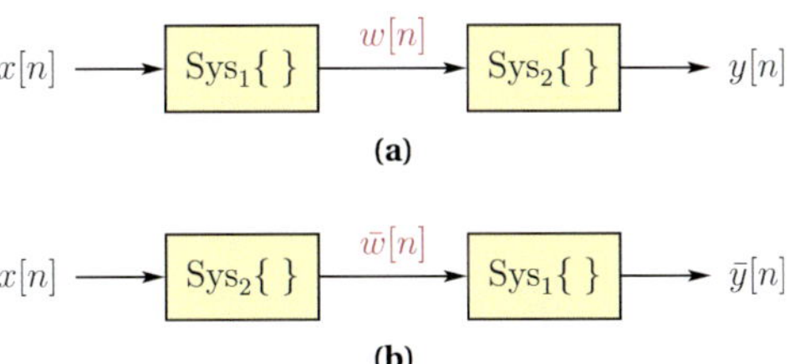

Figure P. 2.3

a. Let the input-output relationships of the two subsystems be given as

$$\text{Sys}_1\left\{x[n]\right\} = 3\,x[n] \quad \text{and} \quad \text{Sys}_2\left\{w[n]\right\} = w[n-2]$$

Write the relationship between $x[n]$ and $y[n]$.

b. Let the order of the two subsystems be changed as shown in Fig. P.2.3(b). Write the relationship between $x[n]$ and $\bar{y}[n]$. Does changing the order of two subsystems change the overall input-output relationship of the system?

2.4. Repeat Problem 2.3 with the following sets of subsystems:

a. $\text{Sys}_1\left\{x[n]\right\} = 3\,x[n]$ and $\text{Sys}_2\left\{w[n]\right\} = n\,w[n]$

b. $\text{Sys}_1\left\{x[n]\right\} = 3\,x[n]$ and $\text{Sys}_2\left\{w[n]\right\} = w[n] + 5$

2.5. The response of a linear and time-invariant system to the input signal $x[n] = \delta[n]$ is given by

$$\text{Sys}\left\{\delta[n]\right\} = \{\, \underset{\substack{\uparrow \\ n=0}}{2}, 1, -1 \,\}$$

Determine the response of the system to the following input signals:

a. $x[n] = \delta[n] + \delta[n-1]$

b. $x[n] = \delta[n] - 2\delta[n-1] + \delta[n-2]$

c. $x[n] = u[n] - u[n-5]$

d. $x[n] = n\,\big(u[n] - u[n-5]\big)$

2.6. Consider a system that is known to be linear but not necessarily time-invariant. Its responses to three impulse signals are given below:

$$\text{Sys}\left\{\delta[n]\right\} = \{\underset{\substack{\uparrow \\ n=0}}{1}, 2, 3\}$$

$$\text{Sys}\left\{\delta[n-1]\right\} = \{\underset{\substack{\uparrow \\ n=1}}{3}, 3, 2\}$$

$$\text{Sys}\left\{\delta[n-2]\right\} = \{\underset{\substack{\uparrow \\ n=2}}{3}, 2, 1\}$$

For each of the input signals listed below, state whether the response of the system can be determined from the information given. If the response can be determined, find it. If not, explain why it cannot be done.

a. $x[n] = 5\delta[n-1]$

b. $x[n] = 3\delta[n] + 2\delta[n-1]$

c. $x[n] = \delta[n] - 2\delta[n-1] + 4\delta[n-2]$

d. $x[n] = u[n] - u[n-3]$

e. $x[n] = u[n] - u[n-4]$

2.7. The discrete-time signal

$$x[n] = \{\underset{\substack{\uparrow \\ n=0}}{1.7}, 2.3, 3.1, 3.3, 3.7, 2.9, 2.2, 1.4, 0.6, -0.2, 0.4\}$$

is used as input to a length-2 moving average filter. Determine the response $y[n]$ for $n = 0,\ldots,9$. Use $x[-1] = 0$.

2.8. Consider again the signal $x[n]$ specified in Problem 2.7. If it is applied to a length-4 moving average filter, determine the output signal $y[n]$ for $n = 0,\ldots,9$. Use $x[-1] = x[-2] = x[-3] = 0$.

2.9. The signal $x[n]$ specified in Problem 2.7 is used as input to an exponential smoother with $\alpha = 0.2$. The initial value of the output signal is $y[-1] = 1$. Determine $y[n]$ for $n = 0,\ldots,9$.

2.10. Consider the difference equation model for the loan payment system explored in Example 2.7. The balance at the end of month-n is given by

$$y[n] = (1 + c)\; y[n-1] - x[n]$$

where c is the monthly interest rate and $x[n]$ represents the payment made in month-n. Let the borrowed amount be \$10,000 to be paid back at the rate of \$250 per month and with a monthly interest rate of 0.5 percent, that is, $c = 0.005$. Determine the monthly balance for $n = 1,\ldots,12$ by iteratively solving the difference equation.

Hint: Start with the initial balance of $y[0] = 10{,}000$ and model the payments with the input signal

$$x[n] = 250\, u[n-1]$$

2.11. Refer to Example 2.9 in which a nonlinear difference equation was found for computing a root of a function. The idea was adapted to the computation of the square root of a positive number A by searching for a root of the function

$$f(w) = w^2 - A$$

By iteratively solving the difference equation given by Eqn. (2.19), approximate the square root of

a. $A = 5$

b. $A = 17$

c. $A = 132$

In each case carry out the iterations until the result is accurate up to the fourth digit after the decimal point.

2.12. For each homogeneous difference equation given below, find the characteristic equation and show that it only has simple real roots. Find the homogeneous solution for $n \geq 0$ in each case subject to the initial conditions specified. Hint: For part (e) use MATLAB function `roots()` to find the roots of the characteristic polynomial.

a. $y[n] + 0.2\,y[n-1] - 0.63\,y[n-2] = 0\,,\; y[-1] = 5\,,\; y[-2] = -3$

b. $y[n] + 1.3\,y[n-1] + 0.4\,y[n-2] = 0\,,\; y[-1] = 0\,,\; y[-2] = 5$

c. $y[n] - 1.7\,y[n-1] + 0.72\,y[n-2] = 0\,,\; y[-1] = 1\,,\; y[-2] = 2$

d. $y[n] - 0.49\,y[n-2] = 0\,,\; y[-1] = -3\,,\; y[-2] = -1$

e. $y[n] + 0.6\,y[n-1] - 0.51\,y[n-2] - 0.28\,y[n-3] = 0\,,\; y[-1] = 3\,,\; y[-2] = 2\,,\; y[-3] = 1$

2.13. For each homogeneous difference equation given below, find the characteristic equation and show that at least some of its roots are complex. Find the homogeneous solution for $n \geq 0$ in each case subject to the initial conditions specified. Hint: For part (d) use MATLAB function `roots()` to find the roots of the characteristic polynomial.

a. $y[n] - 1.4\,y[n-1] + 0.85\,y[n-2] = 0\,,\; y[-1] = 2\,,\; y[-2] = -2$

b. $y[n] - 1.6\,y[n-1] + y[n-2] = 0\,,\; y[-1] = 0\,,\; y[-2] = 3$

c. $y[n] + y[n-2] = 0\,,\; y[-1] = 3\,,\; y[-2] = 2$

d. $y[n] - 2.5\,y[n-1] + 2.44\,y[n-2] - 0.9\,y[n-3] = 0\,,\; y[-1] = 1\,,\; y[-2] = 2\,,\; y[-3] = 3$

2.14. For each homogeneous difference equation given below, find the characteristic equation and show that it has multiple-order roots. Find the homogeneous solution for $n \geq 0$ in each case subject to the initial conditions specified. Hint: For parts (c) and (d) use MATLAB function `roots()` to find the roots of the characteristic polynomial.

a. $y[n] - 1.4\,y[n-1] + 0.49\,y[n-2] = 0\,,\; y[-1] = 1\,,\; y[-2] = 1$

b. $y[n] + 1.8\,y[n-1] + 0.81\,y[n-2] = 0\,,\; y[-1] = 0\,,\; y[-2] = 2$

c. $y[n] - 0.8\,y[n-1] - 0.64\,y[n-2] + 0.512\,y[n-3] = 0\,,\; y[-1] = 1\,,\; y[-2] = 1\,,\; y[-3] = 2$

d. $y[n] + 1.7\,y[n-1] + 0.4\,y[n-2] - 0.3\,y[n-3] = 0\,,\; y[-1] = 1\,,\; y[-2] = 2\,,\; y[-3] = 1$

2.15. Solve each difference equation given below for the specified input signal and initial conditions. Use the general solution technique outlined in Section 2.5.2.

a. $y[n] = 0.6\,y[n-1] + x[n]\,,\quad x[n] = u[n]\,,\quad y[-1] = 2$

b. $y[n] = 0.8\,y[n-1] + x[n]\,,\quad x[n] = 2\sin(0.2n)\,,\quad y[-1] = 1$

c. $y[n] - 0.2\,y[n-1] - 0.63\,y[n-2] = x[n]\,,\quad x[n] = e^{-0.2n}\,,\quad y[-1] = 0\,,\quad y[-2] = 3$

d. $y[n] + 1.4\,y[n-1] + 0.85\,y[n-2] = x[n]\,,\quad x[n] = u[n]\,,\quad y[-1] = -2\,,\quad y[-2] = 0$

e. $y[n] + 1.6\,y[n-1] + 0.64\,y[n-2] = x[n]\,,\quad x[n] = u[n]\,,\quad y[-1] = 0\,,\quad y[-2] = 1$

2.16. Consider the exponential smoother explored in Examples 2.6 and 2.14. It is modeled with the difference equation

$$y[n] = (1-\alpha)\,y[n-1] + \alpha\,x[n]$$

Let $y[-1] = 0$ so that the system is linear.

a. Let the input signal be a unit step, that is, $x[n] = u[n]$. Determine the response of the linear exponential smoother as a function of α.

b. Let the input signal be a unit ramp, that is, $x[n] = n\,u[n]$. Determine the response of the linear exponential smoother as a function of α.

2.17. Consider a first-order differential equation in the form

$$\frac{dy_a(t)}{dt} + A\,y_a(t) = A\,x_a(t)$$

The derivative can be approximated using the backward difference

$$\frac{dy_a(t)}{dt} \approx \frac{y_a(t) - y_a(t-T)}{T}$$

where T is the step size.

a. Using the approximation for the derivative in the differential equation, express $y_a(t)$ in terms of $y_a(t-T)$ and $x_a(t)$.

b. Convert the differential equation to a difference equation by defining discrete-time signals

$$x[n] = x_a(nT)$$
$$y[n] = y_a(nT)$$
$$y[n-1] = y_a(nT-T)$$

Show that the resulting difference equation corresponds to an exponential smoother. Determine its parameter α in terms of A and T.

2.18. Construct a block diagram for each difference equation given below.

a. $y[n] + 0.2\,y[n-1] - 0.63\,y[n-2] = x[n] + x[n-2]$

b. $y[n] - 2.5\,y[n-1] + 2.44\,y[n-2] - 0.9\,y[n-3] = x[n] - 3\,x[n-1] + 2\,x[n-2]$

c. $y[n] - 0.49\,y[n-2] = x[n] - x[n-1]$

d. $y[n] + 0.6\,y[n-1] - 0.51\,y[n-2] - 0.28\,y[n-3] = x[n] - 2\,x[n-2]$

2.19. A discrete-time system is described by the block diagram shown in Fig. P.2.19.

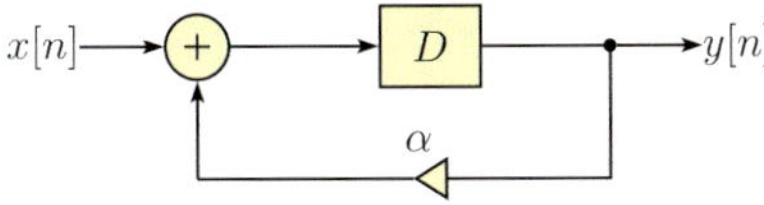

Figure P. 2.19

a. Write a difference equation between the signals $x[n]$ and $y[n]$.

b. Let $\alpha = 0.9$. Using the iterative solution method outlined in Section 2.3, determine the output signal $y[n]$ for $n = 0, \ldots, 9$ if the input signal is a unit impulse, that is, if $x[n] = \delta[n]$.

c. Determine the output signal $y[n]$ for $n = 0, \ldots, 9$ if the input signal is a unit step, that is, if $x[n] = u[n]$.

2.20. A discrete-time system is described by the block diagram shown in Fig. P.2.20.

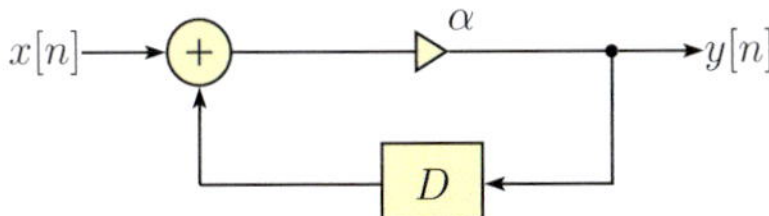

Figure P. 2.20

a. Write a difference equation between the signals $x[n]$ and $y[n]$.

b. Let $\alpha = 0.9$. Using the iterative solution method outlined in Section 2.3, determine the output signal $y[n]$ for $n = 0, \ldots, 9$ if the input signal is a unit impulse, that is, if $x[n] = \delta[n]$.

c. Determine the output signal $y[n]$ for $n = 0, \ldots, 9$ if the input signal is a unit step, that is, if $x[n] = u[n]$.

2.21. Two DTLTI systems with impulse responses $h_1[n]$ and $h_2[n]$ are connected in cascade as shown in Fig. P.2.21(a).

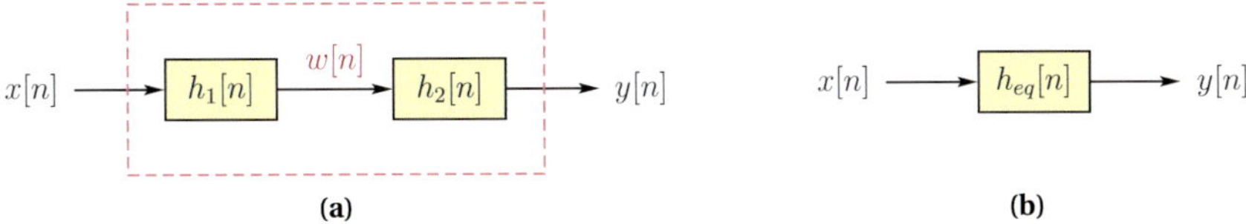

Figure P. 2.21

a. Determine the impulse response $h_{eq}[n]$ of the equivalent system, as shown in Fig. P.2.21(b) in terms of $h_1[n]$ and $h_2[n]$.

Hint: Use convolution to express $w[n]$ in terms of $x[n]$. Afterward use convolution again to express $y[n]$ in terms of $w[n]$.

b. Let $h_1[n] = h_2[n] = u[n] - u[n-5]$. Determine and sketch $h_{eq}[n]$ for the equivalent system.

c. With $h_1[n]$ and $h_2[n]$ as specified in part (b), let the input signal be a unit step, that is, $x[n] = u[n]$. Determine and sketch the signals $w[n]$ and $y[n]$.

2.22. Two DTLTI systems with impulse responses $h_1[n]$ and $h_2[n]$ are connected in parallel as shown in Fig. P.2.22(a).

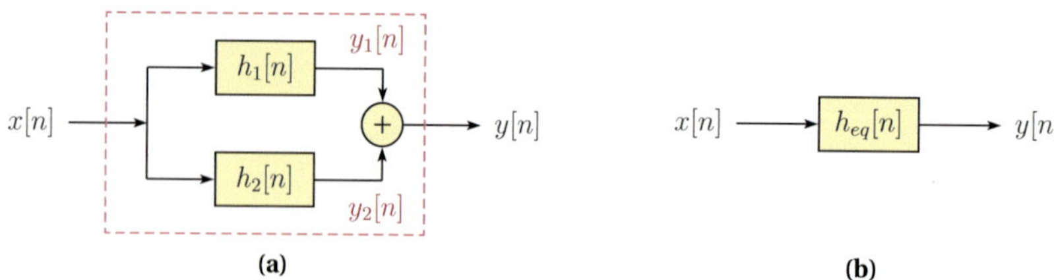

Figure P. 2.22

a. Determine the impulse response $h_{eq}[n]$ of the equivalent system, as shown in Fig. P.2.22(b) in terms of $h_1[n]$ and $h_2[n]$. Hint: Use convolution to express the signals $y_1[n]$ and $y_2[n]$ in terms of $x[n]$. Afterward express $y[n]$ in terms of $y_1[n]$ and $y_2[n]$.

b. Let $h_1[n] = (0.9)^n\, u[n]$ and $h_2[n] = (-0.7)^n\, u[n]$. Determine and sketch $h_{eq}[n]$ for the equivalent system.

c. With $h_1[n]$ and $h_2[n]$ as specified in part (b), let the input signal be a unit step, that is, $x[n] = u[n]$. Determine and sketch the signals $y_1[n]$, $y_2[n]$, and $y[n]$.

2.23. Three DTLTI systems with impulse responses $h_1[n]$, $h_2[n]$, and $h_3[n]$ are connected as shown in Fig. P.2.23(a).

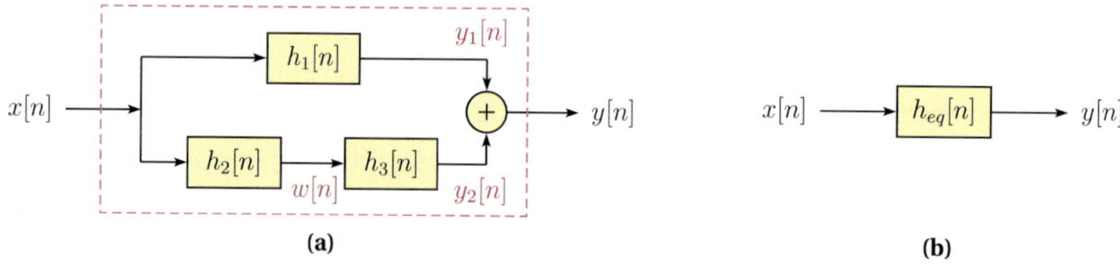

Figure P. 2.23

a. Determine the impulse response $h_{eq}[n]$ of the equivalent system, as shown in Fig. 2.23(b) in terms of $h_1[n]$, $h_2[n]$, and $h_3[n]$.

b. Let $h_1[n] = e^{-0.1n}\, u[n]$, $h_2[n] = \delta[n-2]$, and $h_3[n] = e^{-0.2n}\, u[n]$. Determine and sketch $h_{eq}[n]$ for the equivalent system.

c. With $h_1[n]$, $h_2[n]$, and $h_3[n]$ as specified in part (b), let the input signal be a unit step, that is, $x[n] = u[n]$. Determine and sketch the signals $w[n]$, $y_1[n]$, $y_2[n]$, and $y[n]$.

2.24. Consider the DTLTI system shown in Fig. P.2.24(a).

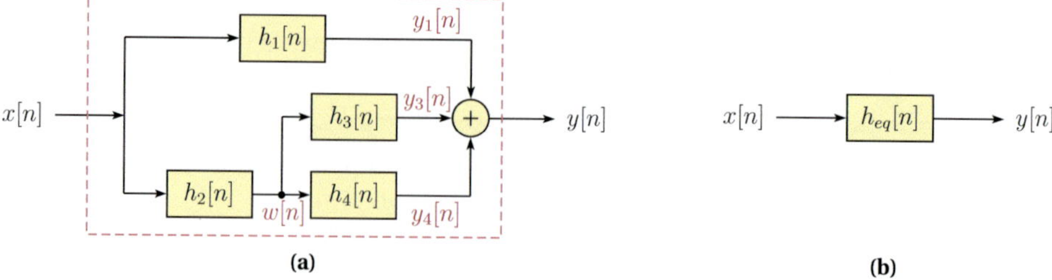

Figure P. 2.24

a. Express the impulse response of the system as a function of the impulse responses of the subsystems.

b. Let

$$h_1[n] = e^{-0.1n}\, u[n]$$

$$h_2[n] = h_3[n] = u[n] - u[n-3]$$

and

$$h_4[n] = \delta[n-2]$$

Determine the impulse response $h_{eq}[n]$ of the equivalent system.

c. Let the input signal be a unit step, that is, $x[n] = u[n]$. Determine and sketch the signals $w[n]$, $y_1[n]$, $y_3[n]$ and $y_4[n]$.

2.25. Consider two finite-length signals $x[n]$ and $h[n]$ that are equal to zero outside the intervals indicated below:

$$x[n] = 0 \quad \text{if } n < N_{x1} \text{ or } n > N_{x2}$$

$$h[n] = 0 \quad \text{if } n < N_{h1} \text{ or } n > N_{h2}$$

In other words, significant samples of $x[n]$ are in the index range $N_{x1}, \ldots, N_{x2}$, and the significant samples of $h[n]$ are in the index range $N_{h1}, \ldots, N_{h2}$.

Let $y[n]$ be the convolution of $x[n]$ and $h[n]$. Starting with the convolution sum given by Eqn. (2.109), determine the index range of significant samples $N_{y1}, \ldots, N_{y2}$ for $y[n]$.

2.26. Let $y[n]$ be the convolution of two discrete-time signals $x[n]$ and $h[n]$, that is

$$y[n] = x[n] * h[n]$$

Show that time shifting either $x[n]$ or $h[n]$ by m samples causes the $y[n]$ to be time shifted by m samples also. Mathematically prove that

$$x[n-m] * h[n] = y[n-m]$$

and

$$x[n] * h[n-m] = y[n-m]$$

2.27. Consider a causal DTLTI system with impulse response $h[n]$. Let $y_u[n]$ and $y_r[n]$ represent the responses of this system to unit step and unit ramp signals respectively, i.e.,

$$y_u[n] = \text{Sys}\left\{u[n]\right\} \quad \text{and} \quad y_r[n] = \text{Sys}\left\{r[n]\right\}$$

a. Show that the unit step response of the system is

$$y_u[n] = \sum_{k=0}^{n} h[k]\,, \qquad n \ge 0$$

b. Show that the unit ramp response of the system is

$$y_r[n] = n\, y_u[n] - \sum_{k=0}^{n} k\, h[k]\,, \qquad n \ge 0$$

2.28. Determine the impulse response of each system described below. Afterward determine whether the system is causal and/or stable.

a. $y[n] = \text{Sys}\{x[n]\} = \sum_{k=-\infty}^{n} x[k]$

b. $y[n] = \text{Sys}\{x[n]\} = \sum_{k=-\infty}^{n} e^{-0.1(n-k)} x[k]$

c. $y[n] = \text{Sys}\{x[n]\} = \sum_{k=0}^{n} x[k] \quad \text{for } n \geq 0$

d. $y[n] = \text{Sys}\{x[n]\} = \sum_{k=n-10}^{n} x[k]$

e. $y[n] = \text{Sys}\{x[n]\} = \sum_{k=n-10}^{n+10} x[k]$

MATLAB Problems

2.29. Refer to the homogeneous difference equations in Problem 2.12. For each one, develop a MATLAB script to iteratively solve it for $n = 0, \ldots, 19$ using the initial conditions specified. Compare the results of iterative solutions to the results obtained analytically in Problem 2.12.

2.30. Refer to Problem 2.3. The cascade combination of the two subsystems will be tested using the input signal

$$x[n] = (0.95)^n \cos(0.1\pi n)\, u[n]$$

Write a script to do the following:

a. Create an anonymous function `x` to compute the input signal at all index values specified in a vector `n`.

b. Implement the cascade system as shown in Fig. P.2.3(a). Compute the signals $w[n]$ and $y[n]$ for $n = 0, \ldots, 29$.

c. Implement the cascade system in the alternative form shown in Fig. P.2.3(b). Compute the signals $\bar{w}[n]$ and $\bar{y}[n]$ for $n = 0, \ldots, 29$. Compare $y[n]$ and $\bar{y}[n]$.

2.31. Repeat the steps of Problem 2.30 using the subsystems described in Problem 2.4.

2.32. Refer to Problem 2.1. Linearity and time invariance of the systems listed will be tested using the two input signals

$$x_1[n] = n\, e^{-0.2n} \left(u[n] - u[n-20]\right);$$

and

$$x_2[n] = \cos(0.05\pi n) \left(u[n] - u[n-20]\right);$$

Develop a script to do the following:

a. Express the two input signals by means of two anonymous functions `x1` and `x2`. Each anonymous function should take a single argument `n` which could either be a scalar or a vector of index values.

b. Express each of the systems described in Problem 2.1 as an anonymous function. Name the anonymous functions `sys1` through `sys5`. Each should take two arguments `n` and `x`.

c. Compute the response of each system to

$$x[n] = x_1[n]$$

$$x[n] = x_2[n]$$

$$x[n] = x_1[n] + x_2[n]$$

$$x[n] = 5\,x_1[n] - 3\,x_2[n]$$

Identify which systems fail the linearity test.

d. Compute the response of each system to

$$x[n] = x_1[n-1]$$

$$x[n] = x_2[n-3]$$

Identify which systems fail the time invariance test.

2.33.

a. Develop a function `ss_loanbal()` to iteratively solve the loan balance difference equation explored in Problem 2.10. It should have the following interface:

```
bal = ss_loanbal(A,B,c,n)
```

The arguments are:

`A`:	Amount of the loan; this is also the balance at $n = 0$
`B`:	Monthly payment amount
`c`:	Monthly interest rate
`n`:	Index of the month in which the balance is sought
`bal`:	Computed balance after `n` months

b. Write a script to test the function with the values specified in Problem 2.10.

2.34. Refer to Problem 2.11.

a. Develop a MATLAB function `ss_sqrt()` that iteratively solves the nonlinear difference equation given by Eqn. (2.19) to approximate the square root of a positive number A. The syntax of the function should be

```
y = ss_sqrt(A,y_init,tol)
```

The argument `A` represents the positive number the square-root of which is being sought, `y_init` is the initial value $y[-1]$, and `tol` is the tolerance limit ε. The function should return when the difference between two consecutive output samples is less than the tolerance limit, that is,

$$\big|y[n] - y[n-1]\big| \le \varepsilon$$

b. Write a script to test the function `ss_sqrt()` with the values $A = 5$, $A = 17$ and $A = 132$.

2.35. Refer to Problem 2.16 in which the response of the linear exponential smoother to unit step and unit ramp signals were found as functions of the parameter α. Consider the input signal shown in Fig. P.2.35.

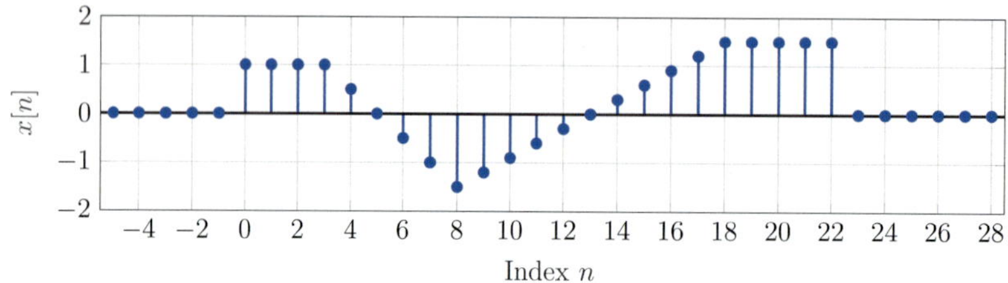

Figure P. 2.35

Develop a script to do the following:

a. Create an anonymous function `yu()` to compute the unit step response of the linear exponential smoother. It should take two arguments, `alpha` and `n`, and should return the unit step response evaluated at each index in the vector `n`.

b. Create an anonymous function `yr()` to compute the unit ramp response of the linear exponential smoother. Like the function `yu()`, the function `yr()` should also take two arguments, `alpha` and `n`, and should return the unit ramp response evaluated at each index in the vector `n`.

c. Express the input signal $x[n]$ using a combination of scaled and time shifted unit step and unit ramp functions.

d. Use the anonymous functions `yu()` and `yr()` to compute and graph the output signal $y[n]$. Try with $\alpha = 0.1, 0.2, 0.3$ and compare.

2.36. Consider the discrete-time system described in Problem 2.19 by the block diagram shown in Fig. P.2.19.

a. Develop a MATLAB script to compute and graph the impulse response of this system for $n = 0, \ldots, 99$ using values $\alpha = 0.7$, $\alpha = 0.9$, and $\alpha = 1.1$.

b. Modify the script developed in part (a) to compute and graph the unit step response of the system for $n = 0, \ldots, 99$ using values $\alpha = 0.7$, $\alpha = 0.9$, and Let $\alpha = 0.9$.

2.37. Consider the discrete-time system described in Problem 2.20 by the block diagram shown in Fig. P.2.20.

a. Develop a MATLAB script to compute and graph the impulse response of this system for $n = 0, \ldots, 99$ using values $\alpha = 0.7$, $\alpha = 0.9$, and $\alpha = 1.1$.

b. Modify the script developed in part (a) to compute and graph the unit step response of the system for $n = 0, \ldots, 99$ using values $\alpha = 0.7$, $\alpha = 0.9$ and Let $\alpha = 0.9$.

MATLAB Projects

2.38. Consider the logistic growth model explored in Example 2.8. The normalized value of the population at index n is modeled using the nonlinear difference equation

$$y[n] = r\,\big(1 - y[n-1]\big)\; y[n-1]$$

where r is the growth rate. This is an example of a *chaotic system.* Our goal in this project is to simulate the behavior of a system modeled by the nonlinear difference equation given, and to determine the dependency of the population growth on the initial value $y[0]$ and the growth rate r.

a. Develop a function called `ss_growth()` with the syntax

```
y = ss_growth(y_init,r,N)
```

where `y_init` is the initial value of the population at $n = 0$, `r` is the growth rate, and `N` is the number of iterations to be carried out. The vector returned by the function should have the values

$$\left[\, y[0], y[1], y[2], \ldots, y[N] \,\right]$$

b. Write a script to compute $y[n]$ for $n = 1, \ldots, 30$ with specified values $y[0]$ and r.

c. With the growth rate set at $r = 1.5$, compute and graph $y[n]$ for the cases of $y[0] = 0.1$, $y[0] = 0.3$, $y[0] = 0.5$, $y[0] = 0.7$. Does the population reach equilibrium? Comment on the dependency of the population at equilibrium on the initial value $y[0]$.

d. Repeat the experiment with $r = 2, 2.5, 2.75, 3$. For each value of r compute and graph $y[n]$ for the cases of $y[0] = 0.1$, $y[0] = 0.3$, $y[0] = 0.5$, $y[0] = 0.7$. Does the population still reach equilibrium? You should find that beyond a certain value of r population behavior should become oscillatory. What is the critical value of r?

2.39. Consider the second-order homogeneous difference equation

$$y[n] + a_1\, y[n-1] + a_2\, y[n-2] = 0$$

with initial conditions $y[-1] = p_1$ and $y[-2] = p_2$. The coefficients a_1 and a_2 of the difference equation and the initial values p_1 and p_2 are all real-valued, and will be left as parameters.

a. Let z_1 and z_2 be the roots of the characteristic polynomial. On paper, solve the homogeneous difference equation for the three possibilities for the roots:

a.1. The roots are real and distinct.

a.2. The roots are a complex conjugate pair.

a.3. The two roots are real and equal.

Find the solution $y[n]$ as a function of the roots z_1 and z_2 as well as the initial values p_1 and p_2.

b. Develop a MATLAB function `ss_de2solve()` with the syntax

```
y = ss_de2solve(a1,a2,p1,p2,n)
```

The vector `n` contains the index values for which the solution should be computed. The returned vector `y` holds the solution at the index values in vector `n` so that the solution can be graphed with

```
stem(n,y)
```

Your function should perform the following steps:

b.1. Form the characteristic polynomial and find its roots.

b.2. Determine which of the three categories the roots fit (simple real roots, complex conjugate pair or multiple roots).

b.3. Compute the solution accordingly.

c. Test the function `ss_de2solve()` with the homogeneous difference equations in Problem 2.12a,b,c,d, Problem 2.13a,b,c, and Problem 2.14a,b.

2.40. One method of numerically approximating the integral of a function is the *trapezoidal integration* method. Consider the function $x_a(t)$ as shown in Fig. P.2.40.

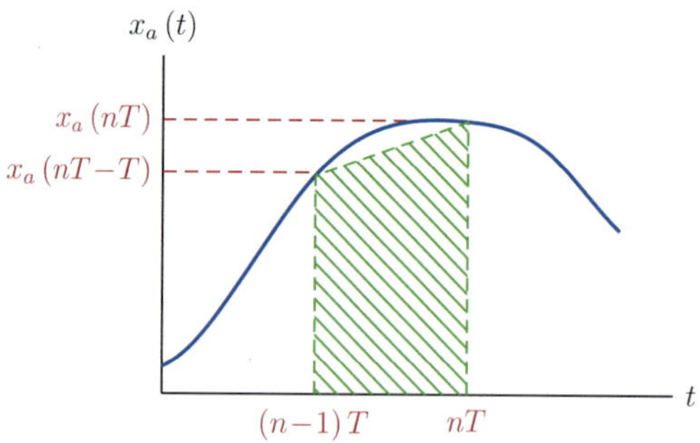

Figure P. 2.40

The area under the function from $t = (n-1)T$ to $t = nT$ may be approximated with the area of the shaded trapezoid:

$$\int_{(n-1)T}^{nT} x_a(t)\,dt \approx \frac{T}{2}\left[x_a(nT-T) + x_a(nT)\right]$$

Let

$$y_a(t) = \int_{-\infty}^{t} x_a(\lambda)\,d\lambda$$

It follows that

$$y_a(nT) \approx y_a(nT-T) + \frac{T}{2}\left[x_a(nT-T) + x_a(nT)\right]$$

If we define discrete-time signals $x[n]$ and $y[n]$ as

$$x[n] = x_a(nT) \qquad \text{and} \qquad y[n] = y_a(nT)$$

then $y[n-1] = y_a(nT-T)$, and we obtain the difference equation

$$y[n] = y[n-1] + \frac{T}{2}\left(x[n] + x[n-1]\right)$$

which is the basis of a trapezoidal integrator.

a. Develop a function `ss_integ()` with the following interface:

```
val = ss_integ(xa,lim1,lim2,y_init,k)
```

The arguments are:

`xa`:	Handle to an anonymous function for $x_a(t)$
`lim1`:	Lower limit of the integral
`lim2`:	Upper limit of the integral
`y_init`:	Initial value of the integral
`k`:	Number of time steps from `lim1` to `lim2`
`val`:	Trapezoidal approximation to the integral

b. Write a script to test the function with the integrand

$$x_a(t) = e^{-t}\,\sin(t)$$

and compare its approximate integral with the correct result given by

$$\int x_a(t)\,dt = \frac{1}{2} - \frac{1}{2}e^{-t}\,\cos(t) - \frac{1}{2}e^{-t}\,\sin(t)$$

Hint: Set the lower limit of the integral to $t = 0$.

CHAPTER 3

FOURIER ANALYSIS FOR DISCRETE-TIME SIGNALS AND SYSTEMS

Chapter Objectives

- Understand the significance of analyzing discrete-time signals in terms of their frequency content.
- Learn the use of discrete-time Fourier series (DTFS) for representing periodic signals using orthogonal basis functions. Develop analysis and synthesis equations for DTFS representation of signals.
- Learn important properties of the DTFS representation such as periodicity, linearity, time shifting and reversal, symmetry properties, and periodic convolution.
- Learn the discrete-time Fourier transform (DTFT) for non-periodic signals as an extension of DTFS for periodic signals.
- Study properties of the DTFT. Understand energy and power spectral density concepts. Learn Parseval's theorem for relating time- and frequency-domain representations of signals to each other from an energy or power perspective.
- Explore frequency-domain characteristics of discrete-time linear and time-invariant (DTLTI) systems. Understand the system function concept for DTLTI systems.
- Learn the autocorrelation function and its properties.
- Learn the use of frequency-domain analysis methods for solving signal-system interaction problems with periodic and non-periodic input signals.

DOI: 10.1201/9781003570462-3

3.1 Introduction

In Chapters 1 and 2 we have developed techniques for analyzing discrete-time signals and systems from a time-domain perspective. A discrete-time signal can be modeled as a function of the sample index. A DTLTI system can be represented by means of a constant-coefficient linear difference equation, or alternatively by means of an impulse response. The output signal of a DTLTI system can be determined by solving the corresponding difference equation or by using the convolution operation.

In this chapter frequency-domain analysis methods are developed for discrete-time signals and systems. Section 3.2 focuses on analyzing periodic discrete-time signals in terms of their frequency content. Discrete-time Fourier series (DTFS) is presented as the counterpart of the exponential Fourier series (EFS) for continuous-time signals. Periodic signals are expressed in terms of orthogonal basis functions. Analysis and synthesis equations are derived. The properties of the DTFS are explored. The concept of periodic convolution is introduced. Frequency-domain analysis methods for non-periodic discrete-time signals are presented in Section 3.3 through the use of the discrete-time Fourier transform (DTFT). Energy and power spectral density concepts for discrete-time signals are introduced in Section 3.4 along with Parseval's theorem. In addition, autocorrelation function is defined, and its fundamental properties are discussed. Section 3.5 introduces the DTFT-based system function concept for describing the behavior of a DTLTI system. In sections 3.6 and 3.7, we cover the use of the DTFT and the system function for finding the output signal of a DTLTI system driven by a periodic or non-periodic input signal.

3.2 Analysis of Periodic Discrete-Time Signals

It is possible to express a discrete-time periodic signal as a linear combination of periodic basis functions. In this section we will explore one such periodic expansion referred to as the *discrete-time Fourier series (DTFS)*. In the process we will discover interesting similarities and differences between DTFS and its continuous-time counterpart, the exponential Fourier series (EFS). One fundamental difference is regarding the number of series terms needed. A continuous-time periodic signal may have an infinite range of frequencies and therefore may require an infinite number of harmonically related basis functions. In contrast, discrete-time periodic signals contain a finite range of angular frequencies, and thus require a finite number of harmonically related basis functions. We will see in later parts of this section that, a discrete-time signal with a period of N samples will require at most N basis functions.

3.2.1 Discrete-time Fourier series (DTFS)

Consider a discrete-time signal $\tilde{x}[n]$ periodic with a period of N samples, that is, $\tilde{x}[n]=\tilde{x}[n+N]$ for all n. Periodic signals will be distinguished through the use of the tilde (~) character over the name of the signal. We would like to explore the possibility of writing $\tilde{x}[n]$ as a linear combination of complex exponential basis functions in the form

$$\phi_k[n] = e^{j\Omega_k n} \tag{3.1}$$

using a series expansion

$$\tilde{x}[n] = \sum_k \tilde{c}_k\, \phi_k[n] = \sum_k \tilde{c}_k\, e^{j\Omega_k n} \tag{3.2}$$

Two important questions need to be answered:

1. How should the angular frequencies Ω_k be chosen?
2. How many basis functions are needed? In other words, what should be the limits of the summation index k in Eqn. (3.2)?

Intuitively, since the period of $\tilde{x}[n]$ is N samples long, the basis functions used in constructing the signal must also be periodic with N. Therefore we require

$$\phi_k[n+N] = \phi_k[n] \tag{3.3}$$

for all n. Substituting Eqn. (3.1) into Eqn. (3.3) leads to

$$e^{j\Omega_k(n+N)} = e^{j\Omega_k n} \tag{3.4}$$

For Eqn. (3.4) to be satisfied we need $e^{j\Omega_k N} = 1$, and consequently $\Omega_k N = 2\pi k$. The angular frequency of the basis function $\phi_k[n]$ must be

$$\Omega_k = \frac{2\pi k}{N} \tag{3.5}$$

leading to the set of basis functions

$$\phi_k[n] = e^{j(2\pi/N)kn} \tag{3.6}$$

Using $\phi_k[n]$ found, the series expansion of the signal $\tilde{x}[n]$ is in the form

$$\tilde{x}[n] = \sum_k \tilde{c}_k\, e^{j(2\pi/N)kn} \tag{3.7}$$

To address the second question, it can easily be shown that only the first N basis functions $\phi_0[n]$, $\phi_1[n], \ldots, \phi_{N-1}[n]$ are unique; all other basis functions, i.e., the ones obtained for $k<0$ or $k \ge N$, are duplicates of the basis functions in this set. To see why this is the case, let us write $\phi_{k+N}[n]$:

$$\phi_{k+N}[n] = e^{j(2\pi/N)(k+N)n} \tag{3.8}$$

Factoring Eqn. (3.8) into two exponential terms and realizing that $e^{j2\pi n} = 1$ for all integer n we obtain

$$\phi_{k+N}[n] = e^{j(2\pi/N)kn}\, e^{j2\pi n} = e^{j(2\pi/N)kn} = \phi_k[n] \tag{3.9}$$

Since $\phi_{k+N}[n]$ is equal to $\phi_k[n]$, we only need to include N terms in the summation of Eqn. (3.7).

> The periodic discrete-time signal $\tilde{x}[n]$ can be constructed as
>
> $$\tilde{x}[n] = \sum_{k=0}^{N-1} \tilde{c}_k\, e^{j(2\pi/N)kn} \tag{3.10}$$

As a specific example, if the period of the signal $\tilde{x}[n]$ is $N=5$, then the only basis functions that are unique would be

$$\phi_k[n]\,, \quad \text{for} \quad k = 0,1,2,3,4$$

Increasing the summation index k beyond $k=4$ would not create any additional unique terms since $\phi_5[n] = \phi_0[n]$, $\phi_6[n] = \phi_1[n]$, and so on. Eqn. (3.10) is referred to as the *discrete-time Fourier series (DTFS)* expansion of the periodic signal $\tilde{x}[n]$. The coefficients $\tilde{c}_k$ used in the summation of Eqn. (3.10) are the DTFS coefficients of the signal $\tilde{x}[n]$.

Example 3.1: DTFS for a discrete-time sinusoidal signal

Determine the DTFS representation of the signal $\tilde{x}[n] = \cos\left(\sqrt{2}\pi n\right)$.

Solution: The angular frequency of the signal is

$$\Omega_0 = \sqrt{2}\pi \text{ rad}$$

and it corresponds to the normalized frequency

$$F_0 = \frac{\Omega_0}{2\pi} = \frac{1}{\sqrt{2}}$$

Since normalized frequency F_0 is an irrational number, the signal specified is not periodic (refer to the discussion in Section 1.5 of Chapter 1). Therefore it cannot be represented in series form using periodic basis functions. It can still be analyzed in the frequency domain, however, using the *discrete-time Fourier transform (DTFT)* which will be explored in Section 3.3.

Example 3.2: DTFS for a discrete-time sinusoidal signal revisited

Determine the DTFS representation of the signal $\tilde{x}[n] = \cos(0.2\pi n)$.

Solution: The angular frequency of the signal is

$$\Omega_0 = 0.2\pi \text{ rad}$$

and the corresponding normalized frequency is

$$F_0 = \frac{\Omega_0}{2\pi} = \frac{1}{10}$$

Based on the normalized frequency, the signal is periodic with a period of $N = 1/F_0 = 10$ samples. A general formula for obtaining the DTFS coefficients will be derived later in this section. For the purpose of this example, however, we will take a shortcut afforded by the sinusoidal nature of the signal $\tilde{x}[n]$ and express it using Euler's formula:

$$\begin{aligned} \tilde{x}[n] =& \frac{1}{2}\, e^{j0.2\pi n} + \frac{1}{2}\, e^{-j0.2\pi n} \\ =& \frac{1}{2}\, e^{j(2\pi/10)n} + \frac{1}{2}\, e^{-j(2\pi/10)n} \end{aligned} \tag{3.11}$$

The two complex exponential terms in Eqn. (3.11) correspond to $\phi_1[n]$ and $\phi_{-1}[n]$, therefore their coefficients must be $\tilde{c}_1$ and $\tilde{c}_{-1}$ respectively. As a result we have

$$\tilde{c}_1 = \frac{1}{2}, \quad \text{and} \quad \tilde{c}_{-1} = \frac{1}{2}$$

As discussed in the previous section we would like to see the series coefficients $\tilde{c}_k$ in the index range $k = 0, \ldots, N-1$ where N is the period. In this case we need to obtain $\tilde{c}_k$ for $k = 0, \ldots, 9$. The basis functions have the property

$$\phi_k[n] = \phi_{k+N}[n]$$

The term $\phi_{-1}[n]$ in Eqn. (3.11) can be written as

$$\phi_{-1}[n] = e^{-j(2\pi/10)n} = e^{-j(2\pi/10)n}\, e^{j2\pi n} = e^{j(18\pi/10)n} = \phi_9[n]$$

Eqn. (3.11) becomes

$$\tilde{x}[n] = \cos(0.2\pi n) = \frac{1}{2}\, e^{j(2\pi/10)n} + \frac{1}{2}\, e^{j(18\pi/10)n} = \tilde{c}_1\, e^{j(2\pi/10)n} + \tilde{c}_9\, e^{j(18\pi/10)n}$$

DTFS coefficients are

$$\tilde{c}_k = \begin{cases} \dfrac{1}{2}, & k = 1 \text{ or } k = 9 \\ 0, & \text{otherwise} \end{cases}$$

The signal $\tilde{x}[n]$ and its DTFS spectrum $\tilde{c}_k$ are shown in Fig. 3.1.

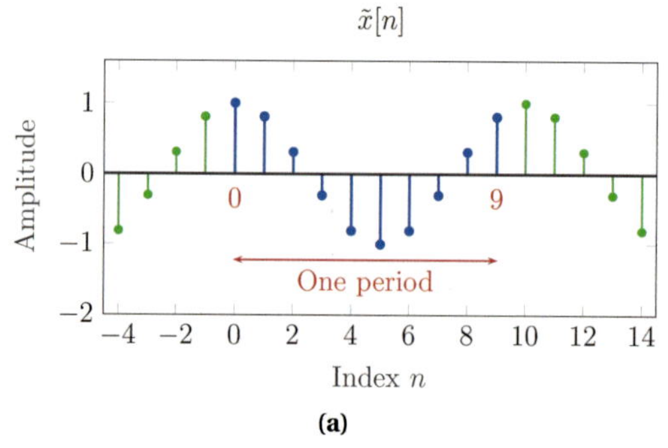

(a)

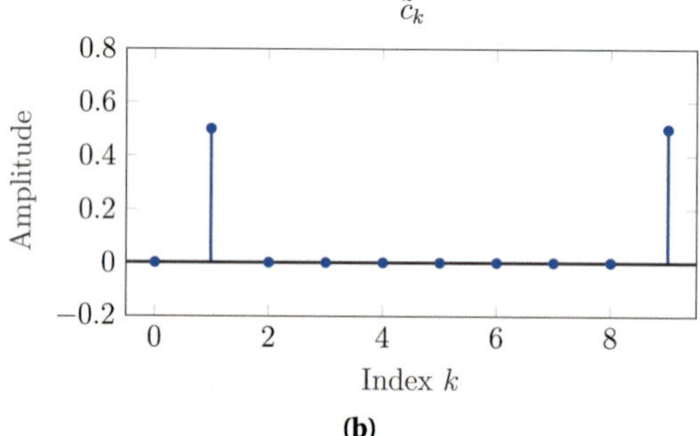

(b)

Figure 3.1 – **(a)** The signal $\tilde{x}[n]$ for Example 3.2 and **(b)** its DTFS spectrum.

Software resource: exdt_3_2.m

Example 3.3: DTFS for a multi-tone signal

Determine the DTFS coefficients for the signal

$$\tilde{x}[n] = 1 + \cos(0.2\pi n) + 2\sin(0.3\pi n)$$

Solution: The two angular frequencies $\Omega_1 = 0.2\pi$ and $\Omega_2 = 0.3\pi$ radians correspond to normalized frequencies $F_1 = 0.1$ and $F_2 = 0.15$, respectively. The normalized fundamental frequency of $\tilde{x}[n]$ is $F_0 = 0.05$, and it corresponds to a period of $N = 20$ samples. Using this value of N, the angular frequencies of the two sinusoidal terms of $\tilde{x}[n]$ are

$$\Omega_1 = 2\Omega_0 = \frac{2\pi}{20}(2) \quad \text{and} \quad \Omega_2 = 3\Omega_0 = \frac{2\pi}{20}(3)$$

We will use Euler's formula to write $\tilde{x}[n]$ in the form

$$\begin{aligned}\tilde{x}[n] =& 1 + \frac{1}{2}\, e^{j(2\pi/20)2n} + \frac{1}{2}\, e^{-j(2\pi/20)2n} + \frac{1}{j}\, e^{j(2\pi/20)3n} - \frac{1}{j}\, e^{-j(2\pi/20)3n} \\ =& 1 + \frac{1}{2}\,\phi_2[n] + \frac{1}{2}\,\phi_{-2}[n] + \frac{1}{j}\,\phi_3[n] - \frac{1}{j}\,\phi_{-3}[n] \\ =& 1 + \frac{1}{2}\,\phi_2[n] + \frac{1}{2}\,\phi_{-2}[n] + e^{-j\pi/2}\,\phi_3[n] - e^{j\pi/2}\,\phi_{-3}[n]\end{aligned}$$

The DTFS coefficients are

$$\begin{aligned}\tilde{c}_0 =& 1 \\ \tilde{c}_2 =& \frac{1}{2}, \qquad \tilde{c}_{-2} = \frac{1}{2} \\ \tilde{c}_3 =& e^{-j\pi/2}, \qquad \tilde{c}_{-3} = e^{j\pi/2}\end{aligned}$$

We know from the periodicity of the DTFS representation that

$$\tilde{c}_{-2} = \tilde{c}_{18} \qquad \text{and} \qquad \tilde{c}_{-3} = \tilde{c}_{17}$$

The signal $\tilde{x}[n]$ and its DTFS spectrum $\tilde{c}_k$ are shown in Fig. 3.2.

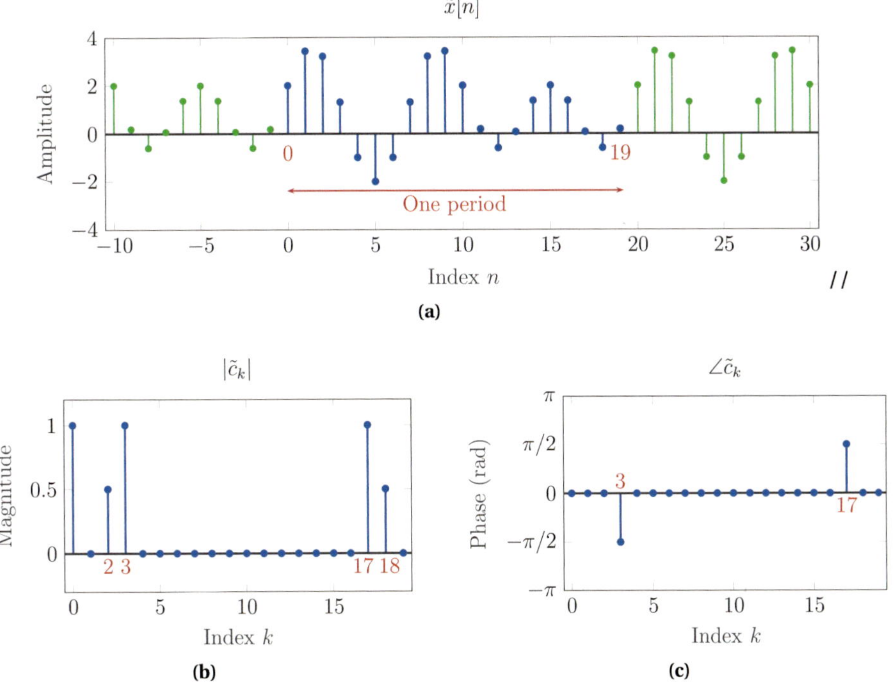

Figure 3.2 – **(a)** The signal $\tilde{x}[n]$ for Example 3.3, **(b)** magnitude of the DTFS spectrum, and **(c)** phase of the DTFS spectrum.

Software resource: exdt_3_3.m

3.2.2 Finding DTFS coefficients

In Examples 3.2 and 3.3 the DTFS coefficients for discrete-time sinusoidal signals were easily determined since the use of Euler's formula gave us the ability to express those signals in a form very similar to the DTFS expansion given by Eqn. (3.7). In order to determine the DTFS coefficients for a general discrete-time signal we will take advantage of the orthogonality of the basis function set $\{\phi_k[n],\ k=0,\ldots,N-1\}$. It can be shown that (see Appendix D)

$$\sum_{n=0}^{N-1} e^{j(2\pi/N)(m-k)n} = \begin{cases} N\,, & (m-k)=rN\,,\ r \text{ integer} \\ 0\,, & \text{otherwise} \end{cases} \tag{3.12}$$

Derivation of DTFS analysis equation:

To derive the expression for the DTFS coefficients, let us first write Eqn. (3.10) using m instead of k as the summation index:

$$\tilde{x}[n] = \sum_{m=0}^{N-1} \tilde{c}_m\, e^{j(2\pi/N)mn} \tag{3.13}$$

Multiplication of both sides of Eqn. (3.7) by $e^{-j(2\pi/N)kn}$ leads to

$$\tilde{x}[n]\, e^{-j(2\pi/N)kn} = \sum_{m=0}^{N-1} \tilde{c}_m\, e^{j(2\pi/N)mn}\, e^{-j(2\pi/N)kn} \tag{3.14}$$

Summing the terms on both sides of Eqn. (3.14) for $n=0,\ldots,N-1$, rearranging the summations, and using the orthogonality property yields

$$\begin{aligned}\sum_{n=0}^{N-1} \tilde{x}[n]\, e^{-j(2\pi/N)kn} &= \sum_{n=0}^{N-1}\sum_{m=0}^{N-1} \tilde{c}_m\, e^{j(2\pi/N)mn}\, e^{-j(2\pi/N)kn} \\ &= \sum_{m=0}^{N-1} \tilde{c}_m \sum_{n=0}^{N-1} e^{j(2\pi/N)(m-k)n} \\ &= \tilde{c}_k N \end{aligned} \tag{3.15}$$

The DTFS coefficients are computed from Eqn. (3.15) as

$$\tilde{c}_k = \frac{1}{N} \sum_{n=0}^{N-1} \tilde{x}[n]\, e^{-j(2\pi/N)kn}\,, \qquad k=0,\ldots,N-1 \tag{3.16}$$

In Eqn. (3.16) the DTFS coefficients are computed for the index range $k=0,\ldots,N-1$ since those are the only coefficient indices that are needed in the DTFS expansion in Eqn. (3.10). If we were to use Eqn. (3.16) outside the specified index range we would discover that

$$\tilde{c}_{k+rN} = \tilde{c}_k \tag{3.17}$$

for all integer r. The DTFS coefficients evaluated outside the range $k = 0,\ldots,N-1$ exhibit periodic behavior with period N. This was evident in Examples 3.2 and 3.3 as well. The development so far can be summarized as follows:

Discrete-Time Fourier Series (DTFS):

1. Synthesis equation:

$$\tilde{x}[n] = \sum_{k=0}^{N-1} \tilde{c}_k \, e^{j(2\pi/N)kn}\,, \qquad \text{all } n \tag{3.18}$$

2. Analysis equation:

$$\tilde{c}_k = \frac{1}{N} \sum_{n=0}^{N-1} \tilde{x}[n] \, e^{-j(2\pi/N)kn} \tag{3.19}$$

Note that the coefficients $\tilde{c}_k$ are computed for all indices k in Eqn. (3.19), however, only the ones in the range $k = 0,\ldots,N-1$ are needed in constructing the signal $\tilde{x}[n]$ in Eqn. (3.18). Due to the periodic nature of the DTFS coefficients $\tilde{c}_k$, the summation in the synthesis equation can be started at any arbitrary index, provided that the summation includes exactly N terms. In other words, Eqn. (3.18) can be written in the alternative form

$$\tilde{x}[n] = \sum_{k=N_0}^{N_0+N-1} \tilde{c}_k \, e^{j(2\pi/N)kn}\,, \qquad \text{all } n \tag{3.20}$$

Example 3.4: Finding DTFS representation

Consider the periodic signal $\tilde{x}[n]$ defined as

$$\tilde{x}[n] = n \text{ for } n = 0,1,2,3,4 \quad \text{and} \quad \tilde{x}[n+5] = \tilde{x}[n]$$

and shown in Fig. 3.3. Determine the DTFS coefficients for $\tilde{x}[n]$. Afterward, verify the synthesis equation in Eqn. (3.18).

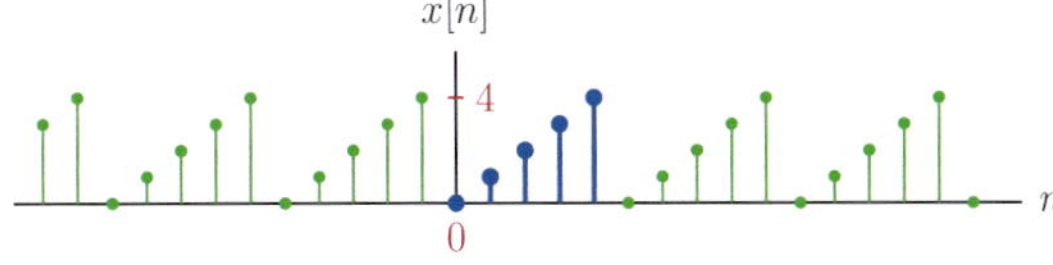

Figure 3.3 – The signal $\tilde{x}[n]$ for Example 3.4.

Solution: Using the analysis equation given by Eqn. (3.19) the DTFS coefficients are found as

$$\begin{aligned}\tilde{c}_k =& \frac{1}{5} \sum_{n=0}^{4} \tilde{x}[n] \, e^{-j(2\pi/5)kn} \\ =& \frac{1}{5} e^{-j2\pi k/5} + \frac{2}{5} e^{-j4\pi k/5} + \frac{3}{5} e^{-j6\pi k/5} + \frac{4}{5} e^{-j8\pi k/5}\end{aligned} \tag{3.21}$$

Evaluating Eqn. (3.21) for $k = 0,\ldots,4$ yields the following values for the coefficients:

$$\begin{aligned}\tilde{c}_0 &= 2\\ \tilde{c}_1 &= -0.5 + j\,0.6882\\ \tilde{c}_2 &= -0.5 + j\,0.1625\\ \tilde{c}_3 &= -0.5 - j\,0.1625\\ \tilde{c}_4 &= -0.5 - j\,0.6882\end{aligned}$$

The signal $\tilde{x}[n]$ can be constructed from the DTFS coefficients as

$$\begin{aligned}\tilde{x}[n] =& \tilde{c}_0 + \tilde{c}_1\, e^{j2\pi n/5} + \tilde{c}_2\, e^{j4\pi n/5} + \tilde{c}_3\, e^{j6\pi n/5} + \tilde{c}_4\, e^{j8\pi n/5}\\ =& 2 + \left(-0.5 + j\,0.6882\right)\, e^{j2\pi n/5} + \left(-0.5 + j\,0.1625\right)\, e^{j4\pi n/5}\\ &+ \left(-0.5 - j\,0.1625\right)\, e^{j6\pi n/5} + \left(-0.5 - j\,0.6882\right)\, e^{j8\pi n/5}\end{aligned}$$

Software resource: exdt_3_4.m

Example 3.5: DTFS for discrete-time pulse train

Consider the periodic pulse train $\tilde{x}[n]$ defined by

$$\tilde{x}[n] = \begin{cases} 1\,, & -L \le n \le L \\ 0\,, & L < n < N - L \end{cases}\,, \qquad \text{and} \quad \tilde{x}[n+N] = \tilde{x}[n]$$

where $N > 2L + 1$ as shown in Fig. 3.4. Determine the DTFS coefficients of the signal $\tilde{x}[n]$ in terms of L and N.

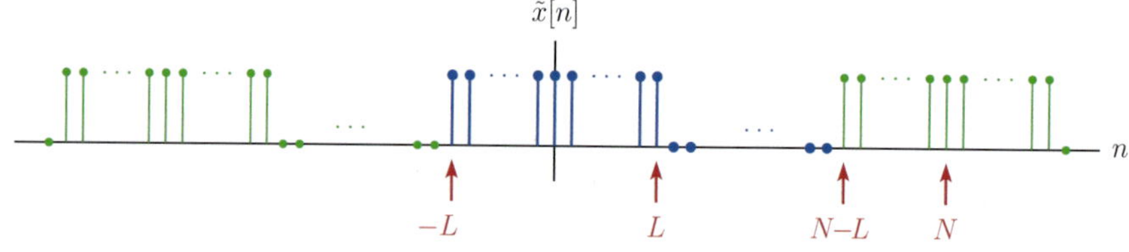

Figure 3.4 – The signal $\tilde{x}[n]$ for Example 3.5.

Solution: Using Eqn. (3.19) the DTFS coefficients are

$$\tilde{c}_k = \frac{1}{N} \sum_{n=-L}^{L} (1)\, e^{-j(2\pi/N)kn}$$

The closed form expression for a finite-length geometric series is (see Appendix C for derivation)

$$\text{Eqn. (C.13):} \qquad \sum_{n=L_1}^{L_2} a^n = \frac{a^{L_1} - a^{L_2+1}}{1-a}$$

Using Eqn. (C.13) with $a = e^{-j(2\pi/N)k}$, $L_1 = -L$ and $L_2 = L$, the closed form expression for $\tilde{c}_k$ is

$$\tilde{c}_k = \frac{1}{N}\,\frac{e^{j(2\pi/N)Lk} - e^{-j(2\pi/N)(L+1)k}}{1 - e^{-j(2\pi/N)k}} \tag{3.22}$$

In order to get symmetric complex exponentials in Eqn. (3.22) we will multiply both the numerator and the denominator of the fraction on the right side of the equal sign with $e^{j\pi k/N}$ resulting in

$$\tilde{c}_k = \frac{1}{N}\,\frac{e^{j(2\pi/N)(L+1/2)k} - e^{-j(2\pi/N)(L+1/2)k}}{e^{j(2\pi/N)(k/2)} - e^{-j(2\pi/N)(k/2)}}$$

which, using Euler's formula, can be simplified to

$$\tilde{c}_k = \frac{\sin\left(\dfrac{\pi k}{N}\,(2L+1)\right)}{N\sin\left(\dfrac{\pi k}{N}\right)}, \quad k = 0,\ldots,N-1$$

The coefficient $\tilde{c}_0$ needs special attention. Using L'Hospital's rule we obtain

$$\tilde{c}_0 = \frac{2L+1}{N}$$

The DTFS representation of the signal $\tilde{x}[n]$ is

$$\tilde{x}[n] = \sum_{k=0}^{N-1}\left[\frac{\sin\left(\dfrac{\pi k}{N}\,(2L+1)\right)}{N\sin\left(\dfrac{\pi k}{N}\right)}\right] e^{j(2\pi/N)kn}$$

The DTFS coefficients are graphed in Figs. 3.5–3.7 for $N = 40$, and $L = 3,5,7$.

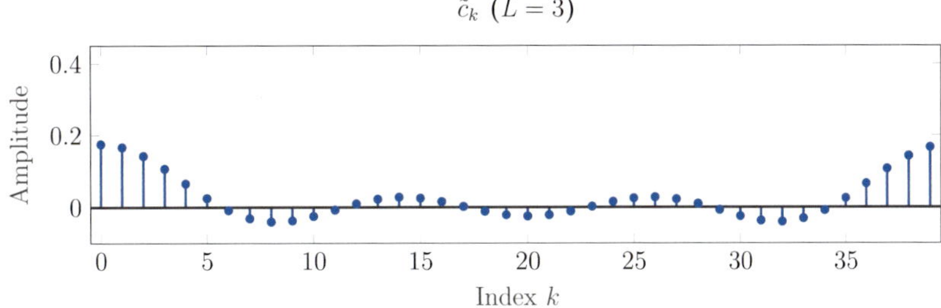

Figure 3.5 – The DTFS coefficients of the signal $\tilde{x}[n]$ of Example 3.5 for $N = 40$ and $L = 3$.

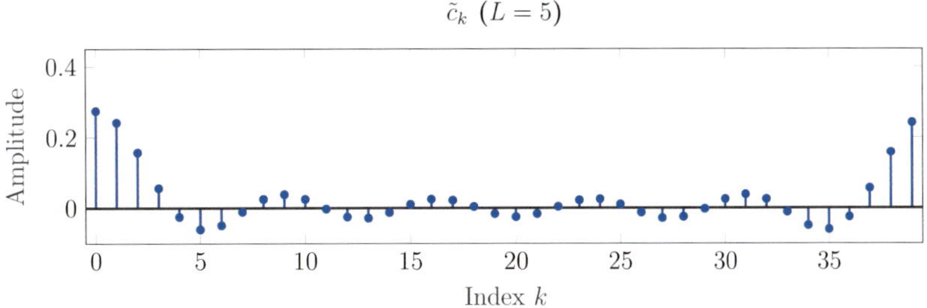

Figure 3.6 – The DTFS coefficients of the signal $\tilde{x}[n]$ of Example 3.5 for $N = 40$ and $L = 5$.

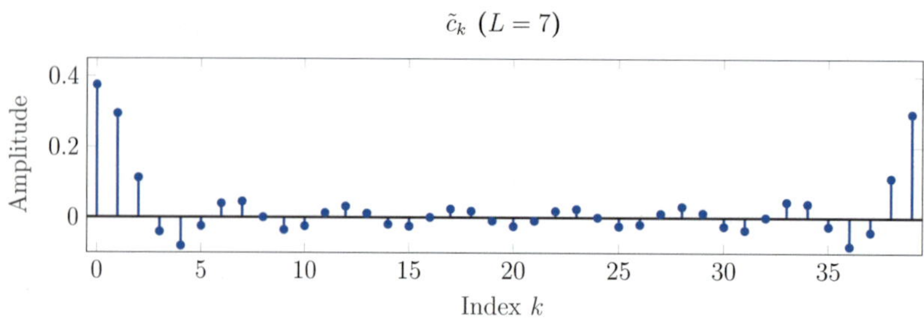

Figure 3.7 – The DTFS coefficients of the signal $\tilde{x}[n]$ of Example 3.5 for $N = 40$ and $L = 7$.

Software resources: exdt_3_5.m , PulseTrainDTFS.mlx

Interactive App: DTFS for discrete-time pulse train

The interactive app in `appDTFS.m` provides a graphical user interface for computing the DTFS representation of the periodic discrete-time pulse train of Example 3.5. The period is fixed at $N = 40$ samples. The parameter L can be varied from 0 to 19. The signal $\tilde{x}[n]$ and its DTFS coefficients $\tilde{c}_k$ are displayed.

a. Set $L = 0$. The signal $x[n]$ becomes a periodic train of unit impulses. Observe the DTFS coefficients for this case, and comment.

b. Increment L to $2, 3, 4, \ldots$ and observe the changes to DTFS coefficients. Pay attention to the outline (or the envelope) of the coefficients. Can you identify a pattern that emerges as L is incremented?

Software resource: `appDTFS.m`

Software resource: See MATLAB Exercises 3.1, 3.2 and 3.3.

3.2.3 Properties of the DTFS

In this section we will summarize a few important properties of the DTFS representation of a periodic signal. To keep the notation compact, we will denote the relationship between the periodic signal $\tilde{x}[n]$ and its DTFS coefficients $\tilde{c}_k$ as

$$\tilde{x}[n] \overset{\text{DTFS}}{\longleftrightarrow} \tilde{c}_k \tag{3.23}$$

Periodicity

DTFS coefficients are periodic with the same period N as the signal $\tilde{x}[n]$. Periodicity of DTFS coefficients follows easily from the analysis equation given by Eqn. (3.19). Both factors inside the summation, namely $\tilde{x}[n]$ and $e^{-j(2\pi/N)kn}$ are periodic with period N. Therefore, the result must be periodic with the same period.

Periodicity of the DTFS:

Given

$$\tilde{x}[n] = \tilde{x}[n+rN]\,, \quad \text{all integer } r$$

it can be shown that

$$\tilde{c}_k = \tilde{c}_{k+rN}\,, \quad \text{all integer } r \tag{3.24}$$

Since the DTFS coefficients are periodic with period N, there is no reason to use more than N terms in the synthesis equation given by Eqn. (3.18).

Linearity

Linearity of the DTFS:

Let $\tilde{x}_1[n]$ and $\tilde{x}_2[n]$ be two signals both periodic with the same period, and with DTFS representations

$$\tilde{x}_1[n] \overset{\text{DTFS}}{\longleftrightarrow} \tilde{c}_k \qquad \text{and} \qquad \tilde{x}_2[n] \overset{\text{DTFS}}{\longleftrightarrow} \tilde{d}_k$$

It can be shown that

$$\alpha_1\, \tilde{x}_1[n] + \alpha_2\, \tilde{x}_2[n] \overset{\text{DTFS}}{\longleftrightarrow} \alpha_1\, \tilde{c}_k + \alpha_2\, \tilde{d}_k \tag{3.25}$$

for any two arbitrary constants α_1 and α_2.

Linearity property is easily proven starting with the synthesis equation in Eqn. (3.18). Start with the coefficient set $\alpha_1\, \tilde{c}_k + \alpha_2\, \tilde{d}_k$ and show that it leads to the signal $\alpha_1\, \tilde{x}_1[n] + \alpha_2\, \tilde{x}_2[n]$.

Time shifting

Time shifting the signal $\tilde{x}[n]$ causes the DTFS coefficients to be multiplied by a complex exponential function.

Time shifting property of the DTFS:

Given that

$$\tilde{x}[n] \overset{\text{DTFS}}{\longleftrightarrow} \tilde{c}_k$$

it can be shown that

$$\tilde{x}[n-m] \overset{\text{DTFS}}{\longleftrightarrow} e^{-j(2\pi/N)km}\, \tilde{c}_k \tag{3.26}$$

Consistency check: Let the signal be time shifted by exactly one period, that is, $m = N$. We know that $\tilde{x}[n-N] = \tilde{x}[n]$ due to the periodicity of $\tilde{x}[n]$. The exponential function on the right side of Eqn. (3.26) would be $e^{-j(2\pi/N)kN} = 1$, and the DTFS coefficients remain unchanged, as expected.

Time reversal

Time reversal property of the DTFS:

Given that

$$\tilde{x}[n] \overset{\text{DTFS}}{\longleftrightarrow} \tilde{c}_k$$

it can be shown that

$$\tilde{x}[-n] \overset{\text{DTFS}}{\longleftrightarrow} \tilde{c}_{-k} = \tilde{c}_{N-k} \tag{3.27}$$

Time reversal of the periodic signal corresponds to an index reversal of its DTFS coeffiecients. The proof of this property is instructive, and will be given below.

Proof of Eqn. (3.27):

Let the DTFS coefficients of the signal $\tilde{x}[-n]$ be $\tilde{d}_k$. Using Eqn. (3.19) we have

$$\tilde{d}_k = \frac{1}{N} \sum_{n=0}^{N-1} \tilde{x}[-n]\, e^{-j(2\pi/N)kn}$$

We will introduce the variable change $m = -n$ so that Eqn. (3.28) can be written as

$$\tilde{d}_k = \frac{1}{N} \sum_{m=0}^{-N+1} \tilde{x}[m]\, e^{j(2\pi/N)km} = \frac{1}{N} \sum_{m=-N+1}^{0} \tilde{x}[m]\, e^{j(2\pi/N)km} \tag{3.28}$$

Since the term inside the summation is periodic with N, the starting index of the summation is not important as long as we add N consecutive terms. Therefore summation limits can be adjusted to yield

$$\tilde{d}_k = \frac{1}{N} \sum_{m=0}^{N-1} \tilde{x}[m]\, e^{j(2\pi/N)km} \tag{3.29}$$

Comparing the right side of Eqn. (3.29) to that of Eqn. (3.19), we conclude that

$$\tilde{d}_k = \tilde{c}_{-k}$$

Symmetry of DTFS coefficients

Conjugate symmetry and conjugate antisymmetry properties were defined for discrete-time signals in Section 1.7 of Chapter 1. Same definitions apply to DTFS coefficients as well. A *conjugate symmetric* set of coefficients satisfy

$$\tilde{c}_k^* = \tilde{c}_{-k} \tag{3.30}$$

for all k. Similarly, the coefficients form a *conjugate antisymmetric* set if they satisfy

$$\tilde{c}_k^* = -\tilde{c}_{-k} \tag{3.31}$$

for all k. For a signal $\tilde{x}[n]$ which is periodic with N samples it is customary to use the DTFS coefficients in the index range $k = 0, \ldots, N-1$. The definitions in Eqns. (3.30) and (3.31) can be adjusted in terms of their indices using the periodicity of the DTFS coefficients. Since $\tilde{c}_{-k} = \tilde{c}_{N-k}$, a conjugate symmetric set of DTFS coefficients have the property

$$\tilde{c}_k^* = \tilde{c}_{N-k}\,, \qquad k = 0, \ldots, N-1 \tag{3.32}$$

Similarly, a conjugate antisymmetric set of DTFS coefficients have the property

$$\tilde{c}_k^* = -\tilde{c}_{N-k}\,, \qquad k = 0,\ldots,N-1 \tag{3.33}$$

If the signal $\tilde{x}[n]$ is real-valued, it can be shown that its DTFS coefficients form a conjugate symmetric set. Conversely, if the signal $\tilde{x}[n]$ is purely imaginary, its DTFS coefficients form a conjugate antisymmetric set.

Symmetry of DTFS coefficients:

$$\tilde{x}[n]\text{: Real,} \quad \text{Im}\left\{\tilde{x}[n]\right\} = 0 \quad \text{implies that} \quad \tilde{c}_k^* = \tilde{c}_{N-k} \tag{3.34}$$

$$\tilde{x}[n]\text{: Imag,} \quad \text{Re}\left\{\tilde{x}[n]\right\} = 0 \quad \text{implies that} \quad \tilde{c}_k^* = -\tilde{c}_{N-k} \tag{3.35}$$

Polar form of DTFS coefficients

DTFS coefficients can be written in polar form as

$$\tilde{c}_k = \left|\tilde{c}_k\right| e^{j\tilde{\theta}_k} \tag{3.36}$$

If the set $\left\{\tilde{c}_k\right\}$ is conjugate symmetric, the relationship in Eqn. (3.32) leads to

$$\left|\tilde{c}_k\right| e^{-j\tilde{\theta}_k} = \left|\tilde{c}_{N-k}\right| e^{j\tilde{\theta}_{N-k}} \tag{3.37}$$

using the polar form of the coefficients. The consequences of Eqn. (3.37) are obtained by equating the magnitudes and the phases on both sides.

Conjugate symmetric coefficients: $\tilde{c}_k^* = \tilde{c}_{N-k}$

$$\text{Magnitude:} \quad \left|\tilde{c}_k\right| = \left|\tilde{c}_{N-k}\right| = \left|\tilde{c}_{-k}\right| \tag{3.38}$$

$$\text{Phase:} \quad \tilde{\theta}_k = -\tilde{\theta}_{N-k} = -\tilde{\theta}_{-k} \tag{3.39}$$

It was established in Eqn. (3.34) that the DTFS coefficients of a real-valued $\tilde{x}[n]$ are conjugate symmetric. Based on the results in Eqns. (3.38) and (3.39) the magnitude spectrum is an even function of k, and the phase spectrum is an odd function of k.

Similarly, if the set $\left\{\tilde{c}_k\right\}$ is conjugate antisymmetric, the relationship in Eqn. (3.33) reflects on polar form of $\tilde{c}_k$ as

$$\left|\tilde{c}_k\right| e^{-j\tilde{\theta}_k} = -\left|\tilde{c}_{N-k}\right| e^{j\tilde{\theta}_{N-k}} \tag{3.40}$$

The negative sign on the right side of Eqn. (3.40) needs to be incorporated into the phase since we could not write $\left|\tilde{c}_k\right| = -\left|\tilde{c}_{N-k}\right|$ (recall that magnitude must to be non-negative). Using $e^{\mp j\pi} = -1$, Eqn. (3.40) becomes

$$\begin{aligned}\left|\tilde{c}_k\right| e^{-j\tilde{\theta}_k} &= \left|\tilde{c}_{N-k}\right| e^{j\tilde{\theta}_{N-k}}\, e^{\mp j\pi} \\ &= \left|\tilde{c}_{N-k}\right| e^{j\left(\tilde{\theta}_{N-k} \mp \pi\right)}\end{aligned} \tag{3.41}$$

The consequences of Eqn. (3.41) are summarized below.

Conjugate antisymmetric coefficients: $\tilde{c}_k^* = -\tilde{c}_{N-k}$

$$\text{Magnitude:} \quad \left|\tilde{c}_k\right| = \left|\tilde{c}_{N-k}\right| = \left|\tilde{c}_{-k}\right| \tag{3.42}$$

$$\text{Phase:} \quad \tilde{\theta}_k = -\tilde{\theta}_{N-k} \pm \pi = -\tilde{\theta}_{-k} \pm \pi \tag{3.43}$$

A purely-imaginary signal $\tilde{x}[n]$ leads to a set of DTFS coefficients with conjugate antisymmetry. The corresponding magnitude spectrum is even function of k as suggested by Eqn. (3.42). The phase spectrum is neither even nor odd.

Example 3.6: Symmetry of DTFS coefficients

Recall the real-valued periodic signal $\tilde{x}[n]$ of Example 3.4 shown in Fig. 3.3. Its DTFS coefficients were found as

$$\begin{aligned}
\tilde{c}_0 =& 2 \\
\tilde{c}_1 =& -0.5 + j\,0.6882 \\
\tilde{c}_2 =& -0.5 + j\,0.1625 \\
\tilde{c}_3 =& -0.5 - j\,0.1625 \\
\tilde{c}_4 =& -0.5 - j\,0.6882
\end{aligned}$$

It can easily be verified that coefficients form a conjugate symmetric set. With $N = 5$ we have

$$\begin{aligned}
k =& 1 \quad \Longrightarrow \quad \tilde{c}_1^* = \tilde{c}_4 \\
k =& 2 \quad \Longrightarrow \quad \tilde{c}_2^* = \tilde{c}_3
\end{aligned}$$

DTFS spectra of even and odd signals

If the real-valued signal $\tilde{x}[n]$ is an even function of index n, the resulting DTFS spectrum $\tilde{c}_k$ is real-valued for all k.

Signal $x[n]$ with even symmetry:

$$\tilde{x}[n] = \tilde{x}[n]\,,\text{all } n \quad \text{implies that} \quad \text{Im}\left\{\tilde{c}_k\right\} = 0\,,\text{all } k \tag{3.44}$$

Conversely it can also be proven that, if the real-valued signal $\tilde{x}[n]$ has odd-symmetry, the resulting DTFS spectrum is purely imaginary.

Signal $x[n]$ with odd symmetry:

$$\tilde{x}[n] = -\tilde{x}[n]\,,\text{all } n \quad \text{implies that} \quad \text{Re}\left\{\tilde{c}_k\right\} = 0\,,\text{all } k \tag{3.45}$$

Example 3.7: DTFS symmetry for periodic waveform

Explore the symmetry properties of the periodic waveform $\tilde{x}[n]$ shown in Fig. 3.8. One period of $\tilde{x}[n]$ has the sample amplitudes

$$\tilde{x}[n] = \{\ \ldots,\ \underset{\substack{\uparrow \\ n=0}}{0}, \tfrac{1}{2}, 1, \tfrac{3}{4}, 0, -\tfrac{3}{4}, -1, -\tfrac{1}{2}, \ldots\ \}$$

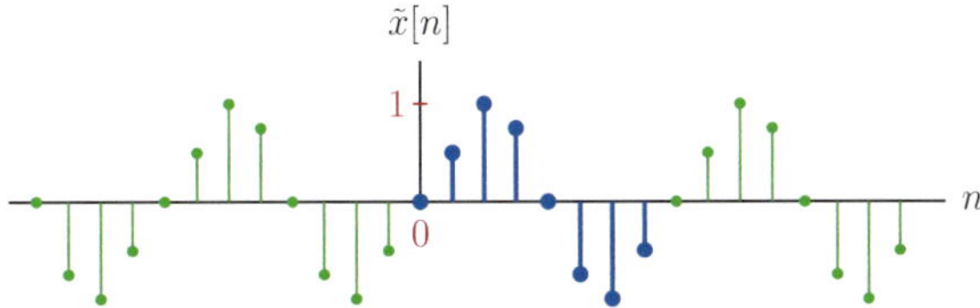

Figure 3.8 – The signal $\tilde{x}[n]$ for Example 3.7.

Solution: The DTFS coefficients for $\tilde{x}[n]$ are computed as

$$\tilde{c}_k = \sum_{n=0}^{7} \tilde{x}[n]\, e^{-j(2\pi/8)kn}$$

and are listed for $k = 0, \ldots, 8$ in Table 3.1 along with magnitude and phase values for each.

Table 3.1 – DTFS coefficients for the pulse train of Example 3.7.

k	$\tilde{c}_k$	$\lvert\tilde{c}_k\rvert$	$\tilde{\theta}_k$
0	0.0000	0.0000	
1	$-j0.4710$	0.4710	$-\pi/2$
2	$j0.0625$	0.0625	$\pi/2$
3	$j0.0290$	0.0290	$\pi/2$
4	0.0000	0.0000	
5	$-j0.0290$	0.0290	$-\pi/2$
6	$-j0.0625$	0.0625	$-\pi/2$
7	$j0.4710$	0.4710	$\pi/2$

Symmetry properties can be easily observed. The signal is real-valued, therefore the DTFS spectrum is conjugate symmetric:

$$\begin{aligned} k=1 &\quad\Longrightarrow\quad \tilde{c}_1^* = \tilde{c}_7 \\ k=2 &\quad\Longrightarrow\quad \tilde{c}_2^* = \tilde{c}_6 \\ k=2 &\quad\Longrightarrow\quad \tilde{c}_3^* = \tilde{c}_5 \end{aligned}$$

Furthermore, odd symmetry of $\tilde{x}[n]$ causes coefficients to be purely imaginary:

$$\text{Re}\{\tilde{c}_k\} = 0\,, \qquad k = 0, \ldots, 7$$

In terms of the magnitude values we have

$$\begin{aligned} k=1 \quad &\Longrightarrow \quad |\tilde{c}_1| = |\tilde{c}_7| \\ k=2 \quad &\Longrightarrow \quad |\tilde{c}_2| = |\tilde{c}_6| \\ k=2 \quad &\Longrightarrow \quad |\tilde{c}_3| = |\tilde{c}_5| \end{aligned}$$

For the phase angles the following relationships hold:

$$\begin{aligned} k=1 \quad &\Longrightarrow \quad \tilde{\theta}_1 = -\tilde{\theta}_7 \\ k=2 \quad &\Longrightarrow \quad \tilde{\theta}_2 = -\tilde{\theta}_6 \\ k=2 \quad &\Longrightarrow \quad \tilde{\theta}_3 = -\tilde{\theta}_5 \end{aligned}$$

The phase values for $\tilde{\theta}_0$ and $\tilde{\theta}_4$ are insignificant since the corresponding magnitude values $|\tilde{c}_0|$ and $|\tilde{c}_4|$ are equal to zero.

Software resource: exdt_3_7.m

Periodic convolution

Consider the convolution of two discrete-time signals defined by Eqn. (2.109) in Chapter 2, and repeated here for convenience:

$$\text{Eqn. (2.109):} \qquad y[n] = x[n] * h[n] = \sum_{k=-\infty}^{\infty} x[k]\, h[n-k]$$

This summation would obviously fail to converge if both signals $x[n]$ and $h[n]$ happen to be periodic with period of N. For such a case, a *periodic convolution* operator can be defined as

$$\tilde{y}[n] = \tilde{x}[n] \otimes \tilde{h}[n] = \sum_{k=0}^{N-1} \tilde{x}[k]\, \tilde{h}[n-k]\,, \quad \text{all } n \tag{3.46}$$

Eqn. (3.46) is essentially an adaptation of the convolution sum to periodic signals where the limits of the summation are modified to cover only one period (we are assuming that both $\tilde{x}[n]$ and $\tilde{h}[n]$ have the same period N). It can be shown that (see Problem 3.10) the periodic convolution of two signals $\tilde{x}[n]$ and $\tilde{h}[n]$ that are both periodic with N is also periodic with the same period.

Example 3.8: Periodic convolution

Two signals $\tilde{x}[n]$ and $\tilde{h}[n]$, each periodic with $N = 5$ samples, are shown in Figs. 3.9 and 3.10. Determine the periodic convolution

$$\tilde{y}[n] = \tilde{x}[n] \otimes \tilde{h}[n]$$

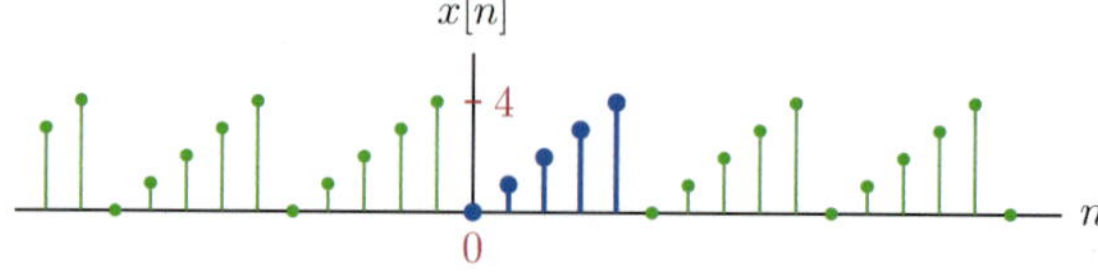

Figure 3.9 – The signal $\tilde{x}[n]$ for Example 3.8.

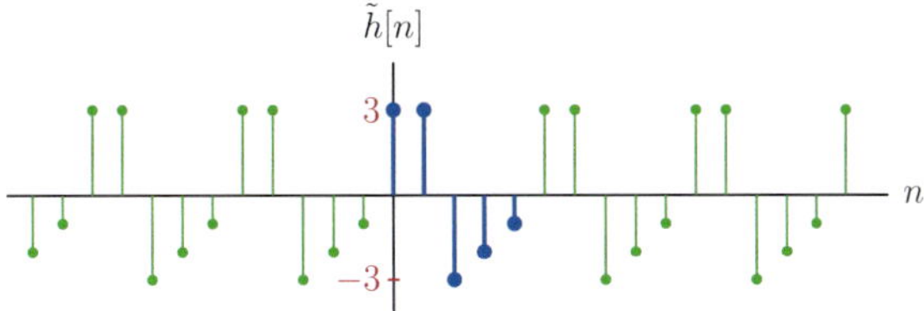

Figure 3.10 – The signal $\tilde{h}[n]$ for Example 3.8.

Solution: Sample amplitudes for one period are

$$\tilde{x}[n] = \{\ \ldots,\ \underset{\substack{\uparrow \\ n=0}}{0}, 1, 2, 3, 4, \ldots\ \}$$

and

$$\tilde{h}[n] = \{\ \ldots,\ \underset{\substack{\uparrow \\ n=0}}{3}, 3, -3, -2, -1, \ldots\ \}$$

The periodic convolution is given by

$$\tilde{y}[n] = \sum_{k=0}^{4} \tilde{x}[k]\, \tilde{h}[n-k]$$

To start, let $n = 0$. The terms $\tilde{x}[k]$ and $\tilde{h}[0-k]$ are shown in Fig. 3.11.

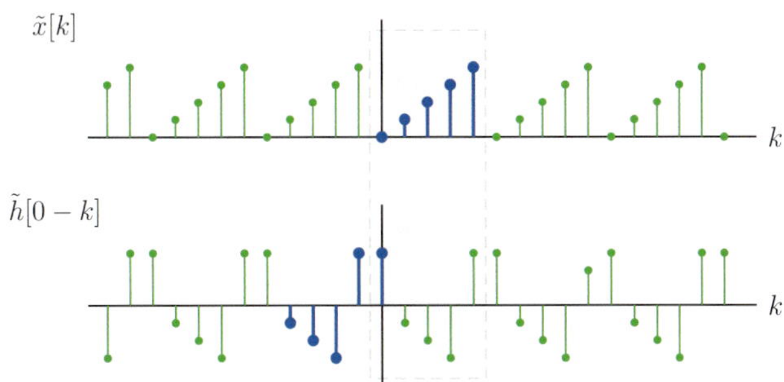

Figure 3.11 – The terms $\tilde{x}[k]$ and $\tilde{h}[0-k]$ in the convolution sum.

In Fig. 3.11 the main period of each signal is indicated with sample amplitudes colored blue. The shaded area contains the terms included in the summation for periodic convolution. The sample $\tilde{y}[0]$ is computed as

$$\begin{aligned}\tilde{y}[0] =& \tilde{x}[0]\, \tilde{h}[0] + \tilde{x}[1]\, \tilde{h}[4] + \tilde{x}[2]\, \tilde{h}[3] + \tilde{x}[3]\, \tilde{h}[2] + \tilde{x}[4]\, \tilde{h}[1] \\ =& (0)\ (3) + (1)\ (-1) + (2)\ (-3) + (3)\ (-3) + (4)\ (3) = -2\end{aligned}$$

Next we will set $n = 1$. The terms $\tilde{x}[k]$ and $\tilde{h}[1-k]$ are shown in Fig. 3.12. The sample $\tilde{y}[1]$ is computed as

$$\begin{aligned}\tilde{y}[1] =& \tilde{x}[0]\, \tilde{h}[1] + \tilde{x}[1]\, \tilde{h}[0] + \tilde{x}[2]\, \tilde{h}[4] + \tilde{x}[3]\, \tilde{h}[3] + \tilde{x}[4]\, \tilde{h}[2] \\ =& (0)\ (3) + (1)\ (3) + (2)\ (-1) + (3)\ (-2) + (4)\ (-3) = -17\end{aligned}$$

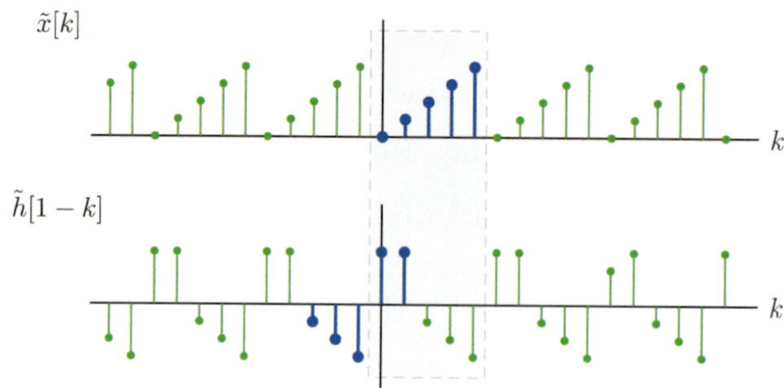

Figure 3.12 – The terms $\tilde{x}[k]$ and $\tilde{h}[1-k]$ in the convolution sum.

Finally, for $n = 2$ the terms involved in the summation are shown in Fig. 3.13.

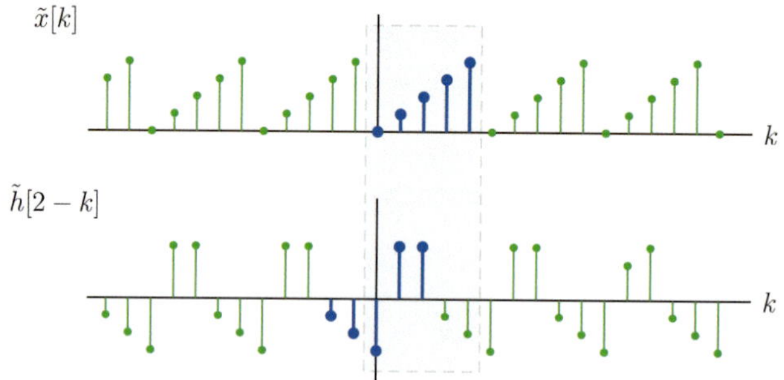

Figure 3.13 – The terms $\tilde{x}[k]$ and $\tilde{h}[2-k]$ in the convolution sum.

The sample $\tilde{y}[2]$ is computed as

$$\begin{aligned}\tilde{y}[2] =& \tilde{x}[0]\,\tilde{h}[2] + \tilde{x}[1]\,\tilde{h}[1] + \tilde{x}[2]\,\tilde{h}[0] + \tilde{x}[3]\,\tilde{h}[4] + \tilde{x}[4]\,\tilde{h}[3] \\ =& (0)\;(-3) + (1)\;(3) + (2)\;(3) + (3)\;(-1) + (4)\;(-2) = -2\end{aligned}$$

Continuing in this fashion, it can be shown that $\tilde{y}[3] = 8$ and $\tilde{y}[4] = 13$. Thus, one period of the signal $\tilde{y}[n]$ is

$$\tilde{y}[n] = \{\ \ldots, \underset{\substack{\uparrow \\ n=0}}{-2}, -17, -2, 8, 13, \ldots\ \}$$

The signal $\tilde{y}[n]$ is shown in Fig. 3.14.

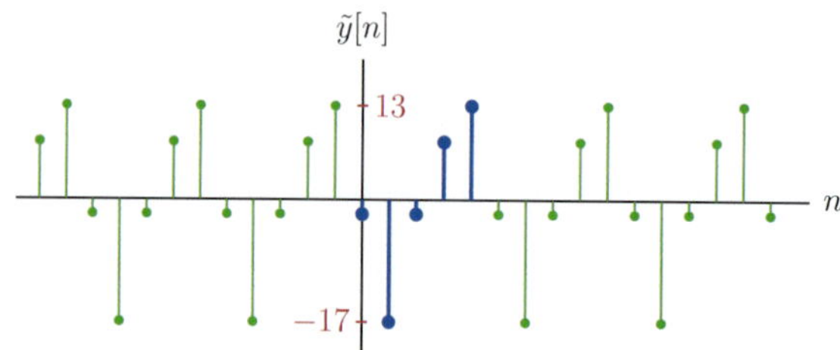

Figure 3.14 – Periodic convolution result $\tilde{y}[n]$ for Example 3.8.

The periodic convolution property of the discrete-time Fourier series can be stated as follows:

Let $\tilde{x}[n]$ and $\tilde{h}[n]$ be two signals both periodic with the same period and with DTFS representations

$$\tilde{x}[n] \overset{\text{DTFS}}{\longleftrightarrow} \tilde{c}_k \qquad \text{and} \qquad \tilde{h}[n] \overset{\text{DTFS}}{\longleftrightarrow} \tilde{d}_k$$

It can be shown that

$$\tilde{x}[n] \otimes \tilde{h}[n] \overset{\text{DTFS}}{\longleftrightarrow} N\,\tilde{c}_k\,\tilde{d}_k \tag{3.47}$$

The DTFS coefficients of the periodic convolution result is equal to N times the product of the DTFS coefficients of the two signals.

Proof for Eqn. (3.47):

Let $\tilde{y}[n] = \tilde{x}[n] \otimes \tilde{h}[n]$, and let the DTFS coefficients of $\tilde{y}[n]$ be $\tilde{e}_k$. Using the DTFS analysis equation, we can write

$$\tilde{e}_k = \frac{1}{N} \sum_{n=0}^{N-1} \tilde{y}[n]\, e^{-j(2\pi/N)kn} \tag{3.48}$$

Substituting

$$\tilde{y}[n] = \sum_{m=0}^{N-1} \tilde{x}[m]\, \tilde{h}[n-m] \tag{3.49}$$

into Eqn. (3.48) we obtain

$$\tilde{e}_k = \frac{1}{N} \sum_{n=0}^{N-1} \left[\sum_{m=0}^{N-1} \tilde{x}[m]\, \tilde{h}[n-m] \right] e^{-j(2\pi/N)kn} \tag{3.50}$$

Changing the order of the two summations and rearranging terms leads to

$$\tilde{e}_k = \sum_{m=0}^{N-1} \tilde{x}[m] \left[\frac{1}{N} \sum_{n=0}^{N-1} \tilde{h}[n-m]\, e^{-j(2\pi/N)kn} \right] \tag{3.51}$$

In Eqn. (3.51) the term in square brackets represents the DTFS coefficients for the time shifted periodic signal $\tilde{h}[n-m]$, and is evaluated as

$$\frac{1}{N} \sum_{n=0}^{N-1} \tilde{h}[n-m]\, e^{-j(2\pi/N)kn} = e^{-j(2\pi/N)km}\, \tilde{d}_k \tag{3.52}$$

Using this result Eqn. (3.51) becomes

$$\tilde{e}_k = \tilde{d}_k \sum_{m=0}^{N-1} \tilde{x}[m]\, e^{-j(2\pi/N)km} = N\,\tilde{c}_k\,\tilde{d}_k \tag{3.53}$$

completing the proof.

Example 3.9: Periodic convolution revisited

Refer to the signals $\tilde{x}[n]$, $\tilde{h}[n]$, and $\tilde{y}[n]$ of Example 3.8. The DTFS coefficients of $\tilde{x}[n]$ were determined in Example 3.4. Find the DTFS coefficients of $\tilde{h}[n]$ and $\tilde{y}[n]$. Afterward verify that the convolution property given by Eqn. (3.47) holds.

Solution: Let $\tilde{c}_k$, $\tilde{d}_k$, and $\tilde{e}_k$ represent the DTFS coefficients of $\tilde{x}[n]$, $\tilde{h}[n]$, and $\tilde{y}[n]$, respectively. Table 3.2 lists the DTFS coefficients for the three signals. It can easily be verified that

$$\tilde{e}_k = 5\,\tilde{c}_k\,\tilde{d}_k\,, \qquad k = 0,\ldots,4$$

Table 3.2 – DTFS coefficients for Example 3.9.

k	$\tilde{c}_k$	$\tilde{d}_k$	$\tilde{e}_k$
0	$2.0000+j\,0.0000$	$0.0000+j\,0.0000$	$0.0000+j\,0.0000$
1	$-0.5000+j\,0.6882$	$1.5326-j\,0.6433$	$-1.6180+j\,6.8819$
2	$-0.5000+j\,0.1625$	$-0.0326-j\,0.6604$	$0.6180+j\,1.6246$
3	$-0.5000-j\,0.1625$	$-0.0326+j\,0.6604$	$0.6180-j\,1.6246$
4	$-0.5000-j\,0.6882$	$1.5326+j\,0.6433$	$-1.6180-j\,6.8819$

Software resource: exdt_3_9.m

Software resource: See MATLAB Exercise 3.4.

3.3 Analysis of Non-Periodic Discrete-Time Signals

In the previous section we have focused on representing periodic discrete-time signals using complex exponential basis functions. The end result was the discrete-time Fourier series (DTFS) that allowed a signal periodic with a period of N samples to be constructed using N harmonically related exponential basis functions. In this section we extend the DTFS concept for use in non-periodic signals.

3.3.1 Discrete-time Fourier transform (DTFT)

In the derivation of the Fourier transform for non-periodic discrete-time signals we will make use of the DTFS formulation of Section 3.2. We will treat a non-periodic discrete-time signal as a limit case of a periodic discrete-time signal, and will derive its Fourier transform from the DTFS. The resulting development is not a mathematically rigorous derivation of the Fourier transform, but it is intuitive.

Let us begin by considering a non-periodic discrete-time signal $x[n]$ as shown in Fig. 3.15. Initially we will assume that $x[n]$ is finite-length with its significant samples confined into the range $-M \le n \le M$ of the index, that is, $x[n] = 0$ for $n < -M$ and for $n > M$. A periodic extension $\tilde{x}[n]$

can be constructed by taking $x[n]$ as one period in $-M \le n \le M$, and repeating it at intervals of $2M+1$ samples. This is illustrated in Fig. 3.16.

$$\tilde{x}[n] = \sum_{k=-\infty}^{\infty} x[n+k(2M+1)] \tag{3.54}$$

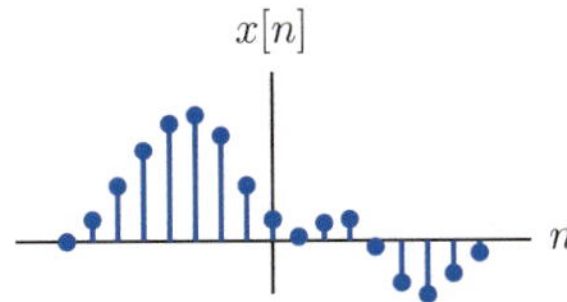

Figure 3.15 – A non-periodic signal $x[n]$.

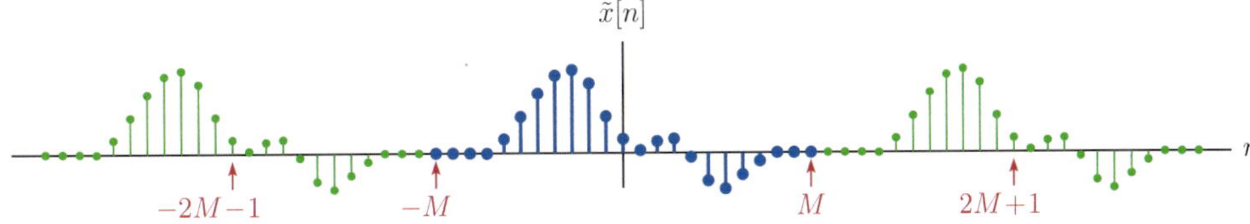

Figure 3.16 – Periodic extension $\tilde{x}[n]$ of the signal $x[n]$.

The periodic extension $\tilde{x}[n]$ can be expressed in terms of its DTFS coefficients. Using Eqn. (3.18) with $N = 2M+1$ we obtain

$$\tilde{x}[n] = \sum_{k=0}^{2M} \tilde{c}_k \, e^{j(2\pi/(2M+1))kn} = \sum_{k=-M}^{M} \tilde{c}_k \, e^{j(2\pi/(2M+1))kn} \tag{3.55}$$

The coefficients are computed through the use of Eqn. (3.19) as

$$\tilde{c}_k = \frac{1}{2M+1} \sum_{n=-M}^{M} \tilde{x}[n] \, e^{-j(2\pi/(2M+1))kn} , \quad k = -M, \ldots, M \tag{3.56}$$

Fundamental angular frequency is

$$\Omega_0 = \frac{2\pi}{2M+1} \quad \text{radians.}$$

The k-th DTFS coefficient is associated with the angular frequency $\Omega_k = k\Omega_0 = 2\pi k/(2M+1)$. The set of DTFS coefficients span the range of angular frequencies

$$\Omega_k\colon \quad \frac{-2M}{2M+1}\pi, \ldots, 0, \ldots, \frac{2M}{2M+1}\pi \tag{3.57}$$

It is worth noting that the set of coefficients in Eqn. (3.57) are roughly in the interval $(-\pi, \pi)$, just slightly short of either end of the interval. Realizing that $\tilde{x}[n] = x[n]$ within the range $-M \le n \le M$, Eqn. (3.56) can be written using $x[n]$ instead of $\tilde{x}[n]$ to yield

$$\tilde{c}_k = \frac{1}{2M+1} \sum_{n=-M}^{M} x[n] \, e^{-j(2\pi/(2M+1))kn} , \quad k = -M, \ldots, M \tag{3.58}$$

If we were to stretch out the period of the signal by increasing the value of M, then $\tilde{x}[n]$ would start to resemble $x[n]$ more and more. Other effects of increasing M would be an increase in the coefficient count and a decrease in the magnitudes of the coefficients $\tilde{c}_k$ due to the $1/(2M+1)$ factor in front of the summation. Let us multiply both sides of Eqn. (3.58) by $2M+1$ to obtain

$$(2M+1)\,\tilde{c}_k = \sum_{n=-M}^{M} x[n]\,e^{-j(2\pi/(2M+1))kn} \tag{3.59}$$

As M becomes very large, the fundamental angular frequency Ω_0 becomes very small, and the spectral lines get closer to each other in the frequency domain, resembling a continuous transform.

$$M \to \infty \quad \text{implies that} \quad \Omega_0 = \frac{2\pi}{2M+1} \to \Delta\Omega \quad \text{and} \quad k\Delta\Omega \to \Omega \tag{3.60}$$

In Eqn. (3.60) we have switched the notation from Ω_0 to $\Delta\Omega$ due to the infinitesimal nature of the fundamental angular frequency. In the limit we have

$$\lim_{M\to\infty} \left[\tilde{x}[n]\right] = x[n] \tag{3.61}$$

in the time domain. Using the substitutions $2\pi k/(2M+1) \to \Omega$ and $(2M+1)\,\tilde{c}_k \to X(\Omega)$ Eqn. (3.59) becomes

$$X(\Omega) = \sum_{n=-\infty}^{\infty} x[n]\,e^{-j\Omega n} \tag{3.62}$$

The result in Eqn. (3.62) is referred to as the *discrete-time Fourier transform (DTFT)* of the signal $x[n]$. In deriving this result we assumed a finite-length signal $x[n]$ the samples of which are confined into the range $-M \le n \le M$, and then took the limit as $M \to \infty$. Would Eqn. (3.62) still be valid for an infinite-length $x[n]$? The answer is yes, provided that the summation in Eqn. (3.62) converges.

Next we will try to develop some insight about the meaning of the transform $X(\Omega)$. Let us apply the limit operation to the periodic extension signal $\tilde{x}[n]$ defined in Eqn. (3.55).

$$x[n] = \lim_{M\to\infty} \left[\tilde{x}[n]\right] = \lim_{M\to\infty} \left[\sum_{k=-M}^{M} \tilde{c}_k\,e^{j(2\pi/(2M+1))kn}\right] \tag{3.63}$$

For large M we have from Eqn. (3.60)

$$\frac{(2M+1)\,\Delta\Omega}{2\pi} \to 1 \tag{3.64}$$

Using this result in Eqn. (3.63) leads to

$$x[n] = \lim_{M\to\infty} \left[\sum_{k=-M}^{M} (2M+1)\,\tilde{c}_k\,e^{jk\Delta\Omega n}\,\frac{\Delta\Omega}{2\pi}\right] \tag{3.65}$$

In the limit we have

$$(2M+1)\,\tilde{c}_k \to X(\Omega)\,, \quad k\Delta\Omega \to \Omega \quad \text{and} \quad \Delta\Omega \to d\Omega \tag{3.66}$$

Furthermore, the summation turns into an integral to yield

$$x[n] = \frac{1}{2\pi} \int_{-\pi}^{\pi} X(\Omega)\,e^{j\Omega n}\,d\Omega \tag{3.67}$$

This result explains how the transform $X(\Omega)$ can be used for constructing the signal $x[n]$. We can interpret the integral in Eqn. (3.67) as a continuous sum of complex exponentials at harmonic frequencies that are infinitesimally close to each other.

In summary, we have derived a transform relationship between $x[n]$ and $X(\Omega)$ through the following equations:

Discrete-Time Fourier Transform (DTFT):

1. Synthesis equation:

$$x[n] = \frac{1}{2\pi} \int_{-\pi}^{\pi} X(\Omega)\, e^{j\Omega n}\, d\Omega \tag{3.68}$$

2. Analysis equation:

$$X(\Omega) = \sum_{n=-\infty}^{\infty} x[n]\, e^{-j\Omega n} \tag{3.69}$$

Often we will use the Fourier transform operator $\mathscr{F}$ and its inverse $\mathscr{F}^{-1}$ in a shorthand notation. The *forward transform* is expressed as

$$X(\Omega) = \mathscr{F}\left\{x[n]\right\} \tag{3.70}$$

In contrast, the *inverse transform* is expressed as

$$x[n] = \mathscr{F}^{-1}\left\{X(\Omega)\right\} \tag{3.71}$$

Sometimes we will use an even more compact notation to express the relationship between $x[n]$ and $X(\Omega)$ by

$$x[n] \overset{\mathscr{F}}{\longleftrightarrow} X(\Omega) \tag{3.72}$$

In general, the Fourier transform, as computed by Eqn. (3.69), is a complex function of Ω. It can be written in Cartesian form as

$$X(\Omega) = X_r(\Omega) + X_i(\Omega) \tag{3.73}$$

or in polar form as

$$X(\Omega) = \left|X(\Omega)\right| e^{j\measuredangle X(\Omega)} \tag{3.74}$$

3.3.2 Developing further insight

In this section we will build on the idea of obtaining the DTFT as the limit case of DTFS coefficients when the signal period is made very large. Consider a discrete-time pulse with 7 unit-amplitude samples as shown in Fig. 3.17.

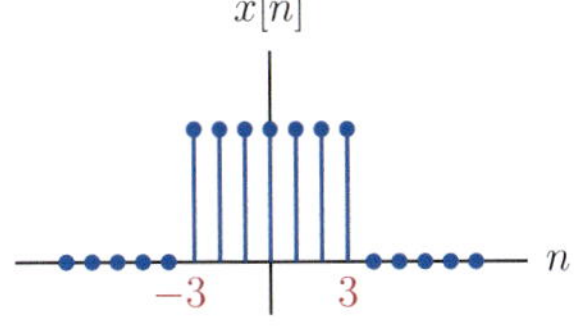

Figure 3.17 – Discrete-time rectangular pulse.

The analytical definition of $x[n]$ is

$$x[n]=\begin{cases}1\,, & n=-3,\ldots,3\\ 0\,, & \text{otherwise}\end{cases} \tag{3.75}$$

Let the signal $\tilde{x}[n]$ be defined as the periodic extension of $x[n]$ with a period of $2M+1$ samples so that

$$\tilde{x}[n]=\sum_{r=-\infty}^{\infty} x[n+r\,(2M+1)] \tag{3.76}$$

as shown in Fig. 3.18.

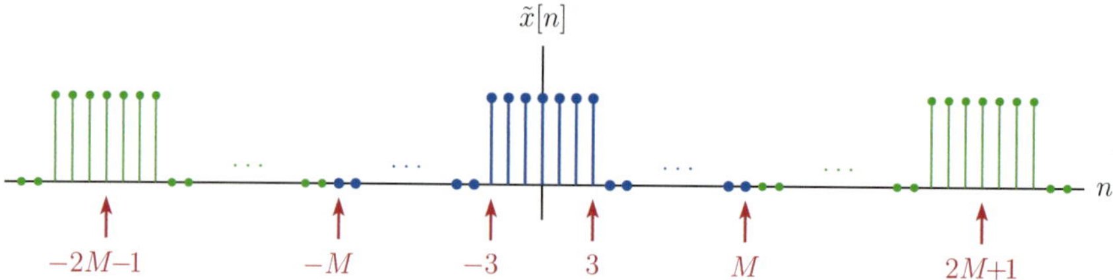

Figure 3.18 – Periodic extension of discrete-time pulse into a pulse train.

One period of $\tilde{x}[n]$ extends from $n=-M$ to $n=M$ for a total of $2M+1$ samples. The general expression for DTFS coefficients can be found by adapting the result obtained in Example 3.5 to the signal $\tilde{x}[n]$ with $L=3$ and $N=2M+1$:

$$\tilde{c}_k=\frac{\sin(7k\Omega_0/2)}{(2M+1)\,\sin(k\Omega_0/2)} \tag{3.77}$$

The parameter Ω_0 is the fundamental angular frequency given by

$$\Omega_0=\frac{2\pi}{2M+1}\quad\text{rad} \tag{3.78}$$

Let us multiply both sides of Eqn. (3.77) by $(2M+1)$ and write the scaled DTFS coefficients as

$$(2M+1)\,\tilde{c}_k=\frac{\sin(7k\Omega_0/2)}{\sin(k\Omega_0/2)} \tag{3.79}$$

Let $M=8$ corresponding to a period length of $2M+1=17$. The scaled DTFS coefficients are shown in Fig. 3.19(a) for the index range $k=-8,\ldots,8$. In addition, the outline (or the envelope) of the scaled DTFS coefficients is also shown. The leftmost coefficient in the figure has the index $k=-8$, and is associated with the angular frequency $-8\Omega_0=-16\pi/17$. Similarly, the rightmost coefficient is at $k=8$, and is associated with the angular frequency $8\Omega_0=16\pi/17$. Fig. 3.19(b) shows the same coefficients and envelope as functions of the angular frequency Ω instead of the integer index k.

It is interesting to check the locations for the zero crossings of the envelope. The first positive zero crossing of the envelope occurs at the index value

$$\frac{7k\Omega_0}{2}=\pi \quad\Rightarrow\quad k=\frac{2\pi}{7\Omega_0}=\frac{2M+1}{7} \tag{3.80}$$

which may or may not be an integer. For $M=8$ the first positive zero crossing is at index value $k=17/7=2.43$ as shown in Fig. 3.19(a). This corresponds to the angular frequency $(17/7)\,\Omega_0=2\pi/7$ radians, independent of the value of M.

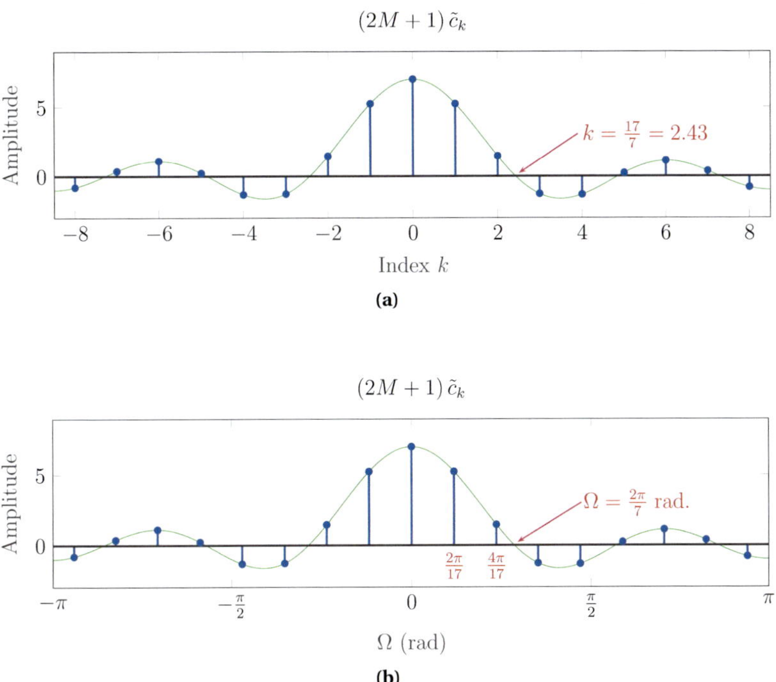

Figure 3.19 – **(a)** The scaled DTFS spectrum $(2M+1)\,\tilde{c}_k$ for $M = 8$ for the signal $\tilde{x}[n]$ and **(b)** the scaled DTFS spectrum as a function of angular frequency Ω.

If we increase the period M, the following changes occur in the scaled DTFS spectrum:

1. The number of DTFS coefficients increases since the total number of unique DTFS coefficients is the same as the period length of $\tilde{x}[n]$ which, in this case, is $2M+1$.
2. The fundamental angular frequency decreases since it is inversely proportional to the period of $\tilde{x}[n]$. The spectral lines in Fig. 3.19(b) move in closer to each other.
3. The leftmost and the rightmost spectral lines get closer to $\pm\pi$ since they are $\pm M\Omega_0 = \pm 2M\pi/(2M+1)$.
4. As $M \to \infty$, the fundamental angular frequency becomes infinitesimally small and the spectral lines come together to form a continuous transform.

We can conclude from the foregoing development that, as the period becomes infinitely large, spectral lines of the DTFS representation of $\tilde{x}[n]$ converge to a continuous function of Ω to form the DTFT of $x[n]$. Taking the limit of Eqn. (3.79) we get

$$X(\Omega) = \lim_{M\to\infty}\Big[(2M+1)\,\tilde{c}_k\Big] = \frac{\sin(7\Omega/2)}{\sin(\Omega/2)} \tag{3.81}$$

We will also obtain this result in Example 3.12 through direct application of the DTFT equation.

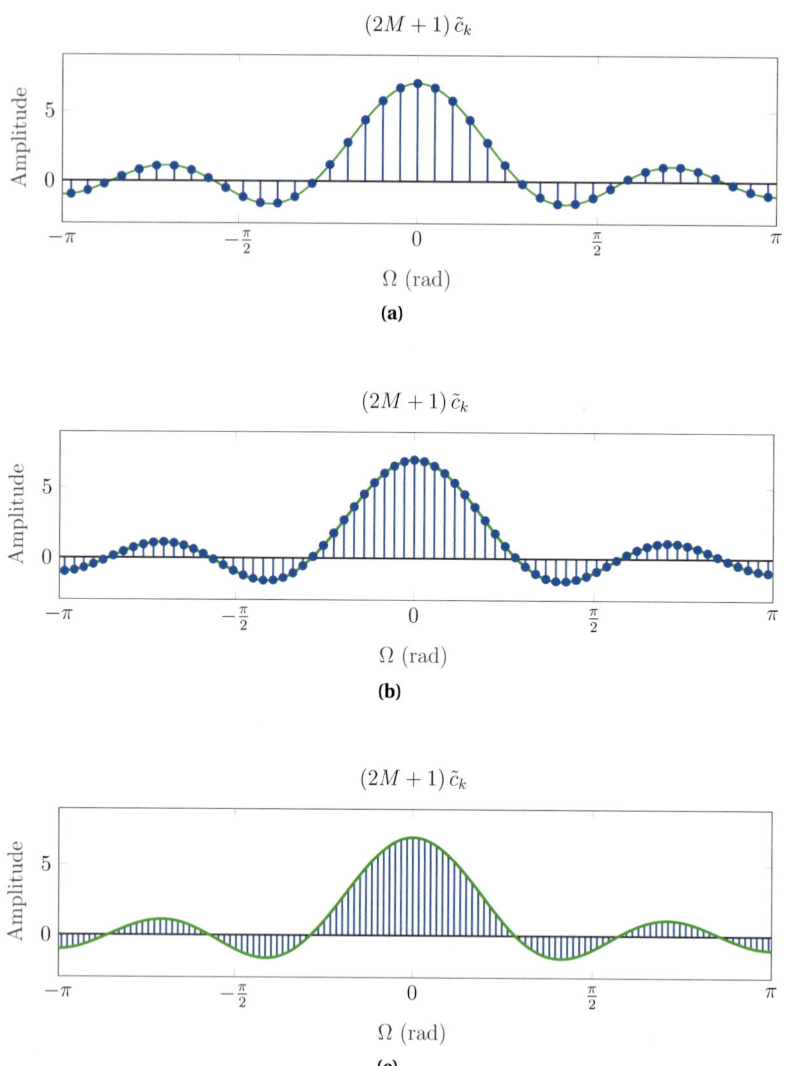

Figure 3.20 – The scaled DTFS spectrum for the signal $\tilde{x}[n]$ for **(a)** $M = 20$, **(b)** for $M = 35$, and **(c)** for $M = 60$.

Interactive App: Experimenting with relationship between DTFS and DTFT

The interactive app in `appDTFT1.m` provides a graphical user interface for experimenting with the development in Section 3.3.2. Refer to Figs. 3.17 through 3.20, Eqns. (3.75) through (3.79). The discrete-time pulse train has a period of $2M+1$. In each period $2L+1$ contiguous samples have unit amplitude. (Keep in mind that $M > L$). Parameters L and M can be adjusted and the resulting scaled DTFS spectrum can be observed. As the period $2M+1$ is increased the DTFS coefficients move in closer due to the fundamental angular frequency Ω_0 becoming smaller. Consequently, the DTFS spectrum of the periodic pulse train approaches the DTFT spectrum of the non-periodic discrete-time pulse with $2L+1$ samples.

a. Set $L = 3$ and $M = 8$. This should duplicate Fig. 3.19. Observe the scaled DTFS coefficients. Pay attention to the envelope of the DTFS coefficients.

b. While keeping $L = 3$ increment M and observe the changes in the scaled DTFS coefficients. Pay attention to the fundamental angular frequency Ω_0 change, causing the coefficients to move in closer together. Observe that the envelope does not change as M is increased.

Software resource: `appDTFT1.m`

3.3.3 Existence of the DTFT

A mathematically thorough treatment of the conditions for the existence of the DTFT is beyond the scope of this text. It will suffice to say, however, that the question of existence is a simple one for the types of signals we encounter in engineering practice. A sufficient condition for the convergence of Eqn. (3.69) is that the signal $x[n]$ be absolute summable, that is,

$$\sum_{n=-\infty}^{\infty} \left|x[n]\right| < \infty \tag{3.82}$$

Alternatively, it is also sufficient for the signal $x[n]$ to be square-summable:

$$\sum_{n=-\infty}^{\infty} \left|x[n]\right|^2 < \infty \tag{3.83}$$

In addition, we will see in the next section that some signals that do not satisfy either condition may still have a DTFT if we are willing to resort to the use of singularity functions in the transform.

3.3.4 DTFT of some signals

In this section we present examples of determining the DTFT for a variety of discrete-time signals.

Example 3.10: DTFT of right-sided exponential signal

Determine the DTFT of the signal $x[n] = \alpha^n\, u[n]$ as shown in Fig. 3.21. Assume $|\alpha| < 1$.

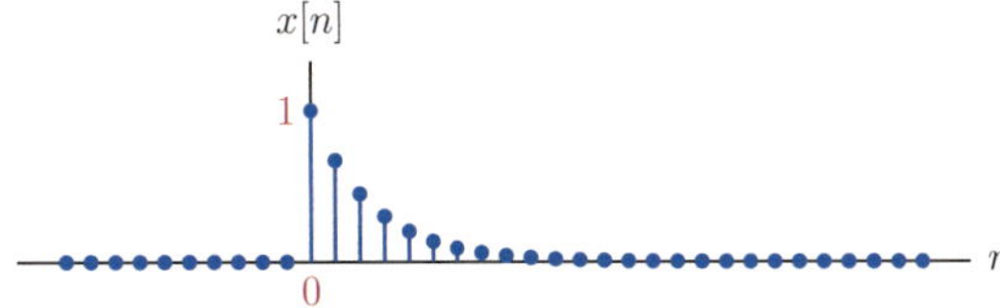

Figure 3.21 – The signal $x[n]$ for Example 3.10.

Solution: The use of the DTFT analysis equation given by Eqn. (3.69) yields

$$X(\Omega) = \sum_{n=-\infty}^{\infty} \alpha^n\, u[n]\, e^{-j\Omega n}$$

The factor $u[n]$ causes all terms of the summation to equal zero for $n<0$. Consequently, we can start the summation at $n=0$ and drop the term $u[n]$ to write

$$X(\Omega)=\sum_{n=0}^{\infty}\alpha^n\,e^{-j\Omega n}=\frac{1}{1-\alpha\,e^{-j\Omega}} \tag{3.84}$$

provided that $|\alpha|<1$. In obtaining the result in Eqn. (3.84) we have used the closed form of the sum of infinite-length geometric series (see Appendix C). The magnitude of the transform is

$$\left|X(\Omega)\right|=\frac{1}{\left|1-\alpha\,e^{-j\Omega}\right|}=\frac{1}{\sqrt{1+\alpha^2-2\alpha\cos(\Omega)}}$$

The phase of the transform is found as the difference of the phases of numerator and denominator of the result in Eqn. (3.84):

$$\measuredangle X(\Omega)=0-\measuredangle\left(1-\alpha\,e^{-j\Omega}\right)=-\tan^{-1}\left[\frac{\alpha\sin(\Omega)}{1-\alpha\cos(\Omega)}\right]$$

The magnitude and the phase of the transform are shown in Fig. 3.22(a),(b) for the case $\alpha=0.4$.

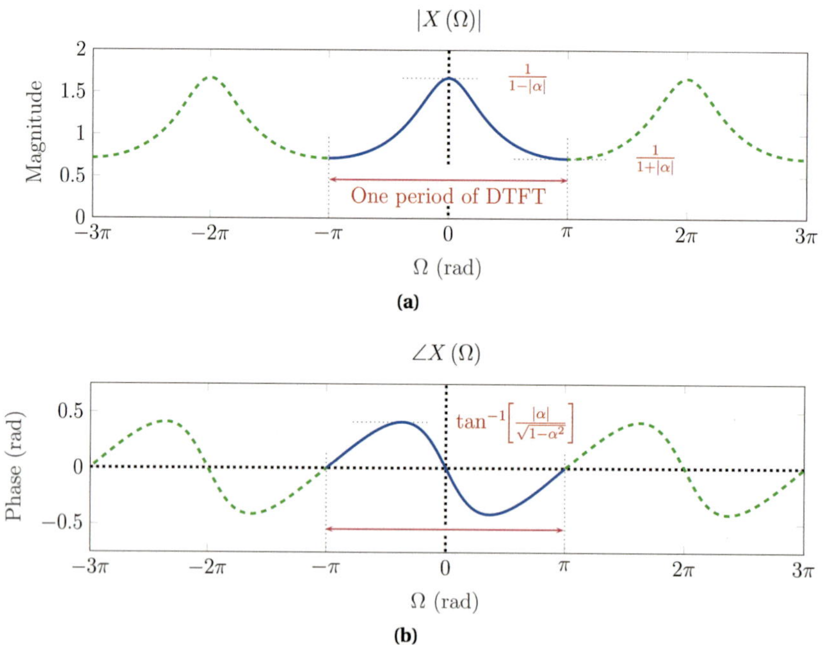

Figure 3.22 – The DTFT of the signal $x[n]$ for Example 3.10 for $\alpha=0.4$: **(a)** the magnitude and **(b)** the phase.

Software resource: exdt_3_10.m

Interactive App: DTFT of right-sided exponential signal

The interactive app in `appDTFT2.m` is based on Example 3.10. The signal $x[n]$ is graphed along with the magnitude and the phase of its DTFT $X(\Omega)$. The parameter α may be varied and its effects on the spectrum may be observed.

a. Set $\alpha = 0.4$ to match the DTFT spectrum in Fig. 3.22.

b. Slowly reduce the value of α. Observe the changes in the exponential decay of the signal $x[n]$ and the corresponding changes in the magnitude $|X(\Omega)|$ of the spectrum. Watch how the spectrum becomes narrower as the decay of the time-domain signal is made slower.

c. Now vary the parameter α in the opposite direction, increasing its value. Observe the changes in the signal and its spectrum.

Software resource: `appDTFT2.m`

Example 3.11: DTFT of unit impulse signal

Determine the DTFT of the unit impulse signal $x[n] = \delta[n]$.

Solution: Direct application of Eqn. (3.69) yields

$$\mathcal{F}\{\delta[n]\} = \sum_{n=-\infty}^{\infty} \delta[n]\, e^{-j\Omega n}$$

Using the sifting property of the impulse function, this can be reduced to

$$\mathcal{F}\{\delta[n]\} = 1\,, \quad \text{all } \Omega$$

Example 3.12: DTFT for discrete-time pulse

Determine the DTFT of the discrete-time pulse signal $x[n]$ given by

$$x[n] = \begin{cases} 1\,, & -L \le n \le L \\ 0\,, & \text{otherwise} \end{cases}$$

Solution: Using Eqn. (3.69) the transform is

$$X(\Omega) = \sum_{n=-L}^{L} (1)\, e^{-j\Omega n}$$

The closed form expression for a finite-length geometric series is (see Appendix C for derivation)

$$\text{Eqn. (C.13):} \qquad \sum_{n=L_1}^{L_2} a^n = \frac{a^{L_1} - a^{L_2+1}}{1-a}$$

Using Eqn. (C.13) with $a = e^{-j\Omega}$, $L_1 = -L$, and $L_2 = L$, the closed form expression for $X(\Omega)$ is

$$X(\Omega) = \frac{e^{j\Omega L} - e^{-j\Omega(L+1)}}{1 - e^{-j\Omega}} \tag{3.85}$$

In order to get symmetric complex exponentials in Eqn. (3.85) we will multiply both the numerator and the denominator of the fraction on the right side of the equal sign with $e^{j\Omega/2}$. The result is

$$X(\Omega) = \frac{e^{j\Omega(L+1/2)} - e^{-j\Omega(L+1/2)}}{e^{j\Omega/2} - e^{-j\Omega/2}}$$

which, using Euler's formula, can be simplified to

$$X(\Omega) = \frac{\sin\left(\dfrac{\Omega}{2}(2L+1)\right)}{\sin\left(\dfrac{\Omega}{2}\right)} \tag{3.86}$$

The value of the transform at $\Omega = 0$ must be resolved through the use of L'Hospital's rule:

$$X(0) = 2L + 1$$

The transform $X(\Omega)$ is graphed in Figs. 3.23–3.25 for $L = 3, 4, 5$.

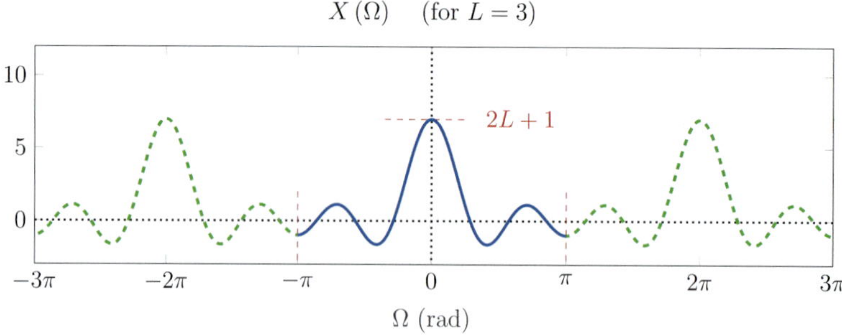

Figure 3.23 – The transform of the pulse signal $x[n]$ of Example 3.12 for $L = 3$.

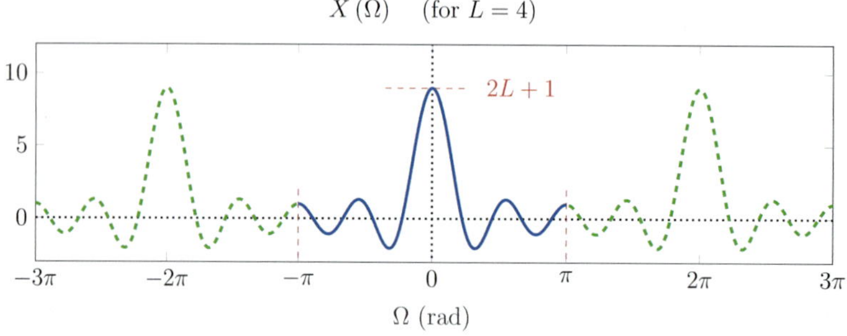

Figure 3.24 – The transform of the pulse signal $x[n]$ of Example 3.12 for $L = 5$.

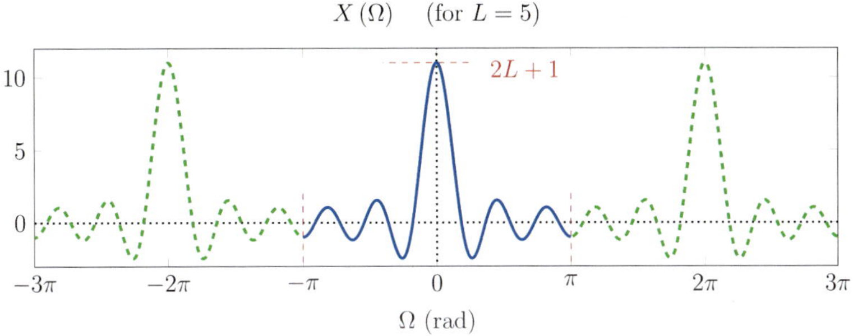

Figure 3.25 – The transform of the pulse signal $x[n]$ of Example 3.12 for $L=7$.

Software resources: exdt_3_12a.m , exdt_3_12b.m , exdt_3_12c.m

Interactive App: DTFT of discrete-time pulse

The interactive app in `appDTFT3.m` is based on computing the DTFT of a discrete-time pulse explored in Example 3.12. The pulse with $2L+1$ unit-amplitude samples centered around $n=0$ leads to the DTFT shown in Figs. 3.23 through 3.25.

a. Set $L=3$ to match Fig. 3.23. Observe the peak magnitude of the spectrum as well as the locations of zero crossings.

b. Set $L=4$ to match Fig. 3.24. Observe how a wider pulse leads to a narrower spectrum. How do the zero crossings of the spectrum relate to the width of the pulse?

c. Repeat with $L=5$ to match Fig. 3.25.

Software resource: `appDTFT3.m`

Example 3.13: Inverse DTFT of rectangular spectrum

Determine the inverse DTFT of the transform $X(\Omega)$ defined in the angular frequency range $-\pi<\Omega<\pi$ by

$$X(\Omega)=\begin{cases}1\,, & -\Omega_c<\Omega<\Omega_c\\ 0\,, & \text{otherwise}\end{cases}$$

Solution: To be a valid transform, $X(\Omega)$ must be 2π-periodic, therefore we need $X(\Omega)=X(\Omega+2\pi)$. The resulting transform is shown in Fig. 3.26. The inverse $x[n]$ is found by application of the DTFT synthesis equation given by Eqn. (3.68).

$$x[n]=\frac{1}{2\pi}\int_{-\Omega_c}^{\Omega_c}(1)\,e^{j\Omega n}\,d\Omega=\frac{\sin(\Omega_c n)}{\pi n} \tag{3.87}$$

It should be noted that the expression found in Eqn. (3.87) for the signal $x[n]$ is for all n, therefore, $x[n]$ is non-causal. For convenience let us use the normalized frequency F_c related to the angular frequency Ω_c by

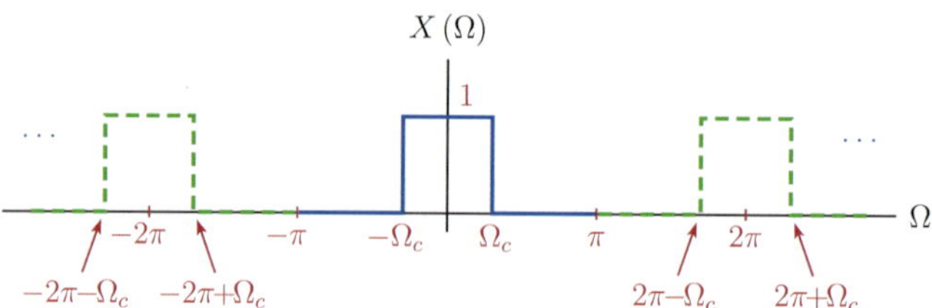

Figure 3.26 – The transform $X(\Omega)$ for Example 3.13.

$$\Omega_c = 2\pi F_c$$

and the signal $x[n]$ becomes

$$x[n] = 2F_c \operatorname{sinc}(2F_c n) \tag{3.88}$$

The result is shown in Fig. 3.27 for the case $F_c = 1/9$. Zero crossings of the sinc function in Eqn. (3.88) are spaced $1/(2F_c)$ apart which has the non-integer value of 4.5 in this case. Therefore, the envelope of the signal $x[n]$ crosses the axis at the midpoint of the samples for $n = 4$ and $n = 5$.

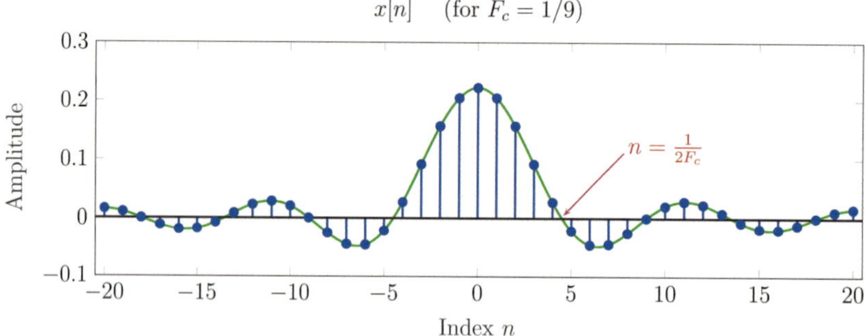

Figure 3.27 – The signal $x[n]$ for Example 3.13.

Software resource: exdt_3_13.m

Interactive App: Inverse DTFT of rectangular spectrum

The interactive app in appDTFT4.m is based on Example 3.13 where the inverse DTFT of the rectangular spectrum shown in Fig. 3.26 was determined using the DTFT synthesis equation. The app displays graphs of the spectrum and the corresponding time-domain signal. The normalized cutoff frequency F_c of the spectrum may be varied through the use of the slider control, and its effects on the signal $x[n]$ may be observed. Recall that $\Omega_c = 2\pi F_c$.

a. Gradually lower the cutoff frequency F_c until it reaches $F_c = 0.02$. Notice how the spectrum starts looking like a periodic series of impulses. Observe the changes in the signal $x[n]$. Comment on what the signal $x[n]$ seems to be approaching.

b. Now gradually increase the cutoff frequency F_c until it reaches $F_c = 0.48$. Notice how the spectrum starts looking like a constant. Observe the changes in the signal $x[n]$. Comment on what the signal $x[n]$ seems to be approaching in this case.

Software resource: `appDTFT4.m`

Example 3.14: Inverse DTFT of the unit impulse function

Find the signal the DTFT of which is $X(\Omega) = \delta(\Omega)$ in the range $-\pi < \Omega < \pi$.

Solution: To be a valid DTFT, the transform $X(\Omega)$ must be 2π-periodic. Therefore, the full expression for $X(\Omega)$ must be

$$X(\Omega) = \sum_{m=-\infty}^{\infty} \delta(\Omega - 2\pi m)$$

as shown in Fig. 3.28.

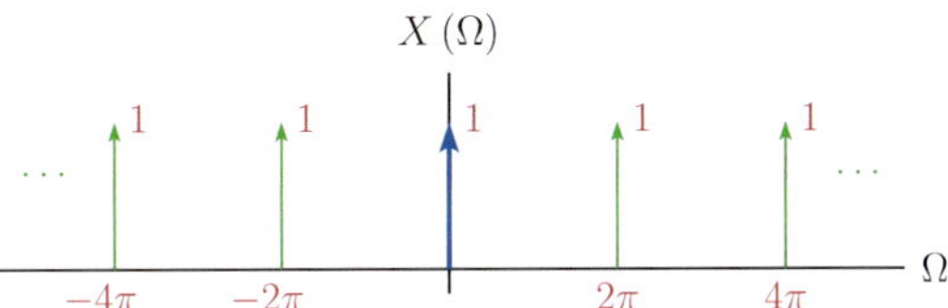

Figure 3.28 – The transform $X(\Omega)$ for Example 3.14.

The inverse transform is

$$x[n] = \frac{1}{2\pi} \int_{-\pi}^{\pi} \delta(\Omega)\, e^{j\Omega n}\, d\Omega$$

Using the sifting property of the impulse function we get

$$x[n] = \frac{1}{2\pi}, \quad \text{all } n$$

Thus we have the DTFT pair

$$\frac{1}{2\pi} \overset{\mathcal{F}}{\longleftrightarrow} \sum_{m=-\infty}^{\infty} \delta(\Omega - 2\pi m) \tag{3.89}$$

An important observation is in order: A signal that has constant amplitude for all index values is a power signal; it is neither absolute summable nor square summable. Strictly speaking, its DTFT does not converge. The transform relationship found in Eqn. (3.89) is a compromise made possible by our willingness to allow the use of singularity functions in $X(\Omega)$. Nevertheless, it is a useful relationship since it can be used in solving problems in the frequency domain. Multiplying both sides of the relationship in Eqn. (3.89) by 2π results in

$$\mathcal{F}\{1\} = 2\pi \sum_{m=-\infty}^{\infty} \delta(\Omega - 2\pi m) \tag{3.90}$$

This relationship is fundamental, and will be explored further in the next example.

Example 3.15: Inverse DTFT of unit impulse function revisited

Consider again the DTFT transform pair found in Eqn. (3.90) of Example 3.14. The unit-amplitude signal $x[n] = 1$ can be thought of as the limit of the rectangular pulse signal explored in Example 3.12. Let

$$x[n] = 1\,,\ \text{all } n \quad \text{and} \quad w[n] = \begin{cases} 1\,, & -L \le n \le L \\ 0\,, & \text{otherwise} \end{cases}$$

so that

$$x[n] = \lim_{L\to\infty} \big[w[n]\big]$$

Adapting from Example 3.12, the transform of $w[n]$ is

$$W(\Omega) = \frac{\sin\left(\dfrac{\Omega}{2}(2L+1)\right)}{\sin\left(\dfrac{\Omega}{2}\right)}$$

and the transform of $x[n]$ was found in Example 3.14 to be

$$X(\Omega) = 2\pi \sum_{m=-\infty}^{\infty} \delta(\Omega - 2\pi m)$$

Intuitively we would expect $W(\Omega)$ to resemble $X(\Omega)$ more and more closely as L is increased. Figs. 3.29–3.31 illustrate shows the transform $W(\Omega)$ for $L = 10, 20, 50$. Observe the transition from $W(\Omega)$ to $X(\Omega)$ as L is increased.

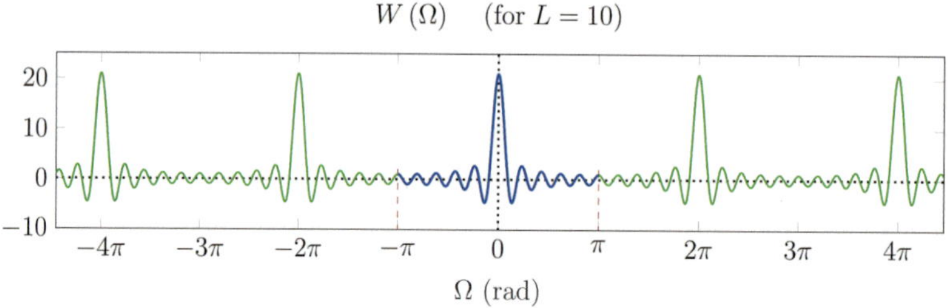

Figure 3.29 – The transform $W(\Omega)$ of Example 3.15 for $L = 10$.

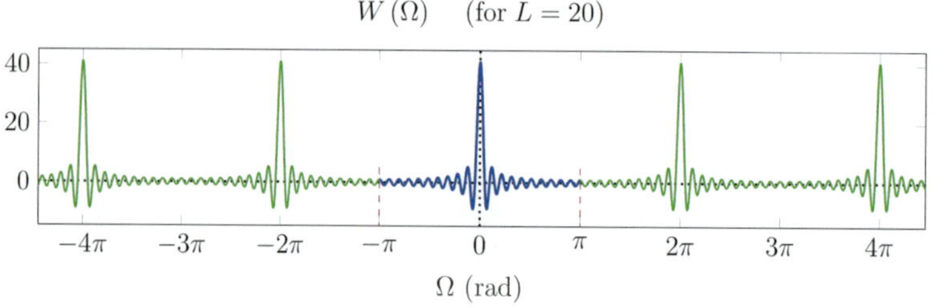

Figure 3.30 – The transform $W(\Omega)$ of Example 3.15 for $L = 20$.

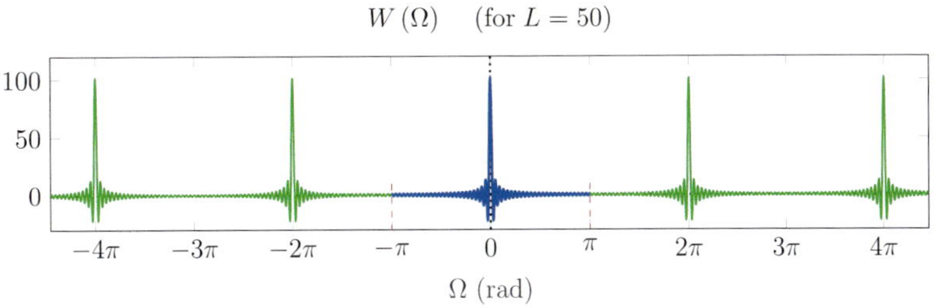

Figure 3.31 – The transform $W(\Omega)$ of Example 3.15 for $L = 50$.

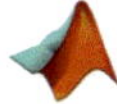

Software resources: exdt_3_15a.m , exdt_3_15b.m , exdt_3_15c.m

Interactive App: Inverse DTFT of the unit impulse function

The interactive app in `appDTFT5.m` is based on Examples 3.14 and 3.15. The signal $w[n]$ is a discrete-time pulse with $2L+1$ unit-amplitude samples centered around $n = 0$. The parameter L may be varied. The DTFT spectrum $W(\Omega)$ has a peak magnitude of $2L+1$ (see Example 3.12). In order to avoid scaling issues as L is increased, we graph $W(\Omega)\,/\,(2L+1)$ instead of $W(\Omega)$. As the value of L is increased, the signal $w[n]$ becomes more and more similar to the constant amplitude signal of Example 3.14, and its spectrum approaches the spectrum shown in Fig. 3.28.

a. Set $L = 10$. Observe the shape of the spectrum, and compare to Fig. 3.29. Observe the peak magnitude of the spectrum as well as the locations of zero crossings.

b. Set $L = 20$ to match Fig. 3.30.

c. Set $L = 50$ to match Fig. 3.31.

Software resource: `appDTFT5.m`

3.3.5 Properties of the DTFT

In this section we will explore some of the fundamental properties of the DTFT. As in the case of the Fourier transform for continuous-time signals, careful use of DTFT properties simplifies the solution of many types of problems.

Periodicity

DTFT is periodic:

$$X(\Omega + 2\pi r) = X(\Omega)\,, \quad \text{all integer } r \tag{3.91}$$

Periodicity of the DTFT is a direct consequence of the analysis equation given by Eqn. (3.69).

Linearity

DTFT is a linear transform:

For two transform pairs

$$x_1[n] \stackrel{\mathcal{F}}{\longleftrightarrow} X_1(\Omega)$$

$$x_2[n] \stackrel{\mathcal{F}}{\longleftrightarrow} X_2(\Omega)$$

and two arbitrary constants α_1 and α_2 it can be shown that

$$\alpha_1\, x_1[n] + \alpha_2\, x_2[n] \stackrel{\mathcal{F}}{\longleftrightarrow} \alpha_1\, X_1(\Omega) + \alpha_2\, X_2(\Omega) \tag{3.92}$$

Proof of linearity of the DTFT:

Use of the forward transform given by Eqn. (3.69) with the signal $(\alpha_1\, x_1[n] + \alpha_2\, x_2[n])$ yields

$$\begin{aligned}\mathcal{F}\left\{\alpha_1\, x_1[n] + \alpha_2\, x_2[n]\right\} &= \sum_{n=-\infty}^{\infty} \left(\alpha_1\, x_1[n] + \alpha_2\, x_2[n]\right)\, e^{-j\Omega n}\\ &= \sum_{n=-\infty}^{\infty} \alpha_1\, x_1[n]\, e^{-j\Omega n} + \sum_{n=-\infty}^{\infty} \alpha_2\, x_2[n]\, e^{-j\Omega n}\\ &= \alpha_1 \sum_{n=-\infty}^{\infty} x_1[n]\, e^{-j\Omega n} + \alpha_2 \sum_{n=-\infty}^{\infty} x_2[n]\, e^{-j\Omega n}\\ &= \alpha_1\, \mathcal{F}\left\{x_1[n]\right\} + \alpha_2\, \mathcal{F}\left\{x_2[n]\right\}\end{aligned} \tag{3.93}$$

Time shifting

Time shifting property of the DTFT:

For a transform pair

$$x[n] \stackrel{\mathcal{F}}{\longleftrightarrow} X(\Omega)$$

it can be shown that

$$x[n-m] \stackrel{\mathcal{F}}{\longleftrightarrow} X(\Omega)\, e^{-j\Omega m} \tag{3.94}$$

The consequence of shifting, or delaying, a signal in time is multiplication of its DTFT by a complex exponential function of angular frequency.

Proof of Eqn. (3.94):

Applying the forward transform in Eqn. (3.69) to the signal $x[n-m]$ we obtain

$$\mathcal{F}\left\{x[n-m]\right\} = \sum_{n=-\infty}^{\infty} x[n-m]\, e^{-j\Omega n} \tag{3.95}$$

Let us apply the variable change $k = n - m$ to the summation on the right side of Eqn. (3.95) so that

$$\mathcal{F}\left\{x[n-m]\right\} = \sum_{k=-\infty}^{\infty} x(k)\, e^{-j\Omega(k+m)} \tag{3.96}$$

The exponential function in the summation of Eqn. (3.96) can be written as the product of two exponential functions to obtain

$$\begin{aligned}\mathcal{F}\left\{x[n-m]\right\} &= \sum_{k=-\infty}^{\infty} x[k]\, e^{-j\Omega k}\, e^{-j\Omega m} \\ &= e^{-j\Omega m} \sum_{k=-\infty}^{\infty} x[k]\, e^{-j\Omega k} \\ &= e^{-j\Omega m}\, X(\Omega)\end{aligned} \tag{3.97}$$

Example 3.16: DTFT of a time shifted signal

Determine the DTFT of the signal $x[n] = e^{-\alpha(n-1)}\, u[n-1]$ as shown in Fig. 3.32. (Assume $|\alpha| < 1$).

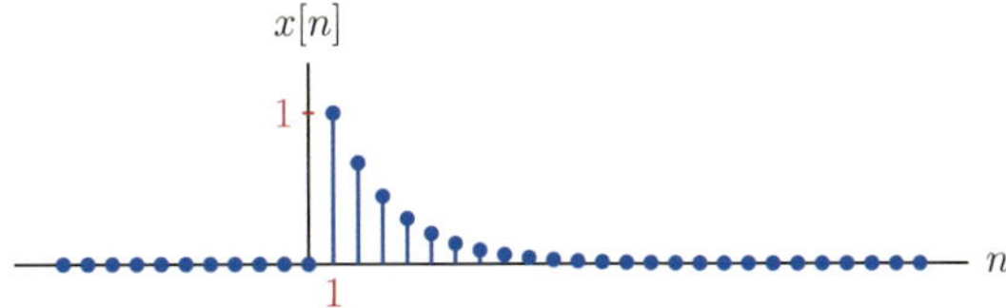

Figure 3.32 – The signal $x[n]$ for Example 3.16.

Solution: The transform of a right-sided exponential signal was determined in Example 3.10. As a result we have the transform pair

$$e^{-\alpha n}\, u[n] \overset{\mathcal{F}}{\longleftrightarrow} \frac{1}{1-\alpha\, e^{-j\Omega}}$$

Applying the time shifting property of the DTFT given by Eqn. (3.94) with $m = 1$ leads to the result

$$X(\Omega) = \mathcal{F}\left\{e^{-\alpha(n-1)}\, u[n-1]\right\} = \frac{e^{-j\Omega}}{1-\alpha\, e^{-j\Omega}}$$

Since $\left|e^{-j\Omega}\right| = 1$, the magnitude of $X(\Omega)$ is the same as that obtained in Example 3.10. Time shifting a signal does not change the magnitude of its transform.

$$\left|X(\Omega)\right| = \frac{1}{\sqrt{1+\alpha^2 - 2\alpha\cos(\Omega)}}$$

The magnitude $|X(\Omega)|$ is shown in Fig. 3.33 for $\alpha = 0.4$. The phase of $X(\Omega)$ is found by first determining the phase angles of the numerator and the denominator, and then subtracting the latter from the former. In Example 3.10 the numerator of the transform was a constant equal to unity; in this case it is a complex exponential. Therefore

$$\begin{aligned}\measuredangle X(\Omega) &= \measuredangle\left(e^{-j\Omega}\right) - \measuredangle\left(1 - \alpha\, e^{-j\Omega}\right)\\ &= -\Omega - \tan^{-1}\left[\frac{\alpha \sin(\Omega)}{1 - \alpha\cos(\Omega)}\right]\end{aligned} \tag{3.98}$$

Thus, the phase of the transform differs from that obtained in Example 3.10 by $-\Omega$, a ramp with a slope of -1. The phase, as computed by Eqn. (3.98), is shown in Fig. 3.34. In graphing the phase of the transform it is customary to fit phase values to be the range $(-\pi, \pi)$ radians. Fig. 3.35 depicts the phase of the transform normalized through phase wrapping.

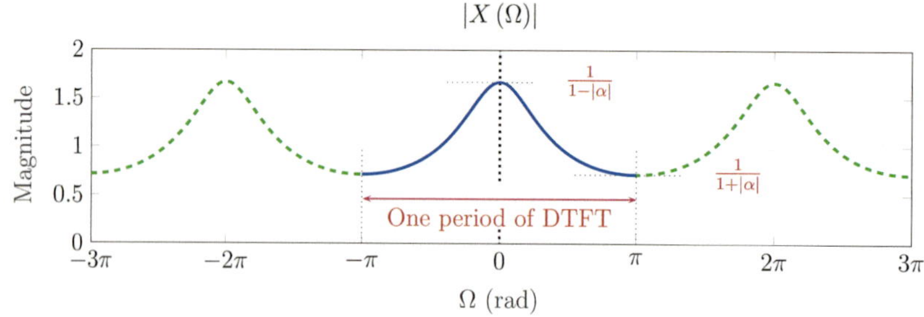

Figure 3.33 – The magnitude of the DTFT for the signal $x[n]$ of Example 3.16 with $\alpha = 0.4$.

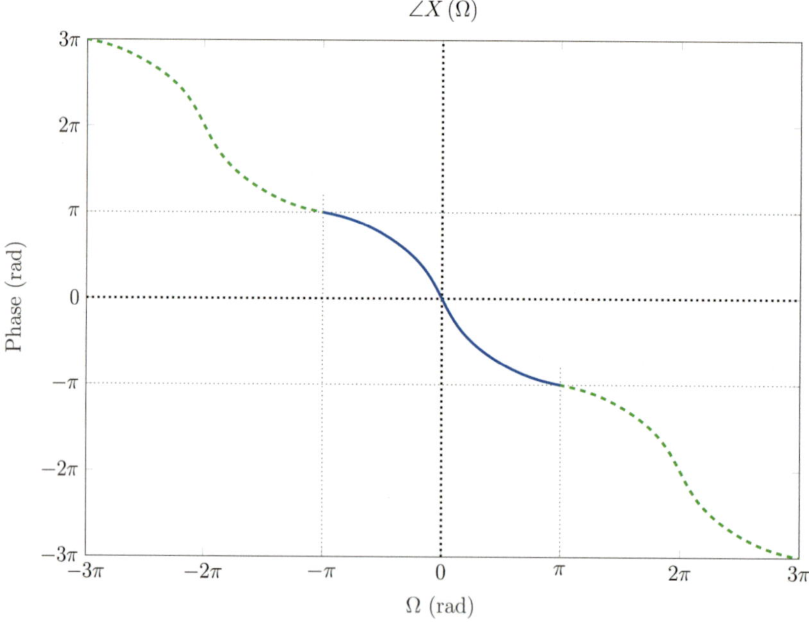

Figure 3.34 – The phase of the DTFT for the signal $x[n]$ of Example 3.16 with $\alpha = 0.4$.

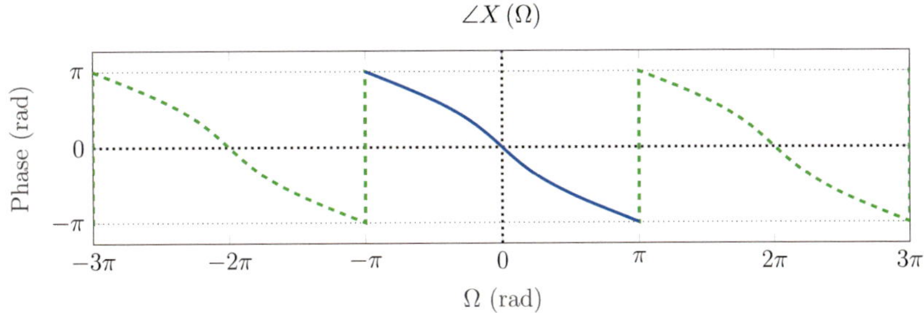

Figure 3.35 – The phase of the DTFT for the signal $x[n]$ of Example 3.16 with $\alpha = 0.4$. Phase wrapping was applied to keep values in the interval $(-\pi, \pi)$.

Software resource: exdt_3_16.m

Interactive App: DTFT of a time-shifted signal

The interactive app in appDTFT6.m is based on Example 3.16. The signal

$$x[n] = e^{-\alpha(n-m)}\, u[n-m]$$

is shown along with the magnitude and the phase of its transform $X(\Omega)$. The time delay m can be varied, and its effect on the spectrum can be observed.

a. Set $\alpha = 0.4$ and $m = 0$. Vary the value of alpha while keeping m fixed. Observe its effects on the magnitude and the phase characteristics.

b. In Example 3.16 we have used $m = 1$ and obtained the expression given by Eqn. (3.98) for the phase characteristic $\measuredangle X(\Omega)$. When the signal is delayed by m samples the corresponding phase is

$$\measuredangle X(\Omega) = -m\Omega - \tan^{-1}\left[\frac{\alpha \sin(\Omega)}{1-\alpha\cos(\Omega)}\right] \tag{3.99}$$

Set $\alpha = 0.4$, and $m = 0$. Increase the delay amount to $m = 1, 2, 3, \ldots$. Observe the changes in the phase characteristic, and try to explain them in light of Eqn. (3.99).

Software resource: appDTFT6.m

Time reversal

For a transform pair

$$x[n] \overset{\mathcal{F}}{\longleftrightarrow} X(\Omega)$$

it can be shown that

$$x[-n] \overset{\mathcal{F}}{\longleftrightarrow} X(-\Omega) \tag{3.100}$$

Time reversal of the signal causes angular frequency reversal of the transform. This property will be useful when we consider symmetry properties of the DTFT.

Proof of Eqn. (3.100):

Direct application of the forward transform in Eqn. (3.69) to the signal $x[-n]$ yields

$$\mathcal{F}\{x[-n]\} = \sum_{n=-\infty}^{\infty} x[-n]\, e^{-j\Omega n} \tag{3.101}$$

Let us apply the variable change $k = -n$ to the summation on the right side of Eqn. (3.101) to obtain

$$\mathcal{F}\{x[-n]\} = \sum_{k=\infty}^{-\infty} x[k]\, e^{j\Omega k} \tag{3.102}$$

which can be written as

$$\mathcal{F}\{x[-n]\} = \sum_{k=-\infty}^{\infty} x[k]\, e^{-j(-\Omega)k} = X(-\Omega) \tag{3.103}$$

Example 3.17: DTFT of two-sided exponential signal

Determine the DTFT of the signal $x[n] = \alpha^{|n|}$ with $|\alpha| < 1$.

Solution: The signal $x[n]$ can be written as

$$x[n] = \begin{cases} \alpha^{n}, & n \ge 0 \\ \alpha^{-n}, & n < 0 \end{cases}$$

It can also be expressed as

$$x[n] = x_1[n] + x_2[n]$$

where $x_1[n]$ is a causal signal and $x_2[n]$ is an anti-causal signal defined as

$$x_1[n] = \alpha^n\, u[n] \quad \text{and} \quad x_2[n] = \alpha^{-n}\, u[-n-1]$$

This is illustrated in Fig. 3.36. Based on the linearity property of the DTFT, the transform of $x[n]$ is the sum of the transforms of $x_1[n]$ and $x_2[n]$.

$$X(\Omega) = X_1(\Omega) + X_2(\Omega)$$

For the transform of $x_1 = \alpha^n\, u[n]$ we will make use of the result obtained in Example 3.10.

$$X_1(\Omega) = \frac{1}{1 - \alpha\, e^{-j\Omega}}$$

Time shifting and time-reversal properties of the DTFT will be used for obtaining the transform of $x_2[n]$. Let a new signal $g[n]$ be defined as a scaled and time-shifted version of $x_1[n]$:

$$g[n] = \alpha\, x_1[n-1] = \alpha\left(\alpha^{n-1}\, u[n-1]\right) = \alpha^n\, u[n-1]$$

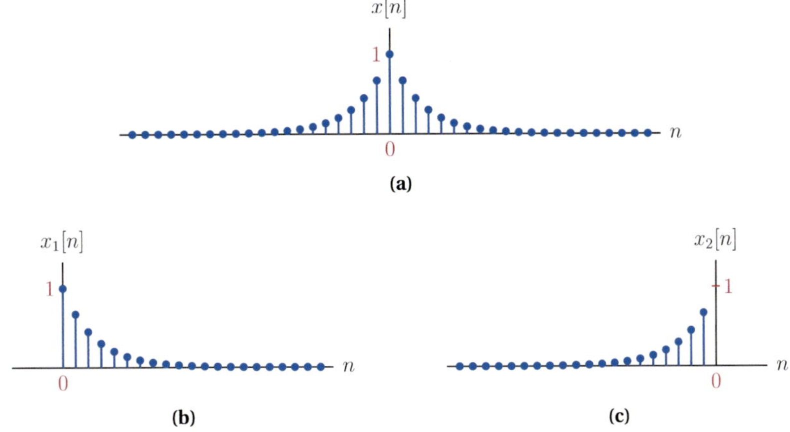

Figure 3.36 – **(a)** Two-sided exponential signal of Example 3.17, **(b)** its causal component, and **(c)** its anti-causal component.

Using the time shifting property, the transform of $g[n]$ is

$$G(\Omega) = \alpha\, X(\Omega)\, e^{-j\Omega} = \frac{\alpha\, e^{-j\Omega}}{1-\alpha\, e^{-j\Omega}}$$

The signal $x_2[n]$ is a time-reversed version of $g[n]$, that is,

$$x_2[n] = g[-n]$$

Applying the time-reversal property, the transform of $x_2[n]$ is found as

$$X_2(\Omega) = G(-\Omega) = \frac{\alpha\, e^{j\Omega}}{1-\alpha\, e^{j\Omega}}$$

Finally, $X(\Omega)$ is found by adding the two transforms:

$$\begin{aligned} X(\Omega) &= \frac{1}{1-\alpha\, e^{-j\Omega}} + \frac{\alpha\, e^{j\Omega}}{1-\alpha\, e^{j\Omega}} \\ &= \frac{1-\alpha^2}{1-2\alpha\cos(\Omega)+\alpha^2} \end{aligned} \tag{3.104}$$

The transform $X(\Omega)$ obtained in Eqn. (3.104) is real-valued, and is shown in Fig. 3.37 for $\alpha = 0.4$.

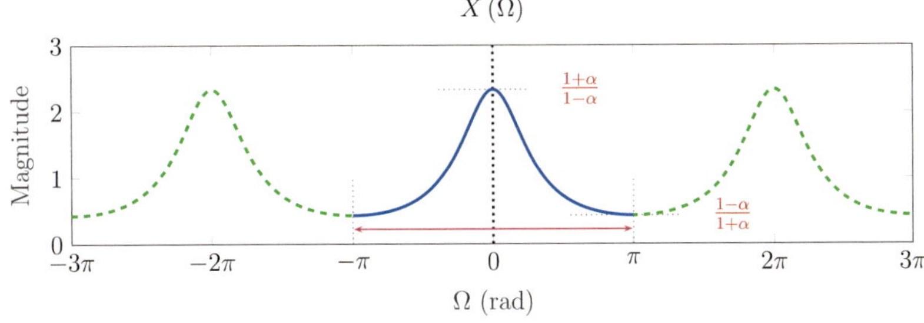

Figure 3.37 – The DTFT of the signal $x[n]$ of Example 3.17 for $a = 0.4$.

Software resource: `exdt_3_17.m`

Interactive App: DTFT of two-sided exponential signal

The interactive app in `appDTFT7.m` is based on Example 3.17. The signal $x[n]$ of Example 3.17 is graphed along with its DTFT spectrum $X(\Omega)$ which is purely real due to the even symmetry of the signal. Parameter α can be varied in order to observe its effect on the spectrum.

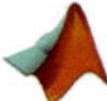

Software resource: `appDTFT7.m`

Conjugation property

Conjugation property of the DTFT:

For a transform pair

$$x[n] \stackrel{\mathcal{F}}{\longleftrightarrow} X(\Omega)$$

it can be shown that

$$x^*[n] \stackrel{\mathcal{F}}{\longleftrightarrow} X^*(-\Omega) \tag{3.105}$$

Conjugation of the signal causes both conjugation and angular frequency reversal of the transform. This property will also be useful when we consider symmetry properties of the DTFT.

Proof of Eqn. (3.105):

Using the signal $x^*[n]$ in the forward transform equation results in

$$\mathcal{F}\left\{x^*[n]\right\} = \sum_{n=-\infty}^{\infty} x^*[n]\, e^{-j\Omega n} \tag{3.106}$$

Writing the right side of Eqn. (3.106) in a slightly modified form we obtain

$$\mathcal{F}\left\{x^*[n]\right\} = \left[\sum_{n=\infty}^{-\infty} x[n]\, e^{j\Omega n}\right]^* = X^*(-\Omega) \tag{3.107}$$

Symmetry of the DTFT

If the signal $x[n]$ is real-valued, it can be shown that its DTFT is conjugate symmetric. Conversely, if the signal $x[n]$ is purely imaginary, its transform is conjugate antisymmetric.

Symmetry of the DTFT:

$$x[n]\text{: Real, } \operatorname{Im}\{x[n]\}=0 \quad \text{implies that} \quad X^*(\Omega)=X(-\Omega) \tag{3.108}$$

$$x[n]\text{: Imag, } \operatorname{Re}\{x[n]\}=0 \quad \text{implies that} \quad X^*(\Omega)=-X(-\Omega) \tag{3.109}$$

Proof of symmetry properties:

1. Real $x(t)$:
   Any real-valued signal is equal to its own conjugate, therefore we have

$$x^*[n]=x[n] \tag{3.110}$$

   Taking the transform of each side of Eqn. (3.110) and using the conjugation property stated in Eqn. (3.105) we obtain

$$X^*(-\Omega)=X(\Omega) \tag{3.111}$$

   which is equivalent to Eqn. (3.108).
2. Imaginary $x(t)$:
   A purely imaginary signal is equal to the negative of its conjugate, i.e.,

$$x^*[n]=-x[n] \tag{3.112}$$

   Taking the transform of each side of Eqn. (3.112) and using the conjugation property given by Eqn. (3.105) yields

$$X^*(-\Omega)=-X(\Omega) \tag{3.113}$$

   This is equivalent to Eqn. (3.109). Therefore the transform is conjugate antisymmetric.

Cartesian and polar forms of the transform

A complex transform $X(\Omega)$ can be written in polar form as

$$X(\Omega)=\left|X(\Omega)\right| e^{j\Theta(\Omega)} \tag{3.114}$$

and in Cartesian form as

$$X(\Omega)=X_r(\Omega)+jX_i(\Omega) \tag{3.115}$$

Case 1: If $X(\Omega)$ is conjugate symmetric, the relationship in Eqn. (3.108) can be written as

$$\left|X(-\Omega)\right| e^{j\Theta(-\Omega)}=\left|X(\Omega)\right| e^{-j\Theta(\Omega)} \tag{3.116}$$

using the polar form of the transform, and

$$X_r(-\Omega)+jX_i(-\Omega)=X_r(\Omega)-jX_i(\Omega) \tag{3.117}$$

using its Cartesian form. The consequences of Eqns. (3.116) and (3.117) can be obtained by equating the magnitudes and the phases on both sides of Eqn. (3.116) and by equating real and imaginary part on both sides of Eqn. (3.117). The results are summarized below:

Conjugate symmetric transform: $X(-\Omega) = X^*(\Omega)$

$$\text{Magnitude:} \quad |X(-\Omega)| = |X(\Omega)| \tag{3.118}$$
$$\text{Phase:} \quad \Theta(-\Omega) = -\Theta(\Omega) \tag{3.119}$$
$$\text{Real part:} \quad X_r(-\Omega) = X_r(\Omega) \tag{3.120}$$
$$\text{Imag. part:} \quad X_i(-\Omega) = -X_i(\Omega) \tag{3.121}$$

The transform of a real-valued $x[n]$ is conjugate symmetric. For such a transform, the magnitude is an even function of Ω and the phase is an odd function. Furthermore, the real part of the transform has even symmetry and its imaginary part has odd symmetry.

Case 2: If $X(\Omega)$ is conjugate antisymmetric, the polar form of $X(\Omega)$ has the property

$$|X(-\Omega)|\, e^{j\Theta(-\Omega)} = -|X(\Omega)|\, e^{-j\Theta(\Omega)} \tag{3.122}$$

The negative sign on the right side of Eqn. (3.122) needs to be incorporated into the phase since we could not write $|X(-\Omega)| = -|X(\Omega)|$ (recall that magnitude needs to be a non-negative function for all Ω). Using $e^{\mp j\pi} = -1$, Eqn. (3.122) can be written as

$$\begin{aligned} |X(-\Omega)|\, e^{j\Theta(-\Omega)} &= |X(\Omega)|\, e^{-j\Theta(\Omega)}\, e^{\mp j\pi} \\ &= |X(\Omega)|\, e^{-j[\Theta(\Omega)\mp\pi]} \end{aligned} \tag{3.123}$$

Conjugate antisymmetry property of the transform can also be expressed in Cartesian form as

$$X_r(-\Omega) + jX_i(-\Omega) = -X_r(\Omega) + jX_i(\Omega) \tag{3.124}$$

The consequences of Eqns. (3.123) and (3.124) are given below.

Conjugate antisymmetric transform: $X(-\Omega) = -X^*(\Omega)$

$$\text{Magnitude:} \quad |X(-\Omega)| = |X(\Omega)| \tag{3.125}$$
$$\text{Phase:} \quad \Theta(-\Omega) = -\Theta(\Omega) \mp \pi \tag{3.126}$$
$$\text{Real part:} \quad X_r(-\Omega) = -X_r(\Omega) \tag{3.127}$$
$$\text{Imag. part:} \quad X_i(-\Omega) = X_i(\Omega) \tag{3.128}$$

We know that a purely-imaginary signal leads to a DTFT that is conjugate antisymmetric. For such a transform the magnitude is still an even function of Ω as suggested by Eqn. (3.125). The phase is neither even nor odd. The real part is an odd function of Ω and the imaginary part is an even function.

Transforms of even and odd signals

If the real-valued signal $x[n]$ is an even function of time, the resulting transform $X(\Omega)$ is real-valued for all Ω.

Real-valued signal $x[n]$ with even symmetry:

$$x[-n] = x[n]\,, \text{ all } n \quad \text{implies that} \quad \text{Im}\left\{X(\Omega)\right\} = 0\,, \text{ all } \Omega \tag{3.129}$$

Proof of Eqn. (3.129):

Using the time reversal property of the DTFT, Eqn. (3.129) implies that

$$X(-\Omega) = X(\Omega) \tag{3.130}$$

Furthermore, since $x[n]$ is real-valued, the transform is conjugate symmetric, therefore

$$X^*(\Omega) = X(-\Omega) \tag{3.131}$$

Combining Eqns. (3.130) and (3.131) we reach the conclusion

$$X^*(\Omega) = X(\Omega) \tag{3.132}$$

Therefore, $X(\Omega)$ must be real.

Conversely it can also be proven that, if the real-valued signal $x[n]$ has odd-symmetry, the resulting transform is purely imaginary. This can be stated mathematically as follows:

Real-valued signal $x[n]$ with odd symmetry:

$$x[-n] = -x[n]\,, \text{ all } n \quad \text{implies that} \quad \text{Re}\left\{X(\Omega)\right\} = 0\,, \text{ all } \Omega \tag{3.133}$$

Conjugating a purely imaginary transform is equivalent to negating it, so an alternative method of expressing the relationship in Eqn. (3.133) is

$$X^*(\Omega) = -X(\Omega) \tag{3.134}$$

Proof of Eqn. (3.134) is similar to the procedure used above for proving Eqn. (3.129) (see Problem 3.20 at the end of this chapter).

Frequency shifting

Frequency shifting property of the DTFT:

For a transform pair

$$x[n] \stackrel{\mathcal{F}}{\longleftrightarrow} X(\Omega)$$

it can be shown that

$$x[n]\,e^{j\Omega_0 n} \stackrel{\mathcal{F}}{\longleftrightarrow} X(\Omega - \Omega_0) \tag{3.135}$$

Proof of Eqn. (3.135):

Applying the DTFT definition given by Eqn. (3.69) to the signal $x[n]\,e^{j\Omega_0 n}$ we obtain

$$\begin{aligned}\mathcal{F}\left\{x[n]\,e^{j\Omega_0 n}\right\} &= \sum_{n=-\infty}^{\infty} x[n]\,e^{j\Omega_0 n}\,e^{-j\Omega n}\\ &= \sum_{n=-\infty}^{\infty} x[n]\,e^{-j(\Omega-\Omega_0)n}\\ &= X(\Omega-\Omega_0)\end{aligned} \tag{3.136}$$

Modulation property

Modulation property of the DTFT:

For a transform pair

$$x[n] \overset{\mathcal{F}}{\longleftrightarrow} X(\Omega)$$

it can be shown that

$$x[n]\cos(\Omega_0 n) \overset{\mathcal{F}}{\longleftrightarrow} \frac{1}{2}\left[X(\Omega-\Omega_0)+X(\Omega+\Omega_0)\right] \tag{3.137}$$

and

$$x[n]\sin(\Omega_0 n) \overset{\mathcal{F}}{\longleftrightarrow} \frac{1}{2}\left[X(\Omega-\Omega_0)\,e^{-j\pi/2}+X(\Omega+\Omega_0)\,e^{j\pi/2}\right] \tag{3.138}$$

Modulation property is an interesting consequence of the frequency shifting property combined with the linearity of the Fourier transform. Multiplication of a signal by a cosine waveform causes its spectrum to be shifted in both directions by the angular frequency of the cosine waveform, and to be scaled by $\frac{1}{2}$. Multiplication of the signal by a sine waveform causes a similar effect with an added phase shift of $-\pi/2$ radians for positive frequencies and $\pi/2$ radians for negative frequencies.

Proofs of Eqns. (3.137) and (3.138):

Using Euler's formula, the left side of the relationship in Eqn. (3.137) can be written as

$$x[n]\cos(\Omega_0 n) = \frac{1}{2}\,x[n]\,e^{\Omega_0 n} + \frac{1}{2}\,x[n]\,e^{-\Omega_0 n} \tag{3.139}$$

The desired proof is obtained by applying the frequency shifting theorem to the terms on the right side of Eqn. (3.139). Using Eqn. (3.135) we obtain

$$x[n]\,e^{j\Omega_0 n} \overset{\mathcal{F}}{\longleftrightarrow} X(\Omega-\Omega_0) \tag{3.140}$$

and

$$x[n]\,e^{-j\Omega_0 n} \overset{\mathcal{F}}{\longleftrightarrow} X(\Omega+\Omega_0), \tag{3.141}$$

which could be used together in Eqn. (3.139) to arrive at the result in Eqn. (3.137). The proof of Eqn. (3.138) is similar, but requires one additional step. Again using Euler's formula, let us write the left side of Eqn. (3.138) as

$$x[n]\,\sin(\Omega_0 n) = \frac{1}{2j}\,x[n]\,e^{j\Omega_0 n} - \frac{1}{2j}\,x[n]\,e^{-j\Omega_0 n} \tag{3.142}$$

Realizing that

$$\frac{1}{j} = -j = e^{-j\pi/2} \quad \text{and} \quad \frac{-1}{j} = j = e^{j\pi/2}$$

Eqn. (3.142) can be rewritten as

$$x[n]\,\sin(\Omega_0 n) = \frac{1}{2}\,x[n]\,e^{j\Omega_0 n}\,e^{-j\pi/2} + \frac{1}{2}\,x[n]\,e^{-j\Omega_0 n}\,e^{j\pi/2} \tag{3.143}$$

The proof of Eqn. (3.138) can be completed by using Eqns. (3.140) and (3.141) on the right side of Eqn. (3.143).

Differentiation in the frequency domain

Differentiation of the DTFT in the frequency domain:

For a transform pair

$$x[n] \overset{\mathcal{F}}{\longleftrightarrow} X(\Omega)$$

it can be shown that

$$n\,x[n] \overset{\mathcal{F}}{\longleftrightarrow} j\,\frac{dX(\Omega)}{d\Omega} \tag{3.144}$$

and

$$n^m\,x[n] \overset{\mathcal{F}}{\longleftrightarrow} j^m\,\frac{d^m X(\Omega)}{d\Omega^m} \tag{3.145}$$

Proofs of Eqns. (3.144) and (3.145):

Differentiating both sides of Eqn. (3.69) with respect to Ω yields

$$\frac{dX(\Omega)}{d\Omega} = \frac{d}{d\Omega}\left[\sum_{n=-\infty}^{\infty} x[n]\,e^{-j\Omega n}\right]$$
$$\sum_{n=-\infty}^{\infty} -j\,n\,x[n]\,e^{-j\Omega n} \tag{3.146}$$

The summation on the right side of Eqn. (3.146) is the DTFT of the signal $-j\,n\,x[n]$. Multiplying both sides of Eqn. (3.146) by j results in the transform pair in Eqn. (3.144). Eqn. (3.145) is proven by repeated use of Eqn. (3.144).

Example 3.18: Use of differentiation in frequency property

Determine the DTFT of the signal $x[n] = n\,e^{-\alpha n}\,u[n]$ shown in Fig. 3.38. Assume $|\alpha| < 1$.

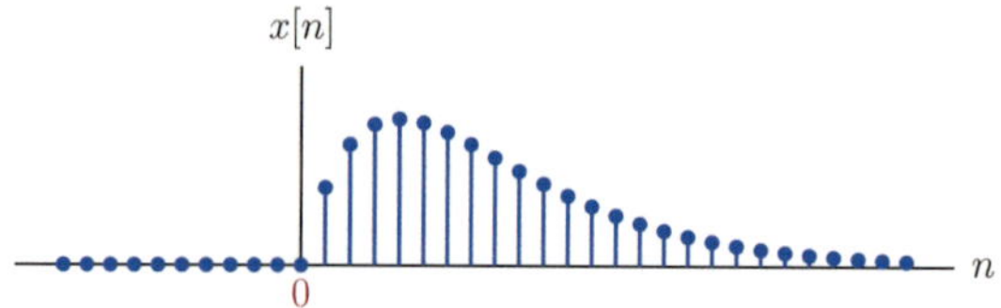

Figure 3.38 – The signal $x[n]$ for Example 3.18.

Solution: In Example 3.10 we have established the following transform pair:

$$e^{-\alpha n}\,u[n] \overset{\mathcal{F}}{\longleftrightarrow} \frac{1}{1-\alpha\,e^{-j\Omega}}$$

Applying the differentiation in frequency property of the DTFT leads to

$$n\,e^{-\alpha n}\,u[n] \overset{\mathcal{F}}{\longleftrightarrow} j\,\frac{d}{d\Omega}\left[\frac{1}{1-\alpha\,e^{-j\Omega}}\right]$$

Differentiation is carried out easily:

$$\frac{d}{d\Omega}\left[\frac{1}{1-\alpha\,e^{-j\Omega}}\right] = \frac{d}{d\Omega}\left[\left(1-\alpha\,e^{-j\Omega}\right)^{-1}\right]$$
$$= -\left(1-\alpha\,e^{-j\Omega}\right)^{-2}\left(j\alpha\,e^{-j\Omega}\right)$$

and the transform we seek is

$$X(\Omega) = \frac{\alpha\,e^{-j\Omega}}{\left(1-\alpha\,e^{-j\Omega}\right)^2}$$

The magnitude and the phase of the transform $X(\Omega)$ are shown in Figs. 3.39 and 3.40 for the case $\alpha = 0.4$.

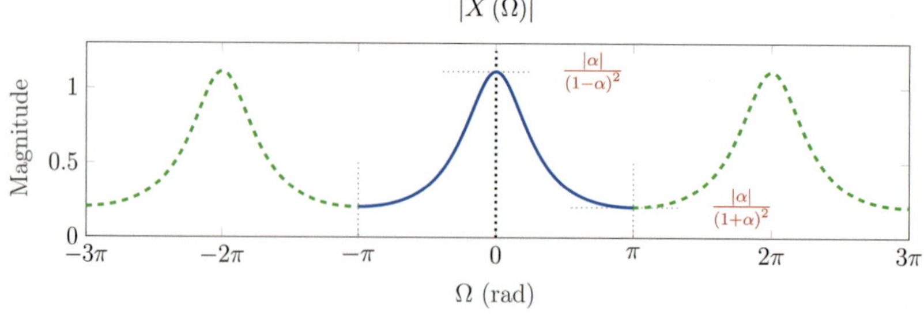

Figure 3.39 – The magnitude of the transform $X(\Omega)$ found in Example 3.18.

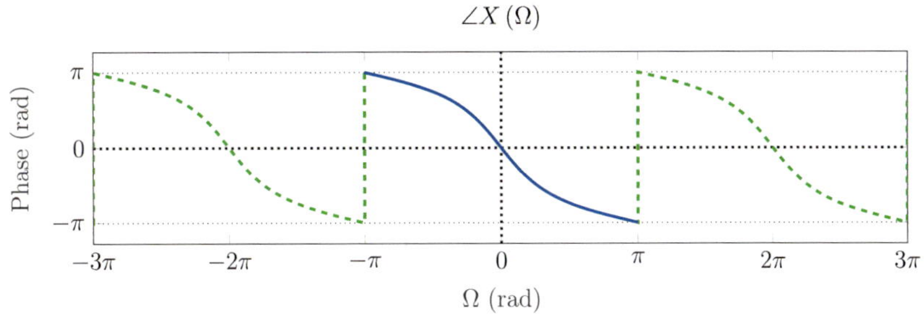

Figure 3.40 – The phase of the transform $X(\Omega)$ found in Example 3.18.

Software resource: exdt_3_18.m

Convolution property

Convolution property of the DTFT:

For two transform pairs

$$x_1[n] \overset{\mathcal{F}}{\longleftrightarrow} X_1(\Omega) \quad \text{and} \quad x_2[n] \overset{\mathcal{F}}{\longleftrightarrow} X_2(\Omega)$$

it can be shown that

$$x_1[n] * x_2[n] \overset{\mathcal{F}}{\longleftrightarrow} X_1(\Omega)\, X_2(\Omega) \tag{3.147}$$

Convolving two signals in the time domain corresponds to multiplying the corresponding transforms in the frequency domain.

Proof of Eqn. (3.147):

The convolution of signals $x_1[n]$ and $x_2[n]$ is given by

$$x_1[n] * x_2[n] = \sum_{k=-\infty}^{\infty} x_1[k]\, x_2[n-k] \tag{3.148}$$

Using Eqn. (3.148) in the DTFT definition yields

$$\mathcal{F}\left\{x_1[n] * x_2[n]\right\} = \sum_{n=-\infty}^{\infty} \left[\sum_{k=-\infty}^{\infty} x_1[k]\, x_2[n-k] \right] e^{-j\Omega n} \tag{3.149}$$

Changing the order of the two summations in Eqn. (3.149) and rearranging terms we obtain

$$\mathcal{F}\left\{x_1[n] * x_2[n]\right\} = \sum_{k=-\infty}^{\infty} x_1[k] \left[\sum_{n=-\infty}^{\infty} x_2[n-k]\, e^{-j\Omega n} \right] \tag{3.150}$$

In Eqn. (3.150) the expression in square brackets should be recognized as the transform of time shifted signal $x_2[n-k]$, and is equal to $X_2(\Omega)\, e^{-j\Omega k}$. Using this result in Eqn. (3.150) leads to

$$\begin{aligned}\mathscr{F}\left\{x_1[n] * x_2[n]\right\} =& \sum_{k=-\infty}^{\infty} x_1[k] \left[X_2(\Omega)\, e^{-j\Omega k}\right] \\ =& X_2(\Omega) \sum_{k=-\infty}^{\infty} x_1[k]\, e^{-j\Omega k} \\ =& X_1(\Omega)\, X_2(\Omega)\end{aligned} \tag{3.151}$$

Example 3.19: Convolution using the DTFT

Two signals are given as

$$h[n] = \left(\frac{2}{3}\right)^n u[n] \quad \text{and} \quad x[n] = \left(\frac{3}{4}\right)^n u[n]$$

Determine the convolution $y[n] = h[n] * x[n]$ of these two signals using the DTFT.

Solution: Transforms of $H(\Omega)$ and $X(\Omega)$ can easily be found using the result obtained in Example 3.10:

$$H(\Omega) = \frac{1}{1-\frac{2}{3}\,e^{-j\Omega}}, \qquad X(\Omega) = \frac{1}{1-\frac{3}{4}\,e^{-j\Omega}}$$

Using the convolution property, the transform of $y[n]$ is

$$\begin{aligned}Y(\Omega) =& H(\Omega)\, X(\Omega) \\ =& \frac{1}{\left(1-\frac{2}{3}\,e^{-j\Omega}\right)\left(1-\frac{3}{4}\,e^{-j\Omega}\right)}\end{aligned} \tag{3.152}$$

The transform $Y(\Omega)$ found in Eqn. (3.152) can be written in the form

$$Y(\Omega) = \frac{-8}{1-\frac{2}{3}\,e^{-j\Omega}} + \frac{9}{1-\frac{3}{4}\,e^{-j\Omega}}$$

The convolution result $y[n]$ is the inverse DTFT of $Y(\Omega)$ which can be obtained using the linearity property of the transform:

$$y[n] = -8\left(\frac{2}{3}\right)^n u[n] + 9\left(\frac{3}{4}\right)^n u[n]$$

Solution of problems of this type will be more practical through the use of z-transform in Chapter 5.

Software resource: exdt_3_19.m

Multiplication of two signals

Multiplication property of the DTFT:

For two transform pairs

$$x_1[n] \overset{\mathcal{F}}{\longleftrightarrow} X_1(\Omega) \quad \text{and} \quad x_2[n] \overset{\mathcal{F}}{\longleftrightarrow} X_2(\Omega) \tag{3.153}$$

it can be shown that

$$x_1[n]\,x_2[n] \overset{\mathcal{F}}{\longleftrightarrow} \frac{1}{2\pi}\int_{-\pi}^{\pi} X_1(\lambda)\,X_2(\Omega-\lambda)\,d\lambda \tag{3.154}$$

The DTFT of the product of two signals $x_1[n]$ and $x_2[n]$ is equal the 2π-periodic convolution of the individual transforms $X_1(\Omega)$ and $X_2(\Omega)$.

Proof of Eqn. (3.154):

Applying the DTFT definition to the product $x_1[n]\,x_2[n]$ leads to

$$\mathcal{F}\left\{x_1[n]\,x_2[n]\right\} = \sum_{n=-\infty}^{\infty} x_1[n]\,x_2[n]\,e^{-j\Omega n} \tag{3.155}$$

Using the DTFT synthesis equation given by Eqn. (3.68) we get

$$x_1[n] = \frac{1}{2\pi}\int_{-\pi}^{\pi} X_1(\lambda)\,e^{j\lambda n}\,d\lambda \tag{3.156}$$

Substituting Eqn. (3.156) into Eqn. (3.155)

$$\mathcal{F}\left\{x_1[n]\,x_2[n]\right\} = \sum_{n=-\infty}^{\infty}\left[\frac{1}{2\pi}\int_{-\pi}^{\pi} X_1(\lambda)\,e^{j\lambda n}\,d\lambda\right] x_2[n]\,e^{-j\Omega n} \tag{3.157}$$

Interchanging the order of summation and integration, and rearranging terms yields

$$\begin{aligned}\mathcal{F}\left\{x_1[n]\,x_2[n]\right\} &= \frac{1}{2\pi}\int_{-\pi}^{\pi} X_1(\lambda)\left[\sum_{n=-\infty}^{\infty} x_2[n]\,e^{-j(\Omega-\lambda)n}\right] d\lambda \\ &= \frac{1}{2\pi}\int_{-\pi}^{\pi} X_1(\lambda)\,X_2(\Omega-\lambda)\,d\lambda \end{aligned} \tag{3.158}$$

Table 3.3 contains a summary of key properties of the DTFT. Table 3.4 lists some of the fundamental DTFT pairs.

3.3.6 Applying DTFT to periodic signals

In previous sections of this chapter we have distinguished between two types of discrete-time signals: periodic and non-periodic. For periodic discrete-time signals we have the DTFS as an analysis and problem solving tool; for non-periodic discrete-time signals the DTFT serves a similar purpose. This arrangement will serve us adequately in cases where we work with one type of signal or the other. There may be times, however, when we need to mix periodic and non-periodic signals within one system. For example, in amplitude modulation, a non-periodic signal may be multiplied with a periodic signal, and we may need to analyze the resulting product in the frequency domain. Alternately, a periodic signal may be used as input to a system the impulse response of which is non-periodic. In these types of scenarios it would be convenient to find a way to use the DTFT for periodic signals as well. We have seen in Example 3.14 that a DTFT can be found for a constant-amplitude signal that does not satisfy the existence conditions, as long as we are willing to accept singularity functions in the transform. The next two examples will expand on this idea. Afterward we will develop a technique for converting the DTFS of any periodic discrete-time signal to a DTFT.

Table 3.3 – DTFT properties.

Theorem	Signal	Transform	
Linearity	$\alpha\, x_1[n]+\beta\, x_2[n]$	$\alpha\, X_1(\Omega)+\beta\, X_2(\Omega)$	
Periodicity	$x[n]$	$X(\Omega)=X(\Omega+2\pi r)\,,$	all integer r
Conjugate symmetry	$x[n]$ real	$X^*(\Omega)=X(-\Omega)$	
		Magnitude:	$\lvert X(-\Omega)\rvert=\lvert X(\Omega)\rvert$
		Phase:	$\Theta(-\Omega)=-\Theta(\Omega)$
		Real part:	$X_r(-\Omega)=X_r(\Omega)$
		Imaginary part:	$X_i(-\Omega)=-X_i(\Omega)$
Conjugate antisymmetry	$x[n]$ imaginary	$X^*(\Omega)=-X(-\Omega)$	
		Magnitude:	$\lvert X(-\Omega)\rvert=\lvert X(\Omega)\rvert$
		Phase:	$\Theta(-\Omega)=-\Theta(\Omega)\mp\pi$
		Real part:	$X_r(-\Omega)=-X_r(\Omega)$
		Imaginary part:	$X_i(-\Omega)=X_i(\Omega)$
Even signal	$x[n]=x[-n]$	$\text{Im}\{X(\Omega)\}=0$	
Odd signal	$x[n]=-x[-n]$	$\text{Re}\{X(\Omega)\}=0$	
Time shifting	$x[n-m]$	$X(\Omega)\, e^{-j\Omega m}$	
Time reversal	$x[-n]$	$X(-\Omega)$	
Conjugation	$x^*[n]$	$X^*(-\Omega)$	
Frequency shifting	$x[n]\, e^{j\Omega_0 n}$	$X(\Omega-\Omega_0)$	
Modulation	$x[n]\,\cos(\Omega_0 n)$	$\frac{1}{2}\left[X(\Omega-\Omega_0)+X(\Omega+\Omega_0)\right]$	
	$x[n]\,\sin(\Omega_0 n)$	$\frac{1}{2}\left[X(\Omega-\Omega_0)\, e^{-j\pi/2}-X(\Omega+\Omega_0)\, e^{j\pi/2}\right]$	
Differentiation in frequency	$n^m\, x[n]$	$j^m\,\dfrac{d^m}{d\Omega^m}\left[X(\Omega)\right]$	
Convolution	$x_1[n]*x_2[n]$	$X_1(\Omega)\, X_2(\Omega)$	
Multiplication	$x_1[n]\, x_2[n]$	$\dfrac{1}{2\pi}\displaystyle\int_{-\pi}^{\pi} X_1(\lambda)\, X_2(\Omega-\lambda)\, d\lambda$	
Parseval's theorem	$\displaystyle\sum_{n=-\infty}^{\infty}\lvert x[n]\rvert^2=\dfrac{1}{2\pi}\int_{-\pi}^{\pi}\lvert X(\Omega)\rvert^2\, d\Omega$		

Table 3.4 – Some DTFT transform pairs.

Name	Signal	Transform
Discrete-time pulse	$x[n]=\begin{cases}1, & \lvert n\rvert\le L\\ 0 & \text{otherwise}\end{cases}$	$X(\Omega)=\dfrac{\sin\left(\dfrac{(2L+1)\,\Omega}{2}\right)}{\sin\left(\dfrac{\Omega}{2}\right)}$
Unit impulse signal	$x[n]=\delta[n]$	$X(\Omega)=1$
Constant-amplitude signal	$x[n]=1\,,\;$ all n	$X(\Omega)=2\pi\sum\limits_{m=-\infty}^{\infty}\delta(\Omega-2\pi m)$
Sinc function	$x[n]=\dfrac{\Omega_c}{\pi}\,\text{sinc}\left(\dfrac{\Omega_c n}{\pi}\right)$	$X(\Omega)=\begin{cases}1, & \lvert\Omega\rvert<\Omega_c\\ 0, & \text{otherwise}\end{cases}$
Right-sided exponential	$x[n]=\alpha^n\,u[n]\,,\;\lvert\alpha\rvert<1$	$X(\Omega)=\dfrac{1}{1-\alpha\,e^{-j\Omega}}$
Complex exponential	$x[n]=e^{j\Omega_0 n}$	$X(\Omega)=2\pi\sum\limits_{m=-\infty}^{\infty}\delta(\Omega-\Omega_0-2\pi m)$

Example 3.20: DTFT of complex exponential signal

Determine the transform of the complex exponential signal $x[n]=e^{j\Omega_0 n}$ with $-\pi<\Omega_0<\pi$.

Solution: The transform of the constant unit-amplitude signal was found in Example 3.14 to be

$$\mathcal{F}\{1\}=2\pi\sum_{m=-\infty}^{\infty}\delta(\Omega-2\pi m)$$

Using this result along with the frequency shifting property of the DTFT given by Eqn. (3.135), the transform of the complex exponential signal is

$$\mathcal{F}\left\{e^{j\Omega_0 n}\right\}=2\pi\sum_{m=-\infty}^{\infty}\delta(\Omega-\Omega_0-2\pi m) \tag{3.159}$$

This is illustrated in Fig. 3.41.

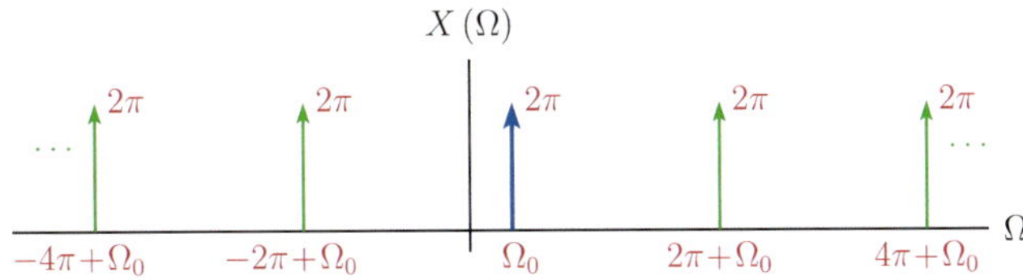

Figure 3.41 – The transform of complex exponential signal $x[n]=e^{j\Omega_0 n}$.

Example 3.21: DTFT of sinusoidal signal

Determine the transform of the sinusoidal signal $x[n] = \cos(\Omega_0 n)$.

Solution: The transform of the constant unit-amplitude signal was found in Example 3.14 to be

$$\mathscr{F}\{1\} = 2\pi \sum_{m=-\infty}^{\infty} \delta(\Omega - 2\pi m)$$

Using this result along with the modulation property of the DTFT given by Eqn. (3.137), the transform of the sinusoidal signal is

$$\mathscr{F}\{\cos(\Omega_0 n)\} = \pi \sum_{m=-\infty}^{\infty} \delta(\Omega - \Omega_0 - 2\pi m) + \pi \sum_{m=-\infty}^{\infty} \delta(\Omega + \Omega_0 - 2\pi m)$$

Let $\bar{X}(\Omega)$ represent the part of the transform in the range $-\pi < \Omega < \pi$.

$$\bar{X}(\Omega) = \pi\,\delta(\Omega - \Omega_0) + \pi\,\delta(\Omega + \Omega_0) \tag{3.160}$$

The DTFT can now be expressed as

$$X(\Omega) = \sum_{m=-\infty}^{\infty} \bar{X}(\Omega - 2\pi m)$$

This is illustrated in Fig. 3.42.

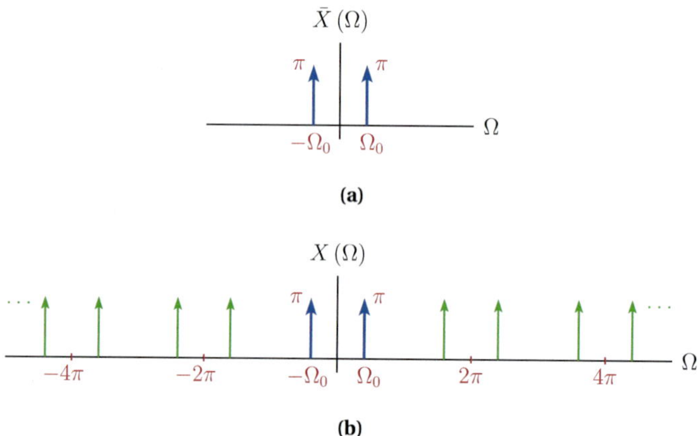

Figure 3.42 – The transform of complex exponential signal $x[n] = \cos(\Omega_0 n)$: **(a)** the middle part $\bar{X}(\Omega)$ and **(b)** the complete transform $X(\Omega)$.

In Examples 3.20 and 3.21, we were able to obtain the DTFT of two periodic signals, namely a complex exponential signal and a sinusoidal signal. The idea can be generalized to apply to any periodic discrete-time signal. The DTFS synthesis equation for a periodic signal $\tilde{x}[n]$ was given by Eqn. (3.18) which is repeated here for convenience:

$$\text{Eqn. (3.18):} \qquad \tilde{x}[n] = \sum_{k=0}^{N-1} \tilde{c}_k\, e^{j(2\pi/N)kn}, \qquad \text{all } n$$

If we were to attempt to find the DTFT of the signal $\tilde{x}[n]$ by direct application of the DTFT analysis equation given by Eqn. (3.69) we would need to evaluate

$$X(\Omega) = \sum_{n=-\infty}^{\infty} \tilde{x}[n]\, e^{-j\Omega n} \tag{3.161}$$

Substituting Eqn. (3.18) into Eqn. (3.161) leads to

$$X(\Omega) = \sum_{n=-\infty}^{\infty} \left[\sum_{k=0}^{N-1} \tilde{c}_k\, e^{j(2\pi/N)kn} \right] e^{-j\Omega n} \tag{3.162}$$

Interchanging the order of the two summations in Eqn. (3.162) and rearranging terms we obtain

$$X(\Omega) = \sum_{k=0}^{N-1} \tilde{c}_k \left[\sum_{n=-\infty}^{\infty} e^{j(2\pi/N)kn}\, e^{-j\Omega n} \right] \tag{3.163}$$

In Eqn. (3.163) the expression in square brackets is the DTFT of the signal $e^{j(2\pi/N)kn}$. Using the result obtained in Example 3.15, and remembering that $2\pi/N = \Omega_0$ is the fundamental angular frequency for the periodic signal $\tilde{x}[n]$, we get

$$\begin{aligned} \sum_{n=-\infty}^{\infty} e^{j(2\pi/N)kn}\, e^{-j\Omega n} &= 2\pi \sum_{m=-\infty}^{\infty} \delta\left(\Omega - \frac{2\pi k}{N} - 2\pi m\right) \\ &= 2\pi \sum_{m=-\infty}^{\infty} \delta(\Omega - k\Omega_0 - 2\pi m) \end{aligned} \tag{3.164}$$

and

$$X(\Omega) = 2\pi \sum_{k=0}^{N-1} \tilde{c}_k \sum_{m=-\infty}^{\infty} \delta(\Omega - k\Omega_0 - 2\pi m) \tag{3.165}$$

The part of the transform in the range $0 < \Omega < 2\pi$ is found by setting $m = 0$ in Eqn. (3.165):

$$\bar{X}(\Omega) = 2\pi \sum_{k=0}^{N-1} \tilde{c}_k\, \delta(\Omega - k\Omega_0) \tag{3.166}$$

Thus, $\bar{X}(\Omega)$ for a periodic discrete-time signal is obtained by converting each DTFS coefficient $\tilde{c}_k$ to an impulse with area equal to $2\pi\tilde{c}_k$ and placing it at angular frequency $\Omega = k\Omega_0$. The DTFT for the signal is then obtained as

$$X(\Omega) = \sum_{m=-\infty}^{\infty} \bar{X}(\Omega - 2\pi m) \tag{3.167}$$

This process is illustrated in Fig. 3.43.

Note: In Eqn. (3.160) of Example 3.21 we have used $\bar{X}(\Omega)$ to represent the part of the transform in the range $-\pi < \Omega < \pi$. On the other hand, in Eqn. (3.166) above, $\bar{X}(\Omega)$ was used as the part of the transform in the range $0 < \Omega < 2\pi$. This should not cause any confusion. In general, $\bar{X}(\Omega)$ represents one period of the 2π-periodic transform, and the starting value of Ω is not important.

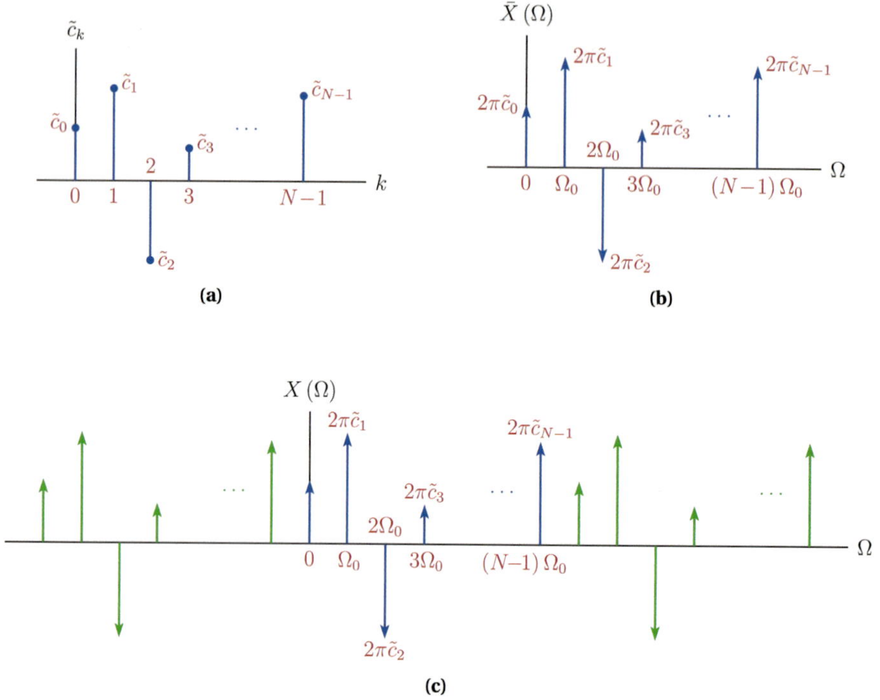

Figure 3.43 – **(a)** DTFS coefficients for a signal periodic with N, **(b)** DTFT for $-\pi < \Omega < \pi$, and **(c)** complete DTFT.

Example 3.22: DTFT of the periodic signal of Example 3.4

Determine the DTFT of the signal $x[n]$ used in Example 3.4 and shown in Fig. 3.3.

Solution: We will first write the transform in the interval $0 < \Omega < 2\pi$.

$$\bar{X}(\Omega) = 2\pi\left[\tilde{c}_0\,\delta(\Omega) + \tilde{c}_1\,\delta(\Omega-\Omega_0) + \tilde{c}_2\,\delta(\Omega-2\Omega_0) + \tilde{c}_3\,\delta(\Omega-3\Omega_0) + \tilde{c}_4\,\delta(\Omega-4\Omega_0)\right]$$

The period of the signal $\tilde{x}[n]$ is $N = 5$, therefore the fundamental angular frequency is $\Omega_0 = 2\pi/5$. Using the values of DTFS coefficients found in Example 3.4 we obtain

$$\begin{aligned}\bar{X}(\Omega) =& 12.566\,\delta(\Omega) + \left(-3.142-j4.324\right)\delta(\Omega-2\pi/5)\\ &+\left(-3.142-j1.021\right)\delta(\Omega-4\pi/5)\\ &+\left(-3.142+j1.021\right)\delta(\Omega-6\pi/5)\\ &+\left(-3.142+j4.324\right)\delta(\Omega-8\pi/5)\end{aligned}$$

The complete transform $X(\Omega)$ is found by periodically extending $\bar{X}(\Omega)$.

$$X(\Omega) = \sum_{m=-\infty}^{\infty} \Big[12.566\,\delta(\Omega-2\pi m) + \left(-3.142-j4.324\right)\delta(\Omega-2\pi/5-2\pi m)$$
$$+\left(-3.142-j1.021\right)\delta(\Omega-4\pi/5-2\pi m)$$
$$+\left(-3.142+j1.021\right)\delta(\Omega-6\pi/5-2\pi m)$$
$$+\left(-3.142+j4.324\right)\delta(\Omega-8\pi/5-2\pi m)\Big]$$

3.4 Energy and Power in the Frequency Domain

We will begin this section with a discussion of an important theorem of Fourier series and transform known as *Parseval's theorem*. We have already defined signal energy and power from a time domain perspective in Chapter 1. Parseval's theorem, named after the French mathematician Marc-Antoine Parseval (1755–1836), can be used as the basis of computing energy or power of a signal from its frequency domain representation. Versions of the theorem for periodic and non-periodic discrete-time signals lead to the definitions of energy and power spectral density concepts.

3.4.1 Parseval's theorem

Parseval's theorem for periodic and non-periodic signals:

For a periodic power signal $\tilde{x}[n]$ with period N and DTFS coefficients $\{\tilde{c}_k\,,\; k=0,\ldots,N-1\}$, it can be shown that

$$\frac{1}{N}\sum_{n=0}^{N-1}\left|\tilde{x}[n]\right|^2 = \sum_{k=0}^{N-1}\left|\tilde{c}_k\right|^2 \tag{3.168}$$

For a non-periodic energy signal $x[n]$ with DTFT $X(\Omega)$, the following holds true:

$$\sum_{n=-\infty}^{\infty}\left|\tilde{x}[n]\right|^2 = \frac{1}{2\pi}\int_{-\pi}^{\pi}\left|X(\Omega)\right|^2\,d\Omega \tag{3.169}$$

The left side of Eqn. (3.168) represents the normalized average power in a periodic signal which was derived in Chapter 1 Eqn. (1.56). The left side of Eqn. (3.169) represents the normalized signal energy as derived in Eqn. (1.46). The relationships given by Eqns. (3.168) and (3.169) relate signal energy or signal power to the frequency-domain representation of the signal. They are two forms of *Parseval's theorem.*

Proofs of Eqns. (3.168) and (3.169):

First we will prove Eqn. (3.168). DTFS representation of a periodic discrete-time signal with period N was given by Eqn. (3.18), and is repeated here for convenience:

$$\text{Eqn. (3.18):}\qquad \tilde{x}[n] = \sum_{k=0}^{N-1}\tilde{c}_k\,e^{j(2\pi/N)kn}\,,\qquad \text{all } n$$

Using $\left|x[n]\right|^2 = x[n]\,x^*[n]$, the left side of Eqn. (3.168) can be written as

$$\begin{aligned}\frac{1}{N}\sum_{n=0}^{N-1}\left|\tilde{x}[n]\right|^2 &= \frac{1}{N}\sum_{n=0}^{N-1}\left[\sum_{k=0}^{N-1}\tilde{c}_k\,e^{j(2\pi/N)kn}\right]\left[\sum_{m=0}^{N-1}\tilde{c}_m\,e^{j(2\pi/N)mn}\right]^* \\ &= \frac{1}{N}\sum_{n=0}^{N-1}\left[\sum_{k=0}^{N-1}\tilde{c}_k\,e^{j(2\pi/N)kn}\right]\left[\sum_{m=0}^{N-1}\tilde{c}_m^*\,e^{-j(2\pi/N)mn}\right]\end{aligned} \tag{3.170}$$

Rearranging the order of the summations in Eqn. (3.170) we get

$$\frac{1}{N}\sum_{n=0}^{N-1}\left|\tilde{x}[n]\right|^2 = \frac{1}{N}\sum_{k=0}^{N-1}\tilde{c}_k\left[\sum_{m=0}^{N-1}\tilde{c}_m^*\left[\sum_{n=0}^{N-1}e^{j(2\pi/N)(k-m)n}\right]\right] \tag{3.171}$$

Using orthogonality of the basis function set $\left\{e^{j(2\pi/N)kn}\,,\ k=0,\ldots,N-1\right\}$, it can be shown that (see Appendix D)

$$\sum_{n=0}^{N-1}e^{j(2\pi/N)(k-m)n} = \begin{cases} N\,, & k-m = 0, \mp N, \mp 2N, \ldots \\ 0\,, & \text{otherwise}\end{cases} \tag{3.172}$$

Using Eqn. (3.172) in Eqn. (3.171) leads to the desired result:

$$\frac{1}{N}\sum_{n=0}^{N-1}\left|\tilde{x}[n]\right|^2 = \sum_{k=0}^{N-1}\tilde{c}_k\,\tilde{c}_k^* = \sum_{k=-0}^{N-1}\left|\tilde{c}_k\right|^2 \tag{3.173}$$

The proof for Eqn. (3.169) is similar. The DTFT synthesis equation was given by Eqn. (3.68) and is repeated here for convenience.

$$\text{Eqn. (3.68):}\qquad x[n] = \frac{1}{2\pi}\int_{-\pi}^{\pi}X(\Omega)\,e^{j\Omega n}\,d\Omega$$

The left side of Eqn. (3.169) can be written as

$$\begin{aligned}\sum_{n=-\infty}^{\infty}\left|\tilde{x}[n]\right|^2 &= \sum_{n=-\infty}^{\infty}x[n]\left[\frac{1}{2\pi}\int_{-\pi}^{\pi}X(\Omega)\,e^{j\Omega n}\,d\Omega\right]^* \\ &= \sum_{n=-\infty}^{\infty}x[n]\left[\frac{1}{2\pi}\int_{-\pi}^{\pi}X^*(\Omega)\,e^{-j\Omega n}\,d\Omega\right]\end{aligned} \tag{3.174}$$

Interchanging the order of the integral and the summation in Eqn. (3.174) and rearranging terms yields

$$\begin{aligned}\sum_{n=-\infty}^{\infty}\left|\tilde{x}[n]\right|^2 &= \frac{1}{2\pi}\int_{-\pi}^{\pi}X^*(\Omega)\left[\sum_{n=-\infty}^{\infty}x[n]\,e^{-j\Omega n}\right]d\Omega \\ &= \frac{1}{2\pi}\int_{-\pi}^{\pi}\left|X(\Omega)\right|^2\,d\Omega\end{aligned} \tag{3.175}$$

3.4.2 Energy and power spectral density

The two statements of Parseval's theorem given by Eqns. (3.168) and (3.169) lead us to the following conclusions:

1. In Eqn. (3.168) the left side corresponds to the normalized average power of the periodic signal $\tilde{x}[n]$, and therefore the summation on the right side must also represent normalized average power. The term $|\tilde{c}_k|^2$ corresponds to the power of the signal component at angular frequency $\Omega = k\Omega_0$. (Remember that $\Omega_0 = 2\pi/N$.) Let us construct a new function $S_x(\Omega)$ as follows:

$$S_x(\Omega) = \sum_{k=-\infty}^{\infty} 2\pi\, |\tilde{c}_k|^2\, \delta(\Omega - k\Omega_0) \tag{3.176}$$

The function $S_x(\Omega)$ consists of impulses placed at angular frequencies $k\Omega_0$ as illustrated in Fig. 3.44. Since the DTFS coefficients are periodic with period N, the function $S_x(\Omega)$ is 2π-periodic.

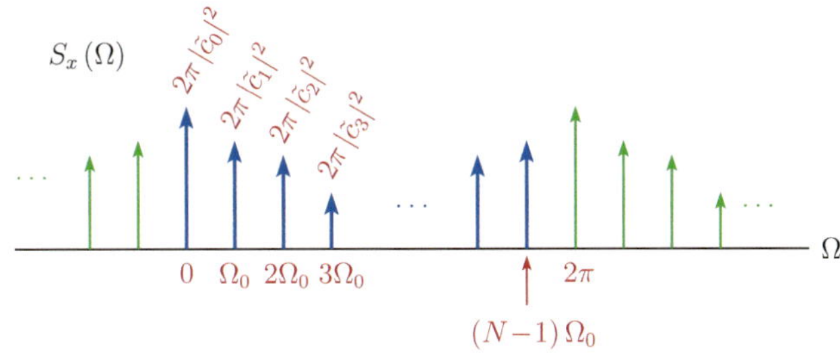

Figure 3.44 – The function $S_x(\Omega)$ constructed using the DTFS coefficients of the signal $\tilde{x}[n]$.

Integrating $S_x(\Omega)$ over an interval of 2π radians leads to

$$\int_0^{2\pi} S_x(\Omega)\, d\Omega = \int_0^{2\pi} \left[\sum_{k=-\infty}^{\infty} 2\pi\, |\tilde{c}_k|^2\, \delta(\Omega - k\Omega_0)\right] d\Omega \tag{3.177}$$

Interchanging the order of integration and summation in Eqn. (3.177) and rearranging terms we obtain

$$\int_0^{2\pi} S_x(\Omega)\, d\Omega = 2\pi \sum_{k=-\infty}^{\infty} |\tilde{c}_k|^2 \left[\int_0^{2\pi} \delta(\Omega - k\Omega_0)\, d\Omega\right] \tag{3.178}$$

Recall that $\Omega_0 = 2\pi/N$. Using the sifting property of the impulse function, the integral between square brackets in Eqn. (3.178) is evaluated as

$$\int_0^{2\pi} \delta(\Omega - k\Omega_0)\, d\Omega = \begin{cases} 1\,, & k = 0, \ldots, N-1 \\ 0\,, & \text{otherwise} \end{cases} \tag{3.179}$$

and Eqn. (3.178) becomes

$$\int_0^{2\pi} S_x(\Omega)\, d\Omega = 2\pi \sum_{k=0}^{N-1} |\tilde{c}_k|^2 \tag{3.180}$$

The normalized average power of the signal $\tilde{x}[n]$ is therefore found as

$$P_x = \sum_{k=0}^{N-1} |\tilde{c}_k|^2 = \frac{1}{2\pi} \int_0^{2\pi} S_x(\Omega)\, d\Omega \tag{3.181}$$

Consequently, the function $S_x(\Omega)$ is the *power spectral density* of the signal $\tilde{x}[n]$. Since $S_x(\Omega)$ is 2π-periodic, the integral in Eqn. (3.181) can be started at any value of Ω as long as it covers a span of 2π radians. It is usually more convenient to write the integral in Eqn. (3.181) to start at $-\pi$.

$$P_x = \frac{1}{2\pi} \int_{-\pi}^{\pi} S_x(\Omega)\, d\Omega \tag{3.182}$$

As an interesting by-product of Eqn. (3.181), the normalized average power of $\tilde{x}[n]$ that is within a specific angular frequency range can be determined by integrating $S_x(\Omega)$ over that range. For example, the power contained at angular frequencies in the range $(-\Omega_c, \Omega_c)$ is

$$P_x \text{ in } (-\Omega_c, \Omega_c) = \frac{1}{2\pi} \int_{-\Omega_c}^{\Omega_c} S_x(\Omega)\, d\Omega \tag{3.183}$$

2. In Eqn. (3.169) the left side is the normalized signal energy for the signal $x[n]$ and the right side must be the same. The integrand $\left|X(\Omega)\right|^2$ is therefore the *energy spectral density* of the signal $x[n]$. Let the function $G_x(\Omega)$ be defined as

$$G_x(\Omega) = \left|X(\Omega)\right|^2 \tag{3.184}$$

Substituting Eqn. (3.184) into Eqn. (3.169), the normalized energy in the signal $x[n]$ can be expressed as

$$E_x = \sum_{n=-\infty}^{\infty} \left|\tilde{x}[n]\right|^2 = \frac{1}{2\pi} \int_{-\pi}^{\pi} G(\Omega)\, d\Omega \tag{3.185}$$

The energy contained at angular frequencies in the range $(-\Omega_0, \Omega_0)$ is found by integrating $G_x(\Omega)$ in the frequency range of interest:

$$E_x \text{ in } (-\Omega_0, \Omega_0) = \frac{1}{2\pi} \int_{-\Omega_0}^{\Omega_0} G_x(\Omega)\, d\Omega \tag{3.186}$$

3. In Eqn. (3.169) we have expressed Parseval's theorem for an energy signal, and used it to lead to the derivation of the energy spectral density in Eqn. (3.184). We know that some non-periodic signals are power signals, therefore their energy cannot be computed. The example of one such signal is the unit step function. The power of a non-periodic signal was defined in Eqn. (1.57) in Chapter 1 which will be repeated here:

$$\text{Eqn. (1.57):} \qquad P_x = \lim_{M\to\infty} \left[\frac{1}{2M+1} \sum_{n=-M}^{M} \left|x[n]\right|^2 \right]$$

In order to write the counterpart of Eqn. (3.169) for a power signal, we will first define a *truncated* version of $x[n]$ as

$$x_T[n] = \begin{cases} x[n], & -M \le n \le M \\ 0, & \text{otherwise} \end{cases} \tag{3.187}$$

Let $X_T(\Omega)$ be the DTFT of the truncated signal $x_T[n]$:

$$X_T(\Omega) = \mathcal{F}\left\{x_T[n]\right\} = \sum_{n=-M}^{M} x_T[n]\, e^{-j\Omega n} \tag{3.188}$$

Now Eqn. (3.169) can be written in terms of the truncated signal and its transform as

$$\sum_{n=-M}^{M} \left|x_T[n]\right|^2 = \frac{1}{2\pi} \int_{-\pi}^{\pi} \left|X_T(\Omega)\right|^2 d\Omega \tag{3.189}$$

Scaling both sides of Eqn. (3.189) by $(2M+1)$ and taking the limit as M becomes infinitely large, we obtain

$$\lim_{M\to\infty}\left[\frac{1}{2M+1}\sum_{n=-M}^{M}\left|x_T[n]\right|^2\right]=\lim_{M\to\infty}\left[\frac{1}{2\pi\,(2M+1)}\int_{-\pi}^{\pi}\left|X_T(\Omega)\right|^2\,d\Omega\right] \tag{3.190}$$

The left side of Eqn. (3.190) is the average normalized power in a non-periodic signal as we have established in Eqn. (1.57). Therefore, the power spectral density of $x[n]$ is

$$S_x(\Omega)=\lim_{M\to\infty}\left[\frac{1}{2M+1}\left|X_T(\Omega)\right|^2\right] \tag{3.191}$$

Example 3.23: Normalized average power for waveform of Example 3.7

Consider again the periodic waveform used in Example 3.7 and shown in Fig. 3.8. Using the DTFS coefficients found in Example 3.7 verify Parseval's theorem. Also determine the percentage of signal power in the angular frequency range $-\pi/3<\Omega_0<\pi/3$.

Solution: The average power computed from the signal is

$$\begin{aligned}P_x=&\frac{1}{8}\sum_{n=0}^{7}\left|\tilde{x}[n]\right|^2\\=&\frac{1}{8}\left[0+\left(\frac{1}{2}\right)^2+(1)^2+\left(\frac{3}{4}\right)^2+0+\left(-\frac{3}{4}\right)^2+(-1)^2+\left(-\frac{1}{2}\right)^2\right]=0.4531\end{aligned}$$

Using the DTFS coefficients in Table 3.1 yields

$$\sum_{k=0}^{7}\left|\tilde{c}_k\right|^2=0+(0.4710)^2+(0.0625)^2+(0.0290)^2+0+(0.0290)^2+(0.0625)^2+(0.4710)^2=0.4531$$

As expected, the value found from DTFS coefficients matches that found from the signal. One period of the power spectral density $S_x(\Omega)$ is

$$\begin{aligned}S_x(\Omega)=&0.0017\pi\,\delta(\Omega+3\pi/4)+0.0078\pi\,\delta(\Omega+\pi/2)+0.4436\pi\,\delta(\Omega+\pi/4)\\&+0.4436\pi\,\delta(\Omega-\pi/4)+0.0078\pi\,\delta(\Omega-\pi/2)+0.0017\pi\,\delta(\Omega-3\pi/4)\end{aligned}$$

The power in the angular frequency range of interest is

$$\begin{aligned}P_x \text{ in } (-\pi/3,\pi/3)=&\frac{1}{2\pi}\int_{-\pi/3}^{\pi/3}S_x(\Omega)\,d\Omega\\=&\frac{1}{2\pi}\,(0.4436\pi+0.4436\pi)=0.4436\end{aligned}$$

The ratio of the power in $-\pi/3<\Omega_0<\pi/3$ to the total signal power is

$$\frac{P_x \text{ in } (-\pi/3,\pi/3)}{P_x}=\frac{0.4436}{0.4531}=0.979$$

Thus, 97.9 percent of the power of the signal is in the frequency range $-\pi/3<\Omega_0<\pi/3$.

Software resource: `exdt_3_23.m`

Example 3.24: Power spectral density of a discrete-time sinusoid

Find the power spectral density of the signal

$$\tilde{x}[n] = 3\cos(0.2\pi n)$$

Solution: Using Euler's formula, the signal can be written as

$$\tilde{x}[n] = \frac{3}{2}\, e^{-j0.2\pi n} + \frac{3}{2}\, e^{j0.2\pi n}$$

The normalized frequency of the sinusoidal signal is $F_0 = 0.1$ corresponding to a period length of $N = 10$ samples. Non-zero DTFS coefficients for $k = 0,\ldots,9$ are

$$\tilde{c}_1 = \tilde{c}_9 = \frac{3}{2}$$

Using Eqn. (3.176), the power spectral density is

$$S_x(\Omega) = \sum_{r=-\infty}^{\infty} \left[\frac{9\pi}{2}\,\delta(\Omega - 0.2\pi - 2\pi r) + \frac{9\pi}{2}\,\delta(\Omega - 1.8\pi - 2\pi r) \right]$$

which is shown in Fig. 3.45.

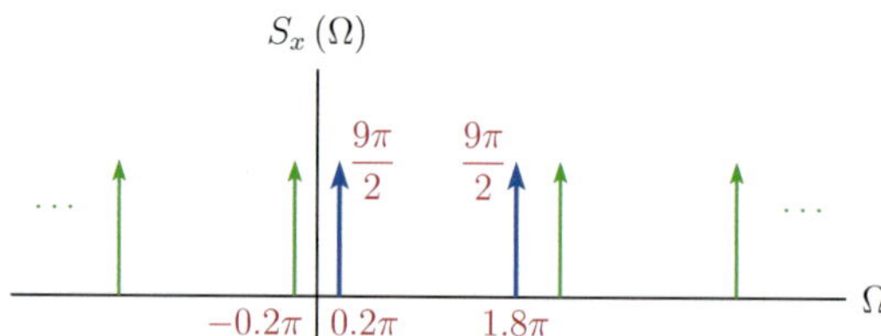

Figure 3.45 – Power spectral density for Example 3.24.

The power in the sinusoidal signal is computed from the power spectral density as

$$P_x = \frac{1}{2\pi}\int_{-\pi}^{\pi} S_x(\Omega)\, d\Omega = \frac{1}{2\pi}\left(\frac{9\pi}{2} + \frac{9\pi}{2}\right) = \frac{9}{2}$$

Example 3.25: Energy spectral density of a discrete-time pulse

Determine the energy spectral density of the rectangular pulse

$$x[n] = \begin{cases} 1, & n = -5,\ldots,5 \\ 0, & \text{otherwise} \end{cases}$$

Also compute the energy of the signal in the frequency interval $-\pi/10 < \Omega < \pi/10$.

Solution: Using the general result obtained in Example 3.12 with $L = 5$, the DTFT of the signal $x[n]$ is

$$X(\Omega) = \frac{\sin\left(\dfrac{11\Omega}{2}\right)}{\sin\left(\dfrac{\Omega}{2}\right)}$$

The energy spectral density for $x[n]$ is found as

$$G_x(\Omega) = \left|X(\Omega)\right|^2 = \frac{\sin^2\left(\dfrac{11\Omega}{2}\right)}{\sin^2\left(\dfrac{\Omega}{2}\right)}$$

which is shown in Fig. 3.46.

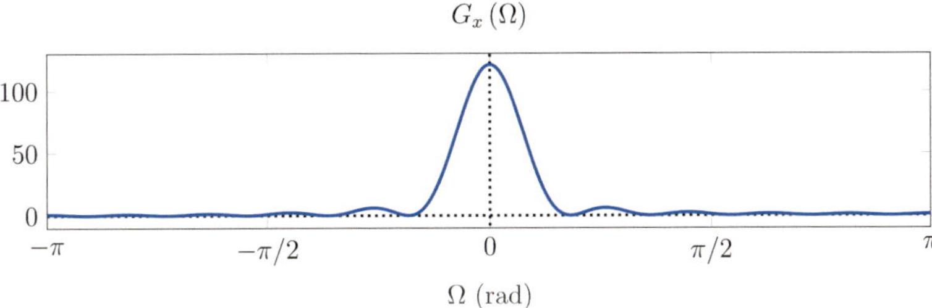

Figure 3.46 – Power spectral density for Example 3.25.

The energy of the signal within the frequency interval $-\pi/10 < \Omega < \pi/10$ is computed as

$$E_x \text{ in } \left(\pi/10, \pi/10\right) = \frac{1}{2\pi} \int_{-\pi/10}^{\pi/10} G_x(\Omega)\, d\Omega \tag{3.192}$$

which is proportional to the shaded area under $G(\Omega)$ in Fig. 3.47.

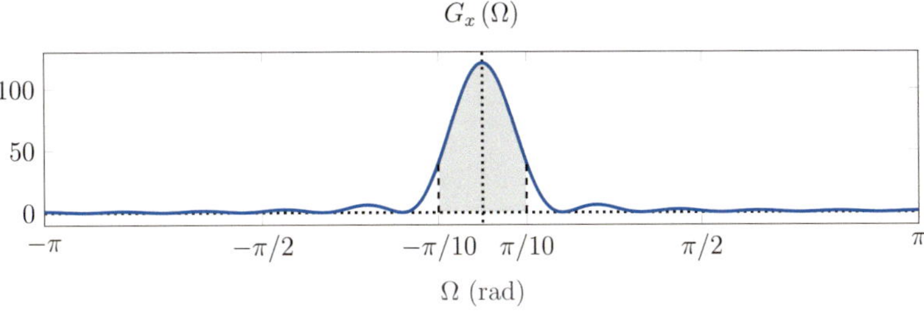

Figure 3.47 – The area under $G_x(\Omega)$ for $-\pi/10 < \Omega/\pi/10$.

Direct evaluation of the integral in Eqn. (3.192) is difficult, however, the result can be obtained by numerical approximation of the integral (see MATLAB Exercise 3.1), and is

$$E_x \text{ in } \left(\pi/10, \pi/10\right) \approx 8.9309$$

Software resource: exdt_3_25.m

3.4.3 Energy or power in a frequency range

Signal power or signal energy that is within a finite range of frequencies is found through the use of Eqns. (3.183) and (3.186) respectively. An interesting interpretation of the result in Eqn. (3.183) is that, for the case of a power signal, the power of $x[n]$ in the frequency range $-\Omega_0 < \Omega < \Omega_0$ is the same as the power of the output signal of a system with system function

$$H(\Omega) = \begin{cases} 1\,, & |\Omega| < \Omega_0 \\ 0\,, & \Omega_0 < |\Omega| < \pi \end{cases} \tag{3.193}$$

driven by the signal $x[n]$. Similarly, for the case of an energy signal, the energy of $x[n]$ in the frequency range $-\Omega_0 < \Omega < \Omega_0$ is the same as the energy of the output signal of a system with the system function in Eqn. (3.193) driven by the signal $x[n]$. These relationships are illustrated in Fig. 3.48(a) and (b).

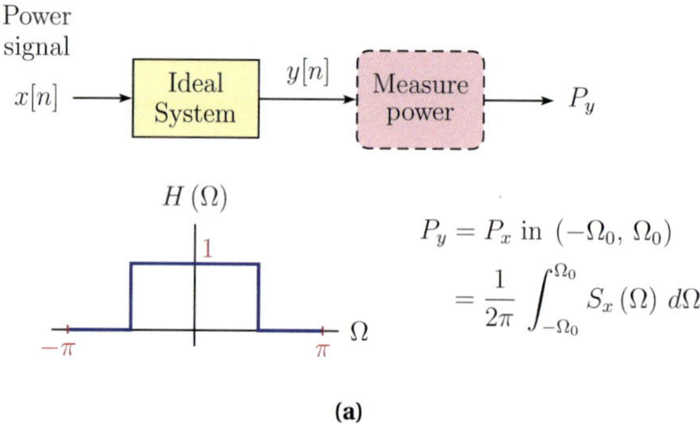

(a)

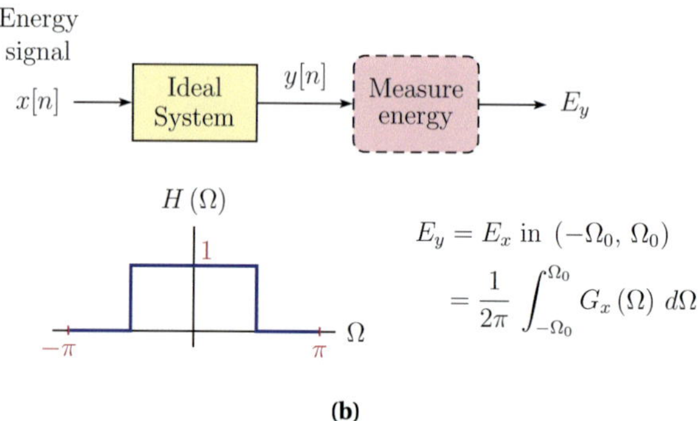

(b)

Figure 3.48 – Computation of signal power or energy in a range of frequencies.

3.4.4 Autocorrelation

The energy spectral density $G_x(\Omega)$ or the power spectral density $S_x(\Omega)$ for a signal can be computed through direct application of the corresponding equations derived in Section 3.4.2; namely Eqn. (3.184) for an energy signal, and either Eqn. (3.176) or Eqn. (3.191) for a power signal depending on its type. In some circumstances it is also possible to compute either from the knowledge of the *autocorrelation function* which will be defined in this section. Let $x[n]$ be a real-valued signal.

Autocorrelation function for various signal types:

For an energy signal $x[n]$ the *autocorrelation function* is defined as

$$r_{xx}[m] = \sum_{n=-\infty}^{\infty} x[n]\,x[n+m] \tag{3.194}$$

For a periodic power signal $\tilde{x}[n]$ with period N, the corresponding definition of the autocorrelation function is

$$\begin{aligned}\tilde{r}_{xx}[m] &= \left\langle \tilde{x}[n]\,\tilde{x}[n+m] \right\rangle \\ &= \frac{1}{N}\sum_{n=0}^{N-1} \tilde{x}[n]\,\tilde{x}[n+m]\end{aligned} \tag{3.195}$$

The triangle brackets indicate time average. The autocorrelation function for a periodic signal is also periodic as signified by the tilde (~) character used over the symbol r_{xx} in Eqn. (3.195). Finally, for a non-periodic power signal, the corresponding definition is

$$\begin{aligned}r_{xx}[m] &= \left\langle x[n]\,x[n+m] \right\rangle \\ &= \lim_{M\to\infty}\left[\frac{1}{2M+1}\sum_{n=-M}^{M} x[n]\,x[n+m]\right]\end{aligned} \tag{3.196}$$

Even though we refer to $r_{xx}[m]$ as a *function,* we will often treat it as if it is a discrete-time signal, albeit one that uses a different index, m, than the signal $x[n]$ for which it is computed. The index m simply corresponds to the time shift between the two copies of $x[n]$ used in the definitions of Eqns. (3.194) through (3.196).

It can be shown that, for an energy signal, the energy spectral density is the DTFT of the autocorrelation function, that is,

$$G_x(\Omega) = \mathcal{F}\{r_{xx}[m]\} \tag{3.197}$$

Proof of Eqn. (3.197):

Let us begin by applying the DTFT definition to the autocorrelation function $r_x[m]$ treated as a discrete-time signal indexed by m:

$$\mathcal{F}\left\{r_{xx}[m]\right\} = \sum_{m=-\infty}^{\infty}\left[\sum_{n=-\infty}^{\infty} x[n]\,x[n+m]\right] e^{-j\Omega m} \tag{3.198}$$

The two summations in Eqn. (3.198) can be rearranged to yield

$$\mathcal{F}\left\{r_{xx}[m]\right\} = \sum_{n=-\infty}^{\infty} x[n]\left[\sum_{m=-\infty}^{\infty} x[n+m]\,e^{-j\Omega m}\right] \tag{3.199}$$

Realizing that the inner summation in Eqn. (3.199) is equal to

$$\sum_{m=-\infty}^{\infty} x[n+m]\,e^{-j\Omega m} = \mathcal{F}\left\{x[m+n]\right\} = e^{j\Omega n}\,X(\Omega) \tag{3.200}$$

Eqn. (3.199) becomes

$$\mathcal{F}\left\{r_{xx}[m]\right\} = X(\Omega)\sum_{n=-\infty}^{\infty} x[n]\,e^{j\Omega n} \tag{3.201}$$

and since $x[n]$ is real, we obtain

$$\mathscr{F}\left\{r_{xx}[m]\right\} = X(\Omega)\, X^*(\Omega) = \left|X(\Omega)\right|^2 \tag{3.202}$$

which completes the proof.

The definition of the autocorrelation function for a periodic signal $\tilde{x}[n]$ can be used in a similar manner in determining the power spectral density of the signal. Let the signal $\tilde{x}[n]$ and the autocorrelation function $\tilde{r}_{xx}[m]$, both periodic with period N, have the DTFS representations given by

$$\tilde{x}[n] = \sum_{k=0}^{N-1} \tilde{c}_k\, e^{j(2\pi/N)kn} \tag{3.203}$$

and

$$\tilde{r}_{xx}[m] = \sum_{k=0}^{N-1} \tilde{d}_k\, e^{j(2\pi/N)km}, \tag{3.204}$$

respectively. It can be shown that the DTFS coefficients $\left\{\tilde{c}_k\right\}$ and $\left\{\tilde{d}_k\right\}$ are related by

$$\tilde{d}_k = \left|\tilde{c}_k\right|^2 \tag{3.205}$$

and the power spectral density is the DTFT of the autocorrelation function, that is,

$$\tilde{S}_x(\Omega) = \mathscr{F}\left\{\tilde{r}_{xx}[m]\right\} \tag{3.206}$$

Proof of Eqn. (3.206):

Using the DTFS analysis equation given by Eqn. (3.19) with the definition of the autocorrelation function in Eqn. (3.195) leads to

$$\begin{aligned}\tilde{d}_k &= \frac{1}{N}\sum_{m=0}^{N-1} \tilde{r}_{xx}[m]\, e^{-j(2\pi/N)km} \\ &= \frac{1}{N}\sum_{m=0}^{N-1} \left[\frac{1}{N}\sum_{n=0}^{N-1} \tilde{x}[n]\,\tilde{x}[n+m]\right] e^{-j(2\pi/N)km}\end{aligned} \tag{3.207}$$

Rearranging the order of the two summations Eqn. (3.207) can be written as

$$\tilde{d}_k = \frac{1}{N}\sum_{n=0}^{N-1} \tilde{x}[n] \left[\frac{1}{N}\sum_{m=0}^{N-1} \tilde{x}[n+m]\, e^{-j(2\pi/N)km}\right] \tag{3.208}$$

Using the time shifting property of the DTFS given by Eqn. (3.26) the expression in square brackets in Eqn. (3.208) is

$$\frac{1}{N}\sum_{m=0}^{N-1} \tilde{x}[n+m]\, e^{-j(2\pi/N)km} = e^{j(2\pi/N)kn}\, \tilde{c}_k \tag{3.209}$$

Substituting Eqn. (3.209) into Eqn. (3.208) yields

$$\begin{aligned}\tilde{d}_k =& \frac{1}{N}\sum_{n=0}^{N-1}\tilde{x}[n]\,e^{j(2\pi/N)kn}\,\tilde{c}_k \\ =& \tilde{c}_k\,\frac{1}{N}\sum_{n=0}^{N-1}\tilde{x}[n]\,e^{j(2\pi/N)kn} \\ =& \tilde{c}_k\,\tilde{c}_k^* = \left|\tilde{c}_k\right|^2 \end{aligned} \tag{3.210}$$

Using the technique developed in Section 3.3.6 the DTFT of $\tilde{r}_{xx}[m]$ is

$$\mathcal{F}\left\{\tilde{r}_{xx}[m]\right\} = 2\pi\sum_{k=-\infty}^{\infty}\tilde{d}_k\,\delta\left(\Omega - k\Omega_0\right) \tag{3.211}$$

Using Eqn. (3.210) in Eqn. (3.211) results in

$$\mathcal{F}\left\{\tilde{r}_{xx}[m]\right\} = 2\pi\sum_{k=-\infty}^{\infty}\left|\tilde{c}_k\right|^2\,\delta\left(\Omega - k\Omega_0\right) \tag{3.212}$$

which is identical to the expression given by Eqn. (3.176) for the power spectral density of a periodic signal.

A relationship similar to the one expressed by Eqn. (3.206) applies to random processes that are *wide-sense stationary*, and is known as the Wiener-Khinchin theorem. It is one of the fundamental theorems of random signal processing.

Example 3.26: Power spectral density of a discrete-time sinusoid revisited

Consider again the discrete-time sinusoidal signal

$$\tilde{x}[n] = 3\,\cos\left(0.2\pi n\right)$$

the power spectral density of which was determined in Example 3.24. Determine the autocorrelation function $R_x[m]$ for this signal, and then find the power spectral density $S_x(\Omega)$ from the autocorrelation function.

Solution: The period of $\tilde{x}[n]$ is $N = 10$ samples. Using the definition of the autocorrelation function for a periodic signal

$$\begin{aligned}r_{xx}[m] =& \frac{1}{10}\sum_{n=0}^{9}\tilde{x}[n]\,\tilde{x}[n+m] \\ =& \frac{1}{10}\sum_{n=0}^{9}\left[3\cos\left(0.2\pi n\right)\right]\left[3\cos\left(0.2\pi\left[n+m\right]\right)\right]\end{aligned} \tag{3.213}$$

By using appropriate trigonometric identity, the result in Eqn. (3.213) is simplified to

$$\begin{aligned}r_{xx}[m] =& \frac{1}{10}\sum_{n=0}^{9}\left[\frac{9}{2}\cos\left(0.4\pi n + 0.2\pi m\right) + \frac{9}{2}\cos\left(0.2\pi m\right)\right] \\ =& \frac{9}{20}\sum_{n=0}^{9}\cos\left(0.4\pi n + 0.2\pi m\right) + \frac{9}{20}\sum_{n=0}^{9}\cos\left(0.2\pi m\right)\end{aligned} \tag{3.214}$$

In Eqn. (3.214) the first summation is equal to zero since its term $\cos(0.4\pi n + 0.2\pi m)$ is periodic with a period of five samples, and the summation is over two full periods. The second summation yields

$$\begin{aligned} r_{xx}[m] &= \frac{9}{20} \sum_{n=0}^{9} \cos(0.2\pi m) \\ &= \frac{9}{20} \cos(0.2\pi m) \sum_{n=0}^{9} (1) \\ &= \frac{9}{2} \cos(0.2\pi m) \end{aligned}$$

for the autocorrelation function. The power spectral density is found as the DTFT of the autocorrelation function. Application of the DTFT was discussed in Section 3.3.6. Using the technique developed in that section (specifically, see Example 3.21) the transform is found as

$$\begin{aligned} S_x(\Omega) &= \mathscr{F}\left\{r_{xx}[m]\right\} = \mathscr{F}\left\{\frac{9}{2}\cos(0.2\pi m)\right\} \\ &= \sum_{r=-\infty}^{\infty} \left[\frac{9\pi}{2}\,\delta(\Omega + 0.2\pi - 2\pi r) + \frac{9\pi}{2}\,\delta(\Omega - 0.2\pi - 2\pi r)\right] \end{aligned}$$

which matches the answer found earlier in Example 3.24.

Software resource: exdt_3_26.m

3.4.5 Properties of the autocorrelation function

The autocorrelation function as defined by Eqns. (3.194), (3.195), and (3.196) has a number of important properties that will be summarized here:

1. $r_{xx}[0] \ge \left|r_{xx}[m]\right|$ for all m.

   To see why this is the case, we will consider the non-negative function $(x[n] \mp x[n+m])^2$. The time average of this function must also be non-negative, therefore

   $$\left\langle (x[n] \mp x[n+m])^2 \right\rangle \ge 0$$

   or equivalently

   $$\left\langle x^2[n] \right\rangle \mp 2\left\langle x[n]\,x[n+m] \right\rangle + \left\langle x^2[n+m] \right\rangle \ge 0$$

   which implies that

   $$r_{xx}[0] \mp 2\,r_{xx}[m] + r_{xx}[0] \ge 0$$

   which is the same as property 1.

2. $r_{xx}[-m] = r_{xx}[m]$ for all m, that is, the autocorrelation function has even symmetry.

   Recall that the autocorrelation function is the inverse Fourier transform of either the energy spectral density or the power spectral density. Since $G_x(\Omega)$ and $S_x(\Omega)$ are purely real, $r_{xx}[m]$ must be an even function of m.

3. If the signal $\tilde{x}[n]$ is periodic with period N, then its autocorrelation function $\tilde{r}_{xx}[m]$ is also periodic with the same period. This property easily follows from the time average based definition of the autocorrelation function given by Eqn. (3.195).

3.5 System Function Concept

In time domain analysis of systems in Chapter 2 two distinct description forms were used for DTLTI systems:

1. A linear constant-coefficient difference equation that describes the relationship between the input and the output signals and
2. An impulse response which can be used with the convolution operation for determining the response of the system to an arbitrary input signal.

In this section, the concept of *system function* will be introduced as the third method for describing the characteristics of a system.

The system function is simply the DTFT of the impulse response:

$$H(\Omega) = \mathcal{F}\left\{h[n]\right\} = \sum_{n=-\infty}^{\infty} h[n]\, e^{-j\Omega n} \tag{3.215}$$

Recall that the impulse response is only meaningful for a system that is both linear and time-invariant (since the convolution operator could not be used otherwise). It follows that the system function concept is valid for linear and time-invariant systems only.

In general, $H(\Omega)$ is a complex function of Ω, and can be written in polar form as

$$H(\Omega) = \left|H(\Omega)\right| e^{j\Theta(\Omega)} \tag{3.216}$$

Obtaining the system function from the difference equation

In finding a system function for a DTLTI system described by means of a difference equation, two properties of the DTFT will be useful: the convolution property and the time shifting property.

1. Since the output signal is computed as the convolution of the impulse response and the input signal, that is, $y[n] = h[n] * x[n]$, the corresponding relationship in the frequency domain is $Y(\Omega) = H(\Omega)\, X(\Omega)$. Consequently, the system function is equal to the ratio of the output transform to the input transform:

$$H(\Omega) = \frac{Y(\Omega)}{X(\Omega)} \tag{3.217}$$

2. Using the time shifting property, transforms of the individual terms in the difference equation are found as

$$y[n-m] \overset{\mathcal{F}}{\longleftrightarrow} e^{-j\Omega m}\, Y(\Omega) \qquad m = 0, 1, \ldots \tag{3.218}$$

and

$$x[n-m] \overset{\mathcal{F}}{\longleftrightarrow} e^{-j\Omega m}\, X(\Omega) \qquad m = 0, 1, \ldots \tag{3.219}$$

The system function is obtained from the difference equation by first transforming both sides of the difference equation through the use of Eqns. (3.218) and (3.219), and then using Eqn. (3.217).

Example 3.27: Finding the system function from the difference equation

Determine the system function for a DTLTI system described by the difference equation

$$y[n]-0.9\,y[n-1]+0.36\,y[n-2]=x[n]-0.2\,x[n-1]$$

Solution: Taking the DTFT of both sides of the difference equation we obtain

$$Y(\Omega)-0.9\,e^{-j\Omega}\,Y(\Omega)+0.36\,e^{-j2\Omega}\,Y(\Omega)=X(\Omega)-0.2\,e^{-j\Omega}\,X(\Omega)$$

which can be written as

$$\left[1-0.9\,e^{-j\Omega}+0.36\,e^{-j2\Omega}\right]\,Y(\Omega)=\left[1-0.2\,e^{-j\Omega}\right]\,X(\Omega)$$

The system function is found by using Eqn. (3.217)

$$H(\Omega)=\frac{Y(\Omega)}{X(\Omega)}=\frac{1-0.2\,e^{-j\Omega}}{1-0.9\,e^{-j\Omega}+0.36\,e^{-j2\Omega}}$$

The magnitude and the phase of the system function are shown in Fig. 3.49.

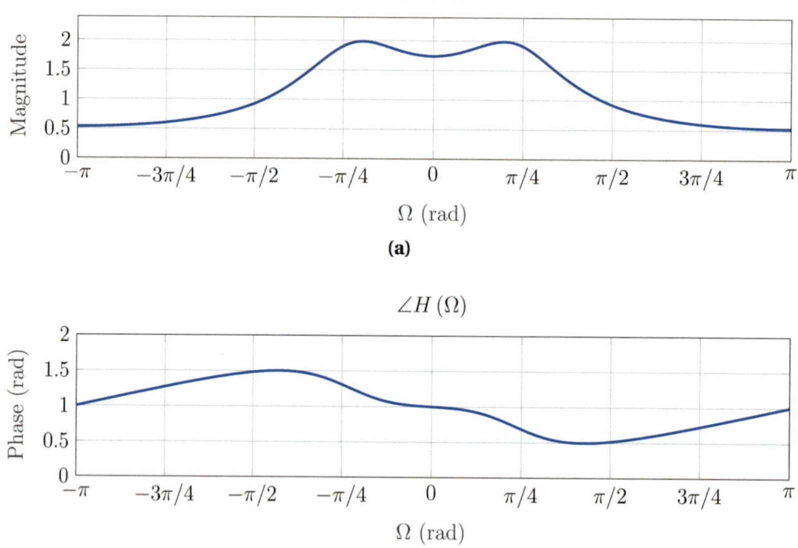

Figure 3.49 – The system function for Example 3.27: **(a)** Magnitude and **(b)** phase.

Software resource: exdt_3_27.m

Example 3.28: System function for length-N moving average filter

Recall the length-N moving average filter with the difference equation

$$y[n] = \frac{1}{N} \sum_{k=0}^{N-1} x[n-k]$$

Determine system function. Graph its magnitude and phase as functions of Ω.

Solution: Taking the DTFT of both sides of the difference equation we obtain

$$Y(\Omega) = \frac{1}{N} \sum_{k=0}^{N-1} e^{-j\Omega k}\, X(\Omega)$$

from which the system function is found as

$$H(\Omega) = \frac{Y(\Omega)}{X(\Omega)} = \frac{1}{N} \sum_{k=0}^{N-1} e^{-j\Omega k} \tag{3.220}$$

The expression in Eqn. (3.220) can be put into closed form as

$$H(\Omega) = \frac{1 - e^{-j\Omega N}}{N\left(1 - e^{-j\Omega}\right)}$$

As the first step in expressing $H(\Omega)$ in polar complex form, we will factor out $e^{-j\Omega N/2}$ from the numerator and $e^{-j\Omega/2}$ from the denominator to obtain

$$\begin{aligned} H(\Omega) &= \frac{e^{-j\Omega N/2}\left(e^{j\Omega N/2} - e^{-j\Omega N/2}\right)}{N e^{-j\Omega/2}\left(e^{j\Omega/2} - e^{-j\Omega/2}\right)} \\ &= \frac{\sin(\Omega N/2)}{N \sin(\Omega/2)}\, e^{-j\Omega(N-1)/2} \end{aligned}$$

The magnitude and the phase of the system function are shown in Fig. 3.50 for $N = 4$.

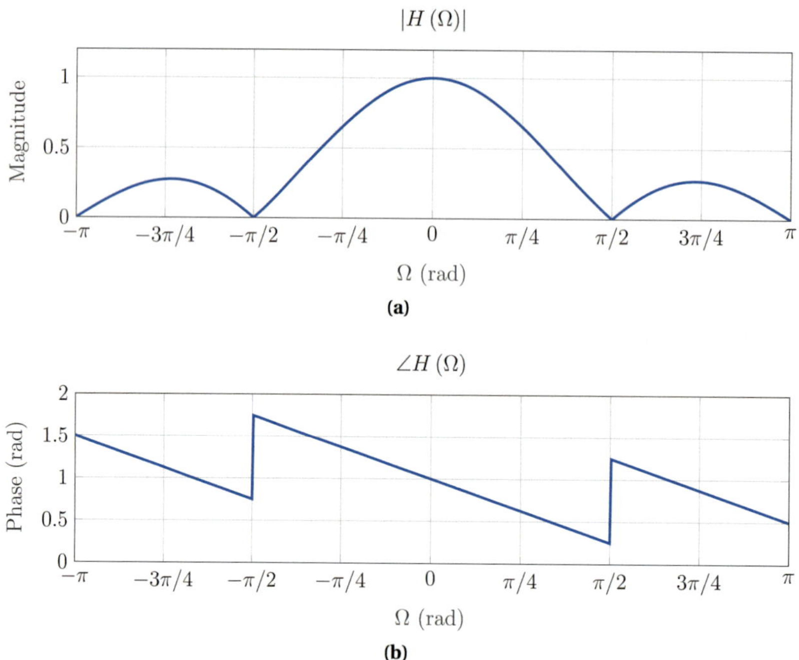

Figure 3.50 – The system function for Example 3.28: **(a)** Magnitude and **(b)** phase.

Software resource: `exdt_3_28.m`

Interactive App: System function for length-N moving average filter

The interactive app in `appSysFuncDT1.m` is based on Example 3.28. It computes and graphs the magnitude and the phase of the system function $H(\Omega)$ for the length-N moving average filter. The filter length N may be varied.

a. Set $N = 4$ to match Fig. 3.50. Afterward try with $N = 5$ and $N = 6$. Observe that zero crossings (dips) of the magnitude spectrum divide the angular frequency interval $0 \le \Omega \le 2\pi$ into N equal segments.

b. Pay attention to the phase of the system function. Its sloped sections all have the same slope, indicating that the phase response is linear. How does the slope of the phase response relate to the filter length N?

Software resource: `appSysFuncDT1.m`

3.6 DTLTI Systems with Periodic Input Signals

In earlier sections of this chapter we have explored the use of the discrete-time Fourier series (DTFS) for representing periodic discrete-time signals. It was shown that a periodic discrete-time signal can be expressed using complex exponential basis functions in the form

$$\text{Eqn. (3.18):} \qquad \tilde{x}[n] = \sum_{k=0}^{N-1} \tilde{c}_k \, e^{j(2\pi/N)kn}\,, \qquad \text{all } n$$

If a periodic signal is used as input to a DTLTI system, the use of the superposition property allows the response of the system to be determined as a linear combination of its responses to individual basis functions

$$\phi_k[n] = e^{j(2\pi/N)kn}$$

3.6.1 Response of a DTLTI system to complex exponential signal

Consider a DTLTI system with impulse response $h[n]$, driven by a complex exponential input signal

$$\tilde{x}[n] = e^{j\Omega_0 n} \tag{3.221}$$

The response $y[n]$ of the system is found by using the convolution relationship that was derived in Section 2.7.2 of Chapter 2.

$$\begin{aligned} y[h] =& h[n] * \tilde{x}[n] \\ =& \sum_{k=-\infty}^{\infty} h[k]\, \tilde{x}[n-k] \end{aligned} \tag{3.222}$$

Derivation of response to complex exponential signal:

Using the signal $\tilde{x}[n]$ given by Eqn. (3.221) in Eqn. (3.222) we get

$$y[n] = \sum_{k=-\infty}^{\infty} h[k]\, e^{j\Omega_0 (n-k)} \tag{3.223}$$

or, equivalently

$$y[n] = e^{j\Omega_0 n} \sum_{k=-\infty}^{\infty} h[k]\, e^{-j\Omega_0 n} \tag{3.224}$$

The summation in Eqn. (3.224) should be recognized as the system function evaluated at the specific angular frequency $\Omega = \Omega_0$. Therefore

$$\tilde{y}[n] = e^{j\Omega_0 n}\, H(\Omega_0) \tag{3.225}$$

We have used the tilde (~) character over the name of the output signal in realization of the fact that it is also periodic. The development in Eqns. (3.223) through (3.225) is based on the inherent assumption that the Fourier transform of $h[n]$ exists. This in turn requires the corresponding DTLTI system to be stable. Any natural response the system may have exhibited at one point would have disappeared a long time ago. Consequently, the response found in Eqn. (3.225) is the *steady-state response* of the system.

Based on the development above, we have the following important relationship for a DTLTI system driven by a complex exponential input signal:

Response to complex exponential input signal:

$$\begin{aligned}\tilde{y}[n] = \text{Sys}\left\{e^{j\Omega_0 n}\right\} &= e^{j\Omega_0 n}\, H(\Omega_0)\\ &= \left|H(\Omega_0)\right|\, e^{j[\Omega_0 t + \Theta(\Omega_0)]}\end{aligned} \tag{3.226}$$

1. The response of a DTLTI system to a complex exponential input signal is a complex exponential output signal with the same angular frequency Ω_0.
2. The effect of the system on the complex exponential input signal is to
   (a) scale its amplitude by an amount equal to the magnitude of the system function at $\Omega = \Omega_0$ and
   (b) shift its phase by an amount equal to the phase of the system function at $\Omega = \Omega_0$.

3.6.2 Response of a DTLTI system to sinusoidal signal

Let the input signal to the DTLTI system under consideration be a sinusoidal signal in the form

$$\tilde{x}[n] = \cos(\Omega_0 n) \tag{3.227}$$

The response of the system in this case will be determined by making use of the results of the previous section.

Derivation of response to cosine signal:

We will use Euler's formula to write the input signal using two complex exponential functions as

$$\tilde{x}[n] = \cos(\Omega_0 n) = \frac{1}{2}\, e^{j\Omega_0 n} + \frac{1}{2}\, e^{-j\Omega_0 n} \tag{3.228}$$

This representation of $\tilde{x}[n]$ allows the results of the previous section to be used. The output signal can be written using superposition:

$$\begin{aligned}\tilde{y}[n] &= \frac{1}{2}\,\text{Sys}\left\{e^{j\Omega_0 n}\right\} + \frac{1}{2}\,\text{Sys}\left\{e^{-j\Omega_0 n}\right\}\\ &= \frac{1}{2}\, e^{j\Omega_0 n}\, H(\Omega_0) + \frac{1}{2}\, e^{-j\Omega_0 n}\, H(-\Omega_0)\\ &= \frac{1}{2}\, e^{j\Omega_0 n}\left|H(\Omega_0)\right| e^{j\Theta(\Omega_0)} + \frac{1}{2}\, e^{-j\Omega_0 n}\left|H(-\Omega_0)\right| e^{j\Theta(-\Omega_0)}\end{aligned} \tag{3.229}$$

If the impulse response $h[n]$ is real-valued, the result in Eqn. (3.229) can be further simplified. Recall from Section 3.3.5 that, for real-valued $h[n]$, the transform $H(\Omega)$ is conjugate symmetric, resulting in

$$\left|H(-\Omega_0)\right| = \left|H(\Omega_0)\right| \quad \text{and} \quad \Theta(-\Omega_0) = -\Theta(\Omega_0) \tag{3.230}$$

Using these relationships, Eqn. (3.229) becomes

$$\begin{aligned}\tilde{y}[n] &= \frac{1}{2}\left|H(\Omega_0)\right| e^{j[\Omega_0 t+\Theta(\Omega_0)]} + \frac{1}{2}\left|H(\Omega_0)\right| e^{-j[\Omega_0 t+\Theta(\Omega_0)]} \\ &= \left|H(\Omega_0)\right| \cos\left(\omega_0 t + \Theta(\Omega_0)\right) \end{aligned} \tag{3.231}$$

For a DTLTI system driven by a cosine input signal, we have the following important relationship:

Response to cosine input signal:

$$\tilde{y}[n] = \text{Sys}\left\{\cos(\Omega_0 n)\right\} = \left|H(\Omega_0)\right| \cos\left(\Omega_0 n + \Theta(\Omega_0)\right) \tag{3.232}$$

1. When a DTLTI system is driven by single-tone input signal at angular frequency Ω_0, its output signal is also a single-tone signal at the same angular frequency.
2. The effect of the system on the input signal is to
   (a) scale its amplitude by an amount equal to the magnitude of the system function at $\Omega = \Omega_0$, and
   (b) shift its phase by an amount equal to the phase of the system function at $\Omega = \Omega_0$.

Example 3.29: Steady-state response of DTLTI system to sinusoidal input

Consider a DTLTI system characterized by the difference equation

$$y[n] - 0.9y[n-1] + 0.36y[n-2] = x[n] - 0.2x[n-1]$$

The system function for this system was determined in Example 3.27 to be

$$H(\Omega) = \frac{Y(\Omega)}{X(\Omega)} = \frac{1 - 0.2\,e^{-j\Omega}}{1 - 0.9\,e^{-j\Omega} + 0.36\,e^{-j2\Omega}}$$

Find the response of the system to the sinusoidal input signal

$$\tilde{x}[n] = 5\cos\left(\frac{\pi n}{5}\right)$$

Solution: Evaluating the system function at the angular frequency $\Omega_0 = \pi/5$ yields

$$H(\pi/5) = \frac{1 - 0.2\,e^{-j\pi/5}}{1 - 0.9\,e^{-j\pi/5} + 0.36\,e^{-j2\pi/5}} = 1.8890 - j0.6133,$$

which can be written in polar form as

$$H(\pi/5) = \left|H(\pi/5)\right| e^{j\Theta(\pi/5)} = 1.9861\,e^{-j0.3139}$$

The magnitude and the phase of the system function are shown in Fig. 3.51. The values of magnitude and phase at the angular frequency of interest are marked on the graphs.

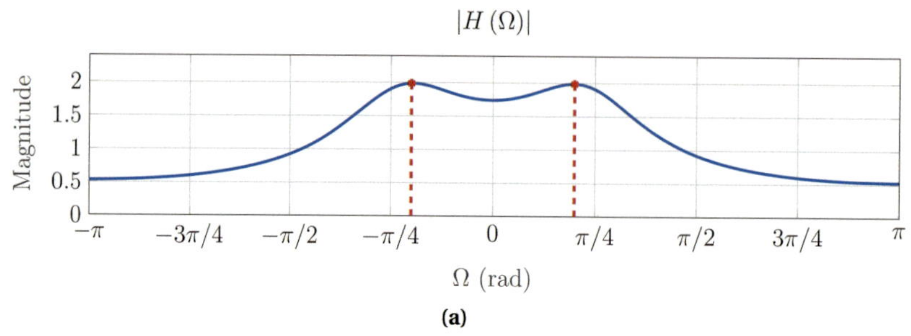

(a)

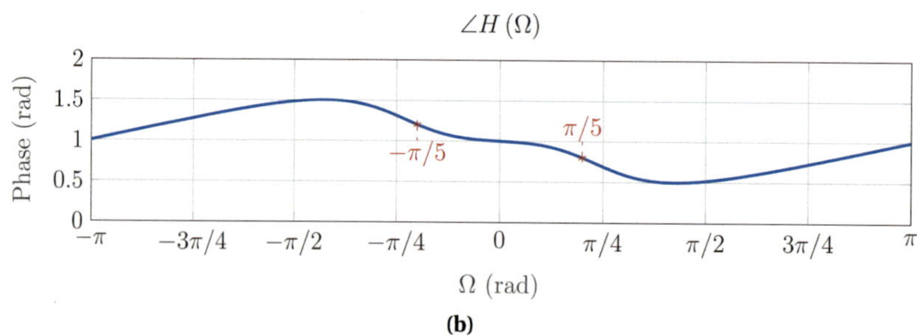

(b)

Figure 3.51 – The system function for Example 3.29: **(a)** Magnitude and **(b)** phase.

The steady-state response of the system to the specified input signal $\tilde{x}[n]$ is

$$\begin{aligned}\tilde{y}[n] =& 5\,\left|H(\pi/5)\right|\,\cos\left(\frac{\pi n}{5}+\Theta(\pi/5)\right)\\ =& 9.9305\,\cos\left(\frac{\pi n}{5}-0.3139\right)\end{aligned}$$

Software resource: exdt_3_29.m

Software resource: See MATLAB Exercise 3.5.

Interactive App: Steady-state response of DTLTI system to sinusoidal input

The interactive app in `appSysFuncDT2.m` is based on Example 3.29. It computes and graphs the steady-state response of the system under consideration to the sinusoidal input signal

$$x[n] = 5\,\cos(\Omega_0 n) = 5\,\cos(2\pi F_0 n)$$

The normalized frequency F_0 can be varied from $F_0 = 0.01$ to $F_0 = 0.49$ using a slider control. As F_0 is varied, the display is updated

a. to show the changes to the steady-state response and

b. to show the critical magnitude and phase and values on the graphs of the system function.

Software resource: `appSysFuncDT2.m`

3.6.3 Response of a DTLTI system to periodic input signal

Using the development in the previous sections, we are now ready to consider the use of a general periodic signal $\tilde{x}[n]$ as input to a DTLTI system. Using the DTFS representation of the periodic input signal, the response of the system is

$$\mathrm{Sys}\left\{\tilde{x}[n]\right\} = \mathrm{Sys}\left\{\sum_{k=0}^{N-1} \tilde{c}_k\, e^{j(2\pi/N)kn}\right\} \tag{3.233}$$

Let us use the linearity of the system to write the response as

$$\mathrm{Sys}\left\{\tilde{x}[n]\right\} = \sum_{k=0}^{N-1} \mathrm{Sys}\left\{\tilde{c}_k\, e^{j(2\pi/N)kn}\right\} = \sum_{k=0}^{N-1} \tilde{c}_k\, \mathrm{Sys}\left\{e^{j(2\pi/N)kn}\right\} \tag{3.234}$$

Based on Eqn. (3.226) the response of the system to an exponential basis function at frequency $\Omega = 2\pi k/N$ is given by

$$\mathrm{Sys}\left\{e^{j(2\pi/N)kn}\right\} = e^{j(2\pi/N)kn}\, H\left(\frac{2\pi k}{N}\right) \tag{3.235}$$

Therefore, the response of the DTLTI system to any periodic signal $\tilde{x}[n]$ with DTFS coefficients $\tilde{c}_k$ can be formulated using Eqn. (3.235) in Eqn. (3.234).

Response to any periodic input signal:

$$\mathrm{Sys}\left\{\tilde{x}[n]\right\} = \sum_{k=0}^{N-1} \tilde{c}_k\, H\left(\frac{2\pi k}{N}\right) e^{j(2\pi/N)kn} \tag{3.236}$$

Two important observations should be made based on Eqn. (3.236):

1. For a DTLTI system driven by an input signal $\tilde{x}[n]$ with period N, the output signal is also periodic with the same period.
2. If the DTFS coefficients of the input signal are $\left\{\tilde{c}_k\,;\, k = 1,\ldots,N-1\right\}$, then the DTFS coefficients of the output signal are $\left\{\tilde{d}_k\,;\, k = 1,\ldots,N-1\right\}$ such that

$$\tilde{d}_k = \tilde{c}_k\, H\left(\frac{2\pi k}{N}\right), \quad k = 0,\ldots,N-1 \tag{3.237}$$

Example 3.30: Response of DTLTI system to discrete-time sawtooth signal

Let the discrete-time sawtooth signal used in Example 3.4 and shown again in Fig. 3.52 be applied to the system with system function

$$H(\Omega) = \frac{1-0.2\,e^{-j\Omega}}{1-0.9\,e^{-j\Omega}+0.36\,e^{-j2\Omega}}$$

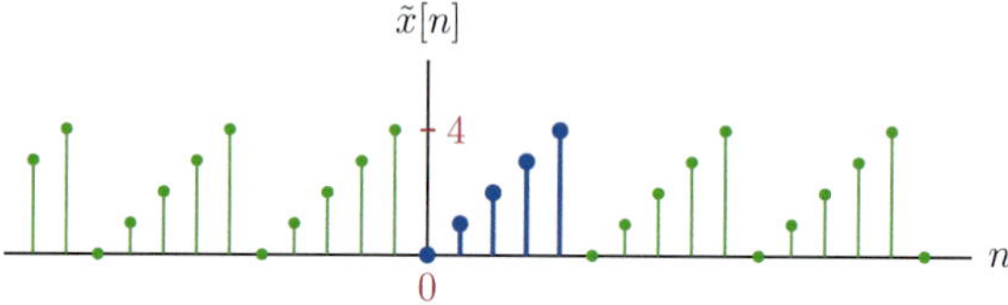

Figure 3.52 – The signal $\tilde{x}[n]$ for Example 3.30.

Find the steady-state response of the system.

Solution: The DTFS coefficients for $\tilde{x}[n]$ were determined in Example 3.4 and are repeated below:

$$\begin{aligned}
\tilde{c}_0 &= 2\\
\tilde{c}_1 &= -0.5+j\,0.6882\\
\tilde{c}_2 &= -0.5+j\,0.1625\\
\tilde{c}_3 &= -0.5-j\,0.1625\\
\tilde{c}_4 &= -0.5-j\,0.6882
\end{aligned}$$

Evaluating the system function at angular frequencies

$$\Omega_k = \frac{2\pi k}{5}\,, \qquad k=0,1,2,3,4$$

we obtain

$$\begin{aligned}
H(0) &= 1.7391\\
H\left(\frac{2\pi}{5}\right) &= 0.8767-j\,0.8701\\
H\left(\frac{4\pi}{5}\right) &= 0.5406-j\,0.1922\\
H\left(\frac{6\pi}{5}\right) &= 0.5406+j\,0.1922\\
H\left(\frac{8\pi}{5}\right) &= 0.8767+j\,0.8701
\end{aligned}$$

The DTFS coefficients for the output signal are found using Eqn. (3.237):

$$\begin{aligned}
\tilde{d}_0 =& \tilde{c}_0\, H(0) = (2)\,(1.7391) = 3.4783 \\
\tilde{d}_1 =& \tilde{c}_1\, H\left(\frac{2\pi}{5}\right) = \left(-0.5000 + j\,0.6882\right)\,\left(0.8767 - j\,0.8701\right) = 0.1604 + j\,1.0384 \\
\tilde{d}_2 =& \tilde{c}_2\, H\left(\frac{4\pi}{5}\right) = \left(-0.5000 + j\,0.1625\right)\,\left(0.5406 - j\,0.1922\right) = -0.2391 + j\,0.1839 \\
\tilde{d}_3 =& \tilde{c}_3\, H\left(\frac{6\pi}{5}\right) = \left(-0.5000 - j\,0.1625\right)\,\left(0.5406 + j\,0.1922\right) = -0.2391 - j\,0.1839 \\
\tilde{d}_4 =& \tilde{c}_4\, H\left(\frac{8\pi}{5}\right) = \left(-0.5000 - j\,0.6882\right)\,\left(0.8767 + j\,0.8701\right) = 0.1604 - j\,1.0384
\end{aligned}$$

The output signal $y[n]$ can now be constructed using the DTFS coefficients $\left\{\,\tilde{d}_k;\ k = 0, 1, 2, 3, 4\,\right\}$:

$$\begin{aligned}
\tilde{y}[n] =& \sum_{k=0}^{4} \tilde{d}_k\, e^{j(2\pi/N)kn} \\
=& 3.4783 + \left(0.1604 + j\,1.0384\right)\, e^{j2\pi n/5} + \left(-0.2391 + j\,0.1839\right)\, e^{j4\pi n/5} \\
&+ \left(-0.2391 - j\,0.1839\right)\, e^{j6\pi n/5} + \left(0.1604 - j\,1.0384\right)\, e^{j8\pi n/5}
\end{aligned}$$

and is shown in Fig. 3.53.

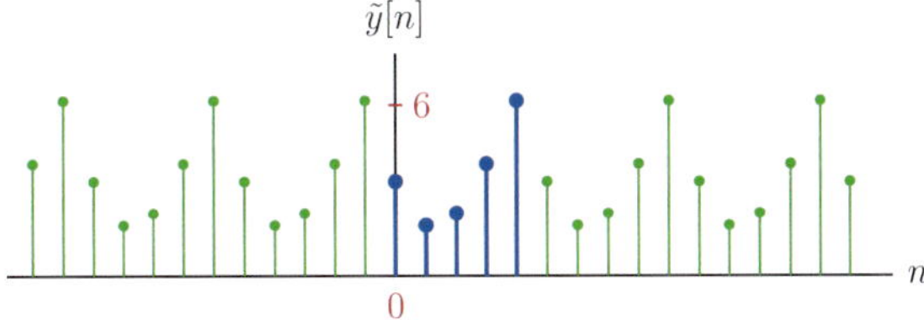

Figure 3.53 – The output signal $\tilde{y}[n]$ for Example 3.30.

Software resource: exdt_3_30.m

3.7 DTLTI Systems with Non-Periodic Input Signals

Let us consider the case of using a non-periodic signal $x[n]$ as input to a DTLTI system. It was established in Section 2.7.2 of Chapter 2 that the output of a DTLTI system is equal to the convolution of the input signal with the impulse response, that is

$$\begin{aligned}
y[n] =& h[n] * x[n] \\
=& \sum_{k=-\infty}^{\infty} h[k]\, x[n-k]
\end{aligned}$$

Let us assume that

1. The system is stable ensuring that $H(\Omega)$ converges and
2. The DTFT of the input signal also converges.

We have seen in Section 3.3.5 of this chapter that the DTFT of the convolution of two signals is equal to the product of individual transforms:

$$Y(\Omega) = H(\Omega)\, X(\Omega) \tag{3.238}$$

The output transform is the product of the input transform and the system function.

Writing each transform involved in Eqn. (3.238) in polar form using its magnitude and phase we obtain the corresponding relationships:

$$\left|Y(\Omega)\right| = \left|H(\Omega)\right| \left|X(\Omega)\right| \tag{3.239}$$

$$\measuredangle Y(\Omega) = \measuredangle X(\Omega) + \Theta(\Omega) \tag{3.240}$$

1. The magnitude of the output spectrum is equal to the product of the magnitudes of the input spectrum and the system function.
2. The phase of the output spectrum is found by adding the phase characteristics of the input spectrum and the system function.

Summary of Key Points

☞ A periodic discrete-time signal $\tilde{x}[n]$ which is periodic with period N can be expressed through its discrete-time Fourier series (DTFS) expansion as

Eqn. 3.18: $$\tilde{x}[n] = \sum_{k=0}^{N-1} \tilde{c}_k\, e^{j(2\pi/N)kn}\,, \qquad \text{all } n$$

The coefficients $\tilde{c}_k$ are also periodic with period N.

☞ DTFS coefficients of the signal $\tilde{x}[n]$ are computed as

Eqn. 3.19: $$\tilde{c}_k = \frac{1}{N} \sum_{n=0}^{N-1} \tilde{x}[n]\, e^{-j(2\pi/N)kn}$$

where N is the period of the signal as well as the set of DTFS coefficients.

☞ Given a DTFS pair

$$\tilde{x}[n] \overset{\text{DTFS}}{\longleftrightarrow} \tilde{c}_k,$$

the following are also valid DTFS pairs:

Time shifting property:

Eqn. 3.26: $$\tilde{x}[n-m] \overset{\text{DTFS}}{\longleftrightarrow} e^{-j(2\pi/N)km}\,\tilde{c}_k$$

Time reversal property:

Eqn. 3.27: $$\tilde{x}[-n] \overset{\text{DTFS}}{\longleftrightarrow} \tilde{c}_{-k} = \tilde{c}_{N-k}$$

☞ Given two DTFS pairs

$$\tilde{x}_1[n] \stackrel{\text{DTFS}}{\longleftrightarrow} \tilde{c}_k \qquad \text{and} \qquad \tilde{x}_2[n] \stackrel{\text{DTFS}}{\longleftrightarrow} \tilde{d}_k$$

and any two arbitrary constants α_1 and α_2, the following are also valid DTFS pairs.

Linearity property:

Eqn. 3.25:
$$\alpha_1\,\tilde{x}_1[n] + \alpha_2\,\tilde{x}_2[n] \stackrel{\text{DTFS}}{\longleftrightarrow} \alpha_1\,\tilde{c}_k + \alpha_2\,\tilde{d}_k$$

Periodic convolution property:

Eqn. 3.47:
$$\tilde{x}[n] \otimes \tilde{h}[n] \stackrel{\text{DTFS}}{\longleftrightarrow} N\,\tilde{c}_k\,\tilde{d}_k$$

where the operator $\otimes$ denotes periodic convolution defined as

Eqn. 3.46:
$$\tilde{y}[n] = \tilde{x}[n] \otimes \tilde{h}[n] = \sum_{k=0}^{N-1} \tilde{x}[k]\,\tilde{h}[n-k]\,, \quad \text{all } n$$

☞ Symmetry properties of the DTFS:

$$\tilde{x}[n] : \text{Real}, \quad \text{Im}\left\{\tilde{x}[n]\right\} = 0 \longrightarrow \tilde{c}_k^* = \tilde{c}_{N-k} \longrightarrow \begin{cases} |\tilde{c}_k| = |\tilde{c}_{N-k}| = |\tilde{c}_{-k}| \\ \tilde{\theta}_k = -\tilde{\theta}_{N-k} = -\tilde{\theta}_{-k} \end{cases}$$

$$\tilde{x}[n] : \text{Imag}, \quad \text{Re}\left\{\tilde{x}[n]\right\} = 0 \longrightarrow \tilde{c}_k^* = -\tilde{c}_{N-k} \longrightarrow \begin{cases} |\tilde{c}_k| = |\tilde{c}_{N-k}| = |\tilde{c}_{-k}| \\ \tilde{\theta}_k = -\tilde{\theta}_{N-k} \pm \pi = -\tilde{\theta}_{-k} \pm \pi \end{cases}$$

☞ Discrete-time Fourier transform (DTFT) of a signal $x[n]$ is

Eqn. 3.69:
$$X(\Omega) = \sum_{n=-\infty}^{\infty} x[n]\,e^{-j\Omega n}$$

The signal $x[n]$ can be obtained from the transform through

Eqn. 3.68:
$$x[n] = \frac{1}{2\pi} \int_{-\pi}^{\pi} X(\Omega)\,e^{j\Omega n}\,d\Omega$$

☞ For the DTFT to exist, a sufficient condition is that the signal $x[n]$ be absolute summable:

Eqn. 3.82:
$$\sum_{n=-\infty}^{\infty} \left|x[n]\right| < \infty$$

Alternatively, it is also sufficient for the signal $x[n]$ to be square-summable:

Eqn. 3.83:
$$\sum_{n=-\infty}^{\infty} \left|x[n]\right|^2 < \infty$$

Some signals that do not satisfy either condition may still have a DTFT because of using of singularity functions in the transform.

☞ DTFT of any signal is periodic with a period of 2π radians.

☞ DTFT is a linear operator. Given two transform pairs

$$x_1[n] \overset{\mathcal{F}}{\longleftrightarrow} X_1(\Omega) \qquad \text{and} \qquad x_2[n] \overset{\mathcal{F}}{\longleftrightarrow} X_2(\Omega)$$

and two arbitrary constants α_1 and α_2, it can be shown that

Eqn. 3.92: $$\alpha_1\, x_1[n] + \alpha_2\, x_2[n] \overset{\mathcal{F}}{\longleftrightarrow} \alpha_1\, X_1(\Omega) + \alpha_2\, X_2(\Omega)$$

☞ Given a transform pair

$$x[n] \overset{\mathcal{F}}{\longleftrightarrow} X(\Omega)$$

the following are also valid transform pairs.

Time shifting property:

Eqn. 3.94: $$x[n-m] \overset{\mathcal{F}}{\longleftrightarrow} X(\Omega)\, e^{-j\Omega m}$$

Time reversal property:

Eqn. 3.100: $$x[-n] \overset{\mathcal{F}}{\longleftrightarrow} X(-\Omega)$$

Conjugation property:

Eqn. 3.105: $$x^*[n] \overset{\mathcal{F}}{\longleftrightarrow} X^*(-\Omega)$$

Frequency shifting property:

Eqn. 3.135: $$x[n]\, e^{j\Omega_0 n} \overset{\mathcal{F}}{\longleftrightarrow} X(\Omega-\Omega_0)$$

Modulation property:

Eqn. 3.137: $$x[n]\cos(\Omega_0 n) \overset{\mathcal{F}}{\longleftrightarrow} \frac{1}{2}\left[X(\Omega-\Omega_0) + X(\Omega+\Omega_0)\right]$$

Eqn. 3.138: $$x[n]\sin(\Omega_0 n) \overset{\mathcal{F}}{\longleftrightarrow} \frac{1}{2}\left[X(\Omega-\Omega_0)\, e^{-j\pi/2} + X(\Omega+\Omega_0)\, e^{j\pi/2}\right]$$

Differentiation in the frequency domain:

Eqn. 3.144: $$n\, x[n] \overset{\mathcal{F}}{\longleftrightarrow} j\,\frac{dX(\Omega)}{d\Omega}$$

Eqn. 3.145: $$n^m\, x[n] \overset{\mathcal{F}}{\longleftrightarrow} j^m\,\frac{d^m X(\Omega)}{d\Omega^m}$$

☞ Given two transform pairs

$$x_1[n] \overset{\mathcal{F}}{\longleftrightarrow} X_1(\Omega) \quad \text{and} \quad x_2[n] \overset{\mathcal{F}}{\longleftrightarrow} X_2(\Omega),$$

the following are also valid transform pairs:

Convolution property:

Eqn. 3.147: $$x_1[n] * x_2[n] \overset{\mathcal{F}}{\longleftrightarrow} X_1(\Omega)\, X_2(\Omega)$$

Multiplication of two signals:

Eqn. 3.154: $$x_1[n]\, x_2[n] \overset{\mathcal{F}}{\longleftrightarrow} \frac{1}{2\pi} \int_{-\pi}^{\pi} X_1(\lambda)\, X_2(\Omega-\lambda)\, d\lambda$$

☞ Symmetry properties of the DTFT:

$$x[n] : \text{Real, } \text{Im}\{x[n]\} = 0 \longrightarrow X^*(\Omega) = X(-\Omega) \longrightarrow \begin{cases} |X(\Omega)| = |X(-\Omega)| \\ \Theta(\Omega) = -\Theta(-\Omega) \end{cases}$$

$$x[n] : \text{Imag, } \text{Re}\{x[n]\} = 0 \longrightarrow X^*(\Omega) = -X(-\Omega) \longrightarrow \begin{cases} |X(\Omega)| = |X(-\Omega)| \\ \Theta(\Omega) = -\Theta(-\Omega) \mp \pi \end{cases}$$

☞ For a periodic power signal $\tilde{x}[n]$ with period N and DTFS coefficients $\{\tilde{c}_k\,,\ k = 0, \ldots, N-1\}$, it can be shown that

Eqn. 3.168: $$\frac{1}{N} \sum_{n=0}^{N-1} \left|\tilde{x}[n]\right|^2 = \sum_{k=0}^{N-1} \left|\tilde{c}_k\right|^2$$

For a non-periodic energy signal $x[n]$ with DTFT $X(\Omega)$, we have

Eqn. 3.169: $$\sum_{n=-\infty}^{\infty} \left|\tilde{x}[n]\right|^2 = \frac{1}{2\pi} \int_{-\pi}^{\pi} \left|X(\Omega)\right|^2 d\Omega$$

These relationships are two forms of Parseval's theorem.

☞ For a periodic signal $\tilde{x}[n]$ with period N and DTFS coefficients $\{\tilde{c}_k\,,\ k = 0, \ldots, N-1\}$, the power spectral density is

Eqn. 3.176: $$S_x(\Omega) = \sum_{k=-\infty}^{\infty} 2\pi\, |\tilde{c}_k|^2\, \delta(\Omega - k\Omega_0)$$

where $\Omega_0 = 2\pi/N$. The normalized average power of the signal $\tilde{x}[n]$ is found as

Eqn. 3.181: $$P_x = \sum_{k=0}^{N-1} |\tilde{c}_k|^2 = \frac{1}{2\pi} \int_0^{2\pi} S_x(\Omega)\, d\Omega$$

☞ For a non-periodic energy signal $x[n]$ with DTFT $X(\Omega)$, the energy spectral density is

Eqn. 3.184: $$G_x(\Omega) = |X(\Omega)|^2$$

The normalized energy in the signal $x[n]$ can be computed as

Eqn. 3.185: $$E_x = \sum_{n=-\infty}^{\infty} \left|\tilde{x}[n]\right|^2 = \frac{1}{2\pi} \int_{-\pi}^{\pi} G(\Omega)\, d\Omega$$

☞ For a non-periodic power signal, the power spectral density is

Eqn. 3.191: $$S_x(\Omega) = \lim_{M\to\infty}\left[\frac{1}{2M+1}\,|X_T(\Omega)|^2\right]$$

where $X_T(\Omega)$ is the DTFT of the truncated signal $x_T[n]$ defined as

Eqn. 3.187: $$x_T[n] = \begin{cases} x[n]\,, & -M \le n \le M \\ 0\,, & \text{otherwise} \end{cases}$$

☞ For an energy signal $x[n]$ the autocorrelation function is defined as

Eqn. 3.194: $$r_{xx}[m] = \sum_{n=-\infty}^{\infty} x[n]\,x[n+m]$$

For a periodic power signal $\tilde{x}[n]$ with period N, the autocorrelation function is

Eqn. 3.195: $$\tilde{r}_{xx}[m] = \frac{1}{N}\sum_{n=0}^{N-1} \tilde{x}[n]\,\tilde{x}[n+m]$$

For a non-periodic power signal, the autocorrelation function is

Eqn. 3.196: $$r_{xx}[m] = \lim_{M\to\infty}\left[\frac{1}{2M+1}\sum_{n=-M}^{M} x[n]\,x[n+m]\right]$$

☞ For an energy signal, the energy spectral density is the DTFT of the autocorrelation function.

Eqn. 3.197: $$G_x(\Omega) = \mathcal{F}\left\{r_{xx}[m]\right\}$$

Similarly, for a power signal, the power spectral density is the DTFT of the autocorrelation function.

Eqn. 3.206: $$S_x(\Omega) = \mathcal{F}\left\{r_{xx}[m]\right\}$$

☞ The autocorrelation function has the following properties:

1. $r_{xx}[0] \ge \left|r_{xx}[m]\right|$ for all m.
2. Autocorrelation function has even symmetry, that is, $r_{xx}[-m] = r_{xx}[m]$ for all m.
3. If the signal $\tilde{x}[n]$ is periodic with period N, then its autocorrelation function $\tilde{r}_{xx}[m]$ is also periodic with the same period.

☞ The system function is the DTFT of the impulse response:

Eqn. 3.215: $$H(\Omega) = \mathcal{F}\left\{h[n]\right\} = \sum_{n=-\infty}^{\infty} h[n]\,e^{-j\Omega n}$$

☞ The response of a DTLTI system with system function $H(\Omega)$ to a complex exponential input signal is

Eqn. 3.226: $$\tilde{y}[n] = \text{Sys}\left\{e^{j\Omega_0 n}\right\} = \left|H(\Omega_0)\right|\,e^{j[\Omega_0 t + \Theta(\Omega_0)]}$$

☞ The response of a DTLTI system with system function $H(\Omega)$ to a cosine input signal is

Eqn. 3.232: $$\tilde{y}[n] = \text{Sys}\left\{\cos(\Omega_0 n)\right\} = \left|H(\Omega_0)\right| \cos\left(\omega_0 n + \Theta(\Omega_0)\right)$$

☞ The response of a DTLTI system with system function $H(\Omega)$ to a periodic signal $\tilde{x}[n]$ with DTFS coefficients $\left\{\tilde{c}_k,\ k=0,\ldots,N-1\right\}$ is

Eqn. 3.236: $$\text{Sys}\left\{\tilde{x}[n]\right\} = \sum_{k=0}^{N-1} \tilde{c}_k\, H\left(\frac{2\pi k}{N}\right) e^{j(2\pi/N)kn}$$

☞ The response of a DTLTI system with system function $H(\Omega)$ to a non-periodic signal $x[n]$ can be determined by finding the output transform $Y(\Omega)$ as

Eqn. 3.238: $$Y(\Omega) = H(\Omega)\, X(\Omega)$$

and then finding the inverse transform $y[n]$.

Further Readings

[1] R.J. Beerends. *Fourier and Laplace Transforms.* Cambridge University Press, 2003.

[2] A. Boggess and F.J. Narcowich. *A First Course in Wavelets with Fourier Analysis.* Wiley, 2011.

[3] R.N. Bracewell. *The Fourier Transform and Its Applications.* Electrical Engineering Series. McGraw-Hill, 2000.

[4] S.C. Chapra. *Applied Numerical Methods with MATLAB for Engineers and Scientists.* McGraw-Hill, 2008.

[5] S.C. Chapra and R.P. Canale. *Numerical Methods for Engineers.* McGraw-Hill, 2010.

[6] E.A. Gonzalez-Velasco. *Fourier Analysis and Boundary Value Problems.* Elsevier Science, 1996.

[7] K.B. Howell. *Principles of Fourier Analysis.* Studies in Advanced Mathematics. Taylor & Francis, 2010.

[8] Ken C. Pohlmann. *Principles of Digital Audio.* McGraw-Hill Professional, 2000.

[9] Kenneth Steiglitz. *Digital Signal Processing Primer.* Courier Dover Publications, 2020.

[10] Elias M. Stein and Rami Shakarchi. *Fourier Analysis: An Introduction.* Vol. 1. Princeton University Press, 2011.

[11] U. Zölzer. *Digital Audio Signal Processing.* Wiley, 2008.

MATLAB Exercises with Solutions

MATLAB Exercise 3.1: Writing functions for DTFS analysis and synthesis

In this exercise we will develop two MATLAB functions to implement DTFS analysis and synthesis equations. Function `ss_dtfs()` given below computes the DTFS coefficients for the periodic signal $\tilde{x}[n]$. The vector `x` holds one period of the signal $\tilde{x}[n]$ for $n = 0, \ldots, N-1$. The vector `idx` holds the values of the index k for which the DTFS coefficients $\tilde{c}_k$ are to be computed. The coefficients are returned in the vector `c`.

```
1  function c = ss_dtfs(x,idx)
2    c = zeros(size(idx));   % Create all-zero vector
3    N = length(x);          % Period of the signal x[n]
4    for kk = 1:length(idx)  % Loop over indices in vector idx
5      k = idx(kk);
6      tmp = 0;              % Reset running sum
7      for nn = 1:length(x)
8        n = nn-1;           % MATLAB array indices start with 1
9        tmp = tmp+x(nn)*exp(-j*2*pi/N*k*n);  % Eqn. (3.19)
10     end
11     c(kk) = tmp/N;
12   end
13 end
```

The inner loop between lines 7 and 10 implements the summation in Eqn. (3.19) for one specific value of k. The outer loop between lines 4 and 12 causes this computation to be performed for all values of k in the vector `idx`.

Function `ss_invdtfs()` implements the DTFS synthesis equation. The vector `c` holds one period of the DTFS coefficients $\tilde{c}_k$ for $k = 0, \ldots, N-1$. The vector `idx` holds the values of the index n for which the signal samples $\tilde{x}[n]$ are to be computed. The synthesized signal $\tilde{x}[n]$ is returned in the vector `x`.

```
function x = ss_invdtfs(c,idx)
  x = zeros(size(idx));   % Create all-zero vector
  N = length(c);          % Period of the coefficient set
  for nn = 1:length(idx)  % Loop over indices in vector idx
    n = idx(nn);
    tmp = 0;              % Reset running sum
    for kk = 1:length(c)
      k = kk-1;           % MATLAB array indices start with 1
      tmp = tmp+c(kk)*exp(j*2*pi/N*k*n);  % Eqn. (3.18)
    end
    x(nn) = tmp;
  end
end
```

Note the similarity in the structures of the two functions. This is due to the similarity of DTFS analysis and synthesis equations.

The functions `ss_dtfs()` and `ss_invdtfs()` are not meant to be computationally efficient or fast. Rather, they are designed to correlate directly with DTFS analysis and synthesis equations (3.19) and (3.18), respectively. A more efficient method of obtaining the same results is to use the FFT (see MATLAB Exercise 7.2) in Chapter 7.

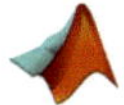

Software resources: `ss_dtfs.m` , `ss_invdtfs.m`

MATLAB Exercise 3.2: Testing DTFS functions

In this exercise we will test the two functions `ss_dtfs()` and `ss_invdtfs()` that were developed in MATLAB Exercise 3.1.

Consider the signal $\tilde{x}[n]$ used in Example 3.4 and shown in Fig. 3.3. It is defined by

$$\tilde{x}[n] = n\,, \quad \text{for } n = 0,\ldots,4\,; \quad \text{and} \quad \tilde{x}[n+5] = \tilde{x}[n]$$

Its DTFS coefficients can be computed using the function `ss_dtfs()` as

```
>> x = [0,1,2,3,4]
>> c = ss_dtfs(x,[0:4])
```

The signal can be reconstructed from its DTFS coefficients using the function `ss_invdtfs()` as

```
>> x = ss_invdtfs(c,[-12:15])
>> stem([-12:15],real(x))
```

The use of the function `real()` is necessary to remove very small imaginary parts due to round-off error. Next, consider the signal of Example 3.5 which is a discrete-time pulse train. We will assume a period of 40 samples and duplicate the DTFS spectra in Fig. 3.5 through 3.7. Let the signal $\tilde{x}_a[n]$ have $L = 3$. Its DTFS coefficients can be computed and graphed with as follows:

```
>> xa=[ones(1,4),zeros(1,33),ones(1,3)]
>> ca = ss_dtfs(xa,[0:39])
>> stem([0:39],real(ca))
```

Note that one period of the signal must be specified using the index range $n = 0,\ldots,39$. The DTFS coefficients for the signal $\tilde{x}_b[n]$ with $L = 5$ are computed and graphed with the following lines:

```
>> xb=[ones(1,6),zeros(1,29),ones(1,5)]
>> cb = ss_dtfs(xb,[0:39])
>> stem([0:39],real(cb))
```

Finally for $\tilde{x}_c[n]$ with $L = 7$ we have

```
>> xc=[ones(1,8),zeros(1,25),ones(1,7)]
>> cc = ss_dtfs(xc,[0:39])
>> stem([0:39],real(cc))
```

Software resources: `mexdt_3_2a.m` , `mexdt_3_2b.m`

MATLAB Exercise 3.3: Improving performance of DTFS functions

Recall that two functions named `ss_dtfs()` and `ss_invdtfs()` were developed in MATLAB Exercise 3.1 for computing forward and inverse DTFS of a periodic signal respectively. While the functions work fine, in writing them we have side-stepped one of the good MATLAB coding principles we established: Avoid looping structures whenever possible and practical. In this exercise we will address the question of what can be done regarding the nested double looping structure within each function.

The DTFS analysis equation was given by Eqn. (3.19) which is repeated here for convenience:

$$\text{Eqn. (3.19):} \qquad \tilde{c}_k = \frac{1}{N} \sum_{n=0}^{N-1} \tilde{x}[n]\, e^{-j(2\pi/N)kn}$$

In order to simplify the notation, let W_N be defined as

$$W_N = e^{-j2\pi/N} \tag{3.241}$$

so that the DTFS analysis equation can be written as

$$\tilde{c}_k = \frac{1}{N} \sum_{n=0}^{N-1} \tilde{x}[n]\, W_N^{kn} \tag{3.242}$$

Let us define a row vector $\mathbf{x}$ and a column vector $\mathbf{A_k}$ as

$$\mathbf{x} = \begin{bmatrix} \tilde{x}[0] & \tilde{x}[1] & \ldots & \tilde{x}[N-1] \end{bmatrix} \quad \text{and} \quad \mathbf{A_k} = \begin{bmatrix} W_N^0 \\ W_N^k \\ \vdots \\ W_N^{(N-1)k} \end{bmatrix} \tag{3.243}$$

Eqn. (3.19) can be written in alternative form as

$$\tilde{c}_k = \frac{1}{N}\, \mathbf{x}\mathbf{A_k} \tag{3.244}$$

This allows us to obtain $\tilde{c}_k$ as the scalar product of two vectors, thus avoiding the need for the inner loop in the code. Consider the function `ss_dtfs2()` listed below.

```
1  function c = ss_dtfs2(x,idx)
2    c = zeros(size(idx));   % Create all-zero vector
3    N = length(x);          % Period of the signal x[n]
4    WN = exp(-j*2*pi/N);    % Eqn. (3.241)
5    n = [0:N-1]';           % Column vector of sample indices for x[n]
6    for kk = 1:length(idx)  % Loop over indices in vector idx
7      k = idx(kk);
8      Ak = WN.^(k*n);       % Eqn. (3.243)
9      c(kk) = x*Ak/N;       % Eqn. (3.244)
10   end
11 end
```

The inverse DTFS function `ss_invdtfs()` could also be modified in a similar way and will not be shown here. An important detail in the use of the function `ss_dtfs2()` is that, for the scalar product in line 9 of the code to work, the vector `x` must be in row format.

It is also possible to eliminate the remaining looping structure over the variable kk by putting vectors $\mathbf{A_k}$ side by side to form a matrix. We will defer discussion of that to Chapter 7 when we develop functions for the discrete Fourier transform (DFT).

Software resources: ss_dtfs2.m , ss_invdtfs2.m

MATLAB Exercise 3.4: A function to implement periodic convolution

In this exercise we will develop a MATLAB function to implement the periodic convolution operation defined by Eqn. (3.46). The function ss_pconv() given below computes the periodic convolution of two length-N signals $\tilde{x}[n]$ and $\tilde{h}[n]$. Vectors x and h hold one period each of signals $\tilde{x}[n]$ and $\tilde{h}[n]$ respectively for $n = 0, \ldots, N-1$. One period of the periodic convolution result $\tilde{y}[n]$ is returned in vector y.

```
1  function y = ss_pconv(x,h)
2    N = length(x);             % Period for all three signals
3    y = zeros(size(x));        % Create all-zero vector
4    for n = 0:N-1
5      tmp = 0;                 % Reset running sum
6      for k = 0:N-1
7        tmp = tmp+ss_per(x,k)*ss_per(h,n-k);  % Eqn. (3.46)
8      end
9      nn = n+1;                % MATLAB array indices start with 1
10     y(nn) = tmp;
11   end
12 end
```

Line 7 of the code is a direct implementation of Eqn. (3.46). It utilizes the function ss_per() which was developed in MATLAB Exercise 1.5 for periodically extending a discrete-time signal. The function ss_pconv() can easily be tested with the signals used in Example 3.8. Recall that the two signals were

$$\tilde{x}[n] = \{\ \ldots,\ \underset{\substack{\uparrow \\ n=0}}{0},\ 1, 2, 3, 4, \ldots\ \}$$

and

$$\tilde{h}[n] = \{\ \ldots,\ \underset{\substack{\uparrow \\ n=0}}{3},\ 3, -3, -2, -1, \ldots\ \}$$

each with a period of $N = 5$. The circular convolution result is obtained as follows:

```
>> x = [0,1,2,3,4]
>> h = [3,3,-3,-2,-1]
>> y = ss_pconv(x,h)
```

Software resources: mexdt_3_4.m , ss_pconv.m

MATLAB Exercise 3.5: Steady-state response of DTLTI system to sinusoidal input

Consider the DTLTI system of Example 3.29 described by the difference equation

$$y[n]-0.9y[n-1]+0.36y[n-2]=x[n]-0.2x[n-1]$$

The steady-state response of the system to the input signal

$$\tilde{x}[n]=5\cos\left(\frac{\pi n}{5}\right)$$

was found to be

$$\tilde{y}[n]=9.9305\cos\left(\frac{\pi n}{5}-0.3139\right)$$

In this exercise we will obtain the response of the system to a sinusoidal signal turned on at $n=0$, that is.

$$x[n]=5\cos\left(\frac{\pi n}{5}\right)u[n]$$

by iteratively solving the difference equation, and compare it to the steady-state response found in Example 3.29. MATLAB function `filter()` will be used in iteratively solving the difference equation. The script listed below computes both responses and compares them.

```
1  % Script: mexdt_3_5.m
2  n = [-10:30];
3  ytilde = 9.9305*cos(pi*n/5-0.3139);     % Steady-state response found in
4                                          %   Example (3.29)
5  x = 5*cos(pi*n/5).*(n>=0);              % Right-sided input signal
6  y = filter([1,-0.2],[1,-0.9,0.36],x);   % Solve the difference equation
7  p1 = stem(n-0.125,ytilde,'b');          % Graph steady-state response
8  hold on
9  p2 = stem(n+0.125,y,'r');               % Graph solution of the diff. eqn.
10 hold off
11 axis([-11,31,-12,12]);
12 xlabel('n');
```

The lines 6 and 8 of the script create two stem plots that are overlaid. In order to observe the two discrete-time signals comparatively, two different colors are used. In addition, horizontal positions of the stems are offset slightly from their integer values, to the left for $\tilde{y}[n]$ and to the right for $y[n]$. The graph produced by the script is shown in Fig. 3.54. The steady-state response is shown in blue and the response to the sinusoidal signal turned on at $n=0$ is shown in red. Notice how the two responses become the same after the transient dies out after about $n=10$.

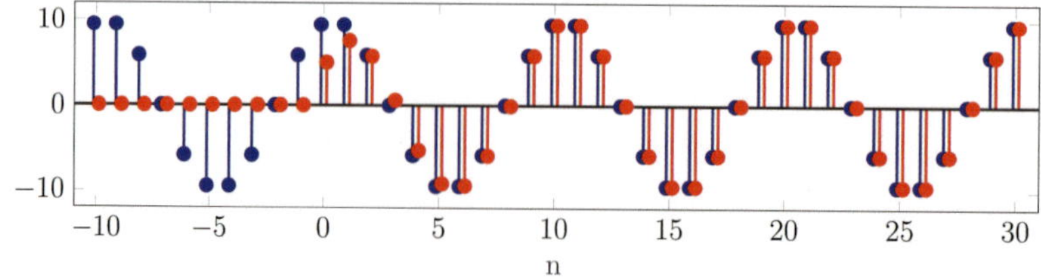

Figure 3.54 – The graph obtained in MATLAB Exercise 3.5.

Software resources: `mexdt_3_5.m`

MATLAB Exercise 3.6: Case study – Echoes and reverberation, part 2

In MATLAB Exercise 2.13 we have used a DTLTI system with the difference equation

$$\text{Eqn. (2.134):} \qquad y[n] = x[n] + r\,x[n-L] + r^2\,x[n-2L] + r^3\,x[n-3L]$$

in an attempt to simulate the reverberation effects of an enclosed space. The impulse response of the feed-forward system is

$$h[n] = \delta[n] + r\,\delta[n-L] + r^2\,\delta[n-2L] + r^3\,\delta[n-3L] \tag{3.245}$$

and the corresponding system function is

$$H_{ff}(\Omega) = 1 + r\,e^{-j\Omega L} + r^2\,e^{-j2\Omega L} + r^3\,e^{-j3\Omega L} \tag{3.246}$$

While this system provided some enhancement for the sound, a good quality simulation of the reverberation effect requires a much larger number of densely packed echoes. The use of a feed-forward system such as the one described would not be very practical for this purpose. As more echoes are added to the difference equation, both the computational cost and the memory requirements would increase significantly.

An alternative for generating the echoes needed for the reverberation effect is to use a system with feedback. Consider the difference equation

$$y[n] = x[n] + r\,y[n-L] \tag{3.247}$$

The main difference is that, instead of using delayed samples of the input signal, we use delayed samples of the output signal to produce echoes. A block diagram that corresponds to the difference equation in Eqn. (3.247) is shown in Fig. 3.55. Note that D_L represents a time delay of L samples.

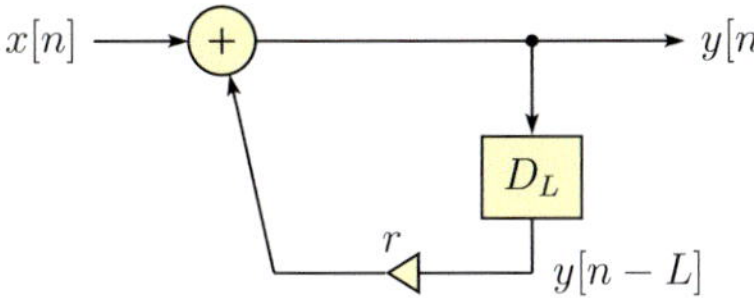

Figure 3.55 – A block diagram for implementing the feedback system with the difference equation in Eqn. (3.247).

It would be interesting to compare the frequency domain characteristics of the two systems derived from the difference equations in Eqns. (2.134) and (3.247). The magnitude responses $|H_{ff}(\Omega)|$ and $|H_{fb}(\Omega)|$ are shown in Fig. 3.56. The system with the difference equation given by Eqn. (3.247) is referred to as a *feedback comb filter* or simply *comb filter* due to the shape of its magnitude spectrum.

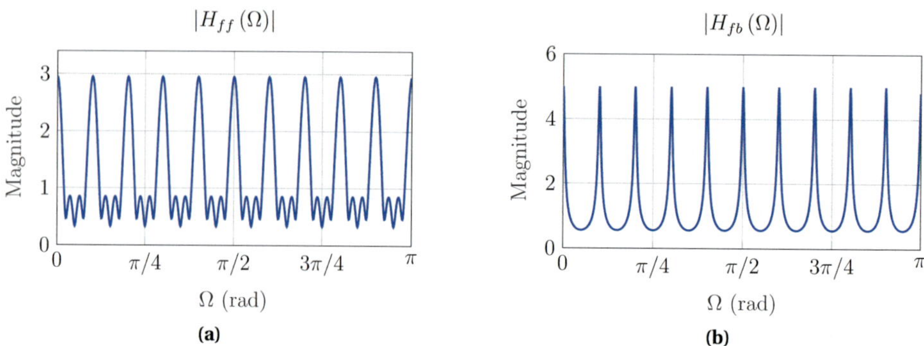

Figure 3.56 – Magnitude of the system function for **(a)** feed-forward system of Eqn. (2.134) and **(b)** feedback system of Eqn. (3.247). In both cases, parameter values $L = 20$ and $r = 0.8$ were used.

The function `ss_comb()` for implementing a comb filter is listed below. Compare the code to that of the function `ss_echo()` developed earlier in MATLAB Exercise 2.13. We only needed to make one small modification to it to obtain the comb filter function `ss_comb()`.

```
function [y,buffer] = ss_comb(x,r,buffer)
  y = x+r*buffer(end);                % Eqn. (3.247)
  buffer = [y;buffer(1:end-1)];  % Update buffer with new output sample
end
```

The first two input arguments `x` and `r` are scalars for the current input sample and the gain factor, respectively. The third argument `buffer` is a length-L column vector that holds the past L samples of the output signal. The script `mexdt_3_6a.m` listed below loads the music signal from the file "`AG_Duet_22050_Hz.flac`" into the vector `x`. We extract the left channel of audio, and then use the function `ss_comb()` to process it. Finally we play back the resulting audio signal.

```
% Script: mexdt_3_6a.m
[x,fs] = audioread('AG_Duet_22050_Hz.flac');
xLeft = x(:,1);                   % Left channel
numSamples = size(xLeft,1);    % Number of samples
yLeft = zeros(size(xLeft));    % Create output vector
buffer = zeros(1024,1);        % L=1024
for i=1:numSamples             % Loop through audio samples
  [yLeft(i),buffer] = ss_comb(xLeft(i),0.8,buffer);
end
sound(yLeft,fs);               % Play back the output vector
```

The function `ss_comb()` takes in one sample of the input signal at a time and computes one sample of the output signal in return. A frame-based version of the same function, called `ss_combf()` is listed below:

```
function [y,buffer] = ss_combf(x,r,buffer)
  y = zeros(size(x));                    % Placeholder for output frame
  frameSize = size(x,1);
  for i=1:frameSize
    y(i,:) = x(i,:)+r*buffer(end,:);       % Eqn. (3.247)
    buffer = [y(i,:);buffer(1:end-1,:)]; % Update buffer
  end
end
```

In this case, the first argument x is a matrix with as many rows as the frame size and as many columns as the number of audio channels. The matrix `buffer` needs to be initialized before calling this function. It needs to have L rows and as many columns as the number of audio channels. A modified script, `mexdt_3_6b.m`, is listed below for real-time processing of an audio file using `ss_combf()`.

```
% Script: mexdt_3_6b.m
% Create an "audio file reader" object
sReader = dsp.AudioFileReader('AG_Duet_22050_Hz.flac','ReadRange',[1,661500]);
% Create an "audio player" object
sPlayer = audioDeviceWriter('SampleRate',sReader.SampleRate);

L = 1024;                              % Corresponds to 46 ms delay
r = 0.8;                               % Gain factor
numChannels = info(sReader).NumChannels;  % Number of channels
buffer = zeros(L,numChannels);         % Initial buffer matrix
while ~isDone(sReader)                 % Repeat until no more frames left
  x = sReader();                       %   Get the next frame of data
  [y,buffer] = ss_combf(x,r,buffer);   %   Compute output frame
  sPlayer(x);                          %   Play back the frame
end

release(sReader);  % We are finished with the input audio file
release(sPlayer);  % We are finished with the audio output device
```

Software resources: `mexdt_3_6a.m` , `mexdt_3_6b.m` , `ss_comb.m` , `ss_combf.m`

Problems

3.1. Determine the DTFS representation of the signal $\tilde{x}[n] = \cos(0.3\pi n)$. Sketch the DTFS spectrum.

3.2. Determine the DTFS representation of the signal $\tilde{x}[n] = 1 + \cos(0.24\pi n) + 3\sin(0.56\pi n)$. Sketch the DTFS spectrum.

3.3. Determine the DTFS coefficients for each of the periodic signals given in Fig. P.3.3.

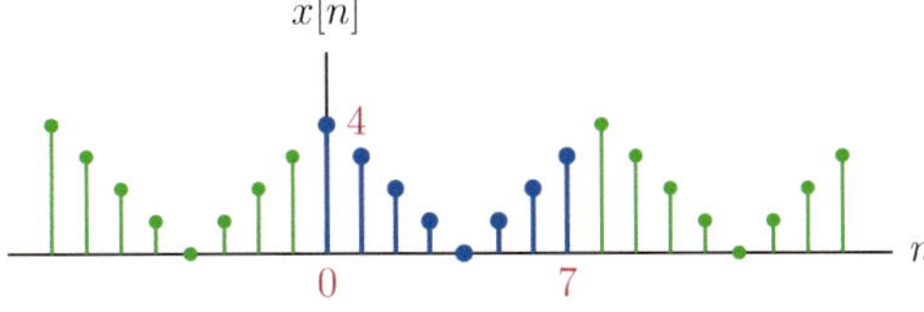

Figure P. 3.3 – (a)

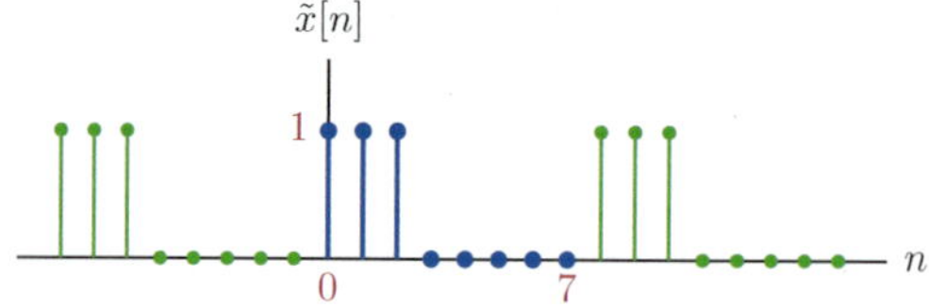

Figure P. 3.3 – (b)

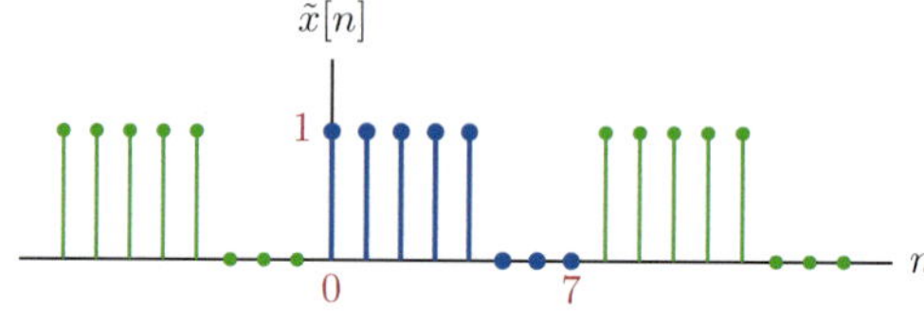

Figure P. 3.3 – (c)

3.4. Consider the periodic signal of Example 3.4. Let $\tilde{g}[n]$ be one sample delayed version of it, that is,

$$\tilde{g}[n] = \tilde{x}[n-1]$$

as shown in Fig. P.3.4.

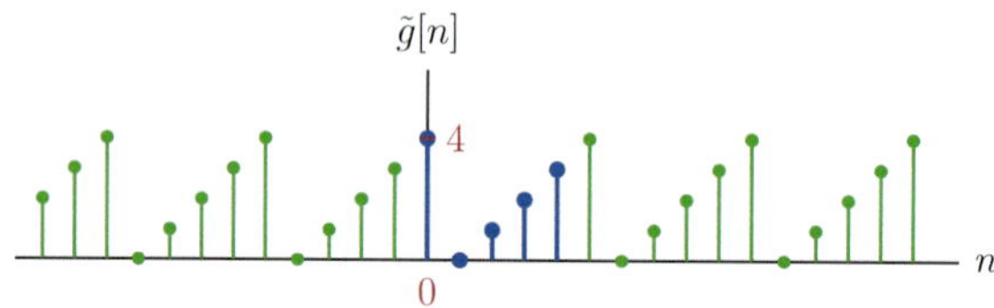

Figure P. 3.4

a. Determine the DTFS coefficients of $\tilde{g}[n]$ directly from the DTFS analysis equation.

b. Determine the DTFS coefficients of $\tilde{g}[n]$ by applying the time shifting property to the coefficients of $\tilde{x}[n]$ that were determined in Example 3.4.

3.5. Using the DTFS result found in part (b) of Problem 3.3 along with linearity and time shifting properties of the DTFS, determine the DTFS coefficients of the periodic signal $\tilde{x}[n]$ shown in Fig. P.3.5.

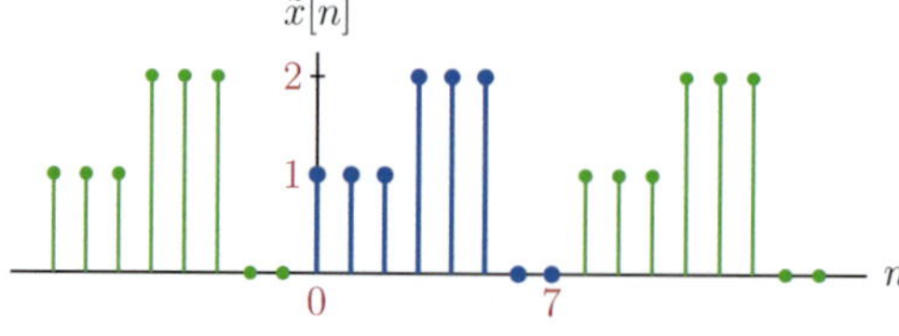

Figure P. 3.5

3.6. Verify that the DTFS coefficients found for the signals in Problem 3.3 satisfy the conjugate symmetry properties outlined in Section 3.2.3. Since the signals are real, DTFS spectra should be conjugate symmetric.

3.7. Consider the periodic signal shown in Fig. P.3.4.

a. Find even and odd components of $\tilde{x}[n]$ so that

$$\tilde{x}[n] = \tilde{x}_e[n] + \tilde{x}_o[n]$$

b. Determine the DTFS coefficients of $\tilde{x}_e[n]$ and $\tilde{x}_o[n]$. Verify the symmetry properties outlined in Section 3.2.3. The spectrum of the even component should be real, and the spectrum of the odd component should be purely imaginary.

3.8. Compute the DTFS coefficients of periodic signal $\tilde{x}[n]$ shown in Fig. P.3.8. What symmetry properties does the signal have? What can you say about the DTFS coefficients of this signal? Are the DTFS coefficients purely real or imaginary? Conjugate symmetric or conjugate antisymmetric? Even or odd? Justify your conclusions.

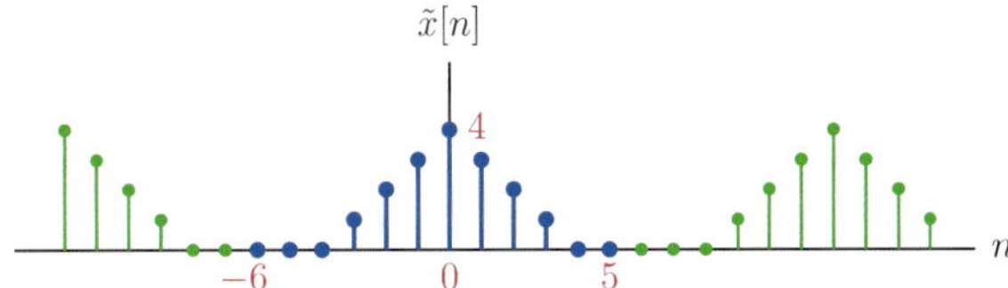

Figure P. 3.8

3.9. Let a N-point signal $x[n]$ be one period of a periodic signal $\tilde{x}[n]$. Let the DTFS coefficients of $x[n]$ be $\tilde{c}_k$. Recall that a signal that is periodic with a period of N is also periodic with a period of $2N$, that is

$$\tilde{x}[n] = \tilde{x}[n+2N]\,, \quad \text{all integer } m$$

a. Suppose a length-$2N$ signal $y[n]$ is defined as two repetitions of $x[n]$:

$$y[n] = x[n] + x[n-N]$$

If $y[n]$ is taken as one period of a periodic signal $\tilde{y}[n]$, express the DTFS coefficients $\tilde{d}_k$ of this signal in terms of the DTFS coefficients $\tilde{c}_k$ of the signal $\tilde{x}[n]$.

b. Since $\tilde{y}[n]$ is essentially the same signal as $\tilde{x}[n]$, the coefficients $\tilde{d}_k$ also represent the latter signal. How can you explain the fact that $\tilde{x}[n]$ has two sets of DTFS coefficients? How does that make sense?

3.10. Let $\tilde{x}[n]$ and $\tilde{h}[n]$ be both periodic with period N. Using the definition of periodic convolution given by Eqn. (3.47), show that the signal

$$\tilde{y}[n] = \tilde{x}[n] \otimes \tilde{h}[n]$$

is also periodic with period N.

3.11. Consider the two periodic signals shown in Fig. P.3.11.

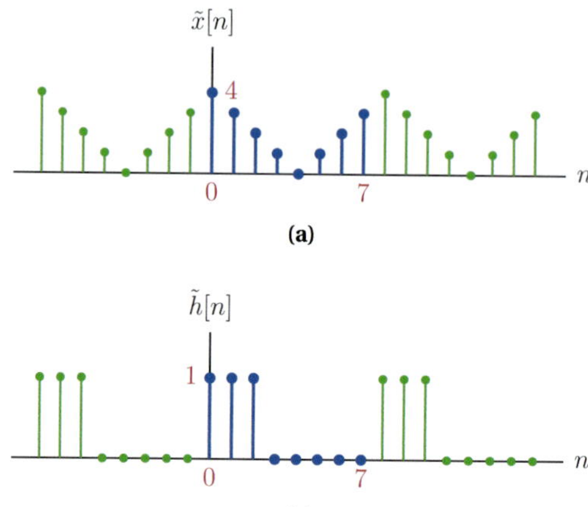

Figure P. 3.11

a. Compute the periodic convolution

$$\tilde{y}[n] = \tilde{x}[n] \otimes \tilde{h}[n]$$

of the two signals.

b. Determine the DTFS coefficients of signals $\tilde{x}[n]$, $\tilde{h}[n]$, and $\tilde{y}[n]$. Verify that the periodic convolution property stated by Eqn. (3.47) holds.

3.12. Repeat Problem 3.11 for the two signals shown in Fig. P.3.12.

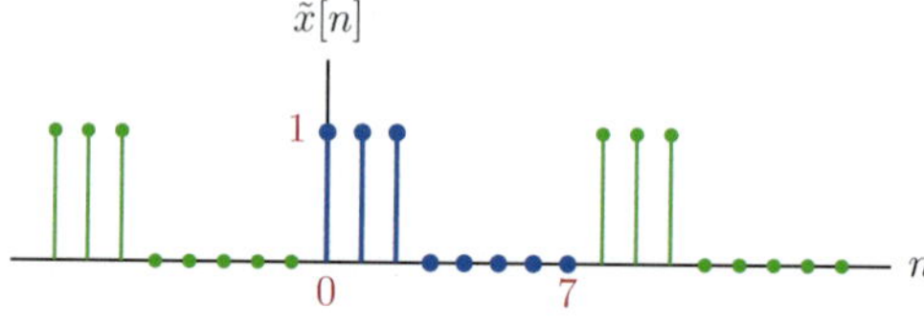

Figure P. 3.12 – (a)

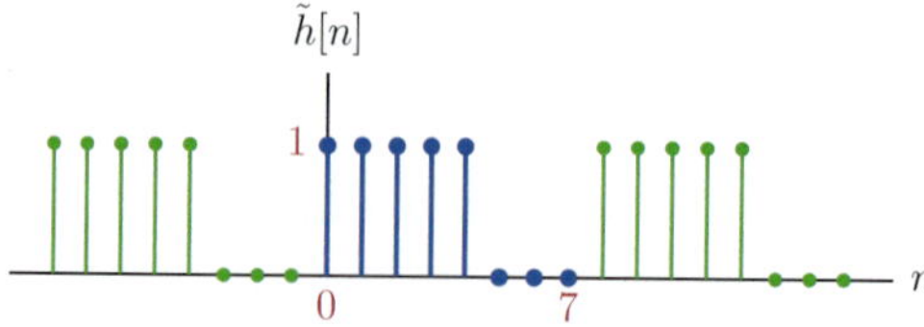

Figure P. 3.12 – (b)

3.13. Find the DTFT of each signal given below. For each, sketch the magnitude and the phase of the transform.

a. $x[n]=\delta[n]+\delta[n-1]$

b. $x[n]=\delta[n]-\delta[n-1]$

c. $x[n]=\begin{cases} 1, & n=0,1,2,3 \\ 0, & \text{otherwise} \end{cases}$

d. $x[n]=\begin{cases} 1, & n=-2,\ldots,2 \\ 0, & \text{otherwise} \end{cases}$

e. $x[n]=(0.7)^n\,u[n]$

f. $x[n]=(0.7)^n\cos(0.2\pi n)\,u[n]$

g. $x[n]=|a|^n\,,\quad |a|<1$

3.14. Find the DTFT of each signal given below. For each, sketch the magnitude and the phase of the transform.

a. $a^n\,u[n]\,,\quad |a|<1$

b. $-a^n\,u[-n-1]\,,\quad |a|>1$

3.15. Find the inverse DTFT of each transform specified below for $-\pi\le\Omega<\pi$.

a. $X(\Omega)=\begin{cases} 1, & |\Omega|<0.2\pi \\ 0, & \text{otherwise} \end{cases}$

b. $X(\Omega)=\begin{cases} 1, & |\Omega|<0.4\pi \\ 0, & \text{otherwise} \end{cases}$

c. $X(\Omega)=\begin{cases} 0, & |\Omega|<0.2\pi \\ 1, & \text{otherwise} \end{cases}$

d. $X(\Omega)=\begin{cases} 1, & 0.1\pi<|\Omega|<0.2\pi \\ 0, & \text{otherwise} \end{cases}$

3.16. Use linearity and time shifting properties of the DTFT to find the transform of each signal given below.

a. $x[n]=(0.5)^n\,u[n-2]$

b. $x[n]=(0.8)^n\,\big(u[n]-u[n-10]\big)$

c. $x[n]=(0.8)^n\,\big(u[n+5]-u[n-5]\big)$

3.17. Use linearity and time reversal properties of the DTFT to find the transform of the signals listed below:

a. $x[n]=(2)^n\,u[-n-1]$

b. $x[n]=(1.25)^n\,u[-n]$

c. $x[n]=(0.8)^{|n|}$

d. $x[n]=(0.8)^{|n|}\,(u[n+5]+u[n-5])$

e. $x[n]=(0.8)^{n}\,\big(u[n+5]-u[n-5]\big)$

3.18.

a. Show that the DTFT of a real-valued and even signal $x[n]$ can be computed as

$$X(\Omega)=x[0]+2\sum_{n=1}^{\infty}x[n]\,\cos(\Omega n)$$

b. If the signal $x[n]$ is real-valued and odd, show that its DTFT can be computed as

$$X(\Omega)=-j2\sum_{n=1}^{\infty}x[n]\,\sin(\Omega n)$$

3.19. Given that $x[n]$ and $X(\Omega)$ are a transform pair, express the DTFT's of the following signals in terms of $X(\Omega)$.

a. $x[-n]$

b. $x^*[n]$

c. $x^*[-n]$

d. $x^*[n-3]$

e. $x[-n+5]$

3.20. Prove that the DTFT of a signal with odd symmetry is purely imaginary. In mathematical terms, prove that if

$$x[-n]=-x[n],$$

then

$$X^*(\Omega)=-X(\Omega)$$

3.21. Signals listed below have even symmetry. Determine the DTFT of each, and graph the magnitude of the transform.

a. $x[n]=\begin{cases}1\,, & n=0\\ 1/2\,, & n=\pm1\\ 0\,, & \text{otherwise}\end{cases}$

b. $x[n]=\begin{cases}5-|n|\,, & |n|\le 4\\ 0\,, & \text{otherwise}\end{cases}$

c. $x[n]=\begin{cases}\cos(0.2\pi n)\,, & n=-4,\ldots,4\\ 0\,, & \text{otherwise}\end{cases}$

3.22. Signals listed below have odd symmetry. For each signal determine the DTFT. Graph the magnitude and the phase of the transform.

a. $x[n] = \begin{cases} -1/4\,, & n=-2 \\ -1/2\,, & n=-1 \\ 1/2\,, & n=1 \\ 1/4\,, & n=2 \\ 0\,, & \text{otherwise} \end{cases}$

b. $x[n] = \begin{cases} n\,, & n=-5,\ldots,5 \\ 0\,, & \text{otherwise} \end{cases}$

c. $x[n] = \begin{cases} \sin(0.2\pi n)\,, & n=-4,\ldots,4 \\ 0\,, & \text{otherwise} \end{cases}$

3.23. Determine the transforms of the signals listed below using the modulation property of the DTFT. Sketch the magnitude and the phase of the transform for each.

a. $x[n] = (0.7)^n \cos(0.2\pi n)\, u[n]$

b. $x[n] = (0.7)^n \sin(0.2\pi n)\, u[n]$

c. $x[n] = \cos(\pi n/5)\, \big(u[n] - u[n-10]\big)$

3.24. Use the differentiation in frequency property of the DTFT to find the transforms of the signals listed below.

a. $x[n] = n\,(0.7)^n\, u[n]$

b. $x[n] = n\,(n+1)\,(0.7)^n\, u[n]$

c. $x[n] = n\,(0.7)^n\, \big(u[n] - u[n-10]\big)$

3.25. Two signals $x[n]$ and $h[n]$ are given by

$$x[n] = u[n] - u[n-10] \qquad \text{and} \qquad h[n] = (0.8)^n\, u[n]$$

a. Determine the DTFT for each signal.

b. Let $y[n]$ be the convolution of these two signals, that is, $y[n] = x[n] * h[n]$. Compute $y[n]$ by direct application of the convolution sum.

c. Determine the DTFT of $y[n]$ by direct application of the DTFT analysis equation. Verify that it is equal to the product of the individual transforms of $x[n]$ and $h[n]$:

$$Y(\Omega) = X(\Omega)\, H(\Omega)$$

3.26. The signal x[n] is defined by

$$x[n] = (0.8)^n\, u[n]$$

Without computing its transform $X(\Omega)$, determine the following:

a. $X(0)$

b. $X(\pi)$

c. $X(\pi/2)$

d. $\dfrac{1}{2\pi}\displaystyle\int_{-\pi}^{\pi} \text{Re}\left\{X(\Omega)\right\} d\Omega$

e. $\dfrac{1}{2\pi}\displaystyle\int_{-\pi}^{\pi} X(\Omega)\, e^{j5\Omega}\, d\Omega$

f. $\displaystyle\int_{-\pi}^{\pi} X(\Omega)\, d\Omega$

3.27. Determine and sketch the DTFT of the following periodic signals.

a. $x[n] = e^{j0.3\pi n}$

b. $x[n] = e^{j0.2\pi n} + 3\, e^{j0.4\pi n}$

c. $x[n] = \cos(\pi n/5)$

d. $x[n] = 2\cos(\pi n/5) + 3\cos(2\pi n/5)$

3.28. Consider the rectangular pulse signal

$$w_R[n] = \begin{cases} 1\,, & -L \le n \le L \\ 0\,, & \text{otherwise} \end{cases}$$

the DTFT of which was determined in Example 3.12, Eqn. (3.86). This is one form of a *rectangular window* sequence that is used in filter design and detection applications of signal processing. Now consider a truncated sinusoidal signal defined by

$$x_T[n] = \begin{cases} \cos(0.2\pi n)\,, & -L \le n \le L \\ 0\,, & \text{otherwise} \end{cases}$$

The signal $x_T[n]$ can be expressed as $x_T[n] = w_R[n]\,\cos(0.2\pi n)$.

a. Determine the DTFT of the signal $x_T[n]$ using the DTFT of the rectangular window along with the modulation property of the DTFT.

b. Sketch the magnitude of the transform $X_T(\Omega)$ found in part (a) in the angular frequency range $-\pi \le \Omega \le \pi$.

c. How does the spectrum $X_T(\Omega)$ found in part (b) compare to the spectrum of the non-truncated sinusoidal signal at the same frequency? Comment.

3.29. An alternative to the rectangular window sequence introduced in Problem 3.28 is the *Hanning window* (also known as the *von Hann window*) given by

$$w_H[n] = \begin{cases} \dfrac{1}{2} + \dfrac{1}{2}\cos\left(\dfrac{\pi n}{L}\right), & -L \le n \le L \\ 0\,, & \text{otherwise} \end{cases}$$

a. Sketch the sequence $w_H[n]$ for $-L \le n \le L$.

b. The rectangular window spectrum $W_R(\Omega)$ was determined in Example 3.12. Making use of that result along with linearity and modularity properties of the DTFT, express the spectrum of the Hanning window sequence in terms of $W_R(\Omega)$. Sketch the magnitude of the spectrum in the angular frequency range $-\pi \le \Omega \le \pi$.

c. Suppose a truncated sinusoidal signal is obtained by multiplying a pure sinusoidal signal with the Hanning window sequence, i.e.,

$$x_T[n] = w_H[n] \cos(0.2\pi n)$$

Determine the spectrum $X_T(\Omega)$ of this signal. How does it compare to the spectrum of the non-truncated sinusoidal signal at the same frequency? Comment.

3.30. Determine and sketch the power spectral density for each signal shown in Fig. P.3.3.

3.31. Consider the periodic signal

$$\tilde{x}[n] = \begin{cases} 1, & -L \le n \le L \\ 0, & L < n < N - L \end{cases} \quad \text{and} \quad \tilde{x}[n+N] = \tilde{x}[n]$$

Recall that the DTFS representation of this signal was found in Example 3.5.

a. Find the power spectral density $S_x(\Omega)$.

b. Let $N = 40$ and $L = 3$. Determine what percentage of signal power is preserved if only three harmonics are kept and the others are discarded.

c. Repeat part (b) with $N = 40$ and $L = 6$.

3.32. Determine the autocorrelation function $\tilde{r}_{xx}[m]$ for each of the signals given below.

a. $x[n] = \cos(0.4\pi n)$

b. $x[n] = \cos(0.15\pi n) + \cos(0.2\pi n)$

c. $x[n] = \sin(0.3\pi n + \pi/4)$

3.33. Determine the autocorrelation function $r_{xx}[m]$ for each of the signals given below.

a. $x[n] = u[n] - u[n-10]$

b. $x[n] = a^n u[n]\,, \quad |a| < 1$

3.34. A DTLTI system is characterized by the difference equation

$$y[n] + y[n-1] + 0.89\, y[n-2] = x[n] + 2\, x[n-1]$$

Determine the steady-state response of the system to the following input signals:

a. $x[n] = e^{j0.2\pi n}$

b. $x[n] = \cos(0.2\pi n)$

c. $x[n] = 2\sin(0.3\pi n)$

d. $x[n] = 3\cos(0.1\pi n) - 5\sin(0.2\pi n)$

Hint: First find the system function $H(\Omega)$.

3.35. Determine the steady-state response of a length-4 moving average filter to the following signals.

a. $x[n]=e^{j0.2\pi n}$

b. $x[n]=\cos(0.2\pi n)$

c. $x[n]=2\sin(0.3\pi n)$

d. $x[n]=3\cos(0.1\pi n)-5\sin(0.2\pi n)$

Hint: First find the system function $H(\Omega)$.

3.36. A DTLTI system is characterized by the difference equation

$$y[n]+y[n-1]+0.89\,y[n-2]=x[n]+2\,x[n-1]$$

Using the technique outlined in Section 3.6.3, determine the steady-state response of the system to each of the periodic signals shown in Fig. P.3.3.

MATLAB Problems

3.37. Develop a MATLAB script to compute the DTFS coefficients of each of the signals considered in Problem 3.3 and shown in Figs. 3.3a–c. Compare the results to your pencil-and-paper solutions. You may want to make use of the function `ss_dtfs()` which was developed in MATLAB Exercise 3.1.

3.38. Develop a MATLAB script to compute the DTFS coefficients of the signal considered in Problem 3.8 and shown in Fig. 3.8. Compare the results to your pencil-and-paper solution. You may want to make use of the function `ss_dtfs()` which was developed in MATLAB Exercise 3.1.

3.39. Consider the periodic signals $\tilde{x}[n]$ and $\tilde{g}[n]$ shown in Fig. P.3.39.

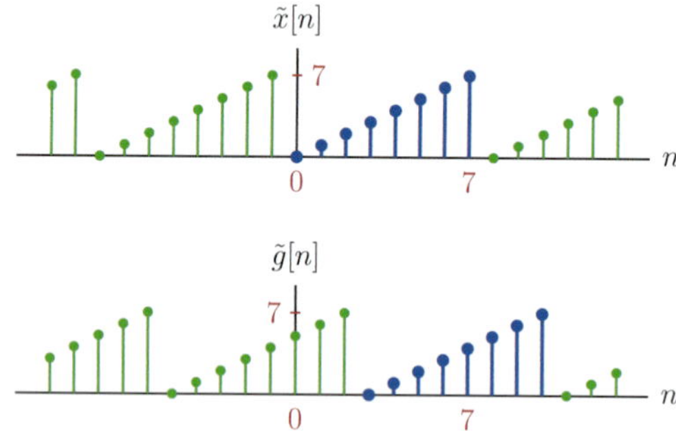

Figure P. 3.39

Write a MATLAB script to do the following:

a. Compute the DTFS coefficients of each signal using the function `ss_dtfs()` developed in MATLAB Exercise 3.1.

b. Compute the DTFS coefficients of $\tilde{g}[n]$ by multiplying the DTFS coefficients of $\tilde{x}[n]$ with the proper exponential sequence as dictated by the time shifting property.

c. Compare the results found in parts (a) and (b).

3.40. Consider again the periodic signals used in Problem 3.39 and shown in Fig. P.3.39. Both signals are real-valued. Consequently, their DTFS coefficients must satisfy the conjugate symmetry property as stated by Eqn. (3.34). Verify this with the values obtained in Problem 3.39.

3.41. Refer to the periodic pulse train used in Example 3.5. Write a MATLAB script to compute and graph its DTFS spectrum. Use the script to obtain graphs for the following parameter configurations:

a. $N = 30\,, \quad L = 5$

b. $N = 30\,, \quad L = 8$

c. $N = 40\,, \quad L = 10$

d. $N = 40\,, \quad L = 15$

3.42. Use the MATLAB script developed in Problem 3.41 to produce graphs that duplicate the three cases shown in Fig. 3.20 parts (a),(b), and (c).

3.43. Refer to Problem 3.31.

a. Let $N = 40$ and $L = 3$. Develop a script to compute and graph a finite-harmonic approximation to $x[n]$ using 3 harmonics.

b. Repeat part (a) using a signal with $N = 40$ and $L = 6$.

3.44. Refer to Problem 3.35. Write a script to compute the response of the length-4 moving average filter to each of the signals specified in the problem 3.35. Use the function `ss_movavg4()` that was developed in MATLAB Exercise 2.1 in Chapter 2. For each input signal compute the response for $n = 0, \ldots, 49$. Compare the steady-state part of the response (in this case it will be after the first three samples) to the theoretical result obtained in Problem 3.35.

3.45. Refer to Problem 3.36.

a. Develop a script to iteratively solve the difference equation given. The system should be assumed initially relaxed, that is, $y[-1] = y[-2] = 0$.

b. Find the response of the system to each of the input signals in Fig. P.3.3 for $n = 0, \ldots, 49$.

c. In each of the responses, disregard the samples for $n = 0, \ldots, 24$, and compare samples $n = 25, \ldots, 49$ to the theoretical steady-state response computed in Problem 3.36.

Hint: You may wish to use the function `ss_dper()` that was developed in MATLAB Exercise 1.5 for creating the periodic signals needed.

3.46. Consider the comb filter discussed in MATLAB Exercise 3.6. Refer to Eqn. (3.247) and Fig. 3.55.

a. Develop a MATLAB script to compute and graph the impulse response of the comb filter. You may want to use functions `ss_dimpulse()` and `ss_comb()` in your script to generate an impulse signal and then process it through a comb filter. Use parameter values $L = 1024$ and $r = 0.8$ for the time delay and the gain factor, respectively. Graph the impulse response for the sample index range $n = 0, \ldots, 4999$.

b. In your script, modify the time delay L and observe its effect on the impulse response. Try with values $L = 64$, 128, 256. Keep the gain factor fixed at $r = 0.8$.

c. Set $L = 1024$ and modify the gain factor r. How does the parameter r affect the impulse response? Try with $r = 0.5$, 0.7, 0.9.

MATLAB Project

3.47. Refer to Problems 3.28 and 3.29. In Problem 3.28 the rectangular window sequence $w_R[n]$ was defined, and its use for representing a truncated sinusoidal signal was discussed. In Problem 3.29, the Hanning window was introduced, and a method for obtaining its DTFT $W_H(\Omega)$ through the use of the rectangular window spectrum $W_R(\Omega)$ was explored. This project is about graphing the spectra of the two window sequences and comparing their characteristics.

Develop a MATLAB script to study the DTFT spectra of rectangular and Hanning window sequences. Follow the specific steps listed below:

a. Within your script, write a local function to compute and return the DTFT of the rectangular window sequence. Your local function may have the following syntax:

```
function WOmg = rwSpectrum(Omega,L)
```

The argument `Omega` is a vector of angular frequency values. The argument L is the parameter of the rectangular window. The function should return the DTFT spectrum computed at specified frequencies. See Eqn. (3.86).

b. Write a second local function that computes the spectrum of the Hanning window sequence at the specified set of frequencies. This local function may have the syntax

```
function WOmg = hanningSpectrum(Omega,L)
```

and may call the local function `rwSpectrum()`.

c. Using the two local functions, compute the spectra $W_R(\Omega)$ and $W_H(\Omega)$ for $L = 20$ and the angular frequency range $-\pi \le \Omega \le \pi$.

d. Compute dB magnitude of each spectrum with the peak value normalized to 0 dB. For a spectrum $W(\Omega)$, the normalized dB magnitude is

$$|W(\Omega)|_{dB} = 20\log_{10}\left(\frac{|W(\Omega)|}{\max\{|W(\Omega)|\}}\right)$$

where $\max\{|W(\Omega)|\}$ is the largest magnitude in the spectrum.

e. Graph the dB magnitudes $|W_R(\Omega)|_{dB}$ and $|W_H(\Omega)|_{dB}$ on the same coordinate system and compare. In each spectrum, identify the main lobe (centered around $\Omega = 0$), and the first sidelobe on either side of the main lobe. How wide is the main lobe for each window? What is the dB difference between the peak of the main lobe and the peak of the first sidelobe for each window?

f. Repeat part (e) with $L = 40$. Does the width of the main lobe change? Does the dB difference between the peak of the main lobe and the peak of the first sidelobe change? Check for each window sequence.

CHAPTER 4

SAMPLING AND RECONSTRUCTION

Chapter Objectives

- Understand the concept of sampling for converting a continuous-time signal to a discrete-time signal.
- Learn how sampling affects the frequency-domain characteristics of the signal, and what precautions must be taken to ensure that the signal obtained through sampling is an accurate representation of the original.
- Consider the issue of reconstructing an analog signal from its sampled version. Understand various interpolation methods used and the spectral relationships involved in reconstruction.
- Discuss methods for changing the sampling rate of a discrete-time signal. Understand upsampling and downsampling operations, interpolation, and decimation.

4.1 Introduction

The term *sampling* refers to the act of periodically measuring the amplitude of a continuous-time signal and constructing a discrete-time signal with the measurements. If certain conditions are satisfied, a continuous-time signal can be completely represented by measurements (samples) taken from it at uniform intervals. This allows us to store and manipulate continuous-time signals on a digital computer.

Consider, for example, the problem of keeping track of temperature variations in a classroom. The temperature of the room can be measured at any time instant, and can therefore be modeled

DOI: 10.1201/9781003570462-4

as a continuous-time signal $x_a(t)$. Alternatively, we may choose to check the room temperature once every 10 minutes and construct a table similar to that shown in Table 4.1.

Table 4.1 – Sampling the temperature signal.

Time	8:30	8:40	8:50	9:00	9:10	9:20	9:30	9.40
Temp. (°C)	22.4	22.5	22.8	21.6	21.7	21.7	21.9	22.2
Index n	0	1	2	3	4	5	6	7

If we choose to index the temperature values with integers as shown in the third row of Table 4.1 then we could view the result as a discrete time signal $x[n]$ in the form

$$x[n] = \{\underset{\substack{\uparrow \\ n=0}}{22.4}, 22.5, 22.8, 21.6, 21.7, 21.7, 21.9, 22.2, \ldots\} \tag{4.1}$$

Thus the act of sampling allows us to obtain a discrete-time signal $x[n]$ from the continuous-time signal $x_a(t)$. While any signal can be sampled with any time interval between consecutive samples, there are certain questions that need to be addressed before we can be confident that $x[n]$ provides an accurate representation of $x_a(t)$. We may question, for example, the decision to wait for 10 minutes between temperature measurements. Do measurements taken 10 minutes apart provide enough information about the variations in temperature? Are we confident that no significant variations occur between consecutive measurements? If that is the case, then could we have waited for 15 minutes between measurements instead of 10 minutes?

Generalizing the temperature example used above, the relationship between the continuous-time signal $x_a(t)$ and its discrete-time counterpart $x[n]$ is

$$x[n] = x_a(t)\Big|_{t=nT_s} = x_a(nT_s) \tag{4.2}$$

where T_s is the *sampling interval*, that is, the time interval between consecutive samples. It is also referred to as the *sampling period*. The reciprocal of the sampling interval is called the *sampling rate* or the *sampling frequency*:

$$f_s = \frac{1}{T_s} \tag{4.3}$$

The relationship between a continuous-time signal and its discrete-time version is illustrated in Fig. 4.1.

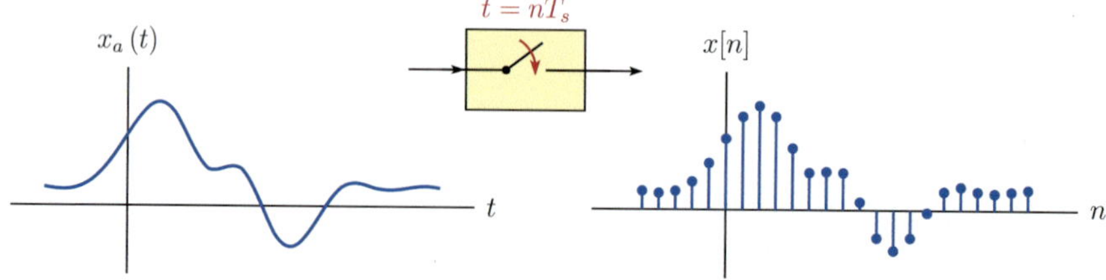

Figure 4.1 – Graphical representation of sampling relationship.

The claim that it may be possible to represent a continuous-time signal without any loss of information by a discrete set of amplitude values measured at uniformly-spaced time intervals may be a bit counter-intuitive at first: How is it possible that we do not lose any information by merely measuring the signal at a discrete set of time instants and ignoring what takes place between those measurements? This question is perhaps best answered by posing another question: Does the behavior of the signal between measurement instants constitute worthwhile information, or is it just redundant behavior that could be completely predicted from the set of measurements? If it is the latter, then we will see that the measurements (samples) taken at intervals of T_s will be sufficient to reconstruct the continuous-time signal $x_a(t)$.

Sampling forms the basis of digital signals we encounter everyday in our lives. For example, an audio signal played back through online streaming, from a file on a computer, or from a compact disc is a signal that has been captured and recorded at discrete time instants. When we look at the amplitude values of the signal, we only see values taken at equally-spaced time instants (often at a rate of 44,100 times per second) with missing amplitude values in between these instants. This is perfectly fine since all the information contained in the original audio signal in the studio can be accounted for in these samples. An image captured by a digital camera is stored in the form of a dense rectangular grid of colored dots (known as pixels). When printed and viewed from an appropriate distance, we cannot tell the individual pixels apart. Similarly, a movie stored in a computer file or on a video disc is stored in the form of consecutive snapshots, taken at equal time intervals. If enough snapshots are taken from the scene and are played back in sequence with the right timing, we perceive motion.

We begin by considering the sampling of continuous-time signals in Section 4.2. The idea of impulse sampling and its implications on the frequency spectrum are studied. Nyquist sampling criterion is introduced. Conversion of the impulse-sampled signal to a discrete-time signal is discussed along with the effect of this conversion on the frequency spectrum. Practical issues in sampling applications are also briefly discussed. The issue of reconstructing a continuous-time signal from its sampled version is the topic of Section 4.3.

4.2 Sampling of a Continuous-Time Signal

Consider a periodic impulse train $\tilde{p}(t)$ with period T_s:

$$\tilde{p}(t) = \sum_{n=-\infty}^{\infty} \delta(t - nT_s) \tag{4.4}$$

Multiplication of any signal $x(t)$ with this impulse train $\tilde{p}(t)$ would result in amplitude information for $x(t)$ being retained only at integer multiples of the period T_s. Let the signal $x_s(t)$ be defined as the product of the original signal and the impulse train, i.e.,

$$\begin{aligned} x_s(t) &= x_a(t)\,\tilde{p}(t) \\ &= x_a(t) \sum_{n=-\infty}^{\infty} \delta(t - nT_s) \\ &= \sum_{n=-\infty}^{\infty} x_a(nT_s)\,\delta(t - nT_s) \end{aligned} \tag{4.5}$$

We will refer to the signal $x_s(t)$ as the *impulse-sampled* version of $x(t)$. Fig. 4.2 illustrates the relationship between the signals involved in impulse sampling. It is important to understand that the impulse-sampled signal $x_s(t)$ is still a continuous-time signal. The subject of converting $x_s(t)$ to a discrete-time signal will be discussed in Section 4.2.2.

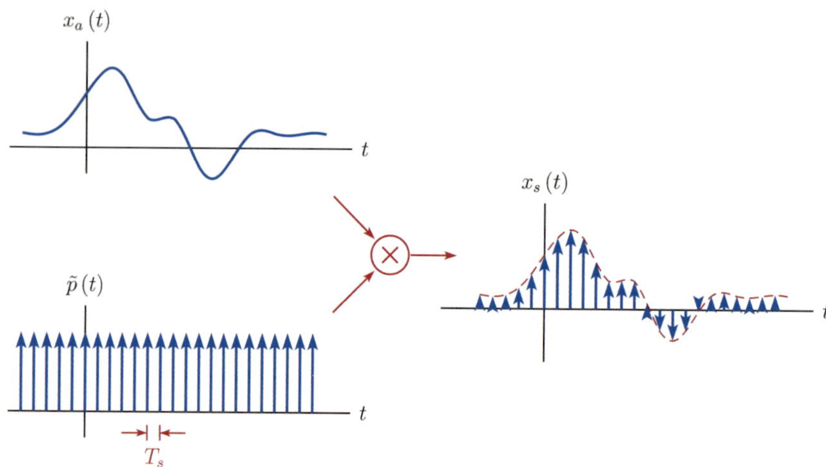

Figure 4.2 – Impulse-sampling a signal: **(a)** Continuous-time signal $x(t)$, **(b)** the pulse train $p(t)$, and **(c)** impulse-sampled signal $x_s(t)$.

Consider the periodic impulse train $\tilde{p}(t)$ which is shown in detail in Fig. 4.3. At this point, we need to pose a critical question: How dense must the impulse train $\tilde{p}(t)$ be so that the impulse-sampled signal $x_s(t)$ is an accurate and complete representation of the original signal $x_a(t)$? In other words, what are the restrictions on the sampling interval T_s or, equivalently, the sampling rate f_s? In order to answer this question, we need to develop some insight into how the frequency spectrum of the impulse-sampled signal $x_s(t)$ relates to the spectrum of the original signal $x_a(t)$.

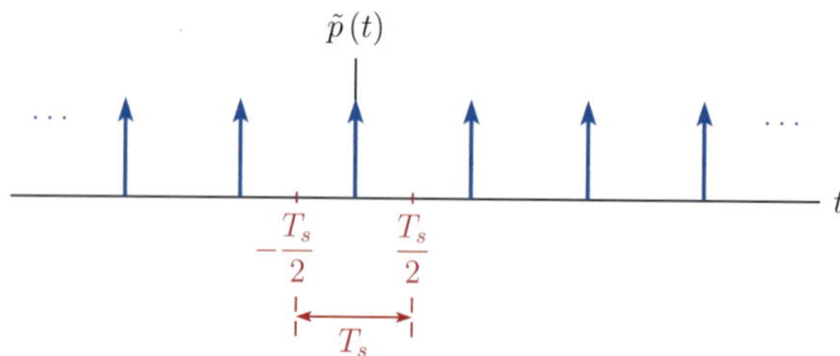

Figure 4.3 – Periodic impulse train $\tilde{p}(t)$.

Spectral relationships in impulse sampling:

The periodic signal $\tilde{p}(t)$ can be represented in an exponential Fourier series expansion in the form

$$\tilde{p}(t) = \sum_{k=-\infty}^{\infty} c_k \, e^{jk\omega_s t} \tag{4.6}$$

where ω_s is both the sampling rate in rad/s and the fundamental frequency of the impulse train. It is computed as $\omega_s = 2\pi f_s = 2\pi / T_s$. The EFS coefficients for $\tilde{p}(t)$ are found as

$$c_k = \frac{1}{T_s} \int_{-T_s/2}^{T_s/2} \tilde{p}(t)\, e^{-jk\omega_s t}\, dt = \frac{1}{T_s}\,, \qquad \text{all } k \tag{4.7}$$

Substituting the EFS coefficients found in Eqn. (4.7) into Eqn. (4.6), the impulse train $p(t)$ becomes

$$\tilde{p}(t) = \frac{1}{T_s} \sum_{k=-\infty}^{\infty} e^{jk\omega_s t} \tag{4.8}$$

Finally, using Eqn. (4.8) in Eqn. (4.5) we get

$$x_s(t) = \frac{1}{T_s} \sum_{k=-\infty}^{\infty} x_a(t)\, e^{jk\omega_s t} \tag{4.9}$$

for the sampled signal $x_s(t)$. In order to determine the frequency spectrum of the impulse sampled signal $x_s(t)$ let us take the Fourier transform of both sides of Eqn. (4.9).

$$\begin{aligned} \mathcal{F}\left\{x_s(t)\right\} &= \mathcal{F}\left\{\frac{1}{T_s} \sum_{k=-\infty}^{\infty} x_a(t)\, e^{jk\omega_s t}\right\} \\ &= \frac{1}{T_s} \sum_{k=-\infty}^{\infty} \mathcal{F}\left\{x_a(t)\, e^{jk\omega_s t}\right\} \end{aligned} \tag{4.10}$$

Linearity property of the Fourier transform was used in obtaining the result in Eqn. (4.10). Furthermore, using the frequency shifting property of the Fourier transform, the term inside the summation becomes

$$\mathcal{F}\left\{x_a(t)\, e^{jk\omega_s t}\right\} = X_a(\omega - k\omega_s) \tag{4.11}$$

The frequency-domain relationship between the signal $x_a(t)$ and its impulse-sampled version $x_s(t)$ follows from Eqns. (4.10) and (4.11).

Fourier transform of impulse-sampled signal:

The Fourier transform of the impulse sampled signal is related to the Fourier transform of the original signal by

$$X_s(\omega) = \frac{1}{T_s} \sum_{k=-\infty}^{\infty} X_a(\omega - k\omega_s) \tag{4.12}$$

This relationship can also be written using frequencies in Hz as

$$X_s(f) = \frac{1}{T_s} \sum_{k=-\infty}^{\infty} X_a(f - kf_s) \tag{4.13}$$

This is a very significant result. The spectrum of the impulse-sampled signal is obtained by adding frequency-shifted versions of the spectrum of the original signal, and then scaling the sum by $1/T_s$. The terms of the summation in Eqn. (4.12) are shifted by all integer multiples of the sampling rate ω_s. Fig. 4.4 illustrates this.

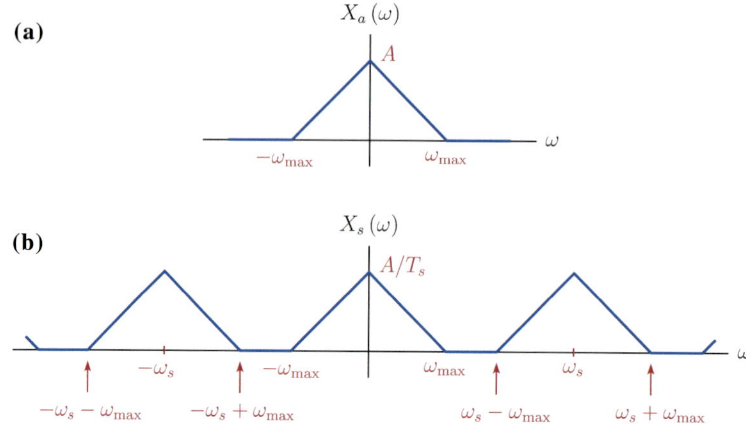

Figure 4.4 – Effects of impulse-sampling on the frequency spectrum: **(a)** The example spectrum $X_a(\omega)$ of the original signal $x_a(t)$ and **(b)** the spectrum $X_s(\omega)$ of the impulse-sampled signal $x_s(t)$.

For the impulse-sampled signal to be an accurate and complete representation of the original signal, $x_a(t)$ should be recoverable from $x_s(t)$. This in turn requires that the frequency spectrum $X_a(\omega)$ be recoverable from the frequency spectrum $X_s(\omega)$. In Fig. 4.4 the example spectrum $X_a(\omega)$ used for the original signal is bandlimited to the frequency range $|\omega| \le \omega_{\text{max}}$. Sampling rate ω_s is chosen such that the repetitions of $X_a(\omega)$ do not overlap with each other in the construction of $X_s(\omega)$. As a result, the shape of the original spectrum $X_a(\omega)$ is preserved within the sampled spectrum $X_s(\omega)$. This ensures that $x_a(t)$ is recoverable from $x_s(t)$.

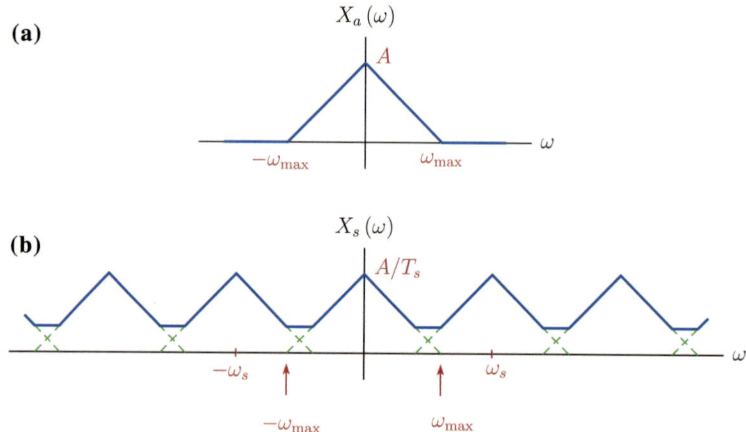

Figure 4.5 – Effects of impulse-sampling on the frequency spectrum when the sampling rate chosen is too low: **(a)** The example spectrum $X_a(\omega)$ of the original signal $x_a(t)$ and **(b)** the spectrum $X_s(\omega)$ of the impulse-sampled signal $x_s(t)$.

Alternatively, consider the scenario illustrated by Fig. 4.5 where the sampling rate chosen causes overlaps to occur between the repetitions of the spectrum. In this case $X_a(\omega)$ cannot be recovered from $X_s(\omega)$. Consequently, the original signal $x_a(t)$ cannot be recovered from its sampled version. Under this scenario, replacing the signal with its sampled version represents an irrecoverable loss of information.

Interactive App: Frequency spectrum of the sampled signal

The interactive app in `appSampling1.m` illustrates the process of obtaining the spectrum $X_s(\omega)$ from the original spectrum $X_a(\omega)$ based on Eqn. (4.12) and Figs. 4.4 and 4.5. Sampling rate f_s and the bandwidth f_{max} of the signal to be sampled can be varied using slider controls. (In the preceding development we have used radian frequencies ω_s and ω_{max}. They are related to related to frequencies in Hz used by the app through $\omega_s = 2\pi f_s$ and $\omega_{\text{max}} = 2\pi f_{\text{max}}$.)

Spectra $X_a(f)$ and $T_s X_s(f)$ are computed and graphed. Note that, in Fig. 4.4(b), the peak magnitude of $X_s(f)$ is proportional to $f_s = 1/T_s$. Same can be observed from the $1/T_s$ factor in Eqn. (4.12). As a result, graphing $X_s(f)$ directly would have required a graph window tall enough to accommodate the necessary magnitude changes as the sampling rate $f_s = 1/T_s$ is varied, and still show sufficient detail. Instead, we opt to graph

$$T_s X_s(f) = \sum_{k=-\infty}^{\infty} X_a(f - kf_s)$$

to avoid the need to deal with scaling issues. Individual terms $X_a(f - kf_s)$ in Eqn. (4.12) are also shown, although they may be under the sum $T_s X_s(f)$, and thus invisible, when the spectral sections do not overlap. When there is an overlap of spectral sections as in Fig. 4.5(b), part of each individual term becomes visible in red dashed lines. They may also be made visible by unchecking the box "`Show sum of terms`". When spectral sections overlap, the word "Aliasing" is displayed, indicating that the spectrum is being corrupted through the sampling process.

a. Set the sampling rate at $f_s = 250$ Hz, and the maximum frequency in the signal $x_a(t)$ to $f_{max} = 100$ Hz. Uncheck the box labeled "`Show sum of terms`". Figure out the frequency gap between the repetitions of spectral segments in the graph for $T_s X_s(f)$.

b. Gradually increase f_{max}. Observe the changes in the lower graph. At what value of f_{max} do spectral segments touch each other?

c. Keep increasing f_{max} and observe spectral segments overlap. Check the box labeled "`Show sum of terms`" and observe the sum of overlapping spectral segments.

d. Set $f_s = 250$ Hz and $f_{max} = 100$ Hz again. Now gradually reduce the sampling rate and observe the changes in the lower graph.

Software resource: `appSampling1.m`

Example 4.1: Impulse-sampling a right-sided exponential

Consider a right-sided exponential signal

$$x_a(t) = e^{-100t}\, u(t)$$

This signal is to be impulse sampled. Determine and graph the spectrum of the impulse sampled signal $x_s(t)$ for sampling rates $f_s = 200$ Hz, $f_s = 400$ Hz, and $f_s = 600$ Hz.

Solution: The frequency spectrum of the signal $x_a(t)$ is

$$X_a(f) = \frac{1}{100 + j2\pi f}$$

which is graphed in Fig. 4.6.

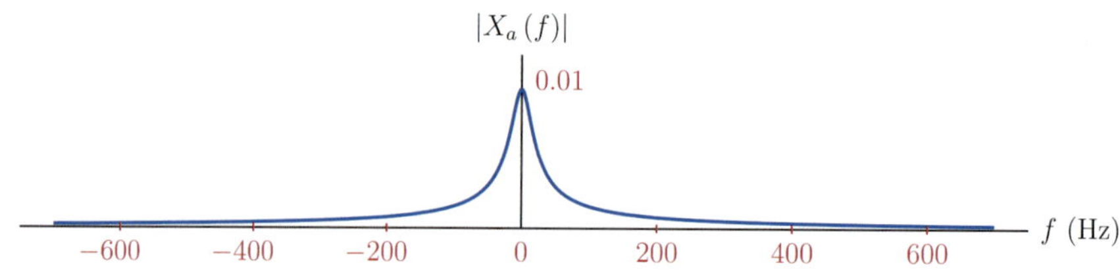

Figure 4.6 – Frequency spectrum of the signal $x_a(t)$ for Example 4.1.

Impulse-sampling $x_a(t)$ at a sampling rate of $f_s = /1/T_s$ yields the signal

$$x_s(t) = \sum_{n=-\infty}^{\infty} e^{-100nT_s}\, u(nT_s)\, \delta(t - nT_s)$$

$$= \sum_{n=0}^{\infty} e^{-100nT_s}\, \delta(t - nT_s)$$

The frequency spectrum of this impulse sampled signal is

$$X_s(f) = \int_{-\infty}^{\infty} \left[\sum_{n=0}^{\infty} e^{-100nT_s}\, \delta(t - nT_s) \right] e^{-j2\pi f t}\, dt$$

$$= \sum_{n=0}^{\infty} e^{-100nT_s} \left[\int_{-\infty}^{\infty} \delta(t - nT_s)\, e^{-j2\pi f t}\, dt \right]$$

$$= \sum_{n=0}^{\infty} e^{-100nT_s}\, e^{-j2\pi f nT_s}$$

which can be put into a closed form through the use of the geometric series formula (see Appendix C) as

$$X_s(f) = \frac{1}{1 - e^{-100T_s}\, e^{-j2\pi f T_s}}$$

The resulting spectrum is shown in Fig. 4.7(a),(b),(c) for $f_s = 200$ Hz, $f_s = 400$ Hz, and $f_s =$ 600 Hz, respectively. Note that the overlap of spectral segments $(1/T_s)\, X_a(f - kf_s)$ causes the shape of the spectrum $X_s(f)$ to be different than the shape of $X(f)$ since the right-sided exponential is not band-limited. This distortion of the spectrum is present in all three cases, but seems to be more pronounced for $f_s = 200$ Hz than it is for the other two choices.

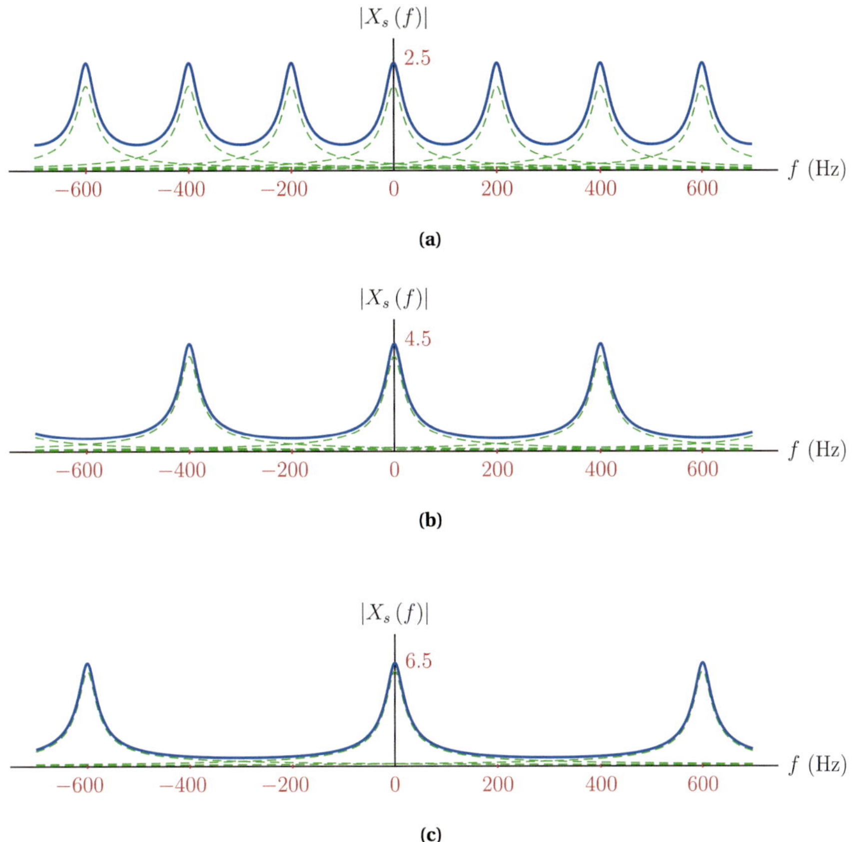

Figure 4.7 – The spectrum of impulse sampled signal for Example 4.1 **(a)** for $f_s = 200$ Hz, **(b)** for $f_s = 400$ Hz, and **(c)** for $f_s = 600$ Hz.

Software resource: `exdt_4_1.m`

Interactive App: Impulse-sampling a right-sided exponential signal

The interactive app in `appSampling2.m` is based on Example 4.2. The continuous-time signal $x_a(t) = \exp(-100t)\, u(t)$ and its impulse sampled version $x_s(t)$ are shown at the top. The bottom graph displays the magnitude of the spectrum

$$T_s X_s(f) = \sum_{k=-\infty}^{\infty} X\left(f - kf_s\right)$$

as well as the magnitudes of contributing terms $X\left(f - kf_s\right)$. Our logic in graphing the magnitude of $T_s X_s(t)$ instead of $X_s(t)$ is the same as in the previous interactive app `appSampling1.m`. Sampling rate f_s may be adjusted using a slider control. Time and frequency domain plots are simultaneously updated to show the effect of sampling rate adjustment.

a. Set the sampling rate at $f_s = 600$ Hz, and observe the spectrum of the impulse-sampled signal. Notice the overlap of spectral segments.

b. Gradually decrease the sampling rate. Observe how the effect of aliasing on the spectrum of $x_s(t)$ increases.

Software resource: `appSampling2.m`

Software resource: See MATLAB Exercise 4.1.

4.2.1 Nyquist sampling criterion

As illustrated in Example 4.1, if the range of frequencies in the signal $x_a(t)$ is not limited, then the periodic repetition of spectral components dictated by Eqn. (4.12) creates overlapped regions. This effect is known as *aliasing* and it results in the shape of the spectrum $X_s(f)$ being different than the original spectrum $X_a(f)$. Once the spectrum is aliased, the original signal is no longer recoverable from its sampled version. Aliasing could also occur when sampling signals that contain a finite range of frequencies if the sampling rate is not chosen carefully.

Let $x_a(t)$ be a signal the spectrum $X_a(f)$ of which is band-limited to f_{max} meaning it exhibits no frequency content for $|f| > f_{\text{max}}$. If $x_a(t)$ is impulse-sampled to obtain the signal $x_s(t)$, the frequency spectrum of the resulting signal is given by Eqn. (4.12). If we want to be able to recover the signal from its impulse-sampled version, then the spectrum $X_a(f)$ must also be recoverable from the spectrum $X_s(f)$. This in turn requires that no overlaps occur between periodic repetitions of spectral segments. Refer to Fig. 4.8.

In order to keep the left edge of the spectral segment centered at $f = f_s$ from interfering with the right edge of the spectral segment centered at $f = 0$, we need

$$f_s - f_{\text{max}} \ge f_{\text{max}} \tag{4.14}$$

and therefore

$$f_s \ge 2 f_{\text{max}} \tag{4.15}$$

Nyquist sampling criterion:

For the impulse-sampled signal to form an accurate representation of the original signal, the sampling rate must be at least twice the highest frequency in the spectrum of the original signal. This is known as the *Nyquist sampling criterion.* It was named after Harry Nyquist (1889–1976) who first introduced the idea in his work on telegraph transmission. Later it was formally proven by his colleague Claude Shannon (1916–2001) in his work that formed the foundations of *information theory.*

In practice, the condition in Eqn. (4.15) is usually met with inequality and with sufficient margin between the two terms to allow for the imperfections of practical samplers and reconstruction systems. In practical implementations of samplers, the sampling rate f_s is typically fixed by the constraints of the hardware used. On the other hand, the highest frequency of the actual signal to be sampled is not always known a priori. One example of this is the sampling of speech signals where the highest frequency in the signal depends on the speaker, and may vary. In order to ensure that the Nyquist sampling criterion in Eqn. (4.15) is met regardless, the signal is processed through an *anti-aliasing filter* before it is sampled, effectively removing all frequencies that are greater than half the sampling rate. This is illustrated in Fig. 4.9.

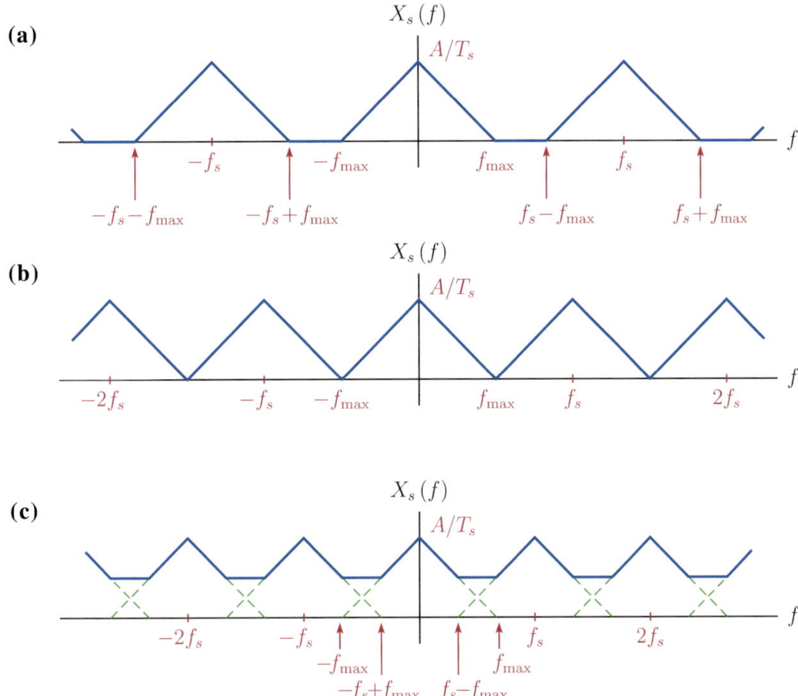

Figure 4.8 – Determination of an appropriate sampling rate: **(a)** The spectrum of a properly impulse-sampled signal, **(b)** the spectrum of the signal with critical impulse-sampling, and **(c)** the spectrum of the signal after improper impulse-sampling.

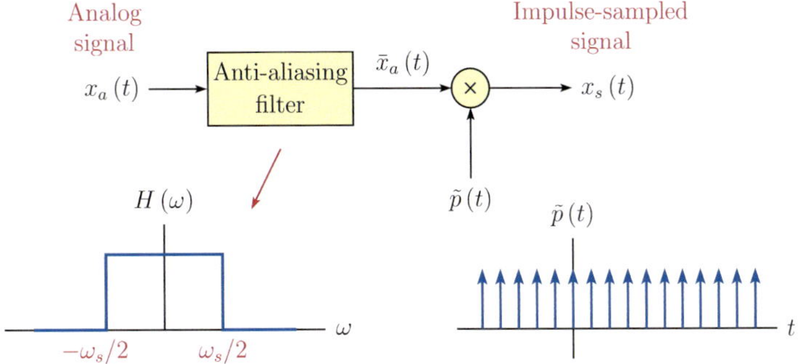

Figure 4.9 – Use of an anti-aliasing filter to ensure compliance with the requirements of Nyquist sampling criterion.

4.2.2 DTFT of sampled signal

The relationship between the Fourier transforms of the continuous-time signal and its impulse-sampled version is given by Eqns. (4.12) and (4.13). As discussed in Section 4.1, the purpose of sampling is to ultimately create a discrete-time signal $x[n]$ from a continuous-time signal $x_a(t)$. The discrete-time signal can then be converted to a digital signal suitable for storage and manipulation on digital computers. Let $x[n]$ be defined in terms of $x_a(t)$ as

$$x[n] = x_a(nT_s) \tag{4.16}$$

The Fourier transform of $x_a(t)$ is defined by

$$X_a(\omega) = \int_{-\infty}^{\infty} x_a(t)\, e^{-j\omega t}\, dt \tag{4.17}$$

Similarly, the DTFT of the discrete-time signal $x[n]$ is

$$X(\Omega) = \sum_{n=-\infty}^{\infty} x[n]\, e^{-j\Omega n} \tag{4.18}$$

We would like to understand the relationship between the two transforms in Eqns. (4.17) and (4.18).

DTFT of sampled signal:

The impulse sampled signal, given by Eqn. (4.5) can be written as

$$x_s(t) = \sum_{n=-\infty}^{\infty} x_a(nT_s)\, \delta(t - nT_s) \tag{4.19}$$

making use of the *sampling property* of the impulse function. The Fourier transform of the impulse-sampled signal is

$$\begin{aligned} X_s(\omega) &= \int_{-\infty}^{\infty} x_s(t)\, e^{-j\omega t}\, dt \\ &= \int_{-\infty}^{\infty} \left[\sum_{n=-\infty}^{\infty} x_a(nT_s)\, \delta(t - nT_s) \right] e^{-j\omega t}\, dt \end{aligned} \tag{4.20}$$

Interchanging the order of integration and summation, and rearranging terms, Eqn. (4.20) can be written as

$$X_s(\omega) = \sum_{n=-\infty}^{\infty} x_a(nT_s) \left[\int_{-\infty}^{\infty} \delta(t - nT_s)\, e^{-j\omega t}\, dt \right] \tag{4.21}$$

Using the sifting property of the impulse function, Eqn. (4.21) becomes

$$X_s(\omega) = \sum_{n=-\infty}^{\infty} x_a(nT_s)\, e^{-j\omega n T_s} \tag{4.22}$$

Compare Eqn. (4.22) to Eqn. (4.18). The two equations would become identical with the adjustment

$$\Omega = \omega\, T_s \tag{4.23}$$

This leads us to the conclusion

$$X(\Omega) = X_s\left(\frac{\Omega}{T_s}\right) \tag{4.24}$$

Frequency spectrum of the sampled signal:

Using Eqn. (4.24) with Eqn. (4.12) yields the relationship between the spectrum of the original continuous-time signal and the DTFT of the discrete-time signal obtained by sampling it:

$$X(\Omega) = \frac{1}{T_s} \sum_{k=-\infty}^{\infty} X_a\left(\frac{\Omega - 2\pi k}{T_s}\right) \tag{4.25}$$

This relationship is illustrated in Fig. 4.10.

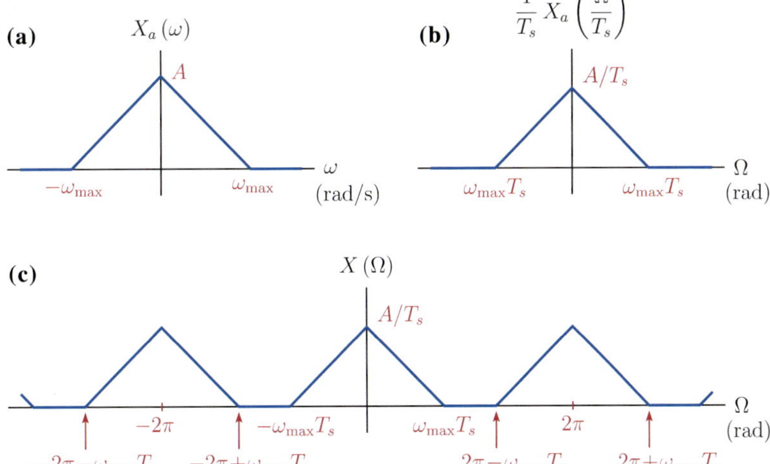

Figure 4.10 – Relationships between the frequency spectra of the original signal and the discrete-time signal obtained from it: **(a)** The example spectrum $X_a(\omega)$ of the original signal $x_a(t)$, **(b)** the term in the summation of Eqn. (4.25) for $k = 0$, and **(c)** the spectrum $X(\Omega)$ of the sampled signal $x[n]$.

It is evident from Fig. 4.10 that, in order to avoid overlaps between repetitions of the segments of the spectrum, we need

$$\omega_{\max} T_s \le \pi \quad \Longrightarrow \quad \omega_{\max} \le \frac{\pi}{T_s} \quad \Longrightarrow \quad f_{\max} \le \frac{f_s}{2} \tag{4.26}$$

consistent with the earlier conclusions.

Software resource: See MATLAB Exercise 4.2.

4.2.3 Sampling of sinusoidal signals

In this section we consider the problem of obtaining a discrete-time sinusoidal signal by sampling a continuous-time sinusoidal signal. Let $x_a(t)$ be defined as

$$x_a(t) = \cos\left(2\pi f_0 t\right) \tag{4.27}$$

and let $x[n]$ be obtained by sampling $x_a(t)$ as

$$x[n] = x_a(nT_s) = \cos\left(2\pi f_0 n T_s\right) \tag{4.28}$$

Using the normalized frequency $F_0 = f_0\, T_s = f_0 / f_s$ the signal $x[n]$ becomes

$$x[n] = \cos\left(2\pi F_0 n\right) \tag{4.29}$$

The DTFT of a discrete-time sinusoidal signal was derived in Section 3.3.6 of Chapter 3, and can be applied to $x[n]$ to yield

$$X(\Omega) = \sum_{k=-\infty}^{\infty} \left[\pi\,\delta\left(\Omega + 2\pi F_0 - 2\pi k\right) + \pi\,\delta\left(\Omega - 2\pi F_0 - 2\pi k\right)\right] \tag{4.30}$$

For the continuous-time signal $x_a(t)$ to be recoverable from $x[n]$, the sampling rate must be chosen properly. In terms of the normalized frequency F_0 we need $|F_0| < 0.5$. Fig. 4.11 illustrates the spectrum of $x[n]$ for proper and improper choices of the sampling rate and the corresponding normalized frequency.

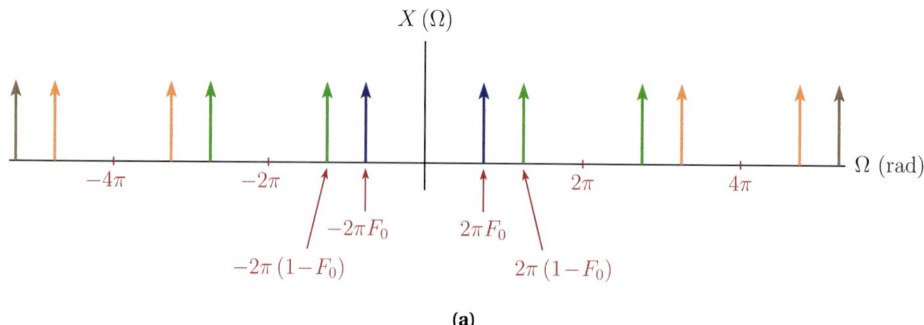

(a)

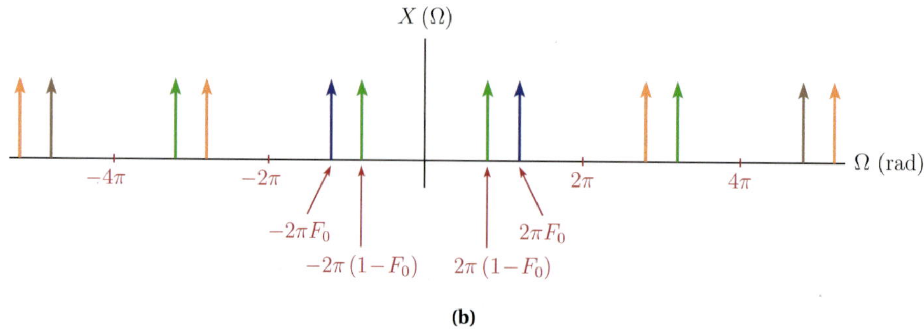

(b)

Figure 4.11 – The spectrum of sampled sinusoidal signal with **(a)** proper sampling rate and **(b)** improper sampling rate.

Example 4.2: Sampling a sinusoidal signal

The sinusoidal signals

$$x_{1a}(t) = \cos(12\pi t) , \qquad x_{2a}(t) = \cos(20\pi t) , \qquad x_{3a}(t) = \cos(44\pi t)$$

are each sampled using the sampling rate $f_s = 16$ Hz and $T_s = 1/f_s = 0.0625$ seconds to obtain the discrete-time signals

$$x_1[n] = x_{1a}(0.0625\,n) , \qquad x_2[n] = x_{2a}(0.0625\,n) , \qquad x_3[n] = x_{3a}(0.0625\,n)$$

Show that the three discrete-time signals are identical, that is,

$$x_1[n] = x_2[n] = x_3[n] , \qquad \text{all } n$$

Solution: Using the specified value of T_s the signals can be written as

$$x_1[n] = \cos(0.75\pi n) , \qquad x_2[n] = \cos(1.25\pi n) , \qquad x_3[n] = \cos(2.75\pi n)$$

Incrementing the phase of a sinusoidal function by any integer multiple of 2π radians does not affect the result. Therefore, $x_2[n]$ can be written as

$$x_2[n] = \cos(1.25\pi n - 2\pi n) = \cos(-0.75\pi n)$$

Since cosine is an even function, we have

$$x_2[n] = \cos(0.75\pi n) = x_1[n]$$

Similarly, $x_3[n]$ can be written as

$$x_3[n] = \cos(2.75\pi n - 2\pi n) = \cos(0.75\pi n) = x_1[n]$$

Thus, three different continuous-time signals correspond to the same discrete-time signal. Fig. 4.12 shows the three signals $x_{1a}(t)$, $x_{2a}(t)$, $x_{3a}(t)$ and their values at the sampling instants.

An important issue in sampling is the *reconstruction* of the original signal from its sampled version. Given the discrete-time signal $x[n]$, how do we determine the continuous-time signal it represents? In this particular case we have at least three possible candidates, $x_{1a}(t)$, $x_{2a}(t)$, and $x_{3a}(t)$, from which $x[n]$ could have been obtained by sampling. Other possible answers could also be found. In fact it can be shown that an infinite number of different sinusoidal signals can be passed through the points shown with red dots in Fig. 4.12(a),(b),(c). Which one is the right answer?

Let us determine the actual and the normalized frequencies for the three signals. For $x_{1a}(t)$ we have

$$f_1 = 6 \text{ Hz} , \qquad F_1 = \frac{6}{16} = 0.375$$

Similarly for $x_{2a}(t)$

$$f_2 = 10 \text{ Hz} , \qquad F_2 = \frac{10}{16} = 0.625$$

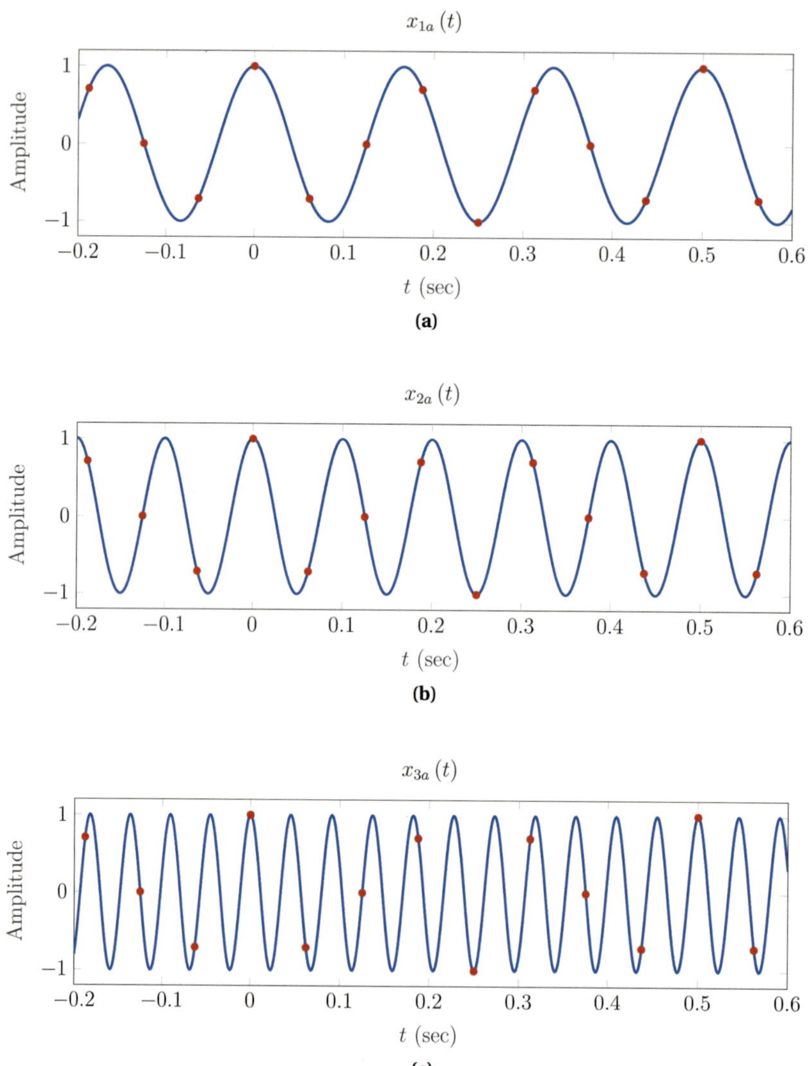

Figure 4.12 – The signals for Example 4.2 and their values at sampling instants.

and for $x_{3a}(t)$

$$f_3 = 22 \text{ Hz}, \qquad F_3 = \frac{22}{16} = 1.375$$

Of the three normalized frequencies only F_1 satisfies the condition $|F| \leq 0.5$, and the other two violate it. In terms of the Nyquist sampling theorem, the signal $x_{1a}(t)$ is sampled using a proper sampling rate, that is, $f_s > 2f_1$. The other two signals are sampled improperly since $f_s < 2f_2$ and $f_s < 2f_3$. Therefore, in the reconstruction process, we would pick $x_{1a}(t)$ based on the assumption that $x[n]$ is a properly sampled signal. Fig. 4.13 shows the signal $x_{2a}(t)$ being improperly sampled with a sampling rate of $f_s = 16$ Hz to obtain $x[n]$. In reconstructing the continuous-time signal from its sampled version, the signal $x_{1a}(t)$ is incorrectly identified as the signal that led to $x[n]$. This is referred to as *aliasing*. In this sampling scheme, $x_{1a}(t)$ is an *alias* for $x_{2a}(t)$.

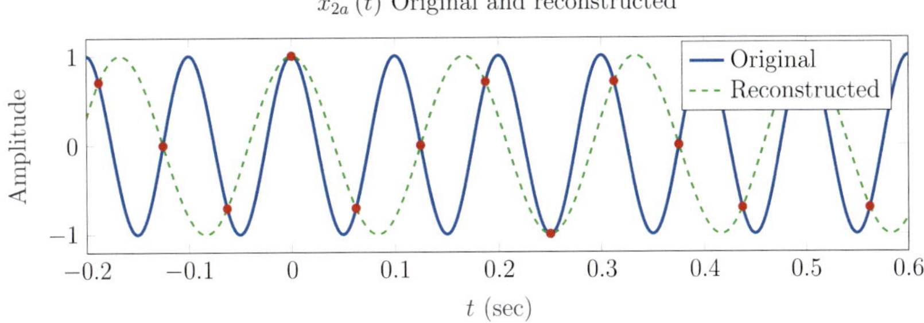

Figure 4.13 – Sampling a 10 Hz sinusoidal signal with sampling rate $f_s = 16$ Hz and attempting to reconstruct it from its sampled version.

Software resources: exdt_4_2a.m , exdt_4_2b.m

Example 4.3: Spectral relationships in sampling a sinusoidal signal

Refer to the signals in Example 4.2. Sketch the frequency spectrum for each, and the use it in obtaining the DTFT spectrum of the sampled signal.

Solution: The Fourier transform of $x_{1a}(t)$ is

$$X_{1a}(\omega) = \pi\,\delta(\omega + 12\pi) + \pi\,\delta(\omega - 12\pi)$$

and is shown in Fig. 4.14(a). Referring to Eqn. (4.25) the term for $k = 0$ is

$$X_{1a}\left(\frac{\Omega}{T_s}\right) = \pi\,\delta\left(\frac{\Omega + 0.75\pi}{0.0625}\right) + \pi\,\delta\left(\frac{\Omega - 0.75\pi}{0.0625}\right)$$

shown in Fig. 4.14(b). The spectrum of the sampled signal $x_1[n]$ is obtained as

$$\begin{aligned} X_1(\Omega) =& \frac{1}{T_s}\sum_{k=-\infty}^{\infty} X_{1a}\left(\frac{\Omega - 2\pi k}{T_s}\right) \\ =& 16\sum_{k=-\infty}^{\infty}\left[\pi\,\delta\left(\frac{\Omega + 0.75\pi - 2\pi k}{0.0625}\right) + \pi\,\delta\left(\frac{\Omega - 0.75\pi - 2\pi k}{0.0625}\right)\right] \end{aligned}$$

which is shown in Fig. 4.14(c). Each impulse in $X_1(\Omega)$ has an area of 16π. The term for $k = 0$ is shown in blue whereas the terms for $k = \mp 1$, $k = \mp 2$, $k = \mp 3$ are shown in green, orange, and brown, respectively. In the reconstruction process, the assumption that $x_1[n]$ is a properly sampled signal would lead us to correctly picking the two blue colored impulses at $\Omega = \pm 0.75\pi$ and ignoring the others.

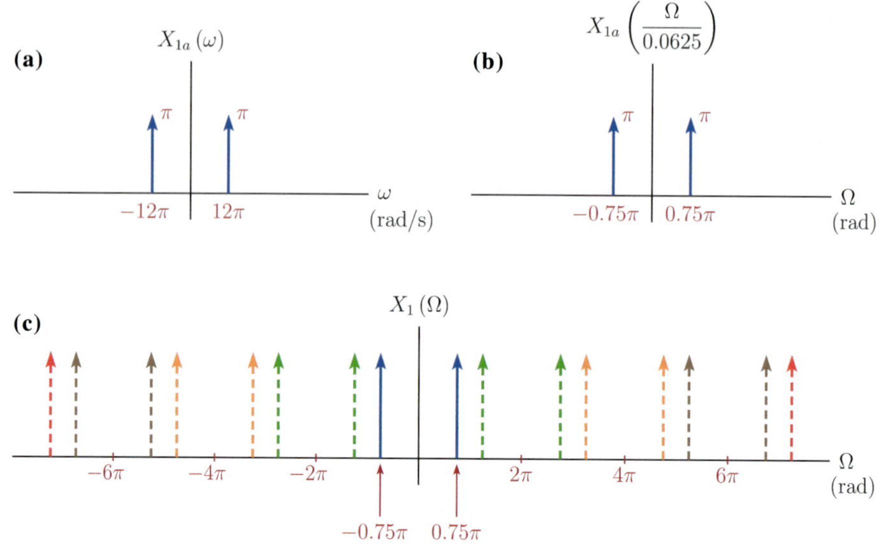

Figure 4.14 – Obtaining the spectrum $X_1(\Omega)$ from the spectrum $X_{1a}(\omega)$ for Example 4.3.

Similar analyses can be carried out for the spectra of the other two continuous-time signals $x_{2a}(t)$ and $x_{3a}(t)$ as well as the discrete-time signals that result from sampling them. Spectral relationships for these cases are shown in Figs. 4.15 and 4.16. Even though $X_1(\Omega) = X_2(\Omega) = X_3(\Omega)$ as expected, notice the differences in the contributions of $k = 0$ term of Eqn. (4.25) and the others in obtaining the DTFT spectrum. In reconstructing the continuous-time signals from $x_2[n]$ and $x_3[n]$ we would also work with the assumption that the signals have been sampled properly, and incorrectly pick the green colored impulses at $\Omega = \pm 0.75\pi$.

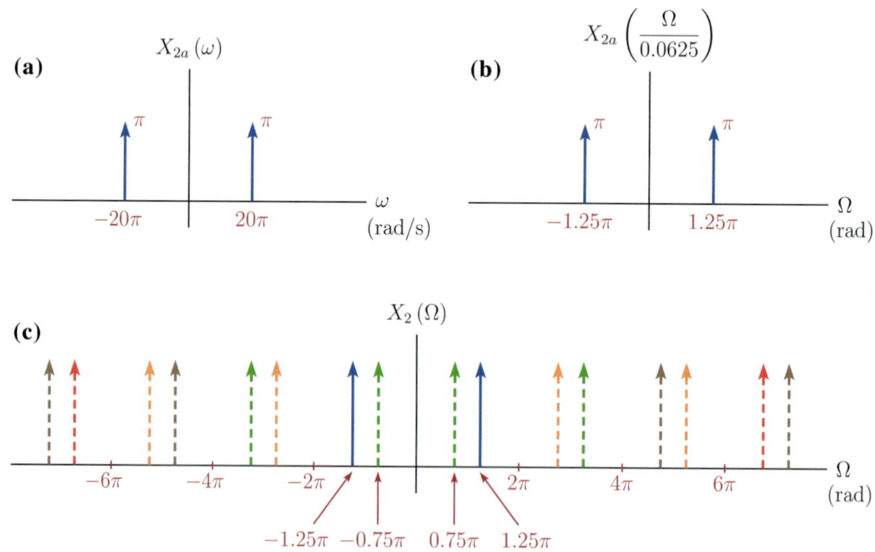

Figure 4.15 – Obtaining the spectrum $X_2(\Omega)$ from the spectrum $X_{2a}(\omega)$ for Example 4.3.

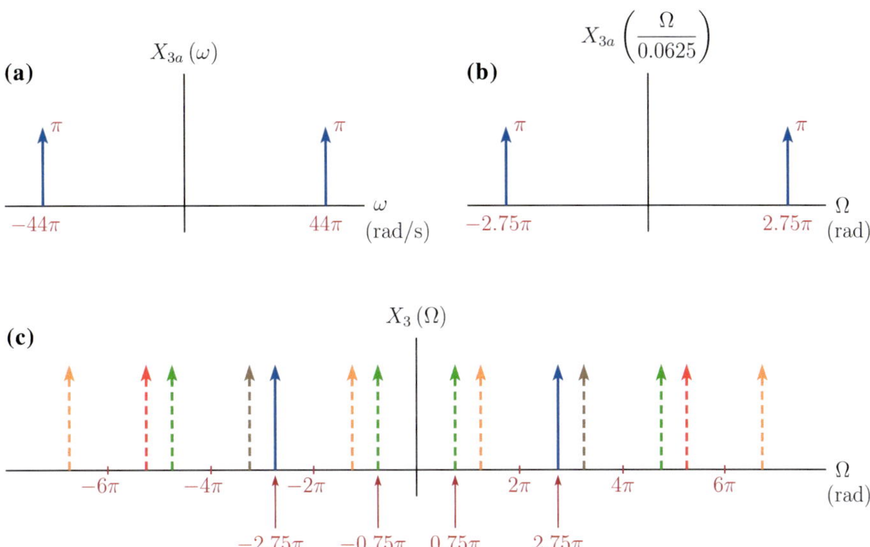

Figure 4.16 – Obtaining the spectrum $X_3(\Omega)$ from the spectrum $X_{3a}(\omega)$ for Example 4.3.

Interactive App: Spectral relationships in sampling sinusoidal signals

The interactive app in `appSampling3.m` is based on Examples 4.2 and 4.3. It allows experimentation with sampling the sinusoidal signal $x_a(t) = \cos(2\pi f_a t)$ to obtain a discrete-time signal $x[n]$. The two signals and their frequency spectra are displayed. The signal $\hat{x}_a(t)$ reconstructed from $x[n]$ may also be displayed, if desired. The signal frequency f_a and the sampling rate f_s may be varied through the use of slider controls.

a. Verify the results of Example 4.2. Begin by setting the signal frequency to $f_a = 6$ Hz and sampling rate to $f_s = 16$ Hz. Uncheck the box "`Show reconstructed signal`" and then check it. Confirm that the reconstructed signal is the same as the original.

b. Slowly increase the signal frequency while observing the changes in the spectra of $x_a(t)$ and $x[n]$.

c. Set the signal frequency to $f_a = 10$ Hz and sampling rate to $f_s = 16$ Hz. Observe the aliasing effect. What is the frequency of the reconstructed signal, and how does it relate to the frequency of the original signal?

d. Set the signal frequency to $f_a = 22$ Hz and sampling rate to $f_s = 16$ Hz. Observe the aliasing effect. Determine the frequency of the reconstructed signal.

e. With $f_a = 22$ Hz, slowly increase the sampling rate until the aliasing disappears, that is, the reconstructed signal is the same as the original.

 **Software resource:** `appSampling3.m`

 **Software resource:** See MATLAB Exercise 4.3.

4.2.4 Practical issues in sampling

In previous sections the issue of sampling a continuous-time signal through multiplication by an impulse-train was discussed. A practical consideration in the design of samplers is that we do not have ideal impulse trains, and must therefore approximate them with pulse trains. Two important questions that arise in this context are:

1. What would happen to the spectrum if we used a pulse train instead of an impulse train in Eqn. (4.25)?
2. How would the use of pulses affect the methods used in recovering the original signal from its sampled version?

When pulses are used instead of impulses, there are two variations of the sampling operation that can be used, namely *natural sampling* and *zero-order-hold sampling*. The former is easier to generate electronically while the latter lends itself better to digital coding through techniques known as *pulse-code modulation* and *delta modulation*. We will review each sampling technique briefly.

Natural sampling

Instead of using the periodic impulse-train of Eqn. (4.5) let the multiplying signal $\tilde{p}(t)$ be defined as a periodic pulse train with a duty cycle of d:

$$\tilde{p}(t) = \sum_{n=-\infty}^{\infty} \Pi\left(\frac{t-nT_s}{dT_s}\right) \tag{4.31}$$

where $\Pi(t)$ represents a unit pulse signal, that is, a pulse with unit amplitude and unit width centered around the time origin $t=0$. The period of the pulse-train is T_s, the same as the sampling interval. The width of each pulse is dT_s as shown in Fig. 4.17.

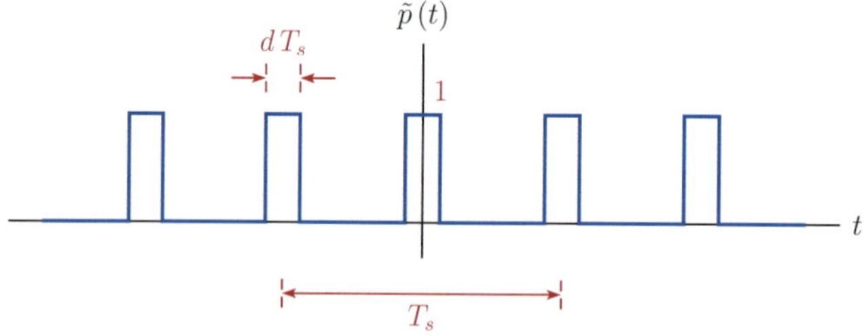

Figure 4.17 – Periodic pulse-train for natural sampling.

Multiplication of the signal $x_a(t)$ with $\tilde{p}(t)$ yields a *natural sampled* version of the signal $x_a(t)$:

$$\bar{x}_s(t) = x_a(t)\,\tilde{p}(t) = x_a(t) \sum_{n=-\infty}^{\infty} \Pi\left(\frac{t-nT_s}{dT_s}\right) \tag{4.32}$$

This is illustrated in Fig. 4.18.

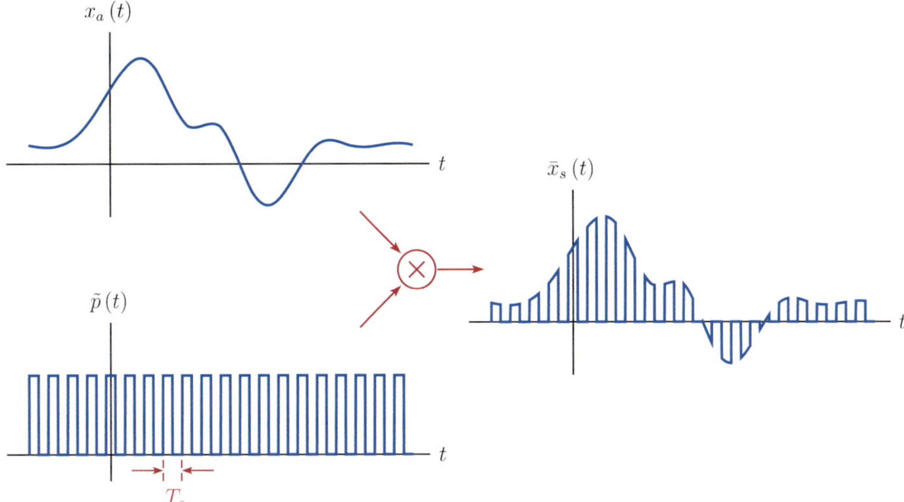

Figure 4.18 – Illustration of natural sampling.

An alternative way to visualize the natural sampling operation is to view the naturally sampled signal as the output of an electronic switch which is controlled by the pulse train $\tilde{p}(t)$. This implementation is shown in Fig. 4.19. The switch is closed when the pulse is present and is opened when the pulse is absent.

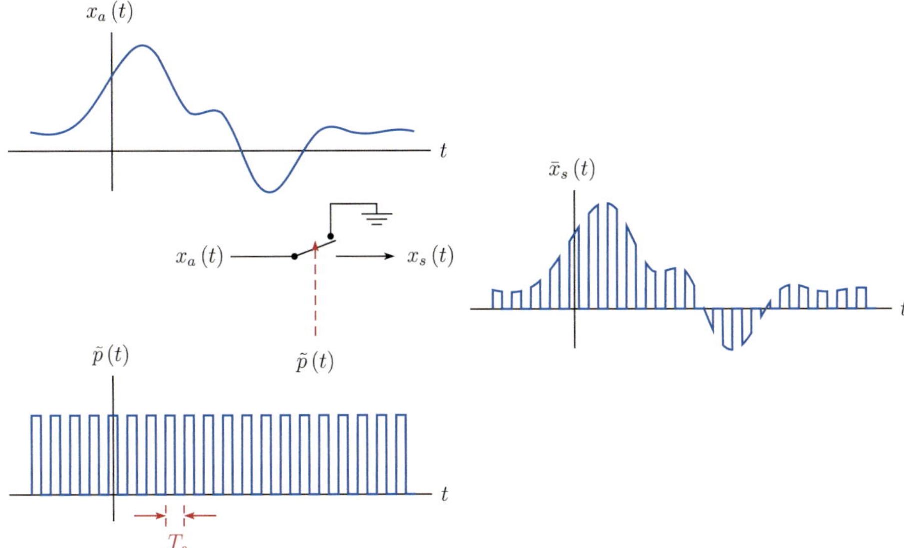

Figure 4.19 – Implementing a natural sampler using an electronically controlled switch.

In order to derive the relationship between frequency spectra of the signal $x_a(t)$ and its naturally sampled version $\bar{x}_s(t)$ we will make use of the exponential Fourier series representation of $\tilde{p}(t)$.

Spectral relationships in natural sampling:

The EFS coefficients for a pulse-train with duty cycle d are

$$c_k = d\,\text{sinc}\,(kd)$$

Therefore the EFS representation of $\tilde{p}\,(t)$ is

$$\tilde{p}\,(t) = \sum_{k=-\infty}^{\infty} c_k\, e^{jk\omega_s t} = \sum_{k=-\infty}^{\infty} d\,\text{sinc}\,(kd)\, e^{jk\omega_s t} \tag{4.33}$$

Fundamental frequency is the same as the sampling rate $\omega_s = 2\pi / T_s$. Using Eqn. (4.33) in Eqn. (4.32) the naturally sampled signal is

$$\bar{x}_s\,(t) = x_a\,(t) \sum_{k=-\infty}^{\infty} d\,\text{sinc}\,(kd)\, e^{jk\omega_s t} \tag{4.34}$$

The Fourier transform of the naturally sampled signal is

$$\bar{X}_s\,(\omega) = \mathcal{F}\left\{\bar{x}_s\,(t)\right\} = \int_{-\infty}^{\infty} \bar{x}_s\,(t)\, e^{-j\omega t}\, dt$$

Using Eqn. (4.34) in Eqn. (4.35) yields

$$\bar{X}_s\,(\omega) = \int_{-\infty}^{\infty} \left[x_a\,(t) \sum_{k=-\infty}^{\infty} d\,\text{sinc}\,(kd)\, e^{jk\omega_s t} \right] e^{-j\omega t}\, dt \tag{4.35}$$

Interchanging the order of integration and summation, and rearranging terms we obtain

$$\bar{X}_s\,(\omega) = d \sum_{k=-\infty}^{\infty} \text{sinc}\,(kd) \left[\int_{-\infty}^{\infty} x_a\,(t)\, e^{-j(\omega - k\omega_s)t}\, dt \right] \tag{4.36}$$

The expression in square brackets is the Fourier transform of $x_a\,(t)$ evaluated for $\omega - k\omega_s$, that is,

$$\int_{-\infty}^{\infty} x_a\,(t)\, e^{-j(\omega - k\omega_s)t}\, dt = X_a\,(\omega - k\omega_s) \tag{4.37}$$

Substituting Eqn. (4.37) into Eqn. (4.36) yields the desired result.

Spectrum of the signal obtained through natural sampling:

$$\bar{X}_s\,(\omega) = d \sum_{k=-\infty}^{\infty} \text{sinc}\,(kd)\, X_a\,(\omega - k\omega_s) \tag{4.38}$$

The effect of natural sampling on the frequency spectrum is shown in Fig. 4.20. It is interesting to compare the spectrum $\bar{X}_s(\omega)$ obtained in Eqn. (4.38) to the spectrum $X_s(\omega)$ given by Eqn. (4.12) for the impulse-sampled signal:

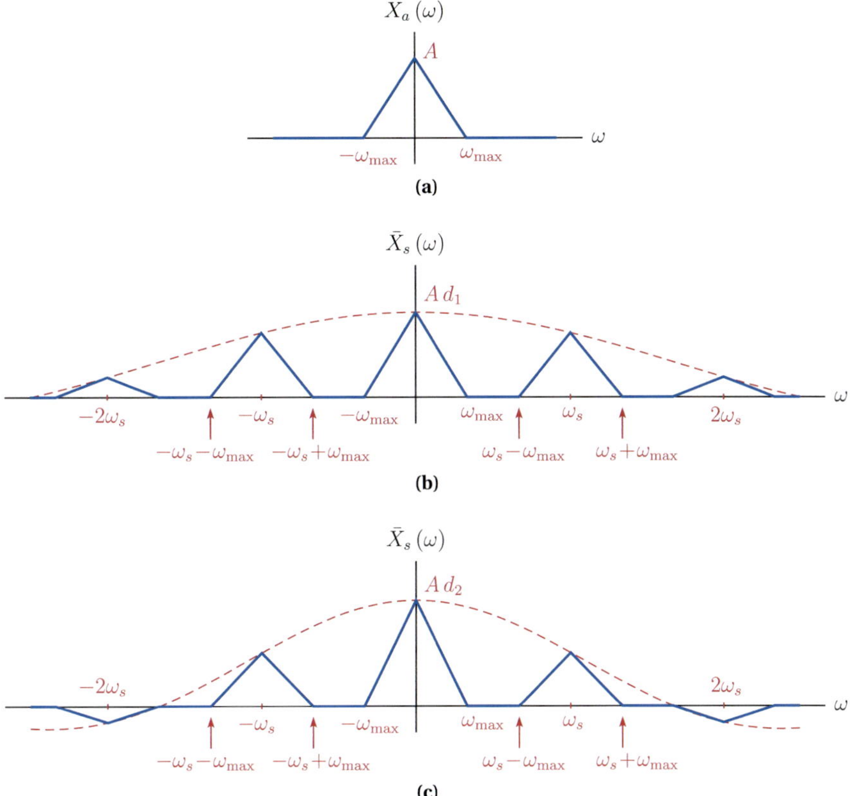

Figure 4.20 – Effects of natural-sampling on the frequency spectrum: **(a)** The example spectrum $X_a(\omega)$ of the original signal $x_a(t)$, **(b)** the spectrum $\bar{X}_s(\omega)$ of the natural-sampled signal $\bar{x}_s(t)$ obtained using a pulse train with duty cycle d_1, and **(c)** the spectrum obtained using a pulse train with duty cycle $d_2 > d_1$.

1. The spectrum of the impulse sampled signal has a scale factor of $1/T_s$ in front of the summation while the spectrum of the naturally sampled signal has a scale factor of d. This is not a fundamental difference. Recall that, in the multiplying signal $p(t)$ of the impulse-sampled signal, each impulse has unit area. On the other hand, in the pulse train $\tilde{p}(t)$ used for natural sampling, each pulse has a width of dT_s and a unit amplitude, corresponding to an area of dT_s. If we were to scale the pulse train $\tilde{p}(t)$ so that each of its pulses has unit area under it, then the amplitude scale factor would have to be $1/dT_s$. Using $(1/dT_s)\,\tilde{p}(t)$ in natural sampling would cause the spectrum $\bar{X}_s(\omega)$ in Eqn. (4.38) to be scaled by $1/dT_s$ as well, matching the scaling of $X_s(\omega)$.
2. As in the case of impulse sampling, frequency-shifted versions of the spectrum $X_a(\omega)$ are added together to construct the spectrum of the sampled signal. A key difference is that each term $X_a(\omega - k\omega_s)$ is scaled by $\text{sinc}(kd)$ as illustrated in Fig. 4.20.

Zero-order hold sampling

In natural sampling the tops of the pulses are not flat, but are rather shaped by the signal $x_a(t)$. This behavior is not always desired, especially when the sampling operation is to be followed by conversion of each pulse to digital format. An alternative is to hold the amplitude of each pulse constant, equal to the value of the signal at the left edge of the pulse. This is referred to as *zero-order hold sampling* or *flat-top sampling*, and is illustrated in Fig. 4.21.

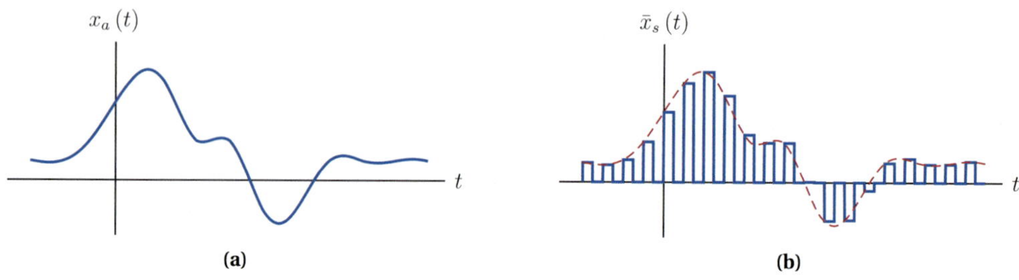

Figure 4.21 – Illustration of zero-order hold sampling.

Conceptually the signal $\bar{x}_s(t)$ can be modeled as the convolution of the impulse sampled signal $x_s(t)$ and a rectangular pulse with unit amplitude and a duration of dT_s as shown in Fig. 4.22.

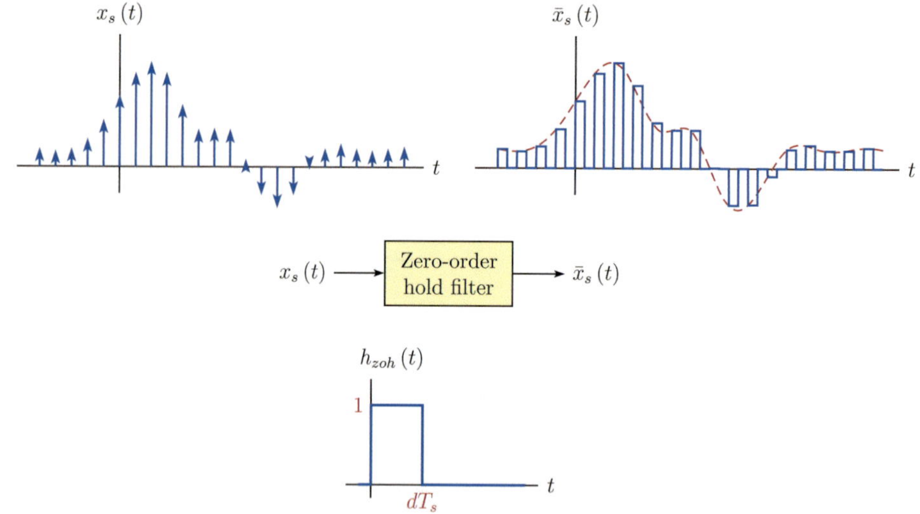

Figure 4.22 – Modeling sampling operation with flat-top pulses.

The impulse response of the zero-order hold filter in Fig. 4.22 is

$$h_{zoh}(t) = \Pi\left(\frac{t-0.5\,dT_s}{dT_s}\right) = u(t) - u(t-dT_s) \tag{4.39}$$

Zero-order hold sampled signal $\bar{x}_s(t)$ can be written as

$$\bar{x}_s(t) = h_{zoh}(t) * x_s(t) \tag{4.40}$$

where $x_s(t)$ represents the impulse sampled signal given by Eqn. (4.5). The frequency spectrum of the zero-order hold sampled signal is found as

$$\bar{X}_s(\omega) = H_{zoh}(\omega)\, X_s(\omega) \tag{4.41}$$

The system function for the zero-order hold filter is

$$H_{zoh}(\omega) = dT_s\,\text{sinc}\left(\frac{\omega dT_s}{2\pi}\right) e^{-j\omega dT_s/2} \tag{4.42}$$

The spectrum of the zero-order hold sampled signal is found by using Eqns. (4.12) and (4.42) in Eqn. (4.41).

Spectrum of the signal obtained through zero-order hold sampling:

$$\bar{X}_s(\omega) = d\,\text{sinc}\left(\frac{\omega d T_s}{2\pi}\right) e^{-j\omega d T_s/2} \sum_{k=-\infty}^{\infty} X_a(\omega - k\omega_s) \tag{4.43}$$

Software resource: See MATLAB Exercises 4.4, 4.5 and 4.4.

4.3 Reconstruction of a Signal from Its Sampled Version

In previous sections of this chapter we explored the issue of sampling an analog signal to obtain a discrete-time signal. Ideally, a discrete-time signal is obtained by multiplying the analog signal with a periodic impulse train. Approximations to the ideal scenario can be obtained through the use of a pulse train instead of an impulse-train.

Often the purpose of sampling an analog signal is to store, process, and/or transmit it digitally, and to later convert it back to analog format. To that end, one question still remains: How can the original analog signal be reconstructed from its sampled version? Given the discrete-time signal $x[n]$ or the impulse-sampled signal $x_s(t)$, how can we obtain a signal identical, or at least reasonably similar, to $x_a(t)$? Obviously we need a way to "fill the gaps" between the impulses of the signal $x_s(t)$ in some meaningful way. In more technical terms, signal amplitudes between sampling instants need to be computed by some form of interpolation.

Let us first consider the possibility of obtaining a signal similar to $x_a(t)$ using rather simple methods. One such method would be to start with the impulse sampled signal $x_s(t)$ given by Eqn. (4.5) and repeated here

$$\text{Eqn. (4.5):} \qquad x_s(t) = \sum_{n=-\infty}^{\infty} x_a(nT_s)\,\delta(t - nT_s)$$

and to hold the amplitude of the signal equal to the value of each sample for the duration of T_s immediately following each sample. The result is a "staircase" type of approximation to the original signal. Interpolation is performed using horizontal lines, or polynomials of order zero, between sampling instants. This is referred to as *zero-order hold* interpolation, and is illustrated in Fig. 4.23(a),(b).

Zero-order-hold interpolation can be achieved by processing the impulse sampled signal $x_s(t)$ through *zero-order hold reconstruction filter*, a linear system the impulse response of which is a rectangle with unit amplitude and a duration of T_s.

$$h_{\text{zoh}}(t) = \Pi\left(\frac{t - T_s/2}{T_s}\right) \tag{4.44}$$

This is illustrated in Fig. 4.24. The linear system that performs the interpolation is called a *reconstruction filter.*

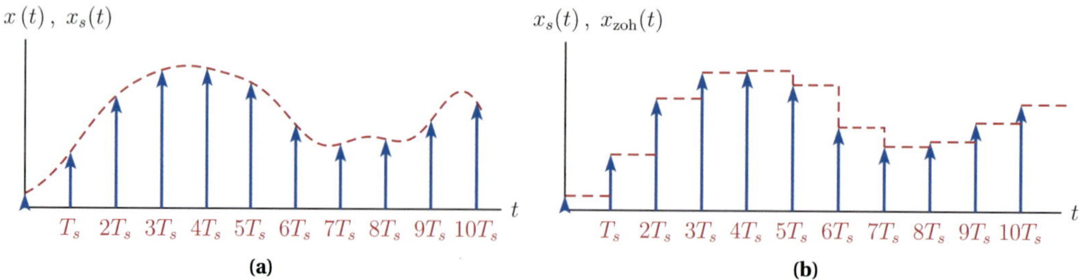

Figure 4.23 – **(a)** Impulse sampling an analog signal $x(t)$ to obtain $x_s(t)$ and **(b)** reconstruction using zero-order-hold interpolation.

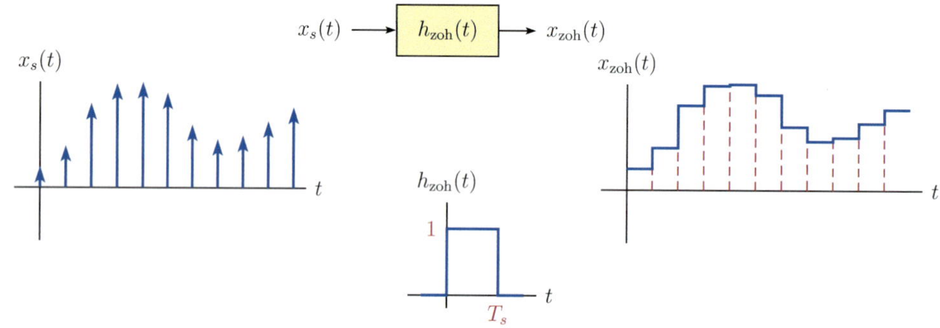

Figure 4.24 – Zero-order hold interpolation using an interpolation filter.

Notice the similarity between Eqn. (4.44) for zero-order hold interpolation and Eqn. (4.39) derived in the discussion of zero-order hold sampling. The two become the same if the duty cycle is set equal to $d = 1$. Therefore, the spectral relationship between the analog signal and its naturally sampled version given by Eqn. (4.43) can be used for obtaining the relationship between the analog signal the signal reconstructed from samples using zero-order hold:

$$X_{zoh}(\omega) = \text{sinc}\left(\frac{\omega T_s}{2\pi}\right) e^{-j\omega T_s/2} \sum_{k=-\infty}^{\infty} X_a(\omega - k\omega_s) \tag{4.45}$$

As an alternative to zero-order hold, the gaps between the sampling instants can be filled by linear interpolation, that is, by connecting the tips of the samples with straight lines as shown in Fig. 4.25. This is also known as *first-order hold* interpolation since the straight line segments used in the interpolation correspond to first order polynomials.

First-order-hold interpolation can also be implemented using a *first-order hold reconstruction filter*. The impulse response of such a filter is a triangle in the form

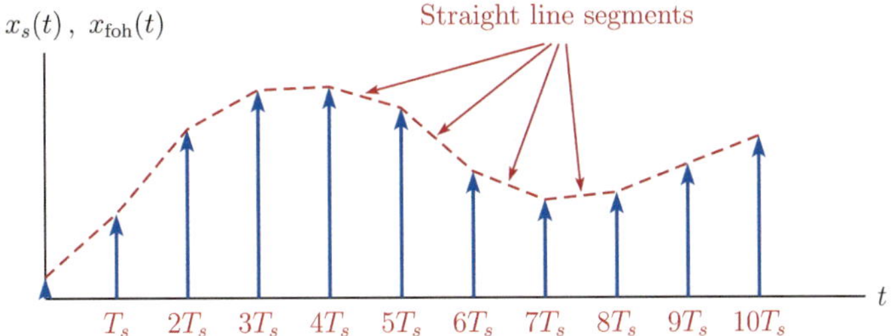

Figure 4.25 – Reconstruction using first-order hold interpolation.

$$h_{\text{foh}}(t) = \begin{cases} 1 + t/T_s\,, & -T_s < t < 0 \\ 1 - t/T_s\,, & 0 < t < T_s \\ 0\,, & |t| \geq T_s \end{cases} \tag{4.46}$$

as illustrated in Fig. 4.26.

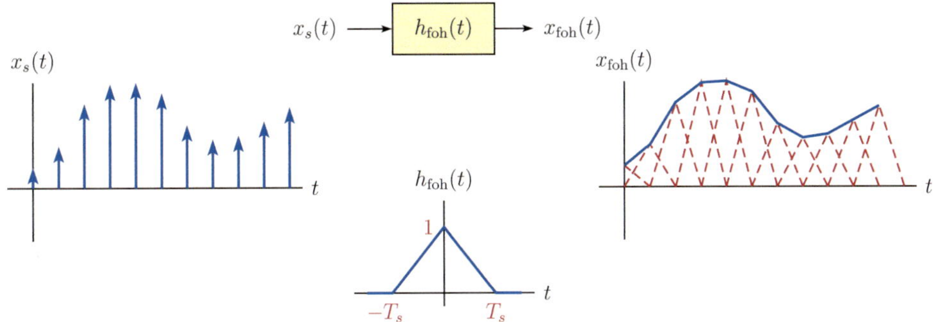

Figure 4.26 – First-order hold interpolation using an interpolation filter.

The impulse response $h_{\text{foh}}(t)$ of the first-order-hold interpolation filter is non-causal since it starts at $t = -T_s$. If a practically realizable interpolator is desired, it would be a simple matter to achieve causality by using a delayed version of $h_{\text{foh}}(t)$ for the impulse response:

$$\bar{h}_{\text{foh}}(t) = h_{\text{foh}}(t - T_s) \tag{4.47}$$

In this case, the reconstructed signal would naturally lag behind the sampled signal by T_s. It is insightful to derive the frequency spectra of the reconstructed signals obtained through zero-order hold and first-order hold interpolation. For convenience we will use f rather than ω in this derivation. The system function for the zero-order-hold filter is

$$H_{\text{zoh}}(f) = T_s\,\text{sinc}(fT_s)\,e^{-j\pi T_s} \tag{4.48}$$

and the spectrum of the analog signal constructed using the zero-order-hold filter is

$$X_{\text{zoh}}(f) = H_{\text{zoh}}(f)\,X_s(f) \tag{4.49}$$

Similarly, the system function for first-order-hold filter is

$$H_{\text{foh}}(f) = T_s\,\text{sinc}^2(fT_s) \tag{4.50}$$

and the spectrum of the analog signal constructed using the first-order-hold filter is

$$X_{\text{foh}}(f) = H_{\text{foh}}(f)\, X_s(f) \tag{4.51}$$

Fig. 4.27 illustrates the process of obtaining the magnitude spectrum of the output of the zero-order-hold interpolation filter for the sample input spectrum used earlier in Section 4.2. Fig. 4.28 illustrates the same concept for the first-order-hold interpolation filter.

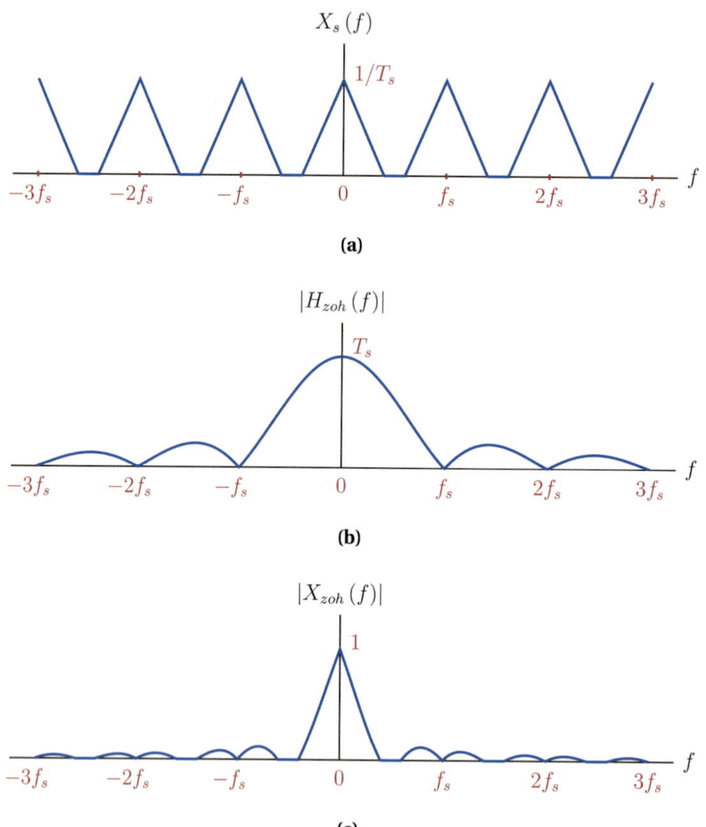

Figure 4.27 – **(a)** Sample spectrum $X_s(f)$ for an impulse-sampled signal, **(b)** magnitude spectrum $|H_{\text{zoh}}(f)|$ for the zero-order-hold interpolation filter, and **(c)** magnitude spectrum $|X_{\text{zoh}}(f)|$ for the signal reconstructed using zero-order-hold interpolation.

Interestingly, both zero-order-hold and first-order-hold filters exhibit lowpass characteristics. A comparison of Figs. 4.27(c) and 4.28(c) reveals that the first-order-hold interpolation filter does a better job in isolating the main section of the signal spectrum around $f = 0$ and suppressing spectral repetitions in $X_s(f)$ compared to the zero-order-hold interpolation filter. A comparison of the time-domain signals obtained through zero-order hold and first-order hold interpolation in Figs. 4.23(b) and 4.25 supports this conclusion as well: The reconstructed signal $x_{\text{foh}}(t)$ is closer to the original signal $x_a(t)$ than $x_{\text{zoh}}(t)$ is.

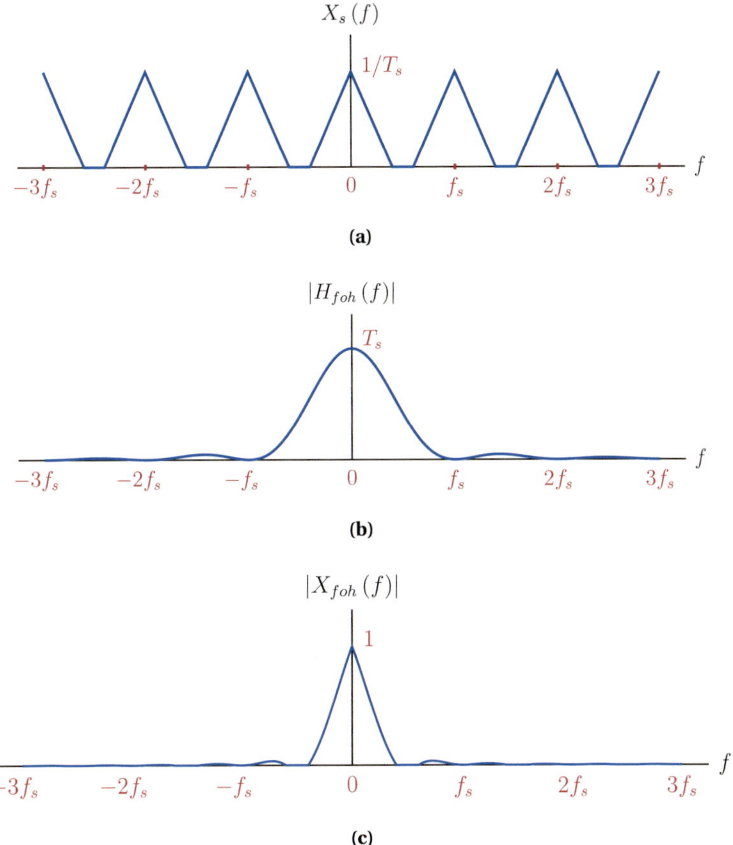

Figure 4.28 – **(a)** Sample spectrum $X_s(f)$ for an impulse-sampled signal, **(b)** magnitude spectrum $|H_{\text{foh}}(f)|$ for the first-order-hold interpolation filter, and **(c)** magnitude spectrum $|X_{\text{foh}}(f)|$ for the signal reconstructed using first-order-hold interpolation.

The fact that both interpolation filters have lowpass characteristics warrants further exploration. The Nyquist sampling theorem states that a properly sampled signal can be recovered perfectly from its sampled version. What kind of interpolation is needed for perfect reconstruction of the analog signal from its impulse-sampled version? The answer must be found through the frequency spectrum of the sampled signal. Recall that the relationship between $X_s(f)$, the spectrum of the impulse-sampled signal, and $X_a(f)$, the frequency spectrum of the original signal, was found in Eqn. (4.13) to be

$$\text{Eqn. (4.13):}\qquad X_s(f) = \frac{1}{T_s}\sum_{k=-\infty}^{\infty} X_a(f - kf_s)$$

As long as the choice of the sampling rate satisfies the Nyquist sampling criterion, the spectrum of the impulse-sampled signal is simply a sum of frequency shifted versions of the original spectrum, shifted by every integer multiple of the sampling rate. An ideal lowpass filter that extracts the term for $k = 0$ from the summation in Eqn. (4.13) and suppresses all other terms for $k = \pm 1, \ldots, \pm\infty$ would recover the original spectrum $X_a(f)$, and therefore the original signal $x_a(t)$. Since the highest frequency in a properly sampled signal would be equal to or less than half the sampling rate, an ideal *lowpass filter* with cutoff frequency set equal to $f_s/2$ is needed. The system function for such a reconstruction filter is

$$H_r(f) = T_s \Pi\left(\frac{f}{f_s}\right) \tag{4.52}$$

where we have also included a magnitude scaling by a factor of T_s within the system function of the lowpass filter in order to compensate for the $1/T_s$ term in Eqn. (4.13).

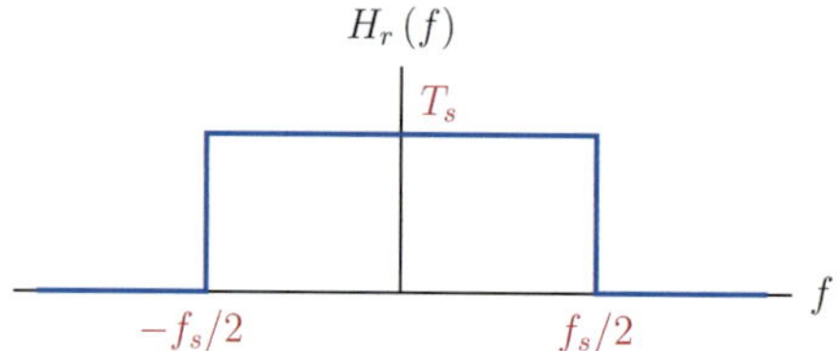

Figure 4.29 – Ideal lowpass reconstruction filter.

The output of the filter defined by Eqn. (4.52) is

$$\begin{aligned} X_r(f) &= H_r(f)\, X_s(f) \\ &= T_s \Pi\left(\frac{f}{f_s}\right) \frac{1}{T_s} \sum_{k=-\infty}^{\infty} X_a(f - kf_s) = X_a(f) \end{aligned} \tag{4.53}$$

This is illustrated in Fig. 4.30. Since $X_r(f) = X_a(f)$ we have $x_r(t) = x_a(t)$. It is also interesting to determine what type of interpolation is implied in the time-domain by the ideal lowpass filter of Eqn. (4.52). The impulse response of the filter is

$$h_r(t) = \text{sinc}(t f_s) = \text{sinc}\left(\frac{t}{T_s}\right) \tag{4.54}$$

which, due to the sinc function, has equally-spaced zero-crossings that coincide with the sampling instants. The signal at the output of the ideal lowpass filter is obtained by convolving $h_r(t)$ with the impulse-sampled signal given by Eqn. (4.5):

$$x_r(t) = \sum_{n=-\infty}^{\infty} x_a(nT_s)\, \text{sinc}\left(\frac{t - nT_s}{T_s}\right) \tag{4.55}$$

The nature of interpolation performed by the ideal lowpass reconstruction filter is evident from Eqn. (4.55). Let us consider the output of the filter at one of the sampling instants, say $t = kT_s$:

$$\begin{aligned} x_r(kT_s) &= \sum_{n=-\infty}^{\infty} x_a(nT_s)\, \text{sinc}\left(\frac{kT_s - nT_s}{T_s}\right) \\ &= \sum_{n=-\infty}^{\infty} x_a(nT_s)\, \text{sinc}(k - n) \end{aligned} \tag{4.56}$$

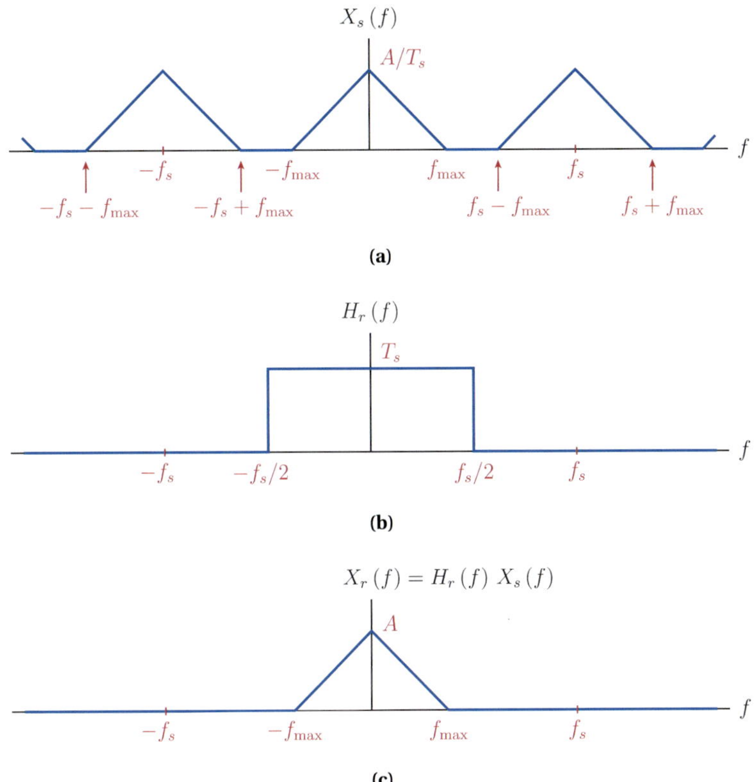

Figure 4.30 – Reconstruction using an ideal lowpass reconstruction filter: **(a)** Sample spectrum $X_s\left(f\right)$ for an impulse-sampled signal, **(b)** system function $H_r\left(f\right)$ for the ideal lowpass filter with cutoff frequency $f_s/2$, and **(c)** spectrum $X_r\left(f\right)$ for the signal at the output of the ideal lowpass filter.

We also know that

$$\text{sinc}\,(k-n) = \begin{cases} 1\,, & n=k \\ 0\,, & n\neq k \end{cases}$$

and Eqn. (4.56) becomes $x_r\,(kT_s) = x_a\,(kT_s)$.

1. The output $x_r\,(t)$ of the ideal lowpass reconstruction filter is equal to the sampled signal at each sampling instant.
2. In between sampling instants $x_r\,(t)$ is obtained by interpolation through the use of sinc functions. This is referred to as *bandlimited interpolation* and is illustrated in Fig. 4.31.

Up to this point we have discussed three possible methods of reconstruction by interpolating between the amplitudes of the sampled signal, namely zero-order-hold, first-order-hold, and bandlimited interpolation. All three methods result in reconstructed signals that have the correct amplitude values at the sampling instants, and interpolated amplitude values in between. Some interesting questions might be raised at this point:

1. What makes the signal obtained by bandlimited interpolation more accurate than the other two?
2. In a practical situation we would only have the samples $x_a\,(nT_s)$ and would not know what the original signal $x_a\,(t)$ looked like in between sampling instants. What if $x_a\,(t)$ had been identical to the zero-order hold result $x_{\text{zoh}}\,(t)$ or the first-order hold result $x_{\text{foh}}\,(t)$ before sampling?

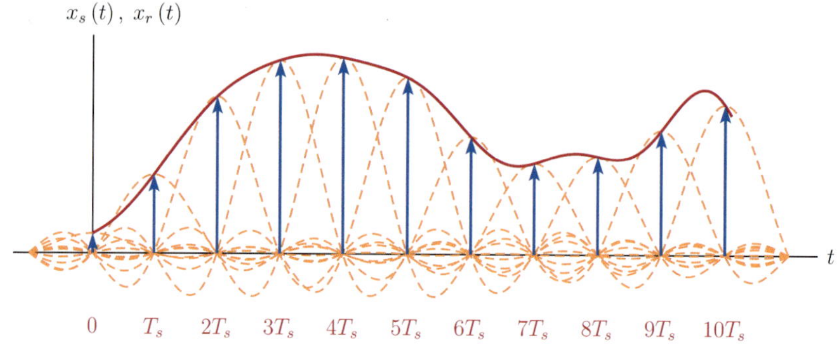

Figure 4.31 – Reconstruction of a signal from its sampled version through bandlimited interpolation.

The answer to both questions lies in the fact that, the signal obtained by bandlimited interpolation is the only signal among the three that is limited to a bandwidth of $f_s/2$. The bandwidth of each of the other two signals is greater than $f_s/2$, therefore, neither of them could have been the signal that produced a properly-sampled $x_s(t)$. From a practical perspective, however, both zero-order-hold and first-order-hold interpolation techniques are occasionally utilized in cases where exact or very accurate interpolation is not needed and simple approximate reconstruction may suffice.

Software resource: See MATLAB Exercises 4.7 and 4.8.

4.4 Resampling Discrete-Time Signals

Sometimes we have the need to change the sampling rate of a discrete-time signal. Subsystems of a large scale system may operate at different sampling rates, and the ability to convert from one sampling rate to another may be necessary to get the subsystems to work together.

Consider a signal $x_1[n]$ that may have been obtained from an analog signal $x_a(t)$ by means of sampling with a sampling rate $f_{s1} = 1/T_1$.

$$x_1[n] = x_a(nT_1) \tag{4.57}$$

Suppose an alternative version of the signal is needed, one that corresponds to sampling $x_a(t)$ with a different sampling rate $f_{s2} = 1/T_2$.

$$x_2[n] = x_a(nT_2) \tag{4.58}$$

The question is: *How can $x_2[n]$ be obtained from $x_1[n]$?*

If $x_1[n]$ is a properly sampled signal, that is, if the conditions of the Nyquist sampling theorem have been satisfied, the analog signal $x_a(t)$ may be reconstructed from it and then resampled at the new rate to obtain $x_2[n]$. This approach may not always be desirable or practical. Realizable reconstruction filters are far from the ideal filters called for in perfect reconstruction of the analog signal $x_a(t)$, and a loss of signal quality would occur in the conversion. We prefer to obtain $x_2[n]$ from $x_1[n]$ using discrete-time processing methods without the need to convert $x_1[n]$ to an intermediate analog signal.

4.4.1 Reducing the sampling rate by an integer factor

Reduction of sampling rate by an integer factor D is easily accomplished by defining the signal $x_d[n]$ as

$$x_d[n] = x[nD] \tag{4.59}$$

This operation is known as *downsampling*. The parameter D is the *downsampling rate*. Graphical representation of a downsampler is shown in Fig. 4.32.

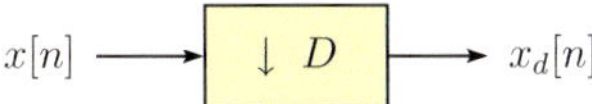

Figure 4.32 – Graphical representation of a downsampler.

Downsampling operation was briefly discussed in Section 1.2 of Chapter 1 in the context of time scaling for discrete-time signals. Figs. 4.33, 4.34, and 4.35 illustrate the downsampling of a signal using downsampling rates of $D = 2$ and $D = 3$.

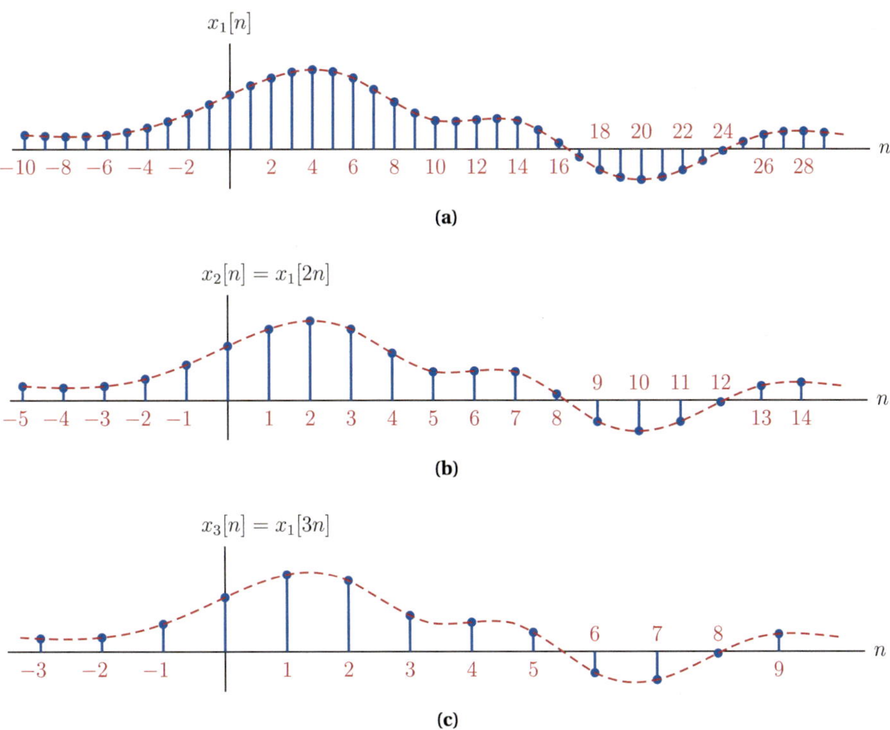

Figure 4.33 – Downsampling a signal $x_1[n]$: **(a)** Original signal $x_1[n]$, and **(b)** $x_2[n] = x_1[2n]$, **(c)** $x_3[n] = x_1[3n]$.

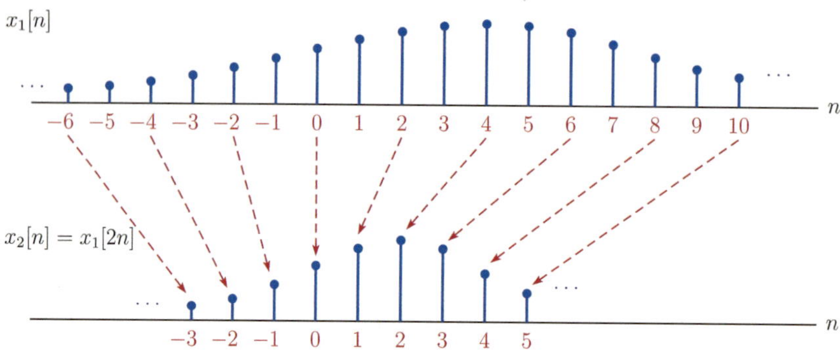

Figure 4.34 – Downsampling by a factor of $D = 2$.

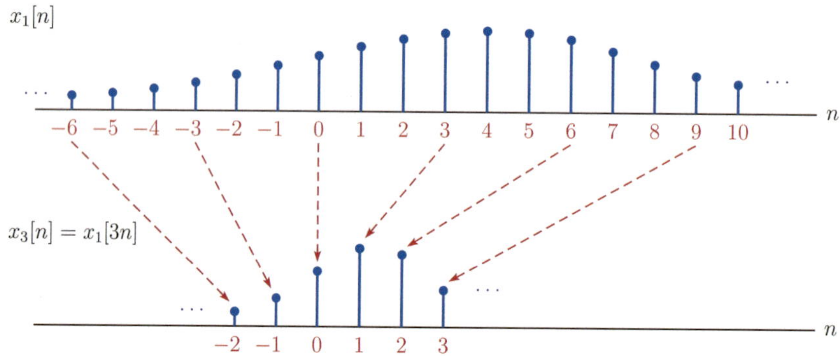

Figure 4.35 – Downsampling by a factor of $D = 3$.

Let us consider the general downsampling relationship in Eqn. (4.59). The signal $x_d[n]$ retains one sample out of each set of D samples of the original signal. For each sample retained $(D-1)$ samples are discarded. The natural question that must be raised is: *Are we losing information by discarding samples, or were those discarded samples just redundant/unnecessary?* In order to answer this question we need to focus on the frequency spectra of the signals involved.

Assume that the original signal was obtained by sampling an analog signal $x_a(t)$ with a sampling rate f_s so that

$$x[n] = x_a\left(\frac{n}{f_s}\right) \tag{4.60}$$

For sampling to be appropriate, the highest frequency of the signal $x_a(s)$ must not exceed $f_s/2$. The downsampled signal $x_d[n]$ may be obtained by sampling $x_a(t)$ with a sampling rate f_s/D:

$$x_d[n] = x_a\left(\frac{nD}{f_s}\right) \tag{4.61}$$

For $x_d[n]$ to represent an appropriately sampled signal the highest frequency in $x_a(t)$ must not exceed $f_s/2D$. This is the more restricting of the two conditions on the bandwidth of $x_a(n)$ as illustrated in Fig. 4.36.

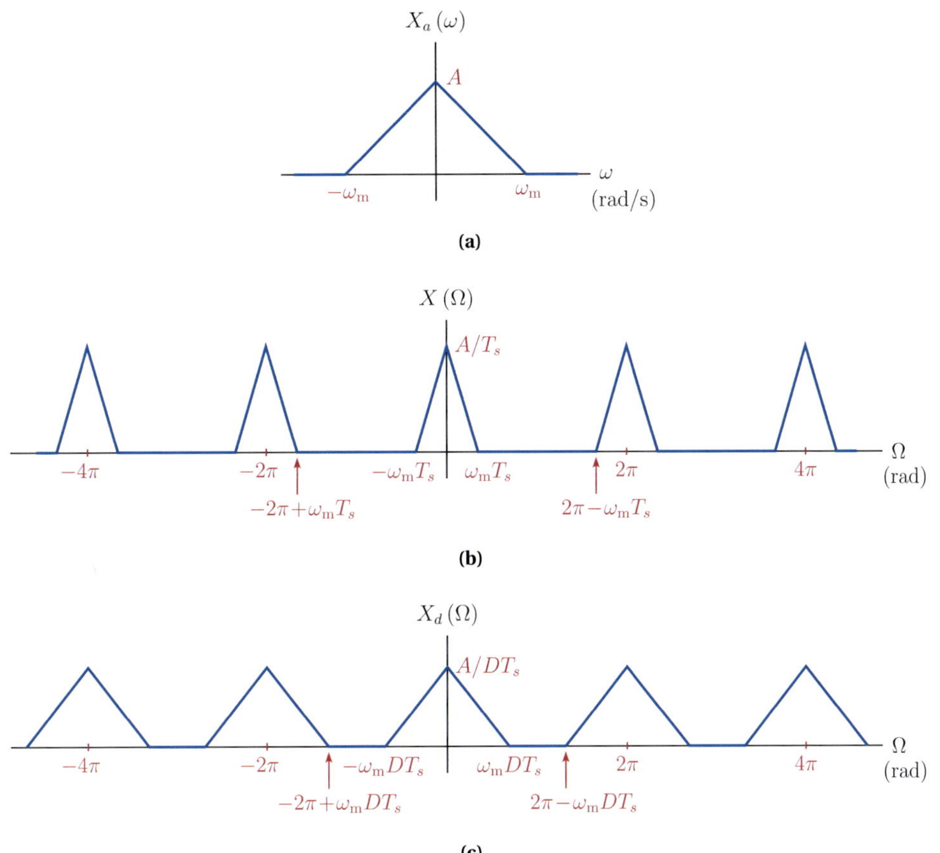

Figure 4.36 – Spectral relationships in downsampling: **(a)** Sample spectrum of analog signal $x_a(t)$, **(b)** corresponding spectrum for $x[n] = x_a\left(n/f_s\right)$, and **(c)** corresponding spectrum for $x_d[n] = x_a\left(nD/f_s\right)$.

The relationship between $X(\Omega)$ and $X_d(\Omega)$ could also be derived without resorting to the analog signal $x_a(t)$.

Expressing $X_d(\Omega)$ in terms of $X(\Omega)$:

The spectrum of $x_d[n]$ is computed as

$$X_d(\Omega) = \sum_{n=-\infty}^{\infty} x_d[n]\, e^{-j\Omega n} \tag{4.62}$$

Substituting Eqn. (4.59) into Eqn. (4.62) yields

$$X_d(\Omega) = \sum_{n=-\infty}^{\infty} x[nD]\, e^{-j\Omega n} \tag{4.63}$$

Using the variable change $m = nD$, Eqn. (4.63) becomes

$$X_d(\Omega) = \sum_{\substack{m=-\infty \\ m=nD}}^{\infty} x[m]\, e^{-j\Omega m/D} \tag{4.64}$$

The restriction on the index of the summation in Eqn. (4.64) makes it difficult for us to put the summation into a closed form. If a signal $w[m]$ with the definition

$$w[m] = \begin{cases} 1\,, & m = nD\,, \quad n: \text{integer} \\ 0\,, & \text{otherwise} \end{cases} \tag{4.65}$$

is available, it can be used in Eqn. (4.65) to obtain

$$X_d(\Omega) = \sum_{m=-\infty}^{\infty} x[m]\, w[m] e^{-j\Omega m/D} \tag{4.66}$$

with no restrictions on the summation index m. Since $w[m]$ is equal to zero for index values that are not integer multiples of D, the restriction that m be an integer multiple of D may be safely removed from the summation. It can be shown (see Problem 4.16 at the end of this chapter) that the signal $w[m]$ can be expressed as

$$w[m] = \frac{1}{D} \sum_{k=0}^{D-1} e^{j2\pi mk/D} \tag{4.67}$$

Substituting Eqn. (4.67) into Eqn. (4.66) leads to

$$\begin{aligned} X_d(\Omega) &= \sum_{m=-\infty}^{\infty} \left[\frac{1}{D} \sum_{k=0}^{D-1} e^{j2\pi mk/D} \right] e^{-j\Omega m/D} \\ &= \frac{1}{D} \sum_{k=0}^{D-1} \left[\sum_{m=-\infty}^{\infty} x[m]\, e^{-j(\Omega-2\pi k)\, m/D} \right] \end{aligned} \tag{4.68}$$

Using Eqn. (4.68) the spectrum of the downsampled signal $x_d[n]$ is related to the spectrum of the original signal $x[n]$ by the following:

Spectrum of downsampled signal:

$$X_d(\Omega) = \frac{1}{D} \sum_{k=0}^{D-1} X\left(\frac{\Omega - 2\pi k}{D} \right) \tag{4.69}$$

This result along with a careful comparison of Fig. 4.36b,c reveals that the downsampling operation could lead to aliasing through overlapping of spectral segments if care is not exercised. Spectral overlap occurs if the highest angular frequency in $x[n]$ exceeds π/D. It is therefore customary to process the signal $x[n]$ through a lowpass anti-aliasing filter before it is downsampled. The combination of the anti-aliasing filter and the downsampler is referred to as a *decimator.*

4.4.2 Increasing the sampling rate by an integer factor

In contrast with downsampling to reduce the sampling rate, the *upsampling* operation is used as the first step in increasing the sampling rate of a discrete-time signal. Upsampled version of a signal $x[n]$ is defined as

$$x_u[n] = \begin{cases} x[n/L]\,, & n = kL\,, \quad k: \text{integer} \\ 0\,, & \text{otherwise} \end{cases} \tag{4.70}$$

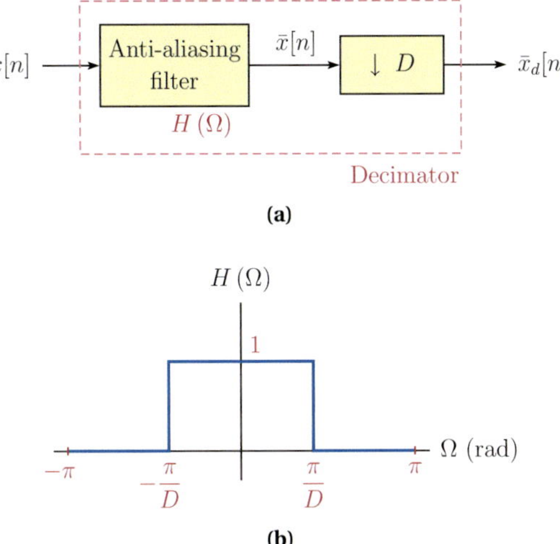

Figure 4.37 – Decimator structure.

where L is the upsampling rate. Graphical representation of an upsampler is shown in Fig. 4.38.

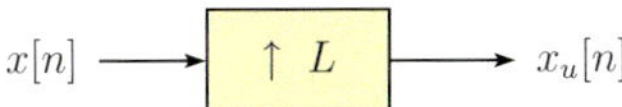

Figure 4.38 – Graphical representation of an upsampler.

Upsampling operation was also briefly discussed in Section 1.2 of Chapter 1 in the context of time scaling for discrete-time signals. Fig. 4.39, illustrates the upsampling of a signal using $L = 2$.

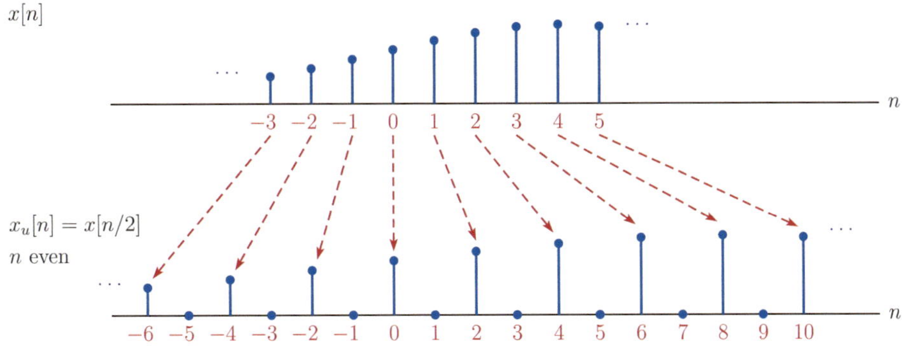

Figure 4.39 – Upsampling by a factor of $L = 2$.

Upsampling operation produces the additional samples needed for increasing the sampling rate by an integer factor, however, the new samples inserted into the signal all have zero amplitudes. Further processing is needed to change the zero-amplitude samples to more meaningful values. In order to understand what type of processing is necessary, we will again use spectral relationships. The DTFT of $x_u[n]$ is

$$X_u(\Omega) = \sum_{n=-\infty}^{\infty} x_u[n]\, e^{-j\Omega n} = \sum_{\substack{n=-\infty \\ n=mL}}^{\infty} x[n/L]\, e^{-j\Omega n} \tag{4.71}$$

Using the variable change $n = mL$, Eqn. (4.71) becomes

$$X_u(\Omega) = \sum_{m=-\infty}^{\infty} x[m]\, e^{-j\Omega Lm} = X(\Omega L) \tag{4.72}$$

This result is illustrated in Fig. 4.40.

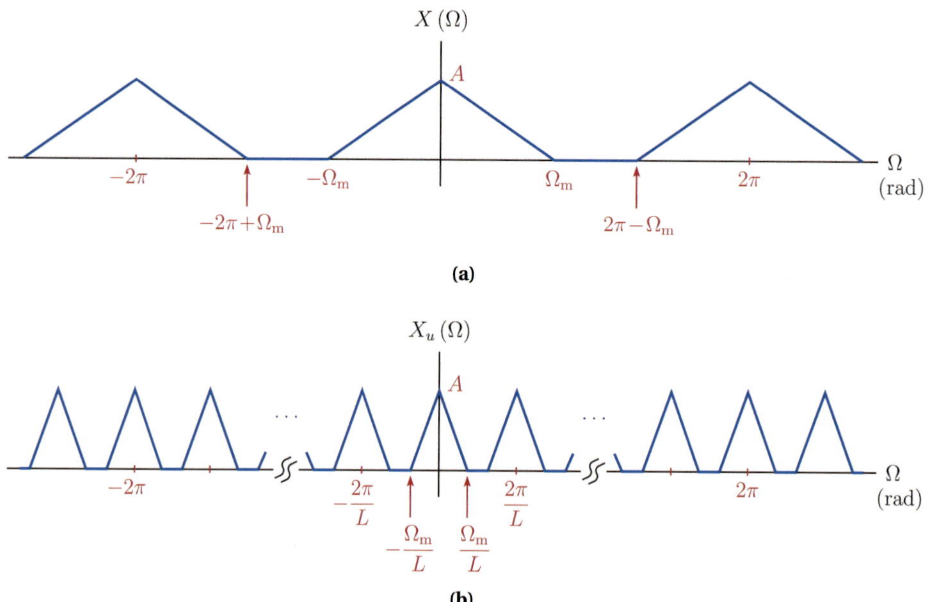

Figure 4.40 – Spectral relationships in upsampling a signal: **(a)** Sample spectrum of signal $x[n]$ and **(b)** corresponding spectrum for $x_u[n] = x[n/L]$.

Spectrum of upsampled signal:

Eqn. (4.72): $X_u(\Omega) = X(\Omega L)$

A lowpass *interpolation filter* is needed to make the zero-amplitude samples "blend-in" with the rest of the signal. The combination of an upsampler and a lowpass interpolation filter is referred to as an *interpolator*. Ideally, the interpolation filter should remove the extraneous spectral segments in $X_u(\Omega)$ as shown in Fig. 4.42.

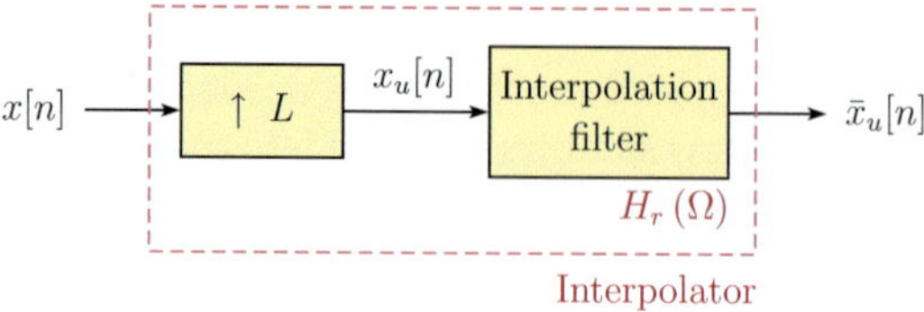

Figure 4.41 – Practical interpolator.

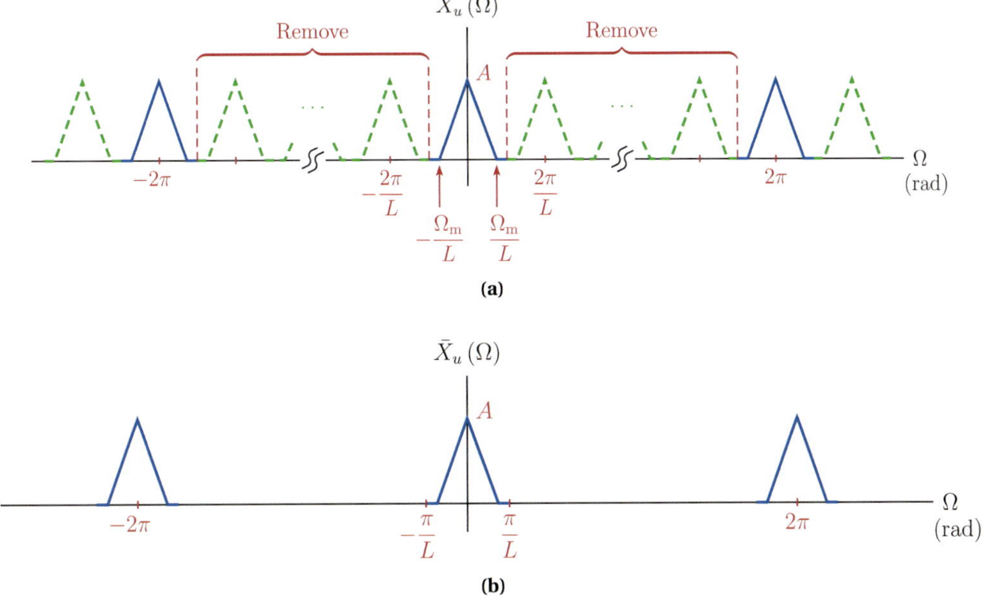

Figure 4.42 – Spectral relationships in interpolating the upsampled signal: **(a)** Sample spectrum of signal $x_u[n]$ and **(b)** corresponding spectrum for $\bar{x}_u[n]$ at filter output.

Thus the ideal interpolation filter is a discrete-time lowpass filter with an angular cutoff frequency of $\Omega_c = \pi/L$ and a gain of L as shown in Fig. 4.43(a). The role of the interpolation filter is similar to that of the ideal reconstruction filter discussed in Section 4.3. The impulse response of the filter is (see Problem 4.17 at the end of this chapter)

$$h_r[n] = \text{sinc}(n/L) \tag{4.73}$$

and is shown in Fig. 4.43b for the sample case of $L = 5$. Notice how every fifth sample has zero amplitude so that convolution of $x_u[n]$ with this filter causes interpolation between existing nonzero samples without changing their values. In practical situations simpler interpolation filters may be used as well. A zero-order hold interpolation filter has the impulse response

$$h_{zoh}[n] = u[n] - u[n-L] \tag{4.74}$$

whereas the impulse response of a first-order hold interpolation filter is

$$h_r[n] = \begin{cases} 1 - \dfrac{|n|}{L}, & n = -L, \ldots, L \\ 0, & \text{otherwise} \end{cases} \tag{4.75}$$

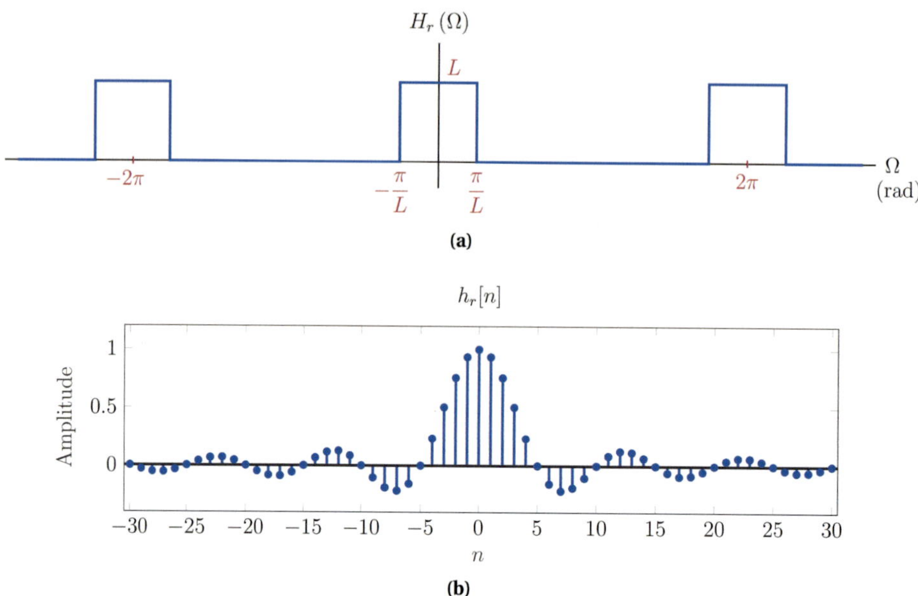

Figure 4.43 – **(a)** Spectrum of ideal interpolation filter and **(b)** impulse response of ideal interpolation filter.

Summary of Key Points

☞ The term *sampling* refers to the act of periodically measuring the amplitude of a continuous-time signal and constructing a discrete-time signal with the measurements. In the simplest form, a continuous-time signal is converted to a discrete-time signal by

Eqn. 4.2: $$x[n] = x_a(nT_s)$$

where T_s is the *sampling interval* or the *sampling period.* Its reciprocal $f_s = 1/T_s$ is the *sampling rate* or the *sampling frequency.*

☞ Impulse sampling is the act of multiplying a continuous-time signal with a periodic train of unit impulses.

Eqn. 4.5: $$x_s(t) = x_a(t) \sum_{n=-\infty}^{\infty} \delta(t - nT_s) = \sum_{n=-\infty}^{\infty} x_a(nT_s)\,\delta(t - nT_s)$$

☞ The frequency spectrum of the impulse-sampled signal can be expressed in terms of the spectrum of the original signal as

Eqn. 4.12: $$X_s(\omega) = \frac{1}{T_s} \sum_{k=-\infty}^{\infty} X_a(\omega - k\omega_s)$$

☞ For the impulse-sampled signal to form an accurate representation of the original signal, the sampling rate must be at least twice the highest frequency in the spectrum of the original signal. This is known as the *Nyquist sampling criterion.*

Eqn. 4.15: $$f_s \geq 2 f_{\max}$$

☞ The DTFT $X(\Omega)$ of a discrete-time signal obtained through sampling a continuous-time signal with Fourier transform $X_a(\omega)$ is

Eqn. 4.25: $$X(\Omega) = \frac{1}{T_s} \sum_{k=-\infty}^{\infty} X_a\left(\frac{\Omega - 2\pi k}{T_s}\right)$$

☞ Natural sampling is the act of sampling a continuous-time signal through multiplication by a pulse train with duty cycle d.

Eqn. 4.32: $$\bar{x}_s(t) = x_a(t) \sum_{n=-\infty}^{\infty} \Pi\left(\frac{t - nT_s}{dT_s}\right)$$

☞ The spectrum of the signal $\bar{x}_s(t)$ obtained through natural sampling a signal $x_a(t)$ is related to the spectrum of the original signal by

Eqn. 4.38: $$\bar{X}_s(\omega) = d \sum_{k=-\infty}^{\infty} \operatorname{sinc}(kd)\, X_a(\omega - k\omega_s)$$

☞ An alternative to natural sampling is *zero-order hold sampling* where the amplitude of each pulse is held constant, equal to the value of the signal at the left edge of the pulse. The sampled signal $\bar{x}_s(t)$ can be modeled as the convolution of the impulse-sampled signal $x_s(t)$ and a rectangular pulse with unit amplitude and a duration of dT_s.

Eqn. 4.39: $$h_{zoh}(t) = \Pi\left(\frac{t - 0.5\,dT_s}{dT_s}\right) = u(t) - u(t - dT_s)$$

☞ The frequency spectrum of the signal obtained through zero-order hold sampling is

Eqn. 4.43: $$\bar{X}_s(\omega) = d \operatorname{sinc}\left(\frac{\omega d T_s}{2\pi}\right) e^{-j\omega d T_s/2} \sum_{k=-\infty}^{\infty} X_a(\omega - k\omega_s)$$

☞ Reconstruction is the act of converting a sampled signal back to its analog form. The simplest method of doing this is *zero-order hold interpolation.* It can be achieved by processing the impulse sampled signal $x_s(t)$ through a zero-order hold reconstruction filter with impulse response

Eqn. 4.44: $$h_{\text{zoh}}(t) = \Pi\left(\frac{t - T_s/2}{T_s}\right)$$

☞ First-order hold interpolation fills the gaps between sampling instants with linear segments. It can be implemented by using a first-order hold reconstruction filter with impulse response in the form of a triangle:

Eqn. 4.46: $$h_{\text{foh}}(t) = \begin{cases} 1 + t/T_s\,, & -T_s < t < 0 \\ 1 - t/T_s\,, & 0 < t < T_s \\ 0\,, & |t| \geq T_s \end{cases}$$

☞ Perfect recovery of an analog signal from its properly sampled version is theoretically possible through *bandlimited interpolation.* It involves processing the impulse-sampled signal through an ideal lowpass filter that only passes frequencies in the range $-f_s/2 \le f \le f_s/2$, and blocks all other frequencies. The impulse response of the ideal reconstruction filter is

Eqn. 4.54:
$$h_r(t) = \text{sinc}\left(t f_s\right) = \text{sinc}\left(\frac{t}{T_s}\right)$$

and the reconstructed signal is

Eqn. 4.55:
$$x_r(t) = \sum_{n=-\infty}^{\infty} x_a(nT_s)\,\text{sinc}\left(\frac{t-nT_s}{T_s}\right)$$

☞ *Downsampling* operation allows the sampling rate of a discrete-time signal to be reduced by an integer factor.

Eqn. 4.59:
$$x_d[n] = x[nD]$$

The parameter D is the *downsampling rate.*

☞ The spectrum of the downsampled signal is

Eqn. 4.69:
$$X_d(\Omega) = \frac{1}{D}\sum_{k=0}^{D-1} X\left(\frac{\Omega - 2\pi k}{D}\right)$$

☞ Downsampling operation could lead to aliasing if the highest angular frequency in the signal $x[n]$ exceeds π/D. A lowpass anti-aliasing filter is typically used on the signal before it is downsampled. The combination of the anti-aliasing filter and the downsampler is referred to as a *decimator.*

☞ *Upsampling* operation provides the first step in increasing the sampling rate of a discrete-time signal by an integer factor.

Eqn. 4.70:
$$x_u[n] = \begin{cases} x[n/L], & n = kL, \quad k: \text{integer} \\ 0, & \text{otherwise} \end{cases}$$

where L is the *upsampling rate.*

☞ The spectrum of the upsampled signal is

Eqn. 4.72:
$$X_u(\Omega) = X(\Omega L)$$

☞ Upsampling creates zero-amplitude samples in between existing samples of the input signal. A lowpass *interpolation filter* is needed to make the zero-amplitude samples "blend-in" with the rest of the signal. The combination of an upsampler and a lowpass interpolation filter is referred to as an *interpolator.*

Further Reading

[1] U. Graf. *Applied Laplace Transforms and Z-Transforms for Scientists and Engineers: A Computational Approach Using a Mathematica Package.* Birkhäuser, 2004.

[2] D.G. Manolakis and V.K. Ingle. *Applied Digital Signal Processing: Theory and Practice.* Cambridge University Press, 2011.

[3] A.V. Oppenheim and R.W. Schafer. *Discrete-Time Signal Processing.* Prentice Hall, 2010.

[4] R.A. Schilling, R.J. Schilling, and P.D. Sandra L. Harris. *Fundamentals of Digital Signal Processing Using MATLAB.* Cengage Learning, 2010.

[5] L. Tan and J. Jiang. *Digital Signal Processing: Fundamentals and Applications.* Elsevier Science, 2013.

MATLAB Exercises with Solutions

MATLAB Exercise 4.1: Spectral relations in impulse sampling

Consider the continuous-time signal

$$x_a(t) = e^{-|t|}$$

Its Fourier transform is

$$X_a(f) = \frac{2}{1+4\pi^2 f^2}$$

Compute and graph the spectrum of $x_a(t)$. If the signal is impulse-sampled using a sampling rate of $f_s = 1$ Hz to obtain the signal $x_s(t)$, compute and graph the spectrum of the impulse-sampled signal. Afterward repeat with $f_s = 2$ Hz.

Solution: The script `mexdt_4_1a.m` listed below utilizes an anonymous function to define the transform $X_a(f)$. It then uses Eqn. (4.13) to compute and graph $X_s(f)$ superimposed with the contributing terms.

```
% Script: mexdt_4_1a.m
Xa = @(f) 2./(1+4*pi*pi*f.*f);     % Original spectrum
f = [-3:0.01:3];
fs = 1;          % Sampling rate
Ts = 1/fs;       % Sampling interval
% Approximate spectrum of impulse-sampled signal
Xs = zeros(size(Xa(f)));
for k=-5:5,
  Xs = Xs+fs*Xa(f-k*fs);  % Eqn. (4.12)
end;
tiledlayout(2,1);
nexttile;        % Graph the original spectrum
plot(f,Xa(f)); grid;
axis([-3,3,-0.5,2.5]);
title('X_{a}(f)');
nexttile;        % Graph spectrum of impulse-sampled signal
plot(f,Xs); grid;
axis([-3,3,-0.5,2.5]);
hold on;
for k=-5:5,
  tmp = plot(f,fs*Xa(f-k*fs),'r--');
end;
```

```
23 hold off;
24 title('X_{s}(f)');
25 xlabel('f (Hz)');
```

Lines 7 through 10 of the code approximate the spectrum $X_s(f)$ using terms for $k = -5, \ldots, 5$ of the infinite summation in Eqn. (4.13). The graphs produced are shown in Fig. 4.44.

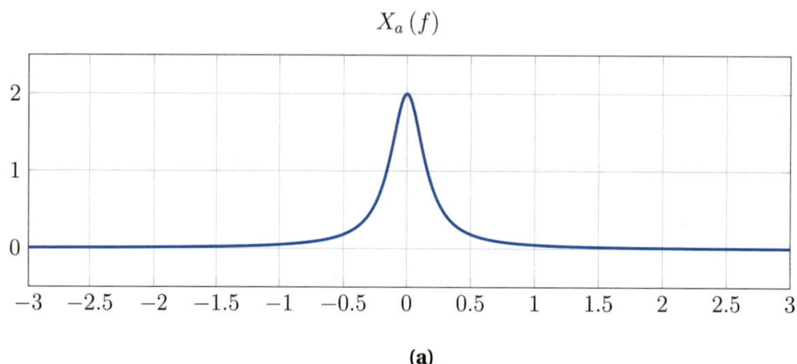

(a)

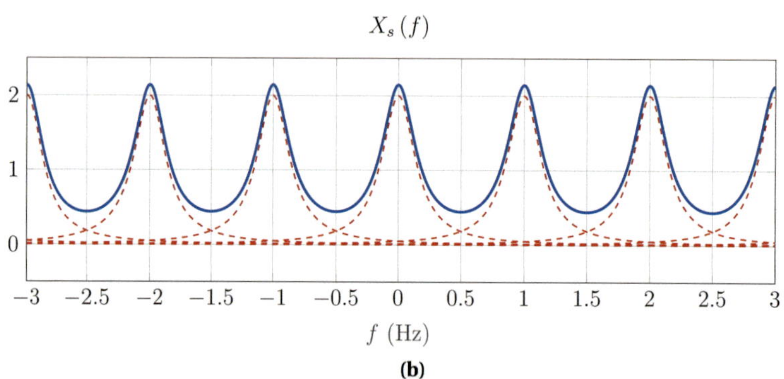

(b)

Figure 4.44 – MATLAB graphs for the script `mexdt_4_1a.m`.

Different sampling rates may be tried easily by changing the value of the variable `fs` in line 4. The spectrum $X_s(f)$ for $f_s = 2$ Hz is shown in Fig. 4.45.

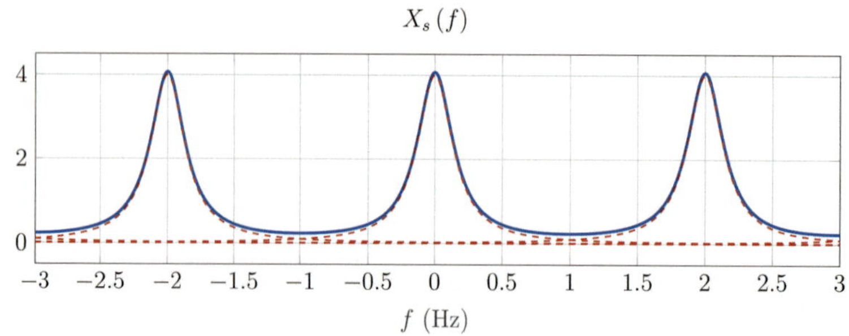

Figure 4.45 – The spectrum $X_s(f)$ for $f_s = 2$ Hz.

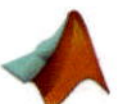

Software resources: `mexdt_4_1a.m` , `mexdt_4_1b.m`

MATLAB Exercise 4.2: DTFT of discrete-time signal obtained through sampling

The two-sided exponential signal

$$x_a(t) = e^{-|t|}$$

which was also used in MATLAB Exercise 4.1 is sampled using a sampling rate of $f_s = 1$ Hz to obtain a discrete-time signal $x[n]$. Compute and graph $X(\Omega)$, the DTFT spectrum of $x[n]$ for the angular frequency range $-\pi \le \Omega \le \pi$. Afterward repeat with $f_s = 2$ Hz.

Solution: The script `mexdt_4_2a.m` listed below graphs the signal $x_a(t)$ and its sampled version $x[n]$.

```
% Script: mexdt_4_2a.m
t = [-5:0.01:5];
xa = @(t) exp(-abs(t));
fs = 2;
Ts = 1/fs;
n = [-15:15];
xn = xa(n*Ts);
tiledlayout(2,1);
nexttile;   % Graph analog signal
plot(t,xa(t)); grid;
title('Signal x_{a}(t)');
nexttile;   % Graph sampled signal
stem(n,xn);
title('Signal x[n]');
```

Recall that the spectrum of the analog signal is

$$X_a(\omega) = \frac{2}{1+\omega^2}$$

The DTFT spectrum $X(\Omega)$ is computed using Eqn. (4.25). The script `mexdt_4_2b.m` is listed below:

```
% Script: mexdt_4_2b.m
Xa = @(omg) 2./(1+omg.*omg);
fs = 1;
Ts = 1/fs;
Omg = [-1:0.001:1]*pi;
XDTFT = zeros(size(Xa(Omg/Ts)));  % Eqn. (4.25)
for k=-5:5,
  XDTFT = XDTFT+fs*Xa((Omg-2*pi*k)/Ts);
end;
plot(Omg,XDTFT); grid;
axis([-pi,pi,0,2.5]);
title('X(\Omega)');
xlabel('\Omega (rad)');
```

The graph produced is shown in Fig. 4.46.

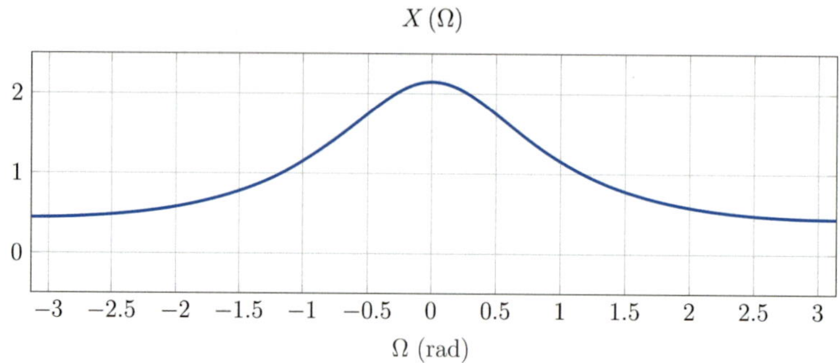

Figure 4.46 – The spectrum $X(\Omega)$ for $f_s = 1$ Hz.

The sampling rate $f_s = 2$ Hz can be obtained by changing the value of the variable `fs` in line 3.

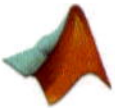

Software resources: `mexdt_4_2a.m` , `mexdt_4_2b.m`

MATLAB Exercise 4.3: Sampling a sinusoidal signal

In Example 4.2 three sinusoidal signals

$$x_{1a}(t) = \cos(12\pi t)$$
$$x_{2a}(t) = \cos(20\pi t)$$
$$x_{3a}(t) = \cos(44\pi t)$$

were each sampled using the sampling rate $f_s = 16$ Hz to obtain three discrete-time signals. It was shown that the resulting signals $x_1[n]$, $x_2[n]$, and $x_3[n]$ were identical. In this exercise we will verify this result using MATLAB. The script listed below computes and displays the first few samples of each discrete-time signal:

```
% Script: mexdt_4_3a.m
x1a = @(t) cos(12*pi*t);
x2a = @(t) cos(20*pi*t);
x3a = @(t) cos(44*pi*t);
t = [0:0.001:0.5];
fs = 16;
Ts = 1/fs;
n = [0:5];
x1n = x1a(n*Ts)
x2n = x2a(n*Ts)
x3n = x3a(n*Ts)
```

The script listed below computes and graphs the three signals and the discrete-time samples obtained by sampling them.

```
% Script: mexdt_4_3b.m
x1a = @(t) cos(12*pi*t);
x2a = @(t) cos(20*pi*t);
x3a = @(t) cos(44*pi*t);
t = [0:0.001:0.5];
fs = 16;
Ts = 1/fs;
n = [0:20];
x1n = x1a(n*Ts);
plot(t,x1a(t),t,x2a(t),t,x3a(t));
hold on;
plot(n*Ts,x1n,'ro');
hold off;
axis([0,0.5,-1.2,1.2]);
```

The graph produced is shown in Fig. 4.47.

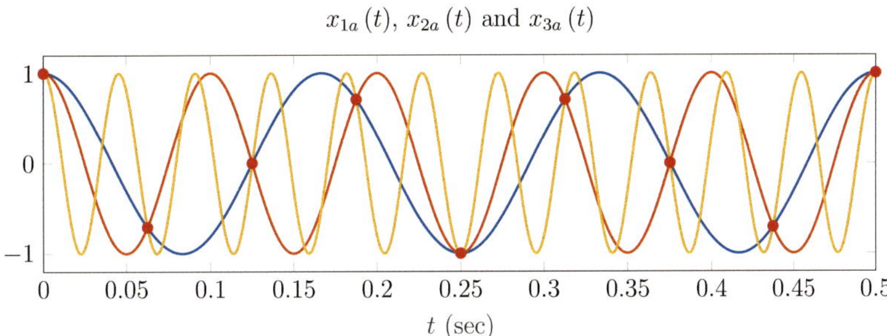

Figure 4.47 – Three sinusoidal signals that produce the same samples.

Software resources: mexdt_4_3a.m , mexdt_4_3b.m

MATLAB Exercise 4.4: Natural sampling

The two-sided exponential signal

$$x_a(t) = e^{-|t|}$$

is sampled using a natural sampler with a sampling rate of $f_s = 4$ Hz and a duty cycle of $d = 0.6$. Compute and graph $X_s(f)$ in the frequency interval $-12 \le f \le 12$ Hz.

Solution: The spectrum given by Eqn. (4.38) may be written using f instead of ω as

$$\bar{X}_s(f) = d \sum_{k=-\infty}^{\infty} \text{sinc}(kd)\, X_a(f - kf_s)$$

The script to compute and graph $\bar{X}_s(f)$ is listed below. It is obtained by modifying the script mexdt_6_1a.m.m" developed in MATLAB Exercise 4.1. The sinc envelope is also shown.

```
% Script: mexdt_4_4.m
Xa = @(f) 2./(1+4*pi*pi*f.*f);
f = [-12:0.01:12];
fs = 4;          % Sampling rate.
Ts = 1/fs;       % sampling interval.
d = 0.6;         % Duty cycle
Xs = zeros(size(Xa(f)));
for k=-5:5,
  Xs = Xs+d*sinc(k*d)*Xa(f-k*fs);  % Eqn. (4.38)
end;
plot(f,Xs,'b-',f,2*d*sinc(f*d/fs),'r--'); grid;
axis([-12,12,-0.5,1.5]);
title('X_{s}(f)');
xlabel('f (Hz)');
```

The spectrum $\bar{X}_s(f)$ is shown in Fig. 4.48.

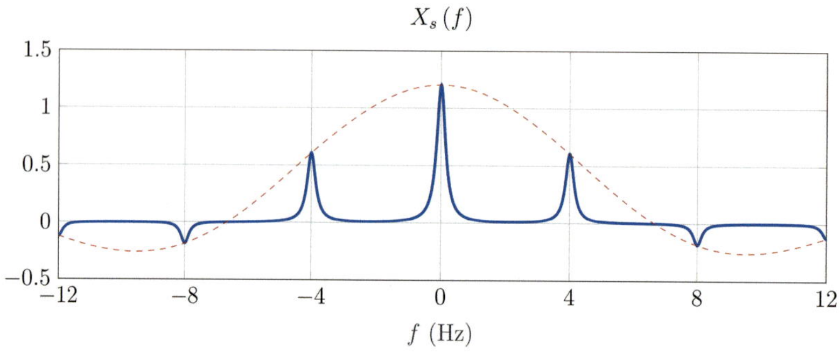

Figure 4.48 – The spectrum $X_s(f)$ for MATLAB Exercise 4.4.

Software resource: `mexdt_4_4.m`

MATLAB Exercise 4.5: Zero-order hold sampling

The two-sided exponential signal

$$x_a(t) = e^{-|t|}$$

is sampled using a zero-order hold sampler with a sampling rate of $f_s = 3$ Hz and a duty cycle of $d = 0.3$. Compute and graph $|X_s(f)|$ in the frequency interval $-12 \le f \le 12$ Hz.

Solution: The spectrum given by Eqn. (4.43) may be written using f instead of ω as

$$\bar{X}_s(f) = d\,\text{sinc}(f d T_s)\, e^{-j\pi f d T_s} \sum_{k=-\infty}^{\infty} X_a(f - k f_s)$$

The script to compute and graph $|\bar{X}_s(f)|$ is listed below. It is obtained by modifying the script `matex_6_1a.m` developed in MATLAB Exercise 4.1.

```
% Script: mexdt_4_5.m
Xa = @(f) 2./(1+4*pi*pi*f.*f);
f = [-12:0.01:12];
fs = 3;         % Sampling rate
Ts = 1/fs;      % Sampling interval
d = 0.3;        % Duty cycle
Xs = zeros(size(Xa(f)));
for k=-5:5,
  Xs = Xs+fs*Xa(f-k*fs);
end;
Xs = d*Ts*sinc(f*d*Ts).*exp(-j*pi*f*d*Ts).*Xs;  % Eqn. (4.43)
plot(f,abs(Xs)); grid;
axis([-12,12,-0.1,0.7]);
title('|X_{s}(f)|');
xlabel('f (Hz)');
```

The magnitude spectrum $|\bar{X}_s(f)|$ is shown in Fig. 4.49.

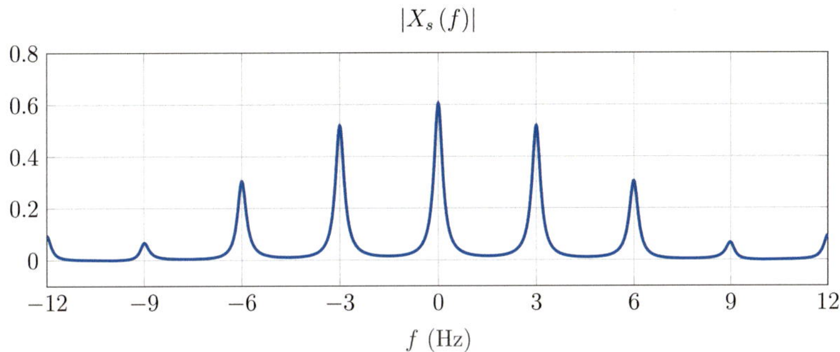

Figure 4.49 – The magnitude spectrum $|X_s(f)|$ for MATLAB Exercise 4.5.

Software resource: mexdt_4_5.m

MATLAB Exercise 4.6: Graphing signals for natural and zero-order hold sampling

In this exercise we will develop and test two functions `ss_natsamp()` and `ss_zohsamp()` for obtaining graphical representation of signals sampled using natural sampling and zero-order sampling respectively. The function `ss_natsamp()` evaluates a naturally sampled signal at a specified set of time instants.

```
function xnat = ss_natsamp(xa,Ts,d,t)
  t1 = (mod(t,Ts)<=d*Ts);
  xnat = xa(t).*t1;
end
```

The function `ss_zohsamp()` evaluates and returns a zero-order hold sampled version of the signal.

```
function xzoh = ss_zohsamp(xa,Ts,d,t)
  t1 = (mod(t,Ts)<=d*Ts);
  xzoh = xa(t).*t1;
  flg = 0;
  for i=1:length(t)
    if not(t1(i))
      flg = 0;
    elseif (t1(i) & (flg==0))
      flg = 1;
      value = xzoh(i);
    end
    if (flg==1)
      xzoh(i) = value;
    end
  end
end
```

For both functions the input arguments are as follows:

`xa`: Name of an anonymous function that can be used for evaluating the analog signal $x_a(t)$ at any specified time instant.

`Ts`: The sampling interval in seconds.

`d`: The duty cycle. Should be $0 < d \leq 1$.

`t`: Vector of time instants at which the sampled signal should be evaluated. For a detailed graph, choose the time increment for the values in vector `t` to be significantly smaller than T_s.

The function `ss_natsamp()` can be tested with the double sided exponential signal using the following statements:

```
>> x = @(t) exp(-abs(t));
>> t = [-4:0.001:4];
>> xnat = ss_natsamp(x,0.2,0.5,t);
>> plot(t,xnat);
```

The function `ss_zohsamp()` can be tested with the following:

```
>> xzoh = ss_zohsamp(x,0.2,0.5,t);
>> plot(t,xzoh);
```

Graphs produced for the naturally-sampled signal and the zero-order hold signal are shown in Fig. 4.50.

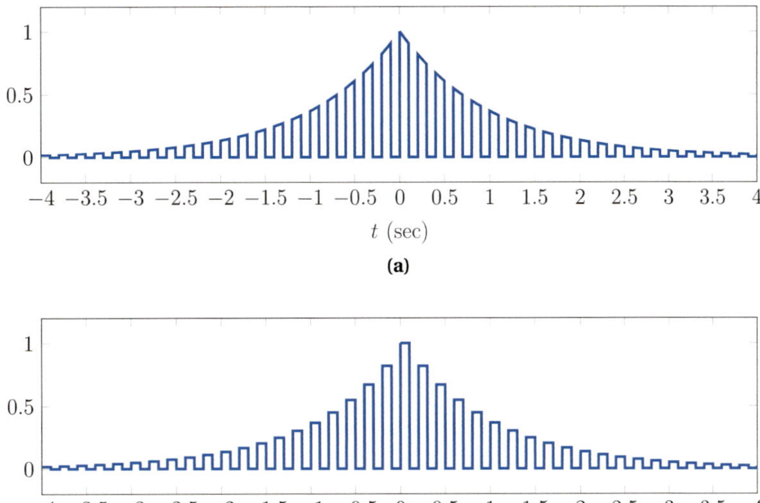

Figure 4.50 – Graphs for MATLAB Exercise 4.6: **(a)** natural sampling and **(b)** zero-order hold sampling.

Software resources: mexdt_4_6a.m , mexdt_4_6b.m , ss_natsamp.m , ss_zohsamp.m

MATLAB Exercise 4.7: Reconstruction of right-sided exponential

Recall that in Example 4.1 we considered impulse-sampling the right-sided exponential signal

$$x_a(t) = e^{-100t}\,u(t)$$

The spectrum $X(f)$ shown in Fig. 4.6 indicates that the right-sided exponential signal is not bandlimited, and therefore there is no sampling rate that would satisfy the requirements of the Nyquist sampling theorem. As a result, aliasing will be present in the spectrum regardless of the sampling rate used. In Example 4.1 three sampling rates, $f_s = 200$ Hz, $f_s = 400$ Hz, and $f_s = 600$ Hz, were used. The aliasing effect is most noticeable for $f_s = 200$ Hz, and less so for $f_s = 600$ Hz.

In a practical application of sampling, we would have processed the signal $x_a(t)$ through an anti-aliasing filter prior to sampling it, as shown in Fig. 4.9. However, in this exercise we will omit the anti-aliasing filter, and attempt to reconstruct the signal from its sampled version using the three techniques we have discussed. The script mexdt_4_7a.m given below produces a graph of the impulse-sampled signal and the zero-order-hold approximation to the analog signal $x_a(t)$.

```
% Script: mexdt_4_7a.m
fs = 200;      % Sampling rate
Ts = 1/fs;     % Sampling interval
% Set index limits "n1" and "n2" to cover the time interval
```

```
5  % from -25 ms to +75 ms
6  n1 = -fs/40;
7  n2 = -3*n1;
8  n = [n1:n2];
9  t = n*Ts;                    % Vector of time instants
10 xs = exp(-100*t).*(n>=0);    % Samples of the signal
11 stem(t,xs,'^'); grid;
12 hold on;
13 stairs(t,xs,'r-');
14 hold off;
15 axis([-0.030,0.080,-0.2,1.1]);
16 title('Reconstruction using zero-order-hold');
17 xlabel('t (sec)');
18 ylabel('Amplitude');
19 text(0.015,0.7,sprintf('Sampling rate = %.3g Hz',fs));
```

The sampling rate can be modified by editing line 2 of the code. The graph generated by this function is shown in Fig. 4.51 for sampling rates 200 Hz and 400 Hz.

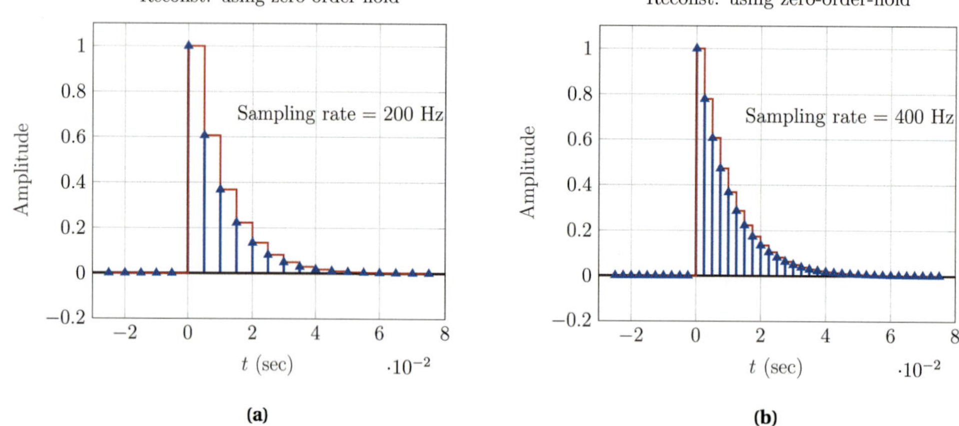

Figure 4.51 – Impulse-sampled right-sided exponential signal and zero-order-hold reconstruction for **(a)** $f_s = 200$ Hz and **(b)** $f_s = 400$ Hz.

Modifying this script to produce first-order-hold interpolation is almost trivial. The modified script `mexdt_4_7b.m` is given below.

```
% Script: mexdt_4_7b.m
fs = 200;      % Sampling rate
Ts = 1/fs;     % Sampling interval
% Set index limits "n1" and "n2" to cover the time interval
% from -25 ms to +75 ms
n1 = -fs/40;
n2 = -3*n1;
n = [n1:n2];
t = n*Ts;                    % Vector of time instants
xs = exp(-100*t).*(n>=0);    % Samples of the signal
stem(t,xs,'^'); grid;
hold on;
```

```
13 plot(t,xs,'r-');
14 hold off;
15 axis([-0.030,0.080,-0.2,1.1]);
16 title('Reconstruction using first-order-hold');
17 xlabel('t (sec)');
18 ylabel('Amplitude');
19 text(0.015,0.7,sprintf('Sampling rate = %.3g Hz',fs));
```

The only functional change is in line 15 where we use the function `plot()` instead of the function `stairs()`. The graph generated by this modified script is shown in Fig. 4.52 for sampling rates 200 Hz and 400 Hz.

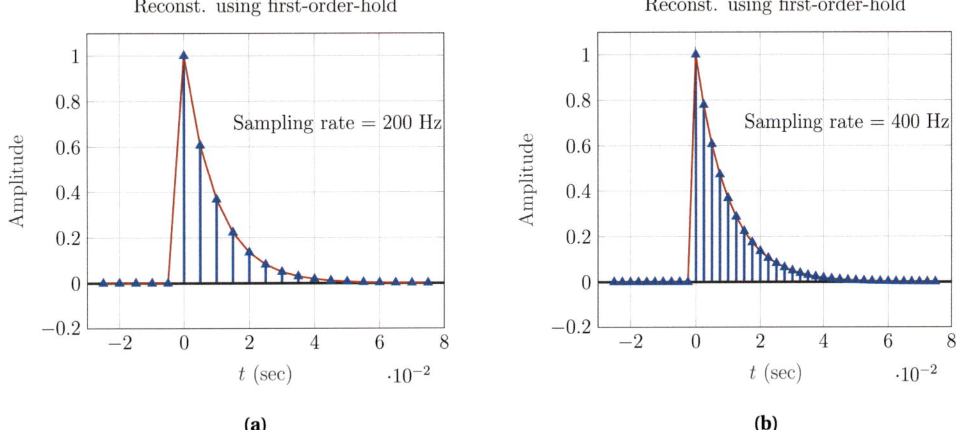

Figure 4.52 – Impulse-sampled right-sided exponential signal and first-order-hold reconstruction for **(a)** $f_s = 200$ Hz and **(b)** $f_s = 400$ Hz.

Reconstruction through bandlimited interpolation requires a bit more work. The script `mexdt_4_7c.m` for this purpose is given below. Note that we have added a new section between lines 13 and 19 to compute the shifted sinc functions called for in Eqn. (4.55).

```
1  % Script: mexdt_4_7c.m
2  fs = 200;      % Sampling rate
3  Ts = 1/fs;     % Sampling interval
4  % Set index limits "n1" and "n2" to cover the time interval
5  % from -25 ms to +75 ms
6  n1 = -fs/40;
7  n2 = -3*n1;
8  n = [n1:n2];
9  t = n*Ts;                    % Vector of time instants
10 xs = exp(-100*t).*(n>=0);    % Samples of the signal
11 % Generate a new, more dense, set of time values for the
12 % sinc interpolating functions
13 t2 = [-0.025:0.0001:0.1];
14 xr = zeros(size(t2));
15 for n=n1:n2,
16   nn = n-n1+1;               % Because MATLAB indices start at 1
17   xr = xr+xs(nn)*sinc((t2-n*Ts)/Ts);
```

```
end;
stem(t,xs,'^'); grid;
hold on;
plot(t2,xr,'r-');
hold off;
axis([-0.030,0.08,-0.2,1.1]);
title('Reconstruction using bandlimited interpolation');
xlabel('t (sec)');
ylabel('Amplitude');
text(0.015,0.7,sprintf('Sampling rate = %.3g Hz',fs));
```

The graph generated by this script is shown in Fig. 4.53 for sampling rates 200 Hz and 400 Hz.

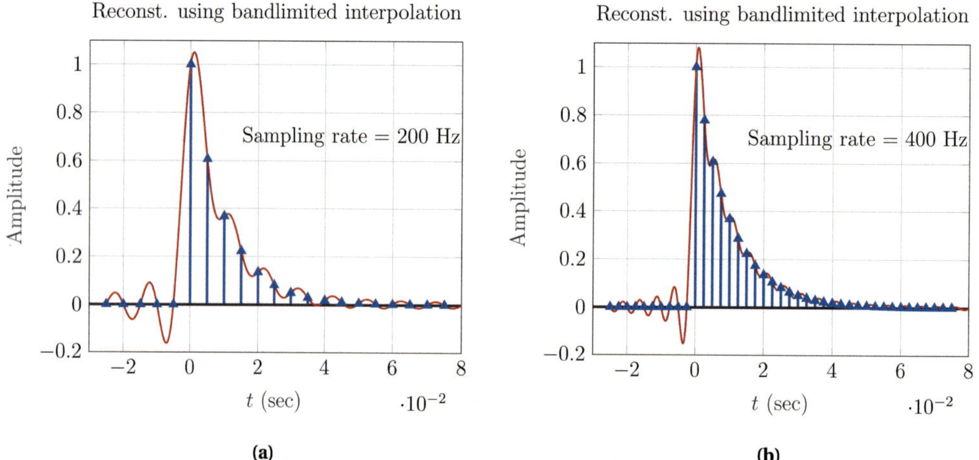

Figure 4.53 – Impulse-sampled right-sided exponential signal and its reconstruction based on bandlimited interpolation for **(a)** $f_s = 200$ Hz and **(b)** $f_s = 400$ Hz.

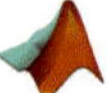

Software resources: mexdt_4_7a.m , mexdt_4_7b.m , mexdt_4_7c.m

MATLAB Exercise 4.8: Frequency spectrum of reconstructed signal

In this exercise we will compute and graph the frequency spectra for the reconstructed signals $x_{zoh}(t)$, $x_{foh}(t)$ and $x_r(t)$ obtained in MATLAB Exercise 4.7. Recall that the frequency spectra for the original right-sided exponential signal and its impulse-sampled version were found in Example 4.1. System functions for zero-order-hold and first-order-hold interpolation filters are given by Eqns. (4.48) and (4.50), respectively. Spectra for reconstructed signals are found through Eqns. (4.49) and (4.51). In the case of bandlimited interpolation, the spectrum of the reconstructed signal can be found by simply truncating the spectrum $X_s(f)$ to retain only the part of it in the frequency range $-f_s/2 \le f \le f_s/2$. The script `mexdt_4_8a.m` listed below computes and graphs each spectrum along with the original spectrum $X_a(f)$ for comparison. The sampling rate used in each case is $f_s = 200$ Hz, and may be modified by editing line 3 of the code.

```
% Script: mexdt_4_8.m
fs = 200;      % Sampling rate
Ts = 1/fs;     % Sampling interval
f = [-700:0.5:700];      % Vector of frequencies
Xa = 1./(100+j*2*pi*f); % Original spectrum
% Compute the spectrum of the impulse-sampled signal.
Xs = 1./(1-exp(-100*Ts)*exp(-j*2*pi*f*Ts));
% Compute system functions of reconstruction filters.
Hzoh = Ts*sinc(f*Ts).*exp(-j*pi*Ts);  % Eqn. (4.48)
Hfoh = Ts*sinc(f*Ts).*sinc(f*Ts);     % Eqn. (4.50)
Hr = Ts*((f>=-0.5*fs)&(f<=0.5*fs));   % Eqn. (4.52)
% Compute spectra of reconstructed signals.
Xzoh = Xs.*Hzoh;  % Eqn. (4.49)
Xfoh = Xs.*Hfoh;  % Eqn. (4.51)
Xr = Xs.*Hr;      % Eqn. (4.53)
% Graph the results.
tiledlayout(3,1);
nexttile;
plot(f,abs(Xzoh),'-',f,abs(Xa),'--'); grid;
title('Spectr. of signal reconstr. through zero-order hold');
xlabel('f (Hz)');
ylabel('Magnitude');
legend('|X_{zoh}(f)|','|X(f)|');
nexttile;
plot(f,abs(Xfoh),'-',f,abs(Xa),'--'); grid;
title('Spectr. of signal reconstr. through first-order hold');
xlabel('f (Hz)');
ylabel('Magnitude');
legend('|X_{foh}(f)|','|X(f)|');
nexttile;
plot(f,abs(Xr),'-',f,abs(Xa),'--'); grid;
title('Spectr. of signal reconstr. through bandlimited interp.');
xlabel('f (Hz)');
ylabel('Magnitude');
legend('|X_{r}(f)|','|X(f)|');
```

The graphs generated by the script are shown in Fig. 4.54. Observe the effect of aliasing on each spectrum.

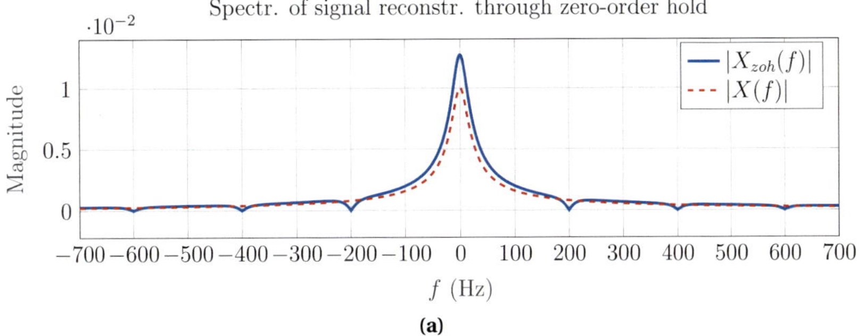

(a)

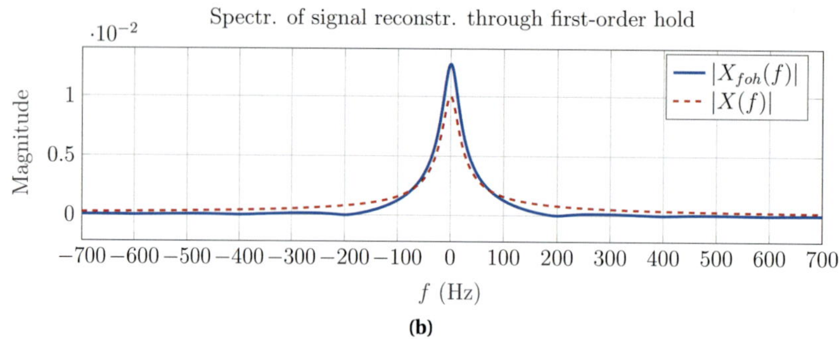

(b)

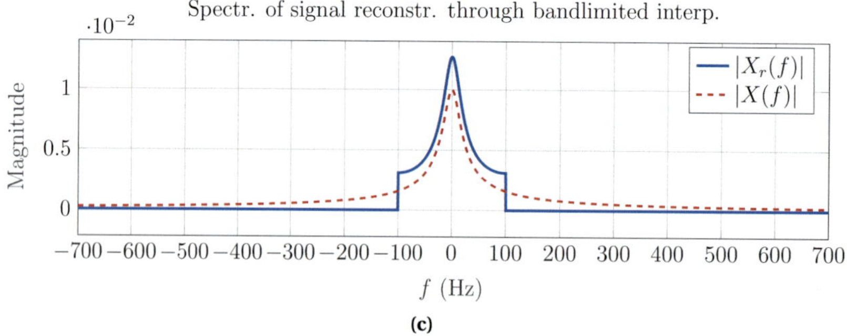

(c)

Figure 4.54 – Frequency spectra of reconstructed signals obtained in MATLAB Exercise 4.8 through **(a)** zero-order-hold, **(b)** first-order-hold, and **(c)** bandlimited interpolation.

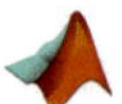

Software resource: `mexdt_4_8.m`

MATLAB Exercise 4.9: Resampling discrete-time signals

Consider the system shown in Fig. 4.55.

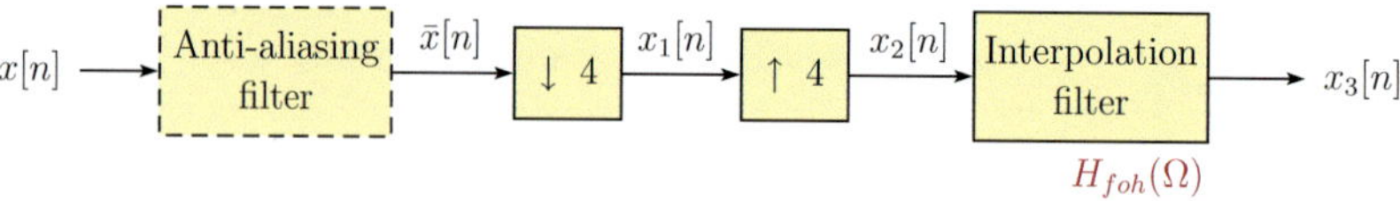

Figure 4.55 – The system for MATLAB Exercise 4.9.

The input signal is

$$x[n] = \cos(0.1\pi n) + 0.7\sin(0.2\pi n) + \cos(0.4\pi n)$$

Simulate the system first without the anti-aliasing filter and then with the anti-aliasing filter. Use first-order hold interpolation after the upsampler.

Solution: The impulse response of the interpolation filter for first-order hold interpolation is a discrete-time triangle peak amplitude of 1 and the two corners at $n = \mp L = \mp 4$.

$$h_r[n] = \{ 0,\ 0.25,\ 0.5,\ 0.75,\ \underset{\substack{\uparrow \\ n=0}}{1.0},\ 0.75,\ 0.5,\ 0.25,\ 0 \} \tag{4.76}$$

The script `mexdt_4_9a.m` listed below implements this system without the anti-aliasing filter.

```
1  % Script: mexdt_4_9a.m
2  n = [0:99];
3  x = cos(0.1*pi*n)+0.7*sin(0.2*pi*n)+cos(0.4*pi*n);
4  x1 = downsample(x,4);
5  x2 = upsample(x1,4);
6  hr = [0,0.25,0.5,0.75,1,0.75,0.5,0.25,0];   % FOH filter
7  x3 = conv(x2,hr);
8  x3 = x3(5:104);     % Compensate for 4 samples of delay
9  n = [0:99];
10 stem(n,x);
11 hold on;
12 plot(n,x3,'r');
13 hold off;
14 xlabel('n');
```

The vector `x3` obtained through convolution in line 7 has 108 samples. In line 8 of the code we discard the first and the last 4 elements of `x3`. The first 4 samples discarded correspond to sample indices $n = -4, \ldots, -1$ since the impulse response $h_r[n]$ starts at index $n = -4$. The last 4 samples discarded are the tail end of the convolution result. We are left with 100 samples of the signal $x_3[n]$ suitable for direct comparison with the 100-sample input signal $x[n]$. Input and output signals are shown in Fig. 4.56. For display purposes the output signal is shown in red with tips of samples connected by straight lines. The effect of aliasing due to the missing anti-aliasing filter is evident.

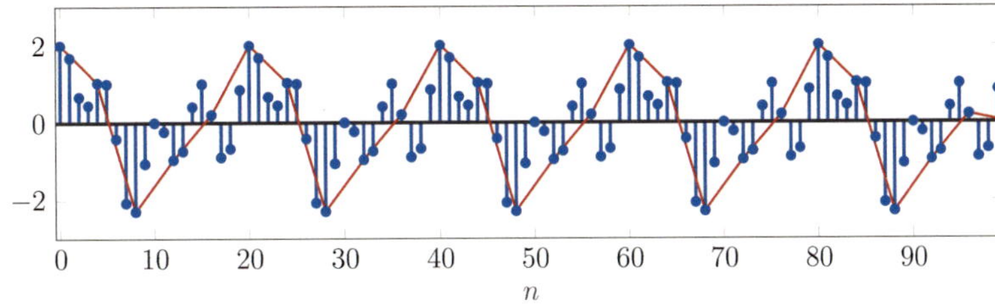

Figure 4.56 – Input and output signals of the system in Fig. 4.55 without the anti-aliasing filter.

For this particular signal the implementation of an anti-aliasing filter is almost trivial. Recall from the discussion in Section 4.4.1 that the signal to be downsampled must not have any frequencies greater than $\Omega_c = \pi/D = \pi/4$. An ideal anti-aliasing filter would simply remove the third term in $x[n]$, the term that has an angular frequency of 0.4π. In the script code we simply modify line 3 to that effect:

```
1  % Script: mexdt_4_9b.m
2  n = [0:99];
3  x = cos(0.1*pi*n)+0.7*sin(0.2*pi*n);
4  x1 = downsample(x,4);
```

```
x2 = upsample(x1,4);
hr = [0,0.25,0.5,0.75,1,0.75,0.5,0.25,0];    % FOH filter
x3 = conv(x2,hr);
x3 = x3(5:104);    % Compensate for 4 samples of delay
n = [0:99];
stem(n,x);
hold on;
plot(n,x3,'r');
hold off;
xlabel('n');
```

Input and output signals for this case are shown in Fig. 4.57.

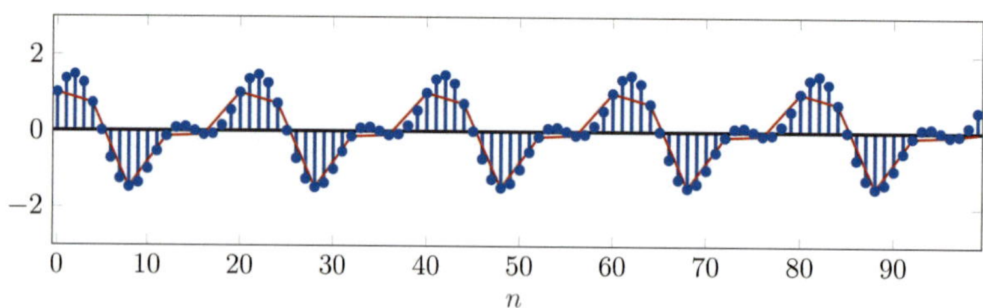

Figure 4.57 – Input and output signals of the system in Fig. 4.55 with the anti-aliasing filter.

Software resources: mexdt_4_9a.m , mexdt_4_9b.m

Problems

4.1. Consider the triangular waveform shown in Fig. P.4.1.

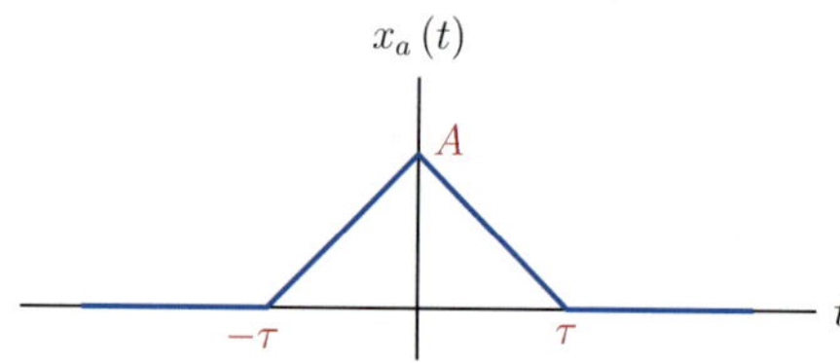

Figure P. 4.1

Its Fourier transform is

$$X_a(f) = A\tau \operatorname{sinc}^2(f\tau)$$

Let $A = 1$ and $\tau = 1$ s. The signal $x_a(t)$ is impulse-sampled using a sampling rate of $f_s = 5$ Hz.

a. Sketch the impulse-sampled signal $x_s(t)$.

b. Find an expression for $X_s(f)$.

c. Sketch $X_s(f)$ for $-10 \le f \le 10$ Hz.

4.2. An analog signal $x_a(t)$ has the Fourier transform shown in Fig. P.4.2.

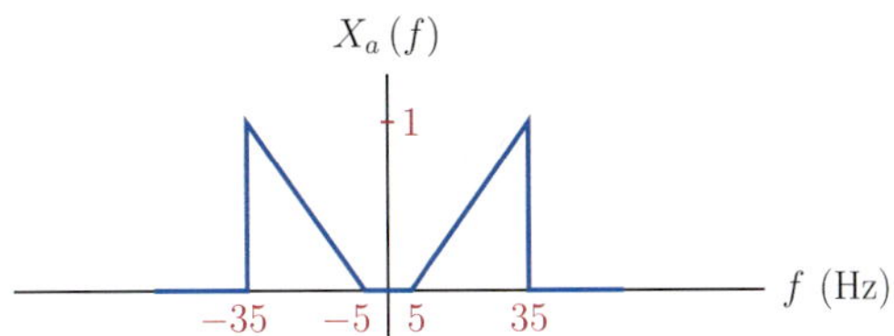

Figure P. 4.2

The signal is impulse sampled using a sampling rate of $f_s = 100$ Hz to obtain the signal $x_s(t)$. Sketch the spectrum $X_s(\omega)$.

4.3. The analog signal $x_a(t)$ with the Fourier transform shown in Fig. P.4.2 is sampled using a sampling rate of $f_s = 100$ Hz to obtain a discrete-time signal $x[n]$. Sketch the spectrum $X(\Omega)$.

4.4. If the analog signal $x_a(t)$ with the Fourier transform shown in Fig. P.4.2 is sampled using a sampling rate that is 10 percent less than the minimum requirement, sketch the spectrum $X(\Omega)$.

4.5. Consider the triangular waveform $x_a(t)$ shown in Fig. P.4.1. Let $A = 1$ and $\tau = 1$ s. This signal is sampled with a sampling rate of $f_s = 12$ Hz to obtain a discrete-time signal $x[n]$.

a. Sketch the signal $x[n]$.

b. Determine and sketch the DTFT spectrum of $x[n]$.

4.6. The signal

$$x_a(t) = \begin{cases} \sin(\pi t), & 0 \le t \le 1 \\ 0, & \text{otherwise} \end{cases}$$

is sampled with a sampling rate of $f_s = 15$ Hz to obtain a discrete-time signal $x[n]$. Determine and sketch the DTFT spectrum of $x[n]$.

4.7. Indicate which of the following signals can be sampled without any loss of information? For signals that can be sampled properly, determine the minimum sampling rate that can be used.

a. $x_a(t) = u(t) - u(t-3)$

b. $x_a(t) = e^{-2t}\, u(t)$

c. $x_a(t) = \cos(100\pi t) + 2\sin(150\pi t)$

d. $x_a(t) = \cos(100\pi t) + 2\sin(150\pi t)\sin(200\pi t)$

e. $x_a(t) = e^{-t}\cos(100\pi t)$

4.8. A sinusoidal signal $x_a(t) = \sin(2\pi f_a t)$ with a frequency of $f_a = 1$ kHz is sampled using a sampling rate of $f_s = 2.4$ kHz to obtain a discrete-time signal $x[n]$.

a. Manually sketch the signal $x_a(t)$ for the time interval $0 < t < 5$ ms.

b. Show the sample amplitudes of the discrete-time signal $x[n]$ on the sketch of $x_a(t)$.

c. Find three alternative frequencies for the analog signal that result in the same discrete-time signal $x[n]$ when sampled with the sampling rate $f_s = 2.4$ kHz.

4.9. Refer to Problem 4.8.

a. Sketch the frequency spectrum of the analog signal for the original sinusoid and each of the three alternative frequencies.

b. For each of the signals and corresponding spectra in part (a), determine the DTFT spectrum of the discrete-time signal that results from sampling with a sampling rate of $f_s = 2.4$ kHz.

4.10. The analog sinusoidal signal $x_a(t) = \sin(500\pi t)$ is sampled to obtain a discrete-time signal $x[n] = \sin(0.4\pi n)$.

a. Assuming that the signal is sampled properly, determine the sampling interval T and the sampling rate f_s.

b. Find two other sampling rates that would produce the same discrete-time signal $x[n]$.

c. Using the sampling rate found in part (a), find two other analog signals that could be sampled to produce the same discrete-time signal $x[n]$.

4.11. The sinusoidal signal

$$x_a(t) = 3\cos(100\pi t) + 5\sin(250\pi t)$$

is sampled at the rate of 100 times per second to obtain a discrete-time signal $x[n]$.

a. Sketch the spectrum $X(\Omega)$ of the signal $x[n]$.

b. If we assume that $x[n]$ is the result of properly sampling an analog signal and try to reconstruct that analog signal, what signal would we get?

4.12. The analog signal

$$x_a(t) = \cos(100\pi t) + \cos(120\pi t)$$

is sampled using natural sampling as shown in Fig. 4.18. The sampling rate used is $f_s = 400$ Hz, and the width of each pulse is $\tau = 0.5$ ms.

a. Write an analytical expression for the Fourier transform $X_a(\omega)$ and sketch it.

b. Find an analytical expression for $\bar{X}_s(\omega)$ the Fourier transform of the naturally-sampled signal $\bar{x}_s(t)$.

c. Sketch the transform $\bar{X}_s(\omega)$.

4.13. Repeat Problem 4.12 if zero-order sampling is used instead of natural sampling.

4.14. The signal $x_a(t) = \cos(150\pi n)$ is impulse-sampled with a sampling rate of $f_s = 200$ Hz and applied to a zero-order-hold reconstruction filter as shown in Fig. P.4.14.

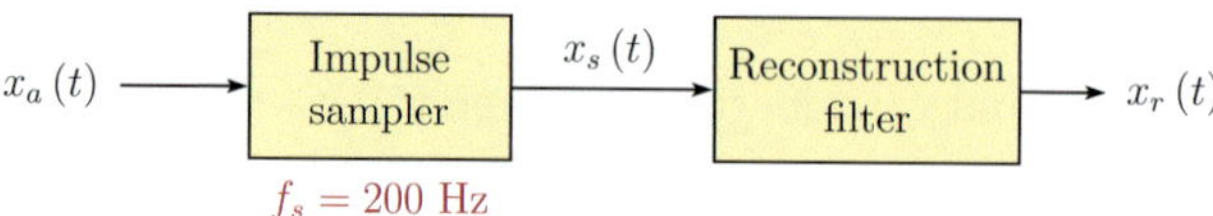

Figure P. 4.14

Sketch the signal at the output of the reconstruction filter.

4.15. Repeat Problem 4.14 if the reconstruction filter is a delayed first-order hold filter with the impulse response shown in Fig. P.4.15.

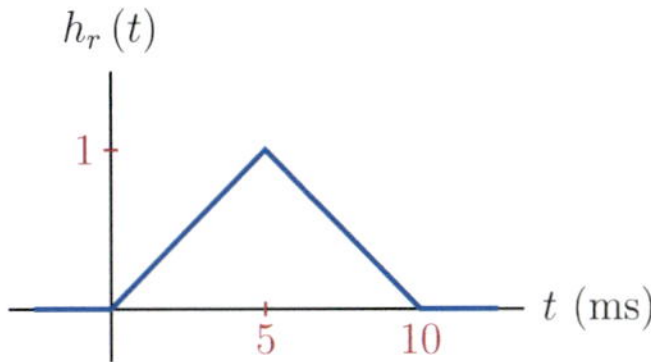

Figure P. 4.15

4.16. Show that the signal $w[m]$ defined as

$$w[m] = \frac{1}{D} \sum_{k=0}^{D-1} e^{j2\pi mk/D}$$

and used in the derivation of the spectrum of a downsampled signal in Section 4.4 satisfies the condition

$$w[m] = \begin{cases} 1, & m = nD, \quad n: \text{integer} \\ 0, & \text{otherwise} \end{cases}$$

4.17. Refer to the ideal interpolation filter spectrum $H_r(\Omega)$ shown in Fig. 4.43a. Show that the impulse response of the ideal interpolation filter is

$$h_r[n] = \text{sinc}(n/L)$$

4.18. Indicate which of the following discrete-time signals can be downsampled without any loss of information? For signals that can be downsampled properly, determine the maximum rate D that can be used.

a. $x[n] = u[n] - u[n-25]$

b. $x[n] = n\,u[n] - (n-10)\,u[n-10]\,, \qquad n = 0, \ldots, 19$

c. $x[n] = \cos\left(\dfrac{\pi n}{3}\right)$

d. $x[n] = \cos\left(\dfrac{\pi n}{10}\right) + 3\sin\left(\dfrac{2\pi n}{7}\right)$

4.19. The signal $x[n] = \sin(0.1\pi n)$ is applied to the system shown in Fig. P.4.19.

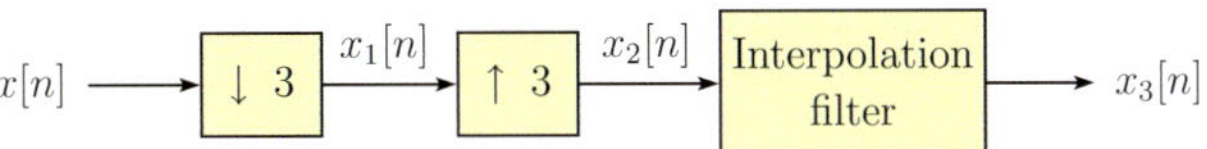

Figure P. 4.19

The interpolation filter is a zero-order hold filter with impulse response

$$h_{zoh}[n] = u[n] - u[n-3]$$

a. Find the spectrum of the zero-order hold filter.

b. Determine and sketch the frequency spectra $X(\Omega)$, $X_1(\Omega)$, $X_2(\Omega)$, and $X_3(\Omega)$.

4.20. Rework Problem 4.19 if the interpolation filter is changed to a first-order hold filter with impulse response

$$h_{foh}[n] = \begin{cases} 1 - \dfrac{|n|}{3}, & n = -3, \ldots, 3 \\ 0, & \text{otherwise} \end{cases}$$

MATLAB Problems

4.21. Refer to the problem described in MATLAB Exercise 4.1.

a. Set the sampling rate as $f_s = 3$ Hz. Compute and graph the spectrum $X_s(f)$ in the interval $-7 \le f \le 7$ Hz.

b. Repeat part (a) with the sampling rate set equal to $f_s = 4$ Hz.

c. Comment on the amount of aliasing for each case.

4.22. Refer to Problem 4.1. Write a script to

- compute and graph the spectrum $X_a(f)$, and
- compute and graph the spectrum $X_s(s)$ for sampling rates $f_s = 5, 3, 2$ Hz.

Use the frequency range for $-10 \le f \le 10$ Hz. for graphs.

4.23. Write a script to compute and graph the DTFT spectrum of the discrete-time signal obtained in Problem 4.5.

4.24. Write a script to compute and graph the DTFT spectrum of the discrete-time signal obtained in Problem 4.6.

4.25. Refer to Problem 4.8.

a. Write a script to graph the signal $x_a(t)$ and the samples of the signal $x[n]$ simultaneously for the time interval $0 < t < 5$ ms.

b. Three alternative frequencies were found that produce the same discrete-time signal. Modify the script in part (a) so that these alternative analog signals are also computed and graphed simultaneously with the original graph.

4.26. Refer to the function `ss_zohsamp()` that was developed in MATLAB Exercise 4.6.

a. Explain how the function works. Pay special attention to the conditional statements between lines 6 and 14 of the code.

b. The function does not work properly when the duty cycle is $d = 1$. Modify it so it also works when $d = 1$. Assume that the sampling rate T_s is always an integer multiple of the time increment used in vector `t`, and explore the use of MATLAB function `kron()` to implement the zero-order hold effect.

4.27. The two-sided exponential signal

$$x_a(t) = e^{-|t|}$$

is sampled using a zero-order hold sampler using a sampling rate of $f_s = 5$ Hz and a duty-cycle of $d = 0.9$.

a. Generate samples of the signal $x_{zoh}(t)$ for $-4 < t < 4$ s using the function `ss_zohsamp(..)`. Use a time vector with increments of 1 ms.

b. We will explore the possibility of obtaining a smoothly reconstructed signal from the zero-order held signal by filtering $x_{zoh}(t)$. A system with system function

$$H(\omega) = \frac{a}{j\omega + a}$$

may be simulated with statements

```
>> sys = tf([a],[1,a]);
>> y = lsim(sys,xzoh,t);
```

Compute and graph the filter output with parameter values $a = 1, 2, 3$ and comment on the results.

4.28. Write a script to simulate the system shown in Fig. P.4.19 with the input signal $x[n] = \sin(0.1\pi n)$ and using a zero-order hold interpolation filter. Generate 200 samples of the signal $x[n]$. Compute and graph the signals $x_1[n]$, $x_2[n]$ and $x_3[n]$.

4.29. Write a script to repeat the simulation in Problem 4.28 using a first-order hold interpolation filter instead of the zero-order hold filter.

MATLAB Projects

4.30. In this project we will explore the concept of aliasing especially in the way it exhibits itself in an audio waveform. MATLAB has a built-in sound file named "`handel`" which contains a recording of Handel's Hallelujah Chorus. It was recorded with a sampling rate of $f_s = 8192$ Hz. Use the statement

```
>> load handel
```

which loads two new variables named `Fs` and `y` into the workspace. The scalar variable `Fs` holds the sampling rate, and the vector `y` holds 73113 samples of the recording that corresponds to about 9 seconds of audio. Once loaded, the audio waveform may be played back with the statement

```
>> sound(y,Fs)
```

a. Graph the audio signal as a function of time. Create a continuous time variable that is correctly scaled for this purpose.

b. Compute and graph the frequency spectrum of the signal using the function `fft()`. Create an appropriate frequency variable to display frequencies in Hz, and use it in graphing the spectrum of the signal. Refer to the discussion in Section 7.2.6 of Chapter 3 for using the DFT to approximate the continuous Fourier transform.

c. Downsample the audio signal in vector y using a downsampling rate of $D = 2$. Do not use an anti-aliasing filter for this part. Play back the resulting signal using the function `sound()`. Be careful to adjust the sampling rate to reflect the act of downsampling or it will play too fast.

d. Repeat part (c) this time using an anti-aliasing filter. A simple Chebyshev type-I lowpass filter may be designed with the following statements:

```
>> [num,den] = cheby1(5,1,0.45)
```

Process the audio signal through the anti-aliasing filter using the statement

```
>> yfilt = filter(num,den,y);
```

The vector `yfilt` represents the signal at the output of the anti-aliasing filter. Downsample this output signal using $D = 2$ and listen to it. How does it compare to the sound obtained in part (c)? How would you explain the difference between the two sounds?

e. Repeat parts (c) and (d) using a downsampling rate of $D = 4$. The anti-aliasing filter for this case should be obtained by

```
>> [num,den] = cheby1(5,1,0.23)
```

4.31. On a computer with a microphone and a sound processor, the following MATLAB code allows 2 seconds of audio signal to be recorded, and a vector "x" to be created with the recording:

```
hRec = audiorecorder;
disp('Press a key to start recording');
pause;
recordblocking(hRec, 2);
disp('Finished recording');
x = getaudiodata(hRec);
```

By default the analog signal $x_a(t)$ captured by the microphone and the sound device is sampled at the rate of 8000 times per second, corresponding to $T = 125$ μs. For a 2-second recording the vector "x" contains 16,000 samples that represent

$$x[n] = x_a\left(\frac{n}{8000}\right), \qquad n = 0, \ldots, 15{,}999$$

Develop a MATLAB script to perform the following steps:

a. Extract 8000 samples of the vector $x[n]$ into a new vector.

$$r[n] = x[n+8000], \qquad n = 0, \ldots, 7999$$

We skip the first 8000 samples so that we do not get a blank period before the person begins to speak. This should create a vector with 8000 elements representing one full second of speech.

b. Convert the sampling rate of the 1-second speech waveform to 12 kHz. Use a combination of downsampling and upsampling along with any necessary filters to achieve this. Justify your choice of which operation should be carried out first.

c. Play back the resulting 12,000-sample waveform using the function `sound()`.

d. Use the function `fft()` to compute and graph the frequency spectra of the signals involved at each step of sampling rate conversion.

CHAPTER 5

THE z-TRANSFORM

Chapter Objectives

- Learn the z-transform as a more generalized version of the discrete-time Fourier transform (DTFT) studied in Chapter 3. Understand how it relates to the DTFT, its similarities and differences.
- Understand the convergence characteristics of the z-transform and the concept of region of convergence.
- Explore the properties of the z-transform such as linearity, time shifting and reversal, differentiation in the z-domain, convolution, correlation, initial value, and summation.
- Understand the use of the z-transform for modeling DTLTI systems. Learn the z-domain system function and its use for solving signal-system interaction problems.
- Learn techniques for obtaining block diagrams for DTLTI systems based on the z-domain system function.
- Learn the use of the unilateral z-transform for solving difference equations with specified initial conditions.

5.1 Introduction

In this chapter we will consider the z-transform which, for discrete-time signals, plays the same role that the Laplace transform plays for continuous-time signals. In the process, we will observe

DOI: 10.1201/9781003570462-5

that the development of the z-transform techniques for discrete-time signals and systems parallels the development of the Laplace transform for continuous-time signals and systems.

For continuous-time signals, the Laplace transform is an extension of the Fourier transform such that the latter can be thought of as a special case, or a limited-view, of the former. The relationship between the z-transform and the DTFT is similar. In Chapter 3 we have discussed the use of the DTFT for analyzing discrete-time signals. The use of the DTFT was adapted to the analysis of DTLTI systems through the concept of the system function. The z-transform is an extension of the DTFT that can be used for the same purposes.

Consider, for example, a discrete-time unit ramp signal $x[n] = n\,u[n]$. Its DTFT does not exist since the signal is not absolute summable. We will, however, be able to use z-transform techniques for analyzing the unit ramp signal. Similarly, the DTFT-based system function is only usable for stable systems since the impulse response of a system must be absolute summable for its DTFT-based system function to exist. The z-transform, on the other hand, can be used with systems that are unstable as well. The DTFT may or may not exist for a particular signal or system while the z-transform will generally exist subject to some constraints.

We will see that the DTFT is a restricted cross-section of the much more general z-transform. The DTFT of a signal, if it exists, can be obtained from its z-transform while the reverse is not generally true. In addition to analysis of signals and systems, implementation structures for discrete-time systems can be developed using z-transform based techniques. Using the unilateral variant of the z-transform we will be able to solve difference equations with non-zero initial conditions.

We begin our discussion with the basic definition of the z-transform and its application to some simple signals. In the process, the significance of the issue of convergence of the transform is highlighted. In Section 5.2 the convergence properties of the z-transform, and specifically the region of convergence concept, are studied. Section 5.3 covers the fundamental properties of the transform that are useful in computing the transforms of signals with more complicated definitions. Proofs of the significant properties of the z-transform are presented to provide further insight and experience on working with transforms. Methods for computing the inverse z-transform are discussed in Section 5.4. Application of the z-transform to the analysis of linear and time-invariant systems is the subject of Section 5.5 where the z-domain system function concept is introduced. Derivation of implementation structures for discrete-time systems based on z-domain system functions are covered in Section 5.6. In Section 5.7 we discuss the unilateral variant of the z-transform that is useful in solving difference equations with specified initial conditions.

z-Transform of a discrete-time signal:

The z-transform of a discrete-time signal $x[n]$ is defined as

$$X(z) = \sum_{n=-\infty}^{\infty} x[n]\, z^{-n} \tag{5.1}$$

where z, the independent variable of the transform, is a complex variable.

If we were to expand the summation in Eqn. (5.1) we would get an expression in the form

$$X(z) = \ldots + x[-2]\, z^{2} + x[-1]\, z^{1} + z[0] + x[1]\, z^{-1} + x[2]\, z^{-2} + \ldots \tag{5.2}$$

which suggests that the transform is a polynomial that contains terms with both positive and negative powers of z. Since the independent variable z is complex, the resulting transform $X(z)$

is also complex-valued in general. Notationally, the transform relationship of Eqn. (5.1) can be expressed in the compact form

$$X(z) = \mathcal{Z}\{x[n]\}$$

which is read "$X(z)$ is the z-transform of $x[n]$". An alternative way to express the same relationship is through the use of the notation

$$x[n] \stackrel{\mathcal{Z}}{\longleftrightarrow} X(z)$$

The derivation of the relationship between the z-transform and the DTFT is straightforward. Let the complex variable z be expressed in polar form as

$$z = r\,e^{j\Omega} \tag{5.3}$$

Since z is a complex variable, we will represent it with a point in the complex z-plane as shown in Fig. 5.1. The parameter r indicates the distance of the point z from the origin. The parameter Ω is the angle, in radians, of the line drawn from the origin to the point z, measured counterclockwise starting from the positive real axis.

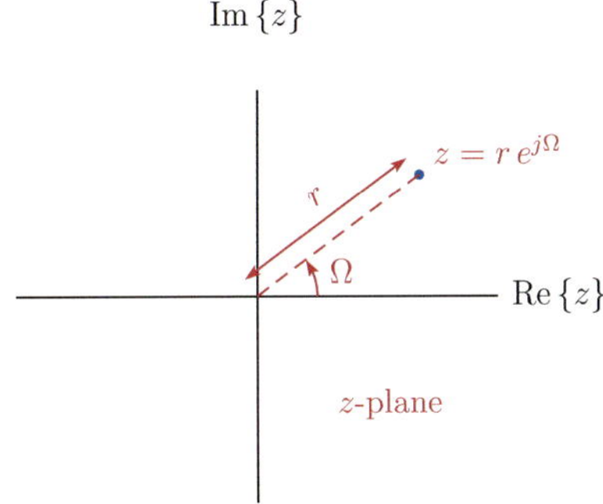

Figure 5.1 – The point $z = r\,e^{j\Omega}$ in the complex plane.

Obviously, any point in the complex z-plane can be expressed by properly choosing the parameters r and Ω. Substituting Eqn. (5.3) into the z-transform definition in Eqn. (5.1) we have

$$\begin{aligned} X(r,\Omega) &= X(z)\Big|_{z=r\,e^{j\Omega}} \\ &= \sum_{n=-\infty}^{\infty} x[n]\left(r\,e^{j\Omega}\right)^{-n} \\ &= \sum_{n=-\infty}^{\infty} x[n]\,r^{-n}\,e^{-j\Omega n} \end{aligned} \tag{5.4}$$

The result is a function of parameters r and Ω as well as the signal $x[n]$. We observe from Eqn. (5.4) that $X(r,\Omega)$ represents the DTFT of the signal $x[n]$ multiplied by an exponential signal r^{-n}:

$$X(r,\Omega) = X(z)\Big|_{z=r\,e^{j\Omega}} = \mathcal{F}\left\{x[n]\,r^{-n}\right\} \tag{5.5}$$

If we choose a fixed value of $r = r_1$ in Eqn. (5.3) and allow the angle Ω to vary from 0 to 2π radians, the resulting trajectory of the complex variable z in the z-plane would be a circle with its center at the origin and with its radius equal to r_1 as shown in Fig. 5.2.

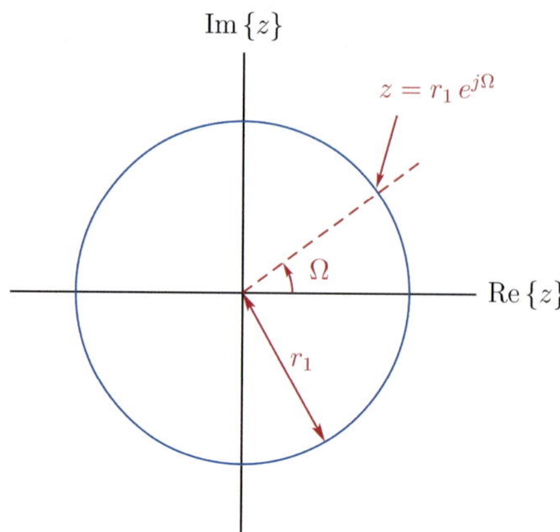

Figure 5.2 – Trajectory of $z = r_1\, e^{j\Omega}$ on the complex plane.

This allows us to make an important observation:

z-Transform evaluated on a circle with radius r_1:

Consider a circle in the z-plane with radius equal to $r = r_1$ and center at the origin. The z-transform of a signal $x[n]$ evaluated on this circle starting at angle $\Omega = 0$ and ending at angle $\Omega = 2\pi$ is equal to the DTFT of the modified signal $x[n]\, r_1^{-n}$.

If the parameter r is chosen to be equal to unity, then we have $z = e^{j\Omega}$ and

$$X(z)\Big|_{z=e^{j\Omega}} = \sum_{n=-\infty}^{\infty} x[n]\, z^{-n} = \mathscr{F}\left\{x[n]\right\} \tag{5.6}$$

This is a significant observation. The z-transform of a signal $x[n]$ evaluated for $z = e^{j\Omega}$ produces the same result as the DTFT of the signal. The trajectory defined by $z = e^{j\Omega}$ in the z-plane is a circle with unit radius, centered at the origin. This circle is referred to as the *unit circle* of the z-plane. Thus we can state an important conclusion:

z-Transform evaluated on the unit circle:

The DTFT of a signal $x[n]$ is equal to its z-transform evaluated at each point on the unit circle of the z-plane described by the trajectory $z = e^{j\Omega}$.

An easy way to visualize the relationship between the z-transform and the DTFT is the following: Imagine that the unit circle of the z-plane is made of a piece of string. Suppose the z-transform is computed at every point on the string. Let each point on the string be identified by the angle Ω so that the range $-\pi \le \Omega < \pi$ covers the entire piece. If we now remove the piece of string from the z-plane and straighten it up, it becomes the Ω axis for the DTFT. The values of the z-transform computed at points on the string would be the values of one period of the DTFT.

Next we would like to develop a graphical representation of the z-transform in order to illustrate its relationship with the DTFT. Suppose the z-transform of some signal $x[n]$ is given by

$$X(z) = \frac{z\,(z-0.7686)}{z^2 - 1.5371z + 0.9025} \tag{5.7}$$

In later sections of this chapter we will study techniques for computing the z-transform of a signal. At this point, however, our interest is in the graphical representation of the z-transform, the DTFT, and the relationship between them. Using the substitution $z = r\,e^{j\Omega}$, the transform in Eqn. (5.7) becomes

$$X(r,\Omega) = \frac{re^{j\Omega}\left(re^{j\Omega} - 0.7686\right)}{r^2 e^{j2\Omega} - 1.5371re^{j\Omega} + 0.9025}$$

Using this result, the transform at a particular point in the z-plane can be computed using r, its distance from the origin, and Ω, the angle it makes with the positive real axis. We would like to graph this transform, however, there are two issues that must first be addressed:

1. The z-transform is complex-valued. We must graph it either through its polar representation (magnitude and phase graphed separately) or its Cartesian representation (real and imaginary parts graphed separately). In this case we will choose to graph only its magnitude $|X(z)|$ as a three-dimensional surface. The magnitude of the transform can be computed by

$$\left|X(r,\Omega)\right| = \left[X(r,\Omega)\; X^*(r,\Omega)\right]^{1/2}$$

where $X^*(r,\Omega)$ is the complex conjugate of the transform $X(r,\Omega)$, and is computed as

$$X^*(r,\Omega) = \frac{re^{-j\Omega}\left(re^{-j\Omega} - 0.7686\right)}{r^2 e^{-j2\Omega} - 1.5371re^{-j\Omega} + 0.9025}$$

2. Depending on the signal $x[n]$, the transform expression given by Eqn. (5.7) is valid either for $r < 0.95$ only, or for $r > 0.95$ only. This notion will elaborated upon when we discuss the "region of convergence" concept later in this chapter. In graphically illustrating the relationship between the z-transform and the DTFT at this point, we will assume that the sample transform in Eqn. (5.7) is valid for $r > 0.95$. The three-dimensional magnitude surface will be graphed for all values of r, however, we will keep in mind that it is only meaningful for values of r greater than 0.95.

The magnitude $|X(z)|$ is shown as a mesh in part (a) of 5.3. The unit circle is shown in the (x,y) plane. Also shown in part (a) of the figure is the set of values of $|X(z)|$ computed at points on the unit circle. In part (b) of Fig. 5.3 the DTFT $X(\Omega)$ of the same signal is graphed. Notice how the unit circle of the z-plane is equivalent to the horizontal axis for the DTFT.

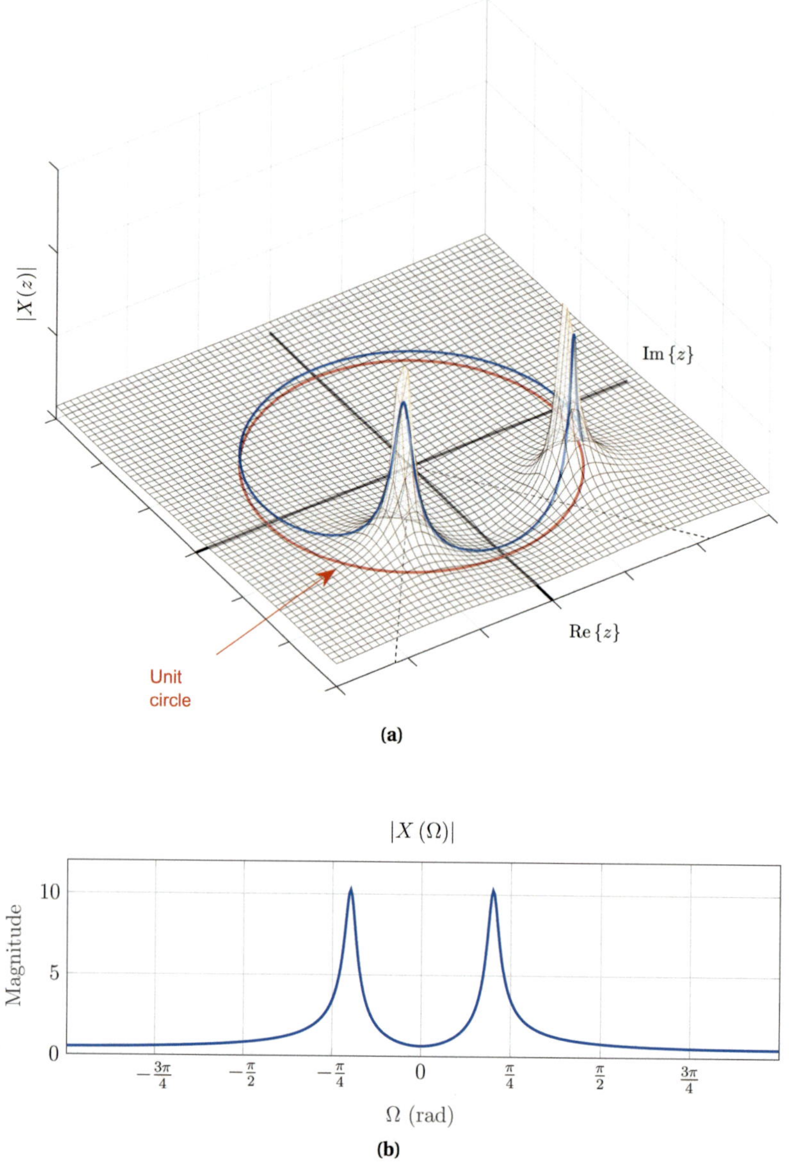

Figure 5.3 – **(a)** The magnitude $|X(z)|$ shown as a surface plot along with the unit circle and magnitude computed on the unit circle and **(b)** the magnitude of the DTFT as a function of Ω.

The z-transform defined by Eqn. (5.1) is sometimes referred to as the *bilateral* (two-sided) z-transform. A simplified variant of the transform termed the *unilateral* (one-sided) z-transform is introduced as an alternative analysis tool. The bilateral z-transform is useful for understanding signal characteristics, signal-system interaction, and fundamental characteristics of systems such as causality and stability. The unilateral z-transform is used for solving a linear constant-coefficient difference equation with specified initial conditions. We will briefly discuss the unilateral z-transform in Section 5.7, however, when we refer to z-transform without the qualifier word "bilateral" or "unilateral", we will always imply the more general bilateral z-transform as defined in Eqn. (5.1).

Interactive App: Three-dimensional view of the z-transform

The interactive app in `appzTrans3D.m` is based on the z-transform

$$X(z) = \frac{z\,(z-0.7686)}{z^2-1.5371z+0.9025}$$

the magnitude of which was shown in Fig. 5.3. The interactive app allows the 3D view to be rotated freely and observed from different angles. The magnitude of the transform $X(z)$ is also evaluated on a circle with radius r. This corresponds to DTFT magnitude for the modified signal $x[n]\,r^{-n}$. It is shown superimposed on the 3D mesh plot, and is also graphed separately as a stand-alone 2D plot. The radius r may be varied through the use of a slider control.

Start with $r=1$. Slowly increase the value of r, and observe the changes in both graphs. Pay attention to the relationship between the 3D mesh plot of $|X(z)|$ and the DTFT magnitude $\left|\mathcal{F}\left\{x[n]\,r^{-n}\right\}\right|$.

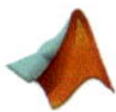

Software resource: `appzTrans3D.m`

Software resource: See MATLAB Exercises 5.1 and 5.2.

Example 5.1: A simple z-transform example

Find the z-transform of the finite-length signal

$$x[n] = \{\, \underset{\substack{\uparrow \\ n=0}}{3.7},\, 1.3,\, -1.5,\, 3.4,\, 5.2 \,\}$$

Solution: For the specified signal non-trivial samples occur for the index range $n=0,\ldots,4$. Therefore, the z-transform is

$$\begin{aligned} X(z) = &\sum_{n=-\infty}^{\infty} x[n]\,z^{-n} \\ =&x[0]+x[1]\,z^{-1}+x[2]\,z^{-2}+x[3]\,z^{-3}+x[4]\,z^{-4} \\ =&3.7+1.3\,z^{-1}-1.5\,z^{-2}+3.4\,z^{-3}+5.2\,z^{-4} \end{aligned}$$

The result is a polynomial of z^{-1}. The samples of the signal $x[n]$ become the coefficients of the corresponding powers of z^{-1}. Essentially, information about time-domain sample amplitudes of the signal is contained in the coefficients of the polynomial $X(z)$. The z-transform, therefore, is just an alternative way representing the signal $x[n]$. The value of the transform at a specific point in the complex plane can be evaluated from the polynomial. For example, if we wanted to know the value of the transform at $z_1 = 1+j2$, we would compute it as

$$
\begin{aligned}
X(1+j2) =& 3.7+1.3\,(1+j2)^{-1}-1.5\,(1+j2)^{-2}+3.4\,(1+j2)^{-3}+5.2\,(1+j2)^{-4} \\
=& 3.7+\frac{1.3}{(1+j2)}-\frac{1.5}{(1+j2)^2}+\frac{3.4}{(1+j2)^3}+\frac{5.2}{(1+j2)^4} \\
=& 3.7826-j0.0259
\end{aligned}
$$

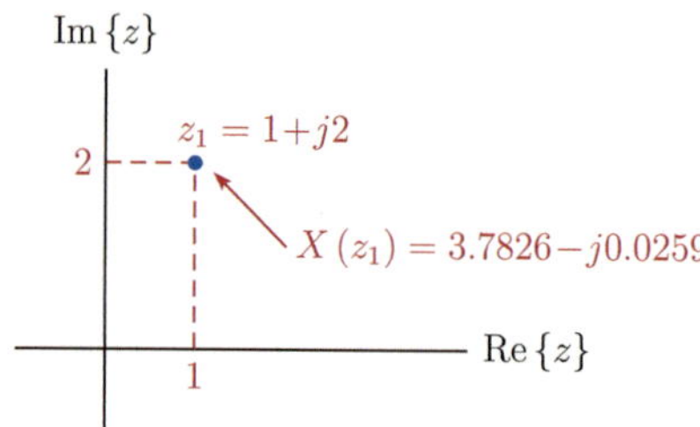

Figure 5.4 – The point z_1 in the z-plane.

The z-transform of a signal does not necessarily exist at every value of z. For example, the transform result we have found above could not be evaluated at the point $z = 0 + j0$ since the transform includes negative powers of z. In this case we conclude that the transform converges at all points in the complex z-plane with the exception of the origin.

 **Software resource:** exdt_5_1.m

Example 5.2: z-transform of a non-causal signal

Find the z-transform of the signal

$$x[n] = \{\ 3.7,\ 1.3,\ \underset{\substack{\uparrow \\ n=0}}{-1.5},\ 3.4,\ 5.2\ \}$$

Solution: This is essentially the same signal the z-transform of which we have computed in Example 5.1 with one difference: It has been advanced by two samples so it starts with index $n = -2$. Applying the z-transform definition given by Eqn. (5.1) we obtain

$$
\begin{aligned}
X(z) =& \sum_{n=-\infty}^{\infty} x[n]\,z^{-n} \\
=& x[-2]\,z^2+x[-1]\,z^1+x[0]+x[1]\,z^{-1}+x[2]\,z^{-2} \\
=& 3.7\,z^2+1.3\,z^1-1.5+3.4\,z^{-1}+5.2\,z^{-2}
\end{aligned}
$$

As in the previous example, the transform fails to converge to a finite value at the origin of the z-plane. In addition, the transform does not converge for infinitely large values of $|z|$ because of the z^1 and z^2 terms included in $X(z)$. It converges at every point in the z-plane with the two exceptions, namely the origin and infinity.

Example 5.3: z-transform of the unit impulse

Find the z-transform of the unit impulse signal

$$x[n] = \delta[n] = \begin{cases} 1\,, & n = 0 \\ 0\,, & n \neq 0 \end{cases}$$

Solution: Since the only nontrivial sample of the signal occurs at index $n = 0$, the z-transform is

$$X(z) = \mathcal{Z}\left\{\delta[n]\right\} = \sum_{n=-\infty}^{\infty} x[n]\, z^{-n} = x[0]\, z^{0} = 1$$

The transform of the unit impulse signal is constant and equal to unity. Since it does not contain any positive or negative powers of z, it converges at every point in the z-plane with no exceptions.

Example 5.4: z-transform of a time shifted unit impulse

Find the z-transform of the time shifted unit impulse signal

$$x[n] = \delta[n-k] = \begin{cases} 1\,, & n = k \\ 0\,, & n \neq k \end{cases}$$

where $k \neq 0$ is a positive or negative integer.

Solution: Applying the z-transform definition we obtain

$$X(z) = \mathcal{Z}\left\{\delta[n-k]\right\} = \sum_{n=-\infty}^{\infty} x[n]\, z^{-n} = z^{-k}$$

The transform converges at every point in the z-plane with one exception:

1. If $k > 0$, then the transform does not converge at the origin $z = 0 + j0$.
2. If $k < 0$, then the transform does not converge at points with infinite radius, that is, at points for which $|z| \to \infty$.

It becomes apparent from the examples above that the z-transform must be considered in conjunction with the criteria for the convergence of the resulting polynomial $X(z)$. The collection of all points in the z-plane for which the transform converges is called the *region of convergence (ROC)*. We conclude from Examples 5.1 and 5.2 that the region of convergence for a finite-length signal is the entire z-plane with the possible exception of $z = 0 + j0$, or $|z| \to \infty$ or both.

1. The origin $z = 0 + j0$ is excluded from the region of convergence if any negative powers of z appear in the polynomial $X(z)$. Negative powers of z are associated with samples of the signal $x[n]$ with positive indices. Therefore, if the signal $x[n]$ has any non-zero valued samples with positive indices, the z-transform does not converge at the origin of the z-plane.
2. Values of z with infinite radius must be excluded from the region of convergence if the polynomial $X(z)$ includes positive powers of z. Positive powers of z are associated with

samples of $x[n]$ that have negative indices. Therefore, if the signal $x[n]$ has any non-zero valued samples with negative indices, that is, if the signal is non-causal, its z-transform does not converge at $|z| \to \infty$.

For finite-length signals, the ROC can be determined based on the sample indices of the leftmost and the rightmost significant samples of the signal. If the index of the leftmost significant sample is negative, then the transform does not converge for $|z| \to \infty$. If the index of the rightmost significant sample is positive, then the transform does not converge at the origin $z = 0 + j0$.

How would we determine the region of convergence for a signal that is not finite-length? Recall that in Eqn. (5.5) we have represented the z-transform of a signal $x[n]$ as equivalent to the DTFT of the modified signal $x[n]\,r^{-n}$. As a result, the convergence of the z-transform of $x[n]$ can be linked to the convergence of the DTFT of the modified signal $x[n]\,r^{-n}$. Using Dirichlet conditions discussed in Chapter 3, this requires that $x[n]\,r^{-n}$ be absolute-summable, that is,

$$\sum_{n=-\infty}^{\infty} \left|x[n]\,r^{-n}\right| < \infty \tag{5.8}$$

The condition stated in Eqn. (5.8) highlights the versatility of the z-transform over the DTFT. Recall that, for the DTFT of a signal $x[n]$ to exist, the signal has to be absolute summable. Therefore, the existence of the DTFT is a binary question; the answer to it is either yes or no. In contrast, if $x[n]$ is not absolute summable, we may still be able to find values of r for which the modified signal $x[n]r^{-n}$ is absolute summable. Therefore, the z-transform of the signal $x[n]$ may exist for some values of r. The question of existence for the z-transform is not a binary one; it is a question of which values of r allow the transform to converge.

The region of convergence concept will be discussed in detail in Section 5.2. The next few examples further highlight the need for a detailed discussion of the region of convergence.

Example 5.5: z-transform of the unit step signal

Find the z-transform of the unit step signal

$$x[n] = u[n] = \begin{cases} 1\,, & n \ge 0 \\ 0\,, & n < 0 \end{cases}$$

Solution: We will again apply the z-transform definition given by Eqn. (5.1).

$$X(z) = \sum_{n=-\infty}^{\infty} u[n]\,z^{-n}$$

Since $u[n] = 0$ for $n < 0$, the lower limit of the summation can be changed to $n = 0$ without affecting the result. Furthermore, $u[n] = 1$ for $n \ge 0$, so $X(z)$ becomes

$$X(z) = \sum_{n=0}^{\infty} z^{-n} \tag{5.9}$$

Eqn. (5.9) represents the sum of an infinite-length geometric series and can be computed in closed form using Eqn. (C.5) in Appendix C to yield

$$X(z) = \frac{1}{1 - z^{-1}} = \frac{z}{z - 1}$$

which converges only for values of z for which

$$\left|z^{-1}\right| < 1 \qquad \Rightarrow \qquad |z| > 1$$

The ROC for the transform of the unit step signal is the collection of points outside a circle with radius equal to unity. This region is shown shaded in Fig. 5.5. Note that the unit circle itself is not part of the ROC since the transform does not converge for values of z on the circle.

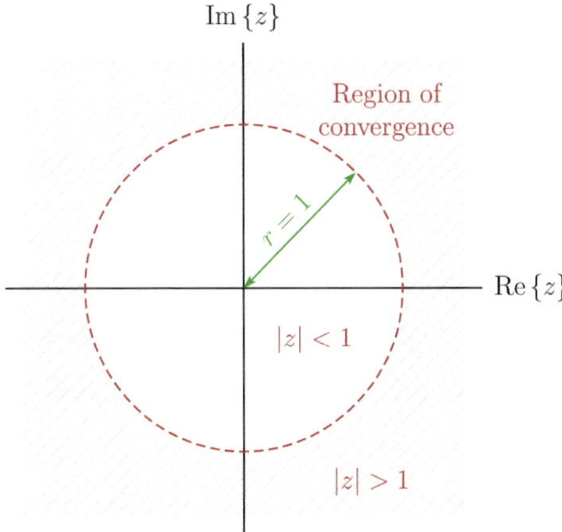

Figure 5.5 – The region of convergence for the z-transform of $x[n] = u[n]$.

Example 5.6: *z*-transform of a causal exponential signal

Find the z-transform of the signal

$$x[n] = a^n\, u[n] = \begin{cases} a^n\,, & n \ge 0 \\ 0\,, & n < 0 \end{cases}$$

where a is any real or complex constant.

Solution: The signal $x[n]$ is causal since $x[n] = 0$ for $n < 0$. Applying the z-transform definition given by Eqn. (5.1) we obtain

$$X(z) = \sum_{n=-\infty}^{\infty} a^n\, u[n]\, z^{-n} \tag{5.10}$$

Let us change the lower limit of the summation to $n = 0$ and drop the factor $u[n]$. Eqn. (5.10) becomes

$$X(z) = \sum_{n=0}^{\infty} a^n\, z^{-n} = \sum_{n=0}^{\infty} \left(a\, z^{-1}\right)^n$$

This is the sum of an infinite-length geometric series and can be put into a closed form as

$$X(z) = \frac{1}{1 - a\,z^{-1}} = \frac{z}{z-a}$$

which is valid only for values of z for which

$$\left|a\,z^{-1}\right| < 1 \qquad \Rightarrow \qquad \frac{|a|}{|z|} < 1 \qquad \Rightarrow \qquad |z| > |a|$$

Four possible forms of the signal $x[n]$ are shown in Fig. 5.6a,b,c,d corresponding to possible real values of the parameter a. It should be noted that there are other possibilities; the parameter a could also have a complex-value. The region of convergence for $X(z)$ is the region outside a circle with radius equal to $|a|$. This is shown in Fig. 5.7a,b for $|a|<1$ and $|a|>1$.

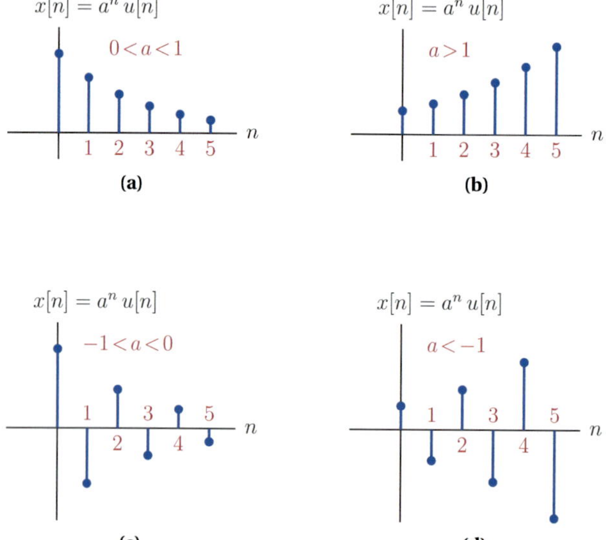

Figure 5.6 – The signal $x[n] = a^n\, u[n]$ for **(a)** $0<a<1$, **(b)** $a>1$, **(c)** $-1<a<0$, and **(d)** $a<-1$.

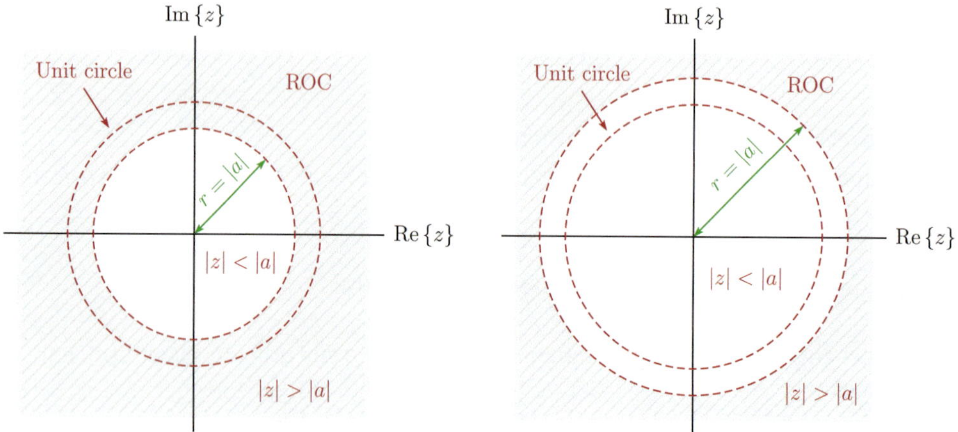

Figure 5.7 – The region of convergence for the z-transform of $x[n] = a^n\, u[n]$ for $|a|<1$ and for $|a|>1$.

For the case $|a|<1$ the unit circle is part of the region of convergence whereas, for the case $|a|>1$, it is not. Recall from the previous discussion that the DTFT is equal to the z-transform evaluated on the unit circle. Consequently, the DTFT of the signal exists only if the region of convergence includes the unit circle. We conclude that, for the DTFT of the signal $x[n]$ to exist, we need $|a|<1$. This is consistent with the existence conditions discussed in Chapter 3.

Example 5.7: z-transform of an anti-causal exponential signal

Find the z-transform of the signal

$$x[n] = -a^n\, u[-n-1] = \begin{cases} -a^n\,, & n<0 \\ 0\,, & n\ge 0 \end{cases}$$

where a is any real or complex constant.

Solution: In this case the signal $x[n]$ is anti-causal since it is equal to zero for $n \ge 0$. Applying the z-transform definition given by Eqn. (5.1) we obtain

$$X(z) = \sum_{n=-\infty}^{\infty} -a^n\, u[-n-1]\, z^{-n} \tag{5.11}$$

Since

$$u[-n-1] = \begin{cases} 1\,, & n<0 \\ 0\,, & n\ge 0 \end{cases}$$

changing the upper limit of the summation to $n = -1$ and dropping the factor $u[-n-1]$ would have no effect on the result. Eqn. (5.11) simplifies to

$$X(z) = -\sum_{n=-\infty}^{-1} a^n\, z^{-n} \tag{5.12}$$

Lower and upper summation limits are both negative. This can be fixed by employing a variable change $m = -n$, and the result found in Eqn. (5.12) can be written as

$$X(z) = -\sum_{m=\infty}^{1} a^{-m}\, z^{m} = -\sum_{m=1}^{\infty} \left(a^{-1}\, z\right)^m$$

which has the familiar sum of geometric series, however, the lower limit of the summation is $m = 1$ rather than $m = 0$ as required for the use of the closed-form formula. Let us apply another variable change to the summation, this time in the form $k = m-1$, to get the lower limit of the summation to start at $k = 0$:

$$X(z) = -\sum_{k=0}^{\infty} \left(a^{-1}\, z\right)^{k+1} = -a^{-1}\, z \sum_{k=0}^{\infty} \left(a^{-1}\, z\right)^k \tag{5.13}$$

The summation can be put into a closed form to yield

$$X(z) = -a^{-1}\, z \left(\frac{1}{1-a^{-1}\, z}\right) = \frac{z}{z-a}\,, \qquad \left|a^{-1} z\right| < 1 \tag{5.14}$$

Notice how the closed-form expression found for the signal $x[n] = -a^n\, u[-n-1]$ in Eqn. (5.14) is identical to the one found in Example 5.6 for a different signal, namely $x[n] = a^n\, u[n]$. The transform expressions are identical, however, the regions in which those expressions are valid are different. The closed-form formula found in Eqn. (5.13) is valid only for values of z for which

$$\left|a^{-1}\, z\right| < 1 \qquad \Rightarrow \qquad \frac{|z|}{|a|} < 1 \qquad \Rightarrow \qquad |z| < |a|$$

Four possible forms of the signal $x[n]$ are shown in Fig. 5.8 corresponding to different ranges of the real-valued parameter a (keep in mind that a could also be complex). The region of convergence for the transform $X(z)$ found in this case is the region inside a circle with radius equal to $|a|$. This is shown in Fig. 5.9a,b for $|a|<1$ and $|a|>1$.

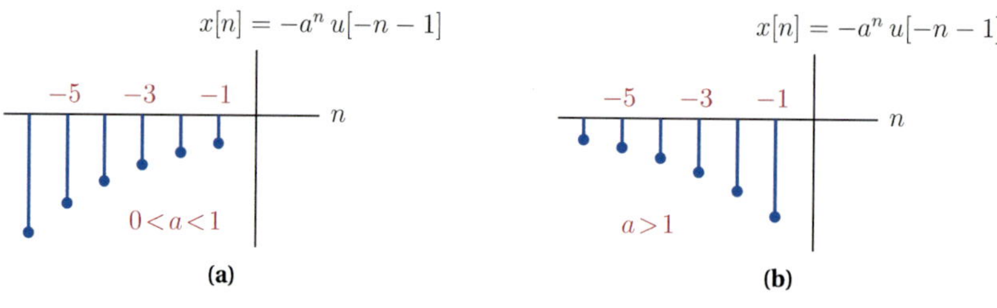

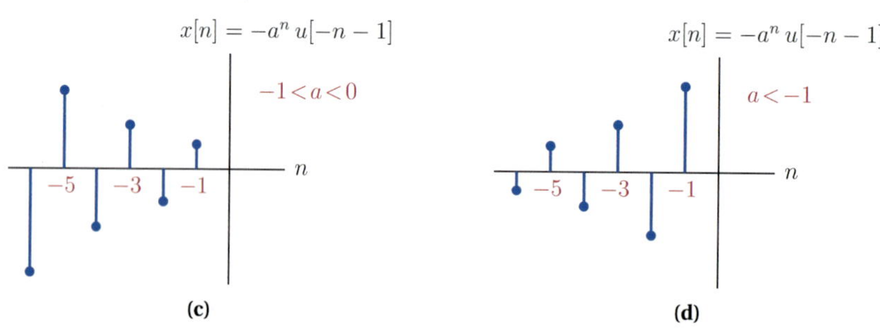

Figure 5.8 – The signal $x[n] = -a^n\, u[-n-1]$ for **(a)** $0<a<1$, **(b)** $a>1$, **(c)** $-1<a<0$, and **(d)** $a<-1$.

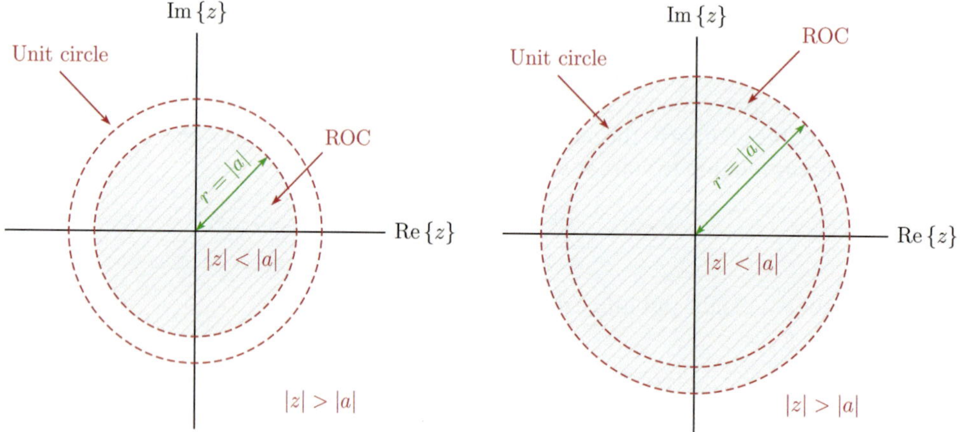

Figure 5.9 – The region of convergence for the z-transform of $x[n] = -a^n\, u[-n-1]$ for $|a|<1$ and for $|a|>1$.

For the case $|a|>1$ the unit circle is part of the region of convergence whereas, for the case $|a|<1$, it is not. Thus, the DTFT of the signal $x[n]$ exists only when $|a|>1$.

Examples 5.6 and 5.7 demonstrate a very important concept: It is possible for two different signals to have the same transform expression for the z-transform $X(z)$. In order for us to uniquely identify which signal among the two led to a particular transform, the region of convergence must be specified along with the transform. The following two transform pairs are fundamental in determining to which of the possible signals a given transform corresponds:

Importance of the region of convergence:

$$a^n\,u[n] \stackrel{\mathcal{Z}}{\longleftrightarrow} \frac{z}{z-a}\,, \qquad \text{ROC: } |z| > |a| \tag{5.15}$$

$$-a^n\,u[-n-1] \stackrel{\mathcal{Z}}{\longleftrightarrow} \frac{z}{z-a}\,, \qquad \text{ROC: } |z| < |a| \tag{5.16}$$

Sometimes we need to deal with the inverse problem of determining the signal $x[n]$ that has a particular z-transform $X(z)$. Given a transform

$$X(z) = \frac{z}{z-a}$$

we need to know the region of convergence to determine if the signal $x[n]$ is the one in Eqn. (5.15) or the one in Eqn. (5.16). This will be very important when we work with inverse z-transform in Section 5.4 later in this chapter. In order to avoid ambiguity, we will adopt the convention that the region of convergence is an integral part of the z-transform, and must be specified explicitly, or must be implied by means of another property of the underlying signal or system, every time a transform is given.

In each of the Examples 5.5, 5.6, and 5.7, we have obtained rational functions of z for the transform $X(z)$. In the general case, a rational transform $X(z)$ is expressed in the form

$$X(z) = K\,\frac{B(z)}{A(z)} \tag{5.17}$$

where the numerator $B(z)$ and the denominator $A(z)$ are polynomials of z. The parameter K is a constant gain factor. Let the numerator polynomial be written in factored form as

$$B(z) = (z-z_1)\,(z-z_2)\,\ldots,(z-z_M) \tag{5.18}$$

and let $p_1, p_2, \ldots, p_N$ be the roots of the denominator polynomial so that

$$A(z) = \left(z-p_1\right)\left(z-p_2\right)\ldots,\left(z-p_N\right) \tag{5.19}$$

The transform can be written as

$$X(z) = K\,\frac{(z-z_1)\,(z-z_2)\,\ldots,(z-z_M)}{\left(z-p_1\right)\left(z-p_2\right)\ldots,\left(z-p_N\right)} \tag{5.20}$$

In Eqns. (5.18) and (5.19) the parameters M and N are the numerator order and the denominator order, respectively. The larger of M and N is the order of the transform $X(z)$. The roots of the numerator polynomial are referred to as the *zeros* of the transform $X(z)$. In contrast, the roots of the denominator polynomial are the *poles*. The transform does not converge at a pole, therefore, the region of convergence cannot contain any poles. In the next section we will see that the boundaries of the region of convergence are determined by the poles of the transform.

Example 5.8: *z*-transform of a discrete-time pulse signal

Determine the z-transform of the discrete-time pulse signal

$$x[n] = \begin{cases} 1\,, & 0 \le n \le N-1 \\ 0\,, & n<0 \quad \text{or} \quad n>N-1 \end{cases}$$

shown in Fig. 5.10.

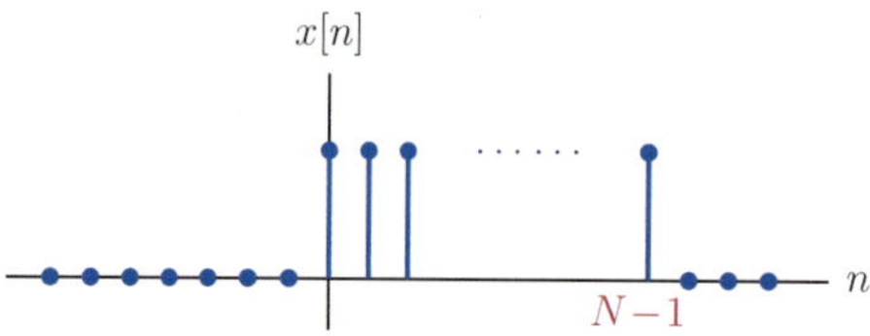

Figure 5.10 – The signal $x[n]$ for Example 5.8.

Solution: The transform in question is computed as

$$X(z) = \sum_{n=0}^{N-1} (1)\, z^{-n} = 1 + z^{-1} + \ldots + z^{-N} = \frac{1-z^{-N}}{1-z^{-1}} \tag{5.21}$$

In computing the closed-form result in Eqn. (5.21) we have used the finite-length geometric series sum formula derived in Appendix C. Since $x[n]$ is finite-length and causal, the transform converges at every point in the z-plane except at the origin $z = 0 + j0$. Thus we have

$$\text{ROC:} \quad |z| > 0$$

Often we find it more convenient to write the transform using non-negative powers of z. Multiplying both the numerator and the denominator polynomials in Eqn. (5.21) by bz^N leads to

$$X(z) = \frac{z^N - 1}{z^{N-1}\,(z-1)} \tag{5.22}$$

which is in the general factored form given by Eqn. (5.20). The appearance of the closed-form result in Eqn. (5.22) may be confusing in the way it relates to the region of convergence. It seems as though $X(z)$ might have a pole at $z = 1$. We need to realize, however, that the numerator polynomial is also equal to zero at $z = 1$:

$$B(z)\Big|_{z=1} = \left(z^N - 1\right)\Big|_{z=1} = 0$$

As a result, the numerator polynomial has a $(z-1)$ factor that effectively cancels the pole at $z = 1$. We will now go through the exercise of determining the zeros and poles of the transform $X(z)$. The roots of the numerator polynomial are found by solving the equation

$$B(z) = 0 \qquad \Longrightarrow \qquad z^N = 1 \tag{5.23}$$

In order to determine the roots of the numerator polynomial that are not immediately obvious, we will use the fact that $e^{j2\pi k} = 1$ for all integer k, and write Eqn. (5.23) as

$$z^N = e^{j2\pi k} \tag{5.24}$$

Taking the N-th root of both sides leads to the set of roots

$$z_k = e^{j2\pi k/N}, \quad k = 0, \ldots, N-1 \tag{5.25}$$

The numerator polynomial of $X(z)$ has N roots that are all on the unit circle of the z-plane. They are equally-spaced at angles that are integer multiples of $2\pi/N$ as shown in Fig. 5.11)(a). Note that in the figure we have used $N = 10$. The roots of the denominator polynomial are found by solving

$$A(z) = 0 \quad \Rightarrow \quad z^{N-1}(z-1) = 0 \tag{5.26}$$

There are $N-1$ solutions at $p_k = 0$ and one solution at $z = 1$ as shown in Fig. 5.11b.

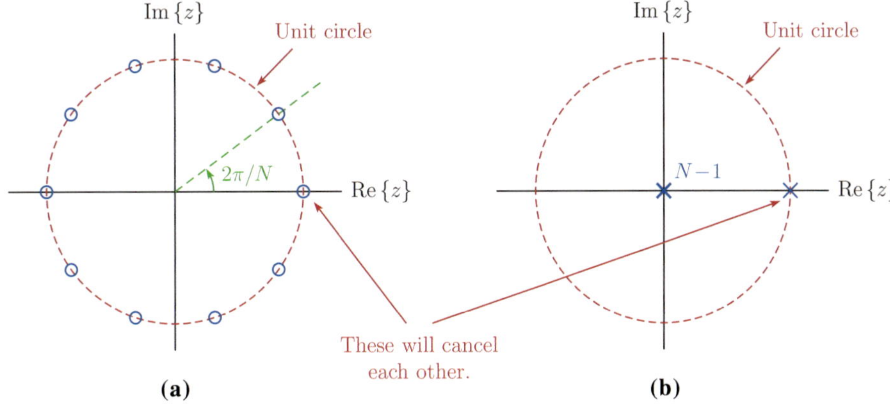

Figure 5.11 – **(a)** Roots of the numerator polynomial $z^N - 1$ and **(b)** roots of the denominator polynomial $z^{N-1}(z-1)$.

The pole-zero diagram for the transform $X(z)$ can now be constructed. The factors $(z-1)$ in numerator and denominator polynomials cancel each other, therefore there is neither a zero nor a pole at $z = 1$. We are left with zeros at

$$z_k = e^{j2\pi k/N}, \quad k = 1, \ldots, N-1$$

and a total of $N-1$ poles all at

$$p_k = 0, \quad k = 1, \ldots, N-1$$

as shown in Fig. 5.12. In light of what we have discovered regarding the poles of $X(z)$ it makes sense that the ROC is the entire z-plane with the exception of a singular point at the origin. Since all $N-1$ poles of the transform are at the origin, that is the only point where the transform does not converge.

The magnitude of the z-transform computed in Eqn. (5.22) is shown in Fig. 5.13a as a surface graph for $N = 10$. Since the origin of the z-plane is excluded from the ROC, the magnitude is not computed at the origin. Also, very large magnitude values in close proximity of the origin are clipped to make the graph fit into a reasonable scale. Note how the zeros equally spaced around the unit circle cause the magnitude surface to dip down. In addition to the surface graph, Fig. 5.13a also shows the unit circle of the z-plane.

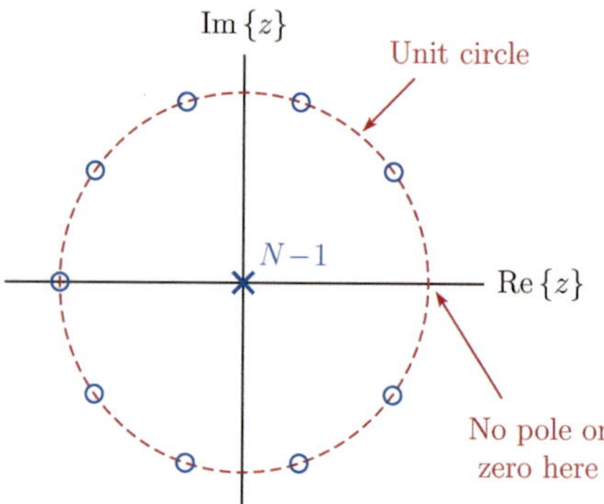

Figure 5.12 – Pole-zero diagram for $X(z)$.

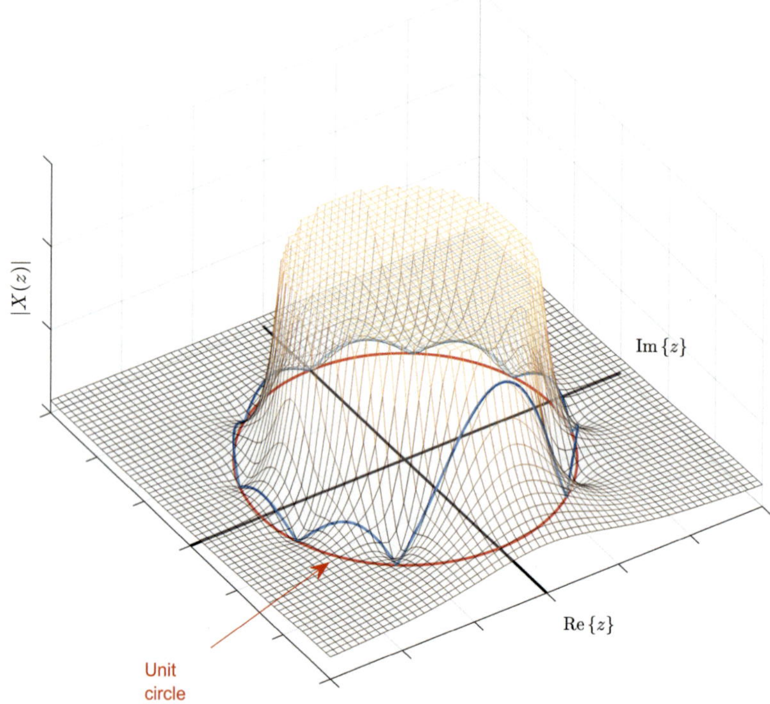

Figure 5.13 – The magnitude $|X(z)|$ shown as a surface plot along with the unit circle and magnitude computed on the unit circle.

Magnitude values computed at points on the unit circle are marked on the surface. Fig. 5.14 shows the magnitude of the DTFT computed for the range of angular frequency $-\pi \leq \Omega < \pi$. It should be compared to the values marked on the surface graph that correspond to the magnitude of the z-transform evaluated on the unit circle.

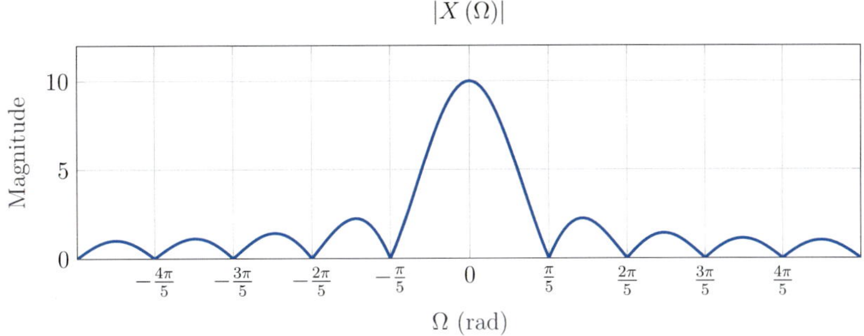

Figure 5.14 – The magnitude of the DTFT of $x[n]$ as a function of Ω.

To provide a slightly different perspective and to help with visualization, imagine that we take a knife and carefully cut through the surface in Fig. 5.13 along the perimeter of the unit circle of the z-plane, as if the surface is a cake that we would like to fit into a cylindrical box. The profile of the cutout would match the DTFT shown in Fig. 5.14. This is illustrated in Fig. 5.15.

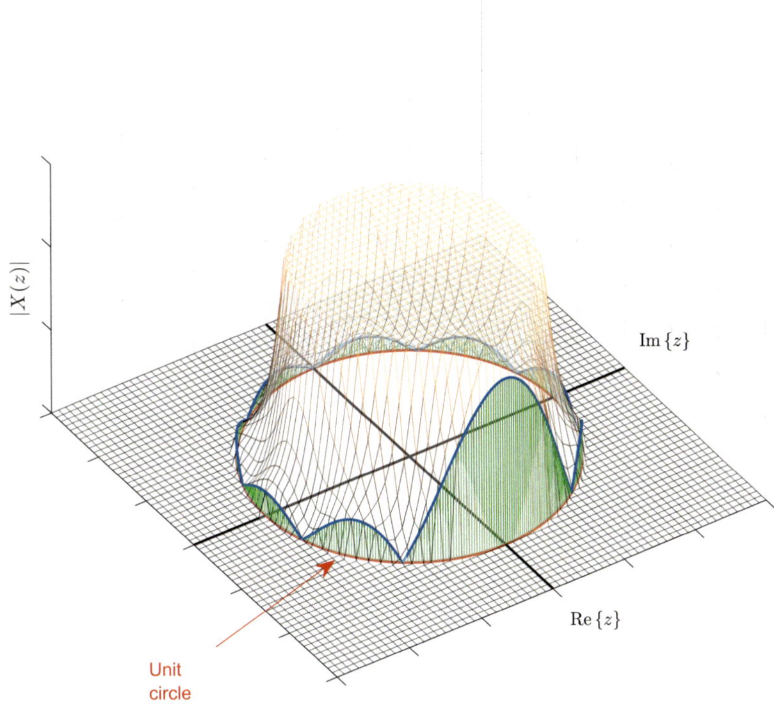

Figure 5.15 – Profile of the cutout along the perimeter of the unit circle for Example 5.8.

Interactive App: z-transform of discrete-time pulse signal

The interactive app in appzTransPulse.m is based on the z-transform of the length-N discrete-time pulse signal analyzed in Example 5.8. The magnitude of the z-transform for this signal was shown in Fig. 5.13 as a 3D mesh plot. The app computes $|X(z)|$ and graphs it as a 3D mesh that may be rotated and viewed from different angles. The length N of the pulse can be varied through the use of a slider control. The magnitude of the transform is also evaluated on a circle with radius r. This corresponds to DTFT magnitude for the modified signal $x[n]\, r^{-n}$. It is shown superimposed on the 3D mesh plot, and is also graphed separately as a stand-alone 2D plot. The radius r may be varied through the use of a slider control.

Start with $r = 1$. Slowly increase the value of r and observe the changes in both graphs. Pay attention to the relationship between the 3D mesh plot of $|X(z)|$ and the DTFT magnitude $\left|\mathcal{F}\left\{x[n]\, r^{-n}\right\}\right|$.

Software resource: appzTransPulse.m

Example 5.9: z-transform of complex exponential signal

Find the z-transform of the signal

$$x[n] = e^{j\Omega_0 n}\, u[n]$$

The parameter Ω_0 is real and positive and is in radians.

Solution: Applying the z-transform definition directly to the signal $x[n]$ leads to

$$X(z) = \sum_{n=-\infty}^{\infty} e^{j\Omega_0 n}\, u[n]\, z^{-n} \tag{5.27}$$

Since $x[n]$ is causal, the lower limit of the summation can be set to $n = 0$, and the $u[n]$ term can be dropped to yield

$$X(z) = \sum_{n=0}^{\infty} e^{j\Omega_0 n}\, z^{-n} \tag{5.28}$$

The expression in Eqn. (5.28) is the sum of an infinite-length geometric series, and can be put into closed form as

$$X(z) = \sum_{n=0}^{\infty} \left(e^{j\Omega_0}\, z^{-1}\right)^n = \frac{1}{1 - e^{j\Omega_0}\, z^{-1}} \tag{5.29}$$

The region of convergence is obtained from the convergence condition for the geometric series:

$$\left|e^{j\Omega_0}\, z^{-1}\right| < 1 \quad \Rightarrow \quad |z| > 1 \tag{5.30}$$

The transform in Eqn. (5.29) can also be written using non-negative powers of z as

$$X(z) = \frac{z}{z - e^{j\Omega_0}} \tag{5.31}$$

It has a zero at the origin and a pole at $z = e^{j\Omega_0}$. The zero and the pole of the transform as well as its ROC are shown in Fig. 5.16.

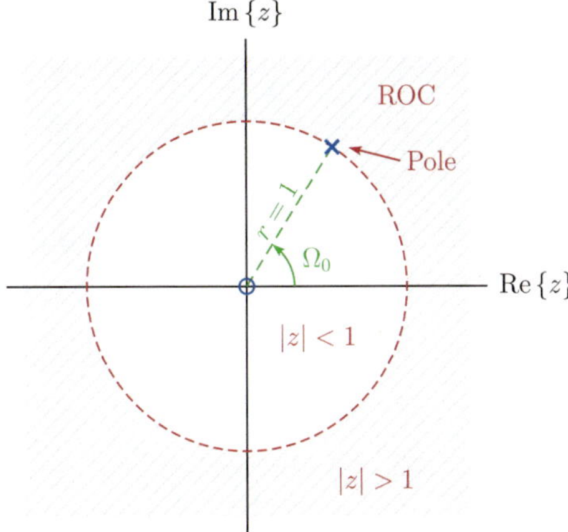

Figure 5.16 – ROC for the transform computed in Example 5.9.

Example 5.10: A more involved z-transform example

Determine the z-transform of the signal $x[n]$ defined by

$$x[n]=\begin{cases}(1/2)^n\,, & n\ge 0 \text{ and even}\\ -(1/3)^n\,, & n>0 \text{ and odd}\\ 0\,, & n<0\end{cases}$$

Solution: This example is somewhat tricky because of the particular way $x[n]$ is defined. Using the definition of the z-transform we can write

$$X(z)=\sum_{\substack{n=0\\ n\text{ even}}}^{\infty}\left(\frac{1}{2}\right)^n z^{-n}+\sum_{\substack{n=1\\ n\text{ odd}}}^{\infty}\left(\frac{1}{3}\right)^n z^{-n} \tag{5.32}$$

The first summation in Eqn. (5.32) will be carried out for only the even values of the index n, and the second summation is for only the odd values of it. To facilitate the goal of limiting the summations to even and odd index values, we will use two variable changes: For the first summation, let $n=2m$ where m is the new integer index. This substitution will ensure that n is always an even value. Similarly, for the second summation, we will use the variable change $n=2m+1$ to ensure that the resulting value of n is always odd. Incorporating these variable changes into the summations, Eqn. (5.32) can be written as

$$X(z)=\sum_{m=0}^{\infty}\left(\frac{1}{2}\right)^{2m} z^{-2m}+\sum_{m=0}^{\infty}\left(\frac{1}{3}\right)^{2m+1} z^{-(2m+1)} \tag{5.33}$$

or in the equivalent form

$$\begin{aligned}X(z)&=\sum_{m=0}^{\infty}\left(\frac{1}{4}\right)^{m}\left(z^2\right)^{-m}+\frac{1}{3}z^{-1}\sum_{m=0}^{\infty}\left(\frac{1}{9}\right)^{m}\left(z^2\right)^{-m}\\ &=\sum_{m=0}^{\infty}\left(\frac{1}{4}z^{-2}\right)^{m}+\frac{1}{3}z^{-1}\sum_{m=0}^{\infty}\left(\frac{1}{9}z^{-2}\right)^{m}\end{aligned} \tag{5.34}$$

The two summations in Eqn. (5.34) can be thought of as two transforms $X_1(z)$ and $X_2(z)$ so that

$$X(z) = X_1(z) + X_2(z)$$

The closed form expression for $X_1(z)$ and its associated ROC can be obtained as

$$X_1(z) = \sum_{m=0}^{\infty} \left(\frac{1}{4} z^{-2}\right)^m = \frac{z^2}{z^2 - \frac{1}{4}}, \qquad \text{ROC:} \quad \left|z^2\right| > \frac{1}{4} \tag{5.35}$$

Similarly, $X_2(z)$ is

$$X_2(z) = \frac{1}{3} z^{-1} \sum_{m=0}^{\infty} \left(\frac{1}{9} z^{-2}\right)^m = \frac{\frac{1}{3} z}{z^2 - \frac{1}{9}}, \qquad \text{ROC:} \quad \left|z^2\right| > \frac{1}{9} \tag{5.36}$$

Combining the closed form expressions for $X_1(z)$ and $X_2(z)$ as given by Eqns. (5.35) and (5.36) under a common denominator, we find the transform $X(z)$ and its ROC as

$$X(z) = \frac{z^4 - \frac{1}{3} z^3 - \frac{1}{9} z^2 + \frac{1}{12} z}{\left(z^2 - \frac{1}{4}\right)\left(z^2 - \frac{1}{9}\right)}, \qquad \text{ROC:} \quad |z| > \frac{1}{2} \tag{5.37}$$

The resulting transform has two real zeros at $z = 0$ and $z = -0.4158$ as well as a complex conjugate pair of zeros at $z = 0.3746 \pm j0.2451$. Its poles are at $z = \pm\frac{1}{2}$ and $z = \pm\frac{1}{3}$. The pole-zero diagram and the ROC for $X(z)$ are shown in Fig. 5.17.

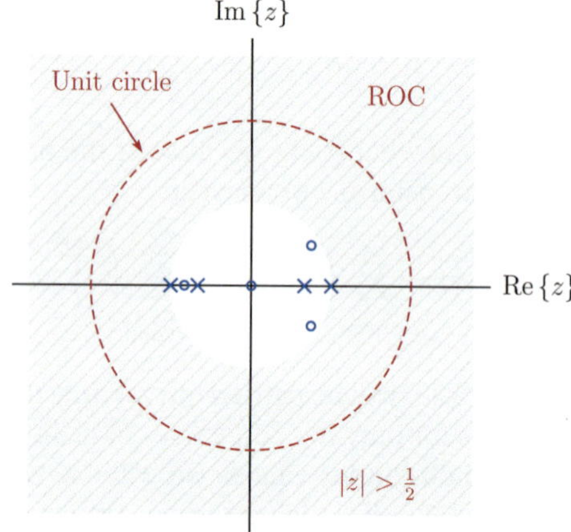

Figure 5.17 – Pole-zero diagram and the ROC for the transform computed in Example 5.10.

Example 5.11: z-transform of a downsampled signal

The downsampling operation was briefly discussed in Chapter 4. Consider a signal $g[n]$ obtained by downsampling another signal $x[n]$ by a factor of 2:

$$g[n] = x[2n]$$

Determine the z-transform of the downsampled signal $g[n]$ in terms of the transform of the original signal $x[n]$.

Solution: Using the z-transform definition, $G(z)$ is

$$G(z) = \sum_{n=-\infty}^{\infty} g[n]\, z^{-n} \tag{5.38}$$

Let us substitute $g[n] = x[2n]$ into Eqn. (5.38) to obtain

$$G(z) = \sum_{n=-\infty}^{\infty} x[2n]\, z^{-n} \tag{5.39}$$

To resolve the summation in Eqn. (5.39) we will use the variable change $m = 2n$ and write the summation in terms of the new index m:

$$G(z) = \sum_{\substack{m=-\infty \\ m \text{ even}}}^{\infty} x[m]\, z^{-m/2} \tag{5.40}$$

Naturally, the summation should only have the terms for which m is even. This restriction on the values of the summation index makes it difficult for us to relate Eqn. (5.40) to the transform $X(z)$ of the original signal. To overcome this hurdle we will use a simple trick. Consider a discrete-time signal $w[n]$ with the following definition:

$$w[m] = \begin{cases} 1\,, & m \text{ even} \\ 0\,, & m \text{ odd} \end{cases}$$

The transform in Eqn. (5.40) can be written as

$$G(z) = \sum_{m=-\infty}^{\infty} x[m]\, w[m]\, z^{-m/2} \tag{5.41}$$

Notice how we removed the restriction on m in Eqn. (5.41) since the factor $w[m]$ ensures that odd-indexed terms do not contribute to the sum. One method of creating the signal $w[m]$ is through

$$w[m] = \frac{1+(-1)^m}{2} \tag{5.42}$$

Substituting Eqn. (5.42) into Eqn. (5.41) we obtain

$$\begin{aligned} G(z) &= \sum_{m=-\infty}^{\infty} x[m] \left[\frac{1+(-1)^m}{2}\right] z^{-m/2} \\ &= \frac{1}{2} \sum_{m=-\infty}^{\infty} x[m]\, z^{-m/2} + \frac{1}{2} \sum_{m=-\infty}^{\infty} x[m]\, (-1)^m\, z^{-m/2} \end{aligned} \tag{5.43}$$

Recognizing that $z^{-m/2} = \left(\sqrt{z}\right)^{-m}$ and $(-1)^m\, z^{-m/2} = \left(-\sqrt{z}\right)^{-m}$, the result in Eqn. (5.43) can be rewritten as

$$\begin{aligned} G(z) &= \frac{1}{2} \sum_{m=-\infty}^{\infty} x[m] \left(\sqrt{z}\right)^{-m} + \frac{1}{2} \sum_{m=-\infty}^{\infty} x[m] \left(-\sqrt{z}\right)^{-m} \\ &= \frac{1}{2} X\left(\sqrt{z}\right) + \frac{1}{2} X\left(-\sqrt{z}\right) \end{aligned} \tag{5.44}$$

Let the ROC for the original transform $X(z)$ be

$$r_1 < |z| < r_2$$

The ROC for $X\left(\sqrt{z}\right)$ and $X\left(-\sqrt{z}\right)$ terms in Eqn. (5.44) is

$$r_1 < \left|\pm\sqrt{z}\right| < r_2 \quad \Rightarrow \quad r_1^2 < |z| < r_2^2$$

and therefore the ROC for $G(z)$ is also

$$r_1^2 < |z| < r_2^2$$

Example 5.12: z-transform of a downsampled signal revisited

Consider the causal exponential signal

$$x[n] = (0.9)^n \, u[n]$$

Let a new signal $g[n]$ be defined as $g[n] = x[2n]$. Find the z-transform of $g[n]$.

Solution: The transform of the signal $x[n]$ is

$$X(z) = \frac{z}{z-0.9}, \qquad \text{ROC:} \quad |z| > 0.9$$

Applying the result found in Eqn. (5.44) of Example 5.11, the transform $G(z)$ of the downsampled signal $g[n]$ is found as

$$G(z) = \frac{\frac{1}{2}\sqrt{z}}{\sqrt{z}-0.9} + \frac{-\frac{1}{2}\sqrt{z}}{-\sqrt{z}-0.9} \tag{5.45}$$

Combining the two terms of Eqn. (5.45) under a common denominator we get

$$G(z) = \frac{z}{z-(0.9)^2}, \qquad \text{ROC:} \quad |z| > (0.9)^2 \tag{5.46}$$

It is easy to verify the validity of the result in Eqn. (5.46) if we apply the downsampling operation in the time domain and then find the z-transform of the resulting signal $g[n]$ directly. The time-domain expression for the signal $g[n]$ is

$$g[n] = x[2n] = \left(0.9^2\right)^n \, u[2n] = (0.81)^n \, u[n]$$

and its z-transform is

$$G(z) = \frac{z}{z-0.81}, \qquad \text{ROC:} \quad |z| > 0.81$$

Software resource: See MATLAB Exercise 5.3.

5.2 Characteristics of the Region of Convergence

The examples we have worked on so far made it clear that the z-transform of a discrete-time signal always needs to be considered in conjunction with its region of convergence, that is, the collection of points in the z-plane for which the transform converges. In this section we will summarize and justify the fundamental characteristics of the region of convergence.

1. The ROC is circularly shaped. It is either the inside of a circle, the outside of a circle, or between two circles as shown in Fig. 5.18.

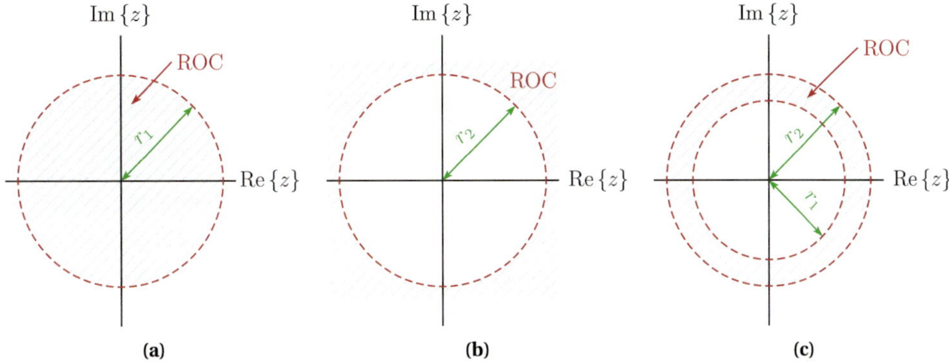

Figure 5.18 – Shape of the region of convergence: **(a)** inside a circle, **(b)** outside a circle, and **(c)** between two circles.

This property is easy to justify when we recall that, for $z = r\,e^{j\Omega}$, the values of the z-transform are identical to the values of the DTFT of the signal $x[n]\,r^{-n}$, as derived in Eqn. (5.5). Therefore, the convergence of the z-transform for $z = r\,e^{j\Omega}$ is equivalent to the convergence of the DTFT of the signal $x[n]\,r^{-n}$ which requires that

$$\sum_{n=-\infty}^{\infty} \left|x[n]\,r^{-n}\right| < \infty \tag{5.47}$$

Thus, the ROC depends only on r and not on Ω, explaining the circular nature of the region. Following are the possibilities for the ROC:

$$\begin{aligned} r < r_1 &: \quad \text{Inside a circle} \\ r > r_2 &: \quad \text{Outside a circle} \\ r_1 < r < r_2 &: \quad \text{Between two circles} \end{aligned}$$

2. The ROC cannot contain any poles.

By definition, poles of $X(z)$ are values of z that make the value of the transform infinitely large. For rational z-transforms, poles are the roots of the denominator polynomial. Since the transform does not converge at a pole, the ROC must naturally exclude all poles of the transform.

3. The ROC for the z-transform of a causal signal is the outside of a circle, and is expressed as

$$|z| > r_1$$

Since a causal signal does not have any significant samples for $n < 0$, its z-transform can be written as

$$X(z) = \sum_{n=0}^{\infty} x[n]\,z^{-n}$$

Writing z as $z = r^{j\Omega}$ and remembering that the convergence of the z-transform is equivalent to the signal $x[n]\,r^{-n}$ being absolute summable leads to the convergence condition

$$\sum_{n=0}^{\infty} \left|x[n]\,r^{-n}\right| < \infty \tag{5.48}$$

which can be expressed in the equivalent form

$$\sum_{n=0}^{\infty} |x[n]| \left(\frac{1}{|r|}\right)^n < \infty \tag{5.49}$$

If we can find a value of r for which Eqn. (5.49) is satisfied, then it is obvious that any larger value of r will satisfy Eqn. (5.49) as well. All we need to do is find the radius r_1 of the bounding circle, and the ROC is the region outside that circle.

A special case of a causal signal is one that is both causal and finite-length. Let N_1 be the largest value of the index for which the signal $x[n]$ has a non-zero value, that is

$$x[n] = 0 \quad \text{for} \quad n < 0 \text{ or } n > N_1$$

In this case the absolute summability condition in Eqn. (5.49) can be written as

$$\sum_{n=0}^{N_1} |x[n]| \left(\frac{1}{|r|}\right)^n < \infty \tag{5.50}$$

The condition in Eqn. (5.50) is satisfied for all values of r with the exception of $r = 0$. Our initial assessment that the ROC is the outside of a circle is still valid if we are willing to consider a circle with radius of $r_1 = 0$, that is, one that shrinks down to a single point at the origin. The ROC for a finite-length causal signal is therefore

$$|z| > 0 \tag{5.51}$$

4. The ROC for the z-transform of an anti-causal signal is the inside of a circle, and is expressed as

$$|z| < r_2$$

The justification of this property will be similar to that of the previous one. An anti-causal signal does not have any significant samples for $n \ge 0$ and its z-transform can be written as

$$X(z) = \sum_{n=-\infty}^{-1} x[n]\, z^{-n}$$

Using $z = r^{j\Omega}$ the condition for the convergence of the z-transform can be expressed through the equivalent condition for the absolute summability of the signal $x[n]\, r^{-n}$ as

$$\sum_{n=-\infty}^{-1} \left| x[n]\, r^{-n} \right| < \infty, \tag{5.52}$$

which can be expressed in the equivalent form

$$\sum_{n=-\infty}^{-1} |x[n]| . |r|^{-n} < \infty \tag{5.53}$$

Let us apply the variable change $n = -m$ to the summation in Eqn. (5.53) to write it as

$$\sum_{m=1}^{\infty} |x[-m]| . |r|^{m} < \infty \tag{5.54}$$

If we can find a value of r for which Eqn. (5.54) is satisfied, then it is obvious that any smaller value of r will satisfy Eqn. (5.54) as well. If r_2 is the radius r_1 of the bounding circle, then the ROC is the region inside that circle.

A special case of an anti-causal signal is one that is both anti-causal and finite-length. Let $-N_2$ be the most negative value of the index for which the signal $x[n]$ has a non-zero value, that is

$$x[n] = 0 \quad \text{for} \quad n \geq 0 \text{ or } n < -N_2$$

In this case the absolute summability condition in Eqn. (5.53) can be written as

$$\sum_{n=-N_2}^{-1} |x[n]|\,.\,|r|^{-n} < \infty \quad \Rightarrow \quad \sum_{m=1}^{N_2} |x[-m]|\,.\,|r|^{m} < \infty \tag{5.55}$$

The condition in Eqn. (5.55) is satisfied for all values of r that are finite; the only exception would be $r \to \infty$. If we simply take the bounding circle to be one with infinite radius $r_2 = \infty$, then our conclusion that the ROC is the inside of a circle is still valid. The ROC for a finite-length anti-causal signal is therefore

$$|z| < \infty \tag{5.56}$$

5. The region of convergence for the z-transform of a signal that is neither causal nor anti-causal is a ring-shaped region between two circles, and can be expressed as

$$r_1 < |z| < r_2$$

Any signal $x[n]$ can be written as the sum of a causal signal and an anti-causal signal. Let the two signals $x_R[n]$ and $x_L[n]$ be defined in terms of the signal $x[n]$ as

$$x_R[n] = \begin{cases} x[n]\,, & n \geq 0 \\ 0\,, & n < 0 \end{cases} \tag{5.57}$$

and

$$x_L[n] = \begin{cases} x[n]\,, & n < 0 \\ 0\,, & n \geq 0 \end{cases} \tag{5.58}$$

so that $x_R[n]$ is causal, and $x_L[n]$ is anti-causal, and they add up to $x[n]$:

$$x[n] = x_R[n] + x_L[n] \tag{5.59}$$

Let the z-transforms of these two signals be

$$X_R(z) = \mathcal{Z}\left\{x_R[n]\right\} \quad \text{ROC: } \left|z\right| > r_1$$

$$X_L(z) = \mathcal{Z}\left\{x_L[n]\right\} \quad \text{ROC: } \left|z\right| < r_2$$

The z-transform of the signal $x[n]$ is

$$X(z) = X_R(z) + X_L(z) \tag{5.60}$$

The ROC for $X(z)$ is at least the overlap of the two regions, that is,

$$r_1 < |z| < r_2 \tag{5.61}$$

provided that $r_2 > r_1$ (otherwise there may be no overlap, and the z-transform may not exist at any point in the z-plane). In some cases the ROC may actually be larger than the overlap in Eqn. (5.61) if the addition of the two transforms in Eqn. (5.60) results in the cancelation of a pole that sets the boundary for either $X_R(z)$ or $X_L(z)$.

As a special case, if the causal term $x_R[n]$ is of finite-length, the inner circle of the ROC may shrink down to a single point at the origin, resulting in $r_1 = 0$. Similarly, if the anti-causal term $x_L[n]$ is of finite-length, the outer circle of the ROC may grow to have infinite radius, resulting in $r_2 \to \infty$.

Example 5.13: z-transform of two-sided exponential signal

Find the z-transform of the signal

$$x[n] = a^{|n|}$$

Solution: The specified signal exists for all values of the index n and can be written as

$$\begin{aligned} x[n] =& \begin{cases} a^n , & n \geq 0 \\ a^{-n} , & n < 0 \end{cases} \\ =& a^n\, u[n] + a^{-n}\, u[-n-1] \end{aligned}$$

or, equivalently as

$$x[n] = a^n\, u[n] + (1/a)^n\, u[-n-1]$$

We will think of $x[n]$ as the sum of a causal signal $x_R[n]$ and an anti-causal signal $x_L[n]$ in the form

$$x[n] = x_R[n] + x_L[n]$$

with the two components given by

$$\begin{aligned} x_R[n] =& a^n\, u[n] \\ x_L[n] =& (1/a)^n\, u[-n-1] \end{aligned}$$

When we discuss the properties of the z-transform later in Section 5.3 we will show that the z-transform is linear, and therefore, the transform of the sum of two signals is equal to the sum of their respective transforms. Therefore $X(z)$ can be written as

$$\begin{aligned} X(z) =& X_R(z) + X_L(z) \\ =& \mathcal{Z}\left\{a^n\, u[n]\right\} + \mathcal{Z}\left\{(1/a)^n\, u[-n-1]\right\} \end{aligned} \tag{5.62}$$

The individual transforms that make up $X(z)$ in Eqn. (5.62) can be determined by adapting the results obtained earlier in Eqns. (5.15) and (5.15) as:

$$\mathcal{Z}\left\{a^n\, u[n]\right\} = \frac{z}{z-a}\,, \qquad \text{ROC: } |z| > |a|$$

$$\mathcal{Z}\left\{(1/a)^n\, u[-n-1]\right\} = -\frac{z}{z-1/a}\,, \qquad \text{ROC: } |z| < \frac{1}{|a|}$$

The regions of convergence for $X_R(z)$ and $X_L(z)$ are shown in Fig. 5.19.

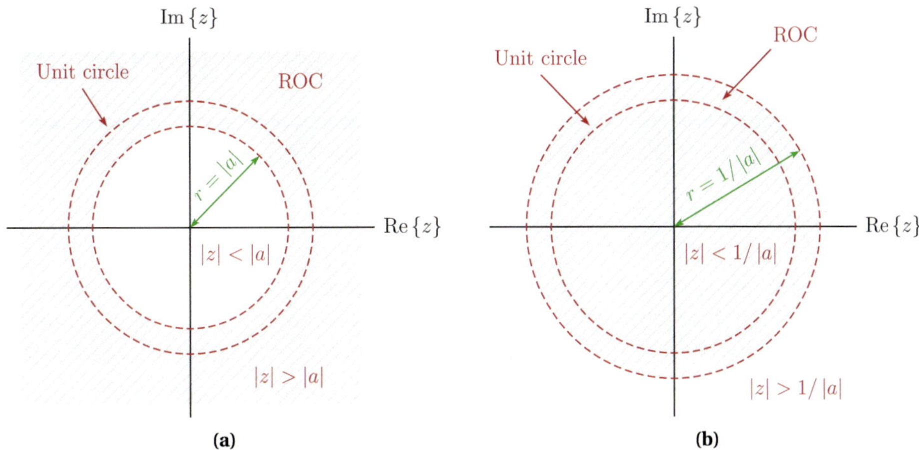

Figure 5.19 – The region of convergence for **(a)** $X_R(z) = \mathcal{Z}\{a^n u[n]\}$ and **(b)** $X_L(z) = \mathcal{Z}\{(1/a)^n u[-n-1]\}$.

The transform $X(z)$ can now be computed as

$$X(z) = \frac{z}{z-a} - \frac{z}{z-1/a} = \frac{\left(a^2-1\right) z}{a z^2 - \left(a^2+1\right) z + a} \tag{5.63}$$

The transform has a zero at $z = 0$, and poles at $z = a$ and $z = 1/a$. For it to converge, both $X_R(z)$ and $X_L(z)$ must converge. Therefore, the ROC for $X(z)$ is the overlap of the ROCs of $X_R(z)$ and $X_L(z)$ if such an overlap exists:

$$\text{ROC: } |a| < |z| < \frac{1}{|a|} \tag{5.64}$$

If $|a| < 1$, then $|a| < 1/|a|$, and the two regions shown in Fig. 5.19 do indeed overlap, creating a ring-shaped region between two circles with radii $|a|$ and $1/|a|$. This is shown in Fig. 5.20.

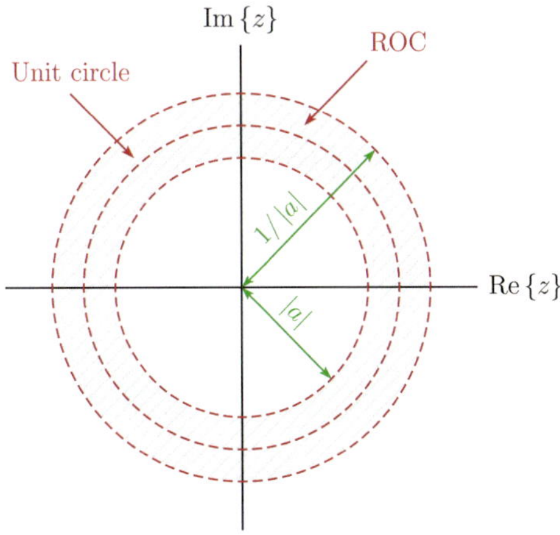

Figure 5.20 – The region of convergence for the transform $X(z)$.

If $|a| \ge 1$, then $|a| \ge 1/|a|$, and therefore the inequality in Eqn. (5.64) cannot be satisfied for any value of z. In that case, the transform found in Eqn. (5.63) does not converge at any point in the z-plane.

5.3 Properties of the z-Transform

In this section we focus on some of the important properties of the z-transform that will help us later in using the z-transform effectively for the analysis and design of discrete-time systems. The proofs of the z-transform properties are also be given. The motivation behind discussing the proofs of various properties of the z-transform is twofold:

1. The techniques used in proving various properties of the z-transform are also useful in working with z-transform problems in general, and
2. The proofs provide further insight into the z-transform and allow us to later identify opportunities for the effective use of z-transform properties in solving problems.

5.3.1 Linearity

The z-transform of a weighted sum of two signals is equal to the same weighted sum of their respective transforms $X_1(z)$ and $X_2(z)$. The ROC for the resulting transform is at least the overlap of the two individual transforms, if such an overlap exists. The ROC may even be greater than the overlap of the two regions if the addition of the two transforms results in the cancelation of a pole that sets the boundary of one of the two regions.

Linearity of the z-transform:

For any two signals $x_1[n]$ and $x_2[n]$ with their respective transforms

$$x_1[n] \stackrel{\mathcal{Z}}{\longleftrightarrow} X_1(z)$$

and

$$x_2[n] \stackrel{\mathcal{Z}}{\longleftrightarrow} X_2(z)$$

and any two constants α_1 and α_2, it can be shown that the following relationship holds:

$$\alpha_1 x_1[n] + \alpha_2 x_2[n] \stackrel{\mathcal{Z}}{\longleftrightarrow} \alpha_1 X_1(z) + \alpha_2 X_2(z) \tag{5.65}$$

Proof of Eqn. (5.65):

We will prove the linearity property in a straightforward manner by using the z-transform definition given by Eqn. (5.1) with the signal $\alpha_1 x_1[n] + \alpha_2 x_2[n]$:

$$\mathcal{Z}\left\{\alpha_1 x_1[n] + \alpha_2 x_2[n]\right\} = \sum_{n=-\infty}^{\infty} \left(\alpha_1 x_1[n] + \alpha_2 x_2[n]\right) z^{-n} \tag{5.66}$$

The summation in Eqn. (5.66) can be separated into two summations as

$$\begin{aligned}
\mathcal{Z}\{\alpha_1 x_1[n] + \alpha_2 x_2[n]\} &= \sum_{n=-\infty}^{\infty} \alpha_1 x_1[n]\, z^{-n} + \sum_{n=-\infty}^{\infty} \alpha_2 x_2[n]\, z^{-n} \\
&= \alpha_1 \sum_{n=-\infty}^{\infty} x_1[n]\, z^{-n} + \alpha_2 \sum_{n=-\infty}^{\infty} x_2[n]\, z^{-n} \\
&= \alpha_1\, \mathcal{Z}\{x_1[n]\} + \alpha_2\, \mathcal{Z}\{x_2[n]\}
\end{aligned}$$

The linearity property proven above for two signals can be generalized to any arbitrary number of signals. The z-transform of a weighted sum of any number of signals is equal to the same weighted sum of their respective transforms.

Example 5.14: Using the linearity property of the z-transform

Determine the z-transform of the signal

$$x[n] = 3\left(\frac{1}{2}\right)^n u[n] - 5\left(\frac{1}{3}\right)^n u[n]$$

Solution: A general expression for the z-transform of the causal exponential signal was found in Example 5.6 as

$$\mathcal{Z}\{a^n u[n]\} = \frac{z}{z-a}\,, \qquad \text{ROC: } |z| > |a|$$

Applying this result to the exponential terms in $x[n]$ we get

$$\mathcal{Z}\left\{\left(\frac{1}{2}\right)^n u[n]\right\} = \frac{z}{z-\frac{1}{2}}\,, \qquad \text{ROC: } |z| > \frac{1}{2} \tag{5.67}$$

and

$$\mathcal{Z}\left\{\left(\frac{1}{3}\right)^n u[n]\right\} = \frac{z}{z-\frac{1}{3}}\,, \qquad \text{ROC: } |z| > \frac{1}{3} \tag{5.68}$$

Combining the results in Eqns. (5.67) and (5.68) using the linearity property we arrive at the desired result:

$$\begin{aligned}
X(z) &= 3\left(\frac{z}{z-\frac{1}{2}}\right) - 5\left(\frac{z}{z-\frac{1}{3}}\right) \\
&= \frac{-2z\left(z-\frac{3}{4}\right)}{\left(z-\frac{1}{2}\right)\left(z-\frac{1}{3}\right)}
\end{aligned} \tag{5.69}$$

The ROC for $X(z)$ is the overlap of the two regions in Eqns. (5.67) and (5.68), namely

$$|z| > \frac{1}{2}$$

The ROC for the transform $X(z)$ is shown in Fig. 5.21 along with poles and zeros of the transform.

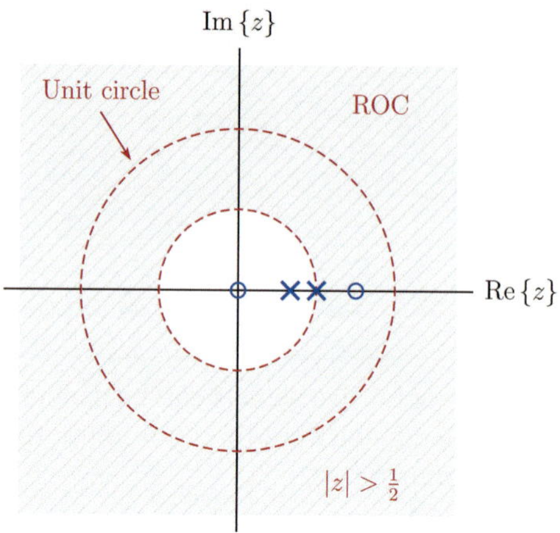

Figure 5.21 – The ROC for the transform in Eqn. (5.69).

Example 5.15: *z*-transform of a cosine signal

Find the z-transform of the signal

$$x[n] = \cos(\Omega_0 n)\; u[n]$$

Solution: Using Euler's formula for the $\cos(\Omega_0 n)$ term, the signal $x[n]$ can be written as

$$\cos(\Omega_0 n)\; u[n] = \frac{1}{2}\, e^{j\Omega_0 n}\, u[n] + \frac{1}{2}\, e^{-j\Omega_0 n}\, u[n]$$

and its z-transform is

$$\mathcal{Z}\left\{\cos(\Omega_0 n)\; u[n]\right\} = \mathcal{Z}\left\{\frac{1}{2}\, e^{j\Omega_0 n}\, u[n] + \frac{1}{2}\, e^{-j\Omega_0 n}\, u[n]\right\} \tag{5.70}$$

Applying the linearity property of the z-transform, Eqn. (5.70) becomes

$$\begin{aligned}\mathcal{Z}\left\{\cos(\Omega_0 n)\; u[n]\right\} &= \frac{1}{2}\, \mathcal{Z}\left\{e^{j\Omega_0 n}\, u[n]\right\} + \frac{1}{2}\, \mathcal{Z}\left\{e^{-j\Omega_0 n}\, u[n]\right\} \\ &= \frac{1/2}{1 - e^{j\Omega_0}\, z^{-1}} + \frac{1/2}{1 - e^{-j\Omega_0}\, z^{-1}}\end{aligned} \tag{5.71}$$

Combining the terms of Eqn. (5.71) under a common denominator we have

$$X(z) = \frac{1 - \cos(\Omega_0)\, z^{-1}}{1 - 2\cos(\Omega_0)\, z^{-1} + z^{-2}} \tag{5.72}$$

Let us multiply both the numerator and the denominator by z^2 to eliminate negative powers of z:

$$X(z) = \frac{z\,[z - \cos(\Omega_0)]}{z^2 - 2\cos(\Omega_0)\, z + 1}$$

Poles of $X(z)$ are both on the unit circle of the z-plane at $z = e^{\pm j\Omega_0}$. Since $x[n]$ is causal, the ROC is the outside of the unit circle. The unit circle itself is not included in the ROC. This is illustrated in Fig. 5.22.

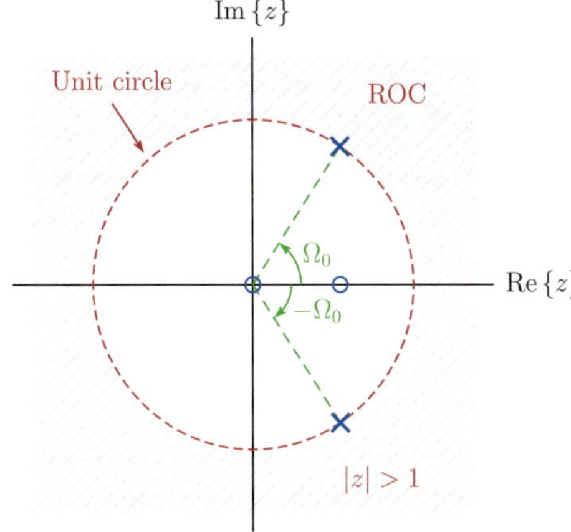

Figure 5.22 – The ROC for the transform in Eqn. (5.72).

Example 5.16: z-transform of a sine signal

Find the z-transform of the signal

$$x[n] = \sin(\Omega_0 n)\, u[n]$$

Solution: This problem is quite similar to that in Example 5.15. Using Euler's formula for the signal $x[n]$ we get

$$\sin(\Omega_0 n)\, u[n] = \frac{1}{2j}\, e^{j\Omega_0 n}\, u[n] - \frac{1}{2j}\, e^{-j\Omega_0 n}\, u[n]$$

and its z-transform is

$$\begin{aligned}
\mathcal{Z}\left\{\sin(\Omega_0 n)\, u[n]\right\} &= \mathcal{Z}\left\{\frac{1}{2j}\, e^{j\Omega_0 n}\, u[n] - \frac{1}{2j}\, e^{-j\Omega_0 n}\, u[n]\right\} \\
&= \frac{1}{2j}\, \mathcal{Z}\left\{e^{j\Omega_0 n}\, u[n]\right\} + \frac{1}{2j}\, \mathcal{Z}\left\{e^{-j\Omega_0 n}\, u[n]\right\} \\
&= \frac{1/2j}{1 - e^{j\Omega_0}\, z^{-1}} - \frac{1/2j}{1 - e^{-j\Omega_0}\, z^{-1}}
\end{aligned} \tag{5.73}$$

Combining the terms of Eqn. (5.73) under a common denominator we have

$$X(z) = \frac{\sin(\Omega_0)\, z^{-1}}{1 - 2\cos(\Omega_0)\, z^{-1} + z^{-2}}$$

or, using non-negative powers of z,

$$X(z) = \frac{\sin(\Omega_0)\, z}{z^2 - 2\cos(\Omega_0)\, z + 1}$$

The ROC of the transform is $|z| > 1$ as in the previous example.

5.3.2 Time shifting

Time shifting property of the z-transform:

Given the transform pair

$$x[n] \overset{\mathcal{Z}}{\longleftrightarrow} X(z)$$

the following is also a valid transform pair:

$$x[n-k] \overset{\mathcal{Z}}{\longleftrightarrow} z^{-k}\,X(z) \tag{5.74}$$

Thus, shifting the signal $x[n]$ in the time domain by k samples corresponds to multiplication of the transform $X(z)$ by z^{-k}. The ability to express the z-transform of a time-shifted version of a signal in terms of the z-transform of the original signal will be very useful in working with difference equations.

Proof of Eqn. (5.74):

The z-transform of $x[n-k]$ is

$$\mathcal{Z}\left\{x[n-k]\right\} = \sum_{n=-\infty}^{\infty} x[n-k]\,z^{-n} \tag{5.75}$$

Let us define a new variable $m = n-k$ and write the summation in Eqn. (5.75) in terms of this new variable:

$$\begin{aligned}\mathcal{Z}\left\{x[n-k]\right\} = &\sum_{m=-\infty}^{\infty} x[m]\,z^{-(m+k)}\\ =&z^{-k}\sum_{m=-\infty}^{\infty} x[m]\,z^{-m}\\ =&z^{-k}\,X(z)\end{aligned}$$

The region of convergence for the resulting transform $z^{-k}\,X(z)$ is the same as that of $X(z)$ with some possible exceptions:

1. If time shifting the signal for $k<0$ samples (left shift) causes some negative indexed samples to appear in $x[n-k]$, then points at $|z|\to\infty$ need to be excluded from the ROC.
2. If time shifting the signal for $k>0$ samples (right shift) causes some positive indexed samples to appear in $x[n-k]$, then the origin $z = 0 + j0$ of the z-plane needs to be excluded from the ROC.

Example 5.17: z-Transform of discrete-time pulse signal revisited

The z-transform of the discrete-time pulse signal

$$x[n] = \begin{cases} 1\,, & 0 \le n \le N-1 \\ 0\,, & n<0 \quad \text{or} \quad n>N-1 \end{cases}$$

was determined earlier in Example 5.8. Find the same transform through the use of linearity and time shifting properties.

Solution: The signal $x[n]$ can be expressed as the difference of a unit step signal and a time shifted unit step signal, that is,

$$x[n] = u[n] - u[n-N]$$

We know from the linearity property of the z-transform that

$$X(z) = \mathcal{Z}\left\{u[n]\right\} - \mathcal{Z}\left\{u[n-N]\right\}$$

The z-transform of the unit step function was found in Example 5.5 as

$$\mathcal{Z}\left\{u[n]\right\} = \frac{z}{z-1}\,, \qquad \text{ROC: } |z|>1 \tag{5.76}$$

Using the time shifting property of the z-transform, we find the transform of the shifted unit step signal as

$$\mathcal{Z}\left\{u[n-N]\right\} = z^{-N}\,\frac{z}{z-1}\,, \qquad \text{ROC: } |z|>1 \tag{5.77}$$

Adding the two transforms in Eqns. (5.76) and (5.77) yields

$$X(z) = \frac{z}{z-1} - z^{-N}\,\frac{z}{z-1} = \left(1-z^{-N}\right)\frac{z}{z-1} \tag{5.78}$$

The result in Eqn. (5.78) can be simplified to

$$X(z) = \frac{z^N-1}{z^{N-1}\,(z-1)}$$

which matches the earlier result found in Example 5.8. Since $x[n]$ is a finite-length causal signal, the transform converges everywhere except at the origin. Thus, the region of convergence is

$$|z|>0$$

This is one example of the possibility mentioned earlier regarding the region of convergence: The ROC for $X(z)$ is larger than the overlap of the two regions given by Eqns. (5.76) and (5.77). The individual transforms $\mathcal{Z}\{u[n]\}$ and $\mathcal{Z}\{u[n-N]\}$ each have a pole at $z=1$ causing the individual ROCs to be the outside of a circle with unit radius. When the two terms are added to construct $X(z)$, however, the pole at $z=1$ is canceled, resulting in the region of convergence for $X(z)$ being larger than the overlap of the individual regions.

5.3.3 Time reversal

Time reversal property of the z-transform:

Given the transform pair

$$x[n] \stackrel{\mathcal{Z}}{\longleftrightarrow} X(z)$$

the following is also a valid transform pair:

$$x[-n] \stackrel{\mathcal{Z}}{\longleftrightarrow} X\left(z^{-1}\right) \tag{5.79}$$

Proof of Eqn. (5.79):

The z-transform of the time reversed signal $x[-n]$ is found by

$$\mathcal{Z}\left\{x[-n]\right\} = \sum_{n=-\infty}^{\infty} x[-n]\, z^{-n} \tag{5.80}$$

We will employ a variable change $m = -n$ on the summation of Eqn. (5.80) to obtain

$$\mathcal{Z}\left\{x[-n]\right\} = \sum_{m=+\infty}^{-\infty} x[m]\, z^{m} \tag{5.81}$$

The summation in Eqn. (5.81) starts at $m = +\infty$ and moves toward $m = -\infty$, adding terms from right to left. The two limits can be swapped without affecting the result since it does not matter whether we add terms from right to left or the other way around. We will also use $z^m = \left(z^{-1}\right)^{-m}$ to write the relationship in Eqn. (5.81) as

$$\mathcal{Z}\left\{x[-n]\right\} = \sum_{m=-\infty}^{\infty} x[m]\, \left(z^{-1}\right)^{-m} = X\left(z^{-1}\right) \tag{5.82}$$

to prove the time reversal property.

Since we have replaced z by z^{-1} in the transform, the ROC must be adjusted for this change. Let the ROC of the original transform $X(z)$ be

$$r_1 < |z| < r_2$$

The ROC for $X\left(z^{-1}\right)$ is

$$r_1 < \left|\frac{1}{z}\right| < r_2 \quad\Longrightarrow\quad \frac{1}{r_2} < |z| < \frac{1}{r_1}$$

Example 5.18: z-Transform of anti-causal exponential signal revisited

Consider the anti-causal exponential signal

$$x[n] = -a^n\, u[-n-1]$$

Its z-transform was found in Example 5.7 as

$$X(z) = \frac{z}{z-a}\,, \quad \text{ROC: } |z| < |a|$$

Obtain the same result from the z-transform of the causal exponential signal using the time reversal and time shifting properties.

Solution: We will start with the known transform relationship and try to convert it into the relationship that we seek, applying appropriate properties of the z-transform at each step. The causal exponential signal and its z-transform are

$$b^n\, u[n] \overset{\mathcal{Z}}{\longleftrightarrow} \frac{z}{z-b}\,, \qquad \text{ROC:} \quad |z| > |b|$$

Let us begin by applying the time reversal operation to the signal on the left, and adjusting the transform on the right according to the time reversal property of the z-transform:

$$b^{-n}\, u[-n] \overset{\mathcal{Z}}{\longleftrightarrow} \frac{z^{-1}}{z^{-1}-b}\,, \qquad \text{ROC:} \quad |z| < \frac{1}{|b|}$$

This transform relationship can be written in the equivalent form

$$\left(\frac{1}{b}\right)^n u[-n] \overset{\mathcal{Z}}{\longleftrightarrow} \frac{-1/b}{z-1/b}\,, \qquad \text{ROC:} \quad |z| < \frac{1}{|b|}$$

We will now apply a time shift to the signal by one sample to the left through the substitution $n \to n+1$ which causes the transform to be multiplied by z, resulting in the relationship

$$\left(\frac{1}{b}\right)^{n+1} u[-n-1] \overset{\mathcal{Z}}{\longleftrightarrow} \frac{(-1/b)\,z}{z-1/b}\,, \qquad \text{ROC:} \quad |z| < \frac{1}{|b|}$$

Multiplying both sides by $-b$ and then choosing $b = 1/a$ leads to

$$-a^n\, u[-n-1] \overset{\mathcal{Z}}{\longleftrightarrow} \frac{z}{z-a}\,, \qquad \text{ROC:} \quad |z| < |a|$$

5.3.4 Multiplication by an exponential signal

Multiplication by an exponential signal:

Given the transform pair

$$x[n] \overset{\mathcal{Z}}{\longleftrightarrow} X(z)$$

the following is also a valid transform pair:

$$a^n\, x[n] \overset{\mathcal{Z}}{\longleftrightarrow} X(z/a) \tag{5.83}$$

Multiplication of $x[n]$ by an exponential signal corresponds to scaling of the z-transform. The ROC must be adjusted for the new transform variable (z/a). Let the ROC of the original transform $X(z)$ be

$$r_1 < |z| < r_2$$

The ROC for $X(z/a)$ is

$$r_1 < \left|\frac{z}{a}\right| < r_2 \qquad \Longrightarrow \qquad |a|\, r_1 < |z| < |a|\, r_2$$

Proof of Eqn. (5.83):

The z-transform of $a^n\, x[n]$ is

$$\mathcal{Z}\left\{a^n\, x[n]\right\} = \sum_{n=-\infty}^{\infty} a^n\, x[n]\, z^{-n} = \sum_{n=-\infty}^{\infty} x[n] \left(\frac{z}{a}\right)^{-n} = X(z/a)$$

Example 5.19: Multiplication by an exponential signal

Determine the z-transform of the signal

$$x[n] = a^n \cos(\omega_0)\; u[n]$$

Assume a is real.

Solution: Let the signal $x_1[n]$ be defined as

$$x_1[n] = \cos(\Omega_0 n)\; u[n]$$

so that

$$x[n] = a^n\, x_1[n]$$

The z-transform of the signal $x_1[n]$ was found in Example 5.15 to be

$$X_1(z) = \mathcal{Z}\left\{\cos(\Omega_0 n)\; u[n]\right\} = \frac{z\left[z - \cos(\Omega_0)\right]}{z^2 - 2\cos(\Omega_0)\; z + 1}$$

Multiplication of the time-domain signal by a^n causes the transform to be evaluated for $z \to z/a$, resulting in the transform relationship

$$\begin{aligned} X(z) =& X_1(z/a) \\ =& \frac{(z/a)\;\left[(z/a) - \cos(\Omega_0)\right]}{(z/a)^2 - 2\cos(\Omega_0)\;(z/a) + 1} \\ =& \frac{z\left[z - a\cos(\Omega_0)\right]}{z^2 - 2a\cos(\Omega_0)\; z + a^2} \end{aligned}$$

The transform $X(z)$ has two poles at $z = a\, e^{\pm j\Omega_0}$. Its ROC is

$$\text{ROC:} \quad |z| > |a|$$

The pole-zero diagram and the ROC are shown in Fig. 5.23.

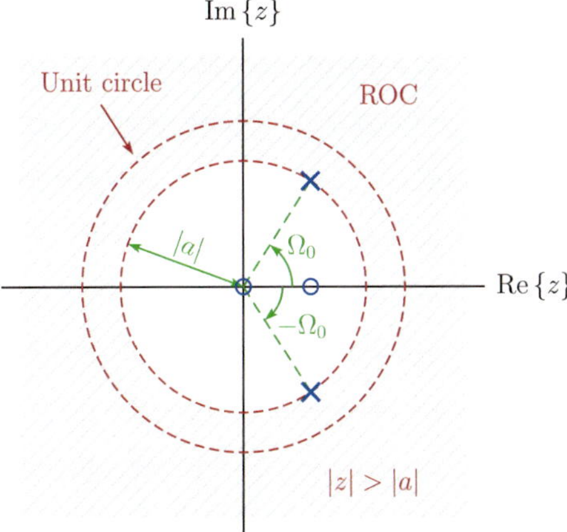

Figure 5.23 – The pole-zero diagram and the ROC for the transform in Example 5.19.

5.3.5 Differentiation in the z-domain

Given the transform pair

$$x[n] \overset{\mathcal{Z}}{\longleftrightarrow} X(z)$$

the following is also a valid transform pair:

$$n\,x[n] \overset{\mathcal{Z}}{\longleftrightarrow} (-z)\,\frac{dX(z)}{dz} \tag{5.84}$$

In general, the ROC of the new transform is the same as that of $X(z)$. Exceptions may occur at $z = 0 + j0$ and/or $|z| \to \infty$. If multiplication by the factor $(-z)$ causes the cancelation of a pole at the origin, then the origin needs to be added to the ROC. If the factor $(-z)$ introduces a new zero, then infinity must be excluded from the ROC.

Proof of Eqn. (5.84):

Let us start with the z-transform definition

$$X(z) = \sum_{n=-\infty}^{\infty} x[n]\,z^{-n}$$

Differentiating both sides with respect to z we obtain

$$\frac{d}{dz}\left[X(z)\right] = \frac{d}{dz}\left[\sum_{n=-\infty}^{\infty} x[n]\,z^{-n}\right]$$

Since summation and differentiation are both linear operators, their order can be changed: The derivative of a sum is equal to the sum of derivatives.

$$\frac{d}{dz}\left[X(z)\right] = \sum_{n=-\infty}^{\infty} \frac{d}{dz}\left[x[n]\,z^{-n}\right]$$

Carrying out the differentiation on each term inside the summation we obtain

$$\frac{d}{dz}\left[X(z)\right] = \sum_{n=-\infty}^{\infty} -n\,x[n]\,z^{-n-1}$$

$$= -z^{-1} \sum_{n=-\infty}^{\infty} n\,x[n]\,z^{-n}$$

from which the z-transform of $n\,x[n]$ can be obtained as

$$\mathcal{Z}\left\{n\,x[n]\right\} = (-z)\,\frac{dX(z)}{dz}$$

Thus we prove Eqn. (5.84).

Example 5.20: Using the differentiation property

Determine the z-transform of the signal

$$x[n] = n\,a^n\,u[n]$$

Solution: In Example 5.6 we found the z-transform of a causal exponential signal to be

$$\mathcal{Z}\left\{a^n\,u[n]\right\} = \frac{z}{z-a}\,, \qquad \text{ROC:} \quad |z| > |a|$$

Applying the differentiation property given by Eqn. (5.84) we get

$$X(z) = (-z)\,\frac{d}{dz}\left[\frac{z}{z-a}\right] = (-z)\,\frac{(-a)}{(z-a)^2} = \frac{az}{(z-a)^2}$$

and the ROC is

$$|z| > |a|$$

Example 5.21: *z*-transform of a unit ramp signal

Determine the z-transform of the unit ramp signal

$$x[n] = n\,u[n]$$

Solution: The solution is straightforward if we use the result obtained in Example 5.20. The signal $n\,a^n\,u[n]$ becomes the unit ramp signal we are considering if the parameter a is set equal to unity. Setting $a = 1$ in the transform found in Example 5.20 we have

$$X(z) = \left.\frac{az}{(z-a)^2}\right|_{a=1} = \frac{z}{(z-1)^2}$$

with the ROC

$$|z| > 1$$

Example 5.22: More on the use of the differentiation property

Determine the z-transform of the signal

$$x[n] = n\,(n+2)\,u[n]$$

Solution: The signal $x[n]$ can be written as

$$x[n] = n^2\,u[n] + 2n\,u[n]$$

The z-transform of the unit ramp function was found in Example 5.21 by using the differentiation property as

$$\mathcal{Z}\{n\,u[n]\} = \frac{z}{(z-1)^2}\,, \qquad \text{ROC:} \quad |z| > 1 \tag{5.85}$$

We will now apply the differentiation property to the transform pair in Eqn. (5.85):

$$\begin{aligned}\mathcal{Z}\{n^2\,u[n]\} &= (-z)\,\frac{d}{dz}\left[\frac{z}{(z-1)^2}\right]\\ &= \frac{z\,(z+1)}{(z-1)^3}\,, \qquad \text{ROC:} \quad |z| > 1\end{aligned} \tag{5.86}$$

Finally, combining the results in Eqns. (5.85) and (5.86) we construct the z-transform of $x[n]$ as

$$\begin{aligned}X(z) &= \mathcal{Z}\{n^2\,u[n]\} + 2\,\mathcal{Z}\{n\,u[n]\}\\ &= \frac{z\,(z+1)}{(z-1)^3} + 2\,\frac{z}{(z-1)^2} = \frac{3z\,(z-1/3)}{(z-1)^3}\,, \qquad \text{ROC:} \quad |z| > 1\end{aligned} \tag{5.87}$$

The transform found in Eqn. (5.87) has three poles at $z = 1$. Its zeros are at $z = 0$ and $z = \frac{1}{3}$. The pole-zero diagram and the ROC are shown in Fig. 5.24.

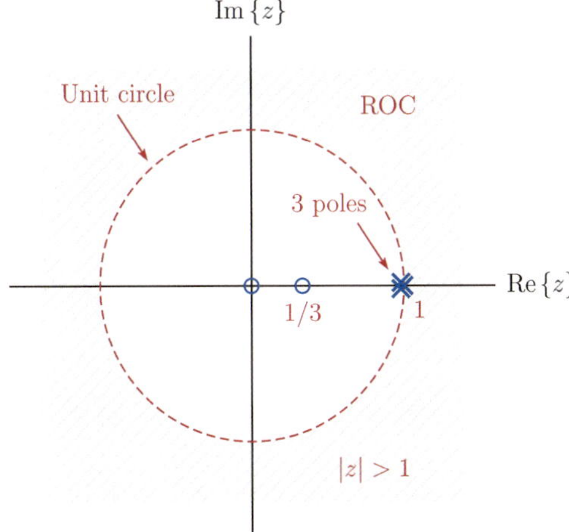

Figure 5.24 – The pole-zero diagram and the ROC for the transform in Example 5.22.

5.3.6 Convolution property

Convolution property of the z-transform is fundamental in its application to DTLTI systems. It forms the basis of the z-domain system function concept that will be explored in detail in Section 5.5.

Convolution property of the z-transform:

For any two signals $x_1[n]$ and $x_2[n]$ with their respective transforms

$$x_1[n] \stackrel{\mathcal{Z}}{\longleftrightarrow} X_1(z)$$

$$x_2[n] \stackrel{\mathcal{Z}}{\longleftrightarrow} X_2(z)$$

it can be shown that the following transform relationship holds:

$$x_1[n] * x_2[n] \stackrel{\mathcal{Z}}{\longleftrightarrow} X_1(z)\, X_2(z) \tag{5.88}$$

As before, the ROC for the resulting transform is at least the overlap of the two individual transforms, if such an overlap exists. It may be greater than the overlap of the two regions if the multiplication of the two transforms $X_1(z)$ and $X_2(z)$ results in the cancelation of a pole that sets a boundary for either $X_1(z)$ or $X_2(z)$.

Proof of Eqn. (5.88):

We will carry out the proof by using the convolution result inside the z-transform definition. Recall that the convolution of two discrete-time signals $x_1[n]$ and $x_2[n]$ is given by

$$x_1[n] * x_2[n] = \sum_{k=-\infty}^{\infty} x_1[k]\, x_2[n-k]$$

Inserting this into the z-transform definition we obtain

$$\begin{aligned}\mathcal{Z}\left\{x_1[n] * x_2[n]\right\} &= \mathcal{Z}\left\{\sum_{k=-\infty}^{\infty} x_1[k]\, x_2[n-k]\right\} \\ &= \sum_{n=-\infty}^{\infty}\left[\sum_{k=-\infty}^{\infty} x_1[k]\, x_2[n-k]\right] z^{-n}\end{aligned} \tag{5.89}$$

Interchanging the order of the two summations in Eqn. (5.89) leads to

$$\begin{aligned}\mathcal{Z}\left\{x_1[n] * x_2[n]\right\} &= \sum_{k=-\infty}^{\infty}\sum_{n=-\infty}^{\infty} x_1[k]\, x_2[n-k]\, z^{-n} \\ &= \sum_{k=-\infty}^{\infty} x_1[k] \sum_{n=-\infty}^{\infty} x_2[n-k]\, z^{-n}\end{aligned} \tag{5.90}$$

Recognizing that the inner summation on the right side of Eqn. (5.90) represents the z-transform of $x_2[n-k]$ and is therefore equal to

$$\sum_{n=-\infty}^{\infty} x_2[n-k]\, z^{-n} = \mathcal{Z}\left\{x_2[n-k]\right\} = z^{-k} X_2(z)$$

we obtain

$$\begin{aligned}\mathcal{Z}\left\{x_1[n] * x_2[n]\right\} &= \sum_{k=-\infty}^{\infty} x_1[k]\, z^{-k}\, X_2(z) \\ &= X_2(z) \sum_{k=-\infty}^{\infty} x_1[k]\, z^{-k} \\ &= X_1(z)\, X_2(z) \qquad (5.91)\end{aligned}$$

Convolution property of the z-transform is very useful in the sense that it provides an alternative to actually computing the convolution of two signals in the time domain. Instead, we can obtain $x[n] = x_1[n] * x_2[n]$ using the procedure outlined below:

Finding convolution result through z-transform:

To compute $x[n] = x_1[n] * x_2[n]$:

1. Find the z-transforms of the two signals.

$$X_1(z) = \mathcal{Z}\left\{x_1[n]\right\}$$
$$X_2(z) = \mathcal{Z}\left\{x_2[n]\right\}$$

2. Multiply the two transforms to obtain $X(z)$.

$$X(z) = X_1(z)\, X_2(z)$$

3. Compute $x[n]$ as the inverse z-transform of $X(z)$.

$$x[n] = \mathcal{Z}^{-1}\left\{X(z)\right\}$$

Example 5.23: Using the convolution property

Consider two signals $x_1[n]$ and $x_2[n]$ given by

$$x_1[n] = \{ \underset{\substack{\uparrow \\ n=0}}{4}, 3, 2, 1 \}$$

and

$$x_2[n] = \{ \underset{\substack{\uparrow \\ n=0}}{3}, 7, 4 \}$$

Let $x[n]$ be the convolution of these two signals, i.e.,

$$x[n] = x_1[n] * x_2[n]$$

Determine $x[n]$ using z-transform techniques.

Solution: Convolution of these two signals was computed in Example 2.19 of Chapter 1 using the time-domain method. In this exercise we will use the z-transform to obtain the same result. Individual transforms of the signals $x_1[n]$ and $x_2[n]$ are

$$X_1(z) = 4 + 3z^{-1} + 2z^{-2} + z^{-3}$$

and

$$X_2(z) = 3 + 7z^{-1} + 4z^{-2}$$

Applying the convolution property, the transform of $x[n]$ can be written as the product of the two transforms:

$$\begin{aligned} X(z) =& X_1(z)\, X_2(z) \\ =& \left(4 + 3z^{-1} + 2z^{-2} + z^{-3}\right)\left(3 + 7z^{-1} + 4z^{-2}\right) \\ =& 12 + 37z^{-1} + 43z^{-2} + 29z^{-3} + 15z^{-4} + 4z^{-5} \end{aligned} \tag{5.92}$$

Comparing the result obtained in Eqn. (5.92) with the definition of the z-transform we conclude that

$$x[0] = 12 \quad x[1] = 37 \quad x[2] = 43 \quad x[3] = 29 \quad x[4] = 15 \quad x[5] = 4$$

Equivalently, the signal $x[n]$ can be written as

$$x[n] = \{ \underset{\substack{\uparrow \\ n=0}}{12}, 37, 43, 29, 15, 4 \}$$

The solution of the convolution problem above using z-transform techniques leads to an interesting result: Convolution operation can be used for multiplying two polynomials. Suppose we need to multiply two polynomials

$$A(v) = a_k v^k + a_{k-1} v^{k-1} + \ldots + a_1 v + a_0$$

and

$$B(v) = b_m v^m + b_{m-1} v^{m-1} + \ldots + b_1 v + b_0$$

and we have a convolution function available in software. To find the polynomial $C(v) = A(v)\, B(v)$, we can first create a discrete-time signal with the coefficients of each polynomial:

$$x_a[n] = \{ \underset{\substack{\uparrow \\ n=0}}{a_0}, a_1, \ldots, a_{k-1}, a_k \}$$

and

$$x_b[n] = \{ \underset{\substack{\uparrow \\ n=0}}{b_0}, b_1, \ldots, b_{m-1}, b_m \}$$

It is important to ensure that the polynomial coefficients are listed in ascending order of powers of v in the signals $x_a[n]$ and $x_b[n]$, and any missing coefficients are accounted for in the form of zero-valued samples. If we now compute the convolution of the two signals as

$$x_c[n] = x_a[n] * x_b[n],$$

the resulting signal $x_c[n]$ holds the coefficients of the product polynomial $C(v)$, also in ascending order of powers of the independent variable. This will be demonstrated in MATLAB Exercise 5.4.

Software resource: See MATLAB Exercise 5.4.

Example 5.24: Finding the output signal of a DTLTI system using inverse z-transform

Consider a DTLTI system described by the impulse response

$$h[n] = (0.9)^n\, u[n]$$

driven by the input signal

$$x[n] = u[n] - u[n-7]$$

Compute the output signal $y[n]$ through the use of z-transform techniques. Recall that this problem was solved in Example 2.22 of Chapter 2 using the convolution operation.

Solution: The z-transform of the impulse response is

$$H(z) = \mathcal{Z}\left\{h[n]\right\} = \frac{z}{z-0.9}\,, \qquad \text{ROC:} \quad |z| > 0.9$$

The z-transform of the input signal $x[n]$ is

$$\begin{aligned} X(z) =& \sum_{n=0}^{6} z^{-n} \\ =& 1 + z^{-1} + z^{-2} + z^{-3} + z^{-4} + z^{-5} + z^{-6} \\ =& \frac{z^7 - 1}{z^6\,(z-1)} \end{aligned}$$

The finite-length geometric series form of the transform $X(z)$ will prove to be more convenient for use in this case. The z-transform of the output signal is found by multiplying the z-transforms of the input signal and the impulse response:

$$\begin{aligned} Y(z) =& X(z)\, H(z) \\ =& \left(1 + z^{-1} + z^{-2} + z^{-3} + z^{-4} + z^{-5} + z^{-6}\right) H(z) \\ =& H(z) + z^{-1} H(z) + z^{-2} H(z) + z^{-3} H(z) + z^{-4} H(z) \\ & + z^{-5} H(z) + z^{-6} H(z) \end{aligned}$$

The output signal $y[n]$ can now be found as the inverse z-transform of $Y(z)$ with the use of linearity and time shifting properties of the z-transform.

$$\begin{aligned} y[n] =& h[n] + h[n-1] + h[n-2] + h[n-3] + h[n-4] \\ & + h[n-5] + h[n-6] \\ =& (0.9)^n\, u[n] + (0.9)^{n-1}\, u[n-1] + (0.9)^{n-2}\, u[n-2] \\ & + (0.9)^{n-3}\, u[n-3] + (0.9)^{n-4}\, u[n-4] \\ & + (0.9)^{n-5}\, u[n-5] + (0.9)^{n-6}\, u[n-6] \end{aligned}$$

It can be shown that this result is identical to the one found in Example 2.22 when the two results are compared on a sample-by-sample basis.

Software resource: `exdt_5_24.m`

5.3.7 Initial value

Initial value property of the z-transform applies to causal signals only.

Initial value property of the z-transform:

Given the transform pair

$$x[n] \overset{\mathcal{Z}}{\longleftrightarrow} X(z)$$

where $x[n]=0$ for $n<0$, it can be shown that

$$x[0] = \lim_{z\to\infty} \left[X(z)\right] \tag{5.93}$$

Proof of Eqn. (5.93):

Consider the z-transform definition given by Eqn. (5.1) applied to a causal signal:

$$X(z) = \sum_{n=0}^{\infty} x[n]\, z^{-n} = x[0] + x[1]\, z^{-1} + x[2]\, z^{-2} + \ldots \tag{5.94}$$

It is obvious from Eqn. (5.94) that, as z becomes infinitely large, all terms that contain negative powers of z tend to zero, leaving behind only $x[0]$.

It is interesting to note that the limit in Eqn. (5.93) exists only for a causal signal, provided that $x[0]$ is finite. Consequently, the convergence of $\lim_{z\to\infty}[X(z)]$ can be used as a test of the causality of the signal. For a causal signal $x[n]$, the ROC of the z-transform must include infinitely large values of z. If the transform $X(z)$ is a rational function of z, the numerator order must not be greater than the denominator order.

Example 5.25: Using the initial value property

The z-transform of a causal signal $x[n]$ is given by

$$X(z) = \frac{3z^3+2z+5}{2z^3-7z^2+z-4}$$

Determine the initial value $x[0]$ of the signal.

Solution: Using Eqn. (5.93) we get

$$\lim_{z\to\infty}\left[X(z)\right] = \lim_{z\to\infty}\left[\frac{3z^3+2z+5}{2z^3-7z^2+z-4}\right] = \lim_{z\to\infty}\left[\frac{3z^3}{2z^3}\right] = \frac{3}{2}$$

The result can be easily justified by computing the inverse z-transform of $X(z)$ using either partial fraction expansion or long division. These techniques will be explored in Section 5.4.

5.3.8 Correlation property

Cross-correlation of two discrete-time signals $x[n]$ and $y[n]$ is defined as

$$r_{xy}[m] = \sum_{n=-\infty}^{\infty} x[n]\, y[n-m] \tag{5.95}$$

Correlation property of the z-transform:

Given two signals $x[n]$ and $y[n]$ with their respective transforms

$$x[n] \overset{\mathcal{Z}}{\longleftrightarrow} X(z)$$

$$y[n] \overset{\mathcal{Z}}{\longleftrightarrow} Y(z),$$

it can be shown that

$$R_{xy}(z) = \mathcal{Z}\left\{r_{XY}[m]\right\} = X(z)\, Y\left(z^{-1}\right) \tag{5.96}$$

The argument for the ROC of the transform uses the same reasoning we have employed before: The ROC for the resulting transform $R_{xy}(z)$ is at least the overlap of the two individual transforms, if such an overlap exists. It may be greater than the overlap of the two regions if the multiplication of the two transforms $X(z)$ and $Y\left(z^{-1}\right)$ results in the cancelation of a pole that sets a boundary for either $X(z)$ or $Y\left(z^{-1}\right)$.

Proof of Eqn. (5.96)

We will first prove this property by applying the z-transform definition directly to the cross-correlation of the two signals.

$$R_{xy}(z) = \sum_{m=-\infty}^{\infty} r_{xy}[m]\, z^{-m} = \sum_{m=-\infty}^{\infty} \left[\sum_{n=-\infty}^{\infty} x[n]\, y[n-m]\right] z^{-m} \tag{5.97}$$

By interchanging the order of the two summations in Eqn. (5.97) and rearranging the terms we get

$$R_{xy}(z) = \sum_{n=-\infty}^{\infty} x[n] \sum_{m=-\infty}^{\infty} y[n-m]\, z^{-m} \tag{5.98}$$

Applying the variable change $n-m=k$ to the inner summation in Eqn. (5.98) yields

$$R_{xy}(z) = \sum_{n=-\infty}^{\infty} x[n] \sum_{k=-\infty}^{\infty} y[k]\, z^{k-n}$$

which can be written in the equivalent form

$$\begin{aligned} R_{xy}(z) &= \left[\sum_{n=-\infty}^{\infty} x[n]\, z^{-n}\right] \left[\sum_{k=-\infty}^{\infty} y[k]\, z^{k}\right] \\ &= \left[\sum_{n=-\infty}^{\infty} x[n]\, z^{-n}\right] \left[\sum_{k=-\infty}^{\infty} y[k]\, \left(z^{-1}\right)^{-k}\right] \\ &= X(z)\, Y\left(z^{-1}\right) \end{aligned}$$

Alternatively, we could have proven this property with less work by using other properties of the z-transform. It is obvious from the definition of the cross-correlation in Eqn. (5.95) that $r_{xy}[n]$ is the convolution of $x[n]$ with $y[-n]$, that is,

$$r_{xy}[n] = x[n] * y[-n]$$

From the convolution property of the z-transform we know that the z-transform of $r_{xy}[n]$ must be the product of the individual transforms of $x[n]$ and $y[-n]$:

$$R_{XY}(z) = \mathcal{Z}\left\{r_{XY}[m]\right\} = \mathcal{Z}\left\{x[n]\right\} \mathcal{Z}\left\{y[-n]\right\}$$

In addition, we know from the time reversal property that

$$\mathcal{Z}\left\{y[-n]\right\} = Y\left(z^{-1}\right)$$

therefore

$$R_{xy}(z) = X(z)\, Y\left(z^{-1}\right) \tag{5.99}$$

As a special case of Eqn. (5.99), the transform of the autocorrelation of a signal $x[n]$ can be found as

$$R_{xx}(z) = X(z)\, X\left(z^{-1}\right)$$

Example 5.26: Using the correlation property

Determine the cross-correlation of the two signals

$$x[n] = a^n\, u[n]$$

and

$$y[n] = u[n] - u[n-3]$$

using the correlation property of the z-transform.

Solution: The z-transform of $x[n]$ is

$$X(z) = \frac{z}{z-a}\,, \qquad \text{ROC:} \quad |z| > |a| \tag{5.100}$$

Next we will determine the z-transform of $y[n]$:

$$Y(z) = 1 + z^{-1} + z^{-2} \qquad \text{ROC:} \quad |z| > 0 \tag{5.101}$$

Substituting $z \to z^{-1}$ in this last result we obtain

$$Y\left(z^{-1}\right) = 1 + z + z^2 \qquad \text{ROC:} \quad |z| < \infty \tag{5.102}$$

which is the z-transform of the time reversed signal $y[-n]$. We have modified the ROC in Eqn. (5.102) to account for the fact that the transform variable z has been reciprocated. Now we obtain $R_{xy}(z)$ by multiplying the two results in Eqns. (5.100) and (5.102):

$$\begin{aligned} R_{XY}(z) =& X(z)\, Y\left(z^{-1}\right) \\ =& \frac{z}{z-a}\left(1+z+z^2\right) \\ =& \frac{z}{z-a} + z\,\frac{z}{z-a} + z^2\,\frac{z}{z-a}\,, \qquad \text{ROC:} \quad |a| < |z| < \infty \end{aligned} \tag{5.103}$$

The cross-correlation $r_{XY}[n]$ is the inverse z-transform of the result in Eqn. (5.103) which, using the time shifting property of the z-transform, can be found as

$$r_{XY}[n] = a^n\, u[n] + a^{n+1}\, u[n+1] + a^{n+2}\, u[n+2]$$

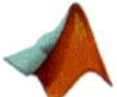

Software resource: exdt_5_26.m

Example 5.27: Finding auto-correlation of a signal

Using the correlation property of the z-transform, determine the auto-correlation of the signal

$$x[n] = a^n\, u[n]$$

Solution: The z-transform of the signal is

$$X(z) = \frac{z}{z-a}\,, \qquad \text{ROC:} \quad |z| > |a| \tag{5.104}$$

Substituting z^{-1} for z in Eqn. (5.104) we obtain

$$X\left(z^{-1}\right) = \frac{z^{-1}}{z^{-1}-a} = \frac{-1/a}{z-1/a}\,, \qquad \text{ROC:} \quad |z| < \frac{1}{|a|}$$

The z-transform of the autocorrelation $r_{xx}[n]$ is

$$\begin{aligned} R_{XX}(z) &= X(z)\, X\left(z^{-1}\right) \\ &= \frac{-z}{a\,z^2 - \left(a^2+1\right)\, z + a}\,, \qquad \text{ROC:} \quad |a| < |z| < \frac{1}{|a|} \end{aligned} \tag{5.105}$$

provided that $|a| < 1$. Recall that in Example 5.13 we found the z-transform of the two-sided exponential signal to be

$$\mathcal{Z}\left\{a^{|n|}\right\} = \frac{\left(a^2-1\right)\, z}{a\,z^2 - \left(a^2+1\right)\, z + a}$$

Comparing this earlier result to what we have in Eqn. (5.105) leads to the conclusion

$$r_{XX}[n] = \left(\frac{1}{1-a^2}\right)\, a^{|n|}$$

5.3.9 Summation property

Summation property of the z-transform:

Given the transform pair

$$x[n] \overset{\mathcal{Z}}{\longleftrightarrow} X(z)$$

the following is also a valid transform pair:

$$\sum_{k=-\infty}^{n} x[k] \overset{\mathcal{Z}}{\longleftrightarrow} \frac{z}{z-1}\, X(z) \tag{5.106}$$

Proof of Eqn. (5.106)

Let a new signal be defined as

$$w[n] = \sum_{k=-\infty}^{n} x[k]$$

Using the definition of the z-transform given by Eqn. (5.1) in conjunction with $w[n]$ we get

$$W(z) = \sum_{n=-\infty}^{\infty} w[n]\, z^{-n} = \sum_{n=-\infty}^{\infty} \left[\sum_{k=-\infty}^{n} x[k] \right] z^{-n}$$

which would be difficult to put into a closed form directly. Instead, we will make use of the other properties of the z-transform discussed earlier, and employ a simple trick of writing $w[n]$ as

$$\begin{aligned} w[n] &= \sum_{k=-\infty}^{n-1} x[k] + x[n] \\ &= w[n-1] + x[n] \end{aligned} \tag{5.107}$$

Eqn. (5.107) provides us with a difference equation between $w[n]$ and $x[n]$. Since the z-transform is linear, the transform of $w[n]$ can be written as

$$\mathcal{Z}\{w[n]\} = \mathcal{Z}\{w[n-1]\} + \mathcal{Z}\{x[n]\}$$

and using the time shifting property we have

$$W(z) = z^{1}\, W(z) + X(z)$$

which we can solve for $W(z)$ to obtain

$$W(z) = \frac{1}{1-z^{1}}\, X(z) = \frac{z}{z-1}\, X(z)$$

An alternative method of justifying the summation property is to think of $w[n]$ as the convolution of the signal $x[n]$ with the unit step function. Using the convolution sum we can write

$$x[n] * u[n] = \sum_{k=-\infty}^{\infty} x[k]\, u[n-k] \tag{5.108}$$

Consider the term $u[n-k]$:

$$u[n-k] = \begin{cases} 1, & n-k \ge 0 \quad \Rightarrow k \le n \\ 0, & n-k < 0 \quad \Rightarrow k > n \end{cases}$$

Therefore, dropping the $u[n-k]$ term from the summation in Eqn. (5.108) and adjusting the summation limits to compensate for it we have

$$x[n] * u[n] = \sum_{k=-\infty}^{n} x[k] = w[n] \tag{5.109}$$

Let us take the z-transform of both sides of Eqn. (5.109) using the convolution property of the z-transform:

$$W(z) = \mathcal{Z}\{x[n] * u[n]\} = \mathcal{Z}\{u[n]\}\, X(z) = \frac{z}{z-1}\, X(z)$$

which provides us with the alternative proof we seek.

Example 5.28: z-transform of a unit ramp signal revisited

The transform of the unit ramp signal

$$x[n] = n\,u[n]$$

was found in Example 5.21 through the use of the differentiation property. In this example we will use it as an opportunity to apply the summation property of the z-transform. The ramp signal $x[n]$ can be expressed as the running sum of a time shifted unit step signal as

$$x[n] = \sum_{k=-\infty}^{n} u[n-1] \tag{5.110}$$

It can easily be verified that the definition in Eqn. (5.110) produces $x[n] = 0$ for $n \le 0$, and $x[1] = 1$, $x[2] = 2$, and so on, consistent with the ramp signal. The z-transform of the time shifted unit step signal is

$$\mathcal{Z}\left\{u[n-1]\right\} = z^{-1}\,\mathcal{Z}\left\{u[n]\right\} = z^{-1}\left(\frac{z}{z-1}\right) = \frac{1}{z-1}$$

Using the summation property we find the transform $X(z)$ as

$$X(z) = \frac{z}{z-1}\,\mathcal{Z}\left\{u[n-1]\right\} = \left(\frac{z}{z-1}\right)\left(\frac{1}{z-1}\right) = \frac{z}{(z-1)^2}$$

which matches the answer found in Example 5.20.

5.4 Inverse z-Transform

Inverse z-transform is the problem of finding $x[n]$ from the knowledge of $X(z)$. Often we need to determine the signal $x[n]$ that has a specified z-transform. There are three basic techniques for computing the inverse z-transform:

1. Direct evaluation of the inversion integral,
2. Partial fraction expansion technique for a rational transform, and
3. Expansion of the rational transform into a power series through long division.

We will focus our attention on the last two methods. The inversion integral method will be briefly mentioned, but will not be considered further due to its complexity. We will find that methods 2 and 3 will be sufficient for most problems encountered in the analysis of signals and linear systems.

5.4.1 Inversion integral

Let $X(r,\Omega)$ be the function obtained by evaluating the z-transform of a signal $x[n]$ for $z = r\,e^{j\Omega}$, that is, at points on a circle with radius equal to r. We have established in Section 5.1 that $X(r,\Omega)$ is the same as the DTFT of the signal $x[n]\,r^{-n}$, that is

$$X(r,\Omega) = X(z)\Big|_{z=r\,e^{j\Omega}} = \mathscr{F}\left\{x[n]\,r^{-n}\right\} \tag{5.111}$$

Consequently, the inverse DTFT equation given by Eqn. (3.68) should yield the signal $x[n]\,r^{-n}$:

$$x[n]\,r^{-n} = \mathscr{F}^{-1}\left\{X(r,\Omega)\right\} = \frac{1}{2\pi}\int_{-\pi}^{\pi} X(r,\Omega)\,e^{j\Omega}\,d\Omega \tag{5.112}$$

Multiplying both sides of Eqn. (5.112) with r^n we obtain

$$\begin{aligned} x[n] =& r^n\,\frac{1}{2\pi}\int_{-\pi}^{\pi} X(r,\Omega)\,e^{j\Omega}\,d\Omega \\ =& \frac{1}{2\pi}\int_{-\pi}^{\pi} X(r,\Omega)\left(re^{j\Omega}\right)^n\,d\Omega \end{aligned} \tag{5.113}$$

Eqns. (5.111) through (5.113) provide us with a method for finding $x[n]$ from its z-transform:

Inverse z-transform through inversion integral:

1. Choose a value for r so that the circle with radius equal to r would be in the ROC of the transform $X(z)$.
2. For the chosen value of the radius r, determine the function $X(r,\Omega)$ by setting $z = r\,e^{j\Omega}$ in the transform $X(z)$:

$$X(r,\Omega) = X(z)\Big|_{z=r\,e^{j\Omega}} = \mathscr{F}\left\{x[n]\,r^{-n}\right\} \tag{5.114}$$

3. Evaluate the integral

$$x[n] = \frac{1}{2\pi}\int_{-\pi}^{\pi} X(r,\Omega)\left(re^{j\Omega}\right)^n\,d\Omega \tag{5.115}$$

to find the signal $x[n]$.

The three-step procedure outlined above can be reduced to a contour integral. Since $z = re^{j\Omega}$ and r is a constant in the evaluation of the integral in Eqn. (5.115), we can write

$$dz = j\,r\,e^{j\Omega}\,d\Omega = j\,z\,d\Omega \tag{5.116}$$

and therefore

$$d\Omega = \frac{1}{jz}\,dz \tag{5.117}$$

In the next step we will substitute Eqn. (5.117) into Eqn. (5.115), change $re^{j\Omega}$ to z, and change $X(r,\Omega)$ to $X(z)$ to arrive at the result

$$x[n] = \frac{1}{2\pi j}\oint X(z)\,z^{n-1}\,dz \tag{5.118}$$

where we have used the contour integral symbol $\oint$ to indicate that the integral should be evaluated by traveling counterclockwise on a closed contour in the z-plane within the ROC of the transform. The values of the transform $X(z)$ on the closed contour are multiplied by z^{n-1} and integrated. Integration can start at an arbitrary point on the contour, but it must end at the same point. Notice the similarity of the inversion integral in Eqn. (5.118) to the contour integral used for computing the inverse Laplace transform for continuous-time signals:

$$x_a(t) = \mathscr{L}^{-1}\left\{X(s)\right\} = \frac{1}{2\pi j}\int_C X(s)\, e^{st}\, ds \tag{5.119}$$

In the case of the Laplace transform, the contour C is a vertical line within the ROC.

In general, direct evaluation of the contour integral is difficult when $X(z)$ is a rational function of z. An indirect method of evaluating the integral in Eqn. (5.118) is to rely on the *Cauchy residue theorem* named after Augustin-Louis Cauchy (1789–1857). In this text we will not consider the inversion integral further for computing the inverse z-transform since more practical methods exist for accomplishing the same task.

5.4.2 Partial fraction expansion

It was established in Examples 5.6 and 5.7 that the z-transform of a causal exponential signal is

$$\mathcal{Z}\left\{a^n\, u[n]\right\} = \frac{z}{z-a}\,, \qquad \text{ROC:}\quad |z| > |a| \tag{5.120}$$

and the z-transform of an anti-causal exponential signal is

$$\mathcal{Z}\left\{-a^n\, u[-n-1]\right\} = \frac{z}{z-a}\,, \qquad \text{ROC:}\quad |z| < |a| \tag{5.121}$$

The two signals in Eqns. (5.120) and (5.121) lead to the same rational function for the transform, albeit with different ROC's. These two transform pairs can be used as the basis for determining the inverse z-transform of rational functions expressed using partial fractions. Consider a transform $X(z)$ given with its denominator factored out as

$$X(z) = \frac{B(z)}{(z-z_1)\,(z-z_2)\ldots(z-z_N)} \tag{5.122}$$

Provided that the order of the numerator polynomial $B(z)$ is not greater than N, expanding the transform into partial fractions in the form

$$X(z) = \frac{k_1\, z}{z-z_1} + \frac{k_2\, z}{z-z_2} + \ldots + \frac{k_N\, z}{z-z_N} \tag{5.123}$$

would allow us to use the standard forms in Eqns. (5.120) and (5.121) for finding the inverse transform of $X(z)$. Let individual terms in the partial fraction expansion be

$$X_i(z) = \frac{k_i\, z}{z-z_i} \quad \text{for} \quad i = 1,\ldots,N \tag{5.124}$$

so that

$$X(z) = X_1(z) + X_2(z) + \ldots + X_N(z) \tag{5.125}$$

The inverse transform is therefore computed as

$$x[n] = x_1[n] + x_2[n] + \ldots + x_N[n] \tag{5.126}$$

where each contributing term $x_i[n]$ represents the inverse transform of the corresponding term on the right side of Eqn. (5.125), that is,

$$x_i[n] = \mathcal{Z}^{-1}\{X_i(z)\} = \mathcal{Z}^{-1}\left\{\frac{k_i\, z}{z - z_i}\right\}, \quad \text{for } i = 1, \ldots, N \tag{5.127}$$

In order to find the correct terms $x_i[n]$ for Eqn. (5.126), we need to determine for each $X_i(z)$ whether it is the transform of a causal signal as in Eqn. (5.120) or an anti-causal signal as in Eqn. (5.121). These decisions must be made by looking at the ROC for $X(z)$ and reasoning what the contribution from the ROC of each individual term $X_i(z)$ must be in order to get the overlap that we have. Since each term in the partial fraction expansion has only one pole, we will adopt the following simple rules:

Determining the inverse of each partial fraction:

1. If the ROC for $X(z)$ is inside the circle that passes through the pole at z_i, then the contribution of $X_i(z)$ to the ROC is in the form $|z| < |z_i|$, and therefore the term $x_i[n]$ is an anti-causal signal in the form

$$x_i[n] = -k_i\, z_i^n\, u[-n-1] \tag{5.128}$$

2. If the ROC for $X(z)$ is outside the circle that passes through the pole at z_i, then the contribution of $X_i(z)$ to the ROC is in the form $|z| > |z_i|$, and therefore the term $x_i[n]$ is a causal signal in the form

$$x_i[n] = k_i\, z_i^n\, u[n] \tag{5.129}$$

These two rules are illustrated in Fig. 5.25. The next two examples will serve to clarify the process explained above.

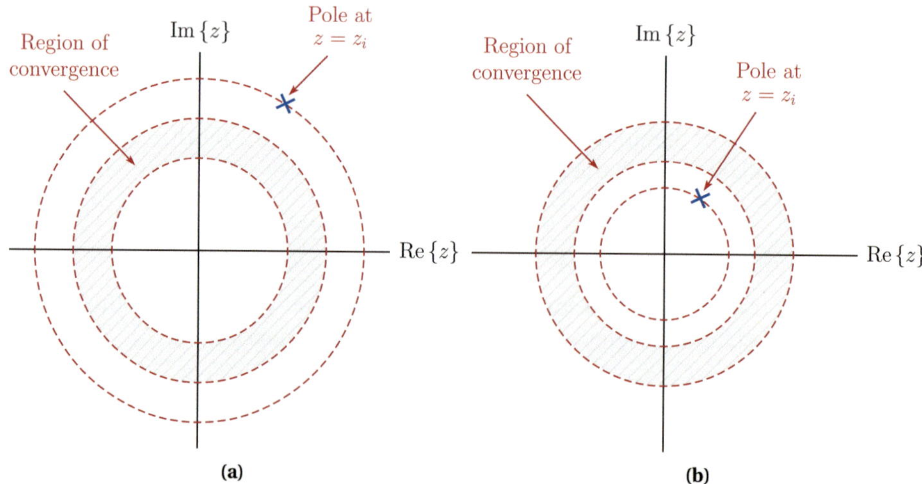

Figure 5.25 – Determining the inverse of each term in partial fraction expansion: **(a)** ROC is inside the circle that passes through the pole at z_i and **(b)** ROC is outside the circle that passes through the pole at z_i.

Example 5.29: Expanding a rational z-transform into partial fractions

Express the z-transform

$$X(z)=\frac{(z-1)(z+2)}{\left(z-\frac{1}{2}\right)(z-2)}$$

using partial fractions.

Solution: The first step is to divide $X(z)$ by z to obtain

$$\frac{X(z)}{z}=\frac{(z-1)(z+2)}{z\left(z-\frac{1}{2}\right)(z-2)} \tag{5.130}$$

The function $X(z)/z$ has three single poles at $z=0,\frac{1}{2}$, and 2, and can be expressed as

$$\frac{X(z)}{z}=\frac{k_1}{z}+\frac{k_2}{z-\frac{1}{2}}+\frac{k_3}{z-2} \tag{5.131}$$

Multiplying both sides of Eqn. (5.131) by z allows us to express $X(z)$ using familiar terms:

$$X(z)=k_1+\frac{k_2 z}{z-\frac{1}{2}}+\frac{k_3 z}{z-2} \tag{5.132}$$

The transition from Eqn. (5.131) to Eqn. (5.132) explains the motivation for dividing $X(z)$ by z in Eqn. (5.130) before expanding the result into partial fractions in Eqn. (5.131). Had we not divided $X(z)$ by z in Eqn. (5.131), we would not have obtained partial fractions in the standard form $z/\left(z-p_i\right)$ in Eqn. (5.132).

The residues in Eqn. (5.131) can be found by using the residue formulas derived in Appendix E:

$$k_1=z\,\frac{X(z)}{z}\bigg|_{z=0}=\frac{(z-1)(z+2)}{\left(z-\frac{1}{2}\right)(z-2)}\bigg|_{z=0}=\frac{(0-1)(0+2)}{\left(0-\frac{1}{2}\right)(0-2)}=-2$$

$$k_2=\left(z-\frac{1}{2}\right)\frac{X(z)}{z}\bigg|_{z=\frac{1}{2}}=\frac{(z-1)(z+2)}{z(z-2)}\bigg|_{z=0}=\frac{\left(\frac{1}{2}-1\right)\left(\frac{1}{2}+2\right)}{\frac{1}{2}\left(\frac{1}{2}-2\right)}=\frac{5}{3}$$

$$k_3=(z-2)\,\frac{X(z)}{z}\bigg|_{z=2}=\frac{(z-1)(z+2)}{z\left(z-\frac{1}{2}\right)}\bigg|_{z=0}=\frac{(2-1)(2+2)}{2\left(2-\frac{1}{2}\right)}=\frac{4}{3}$$

Using the residues computed, the transform $X(z)$ is

$$X(z)=-2+\frac{\frac{5}{3}z}{z-\frac{1}{2}}+\frac{\frac{4}{3}z}{z-2} \tag{5.133}$$

Software resource: See MATLAB Exercise 5.5.

Example 5.30: Finding the inverse z-transform using partial fractions

Consider again the transform $X(z)$ expanded into partial fractions in Example 5.29. The region of convergence was not specified in that example, and therefore, we have only determined the partial fraction expansion for $X(z)$.

a. How many possibilities are there for the ROC?

b. For each possible choice of ROC, determine the inverse transform.

Solution: A pole-zero diagram for $X(z)$ is shown in Fig. 5.26. We know from previous discussion that there can be no poles inside the ROC, and the boundaries of ROC must be determined by poles.

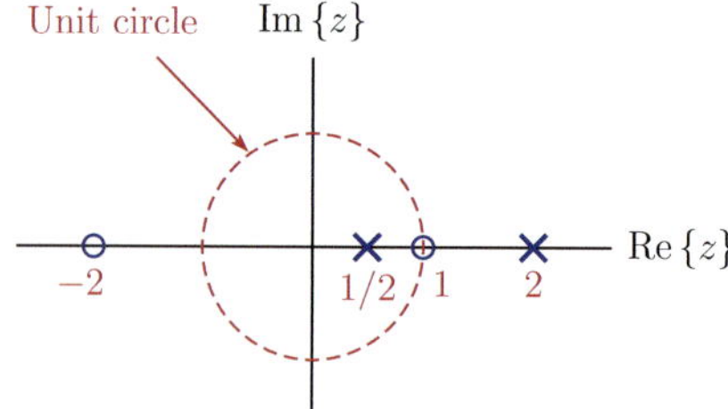

Figure 5.26 – The pole-zero diagram for $X(z)$ in Example 5.30.

We will begin by defining the three components of $X(z)$ in the partial fraction expansion given by Eqn. (5.133). Let

$$X_1(z) = -2$$

$$X_2(z) = \frac{\frac{5}{3}\,z}{z-\frac{1}{2}}$$

and

$$X_3(z) = \frac{\frac{4}{3}\,z}{z-2}$$

The first component, $X_1(z)$, is a constant, and its ROC is always the entire z-plane. Therefore it has no effect on the overall ROC for $X(z)$. The inverse transform of $X_1(z)$ is

$$x_1[n] = \mathcal{Z}^{-1}\{-2\} = -2\,\delta[n]$$

The ROC of $X(z)$ will be determined based on the individual ROCs of the terms $X_2(z)$ and $X_3(z)$. The term $X_2(z)$ has a zero at $z=0$ and a pole at $z=\frac{1}{2}$. Its region of convergence is either the inside or the outside of the circle with a radius of $\frac{1}{2}$. Similarly, the term $X_3(z)$ has a zero at $z=0$ and a pole at $z=2$. Its region of convergence is either the inside or the outside of the circle with a radius of 2. Applying the rules of the ROC in conjunction with the pole-zero diagram shown in Fig. 5.26 we obtain the following possibilities for the ROC of $X(z)$:

Possibility 1: ROC: $|z| < \frac{1}{2}$

In this case both $X_2(z)$ and $X_3(z)$ must correspond to anti-causal signals, so that the overlap of individual ROCs yields the region chosen. Thus we need

$$\text{ROC for } X_2(z): \quad |z| < \frac{1}{2}$$

$$\text{ROC for } X_3(z): \quad |z| < 2$$

The resulting ROC for $X(z)$ is the overlap of the two individual ROC's, and is therefore the inside of the circle with radius $\frac{1}{2}$. This is illustrated in Fig. 5.27.

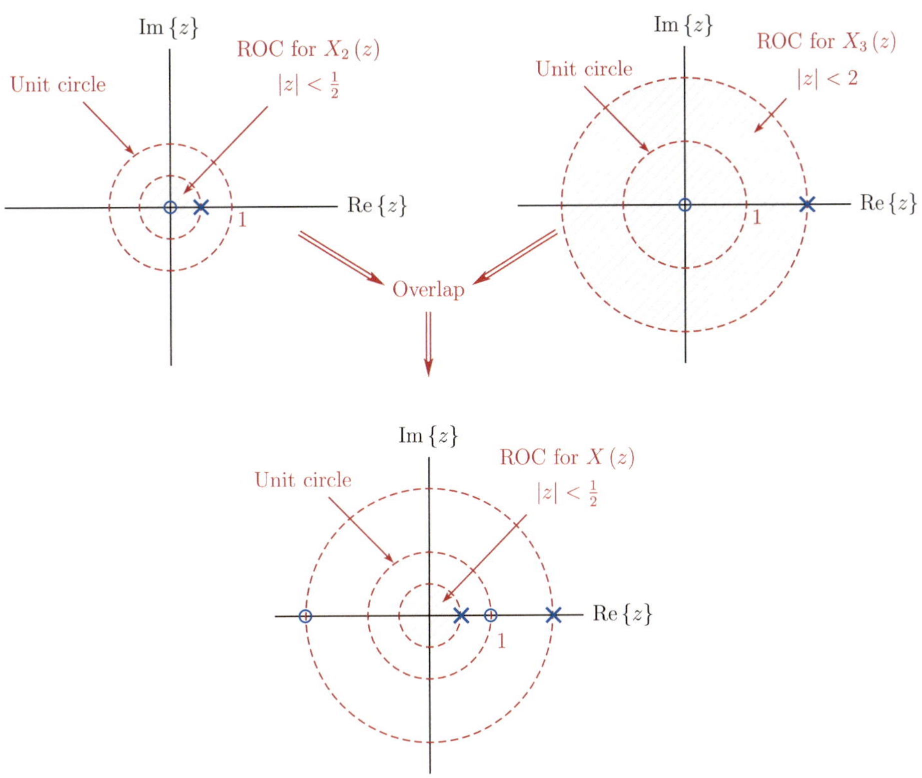

Figure 5.27 – ROCs involved in possibility 1 of Example 5.30.

The inverse transforms of $X_2(z)$ and $X_3(z)$ are determined using the anti-causal signal given by Eqn. (5.121):

$$x_2[n] = \mathcal{Z}^{-1}\left\{\frac{\frac{5}{3}z}{z-\frac{1}{2}}\right\} = -\frac{5}{3}\left(\frac{1}{2}\right)^n u[-n-1]$$

$$x_3[n] = \mathcal{Z}^{-1}\left\{\frac{\frac{4}{3}z}{z-2}\right\} = -\frac{4}{3}(2)^n u[-n-1]$$

Combining $x_1[n]$, $x_2[n]$, and $x_3[n]$, we find the signal $x[n]$ to be

$$x[n] = -2\delta[n] - \frac{5}{3}\left(\frac{1}{2}\right)^n u[-n-1] - \frac{4}{3}(2)^n u[-n-1]$$

Numerical evaluation of $x[n]$ for a few values of the index n results in

$$x[n] = \{\ \ldots, -53.375, -26.75, -13.5, -7, -4, \underset{\substack{\uparrow \\ n=0}}{-2}\ \}$$

Possibility 2: ROC: $|z| > 2$

This ROC is only possible as the overlap of individual ROCs if both $X_2(z)$ and $X_3(z)$ correspond to causal signals, that is,

$$\text{ROC for } X_2(z): \quad |z| > \frac{1}{2}$$

$$\text{ROC for } X_3(z): \quad |z| > 2$$

The resulting ROC for $X(z)$ in this case is the overlap of the two individual ROCs which is the outside of the circle with a radius of 2. This is illustrated in Fig. 5.28.

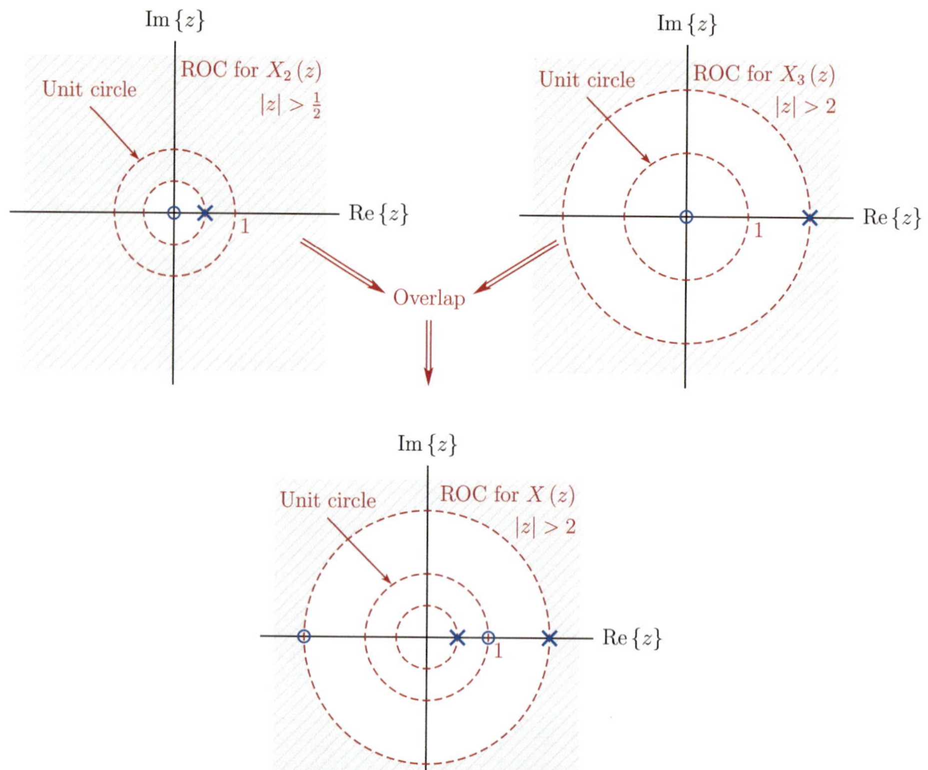

Figure 5.28 – ROCs involved in possibility 2 of Example 5.30.

In this case, the inverse transforms of both $X_2(z)$ and $X_3(z)$ are causal signals, and they are computed from Eqn. (5.120) as

$$x_2[n] = \mathcal{Z}^{-1}\left\{\frac{\frac{5}{3}z}{z-\frac{1}{2}}\right\} = \frac{5}{3}\left(\frac{1}{2}\right)^n u[n]$$

$$x_3[n] = \mathcal{Z}^{-1}\left\{\frac{\frac{4}{3}z}{z-2}\right\} = \frac{4}{3}\,(2)^n\, u[n]$$

Using $x_1[n]$, $x_2[n]$ and $x_3[n]$ found, the inverse transform for $X(z)$ is

$$x[n] = -2\delta[n] + \frac{5}{3}\left(\frac{1}{2}\right)^n u[n] + \frac{4}{3}\,(2)^n\, u[n]$$

In this case the first few samples of $x[n]$ are computed as

$$x[n] = \{ \underset{\substack{\uparrow \\ n=0}}{1}, 3.5, 5.75, 8.2083, 10.7708, 13.3854, \ldots \}$$

Possibility 3: ROC: $\frac{1}{2} < |z| < 2$

For this ROC to be the overlap of the two individual ROC's, $X_2(z)$ must correspond to a causal signal while $X_3(z)$ corresponds to an anti-causal signal. We need

$$\text{ROC for } X_2(z): \quad |z| > \frac{1}{2}$$

$$\text{ROC for } X_3(z): \quad |z| < 2$$

Thus the ROC for $X_2(z)$ is the outside of a circle, and the ROC for $X_3(z)$ is the inside of a circle. The overlap of the two ROCs is the region between the two circles. Figure 5.29 illustrates this possibility.

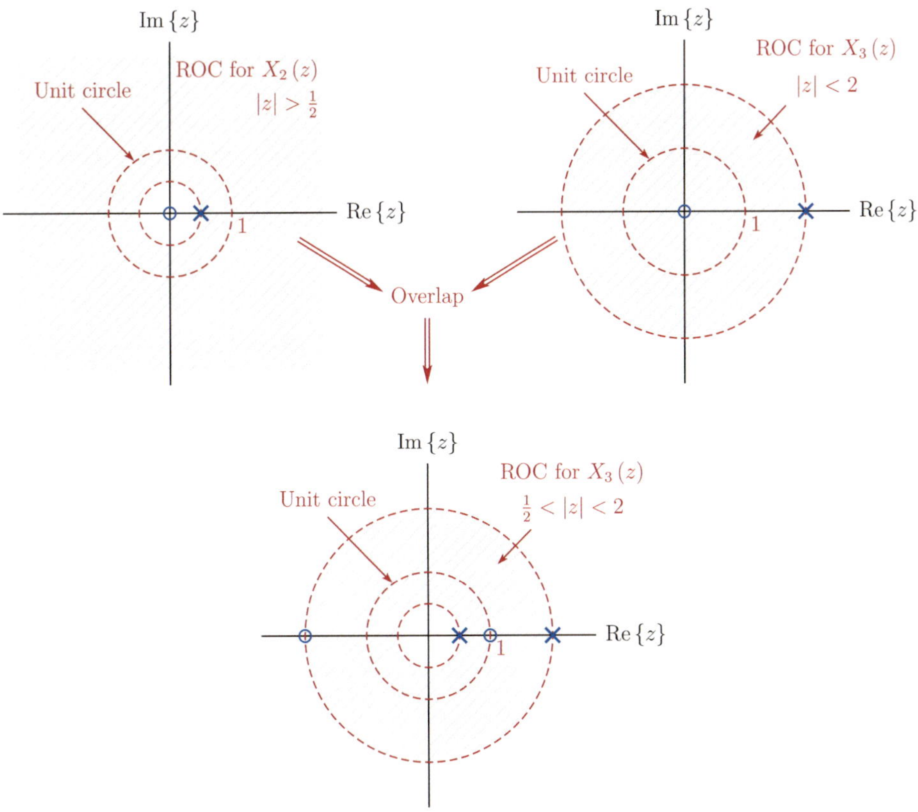

Figure 5.29 – ROCs involved in possibility 3 of Example 5.30.

Inverting each component accordingly, we obtain

$$x_2[n] = \mathcal{Z}^{-1}\left\{ \frac{\frac{5}{3}z}{z - \frac{1}{2}} \right\} = \frac{5}{3}\left(\frac{1}{2}\right)^n u[n]$$

and

$$x_3[n] = \mathcal{Z}^{-1}\left\{\frac{\frac{4}{3}z}{z-2}\right\} = -\frac{4}{3}(2)^n\, u[-n-1]$$

Using $x_1[n]$, $x_2[n]$, and $x_3[n]$ found, the inverse transform for $X(z)$ can be constructed as

$$x[n] = -2\delta[n] + \frac{5}{3}\left(\frac{1}{2}\right)^n u[n] - \frac{4}{3}(2)^n\, u[-n-1]$$

In this case the first few samples of $x[n]$ are computed as

$$x[n] = \{\ \ldots, -0.1667, -0.3333, -0.6667, \underset{\substack{\uparrow \\ n=0}}{-0.3333}, 0.8333, 0.4167, 0.2083, \ldots\ \}$$

5.4.3 Long division

An alternative method for computing the inverse z-transform is the long division technique. Recall that the definition of the z-transform given by Eqns. (5.1) and (5.2) is essentially in the form of a power series involving powers of z and z^{-1}. The long division idea is based on converting a rational transform $X(z)$ back into its power series form, and associating the coefficients of the power series with the sample amplitudes of the signal $x[n]$. In contrast with the partial fraction expansion method discussed in the previous section, the long division method does not produce an analytical solution for the signal $x[n]$. Instead, it allows us to obtain the signal one sample at a time. Its main advantage over the partial fraction expansion method is that it is suitable for use on a computer.

Consider a rational transform in the general form

$$X(z) = \frac{B(z)}{A(z)} = \frac{b_M z^M + b_{M-1} z^{M-1} + \ldots + b_1 z + b_0}{a_N z^N + a_{N-1} z^{N-1} + \ldots + a_1 z + a_0} \tag{5.134}$$

where the M-th order denominator polynomial $B(z)$ is

$$B(z) = b_M z^M + b_{M-1} z^{M-1} + \ldots + b_1 z + b_0$$

and the N-th order numerator polynomial $A(z)$ is

$$A(z) = a_N z^N + a_{N-1} z^{N-1} + \ldots + a_1 z + a_0$$

Initially we will focus on the case where the signal $x[n]$ that led to the transform $X(z)$ is causal. This implies the following:

1. The ROC is in the form $|z| > r_1$.
2. The transform $X(z)$ converges at $z \to \infty$.
3. The order of the numerator is less than or equal to the order of the denominator, that is, $M \le N$.

Dividing the numerator polynomial by the denominator polynomial, the transform $X(z)$ can be written in the alternative form

$$X(z) = \left(\frac{b_M}{a_N}\right) z^{-(N-M)} + \frac{\bar{B}(z)}{A(z)} \tag{5.135}$$

where $(b_M / a_N)\, z^{-(N-M)}$ is the quotient of the division. The remainder is a polynomial $\bar{B}(z)$ in the form

$$\bar{B}(z) = \bar{b}_{M-1} z^{M-1} + \ldots + \bar{b}_{M-N} z^{M-N} \tag{5.136}$$

Associating the expression in Eqn. (5.135) with the power series form of the z-transform we recognize that

$$x[N-M] = \frac{b_M}{a_N} \tag{5.137}$$

providing us with one sample of the signal $x[n]$. We can now take the function

$$\bar{X}(z) = \frac{\bar{B}(z)}{A(z)}$$

and repeat the process, obtaining

$$\bar{X}(z) = \left(\frac{\bar{b}_{M-1}}{a_N}\right) z^{-(N-M+1)} + \frac{\hat{B}(z)}{A(z)} \tag{5.138}$$

which produces another sample of the signal $x[n]$ as

$$x[N-M+1] = \frac{\bar{b}_{M-1}}{a_N} \tag{5.139}$$

and a new remainder polynomial

$$\hat{B}(z) = \hat{b}_{M-2} z^{M-2} + \ldots + \hat{b}_{M-N-1} z^{M-N-1}$$

The next example will illustrate this process.

Example 5.31: Using long division with right-sided signal

Use the long division method to determine the first few samples of the signal $x[n]$ of Example 5.22 from its z-transform which was determined to be

$$X(z) = \frac{3z^2 - z}{(z-1)^3}\,, \qquad \text{ROC:} \quad |z| > 1$$

Solution: By multiplying out the denominator, the transform $X(z)$ can be written as

$$X(z) = \frac{3z^2 - z}{z^3 - 3z^2 + 3z - 1}\,, \qquad \text{ROC:} \quad |z| > 1$$

The numerator polynomial is

$$B(z) = 3z^2 - z$$

and the denominator polynomial is

$$A(z) = z^3 - 3z^2 + 3z - 1$$

Dividing the numerator polynomial by the denominator polynomial we obtain an alternative form of $X(z)$ as

$$X(z) = 3z^{-1} + \frac{8z - 9 + 3z^1}{z^3 - 3z^2 + 3z - 1} = 3z^{-1} + \bar{X}(z) \tag{5.140}$$

Thus we have extracted one term of the power series representation of $X(z)$, and it indicates that $x[1]=3$. Repeating the process with $\bar{X}(z)$, we obtain

$$\bar{X}(z)=8z^{-2}+\frac{15-21z^{-1}+8z^{-2}}{z^3-3z^2+3z-1}=8z^{-2}+\hat{X}(z) \tag{5.141}$$

indicating that $x[2]=8$. The results in Eqns. (5.140) and (5.141) can be combined, yielding

$$X(z)=3z^{-1}+8z^{-2}+\frac{15-21z^{-1}+8z^{-2}}{z^3-3z^2+3z-1}$$

Thus, each iteration through the long division operation produces one more sample of the signal $x[n]$. We are now ready to set up the long division:

$$\begin{array}{r|l}
 & 3z^{-1}+8z^{-2}+15z^{-3}+24z^{-4} \\ \hline
z^3\ -3z^2\ +3z\ -1 & 3z^2\quad -z \\
 & 3z^2\ -9z\ \ +9\ \ -3z^{-1} \\ \cline{2-2}
 & \quad 8z\ \ -9\ \ +3z^{-1} \\
 & \quad 8z\ -24\ +24z^{-1}\ \ -8z^{-2} \\ \cline{2-2}
 & \qquad 15\ -21z^{-1}\ \ +8z^{-2} \\
 & \qquad 15\ -45z^{-1}\ +45z^{-2}\ -15z^{-3} \\ \cline{2-2}
 & \qquad\quad 24z^{-1}\ -37z^{-2}\ +15z^{-3}
\end{array}$$

Using the quotient and the remainder of the division, the transform $X(z)$ can be written as

$$X(z)=3z^{-1}+8z^{-2}+15z^{-3}+24z^{-4}+\frac{24z^{-1}-37z^{-2}+15z^{-3}}{z^3-3z^2+3z-1} \tag{5.142}$$

Comparing the result in Eqn. (5.142) with the definition of the z-transform in Eqn. (5.1) we conclude that the signal $x[n]$ is in the form

$$x[n]=\{\underset{\substack{\uparrow\\ n=1}}{3},8,15,24,\ldots\}$$

The sample amplitudes obtained should be in agreement with those obtained by directly evaluating the values of the signal $x[n]=n(n+2)\,u[n]$ that led to the transform in question.

In Example 5.31 the use of the long division technique produced a causal signal as the inverse transform of the specified function $X(z)$. That was fine since the ROC specified for the transform also supported this conclusion. What if we have a transform and associated ROC that indicate an anti-causal or a non-causal signal as the inverse transform? How would we need to modify the long division technique to produce the correct result in such a case?

Consider again a rational transform $X(z)$ with one change from Eqn. (5.134): This time we will order the terms of numerator and denominator polynomials in ascending powers of z. The result is

$$X(z)=\frac{B(z)}{A(z)}=\frac{b_0+b_1z+\ldots+b_{M-1}z^{M-1}+b_Mz^M}{a_0+a_1z+\ldots+a_{N-1}z^{N-1}+a_Nz^N} \tag{5.143}$$

If we now divide the numerator polynomial by the denominator polynomial, we can write $X(z)$ as

$$X(z) = \left(\frac{b_0}{a_0}\right) + \frac{\bar{B}(z)}{A(z)} \tag{5.144}$$

The term (b_0/a_0) is the quotient of the division. The remainder is a polynomial $\bar{B}(z)$ in the form

$$\bar{B}(z) = \bar{b}_1 z + \bar{b}_2 z^2 + \ldots \tag{5.145}$$

Associating the expression in Eqn. (5.144) with the power series form of the z-transform we recognize that

$$x[0] = \frac{b_0}{a_0} \tag{5.146}$$

We can now take the function

$$\bar{X}(z) = \frac{\bar{B}(z)}{A(z)}$$

and repeat the process, obtaining

$$\bar{X}(z) = \left(\frac{\bar{b}_1}{a_0}\right) z + \frac{\hat{B}(z)}{A(z)}$$

which produces another sample of the signal $x[n]$ as

$$x[-1] = \frac{\bar{b}_1}{a_0}$$

and a new remainder polynomial

$$\hat{B}(z) = \hat{b}_2 z^2 + \hat{b}_3 z^3 + \ldots$$

The next example will illustrate the use of long division with different types of ROC's.

Example 5.32: Using long division with specified ROC

In Example 5.29 we have used the partial fraction expansion method to determine the inverse z-transform of the rational function

$$X(z) = \frac{(z-1)(z+2)}{\left(z-\frac{1}{2}\right)(z-2)}$$

for all possible choices of the ROC. Verify the results of that example by determining the first few samples of the inverse z-transform by using the long division method for each possible choice of the ROC.

Solution: Multiplying out the factors of the numerator and the denominator of $X(z)$ we obtain

$$X(z) = \frac{z^2+z-2}{z^2-2.5z+1} \tag{5.147}$$

As we have discussed in Example 5.30, there are three possible choices for the ROC:

Possibility 1: ROC: $|z| < \frac{1}{2}$

In this case the inverse transform $x[n]$ must be anti-causal. In order to obtain an anti-causal solution from the long division, we will rewrite $X(z)$ with its numerator and denominator polynomials arranged in ascending powers of z:

$$X(z) = \frac{-2+z+z^2}{1-2.5z+z^2}$$

and set up the long division:

$$\begin{array}{r|llllll}
 & -2-4z-7z^2-13.5z^3 & & & & & \\
1\ -2.5z\ +z^2 & -2 & +z & +z^2 & & & \\
 & -2 & +5z & -2z^2 & & & \\
\hline
 & & -4z & +3z^2 & & & \\
 & & -4z & +10z^2 & -4z^3 & & \\
\hline
 & & & -7z^2 & +4z^3 & & \\
 & & & -7z^2 & +17.5z^3 & -7z^4 & \\
\hline
 & & & & -13.5z^3 & +7z^4 & \\
 & & & & -13.5z^3 & +33.75z^4 & -13.5z^5 \\
\hline
 & & & & & -26.75z^4 & +13.5z^5
\end{array}$$

Using the quotient and the remainder of the division, the transform $X(z)$ can be written as

$$X(z) = -2-4z-7z^2-13.5z^3 + \frac{-26.75z^4+13.5z^5}{1-2.5z+z^2}$$

Comparing this result with the z-transform definition in Eqn. (5.1) we conclude that the signal $x[n]$ must be in the form

$$x[n] = \{\ \ldots, -13.5, -7, -4, \underset{\substack{\uparrow \\ n=0}}{-2}\ \}$$

consistent with the answer found in Example 5.30.

Possibility 2: ROC: $|z| > 2$

In this case the signal $x[n]$ is causal. To obtain a causal answer we will use the original form of $X(z)$ in Eqn. (5.147) with numerator and denominator polynomials arranged in descending orders of z. The long division is set up as follows:

$$\begin{array}{r|lllll}
 & 1+3.5z^{-1}+5.75z^{-2} & & & & \\
z^2\ -2.5z\ +1 & z^2 & +z & -2 & & \\
 & z^2 & -2.5z & +1 & & \\
\hline
 & & 3.5z & -3 & & \\
 & & 3.5z & -8.75 & +3.5z^{-1} & \\
\hline
 & & & 5.75 & -3.5z^{-1} & \\
 & & & 5.75 & -14.375z^{-1} & +5.75z^{-2} \\
\hline
 & & & & 10.875z^{-1} & -5.75z^{-2}
\end{array}$$

Using the results obtained so far, the transform $X(z)$ is expressed as

$$X(z) = 1+3.5z^{-1}+5.75z^{-2} + \frac{10.875z^{-1}-5.75z^2}{z^2-2.5z+1}$$

The first few samples of the signal $x[n]$ are

$$x[n] = \{ \underset{\substack{\uparrow \\ n=0}}{1}, 3.5, 5.75, \ldots \}$$

identical to the result that was found in Example 5.30.

Possibility 3: ROC: $\frac{1}{2} < |z| < 2$

In this case the inverse transform $x[n]$ has a causal component and an anti-causal component, so it can be written in the form

$$x[n] = x_R[n] + x_L[n]$$

Accordingly, we need to partition the transform $X(z)$ into two parts: One that corresponds to the causal component $x_R[n]$, and the other one that corresponds to the anti-causal component $x_L[n]$. Recall that in Example 5.29 the transform $X(z)$ was expressed through partial fractions as

$$X(z) = X_1(z) + X_2(z) + X_3(z)$$

with

$$X_1(z) = -2\,, \qquad \text{ROC: all } z$$

$$X_2(z) = \frac{\frac{5}{3}z}{z-\frac{1}{2}}\,, \qquad \text{ROC: } |z| > \frac{1}{2}$$

and

$$X_3(z) = \frac{\frac{4}{3}z}{z-2}\,, \qquad \text{ROC: } |z| < 2$$

It is clear that $X_2(z)$ should be included in $X_R(z)$, and $X_3(z)$ should be included in $X_L(z)$. The constant term, $X_1(z) = -2$, could be included with either function; we will choose to include it with $X_R(z)$. Consequently we have

$$X_R(z) = X_1(z) + X_2(z) = -2 + \frac{\frac{5}{3}z}{z-\frac{1}{2}} = \frac{-\frac{1}{3}z+1}{z-\frac{1}{2}}\,, \qquad \text{ROC: } |z| > \frac{1}{2}$$

and

$$X_L(z) = X_3(z) = \frac{\frac{4}{3}z}{z-2}\,, \qquad \text{ROC: } |z| < 2$$

Let us begin with $x_R[n]$. The long division for the causal component is set up using the numerator and the denominator of $X_R(z)$ arranged in descending powers of z:

$$\begin{array}{r|llll}
 & -\frac{1}{3} + \frac{5}{6}z^{-1} + \frac{5}{12}z^{-2} & & & \\ \hline
z-\frac{1}{2} & -\frac{1}{3}z & +1 & & \\
 & -\frac{1}{3}z & +\frac{1}{6} & & \\ \cline{2-3}
 & & \frac{5}{6} & & \\
 & & \frac{5}{6} & -\frac{5}{12}z^{-1} & \\ \cline{3-4}
 & & & \frac{5}{12}z^{-1} & \\
 & & & \frac{5}{12}z^{-1} & -\frac{5}{24}z^{-2} \\ \cline{4-5}
 & & & & \frac{5}{24}z^{-2}
\end{array}$$

We conclude that $x_R[0] = -1/3$, $x_R[1] = 5/6$, and $x_R[2] = 5/12$. Next we will set up the long division for $X_L[z]$ with numerator and denominator polynomials arranged in ascending powers of z:

$$
\begin{array}{r|l}
 & -\frac{2}{3}z - \frac{1}{3}z^2 - \frac{1}{6}z^3 \\
\hline
-2 + z & \frac{4}{3}z \\
 & \frac{4}{3}z - \frac{2}{3}z^2 \\
\cline{2-2}
 & \frac{2}{3}z^2 \\
 & \frac{2}{3}z^2 - \frac{1}{3}z^3 \\
\cline{2-2}
 & \frac{1}{3}z^3 \\
 & \frac{1}{3}z^3 - \frac{1}{6}z^4 \\
\cline{2-2}
 & \frac{1}{6}z^4
\end{array}
$$

Thus, the first few samples of the anti-causal component of $x[n]$ are $x_L[-1] = -2/3$, $x_L[-2] = -1/3$, and $x_L[-3] = -1/6$. Combining the results of the two long divisions we obtain $x[n]$ as

$$x[n] = \{ \ldots, -\frac{1}{6}, -\frac{1}{3}, -\frac{2}{3}, \underset{\substack{\uparrow \\ n=0}}{-\frac{1}{3}}, \frac{5}{6}, \frac{5}{12}, \ldots \}$$

Software resource: See MATLAB Exercise 5.6.

5.5 Using the z-Transform with DTLTI Systems

In Chapter 2 we have demonstrated that the output signal of a DTLTI system is related to its input signal and its impulse response through the convolution sum. Specifically, if the impulse response of a DTLTI system is $h[n]$, and if the signal $x[n]$ is applied to the system as input, the output signal $y[n]$ is computed as

$$y[n] = x[n] * h[n] = \sum_{k=-\infty}^{\infty} x[k]\, h[n-k] \tag{5.148}$$

Based on the convolution property of the z-transform introduced by Eqn. (5.88) in Section 5.3.6, the z-transform of the output signal is equal to the product of the z-transforms of the input signal and the impulse response, that is,

$$Y(z) = X(z)\, H(z) \tag{5.149}$$

If the input signal $x[n]$ and the impulse response $h[n]$ are specified, we could find the z-transforms $X(z)$ and $H(z)$, multiply them to obtain $Y(z)$, and then determine $y[n]$ by means of inverse z-transform. This process provides us an alternative to computing the output by direct application of the convolution sum, and was demonstrated in Examples 5.23 and 5.24.

Alternatively, if the input and the output signals $x[n]$ and $y[n]$ are specified, their z-transforms can be computed, and then the z-transform of the impulse response can be determined from them. Solving Eqn. (5.149) for $H(z)$ we get

$$H(z) = \frac{Y(z)}{X(z)} \tag{5.150}$$

The function $H(z)$ is the *z-domain system function* of the system. We already know that a DTLTI system can be described fully and uniquely by means of its impulse response $h[n]$. Since the system function $H(z)$ is just the z-transform of the impulse response $h[n]$, it also represents a complete description of the DTLTI system.

5.5.1 Relating the system function to the difference equation

In Section 2.4 of Chapter 2 we have established the fact that a DTLTI system can be described by means of a constant-coefficient linear difference equation in the standard form

$$\sum_{k=0}^{N} a_k\, y[n-k] = \sum_{k=0}^{M} b_k\, x[n-k] \tag{5.151}$$

Therefore it must be possible to obtain the other two forms of description of the DTLTI system, namely the system function and the impulse response, from the knowledge of its difference equation. If we take the z-transform of both sides of Eqn. (5.151) the equality would still be valid:

$$\mathcal{Z}\left\{\sum_{k=0}^{N} a_k\, y[n-k]\right\} = \mathcal{Z}\left\{\sum_{k=0}^{M} b_k\, x[n-k]\right\} \tag{5.152}$$

Using the linearity property of the z-transform, Eqn. (5.152) can be written as

$$\sum_{k=0}^{N} a_k\, \mathcal{Z}\left\{y[n-k]\right\} = \sum_{k=0}^{M} b_k\, \mathcal{Z}\left\{x[n-k]\right\} \tag{5.153}$$

and using the time shifting property of the z-transform leads to

$$\sum_{k=0}^{N} a_k z^{-k}\, Y(z) = \sum_{k=0}^{M} b_k z^{-k}\, X(z) \tag{5.154}$$

The transforms $X(z)$ and $X(z)$ do not depend on the summation index k, and can therefore be factored out of the summations in Eqn. (5.154) resulting in

$$Y(z) \sum_{k=0}^{N} a_k z^{-k} = X(z) \sum_{k=0}^{M} b_k z^{-k} \tag{5.155}$$

Finally, the system function can be obtained from Eqn. (5.155) as

$$H(z) = \frac{Y(z)}{X(z)} = \frac{\displaystyle\sum_{k=0}^{M} b_k z^{-k}}{\displaystyle\sum_{k=0}^{N} a_k z^{-k}} \tag{5.156}$$

Finding the system function from the difference equation:

1. Separate the terms of the difference equation so that $y[n]$ and its time-shifted versions are on the left of the equal sign, and $x[n]$ and its time-shifted versions are on the right of the equal sign, as in Eqn. (5.151).
2. Take the z-transforms of both sides of the difference equation, and use the time shifting property of the z-transform as in Eqn. (5.154).
3. Determine the system function as the ratio of $Y(z)$ to $X(z)$ as in Eqn. (5.156).
4. If the impulse response is needed, it can now be determined as the inverse z-transform of $H(z)$.

At this point we will make two important observations:

1. In the above development leading up to the z-domain system function we have relied on the convolution operation in Eqn. (5.148). In Eqn. (5.156) we have used the convolution property of the z-transform. We know from Chapter 2 that the convolution operation is only applicable to problems involving linear and time-invariant systems. Therefore the system function concept is meaningful only for systems that are both linear and time-invariant. This notion was introduced in earlier discussions involving system functions as well.
2. Furthermore, it was justified in Section 2.4 of Chapter 2 that a constant-coefficient difference equation corresponds to a linear and time-invariant system only if all initial conditions are set to zero.

Consequently, we conclude that, in determining the system function from the difference equation, all initial conditions must be assumed to be zero.

If we need to use z-transform based techniques to solve a difference equation subject to non-zero initial conditions, we can do that by using the unilateral z-transform, but not by using the system function. The unilateral z-transform and its use for solving difference equations will be discussed in Section 5.7.

Example 5.33: Finding the system function from the difference equation

Let a DTLTI system be defined by the difference equation

$$y[n]-0.4y[n-1]+0.89y[n-2]=x[n]-x[n-1]$$

Find the system function $H(z)$ for this system.

Solution: We will assume zero initial conditions and take the z-transforms of both sides of the difference equation to get

$$Y(z)-0.4z^{-1}\,Y(z)+0.89z^{-2}\,Y(z)=X(z)-z^{-1}\,X(z)$$

from which the system function is obtained as

$$H(z)=\frac{1-z^{-1}}{1-0.4z^{-1}+0.89z^{-2}}$$

or, using non-negative powers of z

$$H(z)=\frac{z\,(z-1)}{z^2-0.4z+0.89}$$

We will make another important observation based on the result obtained in Example 5.33: The characteristic equation for the system considered in Example 5.33 is

$$z^2 - 0.4z + 0.89 = 0$$

and the solutions of the characteristic equation are the modes of the system as defined in Section 2.5.1 of Chapter 2. When we find the system function $H(z)$ from the difference equation we see that its denominator polynomial is identical to the characteristic polynomial. The roots of the denominator polynomial are the poles of the system function in the z-domain, and consequently, they are identical to the modes of the difference equation of the system.

Recall that in Section 2.5.1 we have reached some conclusions about the relationship between the modes of the difference equation and the natural response of the corresponding system. The same conclusions would apply to the poles of the system function. Specifically, if all poles p_k of the system are real-valued and distinct, then the natural response is in the form

$$y[n] = \sum_{k=1}^{N} c_k p_k^n$$

Complex poles appear in conjugate pairs provided that all coefficients of the system function are real-valued. A pair of complex conjugate poles

$$p_{1a} = r_1 e^{j\Omega_1}, \quad \text{and} \quad p_{1b} = r_1 e^{-j\Omega_1}$$

yields a response of the type

$$y_1[n] = d_1 r_1^n \cos(\Omega_1 n + \theta_1)$$

Finally, a pole of multiplicity m at $z = p_1$ leads to a response in the form

$$y_1[n] = c_{11} p_1^n + c_{12} n p_1^n + \ldots + c_{1m} n^{m-1} p_1^n + \text{other terms}$$

Justifications for these relationships were given in Section 2.5.1 of Chapter 2 by using the modes of the difference equation, and will not be repeated here.

Sometimes we need to reverse the problem represented in Example 5.33 and find the difference equation from the knowledge of the system function. The next example will demonstrate this.

Example 5.34: Finding the difference equation from the system function

Find the difference equation of the DTLTI system defined by the system function

$$H(z) = \frac{z^2 - 5z + 6}{z^3 + 2z^2 - z - 2}$$

Solution: Let us first write $H(z)$ as a rational function of z^{-1} by multiplying both its numerator and denominator with z^{-3}:

$$H(z) = \frac{z^{-1} - 5z^{-2} + 6z^{-3}}{1 + 2z^{-1} - z^{-2} - 2z^{-3}}$$

We know that the system function is the ratio of the output transform to the input transform, that is,

$$H(z) = \frac{Y(z)}{X(z)}$$

so that we can write

$$\left(1+2z^{-1}-z^{-2}-2z^{-3}\right) Y(z) = \left(z^{-1}-5z^{-2}+6z^{-3}\right) X(z)$$

or equivalently

$$y[n]+2y[n-1]-y[n-2]-2y[n-3] = x[n-1]-5x[n-2]+6x[n-3]$$

5.5.2 Response of a DTLTI system to complex exponential signal

An interesting interpretation of the system function concept is obtained when one considers the response of a DTLTI system to a complex exponential signal in the form

$$x[n] = z_0^n$$

where z_0 represents a point on the z-plane, within the ROC of the system function. Let $h[n]$ be the impulse response of the DTLTI system under consideration. The output signal is determined through the use of the convolution sum

$$y[n] = h[n] * x[n] = \sum_{k=-\infty}^{\infty} h[k]\, x[n-k] \tag{5.157}$$

Substituting $x[n-k] = z_0^{n-k}$ into Eqn. (5.157) and simplifying the resulting summation we obtain

$$\begin{aligned} y[n] &= \sum_{k=-\infty}^{\infty} h[k]\, z_0^{n-k} \\ &= z_0^n \sum_{k=-\infty}^{\infty} h[k]\, z_0^{-k} \end{aligned} \tag{5.158}$$

In Eqn. (5.158) the summation corresponds to the system function $H(z)$ evaluated at the point $z = z_0$, so the response of the DTLTI system to the input signal $x[n] = z_0^n$ is

$$y[n] = z_0^n\, H(z_0) \tag{5.159}$$

We conclude that the response of a DTLTI system to the exponential signal $x[n] = z_0^n$ is a scaled version of the input signal. The scale factor is the value of the system function at the point $z = z_0$.

Example 5.35: Response of a DTLTI system to a complex exponential signal

Consider a DTLTI system with the system function

$$H(z) = \frac{z+1}{z^2 - \frac{5}{6}z + \frac{1}{6}}$$

Find the response of the system to the exponential signal

$$x[n] = \left(0.9\, e^{j0.2\pi}\right)^n$$

Solution: The input signal is complex-valued. Using Euler's formula we can write $x[n]$ as

$$x[n] = (0.9)^n \cos(0.2\pi n) + j\,(0.9)^n \sin(0.2\pi n)$$

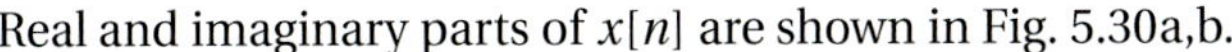

Real and imaginary parts of $x[n]$ are shown in Fig. 5.30a,b.

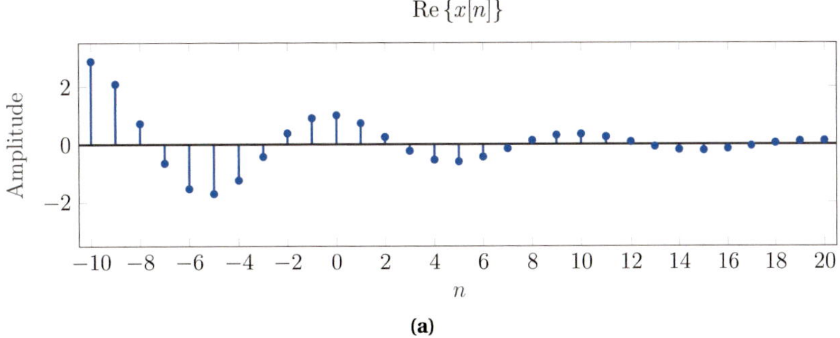

(a)

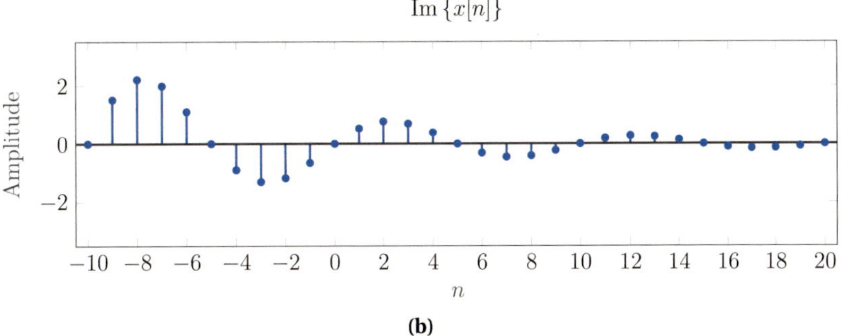

(b)

Figure 5.30 – The signal $x[n]$ for Example 5.35: **(a)** Real part and **(b)** imaginary part.

The value of the system function at $z = 0.9\, e^{j0.2\pi}$ is

$$H\left(0.9\, e^{j0.2\pi}\right) = \frac{0.9\, e^{j0.2\pi} + 1}{\left(0.9\, e^{j0.2\pi}\right)^2 - \frac{5}{6}\left(0.9\, e^{j0.2\pi}\right) + \frac{1}{6}}$$

$$= -1.0627 - j\, 4.6323$$

$$= 4.7526\, e^{-j\, 1.7963}$$

The output signal is found using Eqn. (5.159) as

$$y[n] = \left(0.9\, e^{j0.2\pi}\right)^n \left(4.7526\, e^{-j\, 1.7963}\right)$$

$$= 4.7526\, (0.9)^n\, e^{j\,(0.2\pi n - 1.7963)}$$

or, in Cartesian form

$$y[n] = 4.7526\, (0.9)^n \cos(0.2\pi n - 1.7963) + j\, 4.7526\, (0.9)^n \sin(0.2\pi n - 1.7963)$$

Real and imaginary parts of the output signal $y[n]$ are shown in Fig. 5.31a,b.

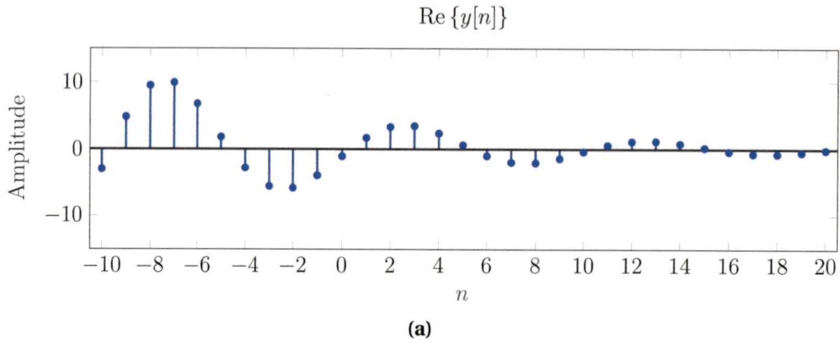

(a)

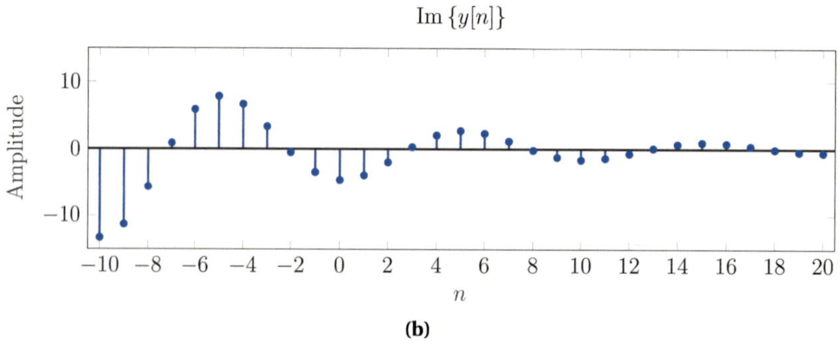

(b)

Figure 5.31 – The signal $y[n]$ for Example 5.35: **a)** Real part and **b)** imaginary part.

Software resources: exdt_5_35a.m , exdt_5_35b.m

5.5.3 Response of a DTLTI system to exponentially damped sinusoid

Next we will consider an input signal in the form of an exponentially damped sinusoid such as

$$x[n] = r_0^n \cos(\Omega_0 n) \tag{5.160}$$

In order to find the response of a DTLTI system to this signal, let us write $x[n]$ using Euler's formula:

$$x[n] = \frac{1}{2}\, r_0^n\, e^{j\Omega_0 n} + \frac{1}{2}\, r_0^n\, e^{-j\Omega_0 n} \tag{5.161}$$

Let z_0 be defined as

$$z_0 = r_0\, e^{j\Omega_0}$$

It follows that

$$z_0^* = r_0\, e^{-j\Omega_0}$$

and Eqn. (5.161) can be written as

$$x[n] = \frac{1}{2}\, z_0^n + \frac{1}{2}\left(z_0^*\right)^n \tag{5.162}$$

The response of the system is determined using the linearity of the system:

$$y[n] = \text{Sys}\left\{\frac{1}{2}\, z_0^n + \frac{1}{2}\left(z_0^*\right)^n\right\} = \frac{1}{2}\,\text{Sys}\left\{z_0^n\right\} + \frac{1}{2}\,\text{Sys}\left\{\left(z_0^*\right)^n\right\}$$

We already know that the response of the system to the term z_0^n is

$$\text{Sys}\left\{z_0^n\right\} = z_0^n\, H(z_0)$$

and its response to the term $\left(z_0^*\right)^n$ is

$$\text{Sys}\left\{\left(z_0^*\right)^n\right\} = \left(z_0^*\right)^n\, H\left(z_0^*\right)$$

so that the output signal $y[n]$ can be written as

$$y[n] = \frac{1}{2}\, z_0^n\, H(z_0) + \frac{1}{2}\left(z_0^*\right)^n\, H\left(z_0^*\right) \tag{5.163}$$

The result obtained in Eqn. (5.163) can be further simplified. Let the value of the system function evaluated at the point $z = z_0$ be written in polar complex form as

$$H(z_0) = H_0\, e^{j\Theta_0} \tag{5.164}$$

where H_0 and Θ_0 represent the magnitude and the phase of the system function at the point $z = z_0$ respectively:

$$H_0 = \left|H(z_0)\right| \tag{5.165}$$

and

$$\Theta_0 = \measuredangle H(z_0) \tag{5.166}$$

It can be shown (see Problem 5.34 at the end of this chapter) that, for a system with a real-valued impulse response, the value of the system function at the point $z = z_0^*$ is the complex conjugate of its value at the point $z = z_0$, that is,

$$H\left(z_0^*\right) = \left[H(z_0)\right]^* = H_0\, e^{-j\Theta_0} \tag{5.167}$$

Using Eqns. (5.164) and (5.167) in Eqn. (5.163), the output signal $y[n]$ becomes

$$\begin{aligned}
y[n] =& \frac{1}{2}\, z_0^n\, H_0\, e^{j\Theta_0} + \frac{1}{2}\left(z_0^*\right)^n\, H_0\, e^{-j\Theta_0} \\
=& \frac{1}{2}\left(r_0\, e^{j\Omega_0}\right)^n\, H_0\, e^{j\Theta_0} + \frac{1}{2}\left(r_0\, e^{-j\Omega_0}\right)^n\, H_0\, e^{-j\Theta_0} \\
=& \frac{1}{2}\, r_0^n\, H_0\left[e^{j(\Omega_0 n+\Theta_0)} + e^{-j(\Omega_0 n+\Theta_0)}\right] \\
=& H_0\, r_0^n\, \cos\left(\Omega_0 n + \Theta_0\right)
\end{aligned} \tag{5.168}$$

Comparison of the input signal in Eqn. (5.160) and the output signal Eqn. (5.168) reveals the following:

1. The amplitude of the signal is multiplied by the magnitude of the system function evaluated at the point $z = r_0\, e^{j\Omega}$.
2. The phase of the cosine function is incremented by an amount equal to the phase of the system function evaluated at the point $z = r_0\, e^{j\Omega}$.

Example 5.36: Response of a DTLTI system to an exponentially damped sinusoid

Consider again the system function used in Example 5.35. determine the response of the system to the input signal

$$x[n] = (0.9)^n \cos(0.2\pi n)$$

Solution: In Example 5.35 we have evaluated the system function at the point $z = 0.9\,e^{j0.2\pi}$ and have obtained

$$H\left(0.9\,e^{j0.2\pi}\right) = H_0\,e^{j\Theta_0} = 4.7526\,e^{-j\,1.7963}$$

Using Eqn. (5.168) with $H_0 = 4.7526$ and $\Theta_0 = -1.7963$ radians, we obtain the output signal as

$$y[n] = 4.7536\,(0.9)^n \cos(0.2\pi n - 1.7963)$$

The input and the output signals are shown in Fig. 5.32a,b.

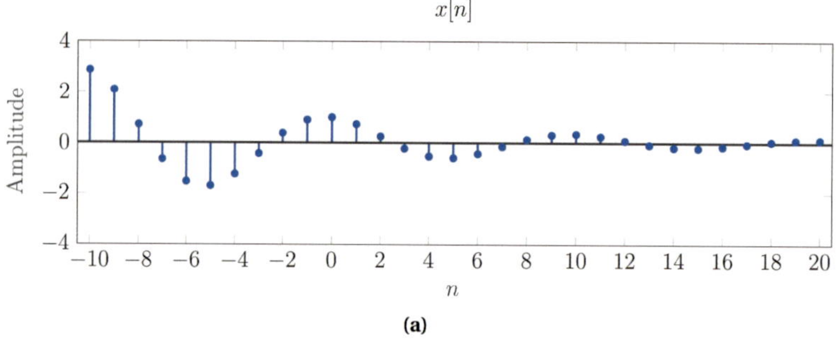

(a)

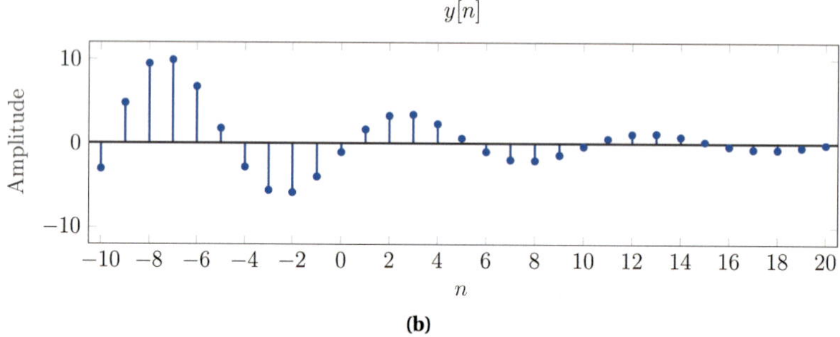

(b)

Figure 5.32 – **(a)** The input signal and **(b)** the output signal for Example 5.36.

Software resource: exdt_5_36.m

5.5.4 Graphical interpretation of the pole-zero plot

It was discussed earlier in this chapter that the complex variable z can be represented as a point in the z-plane. Alternatively, a complex number can be thought of as a vector in the complex plane (see Appendix A), drawn from the origin to the point of interest. Let $z = r\,e^{j\Omega}$. Fig. 5.33 illustrates the use of a vector with norm (or length) equal to r and angle equal to Ω for representing z.

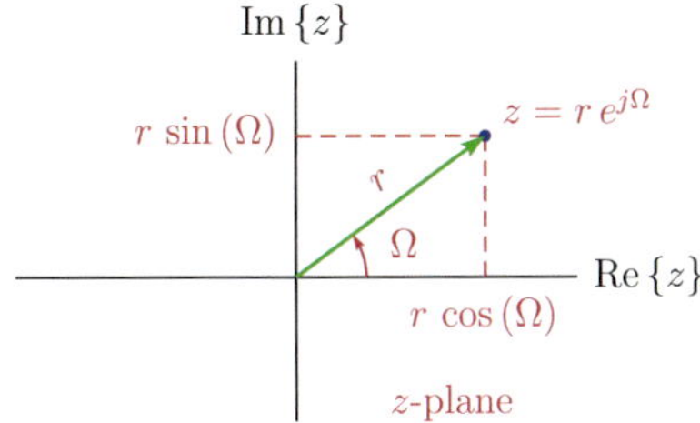

Figure 5.33 – Vector representation of the complex variable $z = r\,e^{j\Omega}$.

Using vector notation we will write

$$\overrightarrow{z} = r\,\overrightarrow{e^{j\Omega}} \tag{5.169}$$

The norm and the angle of the vector $\overrightarrow{z}$ are expressed as

$$\left|\overrightarrow{z}\right| = r \qquad \text{and} \qquad \measuredangle\overrightarrow{z} = \Omega$$

Fixing Ω and changing the value of r causes the tip of the vector to move either toward or away from the origin while the direction of the vector remains unchanged. Fixing r and changing the value of Ω causes the vector to rotate while its length remains unchanged. In this case the tip of the vector stays on a circle with radius r. Consider a fixed point z_1 in the z-plane. It will be interesting to find the vector $(\overrightarrow{z-z_1})$.

$$\begin{aligned}\overrightarrow{z-z_1} &= \overrightarrow{z} - \overrightarrow{z_1}\\ &= \overrightarrow{z} + (\overrightarrow{-z_1})\end{aligned} \tag{5.170}$$

Based on Eqn. (5.170) the vector $(\overrightarrow{z-z_1})$ can be found by adding the two vectors $\overrightarrow{z}$ and $(\overrightarrow{-z_1})$ using the parallelogram rule as shown in Fig. 5.34.

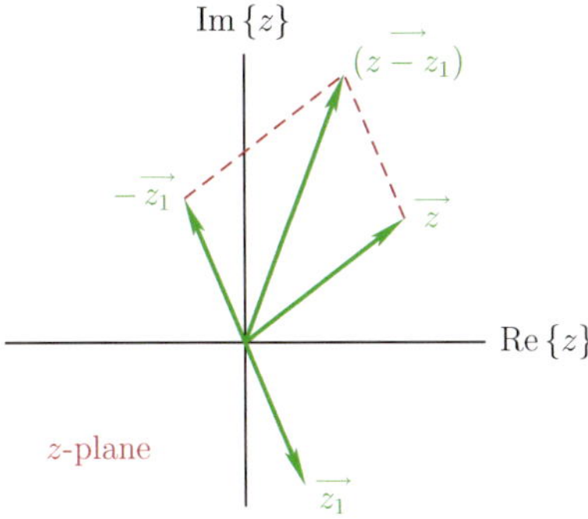

Figure 5.34 – Finding the vector representation for $(\overrightarrow{z-z_1})$.

Any vector is identified by two features: its norm (length) and its direction. The starting point of a vector is not important. Shifting a vector parallel to itself so that it starts at any desired point would be allowed as long as the norm and the direction of the vector are not changed. Accordingly, the vector $(\overrightarrow{z-z_1})$ found through the use of the parallelogram rule in Fig. 5.34 can be shifted so that it originates at the point z_1 and terminates at the point z as shown in Fig. 5.35.

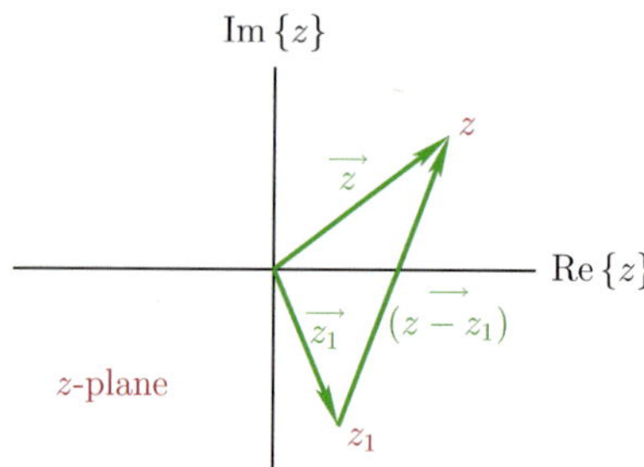

Figure 5.35 – Alternative representation of the vector for $(z - z_1)$.

The alternative representation shown in Fig. 5.35 leads us to an important conclusion:

In the z-plane, the vector $\overrightarrow{(z-z_1)}$ is drawn with an arrow that starts at the point z_1 and ends at the point z.

This conclusion will be critical in graphically interpreting the pole-zero diagram for a system function. We will begin by considering a simple DTLTI system with the impulse response

$$h[n] = a^n\, u[n]$$

and the corresponding z-domain system function

$$H(z) = \frac{z}{z-a}\,, \qquad \text{ROC:} \quad |z| > |a| \tag{5.171}$$

The system has a zero at $z = 0$ and a pole at $z = a$. In vector form, the system function can be written as the ratio of two vectors

$$\overrightarrow{H(z)} = \frac{\overrightarrow{z}}{\overrightarrow{(z-a)}}$$

Suppose we need to evaluate the system function at a specific point $z = z_a$ in the z-plane. Assuming that the point $z = z_a$ is within the ROC of the system function, the vector representation of the system function at the point of interest is

$$\overrightarrow{H(z_a)} = \frac{\overrightarrow{z_a}}{\overrightarrow{(z_a-a)}}$$

with its magnitude and phase computed as

$$\left|\overrightarrow{H(z_a)}\right| = \frac{\left|\overrightarrow{z_a}\right|}{\left|\overrightarrow{(z_a-a)}\right|}$$

and

$$\measuredangle \overrightarrow{H(z_a)} = \measuredangle \overrightarrow{z_a} - \measuredangle \overrightarrow{(z_a-a)}$$

Example 5.37: Evaluating $H(z)$ using vectors

Let a DTLTI system be characterized by the system function

$$H(z) = \frac{z}{z-0.8}\,, \qquad \text{ROC:} \quad |z| > 0.8$$

Discuss how the system function can be graphically evaluated at the points

a. $z_a = 1.5 + j1$

b. $z_b = -0.5 - j1$

Solution: Both points z_a and z_b are within the ROC of the system function. For $H(z_a)$ we need the vectors $\overrightarrow{z_a}$ and $\overrightarrow{(z_a - 0.8)}$ shown in Fig. 5.36.

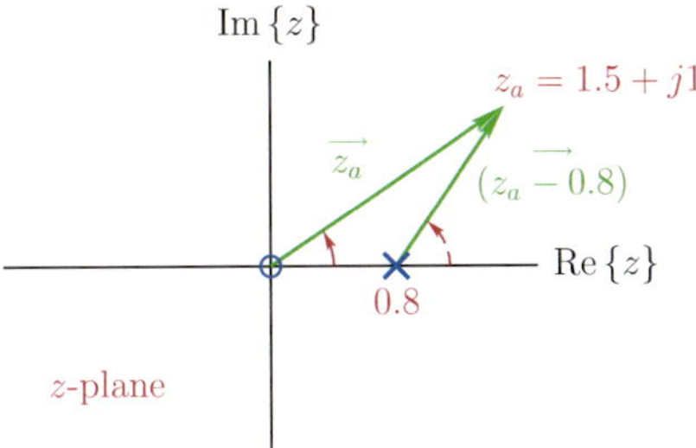

Figure 5.36 – Vectors for graphical representation of $H(z_a)$ in Example 5.37.

If the pole-zero diagram in Fig. 5.36 is drawn to scale, we could simply measure the lengths and the angles of the two vectors, and compute the magnitude and the phase of the system function at $z = z_a$ from those measurements as

$$\left|\overrightarrow{H(z_a)}\right| = \frac{\left|\overrightarrow{z_a}\right|}{\left|\overrightarrow{(z_a - 0.8)}\right|}$$

and

$$\measuredangle\overrightarrow{H(z_a)} = \measuredangle\overrightarrow{z_a} - \measuredangle\overrightarrow{(z_a - 0.8)}$$

It can be shown that

$$\left|\overrightarrow{z_a}\right| = 1.8028\,, \qquad \measuredangle\overrightarrow{z_a} = 0.5880 \text{ rad} = 33.7^o$$

$$\left|\overrightarrow{(z_a - 0.8)}\right| = 1.2207\,, \qquad \measuredangle\overrightarrow{(z_a - 0.8)} = 0.9601 \text{ rad} = 55^o$$

The magnitude and the phase of the system function at $z_a = 1.5 + j1$ are

$$\left|\overrightarrow{H(1.5 + j1)}\right| = \frac{1.8028}{1.2207} = 1.4769$$

and

$$\measuredangle\overrightarrow{H(1.5 + j1)} = 0.5880 - 0.9601 = -0.3721 \text{ rad} = -21.3^o$$

To evaluate $H(z_b)$ we need the vectors $\overrightarrow{z_b}$ and $\overrightarrow{(z_b - 0.8)}$ as shown in Fig. 5.37.

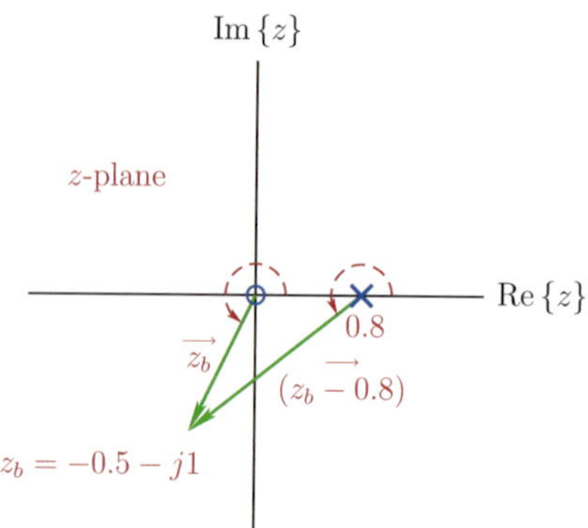

Figure 5.37 – Vectors for graphical representation of $H(z_b)$ in Example 5.37.

With these new vectors it can be shown that

$$\left|\overrightarrow{z_b}\right| = 1.1180\,, \qquad \measuredangle\overrightarrow{z_b} = 4.2487 \text{ rad} = 243.4^o$$

$$\left|\overrightarrow{(z_b - 0.8)}\right| = 1.6401\,, \qquad \measuredangle\overrightarrow{(z_b - 0.8)} = 3.7973 \text{ rad} = 217.6^o$$

The magnitude and the phase of the system function at $z_a = 1.5 + j1$ are

$$\left|\overrightarrow{H(-0.5 - j1)}\right| = \frac{1.1180}{1.6401} = 0.6817$$

and

$$\measuredangle\overrightarrow{H(-0.5 - j1)} = 4.2487 - 3.7973 = 0.4515 \text{ rad} = 25.8^o$$

Vector representation of the factors that make up the system function $H(z)$ is particularly useful for inferring the frequency response of a system from the distribution of its z-domain poles and zeros. It was shown earlier that the DTFT-based frequency response $H(\Omega)$ of a DTLTI system can be obtained from the z-domain system function by evaluating $H(z)$ at each point on the unit circle of the z-plane. Continuing with the first-order system function given by Eqn. (5.171) we obtain

$$H(\Omega) = H(z)\Big|_{z=e^{j\Omega}} = \frac{e^{j\Omega}}{e^{j\Omega} - a} \tag{5.172}$$

In using Eqn. (5.172) we are assuming that $|a| < 1$ so that the ROC for the system function includes the unit circle, and therefore the frequency response $H(\Omega)$ of the system exists. Fig. 5.38 shows the two vectors

$$\overrightarrow{A} = \overrightarrow{e^{j\Omega}}$$

and

$$\overrightarrow{B} = \overrightarrow{\left(e^{j\Omega} - a\right)}$$

that are needed for evaluating $H(\Omega)$. The vectors originate from the zero and the pole and terminate at the point $z = e^{j\Omega}$ on the unit circle.

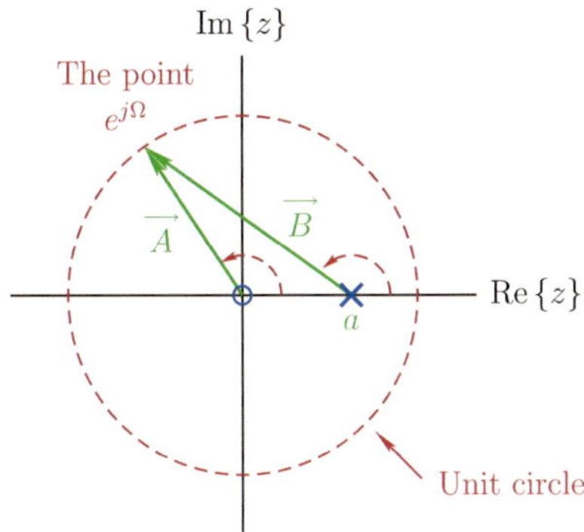

Figure 5.38 – Vectors from the zero and the pole of the system function $H(z) = z/(z-a)$ to a point on the unit circle.

As Ω is varied, the termination point moves on the unit circle, causing the lengths and the angles of the vectors to change, and therefore causing the frequency response $H(\Omega)$ to vary as a function of Ω.

Now consider a more general system function in the form

$$H(z) = K\,\frac{(z-z_1)\,(z-z_2)\ldots(z-z_M)}{(z-p_1)\,(z-p_2)\ldots(z-p_N)} \tag{5.173}$$

The system has M zeros at $z_1, z_2, \ldots, z_M$ and N poles at $p_1, p_2, \ldots, p_N$. Suppose we would like to evaluate the system function at the point $z = z_a$ within the ROC. Vector representation of $H(z_a)$ is

$$\overrightarrow{H(z_a)} = K\,\frac{\overrightarrow{(z_a-z_1)}\,\overrightarrow{(z_a-z_2)}\ldots\overrightarrow{(z_a-z_M)}}{\overrightarrow{(z_a-p_1)}\,\overrightarrow{(z_a-p_2)}\ldots\overrightarrow{(z_a-p_N)}} \tag{5.174}$$

The magnitude of the system function is

$$\left|H(z_a)\right| = K\,\frac{\left|\overrightarrow{(z_a-z_1)}\right|\left|\overrightarrow{(z_a-z_2)}\right|\ldots\left|\overrightarrow{(z_a-z_M)}\right|}{\left|\overrightarrow{(z_a-p_1)}\right|\left|\overrightarrow{(z_a-p_2)}\right|\ldots\left|\overrightarrow{(z_a-p_N)}\right|} \tag{5.175}$$

and its phase is

$$\begin{aligned}\measuredangle H(z_a) =& \measuredangle\overrightarrow{(z_a-z_1)} + \measuredangle\overrightarrow{(z_a-z_2)} + \ldots + \measuredangle\overrightarrow{(z_a-z_M)} \\ & - \measuredangle\overrightarrow{(z_a-p_1)} - \measuredangle\overrightarrow{(z_a-p_2)} - \ldots - \measuredangle\overrightarrow{(z_a-p_N)}\end{aligned} \tag{5.176}$$

The vector-based graphical method described above is useful for understanding the correlation between pole-zero placement and system behavior. In order to build intuition in this regard, Fig. 5.39 illustrates several pole-zero layouts along with the magnitude spectrum that corresponds to each layout.

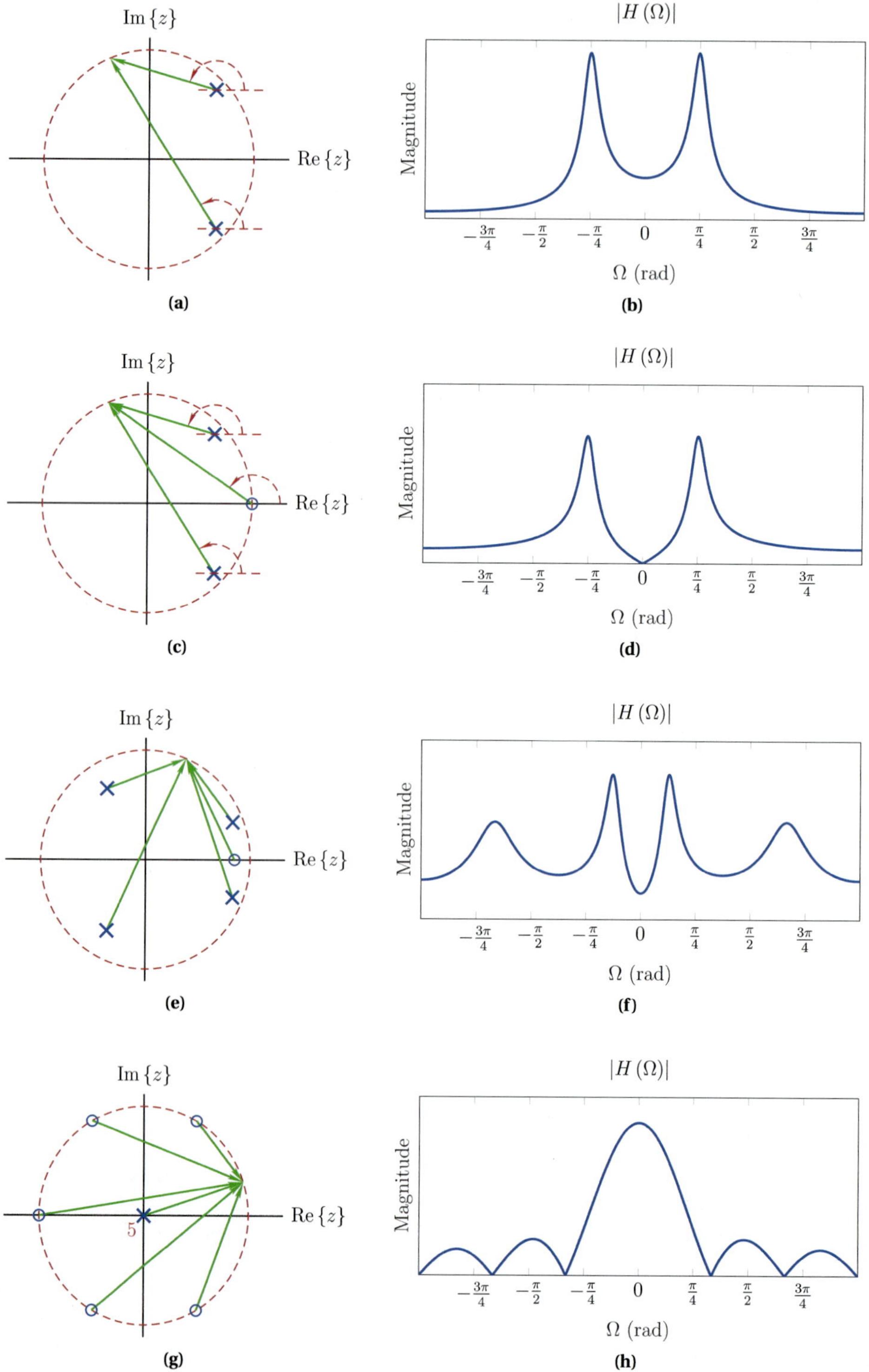
Im {z}
Re {z}
|H (Ω)|
Magnitude
−3π/4
−π/2
−π/4
0
π/4
π/2
3π/4
Ω (rad)
(a)
(b)
(c)
(d)
(e)
(f)
5
(g)
(h)

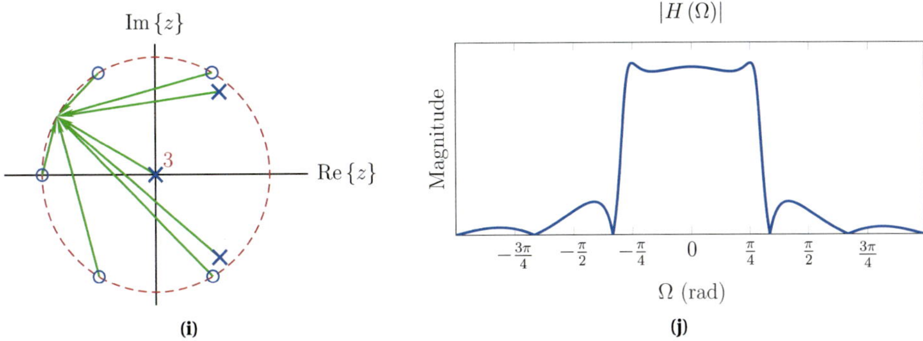

Figure 5.39 – Several pole-zero configurations and the corresponding magnitude responses.

Interactive App: Pole-zero explorer

The interactive app in `appPoleZeroDT.m` allows experimentation with the placement of poles and zeros of the system function. Before using it, two vectors should be created in MATLAB workspace: one containing the poles of the system function and the other one containing its zeros. In the pole-zero explorer user interface, the "import" button is used for importing these vectors. Pole-zero layout in the z-plane is displayed along with the magnitude and the phase of the system function. The vectors from each zero and each pole to a point on the unit circle of the z-plane may optionally be displayed. Individual poles and zeros may be moved, and the effects of movements on magnitude and phase characteristics may be observed. Complex conjugate poles and zeros move together to keep the conjugate relationship.

For example, a system function with zeros at $z_1 = 0$, $z_2 = 1$ and poles at $p_1 = 0.5 + j0.7$, $p_2 = 0.5 - j0.7$ may be studied by creating the following vectors in MATLAB workspace and importing them into the pole-zero explorer program:

```
>> zrs = [0,1];
>> pls = [0.5+j*0.7,0.5-j*0.7];
```

Alternatively, any of the scenarios illustrated in Fig. 5.39 can be recreated using preset layouts #1 through #5.

a. Select preset scenario #1 to match part (a) of Fig. 5.39. Select one of the two poles and change its angle. Observe both poles moving in synchrony to maintain their conjugate relationship. Pay attention to the changes occurring in the magnitude of the system function as the angles of the poles are changed.

b. Reset the angles of the poles back to what they were at the beginning. Now change the radii of each pole, bringing the poles closer to the unit circle of the z-plane. Observe the changes in the magnitude of the system function.

c. Change the radii in the other direction, moving the poles away from the unit circle. Observe the effects on the magnitude response.

d. Repeat above steps with the other parts of Fig. 5.39.

Software resource: `appPoleZeroDT.m`

Software resource: See MATLAB Exercises 5.7 and 5.8.

5.5.5 System function and causality

Causality in linear and time-invariant systems was discussed in Section 2.8 of Chapter 2. For a DTLTI system to be causal, its impulse response $h[n]$ needs to be equal to zero for $n < 0$. Thus, by changing the lower limit of the summation index to $n = 0$ in the definition of the z-transform, the z-domain system function for a causal DTLTI system can be written as

$$H(z) = \sum_{n=-\infty}^{\infty} h[n]\, z^{-n} = \sum_{n=0}^{\infty} h[n]\, z^{-n}$$

As discussed in Section 5.2 of this chapter, the ROC for the system function of a causal system is the outside of a circle in the z-plane. Consequently, the system function must also converge at $|z| \to \infty$. Consider a system function in the form

$$H(z) = \frac{B(z)}{A(z)} = \frac{b_M z^M + b_{M-1} z^{M-1} + \ldots + b_1 z + b_0}{a_N z^N + a_{N-1} z^{N-1} + \ldots + a_1 z + a_0}$$

For the system described by $H(z)$ to be causal we need

$$\lim_{z\to\infty} \left[H(z) \right] = \lim_{z\to\infty} \left[\frac{b_M}{a_N}\, z^{(M-N)} \right] < \infty \tag{5.177}$$

which requires that $M - N \le 0$ and consequently $M \le N$. Thus we arrive at an important conclusion:

Causality condition:

In the z-domain transfer function of a causal DTLTI system the order of the numerator must not be greater than the order of the denominator.

Note that this condition is necessary for a system to be causal, but it is not sufficient. It is also possible for a non-causal system to have a system function with $M \le N$.

5.5.6 System function and stability

In Section 2.9 of Chapter 2 we concluded that for a DTLTI system to be stable its impulse response must be absolute summable, that is,

$$\sum_{n=-\infty}^{\infty} \left| h[n] \right| < \infty$$

Furthermore, we have established in Section 3.3.3 of Chapter 3 that the DTFT of a signal exists if the signal is absolute summable. But the DTFT of the impulse response is equal to the z-domain transfer function evaluated on the unit circle of the z-plane, that is,

$$H(\Omega) = H(z)\Big|_{z=e^{j\Omega}}$$

provided that the unit circle of the z-plane is within the ROC.

Stability condition:

It follows that, for a DTLTI system to be stable, the ROC of its z-domain system function must include the unit circle.

What are the corresponding conditions that must be imposed on the locations of poles and zeros for stability? We will answer this question by taking three-separate cases into account:

1. Causal system:
   The ROC for the system function of a causal system is the outside of a circle with radius r_1, and is expressed in the form
   $$|z| > r_1$$
   For the ROC to include the unit circle, we need $r_1 < 1$. Since the ROC cannot have any poles in it, all the poles of the system function must be on or inside the circle with radius r_1.

   For a causal system to be stable, the system function must not have any poles on or outside the unit circle of the z-plane.

2. Anti-causal system:
   If the system is anti-causal, its impulse response is equal to zero for $n \geq 0$ and the ROC for the system function is the inside of a circle with radius r_2, expressed in the form
   $$|z| < r_2$$
   For the ROC to include the unit circle, we need $r_2 > 1$. Also, all the poles of the system function must reside on or outside the circle with radius r_2 since there can be no poles within the ROC.

   For an anti-causal system to be stable, the system function must not have any poles on or inside the unit circle of the z-plane.

3. Neither causal nor anti-causal system:
   In this case the ROC for the system function, if it exists, is the region between two circles with radii r_1 and r_2, and is expressed in the form
   $$r_1 < |z| < r_2$$
   The poles of the system function may be either

   a. on or inside the circle with radius r_1, or
   b. on or outside the circle with radius r_2.

and the ROC must include the unit circle.

Example 5.38: Impulse response of a stable system

A stable system is characterized by the system function

$$H(z) = \frac{z(z+1)}{(z-0.8)(z+1.2)(z-2)}$$

Determine the impulse response of the system.

Solution: The ROC for the system function is not directly stated, however, we are given enough information to deduce it. The poles of the system are at $p_1 = 0.8$, $p_2 = -1.2$, and $p_3 = 2$. Since the system is known to be stable, its ROC must include the unit circle. The only possible choice is

$$0.8 < |z| < 1.2$$

Partial fraction expansion of $H(z)$ is

$$H(z) = -\frac{0.75\,z}{z-0.8} - \frac{0.0312\,z}{z+1.2} + \frac{0.7813\,z}{z-2}$$

Based on the ROC determined above, the first partial fraction corresponds to a causal signal, and the other two correspond to anti-causal signals. The impulse response of the system is

$$h[n] = -0.75\,(0.8)^n\,u[n] + 0.0312\,(-1.2)^n\,u[-n-1] - 0.7813\,(2)^n\,u[-n-1]$$

which is shown in Fig. 5.40. We observe that $h[n]$ tends to zero as the sample index is increased in both directions, consistent with the fact that $h[n]$ must be absolute summable.

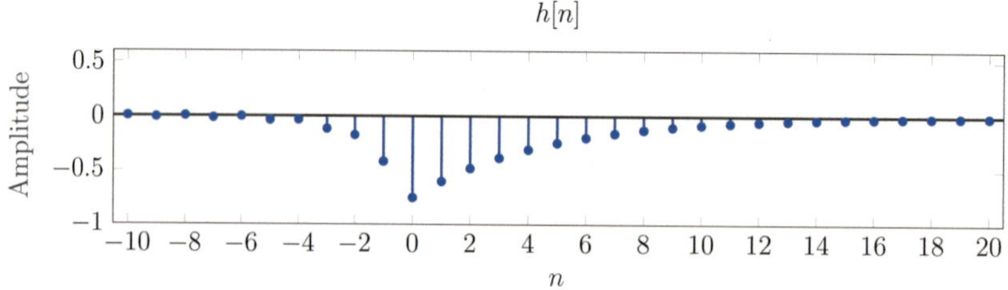

Figure 5.40 – Impulse response $h[n]$ for Example 5.38.

Software resources: exdt_5_38a.m , exdt_5_38b.m

Example 5.39: Stability of a system described by a difference equation

A DTLTI system is characterized by the difference equation

$$y[n] = x[n-1] + 3x[n-2] + 2x[n-3] + 2.3y[n-1] - 2y[n-2] + 1.2y[n-3]$$

Comment on the stability of this system.

Solution: As in the previous example, the ROC for the system function is not directly stated. On the other hand, the difference equation of the system is written with the current output sample $y[n]$ on the left of the equal sign. All terms on the right side of the equal sign are past samples of the input and the output signals. The form of the difference equation suggests that the system is causal. When we find the system function we must choose the ROC in a way that includes infinitely large values of $|z|$.

Rearranging terms of the difference equation we obtain

$$y[n]-2.3y[n-1]+2y[n-2]-1.2y[n-3]=x[n-1]+3x[n-2]+2x[n-3]$$

Taking the z-transform of each side yields

$$\left(1-2.3z^{-1}+2z^{-2}-1.2z^{-3}\right)Y(z)=\left(z^{-1}+3z^{-2}+2z^{-3}\right)X(z)$$

Finally, solving for the ratio of $Y(z)$ and $X(z)$ we find the system function as

$$H(z)=\frac{Y(z)}{X(z)}=\frac{z^{-1}+3z^{-2}+2z^{-3}}{1-2.3z^{-1}+2z^{-2}-1.2z^{-3}}$$

Let us multiply both the numerator and the denominator of $H(z)$ by z^3 so that we express it using powers of z:

$$H(z)=\frac{z^2+3z+2}{z^3-2.3z^2+2z-1.2}$$

Numerator and denominator polynomials can be factored to yield

$$H(z)=\frac{(z+1)(z+2)}{\left(z-0.4+j0.8\right)\left(z-0.4-j0.8\right)(z-1.5)}$$

The pole-zero diagram for the system function is shown in Fig. 5.41. There are two zeros, one at $z=-1$ and one $z=-2$. The system function also has a pole at $z=1.5$ and a conjugate pair of poles at $z=0.4\pm j0.8$.

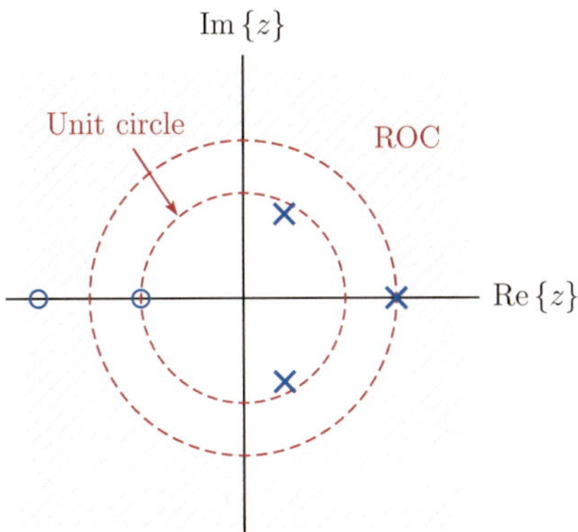

Figure 5.41 – Pole-zero diagram and ROC for the system of Example 5.39.

Since the system is known to be causal, its ROC must be the outside of a circle. The radius of the boundary circle is determined by the outermost pole of the system which happens to be at $z=1.5$. Consequently, the ROC for the system function is

$$|z|>1.5$$

Since the ROC does not include the unit circle, the system is unstable. This can be verified by computing the first few samples of the impulse response $h[n]$ from the system function using long division which yields

$$h[n] = \{ \underset{\underset{n=1}{\uparrow}}{1}, 5.3, 12.19, 18.64, 24.85, 34.5, 52.02, 80.46, 122.42, 183.07, \ldots \}$$

The impulse response keeps growing with increasing values of the index n.

Software resources: exdt_5_39a.m , exdt_5_39b.m , exdt_5_39c.m

5.5.7 Comb filters

Comb filters are used as building blocks in a variety of discrete-time system applications. Two examples of comb filters were used in MATLAB Exercises 2.13 and 3.6 in Chapter 2 and Chapter 3 respectively for generating echoes and reverberation. The *feed-forward comb filter* has the difference equation

$$y[n] = x[n] - r\, x[n-L] \tag{5.178}$$

which can be implemented using the block diagram shown in Fig. 5.42. The notation D_L represents a delay of L samples.

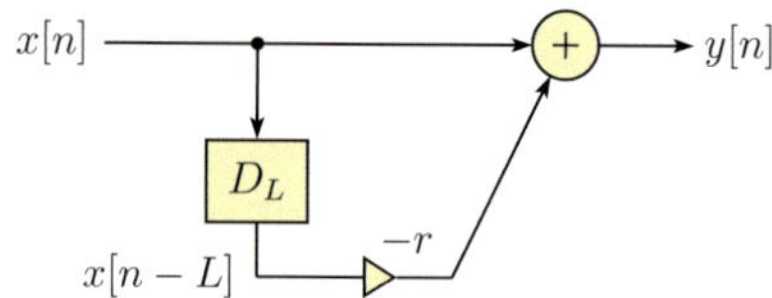

Figure 5.42 – A block diagram for implementing the feed-forward comb filter with the difference equation given by Eqn. (5.178).

System function for the feed-forward comb filter can be found by using the technique detailed in Section 5.5.1:

$$Y(z) = X(z) - r\, z^{-L}\, X(z) \tag{5.179}$$

$$H_{ff}(z) = \frac{Y(z)}{X(z)} = 1 - r\, z^{-L} = \frac{z^L - r}{z^L} \tag{5.180}$$

The ROC for the system function is the entire z-plane with the exception of $z = 0 + j0$. The corresponding impulse response is

$$h_{ff}[n] = \delta[n] - r\, \delta[n-L]$$

The zeros of $H(z)$ are the solutions of

$$z^L = r \tag{5.181}$$

Since a total of L roots are needed, we will write Eqn. (5.181) as

$$z^L = r\, e^{j2\pi k} \tag{5.182}$$

which leads to the set of solutions

$$z_k = r^{1/L}\, e^{j2\pi k/L}, \quad k = 0, \ldots, L-1 \tag{5.183}$$

In addition to the L zeros found, the feed-forward comb filter has L poles, all at the origin. The pole-zero diagram is shown in Fig. 5.43 for the case $L = 10$. The magnitude of the system function is shown in Fig. 5.44.

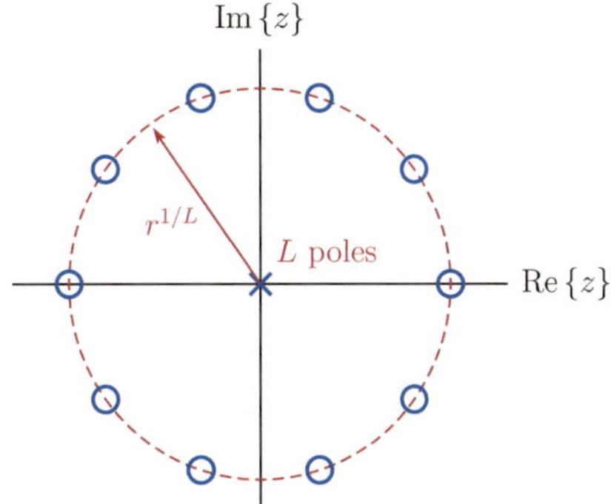

Figure 5.43 – Pole-zero diagram for feed-forward comb filter of order $L = 10$.

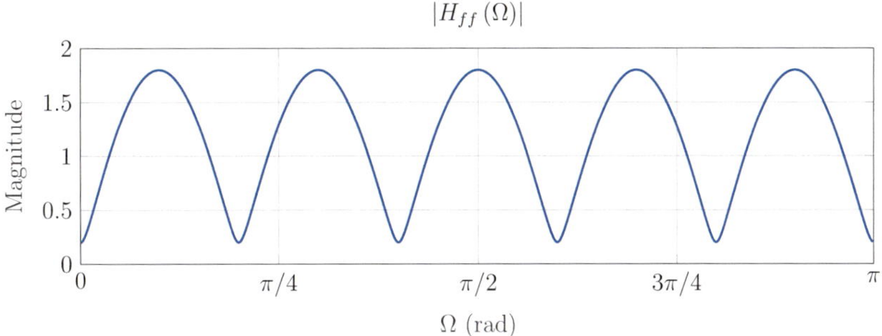

Figure 5.44 – Magnitude of the system function for the feed-forward comb filter of order $L = 10$ with $r = 0.8$.

An alternative to the feed-forward comb filter is the *feedback comb filter* defined by the difference equation

$$y[n] = x[n] + r\,y[n-L] \tag{5.184}$$

which can be implemented using the block diagram shown in Fig. 5.45.

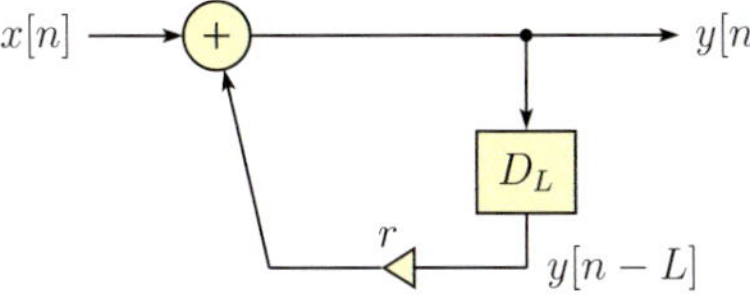

Figure 5.45 – A block diagram for implementing the feedback comb filter with the difference equation given by Eqn. (5.184).

System function for the feedback comb filter is

$$H_{fb}(z) = \frac{1}{1 - r\,z^{-L}} = \frac{z^L}{z^L - r} \tag{5.185}$$

and the ROC is outside the circle with radius $r^{1/L}$. For the system to be stable the ROC should include the unit circle. Thus, we need $|r| < 1$. The system function $H_{fb}(z)$ has L zeros at the origin. Its poles are

$$p_k = r^{1/L}\, e^{j2\pi k/L}, \quad k = 0, \ldots, L-1$$

Pole-zero diagram for the feedback comb filter of order $L = 10$ is shown in Fig. 5.46. The magnitude of the system function is shown in Fig. 5.47.

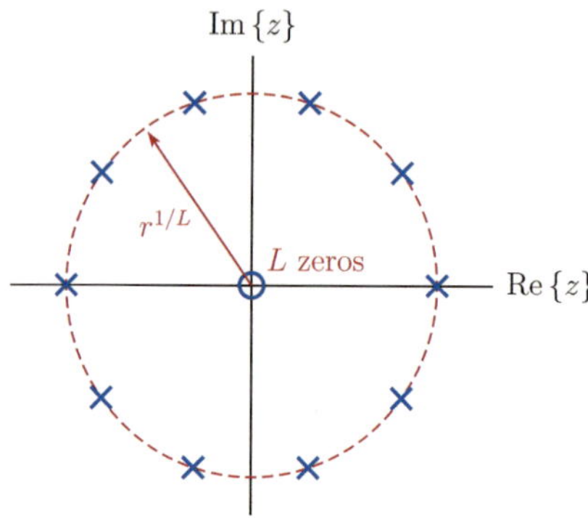

Figure 5.46 – Pole-zero diagram for feedback comb filter of order $L = 10$.

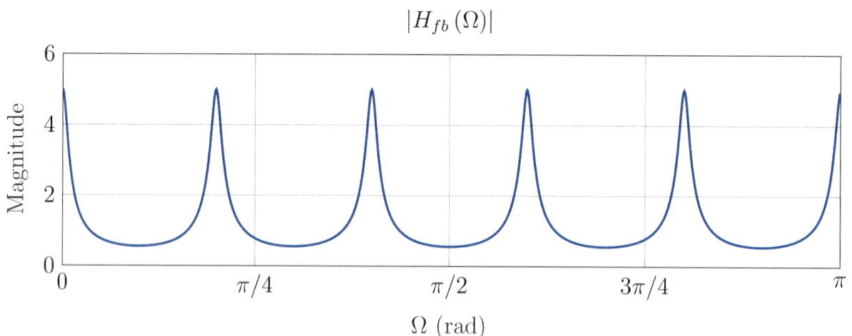

Figure 5.47 – Magnitude of the system function for the feedback comb filter of order $L = 10$ with $r = 0.8$.

5.5.8 All-pass filters

A system the magnitude characteristic of which is constant across all frequencies is called an *all-pass filter*. For a system to be considered an all-pass filter, we need

$$\left|H(\Omega)\right| = C \text{ (constant)} \qquad -\pi \le \Omega \le \pi \tag{5.186}$$

Consider a first-order DTLTI system with a complex zero at

$$z_1 = a = r\, e^{j\Omega_0} \tag{5.187}$$

and a complex pole at

$$p_1 = \frac{1}{a^*} = \frac{1}{r\, e^{-j\Omega_0}} = \frac{1}{r}\, e^{j\Omega_0} \tag{5.188}$$

The resulting system function is

$$H(z) = \frac{z - z_1}{z - p_1} = \frac{z - r\, e^{j\Omega_0}}{z - (1/r)\, e^{j\Omega_0}}$$

The choice made for the zero and the pole ensures that the two lie on a line that goes through the origin. Their distances from the origin are r and $1/r$ respectively. Since it is desirable to have a system that is both causal and stable, we need to choose $r > 1$ so that the resulting pole is inside the unit circle. Naturally this choice causes the zero to be outside the unit circle. This is illustrated in Fig. 5.48.

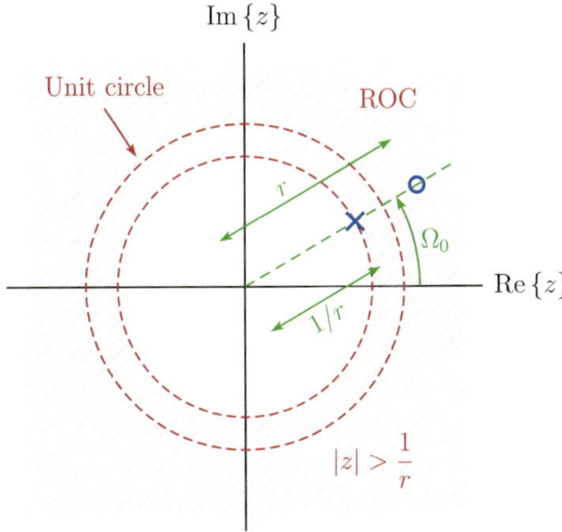

Figure 5.48 – The zero and the pole of a first-order all-pass filter.

A pole and a zero that have the relationship shown in Fig. 5.48 are said to be *mirror images* of each other across the unit circle. To find $H(\Omega)$ we will evaluate the z-domain system function on the unit circle of the z-plane:

$$H(\Omega) = \frac{e^{j\Omega} - r\, e^{j\Omega_0}}{e^{j\Omega} - (1/r)\, e^{j\Omega_0}}$$

For convenience we will determine the squared magnitude of $H(\Omega)$:

$$\begin{aligned} |H(\Omega)|^2 =& H(\Omega)\, H^*(\Omega) \\ =& \left(\frac{e^{j\Omega} - r\, e^{j\Omega_0}}{e^{j\Omega} - (1/r)\, e^{j\Omega_0}}\right)\left(\frac{e^{-j\Omega} - r\, e^{-j\Omega_0}}{e^{-j\Omega} - (1/r)\, e^{-j\Omega_0}}\right) \\ =& \frac{1 + r^2 - 2r\cos(\Omega - \Omega_0)}{1 + (1/r)^2 - 2(1/r)\cos(\Omega - \Omega_0)} \\ =& r^2 \;\; \text{(constant)}. \end{aligned}$$

The magnitude of the system function is $|H(\Omega)| = r$. It can be shown (see Problem 5.40 at the end of this chapter) that the phase of the system function is

$$\measuredangle H(\Omega) = \tan^{-1}\left[\frac{\left(r^2 - 1\right)\sin(\Omega - \Omega_0)}{2r - \left(r^2 + 1\right)\cos(\Omega - \Omega_0)}\right] \tag{5.189}$$

The phase characteristic $\measuredangle H(\Omega)$ is shown in Fig. 5.49 for the case of a pole and a zero both on the real axis, that is, for $\Omega_0 = 0$. The phase is shown for $r = 2.5$, $r = 1.75$ and $r = 1.2$.

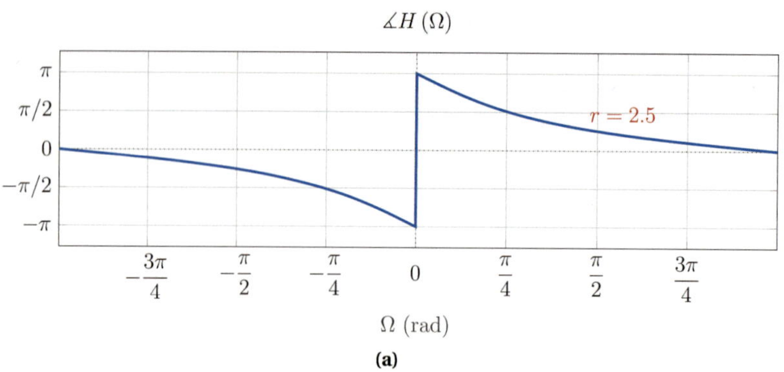

(a)

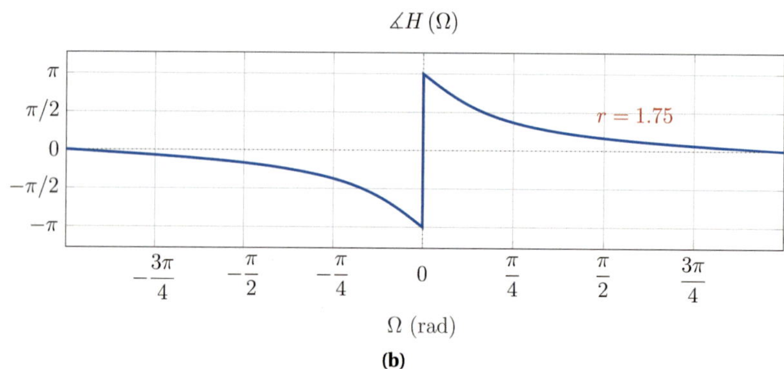

(b)

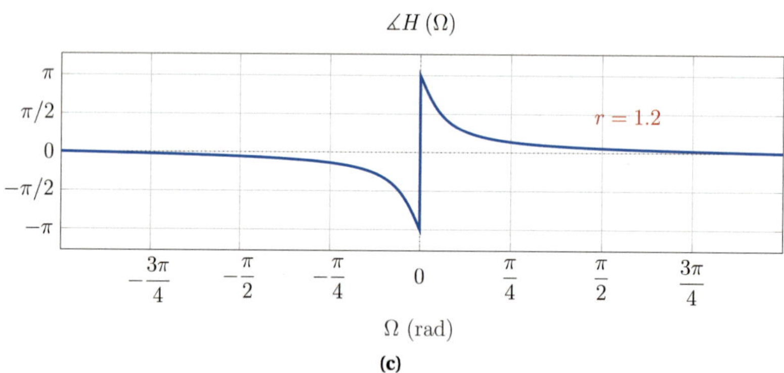

(c)

Figure 5.49 – The phase characteristic of the first-order all-pass filter for $\Omega_0 = 0$ and **(a)** $r = 2.5$, **(b)** $r = 1.75$, and **(c)** $r = 1.2$.

Since the shape of the phase response can be controlled by the choice of r while keeping the magnitude response constant, an all-pass filter is also referred to as a *phase-shifter*. Increased versatility in controlling the phase response can be obtained by choosing the pole and the zero to be complex-valued. This requires $\Omega_0 \neq 0$. Consider a first-order system with a pole at $p_1 = (1/r)\, e^{j\Omega_0}$

and a zero at $z_1 = r\, e^{j\Omega_0}$. Again we choose $r > 1$ to obtain a system that is both causal and stable. Fig. 5.50 illustrates the phase response $\measuredangle H(\Omega)$ is shown for $\Omega_0 = \pi/6$ and $\Omega_0 = \pi/3$. The parameter r is fixed at $r = 1.2$ in both cases.

If a system function with real-valued coefficients is desired, two first-order all-pass sections with complex conjugate zeros and poles can be combined into a second-order all-pass filter. The resulting system function is

$$\begin{aligned} H(z) &= \left(\frac{z - r\, e^{j\Omega_0}}{z - (1/r)\, e^{j\Omega_0}}\right)\left(\frac{z - r\, e^{-j\Omega_0}}{z - (1/r)\, e^{-j\Omega_0}}\right) \\ &= \frac{z^2 - 2r\cos(\Omega_0)\, z + r^2}{z^2 - (2/r)\cos(\Omega_0)\, z + (1/r)^2} \end{aligned} \tag{5.190}$$

Pole-zero diagram for a second-order all-pass filter with parameters r and Ω_0 is shown in Fig. 5.51.

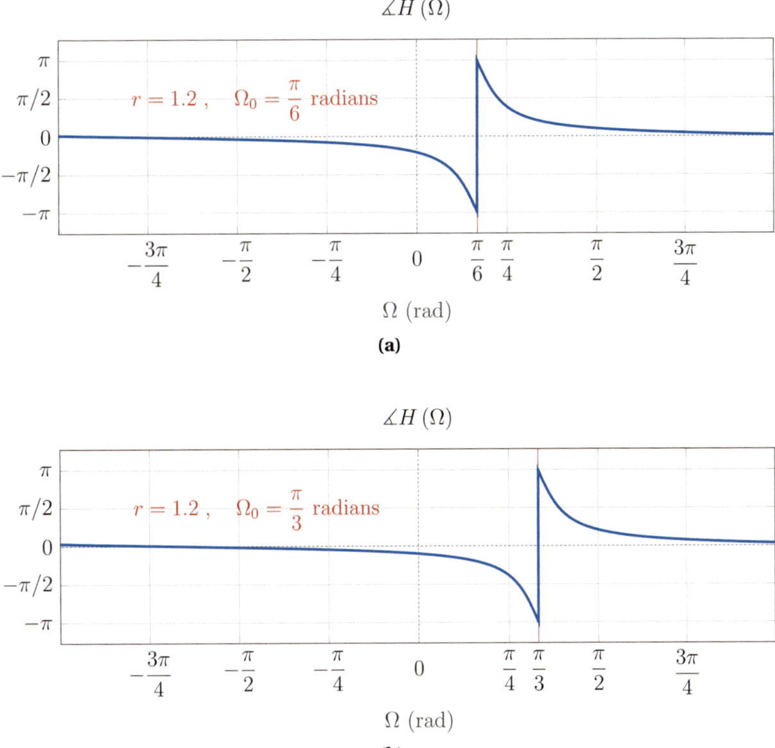

Figure 5.50 – The phase characteristic of the first-order all-pass filter for $r = 1.2$ and **(a)** $\Omega_0 = \pi/6$ and **(b)** $\Omega_0 = \pi/3$.

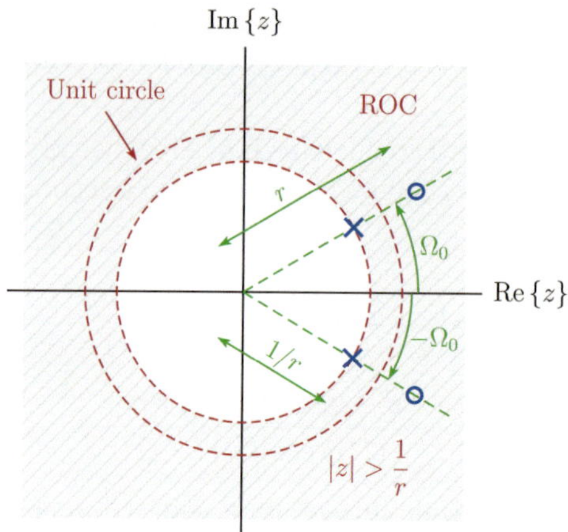

Figure 5.51 – Pole-zero diagram for second-order all-pass filter.

Higher order all-pass filters can be obtained by combining a feedback comb filter and a feed-forward comb filter. An all-pass filter of order L based on this idea is shown in Fig. 5.52.

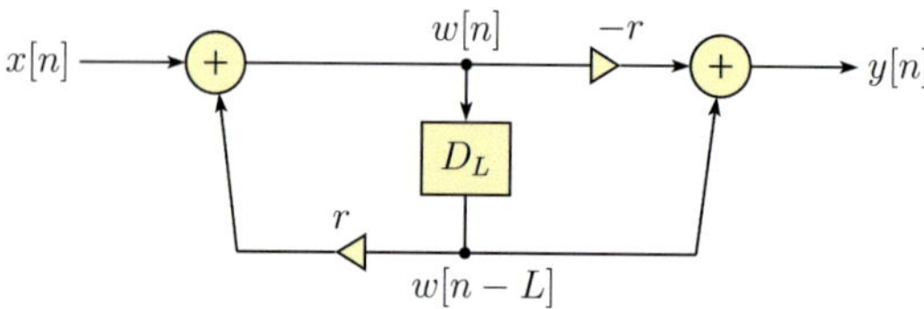

Figure 5.52 – Block diagram for an all-pass filter or order L constructed by combining feed-forward and feedback comb filters.

Its system function is

$$H(z) = \frac{z^{-L} - r}{1 - r\,z^{-L}} = \frac{1 - r\,z^{L}}{z^{L} - r} \tag{5.191}$$

which corresponds to the difference equation

$$y[n] = r\,y[n-L] - r\,x[n] + x[n-L] \tag{5.192}$$

For stability we need $|r| < 1$. Zeros of the system function are at

$$z_k = r^{-1/L}\,e^{j2\pi k/L}, \quad k = 0, \ldots, L-1$$

and its poles are at

$$p_k = r^{1/L}\,e^{j2\pi k/L}, \quad k = 0, \ldots, L-1$$

The pole-zero diagram for $H(z)$ for the all-pass filter of order $L = 10$ is shown in Fig. 5.53. The structure of $H(z)$ guarantees that every pole inside the unit circle is paired with a zero as its mirror image outside the unit circle. Phase response of the all-pass filter with $L = 10$ and $r = 0.8$ is illustrated in Fig. 5.54.

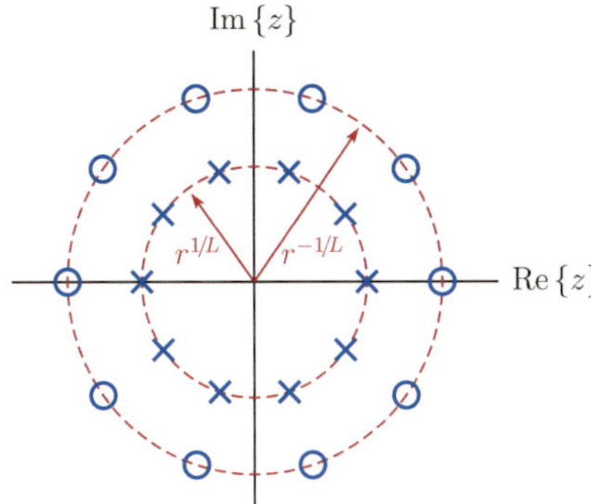

Figure 5.53 – Pole-zero diagram for all-pass filter of order $L = 10$.

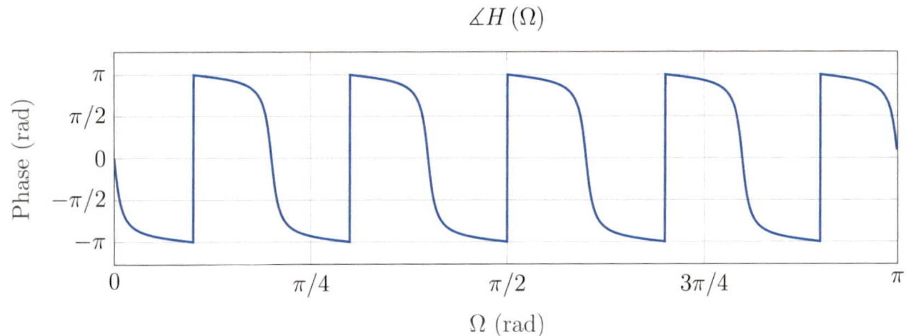

Figure 5.54 – Phase of the system function for an all-pass filter of order $L = 10$ with $r = 0.8$.

5.5.9 Inverse systems

The inverse of the system is another system which, when connected in cascade with the original system, forms an identity system. This relationship is depicted in Fig. 5.55.

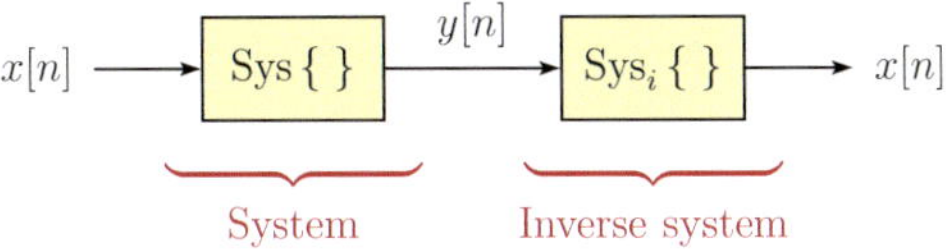

Figure 5.55 – A system and its inverse connected in cascade.

The output signal of the original system is

$$y[n] = \text{Sys}\{x[n]\} \tag{5.193}$$

We require the inverse system to recover the original input signal $x[n]$ from the output signal $y[n]$, therefore

$$x[n] = \text{Sys}_i\{y[n]\} \tag{5.194}$$

Combining Eqns. (5.193) and (5.194) yields

$$\text{Sys}_i\Big\{\text{Sys}\{x[n]\}\Big\} = x[n] \tag{5.195}$$

Let the original system under consideration be a DTLTI system with impulse response $h[n]$, and let the inverse system be also a DTLTI system with impulse response $h_i[n]$ as shown in Fig. 5.194. For the output signal of the inverse system to be identical to the input signal of the original system we need the impulse response of cascade combination to be equal to $\delta[n]$, that is,

$$h_{eq}[n] = h[n] * h_i[n] = \delta[n] \tag{5.196}$$

or, using the convolution summation

$$h_{eq}[n] = \sum_k h[k]\, h_i[n-k] = \delta[n] \tag{5.197}$$

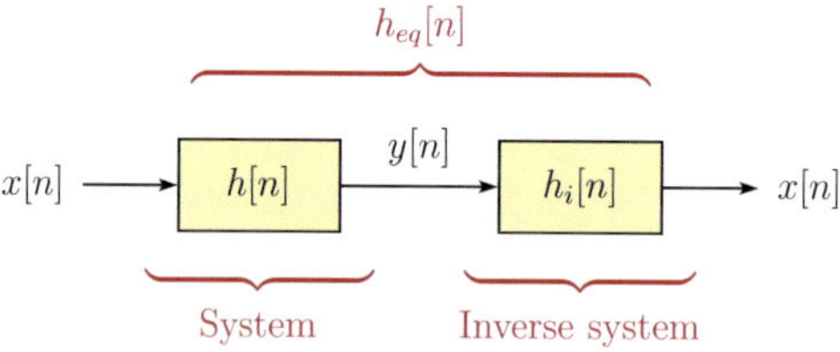

Figure 5.56 – A DTLTI system and its inverse connected in cascade.

Corresponding relationship between the system functions is found by taking the z-transform of Eqn. (5.196):

$$H_{eq}(z) = H(z)\, H_i(z) = 1 \tag{5.198}$$

Consequently, the system function of the inverse system is

$$H_i(z) = \frac{1}{H(z)} \tag{5.199}$$

Two important characteristics of the inverse system are causality and stability. We will first focus on causality. Consider again a rational system function in the standard form

$$H(z) = \frac{B(z)}{A(z)} = \frac{b_M z^M + b_{M-1} z^{M-1} + \ldots + b_1 z + b_0}{a_N z^N + a_{N-1} z^{N-1} + \ldots + a_1 z + a_0}$$

The system function for the inverse system is

$$H_i(z) = \frac{A(z)}{B(z)} = \frac{a_N z^N + a_{N-1} z^{N-1} + \ldots + a_1 z + a_0}{b_M z^M + b_{M-1} z^{M-1} + \ldots + b_1 z + b_0}$$

If the original system with system function $H(z)$ is causal, then $M \le N$ as we have established in Section 5.5.5. By the same token, causality of the inverse system with system function $H_i(z)$ requires $N \le M$. Hence we need $N = M$ if both the original system and its inverse are required to be causal.

To analyze the stability of the inverse system we will find it more convenient to write the system function $H(z)$ in pole zero form. Using $M = N$ we have

$$H(z) = \frac{b_N (z-z_1)(z-z_2)\ldots(z-z_N)}{a_N \left(z-p_1\right)\left(z-p_2\right)\ldots\left(z-p_N\right)}$$

If the original system is both causal and stable, all its poles must be in the unit circle of the z-plane (see Section 5.5.6), therefore

$$\left|p_k\right| < 1\,, \qquad k = 1,\ldots,N \tag{5.200}$$

The system function of the inverse system, written in pole-zero form, is

$$H_i(z) = \frac{a_N(z-p_1)(z-p_2)\dots(z-p_N)}{b_N(z-z_1)(z-z_2)\dots(z-z_N)}$$

For the inverse system to be stable, its poles must also lie in the unit circle. The poles of the inverse system are the zeros of the original system. Therefore, for the inverse system to be stable, both zeros and poles of the original system must be inside the unit circle. In addition to Eqn. (5.200) we also need

$$|z_k| < 1\,, \qquad k = 1,\dots,N \tag{5.201}$$

Minimum-phase system:

A causal DTLTI system that has all of its zeros and poles in the unit circle is referred to as a *minimum-phase system*. A minimum-phase system and its inverse are both causal and stable.

Example 5.40: Inverse of a system described by a difference equation

A DTLTI system is described by a difference equation

$$y[n] = 0.1\,y[n-1] + 0.72\,y[n-2] + x[n] + 0.5\,x[n-1]$$

Determine if a causal and stable inverse can be found for this system. If yes, find a difference equation for the inverse system.

Solution: The system function is found by taking the z-transform of the difference equation and solving for the ratio of the z-transforms of the output signal and the input signal:

$$Y(z)\left[1-0.1\,z^{-1}-0.72\,z^{-2}\right] = X(z)\left[1+0.5\,z^{-1}\right]$$

$$H(z) = \frac{1+0.5\,z^{-1}}{1-0.1\,z^{-1}-0.72\,z^{-2}} = \frac{z\,(z+0.5)}{(z+0.8)\,(z-0.9)}$$

The zeros of the system function are at $z_1 = 0$ and $z_2 = -0.5$. Its two poles are at $p_1 = -0.8$ and $p_2 = 0.9$. Since all poles and zeros are inside the unit circle of the z-plane, $H(z)$ is a minimum phase system. The system function for the inverse system is

$$H_i(z) = \frac{(z+0.8)\,(z-0.9)}{z\,(z+0.5)} = \frac{1-0.1\,z^{-1}-0.72\,z^{-2}}{1+0.5\,z^{-1}}$$

which is clearly causal and stable. Its corresponding difference equation is

$$y[n] = -0.5\,y[n-1] + x[n] - 0.1\,x[n-1] - 0.72\,x[n-2]$$

5.6 Implementation Structures for DTLTI Systems

Block diagram implementation of discrete-time systems was briefly discussed in Section 2.6 of Chapter 2. A method was presented for obtaining a block diagram from a linear constant-coefficient difference equation. Three type of elements were utilized; namely signal adder, constant-gain multiplier, and one-sample delay. In this section we will build on the techniques presented in Section 2.6, this time utilizing the z-domain system function as the starting point.

5.6.1 Direct-form implementations

The general form of the z-domain system function for a DTLTI system is

$$H(z) = \frac{Y(z)}{X(z)} = \frac{b_0 + b_1 z^{-1} + b_2 z^{-2} + \ldots + \ldots b_M z^{-M}}{1 + a_1 z^{-1} + a_2 z^{-2} + \ldots + a_N z^{-N}} \tag{5.202}$$

For the sake of convenience we chose to write $H(z)$ in terms of negative powers of z. $X(z)$ and $Y(z)$ are the z-transforms of the input and the output signals, respectively. The method of obtaining a block diagram from a z-domain system function will be derived using a third-order system, but its generalization to higher-order system functions is quite straightforward. Consider a DTLTI system for which $M = 3$ and $N = 3$. The system function $H(z)$ is

$$H(z) = \frac{Y(z)}{X(z)} = \frac{b_0 + b_1 z^{-1} + b_2 z^{-2} + b_3 z^{-3}}{1 + a_1 z^{-1} + a_2 z^{-2} + a_3 z^{-3}} \tag{5.203}$$

Let us use an intermediate transform $V(z)$ that corresponds to an intermediate signal $v[n]$, and express the system function as

$$H(z) = H_1(z)\, H_2(z) = \frac{Y(z)}{V(z)} \frac{V(z)}{X(z)} \tag{5.204}$$

We will elect to associate $V(z)/X(z)$ with the numerator polynomial, that is,

$$H_1(z) = \frac{V(z)}{X(s)} = b_0 + b_1 z^{-1} + b_2 z^{-2} + b_3 z^{-3} \tag{5.205}$$

To satisfy Eqn. (5.204), the ratio $Y(z)/V(z)$ must be associated with the denominator polynomial.

$$H_2(s) = \frac{Y(z)}{V(z)} = \frac{1}{1 + a_1 z^{-1} + a_2 z^{-2} + a_3 z^{-3}} \tag{5.206}$$

The relationships described by Eqns. (5.203) through (5.206) are illustrated in Fig. 5.57.

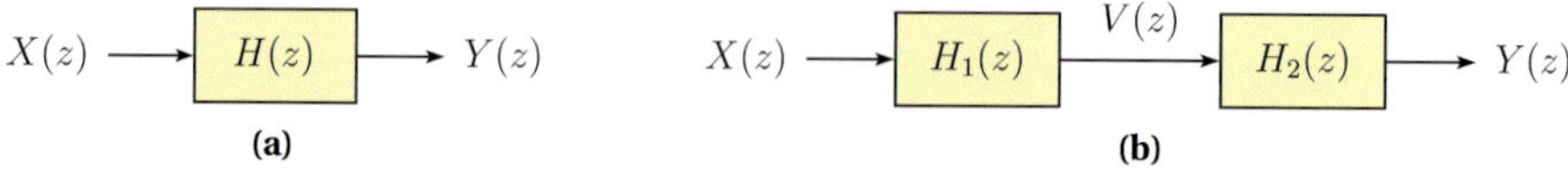

Figure 5.57 – **(a)** DTLTI system with system function $H(z)$ and **(b)** cascade form using an intermediate function $V(z)$.

Solving Eqn. (5.205) for $V(z)$ results in

$$V(z) = b_0 X(z) + b_1 z^{-1} X(z) + b_2 z^{-2} X(z) + b_3 z^{-3} X(z) \tag{5.207}$$

Taking the inverse z-transform of each side of Eqn. (5.207) and remembering that multiplication of the transform by z^{-k} corresponds to a delay of k samples in the time domain, we get

$$v[n] = b_0\, x[n] + b_1\, x[n-1] + b_2\, x[n-2] + b_3\, x[n-3] \tag{5.208}$$

The relationship between $X(z)$ and $V(z)$ can be realized using the block diagram shown in Fig. 5.58(a). The second part of the system will be implemented by expressing the output transform $Y(z)$ in terms of the intermediate transform $V(z)$. From Eqn. (5.206) we write

$$Y(z) = V(z) - a_1 z^{-1}\, Y(z) - a_2 z^{-2}\, Y(z) - a_3 z^{-3}\, Y(z) \tag{5.209}$$

The corresponding time-domain relationship is

$$y[n] = v[n] - a_1\, y[n-1] - a_2\, y[n-2] - a_3\, y[n-3] \tag{5.210}$$

The relationship between $V(z)$ and $Y(z)$ can be realized using the block diagram shown in Fig. 5.58(b).

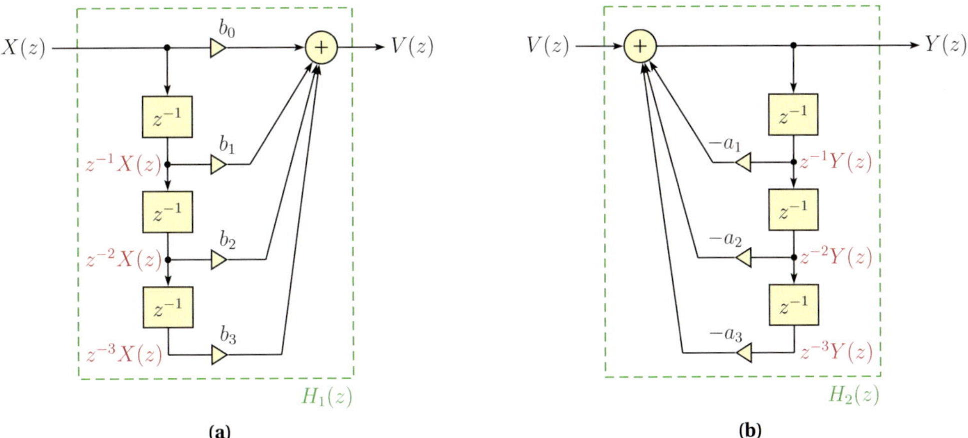

Figure 5.58 – **(a)** Block diagram for $H_1(z)$ and **(b)** block diagram for $H_2(z)$.

The two block diagrams obtained for $H_1(z)$ and $H_2(z)$ can be connected in cascade as shown in Fig. 5.59 to implement the system function $H(z)$. This is the *direct-form I* realization of the system. The block diagram in Fig. 5.59 represents the functional relationship between $X(z)$ and $Y(z)$, the z-transforms of the input and the output signals. If a diagram using time-domain quantities is desired, the conversion is easy using the following steps:

1. Replace each transform $X(z)$, $Y(z)$ and $V(z)$ with its time-domain version $x[n]$, $y[n]$, and $v[n]$.
2. Replace each block with the factor z^{-1} with one-sample time delay.

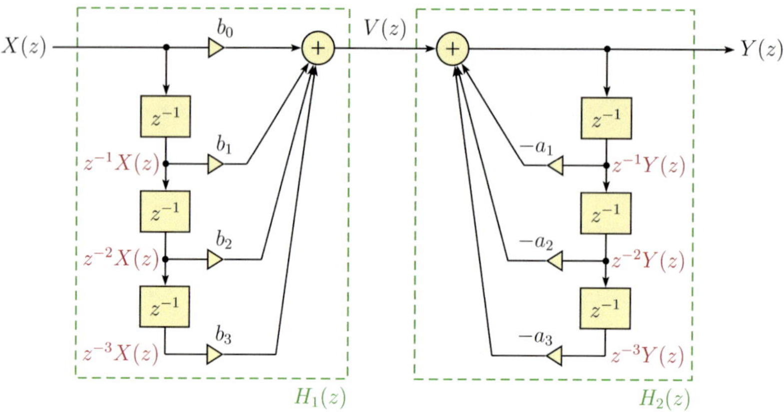

Figure 5.59 – Direct-form I realization of $H(z)$.

The resulting diagram is shown in Fig. 5.60.

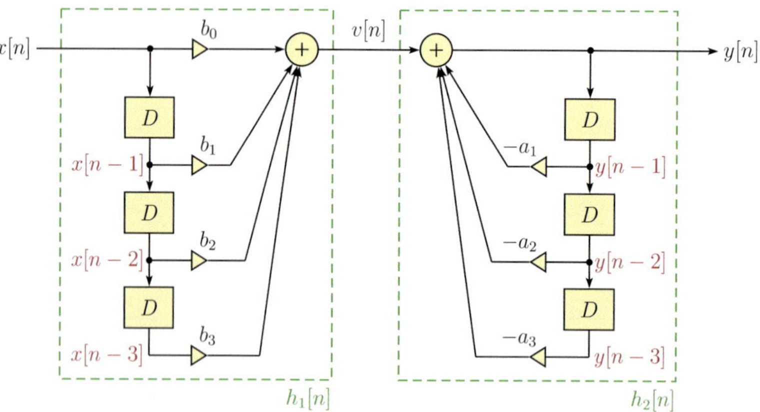

Figure 5.60 – Direct-form I realization of $H(z)$ using time-domain quantities.

Example 5.41: Direct-form type-I implementation of a system

Consider a causal DTLTI system described by the z-domain system function

$$H(z) = \frac{z^3 - 7z + 6}{z^4 - z^3 - 0.34\,z^2 + 0.966\,z - 0.2403}$$

Draw a direct-form I block diagram for implementing this system.

Solution: Let us multiply both the numerator and the denominator by z^{-4} to write the system function in the standard form of Eqn. (5.202):

$$H(z) = \frac{z^{-1} - 7\,z^{-3} + 6\,z^{-4}}{1 - z^{-1} - 0.34\,z^{-2} + 0.966\,z^{-3} - 0.2403\,z^{-4}}$$

We have

$$M = 4\,, \quad N = 4$$
$$b_0 = 0\,, \quad b_1 = 1\,, \quad b_2 = 0\,, \quad b_3 = -7\,, \quad b_4 = 6$$
$$a_0 = 1\,, \quad a_1 = -1; \quad a_2 = -0.34\,, \quad a_3 = 0.966\,, \quad a_4 = -0.2403$$

The block diagram is shown in Fig. 5.61.

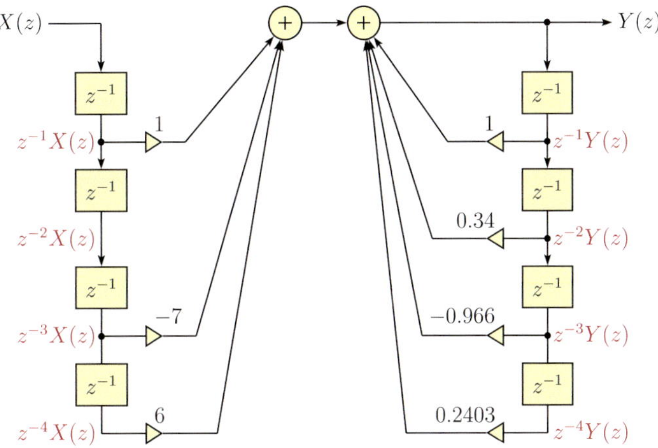

Figure 5.61 – Direct-form type-I block diagram for Example 5.41.

Let us consider the direct-form I diagram in Fig. 5.59 again. The subsystem with system function $H_1(z)$ takes $X(z)$ as input, and produces $V(z)$. The subsystem with system function $H_2(z)$ takes $V(z)$, the output of the first subsystem, and produces $Y(z)$ in response. Since each subsystem is linear [1], it does not matter which one comes first in a cascade connection. An alternative way of connecting the two subsystems is shown in Fig. 5.62, and it leads to the block diagram shown in Fig. 5.63. The middle part of the diagram in Fig. 5.63 has two delay lines running parallel to each other, each set containing four nodes. These two sets of nodes hold identical values $W(z)$, $z^{-1}W(z)$, $z^{-2}W(z)$ and $z^{-3}W(z)$. Consequently, the two sets of nodes can be merged together to result in the diagram shown in Fig. 5.64. This is the *direct-form II* realization of the system.

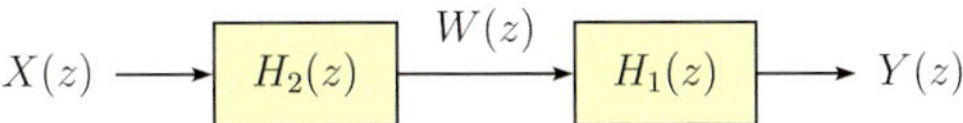

Figure 5.62 – Alternative placement of the two subsystems.

[1] Remember that, for linearity, each subsystem must be initially relaxed. All initial values must be zero. In Fig. 5.60, for example, we need $x[-1] = x[-2] = x[-3] = y[-1] = y[-2] = y[-3] = 0$.

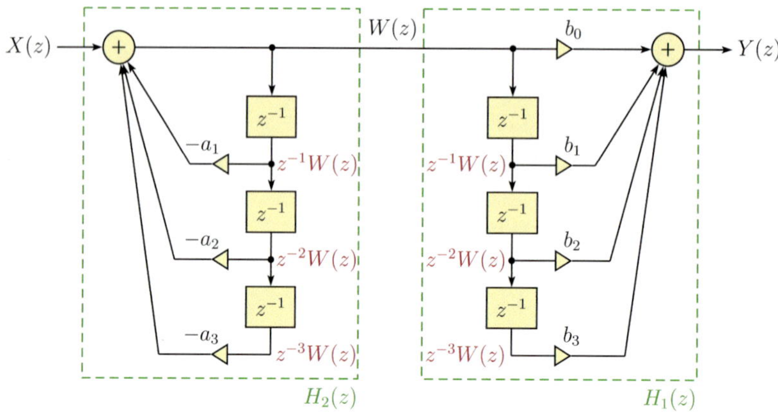

Figure 5.63 – Block diagram obtained by swapping the order of two subsystems.

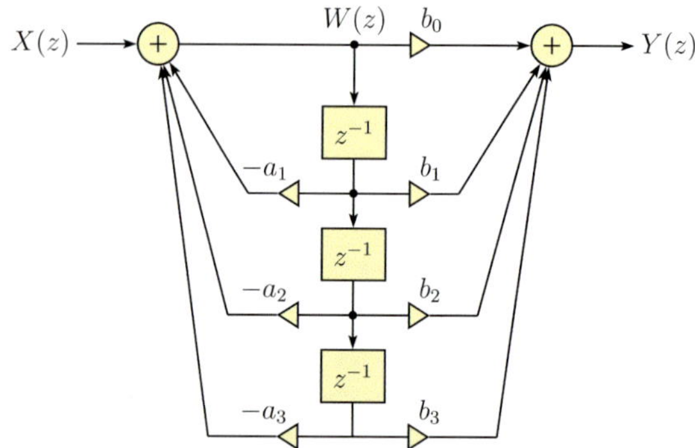

Figure 5.64 – Direct-form II realization of $H(z)$.

The block diagram in Fig. 5.64 could easily have been obtained directly from the system function in Eqn. (5.203) by inspection, using the following set of rules:

Obtaining direct-form II diagram from the system function:

1. Begin by ordering terms of numerator and denominator polynomials in ascending powers of z^{-1}.
2. Ensure that the leading coefficient in the denominator, that is, the constant term, is equal to unity. (If this is not the case, simply scale all coefficients to satisfy this rule.)
3. Set gain factors of feed-forward branches equal to the numerator coefficients.
4. Set gain factors of feedback branches equal to the negatives of the denominator coefficients.
5. Be careful to account for any missing powers of z^{-1} in either polynomial, and treat them as terms with their coefficients equal to zero.

Example 5.42: Direct-form II implementation of a system

Draw a direct-form type-II block diagram for implementing the system used in Example 5.41.

Solution: The diagram obtained by interchanging the order of the two subsystems in Fig. 5.61 and eliminating the nodes that become redundant is shown in Fig. 5.65.

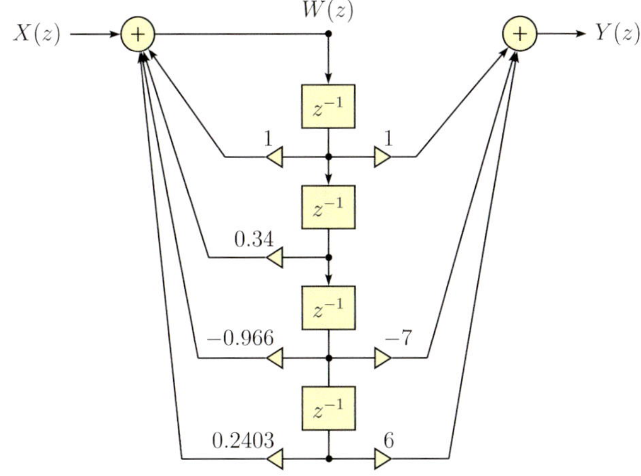

Figure 5.65 – Direct-form type-II block diagram for Example 5.42.

5.6.2 Cascade and parallel forms

Instead of simulating a system with the direct-form block diagram discussed in the previous section, it is also possible to express the system function as either the product or the sum of lower order sections, and base the block diagram on cascade of parallel combination of smaller diagrams. Consider a system function of order M that can be expressed in the form

$$\begin{aligned} H(z) &= H_1(z)\,H_2(z)\ldots H_M(z) \\ &= \frac{W_1(z)}{X(z)}\,\frac{W_2(z)}{W_1(z)}\cdots\frac{Y(z)}{W_{M-1}(z)} \end{aligned} \tag{5.211}$$

One method of simulating this system would be to build a diagram for each of the subsections $H_i(s)$ using the direct-form approach discussed previously, and then to connect those sections in cascade as shown in Fig. 5.66.

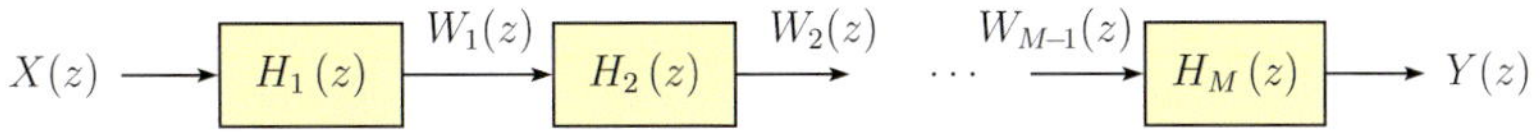

$X(z) \rightarrow H_1(z) \xrightarrow{W_1(z)} H_2(z) \xrightarrow{W_2(z)} \cdots \xrightarrow{W_{M-1}(z)} H_M(z) \rightarrow Y(z)$

Figure 5.66 – Cascade implementation of $H(z)$.

An easy method of sectioning a system function in the style of Eqn. (5.211) is to first find the poles and the zeros of the system function, and then use them for factoring numerator and denominator polynomials. Afterward, each section may be constructed by using one of the poles. Each zero is incorporated into one of the sections, and some sections may have constant numerators.

If some of the poles and zeros are complex-valued, we may choose to keep conjugate pairs together in second-order sections to avoid the need for complex gain factors in the diagram. The next example will illustrate this process.

Example 5.43: Cascade form block diagram

Develop a cascade form block diagram for the DTLTI system used in Example 5.41.

Solution: The system function specified in Example 5.41 has zeros at $z_1 = -3$, $z_2 = 2$, $z_3 = 1$, and poles at $p_1 = -0.9$, $p_2 = 0.3$, $p_{3,4} = 0.8 \pm j0.5$ (see MATLAB Exercise 5.9). It can be written in factored form as

$$H(z) = \frac{(z+3)\,(z-1)\,(z-2)}{(z+0.9)\,(z-0.3)\,\big(z-0.8-j0.5\big)\,\big(z-0.8+j0.5\big)}$$

The system function can be broken down into cascade sections in a number of ways. Let us choose to have two second-order sections by defining

$$H_1(z) = \frac{(z+3)\,(z-1)}{(z+0.9)\,(z-0.3)} = \frac{z^2+2z-3}{z^2+0.6z-0.27}$$

and

$$H_2(z) = \frac{z-2}{\big(z-0.8-j0.5\big)\,\big(z-0.8+j0.5\big)} = \frac{z-2}{z^2-1.6z+0.89}$$

so that

$$H(z) = H_1(z)\,H_2(z)$$

A cascade form block diagram is obtained by using direct-form II for each of $H_1(z)$ and $H_2(z)$, and connecting the resulting diagrams as shown in Fig. 5.67.

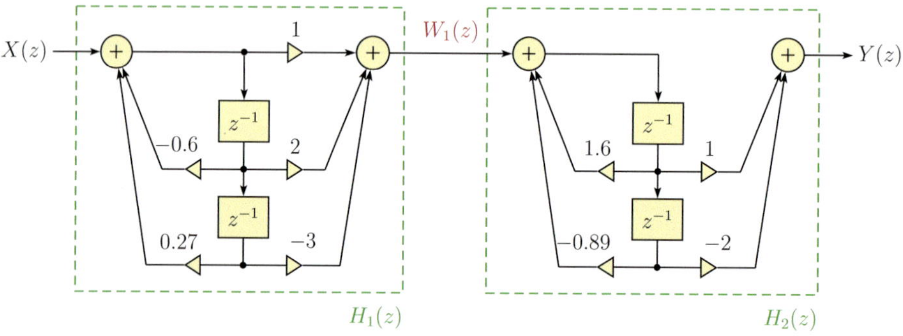

Figure 5.67 – Cascade form block diagram for Example 5.43.

It is also possible to consolidate the neighboring adders in the middle although this would cause the intermediate signal $W_1(z)$ to be lost. The resulting diagram is shown in Fig. 5.68.

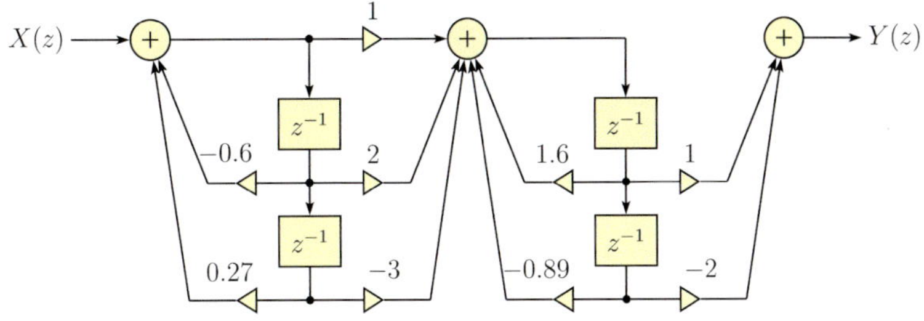

Figure 5.68 – Further simplified cascade form block diagram for Example 5.43.

Software resource: See MATLAB Exercises 5.9 and 5.10.

An alternative to the cascade form simulation diagram is a parallel form diagram which is based on writing the system function as a sum of lower order functions:

$$\begin{aligned} H(s) &= \bar{H}_1(z) + \bar{H}_2(z) + \ldots + \bar{H}_M(z) \\ &= \frac{\bar{W}_1(z)}{X(z)} + \frac{\bar{W}_2(z)}{X(z)} + \ldots \frac{\bar{W}_M(z)}{X(z)} \end{aligned} \tag{5.212}$$

A simulation diagram may be constructed by implementing each term in Eqn. (5.212) using the direct-form approach, and then connecting the resulting subsystems in a parallel configuration as shown in Fig. 5.69.

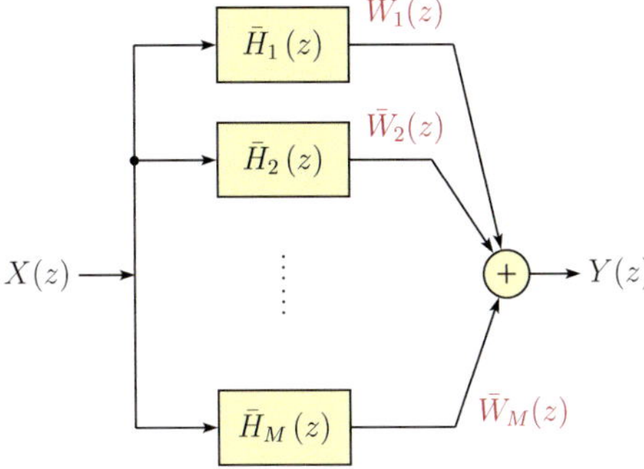

Figure 5.69 – Parallel implementation of $H(z)$.

A rational system function $H(z)$ may be sectioned in the form of Eqn. (5.212) using partial fraction expansion. If some of the poles and zeros are complex-valued, we may choose to keep conjugate pairs together in second-order sections to avoid the need for complex gain factors in the diagram. This process will be illustrated in the next example.

Example 5.44: Parallel form block diagram

Develop a cascade form block diagram for the DTLTI system used in Example 5.41.

Solution: The system function has poles at $p_1 = -0.9$, $p_2 = 0.3$, and $p_{3,4} = 0.8 \pm j0.5$. Let us expand it into partial fractions in the form

$$H(z) = \frac{k_1}{z+0.9} + \frac{k_2}{z-0.3} + \frac{k_3}{z-0.8-j0.5} + \frac{k_4}{z-0.8+j0.5}$$

Note that it is not necessary to divide $H(z)$ by z before expanding it into partial fractions since we are not trying to compute the inverse transform. The residues of the partial fraction expansion are (see MATLAB Exercise 5.11)

$$\begin{aligned} k_1 &= -3.0709\,, & k_2 &= 6.5450 \\ k_3 &= -1.2371 + j1.7480 & k_4 &= -1.2371 - j1.7480 \end{aligned}$$

and $H(s)$ is

$$H(z) = \frac{-3.0709}{z+0.9} + \frac{6.5450}{z-0.3} + \frac{-1.2371+j1.7480}{z-0.8-j0.5} + \frac{-1.2371-j1.7480}{z-0.8+j0.5}$$

Let us combine the first two terms to create a second-order section:

$$\bar{H}_1(z) = \frac{-3.0709}{z+0.9} + \frac{6.5450}{z-0.3} = \frac{3.474\,z+6.812}{z^2+0.6\,z-0.27}$$

The remaining two terms are combined into another second-order section, yielding

$$\bar{H}_2(z) = \frac{-1.2371+j1.7480}{z-0.8-j0.5} + \frac{-1.2371-j1.7480}{z-0.8+j0.5} = \frac{-2.474\,z+0.2314}{z^2-1.6\,z+0.89}$$

so that

$$H(z) = \bar{H}_1(z) + \bar{H}_2(z)$$

The parallel form simulation diagram is shown in Fig. 5.70.

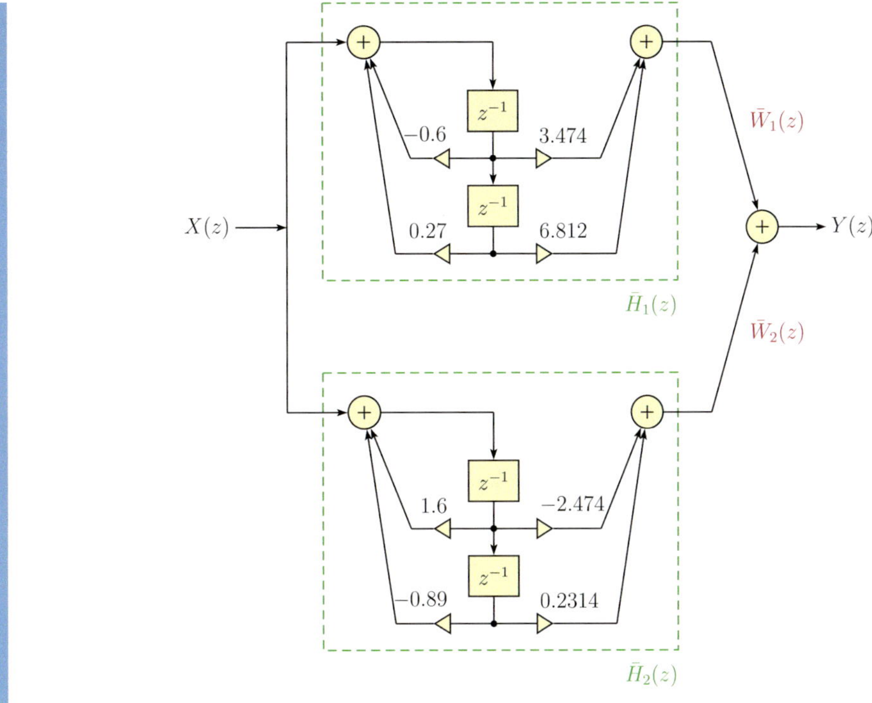

Figure 5.70 – Parallel form block diagram for Example 5.44.

Software resource: See MATLAB Exercise 5.11 and 5.12.

5.7 Unilateral z-Transform

It was mentioned in earlier discussion that the z-transform as defined by Eqn. (5.1) is often referred to as the bilateral z-transform. An alternative version, known as the unilateral z-transform, is defined by

$$X_u(z) = \mathcal{Z}_u\left\{x[n]\right\} = \sum_{n=0}^{\infty} x[n]\, z^{-n} \tag{5.213}$$

We use the subscript "u" to distinguish the unilateral z-transform from its bilateral counterpart. It is clear from a comparison of Eqn. (5.213) with Eqn. (5.1) that the only difference between the two definitions is the lower index of the summation. In Eqn. (5.213) we start the summation at $n = 0$ instead of $n = \infty$. If $x[n]$ is a causal signal, that is, if $x[n] = 0$ for $n < 0$, then the unilateral transform $X_u(z)$ becomes identical to the bilateral transform $X(z)$. On the other hand, for a signal that is $x[n]$ is not necessarily causal, the unilateral transform is essentially the bilateral transform of $x[n]$ multiplied by a unit step function, that is,

$$\begin{aligned}\mathcal{Z}_u\{x[n]\} &= \mathcal{Z}\left\{x[n]\, u[n]\right\} \\ &= \sum_{n=-\infty}^{\infty} x[n]\, u[n]\, z^{-n}\end{aligned} \tag{5.214}$$

Because of the way $X_u(z)$ is defined, its region of convergence is always the outside of a circle and does not have to be explicitly stated.

One property of the unilateral z-transform that differs from its counterpart for the bilateral z-transform is the time shifting property. Consider the time shifted signal $x[n-1]$. In terms of the bilateral z-transform, we have proved the following relationship in earlier discussion:

$$x[n-1] \overset{\mathcal{Z}}{\longleftrightarrow} z^{-1} X(z)$$

The relationship in Eqn. (5.215) does not extend to the unilateral z-transform, that is, we cannot claim that $\mathcal{Z}_u\{x[n-1]\} = z^{-1}\mathcal{Z}_u\{x[n]\}$. The reason for this fundamental difference becomes obvious if we write the unilateral transform in open form:

$$\begin{aligned}\mathcal{Z}_u\{x[n]\} &= \sum_{n=0}^{\infty} x[n]\, z^{-n}\\ &= x[0] + x[1]\, z^{-1} + x[2]\, z^{-2} + x[3]\, z^{-3} + \ldots\end{aligned} \tag{5.215}$$

The transform of the time shifted signal is

$$\begin{aligned}\mathcal{Z}_u\{x[n-1]\} &= \sum_{n=0}^{\infty} x[n-1]\, z^{-n}\\ &= x[-1] + x[0]\, z^{-1} + x[1]\, z^{-2} + x[3]\, z^{-3} + \ldots\end{aligned}$$

The sample $x[-1]$ is not part of $\mathcal{Z}_u\{x[n]\}$ but appears in $\mathcal{Z}_u\{x[n-1]\}$. Let us rewrite the latter by separating $x[-1]$ from the rest of the terms:

$$\begin{aligned}\mathcal{Z}_u\{x[n-1]\} &= x[-1] + \sum_{n=1}^{\infty} x[n-1]\, z^{-n}\\ &= x[-1] + z^{-1} \sum_{n=0}^{\infty} x[n]\, z^{-n}\\ &= x[-1] + z^{-1}\left[x[0] + x[1]\, z^{-1} + x[2]\, z^{-2} + x[3]\, z^{-3} + \ldots\right]\end{aligned}$$

Consequently, the unilateral z-transform of $x[n-1]$ can be expressed in terms of the transform of $x[n]$ as

$$\mathcal{Z}_u\{x[n-1]\} = x[-1] + z^{-1} X_u(z)$$

Through similar reasoning it can be shown that, for the signal $x[n-2]$, we have

$$\mathcal{Z}_u\{x[n-2]\} = x[-2] + x[-1]\, z^{-1} + z^{-2} X_u(z)$$

and in the general case of a signal time shifted by $k > 0$ samples

$$\begin{aligned}\mathcal{Z}_u\{x[n-k]\} &= x[-k] + x[-k+1]\, z^{-1} + \ldots + x[-1]\, z^{-k+1} + z^{-k} X_u(z)\\ &= \sum_{n=-k}^{-1} x[n]\, z^{-n-k} + z^{-k} X_u(z)\end{aligned} \tag{5.216}$$

If the signal is advanced rather than delayed in time, it can be shown that (see Problem 5.45 at the end of this chapter)

$$\mathcal{Z}_u\{x[n+k]\} = z^k X_u(z) - \sum_{n=0}^{k-1} x[n]\, z^{k-n} \tag{5.217}$$

for $k > 0$. The unilateral z-transform is useful in the use of z-transform techniques for solving difference equations with specified initial conditions. The next couple of examples will illustrate this.

Example 5.45: Finding the natural response of a system through z-transform

The homogeneous difference equation for a system is

$$y[n]-\frac{5}{6}\,y[n-1]+\frac{1}{6}\,y[n-2]=0$$

Using z-transform techniques, determine the natural response of the system for the initial conditions

$$y[-1]=19\,,\quad y[-2]=53$$

Recall that this problem was solved in Example 2.12 of Chapter 1 using time-domain solution techniques.

Solution: The first step will be to compute the unilateral z-transform of both sides of the homogeneous difference equation:

$$\mathcal{Z}_u\left\{y[n]-\frac{5}{6}\,y[n-1]+\frac{1}{6}\,y[n-2]\right\}=0$$

Using the linearity of the z-transform we can write

$$\mathcal{Z}_u\{y[n]\}-\frac{5}{6}\,\mathcal{Z}_u\{y[n-1]\}+\frac{1}{6}\,\mathcal{Z}_u\{y[n-2]\}=0 \tag{5.218}$$

The unilateral z-transforms of the time shifted versions of the output signal are

$$\begin{aligned}\mathcal{Z}_u\{y[n-1]\}=&y[-1]+z^{-1}\,Y_u(z)\\=&19+z^{-1}\,Y_u(z)\end{aligned} \tag{5.219}$$

and

$$\begin{aligned}\mathcal{Z}_u\{y[n-2]\}=&y[-2]+y[-1]\,z^{-1}+z^{-2}\,Y_u(z)\\=&53+19\,z^{-1}+z^{-2}\,Y_u(z)\end{aligned} \tag{5.220}$$

Substituting Eqns. (5.219) and (5.220) into Eqn. (5.218) we get

$$Y_u(z)-\frac{5}{6}\left[19+z^{-1}\,Y_u(z)\right]+\frac{1}{6}\left[53+19\,z^{-1}+z^{-2}\,Y_u(z)\right]=0$$

which can be solved for $Y_u(z)$ to yield

$$Y_u(z)=\frac{z\left(7z-\frac{19}{6}\right)}{z^2-\frac{5}{6}z+\frac{1}{6}}=\frac{z\left(7z-\frac{19}{6}\right)}{\left(z-\frac{1}{2}\right)\left(z-\frac{1}{3}\right)}$$

Now we need to find $y[n]$ through inverse z-transform using partial fraction expansion. We will write $Y_u(z)$ as

$$Y_u(z)=\frac{k_1 z}{z-\frac{1}{2}}+\frac{k_2 z}{z-\frac{1}{3}}$$

The residues of the partial fraction expansion are

$$k_1 = \left(z - \frac{1}{2}\right) \frac{Y_u(z)}{z}\bigg|_{z=\frac{1}{2}} = \frac{\left(7z - \frac{19}{6}\right)}{\left(z - \frac{1}{3}\right)}\bigg|_{z=\frac{1}{2}} = \frac{7\left(\frac{1}{2}\right) - \frac{19}{6}}{\left(\frac{1}{2} - \frac{1}{3}\right)} = 2$$

and

$$k_2 = \left(z - \frac{1}{3}\right) \frac{Y_u(z)}{z}\bigg|_{z=\frac{1}{3}} = \frac{\left(7z - \frac{19}{6}\right)}{\left(z - \frac{1}{2}\right)}\bigg|_{z=\frac{1}{3}} = \frac{7\left(\frac{1}{3}\right) - \frac{19}{6}}{\left(\frac{1}{3} - \frac{1}{2}\right)} = 5$$

Thus, the partial fraction expansion for $Y_u(z)$ is

$$Y_u(z) = \frac{2z}{z - \frac{1}{2}} + \frac{5z}{z - \frac{1}{3}}$$

resulting in the homogeneous solution

$$y_h[n] = \mathcal{Z}_u^{-1}\left\{Y_u(z)\right\} = 2\left(\frac{1}{2}\right)^n u[n] + 5\left(\frac{1}{3}\right)^n u[n]$$

This result is in perfect agreement with that found in Example 2.12.

Example 5.46: Finding the forced response of a system through z-transform

Consider a system defined by means of the difference equation

$$y[n] = 0.9\, y[n-1] + 0.1\, x[n]$$

Determine the response of this system for the input signal

$$x[n] = 20\cos(0.2\pi n)$$

if the initial value of the output is $y[-1] = 2.5$. Recall that this problem was solved in Example 2.15 of Chapter 1 using time-domain solution techniques.

Solution: Let us begin by computing the unilateral z-transform of both sides of the difference equation.

$$\begin{aligned}\mathcal{Z}_u\{y[n]\} &= \mathcal{Z}_u\{0.9\, y[n-1]\} + \mathcal{Z}_u\{0.1\, x[n]\} \\ &= 0.9\, \mathcal{Z}_u\{y[n-1]\} + 0.1\, \mathcal{Z}_u\{x[n]\}\end{aligned} \tag{5.221}$$

In Example 5.15 we have found

$$\mathcal{Z}\left\{\cos(\Omega_0 n)\, u[n]\right\} = \frac{z\left[z - \cos(\Omega_0)\right]}{z^2 - 2\cos(\Omega_0)\, z + 1}$$

Adapting that result to the transform of the input signal $x[n]$ given, we get

$$\mathcal{Z}_u\{x[n]\} = \mathcal{Z}_u\{20\cos(0.2\pi n)\} = \frac{20z\left[z - \cos(0.2\pi)\right]}{z^2 - 2\cos(0.2\pi)\, z + 1} \tag{5.222}$$

The unilateral z-transform of the $y[n-1]$ term is

$$\mathcal{Z}_u\{y[n-1]\} = y[-1] + z^{-1}\, Y_u(z) = 2.5 + z^{-1}\, Y_u(z) \tag{5.223}$$

Substituting Eqns. (5.222) and (5.223) into Eqn. (5.221) we obtain

$$\begin{aligned}Y_u(z) =&0.9\left[2.5+z^{-1}\,Y_u(z)\right]+0.1\,\frac{20z\left[z-\cos(0.2\pi)\right]}{z^2-2\cos(0.2\pi)\,z+1}\\ =&0.9z^{-1}\,Y_u(z)+2.25+\frac{2z\left[z-\cos(0.2\pi)\right]}{z^2-2\cos(0.2\pi)\,z+1}\end{aligned}$$

from which the unilateral z-transform of the output signal can be obtained as

$$Y_u(z)=\frac{2z^2\left[z-\cos(0.2\pi)\right]+2.25z\left(z-e^{j0.2\pi}\right)\left(z-e^{-j0.2\pi}\right)}{(z-0.9)\left(z-e^{j0.2\pi}\right)\left(z-e^{-j0.2\pi}\right)}$$

We are now ready to determine $y[n]$ using partial fraction expansion. Let

$$Y_u(z)=\frac{k_1 z}{z-0.9}+\frac{k_2 z}{z-e^{j0.2\pi}}+\frac{k_3 z}{z-e^{-j0.2\pi}}$$

The residues are

$$\begin{aligned}k_1=&(z-0.9)\left.\frac{Y_u(z)}{z}\right|_{z=0.9}\\ =&\left.\frac{2z\left[z-\cos(0.2\pi)\right]+2.25\left(z-e^{j0.2\pi}\right)\left(z-e^{-j0.2\pi}\right)}{\left(z-e^{j0.2\pi}\right)\left(z-e^{-j0.2\pi}\right)}\right|_{z=0.9}\\ =&\frac{2(0.9)\left[0.9-\cos(0.2\pi)\right]+2.25\left(0.9-e^{j0.2\pi}\right)\left(0.9-e^{-j0.2\pi}\right)}{\left(0.9-e^{j0.2\pi}\right)\left(0.9-e^{-j0.2\pi}\right)}\\ =&2.7129\end{aligned}$$

and

$$\begin{aligned}k_2=&\left(z-e^{j0.2\pi}\right)\left.\frac{Y_u(z)}{z}\right|_{z=e^{j0.2\pi}}\\ =&\left.\frac{2z\left[z-\cos(0.2\pi)\right]+2.25\left(z-e^{j0.2\pi}\right)\left(z-e^{-j0.2\pi}\right)}{(z-0.9)\left(z-e^{-j0.2\pi}\right)}\right|_{z=e^{j0.2\pi}}\\ =&\frac{2e^{j0.2\pi}\left[e^{j0.2\pi}-\cos(0.2\pi)\right]}{\left(e^{j0.2\pi}-0.9\right)\left(e^{j0.2\pi}-e^{-j0.2\pi}\right)}\\ =&0.7685-j1.4953\end{aligned}$$

The third residue, k_3, can be computed in a similar manner, however, we will take a shortcut and recognize the fact that complex residues of a rational transform must occur in complex conjugate pairs provided that all coefficients of the denominator polynomial are real-valued. Consequently we have

$$k_3=k_2^*=0.7685+j1.4953$$

It will be more convenient to express k_2 and k_3 in polar complex form as

$$k_2=1.6813\,e^{-j1.096}\,,\quad k_3=1.6813\,e^{j1.096}$$

The forced response of the system is

$$\begin{aligned}y[n] =&2.7129\,(0.9)^n\,u[n]+1.6813\,e^{-j1.096}\,e^{j0.2\pi n}\,u[n]+1.6813\,e^{j1.096}\,e^{-j0.2\pi n}\,u[n]\\ =&2.7129\,(0.9)^n\,u[n]+1.6813\,\left[e^{j(0.2\pi n-1.096)}+e^{-j(0.2\pi n-1.096)}\right]\,u[n]\\ =&2.7129\,(0.9)^n\,u[n]+3.3626\cos\left(0.2\pi n-1.096\right)\,u[n]\end{aligned}$$

or equivalently

$$y[n]=2.7129\,(0.9)^n\,u[n]+1.5371\cos\left(0.2\pi n\right)\,u[n]+2.9907\sin\left(0.2\pi n\right)\,u[n]$$

which is consistent with the answer found earlier in Example 2.15 of Chapter 2.

Software resource: See MATLAB Exercise 5.13.

Summary of Key Points

- The z-transform of a discrete-time signal $x[n]$ is defined as

Eqn. 5.1: $$X(z)=\sum_{n=-\infty}^{\infty}x[n]\,z^{-n}$$

where z, the independent variable of the transform, is a complex variable.

- The z-transform of a signal $x[n]$ evaluated on a circle with radius r_1 starting at angle $\Omega=0$ and ending at angle $\Omega=2\pi$ is equal to the DTFT of the modified signal $x[n]\,r_1^{-n}$.

Eqn. 5.5: $$X(z)\Big|_{z=e^{j\Omega}}=\sum_{n=-\infty}^{\infty}x[n]\,z^{-n}=\mathscr{F}\left\{x[n]\,r^{-n}\right\}$$

- The z-transform of a signal $x[n]$ evaluated on the unit circle of the z-plane starting at angle $\Omega=0$ and ending at angle $\Omega=2\pi$ is the same as the DTFT of the same signal.

Eqn. 5.6: $$X(z)\Big|_{z=e^{j\Omega}}=\sum_{n=-\infty}^{\infty}x[n]\,z^{-n}=\mathscr{F}\left\{x[n]\right\}$$

- The z-transform must be considered in conjunction with the criteria for the convergence of the resulting polynomial $X(z)$. The collection of all points in the z-plane for which the transform converges is called the *region of convergence (ROC).*

- Two different signals may have the same transform with differing regions of convergence:

Eqn. 5.15: $$a^n\,u[n]\;\overset{\mathcal{Z}}{\longleftrightarrow}\;\frac{z}{z-a}\,,\qquad \text{ROC: }|z|>|a|$$

Eqn. 5.16: $$-a^n\,u[-n-1]\;\overset{\mathcal{Z}}{\longleftrightarrow}\;\frac{z}{z-a}\,,\qquad \text{ROC: }|z|<|a|$$

☞ The ROC for a finite-length is the entire z-plane with the possible exception of $z = 0 + j0$, or $|z| \to \infty$ or both.

☞ In general, the ROC is circularly-shaped. It is either the inside of a circle, the outside of a circle, or between two circles. It cannot contain any poles. For the z-transform of a causal signal, the ROC is the outside of a circle. In contrast, for the z-transform of an anti-causal signal, the ROC is the inside of a circle. The ROC for the z-transform of a signal that is neither causal nor anti-causal is a ring-shaped region between two circles, if it exists.

☞ Linearity property:

The z-transform is linear. For any two signals $x_1[n]$ and $x_2[n]$ with their respective transforms

$$x_1[n] \overset{\mathcal{Z}}{\longleftrightarrow} X_1(z)$$

and

$$x_2[n] \overset{\mathcal{Z}}{\longleftrightarrow} X_2(z)$$

and any two constants α_1 and α_2, the following relationship holds:

Eqn. 5.65:
$$\alpha_1\, x_1[n] + \alpha_2\, x_2[n] \overset{\mathcal{Z}}{\longleftrightarrow} \alpha_1\, X_1(z) + \alpha_2\, X_2(z)$$

☞ Given the transform pair

$$x[n] \overset{\mathcal{Z}}{\longleftrightarrow} X(z)$$

the following are also valid transform pairs:

Time shifting property:

Eqn. 5.74:
$$x[n-k] \overset{\mathcal{Z}}{\longleftrightarrow} z^{-k}\, X(z)$$

Time reversal property:

Eqn. 5.79:
$$x[-n] \overset{\mathcal{Z}}{\longleftrightarrow} X\left(z^{-1}\right)$$

Multiplication by an exponential signal:

Eqn. 5.83:
$$a^n\, x[n] \overset{\mathcal{Z}}{\longleftrightarrow} X(z/a)$$

Differentiation in the z-domain:

Eqn. 5.84:
$$n\, x[n] \overset{\mathcal{Z}}{\longleftrightarrow} (-z)\, \frac{dX(z)}{dz}$$

Initial value:

Eqn. 5.93:
$$x[0] = \lim_{z\to\infty} \left[X(z)\right]$$

Summation property:

Eqn. 5.106:
$$\sum_{k=-\infty}^{n} x[k] \overset{\mathcal{Z}}{\longleftrightarrow} \frac{z}{z-1}\, X(z)$$

☞ Convolution Property:

For any two signals $x_1[n]$ and $x_2[n]$ with their respective transforms

$$x_1[n] \stackrel{\mathcal{Z}}{\longleftrightarrow} X_1(z)$$

$$x_2[n] \stackrel{\mathcal{Z}}{\longleftrightarrow} X_2(z)$$

the following transform relationship holds:

Eqn. 5.88:
$$x_1[n] * x_2[n] \stackrel{\mathcal{Z}}{\longleftrightarrow} X_1(z)\, X_2(z)$$

☞ Correlation property:

Cross-correlation of two discrete-time signals $x[n]$ and $y[n]$ is defined as

Eqn. 5.95:
$$r_{xy}[m] = \sum_{n=-\infty}^{\infty} x[n]\, y[n-m]$$

Given two signals $x[n]$ and $y[n]$ with their respective transforms

$$x[n] \stackrel{\mathcal{Z}}{\longleftrightarrow} X(z)$$

$$y[n] \stackrel{\mathcal{Z}}{\longleftrightarrow} Y(z)$$

it can be shown that

Eqn. 5.96:
$$R_{xy}(z) = \mathcal{Z}\left\{r_{XY}[m]\right\} = X(z)\, Y\left(z^{-1}\right)$$

As a special case, the transform of the autocorrelation of a signal $x[n]$ is

$$R_{xx}(z) = X(z)\, X\left(z^{-1}\right)$$

☞ Inverse z-transform can be computed using

1. Direct evaluation of the inversion integral,
2. Partial fraction expansion technique for a rational transform, and
3. Expansion of the rational transform into a power series through long division.

☞ Inversion integral method involves evaluating the integral

Eqn. 5.118:
$$x[n] = \frac{1}{2\pi j} \oint X(z)\, z^{n-1}\, dz$$

on a closed contour in the z-plane within the ROC of the transform. In general, direct evaluation of the contour integral is difficult when $X(z)$ is a rational function of z. An indirect method of evaluating the integral is to rely on the *Cauchy residue theorem.*

☞ Inverse z-transform by partial fraction expansion is based on the idea of expressing a ragtional transform in partial fraction form as

Eqn. 5.123:
$$X(z) = \frac{k_1\, z}{z - z_1} + \frac{k_2\, z}{z - z_2} + \ldots + \frac{k_N\, z}{z - z_N}$$

and determining the residues k_i. Contribution of each term to the inverse transform can be determined based on the relative positioning of its pole with respect to the ROC of the transform.

☞ The long division idea is based on converting a rational transform $X(z)$ back into its power series form, and associating the coefficients of the power series with the sample amplitudes of the signal $x[n]$. It does not produce an analytical solution for the signal $x[n]$. Instead, it allows us to obtain the signal one sample at a time.

☞ System function $H(z)$ of a DTLTI system can be obtained from its difference equation by taking the z-transform of the difference equation, separating terms that have a common factor $X(z)$ from thoise that have a common factor $Y(z)$, and finding the ratio

$$H(z) = \frac{Y(z)}{X(z)}$$

☞ The response of a DTLTI system to a complex exponential signal in the form

$$x[n] = z_0^n$$

where z_0 represents a point on the z-plane within the ROC of the system function is

Eqn. 5.159: $$y[n] = z_0^n \, H(z_0)$$

☞ The response of a DTLTI system to exponentially damped sinusoid in the form

$$x[n] = r_0^n \cos(\Omega_0 n)$$

is computed as

Eqn. 5.168: $$y[n] = H_0 \, r_0^n \cos(\Omega_0 n + \Theta_0)$$

where

$$H_0 = \left|H(z_0)\right| \qquad \text{and} \qquad \Theta_0 = \measuredangle H(z_0)$$

☞ In the z-domain transfer function of a causal DTLTI system the order of the numerator must not be greater than the order of the denominator.

☞ For a DTLTI system to be stable, the ROC of its z-domain system function must include the unit circle. For a causal system, this means that the system function must not have any poles on or outside the unit circle of the z-plane. For an anti-causal system to be stable, the system function must not have any poles on or inside the unit circle of the z-plane.

Further Reading

[1] Sonali Bagchi and Sanjit K Mitra. *The Nonuniform Discrete Fourier Transform and its Applications in Signal Processing.* Vol. 463. Springer Science & Business Media, 2012.

[2] S Allen Broughton and Kurt Bryan. *Discrete Fourier Analysis and Wavelets: Applications to Signal and Image Processing.* John Wiley & Sons, 2018.

[3] J.W. Cooley and J.W. Tukey. "An Algorithm for the Machine Computation of Complex Fourier Series, vol. 19". In: *Mathematics of Computation* (1965), p. 73.

[4] William G Gardner. "Reverberation Algorithms". In: *Applications of Digital Signal Processing to Audio and Acoustics.* Springer, 1998, pp. 85–131.

[5] Tapio Lokki and Jarmo Hiipakka. "A Time-Variant Reverberation Algorithm for Reverberation Enhancement Systems". In: *Proceedings of COST G-6 Conference on Digital Audio Effects (DAFX-01), Limerick, Ireland.* 2001, pp. 28–32.

[6] Huan Mi, Gavin Kearney, and Helena Daffern. "Perceptual Similarities Between Artificial Reverberation Algorithms and Real Reverberation". In: *Applied Sciences* 13.2 (2023), p. 840.

[7] Henri J. Nussbaumer. *The Fast Fourier Transform.* Springer, 1982.

[8] A.V. Oppenheim and R.W. Schafer. *Discrete-Time Signal Processing.* Prentice Hall, 2010.

[9] Manfred R Schroeder and Benjamin F Logan. "Colorless Artificial Reverberation". In: *IRE Transactions on Audio* 6 (1961), pp. 209–214.

[10] Manfred R. Schroeder. "Natural Sounding Artificial Reverberation". In: *Audio Engineering Society Convention 13.* Audio Engineering Society. 1961.

[11] Julius O. Smith. *Mathematics of the Discrete Fourier Transform (DFT): With Audio Applications.* Julius Smith, 2008.

[12] Julius O. Smith. *Physical Audio Signal Processing.* online book, 2010 edition.

[13] Duraisamy Sundararajan. *The Discrete Fourier Transform: Theory, Algorithms and Applications.* World Scientific, 2001.

[14] R. Tolimieri and C. Lu. *Algorithms for Discrete Fourier Transform and Convolution.* Springer, 1989.

[15] Vesa Valimaki et al. "Fifty years of artificial reverberation". In: *IEEE Transactions on Audio, Speech, and Language Processing* 20.5 (2012), pp. 1421–1448.

[16] Man Wah Wong. *Discrete Fourier Analysis.* Vol. 5. Springer Science & Business Media, 2011.

MATLAB Exercises with Solutions

MATLAB Exercise 5.1: Three-dimensional plot of z-transform

In Fig. 5.3 the magnitude of the transform

$$X(z)=\frac{z\,(z-0.7686)}{z^2-1.5371\,z+0.9025}$$

was graphed as a three-dimensional surface. In this exercise we will reproduce that figure using MATLAB, and develop the code to evaluate and display the transform on a circular trajectory $z=r\,e^{j\Omega}$. The first step is to produce a set of complex values of z on a rectangular grid in the z-plane.

```
>> [zr,zi] = meshgrid([-1.5:0.05:1.5],[-1.5:0.05:1.5]);
>> z = zr+j*zi;
```

The next step is to compute the magnitude of the z-transform at each point on the grid. An anonymous function will be used for defining $X(z)$. Additionally, values of magnitude that are greater than 12 will be clipped for graphing purposes.

```
>> Xz = @(z) z.*(z-0.7686)./(z.*z-1.5371*z+0.9025);
>> XzMag = abs(Xz(z));
>> XzMag = XzMag.*(XzMag<=12)+12.*(XzMag>12);
```

A three-dimensional mesh plot of $|X(z)|$ plot can be generated with the following lines:

```
>> mesh(zr,zi,XzMag);
>> axis([-1.5,1.5,-1.5,1.5]);
```

The script listed below produces a mesh plot complete with axis labels and color specifications.

```
1  % Script: mexdt_5_1a.m
2  [zr,zi] = meshgrid([-1.5:0.05:1.5],[-1.5:0.05:1.5]);  % Set up grid
3  z = zr+j*zi;
4  Xz = @(z) z.*(z-0.7686)./(z.*z-1.5371*z+0.9025); % Evaluate X[z] on the grid
5  XzMag = abs(Xz(z));
6  XzMag = XzMag.*(XzMag<=12)+12.*(XzMag>12);        % Clip values greater than 12
7  shading interp;    % Shading method: Interpolated
8  colormap copper;   % Specify the color map used
9  m1 = mesh(zr,zi,XzMag);
10 axis([-1.5,1.5,-1.5,1.5]);
11 % Adjust transparency of surface lines
12 set(m1,'EdgeAlpha',0.5','FaceAlpha',0.5);
13 % Specify x,y,z axis labels
14 xlabel('Re[z]');
15 ylabel('Im[z]');
16 zlabel('|X(z)|');
17 % Specify viewing angles
18 view(gca,[56.5 40]);
```

In line 9 of the script, the handle returned by the function `mesh()` is assigned to the variable `m1` so that it can be used in line 12 for adjusting the transparency of the surface.

The DTFT $X(\Omega)$ is equal to the z-transform evaluated on the unit circle of the z-plane, that is,

$$X(z)\Big|_{z=e^{j\Omega}} = \mathcal{F}\left\{x[n]\right\}$$

Applying this relationship to the magnitudes of the two transforms, the magnitude of the DTFT is obtained by evaluating the magnitude of the z-transform on the unit circle. The script `mexdt_5_1b.m` listed below demonstrates this.

```
% Script: mexdt_5_1b.m
[zr,zi] = meshgrid([-1.5:0.05:1.5],[-1.5:0.05:1.5]);  % Set up grid
z = zr+j*zi;
Xz = @(z) z.*(z-0.7686)./(z.*z-1.5371*z+0.9025); % Evaluate X[z] on the grid
XzMag = abs(Xz(z));
XzMag = XzMag.*(XzMag<=12)+12.*(XzMag>12);        % Clip values greater than 12
% Set the trajectory.
r = 1;                     % Radius of the trajectory
Omg = [0:0.0025:1]*2*pi;   % Vector of angular frequencies
tr = r*exp(j*Omg);         % Circular trajectory on the z plane
% Produce a mesh plot and hold it
shading interp;            % Shading method: Interpolated
colormap copper;           % Specify the color map used
m1 = mesh(zr,zi,XzMag);
set(m1,'EdgeAlpha',0.5','FaceAlpha',0.5);
hold on;
% Superimpose a plot of X(z) magnitude values evaluated
```

```
%   on the trajectory using 'plot3' function
m2 = plot3(real(tr),imag(tr),abs(Xz(tr)),'b-','LineWidth',1.25);
% Show the unit circle in red
m3 = plot3(real(tr),imag(tr),zeros(size(tr)),'r-','LineWidth',1.25);
hold off;
axis([-1.5,1.5,-1.5,1.5]);
% Specify x,y,z axis labels
xlabel('Re[z]');
ylabel('Im[z]');
zlabel('|X(z)|');
% Specify viewing angles
view(gca,[56.5 40]);
```

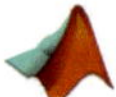

Software resources: mexdt_5_1a.m , mexdt_5_1b.m

MATLAB Exercise 5.2: Computing the DTFT from the z-transform

The DTFT of a signal is equal to its z-transform evaluated on the unit circle of the z-plane.

$$\mathscr{F}\left\{x[n]\right\} = X(z)\Big|_{z=e^{j\Omega}}$$

Consider the z-transform

$$X(z) = \frac{z-0.7686}{z^2-1.5371\,z+0.9025}$$

The first method of computing and graphing the DTFT of the signal is to use an anonymous function for $X(z)$ and evaluate it on the unit circle. The magnitude $|X(\Omega)|$ is graphed using the following statements:

```
>> X = @(z) (z-0.7686)./(z.*z-1.5371*z+0.9025);
>> Omega = [-1:0.001:1]*pi;
>> Xdtft = X(exp(j*Omega));
>> plot(Omega,abs(Xdtft)); grid;
```

If the phase $\measuredangle X(\Omega)$ is needed, it can be graphed using

```
>> plot(Omega,angle(Xdtft)); grid;
```

The second method is to use MATLAB function `freqz()`. Care must be taken to ensure that the function is used with proper arguments, or erroneous results may be obtained. The function `freqz()` requires that the transform $X(z)$ be expressed using negative powers of z in the form

$$X(z) = \frac{0+z^{-1}-0.7686\,z^{-2}}{1-1.5371\,z^{-1}+0.9025\,z^{-2}}$$

Vectors `num` and `den` to hold numerator and denominator coefficients should be entered as

```
>> num = [0,1,-0.7686];
>> den = [1,-1.5371,0.9025];
```

Afterward the magnitude and the phase of the DTFT can be computed and graphed with the statements

```
>> Omega = [-1:0.001:1]*pi;
>> [Xdtft,Omega] = freqz(num,den,Omega);
>> plot(Omega,abs(Xdtft)); grid;
>> plot(Omega,angle(Xdtft)); grid;
```

Common mistake: Pay special attention to the leading 0 in the numerator coefficient vector. Had we omitted it, the function `freqz()` would have incorrectly used the transform

$$X(z) = \frac{1 - 0.7686\,z^{-1}}{1 - 1.5371\,z^{-1} + 0.9025\,z^{-2}}$$

resulting in the correct magnitude $|X(\Omega)|$ but incorrect phase $\measuredangle X(\Omega)$. This example demonstrates the potential for errors when built-in MATLAB functions are used without carefully reading the help information.

Software resources: `mexdt_5_2a.m` , `mexdt_5_2b.m`

MATLAB Exercise 5.3: Graphing poles and zeros

Consider a DTLTI system described by the system function

$$X(z) = \frac{z^3 - 0.5\,z + 0.5}{z^4 - 0.5\,z^3 - 0.5\,z^2 - 0.6\,z + 0.6}$$

Poles and zeros of the system function can be graphed on the z-plane by entering coefficients of numerator and denominator polynomials as vectors and then computing the roots.

```
>> num = [1,0,-0.5,0.5];
>> den = [1,-0.5,-0.5,-0.6,0.6];
>> zrs = roots(num);
>> pls = roots(den);
```

To produce the graph we need

```
>> plot(real(zrs),imag(zrs),'o',real(pls),imag(pls),'x');
```

which uses “o” for a zero and “x” for a pole. The graph produced does not display the unit circle of the z-plane, and the aspect ratio may not be correct. These issues are resolved using the code below:

```
>> Omega = [0:0.001:1]*2*pi;
>> c = exp(j*Omega);   % Unit circle
>> plot(real(zrs),imag(zrs),'o',real(pls),imag(pls),'x',real(c),imag(c),'--');
>> axis equal;
```

Alternatively, built-in function `zplane()` may be used for producing a pole-zero plot, however, caution must be used as in the previous MATLAB exercise. The function `zplane()` expects numerator and denominator polynomials to be specified in terms of negative powers of z. Using the vectors `num` and `den` as entered above and calling the function `zplane()` through

```
>> zplane(num,den);
```

would display an incorrect pole-zero plot with an additional zero at the origin. The correct use for this problem would be

```
>> num = [0,1,0,-0.5,0.5];
>> den = [1,-0.5,-0.5,-0.6,0.6];
>> zplane(num,den);
```

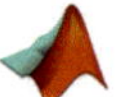

Software resources: mexdt_5_3a.m , mexdt_5_3b.m

MATLAB Exercise 5.4: Using convolution function for polynomial multiplication

The function `conv()` for computing the convolution of two discrete-time signals was used in MATLAB Exercise 2.8 in Chapter 2. In Example 5.23 of this chapter we have discovered another interesting application of the `conv()` function, namely the multiplication of two polynomials. In this exercise we will apply this idea to the multiplication of the two polynomials

$$A(v) = 5v^4 + 7v^3 + 3v - 2$$

and

$$B(v) = 8v^5 + 6v^4 - 3v^3 + v^2 - 4v + 2$$

The problem is to determine the polynomial $C(v) = A(v)\,B(v)$. We will begin by creating a vector with the coefficients of each polynomial ordered from highest power of v down to the constant term, being careful to account for any missing terms:

```
>> A = [5,7,0,3,-2];          % Coefficients a_4, ..., a_0
>> B = [8,6,-3,1,-4,2];       % Coefficients b_5, ..., b_0
```

For the convolution operation the coefficients are needed in ascending order starting with the constant term. MATLAB function `fliplr()` will be used for reversing the order.

```
>> tmpA = fliplr(A);          % Coefficients a_0, ..., a_4
>> tmpB = fliplr(B);          % Coefficients b_0, ..., a_5
>> tmpC = conv(tmpA,tmpB);    % Coefficients c_0, ..., a_9
>> C = fliplr(tmpC)           % Coefficients c_9, ..., c_0

C =
    40   86   27    8  -11  -39   23  -14   14   -4
```

The product polynomial is

$$C(v) = 40v^9 + 86v^8 + 27v^7 + 8v^6 - 11v^5 - 39v^4 + 23v^3 - 14v^2 + 14v - 4$$

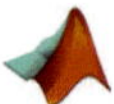

Software resource: mexdt_5_4.m

MATLAB Exercise 5.5: Partial fraction expansion with MATLAB

Find a partial fraction expansion for the transform

$$X(z) = \frac{-0.2\,z^3 + 1.82\,z^2 - 3.39\,z + 3.1826}{z^4 - 1.2\,z^3 + 0.46\,z^2 + 0.452\,z - 0.5607}$$

Solution: We need to create two vectors `num` and `den` with the numerator and the denominator coefficients of $X(z)/z$ and then use the function `residue()`. The statements below compute the z-domain poles and residues:

```
>> num = [-0.2000,1.8200,-3.5900,3.1826];
>> den = [1.0000,-1.2000,0.4600,0.4520,-0.5607,0];
>> [r,p,k] = residue(num,den)

r =
   0.8539 + 0.0337i
   0.8539 - 0.0337i
   1.1111
   2.8571
  -5.6761

p =
   0.5000 + 0.8000i
   0.5000 - 0.8000i
   0.9000
  -0.7000
        0

k =
     []
```

Based on the results obtained, the PFE is in the form

$$\frac{X(z)}{z} = \frac{0.8539 + j0.0337}{z - 0.5 - j0.8} + \frac{0.8539 - j0.0337}{z - 0.5 + j0.8} + \frac{1.1111}{z - 0.9} + \frac{2.8571}{z + 0.7} + \frac{-5.6761}{z}$$

or, equivalently

$$X(z) = \frac{\left(0.8539 + j0.0337\right)\,z}{z - 0.5 - j0.8} + \frac{\left(0.8539 - j0.0337\right)\,z}{z - 0.5 + j0.8} + \frac{1.1111\,z}{z - 0.9} + \frac{2.8571\,z}{z + 0.7} - 5.6761$$

Software resource: `mexdt_5_5.m`

MATLAB Exercise 5.6: Developing a function for long division

In this exercise we will develop a MATLAB function named `ss_longdiv()` to compute the inverse z-transform by long division. MATLAB function `deconv()` will be used for implementing each step of the long division operation. The utility function `ss_longdiv_util()`

listed below implements just one step of the long division.

```
function [q,r] = ss_longdiv_util(num,den)
  M = max(size(num))-1;      % The order of the numerator
  N = max(size(den))-1;      % The order of the denominator
  % Append the shorter vector with zeros
  if (M < N)
    num = [num,zeros(1,N-M)];
  elseif (M > N)
    den = [den,zeros(1,M-N)];
  end
  [q,r] = deconv(num,den);     % Find quotient and remainder
  r = r(2:length(r));          % Drop first element of remainder which is 0
end
```

To test this function, let us try the transform in Example 5.31.

$$X(z) = \frac{3z^2 - z}{z^3 - 3z^2 + 3z - 1}, \qquad \text{ROC:} \quad |z| > 1$$

The first sample of the inverse transform is obtained using the following:

```
>> num = [3,-1,0];
>> den = [1,-3,3,-1];
>> [q,r] = ss_longdiv_util(num,den)

q =
     3

r =
     8    -9     3
```

This result should be compared to Eqn. (5.140). The next step in the long division can be performed with the statement

```
>> [q,r] = ss_longdiv_util(r,den)

q =
     8

r =
    15   -21     8
```

and matches Eqn. (5.141). This process can be repeated as many times as desired. The function ss_longdiv() listed below uses the function ss_longdiv_util() in a loop to compute n samples of the inverse transform.

```
function [x] = ss_longdiv(num,den,n)
  [q,r] = ss_longdiv_util(num,den);
  x = q;
  for nn = 2:n
    [q,r] = ss_longdiv_util(r,den);
    x = [x,q];
  end
end
```

It can be tested using the problem in Example 5.31:

```
>> x = ss_longdiv([3,-1,0],[1,-3,3,-1],6)

x =
     3     8    15    24    35    48
```

If a left-sided result is desired, the function `ss_longdiv()` can be still be used by simply reordering the coefficients in the vectors `num` and `den`. Refer to the transform in Example 5.32.

$$X(z) = \frac{z^2+z-2}{z^2-2.5z+1}\,, \qquad \text{ROC: } |z| < \tfrac{1}{2}$$

The left-sided solution is obtained by

```
>> num = fliplr([1,1,-2]);
>> den = fliplr([1,-2.5,1]);
>> x = ss_longdiv(num,den,6)

x =
   -2.0000   -4.0000   -7.0000  -13.5000  -26.7500  -53.3750
```

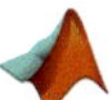

Software resources: `mexdt_5_6.m` , `ss_longdiv.m` , `ss_longdivutil.m`

MATLAB Exercise 5.7: Frequency response of a system from pole-zero layout

The problem of obtaining the frequency response of a DTLTI system from the placement of its poles and zeros in the z-domain was discussed in Section 5.5.4. In this exercise we will use MATLAB to compute the frequency response for a system characterized by the system function

$$H(z) = \frac{z+0.6}{z-0.8}$$

The system has one zero at $z=-0.6$ and one pole at $z=0.8$. Suppose we need the frequency response of this system at the angular frequency of $\Omega = 0.3\pi$ radians, which is equal to the system function evaluated at $z = e^{j0.3\pi}$:

$$H(0.3\pi) = H(z)\Big|_{z=e^{j0.3\pi}} = \frac{e^{j0.3\pi}+0.6}{e^{j0.3\pi}-0.8}$$

Using vector notation, we have

$$H(0.3\pi) = \frac{\vec{B}}{\vec{A}} = \frac{\overrightarrow{(e^{j0.3\pi}+0.6)}}{\overrightarrow{(e^{j0.3\pi}-0.8)}}$$

where we have defined the vectors $\vec{A}$ and $\vec{B}$ for notational convenience. The script listed below computes the vectors $\vec{A}$ and $\vec{B}$ and uses them for computing the magnitude and the phase of the system function at $\omega = 0.3\pi$:

```
1  % Script: mexdt_5_7a.m
2  Omega = 0.3*pi;             % Angular frequency
3  z = exp(j*Omega);           % Set complex variable z
4  B = z+0.6;                  % Numerator value at Omega=0.3*pi
5  A = z-0.8;                  % Denominator value at Omega=0.3*pi
6  mag = abs(B)/abs(A)         % Magnitude response
7  phs = angle(B)-angle(A)     % Phase response
```

This script computes the frequency response of the system at one specific frequency. It would be more interesting if we could use the same idea to compute the frequency response of the system at a large number of angular frequencies so that its magnitude and phase can be graphed as functions of Ω. Let us change the variable `Omega` into a vector by editing line 2 of the script:

```
2  Omega = [-1:0.004:1]*pi;
```

This change causes the variable `z` to become a complex vector with 501 elements. Also, in line 6, the standard division operator "`/`" needs to be changed to the element-by-element division operator "`./`" to read

```
6  mag = abs(B)./abs(A);
```

The script `mexdt_5_7b.m` is listed below with these modifications and the addition of graphing statements:

```
1  % Script: mexdt_5_7b.m
2  Omega = [-1:0.004:1]*pi;   % Vector of angular frequencies
3  z = exp(j*Omega);          % Vector of z values
4  B = z+0.6;                 % Numerator
5  A = z-0.8;                 % Denominator
6  mag = abs(B)./abs(A);      % Magnitude response
7  phs = angle(B)-angle(A);   % Phase response
8  % Graph the magnitude and the phase of the system function
9  tiledlayout(2,1);
10 nexttile;
11 plot(Omega,mag);
12 title('Magnitude of the frequency response');
13 xlabel('\Omega (rad)'); grid;
14 nexttile;
15 plot(Omega,phs);
16 title('Phase of the frequency response');
17 xlabel('\Omega (rad)'); grid;
```

Software resources: mexdt_5_7a.m , mexdt_5_7b.m

MATLAB Exercise 5.8: Frequency response from pole-zero layout revisited

In MATLAB Exercise 5.7 we have explored a method of computing the frequency response of a DTLTI system based on the graphical interpretation of the pole-zero layout of the system function, discussed in Section 5.5.4. The idea can be generalized into the development of a

MATLAB function `ss_freqz()` for computing the frequency response.

```
1  function [mag,phs] = ss_freqz(zrs,pls,gain,Omega)
2    nz = length(zrs);       % Number of zeros.
3    np = length(pls);       % Number of poles.
4    nOmg = length(Omega);   % Number of frequency points.
5    z = exp(j*Omega);       % Get points on the unit circle.
6    mag = ones(1,nOmg);
7    phs = zeros(1,nOmg);
8    if nz>0
9      for n = 1:nz
10       mag = mag.*abs(z-zrs(n));    % See Eqn. (5.175)
11       phs = phs+angle(z-zrs(n));   % See Eqn. (5.176)
12     end
13   end
14   if np>0
15     for n = 1:np
16       mag = mag./abs(z-pls(n));    % See Eqn. (5.175)
17       phs = phs-angle(z-pls(n));   % See Eqn. (5.176)
18     end
19   end
20   mag = mag*gain;
21   phs = wrapToPi(phs);
```

Line 21 of the function causes phase angles to be contained in the interval $(-\pi,\pi)$. The script `mexdt_5_8.m` listed below may be used for testing `ss_freqz()` with the system function

$$H(z) = \frac{z-0.7686}{z^2-1.5371\,z+0.9025}$$

```
% Script: mexdt_5_8.m
num = [1,-0.7686];            % Numerator polynomial
den = [1,-1.5371,0.9025];     % Denominator polynomial
zrs = roots(num);             % Compute zeros
pls = roots(den);             % Compute poles
Omega = [-1:0.001:1]*pi;      % Vector of frequencies
[mag,phs] = ss_freqz(zrs,pls,1,Omega);
tiledlayout(2,1);
nexttile;
plot(Omega,mag); grid;
xlabel('\Omega (rad)');
nexttile;
plot(Omega,phs); grid;
xlabel('\Omega (rad)');
```

Software resources: `mexdt_5_8.m` , `ss_freqz.m`

MATLAB Exercise 5.9: Preliminary calculations for a cascade-form block diagram

In this exercise we will use MATLAB to help with the preliminary calculations for sectioning a system function into cascade subsystems to facilitate the development of a cascade-form block diagram. Consider the system function first used in Example 5.41.

$$H(z) = \frac{z^3 - 7z + 6}{z^4 - z^3 - 0.34\,z^2 + 0.966\,z - 0.2403}$$

A cascade-form block diagram was obtained in Example 5.43. Two vectors `num` and `den` are created with the following statements to hold numerator and denominator coefficients in descending powers of z:

```
>> num=[1,0,-7,6];
>> den=[1,-1,-0.34,0.966,-0.2403];
```

Poles and zeros of the system function are found as

```
>> zrs = roots(num)

zrs =
   -3.0000
    2.0000
    1.0000

>>  pls = roots(den)

pls =
  -0.9000
   0.8000 + 0.5000i
   0.8000 - 0.5000i
   0.3000
```

The returned vector `zrs` holds the three zeros of the system function that are all real-valued. The vector `pls` holds the four poles. The first subsystem $H_1(z)$ will be formed using zeros #1 and #3 along with poles #1 and #4. Using the function `poly()` for polynomial construction from roots, the numerator and denominator coefficients of $H(z)$ can be obtained as follows:

```
>> num1 = poly([zrs(1),zrs(3)])

num1 =
    1.0000    2.0000   -3.0000

>>  den1 = poly([pls(1),pls(4)])

den1 =
    1.0000    0.6000   -0.2700
```

corresponding to

$$H_1(z) = \frac{(z+3)\,(z-1)}{(z+0.9)\,(z-0.3)} = \frac{z^2+2z-3}{z^2+0.6z-0.27}$$

The remaining zero will be combined with poles #2 and #2 to yield the numerator and the denominator of $H_2(z)$:

```
>> num2 = [1,-zrs(2)]

num2 =
    1.0000   -2.0000

>> den2 = poly([pls(2),pls(3)])

den2 =
    1.0000   -1.6000    0.8900
```

The second subsystem is therefore

$$H_2(z) = \frac{z-2}{\left(z-0.8-j0.5\right)\left(z-0.8+j0.5\right)} = \frac{z-2}{z^2-1.6z+0.89}$$

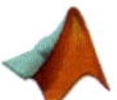

Software resource: mexdt_5_9.m

MATLAB Exercise 5.10: Cascade-form block diagram revisited

An alternative method of breaking up a system function into second-order cascade sections is to use MATLAB functions `tf2zp()` and `zp2sos()`. Working with the same system function as in MATLAB Exercise 5.9 use the following set of statements:

```
>> num = [0,1,0,-7,6];
>> den = [1,-1,-0.34,0.966,-0.2403];
>> [zrs,pls,k] = tf2zp(num,den);
>> [sos,G] = zp2sos(zrs,pls,k)

sos =
         0    1.0000    3.0000    1.0000    0.6000   -0.2700
    1.0000   -3.0000    2.0000    1.0000   -1.6000    0.8900

G =
     1
```

It is important to remember that vectors `num` and `den` hold coefficients of numerator and denominator polynomials in terms of ascending negative powers of z. In general they need to be of the same length, hence requiring a zero-valued element to be added in front of the vector `num`. Second order sections are in the form

$$H_i(z) = \frac{b_{0i} + b_{1i}\,z^{-1} + b_{2i}\,z^{-2}}{1 + a_{1i}\,z^{-1} + a_{2i}\,z^{-2}}$$

Each row of the matrix `sos` holds coefficients of one second-order section in the order

$$b_{0i},\ b_{1i},\ b_{2i},\ 1,\ a_{1i},\ a_{2i}$$

The result obtained above corresponds to second-order sections

$$H_1(z) = \frac{z^{-1} + 3\,z^{-2}}{1 + 0.6\,z^{-1} - 0.27\,z^{-2}}$$

and

$$H_2(z) = \frac{1-3z^{-1}+2z^{-2}}{1-1.6z^{-1}+0.89z^{-2}}$$

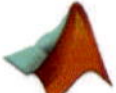

Software resource: mexdt_5_10.m

MATLAB Exercise 5.11: Preliminary calculations for a parallel-form block diagram

This exercise is about using MATLAB to help with the preliminary calculations for sectioning a system function into parallel subsystems to facilitate the development of a parallel-form block diagram. Consider again the system function first used in Example 5.41.

$$H(z) = \frac{z^3-7z+6}{z^4-z^3-0.34z^2+0.966z-0.2403}$$

A parallel-form block diagram was obtained in Example 5.44. In order to obtain the subsystems $\bar{H}_1(z)$ and $\bar{H}_2(z)$ for the parallel-form implementation we will begin by creating vectors `num` and `den` to hold numerator and denominator coefficients in descending powers of z:

```
>> num=[1,0,-7,6];
>> den=[1,-1,-0.34,0.966,-0.2403];
```

Residues are found using the function `residue()`. Recall from the discussion in MATLAB Exercise 5.5 that there is also a function named `residuez()` specifically designed for z-domain residues. However, it utilizes a slightly different format than the conventions we have developed in Section 5.4.2 for the partial fraction expansion, and therefore will not be used here.

```
>> [r,p,k] = residue(num,den)

r =
  -1.2371 + 1.7480i
  -1.2371 - 1.7480i
  -3.0709
   6.5450

p =
   0.8000 + 0.5000i
   0.8000 - 0.5000i
  -0.9000
   0.3000

k =
     []
```

The resulting expansion is

$$H(z) = \frac{-1.2371+j1.7480}{z-0.8-j0.5} + \frac{-1.2371-j1.7480}{z-0.8+j0.5} + \frac{-3.0709}{z+0.9} + \frac{6.5450}{z-0.3}$$

We will recombine the two terms with real poles (terms #3 and #4 in the lists of poles and residues computed by MATLAB) to obtain $\bar{H}_1(z)$:

```
>> [num1,den1] = residue(r(3:4),p(3:4),[]);
>> H1 = tf(num1,den1,-1)

Transfer function:
 3.474 z + 6.812
------------------
z^2 + 0.6 z - 0.27

Sampling time: unspecified
```

The remaining terms (items #1 and #2) will be recombined for the second subsystem $\bar{H}_2(z)$:

```
>> [num2,den2] = residue(r(1:2),p(1:2),[]);
>> H2 = tf(num2,den2,-1)

Transfer function:
-2.474 z + 0.2314
------------------
z^2 - 1.6 z + 0.89

Sampling time: unspecified
```

so that

$$H(z) = \bar{H}_1(z) + \bar{H}_2(z)$$

To see if this is indeed the case, let us add the two system functions:

```
>> H = H1+H2

Transfer function:
    z^3 - 8.882e-016 z^2 - 7 z + 6
---------------------------------------
z^4 - z^3 - 0.34 z^2 + 0.966 z - 0.2403

Sampling time: unspecified
```

This clearly matches the system function under consideration.

Software resource: mexdt_5_11.m

MATLAB Exercise 5.12: Implementing a system using second-order sections

In Section 5.6.2 cascade and parallel implementation forms of a rational system function $H(z)$ were discussed using second-order sections. The second-order rational system function

$$H(z) = \frac{b_0 + b_1 z^{-1} + b_2 z^{-2}}{1 + a_1 z^{-1} + a_2 z^{-2}}$$

may be used as the basis of cascade and parallel implementations, and may be implemented using the direct-form type-II block diagram shown in Fig. 5.71.

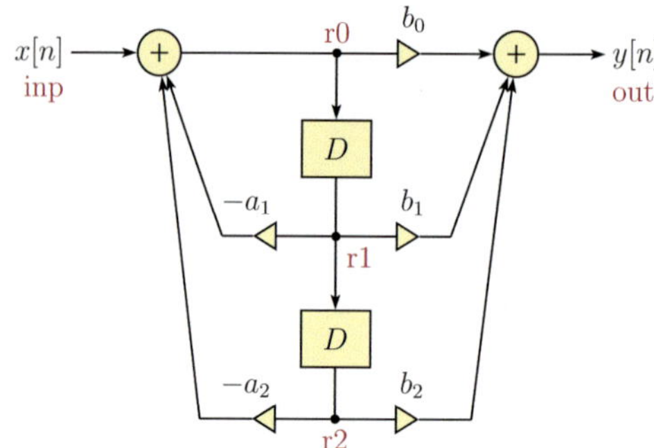

Figure 5.71 – Block diagram for second-order section.

We will develop a MATLAB function `ss_iir2()` for a generic implementation of this second-order system. The focus will be on making the function reusable within a larger block diagram implemented in either cascade or parallel form. For flexibility the function takes in one sample of the input signal at a time, and returns one sample of the output signal. The syntax of the function `ss_iir2()` is

```
[out,states] = ss_iir2(inp,coeffs,states)
```

The scalar `inp` is the current sample of the input signal. The vector `coeffs` holds the coefficients of the second-order IIR filter stage in the following order:

$$\text{coeffs} = \left[b_{0i},\, b_{1i},\, b_{2i},\, 1,\, a_{1i},\, a_{2i}\right]$$

Note that this is the same coefficient order used in MATLAB functions `tf2sos`, `zp2sos`, and `ss2sos`. Finally, the input vector `states` holds values of the internal variables r_1 and r_2 set in the previous step:

$$\text{states} = \left[r_1,\, r_2\right]$$

The return variable `out` is the current output sample computed in response to the current input sample. The function also returns updated values of the internal variables r_1 and r_2 in the vector `states`. These will be needed the for processing the next input sample. The listing for the function `ss_iir2()` is given below:

```
function [out,states] = ss_iir2(inp,coeffs,states)
  % Extract the filter states
  r1 = states(1);
  r2 = states(2);
  % Extract the coefficients
  b0 = coeffs(1);
  b1 = coeffs(2);
  b2 = coeffs(3);
  a1 = coeffs(5);
  a2 = coeffs(6);
  % Compute the output sample
  r0 = inp-a1*r1-a2*r2;
  out = b0*r0+b1*r1+b2*r2;
  % Update the filter states: r1 <-- r0,  r2 <-- r1
  states(1) = r0;
  states(2) = r1;
end
```

To test the function, we will use the DTLTI system used in Examples 5.43 and 5.44. The cascade form block diagram shown in Fig. 5.67 may be implemented with the following script that computes and graphs the unit step response of the system:

```
% Script: mexdt_5_12a.m
states1 = [0,0];                   % Initialize r1 and r2 for stage 1
states2 = [0,0];                   % Initialize r1 and r2 for stage 2
coeffs1 = [1,2,-3,1,0.6,-0.27];   % Coefficients for stage 1
coeffs2 = [0,1,-2,1,-1.6,0.89];   % Coefficients for stage 2
x = ones(1,100);
y = zeros(1,100);
for n=1:100
  inp = x(n);
  [w1,states1] = ss_iir2(inp,coeffs1,states1);
  [y(n),states2] = ss_iir2(w1,coeffs2,states2);
end
stem([0:99],y);
```

Similarly, the script listed below computes and graphs the unit step response of the same system using the parallel form block diagram obtained in Example 5.44 and shown in Fig. 5.70.

```
% Script: mexdt_5_12b.m
states1 = [0,0];                   % Initialize r1 and r2 for stage 1
states2 = [0,0];                   % Initialize r1 and r2 for stage 2
coeffs1 = [0,3.474,6.812,1,0.6,-0.27];     % Coefficients for stage 1
coeffs2 = [0,-2.474,0.2314,1,-1.6,0.89];   % Coefficients for stage 2
x = ones(1,100);
y = zeros(1,100);
for n=1:100
  inp = x(n);
  [w1,states1] = ss_iir2(inp,coeffs1,states1);
  [w2,states2] = ss_iir2(inp,coeffs2,states2);
  y(n) = w1+w2;
end
stem([0:99],y);
```

Software resources: mexdt_5_12a.m , mexdt_5_12b.m , ss_iir2.m

MATLAB Exercise 5.13: Solving a difference equation through z-transform

In Section 5.7 two examples of using the unilateral z-transform for solving difference equations with specified initial conditions were given. In this exercise we will use the symbolic processing capabilities of MATLAB to solve the problem in Example 5.45. The functions `ztrans()` and `iztrans()` are available for symbolic computation of the forward and inverse z-transform. They both assume that the time-domain signal involved is causal. In effect they implement the unilateral variant of the z-transform.

Consider the homogeneous difference equation explored in Example 5.45

$$y[n]-\frac{5}{6}\,y[n-1]+\frac{1}{6}\,y[n-2]=0$$

with the initial conditions

$$y[-1]=19\,,\quad y[-2]=53$$

Unilateral z-transforms of $y[n-1]$ and $y[n-2]$ are

$$Y_1(z)=\mathcal{Z}_u\left\{y[n-1]\right\}=19+z^{-1}\,Y_u(z)$$

and

$$Y_2(z)=\mathcal{Z}_u\left\{y[n-2]\right\}=53+19\,z^{-1}+z^{-2}\,Y_u(z)$$

The following script can be used for solving the problem:

```
1 % Script: mexdt_5_13a.m
2 syms z n Yz
3 Y1 = 19+z^(-1)*Yz;                      % z-transform of y[n-1]
4 Y2 = 53+19*z^(-1)+z^(-2)*Yz;            % z-transform of y[n-2]
5 Yz = solve(Yz-5/6*Y1+1/6*Y2,Yz);        % Solve for Y(z)
6 yn = iztrans(Yz)                        % Inverse z-transform of Y(z)
```

In line 3 we declare three symbolic variables. The variable `Yz` corresponds to the yet unknown transform $Y_u(z)$. Line 5 of the script uses the function `solve()` to solve the equation

$$Y_u(z)-\frac{5}{6}\,Y_1(z)+\;Y_2(z)=0$$

for $Y_u(z)$. Line 7 generates the response

```
yn =
    2*(1/2)^n + 5*(1/3)^n
```

which is in agreement with the solution found in Example 5.45. If desired, the symbolic expression in MATLAB variable yn may be numerically evaluated and graphed with the following statements:

```
>> n = [0:10];
>> y = eval(yn);
>> stem(n,y);
```

Software resources: `mexdt_5_13a.m` , `mexdt_5_13b.m`

MATLAB Exercise 5.14: Case Study – Echoes and reverberation, part 3

In earlier parts of this case study (MATLAB Exercises 2.13 and 3.6) we have used feed-forward and feedback comb filters for adding echoes and simple reverberation effects to audio signals. The feed-forward comb filter used in MATLAB Exercise 2.13 adds a simple echo to the sound. If multiple echoes are needed, perhaps a parallel bank of these filters can be used. Feed-forward comb filters are computationally inefficient and impractical for use when a large number of echoes are needed to simulate the acoustic characteristics of large rooms or halls. The feedback comb filter used in MATLAB Exercise 3.6 is significantly more efficient, and can produce a large number of echoes through the use of feedback.

The problem of simulating the acoustic characteristics of concert halls, churches, or other types of spaces using linear systems can be quite complex. There is a large number of echoes involved. Multiple sound outputs are needed for simulating the spatial distribution of the sound (think of 5-, 7- or 9-speaker sound systems used in home theaters). Assuming the placement of sound sources is fixed, we may be able to come up with a linear system model for what a listener standing at a certain point perceives. If the listener moves to a different point, the system would have to be modified, making the problem even more complex.

M.R. Schroeder suggested that three to four feedback comb filters in parallel may be used for simulating the frequency response of a room that has a reverberation time of 1 s. In the reverberator structure attributed to him, the bank of feedback comb filters is followed by several all-pass filters in series to increase the density of the echoes produced [10].

In this exercise, we will develop MATLAB scripts to implement two Schroeder reverberators for real-time testing. The first reverberator to be tested will be a very simple one, perhaps appropriate for simulating a small room. The second reverberator will be a more complex one, tuned for a larger space. Before we can dive into writing scripts for reverberators, we need to write a function for implementing the all-pass filter of Fig. 5.52. Remember that functions for implementing feedback comb filters were developed in MATLAB Exercise 3.6. The function `ss_comb()` implements a feedback comb filter for processing one sample at a time, and the function `ss_combf()` provides an implementation for frame-by-frame processing.

Function `ss_allpassf()` listed below implements an all-pass filter for frame-based processing. It is a direct implementation of the block diagram in Fig. 5.52 using the following relationships:

$$w[n] = x[n] + r\,w[n-L] \tag{5.224}$$

$$y[n] = -r\,w[n] + w[n-L] \tag{5.225}$$

```
function [y,buffer] = ss_allpassf(x,r,buffer)
  y = zeros(size(x));                    % Placeholder for output frame
  frameSize = size(x,1);
  for i=1:frameSize
    w(i,:) = x(i,:)+r*buffer(end,:);     % Eqn. (5.224)
    y(i,:) = -r*w(i,:)+buffer(end,:);    % Eqn. (5.225)
    buffer = [w(i,:);buffer(1:end-1,:)]; % Update buffer
  end
end
```

The first reverberator structure shown in Fig. 5.72 uses 2 comb filters in parallel followed by 2 all-pass filters in series.

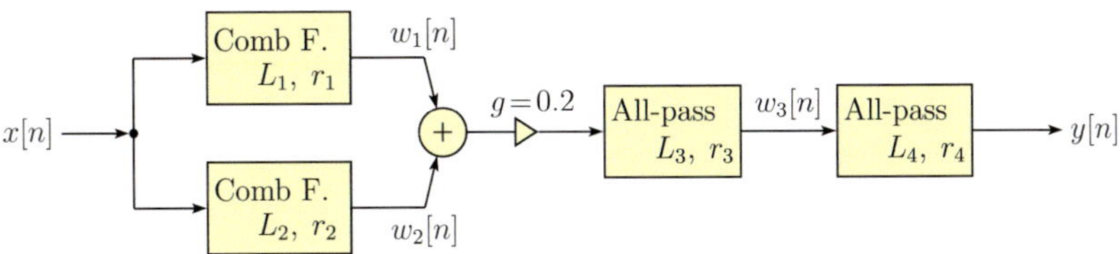

Figure 5.72 – Simple reverberator structure using comb filters and all-pass filters.

The gain factor $g = 0.4$ was added to the diagram in order to prevent sound samples from being clipped. Keep in mind that MATLAB can only play back sample amplitudes in the range $-1 \leq y[n] \leq 1$. Sample amplitudes that fall outside this range are clipped, and they sound distorted. The script file `mexdt_5_14a.m` for implementing this system is shown below.

```
1  % Script: mexdt_5_14a.m
2  % Create an "audio file reader" object
3  sReader = dsp.AudioFileReader('AG_Duet_22050_Hz.flac','ReadRange',[1,661500]);
4  % Create an "audio player" object
5  sPlayer = audioDeviceWriter('SampleRate',sReader.SampleRate);
6  %------------------------------------------------------------
7  L1 = 821;     r1 = 0.85;   % Parameters for comb filter 1
8  L2 = 613;     r2 = 0.75;   % Parameters for comb filter 2
9  L3 = 113;     r3 = 0.7;    % Parameters for allpass filter 1
10 L4 = 41;      r4 = 0.7;    % Parameters for allpass filter 2
11 %------------------------------------------------------------
12 numChannels = info(sReader).NumChannels;  % Number of channels
13 bufferCF1 = zeros(L1,numChannels);         % Buffer for comb filter 1
14 bufferCF2 = zeros(L2,numChannels);         % Buffer for comb filter 2
15 bufferAP1 = zeros(L3,numChannels);         % Buffer for allpass filter 1
16 bufferAP2 = zeros(L4,numChannels);         % Buffer for allpass filter 2
17 % Refer to Fig. 5.72 for the loop
18 while ~isDone(sReader)
19   x = sReader();
20   [w1,bufferCF1] = ss_combf(x,r1,bufferCF1);
21   [w2,bufferCF2] = ss_combf(x,r2,bufferCF2);
22   [w3,bufferAP1] = ss_allpassf((w1+w2)*0.4,r3,bufferAP1);
23   [y,bufferAP2] = ss_allpassf(w3,r4,bufferAP2);
24   sPlayer(y);
25 end
26 release(sReader);  % We are finished with the input audio file
27 release(sPlayer);  % We are finished with the audio output device
```

Parameters of the comb filters and the all-pass filters are shown on the block diagram, and are set in lines 7 through 10 of the code. Lines 13 through 16 initialize the buffer matrices for the filters with all zero amplitudes. The loop between lines 18 and 25 is where frame-by-frame processing of audio takes place. Study lines 20 through 23, and correlate them with the block diagram. Listen to the output sound, and compare it to the original, unprocessed, version.

Caveat: The script was tested with version R2023a of MATLAB running on a Windows-based laptop computer with a 13th generation i7 processor, and real-time playback works fine. On somewhat older computers, or on computers with slower processors, the demands of the algorithm may be too heavy for real-time playback. If that is the case, an alternate version of the script that drops real-time playback and saves the output sound to disc instead is provided among downloadable files. Once the processing is done, open the created file "`Output_sound.flac`" with any audio player, and listen to it.

A more elaborate Schroeder reverberator designed by John Chowning [12] using 4 feedback comb filters and 3 all-pass filters is shown in Fig. 5.73. Again, a gain factor $g = 0.2$ was inserted into the diagram to keep the output from being clipped.

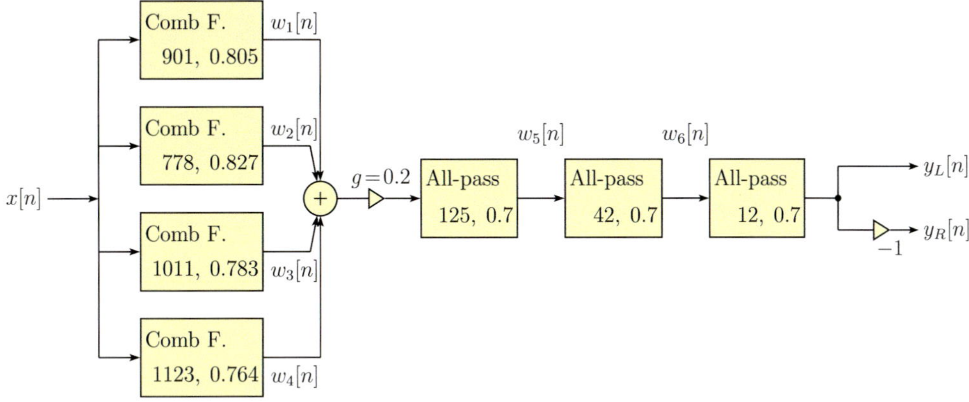

Figure 5.73 – Schroeder reverberator structure by John Chowning [12] using feedback comb filters and all-pass filters.

The script file `mexdt_5_14b.m` for implementing this system is shown below. As in the previous example, an alternate script without playback, but with capability to save the output sound to disc, is provided.

```
% Script: mexdt_5_14b.m
% Create an "audio file reader" object
sReader = dsp.AudioFileReader('AG_Duet_22050_Hz.flac','ReadRange',[1,661500]);
% Create an "audio player" object
sPlayer = audioDeviceWriter('SampleRate',sReader.SampleRate);
%------------------------------------------------------------
L1 = 901;    r1 = 0.805;    % Parameters for comb filter 1
L2 = 778;    r2 = 0.827;    % Parameters for comb filter 2
L3 = 1011;   r3 = 0.783;    % Parameters for allpass filter 1
L4 = 1123;   r4 = 0.764;    % Parameters for allpass filter 2
L5 = 125;    r5 = 0.7;
L6 = 42;     r6 = 0.7;
L7 = 12;     r7 = 0.7;
%------------------------------------------------------------
numChannels = info(sReader).NumChannels;  % Number of channels
bufferCF1 = zeros(L1,numChannels);        % Buffer for comb filter 1
bufferCF2 = zeros(L2,numChannels);        % Buffer for comb filter 2
bufferCF3 = zeros(L3,numChannels);        % Buffer for comb filter 3
bufferCF4 = zeros(L4,numChannels);        % Buffer for comb filter 4
bufferAP1 = zeros(L5,numChannels);        % Buffer for allpass filter 1
bufferAP2 = zeros(L6,numChannels);        % Buffer for allpass filter 2
bufferAP3 = zeros(L7,numChannels);        % Buffer for allpass filter 3
% Refer to Fig. 5.73 for the loop
while ~isDone(sReader)
  x = sReader();
  [w1,bufferCF1] = ss_combf(x,r1,bufferCF1);
  [w2,bufferCF2] = ss_combf(x,r2,bufferCF2);
  [w3,bufferCF3] = ss_combf(x,r3,bufferCF3);
  [w4,bufferCF4] = ss_combf(x,r4,bufferCF4);
```

```
    [w5,bufferAP1] = ss_allpassf((w1+w2+w3+w4)*0.2,r5,bufferAP1);
    [w6,bufferAP2] = ss_allpassf(w5,r6,bufferAP2);
    [y,bufferAP3] = ss_allpassf(w6,r7,bufferAP3);
    y(:,2) = -y(:,2);                        % Negate right channel
    sPlayer(y);
end
release(sReader);  % We are finished with the input audio file
release(sPlayer);  % We are finished with the audio output device
```

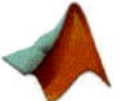

Software resources: mexdt_5_14a.m , mexdt_5_14b.m , mexdt_5_14a_Alt.m , mexdt_5_14b_Alt.m , ss_allpassf.m

Problems

5.1. Using the definition of the z-transform, compute $X(z)$ for the signals listed below. Write each transform using non-negative powers of z. In each case determine the poles and the zeros of the transform, and the region of convergence.

a. $x[n]=\{\underset{\underset{n=0}{\uparrow}}{1},1,1\}$

b. $x[n]=\{\underset{\underset{n=0}{\uparrow}}{1},1,1,1,1\}$

c. $x[n]=\{1,1,\underset{\underset{n=0}{\uparrow}}{1},1,1\}$

d. $x[n]=\{1,1,1,1,\underset{\underset{n=0}{\uparrow}}{1}\}$

5.2. For each transform $X(z)$ listed below, determine if the DTFT of the corresponding signal $x[n]$ exists. If it does, find it.

a. $X(z)=\dfrac{z\,(z-1)}{(z+1)\,(z+2)}\,,\qquad \text{ROC: } \left|z\right|>2$

b. $X(z)=\dfrac{z\,(z+2)}{(z+1/2)\,(z+3/2)}\,,\qquad \text{ROC: } \dfrac{1}{2}\left|z\right|>\dfrac{3}{2}$

c. $X(z)=\dfrac{z^2}{z^2+5z+6}\,,\qquad \text{ROC: } \left|z\right|<2$

d. $X(z)=\dfrac{(z+1)\,(z-1)}{(z+2)\,(z-3)\,(z-4)}\,,\qquad \text{ROC: } \left|z\right|<2$

5.3. Pole-zero diagrams for four transforms are shown in Fig. P.5.3. For each, determine the ROC if it is known that the DTFT of $x[n]$ exists.

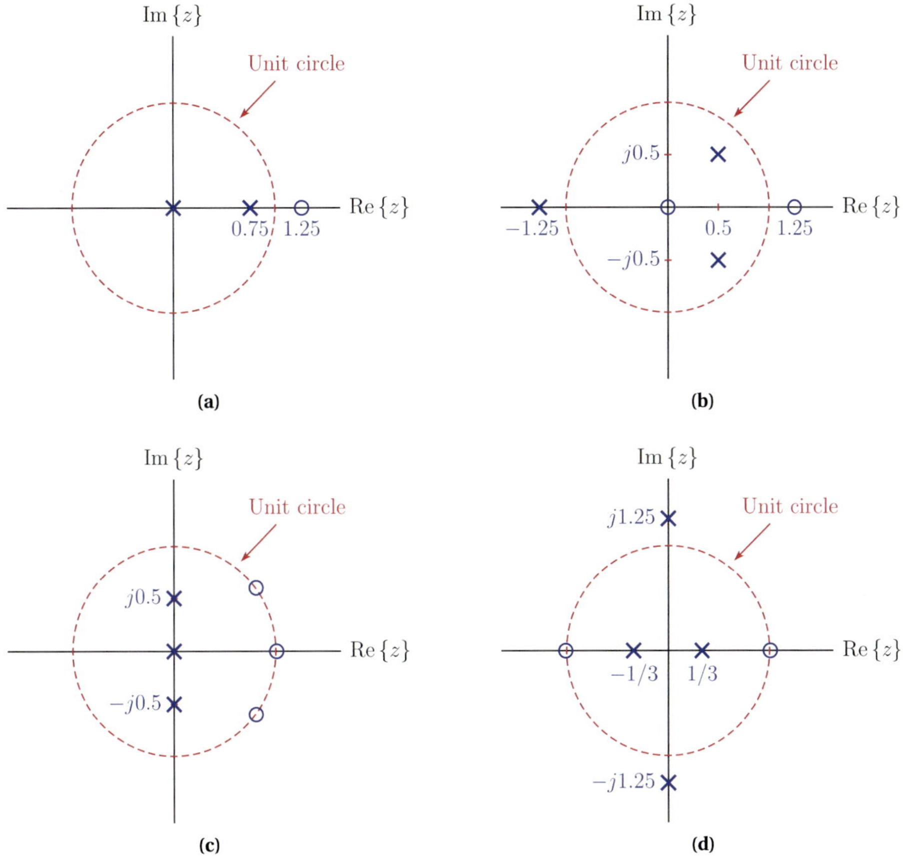

Figure P. 5.3

5.4. Determine the z-transforms of the signals given below. Indicate the ROC for each.

a. $x[n] = \begin{cases} n\,, & n=0,\ldots,9 \\ 0\,, & \text{otherwise} \end{cases}$

b. $x[n] = \begin{cases} n\,, & n=0,\ldots,9 \\ 10\,, & n \ge 10 \\ 0\,, & \text{otherwise} \end{cases}$

c. $x[n] = \begin{cases} n\,, & n=0,\ldots,9 \\ -n+20\,, & n=10,\ldots,19 \\ 0\,, & \text{otherwise} \end{cases}$

5.5. Find the z-transform of the signal

$$x[n] = \begin{cases} 1\,, & n \ge 0 \text{ and even} \\ (2/3)^n\,, & n>0 \text{ and odd} \\ 0\,, & n<0 \end{cases}$$

Also indicate the ROC.

5.6. Consider the signal $g[n]$ specified as

$$g[n] = (0.9)^n \cos(0.3n)\; u[n]$$

a. Determine the transform $G(z)$ and its ROC.

b. Let a new signal $x[n]$ be obtained from $g[n]$ through the relationship $x[n] = g[2n]$. Determine the transform $X(z)$ and its ROC from the transform $G(z)$ using the development outlined in Example 5.11.

c. Determine the transform $X(z)$ and its ROC by direct application of the z-transform definition to the signal $x[n]$. Compare to the result obtained in part (b).

5.7. Determine the z-transform of each signal listed below, and indicate the ROC. Use linearity and time shifting properties of the z-transform when needed.

a. $x[n] = \delta[n+2]$

b. $x[n] = \delta[n-3]$

c. $x[n] = \left(\frac{1}{2}\right)^n u[n]$

d. $x[n] = \left(\frac{1}{2}\right)^n u[n] + \left(\frac{1}{3}\right)^n u[n]$

e. $x[n] = \left(\frac{1}{2}\right)^{n-1} u[n-1]$

f. $x[n] = \left(\frac{1}{2}\right)^{n-1} u[n]$

g. $x[n] = \left(\frac{1}{2}\right)^{n+1} u[n]$

h. $x[n] = (3)^n u[-n-1]$

i. $x[n] = (3)^n u[-n+1]$

5.8. Consider the signal

$$x[n] = (1/2)^{|n|}$$

a. Express $x[n]$ as the sum of a causal signal $x_R[n]$ and an anti-causal signal $x_L[n]$. Find an analytical expression for each component and sketch each component.

b. Determine the transforms $X_R(z)$ and $X_L(z)$ corresponding to the signals $x_R[n]$ and $x_L[n]$, respectively. Also determine the ROC for each transform.

c. Use linearity of the z-transform to express $X(z)$ and to determine its ROC.

5.9. Use the time reversal property of the z-transform to determine $X(z)$ for the signals listed below. Indicate the ROC for each.

a. $x[n] = u[-n]$

b. $x[n] = u[-n-1]$

c. $x[n] = u[-n] - u[-n-5]$

d. $x[n] = n\, u[-n]$

e. $x[n] = \cos(\Omega_0 n)\, u[-n]$

5.10. Use the differentiation property of the z-transform to find the transform of each signal below. Also indicate the ROC for each.

a. $x[n] = \left(n^2 + 3n + 5\right) u[n]$

b. $x[n] = \left(n^2 - 5\right) u[-n-1]$

c. $x[n] = n\cos(\Omega_0 n)\, u[n]$

d. $x[n] = (n+1)\sin(\Omega_0 n)\, u[n]$

5.11. Transforms of several causal signals are listed below. Use the initial value property of the z transform to determine $x[0]$ for each case.

a. $X(z) = \dfrac{z^2 + z + 1}{z^2 + 3z + 2}$

b. $X(z) = \dfrac{z^2 + z + 1}{z^3 + 3z^2 + 3z + 1}$

c. $X(z) = \dfrac{z^{-1} + z^{-2} - z^{-3}}{1 + 0.7z^{-1} + 1.2z^{-2} - 1.5z^{-3}}$

5.12.

a. Use the correlation property of the z-transform to determine the autocorrelation function for

$$x[n] = u[n] - u[n-2]$$

Hint: Write $X(z)$ in polynomial form.

b. Generalize the result in part (a) to find the autocorrelation function for

$$x[n] = u[n] - u[n-N], \qquad N > 0$$

using z-transform techniques.

5.13. Use the correlation property of the z-transform to determine the cross correlation function for the signals

$$x[n] = u[n] - u[n-3]$$

and

$$y[n] = u[n] - u[n-5]$$

Hint: Write $X(z)$ and $Y(z)$ in polynomial form.

5.14. Prove that the cross correlation of signals $x[n]$ and $y[n]$ is equal to the convolution of $x[n]$ and $y[-n]$, that is,

$$r_{xy}[m] = x[n] * y[-n]$$

a. Using direct application of the convolution sum, and

b. Using correlation and time reversal properties of the z-transform.

5.15. Let $x[n] = a^n\, u[n]$. A signal $w[n]$ is defined in terms of $x[n]$ as

$$w[n] = \sum_{k=-\infty}^{n} x[k]$$

a. Determine $X(z)$. Afterwards determine $W(z)$ from $X(z)$ using the summation property of the z-transform.

b. Determine $w[n]$ from the transform $W(z)$ using partial fraction expansion.

c. Determine $w[n]$ directly from the summation relationship using the geometric series formula and compare to the result found in part (b).

5.16. Using the summation property of the z transform prove the well-known formula

$$\sum_{k=0}^{n} k = \frac{n\,(n+1)}{2}$$

Hint: Use $x[n] = n\,u[n]$ and $w[n] = \sum_{k=-\infty}^{n} x[k]$.

5.17. Consider the z-transforms listed below along with their ROC. For each case determine the inverse transform $x[n]$ using partial fraction expansion.

a. $X(z) = \dfrac{z}{(z+1)\,(z+2)}$, ROC: $|z| < 1$

b. $X(z) = \dfrac{z+1}{(z+1/2)\,(z+2/3)}$, ROC: $|z| > \dfrac{2}{3}$

c. $X(z) = \dfrac{z\,(z+1)}{(z-0.4)\,(z+0.7)}$, ROC: $|z| > 0.7$

d. $X(z) = \dfrac{z\,(z+1)}{(z+3/4)\,(z-1/2)\,(z-3/2)}$, ROC: $\dfrac{3}{4} < |z| < \dfrac{3}{2}$

e. $X(z) = \dfrac{z\,(z+1)}{(z+3/4)\,(z-1/2)\,(z-3/2)}$, ROC: $\dfrac{1}{2} < |z| < \dfrac{3}{4}$

5.18. The transform $X(z)$ is given by

$$\frac{(z+1)\,(z-2)}{(z+1/2)\,(z-1)\,(z-2)}$$

The ROC is not specified.

a. Construct a pole-zero plot for $X(z)$ and identify all possibilities for the ROC.

b. Express $X(z)$ using partial fractions and determine the residues.

c. For each choice of the ROC, determine the inverse transform $x[n]$.

5.19. The transforms listed below have complex poles. Each is known to be the transform of a causal signal. Determine the inverse transform $x[n]$ using partial fraction expansion.

a. $X(z) = \dfrac{z^2+3z}{z^2-1.4z+0.85}$

b. $X(z) = \dfrac{z^2}{z^2-1.6z+1}$

c. $X(z) = \dfrac{z^2+3z}{z^2-\sqrt{3}\,z+1}$

5.20. The transforms listed below have multiple poles. Express each transform using partial fractions, and determine the residues.

a. $X(z) = \dfrac{z^2+3z+2}{z^2-2z+1}$

b. $X(z) = \dfrac{z\,(z+1)}{(z+0.9)^2\,(z-1)}$

c. $X(z) = \dfrac{z^2 + 4z - 7}{(z+0.9)^2\,(z-1.2)^2}$

5.21. The difference equation that represents the remaining balance on a car loan was derived in Chapter 2 as

$$y[n] = (1+c)\; y[n-1] - x[n]$$

where c is the monthly interest rate, $x[n]$ is the payment made in month n, and $y[n]$ is the balance remaining.

a. Treat the loan repayment system as a DTLTI system and find its system function $H(z)$.

b. Let A and B represent the initial amount borrowed and the monthly payment amount, respectively. Since we would like to treat the system as DTLTI we cannot use an initial value for the difference equation. Instead, we can view the borrowed amount as a negative payment made at $n = 0$, and monthly payments can be viewed as positive payments starting at $n = 1$. With this approach the input to the system would be

$$x[n] = -A\,\delta[n] + B\,u[n-1]$$

Find the transform $X(z)$.

c. Find the transform $Y(z)$ of the output signal that represents remaining balance each month.

d. We would like to have the loan fully paid back after N payments, that is, $y[N] = 0$. Determine the amount of the monthly payment B in terms of parameters A, c, and N.

5.22. Consider again the transforms listed in Problem 5.19. Find the inverse of each transform using the long division method. In each case carry out the long division for 8 samples.

5.23. Refer to the transform $X(z)$ in Problem 5.18 for which the ROC is not given. For each possible choice of the ROC, determine the inverse transform using long division. Carry out the long division operation to compute significant samples of $x[n]$ in the interval $-5 \le n \le 5$.

5.24. Assume that the transforms listed below correspond to causal signals. Find the inverse of each transform for $n = 0, \ldots, 5$ using long division.

a. $X(z) = \dfrac{z}{(z+1)\,(z+2)}$

b. $X(z) = \dfrac{z+1}{(z+1/2)\,(z+2/3)}$

c. $X(z) = \dfrac{z\,(z+1)}{(z-0.4)\,(z+0.7)}$

5.25. For each transform given below with its ROC, find the inverse using long division. For each, obtain $x[n]$ in the range $-4, \ldots, 4$.

a. $X(z) = \dfrac{z\,(z+1)}{(z+3/4)\,(z-1/2)\,(z-3/2)}\,, \qquad \text{ROC: } \dfrac{3}{4} < |z| < \dfrac{3}{2}$

b. $X(z) = \dfrac{z\,(z+1)}{(z+3/4)\,(z-1/2)\,(z-3/2)}\,, \qquad \text{ROC: } \dfrac{1}{2} < |z| < \dfrac{3}{4}$

Compare with the analytical results found in Problem 5.17 parts (d) and (e).

5.26. Find the system function $H(z)$ for each causal DTLTI system described below by means of a difference equation. Afterward determine the impulse response of the system using partial fraction expansion.

a. $y[n] = 0.9\,y[n-1] + x[n] - x[n-1]$

b. $y[n] = 1.7\,y[n-1] - 0.72\,y[n-2] + x[n] - 2\,x[n-1]$

c. $y[n] = 1.7\,y[n-1] - 0.72\,y[n-2] + x[n] + x[n-1] + x[n-2]$

d. $y[n] = y[n-1] - 0.11\,y[n-2] - 0.07\,y[n-3] + x[n-1]$

5.27. Several causal DTLTI systems are described below by means of their system functions. Find a difference equation for each system.

a. $H(z) = \dfrac{z+1}{z^2+5z+6}$

b. $H(z) = \dfrac{(z-1)^2}{(z+2/3)\,(z+1/2)\,(z-4/5)}$

c. $H(z) = \dfrac{z^2+1}{z^3+1.2\,z^2-1.8}$

5.28. A causal DTLTI system is described by the difference equation

$$y[n] = -0.1\,y[n-1] + 0.56\,y[n-2] + x[n] - 2\,x[n-1]$$

a. Find the system function $H(z)$.

b. Determine the impulse response $h[n]$ of the system.

c. Using z-transform techniques, determine the unit step response of the system.

d. Using z-transform techniques, find the response of the system to a time reversed unit step signal $x[n] = u[-n]$.

5.29. Repeat Problem 5.28 for an anti-causal DTLTI system is described by the difference equation

$$y[n] = -\frac{5}{6}\,y[n+1] - \frac{1}{6}\,y[n+2] + \frac{1}{6}\,x[n+1] + \frac{1}{6}\,x[n+2]$$

5.30. A DTLTI system is characterized by the system function

$$H(z) = \frac{0.04\,z}{z-0.96}$$

a. Using z-transform techniques, determine the response of the system to the causal sinusoidal signal

$$x[n] = \sin(0.01n)\,u[n]$$

b. Determine the response of the system to the non-causal sinusoidal signal

$$x[n] = \sin(0.01n)$$

This is the steady-state response of the system. Refer to the discussion in Section 5.5.3.

c. Compare the responses in parts (a) and (c). Approximately how many samples does it take for the response to the causal sinusoidal signal in part (a) to be almost equal to the steady-state response found in part (b)?

5.31. The impulse response of a DTLTI system is

$$h[n] = (0.8)^n \cos(0.2\pi n)\,u[n]$$

a. Determine the system function $H(z)$ and indicate its ROC.

b. Draw the pole-zero plot for the system function and determine its stability.

c. Write a difference equation for this system.

d. Determine the unit step response of the system using z-transform techniques.

5.32. Input and output signal pairs are listed below for several DTLTI systems. For each case, determine the system function $H(z)$ along with its ROC. Also indicate if the system considered is stable and/or causal.

a. $x[n] = (1/2)^n\, u[n]\,, \qquad y[n] = 3\,(1/2)^n\, u[n] + 2\,(3/4)^n\, u[n]$

b. $x[n] = (1/2)^n\, u[n]\,, \qquad y[n] = (1/2)^{n+3}\, u[n+2] + (1/2)^{n+2}\, u[n+1]$

c. $x[n] = u[n]\,, \qquad y[n] = (n-1)\, u[n-1]$

d. $x[n] = 1.25\,\delta[n] - 0.25\,(0.8)^n\, u[n]\,, \qquad y[n] = (0.8)^n\, u[n]$

5.33. Consider the feedback control system shown in Fig. P.5.33.

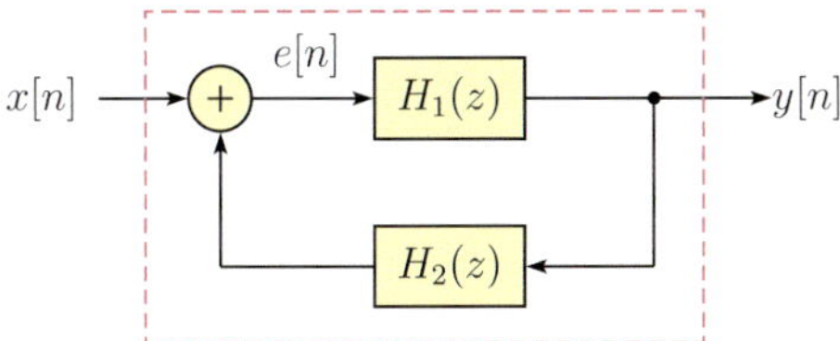

Figure P. 5.33

a. Determine the overall system function $H(z)$ in terms of the system functions $H_1(z)$ and $H_2(z)$ of the two subsystems.

b. Let the impulse responses of the two subsystems be $h_1[n] = u[n]$ and $h_2[n] = K\,\delta[n-1]$ where K is a constant gain parameter. Determine the system functions $H_1(z)$ and $H_2(z)$. Afterward determine the overall system function $H(z)$.

c. Determine the range of K for which the system is stable.

d. Let $K = 3/2$. If the input signal is a unit step function, find the signal $e[n]$ and the output signal $y[n]$.

5.34. The impulse response $h[n]$ of a DTLTI system is real-valued. Let the system function evaluated at some point $z = z_0$ be expressed in the form

$$H(z_0) = H_0\, e^{j\Theta_0}$$

where

$$H_0 = \left|H(z_0)\right|$$

and

$$\Theta_0 = \measuredangle H(z_0)$$

Show that the value of the system function at the point $z = z_0^*$ is the complex conjugate of its value at the point $z = z_0$, that is,

$$H\left(z_0^*\right) = \left[H(z_0)\right]^* = H_0\, e^{-j\Theta_0}$$

Hint: Apply the z-transform definition to $h[n]$, evaluate the resulting expression at $z = z_0^*$ and manipulate the result to complete the proof.

5.35. A DTLTI system has the system function

$$H(z) = \frac{z^2 + 3z}{z^2 - 1.4z + 0.85}$$

Find the steady-state response of the system to each of the following signals:

a. $x[n] = (0.8)^n \, e^{j0.4\pi n}$

b. $x[n] = (0.9)^n \cos(0.3\pi n)$

5.36. A DTLTI system has the system function

$$H(z) = \frac{z - 0.4}{z\left(z^2 - 1.4z + 0.85\right)}$$

a. Draw a pole-zero diagram to show poles and zeros of the system function as well as the unit circle of the z-plane.

b. Using the graphical method discussed in Section 5.5.4 determine the magnitude and the phase of the system function at the angular frequency $\Omega = \pi/6$.

c. Repeat part (b) at $\Omega = \pi/4$.

5.37. For each function $H(z)$ given below, determine if it could be the system function of a *causal and stable* system. In each case provide the reasoning behind your answer.

a. $H(z) = \dfrac{(z-2)(z+2)}{(z+1/2)}$

b. $H(z) = \dfrac{(z+1)^2}{(z-1/2)(z+1/3)}$

c. $H(z) = \dfrac{z(z+1)}{(z+1/2)(z+3/2)}$

5.38. Several causal DTLTI systems are described below by their difference equations. For each system, determine the system function $H(z)$. Draw a pole-zero diagram for the system and indicate the ROC for the system function. Determine if the system is stable.

a. $y[n] = 1.5\,y[n-1] - 0.54\,y[n-2] + x[n] + 3\,x[n-1]$

b. $y[n] = -0.64\,y[n-2] + 2\,x[n]$

c. $y[n] = 0.25\,y[n-1] - 0.125\,y[n-2] - 0.5\,y[n-3] + x[n]$

d. $y[n] = 0.25\,y[n-1] - 0.5\,y[n-2] - 0.75\,y[n-3] + x[n] + x[n-1]$

5.39. Consider again the car loan repayment system discussed in Problem 5.21.

a. Show that the loan repayment system is unstable for any practical setting. You may find it helpful to draw a pole-zero diagram.

b. Find two examples of bounded input signals $|x[n]| < \infty$ that result in unbounded output signals.

c. Can a bounded input signal be found that produces an output signal with a nonzero constant steady-state value? What type of a payment scheme does this represent?

5.40. A first-order all-pass filter section is characterized by the system function

$$H(z) = \frac{z - r\,e^{j\Omega_0}}{z - (1/r)\,e^{j\Omega_0}}$$

Show that the phase characteristic of the system is

$$\measuredangle H(\Omega) = \tan^{-1}\left[\frac{\left(r^2-1\right)\sin(\Omega-\Omega_0)}{2r - \left(r^2+1\right)\cos(\Omega-\Omega_0)}\right]$$

Hint: First find $H(\Omega)$ by substituting $z = e^{j\Omega}$. Afterward express $H(\Omega)$ in Cartesian form as

$$H(\Omega) = H_r(\Omega) + j\,H_i(\Omega)$$

5.41. In Section 5.5.8, the system function for an allpass filter of order L was found to be

$$H(z) = \frac{z^{-L} - r}{1 - r\,z^{-L}}$$

a. Show that the magnitude of the system function is a constant independent of L and r.

b. Find an expression for the phase of the system function.

5.42. Develop a direct-form II block diagram for each system specified below by means of a system function. Assume that each system is causal and initially relaxed.

a. $$H(z) = \frac{z+2}{z-1/2}$$

b. $$H(z) = \frac{z^2+1}{z^3+0.8\,z^2-2.2\,z+0.6}$$

c. $$H(z) = \frac{z^3+2\,z^2-3\,z+4}{2\,z^3+0.8\,z^2+1.8\,z+3.2}$$

5.43. Develop a cascade-form block diagram for each system specified below by means of a system function. Assume that each system is causal and initially relaxed. Use first and second-order cascade sections and ensure that all coefficients are real.

a. $$H(z) = \frac{z+1}{(z+1/2)\,(z+2/3)}$$

b. $$H(z) = \frac{z\,(z+1)}{(z-0.4)\,(z+0.7)}$$

c. $$H(z) = \frac{z\,(z+1)}{\left(z+0.6\right)\left(z^2-1.4\,z+0.85\right)}$$

5.44. Develop a parallel-form block diagram for each system specified in Problem 5.43. Assume that each system is causal and initially relaxed. Use first and second-order parallel sections and ensure that all coefficients are real.

5.45. Let $X_u(z)$ be the unilateral z-transform of $x[n]$.

a. Using the definition of the unilateral z-transform show that the transform of $x[n+1]$ is

$$\mathcal{Z}_u\left\{x[n+1]\right\} = z\,X_u(z) - z\,x[0]$$

b. Show that the transform of $x[n+2]$ is

$$\mathcal{Z}_u\left\{x[n+2]\right\} = z^2\, X_u(z) - z^2\, x[0] - z\, x[1]$$

c. Generalize the results of parts (a) and (b), and show that for $k > 0$

$$\mathcal{Z}_u\left\{x[n+k]\right\} = z^k\, X_u(z) - \sum_{n=0}^{k-1} x[n]\, z^{k-n}$$

5.46. Using the unilateral z-transform determine the natural response of each system with the homogeneous difference equation and initial conditions given below:

a. $y[n] - 1.4y[n-1] + 0.85y[n-2] = 0\,,\quad y[-1] = 5\,,\quad \text{and}\quad y[-2] = 7$

b. $y[n] - 1.6y[n-1] + 0.64y[n-2] = 0\,,\quad y[-1] = 2\,,\quad \text{and}\quad y[-2] = -3$

5.47. Consider the difference equations given below with the specified input signals and initial conditions. Using the unilateral z-transform determine the solution $y[n]$ for each.

a. $y[n] - 2\,y[n-1] = x[n]\,,\quad x[n] = \cos(0.2\pi n)\; u[n]\,,\quad y[-1] = 5$

b. $y[n] + 0.6\,y[n-1] = x[n] + x[n-1]\,,\quad x[n] = u[n]\,,\quad y[-1] = -3$

c. $y[n] - 0.2\,y[n-1] - 0.48 = x[n]\,,\quad x[n] = u[n]\,,\quad y[-1] = -1\,,\quad y[-2] = 2$

5.48. The *Fibonnaci sequence* is a well-known sequence of integers. Fibonnaci numbers start with the pattern

$$0, 1, 1, 2, 3, 5, 8, 13, 21, 34, 55, 89, \ldots$$

They follow the recursion relationship

$$F_{n+2} = F_{n+1} + F_n$$

with $F_0 = 0$ and $F_1 = 1$. In simplest terms, each number in the sequence is the sum of the two numbers before it. In this problem we will use difference equations and the unilateral z-transform to study some properties of the Fibonnaci sequence.

a. Construct a homogeneous difference equation to represent the recurrence relationship of the Fibonnaci sequence. Select $y[n] = F_{n+2}$ with initial conditions $y[-1] = F_1 = 1$ and $y[-2] = F_0 = 0$.

b. Find the solution $y[n]$ of the homogeneous difference equation using the unilateral z-transform.

c. It can be shown that, for large n, the ratio of consecutive Fibonnaci numbers converges to a constant known as the *golden ratio*, that is,

$$\lim_{n\to\infty} \frac{y[n+1]}{y[n]} = \varphi$$

Using the solution $y[n]$ found in part (b) determine the value of φ.

MATLAB Problems

5.49. The signal $x[n]$ is given by

$$x[n] = (0.8)^n \left(u[n] - u[n-8]\right)$$

a. Find $X(z)$. Determine its poles and zeros. Manually construct a pole-zero diagram.

b. Write a MATLAB script to evaluate the magnitude of $X(z)$ at a grid of complex points in the z-plane. Use the function `meshgrid()` to generate the grid of complex points within the ranges $-1.5 < \mathrm{Re}\{z\} < 1.5$ and $-1.5 < \mathrm{Im}\{z\} < 1.5$ with increments of 0.05 in each direction.

c. Use the function `mesh()` to produce a three dimensional mesh plot of $|X(z)|$.

d. Evaluate the z-transform for $z = e^{j\Omega}$ and use the function `plot3()` to plot it over the three-dimensional mesh plot.

5.50.

a. Repeat Problem 5.49 with the signal

$$x[n] = (0.6)^n \left(u[n] - u[n-8]\right)$$

b. Repeat Problem 5.49 with the signal

$$x[n] = (0.4)^n \left(u[n] - u[n-8]\right)$$

5.51. Refer to the system in problem 5.30.

a. Construct a system object for $H(z)$ using the function `zpk()`.

b. Compute the response of the system to the input signal

$$x[n] = \sin(0.01n)\; u[n]$$

for $n = 0, \ldots, 49$ using the function using the function `dlsim()`. Compare the first few samples of the response to the result of hand calculations in Problem 5.30.

5.52. Refer to the system in problem 5.31.

a. Write a MATLAB script to find the impulse response of the system by iteratively solving the difference equation found in Problem 5.31 part (c). Compute $h[n]$ for $n = 0, \ldots, 10$. Does it match the expected answer?

b. Modify the script written in part (a) so that the unit step response of the system is computed by iteratively solving the difference equation. Compare the result to that found in Problem 5.31 part (d).

c. Construct a system object for $H(z)$ using the function `tf()`. Afterwards compute the impulse response and the unit step response of the system using functions `ss_dimpulse()` and `ss_dstep()` respectively. Compare to the answers obtained in previous parts.

5.53. Refer to the DTLTI system in Problem 5.35.

a. Write a MATLAB script to compute and graph the magnitude and the phase of the frequency spectrum for the system. Your script should also mark the points critical for the two input signals used in Problem 5.35.

b. Graph the steady-state response to each of the two input signals specified.

5.54. Develop a script to compute the phase response of a first-order all-pass filter with a real pole at $p = r + j0$ for parameter values $r = 0.2\,, 0.4\,, 0.6,\ 0.8$. Graph the four phase responses on the same coordinate system for comparison.

5.55. Develop a script to compute the phase response of a second-order all-pass filter with a complex conjugate poles at

$$p_{1,2} = r\,e^{j\theta}$$

a. In your script set $\theta = \pi/6$ radians. Compute the phase response for $r = 0.4\,, 0.6\,, 0.8$, and graph the results superimposed for comparison.

b. Set $r = 0.8$. Compute the phase response for $\theta = \pi/6\,, \pi/4\,, \pi/3$ radians, and graph the results superimposed for comparison.

5.56. Refer to Problem 5.46.

a. For each difference equation given, write a script to find the solution for sample indices $n = 0, \ldots, 10$ using the iterative solution method with the specified initial conditions.

b. For each difference equation given, write a script to find the analytical solution using symbolic mathematics capabilities of MATLAB.

MATLAB Project

5.57. In this project the concept of *dual-tone multi-frequency (DTMF)* signaling will be explored. As the name implies, DTMF signals are mixtures of two sinusoids at distinct frequencies. They are used in communications over analog telephone lines. A particular version of DTMF signaling is utilized in dialing a number with push-button telephone handsets, a scheme known as *touch tone dialing.* When the caller dials a number, the DTMF generator produces a dual-tone signal for each digit dialed. The synthesized signal is in the form

$$x_k(t) = \sin\left(2\pi f_1 t\right) + \sin\left(2\pi f_2 t\right)\,, \qquad 0 \le t \le T_d$$

Frequency assignments for the digits on a telephone keypad are shown in Fig. P.5.57a.

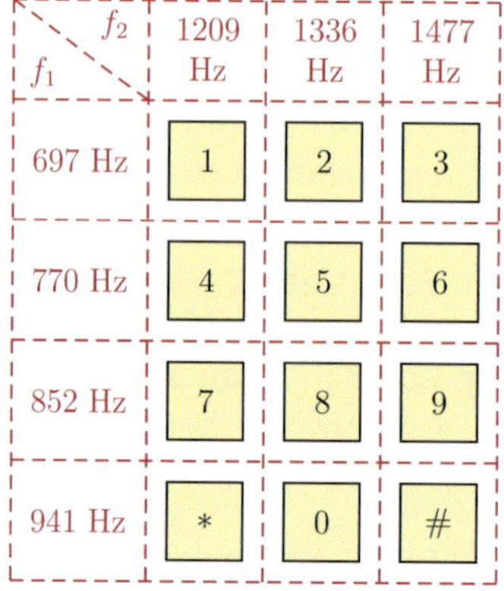

f_2
f_1
1209 Hz
1336 Hz
1477 Hz
697 Hz
770 Hz
852 Hz
941 Hz
1
2
3
4
5
6
7
8
9
*
0
#

Figure P. 5.57a

Although the sinusoidal signals needed for DTMF can be generated using trigonometric functions, in this project we will explore alternative means to generate them.

In some applications we may wish to use a inexpensive processors that may either have no support for trigonometric functions or may be too slow in computing them. An alternative method of computing the samples of a sinusoidal signal is to use a discrete-time system the impulse response of which is sinusoidal, and to apply an impulse signal to its input. Recall that the z-transform of a discrete-time sine signal was found in Example 5.16 as

$$\mathcal{Z}\left\{\sin\left(\Omega_i n\right) u[n]\right\} = \frac{\sin\left(\Omega_i\right) z}{z^2 - 2\cos\left(\Omega_i\right) z + 1}$$

which could be used as the basis of a second-order system as shown in Fig. 5.57b. DTMF signals that correspond to digits on a keypad can be generated by using two systems of this type in parallel as shown in Fig. 5.57b by adjusting the coefficients properly.

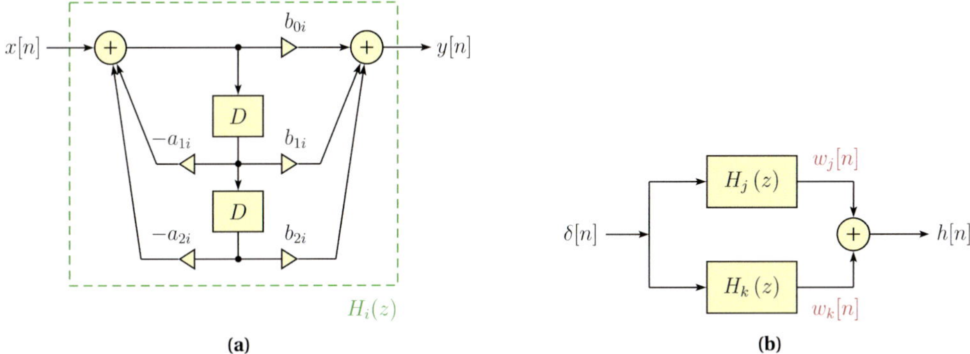

Figure P. 5.57b

a. Develop a MATLAB function `ss_dtmf2()` to generate samples of the DTMF signal for a specified digit for a specified duration. The syntax of the function should be as follows:

```
x = ss_dtmf2(digit,fs,n)
```

The first argument `digit` is the digit for which the DTMF signal is to be generated. Let values digit = 0 through digit = 9 represent the corresponding keys on the keypad. Map the remaining two keys "∗" and "#" to values digit = 10 and digit = 11, respectively. Finally, the value digit = 12 should represent a pause, that is, a silent period. The arguments `fs` and `n` are the sampling rate (in Hz) and the number of samples to be produced, respectively. Internally you may want to use the function `ss_iir2()` that was developed in MATLAB Exercise 5.12.

b. Develop a function named `ss_dtmf()` with the syntax

```
x = ss_dtmf(number,fs,nd,np)
```

The arguments for the function `ss_dtmf()` are defined as follows:

`number`: The phone number to be dialed, entered as a vector. For example, to dial the number 555-1212, the vector `number` would be entered as

```
number = [5,5,5,1,2,1,2]
```

`fs`: The sampling rate (in Hz) used in computing the amplitudes of the DTMF signal.

`nd`: Number of samples for each digit.

`np`: Number of samples for each pause between consecutive digits.

The function `ss_dtmf()` should use the function `ss_dtmf2()` to produce the signals for each digit (and the pauses in between digits) and append them together to create the signal $x[n]$.

c. Write a script to test the function `ss_dtmf()` with the number 555-1212. Use a sampling rate of $f_s = 8000$ Hz. The duration of each digit should be 200 ms with 80 ms pauses between digits.

d. Play back the resulting signal $x[n]$ using the `sound()` function.

CHAPTER 6

STATE-SPACE ANALYSIS OF DISCRETE-TIME SYSTEMS

Chapter Objectives

- Understand the concepts of a *state variable* and *state-space representation* of systems.
- Learn the advantages of state-space modeling over other methods of representing discrete-time systems.
- Learn to derive state-space models for DTLTI systems from the knowledge of other system descriptions such as a system function or a difference equation.
- Learn how to obtain a system function or a difference equation from the state-space model of a system. Explore similarity transformations for obtaining alternative state-space models for a given system.
- Develop techniques for solving state-space models for determining the state variables and the output signal of a system. Learn homogeneous and forced solutions. Understand the significance of the state transition matrix. Learn its fundamental properties and how it is used in solving the state equation.
- Explore methods of converting state space model for a CTLTI system to that of a DTLTI system to allow simulation of CTLTI systems on a computer.

DOI: 10.1201/9781003570462-6

6.1 Introduction

In previous chapters of this text we have considered three different methods for describing a DTLTI system:

1. A linear constant-coefficient difference equation involving the input and the output variables of the system in the time domain,
2. a system function $H(z)$ that describes the frequency-domain behavior characteristics of the system,
3. an impulse response $h[n]$ that describes how the system responds to a unit impulse function $\delta(n)$.

We know that these three descriptions are equivalent. For a DTLTI system, the system function $H(z)$ can be obtained from the difference equation using z-transform techniques, specifically using the linearity and the time shifting properties of the z-transform. Afterward, the impulse response $h[n]$ can be determined from the appropriate system function using inverse transform techniques such as partial fraction expansion. By reversing the steps described above, it is also possible to obtain the system function from the impulse response, or to obtain the difference equation from the system function. Thus, the three definition forms listed above for a linear and time-invariant system are interchangeable since we are able to go back and forth between them freely.

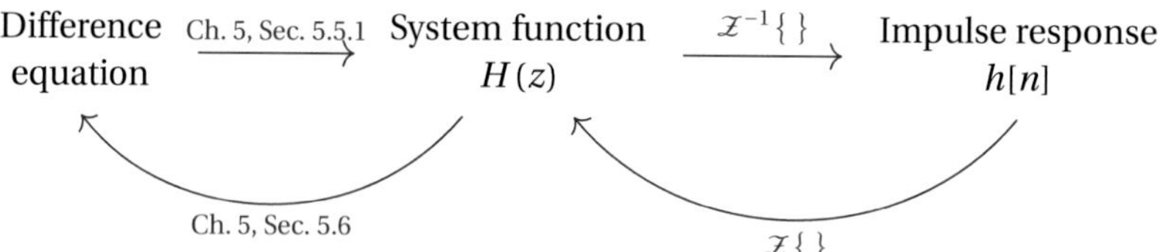

For signal-system interaction problems, that is, for problems involving the actions of a system on an input signal to produce an output signal, the solution of a general N-th order difference equation may be difficult or impossible to carry out analytically for anything other than fairly simple systems and input signals. Sometimes we can avoid the need to solve the difference equation, and attempt to solve the problem using frequency-domain techniques instead. This involves first translating the problem into the frequency domain, solving it in terms of transform-based descriptions of the signals and the systems involved, and finally translating the results back to the time domain. For systems that have nonzero initial conditions, this approach requires the use of the unilateral variant of the z-transform, If the system under consideration is nonlinear or time-varying or both, then the system function based approach cannot be used at all, since system functions are only available for DTLTI systems.

The system descriptions discussed above each represent the system in terms of its input-output relationship. Essentially, the system is treated like a box with input and output ports, and user interaction with the system takes place through those ports only. No attention is paid to what happens internally. Imagine a scenario where an input signal in the sample index range $n = 0, \ldots, L$ drives the system, and we determine the output signal for the same duration. Right after that, a second input signal is applied to the system, this time for the index range $n = L+1, \ldots, 2L$, and we would like to compute the output signal in the same time interval. How would we approach this problem? Clearly, we cannot treat the second part of the problem as a fresh start, since the system has been brought to a certain internal *state* by the first input signal. The solution of the second part of the problem must take this into account.

In this chapter we will introduce yet another method for describing a system, namely *state-space representation* based on a set of variables called the *state variables* of the system. Consider a discrete-time system that is originally described by means of a N-th order difference equation. A state-space model for such a system can be obtained by replacing the N-th order difference equation with N first-order difference equations that involve as many state variables and their one-sample delayed versions. In general, the N first-order difference equations are coupled with each other, and must be solved together as a set rather than as individual equations.

As it should be evident from the foregoing discussion, the state-space model of a system is a time-domain model rather than a frequency-domain one since it is derived from a difference equation. State-space modeling of systems provides a number of advantages over other modeling techniques:

1. State-space modeling provides a standard method of formulating the time-domain input-output relationship of a system independent of the order of the system.
2. State-space model of a system describes the internal structure of the system, in contrast with the other description forms which simply describe the relationship between input and output signals.
3. Systems with multiple inputs and/or outputs can be handled within the standard formulation we will develop in the rest of this chapter without the need to resort to special techniques that may be dependent on the topology of the system or on the number of input or output signals.
4. State-space models for DTLTI systems are naturally suitable for use in computer simulation of these systems.
5. State-space modeling also lends itself to the representation of nonlinear and/or time-varying systems, a task that is not achievable through modeling with the use of the system function or the impulse response. A N-th order difference equation can be used for representing nonlinear and/or time-varying signals, however, computer solutions for these types of systems still necessitate the use of state-space modeling techniques.

In Section 6.2 we discuss general characteristics of state-space models. Section 6.3 focuses on the formulation of state space models for DTLTI systems. The discussion starts with the problem of obtaining a state-space model for a DTLTI system described by means of a difference equation or a system function. The state-space model of a system is not unique; a method for obtaining alternative state-space models for the same system are discussed in Section 6.3.3. Extension to multiple-input multiple-output systems is the subject of Section 6.3.4. Afterward, methods for solving state equations are given. The problem of obtaining the system function from state equations is examined. Conversion of the state-space model of a CTLTI system into that of an approximately equivalent DTLTI system for the purpose of simulation is the subject of Section 6.6.

6.2 State-Space Modeling of Discrete-Time Systems

Consider a discrete-time system with a single input signal $r[n]$ and a single output signal $y[n]$. Such a system can be modeled by means of the relationships between its input and output signals as well as N internal variables $x_i[n]$ for $i = 1,2,\ldots,N$ as shown in Fig. 6.1. Note that in this chapter we will change our notation a bit, and use $r[n]$ for an input signal instead of the usual $x[n]$. This change is needed in order to reserve the use of $x_i[n]$ for state variables as it is customary.

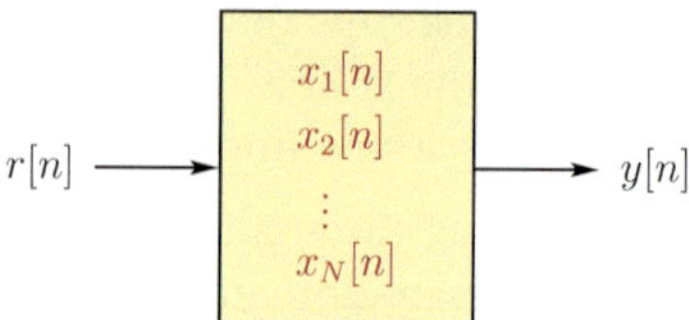

Figure 6.1 – State-space representation of a discrete-time system.

The internal variables $x_1[n], x_2[n], \ldots, x_N[n]$ are the state variables of the discrete-time system. Let us describe the system through a set of first-order difference equations written for each state variable:

$$\begin{aligned}
x_1[n+1] &= f_1\left(x_1[n],\ldots,x_N[n],r[n]\right)\\
x_2[n+1] &= f_2\left(x_1[n],\ldots,x_N[n],r[n]\right)\\
&\vdots\\
x_N[n+1] &= f_N\left(x_1[n],\ldots,x_N[n],r[n]\right)
\end{aligned} \tag{6.1}$$

The equations given in Eqn. (6.1) are the *state equations* of the system. They describe how each state variable changes in the time domain in response to

1. the input signal $r[n]$, and
2. the current values of all state variables.

Observing the time-domain behavior of the state variables provides us with information about the internal operation of the system. Once the N state equations are solved for the N state variables, the output signal can be found through the use of the *output equation*

$$y[n] = g\left(x_1[n],\ldots,x_N[n]\right) \tag{6.2}$$

Eqns. (6.1) and (6.2) together form a *state-space model* for the system.

Systems with multiple inputs and/or outputs can be represented with state-space models as well. Consider a discrete-time system with K input signals $\{r_i[n];\ i = 1,\ldots,K\}$ and M output signals $\{y_j[n];\ i = 1,\ldots,M\}$ as shown in Fig. 6.2.

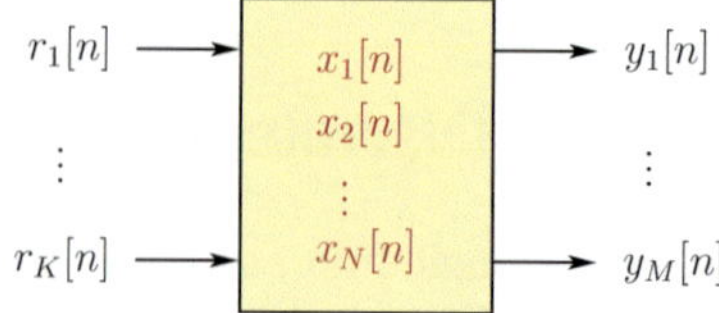

Figure 6.2 – State-space representation of a multiple-input multiple-output discrete-time system.

The state equations given by Eqn. (6.1) can be modified to apply to a multiple-input multiple-output system through a state-space model in the form

$$\begin{aligned}
x_1[n+1] &= f_1\left(x_1[n],\ldots,x_N[n],r_1[n],\ldots,r_K[n]\right)\\
x_2[n+1] &= f_2\left(x_1[n],\ldots,x_N[n],r_1[n],\ldots,r_K[n]\right)\\
&\vdots\\
x_N[n+1] &= f_N\left(x_1[n],\ldots,x_N[n],r_1[n],\ldots,r_K[n]\right)
\end{aligned} \tag{6.3}$$

Each right-side function $f_i(\ldots)$ includes, as arguments, all input signals as well as the state variables. In addition to this change, the output equation given by Eqn. (6.2) needs to be replaced with a set of output equations for computing all output signals:

$$\begin{aligned}
y_1[n] &= g_1\left(x_1[n],\ldots,x_N[n],r_1[n],\ldots,r_K[n]\right)\\
y_2[n] &= g_2\left(x_1[n],\ldots,x_N[n],r_1[n],\ldots,r_K[n]\right)\\
&\vdots\\
y_M[n] &= g_M\left(x_1[n],\ldots,x_N[n],r_1[n],\ldots,r_K[n]\right)
\end{aligned} \tag{6.4}$$

6.3 State-Space Models for DTLTI Systems

If the discrete-time system under consideration is also linear and time-invariant (DTLTI), the right-side functions $f_i(\ldots), i=1,2,\ldots,N$ and $g_i(\ldots), i=1,2,\ldots,M$ in Eqns. (6.3) and (6.4) are linear functions of the state variables and the input signals. For a single-input single-output DTLTI system, the state-space representation is in the form

$$\begin{aligned}
x_1[n+1] &= a_{11}\,x_1[n]+a_{12}\,x_2[n]+\ldots+a_{1N}\,x_N[n]+b_1\,r[n]\\
x_2[n+1] &= a_{21}\,x_1[n]+a_{22}\,x_2[n]+\ldots+a_{2N}\,x_N[n]+b_2\,r[n]\\
&\vdots\\
x_N[n+1] &= a_{N1}\,x_1[n]+a_{N2}\,x_2[n]+\ldots+a_{NN}\,x_N[n]+b_N\,r[n]
\end{aligned} \tag{6.5}$$

and the output equation is

$$y[n] = c_1\,x_1[n]+c_2\,x_2[n]+\ldots,c_N\,x_N[n]+d\,r[n] \tag{6.6}$$

where the coefficients a_{ij}, b_i, c_j and d used in the state equations and the output equation are all constants independent of the time variable. Eqns. (6.5) and (6.6) can be expressed in matrix notation as follows:

State-space representation of a single-input single-output DTLTI system:

State equation: $$\mathbf{x}[n+1] = \mathbf{A}\mathbf{x}[n]+\mathbf{B}\,r[n] \tag{6.7}$$

Output equation: $$y[n] = \mathbf{C}\mathbf{x}[n]+d\,r[n] \tag{6.8}$$

The coefficient matrices used in Eqns. (6.7) and (6.8) are

$$\mathbf{A}=\begin{bmatrix} a_{11} & a_{12} & \dots & a_{1N} \\ a_{21} & a_{22} & \dots & a_{2N} \\ \vdots & & & \\ a_{N1} & a_{N2} & \dots & a_{NN} \end{bmatrix}, \quad \mathbf{B}=\begin{bmatrix} b_1 \\ b_2 \\ \vdots \\ b_N \end{bmatrix}, \quad \mathbf{C}=\begin{bmatrix} c_1 \\ c_2 \\ \vdots \\ c_N \end{bmatrix}^T \tag{6.9}$$

The vector $\mathbf{x}[n]$ is the *state vector* that consists of the N state variables:

$$\mathbf{x}[n]=\begin{bmatrix} x_1[n] \\ x_2[n] \\ \vdots \\ x_N[n] \end{bmatrix} \tag{6.10}$$

The vector $\mathbf{x}[n+1]$ contains versions of the N state variables advanced by one sample:

$$\mathbf{x}[n+1]=\begin{bmatrix} x_1[n+1] \\ x_2[n+1] \\ \vdots \\ x_N[n+1] \end{bmatrix} \tag{6.11}$$

Thus, a single-input single-output DTLTI system can be uniquely described by means of the matrix $\mathbf{A}$, the vectors $\mathbf{B}$ and $\mathbf{C}$, and the scalar d. If the initial value $\mathbf{x}[0]$ of the state vector is known, the state equation given by Eqn. (6.7) can be solved for $\mathbf{x}[n]$. Once $\mathbf{x}[n]$ is determined, the output signal can be found through the use of Eqn. (6.8).

6.3.1 Obtaining state-space model from a difference equation

Converting a N-th order difference equation to a state-space model requires finding N first-order difference equations that represent the same system, and an output equation that links the output signal to the state variables. As in the case of continuous-time state-space models, the solution is not unique. Consider a DTLTI system described by a difference equation in the form

$$y[n]+a_1\,y[n-1]+a_2\,y[n-2]+\ldots+a_{N-1}\,y[n-N+1]+a_N\,y[n-N]=b_0\,r[n] \tag{6.12}$$

Let us rewrite the difference equation so that $y[n]$ is computed from the remaining terms:

$$y[n]=-a_1\,y[n-1]-a_2\,y[n-2]-\ldots-a_{N-1}\,y[n-N+1]-a_N\,y[n-N]+b_0\,r[n] \tag{6.13}$$

We will choose the state variables as follows:

$$\begin{aligned} x_1[n] &= y[n-N] & & \\ x_2[n] &= y[n-N+1] & \Longrightarrow \quad x_1[n+1] &= x_2[n] \\ x_3[n] &= y[n-N+2] & \Longrightarrow \quad x_2[n+1] &= x_3[n] \\ &\vdots & & \\ x_N[n] &= y[n-1] & \Longrightarrow \quad x_{N-1}[n+1] &= x_N[n] \end{aligned}$$

The resulting state equations are

$$
\begin{aligned}
x_1[n+1] &= x_2[n] \\
x_2[n+1] &= x_3[n] \\
&\vdots \\
x_{N-1}[n+1] &= x_N[n] \\
x_N[n+1] &= -a_N\,x_1[n] - a_{N-1}\,x_2[n] - \ldots - a_2\,x_{N-1}[n] - a_1\,x_N[n] + b_0\,r[n]
\end{aligned}
\tag{6.14}
$$

The last state equation is obtained by using the state variables $x_1[n]$ through $x_N[n]$ defined above in the difference equation given by Eqn. (6.13), and writing $y[n] = x_N[n+1]$ in terms of the state variables $x_1[n]$ through $x_N[n]$ and the input signal $r[n]$. The output equation is readily obtained from the last state equation:

$$
\begin{aligned}
y[n] &= x_N[n+1] \\
&= -a_N\,x_1[n] - a_{N-1}\,x_2[n] - \ldots - a_2\,x_{N-1}[n] - a_1\,x_N[n] + b_0\,r[n]
\end{aligned}
\tag{6.15}
$$

In matrix form, the state-space model is

$$
\begin{aligned}
\mathbf{x}[n+1] &= \mathbf{A}\mathbf{x}[n] + \mathbf{B}\,r[n] \\
y[n] &= \mathbf{C}\mathbf{x}[n] + d\,r[n]
\end{aligned}
$$

with

$$
\mathbf{A} = \begin{bmatrix}
0 & 1 & 0 & \ldots & 0 & 0 \\
0 & 0 & 1 & \ldots & 0 & 0 \\
\vdots & \vdots & \vdots & \ldots & \vdots & \vdots \\
0 & 0 & 0 & \ldots & 1 & 0 \\
0 & 0 & 0 & \ldots & 0 & 1 \\
-a_N & -a_{N-1} & -a_{N-2} & \ldots & -a_2 & -a_1
\end{bmatrix}, \qquad
\mathbf{B} = \begin{bmatrix} 0 \\ 0 \\ 0 \\ \vdots \\ 0 \\ b_0 \end{bmatrix}
\tag{6.16}
$$

for the state equation, and

$$
\mathbf{C} = \begin{bmatrix} -a_N & -a_{N-1} & -a_{N-2} & \ldots & -a_2 & -a_1 \end{bmatrix}, \qquad d = b_0
\tag{6.17}
$$

for the output equation. This is the familiar *phase-variable canonical form* of state equations. The coefficient matrix **A** is given below in partitioned form:

$$
\mathbf{A} = \left[\begin{array}{c|ccccc}
0 & 1 & 0 & \ldots & 0 & 0 \\
0 & 0 & 1 & \ldots & 0 & 0 \\
\vdots & \vdots & \vdots & \ldots & \vdots & \vdots \\
0 & 0 & 0 & \ldots & 1 & 0 \\
0 & 0 & 0 & \ldots & 0 & 1 \\
\hline
-a_N & -a_{N-1} & -a_{N-2} & \ldots & -a_2 & -a_1
\end{array}\right]
\tag{6.18}
$$

The top left partition is a $(N-1)\times 1$ vector of zeros; the top right partition is an identity matrix of order $(N-1)$; the bottom row is a vector of difference equation coefficients in reverse order and with a sign change. This is handy for obtaining the state-space model by directly inspecting the difference equation. It is important to remember, however, that the coefficient of $y[n]$ in Eqn. (6.12) must be equal to unity for this to work.

Example 6.1: State-space model from difference equation

Find a state-space model for a DTLTI system described by the difference equation

$$y[n]+1.2\,y[n-1]-0.13\,y[n-2]-0.36\,y[n-3]=2\,r[n]$$

Solution: Expressing $y[n]$ as a function of the other terms leads to

$$y[n]=-1.2\,y[n-1]+0.13\,y[n-2]+0.36\,y[n-3]+2\,r[n]$$

State variables can be defined as

$$\begin{aligned} x_1[n] &= y[n-3] \\ x_2[n] &= y[n-2] \quad \Longrightarrow \quad x_1[n+1]=x_2[n] \\ x_3[n] &= y[n-1] \quad \Longrightarrow \quad x_2[n+1]=x_3[n] \end{aligned}$$

Recognizing that

$$y[n]=x_3[n+1]$$

the difference equation can be written as

$$x_3[n+1]=-1.2\,x_3[n]+0.13\,x_2[n]+0.36\,x_1[n]+2\,r[n]$$

In matrix form, the state-space model is

$$\mathbf{x}[n+1]=\begin{bmatrix} 0 & 1 & 0 \\ 0 & 0 & 1 \\ 0.36 & 0.13 & -1.2 \end{bmatrix}\mathbf{x}[n]+\begin{bmatrix} 0 \\ 0 \\ 2 \end{bmatrix} r[n]$$

$$y[n]=\begin{bmatrix} 0.36 & 0.13 & -1.2 \end{bmatrix}\mathbf{x}[n]+2\,r[n]$$

The inspection method for deriving the state-space model directly from the difference equation would not be as straightforward to use if delayed versions of the input signal $r[n]$ appear in the difference equation. We will, however, generalize the relationship between the coefficients of the difference equation and the state-space model when we consider the derivation of the latter from the system function in the next section.

6.3.2 Obtaining state-space model from a system function

If the DTLTI system under consideration has a system function that is simple in the sense that it has no multiple poles and all of its poles are real-valued, then a state-space model can be easily found using partial fraction expansion. This idea will be explored in the next example.

Example 6.2: State-space model from z-domain system function

A DTLTI system has the system function

$$H(z) = \frac{Y(z)}{R(z)} = \frac{7z}{(z-0.6)\,(z+0.8)}$$

Find a state-space model for this system.

Solution: Expanding $H(z)$ into partial fractions we get

$$H(z) = \frac{3}{z-0.6} + \frac{4}{z+0.8}$$

Note that we did not divide $H(z)$ by z before expanding it into partial fractions, since in this case we do not need the z term in the numerator of each partial fraction. The z-transform of the output signal is

$$Y(z)\,\frac{3}{z-0.6}\,R(z) + \frac{4}{z+0.8}\,R(z)$$

Let us define $X_1(z)$ and $X_2(z)$ as

$$X_1(z) = \frac{3}{z-0.6}\,R(z)$$

and

$$X_2(z) = \frac{4}{z+0.8}\,R(z)$$

so that $Y(z) = X_1(x) + X_2(z)$. Rearranging the terms in these two equations we get

$$z\,X_1(z) = 0.6\,X_1(z) + 3\,R(z)$$

and similarly

$$z\,X_2(z) = -0.8\,X_2(z) + 4\,R(z)$$

Taking inverse z-transforms leads to the two first-order difference equations

$$\begin{aligned} x_1[n+1] &= 0.6\,x_1[n] + 3\,r[n] \\ x_2[n+1] &= -0.8\,x_2[n] + 4\,r[n] \end{aligned}$$

The output equation is

$$y[n] = x_1[n] + x_2[n]$$

Using $x_1[n]$ and $x_2[n]$ as the state variables, the state-space model can be written as

$$\begin{bmatrix} x_1[n+1] \\ x_2[n+1] \end{bmatrix} = \begin{bmatrix} 0.6 & 0 \\ 0 & -0.8 \end{bmatrix} \begin{bmatrix} x_1[n] \\ x_2[n] \end{bmatrix} + \begin{bmatrix} 3 \\ 4 \end{bmatrix} r[n]$$

and

$$y[n] = \begin{bmatrix} 1 & 1 \end{bmatrix} \begin{bmatrix} x_1[n] \\ x_2[n] \end{bmatrix}$$

A state-space model was easily obtained for the system in Example 6.2 owing to the fact that the system function $H(z)$ has only first-order real poles. As in the continuous-time case, the state equations obtained are uncoupled from each other, that is, each state equation involves only one of the state variables. This leads to a state matrix **A** that is diagonal. It is therefore possible to solve the state equations independently of each other.

If the system function does not have the properties that afforded us the simplified solution of Example 6.2, finding a state-space model for a system function involves the use of a block diagram as an intermediate step. Consider the system function

$$H(z) = \frac{Y(z)}{X(z)} = \frac{b_0 + b_1 z^{-1} + b_2 z^{-2} + b_3 z^{-3}}{1 + a_1 z^{-1} + a_2 z^{-2} + a_3 z^{-3}} \tag{6.19}$$

A direct-form II block diagram for implementing this system is shown in Fig. 6.3 (see Section 5.6 of Chapter 5 for the details of obtaining a block diagram).

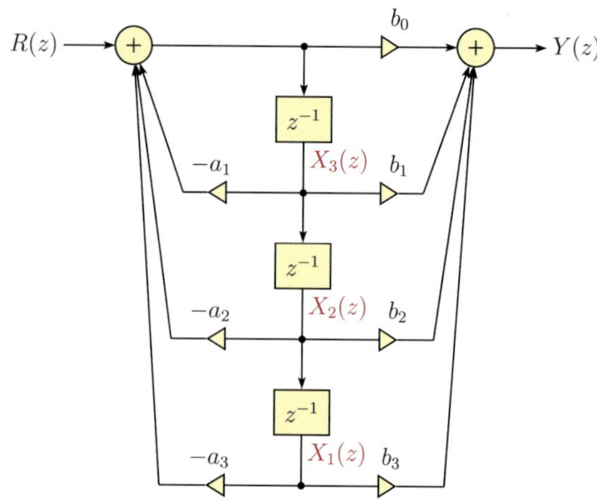

Figure 6.3 – Simulation diagram for $H(z)$ of Eqn. (6.19).

We know from time shifting property of the z-transform (see Section 5.3.2 of Chapter 5) that multiplication of a transform by z^{-1} is equivalent to right shifting the corresponding signal by one sample in the time domain. A time-domain equivalent of the block diagram in Fig. 6.3 is obtained by replacing z^{-1} components with delays and using time-domain versions of input and output quantities. The resulting diagram is shown in Figure 6.4.

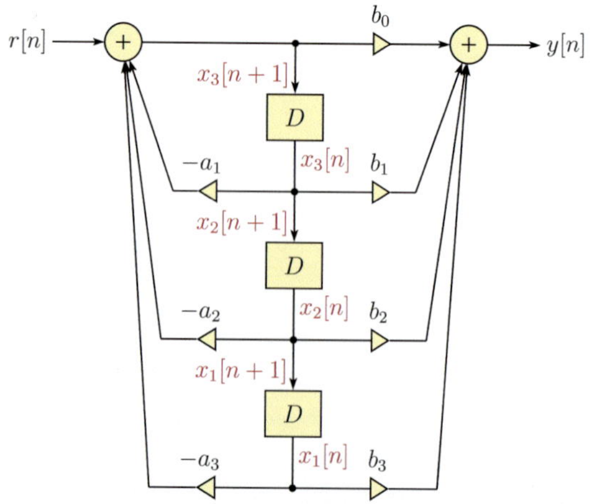

Figure 6.4 – Selecting states from the block diagram for $H(z)$.

Obtaining state-space model from a block diagram:

A state-space model can be obtained from the diagram of Fig. 6.4 as follows:

1. Designate the output signal of each delay element as a state variable. These are marked in Fig. 6.4 as $x_1[n]$, $x_2[n]$, and $x_3[n]$.
2. The input of each delay element is a one sample advanced version of the corresponding state variable. These are marked in Fig. 6.4 as $x_1[n+1]$, $x_2[n+1]$ and $x_3[n+1]$.
3. Express the input signal of each delay element as a function of the state variables and the input signal as dictated by the diagram.

Following state equations are obtained by inspecting the diagram:

$$x_1[n+1] = x_2[n] \tag{6.20}$$

$$x_2[n+1] = x_3[n] \tag{6.21}$$

$$x_3[n+1] = -a_3\, x_1[n] - a_2\, x_2[n] - a_1\, x_3[n] + r(t) \tag{6.22}$$

The output signal can be written as

$$y[n] = b_3\, x_1[n] + b_2\, x_2[n] + b_1\, x_3[n] + b_0\, x_3[n+1] \tag{6.23}$$

Since we do not want the $x_3[n+1]$ term on the right side of the output equation, Eqn. (6.22) needs to be substituted into Eqn. (6.23) to obtain the proper output equation:

$$y[n] = (b_3 - b_0\, a_3)\, x_1[n] + (b_2 - b_0\, a_2)\, x_2[n] + (b_1 - b_0\, a_1)\, x_3[n] + b_0\, r[n] \tag{6.24}$$

Expressed in matrix form, the state-space model is

$$\dot{\mathbf{x}}[n+1] = \begin{bmatrix} 0 & 1 & 0 \\ 0 & 0 & 1 \\ -a_3 & -a_2 & -a_1 \end{bmatrix} \mathbf{x}[n] + \begin{bmatrix} 0 \\ 0 \\ 1 \end{bmatrix} r[n] \tag{6.25}$$

$$y[n] = \begin{bmatrix} (b_3 - b_0\, a_3) & (b_2 - b_0\, a_2) & (b_1 - b_0\, a_1) \end{bmatrix} \mathbf{x}[n] + b_0\, r[n] \tag{6.26}$$

The results can be easily extended to a N-th order system function in the form

$$H(z) = \frac{Y(z)}{X(z)} = \frac{b_0 + b_1 z^{-1} + b_2 z^{-2} + \ldots + \ldots b_M z^{-M}}{1 + a_1 z^{-1} + a_2 z^{-2} + \ldots + a_N z^{-N}} \tag{6.27}$$

to produce the state-space model

$$\mathbf{x}[n+1] = \begin{bmatrix} 0 & 1 & 0 & 0 & \ldots & 0 \\ 0 & 0 & 1 & 0 & \ldots & 0 \\ 0 & 0 & 0 & 1 & \ldots & 0 \\ \vdots & \vdots & \vdots & \vdots & \ldots & \vdots \\ 0 & 0 & 0 & 0 & \ldots & 1 \\ -a_N & -a_{N-1} & -a_{N-2} & -a_{N-3} & \ldots & -a_1 \end{bmatrix} \mathbf{x}[n] + \begin{bmatrix} 0 \\ 0 \\ 0 \\ \vdots \\ 0 \\ 1 \end{bmatrix} r[n] \tag{6.28}$$

$$y[n] = \begin{bmatrix} (b_N - b_0\, a_N) & (b_{N-1} - b_0\, a_{N-1}) & \ldots & (b_1 - b_0\, a_1) \end{bmatrix} \mathbf{x}[n] + b_0\, r[n] \tag{6.29}$$

The structure of the coefficient matrix **A** is the same as in Eqn. (6.16). Similarly, the vector **B** is identical to that found in Section 6.3.1.

Example 6.3: State-space model from block diagram for DTLTI system

Find a state-space model for the DTLTI system described through the system function

$$H(z) = \frac{z^3 - 7z + 6}{z^4 - z^3 - 0.34\,z^2 + 0.966\,z - 0.2403}$$

Solution: A cascade form block diagram for this system was drawn in Example 6.3 and is shown again in Fig. 6.5 with state variable assignments marked.

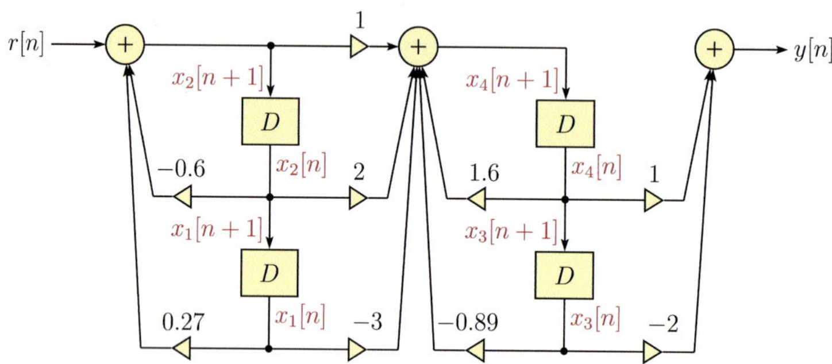

Figure 6.5 – Cascade form block diagram for Example 6.3.

Writing the input signal of each delay element in terms of the other signals in the diagram we obtain

$$\begin{aligned}
x_1[n+1] &= x_2[n] \\
x_2[n+1] &= 0.27\,x_1[n] - 0.6\,x_2[n] + r[n] \\
x_3[n+1] &= x_4[n] \\
x_4[n+1] &= -3\,x_1[n] + 2\,x_2[n] + x_2[n+1] - 0.89\,x_3[n] + 1.6\,x_4[n]
\end{aligned}$$

The equation for $x_4[n+1]$ contains the undesired term $x_2[n+1]$ on the right side of the equal sign. This needs to be resolved by substitution:

$$x_4[n+1] = -2.73\,x_1[n] + 1.4\,x_2[n] - 0.89\,x_3[n] + 1.6\,x_4[n] + r[n]$$

The output equation is

$$y[n] = -2\,x_3[n] + x_4[n]$$

In matrix form, the state-space model is

$$\mathbf{x}[n+1] = \begin{bmatrix} 0 & 1 & 0 & 0 \\ 0.27 & -0.6 & 0 & 0 \\ 0 & 0 & 0 & 1 \\ -2.73 & 1.4 & -0.89 & 1.6 \end{bmatrix} \mathbf{x}[n] + \begin{bmatrix} 0 \\ 1 \\ 0 \\ 1 \end{bmatrix} r[n]$$

$$y[n] = \begin{bmatrix} 0 & 0 & -2 & 1 \end{bmatrix} \mathbf{x}[n]$$

6.3.3 Alternative state-space models

State-space model of a system is not unique; alternative models can be found by means of a transformation. Consider a DTLTI system defined through Eqns. (6.7) and (6.8). Let $\mathbf{q}[n]$ be a length-N column vector of variables $q_1[n],\ldots,q_N[n]$, related to the original state vector $\mathbf{x}[n]$ of the DTLTI system through

$$\mathbf{x}[n] = \mathbf{P}\,\mathbf{q}[n] \tag{6.30}$$

where $\mathbf{P}$ is a constant N by N transformation matrix. The only requirement on the matrix $\mathbf{P}$ is that it be non-singular (invertible). Substituting Eqn. (6.30) into Eqn. (6.7) leads to

$$\mathbf{P}\,\mathbf{q}[n+1] = \mathbf{A}\mathbf{P}\,\mathbf{q}[n] + \mathbf{B}\,r[n] \tag{6.31}$$

Multiplication of both sides of Eqn. (6.31) with the inverse of $\mathbf{P}$ from the left yields

$$\mathbf{q}[n+1] = \mathbf{P}^{-1}\mathbf{A}\mathbf{P}\,\mathbf{q}[n] + \mathbf{P}^{-1}\mathbf{B}\,r[n] \tag{6.32}$$

Finally, using Eqn. (6.30) with Eqn. (6.8) we get

$$y[n] = \mathbf{C}\mathbf{P}\,\mathbf{q}[n] + d\,r[n] \tag{6.33}$$

Eqns. (6.32) and (6.33) provide us with an alternative state-space model for the same DTLTI system, using a new state vector $\mathbf{q}[n]$. They can be written in the standard form of Eqns. (6.7) and (6.8) as

$$\mathbf{q}[n+1] = \tilde{\mathbf{A}}\,\mathbf{q}[n] + \tilde{\mathbf{B}}\,r[n] \tag{6.34}$$

$$y[n] = \tilde{\mathbf{C}}\,\mathbf{q}[n] + d\,r[n] \tag{6.35}$$

with newly defined coefficient matrices

$$\tilde{\mathbf{A}} = \mathbf{P}^{-1}\mathbf{A}\mathbf{P} \tag{6.36}$$

$$\tilde{\mathbf{B}} = \mathbf{P}^{-1}\mathbf{B} \tag{6.37}$$

and

$$\tilde{\mathbf{C}} = \mathbf{C}\mathbf{P} \tag{6.38}$$

The transformation relationship $\tilde{\mathbf{A}} = \mathbf{P}^{-1}\mathbf{A}\mathbf{P}$ is known in control system theory as a *similarity transformation.* It can be used for obtaining alternative state-space models of a system. Each choice of the transformation matrix $\mathbf{P}$ leads to a different, but equivalent, state-space model. In each of these alternative models the input-output relationship remains the same although the state variables are different. Since any non-singular matrix $\mathbf{P}$ can be used in a similarity transformation, and infinite number of different state-space models can be found for a given system. In the next several examples we will explore this concept further.

Example 6.4: Finding an alternative state-space model

Consider the DTLTI system of Example 6.2 for which a state-space model was found in the form

$$\begin{bmatrix} x_1[n+1] \\ x_2[n+1] \end{bmatrix} = \begin{bmatrix} 0.6 & 0 \\ 0 & -0.8 \end{bmatrix} \begin{bmatrix} x_1[n] \\ x_2[n] \end{bmatrix} + \begin{bmatrix} 3 \\ 4 \end{bmatrix} r[n]$$

and

$$y[n] = \begin{bmatrix} 1 & 1 \end{bmatrix} \begin{bmatrix} x_1[n] \\ x_2[n] \end{bmatrix}$$

Find an alternative model for this system using the transformation

$$\mathbf{x}[n] = \begin{bmatrix} 1/2 & 1/2 \\ -1/2 & 1/2 \end{bmatrix} \mathbf{q}[n]$$

Solution: The transformation matrix and its inverse are

$$\mathbf{P} = \begin{bmatrix} 1/2 & 1/2 \\ -1/2 & 1/2 \end{bmatrix} \quad \longrightarrow \quad \mathbf{P}^{-1} = \begin{bmatrix} 1 & -1 \\ 1 & 1 \end{bmatrix}$$

Using Eqns. (6.36) through (6.38) we obtain

$$\tilde{\mathbf{A}} = \mathbf{P}^{-1}\mathbf{A}\mathbf{P} = \begin{bmatrix} 1 & -1 \\ 1 & 1 \end{bmatrix} \begin{bmatrix} 0.6 & 0 \\ 0 & -0.8 \end{bmatrix} \begin{bmatrix} 1/2 & 1/2 \\ -1/2 & 1/2 \end{bmatrix} = \begin{bmatrix} -0.1 & 0.7 \\ 0.7 & -0.1 \end{bmatrix}$$

$$\tilde{\mathbf{B}} = \mathbf{P}^{-1}\mathbf{B} = \begin{bmatrix} 1 & -1 \\ 1 & 1 \end{bmatrix} \begin{bmatrix} 3 \\ 4 \end{bmatrix} = \begin{bmatrix} -1 \\ 7 \end{bmatrix}$$

and

$$\tilde{\mathbf{C}} = \mathbf{C}\mathbf{P} = \begin{bmatrix} 1 & 1 \end{bmatrix} \begin{bmatrix} 1/2 & 1/2 \\ -1/2 & 1/2 \end{bmatrix} = \begin{bmatrix} 0 & 1 \end{bmatrix}$$

The alternative model for the system can be written as

$$\begin{bmatrix} q_1[n+1] \\ q_2[n+1] \end{bmatrix} = \begin{bmatrix} -0.1 & 0.7 \\ 0.7 & -0.1 \end{bmatrix} \begin{bmatrix} q_1[n] \\ q_2[n] \end{bmatrix} + \begin{bmatrix} -1 \\ 7 \end{bmatrix} r[n]$$

and

$$y[n] = \begin{bmatrix} 0 & 1 \end{bmatrix} \begin{bmatrix} q_1[n] \\ q_2[n] \end{bmatrix}$$

6.3.4 DTLTI systems with multiple inputs and/or outputs

A DTLTI system with multiple input and/or output signals can also be represented with a state-space model in the standard form of Eqns. (6.7) and (6.8). Consider a DTLTI system with K input signals $\{\, r_i[n];\ i = 1,2,\ldots,K \,\}$ and M output signals $\{\, y_j[n];\ i = 1,2,\ldots,M \,\}$. Let $\mathbf{r}[n]$ be the $K \times 1$ vector of input signals:

$$\mathbf{r}[n] = \begin{bmatrix} r_1[n] \\ r_2[n] \\ \vdots \\ r_K[n] \end{bmatrix} \tag{6.39}$$

Similarly, $\mathbf{y}[n]$ is the $M \times 1$ vector of output signals.

$$\mathbf{y}[n] = \begin{bmatrix} y_1[n] \\ y_2[n] \\ \vdots \\ y_M[n] \end{bmatrix} \tag{6.40}$$

With these new vectors, the state equation is in the form

$$\mathbf{x}[n+1] = \mathbf{A}\mathbf{x}[n] + \mathbf{B}\mathbf{r}[n] \tag{6.41}$$

The matrix **A** is still the $N \times N$ coefficient matrix, and the vector $\mathbf{x}[n]$ is still the $N \times 1$ state vector as before. Dimensions of the matrix **B** will need to be adjusted, however, to make it compatible with the $K \times 1$ vector $\mathbf{r}[n]$. Therefore, **B** is now a $N \times K$ matrix. The output equation is in the form

$$\mathbf{y}[n] = \mathbf{C}\mathbf{x}[n] + \mathbf{D}\mathbf{r}[n] \tag{6.42}$$

Dimensions of **C** need to be updated as well. Since $\mathbf{y}[n]$ is a $M \times 1$ vector, **C** is a $M \times N$ matrix. Additionally, the scalar d in the output equation in Eqn. (6.8) is replaced with a $M \times K$ matrix **D**.

6.4 Solution of State-Space Model

Easiest method of solving discrete-time state equations is through iteration. Consider the state-space model given by Eqns. (6.7) and (6.8). Given the initial state vector $\mathbf{x}[0]$ and the input signal $r[n]$ the iterative solution proceeds as follows:

Solving discrete-time state equations iteratively:

1. Using $n = 0$, compute $\mathbf{x}[1]$, the state vector at $n+1 = 1$ using

$$\mathbf{x}[1] = \mathbf{A}\mathbf{x}[0] + \mathbf{B}\,r[0]$$

2. Compute the output sample at $n = 0$ using

$$y[0] = \mathbf{C}\mathbf{x}[0] + d\,r[0]$$

3. Increment n by 1, and repeat steps 1 and 2.

The iterative solution method allows as many samples of the output signal to be computed as desired. If an analytical solution is needed, however, we need to follow an alternative approach and obtain a general expression for $y[n]$. We will use z-transform methods for this. Taking the unilateral z-transform of both sides of Eqn. (6.7) we obtain

$$z\mathbf{X}(z) - z\mathbf{x}[0] = \mathbf{A}\mathbf{X}(z) + \mathbf{B}\,R(z) \tag{6.43}$$

Solving Eqn. (6.43) for the vector $\mathbf{X}(z)$ yields

$$\mathbf{X}(z) = z\left[z\mathbf{I} - \mathbf{A}\right]^{-1}\mathbf{x}[0] + \left[z\mathbf{I} - \mathbf{A}\right]^{-1}\mathbf{B}\,R(z) \tag{6.44}$$

The matrix

$$\mathbf{\Phi}(z) = z\left[z\mathbf{I} - \mathbf{A}\right]^{-1} \tag{6.45}$$

is the z-transform version of the *resolvent matrix*. Using it in Eqn. (6.44) we get

$$\mathbf{X}(z) = \mathbf{\Phi}(z)\,\mathbf{x}(0) + z^{-1}\,\mathbf{\Phi}(z)\,\mathbf{B}\,R(z) \tag{6.46}$$

for the solution in the z domain. Taking the inverse z-transform of Eqn. (6.46) and recognizing that the inverse z-transform of the product $\boldsymbol{\Phi}(z)\,R(z)$ is the convolution of the corresponding time-domain functions leads to the solution

$$\mathbf{x}[n] = \boldsymbol{\phi}[n]\,\mathbf{x}[0] + \sum_{m=0}^{n-1} \boldsymbol{\phi}[n-1-m]\,\mathbf{B}\,r[m] \tag{6.47}$$

where $\boldsymbol{\phi}[n]$ is the discrete-time version of the *state transition matrix* computed as

$$\boldsymbol{\phi}[n] = \mathcal{Z}^{-1}\left\{\boldsymbol{\Phi}(z)\right\} = \mathcal{Z}^{-1}\left\{z\left[z\mathbf{I}-\mathbf{A}\right]^{-1}\right\} \tag{6.48}$$

Example 6.5: Solution of discrete-time state-space model

Compute the unit step response of the discrete-time system described by the state-space model

$$\mathbf{x}[n+1] = \begin{bmatrix} 0 & -0.5 \\ 0.25 & 0.75 \end{bmatrix} \mathbf{x}[n] + \begin{bmatrix} 2 \\ 1 \end{bmatrix} r[n]$$

$$y[n] = \begin{bmatrix} 3 & 1 \end{bmatrix} \mathbf{x}[n]$$

using the initial state vector

$$\mathbf{x}[0] = \begin{bmatrix} 2 \\ 0 \end{bmatrix}$$

Solution: We will begin by forming the matrix $[z\mathbf{I}-\mathbf{A}]$:

$$\left[z\mathbf{I}-\mathbf{A}\right] = z\begin{bmatrix} 1 & 0 \\ 0 & 1 \end{bmatrix} - \begin{bmatrix} 0 & -0.5 \\ 0.25 & 0.75 \end{bmatrix} = \begin{bmatrix} z & 0.5 \\ -0.25 & z-0.75 \end{bmatrix}$$

The resolvent matrix is found as

$$\boldsymbol{\Phi}(z) = \frac{z}{(z-0.5)\,(z-0.25)} \begin{bmatrix} z-0.75 & -0.5 \\ 0.25 & z \end{bmatrix}$$

The state transition matrix is found by taking the inverse z-transform of the resolvent matrix:

$$\boldsymbol{\phi}[n] = \begin{bmatrix} -(0.5)^n\,u[n] + 2\,(0.25)^n\,u[n] & -2\,(0.5)^n\,u[n] + 2\,(0.25)^n\,u[n] \\ (0.5)^n\,u[n] - (0.25)^n\,u[n] & 2\,(0.5)^n\,u[n] - (0.25)^n\,u[n] \end{bmatrix}$$

The solution of the state equation is found by using the state transition matrix in Eqn. (6.47):

$$\mathbf{x}[n] = \begin{bmatrix} 6\,(0.5)^n\,u[n] - 4\,(0.25)^n\,u[n] \\ -6\,(0.5)^n\,u[n] + 2\,(0.25)^n\,u[n] + 4\,u[n] \end{bmatrix}$$

The output signal is

$$\begin{aligned} y[n] &= \begin{bmatrix} 3 & 1 \end{bmatrix} \mathbf{x}[n] \\ &= 12\,(0.5)^n\,u[n] - 10\,(0.25)^n\,u[n] + 4\,u[n] \end{aligned}$$

Software resources: exdt_6_5a.m , exdt_6_5b.m

6.5 Obtaining System Function from State-Space Model

The z-domain system function $H(z)$ of a DTLTI system can be easily obtained from its state-space description. Consider Eqn. (6.44) derived in the process of obtaining the solution for the z-transform of the state vector:

$$\text{Eqn. (6.44):} \qquad \mathbf{X}(z) = z\left[z\mathbf{I}-\mathbf{A}\right]^{-1}\mathbf{x}[0] + \left[z\mathbf{I}-\mathbf{A}\right]^{-1}\mathbf{B}\,R(z)$$

Recall that the system function concept is valid only for a system that is both linear and time-invariant. If the system is DTLTI, then the initial value of the state vector is $\mathbf{x}[0] = \mathbf{0}$, and Eqn. (6.44) becomes

$$\mathbf{X}(z) = \left[z\mathbf{I}-\mathbf{A}\right]^{-1}\mathbf{B}\,R(z) \tag{6.49}$$

Using $\mathbf{X}(z)$ in the output equation leads to

$$\begin{aligned} Y(z) =& \mathbf{C}\mathbf{X}(z) + d\,R(z) \\ =& \mathbf{C}\left[z\mathbf{I}-\mathbf{A}\right]^{-1}\mathbf{B}\,R(z) + d\,R(z) \end{aligned} \tag{6.50}$$

from which the system function follows as

$$H(z) = \frac{Y(z)}{R(z)} = \mathbf{C}\left[z\mathbf{I}-\mathbf{A}\right]^{-1}\mathbf{B} + d \tag{6.51}$$

6.6 Discretization of Continuous-Time State-Space Model

As discussed in Section 6.4 of this chapter, discrete-time state-space models have the added advantage that they can be solved iteratively one step at a time. On the other hand, a continuous-time state-space model cannot be solved iteratively unless it is first converted to an approximate discrete-time model. There are times when we may want to simulate an analog system on a digital computer. In this section we will discuss methods for *discretizing* a continuous-time state-space model. Only single-input-single-output systems will discussed although extension to systems with multiple inputs and/or outputs is straightforward. Let an analog system be described by the equations

$$\dot{\mathbf{x}}_{\mathbf{a}}(t) = \mathbf{A}\mathbf{x}_{\mathbf{a}}(t) + \mathbf{B}\,r_a(t) \tag{6.52}$$

$$y_a(t) = \mathbf{C}\mathbf{x}_{\mathbf{a}}(t) + d\,r_a(t) \tag{6.53}$$

In Eqns. (6.52) and (6.52) the subscript "a" is used to indicate analog variables. Our goal is to find a discrete-time system with the state-space model

$$\mathbf{x}[n+1] = \bar{\mathbf{A}}\mathbf{x}[n] + \bar{\mathbf{B}}\,r[n] \tag{6.54}$$

$$y[n] = \bar{\mathbf{C}}\mathbf{x}[n] + \bar{d}\,r[n] \tag{6.55}$$

so that, for a sampling interval T_s we have

$$\mathbf{x}[n] \approx \mathbf{x}_{\mathbf{a}}(nT_s) \qquad \text{and} \qquad y[n] \approx y_a(nT_s) \tag{6.56}$$

when $r[n] = r_a(nT_s)$. The value of T_s must be chosen to satisfy the sampling criteria (see Chapter 4).

Finding an approximate discrete-time state-space model:

The solution for the continuous-time state-space model is

$$\mathbf{x_a}(t) = e^{\mathbf{A}(t-t_0)}\,\mathbf{x_a}(t_0) + \int_{t_0}^{t} e^{\mathbf{A}(t-\tau)}\,\mathbf{B}\,r_a(\tau)\,d\tau \tag{6.57}$$

Eqn. (6.57) starts with the solution at time t_0 and returns the solution at time t. By selecting the limits of the integral to be $t_0 = nT_s$ and $t = nT_s$, Eqn. (6.57) becomes

$$\mathbf{x_a}\big((n+1)\,T_s\big) = e^{\mathbf{A}T_s}\,\mathbf{x_a}\big(nT_s\big) + \int_{nT_s}^{(n+1)T_s} e^{\mathbf{A}((n+1)T_s-\tau)}\,\mathbf{B}\,r_a(\tau)\,d\tau$$

If T_s is small enough the term $r_a(\tau)$ may be assumed to be constant within the span of the integral, that is,

$$r_a(\tau) \approx r_a(nT_s) \qquad \text{for} \quad nT_s \le \tau \le (n+1)\,T_s \tag{6.58}$$

With this assumption Eqn. (6.58) becomes

$$\mathbf{x_a}\big((n+1)\,T_s\big) \approx e^{\mathbf{A}T_s}\,\mathbf{x_a}\big(nT_s\big) + \left[\int_{nT_s}^{(n+1)T_s} e^{\mathbf{A}((n+1)T_s-\tau)}\,d\tau\right]\mathbf{B}\,r_a(nT_s) \tag{6.59}$$

Using the variable change $\lambda = (n+1)\,T_s - \tau$ on the integral in Eqn. (6.59) we obtain

$$\begin{aligned}\mathbf{x_a}\big((n+1)\,T_s\big) &\approx e^{\mathbf{A}T_s}\,\mathbf{x_a}\big(nT_s\big) + \left[\int_0^{T_s} e^{\mathbf{A}\lambda}\,d\lambda\right]\mathbf{B}\,r_a(nT_s)\\ &= e^{\mathbf{A}T_s}\,\mathbf{x_a}\big(nT_s\big) + \mathbf{A}^{-1}\left[e^{\mathbf{A}T_s} - \mathbf{I}\right]\mathbf{B}\,r_a(nT_s)\end{aligned} \tag{6.60}$$

Finally, substitution of $r_a(nT_s) = r[n]$ and $\mathbf{x_a}(nT_s) = \mathbf{x}[n]$ yields

$$\mathbf{x}[n+1] = e^{\mathbf{A}T_s}\,\mathbf{x}[n] + \mathbf{A}^{-1}\left[e^{\mathbf{A}T_s} - \mathbf{I}\right]\mathbf{B}\,r[n]$$

The output equation is simply

$$y[n] = \mathbf{C}\mathbf{x}[n] + d\,r[n] \tag{6.61}$$

In summary, the continuous-time system described by Eqns. (6.52) and (6.53) can be approximated with the discrete-time system in Eqns. (6.54) and (6.55) by choosing the coefficient matrices as follows:

Discretization of state-space model:

$$\bar{\mathbf{A}} = e^{\mathbf{A}T_s}\,, \qquad \bar{\mathbf{B}} = \mathbf{A}^{-1}\left[e^{\mathbf{A}T_s} - \mathbf{I}\right]\mathbf{B}\,, \qquad \bar{\mathbf{C}} = \mathbf{C}\,, \qquad \bar{d} = d \tag{6.62}$$

Example 6.6: Discretization of state-space model for RLC circuit

Consider the continuous-time RLC circuit shown in Fig. 6.6.

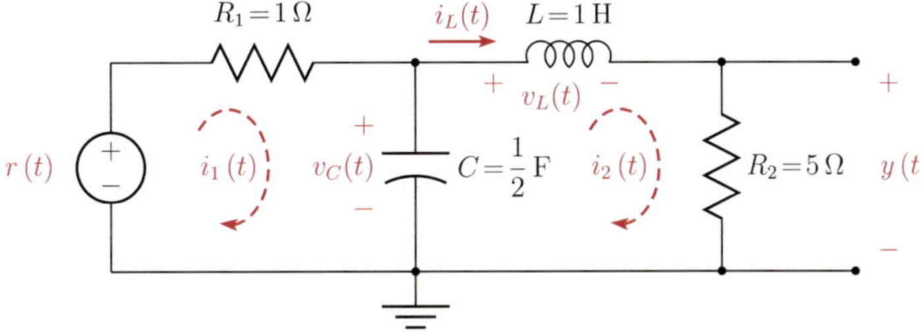

Figure 6.6 – The RLC circuit for Example 6.6.

Designating the capacitor voltage and the inductor current as the two state variables, that is, by choosing

$$x_1(t) = v_C(t) \quad \text{and} \quad x_2(t) = i_L(t)$$

a continuous-time state-space model for this circuit can be found as follows:

$$\dot{\mathbf{x}}(t) = \begin{bmatrix} -2 & -2 \\ 1 & -5 \end{bmatrix} \mathbf{x}(t) + \begin{bmatrix} 1 \\ 0 \end{bmatrix} r(t)$$

$$y(t) = \begin{bmatrix} 0 & 5 \end{bmatrix} \mathbf{x}(t)$$

The state transition matrix corresponding to this model is

$$e^{\mathbf{A}t} = \boldsymbol{\phi}(t) = \begin{bmatrix} \left(2e^{-3t} - e^{-4t}\right) & \left(-2e^{-3t} + 2e^{-4t}\right) \\ \left(e^{-3t} - e^{-4t}\right) & \left(-e^{-3t} + 2e^{-4t}\right) \end{bmatrix}$$

For details, see Examples 5.1 and 5.13 in Chapter 5 of the companion textbook *Continuous-Time Signals and Systems: A MATLAB Integrated Approach.* In this exercise, our interest is in converting this state-space model to an approximate discrete-time model that can be simulated on a computer. Using a step size of $T_s = 0.1$ s, find a discrete-time state-space model to approximate the continuous-time model given above.

Solution: : Let us begin by evaluating the state transition matrix at $t = T_s = 0.1$ s.

$$e^{\mathbf{A}T_s} = \boldsymbol{\phi}(0.1) = \begin{bmatrix} \left(2e^{-0.3} - e^{-0.4}\right) & \left(-2e^{-0.3} + 2e^{-0.4}\right) \\ \left(e^{-0.3} - e^{-0.4}\right) & \left(-e^{-0.3} + 2e^{-0.4}\right) \end{bmatrix} = \begin{bmatrix} 0.8113 & -0.1410 \\ 0.0705 & 0.5998 \end{bmatrix}$$

The state matrix of the discrete-time model is

$$\bar{\mathbf{A}} = e^{\mathbf{A}T_s} = \begin{bmatrix} 0.8113 & -0.1410 \\ 0.0705 & 0.5998 \end{bmatrix}$$

The inverse of state matrix $\mathbf{A}$ is

$$\mathbf{A}^{-1} = \begin{bmatrix} -0.4167 & 0.1667 \\ -0.0833 & -0.1667 \end{bmatrix}$$

which can be used for computing the vector $\bar{\mathbf{B}}$ as

$$\bar{\mathbf{B}} = \mathbf{A}^{-1}\left[e^{\mathbf{A}T_s} - \mathbf{I}\right]\mathbf{B}$$

$$= \begin{bmatrix} -0.4167 & 0.1667 \\ -0.0833 & -0.1667 \end{bmatrix} \left(\begin{bmatrix} 0.8113 & -0.1410 \\ 0.0705 & 0.5998 \end{bmatrix} - \begin{bmatrix} 1 & 0 \\ 0 & 1 \end{bmatrix} \right) \begin{bmatrix} 1 \\ 0 \end{bmatrix} = \begin{bmatrix} 0.0904 \\ 0.0040 \end{bmatrix}$$

Therefore, the discretized state-space model for the RLC circuit in question is

$$\mathbf{x}[n+1] = \begin{bmatrix} 0.8113 & -0.1410 \\ 0.0705 & 0.5998 \end{bmatrix} \mathbf{x}[n] + \begin{bmatrix} 0.0904 \\ 0.0040 \end{bmatrix} r[n]$$

$$y[n] = \begin{bmatrix} 0 & 5 \end{bmatrix} \mathbf{x}[n]$$

Software resource: exdt_6_6.m

Software resource: See MATLAB Exercise 6.2.

The only potential drawback for the discretization technique derived above is that it requires inversion of the matrix **A** and computation of the state transition matrix. Simpler approximations are also available. An example of a simpler technique is the *Euler method.*

Derivation of a discrete-time model using Euler method:

Let us evaluate the continuous-time state equation given by Eqn. (6.52) at $t = nT_s$:

$$\dot{\mathbf{x}}_{\mathbf{a}}(t)\Big|_{t=nT_s} = \mathbf{A}\mathbf{x}_{\mathbf{a}}(nT_s) + \mathbf{B}\,r_a(nT_s) \tag{6.63}$$

The derivative on the left side of Eqn. (6.63) can be approximated using the first difference

$$\dot{\mathbf{x}}_{\mathbf{a}}(t)\Big|_{t=nT_s} \approx \frac{1}{T_s}\left[\mathbf{x}_{\mathbf{a}}\left((n+1)\,T_s\right) - \mathbf{x}_{\mathbf{a}}\left(nT_s\right)\right] \tag{6.64}$$

Substituting Eqn. (6.64) into Eqn. (6.63) and rearranging terms gives

$$\mathbf{x}\left((n+1)\,T_s\right) \approx \left[\mathbf{I} + \mathbf{A}\,T_s\right]\mathbf{x}\left(nT_s\right) + T_s\,\mathbf{B}\,r\left(nT_s\right) \tag{6.65}$$

Through substitutions $r_a(nT_s) = r[n]$ and $\mathbf{x}_{\mathbf{a}}(nT_s) = \mathbf{x}[n]$ we obtain

$$\mathbf{x}[n+1] = \left[\mathbf{I} + \mathbf{A}\,T_s\right]\mathbf{x}[n] + T_s\,\mathbf{B}\,r[n] \tag{6.66}$$

The output equation is

$$y[n] = \mathbf{C}\mathbf{x}[n] + d\,r[n] \tag{6.67}$$

as before.

The results can be summarized as follows:

Discretization using Euler method:

$$\bar{\mathbf{A}} = \mathbf{I} + \mathbf{A}\,T_s\,, \qquad \bar{\mathbf{B}} = T_s\,\mathbf{B}\,, \qquad \bar{\mathbf{C}} = \mathbf{C}\,, \qquad \bar{d} = d \tag{6.68}$$

Since this is a first-order approximation method, the step size (sampling interval) T_s chosen should be significantly small compared to the time constant.

Software resource: See MATLAB Exercise 6.3.

Summary of Key Points

- State-space model of a system describes the internal structure of the system, in contrast with the other description forms which simply describe the relationship between input and output signals. It can also be used for representing systems with multiple inputs and/or outputs.
- State-space representation of a single-input single-output DTLTI system is in the form

  Eqn. 6.7: State equation: $\mathbf{x}[n+1] = \mathbf{A}\mathbf{x}[n] + \mathbf{B}\,r[n]$

  Eqn. 6.8: Output equation: $y[n] = \mathbf{C}\mathbf{x}[n] + d\,r[n]$

- Phase-variable canonical form of the state-space model for a DTLTI system can be obtained from its difference equation using the technique outlined in Section 6.3.1.
- A state-space model for a DTLTI system can also be obtained from a block diagram by designating the output of each delay element as a state variable, and expressing the input signal of each delay element as a linear combination of state variables and system inputs as outlined in Section 6.3.2.
- Alternative state-space descriptions of a system can be obtained by transforming the state variables through the transformation

  Eqn. 6.30: $\mathbf{x}[n] = \mathbf{P}\mathbf{q}[n]$

  where P is a non-singular N by N transformation matrix. New coefficients matrices are

  $$\tilde{\mathbf{A}} = \mathbf{P}^{-1}\mathbf{A}\mathbf{P}\,, \qquad \tilde{\mathbf{B}} = \mathbf{P}^{-1}\mathbf{B}\,, \qquad \tilde{\mathbf{C}} = \mathbf{C}\mathbf{P}$$

- The easiest method of solving discrete-time state equations is through iteration.
- If an analytical solution is needed, z-transform-based solution method can be used to obtain the general solution

  Eqn. 6.47: $\mathbf{x}[n] = \boldsymbol{\phi}[n]\,\mathbf{x}[0] + \sum_{m=0}^{n-1} \boldsymbol{\phi}[n-1-m]\,\mathbf{B}\,r[m]$

  where $\boldsymbol{\phi}[n]$ is the discrete-time *state transition matrix* computed as

  Eqn. 6.48: $\boldsymbol{\phi}[n] = \mathcal{Z}^{-1}\left\{\boldsymbol{\Phi}(z)\right\} = \mathcal{Z}^{-1}\left\{z\left[z\mathbf{I} - \mathbf{A}\right]^{-1}\right\}$

☞ The system function $H(z))$ can be obtained from the state-space model as

Eqn. 6.51: $$H(z) = \frac{Y(z)}{R(z)} = \mathbf{C}\left[z\mathbf{I}-\mathbf{A}\right]^{-1}\mathbf{B}+d$$

☞ A single-input-single-output CTLTI system described by a state-space model

$$\begin{aligned}\dot{\mathbf{x}}_{\mathbf{a}}(t) =& \mathbf{A}\mathbf{x}_{\mathbf{a}}(t)+\mathbf{B}\,r_a(t)\\ y_a(t) =& \mathbf{C}\mathbf{x}_{\mathbf{a}}(t)+d\,r_a(t)\end{aligned}$$

can be converted to a DTLTI system with state-space model

$$\begin{aligned}\mathbf{x}[n+1] =& \bar{\mathbf{A}}\mathbf{x}[n]+\bar{\mathbf{B}}\,r[n]\\ y[n] =& \bar{\mathbf{C}}\mathbf{x}[n]+\bar{d}\,r[n]\end{aligned}$$

so that, for a sampling interval T_s we have

$$\mathbf{x}[n] \approx \mathbf{x}_{\mathbf{a}}(nT_s) \qquad \text{and} \qquad y[n] \approx y_a(nT_s)$$

when $r[n] = r_a(nT_s)$. Conversion equations are

Eqn. 6.62: $$\bar{\mathbf{A}} = e^{\mathbf{A}T_s}\,, \qquad \bar{\mathbf{B}} = \mathbf{A}^{-1}\left[e^{\mathbf{A}T_s}-\mathbf{I}\right]\mathbf{B}\,, \qquad \bar{\mathbf{C}} = \mathbf{C}\,, \qquad \bar{d} = d$$

Further Reading

[1] R.C. Dorf and R.H. Bishop. *Modern Control Systems.* Prentice Hall, 2011.

[2] B. Friedland. *Control System Design: An Introduction to State-Space Methods.* Dover Publications, 2012.

[3] K. Ogata. *Modern Control Engineering.* Instrumentation and Controls Series. Prentice Hall, 2010.

[4] L.A. Zadeh and C.A. Desoer. *Linear System Theory: The State Space Approach.* Dover Civil and Mechanical Engineering Series. Dover Publications, 2008.

MATLAB Exercises with Solutions

MATLAB Exercise 6.1: Obtaining system function from discrete-time state-space model

Determine the system function for a DTLTI sytem described by the state-space model

$$\mathbf{x}[n+1] = \begin{bmatrix} 0 & -0.5 \\ 0.25 & 0.75 \end{bmatrix}\mathbf{x}[n] + \begin{bmatrix} 2 \\ 1 \end{bmatrix} r[n]$$

$$y[n] = \begin{bmatrix} 3 & 1 \end{bmatrix}\mathbf{x}[n]$$

Solution: : The system function $H(z)$ can be found with the following statements:

```
>> A = [0,-0.5;0.25,0.75];
>> B = [2;1];
>> C = [3,1];
>> D = [0];
>> z = sym('z');
>> tmp = z*eye(2)-A;
>> rsm = z*inv(tmp)

rsm =

[ (2*z*(4*z - 3))/(8*z^2 - 6*z + 1),  -(4*z)/(8*z^2 - 6*z + 1)]
[            (2*z)/(8*z^2 - 6*z + 1), (8*z^2)/(8*z^2 - 6*z + 1)]

>> H = C*rsm*B+D

H =

(8*z^2)/(8*z^2-6*z+1) - (8*z)/(8*z^2-6*z+1) + (12*z*(4*z-3))/(8*z^2-6*z+1)
```

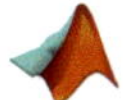

Software resource: mexdt_6_1.m

MATLAB Exercise 6.2: Discretization of state-space model

Consider again the continuous-time state-space model for the RLC circuit used in Example 6.6 and shown in Fig. 6.6.

$$\dot{\mathbf{x}}_{\mathbf{a}}(t) = \begin{bmatrix} -2 & -2 \\ 1 & -5 \end{bmatrix} \mathbf{x}_{\mathbf{a}}(t) + \begin{bmatrix} 1 \\ 0 \end{bmatrix} r_a(t)$$

$$y_a(t) = \begin{bmatrix} 0 & 5 \end{bmatrix} \mathbf{x}_{\mathbf{a}}(t)$$

The unit step response of the circuit obtained analytically from the state-space model is

$$y_a(t) = \frac{5}{12} + \frac{70}{3} e^{-3t} - \frac{135}{4} e^{-4t}, \qquad t \ge 0 \tag{6.69}$$

subject to the initial state vector $\mathbf{x}_{\mathbf{a}}(0) = [3 \ \ -2]^T$. Convert the state-space description to discrete-time, and obtain a numerical solution for the unit step response using a sampling interval of $T_s = 0.1$ s. Compare the approximate solution to the analytical solution.

Solution: : The discrete-time state-space model can be computed with the following statements that are based on Eqn. (6.62):

```
>> A = [-2,-2;1,-5];
>> B = [1;0];
>> C = [0,5];
>> d = 0;
>> Ts = 0.1;
>> A_bar = expm(A*Ts)
```

```
A_bar =
    0.8113   -0.1410
    0.0705    0.5998

>> B_bar = inv(A)*(A_bar-eye(2))*B

B_bar =
    0.0904
    0.0040

>> C_bar = C

C_bar =
     0     5

>> d_bar = d

d_bar =
     0
```

Once coefficient matrix and vectors `A_bar`, `B_bar`, `C_bar`, and `d_bar` are created, the output signal can be approximated by iteratively solving the discrete-time state-space model. The script `mexdt_6_2.m` listed below computes the iterative solution for $n = 0, \ldots, 30$. It also graphs the approximate solution found (red dots) superimposed with the analytical solution given in Eqn. (6.69).

```
% Script: mexdt_6_2.m
A = [-2,-2;1,-5];
B = [1;0];
C = [0,5];
d = 0;
Ts = 0.1;
A_bar = expm(A*Ts);
B_bar = inv(A)*(A_bar-eye(2))*B;
C_bar = C;
d_bar = d;
xn = [3;-2];             % Initial value of state vector
n = [0:30];              % Vector of indices
yn = [];                 % Empty vector to start
for nn=0:30,
  xnp1 = A_bar*xn+B_bar;  % 'xnp1' represents x[n+1]
  yn = [yn,C*xn];         % Append to vector 'yn'
  xn = xnp1;              % New becomes old for next iteration
end;
% Graph correct vs. approximate solution
t = [0:0.01:3];
ya = 5/12+70/3*exp(-3*t)-135/4*exp(-4*t);  % Eqn. (6.69)
plot(t,ya,'b-',n*Ts,yn,'ro'); grid;
title('y_{a}(t) and y[n]');
xlabel('t (sec)');
```

The graph is shown in Fig. 6.7.

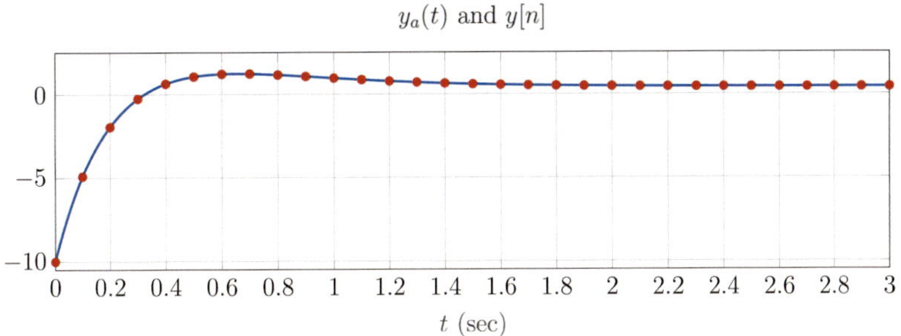

Figure 6.7 – Graph obtained in MATLAB Exercise 6.2.

Software resource: mexdt_6_2.m

MATLAB Exercise 6.3: Discretization using Euler method

Consider again the state-space model for the RLC circuit of Fig. 6.6 which was used in Example 6.6 and MATLAB Exercise 6.2. Convert the state-space description to an approximate discrete-time model using Euler's method, and obtain a numerical solution for the unit step response using a sampling interval of $T_s = 0.1$ s. The initial state vector is $\mathbf{x_a}(0) = [3 \ \ -2]^T$.

Solution: : The discrete-time state-space model can be computed with the following statements that are based on Eqn. (6.68):

```
>> A = [-2,-2;1,-5];
>> B = [1;0];
>> C = [0,5];
>> d = 0;
>> Ts = 0.1;
>> A_bar = eye(2)+A*Ts

A_bar =
    0.8000   -0.2000
    0.1000    0.5000

>> B_bar = B*Ts;

B_bar =
    0.1000
         0

>> C_bar = C

C_bar =
     0     5

>> d_bar = d

d_bar =
     0
```

The script mexdt_6_3.m listed below computes the iterative solution for $n = 0, \ldots$. It also graphs the approximate solution found (red dots) superimposed with the analytical solution given by Eqn. (6.69).

```
% Script: mexdt_6_3.m
A = [-2,-2;1,-5];
B = [1;0];
C = [0,5];
d = 0;
Ts = 0.1;
A_bar = eye(2)+A*Ts;
B_bar = B*Ts;
C_bar = C;
d_bar = d;
xn = [3;-2];          % Initial value of state vector
n = [0:30];           % Vector of indices
yn = [];              % Empty vector to start
for nn=0:30,
  xnp1 = A_bar*xn+B_bar;  % 'xnp1' represents x[n+1]
  yn = [yn,C*xn];         % Append to vector 'yn'
  xn = xnp1;              % New becomes old for next iteration
end;
% Graph correct vs. approximate solution
t = [0:0.01:3];
ya = 5/12+70/3*exp(-3*t)-135/4*exp(-4*t);  % Eqn. (6.69)
plot(t,ya,'b-',n*Ts,yn,'ro'); grid;
title('y_{a}(t) and y[n]');
xlabel('t (sec)');
```

The graph is shown in Fig. 6.8.

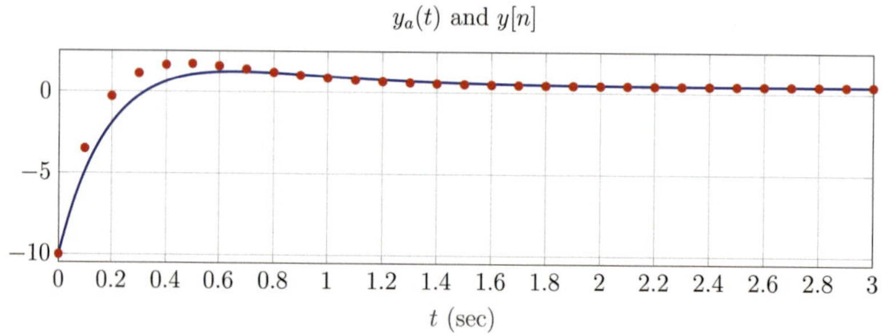

Figure 6.8 – Graph obtained in MATLAB Exercise 6.3.

The quality of approximation is certainly not as good as the one in MATLAB Exercise 6.2, but it can be improved by reducing T_s.

Software resource: mexdt_6_3.m

Problems

6.1. Find a state-space model for each DTLTI system described below by means of a difference equation.

a. $y[n]-0.9\,y[n-1]=r[n]$

b. $y[n]-1.7\,y[n-1]+0.72\,y[n-2]=3\,r[n]$

c. $y[n]-y[n-1]+0.11\,y[n-2]+0.07\,y[n-3]=r[n]$

6.2. System functions for several DTLTI systems are given below. Using the technique outlined in Example 6.2 that is based on partial fraction expansion, find a state-space model for each system. Ensure that the matrix **A** is diagonal in each case.

a. $H(z)=\dfrac{z+1}{(z+1/2)\,(z+2/3)}$

b. $H(z)=\dfrac{z\,(z+1)}{(z-0.4)\,(z+0.7)}$

c. $H(z)=\dfrac{z\,(z+1)}{(z+3/4)\,(z-1/2)\,(z-3/2)}$

6.3. Consider the system functions given in Problem 6.2. For each system draw a direct-form II block diagram and select the state variables. Then find a state-space model based on the block diagram.

6.4. Consider the DTLTI systems described below by means of their difference equations. The direct method of finding a state-space model explained in Example 6.1 would be difficult to use in this case, due to the inclusion of delayed input terms $r[n-1]$, $r[n-2]$, etc. Find a state-space model for each system in two steps: First find the system function $H(z)$, and then find a state-space model in phase-variable canonical form by inspecting the system function.

a. $y[n]-0.9\,y[n-1]=r[n]+r[n-1]$

b. $y[n]-1.7\,y[n-1]+0.72\,y[n-2]=3\,r[n]+2\,r[n-2]$

c. $y[n]-y[n-1]+0.11\,y[n-2]+0.07\,y[n-3]=r[n]-r[n-1]+r[n-2]$

6.5. A discrete-time system has the state-space model

$$\mathbf{x}[n+1]=\begin{bmatrix}0 & -0.5\\ 0.25 & 0.75\end{bmatrix}\mathbf{x}[n]+\begin{bmatrix}2\\ 1\end{bmatrix}r[n]$$

$$y[n]=\begin{bmatrix}3 & 1\end{bmatrix}\mathbf{x}[n]$$

The initial value of the state vector is

$$\mathbf{x}[0]=\begin{bmatrix}2\\ 0\end{bmatrix}$$

The input to the system is a unit step function, that is, $r[n]=u[n]$. Manually solve for the output signal $y[n]$ for $n=0,\ldots,3$.

6.6. A discrete-time system is described by the state-space model

$$\mathbf{x}[n+1]=\begin{bmatrix}-0.1 & -0.7\\ -0.8 & 0\end{bmatrix}\mathbf{x}[n]+\begin{bmatrix}3\\ 1\end{bmatrix}r[n]$$

$$y[n]=\begin{bmatrix}2 & -1\end{bmatrix}\mathbf{x}[n]$$

a. Find the resolvent matrix $\mathbf{\Phi}(z)$.

b. Find the state transition matrix $\boldsymbol{\phi}[n]$.

c. Compute the unit step response of the system using the initial state vector

$$\mathbf{x}[0]=\begin{bmatrix} 2 \\ 0 \end{bmatrix}$$

6.7. A discrete-time system is described by the state-space model

$$\mathbf{x}[n+1]=\begin{bmatrix} -0.1 & -0.7 \\ -0.8 & 0 \end{bmatrix}\mathbf{x}[n]+\begin{bmatrix} 3 \\ 1 \end{bmatrix} r[n]$$

$$y[n]=\begin{bmatrix} 2 & -1 \end{bmatrix}\mathbf{x}[n]$$

Find the system function $H(z)$.

MATLAB Problems

6.8. A DTLTI system has the system function

$$H(z)=\frac{z^3-7z+6}{z^4-0.2\,z^3-0.93\,z^2+0.198\,z+0.1296}$$

Develop a MATLAB script to do the following:

a. Use function `tf2ss()` to find a state-space model.

b. Use function `eig()` to find a transformation matrix **P** that converts the state-space model found in part (a) to one with a diagonal state matrix.

c. Obtain an alternative state-space model using the transformation matrix **P** found in part (b).

Hint: The use of the function `tf2ss()` with discrete-time system functions requires some care. Write the numerator and the denominator of the system function as polynomials of z^{-1}, and then set up vectors for numerator and denominator coefficients for use with the function `tf2ss()`.

6.9. Refer to the system described in Problem 6.5. Write a script to compute and graph the unit step response of the system for $n=0,\ldots,99$ using the iterative method.

6.10. Refer to the system in Problem 6.6. Write a script to determine the unit step response of the system using symbolic mathematics functions of MATLAB. Verify the analytical solution found in Problem 6.6.

6.11. Refer to the system in Problem 6.7. Write a script to determine the system function $H(z)$ using symbolic mathematics functions of MATLAB. Verify that it matches the solution found manually in Problem 6.7.

6.12. Repeat MATLAB Exercise 6.3 using a sampling interval of $T=0.02$ s. Simulate the system for $0\le t\le 3$ s. Graph the approximated output signal against the analytical solution found given by Eqn. (6.69).

6.13. Consider the CTLTI system with a state-space model

$$\dot{\mathbf{x}}(t)=\begin{bmatrix} 0 & 1 & 0 \\ 0 & 0 & 1 \\ -15 & -11 & -5 \end{bmatrix}\mathbf{x}(t)+\begin{bmatrix} 0 \\ 0 \\ 3 \end{bmatrix} r(t)$$

$$y(t)=\begin{bmatrix} 1 & 0 & 0 \end{bmatrix}\mathbf{x}(t)$$

a. Find a discrete-time state-space model to approximate this system using the technique outlined in Eqns. (6.57) through (6.62). Use a sampling interval of $T_s = 0.1$ s.

b. Using the discretized state-space model found, compute an approximation to the unit step response of the system. Assume zero initial conditions.

6.14. Consider the CTLTI system with a state-space model

$$\dot{\mathbf{x}}(t) = \begin{bmatrix} 0 & 1 & 0 \\ 0 & 0 & 1 \\ -15 & -11 & -5 \end{bmatrix} \mathbf{x}(t) + \begin{bmatrix} 0 \\ 7 \\ -32 \end{bmatrix} r(t)$$

$$y(t) = \begin{bmatrix} 1 & 0 & 0 \end{bmatrix} \mathbf{x}(t)$$

a. Find a discrete-time state-space model to approximate this system using Euler's method with the sampling interval of $T_s = 0.1$ s.

b. Using the discretized state-space model found, compute an approximation to the unit step response of the system. Assume zero initial conditions.

CHAPTER 7

DISCRETE FOURIER TRANSFORM

Chapter Objectives

- Learn the fundamentals of the discrete Fourier transform (DFT). Understand its significance in signal processing applications. Learn its relationship to the discrete-time Fourier series (DTFS) and the discrete-time Fourier transform (DTFT).
- Explore the idea of zero padding a signal before computing its DFT. Learn how zero padding affects the transform and how it can be used for approximating the DTFT of a finite-length signal. Understand the challenges faced in the use of the DFT for detecting sinusoidal components in finite-length signals. Learn the difference between frequency spacing and frequency resolution. Understand how zero padding affects each.
- Learn how to use the DFT to approximate exponential Fourier series (EFS) coefficients for continuous-time periodic signals. Understand the conditions that must be satisfied for obtaining a good approximation. Also learn to use the DFT for approximating the Fourier transform for non-periodic continuous-time signals.
- Adapt the formulation of the DFT so that matrix algebra can be used in its computation. Learn how the DFT can be computed as the product of a square coefficient matrix and a column vector representing the signal.
- Learn properties of the DFT such as linearity, time shifting, time reversal, frequency shifting, conjugation, conjugate symmetry and circular convolution. Understand how these properties can be utilized for easier and/or more efficient computation of the transform.
- Learn special uses of the DFT such as the Goertzel algorithm, the chirp transform algorithm (CTA) and the short-time Fourier transform (STFT).
- Learn efficient computation methods for the DFT leading to the fast Fourier transform (FFT). Explore the decimation-in-time and decimation-in-frequency techniques for speeding up the computation of the DFT.

DOI: 10.1201/9781003570462-7

- Learn how to compute linear convolution using the FFT. Explore methods for using the FFT to compute convolution of signals in real-time processing.

7.1 Introduction

In Chapter 3, we studied the discrete-time Fourier series (DTFS) for periodic signals and the discrete-time Fourier transform (DTFT) for non-periodic signals. The result of DTFT analysis of a discrete-time signal $x[n]$ is a transform $X(\Omega)$ which, if it exists, is a 2π-periodic function of the continuous variable Ω. Storing the DTFT of a signal on a digital computer is impractical because of the continuous nature of the transform. On the other hand, the DTFS representation of a signal $\tilde{x}[n]$ that is periodic with N samples is a set of coefficients $\tilde{c}_k$ that is also periodic with N. While this combination would certainly be suitable for computer implementation and storage, it is only valid for periodic signals. We often deal with signals that are not necessarily periodic.

In the analysis of non-periodic discrete-time signals, sometimes it is desirable to have a transform that is also discrete. This can be accomplished through the use of the *discrete Fourier transform (DFT)* provided that the signal under consideration is of finite-length.

In Section 7.2, the concept of DFT is developed starting with the DTFS representation of a periodic signal. The relationship between DFT and DTFT is explored. An important application of the DFT, namely the detection of sinusoidal components in finite-length signals, is discussed. Concepts of zero padding, frequency spacing, and frequency resolution are introduced. The use of the DFT for approximating the continuous Fourier transform and the exponential Fourier series (EFS) is detailed. Properties of the DFT such as linearity, time and frequency shifting, time reversal, conjugation, circular convolution, and symmetry are discussed in Section 7.3. Section 7.4 introduces examples of algorithms that are useful in special types of applications. These algorithms are Goertzel algorithm, chirp transform algorithm (CTA), and short time Fourier transform (STFT). Methods for improving the computational efficiency of the DFT are discussed in Section 7.5. Decimation-in-time and decimation-in-frequency approaches are introduced, leading to the fast Fourier transform (FFT). Finally, the use of FFT for implementation of the convolution operation in real-time systems is discussed in Section 7.6.

7.2 Discrete Fourier Transform (DFT)

Let $x[n]$ be a signal the non-trivial samples of which are limited to the index range $n = 0, 1, \ldots, N-1$, that is,

$$x[n] = 0 \quad \text{for } n < 0 \;\text{ or }\; n \ge N$$

We will refer to $x[n]$ as a *length-N* signal. An easy method of representing $x[n]$ with a transform that is also length-N would be as follows:

1. Consider $x[n]$ as one period of a periodic signal $\tilde{x}[n]$ defined as

$$\tilde{x}[n] = \sum_{m=-\infty}^{\infty} x[n - mN] \tag{7.1}$$

   The signal $\tilde{x}[n]$ is the *periodic extension* of $x[n]$.

2. Determine the DTFS coefficients $\tilde{c}_k$ for the periodic extension $\tilde{x}[n]$.

$$\tilde{x}[n] \overset{\text{DTFS}}{\longleftrightarrow} \tilde{c}_k$$

3. DTFS coefficients of $\tilde{x}[n]$ form a set that is also periodic with N. Let us extract just one period $\{c_k\,;\ k=0,1,\ldots,N-1\}$ from the DTFS coefficients $\{\tilde{c}_k\,;\ \text{all } k\}$:

$$c_k = \begin{cases} \tilde{c}_k\,, & k=0,1,\ldots,N-1 \\ 0\,, & \text{otherwise} \end{cases} \tag{7.2}$$

This gives us the ability to represent the signal $x[n]$ with the set of coefficients c_k for $k = 0,1,\ldots,N-1$. The coefficients can be obtained from the signal using the three steps outlined above. Conversely, the signal can be reconstructed from the coefficients by simply reversing the order of the steps. The DFT will be defined by slightly modifying the idea presented above. The forward transform is

$$X[k] = \sum_{n=0}^{N-1} x[n]\, e^{-j(2\pi/N)kn}\,, \qquad k=0,1,\ldots,N-1 \tag{7.3}$$

Thus, length-N signal $x[n]$ leads to length-N transform $X[k]$. Compare Eqn. (7.3) with DTFS analysis equation given by Eqn. (3.19) in Chapter 3 and repeated here:

$$\text{Eqn. (3.19):} \qquad \tilde{c}_k = \frac{1}{N} \sum_{n=0}^{N-1} \tilde{x}[n]\, e^{-j(2\pi/N)kn}$$

The only difference between Eqns. (7.3) and (3.19) is the scale factor $1/N$:

$$X[k] = N\,c_k\,, \qquad k=0,1,\ldots,N-1 \tag{7.4}$$

It is also possible to obtain the signal $x[n]$ from the transform $X[k]$ using the inverse DFT relationship

$$x[n] = \frac{1}{N} \sum_{k=0}^{N-1} X[k]\, e^{j(2\pi/N)kn}\,, \qquad n=0,1,\ldots,N-1 \tag{7.5}$$

The notation can be made a bit more compact by defining W_N as

$$W_N = e^{-j(2\pi/N)} \tag{7.6}$$

Using W_N the forward DFT equation becomes

$$X[k] = \sum_{n=0}^{N-1} x[n]\, W_N^{kn}\,, \qquad k=0,1,\ldots,N-1 \tag{7.7}$$

and the inverse DFT is found as

$$x[n] = \frac{1}{N} \sum_{k=0}^{N-1} X[k]\, W_N^{-kn}\,, \qquad n=0,1,\ldots,N-1 \tag{7.8}$$

Notationally the DFT relationship between a signal and its transform can be represented as

$$x[n] \overset{\text{DFT}}{\longleftrightarrow} X[k]$$

An alternative method of showing the relationship between $x[n]$ and $X[k]$ is as follows:

$$X[k] = \text{DFT}\{x[n]\}$$

$$x[n] = \text{DFT}^{-1}\{X[k]\}$$

Discrete Fourier transform (DFT):

1. Analysis equation (Forward transform):

   Eqn. (7.7): $$X[k] = \sum_{n=0}^{N-1} x[n]\, W_N^{kn}\,, \qquad k = 0, 1, \ldots, N-1$$

2. Synthesis equation (Inverse transform):

   Eqn. (7.8): $$x[n] = \frac{1}{N} \sum_{k=0}^{N-1} X[k]\, W_N^{-kn} \qquad n = 0, 1, \ldots, N-1$$

The DFT is a very popular tool in a wide variety of engineering applications for a number of reasons:

1. The signal $x[n]$ and its transform $X[k]$ each have N samples, making the DFT practical for computer implementation. N samples of the signal can be replaced in memory with N samples of the transform without losing any information. It does not matter whether we store the signal or the transform since one can always be obtained from the other.
2. Fast and efficient algorithms, known as *fast Fourier transform (FFT)*, are available for the computation of the DFT.
3. DFT can be used for approximating other forms of Fourier series and transforms for both continuous-time and discrete-time systems. It can also be used for fast convolution in filtering applications.
4. Dedicated processors are available for fast and efficient computation of the DFT with minimal or no programming needed.

Example 7.1: DFT of simple signal

Determine the DFT of the signal

$$x[n] = \{ \underset{\substack{\uparrow \\ n=0}}{1}, \; -1, 2 \}$$

Solution: The discrete Fourier transform is

$$\begin{aligned} X[k] =& e^{-j(2\pi/3)k(0)} - e^{-j(2\pi/3)k(1)} + 2\,e^{-j(2\pi/3)k(2)} \\ =& 1 - e^{-j2\pi k/3} + 2\,e^{-j4\pi k/3} \end{aligned}$$

We will evaluate this result for $k = 1, 2, 3$:

$$\begin{aligned} X[0] =& 1 - 1 + 2 = 2 \\ X[1] =& 1 - e^{-j2\pi/3} + 2\,e^{-j4\pi/3} = 0.5 + j\,2.5981 \\ X[2] =& 1 - e^{-j4\pi k/3} + 2\,e^{-j8\pi k/3} = 0.5 - j\,2.5981 \end{aligned}$$

Software resource: exdt_7_1.m

Example 7.2: DFT of discrete-time pulse

Determine the DFT of the discrete-time pulse signal

$$x[n] = u[n] - u[n-10] = \{ \underset{\substack{\uparrow \\ n=0}}{1}, 1, 1, 1, 1, 1, 1, 1, 1, 1 \}$$

Solution: The discrete Fourier transform is

$$X[k] = \sum_{n=0}^{9} e^{-j(2\pi/10)kn}, \qquad k = 0, 1, \ldots, 9$$

which can be put into closed form using the finite-length geometric series formula (see Appendix C)

$$X[k] = \frac{1 - e^{-j2\pi k}}{1 - e^{-j2\pi k/10}} = \begin{cases} 10, & k = 0 \\ 0, & k = 1, \ldots, 9 \end{cases} \tag{7.9}$$

Note that L'Hospital's rule was used for determining the value $X[0]$. The signal $x[n]$ and its DFT are shown in Fig. 7.1.

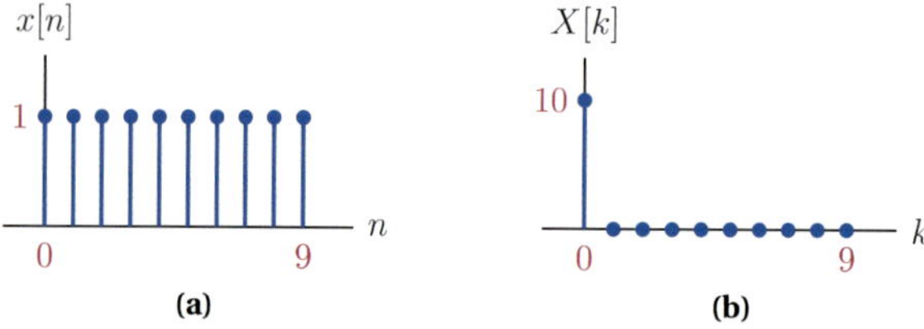

Figure 7.1 – The signal $x[n]$ and the transform $X[k]$ for Example 7.2.

Software resource: exdt_7_2.m

Software resource: See MATLAB Exercise 7.1.

7.2.1 Relationship of the DFT to the DTFT

Consider again a length-N signal $x[n]$. Its DTFT is computed as

$$X(\Omega) = \sum_{n=0}^{N-1} x[n]\, e^{-j\Omega n} \tag{7.10}$$

where the summation limits have been adjusted to account for the fact that the only significant samples of $x[n]$ are in the interval $n = 0, \ldots, N-1$. A comparison of Eqn. (7.10) with Eqn. (3.69) reveals the simple relationship between the DTFT and the DFT.

DFT vs. DTFT:

The DFT of a length-N signal is equal to its DTFT evaluated at a set of N angular frequencies equally spaced in the interval $0 \le \Omega < 2\pi$. Note that the interval includes $\Omega = 0$ but does not include $\Omega = 2\pi$. Let an indexed set of angular frequencies be defined as

$$\Omega_k = \frac{2\pi k}{N}, \quad k = 0, \ldots, N-1$$

The DFT of the signal is written as

$$X[k] = X(\Omega_k) = \sum_{n=0}^{N-1} x[n]\, e^{-j\Omega_k n} \tag{7.11}$$

It is obvious from Eqn. (7.11) that, for a length-N signal, the DFT is very similar to the DTFT with one fundamental difference: In the DTFT, the transform is computed at every value of Ω in the range $0 \le \Omega < 2\pi$. In the DFT, however, the same is computed only at frequencies that are integer multiples of $2\pi/N$. In a way, looking at the DFT is similar to looking at the DTFT placed behind a picket fence with N equally spaced openings. This is referred to as the *picket fence effect*. We will elaborate on this relationship further in the next example.

Example 7.3: DFT of a discrete-time pulse revisited

Consider again the discrete-time pulse used in Example 7.2.

$$x(n) = u[n] - u[n-10]$$

The DFT of this pulse was determined in Example 7.2 and shown graphically in Fig. 7.1(b). The DTFT of $x[n]$ is

$$X(\Omega) = \sum_{n=0}^{9} e^{-j\Omega n}$$

and it can easily be put into a closed form as

$$X(\Omega) = \frac{\sin(5\Omega)}{\sin(0.5\Omega)}\, e^{-j4.5\Omega} \tag{7.12}$$

Recall from earlier discussion (see Eqn. (7.11)) that the transform sample with index k corresponds to the angular frequency $\Omega_k = 2\pi k/N$. If we want to graph the DTFT and the DFT on the same frequency axis, we need to place the DFT samples using an angular frequency spacing of $2\pi/N$ radians. Fig. 7.2 shows both the DTFT and the DFT of the signal $x[n]$, and reveals why we obtained such a trivial looking result in Eqn. (7.9) for the DFT. For $k = 1, \ldots, 9$, the locations of the transform samples in $X[k]$ coincide with the zero-crossings of the DTFT. This is the so-called *picket-fence effect*. It is as though we are looking at the DTFT of the signal $x[n]$ through a picket fence that has narrow openings spaced $2\pi/N$ radians apart. In this case we see mostly the zero-crossings of the DTFT and miss the detail in between. On the other hand, the DFT as given by Eqn. (7.9) is still a complete transform, and the signal $x[n]$ can be obtained from it using the inverse DFT equation given by Eqn. (7.8).

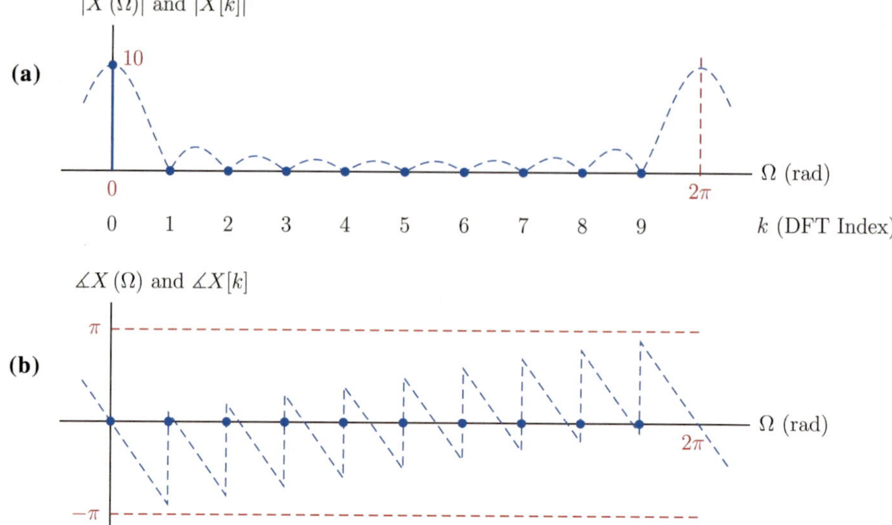

Figure 7.2 – Relationship between the DFT and the DTFT of the signal $x[n]$ used in Example 7.3: **(a)** Magnitudes and **(b)** phases of the two transforms.

Software resource: See MATLAB Exercise 7.3.

Consider again the relationship given by Eqn. (7.11) between DFT and DTFT of a length-N signal $x[n]$. We know that the length-N transform $X[k]$ fully represents the signal $x[n]$ since we can obtain either sequence from the other using Eqns. (7.7) and (7.8). This must mean that the values of the DTFT $X(\Omega)$ between DFT frequencies $\Omega_k = 2\pi k/N$ must be redundant. They must not contain any additional information about the signal $x[n]$. If that is the case, we should be able to obtain $X(\Omega)$ directly from $X[k]$. Let's start with the DTFT of $x[n]$:

$$\text{Eqn. (7.10):} \qquad X(\Omega) = \sum_{n=0}^{N-1} x[n]\, e^{-j\Omega n}$$

We will substitute the inverse transform relationship given by Eqn. (7.5) for $x[n]$ to obtain

$$\begin{aligned} X(\Omega) &= \sum_{n=0}^{N-1} \left[\frac{1}{N} \sum_{k=0}^{N-1} X[k]\, e^{j2\pi kn/N} \right] e^{-j\Omega n} \\ &= \sum_{k=0}^{N-1} X[k] \underbrace{\frac{1}{N} \sum_{n=0}^{N-1} e^{-j(\Omega - 2\pi k/N)n}}_{B_N(\Omega - 2\pi k/N)} \end{aligned} \tag{7.13}$$

Let $B_N(\Omega)$ be defined as

$$B_N(\Omega) = \frac{1}{N} \sum_{n=0}^{N-1} e^{-j\Omega n} = \frac{1 - e^{-j\Omega N}}{N\left(1 - e^{-j\Omega}\right)} \tag{7.14}$$

where we have used the closed form formula for a finite-length geometric series. See Eqn. (C.10) in Appendix C for details. The function $B_N(\Omega)$ can be simplified to

$$B_N(\Omega) = \frac{\sin\left(\frac{\Omega N}{2}\right)}{N \sin\left(\frac{\Omega}{2}\right)}\, e^{-j\Omega(N-1)/2} \tag{7.15}$$

Comparing Eqns. (7.13) and (7.15) we have

$$X(\Omega) = \sum_{k=0}^{N-1} X[k]\, B_N\left(\Omega - \frac{2\pi k}{N}\right) \tag{7.16}$$

Thus, $B_N(\Omega)$ acts as an interpolating function in filling the gaps between samples of the DFT sequence $X[k]$ to produce $X(\Omega)$. For each index k, the amount of frequency shift applied to $B_N(\Omega)$ is $2\pi k/N$.

Example 7.4: Obtaining $X(\Omega)$ from $X[k]$

A length-10 signal $x[n]$ has the DFT

$$X[k] = \{ \underset{\substack{\uparrow \\ k=0}}{1}, 0, 0.5, 0, 2, 0, 0, 0, 0, 0 \}$$

Without finding the signal $x[n]$, determine and graph its DTFT $X(\Omega)$.

Solution:

The only significant samples of $X[k]$ are

$$X[0] = 1\,, \quad X[2] = 0.5 \quad X[4] = 2$$

Using the interpolating function $B_{10}(\Omega)$ given by Eqn. (7.15), $X(\Omega)$ is found as

$$X(\Omega) = B_{10}(\Omega) + 0.5\, B_{10}\left(\Omega - \frac{4\pi}{10}\right) + 2\, B_{10}\left(\Omega - \frac{8\pi}{10}\right)$$

Magnitude and phase of $X(\Omega)$ are shown in Fig. 7.3.

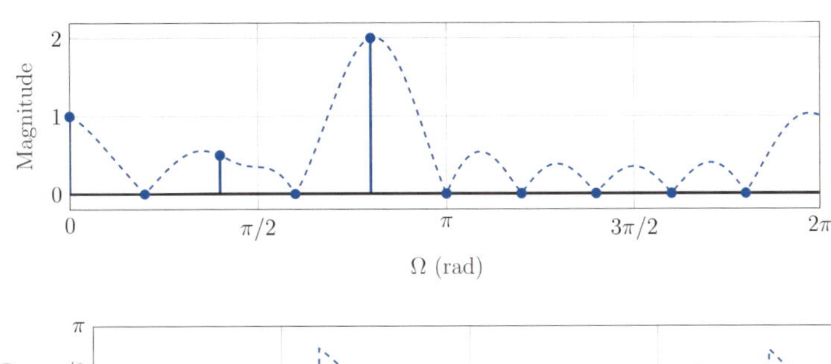

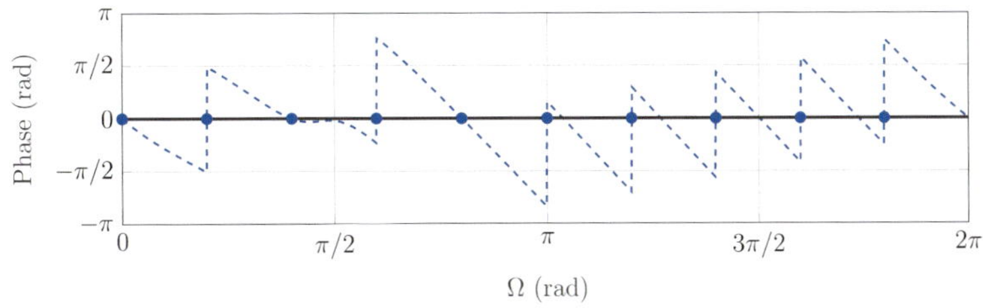

Figure 7.3 – Magnitude and phase of $X(\Omega)$ for the signal of Example 7.4.

Software resources: exdt_7_4.m , DTFTvsDFT.mlx

7.2.2 Zero padding

Even though the DFT computed in Examples 7.2 and 7.3 is a complete and accurate representation of signal $x[n]$, from a visual perspective it does not show the details between frequencies $2\pi k/N$. Sometimes we would like more visual detail than provided by the DFT result. Continuing with the picket fence analogy, we may want to observe the DTFT through a more dense picket fence with more openings in a 2π-radian range of the angular frequency. This can be accomplished by *zero padding* the original signal, that is, by extending it with zero-amplitude samples before computing its DFT.

Consider again the length-N signal $x[n]$. Let us define a length-$(N+M)$ signal $q[n]$ as follows:

$$q[n] = \begin{cases} x[n]\,, & n = 0, 1, \ldots, N-1 \\ 0\,, & n = N, \ldots, N+M-1 \end{cases} \tag{7.17}$$

The DFT of the newly defined signal $q[n]$ is

$$\begin{aligned} Q[k] &= \sum_{n=0}^{N+M-1} q[n]\, e^{-j2\pi kn/(N+M)} \\ &= \sum_{n=0}^{N-1} x[n]\, e^{-j2\pi kn/(N+M)} \end{aligned} \tag{7.18}$$

We have changed the upper limit of the summation back to $N-1$ since sample amplitudes beyond that are all equal to zero. Comparing Eqn. (7.18) with Eqn. (7.10) we conclude that

$$Q[k] = X(\Omega)\Big|_{\Omega = 2\pi k/(N+M)} \tag{7.19}$$

Thus, $Q[k]$ corresponds to observing the DTFT of $x[n]$ through a picket fence with openings spaced $2\pi/(N+M)$ radians as opposed to $2\pi/N$ radians. The number of zeros to be appended to the end of the original signal can be chosen to obtain any desired angular frequency spacing.

Example 7.5: Zero padding the discrete-time pulse

Consider again the length-10 discrete-time pulse used in Example 7.3. Create a new signal $q[n]$ by zero-padding it to 20 samples, and compare the 20-point DFT of $q[n]$ to the DTFT of $x[n]$.

Solution: The new signal $q[n]$ is

$$q[n] = \begin{cases} 1, & n = 0, 1, \ldots, 9 \\ 0, & n = 10, \ldots, 19 \end{cases}$$

The 20-point DFT of $q[n]$ is

$$Q[k] = \sum_{n=0}^{19} q[n]\, e^{-j2\pi nk/20} \tag{7.20}$$

Since $q[n] = 0$ for $n = 10,\ldots,19$, Eqn. (7.20) can be written as

$$Q[k] = \sum_{n=0}^{9} e^{-j2\pi nk/20} = \frac{1-e^{-j\pi k}}{1-e^{-j2\pi k/20}}\,, \qquad k = 0,1,\ldots,19$$

$Q[k]$ is graphed in Fig. 7.4 along with the DTFT $X(\Omega)$. Compare the figure to Fig. 7.3. Notice how 10 new transform samples appear in between the 10 transform samples that were there previously. This is equivalent to obtaining new points between existing ones through some form of interpolation. After the zero padding operation the new angular frequency spacing between the transform samples is $\Omega_k = 2\pi/20$.

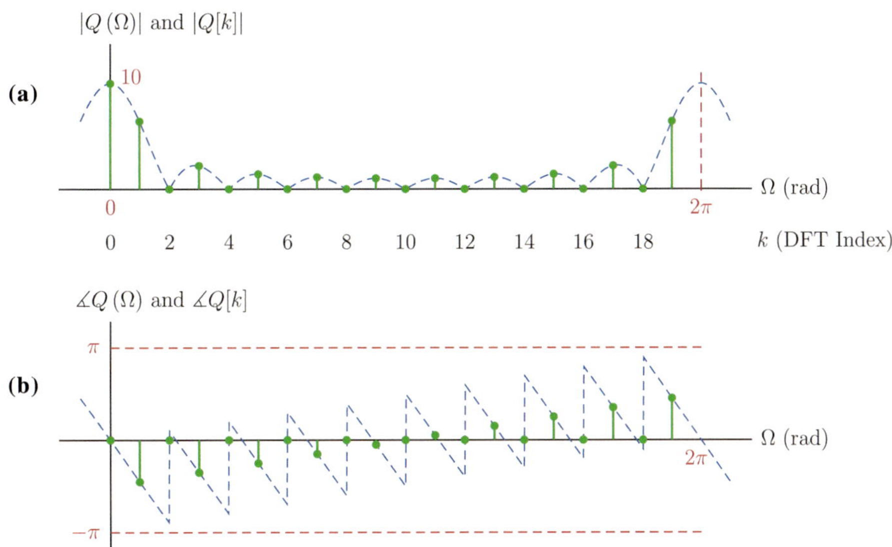

Figure 7.4 – Relationship between the DFT and the DTFT of the zero-padded signal $q[n]$ used in Example 7.5: **(a)** Magnitudes and **(b)** phases of the two transforms.

Software resource: `exdt_7_5.m`

Interactive App: Picket-fence effect of the DFT

The interactive app in `appPicketFence.m` is based on Examples 7.3 and 7.5 as well as Figs. 7.2 and 7.4. The DFT and the DTFT of the 10-sample discrete-time pulse signal are graphed on the same coordinate system to show the picket-fence effect of the DFT. The input signal may be padded with a user-specified number (between 0 and 30) of zero-amplitude samples before the transform is computed. Accordingly, the case in Fig. 7.2 may be duplicated with no additional zero-amplitude samples whereas padding the signal with 10 zero-amplitude samples leads to the situation in Fig. 7.4. The horizontal axis variable for the graphs can be one of three choices: It can display the angular frequency Ω in the range from 0 to 2π, the normalized frequency F in the range from 0 to 1, or the DFT index k in the range from 0 to $N-1$ where N is the size of the DFT. Recall that the relationships between these parameters are $\Omega = 2\pi F$, $\Omega_k = 2\pi k/N$, and $F_k = k/N$.

a. Let L be the number of zero-amplitude data samples with which the 10-sample pulse signal is padded. Compare the graphs for $L=0$ and $L=10$. Going from $L=0$ to $L=10$, observe how the additional DFT samples for $L=10$ interpolate in between the existing DFT samples.

b. Compare the graphs for $L=0$ and $L=20$. In going from $L=0$ to $L=20$, observe that now two additional DFT samples are inserted in between existing DFT samples.

c. Change the horizontal axis to display the DFT index k. Pay attention to the locations of DFT samples in terms of the angular frequency Ω. Explain why the last DFT sample does not coincide with the second peak of the magnitude function, but rather appears slightly to the left of it.

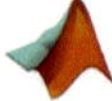

Software resource: `appPicketFence.m`

Software resource: See MATLAB Exercises 7.3 and 7.4.

7.2.3 Detection of sinusoidal signals with the DFT

One of the fundamental uses of the DFT is in detection and estimation. Detection is the act of determining if a particular component is present in the signal being analyzed. Often, the component of interest is a sinusoid or a set of sinusoids. Examples of this are seen in radar and sonar applications. Estimation part of the problem involves estimating values of key parameters of the signal components detected. For sinusoidal components, typical parameters to be estimated might include amplitude, frequency, phase, duration, and so on. In practice, detection problems are complicated by factors such as additive noise, interference and signal distortion. These factors are outside the scope of this textbook, and are typically discussed in texts dealing with communication systems, detection and estimation, and stochastic signal processing. However, one complicating factor will be of interest to us in this discussion, and that is *spectral leakage.*

We will begin by considering a pure sinusoidal signal in the form

$$x[n] = \cos(\lambda n)\,, \quad -\infty < n < \infty \tag{7.21}$$

which may have been obtained by sampling a continuous-time sinusoidal signal

$$x_a(t) = \cos(\omega_0 t)$$

at time instants $t = nT_s$. The DTFT of the signal $x[n]$ was found in Chapter 3 (see Example 3.21) to be

$$X(\Omega) = \sum_{m=-\infty}^{\infty} \Big[\pi\delta(\Omega-\lambda-2\pi m) + \pi\delta(\Omega+\lambda-2\pi m)\Big] \tag{7.22}$$

Fig. 7.5 depicts the DTFT spectrum of the signal $x[n]$.

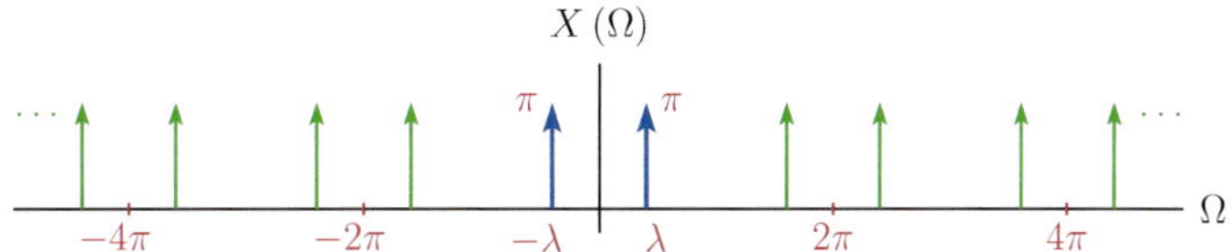

Figure 7.5 – DTFT of the sinusoidal signal in Eqn. (7.21).

Of course, this is a mathematical idealization. In practical situations there are no infinite-length signals. Every signal, no matter how long it might be, is of finite-length. Detection of sinusoids within finite-length signals has its own challenges as we will discover. Consider a truncated, length-N, version of the signal in in Eqn. (7.21):

$$x_T[n] = \cos(\lambda n) \;, \quad n = 0, 1, \ldots, N-1 \tag{7.23}$$

We will explore the problem of finding the transform of this signal using two numerical examples. In the first case, let the angular frequency be $\lambda_1 = 0.2\pi$ for the signal $x_1[n]$. Also, we will choose the length of the signal to be $N = 50$. The DFT of the signal is computed as

$$X_1[k] = \sum_{n=0}^{49} \cos(0.2\pi n)\, e^{-j(2\pi k n/50)}$$

In this case the transform $X_1[k]$ is purely real, and can be written as

$$X_1[k] = 25\,\delta[k-5] + 25\,\delta[k-45] \tag{7.24}$$

The magnitude $|X_1[k]|$ is graphed in Fig. 7.6(a). Discrete-time impulses are clearly visible at sample indices $k = 5$ and $k = 45$. These correspond to angular frequencies as follows:

$$k = 5 \longrightarrow \Omega_k = \frac{10\pi}{50} = 0.2\pi$$

$$k = 45 \longrightarrow \Omega_k = \frac{90\pi}{50} = \frac{-10\pi}{50} = -0.2\pi$$

For the second case, let us change the angular frequency to $\lambda = 0.22\pi$ for the signal $x_2[n]$. The DFT of the new signal is computed as

$$X_2[k] = \sum_{n=0}^{49} \cos(0.22\pi n)\, e^{-j(2\pi k n/50)}$$

The magnitude of $|X_2[k]|$ is shown in Fig. 7.6(b). Unlike the previous case, the spectrum has quite a bit of detail at almost every DFT frequency in the set $\Omega_k = 2\pi k/N$ for $k = 0, \ldots, 49$. This is referred to as *spectral leakage.* Without prior knowledge that $x_2[n]$ has only one sinusoidal component, we may be tempted to conclude that it has two or more sinusoids in its makeup.

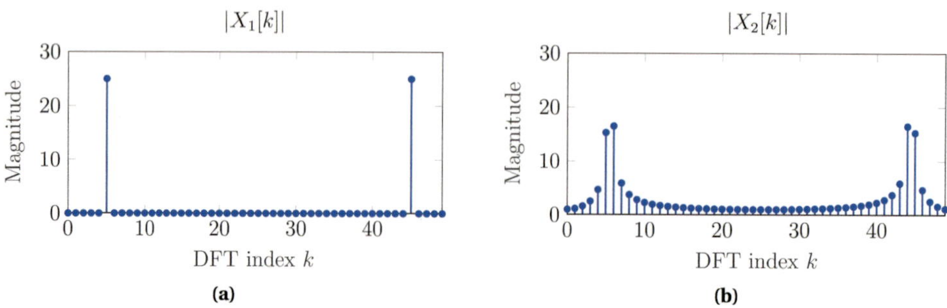

Figure 7.6 – The 50-point DFT of the length-50 cosine signal with **(a)** $\lambda = 0.2\pi$ radians and **(b)** $\lambda = 0.22\pi$ radians.

In order to understand the reason for the disparity between the two results, we need to look at the angular frequencies λ_1 and λ_2 in relation to the length of the signal. For both cases, the DFT frequency spacing is

$$\Delta\Omega = \frac{2\pi}{N} = \frac{2\pi}{50} = 0.04\pi \text{ radians}$$

and the DFT frequencies are

$$\Omega_k = \frac{2\pi k}{50} = 0.04\pi k\,, \quad k = 0, 1, \ldots, 49$$

In the first case, the frequency $\lambda = \lambda_1 = 0.2\pi$ coincides perfectly with the DFT frequency Ω_5. The fundamental period of $x_1[n]$ is

$$N_1 = \frac{2\pi}{\lambda_1} = \frac{2\pi}{0.2\pi} = 10 \text{ samples}$$

Consequently, the 50-sample length of the signal $x_1[n]$ contains 5 full periods of $x[n]$. Recall that, by its definition, DFT has the inherent assumption that the sequence being transformed represents one period of a periodic signal (see Section 7.2, Eqns. (7.1) through (7.4) for details). The periodic extension $\tilde{x}_1[n]$ of the signal $x_1[n]$ is graphed in Fig. 7.7. Successive terms $x_1[n - mN]$ that form the periodic extension fit together perfectly, so we have $\tilde{x}_1[n] = x[n]$. That is the reason why we do not observe spectral leakage in $X_1[k]$ shown in Fig. 7.6(a).

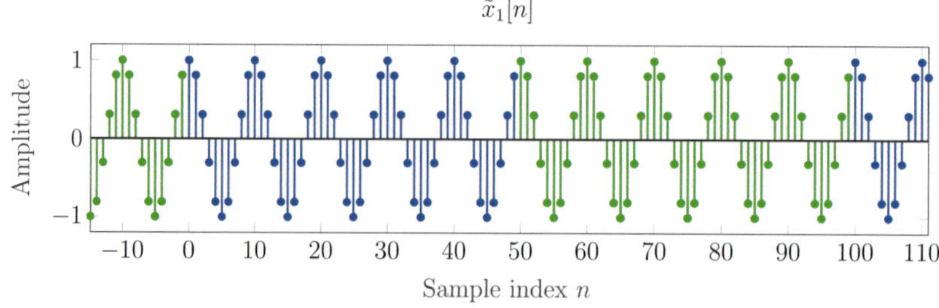

Figure 7.7 – Periodic extension of the signal $x_1[n]$ with $\lambda = 0.2\pi$ radians.

In the second case the frequency, $\lambda = \lambda_2 = 0.22\pi$, falls between $\Omega_5 = 0.2\pi$ and $\Omega_6 = 0.24\pi$. The fundamental period of $x_2[n]$ with $\lambda = \lambda_2$ is

$$N_2 = \frac{2\pi r}{\lambda_2} = \frac{2\pi r}{0.22\pi}$$

which resolves to $N_2 = 100$ samples for $r = 11$ cycles of the sinusoid in $x_a(t)$. The 50-sample length of the signal $x_2[n]$ contains 5.5 cycles of the sinusoid in $x_a(t)$, ending on a half cycle. The periodic extension of $x_2[n]$ is shown in Fig. 7.8. The terms $x_2[n-mN]$ do not fit together perfectly, and give way to discontinuities (see transitions from sample 49 to 50, and from sample 99 to 100). For this case $\tilde{x}_2[n] \neq x[n]$. These discontinuities translate to the frequency spectrum as spectral leakage. A more detailed view of the periodic extension $\tilde{x}_2[n]$ around the transition from one period to the next is shown in Fig 7.9.

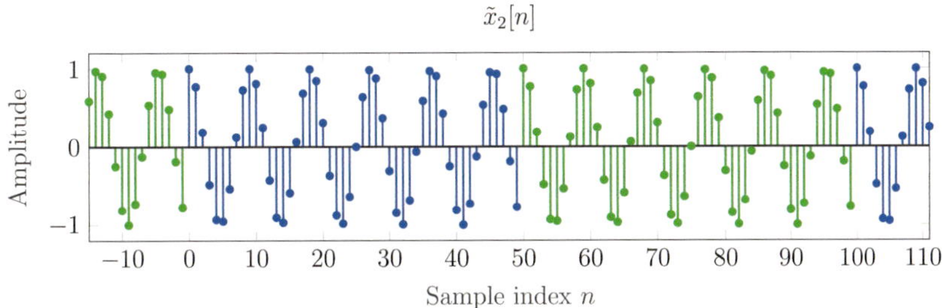

Figure 7.8 – Periodic extension of the signal $x_2[n]$.

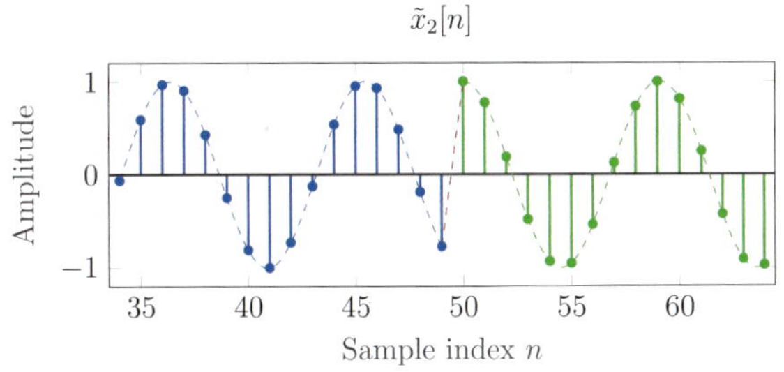

Figure 7.9 – More detailed view of $\tilde{x}_2[n]$.

It would also be interesting to look at the DTFT of a finite-length sinusoid. Magnitudes of transform $X_1(\Omega)$ is shown in Fig. 7.10. Spectral leakage that was not visible in the 50-point DFT $X_1[k]$ is clearly visible in $X_1(\Omega)$.

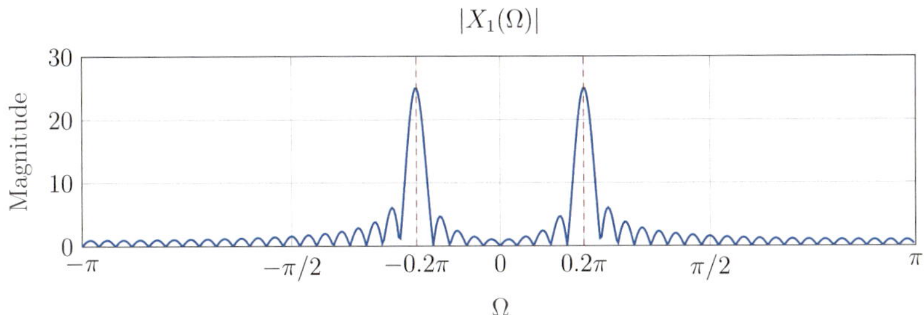

Figure 7.10 – The DTFT of the length-50 cosine signal with frequency $\lambda = 0.2\pi$ radians.

In order to understand spectral leakage from a DTFT perspective, we need to look at the effect of truncating a sinusoidal signal. The length-N signal $x_1[n]$ can be obtained by multiplying $x[n]$

with a *rectangular window* sequence

$$x_1[n] = x[n]\, w_R[n]$$

where $w_R[n]$ is defined as

$$w_R[n] = u[n] - u[n-N] = \begin{cases} 1\,, & n = 0, 1, \ldots, N-1 \\ 0\,, & \text{otherwise} \end{cases} \tag{7.25}$$

Multiplication property of the DTFT given by Eqn. (3.154) is useful in this case. The DTFT of the product $x[n]\, w[n]$ is the periodic convolution of the two transforms $X(\Omega)$ and $W(\Omega)$, that is,

$$X_1(\Omega) = X(\Omega) \otimes W_R(\Omega) = \frac{1}{2\pi} \int_{-\pi}^{\pi} X(\lambda)\, W_R(\Omega - \lambda)\, d\lambda \tag{7.26}$$

The spectrum of the rectangular window sequence is (see Problem 7.11 at the end of this chapter)

$$W_R(\Omega) = \frac{\sin\left(\frac{\Omega N}{2}\right)}{\sin\left(\frac{\Omega}{2}\right)}\, e^{-j\Omega(N-1)/2} \tag{7.27}$$

The magnitude $\left|W_R(\Omega)\right|$ is shown in Fig. 7.11 on both linear and logarithmic (dB) scales. Compare Figs. 7.5, 7.10, and 7.11 to identify the circular convolution relationship between them that is established by Eqn. (7.26).

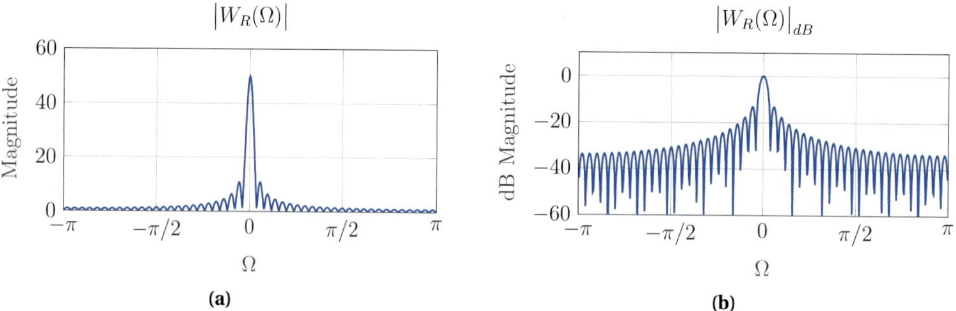

Figure 7.11 – The DTFT spectrum of the rectangular window **(a)** on linear scale and **(b)** on dB scale.

The DTFT spectrum of the rectangular window exhibits a main lobe centered around $\Omega = 0$, and sidelobes on both sides of the main lobe. A reduction in spectral leakage requires narrower main lobe and weaker sidelobes. To determine the width of the main lobe, we need to look at the zero crossings of the numerator of Eqn. (7.27). The zero crossings that are closest to the origin occur for

$$\sin\left(\frac{\Omega N}{2}\right) = \pm\pi \longrightarrow \Omega = \pm\frac{2\pi}{N}$$

The width of the main lobe is therefore $4\pi/N$. It can be made narrower by increasing the length of the window sequence $w_R[n]$ and, consequently, the observation interval for the sinusoidal data. The first sidelobe on either side of the main lobe is only about 13 dB below the peak of the main lobe. Sidelobes in the window spectrum cannot be weakened by simply increasing N. Instead, we need to modify the shape of the window sequence to suppress the sidelobes. We will discover, however, that sidelobe suppression is achieved at the expense of widening the main lobe, so there is a tradeoff involved.

An alternative to the rectangular window is the *Hamming window* given by

$$w_H[n] = \begin{cases} 0.54 - 0.46\cos\left(\dfrac{2\pi n}{N-1}\right), & n = 0, 1, \ldots, N-1 \\ 0, & \text{otherwise} \end{cases} \tag{7.28}$$

The shape of the Hamming window and the magnitude of its DTFT spectrum are shown in Fig. 7.12. Fig. 7.13 depicts a comparison of the dB magnitude spectra for rectangular and Hamming windows. Hamming window does a significantly better job at suppressing the sidelobes, however, the cost of that improvement is a main lobe that is roughly twice as wide compared to the rectangular window.

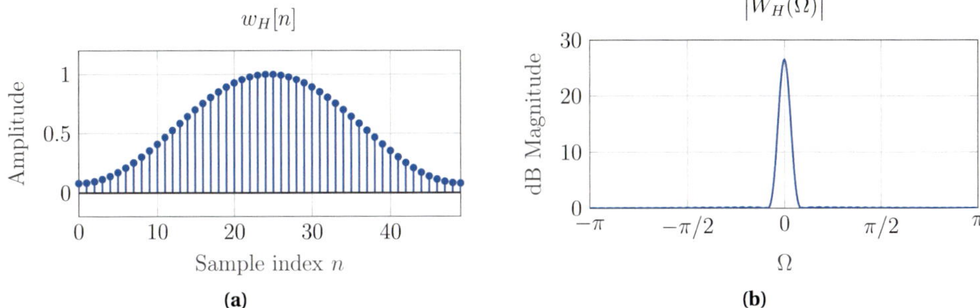

Figure 7.12 – **(a)** Length-50 Hamming window and **(b)** the magnitude of its DTFT spectrum.

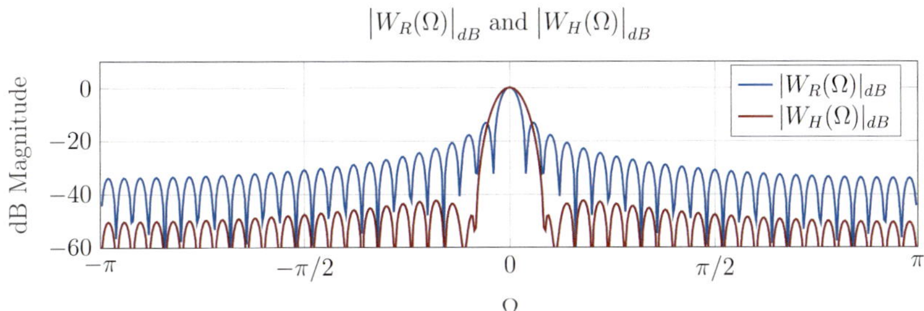

Figure 7.13 – Comparison of dB magnitude spectra for rectangular and Hamming windows of length 50.

7.2.4 Frequency spacing vs. frequency resolution

In comparing the DFT with the DTFT in Section 7.2.1 it was established that the DFT of a signal is essentially a sampled version of its DTFT:

$$X[k] = X(\Omega)\Big|_{\Omega=\Omega_k=\frac{2\pi k}{N}}$$

Thus, the *frequency spacing* between successive samples of the N-point DFT is $\Delta\Omega = 2\pi/N$ radians. In Section 7.2.2 the effects of zero padding a signal before computing its DFT was discussed. Padding a length-N signal with M additional zero-amplitude samples extends its effective length to $(N+M)$ samples, resulting in a new frequency spacing of $\Delta\Omega = 2\pi/(N+M)$ radians. It is possible to achieve any desired frequency spacing by properly choosing the parameter M for zero padding.

In contrast, the frequency resolution Ω_r of the DFT is the smallest frequency difference between two sinusoidal signals that the transform is able to distinguish. Consider the multitone signal $x[n]$ given by

$$x[n] = \cos(\Omega_1 n) + \cos(\Omega_2 n)\ , \quad n = 0, 1, \ldots, N-1$$

Would the N-point DFT of this signal indicate two distinct sinusoidal components? The answer depends on the difference of the two frequencies. If

$$\Omega_2 - \Omega_1 \ge \frac{2\pi}{N},$$

then we would expect the DFT to detect two frequency components. If Ω_1 and Ω_2 are closer to each other than the DFT frequency spacing, the transform would show only one peak. In this particular case, the frequency resolution of the DFT is

$$\Omega_r = \frac{2\pi}{N}$$

same as the frequency spacing. A question that may be raised is: Would it be possible to improve the frequency resolution of the DFT by zero padding the signal $x[n]$? The answer is no. Zero padding does not add any new information to the signal, and we cannot realistically expect it to improve the ability of the transform when it comes to detecting sinusoidal components. For a length-N signal padded with M zero-amplitude samples, the frequency resolution of the DFT is still $\Omega_r = 2\pi/N$. If the multitone signal is multiplied by a window sequence other than the rectangular window, the ability to distinguish sinusoids closely spaced in frequency would be further diminished.

Example 7.6: Frequency resolution of the DFT

a. In this example we will experiment with the frequency resolution of the DFT. Consider the length-32 signal

$$x[n] = \cos(0.8345n) + \cos(0.9327n)\ , \quad n = 0, 1, \ldots, 31$$

The frequency spacing of the 32-point DFT is

$$\Delta\Omega = \frac{2\pi}{32} = 0.1963 \text{ radians}$$

In this case, the frequency resolution of the DFT is $\Omega_r = \Delta\Omega = 0.1963$ radians. The difference of the two frequencies is

$$\Omega_2 - \Omega_1 = 0.0982 \text{ radians}$$

Since $\Omega_2 - \Omega_1 < \Omega_r$, we do not expect the DFT to resolve the two frequencies. Fig. 7.14 depicts the DFT of the signal $x[n]$ under various circumstances, with and without zero padding. It is clear that no amount of zero padding enables the detection of two separate frequencies.

b. In order to improve the frequency resolution of the DFT, we will increase the amount of data. Let the signal $x[n]$ be redefined as a length-64 signal:

$$x[n] = \cos(0.8345n) + \cos(0.9327n)\ , \quad n = 0, 1, \ldots, 64$$

Now the frequency resolution is

$$\Omega_r = \frac{2\pi}{64} = 0.0982 \text{ radians}$$

and the DFT should be able to resolve the two sinusoids. Fig. 7.15 illustrates 64-point and 512-point transforms of the redefined signal. Two sinusoidal peaks are clearly visible in both cases.

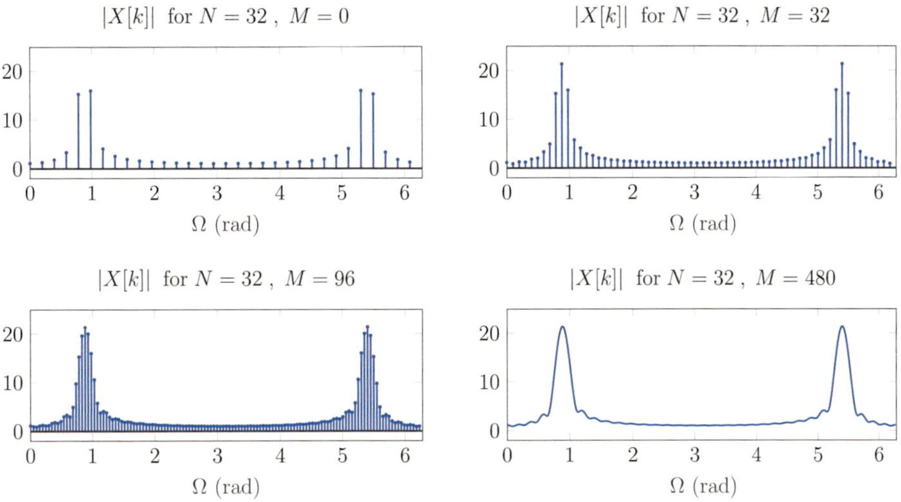

Figure 7.14 – The DFT of the length-32 signal $x[n]$ in Example 7.6.

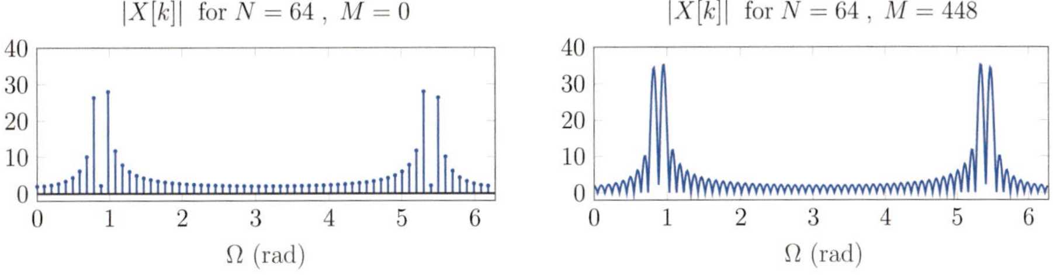

Figure 7.15 – The DFT of the length-64 signal $x[n]$ in Example 7.6.

Software resources: exdt_7_6.m , DFTResolution.mlx

7.2.5 Using the DFT to approximate the EFS coefficients

EFS representation of a continuous-time signal $x_a(t)$ periodic with a period T_0 is given by

$$c_k = \frac{1}{T_0} \int_{t_0}^{t_0+T_0} x_a(t)\, e^{-jk\omega_0 t}\, dt \tag{7.29}$$

One method of approximating the coefficients c_k on a computer is by approximating the integral in Eqn. (7.29) using the rectangular approximation method. More sophisticated methods for approximating integrals exist, and are explained in detail in a number of excellent texts on numerical analysis. The rectangular approximation method is quite simple and will be sufficient for our purposes. Suppose we would like to approximate the following integral:

$$G = \int_0^{T_0} g(t)\, dt \tag{7.30}$$

Since integrating the function $g(t)$ amounts to computing the area under the function, a simple approximation is

$$G \approx \sum_{n=0}^{N-1} g(nT)\, T \tag{7.31}$$

We have used the sampling interval $T = T_0/N$ and assumed that N is sufficiently large for the approximation to be a good one. The area under the function $g(t)$ is approximated using the areas of successive rectangles formed by the samples $g(nT)$. This is depicted graphically in Fig. 7.16.

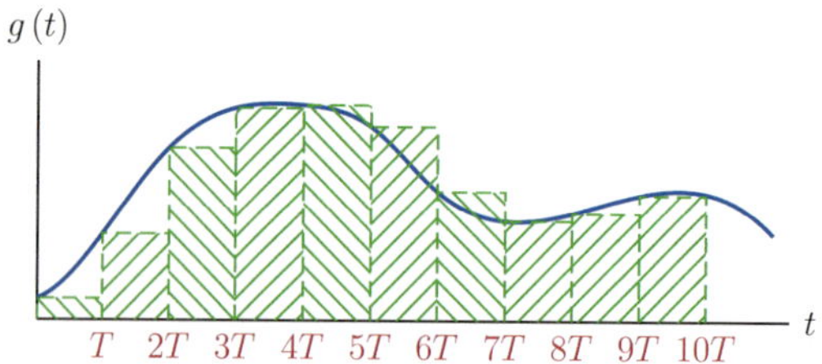

Figure 7.16 – Rectangular approximation to an integral.

For the purpose of approximating Eqn. (7.29) let $g(t)$ be chosen as

$$g(t) = \frac{1}{T_0}\, x_a(t)\, e^{-jk\omega_0 t} \tag{7.32}$$

so that the integral in Eqn. (7.29) is approximated as

$$c_k \approx \frac{1}{T_0} \sum_{n=0}^{N-1} x_a(nT)\, e^{-jk\omega_0 nT}\, T \tag{7.33}$$

Recalling that $T_0 = NT$ and $\omega_0 = 2\pi/T_0$, and using the discrete-time signal $x[n] = x_a(nT)$, we have

$$c_k \approx \frac{1}{N} \sum_{n=0}^{N-1} x[n]\, e^{-j(2\pi/N)kn} = \frac{1}{N}\, X[k] \tag{7.34}$$

The EFS coefficients of a periodic signal can be approximated by sampling the signal at N equally-spaced time instants over one period, computing the DFT of the resulting discrete-time signal $x[n]$, and scaling the DFT result by $1/N$. Some caveats are in order:

1. The number of samples over one period, N, should be sufficiently large. If we want to be more rigorous, the conditions of the Nyquist sampling theorem must be observed, and the sampling rate $f_s = 1/T$ should be at least twice the highest frequency in the periodic signal $x_a(t)$.
2. Only the first half of the DFT values can be used as approximations for positive indexed coefficients. Assuming the DFT size is even, we can write

$$c_k \approx \frac{1}{N}\, X[k], \quad k = 0, \ldots, \frac{N}{2} - 1 \tag{7.35}$$

with negative indexed coefficients obtained from the second half of the DFT by

$$c_{-k} \approx \frac{1}{N}\,X[N-k], \quad k=1,\ldots,\frac{N}{2} \tag{7.36}$$

3. If the conditions of the Nyquist sampling theorem cannot be satisfied in a strict sense (such as when we try to approximate the Fourier series of a pulse train which is not limited in terms of the highest frequency it contains), only the first few terms of each set in Eqns. (7.35) and (7.36) should be used to keep the quality of the approximation acceptable.

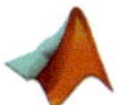

Software resource: See MATLAB Exercises 7.10 and 7.11.

7.2.6 Using the DFT to approximate the continuous Fourier transform

It is also possible to use the DFT for approximating the Fourier transform of a non-periodic continuous-time signal. For a signal $x_a(t)$ the Fourier transform is defined

$$X_a(\omega) = \int_{-\infty}^{\infty} x_a(t)\, e^{-j\omega t}\, dt \tag{7.37}$$

Suppose $x_a(t)$ is a finite-length signal that is zero outside the interval $0 \le t \le t_1$. Adjusting the integration limits, we have

$$X_a(\omega) = \int_{0}^{t_1} x_a(t)\, e^{-j\omega t}\, dt \tag{7.38}$$

The integral can be evaluated in an approximate sense using the rectangular approximation technique outlined in Eqn. (7.31) with

$$g(t) = x_a(t)\, e^{-j\omega t} \tag{7.39}$$

The function $g(t)$ needs to be sampled at N equally spaced time instants in the interval $0 \le t \le t_1$, and thus $NT = t_1$. Rectangular rule approximation to the integral in Eqn. (7.38) is

$$X_a(\omega) \approx \sum_{n=0}^{N-1} x_a(nT)\, e^{-j\omega nT}\, T \tag{7.40}$$

Using the discrete-time signal $x[n] = x_a(nT)$ and evaluating Eqn. (7.40) at a discrete set of frequencies

$$\omega_k = \frac{k\omega_s}{N} = \frac{2\pi k}{NT}, \qquad k=0,\ldots,N-1$$

where ω_s is the sampling rate in rad/s, we obtain

$$X_a(\omega_k) \approx T \sum_{n=0}^{N-1} x[n]\, e^{-j(2\pi/N)kn} = T\,X[k] \tag{7.41}$$

The Fourier transform of a continuous-time signal can be approximated by sampling the signal at N equally-spaced time instants, computing the DFT of the resulting discrete-time signal $x[n]$, and scaling the DFT result by the sampling interval T.

It is also possible to obtain the approximation for $X_a(\omega)$ at a more closely spaced set of frequencies than $\omega_k = k\omega_s/N$ by zero-padding $x[n]$ prior to computing the DFT. Let

$$\bar{\omega}_k = \frac{k\omega_s}{N+M} = \frac{2\pi k}{(N+M)\,T}, \qquad k=0,\ldots,N+M-1$$

where M is an integer. Using $\bar{\omega}_k$ in Eqn. (7.41) leads to

$$X_a(\bar{\omega}_k) \approx T \sum_{n=0}^{N-1} x[n]\, e^{-j2\pi kn(N+M)} \qquad (7.42)$$

The summation on the right side of Eqn. (7.42) is the DFT of the signal $x[n]$ zero-padded with M additional samples. As noted in the discussion of the previous section, the conditions of the Nyquist sampling theorem apply to this case as well. Ideally we would like the sampling rate ω_s to be at least twice the highest frequency of the signal the transform of which is being approximated. On the other hand, it can be shown that a time-limited signal contains an infinite range of frequencies, and strict adherence to the Nyquist sampling theorem is not possible. Approximation errors are due to the aliasing that occurs in sampling a time-limited signal. For the approximation to be acceptable, we need to ensure that the effect of aliasing is negligible.

Based on the Nyquist sampling theorem, only the first half of the DFT samples in Eqns. (7.41) and (7.42) should be used for approximating the transform $X_a(\omega)$ at positive frequencies in the range $0 \le \omega < \omega_s/2$. The second half of the DFT samples represent an approximation to the transform $X_a(\omega)$ in the negative frequency range $-\omega_s/2 \le \omega < 0$.

Software resource: See MATLAB Exercise 7.12.

Interactive App: Using DFT to approximate the continuous Fourier transform

The interactive app in `appApproxCFT.m` illustrates the use of DFT for approximating the Fourier transform of a continuous-time function. It is based on Eqn. (7.41), MATLAB Exercise 7.12, and Fig. 7.36.

The continuous-time function $x_a(t)$ used in MATLAB Exercise 7.12 is graphed on the left side of the screen along with its sampled form $x[n]$. The parameter N represents the number of samples in the time interval $0 \le t \le 1$ s, and the parameter M represents the number of padded zero-amplitude samples, paralleling the development in Eqns. (7.37) through (7.42). Magnitude and phase of the actual spectrum $X_a(f)$ as well as the DFT-based approximation are shown on the right side.

a. The case in Fig. 7.36 may be duplicated with the choices of $N = 8$ and $M = 0$. We know that $NT = 1$ s, so the sampling interval is $T = 1/N = 0.125$ s, and the sampling rate is $f_s = 8$ Hz. The approximated samples appear from $f_1 = -4$ Hz to $f_2 = 3$ Hz with a frequency increment of $\Delta f = 1$ Hz.

b. While keeping N unchanged, set $M = 8$ which causes 8 zero-amplitude samples to be appended to the right of the existing samples before the DFT computation. Now there are 16 estimated samples of the transform $X_a(f)$ starting at $f_1 = -4$ Hz with an increment of $\Delta f = 0.5$ s.

c. Set $N = 16$ and $M = 0$. The sampling rate is $f_s = 16$ Hz now, and therefore the estimated samples start at $f_1 = -8$ Hz and go up to $f_2 = 7$ Hz with an increment of $\Delta f = 1$ Hz.

Understanding these relationships is key to understanding effective use of the DFT as a tool for analyzing continuous-time signals.

Software resource: `appApproxCFT.m`

7.2.7 Matrix Formulation of the DFT

Consider again the DFT analysis equation

$$X[k] = \sum_{n=0}^{N-1} x[n]\, W_N^{kn}, \qquad k = 0,\ldots,N-1$$

For a specific value of the transform index $k = m$, the result $X[m]$ is in the same form as the product of the row vector $\mathbf{A_m}$ and the column vector $\mathbf{x}$ defined as follows:

$$\mathbf{A_m} = \begin{bmatrix} W_N^0 & W_N^m & W_N^{2m} & \ldots & W_N^{(N-1)m} \end{bmatrix} \quad \text{and} \quad \mathbf{x} = \begin{bmatrix} x[0] \\ x[1] \\ x[2] \\ \vdots \\ x[N-1] \end{bmatrix}$$

Thus the m-th sample of the transform can be computed as

$$X[m] = \mathbf{A_m}\,\mathbf{x} \tag{7.43}$$

A similar row vector can be found for each sample of the transform. If we want to compute the entire transform in this fashion, we need to construct a matrix $\mathbf{A}$ each row of which corresponds to the corresponding vector $\mathbf{A_m}$ for $m = 0, 1, \ldots, N-1$.

$$\mathbf{A} = \begin{bmatrix} \mathbf{A_0} \\ \mathbf{A_1} \\ \mathbf{A_2} \\ \vdots \\ \mathbf{A_{N-1}} \end{bmatrix} = \begin{bmatrix} a_{11} & a_{12} & a_{13} & \ldots & a_{1N} \\ a_{21} & a_{22} & a_{23} & \ldots & a_{2N} \\ a_{31} & a_{32} & a_{33} & \ldots & a_{3N} \\ \vdots & \vdots & \vdots & \ldots & \vdots \\ a_{N1} & a_{N2} & a_{N3} & \ldots & a_{NN} \end{bmatrix} \tag{7.44}$$

Individual elements of the matrix $\mathbf{A}$ can be computed as follows:

$$\text{Element at row-}p\text{, column-}r: \quad a_{pr} = W_N^{(p-1)(r-1)} \tag{7.45}$$

Let the transform also be expressed as a column vector:

$$\mathbf{X} = \begin{bmatrix} X[0] \\ X[1] \\ X[2] \\ \vdots \\ X[N-1] \end{bmatrix} \tag{7.46}$$

The DFT synthesis equation can now be written in matrix form as

$$\mathbf{X} = \mathbf{A}\mathbf{x} \tag{7.47}$$

Example 7.7: Matrix formulation of the DFT

Using matrix formulation of the DFT, find the transform of the signal

$$x[n] = \{ \underset{\substack{\uparrow \\ n=0}}{1}, \; -3, 2, 5, -6 \}$$

Solution: The first step is to form the 5 by 5 coefficient matrix **A**. Recall that its row-p column-r element is $a_{pr} = W_5^{(p-1)(r-1)}$. Therefore, we have

$$\mathbf{A} = \begin{bmatrix} W_5^0 & W_5^0 & W_5^0 & W_5^0 & W_5^0 \\ W_5^0 & W_5^1 & W_5^2 & W_5^3 & W_5^4 \\ W_5^0 & W_5^2 & W_5^4 & W_5^6 & W_5^8 \\ W_5^0 & W_5^3 & W_5^6 & W_5^9 & W_5^{12} \\ W_5^0 & W_5^4 & W_5^8 & W_5^{12} & W_5^{16} \end{bmatrix}$$

Using the periodicity of $W_5^k = e^{-j2\pi k/5}$ the matrix **A** can be simplified to the following form:

$$\mathbf{A} = \begin{bmatrix} 1 & 1 & 1 & 1 & 1 \\ 1 & W_5^1 & W_5^2 & W_5^3 & W_5^4 \\ 1 & W_5^2 & W_5^4 & W_5^1 & W_5^3 \\ 1 & W_5^3 & W_5^1 & W_5^4 & W_5^2 \\ 1 & W_5^4 & W_5^3 & W_5^2 & W_5^1 \end{bmatrix}$$

Using numerical values for W_5^k for $k = 1,2,3,4$ we can now populate the matrix:

$$\mathbf{A} = \begin{bmatrix} 1 & 1 & 1 & 1 & 1 \\ 1 & 0.309 - j0.951 & -0.809 - j0.588 & -0.809 + j0.588 & 0.309 + j0.951 \\ 1 & -0.809 - j0.588 & 0.309 + j0.951 & 0.309 - j0.951 & -0.809 + j0.588 \\ 1 & -0.809 + j0.588 & 0.309 - j0.951 & 0.309 + j0.951 & -0.809 - j0.588 \\ 1 & 0.309 + j0.951 & -0.809 + j0.588 & -0.809 - j0.588 & 0.309 - j0.951 \end{bmatrix}$$

Defining the column vector **x** as

$$\mathbf{x} = \begin{bmatrix} 1 & -3 & 2 & 5 & -6 \end{bmatrix}^T$$

the transform is found to be

$$\mathbf{X} = \mathbf{A}\mathbf{x} = \begin{bmatrix} -1 \\ -7.444 - j1.090 \\ 10.444 - j4.617 \\ 10.444 + j4.617 \\ -7.444 + j1.090 \end{bmatrix}$$

Software resource: exdt_7_7.m

Software resource: See MATLAB Exercise 7.13.

7.3 Properties of the DFT

Important properties of the DFT will be summarized in this section. It will become apparent in that process that the properties of the DFT are similar to those of DTFS and DTFT with one significant difference: Any shifts in the time domain or the transform domain are *circular shifts* rather than linear shifts. Also, any time reversals used in conjunction with the DFT are *circular time reversals* rather than linear ones. Therefore, the concepts of circular shift and circular time reversal will be introduced here in preparation for the discussion of DFT properties.

In the derivation leading to forward and inverse DFT relationships in Eqns. (7.7) and (7.8) for a length-N signal $x[n]$ we have relied on the DTFS representation of the periodic extension signal $\tilde{x}[n]$. Let us consider the following scenario:

1. Obtain periodic extension $\tilde{x}[n]$ from $x[n]$ using Eqn. (7.1).
2. Apply a time shift to $\tilde{x}[n]$ to obtain $\tilde{x}[n-m]$. The amount of the time shift may be positive or negative.
3. Obtain an length-N signal $g[n]$ by extracting the main period of $\tilde{x}[n-m]$.

$$g[n] = \begin{cases} \tilde{x}[n-m]\,, & n = 0, 1, \ldots, N-1 \\ 0\,, & \text{otherwise} \end{cases} \tag{7.48}$$

The resulting signal $g[n]$ is a *circularly shifted* version of $x[n]$, that is

$$g[n] = x[n-m]_{\text{mod } N} \tag{7.49}$$

The term on the right side of Eqn. (7.49) uses *modulo indexing*. The signal $x[n]$ has meaningful samples only for $n = 0, 1, \ldots, N-1$. The index $n-m$ for a particular set of n and m values may or may not be in this range. Modulo-N value of the index is found by adding integer multiples of N to the index until the result is within the range $n = 0, 1, \ldots, N-1$. A few examples are given below:

$$x[-3]_{\text{mod } 8} = x[5] \qquad x[12]_{\text{mod } 10} = x[2] \qquad x[-7]_{\text{mod } 25} = x[18]$$

$$x[16]_{\text{mod } 16} = x[0] \qquad x[-3]_{\text{mod } 4} = x[1] \qquad x[95]_{\text{mod } 38} = x[19]$$

The process that led to Eqn. (7.49) is illustrated in Figs. 7.17 and 7.18 for an example length-8 signal. For the example, we are considering in Figs. 7.17 and 7.18 imagine a picture frame that fits samples $0, \ldots, 7$. Right shifting the signal by two samples causes two samples to leave frame from the right edge and re-enter from the left edge, as shown in Fig. 7.17 Left shifting the signal has the opposite effect: Samples leave the frame from the left edge and re-enter from the right edge, as shown in Fig. 7.18. Fig. 7.19 further illustrates the concept of circular shifting.

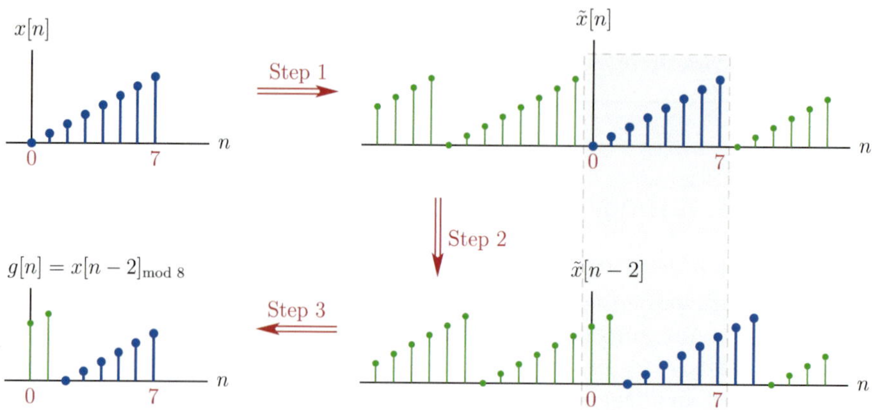

Figure 7.17 – Obtaining a circular shift to the right by 2 samples. In step 1 a periodic extension $\tilde{x}[n]$ is formed. In step 2 the periodic extension is time shifted to obtain $\tilde{x}[n-2]$. In step 3 the main period is extracted to obtain $g[n]$.

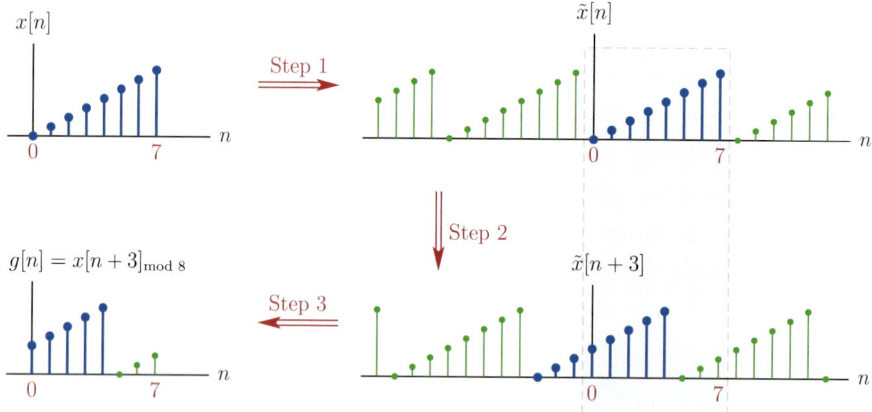

Figure 7.18 – Obtaining a circular shift to the left by 3 samples.

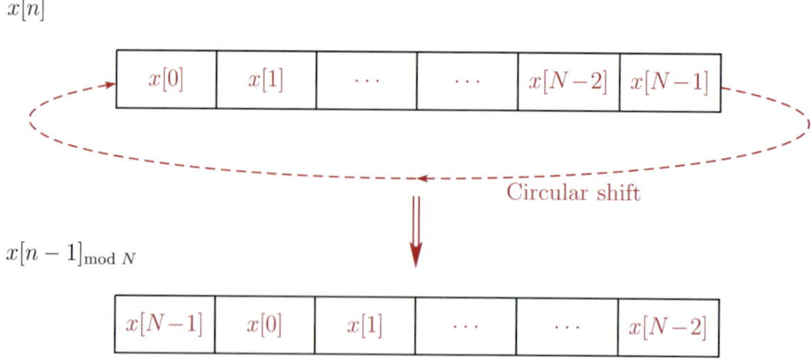

Figure 7.19 – Circular shifting a length-N signal.

For the time reversal operation consider the following steps:

1. Obtain periodic extension $\tilde{x}[n]$ from $x[n]$ using Eqn. (7.1).
2. Apply the time reversal operation to $\tilde{x}[n]$ to obtain $\tilde{x}[-n]$.
3. Obtain an length-N signal $g[n]$ by extracting the main period of $\tilde{x}[-n]$.

$$g[n] = \begin{cases} \tilde{x}[-n]\,, & n = 0,\ldots,N-1 \\ 0\,, & \text{otherwise} \end{cases} \tag{7.50}$$

The resulting signal $g[n]$ is a *circularly time reversed* version of $x[n]$, that is

$$g[n] = x[-n]_{\text{mod } N} \tag{7.51}$$

The process that led to Eqn. (7.51) is illustrated in Fig. 7.20 for an example length-8 signal.

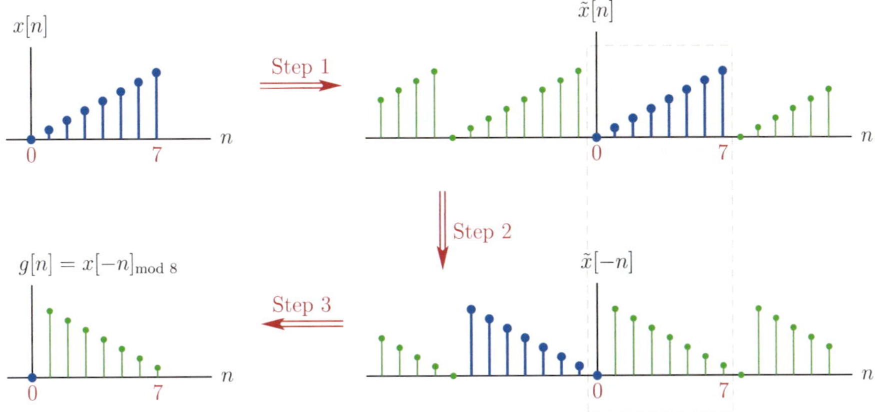

Figure 7.20 – Circular time reversal of a length-8 signal.

Software resource: See MATLAB Exercise 7.5.

For DFT related operations the definitions of conjugate symmetry properties also need to be adjusted so that they utilize circular time reversals. A length-N signal $x[n]$ is *circularly conjugate symmetric* if it satisfies

$$x^*[n] = x[-n]_{\text{mod } N} \tag{7.52}$$

or *circularly conjugate antisymmetric* if it satisfies

$$x^*[n] = -x[-n]_{\text{mod } N} \tag{7.53}$$

A signal that satisfies neither Eqn. (7.52) nor Eqn. (7.53) can still be decomposed into two components such that one is circularly conjugate symmetric and the other is circularly conjugate antisymmetric. The conjugate symmetric component is computed as

$$x_E[n] = \frac{x[n] + x^*[-n]_{\text{mod } N}}{2} \tag{7.54}$$

and the conjugate antisymmetric component is computed as

$$x_O[n] = \frac{x[n] - x^*[-n]_{\text{mod } N}}{2} \tag{7.55}$$

respectively, so that

$$x[n] = x_E[n] + x_O[n] \tag{7.56}$$

Software resource: See MATLAB Exercise 7.6.

We are now ready to explore the properties of the DFT. All properties listed in this section assume length-N signals and transforms.

7.3.1 Linearity

Linearity property of the DFT:

Let $x_1[n]$ and $x_2[n]$ be two length-N signals with discrete Fourier transforms

$$x_1[n] \overset{\text{DFT}}{\longleftrightarrow} X_1[k] \qquad \text{and} \qquad x_2[n] \overset{\text{DFT}}{\longleftrightarrow} X_2[k]$$

It can be shown that

$$\alpha_1\, x_1[n] + \alpha_2\, x_2[n] \overset{\text{DFT}}{\longleftrightarrow} \alpha_1\, X_1[k] + \alpha_2\, X_2[k] \tag{7.57}$$

for any two arbitrary constants α_1 and α_2.

Linearity property is easily proven through direct application of the DFT definition given by Eqn. (7.7) to the signal $\alpha_1\, x_1[n] + \alpha_2\, x_2[n]$.

7.3.2 Time shifting

Time shifting property of the DFT:

Given a transform pair

$$x[n] \overset{\text{DFT}}{\longleftrightarrow} X[k],$$

it can be shown that

$$x[n-m]_{\text{mod } N} \overset{\text{DFT}}{\longleftrightarrow} e^{-j(2\pi/N)km}\, X[k] \tag{7.58}$$

Circular shifting of the signal $x[n]$ causes its DFT $X[k]$ to be multiplied by a complex exponential function.

Consistency check: Let the signal be circularly shifted by exactly one period, that is, $m = N$. We know that

$$x[n-N]_{\text{mod } N} = x[n]$$

In this case the exponential function on the right side of Eqn. (7.58) would be $e^{-j(2\pi/N)kN} = 1$, and the transform remains unchanged, as expected.

Example 7.8: Gaining insight into the time shifting property of DFT

Let a signal $x[n]$ be given by

$$x[n] = \{ \underset{\substack{\uparrow \\ n=0}}{a}, b, c, d \}$$

where a, b, c, d arbitrary signal amplitudes. Write $X[k]$, the DFT of $x[n]$, in terms of the parameters a, b, c, d. Afterward construct the transform

$$G[k] = e^{-2\pi k/4}\, X[k]$$

and determine the signal $g[n]$ to which it corresponds.

Solution: The DFT of $x[n]$ is

$$X[k] = a + b\, e^{-j\pi k/2} + c\, e^{-j\pi k} + d\, e^{-j3\pi k/2} \tag{7.59}$$

The transform $G[k]$ is obtained as

$$G[k] = e^{-2\pi k/4}\, X[k] = a\, e^{-j\pi k/2} + b\, e^{-j\pi k} + c\, e^{-j3\pi k/2} + d\, e^{-j2\pi k}$$

Realizing that $e^{-j2\pi k} = 1$ for any integer value of k, the transform $G[k]$ becomes

$$G[k] = d + a\, e^{-j\pi k/2} + b\, e^{-j\pi k} + c\, e^{-j3\pi k/2} \tag{7.60}$$

Comparing Eqn. (7.60) with Eqn. (7.59) we conclude that $G[k]$ is the DFT of the signal

$$g[n] = \{ \underset{\substack{\uparrow \\ n=0}}{d}, a, b, c \}$$

which is a circularly shifted version of $x[n]$, that is,

$$g[n] = x[n-1]_{\text{mod } 4}$$

7.3.3 Time reversal

Time reversal property of the DFT:

For a transform pair

$$x[n] \overset{\text{DFT}}{\longleftrightarrow} X[k],$$

it can be shown that

$$x[-n]_{\text{mod } N} \overset{\text{DFT}}{\longleftrightarrow} X[-k]_{\text{mod } N} \tag{7.61}$$

Example 7.9: Gaining insight into the time reversal property of DFT

Consider again the signal $x[n]$ used in Example 7.8:

$$x[n] = \{ \underset{\substack{\uparrow \\ n=0}}{a}, b, c, d \}$$

The transform $X[k] = \text{DFT}\{x[n]\}$ was derived in Eqn. (7.59). Construct the transform

$$G[k] = X[-k]_{\text{mod } 4}$$

and determine the signal $g[n]$ to which it corresponds.

Solution: Writing $X[k]$ for each value of the index k we get

$$\begin{aligned} X[0] &= a+b+c+d \\ X[1] &= (a-c) - j\,(b-d) \\ X[2] &= a-b+c-d \\ X[3] &= (a-c) + j\,(b-d) \end{aligned}$$

The transform $G[k]$ is a circularly time reversed version of $X[k]$. Its samples are

$$\begin{aligned} G[0] &= X[0] = a+b+c+d \\ G[1] &= X[3] = (a-c) - j\,(d-b) \\ G[2] &= X[2] = a-b+c-d \\ G[3] &= X[1] = (a-c) + j\,(d-b) \end{aligned}$$

Comparing $G[k]$ with $X[k]$ we conclude that the expressions for $G[0]$ through $G[3]$ can be obtained from those for $X[0]$ through $X[3]$ by simply swapping the roles of the parameters b and d. Consequently the signal $g[n]$ is

$$g[n] = \{ \underset{\substack{\uparrow \\ n=0}}{a}, d, c, b \}$$

which is a circularly reversed version of $x[n]$, that is,

$$g[n] = x[-n]_{\text{mod } 4}$$

7.3.4 Conjugation property

Conjugation property of the DFT:

For a transform pair

$$x[n] \overset{\text{DFT}}{\longleftrightarrow} X[k],$$

it can be shown that

$$x^*[n] \overset{\text{DFT}}{\longleftrightarrow} X^*[-k]_{\text{mod } N} \tag{7.62}$$

7.3.5 Symmetry of the DFT

If the signal $x[n]$ is real-valued, it can be shown that its DFT is circularly conjugate symmetric. Conversely, if the signal $x[n]$ is purely imaginary, its transform is circularly conjugate antisymmetric. When we discuss conjugate symmetry properties in the context of the DFT we will always imply circular conjugate symmetry. If the signal $x[n]$ is conjugate symmetric, its DFT is purely real. In contrast, the DFT of a conjugate antisymmetric signal is purely imaginary.

Symmetry properties of the DFT:

$$x[n]\text{: Real, } \operatorname{Im}\{x[n]\} = 0 \implies X^*[k] = X[-k]_{\text{mod } N} \tag{7.63}$$

$$x[n]\text{: Imag, } \operatorname{Re}\{x[n]\} = 0 \implies X^*[k] = -X[-k]_{\text{mod } N} \tag{7.64}$$

$$x^*[n] = x[-n]_{\text{mod } N} \implies X[k]\text{ : Real} \tag{7.65}$$

$$x^*[n] = -x[-n]_{\text{mod } N} \implies X[k]\text{ : Imag} \tag{7.66}$$

Consider a length-N signal $x[n]$ that is complex-valued. In Cartesian complex form $x[n]$ can be written as

$$x[n] = x_r[n] + j\,x_i[n]$$

Let the discrete Fourier transform, $X[k]$ of the signal $x[n]$ be written in terms of its conjugate symmetric and conjugate antisymmetric components as

$$X[k] = X_E[k] + X_O[k]$$

The transform relationship between $x[n]$ and $X[k]$ is

$$x_r[n] + j\,x_i[n] \overset{\text{DFT}}{\longleftrightarrow} X_E[k] + X_O[k]$$

We know from Eqns. (7.63) and (7.64) that the DFT of a real signal must be conjugate symmetric, and the DFT of a purely imaginary signal must be conjugate antisymmetric. Therefore it follows that the following must be valid transform pairs:

$$x_r[n] \overset{\text{DFT}}{\longleftrightarrow} X_E[k] \tag{7.67}$$

$$j\,x_i[n] \overset{\text{DFT}}{\longleftrightarrow} X_O[k] \tag{7.68}$$

A similar argument can be made by writing the signal $x[n]$ as the sum of a conjugate symmetric signal and a conjugate antisymmetric signal

$$x[n] = x_E[n] + x_O[n]$$

and writing the transform $X[k]$ in Cartesian complex form

$$X[k] = X_r[k] + j\,X_i[k]$$

The transform relationship between the two is

$$x_E[n] + x_O[n] \overset{\text{DFT}}{\longleftrightarrow} X_r[k] + j\,X_i[k],$$

which leads to the following transform pairs:

$$x_E[n] \stackrel{\text{DFT}}{\longleftrightarrow} X_r[k] \tag{7.69}$$

$$x_O[n] \stackrel{\text{DFT}}{\longleftrightarrow} jX_i[k] \tag{7.70}$$

Example 7.10: Using symmetry properties of the DFT

The DFT of a length-4 signal $x[n]$ is given by

$$X[k] = \{ \underset{\substack{\uparrow \\ k=0}}{(2+j3)}, (1+j5), (-2+j4), (-1-j3) \}$$

Without computing $x[n]$ first, determine the DFT of $x_r[n]$, the real part of $x[n]$.

Solution: We know from the symmetry properties of the DFT that the transform of the real part of $x[n]$ is the conjugate symmetric part of $X[k]$:

$$\text{DFT}\{x_r[n]\} = X_E[k] = \frac{X[k] + X^*[-k]_{\text{mod } N}}{2}$$

The complex conjugate of the index reversed transform is

$$X^*[-k]_{\text{mod } 4} = \{ \underset{\substack{\uparrow \\ k=0}}{(2-j3)}, (-1+j3), (-2-j4), (1-j5) \}$$

The conjugate symmetric component of $X[k]$ is

$$X_E[k] = \{ \underset{\substack{\uparrow \\ k=0}}{2}, j4, -2, -j4 \}$$

The real part of $x[n]$ can be found as the inverse transform of $X_E[k]$:

$$x_r[n] = \{ \underset{\substack{\uparrow \\ n=0}}{0}, -1, 0, 3 \}$$

Software resource: `exdt_7_10.m`

Example 7.11: Using symmetry properties of the DFT to increase efficiency

Consider the real-valued signals $g[n]$ and $h[n]$ specified as

$$g[n] = \{ \underset{\substack{\uparrow \\ n=0}}{11}, -2, 7, 9 \}$$

$$h[n] = \{ \underset{\substack{\uparrow \\ n=0}}{6}, 14, -13, 8 \}$$

Devise a method of obtaining the DFT's of $g[h]$ and $h[n]$ by computing only one 4-point DFT and utilizing symmetry properties.

Solution: Let us construct a complex signal $x[n]$ as

$$\begin{aligned} x[n] &= g[n] + j\,h[n] \\ &= \{ \underset{\substack{\uparrow \\ n=0}}{(11+j6)},\ (-2+j14),\ (7-j13),\ (9+j8) \} \end{aligned}$$

The 4-point DFT of $x[n]$ is

$$X[k] = \{ \underset{\substack{\uparrow \\ n=0}}{(25+j15)},\ (10+j30),\ (11-j29),\ (-2+j8) \}$$

The conjugate symmetric component of $X[k]$ is

$$X_E[k] = \frac{X[k] + X^*[-k]_{\text{mod } 4}}{2} = \{ \underset{\substack{\uparrow \\ n=0}}{25},\ (4+j11),\ (11),\ (4-j11) \}$$

and its conjugate antisymmetric component is

$$X_O[k] = \frac{X[k] - X^*[-k]_{\text{mod } 4}}{2} = \{ \underset{\substack{\uparrow \\ n=0}}{j15},\ (6+j19),\ -j29,\ (-6+j19) \}$$

Based on the symmetry properties of the DFT we have $\text{DFT}\{g[n]\} = X_E[k]$ and $\text{DFT}\{j\,h[n]\} = X_O[k]$. Therefore

$$G[k] = X_E[k] = \{ \underset{\substack{\uparrow \\ n=0}}{25},\ (4+j11),\ 11,\ (4-j11) \}$$

and

$$H[k] = -j\,X_O[k] = \{ \underset{\substack{\uparrow \\ n=0}}{15},\ (19-j6),\ -29,\ (19+j6) \}$$

It can easily be verified that $G[k]$ and $H[k]$ found above are indeed the DFT's of the two signals $g[n]$ and $h[n]$.

Software resource: exdt_7_11.m

Software resource: See MATLAB Exercise 7.7.

7.3.6 Frequency shifting

Frequency shifting property of the DFT:

For a transform pair

$$x[n] \stackrel{\text{DFT}}{\longleftrightarrow} X[k],$$

it can be shown that

$$x[n]\,e^{j(2\pi/N)mn} \stackrel{\text{DFT}}{\longleftrightarrow} X[k-m]_{\text{mod } N} \qquad (7.71)$$

Multiplication of the signal $x[n]$ with a complex exponential causes a circular shift in the transform. We will prove this property starting with the DTFS coefficients of periodic extensions of the signals involved.

Proof of Eqn. (7.71):

We will begin by defining the signal $g[n]$ as

$$g[n] = x[n]\, e^{j(2\pi/N)mn}$$

Let $\tilde{x}[n]$ and $\tilde{g}[n]$ be the periodic extensions of the two signals $x[n]$ and $g[n]$, that is

$$\tilde{x}[n] = \sum_{r=-\infty}^{\infty} x[n-rN] \qquad \text{and} \qquad \tilde{g}[n] = \sum_{r=-\infty}^{\infty} g[n-rN]$$

with DTFS coefficients

$$\tilde{x}[n] \overset{\text{DTFS}}{\longleftrightarrow} \tilde{c}[k] \qquad \text{and} \qquad \tilde{g}[n] \overset{\text{DTFS}}{\longleftrightarrow} \tilde{d}[k]$$

Using the DTFS analysis equation given by Eqn. (3.19) with $\tilde{g}[n]$ it follows that

$$\begin{aligned}
\tilde{d}_k &= \sum_{n=0}^{N-1} \tilde{g}[n]\, e^{-j(2\pi/N)kn} \\
&= \sum_{n=0}^{N-1} \tilde{x}[n]\, e^{j(2\pi/N)mn}\, e^{-j(2\pi/N)kn} \\
&= \sum_{n=0}^{N-1} \tilde{x}[n]\, e^{-j(2\pi/N)(k-m)n} = \tilde{c}[k-m]
\end{aligned}$$

which justifies the result in Eqn. (7.71). Recall the relationship between the DFT of a signal $x[n]$ and the DTFS coefficients of its periodic extension, summarized by Eqns. (7.2) and (7.4).

7.3.7 Circular convolution

Periodic convolution of two periodic signals $\tilde{x}[n]$ and $\tilde{h}[n]$ was defined in Section 3.2.3, Eqn. (3.46). In this section we will define *circular convolution* for length-N signals in the context of the discrete Fourier transform. Let $x[n]$ and $h[n]$ be length-N signals. Consider the following set of steps:

1. Obtain periodic signals $\tilde{x}[n]$ and $\tilde{h}[n]$ as periodic extensions of $x[n]$ and $h[n]$:

$$\tilde{x}[n] = \sum_{m=-\infty}^{\infty} x[n+mN]$$

$$\tilde{h}[n] = \sum_{m=-\infty}^{\infty} h[n+mN]$$

2. Compute $\tilde{y}[n]$ as the periodic convolution of $\tilde{x}[n]$ and $\tilde{h}[n]$.

$$\tilde{y}[n] = \tilde{x}[n] \otimes \tilde{h}[n] = \sum_{k=0}^{N-1} \tilde{x}[k]\, \tilde{h}[n-k]$$

3. Let $y[n]$ be the length-N signal that is equal to one period of $\tilde{y}[n]$:

$$y[n] = \tilde{y}[n]\,, \quad \text{for } n = 0, \ldots, N-1$$

The signal $y[n]$ is the *circular convolution* of $x[n]$ and $h[n]$. It can be expressed in compact form as

$$y[n] = x[n] \otimes h[n] = \sum_{k=0}^{N-1} x[k]\, h[n-k]_{\text{mod } N}\,, \qquad n = 0, \ldots, N-1 \tag{7.72}$$

Example 7.12: Circular convolution of two signals

Determine the circular convolution of the length-5 signals

$$x[n] = \{ \underset{\substack{\uparrow \\ n=0}}{1}, 3, 2, -4, 6 \}$$

and

$$h[n] = \{ \underset{\substack{\uparrow \\ n=0}}{5}, 4, 3, 2, 1 \}$$

using the definition of circular convolution given by Eqn. (7.72).

Solution: Adapting Eqn. (7.72) to length-5 signals we have

$$y[n] = x[n] \otimes h[n] = \sum_{k=0}^{4} x[k]\, h[n-k]_{\text{mod } 5}\,, \qquad n = 0, 1, 2, 3, 4$$

Fig. 7.21 illustrates the steps involved in computing the circular convolution.

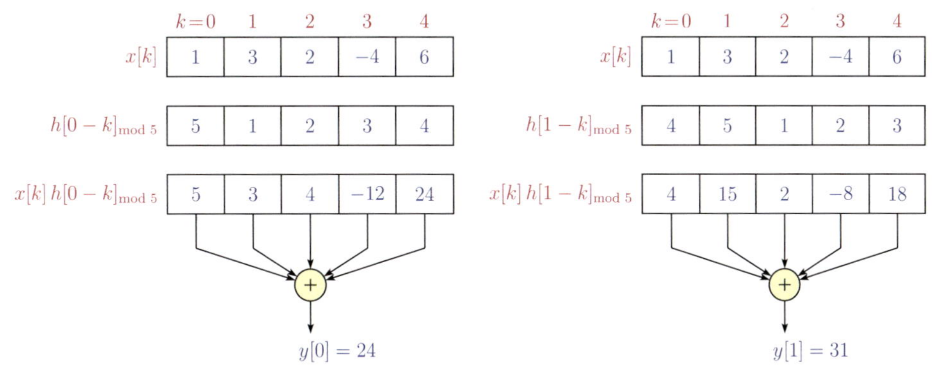

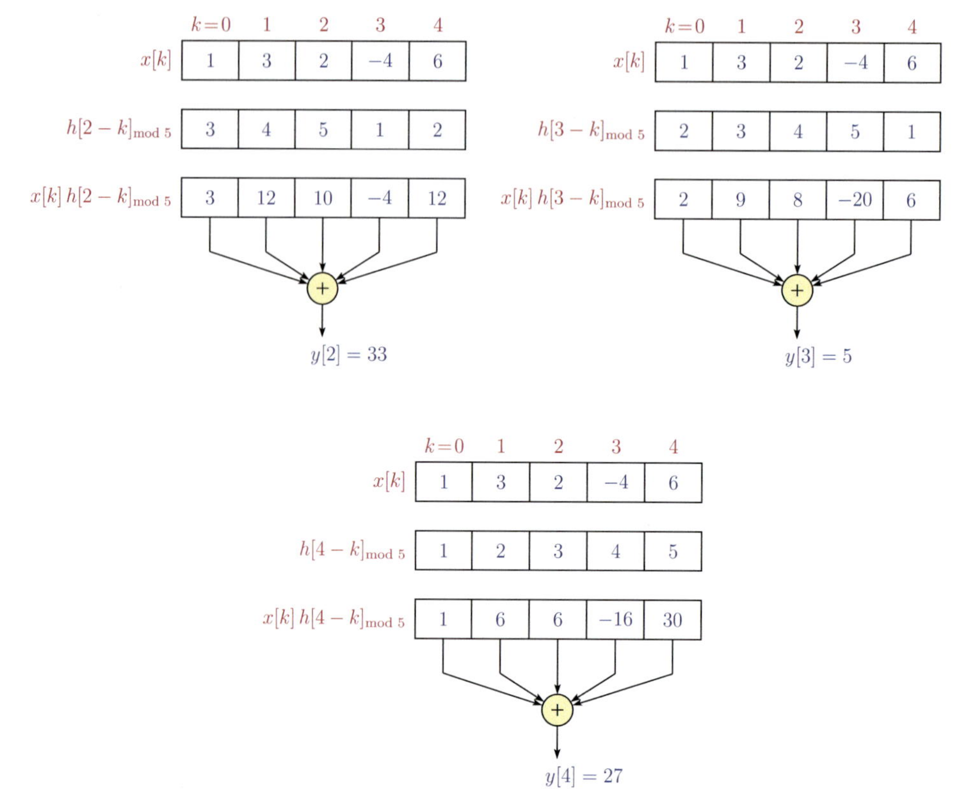

Figure 7.21 – The circular convolution for Example 7.12.

The result is

$$y[n] = \{ \underset{\substack{\uparrow \\ n=0}}{24}, 31, 33, 5, 27 \}$$

Circular convolution property of the DFT:

Let $x[n]$ and $h[n]$ be two length-N signals with discrete-Fourier transforms (DFTs)

$$x[n] \overset{\text{DFT}}{\longleftrightarrow} X[k] \qquad \text{and} \qquad h[n] \overset{\text{DFT}}{\longleftrightarrow} H[k]$$

It can be shown that

$$x[n] \otimes h[n] \overset{\text{DFT}}{\longleftrightarrow} X[k]\, H[k] \tag{7.73}$$

The DFT of the circular convolution of two signals $x[n]$ and $h[n]$ is equal to the product of individual DFT's $X[k]$ and $H[k]$.

This is a very significant result in the use of the DFT for signal processing applications. The proof is straightforward using the periodic convolution property of the discrete-time Fourier series (DTFS), and will be given here.

Proof of Eqn. (7.73):

Let $\tilde{x}[n]$ and $\tilde{h}[n]$ be periodic extensions of the length-N signals $x[n]$ and $h[n]$. Furthermore, let $\tilde{c}_k$ and $\tilde{d}_k$ be the DTFS coefficients for $\tilde{x}[n]$ and $\tilde{h}[n]$, respectively:

$$\tilde{x}[n] \overset{\text{DTFS}}{\longleftrightarrow} \tilde{c}_k \quad \text{and} \quad \tilde{h}[n] \overset{\text{DTFS}}{\longleftrightarrow} \tilde{d}_k$$

The periodic convolution of $\tilde{x}[n]$ and $\tilde{h}[n]$ is

$$\tilde{y}[n] = \tilde{x}[n] \otimes \tilde{h}[n] = \sum_{m=0}^{N-1} \tilde{x}[m]\, \tilde{h}[n-m]$$

and the DTFS coefficients of $\tilde{y}[n]$ are

$$\tilde{e}_k = \frac{1}{N} \sum_{n=0}^{N-1} \tilde{y}[n]\, e^{-j(2\pi/N)kn}$$

We know from Eqn. (3.47) that

$$\tilde{e}_k = N\, \tilde{c}_k\, \tilde{d}_k \tag{7.74}$$

Recall the relationship between the DTFS and the DFT given by Eqn. (7.4). The DFT's of length-N signals $x[n]$, $h[n]$, and $y[n]$ are related to the DTFS coefficients of the periodic extensions by

$$X[k] = N\, \tilde{c}_k\,, \quad k = 0, \ldots, N-1 \tag{7.75}$$

$$H[k] = N\, \tilde{d}_k\,, \quad k = 0, \ldots, N-1 \tag{7.76}$$

$$Y[k] = N\, \tilde{e}_k\,, \quad k = 0, \ldots, N-1 \tag{7.77}$$

Using Eqns. (7.75), (7.76), and (7.77) in Eqn. (7.74) we obtain the desired result:

$$Y[k] = X[k]\, H[k] \tag{7.78}$$

Example 7.13: Circular convolution through DFT

Consider again the length-5 signals $x[n]$ and $h[n]$ of Example 7.12. The circular convolution

$$y[n] = x[n] \otimes h[n]$$

was determined in Example 7.12 in the time-domain. Verify the circular convolution property of the DFT using these signals.

Solution: Table 7.1 lists the DFT for the three signals. It can easily be verified that

$$Y[k] = X[k]\, H[k]$$

Table 7.1 – DFT's of the three signals Example 7.13.

k	$X[k]$	$H[k]$	$Y[k]$
0	8.0000+j0.0000	15.0000+j0.0000	120.0000+j0.0000
1	5.3992+j0.6735	2.5000+j3.4410	11.1803+j20.2622
2	-6.8992-j7.4697	2.5000+j0.8123	-11.1803-j24.2784
3	-6.8992+j7.4697	2.5000-j0.8123	-11.1803+j24.2784
4	5.3992-j0.6735	2.5000-j3.4410	11.1803-j20.2622

Software resource: exdt_7_13.m

If the circular convolution of two length-N signals is desired, the convolution property of the DFT provides an easy and practical method of computing it.

Obtaining circular convolution:

1. Compute the DFT's
$$X[k] = \text{DFT}\{x[n]\}\,, \quad \text{and} \quad H[k] = \text{DFT}\{h[n]\}$$
2. Multiply the two DFT's to obtain $Y[k]$.
$$Y[k] = X[k]\,H[k]$$
3. Compute $y[n]$ through inverse DFT:
$$y[n] = \text{DFT}^{-1}\{Y[k]\}$$

In most applications of signal processing, however, we are interested in computing the *linear convolution* of two signals rather than their circular convolution. The output signal of a DTLTI system is equal to the linear convolution of its impulse response with the input signal. The ability to use the DFT as a tool for the computation of linear convolution is very important due to the availability of fast and efficient algorithms for computing the DFT. Therefore, the following two questions need to be answered:

1. How is the circular convolution of two length-N signals related to their linear convolution?
2. What can be done to ensure that the circular convolution result obtained using the DFT method matches the linear convolution result?

The next example will address the first question.

Example 7.14: Linear vs. circular convolution

Consider again the length-5 signals $x[n]$ and $h[n]$ of Example 7.12.

$$x[n] = \{\underset{\substack{\uparrow\\ n=0}}{1}, 3, 2, -4, 6\}$$

$$h[n] = \{\underset{\substack{\uparrow\\ n=0}}{5}, 4, 3, 2, 1\}$$

The circular convolution of these two signals was determined in Example 7.12 as

$$y[n] = x[n] \otimes h[n] = \{ \underset{\substack{\uparrow \\ n=0}}{24}, 31, 33, 5, 27 \} \tag{7.79}$$

The linear convolution of $x[n]$ and $h[n]$, computed using the convolution sum

$$y_\ell[n] = \sum_{k=-\infty}^{\infty} x[k]\, h[n-k]$$

is found as

$$y_\ell[n] = \{ \underset{\substack{\uparrow \\ n=0}}{5}, 19, 25, -1, 27, 19, 12, 8, 6 \} \tag{7.80}$$

Note that in Eqn. (7.80) we have used the notation $y_\ell[n]$ for linear convolution to differentiate it from the circular convolution result of Eqn. (7.79). The most obvious difference between the two results $y[n]$ and $y_\ell[n]$ is the length of each: The circular convolution result is 5 samples long, however the linear convolution result is 9 samples long. This is the first step toward explaining why the DFT method does not produce the linear convolution result. $X[k]$ and $H[k]$ are length-5 transforms, and the inverse DFT of their product yields a length-5 result for $y[n]$.

How does $y_\ell[n]$ relate to $y[n]$? Imagine filling out a form that has 5 boxes for entering values, yet we have 9 values we must enter. We start from with the leftmost box and enter the first 5 of 9 values. At this point each box has a value in it, and we still have 4 more values not entered into any box. Suppose we decide to go back to the leftmost box, and start entering additional values into each box as needed. This is illustrated in Fig. 7.22 using the samples of the linear convolution result to fill the boxes.

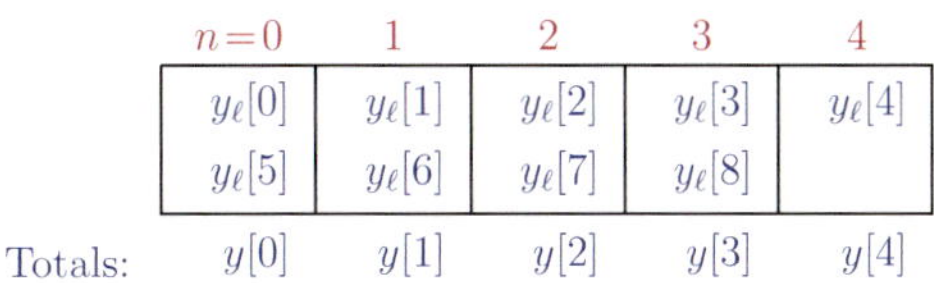

	$n=0$	1	2	3	4
	$y_\ell[0]$	$y_\ell[1]$	$y_\ell[2]$	$y_\ell[3]$	$y_\ell[4]$
	$y_\ell[5]$	$y_\ell[6]$	$y_\ell[7]$	$y_\ell[8]$	
Totals:	$y[0]$	$y[1]$	$y[2]$	$y[3]$	$y[4]$

	$n=0$	1	2	3	4
	5	19	25	-1	27
	19	12	8	6	
Totals:	24	31	33	5	27

Figure 7.22 – Relationship between linear and circular convolution in Example 7.14.

Each sample of the circular convolution result is equal to the sum of values in the corresponding box. For example, $y[0] = y_\ell[0] + y_\ell[5]$ and $y[1] = y_\ell[1] + y_\ell[6]$. If a box has a single value, the circular convolution result is identical to the linear convolution result for the corresponding n. For boxes that have multiple entries, the circular convolution result is a corrupted version of the linear convolution result.

Software resource: `exdt_7_14.m`

Generalizing the results of Example 7.14 the circular convolution of two signals can be expressed in terms of their linear convolution as

$$y[n] = \sum_{m=-\infty}^{\infty} y_\ell[n+mN] \tag{7.81}$$

If the circular convolution result is desired to be identical to the linear convolution result, the length of the circular convolution result must be sufficient to accommodate the number of samples expected from linear convolution. Using the analogy employed in Example 7.14, namely filling out a form with the results, there must be enough "boxes" to accommodate all samples of $y_\ell[k]$ without any overlaps. One method of achieving this is through zero-padding $x[n]$ and $h[n]$ before the computation of the DFT.

Computing linear convolution using the DFT:

Given two finite length signals with N_x and N_h samples respectively

$$x[n]\,, \quad n=0,\ldots,N_x-1 \qquad \text{and} \qquad h[n]\,, \quad n=0,\ldots,N_h-1,$$

the linear convolution $y_\ell[n] = x[n] * h[n]$ can be computed as follows:

1. Anticipating the length of the linear convolution result to be $N_y = N_x + N_h - 1$, extend the length of each signal to N_y through zero padding:

$$x_p[n] = \begin{cases} x[n]\,, & n=0,\ldots,N_x-1 \\ 0\,, & n=N_x,\ldots,N_y-1 \end{cases}$$

$$h_p[n] = \begin{cases} h[n]\,, & n=0,\ldots,N_h-1 \\ 0\,, & n=N_h,\ldots,N_y-1 \end{cases}$$

2. Compute the DFT's of the zero-padded signals $x_p[n]$ and $h_p[n]$.

$$X_p[k] = \text{DFT}\left\{x_p[n]\right\} \quad \text{and} \quad H_p[k] = \text{DFT}\left\{h_p[n]\right\}$$

3. Multiply the two DFT's to obtain $Y_p[k]$.

$$Y_p[k] = X_p[k]\,H_p[k]$$

4. Compute $y_p[n]$ through inverse DFT:

$$y_p[n] = \text{DFT}^{-1}\left\{Y_p[k]\right\}$$

The result $y_p[n]$ is the same as the linear convolution of the signals $x[n]$ and $y[n]$.

$$y_p[n] = y_\ell[n] \quad \text{for } n=0,\ldots,N_y-1$$

Software resource: See MATLAB Exercises 7.8 and 7.9.

7.4 Special Uses of the DFT

Direct computation of the DFT using the analysis equation given by Eqn. (7.7) yields a transform $X[k]$ that has the same length as the signal $x[n]$. In Section 7.5, efficient algorithms will be presented for fast computation of the same result. There may be times, however, when a full transform is not needed. For example, we may be interested in only one sample of the transform. Computing the entire transform just for the purpose of looking at one of its samples would be inefficient. Alternately, we may need a set of samples of the transform at angular frequencies that do not necessarily coincide with DFT frequencies. Yet another example situation is that of working with an audio or video stream where the frequency composition of the signal changes over time. The DFT result $X[k]$ only provides information as to which frequencies are contained in the entirety of the signal, but does not tell us when those frequencies show up or disappear. In this section, several algorithms will be presented to address these situations.

7.4.1 Goertzel algorithm

Goertzel algorithm provides a convolution perspective to the computation of the DFT, and achieves a modest improvement in computational efficiency. This improvement in computational efficiency/speed is not as dramatic as it is with decimation-in-time and decimation-in-frequency algorithms we will study in Section 7.5. It is still useful, however, in situations where we might not need to compute all N transform samples but just one or a few of them.

Let's assume that we need to compute the transform for one specific value of the index k. An example application might be a system to detect the presence of a sinusoidal signal at a particular frequency. Direct computation of the DFT is carried out through the use of Eqn. (7.7) which is repeated here for convenience:

$$\text{Eqn. (7.7):} \qquad X[k] = \sum_{r=0}^{N-1} x[r]\, W_N^{kr}$$

Recall that $W_N = e^{-j2\pi/N}$. We have changed the independent variable of the summation to r for notational convenience in the development that will follow. The complex sequence W_N^{-kr} is periodic with a period of N, therefore we have

$$W_N^{-kN} = e^{2\pi k} = 1 \tag{7.82}$$

The DFT summation in Eqn. (7.7) can be multiplied by W_N^{-kN} without changing the result.

$$\begin{aligned} X[k] &= W_N^{-kN} \sum_{r=0}^{N-1} x[r]\, W_N^{kr} \\ &= \sum_{r=0}^{N-1} x[r]\, W_N^{-k(N-r)} \end{aligned} \tag{7.83}$$

The summation in Eqn. (7.83) looks curiously similar to the convolution sum defined in Eqn. (2.111). In fact, if we had $(n-r)$ in place of $(N-r)$ in the exponent, we would have a convolution sum on the right side. In order to exploit this similarity, let's imagine a system with its impulse response $h_k[n]$ defined as

$$h_k[n] = W_N^{-kn}\, u[n] \tag{7.84}$$

Also, let $y_k[n]$ be the output signal of this system in response to the input signal $x[n]$. We know that the signal $x[n]$ has N significant samples in the sample index range $n = 0, \ldots, N-1$. The input-output relationship of the system can be written as

$$y_k[n] = x[n] * h_k[n] = \sum_{r=0}^{\infty} x[r]\, W_N^{-k(n-r)}\, u[n-r]$$

If we now evaluate the output signal $y_k[n]$ for sample index $n = N$ we get

$$\begin{aligned} y_k[n]\Big|_{n=N} &= \sum_{r=0}^{\infty} x[r]\, W_N^{-k(N-r)}\, u[N-r] \\ &= \sum_{r=0}^{N-1} x[r]\, W_N^{-k(N-r)} = X[k] \end{aligned} \tag{7.85}$$

The transform sample $X[k]$ is numerically equal to the amplitude of the output sample $y_k[N]$. Of course we need to iteratively compute (and later discard) output samples $y_k[0], \ldots, y_k[N-1]$ until we can get to the sample $y_k[N]$. At this point it would be instructive to determine the system function $H_k(\Omega)$ for the system with impulse response $h_k[n]$ given by Eqn. (7.84). Using the result of Example 3.10, the system function is

$$H_k(\Omega) = \frac{1}{1 - W_N^{-k}\, e^{-j\Omega}} \tag{7.86}$$

The difference equation for the system is

$$y_k[n] = x[n] + W_N^{-k}\, y_k[n-1] \tag{7.87}$$

A block diagram can be obtained as shown in Fig. 7.23, and can be used as the basis of implementing Goertzel's algorithm.

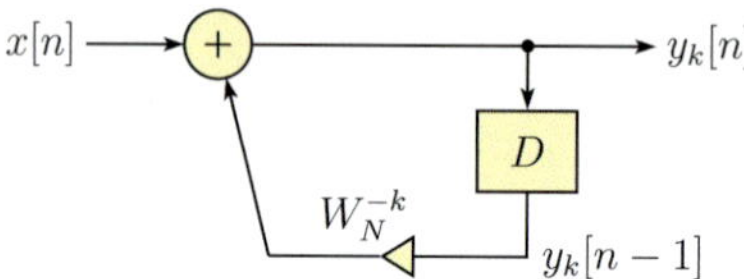

Figure 7.23 – Block diagram for Goertzel algorithm based on the difference equation in Eqn. (7.87).

Let's compare the computational cost of finding $X[k]$ through direct application of Eqn. (7.7) vs. through Goertzel's algorithm.

- Assuming $x[n]$ is a complex-valued signal, direct computation of $X[k]$ requires N complex multiplications and $N-1$ complex additions. In terms of real multiplications and additions, that translates to about $4N$ of each.

- Using the block diagram in Fig. 7.23 for Goertzel's algorithm, computation of each output sample $y_k[n]$ requires one complex multiplication and one complex addition. We need $y_k[n]$, therefore we will need N complex multiplications and as many complex additions to get there. In terms of real operations, again we need about $4N$ of each. The only slight advantage with Goertzel's algorithm is that we don't need to compute and store the exponential terms W_N^{kr}.

It is possible to modify the system function in Eqn. (7.86) to reduce the computational cost of Goertzel's algorithm by about half. Let's multiply and divide $H_k(\Omega)$ by $\left(1-W_N^k e^{-j\Omega}\right)$.

$$H_k(\Omega)=\frac{1-W_N^k e^{-j\Omega}}{\left(1-W_N^k e^{-j\Omega}\right)\left(1-W_N^k e^{-j\Omega}\right)}$$

$$=\frac{1-W_N^k e^{-j\Omega}}{1-\left(W_N^{-k}+W_N^k\right)e^{-j\Omega}+e^{-j2\Omega}}$$

Recognizing that $W_N^{-k}+W_N^k=2\cos\left(\frac{2\pi k}{N}\right)$, we have

$$H_k(\Omega)=\frac{1-W_N^k e^{-j\Omega}}{1-2\cos\left(\frac{2\pi k}{N}\right)e^{-j\Omega}+e^{-j2\Omega}} \tag{7.88}$$

The system function in Eqn. (7.88) has only real coefficients. The corresponding difference equation is

$$y_k[n]=x[n]-W_N^k\,x[n-1]+2\cos(2\pi k/N)\,y_k[n-1]-y_k[n-2] \tag{7.89}$$

This new difference equation can be represented with the block diagram shown in Fig. 7.24.

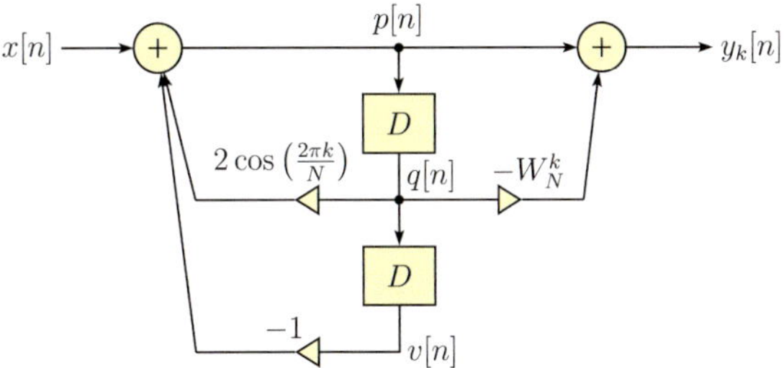

Figure 7.24 – A more efficient block diagram implementation of Goertzel algorithm based on the difference equation in Eqn. (7.89).

In implementing Goertzel algorithm using the diagram of Fig. 7.24, we do not need to compute the output signal $y_k[n]$ for $n=0,\ldots,N-1$. Recall that these output samples would be discarded anyway. For this range of indices we only need to maintain and update node variables $p[n]$, $q[n]$ and $v[n]$.

$$p[n]=x[n]+2\cos\left(\frac{2\pi k}{N}\right)q[n]-v[n]\,, \qquad n=0,\ldots,N \tag{7.90}$$

When we reach index $n=N$, the output $y_k[N]$ can be computed from the knowledge of $p[N]$ and $q[N]$.

$$X[k]=y_k[N]=p[N]-W_N^k\,q[N] \tag{7.91}$$

Assuming $x[n]$ is complex-valued, node variables would also be complex. The computation of the node variable $p[n]$ requires 2 complex additions, translating to 4 real additions. The only product is between complex-valued $q[n]$ and the real-valued cosine term, requiring 2 real multiplications. Thus, iteration through N samples requires $2N$ real multiplications and $4N$ real additions. One complex multiplication and one complex addition is needed at the end to compute $y_k[N]$.

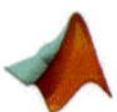

Software resource: See MATLAB Exercise 7.14.

7.4.2 Chirp transform algorithm (CTA)

Chirp transform algorithm (CTA) is used for computing the DTFT of a finite-length signal at an arbitrary set of equally-spaced frequencies. Similar to Goertzel's algorithm, CTA also provides a convolution perspective to the computation of the transform. Since the process for computing the required transform result can be converted to the convolution of two sequences, FFT-based implementation is possible. The main advantage is that we are not limited to FFT-based frequency spacing of $2\pi/N$ radians.

Consider a N-point signal $x[n]$. The DTFT of the signal is given by

$$X(\Omega) = \sum_{n=0}^{N-1} x[n]\, e^{-j\Omega n}$$

and its DFT is

$$X[k] = \sum_{n=0}^{N-1} x[n]\, e^{-j2\pi kn/N}$$

The relationship between the two transforms was explored in Section 7.2.1. Samples of the DFT have a frequency spacing (or resolution) of $2\pi/N$ radians. The DFT spectrum is a sampled version (recall the discussion on *picket-fence effect* in Section 7.2.1) of the DTFT spectrum at angular frequencies

$$\Omega_k = \frac{2\pi k}{N}\,, \qquad k = 0,\ldots,N-1$$

What if we need the values of the DTFT at an arbitrary set of equally-spaced frequencies with frequency spacing not necessarily equal to $2\pi/N$ radians? Consider a set of M angular frequencies of interest that start at $\Omega = \Omega_0$ and continue with increment of $\Delta\Omega$:

$$\Omega_k = \Omega_0 + k\Delta\Omega\,, \qquad k = 0,\ldots,M-1$$

If we wanted to compute the DTFT at this set of frequencies, we would have

$$X(\Omega_k) = \sum_{n=0}^{N-1} x[n]\, e^{-j(\Omega_0 + k\Delta\Omega)n} \tag{7.92}$$

Let $V = e^{-j\Delta\Omega}$. With some simplification, Eqn. (7.92) becomes

$$X(\Omega_k) = \sum_{n=0}^{N-1} x[n]\, e^{-j\Omega_0 n}\, V^{kn} \tag{7.93}$$

Using the identity

$$nk = \frac{1}{2}\left[n^2 + k^2 - (k-n)^2\right]$$

Eqn. (7.93) can be written as

$$\begin{aligned} X(\Omega_k) &= \sum_{n=0}^{N-1} x[n]\, e^{-j\Omega_0 n}\, V^{n^2/2}\, V^{k^2/2}\, V^{(k-n)^2/2} \\ &= V^{k^2/2} \sum_{n=0}^{N-1} \underbrace{x[n]\, e^{-j\Omega_0 n}\, V^{n^2/2}}_{g[n]}\; \underbrace{V^{(k-n)^2/2}}_{h[k-n]} \end{aligned} \tag{7.94}$$

The summation in Eqn. (7.94) is starting to resemble the convolution sum. We will define two new sequences as follows:

$$g[n] = x[n]\,e^{-j\Omega_0 n}\,V^{n^2/2}\,, \qquad n = 0,\ldots,N-1 \tag{7.95}$$

$$h[n] = V^{(k-n)^2/2} \tag{7.96}$$

Now we can write Eqn. (7.94) in simplified form using $g[n]$ and $h[n]$:

$$X(\Omega_k) = V^{k^2/2} \sum_{n=0}^{N-1} g[n]\,h[k-n] \tag{7.97}$$

Thus, the desired transform samples $X(\Omega_k)$ can be obtained by convolving the two sequences $g[n]$ and $h[n]$, and multiplying the convolution result by $V^{k^2/2}$. The sequence $g[n]$ is defined for the index range $0 \le n \le N-1$. The sequence $h[n]$ is non-causal, and exists for all n. Fortunately, we do not need all samples of $h[n]$ in the computations, but only the ones that are needed for the computation of the convolution result for $0 \le k \le M-1$. We will look at the two ends of the frequency range, namely Ω_0 and Ω_{M-1}. For $k = 0$ we have

$$X(\Omega_0) = V^0 \sum_{n=0}^{n-1} g[n]\,h[-n]$$

and samples of $h[n]$ are needed in the range $-N+1 \le n \le 0$. At the high end of the frequency range, for $k = M-1$, Eqn. (7.97) becomes

$$X(\Omega_{M-1}) = V^{(M-1)^2/2} \sum_{n=0}^{n-1} g[n]\,h[M-1-n]$$

Samples of $h[n]$ are needed for computing $X(\Omega_{M-1})$ are in the range $M-N \le n \le M-1$. Combining the two index ranges, it becomes apparent that we need to use samples of $h[n]$ in the range $-N+1 \le n \le M-1$. Based on this, Eqn. (7.96) can be revised as follows:

$$h[n] = \begin{cases} V^{(k-n)^2/2}\,, & -N+1 \le n \le M-1 \\ 0\,, & \text{otherwise} \end{cases} \tag{7.98}$$

With the established limits for $g[n]$ and $h[n]$, the convolution result obtained in Eqn. (7.97) will be in the index range $-N+1 \le k \le M+N-2$ (see Problem 2.25 at the end of Chapter 2). We only need the part in $0 \le k \le M-1$ which means the first and the last $(N-1)$ samples can be discarded.

Software resource: See MATLAB Exercise 7.15.

7.4.3 Short time Fourier transform (STFT)

Previous sections of this chapter focused on analyzing signals in the frequency domain through the use of the DTFS and the DFT. The signal that is being analyzed is thought of as a mixture of sinusoidal components at various frequencies. Analysis methods attempt to determine how much of each frequency component is contained in the entirety of the signal. This kind of analysis may be useful for certain types of signals, however, signals involved in most practical applications are nonstationary, that is, their frequency composition changes over time. Consider, for example, a piece of music played by an instrument. As various notes are played one after the other, the frequency of the signal changes. Amplitudes and phases of sinusoidal components change as well. Similar examples can be found in speech, radar, sonar, satellite communications, and so on.

Analysis of nonstationary signals requires a transform that is a function of both frequency and time. One such transform is the *short time Fourier transform (STFT)*. For the signal $x[n]$, the STFT is defined as

$$X_{STFT}(m,\Omega) = \sum_{n=-\infty}^{\infty} x[n+m]\, w[n]\, e^{-j\Omega n} \tag{7.99}$$

where $w[n]$ is a length-M window sequence that has the effect of selecting a small segment of the signal $x[n]$. The only requirement on the window sequence is that $w[n]=0$ for $n<0$ or $n>M-1$. As the parameter m is incremented, the signal $x[n]$ is time shifted to the left against the stationary window sequence. An alternative form of of Eqn. (7.99) is obtained through the variable change $n'=n+m$:

$$X_{STFT}(m,\Omega) = \sum_{n'=-\infty}^{\infty} x[n']\, w[n'-m]\, e^{-j\Omega(n'-m)} \tag{7.100}$$

We will find working with this form of the STFT equation a bit more convenient. Since $w[n'-m]=0$ for $n'<m$ or $n'>m+M-1$, the summation limits in Eqn. (7.100) can be adjusted to yield

$$X_{STFT}(m,\Omega) = e^{j\Omega m} \sum_{n=m}^{m+M-1} x[n]\, w[n-m]\, e^{-j\Omega n} \tag{7.101}$$

where we changed the summation variable n' back to n. In the alternative form of Eqn. (7.101), as parameter m is incremented, the window sequence is shifted to the right. Signals involved in the summation in Eqn. (7.101) are illustrated in Fig. 7.25. Note that these are discrete-time signals even though they are shown in the form of line graphs to allow for details to be displayed.

The product $x[n+m]\, w[n]$ can be recovered from the transform in Eqn. (7.99) through the application of the inverse DTFT equation given by Eqn. (3.68).

$$x[n+m]\, w[n] = \frac{1}{2\pi}\int_{-\pi}^{\pi} X_{STFT}(m,\Omega)\, e^{j\Omega n}\, d\Omega \tag{7.102}$$

The signal $x[n]$ can be obtained from Eqn. (7.102) as

$$x[n+m] = \frac{1}{2\pi\, w[n]}\int_{-\pi}^{\pi} X_{STFT}(m,\Omega)\, e^{j\Omega n}\, d\Omega \tag{7.103}$$

Any nonzero sample of the window sequence can be used in Eqn. (7.103). For example, if $w[0]\neq 0$, then

$$x[m] = \frac{1}{2\pi\, w[0]}\int_{-\pi}^{\pi} X_{STFT}(m,\Omega)\, d\Omega\,, \quad -\infty<m<\infty \tag{7.104}$$

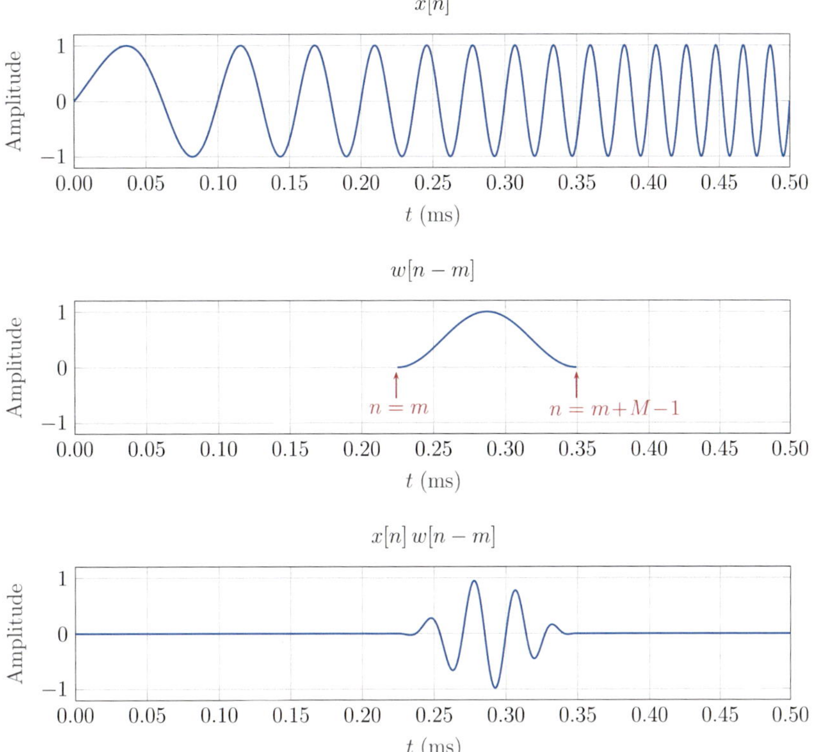

Figure 7.25 – Signals involved in Eqn. (7.100).

If $w[0] = 0$ but another sample of the window sequence can be found so that $w[\ell] \neq 0$, then we can write

$$x[\ell + m] = \frac{1}{2\pi\, w[\ell]} \int_{-\pi}^{\pi} X_{STFT}(m,\Omega)\, e^{j\Omega\ell}\, d\Omega\,, \quad -\infty < m < \infty \tag{7.105}$$

In practical applications we often prefer to use DFT-based computation of the STFT due to the availability of fast algorithms. Let

$$\Omega = \Omega_k = \frac{2\pi k}{N}$$

where $N \geq M$ is the size of the DFT to be used. Substituting this into Eqn. (7.101) yields

$$\begin{aligned} X_{STFT}[m,k] &= e^{j(2\pi/N)km} \sum_{n=m}^{m+M-1} x[n]\, w[n-m]\, e^{-j(2\pi/N)kn} \\ &= e^{j(2\pi/N)km}\, \text{DFT}\left\{ x[n]\, w[n-m] \right\} \end{aligned} \tag{7.106}$$

In practical use of the STFT, we are often interested in just the magnitude $\left|X_{STFT}[m,k]\right|$ and not the phase. The exponential term $e^{-j(2\pi/N)km}$ in Eqn. (7.106) can be dropped in the interest of computational efficiency.

It is also possible to view the STFT as the equivalent of a bank of N filters each designed to detect a particular DFT frequency. Eqn. (7.106) can be slightly manipulated to obtain

$$X_{STFT}[m,k] = \sum_{n=m}^{m+M-1} x[n]\, w[-(m-n)]\, e^{j(2\pi/N)k(m-n)} \tag{7.107}$$

Let $h_k[n]$ be defined as

$$h_k[n] = w[-n]\, e^{j(2\pi/N)kn} \tag{7.108}$$

which allows us to write

$$X_{STFT}[m,k] = \sum_{n=m}^{m+M-1} x[n]\, h_k[m-n] = x[m] * h_k[m]\,, \qquad k = 0,1,\ldots,N-1 \tag{7.109}$$

For each value of k, the transform $X_{STFT}[m,k]$ can be found by processing the signal $x[n]$ through a filter with impulse response $h_k[n]$. As defined by Eqn. (7.108), $h_k[n]$ corresponds to a non-causal filter. It can be made causal by applying a time shift of $M-1$ samples. Fig. 7.26 illustrates the implementation of the STFT as a filter bank.

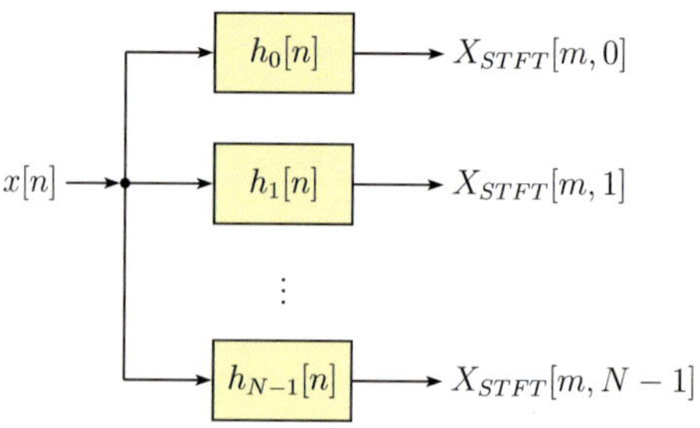

Figure 7.26 – Filter bank implementation of the STFT.

Software resource: See MATLAB Exercise 7.16.

7.5 Improving Computational Efficiency of the DFT

The DFT introduced in Section 7.2 of this chapter is a very important tool in spectral analysis as well as indirect computation of the convolution of two signals. In Section 7.6 we will further elaborate on the use of DFT for real-time processing of audio signals by means of fast convolution techniques. On the other hand, computation of the DFT becomes tedious for large signal and transform sizes. Consider again the DFT analysis equation given by Eqn. (7.8):

$$X[k] = \sum_{n=0}^{N-1} x[n]\, W_N^{kn}, \qquad k = 0,\ldots,N-1$$

where $W_N = e^{-j2\pi/N}$. In general, both the signal $x[n]$ and the transform $X[k]$ may be complex valued. Computation of a single sample of the transform, say for $k = k_1$, involves N complex multiplications and $N-1$ complex additions. (We are assuming that first each term in the summation is computed, and then the remaining $N-1$ terms are added on top of the first term.) The complete transform has N samples for $k = 0,\ldots,N-1$, therefore, it requires N^2 complex multiplications and $N(N-1)$ complex additions. For large N it would be reasonable to say that the computational complexity of the N-point DFT is on the order of N^2 complex multiplications and additions or, equivalently, $4N^2$ real multiplications and additions. (Recall that a complex multiplication requires four real multiplications and two real additions. A complex addition requires two real additions.) This illustrates the daunting task at hand for large values of N. Imagine a situation where an audio signal sampled at a rate of $f_s = 20$ kHz being played back, and we would like to compute and graph its frequency spectrum every 50 ms. This would mean computing a 1000-point DFT each time. It would require about a million complex multiplications and almost as many complex additions to be completed in just 50 ms.

It is clear from the foregoing discussion that, if the DFT is going to be utilized in practical scenarios, its computation needs to be made more efficient. Fortunately, there are ways to do just that. One simple method of computing the DFT a bit faster is to utilize its symmetry properties whenever applicable.

Consider, for example, the problem of computing the DFT of a length-N real-valued signal $x[n]$, that is, $\text{Im}\{x[n]\} = 0$. Since $x[n]$ is purely real, the general term of the DFT summation has the product of a real number with a complex number instead of two complex numbers. This cuts down the amount of arithmetic by half since we can now write

$$X[k] = \sum_{n=0}^{N-1} \left[x[n] \cos\left(\frac{2\pi}{N}kn\right) - j\, x[n] \sin\left(\frac{2\pi}{N}kn\right) \right] \qquad k = 0,\ldots,N-1 \tag{7.110}$$

requiring $2N^2$ real multiplications. In addition, we know from Section 7.3 that the DFT of a real signal is conjugate symmetric, and not all transform samples need to be computed. We only need to compute about half the transform samples. Specifically, if N is even, we need

$$X[k] = \sum_{n=0}^{N-1} \left[x[n] \cos\left(\frac{2\pi}{N}kn\right) - j\, x[n] \sin\left(\frac{2\pi}{N}kn\right) \right] \qquad k = 0,\ldots,\frac{N}{2} \tag{7.111}$$

Remaining transform samples can be computed as

$$X[N-k] = X^*[k], \qquad k = 1,\ldots,\frac{N}{2} - 1 \tag{7.112}$$

using the conjugate symmetry of the transform. This further cuts the number of real multiplications down to about N^2. If N is odd, the same technique can be used by slightly modifying Eqns. (7.111) and (7.112) (see Problem 7.24 at the end of this chapter).

Another method of increasing the efficiency of DFT computation is to combine two real-valued length-N signals into one complex signal. This idea was first explored in Example 7.11 in Section 7.3. Let $g[n]$ and $h[n]$ be two real length-N signals. Define a new signal $x[n]$ as

$$x[n] = g[n] + j\, h[n]$$

At this point we will go ahead and compute the N-point DFT $X[k]$, and obtain the two desired transforms $G[k]$ and $H[k]$ from it. We know from the symmetry properties of the DFT that the transform of $g[n]$ is the conjugate symmetric component of $X[k]$, that is,

$$\text{DFT}\{g[n]\} = X_E[k] \qquad \Longrightarrow \qquad G[k] = X_E[k] = \frac{X[k] + X^*[N-k]}{2} \tag{7.113}$$

and the transform of $j\, h[n]$ is the conjugate antisymmetric component of $X[k]$. With some scaling we conclude that

$$\text{DFT}\{j\, h[n]\} = X_O[k] \qquad \Longrightarrow \qquad H[k] = -j\, X_O[k] = -j\, \frac{X[k] - X^*[N-k]}{2} \tag{7.114}$$

The development above allows us some reduction in the amount of arithmetic required for computing the transform. We can do significantly better, however, by exploiting the redundancy created by the cyclic nature of the term W_N^{nk}. The next few subsections will focus on this to obtain more dramatic improvements in computation speed. We will begin by analyzing the length-4 transform followed by the length-8 transform, and finally generalize the results for longer transform lengths. This will lead to a set of algorithms known as the *fast Fourier transform (FFT).*

7.5.1 Length-4 DFT

Let's begin by writing the DFT of a length-4 signal $x[n]$ in open form:

$$\begin{aligned} X[k] &= \sum_{n=0}^{3} x[n]\, W_4^{nk} \\ &= x[0] + x[1]\, W_4^{k} + x[2]\, W_4^{2k} + x[3]\, W_4^{3k}\,, \qquad k = 0,\ldots,3 \end{aligned} \tag{7.115}$$

Changing the order of the terms on the right side, specifically, placing even indexed terms first and odd indexed terms last, we obtain

$$\begin{aligned} X[k] &= x[0] + x[2]\, W_4^{2k} + x[1]\, W_4^{k} + x[3]\, W_4^{3k} \\ &= x[0] + x[2]\, W_4^{2k} + W_4^{k}\left[x[1] + x[3]\, W_4^{2k}\right], \qquad k = 0,\ldots,3 \end{aligned} \tag{7.116}$$

Since $W_4^{2k} = W_2^{k}$, Eqn. (7.116) can be written as

$$X[k] = x[0] + x[2]\, W_2^{k} + W_4^{k}\left[x[1] + x[3]\, W_2^{k}\right], \qquad k = 0,\ldots,3 \tag{7.117}$$

Let's define two new signals $g[n]$ and $h[n]$ as follows:

$$g[n] = \{\, \underset{\substack{\uparrow \\ n=0}}{x[0]},\, x[2] \,\}$$

$$h[n] = \{\, \underset{\substack{\uparrow \\ n=0}}{x[1]},\, x[3] \,\}$$

The transform $X[k]$ becomes

$$\begin{aligned} X[k] &= \underbrace{g[0] + g[1]\, W_2^{k}}_{G[k],\ k=0,1} + W_4^{k}\, \underbrace{\left[h[0] + h[1]\, W_2^{k}\right]}_{H[k],\ k=0,1} \\ &= G[k] + W_4^{k}\, H[k]\,, \qquad k = 0,\ldots,3 \end{aligned} \tag{7.118}$$

Eqn. (7.118) represents a significant result. We decompose, or split, the time domain signal $x[n]$ into two half-length signals $g[n]$ and $h[n]$. The DFT of each half-length signal is computed, leading to half-length transforms $G[k]$ and $H[k]$. These two are then merged to yield the desired transform $X[k]$. Since the two half-length signals are obtained by downsampling (decimating) the original signal $x[n]$ using a factor of 2, we refer to this technique as *radix-2 decimation-in-time* decomposition.

One question that may be raised is about the sizes of transforms involved in Eqn. (7.118): The transform $X[k]$ should have four samples in the range $k = 0,\ldots,3$. On the other hand, the transforms $G[k]$ and $H[k]$ are each computed for $k = 0,1$. Remembering the cyclic nature of the DFT, we can write

$$G[2] = G[0], \quad G[3] = G[1], \quad H[2] = H[0] \quad \text{and} \quad H[3] = H[1]$$

Eqn. (7.118) for merging the two half-size transforms can now be written in open form as

$$\begin{aligned} X[0] &= G[0] + W_4^0\, H[0] = G[0] + H[0] \\ X[1] &= G[1] + W_4^1\, H[1] = G[1] - j\, H[1] \\ X[2] &= G[2] + W_4^2\, H[2] = G[0] - H[0] \\ X[3] &= G[3] + W_4^3\, H[3] = G[1] + j\, H[1] \end{aligned}$$

where we have also recognized that $W_4^1 = -j$, $W_4^2 = -1$ and $W_4^3 = j$. The process of merging two half-length DFT's into a full size DFT is illustrated in Fig. 7.27.

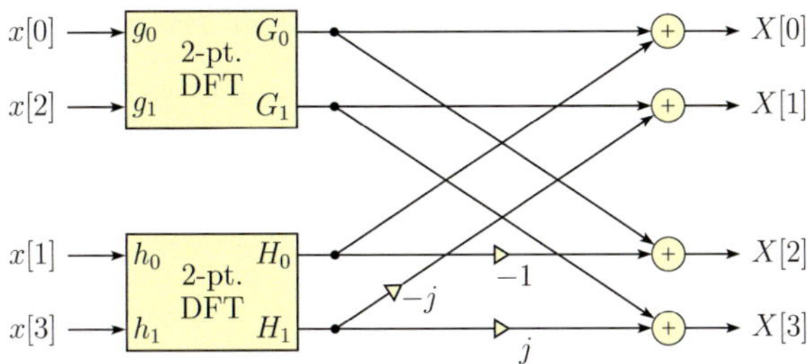

Figure 7.27 – Combining length-2 DFT's into a length-4 DFT.

Computation of length-2 DFT's shown as rectangular boxes in Fig. 7.27 is straightforward:

$$\begin{aligned} G[k] &= g[0] + g[1]\, W_2^k \quad \Longrightarrow \quad G[0] = g[0] + g[1], \quad G[1] = g[0] - g[1] \\ H[k] &= h[0] + h[1]\, W_2^k \quad \Longrightarrow \quad H[0] = h[0] + h[1], \quad H[1] = h[0] - h[1] \end{aligned}$$

A block diagram for the first of the length-2 DFT's is shown in Fig. 7.28. This is often referred to as a *butterfly* due to the shape of the diagram structure.

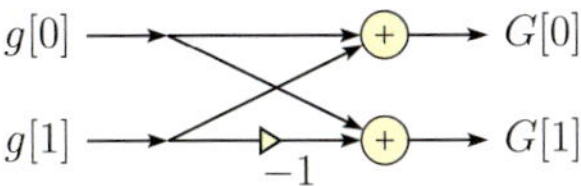

Figure 7.28 – Length-2 DFT block diagram.

Substituting the butterfly structures for length-2 DFT's into the diagram of Fig. 7.27 we obtain the completed diagram in Fig. 7.29.

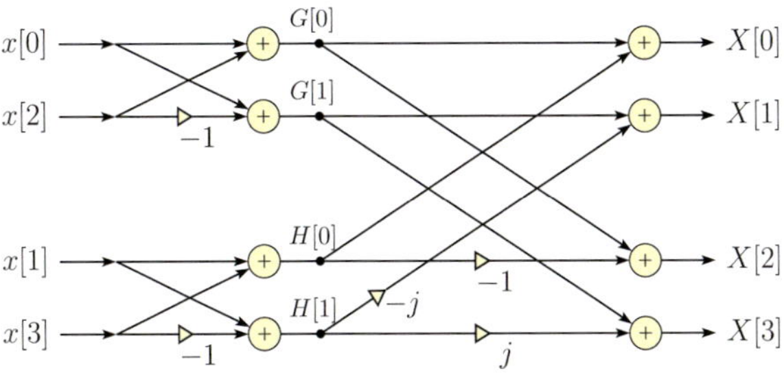

Figure 7.29 – Length-4 FFT block diagram.

The DFT computed using the radix-2 decimation-in-time approach in this fashion is referred to as the fast Fourier transform (FFT).

Example 7.15: Length-4 FFT using radix-2 decimation in time

A length-4 signal $x[n]$ is given by

$$x[n] = \{ \underset{\substack{\uparrow \\ n=0}}{1.7}, 2.4, -1.2, 4.3 \}$$

Compute the FFT $X[k]$ of this signal using the radix-2 decimation in time technique outlined above.

Solution: Following the development in Section 7.5.1 we have

$$g[n] = \{ \underset{\substack{\uparrow \\ n=0}}{1.7}, -1.2 \}$$

and

$$h[n] = \{ \underset{\substack{\uparrow \\ n=0}}{2.4}, 4.3 \}$$

The transforms $G[k]$ and $H[k]$ are computed using the butterfly structure as

$$G[0] = g[0] + g[1] = 0.5$$

$$G[1] = g[0] - g[1] = 2.9$$

and

$$H[0] = h[0] + h[1] = 6.7$$

$$H[1] = h[0] - h[1] = -1.9$$

Finally, using the block diagram in Fig. 7.27

$$X[0] = G[0] + H[0] = 7.2$$

$$X[1] = G[1] - j\,H[1] = 2.9 - j\,1.9$$

$$X[2] = G[0] - H[0] = -6.2$$

$$X[3] = G[1] + j\,H[1] = 2.9 + j\,1.9$$

Software resource: `exdt_7_15.m`

7.5.2 Length-8 DFT

In this section we will apply the decimation-in-time technique to the problem of computing the DFT of a length-8 signal. The DFT analysis equation for this case is

$$X[k] = \sum_{n=0}^{7} x[n]\, W_8^{nk}\,, \qquad k = 0,\ldots,7$$

which can be written in open form as

$$\begin{aligned} X[k] =& x[0] + x[1]\, W_8^{k} + x[2]\, W_8^{2k} + x[3]\, W_8^{3k} + \\ & x[4]\, W_8^{4k} + x[5]\, W_8^{5k} + x[6]\, W_8^{6k} + x[7]\, W_8^{7k}\,, \qquad k = 0,1,\ldots,7 \end{aligned} \tag{7.119}$$

Reordering the terms on the right side so that even indexed terms appear first, followed by odd indexed terms, yields

$$\begin{aligned} X[k] =& x[0] + x[2]\, W_8^{2k} + x[4]\, W_8^{4k} + x[6]\, W_8^{6k} + \\ & x[1]\, W_8^{k} + x[3]\, W_8^{3k} + x[5]\, W_8^{5k} + x[7]\, W_8^{7k}\,, \qquad k = 0,1,\ldots,7 \end{aligned} \tag{7.120}$$

Factoring out W_8^k from odd indexed terms and realizing that $W_8^{2k} = W_4^k$ we have

$$X[k] = x[0] + x[2]\, W_4^{k} + x[4]\, W_4^{2k} + x[6]\, W_4^{3k} + W_8^k \left[x[1] + x[3]\, W_4^{k} + x[5]\, W_4^{2k} + x[7]\, W_4^{3k} \right]$$

We will define two new signals $g[n]$ and $h[n]$ as

$$g[n] = \{ \underset{\substack{\uparrow \\ n=0}}{x[0]},\, x[2],\, x[4],\, x[6] \}$$

$$h[n] = \{ \underset{\substack{\uparrow \\ n=0}}{x[1]},\, x[3],\, x[5],\, x[7] \}$$

The transform $X[k]$ can now be written as

$$\begin{aligned} X[k] &= \underbrace{g[0] + g[1]\, W_4^{k} + g[2]\, W_4^{2k} + g[3]\, W_4^{3k}}_{G[k],\quad k=0,\ldots,3} + W_8^k \underbrace{\left[h[0] + h[1]\, W_4^{k} + h[2]\, W_4^{2k} + h[3]\, W_4^{3k} \right]}_{H[k],\quad k=0,\ldots,3} \\ &= G[k] + W_8^k\, H[k]. \end{aligned} \tag{7.121}$$

Eqn. (7.121) is similar to Eqn. (7.118). It shows us how to merge two length-4 transforms to obtain the length-8 transform. This is depicted in Fig. 7.30. Substituting the radix-2 structure shown in 7.29 into the diagram in Fig. 7.30, we obtain the completed diagram for length-8 radix-2 FFT shown in Fig. 7.31.

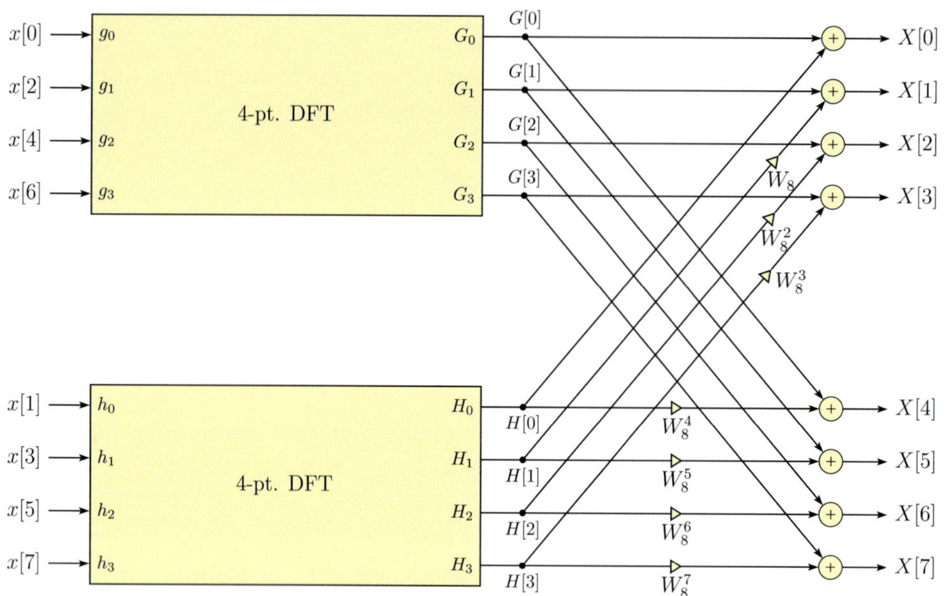

Figure 7.30 – Combining length-4 DFT's into a length-8 DFT.

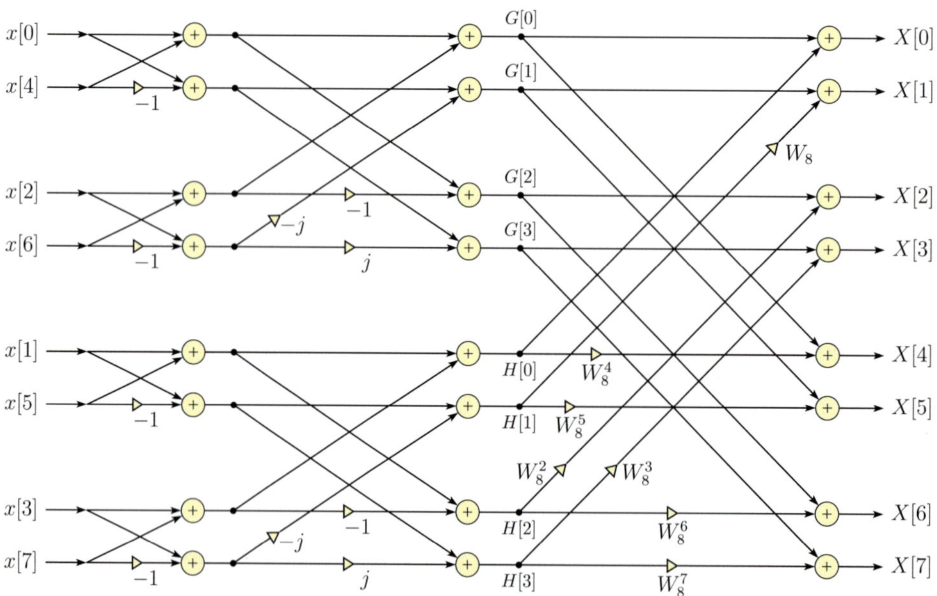

Figure 7.31 – Length-8 FFT block diagram.

7.5.3 Radix-2 Decimation-In-Time Fast Fourier Transform

In previous sections we have applied radix-2 decimation-in-time technique to the computation of length-4 and length-8 transforms. The idea of splitting a large DFT problem into two half-size problems can be generalized for any even transform length N. Consider the general form of the

DFT analysis equation:

$$X[k] = \sum_{n=0}^{N-1} x[n]\, W_N^{nk}\,, \qquad k = 0,\ldots,N-1$$

Provided that N is an even number, the DFT summation can be broken down into two summations, one containing even indexed samples and the other containing odd indexed samples.

$$X[k] = \sum_{n\ \text{even}} x[n]\, W_N^{nk} + \sum_{n\ \text{odd}} x[n]\, W_N^{nk}\,, \qquad k = 0,\ldots,N-1 \tag{7.122}$$

We will employ a change of variables on the two summations. Recognizing that even and odd indices can be expressed as $n = 2m$ and $n = 2m+1$, respectively, Eqn. (7.122) can be written as

$$X[k] = \sum_{m=0}^{N/2-1} x[2m]\, W_N^{2mk} + \sum_{m=0}^{N/2-1} x[2m+1]\, W_N^{(2m+1)k}\,, \qquad k = 0,\ldots,N-1 \tag{7.123}$$

using the new summation index variable m. Factoring out W_N^k from the terms of the second summation and remembering that $W_N^{2m} = W_{N/2}^m$ we obtain

$$X[k] = \sum_{m=0}^{N/2-1} x[2m]\, W_{N/2}^{mk} + W_N^k \sum_{m=0}^{N/2-1} x[2m+1]\, W_{N/2}^{mk}\,, \qquad k = 0,\ldots,N-1 \tag{7.124}$$

Let two signals $g[n]$ and $h[n]$ be defined as

$$g[n] = x[2n], \qquad n = 0,\ldots,\frac{N}{2}-1 \tag{7.125}$$

$$h[n] = x[2n+1], \qquad n = 0,\ldots,\frac{N}{2}-1 \tag{7.126}$$

The transform $X[k]$ becomes

$$X[k] = \underbrace{\sum_{m=0}^{N/2-1} g[m]\, W_{N/2}^{mk}}_{G[k],\ k=0,\ldots,\frac{N}{2}-1} + W_N^k \underbrace{\sum_{m=0}^{N/2-1} h[m]\, W_{N/2}^{mk}}_{H[k],\ k=0,\ldots,\frac{N}{2}-1}$$

$$= G[k] + W_N^k\, H[k]\,, \qquad k = 0,\ldots,N-1 \tag{7.127}$$

Eqn. (7.127) provides us with the algorithm for merging two half-size transforms into a full-size transform.

Let us now consider the question of what we have gained by doing this: Direct computation of the N-point DFT would require about N^2 complex multiplications. In contrast, each $(N/2)$-point transform requires $N^2/4$ complex multiplications. Merging them would require N additional complex multiplications. A comparison is given below:

Direct computation: N^2 comp. mult

Split into two: $2\left(\frac{N^2}{4}\right) + N = \frac{N^2}{2} + N$ comp. mult.

Now imagine that the two length-$(N/2)$ transforms are each further divided into length-$(N/4)$ transforms. Computation of each $(N/4)$-point transform requires $N^2/16$ complex multiplications. Merging two $(N/4)$-point transforms takes $N/2$ complex multiplications, and merging the resulting $(N/2)$-point transforms takes N complex multiplications. Arithmetic complexity for this case is as follows:

Direct computation:	N^2 comp. mult.
Split, and split again:	$4\left(\dfrac{N^2}{16}\right)+\dfrac{N}{2}+\dfrac{N}{2}+N=\dfrac{N^2}{4}+2N$ comp. mult.

If we were to split the transforms yet again, this time down to $(N/8)$-point DFT's, the total number of complex multiplications required would be about $\dfrac{N^2}{8}+3N$. We can see a pattern emerging here for computational complexity. Each additional stage of radix-2 decimation-in-time decomposition cuts the exponential term by half, and adds a linear term. The ideal scenario would be observed if the number of samples N is an integer power of 2. If that is the case, we can keep splitting the transform all the way down to 2-point transforms which can each be computed using the butterfly structures discussed earlier. There would be a total of $\log_2(N)$ stages in the computation. The number of required complex multiplications in this case is on the order of $\left[N\log_2(N)\right]$ as opposed to N^2 for direct computation.

Software resource: See MATLAB Exercise 7.17.

7.5.4 Generalizing Decimation-In-Time Algorithms

In the previous section we have focused on radix-2 decimation-in-time technique for improving the speed of DFT computation. Radix-2 algorithms are based on splitting the signal into two parts at each stage, computing two half-size transforms, and merging them back to obtain the full-size transform. They require the number of samples N to be an integer power of 2 so that the signal can be split over and over until we reach 2-point DFT's easily computable using the butterfly structures. The restriction on N does not pose a significant problem in most cases. Any signal could simply be padded with zero-amplitude samples to increase its length to the next power of 2. There may be cases, however, where it may be desirable to use a radix other than 2. In this section we will formulate radix-3 decimation-in-time FFT as an example. Similarities between the derivarions of radix-2 and radix-3 algorithms will become apparent, and will make it relatively easy for the reader to formulate decimation-in-time FFT with any other radix.

Assume N is an integer multiple of 3. Let's split the DFT summation into three separate summations, each taking a radix-3 downsampled version of the signal $x[n]$ as follows:

$$X[k]=\sum_{m=0}^{N/3-1}x[3m]\,W_N^{3mk}+\sum_{m=0}^{N/3-1}x[3m+1]\,W_N^{(3m+1)k}+\sum_{m=0}^{N/3-1}x[3m+2]\,W_N^{(3m+2)k} \tag{7.128}$$

Factoring out W_N^{mk} and W_N^{2mk} from the second and the third summations respectively, we have

$$X[k]=\sum_{m=0}^{N/3-1}x[3m]\,W_N^{3mk}+W_N^{k}\sum_{m=0}^{N/3-1}x[3m+1]\,W_N^{(3m)k}+W_N^{2k}\sum_{m=0}^{N/3-1}x[3m+2]\,W_N^{(3m)k} \tag{7.129}$$

Let three signals $p[n]$, $q[n]$ and $r[n]$ be defined as

$$p[n]=x[3n],\quad q[n]=x[3n+1]\quad \text{and}\quad r[n]=x[3n+2],\qquad n=0,\ldots,\frac{N}{3}-1$$

Also recognizing that $W_N^{3mk} = W_{N/3}^{mk}$, Eqn. (7.129) can be written as

$$X[k] = \underbrace{\sum_{m=0}^{N/3-1} p[m]\, W_{N/3}^{mk}}_{P[k],\; k=0,\ldots,\frac{N}{3}-1} + W_N^k \underbrace{\sum_{m=0}^{N/3-1} q[m]\, W_{N/3}^{mk}}_{Q[k],\; k=0,\ldots,\frac{N}{3}-1} + W_N^{2k} \underbrace{\sum_{m=0}^{N/3-1} r[m]\, W_{N/3}^{mk}}_{R[k],\; k=0,\ldots,\frac{N}{3}-1}$$

$$= P[k] + W_N^k\, Q[k] + W_N^{2k}\, R[k]\,, \qquad k=0,\ldots,N-1 \tag{7.130}$$

Eqn. (7.130) is the radix-3 equivalent of the decimation-in-time technique that was formulated for radix-2 in Eqn. (7.126).

7.5.5 Radix-2 Decimation-In-Frequency Fast Fourier Transform

An alternative to decimation-in-time decomposition discussed in earlier sections is decimation-in-frequency decomposition. We will begin by writing the general form of the DFT analysis equation separately for even and odd values of the transform index k:

$$k \text{ even:} \qquad X[2\ell] = \sum_{n=0}^{N-1} x[n]\, W_N^{2\ell n}, \qquad \ell = 0,\ldots,\tfrac{N}{2}-1$$

$$k \text{ odd:} \qquad X[2\ell+1] = \sum_{n=0}^{N-1} x[n]\, W_N^{(2\ell+1)n}, \qquad \ell = 0,\ldots,\tfrac{N}{2}-1$$

Recognizing that $W_N^{2\ell n} = W_{N/2}^{\ell n}$, even indexed samples of the transform can be written as

$$X[2\ell] = \sum_{n=0}^{N-1} x[n]\, W_{N/2}^{\ell n}\,, \qquad \ell = 0,\ldots,\frac{N}{2}-1 \tag{7.131}$$

The summation in Eqn. (7.131) has N terms, yet only the first half of the exponential factors $W_{N/2}^{\ell n}$ are unique. At this point it would make sense to break the summation into two parts as follows:

$$X[2\ell] = \sum_{n=0}^{N/2-1} x[n] W_{N/2}^{\ell n} + \sum_{n=N/2}^{N-1} x[n]\, W_{N/2}^{\ell n}\,, \qquad \ell = 0,\ldots,\frac{N}{2}-1 \tag{7.132}$$

Applying a variable change $n \rightarrow n - N/2$ to the second summation in Eqn. (7.132) leads to

$$X[2\ell] = \sum_{n=0}^{N/2-1} x[n]\, W_{N/2}^{\ell n} + \sum_{n=0}^{N/2-1} x[n+N/2]\, W_{N/2}^{\ell(n+N/2)}\,, \qquad \ell = 0,\ldots,\frac{N}{2}-1$$

Recognizing that $W_{N/2}^{\ell(n+N/2)} = W_{N/2}^{\ell n}$, and combining the two summations into one, we have

$$X[2\ell] = \sum_{n=0}^{N/2-1} x[n]\, W_{N/2}^{\ell n} + \sum_{n=0}^{N/2-1} x[n+N/2]\, W_{N/2}^{\ell n}$$

$$= \sum_{n=0}^{N/2-1} \Big(x[n] + x[n+N/2]\Big)\, W_{N/2}^{\ell n}\,, \qquad \ell = 0,\ldots,\frac{N}{2}-1 \tag{7.133}$$

A similar derivation will be used for computing the odd indexed transform samples given by

$$X[2\ell+1] = \sum_{n=0}^{N-1} x[n]\, W_N^n\, W_{N/2}^{\ell n}\,, \qquad \ell = 0,\ldots,\frac{N}{2}-1 \tag{7.134}$$

The first step is to write the summation in Eqn. (7.134) in two parts as

$$X[2\ell+1]=\sum_{n=0}^{N/2-1} x[n]\,W_N^n\,W_{N/2}^{\ell n}+\sum_{n=N/2}^{N-1} x[n]\,W_N^n\,W_{N/2}^{\ell n}\,,\qquad \ell=0,\ldots,\frac{N}{2}-1 \tag{7.135}$$

As before, we will apply the variable change $n\rightarrow n-N/2$ to the second summation in Eqn. (7.135), and combine the two summations into one to obtain the result we seek:

$$\begin{aligned} X[2\ell+1] &= \sum_{n=0}^{N/2-1} x[n]\,W_N^n\,W_{N/2}^{\ell n}+\sum_{n=0}^{N/2-1} x[n+N/2]\,W_N^{n+N/2}\,W_{N/2}^{\ell(n+N/2)n} \\ &= \sum_{n=0}^{N/2-1}\Big(x[n]-x[n+N/2]\Big)\,W_N^n\,W_{N/2}^{\ell n}\,,\qquad \ell=0,\ldots,\frac{N}{2}-1 \end{aligned} \tag{7.136}$$

Now we are ready to examine two parts of the transform $X[k]$ expressed in Eqns. (7.133) and (7.136). The right side of Eqn. (7.133) is identical to the $(N/2)$-point DFT $G[k]$ of a signal $g[n]$ defined as

$$g[n]=x[n]+x[n+N/2]\,,\qquad n=0,\ldots,\frac{N}{2}-1 \tag{7.137}$$

In contrast, the right side of Eqn. (7.136) is identical to the $(N/2)$-point DFT $H[k]$ of a signal $h[n]$ defined as

$$h[n]=\Big(x[n]-x[n+N/2]\Big)\,W_N^n\,,\qquad n=0,\ldots,\frac{N}{2}-1 \tag{7.138}$$

The transform $X[k]$ can now be obtained by merging $G[k]$ and $H[k]$ as follows:

$$X[k]=\begin{cases} G[k/2]\,, & k \text{ even} \\ H[(k-1)/2]\,, & k \text{ odd}\end{cases}\qquad k=0,\ldots,N-1 \tag{7.139}$$

Software resource: See MATLAB Exercise 7.18.

7.6 Using FFT to Implement Convolution in Real Time

In previous sections of this chapter we have discussed the utility of the DFT for implementing linear convolution between an input signal $x[n]$ and an impulse response $h[n]$. The product of two transforms $X[k]$ and $H[k]$ corresponds to circular convolution of the signals $x[n]$ and $h[n]$ involved. With proper use of zero padding and careful selection of the DFT size, the circular convolution result can be made to equal the linear convolution of the two signals (see Section 7.3.7). This allows us to use DFT/FFT algorithms for implementing linear systems. Fast convolution can be implemented through the use of the FFT (see MATLAB Exercise 7.9).

One requirement in the use of convolution for implementing a linear system is that both the input signal and the system impulse response be of finite length. If the impulse response of the system is of infinite length, then we need to change our strategy, and implement the system by iteratively solving the difference equation instead (see MATLAB Exercises 2.4 and 2.6).

It is also possible that the input signal may be of infinite length. An example of this is an audio signal that is being streamed for a long period. Consider the case where a linear system with a finite-length impulse response is used for processing an infinite-length audio signal. For practical purposes, we would like to process the signal while it is being played back. This is called *real-time processing.* It requires the signal to be partitioned into finite-length frames (see MATLAB Exercise 1.11 about frame-based processing).

An infinite-length signal $x[n]$ may be partitioned into frames of length L defined as

$$x_r[n] = \begin{cases} x[n+rL], & n = 0,1,\ldots,L-1 \\ 0, & \text{otherwise} \end{cases}$$

$$= x[n]\,\Big(u[n-rL] - u[n-(r+1)\,L+1]\Big) \tag{7.140}$$

We will refer to L as the *frame size.* The signal $x[n]$ can be written as

$$x[n] = \sum_{r=0}^{\infty} x_r[n-rL] \tag{7.141}$$

Samples of frame 0 occupy sample indices $0 \le n \le L-1$. Frame 1 is time shifted to fill the index range $L \le n \le 2L-1$, and so on. This is depicted in Fig. 7.32.

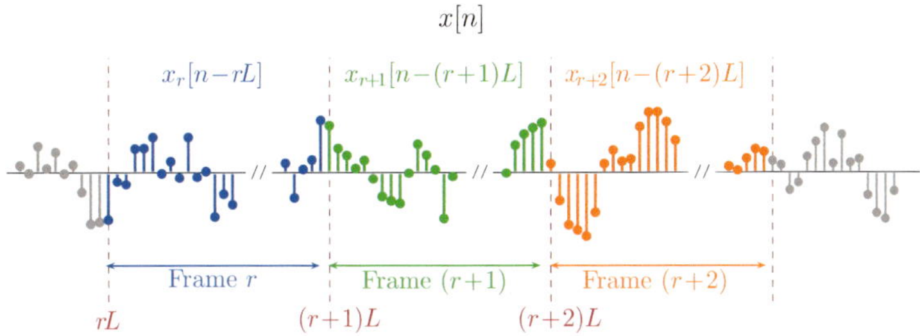

Figure 7.32 – Partitioning signal $x[n]$ into frames.

Let the impulse response $h[n]$ be a length-M sequence that is equal to zero outside the interval $0 \le n \le M-1$. Generally we will have $M < L$ although this is not a strict requirement. The convolution of $h[n]$ with $x[n]$ is

$$y[n] = h[n] * x[n] = \sum_{k=0}^{M-1} h[k]\,x[n-k] \tag{7.142}$$

Substituting Eqn. (7.141) into Eqn. (7.142) yields

$$y[n] = \sum_{k=0}^{M-1} h[k] \sum_{r=0}^{\infty} x_r[n-rL-k]$$

$$= \sum_{r=0}^{\infty} \underbrace{\sum_{k=0}^{M-1} h[k]\,x_r[n-rL-k]}_{y_r[n-rL]} \tag{7.143}$$

where $y_r[n]$ is the convolution of the impulse response of the system with frame r of the input signal:

$$y_r[n] = \sum_{k=0}^{M-1} h[k]\,x_r[n-k]$$

Thus, the output signal is

$$y[n] = \sum_{r=0}^{\infty} y_r[n - rL] \tag{7.144}$$

The sequence $y_r[n]$ is zero outside the range $0 \le n \le L + M - 2$. It has a total of $L + M - 1$ samples compared to L samples for the input frame $x_r[n]$. Comparing Eqns. (7.141) and Eqn. (7.144), we see that both $x_r[n - rL]$ and $y_r[n - rL]$ start with index $n = rL$. The input frame $x_r[n - rL]$ ends at index $(r+1)L - 1$. In contrast, $y_r[n - rL]$ ends at index $(r+1)L + M - 2$. Fig. 7.33 illustrates how output components $y_r[n]$ are fitted together to construct the output signal $y[n]$.

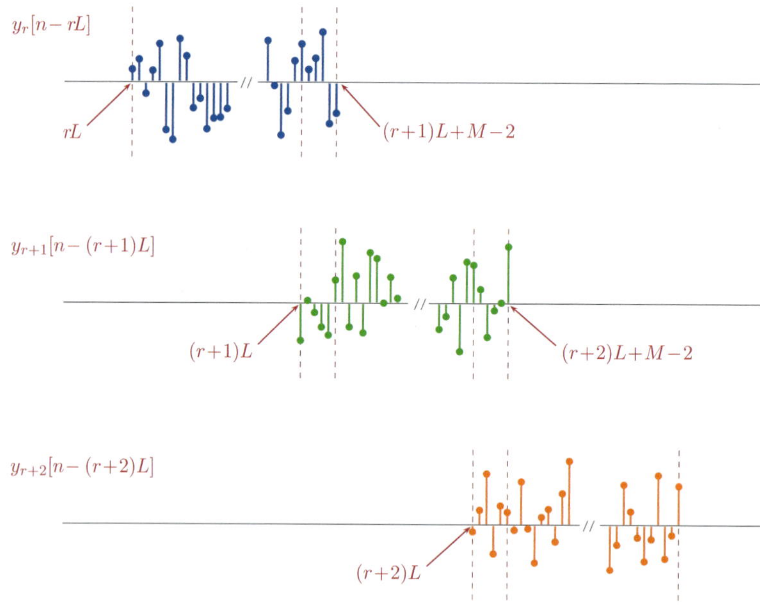

Figure 7.33 – Signals involved in Eqn. (7.100).

The last $(M-1)$ samples of $y_r[n]$ overlap with the first $(M-1)$ samples of $y_{r+1}[n]$. Similarly, the last $(M-1)$ samples of $y_{r+1}[n]$ overlap with the first $(M-1)$ samples of $y_{r+2}[n]$, and so on. Frame by frame processing of an audio signal in a seamless manner requires that, for each frame of data received from the input stream, a frame of the same size must be written to the output stream. The excess samples in each output frame need to be saved and added to the beginning of the following frame. This is referred to as the *overlap-add method.* The steps involved in fast convolution for real-time processing can be summarized as follows:

Real-time convolution:

1. Receive the input frame $x_r[n]$.
2. Obtain $y_r[n]$ using fast convolution. Sequences $x_r[n]$ and $h[n]$ should each be zero padded to extend their lengths to $L+M-1$.

$$H[k] = \text{FFT}\{h[n]\} \searrow$$

$$Y_r[k] = H[k]\,X_r[k] \longrightarrow y_r[n] = \text{FFT}^{-1}\{Y_r[k]\}$$

$$X_r[k] = \text{FFT}\{x_r[n]\} \nearrow$$

Computation of $H[k]$ does not need to be repeated for each frame of data. It can be computed ahead of time, and reused with each frame.

3. Add the contents of the buffer sequence $b_{r-1}[n]$ left over from the previous frame to the beginning of $y_r[n]$.

$$v_r[n] = \begin{cases} y_r[n] + b_{r-1}[n]\,, & n = 0, 1 \ldots, M-2 \\ y_r[n]\,, & n = M-1, \ldots, L+M-2 \end{cases}$$

4. Save the last $(M-1)$ samples of $v_r[n]$ into the buffer sequence $b_r[n]$.

$$b_r[n] = \begin{cases} v_r[n+L]\,, & n = 0, 1, \ldots, M-2 \\ 0\,, & \text{otherwise} \end{cases}$$

5. Construct output frame $w_r[n]$ by truncating $v_r[n]$ to L samples (frame size).

$$w_r[n] = \begin{cases} v_r[n]\,, & n = 0, 1, \ldots, L-1 \\ 0\,, & \text{otherwise} \end{cases}$$

6. Write $w_r[n]$ into the output stream, and go back to step 1.

Software resource: See MATLAB Exercise 7.19.

Summary of Key Points

☞ The discrete Fourier transform (DFT) of a length-N signal $x[n]$ is

Eqn. 7.7:
$$X[k] = \sum_{n=0}^{N-1} x[n]\,W_N^{kn}\,, \qquad k = 0, 1, \ldots, N-1$$

where $W_N = e^{-j(2\pi k/N)}$. The signal can be recovered from its DFT as

Eqn. 7.8:
$$x[n] = \frac{1}{N}\sum_{k=0}^{N-1} X[k]\,W_N^{-kn}\,, \qquad n = 0, 1, \ldots, N-1$$

☞ The DFT of a length-N signal is equal to its DTFT evaluated at a set of N angular frequencies equally spaced in the interval $0 \le \Omega < 2\pi$. This interval includes $\Omega = 0$ but does not include $\Omega = 2\pi$.

Eqn. 7.11:
$$X[k] = X(\Omega)\Big|_{\Omega=\Omega_k=2\pi k/N}$$

Looking at the DFT is similar to looking at the DTFT placed behind a picket fence with N equally spaced openings. This is referred to as the *picket fence effect.*

☞ The DTFT of a length-N signal can be obtained from its N-point DFT through the interpolation formula

Eqn. 7.16:
$$X(\Omega) = \sum_{k=0}^{N-1} X[k]\, B_N\left(\Omega - \frac{2\pi k}{N}\right)$$

where $B_N(\Omega)$ is the interpolating function defined as

Eqn. 7.15:
$$B_N(\Omega) = \frac{\sin\left(\dfrac{\Omega N}{2}\right)}{N \sin\left(\dfrac{\Omega}{2}\right)}\, e^{-j\Omega(N-1)/2}$$

☞ Zero padding is the act of appending a set of zero-amplitude samples to the end of a finite-length signal. It can be used for controlling the frequency spacing of transform samples.

Eqn. 7.17:
$$q[n] = \begin{cases} x[n], & n = 0, 1, \ldots, N-1 \\ 0, & n = N, \ldots, N+M-1 \end{cases}$$

The DFT of the newly defined signal $q[n]$ is

Eqn. 7.19:
$$Q[k] = X(\Omega)\Big|_{\Omega=2\pi k/(N+M)}$$

☞ Spectral leakage is caused by truncation of infinite-length signals. Simply truncating a signal is equivalent to multiplying it with a rectangular window sequence. Alternative window sequences may be used for controlling spectral leakage.

☞ The frequency spacing between successive samples of the N-point DFT is $\Delta\Omega = 2\pi/N$ radians. Frequency resolution Ω_r of the DFT is the smallest frequency difference between two sinusoidal signals that the transform is able to distinguish. It is a function of the genuine length of the signal without taking into account any zero-amplitude samples that may have been added on. The use of window sequences other than the rectangular window may further diminish the frequency resolution of the DFT.

☞ The EFS coefficients of a continuous-time periodic signal can be approximated by sampling the signal at N equally-spaced time instants over one period, computing the DFT of the resulting discrete-time signal $x[n]$, and scaling the DFT result by $1/N$. For accuracy of approximation, the number of samples over one period, N, should be sufficiently large. Only the first half of the DFT values can be used as approximations for positive indexed coefficients. Negative indexed coefficients are obtained form the second half of the DFT. If the conditions of the Nyquist sampling theorem cannot be satisfied in a strict sense, only the first few EFS coefficients should be used.

☞ The Fourier transform of a continuous-time signal can be approximated by sampling the signal at N equally-spaced time instants, computing the DFT of the resulting discrete-time signal $x[n]$, and scaling the DFT result by the sampling interval T.

Eqn. 7.41:
$$X_a(\omega_k) \approx T\,X[k]$$

Ideally, the sampling rate ω_s should be at least twice the highest frequency of the signal the transform of which is being approximated. On the other hand, it can be shown that a time-limited signal contains an infinite range of frequencies, and strict adherence to the Nyquist sampling theorem is not possible. For the approximation to be acceptable, care must be taken to ensure that the effect of aliasing is negligible. Only the first half of the DFT samples should be used for approximating the transform $X_a(\omega)$ at positive frequencies in the range $0 \le \omega < \omega_s/2$. The second half of the DFT samples represent an approximation to the transform $X_a(\omega)$ in the negative frequency range $-\omega_s/2 \le \omega < 0$.

☞ DFT is a linear transform. Given two length-N signals $x_1[n]$ and $x_2[n]$ with discrete Fourier transforms

$$x_1[n] \stackrel{\text{DFT}}{\longleftrightarrow} X_1[k] \qquad \text{and} \qquad x_2[n] \stackrel{\text{DFT}}{\longleftrightarrow} X_2[k],$$

it can be shown that

Eqn. 7.57:
$$\alpha_1\,x_1[n] + \alpha_2\,x_2[n] \stackrel{\text{DFT}}{\longleftrightarrow} \alpha_1\,X_1[k] + \alpha_2\,X_2[k]$$

for any two arbitrary constants α_1 and α_2.

☞ Given a transform pair

$$x[n] \stackrel{\text{DFT}}{\longleftrightarrow} X[k]$$

the following are also valid transform pairs:

Time shifting property:

Eqn. 7.58:
$$x[n-m]_{\text{mod } N} \stackrel{\text{DFT}}{\longleftrightarrow} e^{-j(2\pi/N)km}\,X[k]$$

Time reversal property:

Eqn. 7.61:
$$x[-n]_{\text{mod } N} \stackrel{\text{DFT}}{\longleftrightarrow} X[-k]_{\text{mod } N}$$

Conjugation property:

Eqn. 7.62:
$$x^*[n] \stackrel{\text{DFT}}{\longleftrightarrow} X^*[-k]_{\text{mod } N}$$

Frequency shifting property:

Eqn. 7.71:
$$x[n]\,e^{j(2\pi/N)mn} \stackrel{\text{DFT}}{\longleftrightarrow} X[k-m]_{\text{mod } N}$$

☞ Circular convolution property:

Given two length-N signals $x[n]$ and $h[n]$ with discrete-Fourier transforms

$$x[n] \stackrel{\text{DFT}}{\longleftrightarrow} X[k] \qquad \text{and} \qquad h[n] \stackrel{\text{DFT}}{\longleftrightarrow} H[k],$$

it can be shown that

Eqn. 7.73:
$$x[n] \otimes h[n] \stackrel{\text{DFT}}{\longleftrightarrow} X[k]\,H[k]$$

☞ Symmetry properties of the DFT:

$$\begin{aligned}
&x[n]\text{: Real, } \operatorname{Im}\{x[n]\}=0 &&\Longrightarrow && X^*[k]=X[-k]_{\bmod N} \\
&x[n]\text{: Imag, } \operatorname{Re}\{x[n]\}=0 &&\Longrightarrow && X^*[k]=-X[-k]_{\bmod N} \\
&x^*[n]=x[-n]_{\bmod N} &&\Longrightarrow && X[k]\text{: Real} \\
&x^*[n]=-x[-n]_{\bmod N} &&\Longrightarrow && X[k]\text{: Imag}
\end{aligned}$$

☞ Goertzel algorithm is useful in situations where only one or a few transform samples are needed as opposed to a complete transform. The transform sample $X[k]$ can be computed by iteratively solving the difference equation

Eqn. 7.87:
$$y_k[n]=x[n]+W_N^{-k}\,y_k[n-1]$$

and obtaining the output sample for index N. Then we have

$$X[k]=y_k[N]$$

Alternatively, the difference equation

Eqn. 7.89:
$$y_k[n]=x[n]-W_N^{k}\,x[n-1]+2\,\cos(2\pi k/N)\;y_k[n-1]-y_k[n-2]$$

can be solved to obtain the same result.

☞ Chirp transform algorithm (CTA) is used for computing the DTFT of a finite-length signal at an arbitrary set of equally-spaced frequencies. For a set of frequencies

$$\Omega_k=\Omega_0+k\Delta\Omega\,,\qquad m=0,\ldots,M-1$$

and $V=e^{-j\Delta\Omega}$, we define

Eqn. 7.95:
$$g[n]=x[n]\,e^{-j\Omega_0 n}\,V^{n^2/2}\,,\qquad n=0,1,\ldots,N-1$$

and

Eqn. 7.96:
$$h[n]=V^{(k-n)^2/2}$$

The desired transform samples $X(\Omega_k)$ can be obtained by convolving the two sequences $g[n]$ and $h[n]$, and multiplying the convolution result by $V^{k^2/2}$.

☞ Short time Fourier transform (STFT) is useful for analyzing non-stationary signals. It yields a transform result that is a function of both time and frequency.

☞ For large N, the computational complexity of the N-point DFT is on the order of N^2 complex multiplications and additions or, equivalently, $4N^2$ real multiplications and additions. Symmetry properties of the DFT can be utilized for achieving relatively modest improvements in computational speed especially when the signal $x[n]$ is real-valued. Decimation-in-time and decimation-in-frequency fast Fourier transform (FFT) algorithms can be utilized for more significant improvements in speed and efficiency of computation.

☞ FFT-based fast convolution can be used in real-time processing of signals through the use of the overlap-and-add method. In processing each frame, the excess length of the convolution result is saved in a buffer and added to the beginning of the following frame.

Further Reading

[1] Sonali Bagchi and Sanjit K Mitra. *The Nonuniform Discrete Fourier Transform and its Applications in Signal Processing.* Vol. 463. Springer Science & Business Media, 2012.

[2] S Allen Broughton and Kurt Bryan. *Discrete Fourier Analysis and Wavelets: Applications to Signal and Image Processing.* John Wiley & Sons, 2018.

[3] J.W. Cooley and J.W. Tukey. "An Algorithm for the Machine Computation of Complex Fourier Series, vol. 19". In: *Mathematics of Computation* (1965), p. 73.

[4] Henri J. Nussbaumer. *The Fast Fourier Transform.* Springer, 1982.

[5] Julius O. Smith. *Mathematics of the Discrete Fourier Transform (DFT): With Audio Applications.* Julius Smith, 2008.

[6] Duraisamy Sundararajan. *The Discrete Fourier Transform: Theory, Algorithms and Applications.* World Scientific, 2001.

[7] R. Tolimieri and C. Lu. *Algorithms for Discrete Fourier Transform and Convolution.* Springer, 1989.

[8] Man Wah Wong. *Discrete Fourier Analysis.* Vol. 5. Springer Science & Business Media, 2011.

MATLAB Exercises with Solutions

MATLAB Exercise 7.1: Writing and testing a function for DFT

In this exercise we will develop a MATLAB function for computing the DFT of a signal. A built-in MATLAB function `fft()` is available for fast computation of the DFT, and should be used whenever possible. On the other hand, developing our own version will help us understand the DFT better. Along the process we will have the opportunity to see some pitfalls encountered in writing MATLAB functions, and how to avoid them.

Consider the DFT equation given by Eqn. (7.7) and repeated here:

$$\text{Eqn. (7.7):} \qquad X[k] = \sum_{n=0}^{N-1} x[n]\, W_N^{kn}\,, \quad k = 0, 1, \ldots, N-1$$

For the summation in Eqn. (7.7) we could use a loop over the index `n`. The would give us one sample of $X[k]$. An second, outer, loop over k would then be needed to repeat the process for $k = 0, 1, \ldots, N-1$. In MATLAB, loop structures are slow and inefficient, and we try to avoid them whenever it is possible and practical. The summation result in Eqn. (7.7) can be obtained as a scalar product instead. Let us define a row vector $\mathbf{x}$ and a column vector $\mathbf{A_k}$ as

$$\mathbf{x} = \begin{bmatrix} x[0] & x[1] & \ldots & x[N-1] \end{bmatrix} \quad \text{and} \quad \mathbf{A_k} = \begin{bmatrix} W_N^0 \\ W_N^k \\ \vdots \\ W_N^{(N-1)k} \end{bmatrix} \tag{7.145}$$

Eqn. (7.7) can be written in alternative form as

$$X[k] = \mathbf{x}\mathbf{A_k}\,, \quad k = 0, 1, \ldots, N-1 \tag{7.146}$$

Consider the listing for the function ss_dft1() shown below. We will refer to this as version 1 of our function.

```
1  function Xk = ss_dft1(xn)
2    N = length(xn);             % Number of samples in vector xn
3    Xk = zeros(size(xn));       % Create a compatible vector Xk with all zeros
4    WN = exp(-j*2*pi/N);        % Eqn. (7.6)
5    n = [0:N-1]';               % Column vector of sample indices for x[n]
6    for k=0:N-1                 % Compute transform samples for k=0,...,N-1
7      Ak = WN.^(k*n);           % Eqn. (7.145)
8      Xk(k+1) = xn*Ak;          % Eqn. (7.146)
9    end
10 end
```

On line 5 we create a column vector n of sample indices, and on line 7 we form the column vector Ak for use in the scalar product. The argument xn must be in row format so that the scalar product can be computed in line 8. The transform result is Xk is also returned in row format. The script mexdt_7_1a.m listed below tests the function ss_dft1() against built-in function fft() using a simple pulse signal in both cases, and it should work fine.

```
% Script: mexdt_7_1.m
n = [0:15];
xn = ss_dpulse(n,5);
Xk = ss_dft1(xn);
disp('Xk from ss_dft1()')
Xk
Xk = fft(xn);
disp('Xk from fft()')
Xk
```

What if the argument xn is a column vector? An attempt to use the function ss_dft1() with a vector xn in column format proceeds as follows:

```
>> n = [0:15]';
>> xn = ss_dpulse(n,5);
>> Xk = ss_dft1(xn);

Error using  *
Incorrect dimensions for matrix multiplication. Check that the
number of columns in the first matrix matches..
```

The scalar product on line 8 fails since vectors xn and Ak do not have the proper dimensions for it. This is a typical example of a function that may work well only when certain conditions are met. It may be unreliable when used as part of a larger project or used by someone unfamiliar with its restrictions. A minimalistic solution to this problem would be for the function to check the argument xn, and reject it if it is not in the proper format. The listing below is ss_dft2(), version 2 of our DFT function:

```
function Xk = ss_dft2(xn)
  if ~isrow(xn)
    error('Argument xn must be a row vector.');
  end
  N = length(xn);             % Number of samples in vector xn
  Xk = zeros(size(xn));       % Create a compatible vector Xk with all zeros
```

```
    WN = exp(-j*2*pi/N);        % Eqn. (7.6)
    n = [0:N-1]';               % Column vector of sample indices for x[n]
    for k=0:N-1                 % Compute transform samples for k=0,...,N-1
      Ak = WN.^(k*n);           % Eqn. (7.145)
      Xk(k+1) = xn*Ak;          % Eqn. (7.146)
    end
  end
```

In the final version of the function, we will add a few more lines of code. Instead of simply rejecting an input argument in column format, the function internally converts it to row format to allow the scalar product to work. It also ensures that the returned vector Xk is always in the same format as the input vector xn.

```
1  function Xk = ss_dft(xn)
2    if ~isvector(xn)
3      error('Argument xn must be a vector.');
4    end
5    fixed = false;
6    if iscolumn(xn)             % If the argument xn is in column format
7      xn = xn.';                %   transpose it
8      fixed = true;
9    end
10   N = length(xn);             % Number of samples in vector xn
11   Xk = zeros(size(xn));       % Create a compatible vector Xk with all zeros
12   WN = exp(-j*2*pi/N);        % Eqn. (7.6)
13   n = [0:N-1]';               % Column vector of sample indices for x[n]
14   for k=0:N-1                 % Compute transform samples for k=0,...,N-1
15     Ak = WN.^(k*n);           % Eqn. (7.145)
16     Xk(k+1) = xn*Ak;          % Eqn. (7.146)
17   end
18   if fixed                    % If input argument was in column format
19     Xk = Xk.';                %   then transpose output argument too
20   end
21 end
```

In lines 7 and 19, it is important to use the *transpose* operator and not the *complex conjugate transpose* operator as the latter would also conjugate the results in the case of complex vectors. As a final note, it is also possible to eliminate the loop over k (lines 14 and 17). See the discussion in Section 7.2.7 and in MATLAB Exercise 7.13.

Software resources: mexdt_7_1.m , ss_dft1.m , ss_dft2.m , ss_dft.m

MATLAB Exercise 7.2: Using the DFT to compute DTFS coefficients

The DTFS coefficients of a periodic signal $\tilde{x}(n)$ can be computed using Eqn. (3.19) from Chapter 3 which is repeated here for convenience.

$$\text{Eqn. (3.19):} \qquad \tilde{c}_k = \frac{1}{N} \sum_{n=0}^{N-1} \tilde{x}[n]\, e^{-j(2\pi/N)kn}$$

In the development leading to the DFT, we saw that the DFT and the DTFS are closely related. If $x[n]$ is one period of $\tilde{x}[n]$, its DFT is $X[k] = N\,c_k$ where c_k represent one period of DTFS coefficients $\tilde{c}_k$. This relationship is summarized in Fig. 7.34.

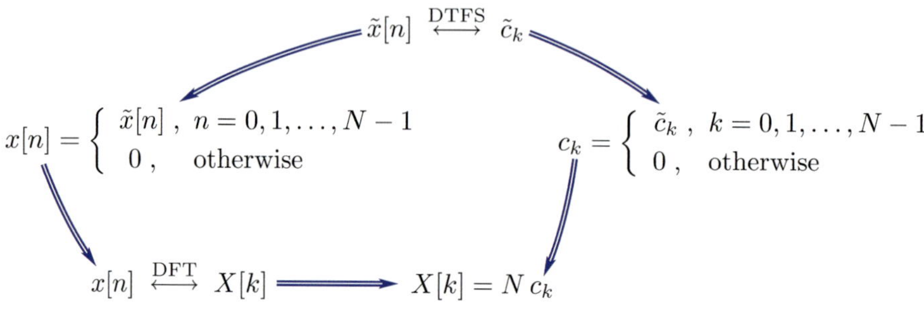

$\tilde{x}[n] \overset{\text{DTFS}}{\longleftrightarrow} \tilde{c}_k$
$x[n] = \begin{cases} \tilde{x}[n]\,, & n = 0, 1, \ldots, N-1 \\ 0\,, & \text{otherwise} \end{cases}$
$c_k = \begin{cases} \tilde{c}_k\,, & k = 0, 1, \ldots, N-1 \\ 0\,, & \text{otherwise} \end{cases}$
$x[n] \overset{\text{DFT}}{\longleftrightarrow} X[k]$
$X[k] = N\,c_k$

Figure 7.34 – Relationship between the DTFS and the DFT.

The script `mexdt_7_2.m` listed below computes the DTFS coefficients of the signal $\tilde{x}[n]$ encountered in Example 3.5 of Chapter 3 for parameter values of $L = 5$ and $N = 40$. In lines 4 through 6 we utilize functions `ss_dpulse()` and `ss_dper()` to create a vector `xPer`. It holds amplitudes of exactly one period of the signal $\tilde{x}[n]$ for $n = 0, 1, \ldots, N-1$. Compare the graph displayed by the script to Fig. 3.6.

```
1  % Script: mexdt_7_2.m
2  N = 40;          % Period of the signal
3  L = 5;           % See Example 3.5 in Chapter
4  %3
5  n = [0:N-1];     % Vector of sample indices
6  xPer = ss_dpulse(n,2*L+1);  % One period of the signal
7  x = ss_dper(xPer,n+L);      % Periodic extension evaluated for n=0,1,...,N-1
8  ck = fft(x)/N;              % See Eqn. (7.4)
9  k = [0:N-1];
10 stem(k,real(ck));           % Graph DTFS coefficients
11 title('c_k (L=5)');
12 xlabel('Index k');
13 ylabel('Amplitude');
```

Software resource: `mexdt_7_2.m`

MATLAB Exercise 7.3: Exploring the relationship between the DFT and the DTFT

The DFT and the DTFT of the length-10 discrete-time pulse were computed in Example 7.3. In this exercise we will duplicate the results of that example in MATLAB. The script listed below computes the 10-point DFT and graphs it on the same coordinate system with the DTFT of the signal.

```
% Script: mexdt_7_3a.m
xn = ones(1,10);    % Length-10 pulse signal
Xk = fft(xn);       % X[k] = DFT of x[n]
```

```
% Create a vector for angular frequencies, and compute the DTFT
Omg = [-0.1:0.01:1.1]*2*pi+eps;
XDTFT = sin(5*Omg)./sin(0.5*Omg).*exp(-j*4.5*Omg);  % Eqn. (7.12)
% Compute frequencies that correspond to DFT samples
k = [0:9];
Omg_k = 2*pi*k/10;
% Graph the DTFT and the DFT on the same coordinate system
tiledlayout(2,1);
nexttile;                 % Graph magnitudes
plot(Omg,abs(XDTFT),'-',Omg_k,abs(Xk),'ro'); grid;
axis([-0.2*pi,2.2*pi,-1,11]);
xlabel('Angular frequency (rad)');
ylabel('Magnitude');
nexttile;                 % Graph phases
plot(Omg,angle(XDTFT),'-',Omg_k,angle(Xk),'ro'); grid;
axis([-0.2*pi,2.2*pi,-pi,pi]);
xlabel('Angular frequency (rad)');
ylabel('Phase (rad)');
```

A slightly modified script is given below. The signal is padded with 10 additional zero-amplitude samples to extend its length to 20. This causes new transform samples to be displayed in between the existing transform samples.

```
% Script: mexdt_7_3b.m
xn = ones(1,10);          % Length-10 pulse signal
Xk = fft(xn,20);          % X[k] = DFT of x[n] zero padded to 20
% Create a vector for angular frequencies, and compute the DTFT
Omg = [-0.1:0.01:1.1]*2*pi+eps;
XDTFT = sin(5*Omg)./sin(0.5*Omg).*exp(-j*4.5*Omg);  % Eqn. (7.12)
% Compute frequencies that correspond to DFT samples
k = [0:19];
Omg_k = 2*pi*k/20;
% Graph the DTFT and the DFT on the same coordinate system
tiledlayout(2,1);
nexttile;    % Graph magnitudes
plot(Omg,abs(XDTFT),'-',Omg_k,abs(Xk),'ro'); grid;
axis([-0.2*pi,2.2*pi,-1,11]);
xlabel('Omega (rad)');
ylabel('Magnitude')
nexttile;    % Graph phases
plot(Omg,angle(XDTFT),'-',Omg_k,angle(Xk),'ro'); grid;
axis([-0.2*pi,2.2*pi,-pi,pi]);
xlabel('Omega (rad)');
ylabel('Phase (rad)');
```

 **Software resources:** mexdt_7_3a.m , mexdt_7_3b.m ,

MATLAB Exercise 7.4: Using the DFT to approximate the DTFT

Recall from the discussion in Section 7.2.2 that zero padding a length-N signal by M additional zero-amplitude samples before the computation of the DFT results in an angular frequency spacing of $\Delta\Omega = 2\pi/(N+M)$ radians. By choosing a large value for M, the angular frequency spacing can be made as small as desired. This allows us to use the DFT as a tool for approximating the DTFT. If sufficient DFT samples are available, they can be graphed in the form of a continuous function of Ω to mimic the appearance of the DTFT graph. The script listed below graphs the approximate DTFT of the length-10 discrete-time pulse

$$x[n] = u[n] - u[n-10]$$

The signal is padded with 490 zero-amplitude samples to extend its length to 500. The DFT result is obtained for $k = 0, 1, \ldots, 499$. The DFT sample for index k corresponds to the angular frequency $\Omega_k = 2\pi k/500$. The last DFT sample is at the angular frequency $\Omega_{499} = 998\pi/500$ which is just slightly less than 2π radians.

```
% Script: mexdt_7_4.m
x = ones(1,10);           % Generate x[n]
Xk = fft(x,500);          % Compute 500-point DFT
k = [0:499];              % DFT indices
Omg_k = 2*pi*k/500;       % Angular frequencies
% Graph the magnitude and the phase of the DTFT
tiledlayout(2,1);
nexttile;                 % Graph magnitudes
plot(Omg_k,abs(Xk)); grid;
title('|X(\Omega)|');
xlabel('\Omega (rad)');
ylabel('Magnitude');
axis([0,2*pi,-1,11]);
nexttile;                 % Graph phases
plot(Omg_k,angle(Xk)); grid;
title('\angle X(\Omega)');
xlabel('\Omega (rad)');
ylabel('Phase (rad)');
axis([0,2*pi,-pi,pi]);
```

Software resource: `mexdt_7_4.m`

MATLAB Exercise 7.5: Writing functions for circular time shifting and time reversal

Circular versions of time shifting and time reversal operations can be implemented in MATLAB using the steps presented in Section 7.3. Recall that the idea behind circularly shifting a length-N signal is to first create its periodic extension with period N, apply time shifting to the periodic extension, and finally extract the main period for $n = 0, 1, \ldots, N-1$. The function `ss_dper()` developed in MATLAB Exercise 1.5 will be utilized for periodic extension.

The function `ss_cshift()` listed below circularly shifts a signal by the specified number of samples. The vector x holds the samples of the signal $x[n]$ for $n = 0, 1, \ldots, N-1$. The second argument m is the amount of circular shift needed. The returned vector g holds samples of the circularly shifted signal for $n = 0, 1, \ldots, N-1$.

```
function g = ss_cshift(x,m)
  N = length(x);          % Length of x[n]
  n = [0:N-1];            % Vector of indices
  g = ss_dper(x,n-m);  % Compute periodic extension, then shift
end
```

The function ss_crev() listed below circularly reverses a signal. Again, the vector x holds the samples of the signal $x[n]$ for $n = 0, 1, \ldots, N-1$.

```
function g = ss_crev(x)
  N = length(x);          % Length of x[n].
  n = [0:N-1];            % Vector of indices.
  g = ss_dper(x,-n);    % Compute periodic extension, then time reverse
end
```

Let us create a length-10 signal to test the functions.

```
>> n = [0:9];
>> x = [0,2,3,4,4.5,3,1,-1,2,1];
```

The signal $g[n] = x[n-3]_{\text{mod } 10}$ is obtained and graphed by typing

```
>> g = ss_cshift(x,3);
>> stem(n,g);
```

A circular shift to the left can be obtained as well. To compute $g[n] = x[n+2]_{\text{mod } 10}$ type the following:

```
>> g = ss_cshift(x,-2);
```

A circularly time reversed version of $x[n]$ is obtained by

```
>> g = ss_crev(x);
```

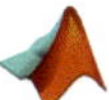

Software resources: mexdt_7_5.m , ss_cshift.m , ss_crev.m

MATLAB Exercise 7.6: Circular conjugate symmetric and antisymmetric components

Any complex length-N signal or transform can be written as the sum of two components one of which is circularly conjugate symmetric and the other circularly conjugate antisymmetric. The two components are computed using Eqns. (7.54) and (7.55) repeated here:

$$\text{Eqn. (7.54):} \qquad x_E[n] = \frac{x[n] + x^*[-n]_{\text{mod } N}}{2}$$

$$\text{Eqn. (7.55):} \qquad x_O[n] = \frac{x[n] - x^*[-n]_{\text{mod } N}}{2}$$

The terms needed in Eqns. (7.54) and (7.55) can easily be computed using the function `ss_crev()` developed in MATLAB Exercise 7.5. Given a vector `x` that holds samples of $x[n]$ for $n = 0, 1, 1, \ldots, N-1$, the signals $x_E[n]$ and $x_O[n]$ are computed as follows:

```
>> xE = 0.5*(x+conj(ss_crev(x)));
>> x0 = 0.5*(x-conj(ss_crev(x)));
```

Software resource: `mexdt_7_6.m`

MATLAB Exercise 7.7: Using the symmetry properties of the DFT

Consider Example 7.11 where two real signals $g[n]$ and $h[n]$ were combined to construct a complex signal $x[n]$. The DFT of the complex signal was then computed and separated into its circularly conjugate symmetric and antisymmetric components, and the individual transforms of $g[n]$ and $h[n]$ were extracted.

If the samples of the two signals are placed into MATLAB vectors `g` and `h`, the following script can be used to compute the transforms $G[k]$ and $H[k]$.

```
% Script: mexdt_7_7.m
x = g+j*h;              % Construct complex signal
Xk = fft(x);            % Compute DFT
% Compute the two components of the DFT
XE = 0.5*(Xk+conj(ss_crev(Xk)));  % Eqn. (7.54)
XO = 0.5*(Xk-conj(ss_crev(Xk)));  % Eqn. (7.55)
% Extract DFT's of the two signals
Gk = XE;                % See Example 7.11
Hk = -j*XO;             % See Example 7.11
disp(ifft(Gk))          % Should be the same as vector 'g'
disp(ifft(Hk))          % Should be the same as vector 'h'
```

Software resource: `mexdt_7_7.m`

MATLAB Exercise 7.8: Circular and linear convolution using the DFT

Consider the two length-5 signals $x[n]$ and $h[n]$ used in Example 7.14. The script listed below computes the circular convolution of the two signals using the DFT method:

```
% Script: mexdt_7_8a.m
x = [1,3,2,-4,6];  % Signal x[n]
h = [5,4,3,2,1];   % Signal h[n]
Xk = fft(x);       % DFT of x[n]
Hk = fft(h);       % DFT of y[n]
Yk = Xk.*Hk;       % Eqn. (7.73)
y = ifft(Yk);      % Compute inverse DFT
```

If linear convolution of the two signals is desired, the signals must be extended to at least 9 samples each prior to computing the DFT's. The script `mexdt_7_8a.m` listed below computes the linear convolution of the two signals using the DFT method.

```
% Script: mexdt_7_8b.m
x = [1,3,2,-4,6];      % Signal x[n]
h = [5,4,3,2,1];       % Signal h[n]
xp = [x,zeros(1,4)];  % Extend by zero padding
hp = [h,zeros(1,4)];  % Extend by zero padding
Xpk = fft(xp);         % DFT of xp[n]
Hpk = fft(hp);         % DFT of hp[n]
Ypk = Xpk.*Hpk;        % Eqn. (7.73)
yp = ifft(Ypk);        % Compute inverse DFT
```

The script `mexdt_7_8b.m` can be further simplified by using an alternative syntax for the function `fft()`. The statement

```
>> fft(x,9)
```

computes the DFT of the signal in vector `x` after internally extending the signal to 9 samples by zero padding. Consider the modified script `mexdt_7_8c.m` listed below:

```
% Script: mexdt_7_8c.m
x = [1,3,2,-4,6];     % Signal x[n]
h = [5,4,3,2,1];      % Signal h[n]
Xpk = fft(x,9);       % 9-point DFT of x[n]
Hpk = fft(h,9);       % 9-point DFT of h[n]
Ypk = Xpk.*Hpk;       % Eqn. (7.73)
yp = ifft(Ypk);       % Compute inverse DFT
```

Software resources: `mexdt_7_8a.m` , `mexdt_7_8b.m` , `mexdt_7_8c.m`

MATLAB Exercise 7.9: Writing a convolution function using the DFT

In this exercise we will develop a function for convolving two signals. Even though the built-in MATLAB function `conv()` accomplishes this task very well, it is still instructive to develop our own function that uses the DFT. As discussed in Section 7.3 linear convolution result can be obtained through the use of the DFT as long as the length of the transform is adjusted carefully. The function `ss_conv2()` given below computes the linear convolution of two finite-length signals $x[n]$ and $h[n]$ the samples of which are stored in vectors `x` and `h`.

```
function y = ss_conv2(x,h)
  N1 = length(x); % Length of signal 'x'
  N2 = length(h); % Length of signal 'h'
  N = N1+N2-1;    % Length of linear convolution result
  Xk = fft(x,N);  % DFT of x[n] zero padded to N
  Hk = fft(h,N);  % DFT of h[n] zero padded to N
  Yk = Xk.*Hk;    % Multiply two DFT's
  y = ifft(Yk,N); % Compute inverse DFT
end
```

The script `mexdt_7_9.m` listed below uses the signals of Example 7.14 to test the function `ss_conv2()`.

```
% Script: mexdt_7_9.m
x = [1,3,2,-4,6];  % Vector for signal x[n]
h = [5,4,3,2,1];   % Vector for signal h[n]
y = ss_conv2(x,h)
```

Caveat: Similar to what we have encountered in writing the function `ss_dft()` (see MATLAB Exercise 7.1), the function `ss_conv2()` fails if its two arguments have different orientations. The multiplication on line 7 is the cause of this failure. The problem can be solved by adding validation and correction code, similar to what was done for `ss_dft()` (see Problem 7.40 at the end of this chapter).

Software resources: mexdt_7_9.m , ss_conv2.m

MATLAB Exercise 7.10: Exponential Fourier series approximation using the DFT

In this example, we will develop a MATLAB function to approximate the EFS coefficients of a periodic signal. The strategy used will parallel the development in Eqns. (7.30) through (7.36). We have concluded in Section 7.2.5 that the DFT can be used for approximating the EFS coefficients through the use of Eqns. (7.34) and (7.35) provided that the largest coefficient index k that is used in approximation is sufficiently small compared to the DFT size N.

The code for the function `ss_efsapprox()` is given below.

```
function c = ss_efsapprox(x,k)
  Nx = length(x);  % Size of vector 'x'
  Nk = length(k);  % Size of vector 'k'
  % Create a return vector same size as 'k'
  c = zeros(1,Nk);
  Xk = fft(x)/Nx;  % Eqn. (7.35)
  % Copy the coefficients requested
  for i = 1:Nk
    kk = k(i);
    if (kk >= 0)
      c(i) = Xk(kk+1);
    else
      c(i) = Xk(Nx+1+kk);
    end
  end
end
```

The input argument `x` is a vector holding samples of one period of the periodic signal. The argument `k` is the index (or a set of indices) at which we wish to approximate the exponential Fourier series coefficients, so it can be specified as either a scalar index or a vector of index values. The return value `c` is either a scalar or a vector of coefficients. It has the same dimensions as `k`.

Suppose that 500 samples representing one period of a periodic signal $\tilde{x}(t)$ have been placed into the vector `x`. Using the statement

```
>> c = ss_efsapprox(x,2)
```

causes an approximate value for the coefficient c_2 to be computed and returned. On the other hand, issuing the statement

```
>> c = ss_efsapprox(x,[-2,-1,0,1,2])
```

results in the approximate values of coefficients $\{c_{-2}, c_{-1}, c_0, c_1, c_2, \}$ to be computed and returned as a vector. To keep the function listing simple, no error checking code has been included. Therefore, it is the user's responsibility to call the function with an appropriate set of arguments. Specifically, the index values in k must not cause an out-of-bounds error with the dimensions of the vector x. Furthermore, for good quality approximate values to be obtained, indices used should satisfy $|k| << N$.

Software resource: `ss_efsapprox.m`

MATLAB Exercise 7.11: Testing the EFS approximation function

In this exercise we will test the function `ss_efsapprox()` that was developed in MATLAB Exercise 7.10. Consider the periodic pulse train shown in Fig. 7.35.

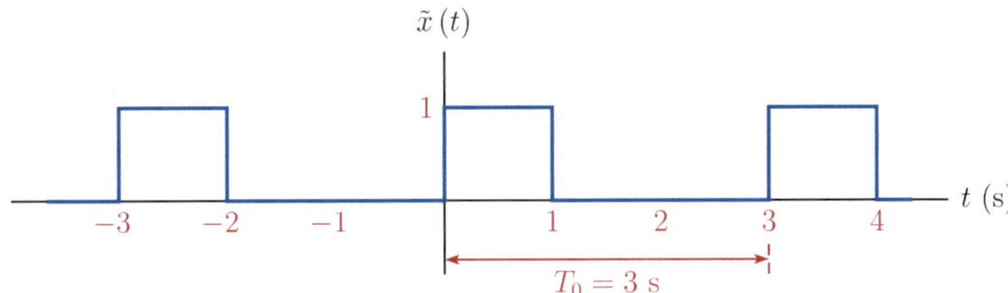

Figure 7.35 – The periodic pulse train for MATLAB Exercise 7.11.

One period is defined as

$$\tilde{x}(t) = \begin{cases} 1, & 0 \le t < 1 \\ 0, & 1 \le t < 3 \end{cases}$$

We will compute 1024 samples of this signal within one period. The test code for approximating c_k for $k = 0, \ldots, 10$ is given below:

```
1  % Script: mexdt_7_11.m
2  t = [0:1023]/1024*3;     % 1024 samples in one period
3  x = (t<=1);              % x(t)=1 if t<=1
4  % Compute and print approximate EFS coefficients for k=0,...,10
5  for k = 0:10
6    coeff = ss_efsapprox(x,k);
7    str = sprintf('k=%3d, magnitude=%0.5f, phase=%0.5f',..
8      k,abs(coeff),angle(coeff));
9    disp(str);
10 end
```

One point needs to be clarified: In line 3 of the code we generate a vector t of time samples spanning one period of $T = 3$ seconds. The first element of the time vector is equal to 0. We might intuitively think that the last element of this vector should be set equal to 3, the length of the period, yet it is equal to $1023(3)/(1024)$ which is slightly less than 3. The reason for this subtle difference becomes obvious when we look at Fig. 7.16. The width of each rectangle used in the numerical approximation of the integral is $T = 3/1024$. The first rectangle starts at $t = 0$ and extends to $t = T = 3/1024$. The second rectangle starts at $t = T = 3/1024$ and extends to $t = 2T = 6/1024$. Continuing to reason in this fashion, the left edge of the last rectangle of the period is at $t = 1023T = 1023(3)/(1024)$, and the right edge of it is at $t = 1$. The vector `t` holds the left edge of each rectangle in the approximation, and therefore its last value is just shy of the length of the period by one rectangle width.

Actual vs. approximate magnitude and phase values of exponential Fourier series coefficients of the signal $\tilde{x}(t)$ are listed in Table 7.2. The same table also includes percent error calculations between actual and approximated values. Note that for index values at which the magnitude of a coefficient is zero or near-zero, phase calculations are meaningless, and percent error values are omitted.

Table 7.2 – Exact vs. approximate exponential Fourier series coefficients for the test case in MATLAB Example 7.11.

k	Magnitude			Phase (rad)		
	Actual	Approx.	% Error	Actual	Approx.	% Error
0	0.3333	0.3340	0.20	0.00	0.000	0.000
1	0.2757	0.2760	0.12	−1.047	−1.046	−0.098
2	0.1378	0.1375	−0.24	−2.094	−2.092	−0.098
3	0.0000	0.0007	0.00	3.142	0.003	
4	0.0689	0.0692	0.47	−1.047	−1.043	−0.391
5	0.0551	0.0548	−0.59	−2.094	−2.089	−0.244
6	0.0000	0.0007	0.00	3.142	0.006	
7	0.0394	0.0397	0.82	−1.047	−1.040	−0.683
8	0.0345	0.0341	−0.94	−2.094	−2.086	−0.390
9	0.0000	0.0007	0.00	3.142	0.009	
10	0.0276	0.0279	1.17	−1.047	−1.037	−0.976

Software resource: `mexdt_7_11.m`

MATLAB Exercise 7.12: Fourier transform approximation using the DFT

The continuous-time signal

$$x_a(t) = \sin(\pi t)\, \Pi\left(t - \frac{1}{2}\right) = \begin{cases} \sin(\pi t)\,, & 0 \le t \le 1 \\ 0\,, & \text{otherwise} \end{cases}$$

has the Fourier transform

$$X_a(f) = \frac{1}{2}\left[\operatorname{sinc}\left(f + \frac{1}{2}\right) + \operatorname{sinc}\left(f - \frac{1}{2}\right)\right] e^{-j\pi f}$$

In this example we will develop the MATLAB code to approximate this transform using the DFT as outlined by Eqns. (7.41) and (7.42). The MATLAB script given below computes and graphs both the actual and the approximated transforms for the signal $x_a(t)$.

1. The actual transform is computed in the frequency range $-10\text{ Hz} \le f \le 10\text{ Hz}$.
2. For the approximate solution, $N = 16$ samples of the signal $x_a(t)$ are taken in the time interval $0 \le t \le 1$ s. This corresponds to a sampling rate of $f_s = 16$ Hz, and to a sampling interval of $T = 0.0625$ s.
3. Initially no zero padding is performed, therefore $M = 0$.
4. Approximations to the continuous Fourier transform are obtained at frequencies that are spaced $f_s/N = 1$ Hz apart.
5. The DFT result has 16 samples as approximations to the continuous Fourier transform at frequencies $f_k = kf_s/N = k$ Hz. Only the first 8 DFT samples are usable for positive frequencies from 0 to 7 Hz. This is due to Nyquist sampling theorem. The second half of the transform represents negative frequencies from -8 Hz to -1 Hz, and must be placed in front of the first half. The function `fftshift()` used in line 20 accomplishes that.

```
1  % Script: mexdt_7_12.m
2  f = [-10:0.01:10];  % Create a vector of frequencies
3  % Compute the actual transform
4  X_actual = 0.5*(sinc(f+0.5)+sinc(f-0.5)).*exp(-j*pi*f);
5  % Set parameters for the approximate transform
6  t1 = 1;                 % Upper limit of the time range
7  N = 16;                 % Number of samples
8  M = 0;                  % Number of samples for zero-padding
9  T = t1/N;               % Sampling interval
10 fs = 1/T;               % Sampling rate
11 n = [0:N-1];            % Index n for the sampled signal x[n]
12 k = [0:N+M-1];          % Index k for the DFT X[k]
13 time = n*T;             % Sampling instants
14 % Sample the signal and compute the DFT
15 xn = sin(pi*time);
16 Xk = T*fft(xn,N+M);
17 fk = k*fs/(N+M);
18 % Use fftshift() function on the DFT result to bring the zero-
19 % frequency to the middle. Also adjust vector fk for it.
20 Xk = fftshift(Xk);
21 fk = fk-0.5*fs;
22 % Graph the results
23 tiledlayout(2,1);
24 nexttile;               % Graph magnitudes
25 plot(f,abs(X_actual),'-',fk,abs(Xk),'r*'); grid;
26 title('Magnitude of actual and approximate transforms');
27 xlabel('f (Hz)');
28 ylabel('Magnitude');
29 nexttile;               % Graph phases
30 plot(f,angle(X_actual),'-',fk,angle(Xk),'r*'); grid;
31 title('Phase of actual and approximate transforms');
32 xlabel('f (Hz)');
33 ylabel('Phase (rad)');
```

The MATLAB graph produced by the script is shown in Fig. 7.36.

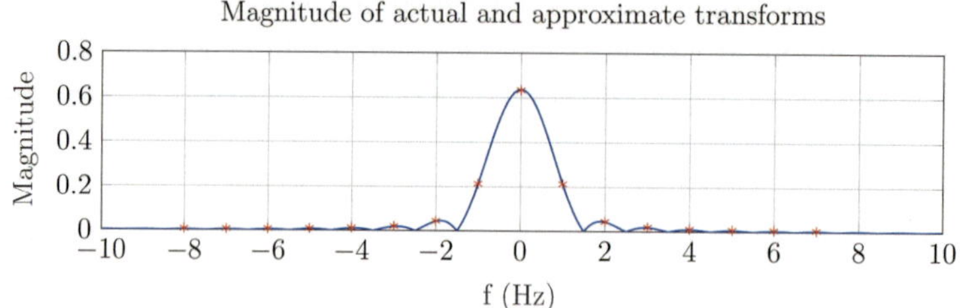

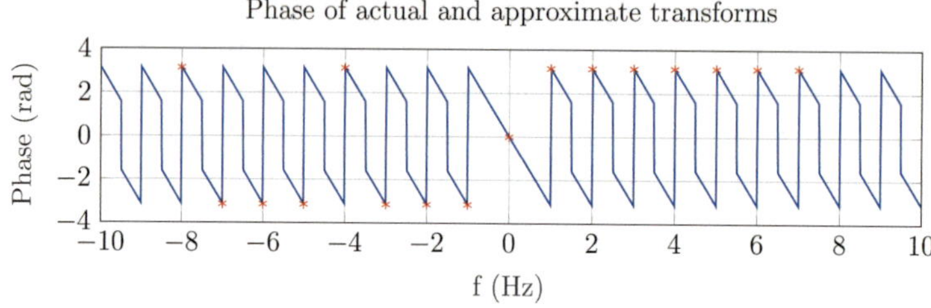

Figure 7.36 – The graph obtained in MATLAB Exercise 7.12. Magnitude and phase of the actual spectrum shown are in blue, and approximated values are shown with red asterisks.

The zero padding parameter M can be used to control the frequency spacing. For example, setting $M = 16$ causes the continuous Fourier transform to be estimated at intervals of 0.5 Hz starting at −8 Hz and ending at 7.5 Hz.

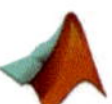

Software resource: `mexdt_7_12.m`

MATLAB Exercise 7.13: Writing a function for DFT using matrix formulation

A matrix formulation of DFT was derived in Section 7.2.7, In this exercise we will develop a function to compute the DFT of a signal using this approach. Consider the function `ss_dftm()` given below.

```
function Xk = ss_dftm(xn)
  if isrow(xn)        % Check if xn is in row format
    xvec = xn.';      % xvec is always in column format
  else
    xvec = xn;
  end
  N = length(xvec);  % Find length of input signal
  a = [0:N-1];        % Create a vector of integers
  b = a'*a;           % N by N matrix of exponent kn
  Wmat = exp(-j*2*pi/N).^b;  % Eqn. (7.45)
  Xk = Wmat*xvec;
  if isrow(xn)        % If xn is in row format, transpose Xk to match it
    Xk = Xk.';
  end
end
```

Lines 8 through 10 facilitate construction of the matrix $\mathbf{W}$ as described in Eqn. (7.44). The transform is computed on line 11. Lines 2 through 6 and 12 through 14 deal with orientation issues for input and output vectors so that the output vector `Xk` has the same orientation as the input vector `xn`.

 **Software resource:** `ss_dftm.m`

MATLAB Exercise 7.14: Implementing Goertzel algorithm

In this exercise, two MATLAB functions will be developed for implementing Goertzel algorithm discussed in Section 7.4.1. Both functions compute the transform $X[k]$ for one specified value of k. The first function `ss_goertzel1()` is based on the difference equation in Eqn. (7.87) and the block diagram shown in Fig. 7.23. The loop between lines 6 and 9 iterates through the difference equation to compute $X[k] = y_k[N]$.

```
1  function Xk = ss_goertzel1(x,k)
2    N = length(x);              % Number of samples in vector x[n]
3    gain = exp(j*2*pi*k/N);
4    yknm1 = 0;                  % Buffer for yk[n-1]
5    x = [x,0];
6    for n=1:N+1                 % Loop through the filter
7      ykn = x(n)+gain*yknm1;    % Eqn. (7.87)
8      yknm1 = ykn;              % Update yk[n-1]
9    end
10   Xk = ykn;
11 end
```

The second function `ss_goertzel2()` is based on the difference equation given by Eqn. (7.89) and its block diagram implementation shown in Fig. 7.24. The loop between lines 7 and 11 implements the feedback portion of the block diagram. The feed-forward portion of the diagram is not implemented inside the loop since intermediate output samples are not needed. The final output sample $y_k[N] = X[k]$ is computed in line 12, after the loop is finished.

```
1  function Xk = ss_goertzel2(x,k)
2    N = length(x);                % Number of samples in vector x[n]
3    gain = 2*cos(2*pi*k/N);
4    q = 0;                        % Node variable q[n] -- see Fig. 7.24
5    v = 0;                        % Node variable v[n] -- see Fig. 7.24
6    x = [x,0];                    % Zero-pad x[n] with one more sample
7    for n=1:N+1                   % Loop through the filter
8      p = x(n)+gain*q-v;          % Eqn. (7.90)
9      v = q;                      % Update node variable q[n]
10     q = p;                      % Update node variable v[n]
11   end
12   Xk = p-exp(-j*2*pi*k/N)*v;    % Eqn. (7.91)
13 end
```

The script `mexdt_7_14.m` listed below allows testing of the two functions discussed above. A length-128 signal $x[n]$ is generated with random amplitude values. The script computes the transform sample $X[47]$ in four different ways:

- Through direct use of the DFT (line 5),
- using convolution as given by Eqn. (7.85) (line 10),
- using function `ss_goertzel1()` (line 14), and
- using function `ss_goertzel2()` (line 16).

```
1  % Script: mexdt_7_14.m
2  N = 128;                    % Number of samples in the signal
3  k = 47;                     % DFT sample needed
4  x = randn(1,N);             % Create a random signal
5  X = fft(x);                 %Compute the FFT
6  disp('DFT result')
7  X(k+1)                      % Display X[k]  (Remember MATLAB indices start with 1)
8  WN = exp(-j*2*pi/N);
9  hk = WN.^(-k*[0:N]);        % Impulse response for Goertzel system
10 yk = conv(x,hk);            % Convolve x[n] and hk[n], Eqn. (7.85)
11 disp('Convolution result')
12 yk(N+1)                     % Display yk[N]  (Remember MATLAB indices start with 1)
13 disp('Use ss_goertzel1()')
14 Xk = ss_goertzel1(x,k)
15 disp('Use ss_goertzel2()')
16 Xk = ss_goertzel2(x,k)
```

Software resources: mexdt_7_14.m , ss_goertzel1.m , ss_goertzel2.m

MATLAB Exercise 7.15: Implementing chirp transform algorithm (CTA)

A MATLAB function `ss_cta()` for implementing the chirp transform algorithm (CTA) based on the development of Section 7.4.2 is given below. Inputs to the function are the signal $x[n]$ as a vector followed by three scalars for Ω_0, $\Delta\Omega$ and M. Lines 7 through 15 of the code correspond directly with Eqns. Eqn. (7.95), (7.98), and (7.97).

```
1  function XDTFT = ss_cta(xn,Omg0,Delta,M)
2    N = length(xn);         % Length of signal x[n]
3    xvec = xn;              % Make a copy of xn in xvec
4    if isrow(xn)            % If xn is in row format, transpose xvec to
5      xvec = xn.';          %   column format
6    end
7    n = [0:N-1]';           % Vector of sample indices
8    V = exp(-j*Delta);
9    gn = xvec.*exp(-j*Omg0*n).*V.^(n.*n/2);  % Eqn. (7.95)
10   nn = [-N+1:M-1]';       % Index range for h[n] -- See section 7.4.2
11   hn = V.^(-nn.*nn/2);    % Eqn. (7.98)
12   tmp = conv(gn,hn);      % Convolve g[n] and h[n]
13   tmp = tmp(N:N+M-1);     % Discard first and last (N-1) samples
14   k = [0:M-1]';
15   XDTFT = tmp.*V.^(k.*k/2);  % Eqn. (7.97)
16   if isrow(xn)            % If xn is in row format, transpose XDTFT to match
17     XDTFT = XDTFT.';      %   it so that XDTFT is in row format too
18 end
```

The script `mexdt_7_15.m` allows the function `ss_cta()` to be tested using a length-32 test signal $x[n]$ with random sample amplitudes.

```
% Script: mexdt_7_15.m
N = 32;                    % Number of samples in the signal x[n]
xn = randn(N,1);           % Generate a random signal
M = 5;                     % Number of transform samples requested
Omg0 = 0.5;                % Initial angular frequency
Delta = 2*pi/17;           % Angular frequency increment
k = [0:M-1];               % Indices of angular frequencies
Omg = Omg0 + k*Delta;      % Vector of angular frequencies
disp('Results obtained from DTFT')
ss_dtft(xn,Omg)
disp('Results obtained using CTA')
ss_cta(xn,Omg0,Delta,M)
```

Software resources: `mexdt_7_15.m` , `ss_cta.m`

MATLAB Exercise 7.16: Short time Fourier transform

Let a signal $x[n]$ be defined as

$$x[n] = \sin(2\pi F_1 n)\Big(u[n]-u[n-100]\Big)+\sin(2\pi F_2 n)\Big(u[n-50]-u[n-150]\Big)$$
$$+\sin(2\pi F_3 n)\Big(u[n-125]-u[n-225]\Big)+\sin(2\pi F_4 n)\Big(u[n-175]-u[n-300]\Big)$$

This signal is graphed in Fig. 7.37.

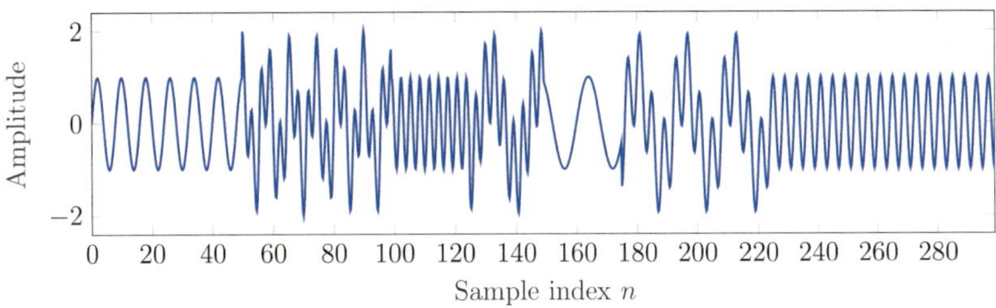

Figure 7.37 – The signal $x[n]$ for Matlab Exercise 7.16.

MATLAB built-in function `spectrogram()` can be used for computing the STFT of this signal. In order to provide further insight into the process, however, we will opt to develop our own script to compute and graph the STFT. The script `mexdt_7_16.m` listed below illustrates the process of computing the STFT. The signal $x[n]$ is generated in lines 2 through 7. In the loop between lines 9 and 13, length-16 segments of the signal are taken one at a time. In the first pass, we take $x[n]$ for $n = 0,\ldots,15$. This is essentially $x[n]\,w[n]$ where $w[n]$ represents the length-16 rectangular window. In the second pass, we have $x[n]$ for $n = 1,\ldots,16$ which is the same as $x[n]\,w[n-1]$. Each segment is zero padded to length 256, its FFT is computed, and multiplied by the proper exponential factor based on Eqn. (7.106). The result is appended to matrix `Xkm` as a new column.

```
% Script: mexdt_7_16.m
n = [0:299];                                   % Vector of sample indices
x1 = sin(2*pi*0.125*n).*(n<=99);
x2 = sin(2*pi*0.325*n).*((n>=50)&(n<=149));
x3 = sin(2*pi*0.0625*n).*((n>=125)&(n<=224));
x4 = sin(2*pi*0.25*n).*((n>=175)&(n<=299));
x = x1+x2+x3+x4;                               % Generate the signal x[n]
Xkm = [];                                      % Start with empty matrix
for m=0:284
  xw = x(m+1:m+16)';          % Get 16 samples of the signal for m<=n<=m+16
  tmp = fft(xw,256).*exp(-j*2*pi/256*m);       % Eqn. (7.106)
  Xkm = [Xkm,tmp];                             % Add new column to matrix
end
k = [127:-1:0];                                % Transform indices
Omg = 2*pi*k/256;                              % Vector of angular frequencies
[M,K] = meshgrid([0:284],[127:-1:0]/256*2*pi);
mesh(M,K,abs(Xkm(128:-1:1,:)),'FaceAlpha',0.5);
axis([0,284,0,pi]);
shading interp;
xlabel('$m$','Interpreter','latex');
ylabel('$\Omega$','Interpreter','latex');
zlabel('$|X_{STFT}\left(m,\Omega\right)|$','Interpreter','latex');
set(gca,'YTick',[0,pi/2,pi,3*pi/2,2*pi]);
set(gca,'YTickLabel',{'$0$','$\frac{\pi}{2}$','$\pi$',..
    '$\frac{3\pi}{2}$','$2\pi$'},'TickLabelInterpreter','latex');
```

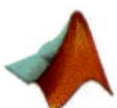

Software resource: mexdt_7_16.m

MATLAB Exercise 7.17: Writing a function for radix-2 decimation in time FFT

Even though MATLAB has a fast and efficient function for computing the FFT of a discrete-time signal, it is instructive to develop our own version to gain insight into the workings of the radix-2 decimation-in-time algorithm. We will approach the development of the function in two stages: The first stage will involve writing a function ss_fftr2dit_() that computes the transform based on the development outlined in Eqns. (7.123) through (7.126), yet is far from being ideal since it contains a set of restrictions that make it impractical for general use. The second stage will be to develop a *wrapper function* ss_fftr2dit() that checks for possible errors, arranges the input and output variables into proper form, and calls the function ss_fftr2dit_() to do the computations.

Consider the first function ss_fftr2dit_() listed below. Lines 9 and 10 implement Eqns. (7.125) and (7.126) to obtain decimated signals $g[n]$ and $h[n]$. In lines 11 and 12, the function calls itself to compute the half-length transforms $G[k]$ and $H[k]$. Such a function is referred to as a *recursive function.* Once half-length transforms are obtained, they are merged in lines 14 through 18 where we simply implement Eqn. (7.127).

```
1 function Xk = ss_fftr2dit_(xn)
2   N = length(xn);
3   if (N==1)
4     % If the length of xn is 1, return Xk = xn (1-point DFT)
5     Xk = xn;
```

```
6     else
7       % Otherwise split x[n] into half-length signals g[n] and h[n],
8       % and compute their transforms
9       gn = xn(1:2:N-1);        % Eqn. (7.125)
10      hn = xn(2:2:N);          % Eqn. (7.126)
11      Gk = ss_fftr2dit_(gn);  % Recursively call the function ss_fftr2dit_()
12      Hk = ss_fftr2dit_(hn);  %   to compute half-length transforms
13      % Prepare the vector S = W.^k needed for merging Gk and Hk
14      k = [0:N-1]';
15      S = exp(-j*2*pi/N).^k;
16      % Combine the two half-length transforms into a full-length
17      % transform
18      Xk = [Gk;Gk]+S.*[Hk;Hk];  % Eqn. (7.127)
19    end;
20  end
```

The function ss_fftr2dit_() will continue calling itself recursively over and over until the length of resulting half-length signals becomes $N = 1$. When that happens, one-point DFT's are computed as shown in line 5, and the function terminates. Every recursive function needs an exit condition.

In order for the function ss_fftr2dit_() to work properly, several conditions must be satisfied. First, the number of samples in the input signal $x[n]$ must be an integer power of 2, otherwise the function will produce runtime errors or results that do not make sense. Furthermore, the input argument xn must be a column vector, and the output transform Xk will also be returned as a column vector. It is conceivable that the user might call ss_fftr2dit_() using a matrix for xn which would break the code. Because of these issues, this function should never be used directly. Instead, we will develop a wrapper function ss_fft2dit() to circumvent violations of these conditions, and then call the function ss_fft2dit_() in a safe manner. The listing for ss_fftr2dit() is given below.

```
function Xk = ss_fftr2dit(xn)
  % Wrapper function for radix-2 decimation-in-time FFT
  if (min(size(xn)) > 1)    % Check to make sure xn is a vector, not a matrix
    error('Input argument ''xn'' must be a vector.');
  else
    N = max(size(xn));
    if ~ss_ispowerof2(N)    % Length of xn must be a power of 2
      error('Length of ''xn'' must be a power of 2')
    end
    xvec = xn;                % Set xvec equal to xn
    if isrow(xn)              % If xn is in row format, transpose xvec to
      xvec = xn.';            %   satisfy the requirements of ss_fftr2dit_()
    end
    Xk = ss_fftr2dit_(xvec);  % Compute the FFT
    if isrow(xn)              % If xn is in row format, transpose Xk to match it
      Xk = Xk.';              %   so that Xk is in row format too
    end
  end
end
```

Most of the code in ss_fftr2dit() is self explanatory. The only part that needs an explanation is the utility function ss_ispowerof2() which returns true or false depending on whether N is an integer power of 2 or not. Its listing is given below, and involves a simple *bitwise and* operation between N and $(N-1)$.

```
function a = ss_ispowerof2(N)
  a = ~bitand(N,N-1);
end
```

Software resources: `ss_fftr2dit.m` , `ss_ispowerof2.m`

MATLAB Exercise 7.18: Writing a function for radix-2 decimation-in-frequency FFT

In this exercise we will write a MATLAB function for computing the FFT of a signal using radix-2 decimation-in-frequency technique. The development will be similar to what was done in MATLAB Exercise 7.17. The function `ss_fftr2dif_()` listed below computes the FFT using the development of Section 7.5.5.

```
1  function Xk = ss_fftr2dif_(xn)
2    N = length(xn);
3    if (N==1)
4      % If the length of xn is 1, return Xk = xn
5      Xk = xn;
6    else
7      % Otherwise split x[n] into half-length signals g[n] and h[n],
8      % and compute their transforms
9      gn = xn(1:N/2)+xn(N/2+1:N);         % Eqn. (7.137)
10     idx = [0:N/2-1]';
11     S = exp(-j*2*pi/N).^idx;            % Compute vector W.^n
12     hn = (xn(1:N/2)-xn(N/2+1:N)).*S;    % Eqn. (7.138)
13     Gk = ss_fftr2dif_(gn);  % Recursively call the function ss_fftr2dif_()
14     Hk = ss_fftr2dif_(hn);  %   to compute half-length transforms
15     % Combine the two half-length transforms into a full-length
16     % transform as detailed in Eqn. (7.139)
17     Xk = [Gk.';Hk.'];
18     Xk = Xk(:);
19   end
20 end
```

Examine the code in lines 9 through 18 to confirm its correspondence to Eqns. (7.131) through (7.139). Also, make sure that you understand how lines 17 and 18 work in implementing Eqn. (7.139). This function has restrictions on its use that are similar to those of function `ss_fftr2dit_()` of the earlier exercise, and therefore should not be used directly. The wrapper function `ss_fftr2dif()` should be used instead, and is given below.

```
function Xk = ss_fftr2dif(xn)
  % Wrapper function for radix-2 decimation-in-time FFT.
  if (min(size(xn)) > 1)       % Check to make sure xn is a vector, not a matrix
    error('Input argument ''xn'' must be a vector.');
  else
    N = max(size(xn));
    if ~ss_ispowerof2(N)       % Length of xn must be a power of 2
      error('Length of ''xn'' must be a power of 2')
    end
```

```
    xvec = xn;                    % Set xvec equal to xn
    if isrow(xn)                  % If xn is in row format, transpose xvec to
      xvec = xn.';                %   satisfy the requirements of ss_fftr2dif_()
    end
    Xk = ss_fftr2dif_(xvec);      % Compute the FFT
    if isrow(xn)                  % If xn is in row format, transpose Xk to match it
      Xk = Xk.';                  %   so that Xk is in row format too
    end
  end
end
```

Software resource: `ss_fftr2dif.m`

MATLAB Exercise 7.19: Convolution in real-time processing using overlap-add method

In this exercise we will use the technique developed in Section 7.6 in processing an audio file using fast convolution while playing the resulting sound back simultaneously. Consider the script `mexdt_7_19.m` listed below.

```
% Script: mexdt_7_19.m
% Design a highpass filter to suppress low frequencies
Frq = [0,0.05,0.08,1];          % MATLAB-normalized critical frequencies
Mag = [0,0,1,1];                % Desired magnitude values
M = 27;
hn = firpm(M-1,Frq,Mag);        % Impulse response of the filter with M=27

% Create an "audio file reader" object
sReader = dsp.AudioFileReader('Ballad_22050_Hz.flac');
sReader.ReadRange = [1,441000];
% Create an "audio player" object
sPlayer = audioDeviceWriter('SampleRate',sReader.SampleRate);
L = sReader.SamplesPerFrame;    % Frame size
Hk = fft(hn',L+M-1);            % Zero pad h[n] to L+M+1 and compute FFT

buffer = zeros(M-1,2);          % Create length M-1 buffer vectors
while ~isDone(sReader)
  xn = sReader();               % Get next frame of data from the reader
  % Implement fast convolution using FFT
  Xk = fft(xn,L+M-1);           % Compute the FFT of zero-padded x[n]
  Yk = Hk.*Xk;                  % Multiply the two FFTs
  yn = real(ifft(Yk));          % Inverse FFT to find y[n]
  yn(1:M-1,:) = yn(1:M-1,:)+buffer;  % Add buffer to beginning of y[n]
  buffer = yn(L+1:L+M-1,:);          % Save last M-1 samples to buffer
  yn = yn(1:L,:);               % Truncate y[n] to L samples
  sPlayer(yn);                  % Play back the frame
end

release(sReader);  % We are finished with the input audio file
release(sPlayer);  % We are finished with the audio output device
```

The first step is to design a highpass filter with a finite-length impulse response $h[n]$. The filter is meant to block normalized frequencies less than $F < 0.025$. In terms of analog frequencies this corresponds to $f < 551.25$ Hz since the sampling rate of the audio file loaded is 22.05 kHz. The design of the highpass filter takes place in lines 3 through 6. The process used for designing the filter will be covered in more detail in Chapter 8.

Object oriented classes for working with audio files were covered in MATLAB exercises 1.11, 2.12, 2.13, and others in earlier chapters. We will work with the default frame size which happens to be $L = 1024$. For the FFT-based convolution result to be identical to linear convolution, the size of each transform needs to be at least $L + M - 1 = 1050$. In line 14, 1050-point FFT of the impulse response $h[n]$ is computed outside the processing loop. Inside the loop, fast convolution is implemented in lines 20 through 22.

Note that in line 16, `buffer` is set as a matrix with $M-1$ rows and two columns to accommodate left and right channels. In line 18, `xn` received from the object `sReader` is also a matrix with L rows and two columns. When the function `fft()` is applied to the matrix `xn` in line 20, it computes the FFT of each column. The result `Xk` has the same dimensions as `xn`. Element by element multiplication in line 21 is also done between `Hk` and each column of `Xk`.

Running the script causes the processed audio signal to be played back. Proper handling of overlapped output segments allows a seamless playback. Listen to both the original and the processed sounds, and pay attention to the difference the highpass filter creates.

Software resources: `mexdt_7_19.m` , `mexdt_7_19_Alt.m`

Problems

7.1. Compute the DFT's of the signals given below.

a. $x[n] = \{ \underset{\substack{\uparrow \\ n=0}}{1}, 1, 1 \}$

b. $x[n] = \{ \underset{\substack{\uparrow \\ n=0}}{1}, 1, 1, 0, 0 \}$

c. $x[n] = \{ \underset{\substack{\uparrow \\ n=0}}{1}, 1, 1, 0, 0, 0, 0 \}$

7.2. Compute the DFT's of the signals given below. Simplify the results and show that each transform is purely real. In each case, explain the cause of this behavior.

a. $x[n] = \{ \underset{\substack{\uparrow \\ n=0}}{1}, 1, 0, 0, 0, 0, 0, 1 \}$

b. $x[n] = \{ \underset{\substack{\uparrow \\ n=0}}{1}, 1, 1, 0, 0, 0, 1, 1 \}$

c. $x[n] = \{ \underset{\substack{\uparrow \\ n=0}}{1}, 1, 1, 1, 0, 1, 1, 1 \}$

7.3. Compute the DFT's of the signals given below. Simplify the results and show that each transform is purely imaginary. In each case, explain the cause of this behavior.

a. $x[n] = \{ \underset{\substack{\uparrow \\ n=0}}{0}, 1, 1, 0, 0, 0, -1, -1 \}$

b. $x[n] = \{ \underset{\substack{\uparrow \\ n=0}}{0}, 1, 2, -3, 0, 3, -2, -1 \}$

c. $x[n] = \{ \underset{\substack{\uparrow \\ n=0}}{0}, -1, 0, -1, 0, 1, 0, 1 \}$

7.4. Given that a function is available for computing the DFT of a length-N signal $x[n]$ as defined by Eqn. (7.7), we would like to devise a method of using the same function for computing the inverse DFT of a length-N transform $X[k]$. The function is to be treated as a closed box, that is, no internal modifications are possible. Explain how the inverse transform can be computed, and draw a block diagram.

7.5.

a. The DFT of a length-N signal is equal to its DTFT sampled at N equally spaced angular frequencies. In this problem we will explore the possibility of sampling a DTFT at fewer than N frequency values. Let $X(\Omega)$ be the DTFT of the length-12 signal

$$x[n] = u[n] - u[n-12]$$

Obtain $S[k]$ by sampling the transform $X(\Omega)$ at 10 equally spaced angular frequencies, that is,

$$S[k] = X\left(\frac{2\pi}{10}\right), \qquad k = 0, 1, \ldots, 9$$

Determine $s[n]$, the inverse DFT of $S[k]$.

b. Repeat part (a) with the length-15 signal

$$x[n] = u[n] - u[n-15]$$

$S[k]$ is still obtained by sampling the transform $X(\Omega)$ at 10 equally spaced angular frequencies. How does the result compare to that found in part (a)?

c. Consider the general case where $x[n]$ may be an infinitely long signal, and its DTFT is

$$X(\Omega) = \sum_{n=-\infty}^{\infty} x[n]\, e^{-j\Omega n}$$

$S[k]$ is obtained by sampling $X(\Omega)$ at M equally spaced angular frequencies:

$$S[k] = X\left(\frac{2\pi}{M}\right), \qquad k = 0, 1, \ldots, M-1$$

$s[n]$ is the length-M inverse DFT of $S[k]$. Show that $s[n]$ is related to $x[n]$ by

$$s[n] = \sum_{r=-\infty}^{\infty} x[n + rM], \qquad n = 0, 1, \ldots, M-1$$

7.6. In this problem we will explore a situation similar to the one in Problem 7.5, with the added assumption that the signal size N is an integer multiple of the DFT size M.

a. Let $x[n]$ be a length-N signal with its DTFT computed as

$$X(\Omega) = \sum_{n=0}^{N-1} x[n]\, e^{-j\Omega n}$$

Let $N = 2M$. A length-M transform sequence $S[k]$ is found by sampling $X(\Omega)$ at M equally-spaced angular frequencies as

$$S[k] = X\left(\frac{2\pi k}{M}\right), \quad k = 0, 1, \ldots, M-1$$

Express the length-M signal $s[n]$ the DFT of which is $S[k]$, in terms of the original signal $x[n]$.

b. Repeat part (a) with the assumption that the length of $x[n]$ is $N = 3M$, and $S[k]$ is found by sampling the DTFT at M equally-spaced angular frequencies.

c. Let $x[n]$ be a length-15 signal. We need the value of its DTFT at angular frequencies $\Omega_a = 2\pi/5$ and $\Omega_b = 4\pi/5$. Utilizing the results obtained in parts (a) and (b), devise a method of computing $X(\Omega_a)$ and $X(\Omega_b)$ with the smallest size DFT possible. Hint: First, determine what the minimum DFT size M is for solving this problem. Afterward, find a way to obtain a length-M sequence $s[n]$ the DFT of which would contain the transform values at desired frequencies.

7.7. Let $x[n]$ be a length-N signal. It is padded with M zero-amplitude samples, and its $(N+M)$-point DFT $X[k]$ is computed. For each combination of parameters below, express the angular frequencies Ω_k to which the transform samples correspond.

a. $N = 35, \quad M = 93$

b. $N = 40, \quad M = 60$

c. $N = 40, \quad M = 472$

7.8. Consider again the scenario in Problem 7.7 where a length-N signal $x[n]$ was padded with M zero-amplitude samples, and its $(N+M)$-point DFT $X[k]$ was computed. This time we will make one change: We will take the second half of the DFT sequence, and place it in front of the first half to bring $\Omega = 0$ into the middle. Repeat Problem 7.7 with this change.

7.9. Consider following multitone signals that each have two sinusoidal components. We want to observe 32 samples of each for $n = 0, 1, \ldots, 31$, and then compute a 512-point DFT for detecting the frequencies contained in the signal. For each signal, speculate whether the DFT would indicate two distinct sinusoidal components or not. Explain your reasoning in each case.

a. $x_a[n] = \cos(0.9817n) + \cos(1.0308n)$

b. $x_b[n] = \cos(0.9817n) + \cos(1.0799n)$

c. $x_c[n] = \cos(0.9817n) + \cos(1.2763n)$

d. $x_d[n] = \cos(0.9817n) + 0.05\cos(1.2763n)$

e. $x_e[n] = \cos(0.9817n) + 0.05\cos(1.6690n)$

7.10. Refer to Problem 7.9. In terms of the ability to detect two distinct sinusoids, which, if any, of the multitone signals would benefit from the use of a Hamming window in place of a rectangular window? Explain your reasoning.

7.11. Rectangular window sequence was discussed in Section 7.2.3, and its characteristics were discussed in the context of truncating infinite-length signals. Length-N rectangular window sequence is defined as

$$w_R[n] = \begin{cases} 1\,, & n = 0, 1, \ldots, N-1 \\ 0\,, & \text{otherwise} \end{cases}$$

a. Determine the spectrum $W_R(\Omega)$.

b. For $N >> 1$, determine the width of the main lobe. Is the width of the main lobe dependent on N? Hint: Find the zero crossings of the numerator term in the expression for $W_R(\Omega)$.

c. Determine the angular frequency for which the peak of the first sidelobe is observed. Compute the approximate peak magnitude of the first sidelobe. Hint: Find the locations of peaks for the numerator term in the expression for $W_R(\Omega)$. Also remember that, for $\alpha << 1$, $\sin(\alpha) \approx \alpha$.

d. Determine the dB difference between the peak of the main lobe and the peak of the first sidelobe. Is the dB difference dependent on N?

e. Determine the angular frequencies for the peaks of second and third sidelobes. Afterward compute approximate peak magnitudes of second and third sidelobes.

7.12. Rectangular window sequence has the narrowest main lobe for a given length N among all window functions, however, its sidelobe behavior may be problematic for detecting weak sinusoids. Alternative window functions are sometimes used. Two popular alternative window sequences are the von Hann window (also referred to as Hanning window) and the Hamming window.

a. The von Hann (or Hanning) window is defined as

$$w[n] = \begin{cases} 0.5 - 0.5\cos\left(\dfrac{2\pi n}{N-1}\right), & n = 0, 1, \ldots, N-1 \\ 0\,, & \text{otherwise} \end{cases}$$

Recall that the spectrum of the rectangular window sequence was given in Eqn. (7.27). Using linearity and modulation properties of the DTFT, express the spectrum of the von Hann window in terms of the spectrum of the rectangular window.

b. The Hamming window is defined as

$$w[n] = \begin{cases} 0.54 - 0.46\cos\left(\dfrac{2\pi n}{N-1}\right), & n = 0, 1, \ldots, N-1 \\ 0\,, & \text{otherwise} \end{cases}$$

Repeat the requirements of part (a) for this case.

7.13. The two window sequences we explored in Problem 7.12, namely von Hann and Hamming windows, are both raised-cosine type functions. They can be expressed using the common format

$$w[n] = \begin{cases} \alpha - (1-\alpha)\cos\left(\dfrac{2\pi n}{N-1}\right), & n = 0, 1, \ldots, N-1 \\ 0\,, & \text{otherwise} \end{cases}$$

Obviously, choosing $\alpha = 0.5$ leads to the von Hann window, whereas the Hamming window can be obtained with $\alpha = 0.54$.

a. Leave α as a parameter. Use linearity and modulation properties of the DTFT to express the spectrum of the general raised-cosine window in terms of the spectrum of the rectangular window.

b. Using the result found in part (a), express the approximate peak magnitude of the first sidelobe as a function of parameter α. You may find that some of the answers found in Problem 7.11 are useful here.

c. Is there a value of α that makes the peak magnitude of the first sidelobe equal zero? If so, what is that value? Which of the two raised-cosine type windows seems optimum in that sense?

7.14. Refer to the matrix formulation of the DFT explored in Section 7.2.7. Using Eqns. 7.44 and 7.45, construct the matrix **A** for the 6-point DFT. Afterward, use it to compute the DFT of the signal

$$x[n] = \{ \underset{\substack{\uparrow \\ n=0}}{2}, 6, -3, -4, 1, 7 \}$$

7.15. For each finite-length signal listed below, find the specified circularly shifted version.

a. $x[n] = \{ \underset{\substack{\uparrow \\ n=0}}{4}, 3, 2, 1 \}$ find $x[n-2]_{\text{mod } 4}$

b. $x[n] = \{ \underset{\substack{\uparrow \\ n=0}}{1}, 1, 1, 0, 0 \}$ find $x[n-4]_{\text{mod } 5}$

c. $x[n] = \{ \underset{\substack{\uparrow \\ n=0}}{1}, 4, 2, 3, 1, -2, -3, 1 \}$ find $x[-n]_{\text{mod } 8}$

d. $x[n] = \{ \underset{\substack{\uparrow \\ n=0}}{1}, 4, 2, 3, 1, -2, -3, 1 \}$ find $x[-n+2]_{\text{mod } 8}$

7.16. Refer to Problem 7.15. Find the circularly conjugate symmetric and antisymmetric components $x_E[n]$ and $x_O[n]$ for each signal listed.

7.17. Consider the finite-length signal

$$x[n] = \{ \underset{\substack{\uparrow \\ n=0}}{1}, 1, 1, 0, 0 \}$$

a. Compute the 5-point DFT $X[k]$ for $k = 0, \ldots, 4$.

b. Multiply the DFT found in part (a) with $e^{-j2\pi k/5}$ to obtain the product

$$R[k] = e^{-j2\pi k/5}\, X[k] \qquad \text{for } n = 0, \ldots, 4$$

c. Compute the signal $r[n]$ as the inverse DFT of $R[k]$ and compare to the original signal $x[n]$.

d. Repeat parts (b) and (c) with

$$S[k] = e^{-j4\pi k/5}\, X[k] \qquad \text{for } n = 0, \ldots, 4$$

e. Provide justification for the answers found using the properties of the DFT.

7.18. Consider the finite-length signal

$$x[n] = \{ \underset{\substack{\uparrow \\ n=0}}{5}, 4, 3, 2, 1 \}$$

a. Compute the 5-point DFT $X[k]$ for $k = 0,\ldots,4$.

b. Reverse the DFT found in part (a) to obtain a new transform $R[k]$ as

$$R[k] = X[-k]_{\text{mod }5} \qquad \text{for } n = 0,\ldots,4$$

c. Compute the signal $r[n]$ as the inverse DFT of $R[k]$ and compare to the original signal $x[n]$. Justify the answer found using the properties of the DFT.

7.19. The DFT of a length-6 signal $x[n]$ is given by

$$X[k] = \{ \underset{\underset{k=0}{\uparrow}}{(2+j3)}, (1+j5), (-2+j4), (-1-j3), (2), (3+j1) \}$$

Without computing $x[n]$ first, determine

a. The DFT of $x_r[n]$, the real part of $x[n]$,

b. The DFT of $x_i[n]$, the imaginary part of $x[n]$.

Explain your reasoning for each part.

7.20. Consider two length-4 signals

$$g[n] = \{ \underset{\underset{n=0}{\uparrow}}{6}, 3, -4, -3 \}$$

$$h[n] = \{ \underset{\underset{n=0}{\uparrow}}{2}, -7, -5, 6 \}$$

Construct a complex signal $x[n] = g[n] + j\,h[n]$. Manually compute its DFT $X[k]$. Afterward obtain the DFT's $G[k]$ and $H[k]$ from $X[k]$ using the appropriate symmetry properties.

7.21. Two signals $x[n]$ and $h[n]$ are given by

$$x[n] = \{ \underset{\underset{n=0}{\uparrow}}{2}, -3, 4, 1, 6 \}$$

$$h[n] = \{ \underset{\underset{n=0}{\uparrow}}{1}, 1, 1, 0, 0 \}$$

a. Compute the circular convolution $y[n] = x[n] \otimes h[n]$ through direct application of the circular convolution sum.

b. Compute the 5-point transforms $X[k]$ and $H[k]$.

c. Compute $Y[k] = X[k]\,H[k]$, and the obtain $y[n]$ as the inverse DFT of $Y[k]$. Verify that the same result is obtained as in part (a).

7.22. Refer to the signals $x[n]$ and $h[n]$ in Problem 7.21. Let $X[k]$ and $H[k]$ represent 7-point DFT's of these signals, and let $Y[k]$ be their product, that is, $Y[k] = X[k]\,H[k]$ for $n = 0,\ldots,6$. The signal $y[n]$ is the inverse DFT of $Y[k]$. Determine $y[n]$ without actually computing any DFT's. Explain your reasoning in detail.

7.23. An analog signal $x_a(t)$ is sampled with a sampling rate of $f_s = 10$ kHz to produce a discrete-time signal $x[n]$. Consecutive segments that are each 100 ms long are received from this signal for analysis. We are interested in detecting the presence of two frequencies, namely $f_1 = 600$ Hz and $f_2 = 1700$ Hz.

a. Explain how Goertzel algorithm can be used for this purpose.
b. Using Fig. 7.24 as a building block, develop a block diagram for a detector. Specify all relevant values in the diagram. Explain how the outputs can be used for detecting the presence of the two frequencies of interest.

7.24. Let $x[n]$ be a real-valued length-N signal where N is an odd number. Modify Eqns. (7.111) and (7.112) so that the DFT $X[k]$ can be computed using about N^2 real multiplications.

7.25. A length-8 real signal $x[n]$ is given by

$$x[n] = \{ \underset{\substack{\uparrow \\ n=0}}{3.7}, 2.4, -1.8, -2.7, 3.2, 4.3, -1.1, -0.5 \}$$

Use Eqn. (7.111) to compute the DFT of this signal for $N = 0, \ldots, 4$. Afterward, obtain the remaining samples of $X[k]$ using the symmetry properties of the DFT.

7.26. A length-7 real-valued signal $x[n]$ is given by

$$x[n] = \{ \underset{\substack{\uparrow \\ n=0}}{3.7}, 2.4, -1.8, -2.7, 3.2, 4.3, -1.1 \}$$

Use Eqn. (7.111) to compute the DFT of this signal for $N = 0, \ldots, 3$. Afterwards, obtain the remaining samples of $X[k]$ using the symmetry properties of the DFT.

7.27. A length-4 signal $x[n]$ is given by

$$x[n] = \{ \underset{\substack{\uparrow \\ n=0}}{3.7}, 2.4, -1.8, -2.7 \}$$

Compute the DFT $X[k]$ of this signal using the radix-2 decimation-in-time technique outlined in Section 7.5.1. Show each step involved in the computation.

7.28. A length-4 signal $x[n]$ is given by

$$x[n] = \{ \underset{\substack{\uparrow \\ n=0}}{3.7}, 2.4, -1.8, -2.7 \}$$

Compute the DFT $X[k]$ of this signal using the radix-2 decimation-in-frequency technique outlined in Section 7.5.5. Show each step involved in the computation.

7.29. A length-9 signal $x[n]$ is given by

$$x[n] = \{ \underset{\substack{\uparrow \\ n=0}}{3.7}, 2.4, -1.8, -2.7, 3.2, 4.3, -1.1, -0.5 - 3.2 \}$$

Compute the FFT $X[k]$ of this signal using the radix-3 decimation-in-time technique outlined in Section 7.5.4. Show each step involved in the computation.

MATLAB Problems

7.30. Refer to Problem 7.4 where we explored the idea of using the same function for computing both forward and inverse DFT. Write a script to perform the following steps:

a. Generate a length-128 real-valued sequence $x[n]$ using the function `randn()`.

b. Compute the 128-point DFT $X[k]$ of the signal $x[n]$ using the function `ss_dft()` that was developed in MATLAB Exercise 7.1.

c. Again using the function `ss_dft()`, compute the inverse DFT of $X[k]$. Make the necessary adjustments before and after the call to the function. Check the result for accuracy by comparing it to the original signal generated in part (a).

7.31. In Problem 7.6, the idea of sampling a DTFT at fewer points than the length of the signal was explored. Develop a MATLAB script to perform the following steps:

a. Generate a length-32 signal $x[n]$ with random amplitudes. Compute its 32-point DFT $X[k]$.

b. Obtain a length-16 sequence $S[k]$ by taking every other sample of $X[k]$. Note that this produces the same result as sampling the DTFT of $x[n]$ at 16 points. Find the inverse transform $s[n]$.

c. Compare signals $x[n]$ and $s[n]$, and verify their relationship found in Problem 7.6.

7.32. In part (c) of Problem 7.6, a method was devised for computing the DTFT of a length-15 signal at frequencies $\Omega_a = 2\pi/5$ and $\Omega_b = 4\pi/5$ using a smaller size DFT.

a. Write a script to test the method devised in Problem 7.6. Your script should begin by generating a length-15 signal $x[n]$ with random amplitudes, and computing its 15-point DFT. Determine which two samples of the 15-point DFT correspond to the frequencies Ω_a and Ω_b.

b. In your script, apply the method of Problem 7.6 to compute the values $X(\Omega_a)$ and $X(\Omega_b)$ using the smallest possible DFT. Let's assume the small size DFT is length-M. Identify the samples of the M-point DFT that correspond to the frequencies Ω_a and Ω_b. Verify that the transform values of interest are identical in both cases.

7.33. A continuous-time signal $x_a(t)$ is observed in terms of segments that are each 10.5 ms long. The signal is sampled with a sampling rate of $f_s = 10$ kHz to create a length-105 discrete-time signal $x_i[n]$ for each segment. We are only interested in detecting possible presence of two frequencies in the signal, namely $f_1 = 2$ kHz and $f_2 = 4$ kHz. We would like to accomplish that in the most computationally efficient way possible, using the smallest size DFT we can employ.

a. Determine M, the size of the smallest DFT that would include the two frequencies of interest.

b. Using the results obtained in Problems 7.5 and 7.6, devise a method of converting $x[n]$ to a length-M signal $s[n]$ so that a M-point DFT provide the desired result.

c. The datafile `Prob_7_33.mat` contains five different segments from the signal. Write a script that would test each one using the technique of part (b), and compute the values of the transform at the frequencies of interest. You may want to write a local function to convert $x_i[n]$ to $s_i[n]$ for each segment and compute the DFT of $s_i[n]$.

d. Your script should also compute a 105-point DFT on each segment data $x_i[n]$ for the purpose of verifying the results found in part (c).

7.34. In Section 7.2.3 the issue of spectral leakage was discussed. Consider part (a) of Fig. 7.6 which represents the transform of the length-50 signal

$$x_1[n] = \cos(0.2\pi n) , \quad n = 0, 1, \ldots, N-1$$

The transform clearly exhibits two peaks with no spectral leakage visible.

a. Write a MATLAB script to generate the signal $x_1[n]$. Compute the 100-point DFT of $x1[n]$. Recall that this is equivalent to zero padding the signal before computing its DFT. Graph the magnitude of this transform. Is spectral leakage visible now? Explain why.

b. Repeat part (a) using a 200-point DFT.

7.35. The use of the DFT for computing approximate values of the exponential Fourier series (EFS) coefficients of a continuous-time signal was discussed in Section 7.2.5. In MATLAB Exercise 7.10, the function `ss_efsapprox()` was developed for this purpose.

Consider the half-wave rectified sinusoidal signal $\tilde{x}_a(t)$ with period T_0 defined by

$$\tilde{x}(t) = \begin{cases} \sin(\omega_0 t), & 0 \le t < T_0/2 \\ 0, & T_0/2 \le t < T_0 \end{cases} \qquad \text{and} \quad \tilde{x}(t+T_0) = \tilde{x}(t)$$

where $\omega_0 = 2\pi/T_0$. The signal is illustrated in Fig. P.7.35. It can be shown that the EFS coefficients of this signal are

$$c_k = \begin{cases} 0, & k \text{ odd and } k \ne \mp 1 \\ -j/4, & k = 1 \\ j/4, & k = -1 \\ \dfrac{-1}{\pi\left(k^2-1\right)}, & k \text{ even} \end{cases}$$

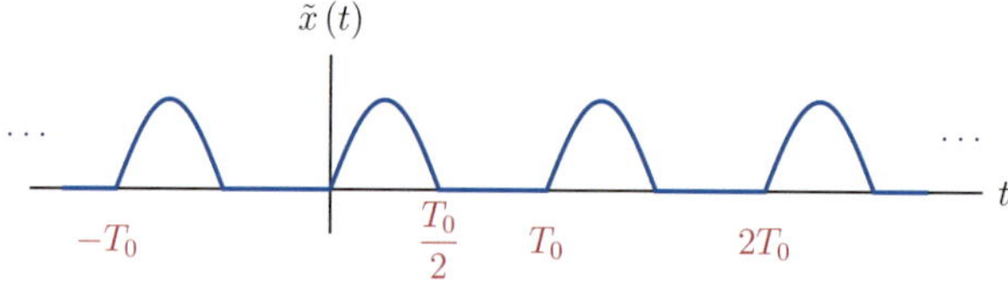

Figure P. 7.35

Develop a MATLAB script to implement the following steps:

a. Generate 250 samples of one period of the signal $\tilde{x}_a(t)$. Use a period length of $T_0 = 1$ s.

b. Using the function `ss_efsapprox()`, compute approximate EFS coefficients $\tilde{c}_k$ for $k = -15,\ldots,15$.

c. Compare the approximate coefficients to their correct values, and determine the percent error for each. Note that the coefficients are complex in general, and their magnitudes need to be used in comparisons.

d. Reconstruct a signal from the approximate coefficient values computed. Graph it for $0 \le t \le 4$ s.

7.36. Matrix formulation of the DFT was discussed in Section 7.2.7. With the signal $x[n]$ represented by a column vector $\mathbf{x}$ of size N, the transform can be obtained in the form of a column vector $\mathbf{X}$ of size N as

$$\mathbf{X} = \mathbf{A}\mathbf{x}$$

where $\mathbf{A}$ is a $N \times N$ matrix of exponential factors in which the element at row-p, column-r is

$$a_{pr} = W_N^{(p-1)(r-1)} = e^{-j\frac{2\pi}{N}(p-1)(r-1)}$$

For an N-point DFT, constructing the matrix **A** this way would require the computation N^2 exponential terms. On the other hand, if we make use of the periodicity of W_N, there can only be N unique values for the elements a_{pr} of the matrix **A**. General term a_{pr} can be written as

$$a_{pr} = W_N^{[(p-1)(r-1)]_{\text{mod } N}}$$

Develop a MATLAB function `ss_dftmatrix()` that takes a single argument N for the size of the DFT, and returns the coefficient matrix **A** for implementing it. In order to make the function as computationally efficient as possible, follow the guidelines given below:

a. Create a vector **W** to hold the N unique exponential factors:

$$\mathbf{W} = [\ W_N^0 \quad W_N^1 \quad W_N^2 \quad \dots \quad W_N^{(N-1)}\ ]$$

b. Create an index matrix **L** as

$$\mathbf{L} = \left[\ell_{pr}\right]_{N\times N}$$

where row-p and column-r element is

$$\ell_{pr} = \left[(p-1)(r-1)\right]_{\text{mod } N} + 1$$

Thus, each element of the matrix **L** is an index into the appropriate exponential factor in vector **W**. Avoid the use of any looping structures in creating the matrix **L**.

c. Construct the matrix **A** by indexing the vector **W** with the matrix **L**. As before, avoid the use of any looping structures.

d. Test the function `ss_dftmatrix()` by computing a 100-point DFT in matrix form and comparing it to the result returned by the function `fft()`.

7.37. Refer to the discussion in Section 7.2.7 about matrix formulation of the DFT. Recall that the coefficient matrix **A** was determined to be

$$\mathbf{A} = \left[a_{pr}\right]_{N\times N}$$

where the element at row-p, column-r is

$$a_{pr} = W_N^{(p-1)(r-1)} = e^{-j\frac{2\pi}{N}(p-1)(r-1)}$$

a. Show that, column r of the matrix **A** is the same as the N-point DFT of $\delta[n-r+1]$.

b. Also show that, based on part (a), the DFT matrix **A** can be constructed by computing the transforms of $\delta[n]$, $\delta[n-1]$, ..., $\delta[n-N+1]$ as column vectors, and then placing them side by side to form a $N \times N$ matrix.

c. Consider the built-in function `fft()`. If it is called with a matrix as its input argument, it computes the transform of each column, and returns a matrix. By this logic, if we use a $N \times N$ identity matrix with the function `fft()`, we should get the DFT matrix **A**. Write a MATLAB script to test this idea. Use $N = 10$ to compute the 10×10 DFT matrix, and check it for accuracy.

7.38. Refer to Problem 7.23. A test signal fitting the description of $x[n]$ in that problem is provided in audio file `Prob_7_38.flac`. It was sampled at $f_s = 10$ kHz, and it contains the two frequencies, 600 Hz and 1700 Hz, turned on and off at random. Develop a script for processing this signal on a frame-by-frame basis. Your script should perform the following steps:

a. Use audio objects to facilitate frame-by-frame processing. Set the frame size appropriately to obtain 100 ms of data in each frame.

b. Create a stem plot with just two samples, showing the strength of the two frequency components at 600 Hz and 1700 Hz. You may want to review part (d) of MATLAB Exercise 1.11 of Chapter 1 for an example of how this can be done.

c. Within the processing loop, play back each frame. Use the function `ss_goertzel2()` with the appropriate parameters to detect the two frequencies of interest. Update the stem plot created in part (b) with the results from Goertzel algorithm.

d. Run the code and make sure the updates to the graph are consistent with what you hear.

7.39. In this problem we will explore the numerical efficiency of the DFT as defined by Eqn. (7.7) versus that of the radix-2 FFT by measuring the time it takes to compute each in MATLAB. An easy way to measure elapsed time in MATLAB is to use the functions `tic` and `toc`. The former starts a timer, and the latter returns the time elapsed since the timer was started. By inserting some code between the calls to these two functions, we can get an idea of how much time it took to execute the code. For example, the lines

```
tic
Xk = ss_dft(xn);
time1 = toc
```

measure the time to compute the DFT. The elapsed time is saved in the variable `time1`. Write a script to perform the following steps:

a. Generate a length-4096 random complex sequence $x[n]$. Use the function `randn()` for real and imaginary parts.

b. Between calls to functions `tic` and `toc`, compute the DFT of $x[n]$ 10 times. Record the elapsed time at the end.

c. Repeat part (b), this time using the function `ss_fftr2dit()` instead of `ss_dft()`.

d. Compute the ratio of elapsed time periods obtained in parts (b) and (c). How does it compare to the expected ratio in terms of computational complexities of DFT vs. FFT? Keep in mind that the result may be a bit skewed due to the recursive nature of the latter function and the overhead associated with it.

e. Repeat steps (b) through (d) above to compare the function `ss_dft()` against the built-in function `fft()`. The results may be further skewed in this comparison since the function `fft()` is compiled, and the function `ss_dft()` is interpreted.

7.40. Refer to the function `ss_conv2()` that was developed in MATLAB Exercise 7.9. It was noted that the function may fail under certain circumstances.

a. Using MATLAB, generate vectors `x` and `h` representing a length-5 signal $x[n]$ and a length-8 signal $h[n]$. Determine the linear convolution $y[n] = x[n] * h[n]$ by using the function `ss_conv2()` with both vectors in row orientation. Print the result and comment.

b. Repeat part (a) with both vectors `x` and `h` in column orientation.

c. Now try the function with `x` in column orientation and `h` in row orientation. What happens? Can you explain the reason for the failure? You may want to trace your steps by typing each line of the function in the command window and observing the intermediate results.

d. Fix the function by adding the necessary validation code (see MATLAB Exercise 7.1 for an example of this). First, ensure that both arguments are vectors, and not matrices. Afterward, if the two vectors have different orientations, change the orientation of `h` to match that of `x`.

MATLAB Projects

7.41. This problem is about increasing the efficiency of DFT computation by recognizing the symmetry properties of the DFT and using them to our advantage. Recall from the symmetry properties discussed in Section 7.3 that for a signal expressed in Cartesian complex form

$$x[n] = x_r[n] + j\,x_i[n]$$

and its transform expressed using circularly conjugate symmetric and antisymmetric components as

$$X[k] = X_E[k] + X_O[k]$$

The following relationships exist:

$$x_r[n] \overset{\text{DFT}}{\longleftrightarrow} X_E[k]$$

$$j\,x_i[n] \overset{\text{DFT}}{\longleftrightarrow} X_O[k]$$

Develop a MATLAB script to compute the DFT's of two real-valued length-N signals $r[n]$ and $s[n]$ with one call to the function `fft()` using the following approach:

- **a.** Construct a complex signal $x[n] = r[n] + j\,s[n]$.
- **b.** Compute the DFT of the complex signal $x[n]$ using the function `fft()`.
- **c.** Find the DFT's of the signals $r[n]$ and $s[n]$ by splitting the transform $X[k]$ into its two components $X_E[k]$ and $X_O[k]$ and making the necessary adjustments. You may wish to use the function `ss_crev()` developed in MATLAB Exercise 7.6.

Test your script using two 128-point signals with random amplitudes. (Use MATLAB function `randn()` to generate them.) Compare the transforms for $R[k]$ and $S[k]$ to results obtained by direct application of the function `fft()` to signals $r[n]$ and $s[n]$.

7.42. One of the popular uses of the DFT is for analyzing the frequency spectrum of a continuous-time signal that has been sampled. The theory of sampling is covered in detail in Chapter 4. In this problem we will focus on computing and graphing the approximate spectrum of a recorded audio signal using the technique discussed in Section 7.2.6.

Provided that a computer with a microphone and a sound processor is available, the following MATLAB code allows 3 seconds of audio signal to be recorded, and a vector `x` to be created with the recording:

```
hRec = audiorecorder;
disp('Press a key to start recording');
pause;
recordblocking(hRec, 3);
disp('Finished recording');
x = getaudiodata(hRec);
```

By default the analog signal $x_a(t)$ captured by the microphone and the sound device is sampled at the rate of 8000 times per second, corresponding to a sampling interval of $T = 125$ μs. For a 3-second recording the vector `x` contains 24,000 samples that represent

$$x[n] = x_a\left(\frac{n}{8000}\right), \qquad n = 0, \ldots, 23999$$

Develop a MATLAB script to perform the following steps to display the approximate spectrum of a speech signal $x_a(t)$:

a. Extract 1024 samples of the vector $x[n]$ into a new vector.

$$r[n] = x[n+8000]\,, \qquad n = 0,\ldots,1023$$

We skip the first 8000 samples so that we do not get a blank period before the person begins to speak.

b. Compute the DFT $R[k]$ of the signal $r[n]$. Scale it as outlined in Section 7.2.6 so that it can be used for approximating the continuous Fourier transform of $x_a(t)$. Also create an approximate vector `f` of frequencies in Hz for use in graphing the approximate spectrum $X_a(f)$.

c. Compute and graph the dB magnitude of the approximate spectrum as a function of frequency.

d. Record your own voice and use it to test the script.

7.43. In this project our focus will be to compare different window sequences in terms of their frequency spectra. Following window sequences will be compared:

Rectangular window:
$$w[n] = \begin{cases} 1\,, & n = 0,1,\ldots,N-1 \\ 0\,, & \text{otherwise} \end{cases}$$

von Hann window:
$$w[n] = \begin{cases} 0.5 - 0.5\cos\left(\dfrac{2\pi n}{N-1}\right), & n = 0,1,\ldots,N-1 \\ 0\,, & \text{otherwise} \end{cases}$$

Hamming window:
$$w[n] = \begin{cases} 0.54 - 0.46\cos\left(\dfrac{2\pi n}{N-1}\right), & n = 0,1,\ldots,N-1 \\ 0\,, & \text{otherwise} \end{cases}$$

Develop a MATLAB script to do the following:

a. Generate samples of each window sequence for $n = 0,1,\ldots,39$.

b. Compute the spectrum of each window sequence through the use of a 512-point FFT. Graph the magnitude of each spectrum.

c. Compute dB magnitude of each spectrum with the peak value normalized to 0 dB. For a spectrum $W(\Omega)$, the normalized dB magnitude is

$$|W(\Omega)|_{dB} = 20\log_{10}\left(\frac{|W(\Omega)|}{\max\{|W(\Omega)|\}}\right)$$

where $\max\{|W(\Omega)|\}$ is the largest magnitude in the spectrum. Graph normalized dB magnitudes of all three window sequences on the same coordinate system. Comment on the results.

7.44.

a. Using the development outlined in MATLAB Exercise 7.17, write a function to compute the DFT of a signal using the radix-3 decimation-in-time algorithm. The computational function should be called `ss_fftr3dit_()`. It should take the vector `xn` with sample amplitudes of the signal $x[n]$ in column format, and return the transform vector `Xk`, also in column format. Obviously, the number of samples in the vector `xn` needs to be a power of 3. Similar to what was done in MATLAB Exercise 7.17, the function `fftr3dit_()` does not do any validation or error checking. It is designed for use only by a wrapper function.

b. Afterward, develop a wrapper function `ss_fftr3dit()` that checks the size and the orientation of the input argument `xn`. Quit if the length of vector `xn` is not a power of 3. If `xn` is in row format, create an appropriate copy of it in column format. Call `ss_fftr3dit_()` in a safe manner. This ensures that argument `xn` for the wrapper function could be in either row or column format. The orientation of the returned output vector `Xk` should match that of `xn`.

c. Write a script to test the function `ss_fftr3dit()`. Generate a length-81 vector `xn` with random amplitude values. Built-in function `rand()` can be used for this purpose. Call the wrapper function with both row-format and column-format versions of this vector. Compare numerical results against the results obtained using the function `fft()`.

CHAPTER 8

ANALYSIS AND DESIGN OF DISCRETE-TIME FILTERS

Chapter Objectives

- Develop the concept of discrete-time filters. Learn similarities and differences between IIR and FIR filters, and reasons for choosing one over the other depending on needs and circumstances.
- Understand how discrete-time filters can be used for processing continuous-time signals through sampling and reconstruction operations. Explore the relationships between analog frequencies for a continuous-time signal versus normalized or angular frequencies for a discrete-time signal.
- Learn pole-zero placement design methods for resonant bandpass filters and notch filters. Understand how to control the bandwidth of each type of filter by careful selection of the pole radius.
- Study the design of IIR filters from analog prototype filters. Briefly review basic analog filter approximation formulas. Understand ways an analog filter can be converted to discrete-time. Learn impulse invariance and bilinear transformation methods.
- Understand the significance of linear phase in FIR filters. Learn the types of symmetries that can be utilized for obtaining linear phase.
- Learn how to design linear-phase FIR filters using various techniques such as Fourier series method with windows, frequency sampling design, least-squares design, and Parks McClellan method.

DOI: 10.1201/9781003570462-8

8.1 Introduction

In many signal processing applications the need arises to change the strength, or the relative significance, of various frequency components in a given signal. Sometimes we may need to eliminate certain frequency components; at other times we may need to boost the strength of a range of frequencies over others. This act of changing the relative amplitudes of frequency components in a signal is referred to as *filtering*, and the system that facilitates this is referred to as a *filter*.

In Chapters 3 and 5 we have developed techniques for transform-domain analysis of DTLTI systems. Specifically, the relationship between the input and the output signals of a DTLTI system is expressed in the transform domain as

$$Y(\Omega) = H(\Omega)\, X(\Omega) \tag{8.1}$$

in terms of the Fourier transforms of the signals involved, or as

$$Y(z) = H(z)\, X(z) \tag{8.2}$$

using the z-transforms of the signals. In either case, the system function $H(\Omega)$ or $H(z)$ serves as a multiplier function that shapes the spectrum of the input signal to create the spectrum of the output signal. In a general sense any DTLTI system can be taken as a filter. A number of well known filter types have already been explored in previous chapters. In Chapter 2 we discussed moving average filters and exponential smoothers. Feed-forward and feedback comb filters were introduced in MATLAB exercises of Chapter 3, and were discussed in more detail in Chapter 5. All-pass filters were also detailed in Chapter 5.

Discrete-time filters are viewed under two broad categories: *infinite impulse response (IIR)* filters, and *finite impulse response (FIR)* filters. The system function of an IIR filter has both poles and zeros. The corresponding difference equation has feedback terms, that is, the current output sample depends on one or more past samples of the output signal in addition to current and past samples of the input signal. Consequently, the impulse response of the filter is of infinite length. It should be noted, however, that the impulse response of a stable IIR filter must also be absolute summable, and must therefore decay over time. In contrast, the behavior of an FIR filter is controlled only by the placement of the zeros of its system function. For causal FIR filters all of the poles are at the origin, and they do not contribute to the magnitude characteristics of the filter. The difference equation of an FIR filter has no feedback terms, that is, the current output sample depends only on current and past samples of the input signal. The resulting impulse response is of finite length.

For a given set of specifications, an IIR filter is generally more efficient than a comparable FIR filter. On the other hand, FIR filters are always stable. Additionally, a linear phase characteristic is possible with FIR filters whereas causal and stable IIR filters cannot have linear phase. The significance of a linear phase characteristic is that the time-delay is constant for all frequencies. Linear phase is desirable in some applications, and requires the use of a FIR filter.

In this chapter we will formalize the design methods for frequency selective discrete-time filters. First, two simple filter types, namely bandpass resonant filters and notch filters, will be considered in Section 8.3. These two types of filters can be designed through pole-zero placement, i.e., choosing locations of poles and zeros of the system function. Section 8.4 will focus on the design of IIR filters from analog prototypes through the application of one of several conversion techniques. Section 8.5 will cover several design methods for obtaining FIR filters with desired behavior. Implementation methods for the designed IIR and FIR filters will be covered through the MATLAB exercises at the end of the chapter.

8.2 Discrete-Time Processing of Continuous-Time Signals

Even though we are discussing analysis and design techniques for discrete-time filters in this chapter, the most common use of discrete-time filters is in processing continuous time signals. This is illustrated in Fig. 8.1. The analog signal $x_a(t)$ is converted to a discrete-time signal $x[n]$ which is then processed by the DTLTI system. The result of that processing is another discrete-time signal $y[n]$ which is converted back to continuous-time to yield the output signal $y_a(t)$. The cascade combination of the three components, namely the continuous-to-discrete converter, the DTLTI system, and the discrete-to-continuous converter, is equivalent to a CTLTI system with system function $G(s)$.

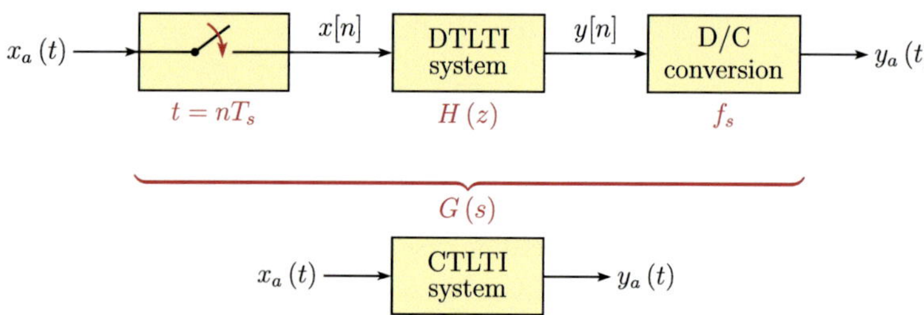

Figure 8.1 – Discrete-time processing of a continuous-time signal through sampling and reconstruction.

In a practical application, some of the parameters may be dictated by external factors. For example, the sampling rate f_s may be chosen based on the bandwidth of the input signal $x_a(t)$, hardware available for implementation, restrictions of the Nyquist sampling theorem, and the level of processing quality desired. A fundamental question that needs to be resolved is the relationship between $G(s)$ and $H(z)$. Often we either know or can figure out what kind of CTLTI system function $G(s)$ is needed. Then the problem becomes one of finding a DTLTI system function $H(z)$ so that the overall system behavior is as similar to an equivalent CTLTI system as possible.

Let us begin by considering a simple sinusoidal signal

$$x_a(t) = \cos\left(2\pi f_0 t\right)$$

Sampling $x_a(t)$ at time instants $t = nT_s$ leads to the discrete-time signal

$$x[n] = x_a(nT_s) = \cos\left(2\pi f_0 n T_s\right)$$

Since $f_s = 1/T_s$, we have

$$x[n] = \cos\left(2\pi\left(\frac{f_0}{f_s}\right)n\right) = \cos\left(2\pi F_0 n\right)$$

Basic relationships between various forms of frequencies can be summarized as follows:

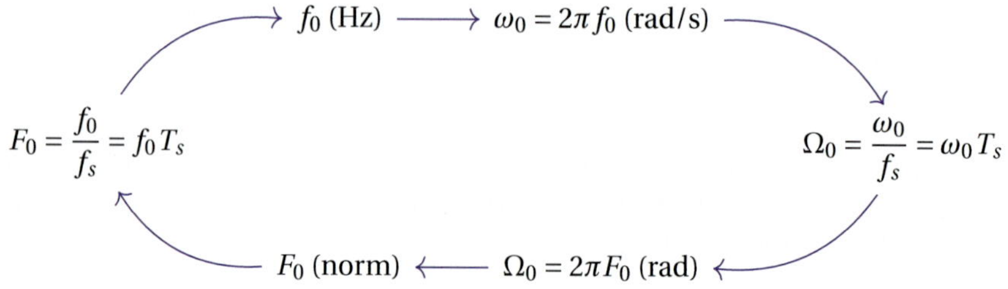

f_0 (Hz)
$\omega_0 = 2\pi f_0$ (rad/s)
$F_0 = \dfrac{f_0}{f_s} = f_0 T_s$
$\Omega_0 = \dfrac{\omega_0}{f_s} = \omega_0 T_s$
F_0 (norm)
$\Omega_0 = 2\pi F_0$ (rad)

Even though these relationships were derived for a sinusoidal signal, they apply to any signal that can be represented as a collection of sinusoids, that is, any signal that has a Fourier series or transform representation.

Example 8.1: Finding DTLTI system specifications

An analog signal $x_a(t)$ bandlimited to 10 kHz is sampled with a sampling rate of $f_s = 22{,}050$ Hz to obtain a discrete-time signal $x[n]$. The goal is to emulate a CTLTI system that keeps the frequencies in the range 2 kHz $\le f \le$ 3 kHz and removes all others, that is,

$$|G(f)| = \begin{cases} 1\,, & 2000 \le |f| \le 3000 \\ 0\,, & \text{otherwise} \end{cases}$$

Determine the specifications for the equivalent DTLTI system.

Solution: First, analog frequency specifications need to be converted to normalized frequencies.

$$F_1 = \frac{2000}{22050} = 0.0907\,, \quad F_2 = \frac{3000}{22050} = 0.1361$$

Critical angular frequencies are

$$\Omega_1 = 2\pi F_1 = 0.5699\,, \quad \Omega_2 = 2\pi F_2 = 0.8549$$

The equivalent DTLTI system magnitude response is

$$|H(\Omega)| = \begin{cases} 1\,, & 0.5699 \le |\Omega| \le 0.8549 \\ 0\,, & \text{otherwise} \end{cases}$$

Software resource: See MATLAB Exercise 8.1.

8.3 Pole-Zero Placement Design of Filters

In this section we will focus on designing some simple but useful filters through careful placement of z-plane poles and zeros of the system function. Of particular interest are resonant bandpass filters for picking a single frequency component from a signal, and notch filters for suppressing a single frequency.

8.3.1 Resonant bandpass filters

In general, a bandpass filter is a system that allows a particular range of frequencies to pass through while blocking frequencies that are outside this range. An ideal bandpass filter would have the system function

$$H_{BP}(\Omega) = \begin{cases} 1\,, & \Omega_1 \le |\Omega| \le \Omega_2 \\ 0\,, & \text{otherwise} \end{cases} \tag{8.3}$$

We will explore methods for the design of practical IIR and FIR bandpass filters in later parts of this chapter. A *resonant bandpass filter*, on the other hand, has a slightly different purpose: Pass a single frequency or a very narrow band of frequencies around it, and block everything else. It is a narrowband filter that is tuned to one particular frequency. Fig. 8.2 depicts the system function of a bandpass filter with a passband center frequency of $\Omega = \Omega_0$ and a narrow bandwidth of Ω_b.

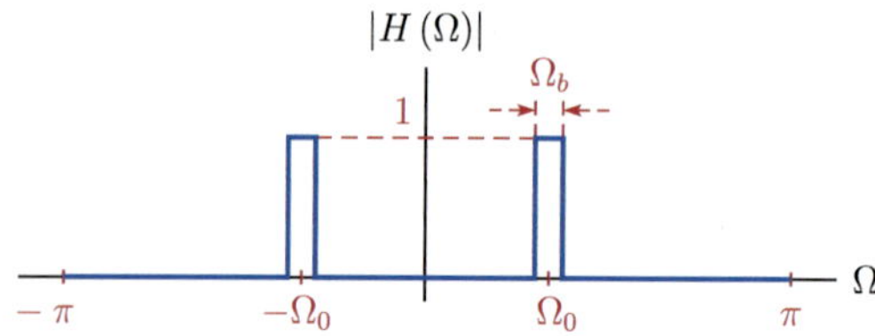

Figure 8.2 – Idealized magnitude response of a resonant bandpass filter with center frequency of Ω_0 and a bandwidth of Ω_b.

For the design of resonant bandpass filters we will use the ideas explored in Section 5.5.4 of Chapter 5. A sharp peak in the magnitude response at angular frequency Ω_0 can be obtained by placing a pole on a line with angle Ω_0 and close to the unit circle. Let a pole be placed at $p_1 = r\,e^{j\Omega_0}$. This results in the system function

$$H(z) = \frac{K}{z - r\,e^{j\Omega_0}} = \frac{K}{z - r\cos(\Omega_0) - j\,r\sin(\Omega_0)} \tag{8.4}$$

where K is a constant gain factor to be adjusted later. The placement of the pole and the magnitude response of the corresponding system function are illustrated in Fig. 8.3.

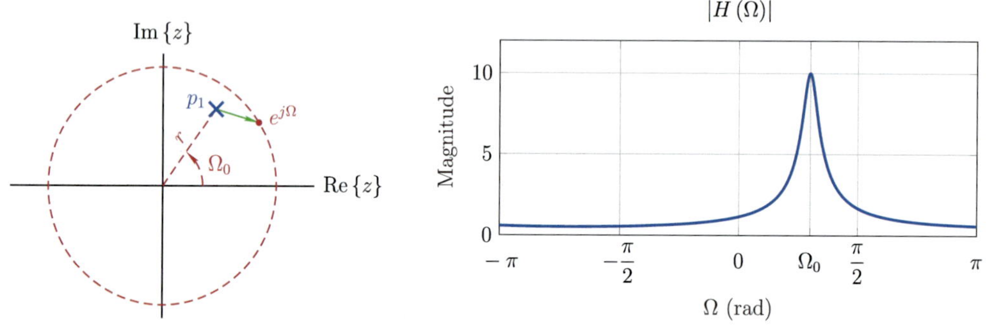

Figure 8.3 – Placement of a pole at $p_1 = r\,e^{j\Omega_0}$ to obtain a peak at angular frequency Ω_0 in the magnitude response. Gain factor used was $K = 1$.

The sharpness of the peak can be controlled by the choice of the pole radius r. Moving the pole closer to the unit circle increases the sharpness of the response around $\Omega = \Omega_0$. Obviously we need to be careful not to place the pole on or outside the unit circle, otherwise we would have an unstable system.

One practical issue with the system function in Eqn. (8.4) is that it has a complex coefficient in the denominator. In most applications we work with real signals, and would prefer to use systems with real coefficients. This requires that the complex poles and zeros of the system function appear in conjugate pairs. Addition of the conjugate pole at $p_2 = r\,e^{-j\Omega_0}$ results in the system function

$$H(z) = \frac{K}{\left(z - r\,e^{j\Omega_0}\right)\left(z - r\,e^{-j\Omega_0}\right)} = \frac{K}{z^2 - 2r\cos(\Omega_0)\,z + r^2} \tag{8.5}$$

in which all coefficients are real. Pole-zero diagram and the magnitude response for this case are shown in Fig. 8.4.

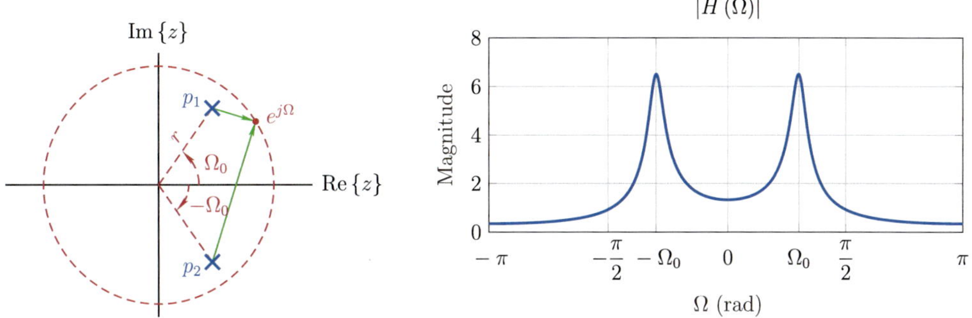

Figure 8.4 – Pole-zero diagram and the corresponding magnitude response after adding the conjugate pole at $p_2 = r\,e^{-j\Omega_0}$. Gain factor used was $K = 1$.

From a casual inspection of the magnitude spectrum $|H(\Omega)|$ in Fig. 8.4, it seems that adding a couple of zeros to the system function would be beneficial for pulling down the magnitude response at $\Omega = 0$ and $\Omega = \pm\pi$. We will therefore add zeros at $z_1 = e^{j0} = 1 + j0$ and $z_2 = e^{j\pi} = -1 + j0$ to obtain the system function

$$H(z) = \frac{K\,(z-1)\,(z+1)}{\left(z - r\,e^{j\Omega_0}\right)\left(z - r\,e^{-j\Omega_0}\right)} = \frac{K\left(z^2 - 1\right)}{z^2 - 2r\cos(\Omega_0)\,z + r^2} \tag{8.6}$$

Completed pole-zero diagram with the added zeros and the updated magnitude response are shown in Fig. 8.5. A block diagram for implementing the filter is shown in Fig. 8.6.

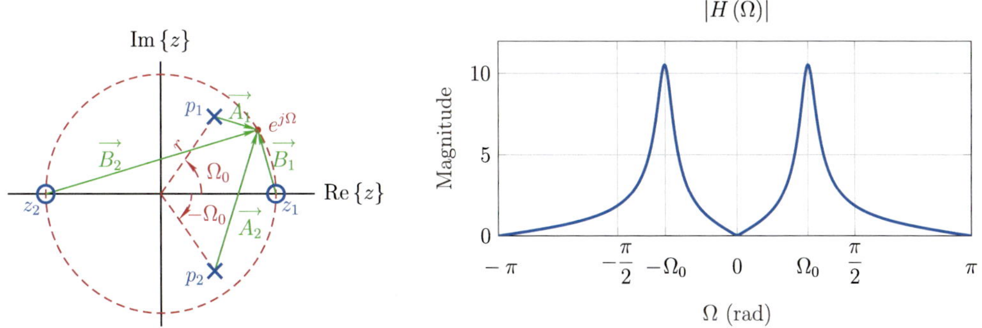

Figure 8.5 – Pole-zero diagram and the corresponding magnitude response after adding the zeros at $z_1 = 1 + j0$ and $z_2 = -1 + j0$.

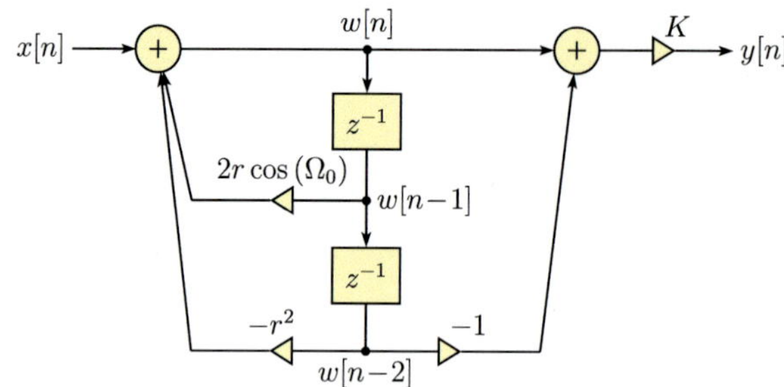

Figure 8.6 – Block diagram for direct-form-II implementation of the resonant bandpass filter system function given by Eqn. (8.6).

There are two parameters still to be determined, namely the pole radius r and the gain factor K. The parameter r controls the sharpness of the peaks in the magnitude response. A meaningful design parameter that can be used in computing an appropriate value for r is the 3-dB bandwidth Ω_b. We would like to adjust r so that the magnitude of the system function at $\Omega = \Omega_0 \pm \Omega_b/2$ is 3-dB below the peak magnitude at $\Omega = \Omega_0$.

$$\left|H\left(\Omega_0 \pm \Omega_b/2\right)\right| = \frac{\left|H\left(\Omega_0\right)\right|}{\sqrt{2}} \tag{8.7}$$

Refer to the set of vectors shown in Fig. 8.5a. The magnitude of the system function at $\Omega = \Omega_0$ is

$$\left|H\left(\Omega_0\right)\right| = \frac{K\left|e^{j\Omega_0}-1\right|.\left|e^{j\Omega_0}+1\right|}{\left|e^{j\Omega_0}-r\,e^{j\Omega_0}\right|.\left|e^{j\Omega_0}-r\,e^{-j\Omega_0}\right|} = \frac{K\left|\overrightarrow{B_1}\right|.\left|\overrightarrow{B_2}\right|}{\left|\overrightarrow{A_1}\right|.\left|\overrightarrow{A_2}\right|} \tag{8.8}$$

The magnitude of the system function at $\Omega = \Omega_0 - \Omega_b/2$, the lower end of the 3-dB bandwidth, is

$$\left|H\left(\Omega_0 - \Omega_b/2\right)\right| = \frac{K\left|e^{(j\Omega_0-\Omega_b/2)}-1\right|.\left|e^{(j\Omega_0-\Omega_b/2)}+1\right|}{\left|e^{(j\Omega_0-\Omega_b/2)}-r\,e^{j\Omega_0}\right|.\left|e^{(j\Omega_0-\Omega_b/2)}-r\,e^{-j\Omega_0}\right|} = \frac{K\left|\overrightarrow{B_1'}\right|.\left|\overrightarrow{B_2'}\right|}{\left|\overrightarrow{A_1'}\right|.\left|\overrightarrow{A_2'}\right|} \tag{8.9}$$

Fig. 8.7 illustrates the bandwidth around the resonant peak. Assuming $\Omega_b << 2\pi$, the following conclusions should be valid:

$$\left|\overrightarrow{A_2'}\right| \approx \left|\overrightarrow{A_2}\right|, \qquad \left|\overrightarrow{B_1'}\right| \approx \left|\overrightarrow{B_1}\right|, \qquad \left|\overrightarrow{B_2'}\right| \approx \left|\overrightarrow{B_2}\right|$$

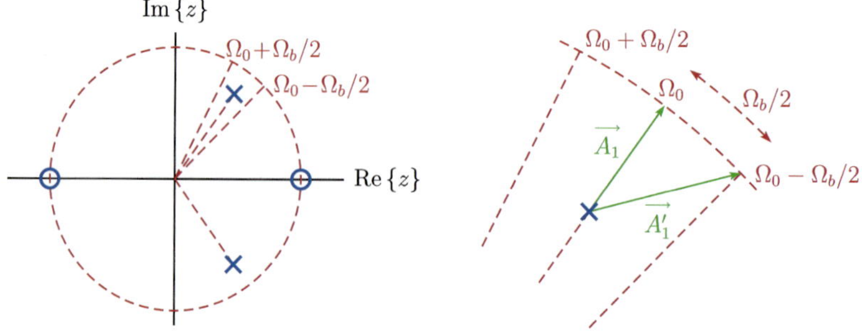

Figure 8.7 – Detail view around the center frequency $\Omega = \Omega_0$.

Vectors $\overrightarrow{A_2}$, $\overrightarrow{B_1}$, and $\overrightarrow{B_1}$ do not change significantly when their tip moves from Ω_0 to $\Omega_0 - \Omega_b/2$ on the unit circle. Consequently, the ratio of the two magnitudes in Eqn. (8.7) is controlled by vectors $\overrightarrow{A_1}$ and $\overrightarrow{A'_1}$ only.

$$\frac{\left|H(\Omega_0 - \Omega_b/2)\right|}{\left|H(\Omega_0)\right|} \approx \frac{\left|\overrightarrow{A_1}\right|}{\left|\overrightarrow{A'_1}\right|} = \frac{1}{\sqrt{2}} \tag{8.10}$$

Since we know that $\left|\overrightarrow{A_1}\right| = (1-r)$, we must have $\left|\overrightarrow{A'_1}\right| = \sqrt{2}\,(1-r)$. Using the right triangle formed by the two vectors and part of the unit circle in Fig. 8.7b, we can write

$$2\,(1-r)^2 = (1-r)^2 + \left(\frac{\Omega_b}{2}\right)^2 \tag{8.11}$$

which leads to the conclusion

$$r = 1 - \frac{\Omega_b}{2} \tag{8.12}$$

The next step is to determine the value of the gain factor K. We would like the peak magnitude of the system function to be equal to 1. Setting $\left|H(\Omega_0)\right| = 1$ in Eqn. (8.8) and solving for K yields

$$K = \frac{\left|e^{j\Omega_0} - r\,e^{j\Omega_0}\right| . \left|e^{j\Omega_0} - r\,e^{-j\Omega_0}\right|}{\left|e^{j\Omega_0} - 1\right| . \left|e^{j\Omega_0} + 1\right|} \tag{8.13}$$

Example 8.2: Design of a resonant bandpass filter

An analog signal $x_a(t)$ is sampled with a sampling rate of $f_s = 22{,}050$ Hz to obtain a discrete-time signal $x[n]$. Design a discrete-time resonant bandpass filter with a center frequency of $f_0 = 660$ Hz and a 3-dB bandwidth of $f_b = 60$ Hz to extract the 660 Hz component from the signal.

Solution: First, analog frequency and bandwidth specifications need to be converted to normalized and angular frequencies. The normalized center frequency is

$$F_0 = \frac{f_0}{f_s} = \frac{660}{22050} = 0.0299$$

and the normalized 3-dB bandwidth is

$$F_b = \frac{f_b}{f_s} = \frac{60}{22050} = 0.0027$$

Converting these values to angular frequencies, we have

$$\Omega_0 = 2\pi F_0 = 0.1881\,, \qquad \Omega_b = 2\pi F_b = 0.0171$$

The pole radius needed to meet the specifications is

$$r = 1 - \frac{\Omega_b}{2} = 1 - \frac{0.0171}{2} = 0.9915$$

Thus, the poles of the system function are at

$$p_{1,2} = r\,e^{\pm j\Omega_0} = 0.9915\,e^{\pm j0.1881} = 0.9740 \pm j0.1854$$

The gain factor is

$$K = \frac{\left|e^{j(2)(0.1881)} - 1.9479\, e^{j0.1881} + 0.9830\right|}{\left|e^{j(2)(0.1881)} - 1\right|} = 0.0085$$

and the system function for the designed filter can be written as

$$H(z) = \frac{0.0085\left(z^2 - 1\right)}{z^2 - 1.9479\, z + 0.9830}$$

The magnitude of the system function is shown in Fig. 8.8.

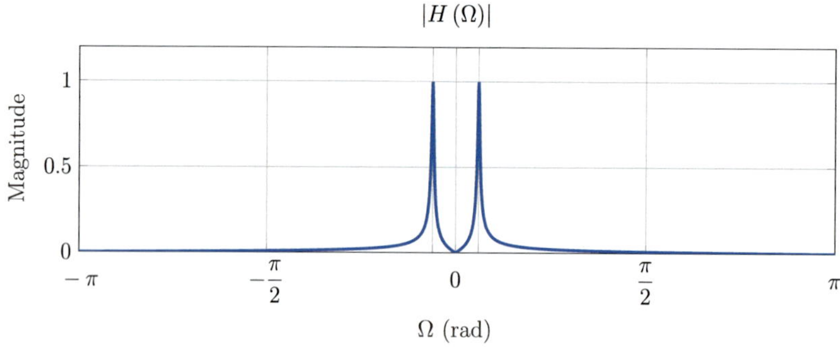

Figure 8.8 – Magnitude response of the resonant bandpass filter designed in Example 8.2.

Software resource: exdt_8_2.m

Software resources: See MATLAB Exercises 8.2 and 8.3, and live script ResonatorDesign.mlx.

8.3.2 Notch filters

A bandstop filter is a system that blocks a particular range of frequencies, and allows everything else to pass through. Design of general bandstop filters will be discussed in later sections of this chapter. In this section, however, we will focus our attention on a special class of bandstop filters designed to block just one particular frequency, or a very narrow band of frequencies. Such a filter is referred to as a *notch filter*.

As in the case of resonant bandpass filters, notch filters can also be designed through strategic placement of poles and zeros of the system function on the z-plane. Let us begin with a system that has a single zero on the unit circle at $z_1 = e^{j\Omega_0}$. In order to keep the system causal, we will also add a pole at the origin, that is, at $p_1 = 0 + j0$. The resulting system function is in the form

$$H(z) = \frac{K\left(z - e^{j\Omega_0}\right)}{z} \tag{8.14}$$

where K is a constant gain factor to be determined. The pole-zero diagram and the resulting magnitude response are shown in Fig. 8.9.

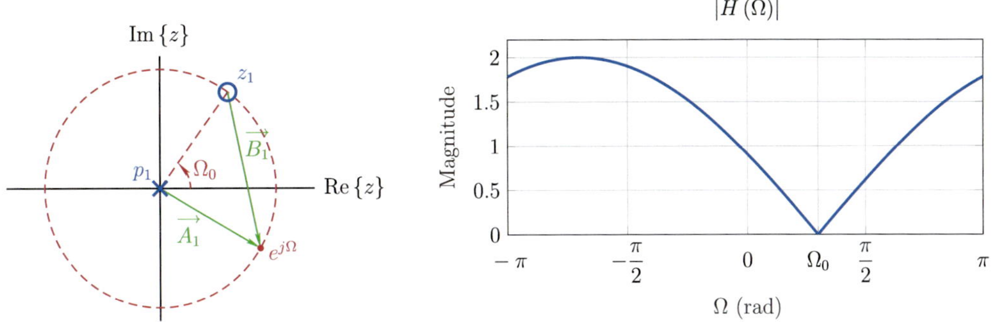

Figure 8.9 – First attempt at a notch filter using a zero on the unit circle at $e^{j\Omega_0}$ and a pole at the origin. Gain factor used was $K = 1$.

While this system would clearly block the frequency $\Omega = \Omega_0$, the shape of the magnitude response in Fig, 8.9b leaves a lot to be desired. In addition to blocking the undesired frequency component, the system would also distort the amplitudes of all remaining frequency components. To understand the reason for this, let us look at the vectors $\overrightarrow{A_1}$ and $\overrightarrow{B_1}$ on Fig. 8.9a originating from the zero and the pole, and pointing to $z = e^{j\Omega}$ on the unit circle. The magnitude of the system function is

$$\left|H(\Omega)\right| = \frac{K\left|e^{j\Omega} - e^{j\Omega_0}\right|}{\left|e^{j\Omega}\right|} = \frac{K\,|\overrightarrow{B_1}|}{|\overrightarrow{A_1}|} \tag{8.15}$$

Since $|\overrightarrow{A_1}| = 1$ independent of the value of Ω, the shape of the magnitude $\left|H(\Omega)\right|$ is determined solely by $|\overrightarrow{B_1}|$ as its tip follows $z = e^{j\Omega}$ on the unit circle. If we could somehow get $|\overrightarrow{A_1}|$ to mimic the behavior of $|\overrightarrow{B_1}|$ as z is moved on the unit circle, the ratio between the norms of the two vectors could be made approximately constant for frequencies away from Ω_0. The key to achieving this is to move the pole p_1 closer to the zero z_1. Let $p_1 = r\,e^{j\Omega_0}$ where $r < 1$ for stability. This is depicted in Fig. 8.10. The system function is

$$H(z) = \frac{K\left(z - e^{j\Omega_0}\right)}{z - r\,e^{j\Omega_0}} \tag{8.16}$$

and its magnitude is

$$\left|H(\Omega)\right| = \frac{K\left|e^{j\Omega} - e^{j\Omega_0}\right|}{\left|e^{j\Omega} - r\,e^{j\Omega_0}\right|} = \frac{K\,|\overrightarrow{B_1}|}{|\overrightarrow{A_1}|} \tag{8.17}$$

An inspection of Fig. 8.10a reveals that vectors $|\overrightarrow{A_1}|$ and $|\overrightarrow{B_1}|$ change in similar ways to keep the ratio close to a constant. The only exception occurs in the vicinity of $\Omega = \Omega_0$ where the zero at z_1 on the unit circle causes the magnitude response to dip regardless of how close the pole at p_1 is.

The system function in Eqn. (8.16) has complex coefficients. A system function with real coefficients can be obtained by adding another pole and zero pair at conjugate locations $p_2 = p_1^*$ and $z_2 = z_1^*$ as shown in Fig. 8.11.

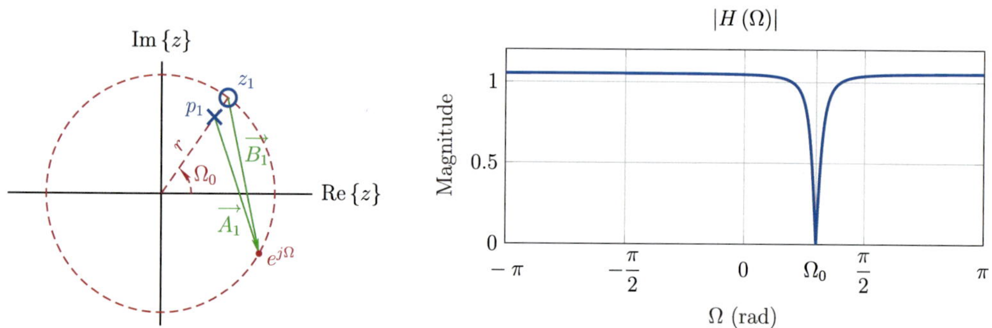

Figure 8.10 – Pole-zero diagram and the corresponding magnitude response after moving the pole from origin to $r\,e^{j\Omega_0}$ to improve the performance.

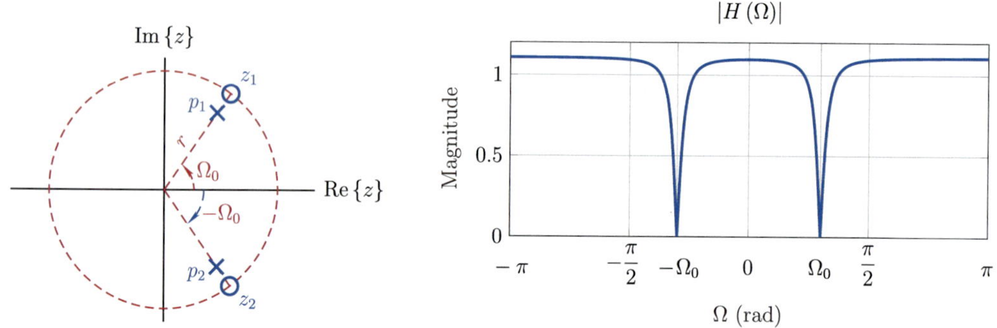

Figure 8.11 – Pole-zero diagram and the corresponding magnitude response after adding conjugate pole and zero to obtain a system function with real coefficients.

The system function for the second-order notch filter is

$$H(z) = \frac{K\left(z - e^{j\Omega_0}\right)\left(z - e^{-j\Omega_0}\right)}{\left(z - r\,e^{j\Omega_0}\right)\left(z - r\,e^{-j\Omega_0}\right)} = \frac{K\left(z^2 - 2\cos(\Omega_0)\,z + 1\right)}{z^2 - 2r\cos(\Omega_0)\,z + r^2} \tag{8.18}$$

A direct-form-II block diagram for $H(z)$ is shown in Fig. 8.12.

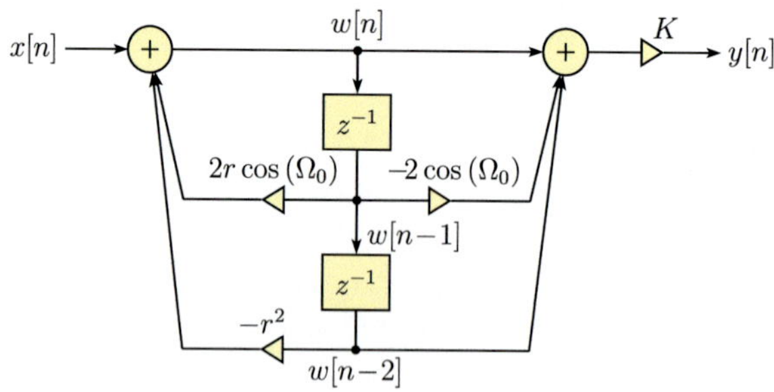

Figure 8.12 – Block diagram for direct-form-II implementation of the notch filter system function given by Eqn. (8.18).

The gain factor can be determined by setting the magnitude at $\Omega = 0$ or, equivalently, at $z = 1$, equal to unity.

$$K = \frac{\left|1 - 2r\,\cos(\Omega_0) + r^2\right|}{\left|2 - 2\,\cos(\Omega_0)\right|} \tag{8.19}$$

It can be shown (see Problem 8.4) that the relationship between the 3-dB width of the stopband and the pole radius r is the same as it was in the case of bandpass resonant filter, that is,

$$r = 1 - \frac{\Omega_b}{2} \tag{8.20}$$

Example 8.3: Design of a notch filter

Consider again a discrete-time signal $x[n]$ obtained by sampling an analog signal $x_a(t)$ with a sampling rate of $f_s = 22{,}050$ Hz. Design a notch filter with the appropriate center frequency and a 3-dB bandwidth of $f_b = 20$ Hz to eliminate the 220 Hz component from the signal.

Solution: We will begin by converting analog frequency and bandwidth specifications to normalized and angular frequencies. The normalized center frequency is

$$F_0 = \frac{f_0}{f_s} = \frac{220}{22050} = 0.01$$

and the normalized 3-dB bandwidth is

$$F_b = \frac{f_b}{f_s} = \frac{20}{22050} = 0.00091$$

Converting these values to angular frequencies, we have

$$\Omega_0 = 2\pi F_0 = 0.0627\,, \qquad \Omega_b = 2\pi F_b = 0.0057$$

The pole radius needed to meet the specifications is

$$r = 1 - \frac{\Omega_b}{2} = 1 - \frac{0.0057}{2} = 0.9972$$

Thus, the zeros of the system function are at

$$z_{1,2} = e^{\pm j\Omega_0} = e^{\pm j0.0627} = 0.9980 \pm j0.0626$$

and the poles are at

$$p_{1,2} = r\,e^{\pm j\Omega_0} = 0.9972\,e^{\pm j0.0627} = 0.9952 \pm j0.0625$$

The gain factor is

$$K = \frac{\left|1 - 2\,(0.9972)\,\cos(0.0627) + (0.9972)^2\right|}{\left|2 - 2\,\cos(0.0627)\right|} = 0.9992$$

and the system function for the designed filter can be written as

$$H(z) = \frac{0.9992\,\left(z^2 - 1.9961\,z + 1\right)}{z^2 - 1.9904\,z + 0.9943}$$

The magnitude of the system function is shown in Fig. 8.13. Fig. 8.14 illustrates a zoomed-in version of the magnitude response around $\Omega = 0$. Also, the frequency axis uses analog frequencies in Hz obtained by multiplying normalized frequencies with the sampling rate, that is, $f = F f_s$.

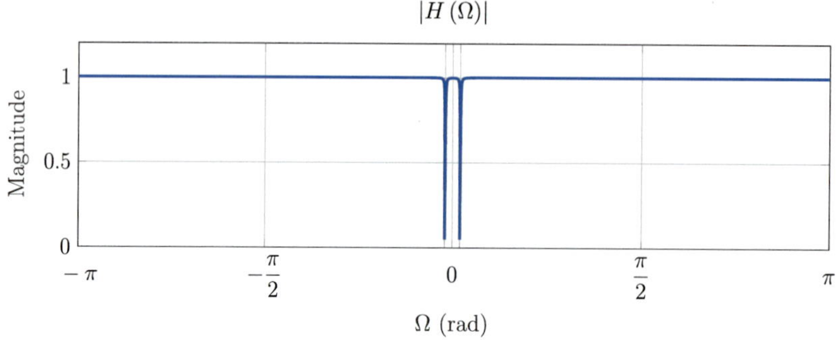

Figure 8.13 – Magnitude response of the notch filter designed in Example 8.2.

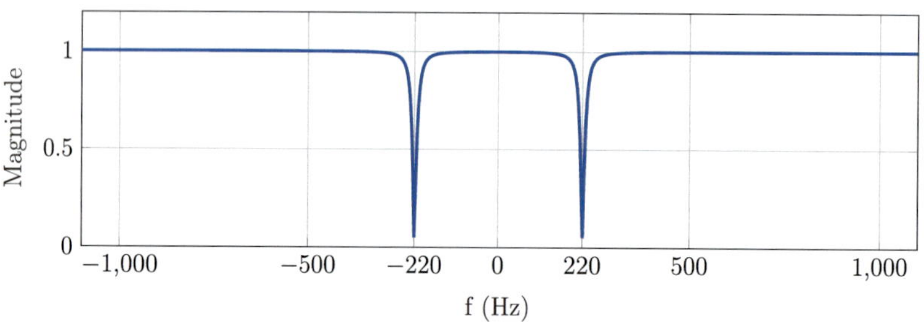

Figure 8.14 – Detail view of the magnitude response of the notch filter designed in Example 8.2 at low frequencies. The frequency axis is marked with analog frequencies rather than angular frequencies.

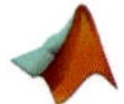

Software resource: exdt_8_3.m

Software resources: See MATLAB Exercise 8.4 and live script NotchFilterDesign.mlx.

8.4 IIR Filters

Recall from earlier discussion that infinite impulse response (IIR) filters are recursive filters. The difference equation of an IIR filter utilizes past samples of the output signal in the computation of the current output sample. In addition, it may also utilize current and past samples of the input signal. The corresponding block diagram for the IIR filter has one or more feedback paths due to the past output samples appearing in the difference equation. When feedback is involved, stability of the IIR filter must always be an issue of concern. An unstable filter would not be usable in practical applications. Therefore, any technique for designing IIR filters must make it a top priority to produce a stable result. In most applications we also require the designed filter to be causal.

IIR filter design problem starts with a set of specifications in the frequency domain that the resulting filter must satisfy. In previous sections of this chapter we discussed pole-zero placement design of some special filter types, namely resonant bandpass filters and notch filters. General frequency selective filters could also be designed using pole-zero placement through trial and error, however, a more systematic approach that works in a predictable way would be desirable. Most common method of IIR design is to find a suitable analog prototype filter, and to apply some kind of transformation to turn it into a discrete-time filter. The motivation for doing this is the fact that analog filter design methods are well established.

There are also some optimization-based methods in the literature for direct design of IIR filters without resorting to the use of an analog prototype. These techniques tend to be more difficult and impractical to use due to nonlinearity of the equations that need to be solved as part of the optimization problem as well as stability concerns for the designed filters.

In the remainder of this chapter we will pursue design of IIR filters from analog prototypes by means of a transformation. The steps used in the design of IIR filters with this approach are as follows:

Procedure for designing an IIR filter:

1. The specifications of the desired discrete-time filter are converted to the specifications of an appropriate analog filter that can be used as a prototype. Let the desired discrete-time filter be specified through critical frequencies Ω_1 and Ω_2 along with tolerance values Δ_1 and Δ_2. Analog prototype filter parameters ω_1 and ω_2 need to be determined. (If the filter type is bandpass or bandstop, two additional frequencies, ω_3 and ω_4, also need to be determined from the specifications for Ω_3 and Ω_4.)
2. An analog prototype filter that satisfies the design specifications in step 1 is designed. Its system function $G(s)$ is constructed.
3. The analog prototype filter is converted to a discrete-time filter by means of a transformation. Specifically, a z-domain system function $H(z)$ is obtained from the analog prototype system function $G(s)$. The designed discrete-time filter can be analyzed using the techniques discussed in Chapters 3 and 5, and implemented using the block diagram methods discussed in Chapter 5.

The methods for specifying frequency domain behavior of the desired filter will be discussed in the next section. A brief review of analog filter design formulas (step 2) will follow. Afterward we will discuss several transformation methods that could be used for converting an analog prototype to a discrete-time filter. This would fulfill step 3 of the procedure described above. Finally step 1, the problem of converting the specifications of the IIR filter to the specifications of a suitable analog prototype will be discussed. The reason why discussion of step 1 is deferred till the end is because how step 1 is performed depends on what transformation is to be used in step 3.

Practical implementation of designed IIR filters is also an important issue. It is often hardware dependent, however, building a foundation for real-time and post processing implementation is still helpful. This will be done in the section on MATLAB exercises.

8.4.1 IIR filter specifications

In general, a frequency selective filter allows a specified band of frequencies to pass through while blocking all other frequencies. Based on the shape of the desired magnitude spectrum, frequency selective filters are categorized into four types, namely lowpass, highpass, bandpass, and band-stop filters. The specification diagram for a lowpass IIR filter is shown in Fig. 8.15. Note that IIR filter specifications are typically given in terms of the desired magnitude response as a function of angular frequency.

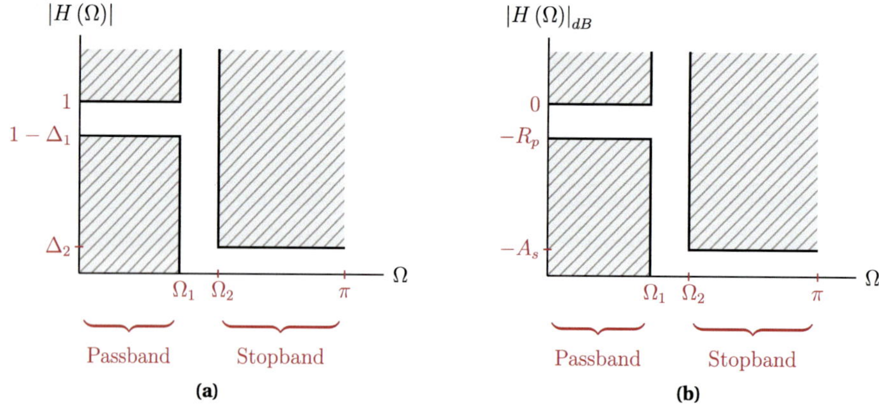

Figure 8.15 – Tolerance specifications for a discrete-time lowpass filter.

The magnitude response of a discrete-time filter is 2π-periodic, and only one period needs to be specified in the angular frequency range $-\pi < \Omega < \pi$. However, in Fig. 8.15, the specifications are given in the range $0 \le \Omega < \pi$. The reason for this is easy to understand if we remember that the system function of the filter to be designed contains only real coefficients. Corresponding impulse response is also real-valued. Consequently, the magnitude is an even function of Ω. The missing left half of the magnitude response in the range $-\pi \le \Omega < 0$ is simply the mirror image of what is shown in Fig. 8.15.

A lowpass filter has a *passband* for $0 < \Omega < \Omega_1$, and a *stopband* for $\Omega_2 < \Omega < \pi$. The frequencies Ω_1 and Ω_2 are termed the *critical frequencies* of the lowpass filter. The range of frequencies from Ω_1 to Ω_2 is called the *transition band*. The magnitude of the desired filter is required to stay in the unshaded regions. This can be expressed as

$$\begin{aligned} 1-\Delta_1 < \left|H(\Omega)\right| < 1 \qquad & \text{for } 0 < \Omega < \Omega_1 \\ \left|H(\Omega)\right| < \Delta_2 \qquad & \text{for } \Omega_2 < \Omega < \pi \end{aligned}$$

Parameters Δ_1 and Δ_2 are referred as the *passband tolerance* and the *stopband tolerance*. Part (b) of Fig. 8.15 has the magnitude specifications on a dB scale. Parameters R_p and A_s are the *maximum passband ripple* and the *minimum stopband attenuation*, respectively. They can be related to Δ_1 and Δ_2 as

$$-R_p = 20\log_{10}(1-\Delta_1), \qquad -A_s = 20\log_{10}(\Delta_2) \tag{8.21}$$

If Δ_1 and Δ_2 are needed, they can be obtained from R_p and A_s as

$$\Delta_1 = 1 - 10^{-R_p/20}, \qquad \Delta_2 = 10^{-A_s/20} \tag{8.22}$$

Specification diagrams for other filter types (highpass, bandpass or bandstop) are shown in Figs. 8.16 and 8.17. A highpass filter has a stopband for low frequencies followed by a passband for higher frequencies. A bandpass filter has a passband sandwiched between two stopbands, and a bandstop filter has the exact opposite combination. Describing the desired magnitude response involves specifying the tolerance values Δ_1 and Δ_2 along with two or four critical frequencies depending on the type of filter being designed.

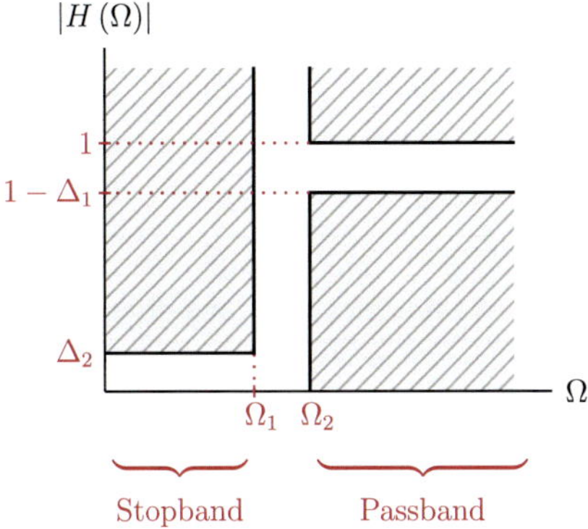

Figure 8.16 – Tolerance specifications for a discrete-time highpass filter.

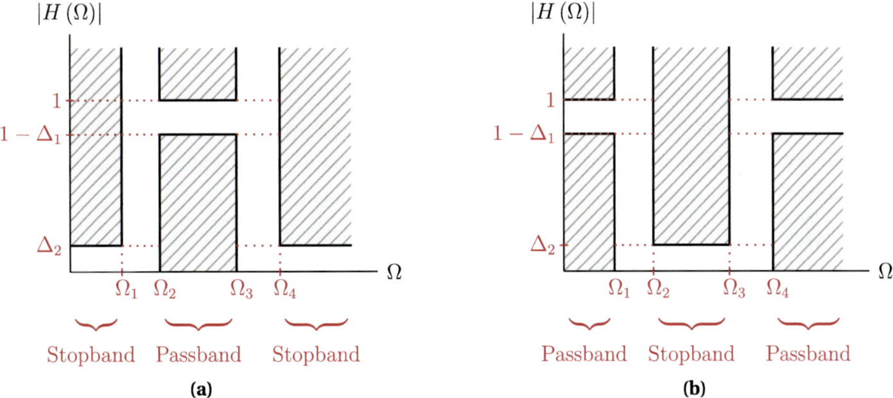

Figure 8.17 – Tolerance specifications for discrete-time bandpass and bandstop filters.

8.4.2 Review of analog filter design formulas

In general, the analog filter design problem begins with the design of a lowpass prototype filter regardless of the type of filter that is ultimately needed. If a highpass, bandpass, or bandstop analog filter is desired, it is obtained subsequently by means of a *frequency transformation* to be applied to the lowpass prototype filter. This approach is motivated by the availability of well established methods for the design of analog lowpass filters. Specification diagram for an analog lowpas filter is illustrated in Fig. 8.18 in both linear and dB scales.

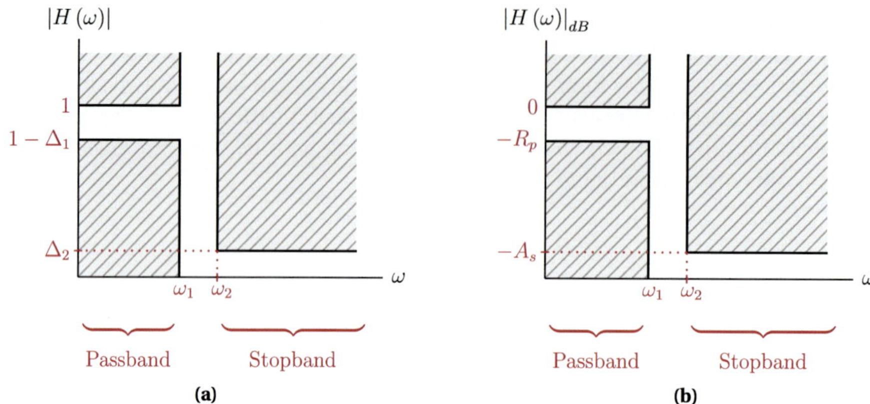

Figure 8.18 – Tolerance specifications for a discrete-time lowpass filter.

Approximation formulas for analog lowpass filters are typically given in terms of the squared-magnitude function $|H(\omega)|^2$. For a system function $H(s)$ with real coefficients, it can be shown that

$$\left|H(\omega)\right|^2 = H(s)\, H(-s)\Big|_{s=j\omega} \tag{8.23}$$

The procedure in Eqn. (8.23) needs to be reversed so that the system function $H(s)$ can be determined from the knowledge of the squared-magnitude function $|H(\omega)|^2$.

$$H(s)\, H(-s) = \left|H(\omega)\right|^2\Big|_{\omega^2=-s^2} \tag{8.24}$$

At this point, poles and zeros of the product $H(s)\,H(-s)$ can be determined. Poles and zeros found need to be divided up between the terms $H(s)$ and $H(-s)$, being careful to assign only stable poles to the former.

Butterworth lowpass filters

Butterworth lowpass filters are characterized by the squared-magnitude function

$$\left|H(\omega)\right|^2 = \frac{1}{1+(\omega/\omega_c)^{2N}} \tag{8.25}$$

where the parameters N and ω_c are the filter-order and the 3-dB cutoff frequency, respectively. The poles of the product $H(s)\,H(-s)$ are at

$$p_k = \begin{cases} \omega_c\, e^{jk\pi/N}, & k=0,\ldots,2N-1 \quad \text{if } N \text{ is odd} \\ \omega_c\, e^{j(2k+1)\pi/2N}, & k=0,\ldots,2N-1 \quad \text{if } N \text{ is even} \end{cases} \tag{8.26}$$

The desired behavior of a lowpass filter to be designed is usually specified in terms of the two critical frequencies ω_1 and ω_2 as well as the dB tolerance values R_p and A_s as shown in Fig. 8.18b. The corresponding parameter values for ω_c and N need to be determined from the provided set of specifications so that the filter can be designed.

$$N = \frac{\log_{10}\sqrt{\left(10^{A_s/10}-1\right)/\left(10^{R_p/10}-1\right)}}{\log_{10}(\omega_2/\omega_1)} \quad \Longrightarrow \quad \text{Round up to next integer} \tag{8.27}$$

After determining N, the frequency ω_c can be computed from one of the following:

$$\left(\frac{\omega_1}{\omega_c}\right)^{2N} = 10^{R_p/10} - 1 \qquad \text{or} \qquad \left(\frac{\omega_2}{\omega_c}\right)^{2N} = 10^{A_s/10} - 1 \tag{8.28}$$

Chebyshev type-I lowpass filters

As an alternative to the Butterworth approximation formula, the Chebyshev type-I approximation formula for the squared-magnitude function of a lowpass filter is

$$\left|H(\omega)\right|^2 = \frac{1}{1+\varepsilon^2\, C_N^2(\omega/\omega_1)} \tag{8.29}$$

where ε is a positive constant. The frequency ω_1 is the passband edge frequency. The function $C_N(v)$ in the denominator represents the *Chebyshev polynomial* of order N.

Poles of the product $H(s)\,H(-s)$ for the Chebyshev type-I lowpass filter are found as

$$p_k = j\omega_1\left[\cos(\alpha_k)\cosh(\beta_k) - j\sin(\alpha_k)\sinh(\beta_k)\right], \qquad k = 0,\ldots,2N-1 \tag{8.30}$$

where

$$\alpha_k = \frac{(2k+1)\,\pi}{2N} \qquad \text{and} \qquad \beta_k = \frac{\sinh^{-1}(1/\varepsilon)}{N}$$

Given dB tolerances R_p and A_s and critical frequencies ω_1 and ω_2, the parameters of the Chebyshev type-I lowpass filter are found as follows:

$$\omega_0 = \frac{\omega_2}{\omega_1}, \qquad \text{and} \qquad F = \sqrt{\frac{10^{A_s/10}-1}{10^{R_p/10}-1}} \tag{8.31}$$

$$N = \frac{\cosh^{-1}(F)}{\cosh^{-1}(\omega_0)} \quad \Longrightarrow \quad \text{Round up to next integer} \tag{8.32}$$

Parameter ε is computed from one of the following:

$$10\log_{10}\left(\frac{1}{1+\varepsilon^2}\right) = -R_p \qquad \text{or} \qquad 10\log_{10}\left(\frac{1}{1+\varepsilon^2\, C_N^2(\omega_0)}\right) = -A_s \tag{8.33}$$

Chebyshev type-II lowpass filters

Chebyshev type-II approximation formula for the squared-magnitude function is similar to the Chebyshev type-I formula.

$$\left|H(\omega)\right|^2 = \frac{\varepsilon^2\, C_N^2(\omega_2/\omega)}{1+\varepsilon^2\, C_N^2(\omega_2/\omega)} \tag{8.34}$$

The parameter ω_2 is the *stopband edge frequency*. Poles of $H(s)\,H(-s)$ for the Chebyshev type-II lowpass filter are found by

$$p_k = \frac{j\omega_2}{\cos(\alpha_k)\cosh(\beta_k) - j\sin(\alpha_k)\sinh(\beta_k)}, \qquad k = 0,\ldots,2N-1 \tag{8.35}$$

where

$$\alpha_k = \frac{(2k+1)\,\pi}{2N} \qquad \text{and} \qquad \beta_k = \frac{\sinh^{-1}(1/\varepsilon)}{N}$$

Zeros of $H(s)$ are

$$z_k = \frac{\pm j\omega_2}{\cos\left(\dfrac{(2k-1)\,\pi}{2N}\right)}\,, \qquad k = 1,\ldots,K \tag{8.36}$$

where

$$K = \begin{cases} (N-1)/2\,, & N \text{ odd} \\ N/2\,, & N \text{ even} \end{cases}$$

Given dB tolerances R_p and A_s and critical frequencies ω_1 and ω_2, the parameters of the Chebyshev type-II lowpass filter are

$$\omega_0 = \frac{\omega_2}{\omega_1}\,, \qquad \text{and} \qquad F = \sqrt{\frac{10^{A_s/10}-1}{10^{R_p/10}-1}} \tag{8.37}$$

and

$$N = \frac{\cosh^{-1}(F)}{\cosh^{-1}(\omega_0)} \qquad \Longrightarrow \qquad \text{Round up to next integer} \tag{8.38}$$

Parameter ε is obtained from one of the following:

$$\varepsilon = \frac{1}{\sqrt{10^{A_s/10}-1}} \qquad \text{or} \qquad \varepsilon = \frac{1}{\sqrt{C_N^2(\omega_0)\left(10^{R_p/10}-1\right)}} \tag{8.39}$$

Elliptic lowpass filters

Finally, an elliptic lowpass filter is characterized by the squared-magnitude function

$$\left|H(\omega)\right|^2 = \frac{1}{1+\varepsilon^2\,\psi_N^2(\omega/\omega_1)} \tag{8.40}$$

The parameter ε is a positive constant. The function $\psi_N(v)$ is called a *Chebyshev rational function,* and is defined in terms of *Jacobi elliptic functions.* A thorough treatment of Jacobi elliptic functions is well beyond the scope of this text, and will not be given here. Details of elliptic functions, related formulas and derivations can be found in references [1],[8].

Analog filter transformations

If a highpass, bandpass, or bandstop analog filter is needed, it is obtained from a lowpass filter, referred to as a *lowpass prototype,* by means of an analog filter transformation. Let $G(s)$ be the system function of an analog lowpass filter, and let $H(\lambda)$ represent the new filter to be obtained from it. For the new filter we use λ as the Laplace transform variable. $H(\lambda)$ is obtained from $G(s)$ through a transformation such that

$$H(\lambda) = G(s)\Big|_{s=f(\lambda)} \tag{8.41}$$

The function $s = f(\lambda)$ is the transformation that converts the lowpass filter into the type of filter desired.

Lowpass to highpass transformation:

Consider the specification diagrams shown in Fig. 8.19.

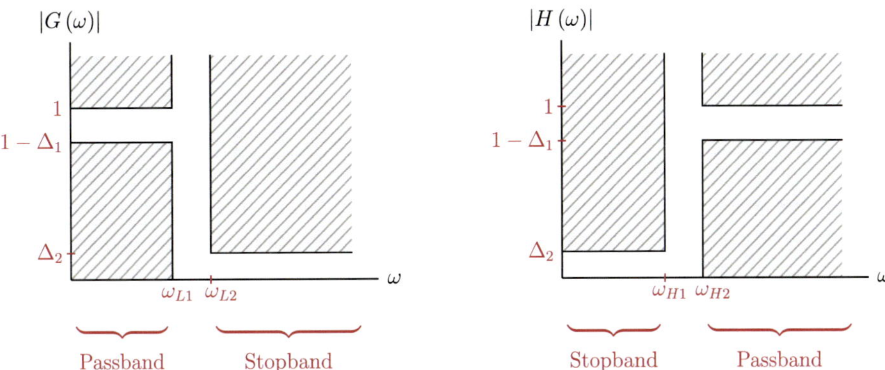

Figure 8.19 – Transforming a lowpass analog filter to highpass.

It is desired to obtain the highpass filter system function $H(\lambda)$ from the lowpass filter system function $G(s)$. The transformation

$$\frac{s}{\omega_{L1}} = \frac{\omega_{H2}}{\lambda} \qquad \Longrightarrow \qquad s = \frac{\omega_{L1}\,\omega_{H2}}{\lambda} = \frac{\omega_0^2}{\lambda} \tag{8.42}$$

can be used for this purpose. Magnitude responses of the two filters are identical at their respective passband edge frequencies, that is,

$$\left|H(\omega_{H2})\right| = \left|G(\omega_{L1})\right| \tag{8.43}$$

The stopband edge of the resulting highpass filter is at

$$\omega_{H1} = \frac{\omega_{L1}\,\omega_{H2}}{\omega_{L2}} \tag{8.44}$$

so that we have

$$\left|H(\omega_{H1})\right| = \left|G(\omega_{L2})\right| \tag{8.45}$$

Lowpass to bandpass transformation:

See the specification diagrams in Fig. 8.20. It is desired to obtain the bandpass filter system function $H(\lambda)$ from the lowpass filter system function $G(s)$. The transformation is

$$s = \frac{\lambda^2 + \omega_0^2}{B\lambda} \tag{8.46}$$

We require

$$\left|H(\omega_{B2})\right| = \left|G(-\omega_{L1})\right| \tag{8.47}$$

and

$$\left|H(\omega_{B3})\right| = \left|G(\omega_{L1})\right| \tag{8.48}$$

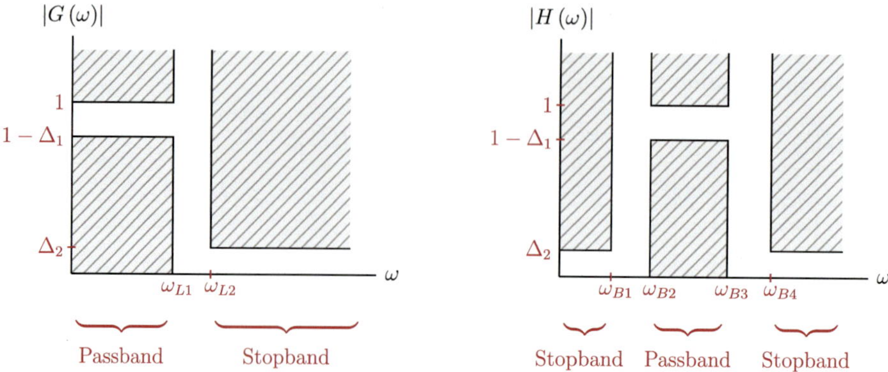

Figure 8.20 – Transforming a lowpass analog filter to bandpass.

These requirements are met if the parameters ω_0 and B are chosen as

$$\omega_0 = \sqrt{\omega_{B2}\,\omega_{B3}} \qquad \text{and} \qquad B = \frac{\omega_{B3} - \omega_{B2}}{\omega_{L1}} \tag{8.49}$$

The parameter ω_0 is the *geometric mean* of the passband edge frequencies of the bandpass filter. The parameter B is the ratio of the bandwidth of the bandpass filter to the bandwidth of the lowpass filter. Alternatively, it is also possible to map the stopband edge of the lowpass filter to the stopband edges of the bandpass filter. Instead of satisfying Eqns. (8.47) and (8.48), we could satisfy

$$\left|H(\omega_{B1})\right| = \left|G(-\omega_{L2})\right| \tag{8.50}$$

and

$$\left|H(\omega_{B4})\right| = \left|G(\omega_{L2})\right| \tag{8.51}$$

This requires that the parameters ω_0 and B of the transformation in Eqn. (8.46) be chosen as

$$\omega_0 = \sqrt{\omega_{B1}\,\omega_{B4}} \qquad \text{and} \qquad B = \frac{\omega_{B4} - \omega_{B1}}{\omega_{L2}} \tag{8.52}$$

This is usually the preferred approach in designing Chebyshev type-II filters.

Lowpass to bandstop transformation:

See the specification diagrams in Fig. 8.21.

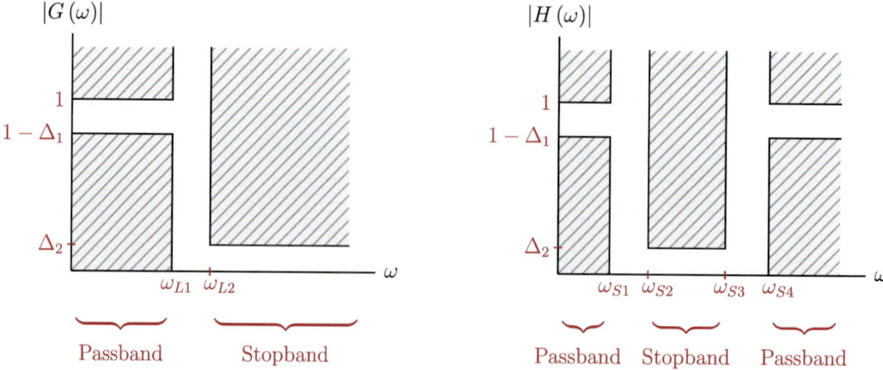

Figure 8.21 – Transforming a lowpass analog filter to bandstop.

In order to obtain the bandstop filter system function $H(\lambda)$ from the lowpass filter system function $G(s)$, the transformation to be used is in the form

$$s = \frac{B\lambda}{\lambda^2 + \omega_0^2} \tag{8.53}$$

We require

$$|H(\omega_{S4})| = |G(-\omega_{L1})| \tag{8.54}$$

and

$$|H(\omega_{S1})| = |G(\omega_{L1})| \tag{8.55}$$

These requirements are met if the parameters ω_0 and B are chosen as

$$\omega_0 = \sqrt{\omega_{S1}\,\omega_{S4}} \qquad \text{and} \qquad B = \frac{\omega_{S4} - \omega_{S1}}{\omega_{L1}} \tag{8.56}$$

Again, an alternative forms of Eqns. (8.54), (8.55), and (8.56) can be used by matching the stopband edges instead. The requirements are set up as

$$|H(\omega_{S3})| = |G(-\omega_{L2})| \tag{8.57}$$

and

$$|H(\omega_{S2})| = |G(\omega_{L2})| \tag{8.58}$$

In this case the parameters ω_0 and B should be chosen using

$$\omega_0 = \sqrt{\omega_{S2}\,\omega_{S3}} \qquad \text{and} \qquad B = \frac{\omega_{S3} - \omega_{S2}}{\omega_{L2}} \tag{8.59}$$

8.4.3 Analog to discrete-time transformation methods

Step 3 of the design procedure discussed above requires a discrete-time system function to be obtained from the analog system function. The goal is to come up with a system function $H(z)$ the characteristics of which resemble those of the analog system function $G(s)$ in some way. The goal is somewhat loosely defined; there is no clear definition of when two filters resemble each other. There is also the question of which characteristics of the two filters (magnitude, phase, time-delay, impulse response, etc.) should exhibit similarity. In most cases analog prototype filters are designed based on their magnitude responses. Therefore, our goal will be to obtain a reasonable degree of similarity between the magnitude characteristics of the two filters.

The frequency response of an analog filter is obtained from the values of the system function $G(s)$ on the $j\omega$ axis of the s-plane. In contrast, the frequency response of a discrete-time filter is obtained by evaluating the system function $H(z)$ on the unit circle of the z-plane. Ideally, for the two filters to exhibit similar characteristics, we need a transformation to map the $j\omega$ axis of the s-plane onto the unit circle of the z-plane. The main challenge in doing this is the fact that $H(\Omega)$ is 2π-periodic whereas $G(\omega)$ is not. How can a frequency axis that is infinitely wide be mapped onto the circumference of a circle with unit radius? We will see two different approaches to doing this when we discuss impulse invariance and bilinear transformation techniques. Both involve some compromises.

An even more serious requirement is that a stable analog prototype should lead to a stable discrete-time filter. Assuming that the analog prototype filter is stable and causal, all the poles of $G(s)$ are located in the left half s-plane. For a stable and causal discrete-time filter, the poles of $H(z)$ should be inside the unit circle of the z-plane. Therefore, a usable transformation should map the left half of the s-plane to the inside of the unit circle. If any points on the left half s-plane are mapped to points on or outside the unit circle of the z-plane, there is the danger of getting an unstable system function $H(z)$.

In the following sections, several transformation methods will be discussed. Some of them are useful. Others do not meet the requirements discussed above, but are still included to show what might go wrong.

8.4.4 Impulse invariance

A possible method of obtaining a discrete-time filter from an analog prototype is to ensure that the impulse response is preserved in the conversion process. Specifically, the discrete-time filter is chosen such that its impulse response is equal to a scaled and sampled version of the impulse response of the analog prototype:

$$h[n] = T\,g(nT) \tag{8.60}$$

The relationship in Eqn. (8.60) is the sampling relationship discussed in Chapter 4. The reason for the scale factor T will become obvious when we consider the magnitude spectrum of the resulting discrete-time filter. Let the system function for a causal analog prototype filter be $G(s)$ which can be expanded into partial fractions in the form

$$G(s) = \sum_{i=1}^{N} \frac{k_i}{s - p_i} \tag{8.61}$$

where p_i are the poles, and k_i are the corresponding residues. We will limit this discussion to the case where the poles of the analog prototype filter are distinct. The impulse response can be written from the partial fraction expansion as

$$g(t) = \sum_{i=1}^{N} k_i\, e^{p_i t}\, u(t) \tag{8.62}$$

Under impulse invariant design the impulse response of the discrete-time system is required to be a sampled version of $g(t)$ in Eqn. (8.62).

$$h[n] = T\,g(nT) = \sum_{i=1}^{N} T\,k_i\, e^{p_i n T}\, u[n] \tag{8.63}$$

The system function of the discrete-time filter is found by taking the z transform of $h[n]$ as

$$H(z) = \sum_{i=1}^{N} \frac{T\,k_i\, z}{z - e^{p_i T}} \tag{8.64}$$

Comparing Eqns. (8.61) and (8.64) we conclude that $H(z)$ can be obtained directly from the partial fraction expansion of $G(z)$ without the intermediate steps of computing $g(t)$ and sampling it. Each residue k_i is multiplied by the sampling interval T. Each pole p_i in Eqn. (8.61) is converted as

$$\bar{p}_i = e^{p_i T}$$

An important issue in the conversion of an analog filter to a discrete-time filter is the requirement that a stable discrete-time filter be obtained from a stable analog filter. The poles of the analog prototype filter are at

$$p_i = \sigma_i + j\omega_i\,, \qquad i = 1, \ldots, N \tag{8.65}$$

The corresponding poles of the discrete-time filter obtained through impulse invariance are at

$$\bar{p}_i = e^{p_i T} = e^{\sigma_i T} e^{j\omega_i T}, \qquad i = 1, \ldots, N \tag{8.66}$$

For a stable analog prototype filter the poles are in the left half s-plane, and therefore $\sigma_i < 0$ for $i = 1, \ldots, N$. This implies that

$$\left|\bar{p}_i\right| = e^{\sigma_i T} < 1, \qquad i = 1, \ldots, N \tag{8.67}$$

Poles in the left half s plane map to the inside of the unit circle on the z-plane. Therefore we are assured that a stable analog prototype filter leads to a stable discrete-time filter when the impulse invariance technique is used. Finally we need to understand the relationship between the magnitude responses of the analog prototype filter and the resulting discrete-time filter. This step is easy owing to the sampling relationship between $g(t)$ and $h[n]$ given by Eqn. (8.60). Using Eqn. (4.25) from Chapter 4 we obtain

$$H(\Omega) = \sum_{k=-\infty}^{\infty} G\left(\frac{\Omega - 2\pi k}{T}\right) \tag{8.68}$$

In order to avoid aliasing, the analog prototype filter must be strictly bandlimited, so that

$$G(\omega) = 0, \qquad |\omega| \geq \frac{\pi}{T} \tag{8.69}$$

and

$$H(\Omega) = G\left(\frac{\Omega}{T}\right) \qquad -\pi < \omega < \pi \tag{8.70}$$

and the spectrum of the resulting discrete-time filter is a frequency-scaled version of the analog prototype filter spectrum in the range $-\pi < \omega < \pi$. Nevertheless, the bandlimiting condition given by Eqn. (8.69) is often not satisfied by the analog prototype filter (none of the analog designs considered in Section 8.4.2 are perfectly bandlimited). Therefore, the spectrum of the resulting discrete-time filter is an aliased version of the spectrum of the analog prototype. This is illustrated in Fig. 8.22.

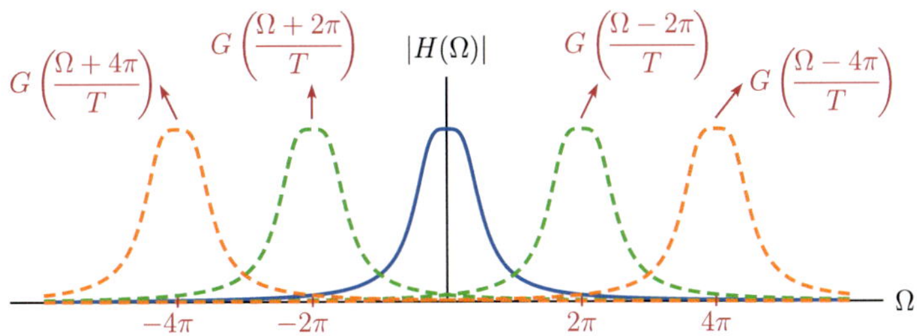

Figure 8.22 – Aliasing effect in impulse invariant design.

Based on Eqn. (8.68) the value of the system function $H(\Omega)$ at a specific frequency $\Omega = \Omega_0$ is determined by contributions from the analog prototype system function $G(\omega)$ at frequencies

$$\omega_k = \frac{\Omega_0 - 2\pi k}{T}, \qquad \text{all } k \tag{8.71}$$

Due to the aliasing of the spectrum, impulse invariant designs are only useful for lowpass filters for which aliasing can be kept at acceptable levels.

Interactive App: Spectral relationships under impulse invariance

This interactive app in `appImpulseInv.m` illustrates the mapping of the $j\omega$-axis of the s-plane to the unit circle of the z-plane for impulse invariant design. Magnitude response of an analog prototype filter with system function

$$G(s) = \frac{2}{s+2}$$

is shown along with the magnitude response of the discrete-time filter derived from it using the impulse invariance technique. Magnitude responses are shown in terms of frequencies f in Hz, and F normalized. The sampling interval T may be adjusted using a slider control. For a selected value of $F = F_0$, the corresponding alias frequencies are shown on the graph for $|G(f)|$.

a. Start with $T = 0.1$ s and $F_0 = 0.1$. Identify the aliased frequencies shown on the graph for $|G(f)|$. They indicate the magnitude values that will be added together to form the magnitude $|H(F)|$ at $F = F_0$.
b. Slowly increase F_0. Again identify the corresponding alias frequencies of $|G(f)|$, and explain how they form $|H(F_0)|$.
c. Fix $F_0 = 0.1$, and gradually increase T. Observe how the severity of the aliasing effect increases with increasing T.

Software resource: `appImpulseInv.m`

Example 8.4: Impulse invariant design of a lowpass filter

Consider the third order Chebyshev type-II analog lowpass filter with system function

$$G(s) = \frac{0.6250}{s^3 + 1.1542\,s^2 + 1.4161\,s + 0.6250}$$

Convert this filter to a discrete-time filter using the impulse invariance technique with $T = 0.2$ s. Afterward compute and graph the magnitude response of the discrete-time filter.

Solution: The system function can be written in partial fraction form as

$$G(s) = \frac{-0.2885 - j\,0.0833}{s + 0.2886 - j\,1} + \frac{-0.2885 + j\,0.0833}{s + 0.2886 + j\,1} + \frac{0.5771}{s + 0.5771}$$

Using Eqn. (8.64) the discrete-time filter system function is written in partial fraction form as

$$H(z) = \frac{-0.0577 - j\,0.0167}{z - 0.9251 - j\,0.1875} + \frac{-0.0577 + j\,0.0167}{z - 0.9251 + j\,0.1875} + \frac{0.1154}{z - 0.8910}$$

The closed form expression for $H(z)$ is

$$H(z) = \frac{0.0023\,z^2 + 0.0021\,z}{z^3 - 2.7412\,z^2 + 2.5395\,z - 0.7939}$$

The magnitude and the phase of $H(z)$ are shown in Fig. 8.23.

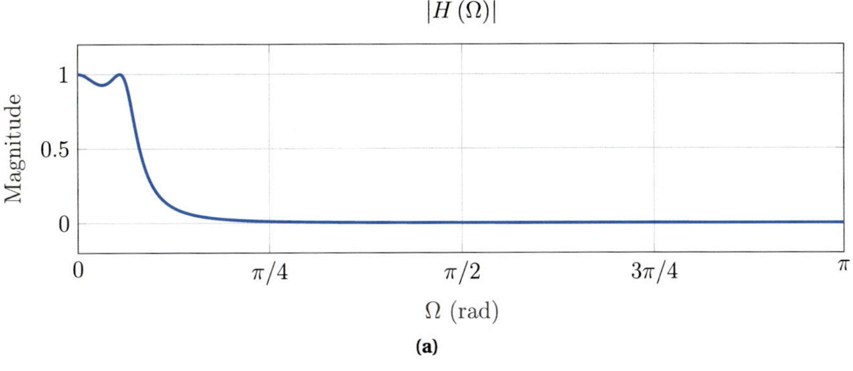

(a)

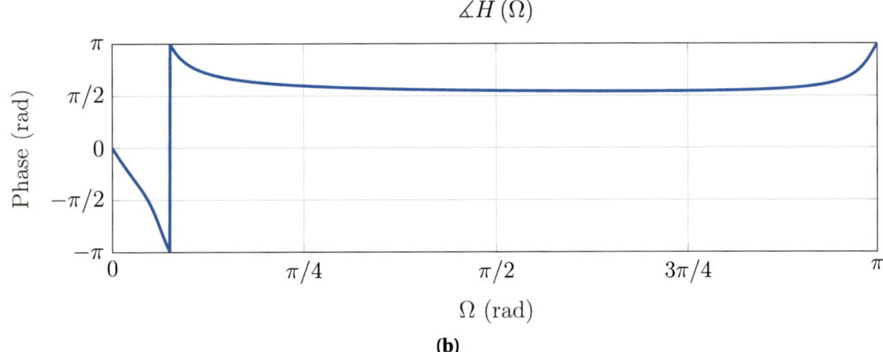

(b)

Figure 8.23 – The magnitude and the phase of the third order Chebyshev lowpass IIR filter designed in Example 8.4.

Note that aliasing can be kept at a negligible level with the choice of T, and the analog frequency $\omega_1 = 1$ rad/s corresponds to the discrete-time frequency $\Omega = 0.2$ radians.

Software resources: exdt_8_4a.m , exdt_8_4b.m

Software resource: See MATLAB Exercise 8.6.

Example 8.4 demonstrates that, if the analog prototype filter system function $G(s)$ is fixed, then the sampling interval should be chosen as small as possible to minimize the effect of aliasing. However, a typical IIR filter design problem begins with the critical frequencies Ω_1 and Ω_2 of the discrete-time filter. We then work backward to determine the critical frequencies of an appropriate analog prototype filter. Let the corresponding critical frequencies of the analog prototype filter to be designed be ω_1 and ω_2 respectively. Recall from the sampling relationship that the analog frequencies and the discrete-time frequencies are related by $\omega = \Omega/T$. If the sampling interval is T, then the sampling rate is $\omega_s = 2\pi/T$, and we obtain

$$\frac{\omega_1}{\omega_s} = \frac{\Omega_1}{2\pi} = \text{constant} \tag{8.72}$$

and

$$\frac{\omega_2}{\omega_s} = \frac{\Omega_2}{2\pi} = \text{constant} \tag{8.73}$$

Thus, if the design problem starts with the discrete-time critical frequencies, then increasing the sampling rate ω_s does not improve the aliasing condition since the bandwidth of the signal $g(t)$ to be sampled also increases proportionally. Therefore the choice of T is irrelevant, and we often use $T = 1$ for simplicity. Example 8.5 will illustrate this.

Example 8.5: Impulse invariant design specifications

A lowpass IIR filter with is to be designed with the following specifications:

$$\Omega_1 = 0.2\pi\,, \qquad \Omega_2 = 0.25\pi\,, \qquad R_p = 1 \text{ dB}\,, \qquad A_s = 30 \text{ dB}$$

Impulse invariance technique is to be used for converting an analog prototype filter to an discrete-time filter. Determine the specifications of the analog prototype if the sampling interval is to be

a. $T = 1$ s,

b. $T = 2$ s.

Solution:

a. Using $T = 1$ s, the critical frequencies of the analog prototype filter are

$$\omega_1 = \frac{\Omega_1}{T} = \frac{\Omega_1}{1} = 0.2\pi\,, \qquad \omega_2 = \frac{\Omega_2}{T} = \frac{\Omega_2}{1} = 0.25\pi$$

The dB tolerance limits are unchanged:

$$R_p = 1 \text{ dB}\,, \qquad A_s = 30 \text{ dB}$$

Let the system function for the analog prototype filter be $G_1(\omega)$ yielding an impulse response $g_1(t)$. The impulse response of the discrete-time filter is

$$h_1[n] = T\, g_1(nT) = g_1(n)$$

b. With $T = 2$ s, the critical frequencies of the analog prototype filter are

$$\omega_1 = \frac{\Omega_1}{T} = \frac{\Omega_1}{2} = 0.1\pi\,, \qquad \omega_2 = \frac{\Omega_2}{T} = \frac{\Omega_2}{2} = 0.125\pi$$

The dB tolerance limits are unchanged as before. Let the system function for the analog prototype filter be $G_2(\omega)$ so that its impulse response is $g_2(t)$. The impulse response of the discrete-time filter is obtained by sampling $g_2(t)$ every 2 seconds, that is,

$$h_2[n] = T\, g_2(nT) = 2\, g_2(2n)$$

We have thus obtained two discrete-time filters with the two choices of the sampling interval T. What is the relationship between these two filters? Let us realize that $G_1(0.2\pi) = G_2(0.1\pi)$ and $G_1(0.25\pi) = G_2(0.125\pi)$. Generalizing these relationships we have

$$G_1(\omega) = G_2\left(\frac{\omega}{2}\right)$$

Based on the scaling property of the Fourier transform, this implies that

$$g_1(n) = 2\,g_2(2n)$$

and therefore

$$h_1[n] = h_2[n]$$

The two IIR filters designed are identical to each other, independent of the choice of T.

8.4.5 Transformations based on rectangular approximation to integrals

In Section 8.4.4 we discussed the impulse invariance method for transforming a continuous-time filter to a discrete-time filter. Its major shortcoming is the aliasing of analog frequencies on the unit circle of the z-plane which practically limits its usefulness to lowpass and bandpass filters only. As an alternative approach, we can convert the differential equation of a CTLTI system to an approximately equivalent difference equation, and thus obtain a DTLTI system. In order to explore this idea we will work with a first-order system, however, the technique developed will be applicable to higher order systems as well. Consider a first-order causal analog prototype filter described by the system function

$$G(s) = \frac{Y_a(s)}{X_a(s)} = \frac{a}{s-b} \tag{8.74}$$

where a and b are arbitrary constants with the only requirement being $\text{Re}\{b\} < 0$ for stability. The relationship in Eqn. (8.74) can also be written as

$$s\,Y_a(s) = b\,Y_a(s) + a\,X_a(s) \tag{8.75}$$

Taking the inverse Laplace transform of each side of Eqn. (8.75) leads to the differential equation for the system:

$$\frac{dy_a(t)}{dt} = b\,y_a(t) + a\,x_a(t) \tag{8.76}$$

Let us define an intermediate variable $w_a(t)$ as

$$w_a(t) = b\,y_a(t) + a\,x_a(t) \tag{8.77}$$

so that the differential equation in Eqn. (8.76) becomes

$$\frac{dy_a(t)}{dt} = w_a(t) \tag{8.78}$$

Integrating both sides of Eqn. (8.78) yields

$$y_a(t) = \int_{t_0}^{t} w_a(t)\,dt + y_a(t_0) \tag{8.79}$$

Letting $t_0 = (n-1)\,T$ and $t = nT$, Eqn. (8.79) becomes

$$y_a(nT) = \int_{(n-1)T}^{nT} w_a(t)\,dt + y_a\big((n-1)\,T\big) \tag{8.80}$$

At this point, let us define discrete-time signals $x[n]$, $y[n]$, and $w[n]$ as sampled versions of their analog counterparts, i.e.,

$$x[n] = x_a(nT)\,, \quad y[n] = y_a(nT)\,, \quad w[n] = w_a(nT)$$

With these definitions, we can write

$$
\begin{aligned}
y[n] &= \int_{(n-1)T}^{nT} w_a(t)\,dt + y[n-1] \\
&= \Delta y + y[n-1]
\end{aligned}
\tag{8.81}
$$

With Eqn. (8.81) we are close to obtaining a difference equation that approximates Eqn. (8.76). We just need to find and an approximation for the integral

$$\Delta y = \int_{(n-1)T}^{nT} w_a(t)\,dt \tag{8.82}$$

Forward rectangular approximation

Let's assume that the sampling interval T is small enough so that $w_a(t)$ does not vary appreciably within the interval. This assumption should be a reasonable one as long as the conditions of Nyquist sampling theorem are observed in choosing the sampling rate $f_s = 1/T$. The integral in Eqn. (8.82) can therefore be approximated using the area of a rectangle with width equal to T. The only question that remains is the height of the rectangle. *Forward rectangular approximation* is obtained by using the left edge of the integrand as the height of the rectangle, and *backward rectangular approximation* is obtained by using the right edge instead. Fig. 8.24 illustrates these two choices.

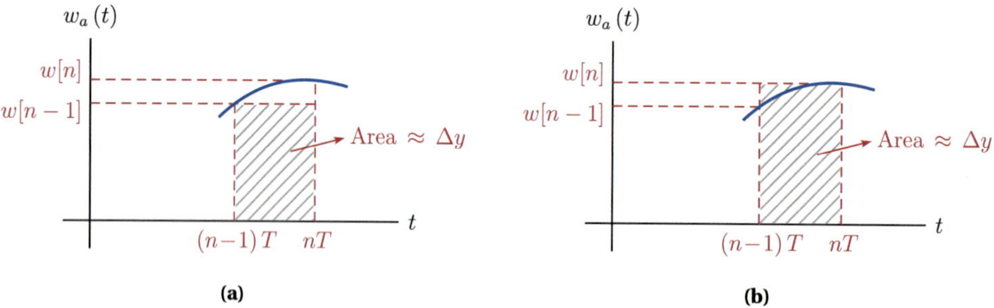

Figure 8.24 – Approximating Δy using **(a)** forward rectangular approximation and **(b)** backward rectangular approximation.

Using forward rectangular approximation, we have

$$\Delta y \approx T\,w[n-1] = bT\,y[n-1] + aT\,x[n-1] \tag{8.83}$$

Incorporating Eqn. (8.83) into Eqn. (8.81) yields the difference equation

$$
\begin{aligned}
y[n] &= bT\,y[n-1] + aT\,x[n-1] + y[n-1] \\
&= (1+bT)\,y[n-1] + aT\,x[n-1]
\end{aligned}
\tag{8.84}
$$

Taking the z-transform of both sides of the difference equation in Eqn. (8.84) we obtain

$$Y(z) = (1+bT)\,z^{-1}\,Y(z) + aT\,z^{-1}\,X(z) \tag{8.85}$$

The system function $H(z)$ for the DTLTI system is

$$H(z) = \frac{Y(z)}{X(z)} = \frac{aT\,z^{-1}}{1-(1+bT)\,z^{-1}} \tag{8.86}$$

Multiplying both the numerator and the denominator by z and simplifying, we get

$$H(z) = \frac{a}{\dfrac{z-1}{T} - b} \tag{8.87}$$

Comparing $H(z)$ with $G(s)$ in Eqn. (8.74) reveals that the former can be obtained from the latter by replacing s with

$$s = \frac{z-1}{T} \tag{8.88}$$

This is the mapping relationship between transform variables s and z. Solving Eqn. (8.88) for z yields

$$z = 1 + T\,s \tag{8.89}$$

Using the Cartesian form of $s = \sigma + j\omega$, the corresponding value of z is

$$z = z_r + j\,z_i = 1 + T\left(\sigma + j\omega\right) \quad \Longrightarrow \quad z_r = 1 + \sigma T\,, \quad z_i = \omega T$$

Earlier, two fundamental questions were raised about any transformation from the s domain to the z domain. The first question concerns the trajectory in the z-plane that corresponds to the set of points on the $j\omega$ axis of the s-plane. Setting $\sigma = 0$ we obtain

$$z_r = 1 \quad \text{and} \quad z_i = j\omega$$

Thus, the $j\omega$ axis of the s-plane maps to a vertical line that passes through $z = 1 + j0$ in the z-plane. This line is tangent to the unit circle of the z-plane at the point $z = 1 + j0$. The second fundamental question concerns the mapping of the left half s plane, that is, the stable zone of the analog prototype, onto the z-plane. The area where $\sigma < 0$ translates to $z_r < 1$, indicating that the left half s-plane maps to the area in the z-plane that is to the left of the vertical line passing through $z = 1 + j0$. These relationships are depicted in Fig. 8.25.

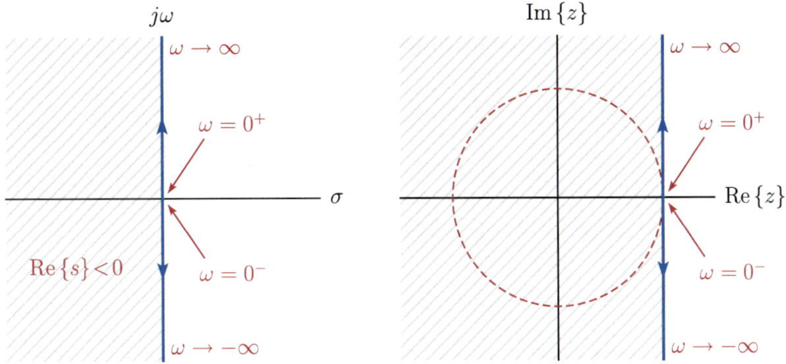

Figure 8.25 – Mapping of the s-plane to the z-plane for forward rectangular integration.

The $j\omega$ axis of the s-plane is not mapped onto the unit circle of the z-plane. Therefore, the frequency domain behavior of the two filters will show similarity only in the vicinity of $\omega = 0$ where the unit circle and the vertical line are tangent to each other. The more serious problem with this transformation method, however, is the fact that some of the left half s-plane points map to points outside the unit circle of the z-plane. As a result, a stable analog prototype may lead to an unstable discrete-time filter.

Transformation using backward rectangular approximation:

1. DTLTI system function is obtained by replacing each occurrence of s in the CTLTI system function with
$$s = \frac{z-1}{T}$$
2. A stable analog prototype filter does not necessarily produce a stable discrete-time filter.
3. Discrete-time filter frequency response resembles that of the analog prototype only in the vicinity of $\omega = 0$ where the unit circle and the vertical line $z = 1 + j\omega$ are tangent to each other.

Backward rectangular approximation

If the right edge of the integrand is used in forming the approximating rectangle (see Fig. 8.24b), then Δy is

$$\Delta y \approx T\, w[n] = bT\, y[n] + aT\, x[n] \tag{8.90}$$

Using Eqn. (8.90) in Eqn. (8.81) leads to the difference equation

$$y[n] = bT\, y[n] + aT\, x[n] + y[n-1] \tag{8.91}$$

Taking the z-transform of both sides of Eqn. (8.91) and rearranging terms, we get

$$(1-bT)\, Y(z) - z^{-1}\, Y(z) = aT\, X(z) \tag{8.92}$$

Now the DTLTI system function can be obtained as

$$H(z) = \frac{aT}{(1-bT) - z^{-1}} \tag{8.93}$$

With a bit of manipulation, $H(z)$ can be written in the form

$$H(z) = \frac{a}{\dfrac{z-1}{Tz} - b} \tag{8.94}$$

Comparing Eqn. (8.94) with Eqn. (8.74) it becomes obvious that $H(z)$ can be obtained from the analog prototype $G(s)$ by replacing every occurrence of s with

$$s = \frac{z-1}{Tz} \tag{8.95}$$

Solving Eqn. (8.95) for z we obtain

$$z = \frac{1}{1-Ts} \tag{8.96}$$

First we need to know how the $j\omega$ axis of the s-plane is mapped onto the z-plane. Using $s = j\omega$ in Eqn. (8.96) we have

$$z = \frac{1}{1-j\omega T} \tag{8.97}$$

The trajectory on the z-plane is a circle with radius of $1/2$, centered at the point $z = 1/2 + j0$. To see that this is indeed the case, let us first use Eqn. (8.97) to write $z - 1/2$ as

$$z - \frac{1}{2} = \frac{1+j\omega T}{2\,(1-j\omega T)} \tag{8.98}$$

Computing the absolute values of both sides of Eqn. (8.98) results in

$$\text{for } \sigma = 0 \quad \Longrightarrow \quad \left| z - \frac{1}{2} \right| = \frac{\sqrt{1+\omega^2 T^2}}{(2)\,\sqrt{1+\omega^2 T^2}} = \frac{1}{2} \tag{8.99}$$

The left side of Eqn. (8.99) represents the distance of z from the point $z = 1/2 + j0$. Thus, all points on the trajectory have the same distance of $1/2$ to that point, therefore they must at least be on a circular arc. Does this arc form a full circle? Evaluating Eqn. (8.97) at $\omega = 0$ and $\omega \to \mp\infty$ yields

$$\omega = 0 \quad \Longrightarrow \quad z = 1$$

$$\omega = \mp\infty \quad \Longrightarrow \quad z = 0\mp$$

Therefore, the trajectory is a complete circle. This circle is completely inside the unit circle, and is also tangent to the unit circle at the point $z = 1 + j0$. The second important question is about how the left half s-plane points map onto the z-plane. Let $s = \sigma + j\omega$ with $\sigma < 0$. The distance of z to the point $z = 1/2 + j0$ is

$$\left| z - \frac{1}{2} \right| = \frac{\sqrt{(1+\sigma T)^2 + \omega^2 T^2}}{(2)\,\sqrt{(1-\sigma T)^2 + \omega^2 T^2}} \tag{8.100}$$

Since $\sigma < 0$, we have $(1 - \sigma T) > (1 + \sigma T)$. Consequently

$$\text{for } \sigma < 0 \quad \Longrightarrow \quad \left| z - \frac{1}{2} \right| < \frac{1}{2}$$

Points on the left half s-plane map to the inside of the small circle in the z-plane. This is illustrated in Figure 8.26.

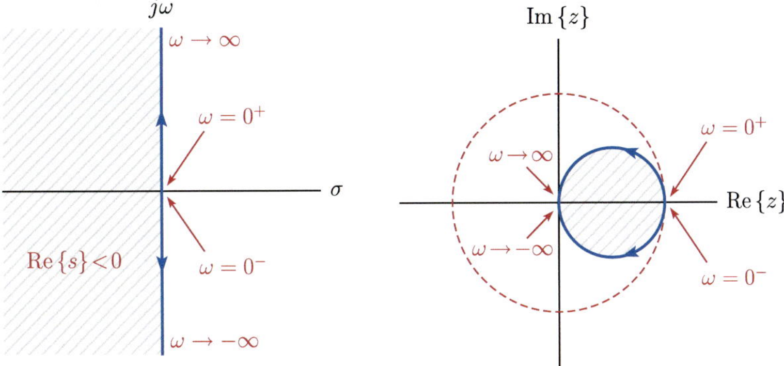

Figure 8.26 – Mapping of the s-plane to the z-plane for backward rectangular integration.

There can be no poles outside the unit circle of the z-plane, provided that the analog prototype filter does not have any poles in the right half of the s-plane. Thus a stable analog prototype filter will always lead to a stable discrete-time filter when this technique is used. The only drawback is that the frequency responses of the two filters will show similar behavior only in the vicinity of $\omega = 0$ where the two circles are tangent to each other. Based on this development, conclusions can be summarized as follows:

Transformation using backward rectangular approximation:

1. DTLTI system function is obtained by replacing each occurrence of s in the CTLTI system function with
$$s = \frac{z-1}{T\,z}$$
2. A stable analog prototype filter is guaranteed to produce a stable discrete-time filter.
3. Discrete-time filter frequency response resembles that of the analog prototype only in the vicinity of $\omega = 0$ where the two circles are tangent to each other.

8.4.6 Bilinear transformation

Among the transformation techniques discussed up to this point, impulse invariance is the only one that maps the $j\omega$ axis of the s-plane to the unit circle of the z-plane. On the other hand, it has a severe shortcoming that prevents its adoption as a general method applicable to the design of all filter types. The aliasing effect that results from sampling the impulse response limits its usefulness to lowpass and some bandpass filters only. An alternative technique known as *bilinear transformation* avoids aliasing, and overcomes this limitation.

Recall that, in Section 8.4.5, two transformation techniques were discussed. Both were based on the idea of converting the differential equation of the analog prototype filter to a difference equation by means of approximating an integral using rectangular approximation. Neither technique turned out to be very useful. Forward rectangular approximation could lead to unstable discrete-time filters, and was therefore deemed unusable. Backward rectangular approximation led to stable filters that just were not very good. What if we use trapezoidal approximation to the integral in Eqn. (8.81) instead?

As before, we will start with a first order analog prototype with system function

$$\text{Eqn. (8.74):} \qquad G(s) = \frac{Y_a(s)}{X_a(s)} = \frac{a}{s-b}$$

where a and b are arbitrary constants with the only requirement being $\text{Re}\{b\} < 0$ for stability. The differential equation for this system was determined in Eqn. (8.81) which is repeated here for convenience:

$$\text{Eqn. (8.81):} \qquad y[n] = \Delta y + y[n-1]$$

with Δy defined as

$$\Delta y = \int_{(n-1)T}^{nT} w_a(t)\,dt\,, \qquad w_a(t) = b\,y_a(t) + a\,x_a(t)$$

Consider approximating Δy with the area of the trapezoid formed between sampling instants $t = (n-1)\,T$ and $t = nT$ as shown in Fig. 8.27.

$$\begin{aligned}\Delta y &\approx \frac{w[n] + w[n-1]}{2}\,T \\ &= \frac{bT}{2}\big(y[n] + y[n-1]\big) + \frac{aT}{2}\big(x[n] + x[n-1]\big) \qquad (8.101)\end{aligned}$$

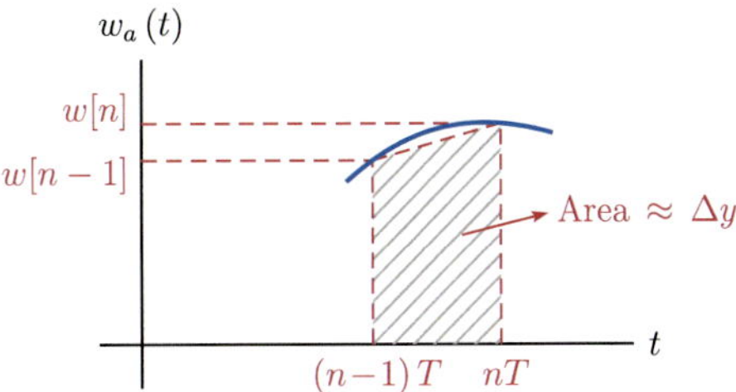

Figure 8.27 – Approximating Δy using trapezoidal approximation for the integral.

Incorporating Eqn. (8.101) into the difference equation in Eqn. (8.81) and simplifying the resulting expression yields

$$y[n] = \frac{2+bT}{2-bT}\, y[n-1] + \frac{aT}{2-bT}\, x[n] + \frac{aT}{2-bT}\, x[n-1] \tag{8.102}$$

The system function $H(z)$ can be obtained by taking the z-transform of the difference equation in Eqn. (8.102) and simplifying the resulting expression.

$$\begin{aligned} H(z) = \frac{Y(z)}{X(z)} &= \frac{aT\left(1+z^{-1}\right)}{2\left(1-z^{-1}\right) - bT\left(1+z^{-1}\right)} \\ &= \frac{a}{\left(\dfrac{2}{T}\,\dfrac{1-z^{-1}}{1+z^{-1}}\right) - b} \end{aligned} \tag{8.103}$$

By comparing the result in Eqn. (8.103) with the analog prototype system function $G(s)$ we conclude that the transformation between the s-plane and the z-plane is given by

$$s = \frac{2}{T}\,\frac{1-z^{-1}}{1+z^{-1}} \tag{8.104}$$

This is referred to as *bilinear transformation.* Solving Eqn. (8.104) for z yields

$$z = \frac{2+sT}{2-sT} \tag{8.105}$$

as the inverse of bilinear transformation. Since the relationship between s and z is invertible, bilinear transformation represents a one-to-one mapping between the s-plane and the z-plane. For each point in the s-plane, there is one and only one point in the z-plane, and vice versa. Contrast this with the impulse invariance technique that maps multiple frequencies on the $j\omega$-axis of the s-plane to a single point on the unit circle of the z-plane, resulting in aliasing.

In order to understand how the $j\omega$-axis of the s-plane is mapped to the z-plane we will evaluate Eqn. (8.105) for $s = j\omega$:

$$z = \frac{2+j\omega T}{2-j\omega T} \tag{8.106}$$

The magnitude of z is

$$|z| = \frac{\sqrt{4+\omega^2 T^2}}{\sqrt{4+\omega^2 T^2}} = 1 \tag{8.107}$$

Setting $z = e^{j\Omega}$ in Eqn. (8.107) we have

$$e^{j\Omega} = \frac{\sqrt{4+\omega^2 T^2}}{\sqrt{4+\omega^2 T^2}} = 1 \tag{8.108}$$

and

$$\Omega = \measuredangle\, e^{j\Omega} = 2\tan^{-1}\left(\frac{\omega T}{2}\right) \tag{8.109}$$

Eqn. (8.109) gives the relationship between analog frequencies ω and discrete-time frequencies Ω. It is shown graphically in fig. 8.28. Since the magnitude of z is equal to unity for all ω, the $j\omega$-axis of the s-plane is mapped onto the unit circle of the z-plane. Furthermore, the entire $j\omega$-axis of the s-plane is mapped onto the unit circle of the z-plane only once, and there is no aliasing. Bilinear transform does not suffer from the limitations of the impulse invariance technique, and can be used for designing all four frequency-selective filter types.

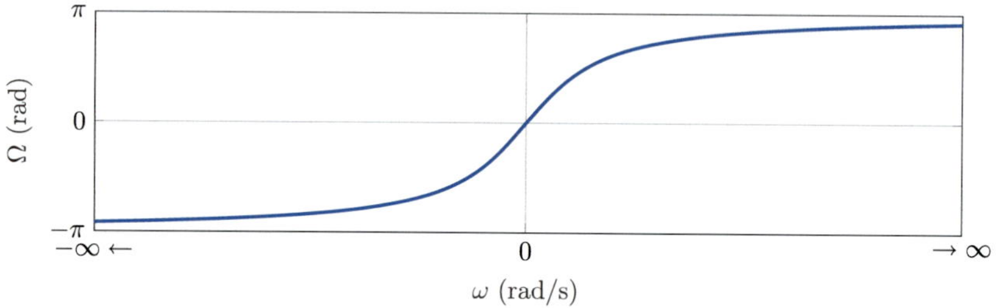

Figure 8.28 – Mapping of frequencies in bilinear transformation.

It is evident from fig. 8.28 that the mapping of frequencies is highly nonlinear. For low frequencies the relationship between ω and Ω is almost linear, however, for high frequencies we see that significant changes in ω trigger minuscule changes in Ω. This behavior is necessary if we want to map an infinite range of analog frequencies to a 2π-radian range of discrete-time frequencies. Fig. 8.29 illustrates the mapping from the $j\omega$-axis of the s-plane and the unit circle of the z-plane.

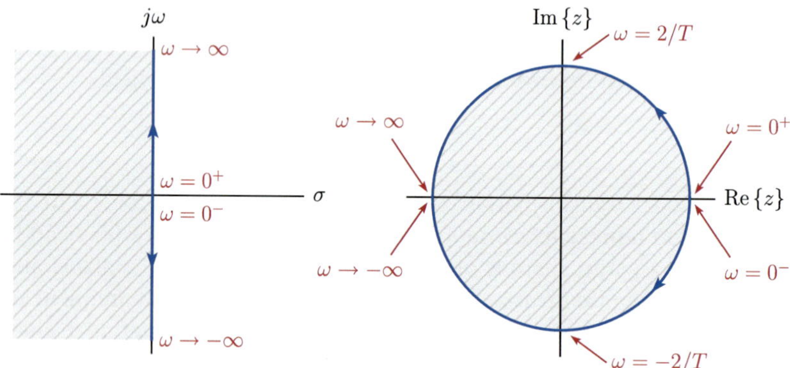

Figure 8.29 – Mapping $j\omega$-axis of the s-plane to the unit circle of the z-plane.

The $j\omega$-axis of the s-plane is mapped onto the unit circle of the z-plane. Each point on the $j\omega$-axis of the s-plane maps to one and only one point on the unit circle of the z-plane. Furthermore, each point in the left half s-plane maps to one and only one point inside the unit circle of the z-plane. This ensures that a stable analog prototype filter leads to a stable discrete-time filter under bilinear transformation.

Interactive App: Spectral relationships under bilinear transformation

The interactive app in `appBilinear.m` illustrates the mapping of the $j\omega$-axis of the s-plane to the unit circle of the z-plane for bilinear transformation. Magnitude response of an analog prototype filter

$$G(s) = \frac{2}{s+2}$$

is shown along with the magnitude response of the discrete-time filter derived from it using bilinear transformation. Magnitude responses are shown in terms of frequencies f in Hz, and F normalized. The sampling interval T may be adjusted using a slider control. For a selected value of f, the corresponding value of F is shown.

a. Start with $T = 0.1$ s and $F = 0.1$. Gradually increase the value of F. Watch the mapped frequencies on $|G(f)|$ and $|H(F)|$ as F is increased. Pay attention to the changes in f as the normalized frequency F is incremented. Initially the increases in f are by small increments, yet they become larger and larger as F is increased. Explain this phenomenon using the nonlinear relationship between f and F.

b. Bring F back down to 0.1. Now increment the sampling interval T, and observe the changes in the relationship between f and F.

Software resource: `appBilinear.m`

Example 8.6: Applying bilinear transformation

A second order analog filter has the system function

$$G(s) = \frac{2}{(s+1)\,(s+2)}$$

Convert this filter to a discrete-time filter using bilinear transformation with $T = 2$ s. Afterward compute and graph the magnitude response of the discrete-time filter.

Solution: Substituting

$$s = \frac{2}{T}\,\frac{1-z^{-1}}{1+z^{-1}} = \frac{1-z^{-1}}{1+z^{-1}}$$

we obtain

$$H(z) = \frac{2}{\left(\dfrac{1-z^{-1}}{1+z^{-1}}+1\right)\left(\dfrac{1-z^{-1}}{1+z^{-1}}+2\right)}$$

which can be simplified to

$$H(z) = \frac{z^2+2z+1}{z\,(3z+1)}$$

The magnitude and the phase of $H(z)$ are shown in Fig. 8.30.

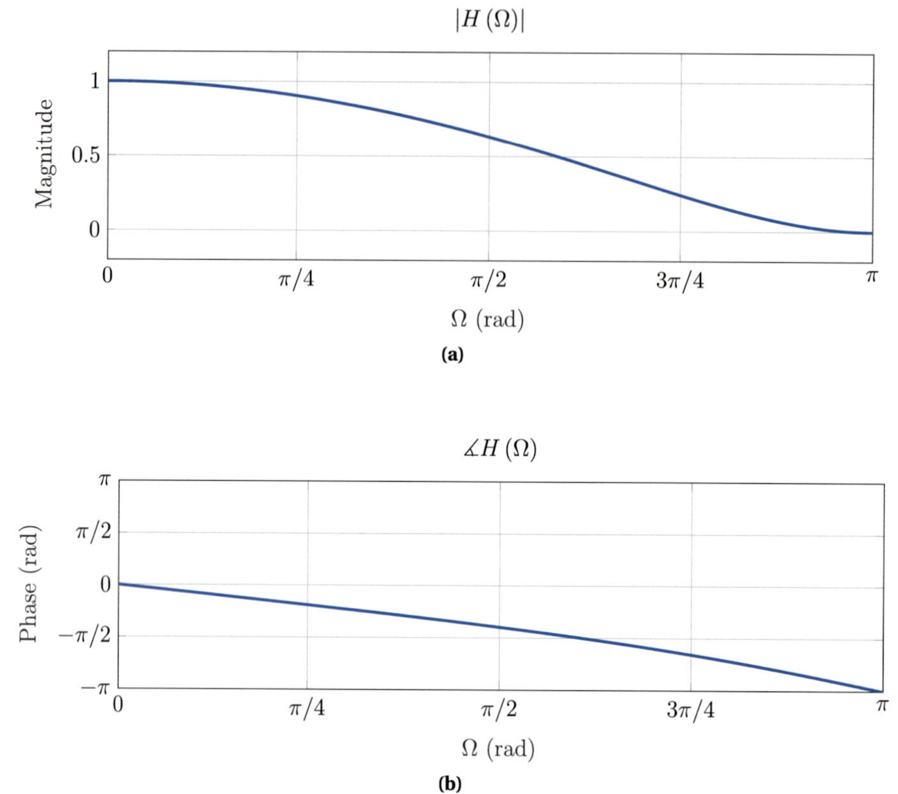

Figure 8.30 – The magnitude and the phase of the second-order IIR filter designed in Example 8.6.

Software resource: exdt_8_6.m

8.4.7 Obtaining analog prototype specifications

As discussed earlier, the first step in designing IIR filters is to convert the specifications of the desired discrete-time filter to the corresponding specifications of an appropriate analog prototype filter. For lowpass and highpass filters two edge frequencies Ω_1 and Ω_2 are given along with the tolerance values Δ_1 and Δ_2, or their decibel equivalents R_p and A_s. The critical frequencies for the discrete-time IIR filter must be translated to the critical frequencies of the analog prototype based on which method is to be used in step 3 of the design process. If impulse invariance will be used in converting the analog prototype to a discrete-time filter, critical frequencies of the analog prototype are computed as

$$\omega_1 = \frac{\Omega_1}{T}\,, \qquad \text{and} \qquad \omega_2 = \frac{\Omega_2}{T} \tag{8.110}$$

since the conversion process is based on sampling the impulse response of the analog prototype filter. If, on the other hand, bilinear transformation is to be used, then the nonlinear warping (distortion) of the frequency axis must be taken into account. In this case the critical frequencies for the analog prototype filter are computed as

$$\omega_1 = \frac{2}{T}\tan\left(\frac{\Omega_1}{2}\right), \qquad \text{and} \qquad \omega_2 = \frac{2}{T}\tan\left(\frac{\Omega_2}{2}\right) \tag{8.111}$$

This is called *prewarping*. Recall that, when bilinear transformation is applied to the analog prototype filter in step 3 of the design process, the frequency axis is distorted as described by Eqn. (8.109). Prewarping counteracts this distortion at the critical frequencies.

Example 8.7: IIR filter design using bilinear transformation

Using bilinear transformation, design a Butterworth lowpass filter with the following specifications:

$$\Omega_1 = 0.2\pi\,, \qquad \Omega_2 = 0.36\pi\,, \qquad R_p = 2\text{ dB}\,, \qquad A_s = 20\text{ dB}$$

Solution: We will use $T = 1$ s. The critical frequencies of the analog prototype are found using the prewarping equation applied to Ω_1 and Ω_2:

$$\omega_1 = 2\tan(0.2\pi/2) = 0.6498 \text{ rad}$$

$$\omega_2 = 2\tan(0.36\pi/2) = 1.2692 \text{ rad}$$

Next the filter order N needs to be determined using Eqn. (8.27).

$$N \ge \frac{\log_{10}\sqrt{\left(10^{2/10}-1\right)/\left(10^{20/10}-1\right)}}{\log_{10}(0.6498/1.2692)} = 3.8326$$

The filter order needs to be chosen as $N = 4$. The 3-dB cutoff frequency is found by setting the magnitude at the stopband edge to $-A_s$ dB by solving

$$\left(\frac{\omega_2}{\omega_c}\right)^{2N} = 10^{A_s/10} - 1$$

so that

$$\left(\frac{1.2692}{\omega_c}\right)^{8} = 10^{20/10} - 1 = 99$$

which yields $\omega_c = 0.7146$ rad/s. The analog prototype filter can now be designed using Butterworth lowpass filter design technique described in Section 8.4.2 with the values of N and ω_c found. Stable poles of the analog prototype are

$$p_{1,2} = -0.2375 \pm j0.6602\,, \qquad p_{3,4} = -0.6602 \pm j0.2735$$

and the system function is

$$G(s) = \frac{0.2608}{s^4 + 1.8675\,s^3 + 1.7437\,s^2 + 0.9537\,s + 0.2608}$$

The system function for the discrete-time filter is found through bilinear transformation using the substitution

$$s = 2\,\frac{1-z^{-1}}{1+z^{-1}}$$

which results in

$$H(z) = \frac{0.0065\,z^4 + 0.0260\,z^3 + 0.0390\,z^2 + 0.0260\,z + 0.0065}{z^4 - 2.2209\,z^3 + 2.0861\,z^2 - 0.9204\,z + 0.1594}$$

The dB magnitude of the resulting filter is shown in Fig. 8.31 with the tolerance limits.

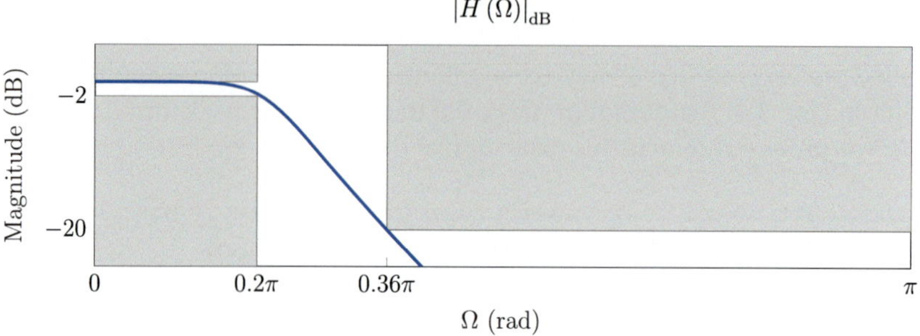

Figure 8.31 – dB Magnitude for the filter designed in Example 8.7.

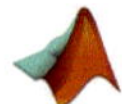

Software resources: exdt_8_7a.m , exdt_8_7b.m

Software resource: See MATLAB Exercises 8.7, 8.8, and 8.9.

8.5 FIR Filters

A length-N FIR filter is completely characterized by its finite-length impulse response $h[n]$ for $n = 0, \ldots, N-1$. The system function for such a filter is computed as

$$\begin{aligned} H(z) &= \sum_{n=0}^{N-1} h[n]\, z^{-n} \\ &= h[0] + h[1]\, z^{-1} + \ldots + h[N-1]\, z^{-(N-1)} \end{aligned} \tag{8.112}$$

The system function can also be written using non-negative powers of z in the form

$$H(z) = \frac{h[0]\, z^{N-1} + h[1]\, z^{N-2} + \ldots + h[N-2]\, z + h[N-1]}{z^{N-1}} \tag{8.113}$$

which leads us to the following conclusions:

FIR filters:

1. Length-N FIR filter has a system function that is of order $N-1$.
2. The system function $H(z)$ has $N-1$ zeros and as many poles.
3. The placement of zeros of the filter is determined by the sample amplitudes of the impulse response $h[n]$.
4. In contrast, all poles of the length-N FIR filter are at the origin, independent of the impulse response. Therefore, FIR filters are inherently stable.
5. The magnitude response of the filter is determined only by the locations of the zeros. Poles at the origin do not affect the magnitude response.

The first point above is important enough to warrant repetition here: A FIR filter of order $N-1$ has N coefficients in its impulse response. Sometimes the design specifications may refer to the length of the impulse response; at other times the order of the desired filter may be specified. Special attention needs to be paid to the distinction between them.

The sample amplitudes of the impulse response $h[n]$ for $n = 0,\ldots,N-1$ are often referred to as the *filter coefficients.* or *filter taps.* In addition to being inherently stable, FIR filters also offer several other advantages over IIR filters. One significant advantage is that FIR filters can be designed to have a linear phase characteristic. To appreciate the significance of this, recall from Section 3.6.2 of Chapter 3 that the steady-state response of a discrete-time system to a sinusoidal signal with angular frequency Ω_0 is

$$\begin{aligned}\text{Sys}\left\{\cos(\Omega_0 n)\right\} &= \left|H(\Omega_0)\right| \cos\left(\Omega_0 n + \Theta(\Omega_0)\right)\\ &= \left|H(\Omega_0)\right| \cos\left(\Omega_0 \left[n + \frac{\Theta(\Omega_0)}{\Omega_0}\right]\right)\end{aligned} \tag{8.114}$$

where $\left|H(\Omega)\right|$ and $\Theta(\Omega)$ are the magnitude and the phase of the system function, respectively. Now suppose the phase of the system is linear, i.e.,

$$\Theta(\Omega) = -M\Omega$$

Using this phase response in Eqn. (8.114), we get

$$\text{Sys}\left\{\cos(\Omega_0 n)\right\} = \left|H(\Omega_0)\right| \cos\left(\Omega_0 (n-M)\right) \tag{8.115}$$

The significance of Eqn. (8.115) is that a linear-phase system delays every frequency component of the input signal by the same amount. It does not distort the time alignment of the frequency components it passes through.

Another advantage of FIR filters over IIR filters is that they offer more choices in terms of what types of filters can be designed. With FIR filters we are not limited to just lowpass, highpass, bandpass, and bandstop filters. Multiband filters with arbitrary magnitude responses can be approximated. Differentiators and Hilbert transform filters can also be designed.

The only drawback FIR filters exhibit is about their computational complexity. With IIR filters, the use of feedback makes it possible to obtain sharp transitions from one frequency band to the next with relatively low filter orders. Achieving similar frequency domain behavior with a FIR filter requires a significantly higher filter order, and thus more arithmetic operations per output sample. This may be especially important in real-time processing where we have a limited time interval to finish processing the current frame of data before the next frame arrives. On the other hand, the use of FFT-based fast convolution algorithms in real-time processing may alleviate the computational burden to a certain degree (see Section 7.6 of Chapter 7).

8.5.1 Linear phase in FIR filters

It can be shown that a length-N FIR filter with impulse response $h[n]$ has a linear phase characteristic if the impulse response exhibits positive symmetry

$$h[n] = h[N-1-n] \qquad \text{for } n = 0,\ldots,N-1 \tag{8.116}$$

or negative symmetry (may be referred to as *antisymmetry*)

$$h[n] = -h[N-1-n] \qquad \text{for } n = 0,\ldots,N-1 \tag{8.117}$$

The conditions given by Eqns. (8.116) (8.117) lead to four possible scenarios:

- Type-I: N: odd, $h[n] = h[N-1-n]$ for $n = 0,\ldots,N-1$
- Type-II: N: even, $h[n] = h[N-1-n]$ for $n = 0,\ldots,N-1$
- Type-III: N: odd, $h[n] = -h[N-1-n]$ for $n = 0,\ldots,N-1$
- Type-IV: N: even, $h[n] = -h[N-1-n]$ for $n = 0,\ldots,N-1$

The constant time delay for a length-N linear-phase FIR filter is

$$M = \frac{N-1}{2} \tag{8.118}$$

for all four possibilities. The parameter M is also the axis of symmetry. For odd values of N, the symmetry axis corresponds with a sample index. It has equal number of samples on either side of it. For even values of N, the symmetry axis is at the midpoint of two sample indices. It is also interesting to see that, for even N, the time-delay M is not an integer, but rather includes a half sample. We will further elaborate what a half-sample delay means for a discrete-time signal later in this chapter. A complete proof of linear-phase conditions must address each of the four possibilities listed above.

Type-I FIR filters

N odd, and $h[n] = h[N-1-n]$ for $n = 0,\ldots,N-1$:

The system function for the FIR filter is

$$H(\Omega) = \sum_{n=0}^{N-1} h[n]\, e^{-j\Omega n}$$

We will partition this expression so that the first $(N-1)/2$ terms form one summation, the last $(N-1)/2$ terms form another, and the middle term at index position $(N-1)/2$ is left by itself. This partitioning leads to

$$H(\Omega) = \sum_{n=0}^{(N-3)/2} h[n]\, e^{-j\Omega n} + h\left[\frac{N-1}{2}\right] e^{-j\Omega(N-1)/2} + \sum_{n=(N+1)/2}^{N-1} h[n]\, e^{-j\Omega n} \tag{8.119}$$

Now let us apply the variable change $n = N-1-n'$ to the second summation in Eqn. (8.119) to obtain

$$H(\Omega) = \sum_{n=0}^{(N-3)/2} h[n]\, e^{-j\Omega n} + h\left[\frac{N-1}{2}\right] e^{-j\Omega(N-1)/2} + \sum_{n'=0}^{(N-3)/2} h[N-1-n']\, e^{-j\Omega(N-1-n')} \tag{8.120}$$

Since $h[N-1-n'] = h[n']$, Eqn. (8.120) can be written as

$$H(\Omega) = \sum_{n=0}^{(N-3)/2} h[n]\, e^{-j\Omega n} + h\left[\frac{N-1}{2}\right] e^{-j\Omega(N-1)/2} + \sum_{n=0}^{(N-3)/2} h[n]\, e^{j\Omega(n-N+1)}$$

$$= \left[h\left[\frac{N-1}{2}\right] + \sum_{n=0}^{(N-3)/2} \left(h[n]\, e^{-j\Omega(n-(N-1)/2)} + h[n]\, e^{j\Omega(n-(N-1)/2)}\right)\right] e^{-j\Omega(N-1)/2} \tag{8.121}$$

From this point on, we will use $M = (N-1)/2$ to keep the notation compact. Eqn. (8.121) can be further simplified to

$$H(\Omega) = \underbrace{\left[h[M] + 2\sum_{n=0}^{M-1} h[n]\, \cos\big(\Omega\,(n-M)\big) \right]}_{H_r(\Omega)} e^{-j\Omega M} \tag{8.122}$$

The expression in square brackets is purely real. We will refer to it as the *amplitude response* of the filter. Let us write the system function in Eqn. (8.122) as

$$H(\Omega) = H_r(\Omega)\, e^{-j\Omega M} \tag{8.123}$$

The phase term in Eqn. (8.123) is linear with a slope of $-M$. We are not quite ready to call this the phase of the FIR filter, however, since the amplitude response could have negative values for some Ω, and therefore is not the magnitude of $H(\Omega)$. Let us write $H_r(\Omega)$ as

$$H_r(\Omega) = \left|H_r(\Omega)\right|\, e^{j\theta_r(\Omega)} \tag{8.124}$$

where

$$\theta_r(\Omega) = \begin{cases} \pm\pi\,, & \text{if } H_r(\Omega) < 0 \\ 0\,, & \text{otherwise} \end{cases} \tag{8.125}$$

The magnitude response of the type-I linear-phase FIR filter is

$$\left|H(\Omega)\right| = \left|H_r(\Omega)\right| = \left| h[M] + 2\sum_{n=0}^{M-1} h[n]\, \cos\big(\Omega\,(n-M)\big) \right| \tag{8.126}$$

and its phase response is

$$\measuredangle H(\Omega) = -\Omega M + \theta_r(\Omega) = \begin{cases} -\Omega M \pm \pi\,, & \text{if } H_r(\Omega) < 0 \\ -\Omega M\,, & \text{otherwise} \end{cases} \tag{8.127}$$

The phase response shows discontinuities of $\pm\pi$ radians at points where $H_r(\Omega)$ changes from positive to negative or vice versa. Its slope is constant, independent of frequency. This translates to a constant time delay of $(N-1)/2$ samples.

Inspection of Eqn. (8.126) reveals no restrictions on what values $\left|H(\Omega)\right|$ can take at any frequency. We will see that each of the other three linear-phase filter types pose restrictions at certain frequencies. Because of this, type-I is the most widely used linear-phase FIR filter type.

Example 8.8: Type-I FIR filter

A length-5 FIR filter has the impulse response

$$h[n] = \{ \underset{\substack{\uparrow \\ n=0}}{3}, 2, 1, 2, 3 \}$$

Compute the magnitude and the phase characteristics of this filter. Show that the phase is a linear function of Ω.

Solution: The DTFT of the impulse response is

$$H(\Omega) = 3 + 2e^{-j\Omega} + e^{-j2\Omega} + 2e^{-j3\Omega} + 3e^{-j4\Omega}$$

Let us factor out $e^{-j2\Omega}$ and write the result as

$$H(\Omega) = \underbrace{\left[3\,e^{j2\Omega} + 2\,e^{j\Omega} + 1 + 2\,e^{-j\Omega} + 3\,e^{-j2\Omega}\right]}_{H_r(\Omega)}\,e^{-j2\Omega}$$

The amplitude response in square brackets contains symmetric exponentials. Using Euler's formula, it can be written as

$$\begin{aligned} H_r(\Omega) &= 3\,e^{j2\Omega} + 2\,e^{j\Omega} + 1 + 2\,e^{-j\Omega} + 3\,e^{-j2\Omega} \\ &= 1 + 4\cos(\Omega) + 6\cos(2\Omega) \end{aligned} \tag{8.128}$$

which is purely real, and is shown in Fig. 8.32. Magnitude and phase of $H_r(\Omega)$ are shown in Fig. 8.33. Note how $\theta_r = \pm\pi$ where $H_r(\Omega) < 0$.

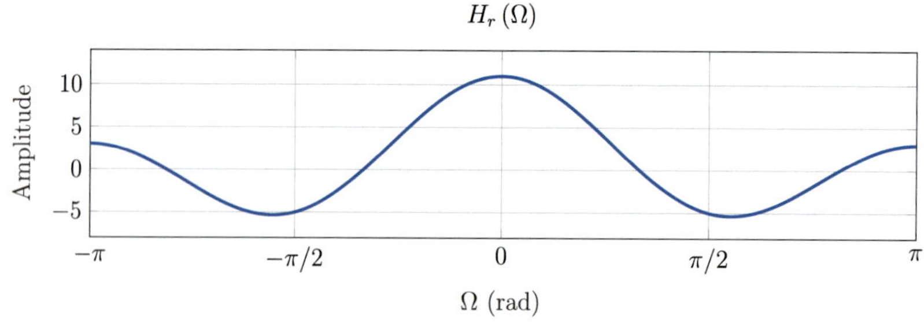

Figure 8.32 – $H_r(\Omega)$ found in Eqn. (8.128).

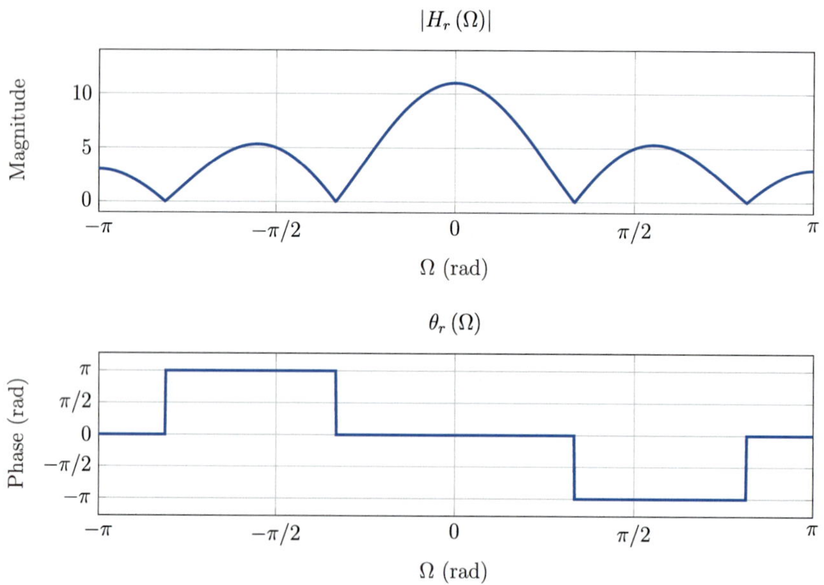

Figure 8.33 – Magnitude and the phase of $H_r(\Omega)$ found in Eqn. (8.128).

The magnitude and the phase of $H(\Omega)$ are

$$\left|H(\Omega)\right| = \left|H_r(\Omega)\right|, \qquad \text{and} \qquad \measuredangle H(\Omega) = -2\Omega + \theta_r(\Omega)$$

$|H_r(\Omega)|$ and $\measuredangle H(\Omega)$ are shown in Fig. 8.34. The phase characteristic is considered linear in spite of the discontinuities it exhibits. Vertical jumps in the phase are due to the fact that $H_r(\Omega)$ is negative for some frequencies. What matters for linear phase is that all non-vertical sections of the phase characteristic have the same slope value.

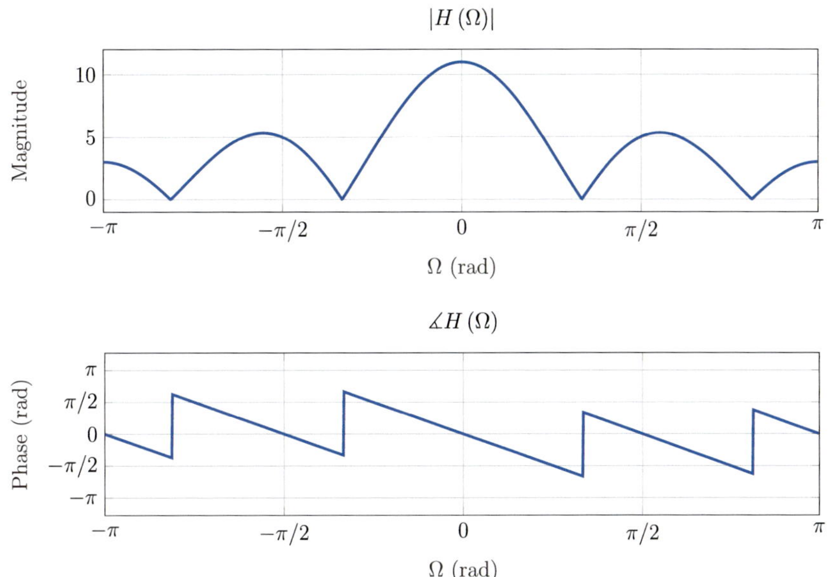

Figure 8.34 – Magnitude and the phase of the FIR filter used in Example 8.8.

Software resource: `exdt_8_8.m`

Type-II FIR filters

N even, and $h[n] = h[N-1-n]$ for $n = 0, \ldots, N-1$:

Similar to the derivation for type-I filters, we will write the system function using two summations:

$$H(\Omega) = \sum_{n=0}^{N/2-1} h[n]\, e^{-j\Omega n} + \sum_{n=N/2}^{N-1} h[n]\, e^{-j\Omega n} \tag{8.129}$$

Application of the variable change $n = N-1-n'$ to the second summation in Eqn. (8.129) yields

$$H(\Omega) = \sum_{n=0}^{N/2-1} h[n]\, e^{-j\Omega n} + \sum_{n'=0}^{N/2-1} h[N-1-n']\, e^{-j\Omega(N-1-n')} \tag{8.130}$$

Using the symmetry condition $h[N-1-n'] = h[n']$ and substituting M for $(N-1)/2$, Eqn. (8.130) becomes

$$H(\Omega) = \underbrace{\left[2 \sum_{n=0}^{M-1/2} h[n] \cos\left(\Omega(n-M)\right) \right]}_{H_r(\Omega)} e^{-j\Omega M} \tag{8.131}$$

and the magnitude of the system function is

$$\left|H(\Omega)\right| = 2\left|\sum_{n=0}^{M-1/2} h[n]\,\cos\left(\Omega\,(n-M)\right)\right| \tag{8.132}$$

As before, the slope of the linear-phase characteristic is also the amount of the time delay caused by the filter which is $M = (N-1)/2$ samples. Since N is even, the time delay is not an integer, but includes a half sample delay.

For type-II filters, $\left|H(\Omega)\right| = 0$ for $\Omega = \pm\pi$. Equivalently, the z-domain system function $H(z)$ always has a zero at $z = -1 + j0$. Therefore, type-II filters cannot be used for highpass and bandstop filters. Their use is limited to lowpass and bandpass filters and differentiators only.

Example 8.9: Type-II FIR filter

A length-6 FIR filter has the impulse response

$$h[n] = \{ \underset{\underset{n=0}{\uparrow}}{3}, 2, 1, 1, 2, 3 \}$$

Compute the magnitude and the phase characteristics of this filter. Show that the phase is a linear function of Ω.

Solution: The DTFT of the impulse response is

$$H(\Omega) = 3 + 2\,e^{-j\Omega} + e^{-j2\Omega} + e^{-j3\Omega} + 2\,e^{-j4\Omega} + 3\,e^{-j5\Omega}$$

We will factor out $e^{-j5\Omega/2}$ and write the result as

$$\begin{aligned} H(\Omega) &= \left[3\,e^{j5\Omega/2} + 2\,e^{j3\Omega/2} + e^{j\Omega/2} + e^{-j\Omega/2} + 2\,e^{-j3\Omega/2} + 3\,e^{-j5\Omega/2}\right] e^{-j5\Omega/2} \\ &= \left[2\,\cos\left(\Omega/2\right) + 4\,\cos\left(3\Omega/2\right) + 6\,\cos\left(5\Omega/2\right)\right] e^{-j5\Omega/2} \end{aligned}$$

The amplitude response is

$$H_r(\Omega) = 2\,\cos\left(\Omega/2\right) + 4\,\cos\left(3\Omega/2\right) + 6\,\cos\left(5\Omega/2\right)$$

The magnitude of $H(\Omega)$ is the same as the magnitude of $H_r(\Omega)$, that is,

$$\left|H(\Omega)\right| = \left|H_r(\Omega)\right| = \left|2\,\cos\left(\Omega/2\right) + 4\,\cos\left(3\Omega/2\right) + 6\,\cos\left(5\Omega/2\right)\right|$$

and the phase response is

$$\measuredangle H(\Omega) = -\frac{5\Omega}{2} + \theta_r(\Omega)$$

Magnitude and phase characteristics of the filter are shown in Fig. 8.35. The magnitude response is equal to zero at $\Omega = \pm\pi$. This is a common restriction for type-II FIR filters.

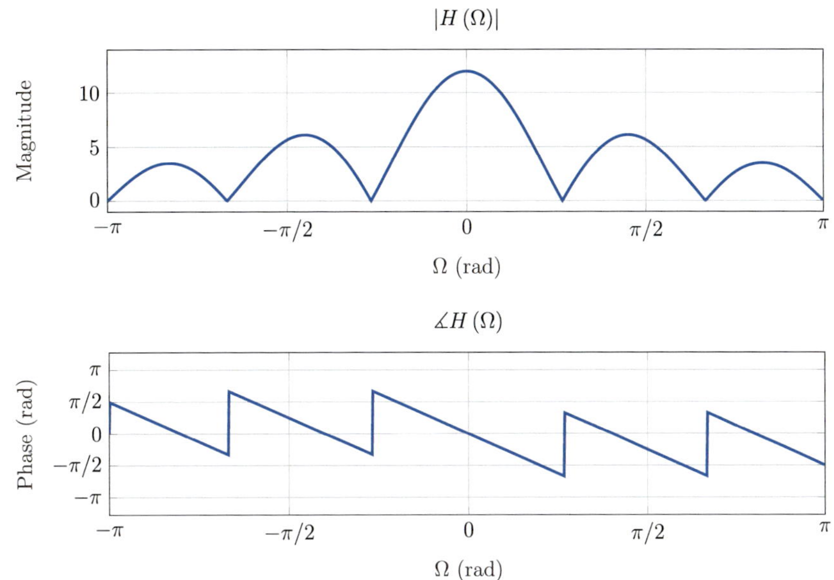

Figure 8.35 – Magnitude and phase responses of the FIR filter of Example 8.9.

Software resource: exdt_8_9.m

Type-III FIR filters

N odd, and $h[n] = -h[N-1-n]$ for $n = 0,\ldots,N-1$:

Derivation of the system function for this case will be similar to that of type-I. Because of the negative symmetry condition, we have $h[M] = 0$ for the middle term. It can be shown that the system function is

$$\begin{aligned} H(\Omega) &= -j2\sum_{n=0}^{M-1} h[n]\,\sin\big(\Omega(n-M)\big)\,e^{-j\Omega M} \\ &= \underbrace{\left[2\sum_{n=0}^{M-1} h[n]\,\sin\big(\Omega(n-M)\big)\right]}_{H_r(\Omega)} e^{-j\Omega M}\,e^{-j\pi/2} \end{aligned} \tag{8.133}$$

The magnitude response of the type-III linear-phase FIR filter is

$$\big|H(\Omega)\big| = 2\left|\sum_{n=0}^{M-1} h[n]\,\sin\big(\Omega(n-M)\big)\right| \tag{8.134}$$

It should be noted that, due to the form of Eqn. (8.126), $\big|H(\Omega)\big| = 0$ for $\Omega = 0$ and $\Omega = \pm\pi$. Consequently, type-III filters cannot be used for designing frequency selective filters other than bandpass filters.

Example 8.10: Type-III FIR filter

A length-5 FIR filter has the impulse response

$$h[n] = \{ \underset{\substack{\uparrow \\ n=0}}{3}, 2, 0, -2, -3 \}$$

Compute the magnitude and the phase characteristics of this filter. Show that the phase is a linear function of Ω.

Solution: The DTFT of the impulse response is

$$H(\Omega) = 3 + 2\,e^{-j\Omega} - 2\,e^{-j3\Omega} - 3\,e^{-j4\Omega}$$

Factoring out $e^{-j2\Omega}$, the result becomes

$$\begin{aligned} H(\Omega) &= \left[3\,e^{j2\Omega} + 2\,e^{j\Omega} - 2\,e^{-j\Omega} - 3\,e^{-j2\Omega} \right] e^{-j2\Omega} \\ &= j\left[4\sin(\Omega) + 6\sin(2\Omega) \right] e^{-j2\Omega} \end{aligned}$$

Since $j = e^{j\pi/2}$, we can write $H(\Omega)$ as

$$H(\Omega) = \underbrace{\left[4\sin(\Omega) + 6\sin(2\Omega) \right]}_{H_r(\Omega)} e^{-j(2\Omega - \pi/2)}$$

The magnitude of $H(\Omega)$ is the same as the magnitude of $H_r(\Omega)$, that is,

$$\left|H(\Omega)\right| = \left|H_r(\Omega)\right| = \left|4\sin(\Omega) + 6\sin(2\Omega)\right|$$

and the phase response is

$$\measuredangle H(\Omega) = -2\Omega + \frac{\pi}{2} + \theta_r(\Omega)$$

Magnitude and phase characteristics of the filter are shown in Fig. 8.36.

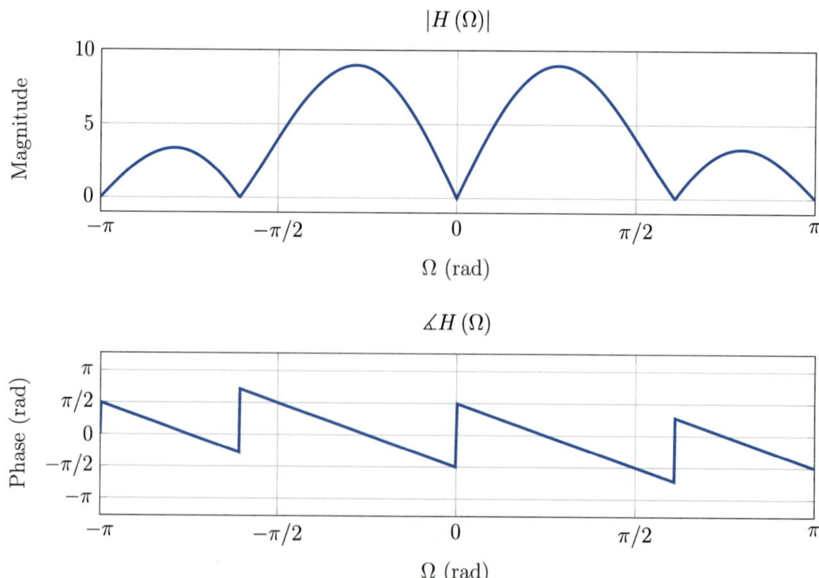

Figure 8.36 – Magnitude and phase responses of the FIR filter of Example 8.10.

The magnitude response is equal to zero at $\Omega = 0$ and $\Omega = \pm\pi$. This is quite restrictive, and limits the use of type-III for bandpass filters only.

Software resource: exdt_8_10.m

Type-IV FIR filters

N even, and $h[n] = -h[N-1-n]$ for $n = 0, \ldots, N-1$:

The derivation in this case is similar to that of type-II filters. It can be shown that the system function is

$$H(\Omega) = -j2 \sum_{n=0}^{M-1/2} h[n] \sin\big(\Omega(n-M)\big)\, e^{-j\Omega M}$$

$$= \underbrace{\left[2 \sum_{n=0}^{M-1/2} h[n] \sin\big(\Omega(n-M)\big)\right]}_{H_r(\Omega)} e^{-j\Omega M}\, e^{-j\pi/2} \qquad (8.135)$$

and the magnitude of the system function is

$$\big|H(\Omega)\big| = 2\left|\sum_{n=0}^{M-1/2} h[n] \sin\big(\Omega(n-M)\big)\right| \qquad (8.136)$$

The magnitude is equal to zero for $\Omega = 0$. The corresponding z-domain system function has a zero at $z = 1 + j0$. Lowpass and bandstop filters cannot be designed as type-IV filters.

Example 8.11: Type-IV FIR filter

A length-6 FIR filter has the impulse response

$$h[n] = \{ \underset{\underset{n=0}{\uparrow}}{3}, 2, 1, -1, -2, -3 \}$$

Compute the magnitude and the phase characteristics of this filter. Show that the phase is a linear function of Ω.

Solution: The DTFT of the impulse response is

$$H(\Omega) = 3 + 2\,e^{-j\Omega} + e^{-j2\Omega} - e^{-j3\Omega} - 2\,e^{-j4\Omega} - 3\,e^{-j5\Omega}$$

Factoring out $e^{-j5\Omega/2}$, the result becomes

$$H(\Omega) = \left[3\,e^{j5\Omega/2} + 2\,e^{j3\Omega/2} + e^{j\Omega/2} - e^{-j\Omega/2} - 2\,e^{-j3\Omega/2} - 3\,e^{-j5\Omega/2}\right] e^{-j5\Omega/2}$$

$$= j\,\big[2\sin(\Omega/2) + 4\sin(3\Omega/2) + 6\sin(5\Omega/2)\big]\, e^{-j5\Omega/2}$$

Susbtituting $j = e^{j\pi/2}$, we can write $H(\Omega)$ as

$$H(\Omega) = \underbrace{\big[2\sin(\Omega/2) + 4\sin(3\Omega/2) + 6\sin(5\Omega/2)\big]}_{H_r(\Omega)}\, e^{-j(5\Omega/2 - \pi/2)}$$

The magnitude of $H(\Omega)$ is the same as the magnitude of $H_r(\Omega)$, that is,

$$|H(\Omega)| = |H_r(\Omega)| = \left|2\sin(\Omega/2) + 4\sin(3\Omega/2) + 6\sin(5\Omega/2)\right|$$

and the phase response is

$$\measuredangle H(\Omega) = -\frac{5\Omega}{2} + \frac{\pi}{2} + \theta_r(\Omega)$$

Magnitude and phase characteristics of the filter are shown in Fig. 8.37.

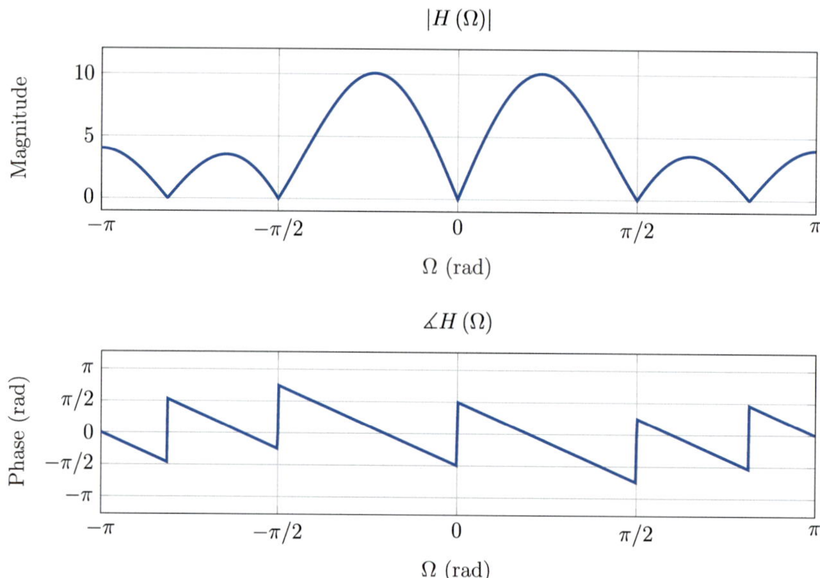

Figure 8.37 – Magnitude and phase responses of the FIR filter of Example 8.11.

As expected, the magnitude response is equal to zero at $\Omega = 0$ which makes it impossible to design lowpass and bandstop filters in type-IV.

 **Software resource:** exdt_8_11.m

8.5.2 Design of FIR filters

In this section we will present two approaches to the problem of designing FIR filters; one very simplistic approach that is nevertheless instructive, and one elegant approach that is based on computer optimization. In a general sense, the design procedure for FIR filters consists of the following steps:

1. Start with the desired frequency response. Select the appropriate length N of the filter. This choice may be an educated guess, or may rely on empirical formulas.
2. Choose a design method that attempts to minimize, in some way, the difference between the desired frequency response and the actual frequency response that results.
3. Determine the filter coefficients $h[n]$ using the design method chosen.
4. Compute the system function and decide if it is satisfactory. If not, repeat the process with a different value of N and/or a different design method.

It was discussed earlier in this chapter that one of the main reasons for preferring FIR filters over IIR filters is the possibility of a linear phase characteristic. Linear phase is desirable since it leads to a time-delay characteristic that is constant independent of frequency. As far as real-time implementation of IIR and FIR filters are concerned, we have seen that IIR filters are mathematically more efficient and computationally less demanding of hardware resources compared to FIR filters. If linear phase is not a significant concern in a particular application, then an IIR filter may be preferred. If linear phase is a requirement, on the other hand, a FIR filter must be chosen even though its implementation may be more costly. Therefore, in the discussion of FIR filter design, we will focus on linear-phase FIR filters only.

A simple example: First-order backward differentiator

Before we start discussing more formal design methods for FIR filters, we will consider a fairly simple problem of designing a differentiator. In analog systems, a building block may be needed the output of which is the time derivative of its input, i.e.,

$$y_a(t) = \frac{dx_a(t)}{dt} \tag{8.137}$$

In radar applications, for example, estimation of the velocity of a moving target requires estimating the time derivative of position data. Transforming both sides of Eqn. (8.137) through the use of the Laplace transform yields

$$Y_a(s) = s\,X_a(s) \tag{8.138}$$

Therefore, the system function of an analog differentiator is $G(s) = s$. Our goal is to design the simplest possible system to compute an estimate of the derivative in discrete-time. Fig. 8.38 depicts the idea of using a discrete-time differentiator for estimating dx_a/dt.

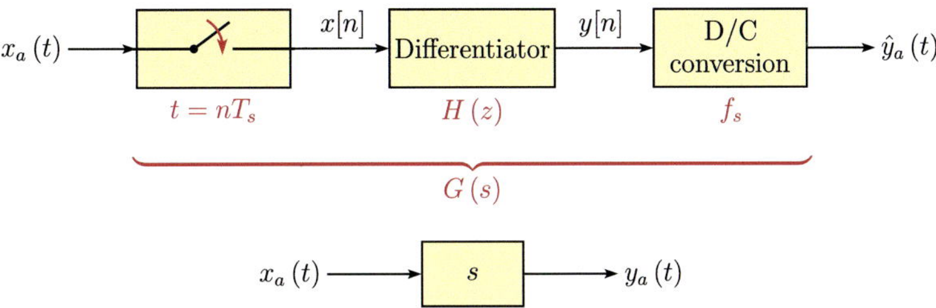

Figure 8.38 – Ideal differentiator and its discrete-time approximation.

We know from basic calculus that one way of approximating a derivative is through the use of finite differences. At the time instant $t = t_0$, the derivative of $x_a(t)$ can be approximated using the *first backward difference* as

$$y_a(t_0) = \left.\frac{dx_a(t)}{dt}\right|_{t=t_0} \approx \frac{x_a(t_0) - x_a(t_0 - T)}{T} \tag{8.139}$$

provided that the time step T is sufficiently small. This is illustrated in Fig. 8.39. The slope of the tangent passing through the point $[t_0, x_a(t_0)]$ represents the true value of the derivative $dx_a(t)/dt$ at time instant $t = t_0$. The slope of the chord passing through the points $[t_0 - T, x_a(t_0 - T)]$ and $[t_0, x_a(t_0)]$ is the approximate value of the same.

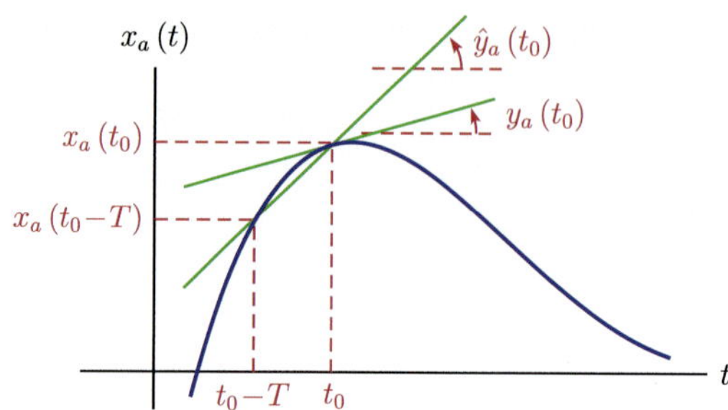

Figure 8.39 – Approximating the first derivative using a finite difference.

Let $t_0 = nT$ where T is the sampling interval. Rewriting the relationship in Eqn. (8.139) with this change we obtain

$$y_a(nT) \approx \frac{x_a(nT) - x_a(nT-T)}{T} \tag{8.140}$$

Defining discrete-time versions $x[n] = x_a(nT)$ and $y[n] = y_a(nT)$ of the signals involved, Eqn. (8.140) becomes

$$y[n] = \frac{1}{T}\,x[n] - \frac{1}{T}\,x[n-1] \tag{8.141}$$

which is the difference equation of a first-order differentiator. It corresponds to a FIR filter with impulse response

$$h[n] = \{\underset{\substack{\uparrow \\ n=0}}{1/T}, -1/T\} \tag{8.142}$$

By inspecting the impulse response we conclude that this is a type-IV linear-phase FIR filter. The z-domain system function is

$$H(z) = \frac{1-z^{-1}}{T} = \frac{z-1}{Tz} \tag{8.143}$$

The frequency response of the differentiator is obtained by evaluating $H(z)$ on the unit circle of the z-plane.

$$H(\Omega) = \frac{1-e^{-j\Omega}}{T} = \frac{2}{T}\sin\left(\Omega/2\right)e^{-j(\Omega/2-\pi/2)} \tag{8.144}$$

It is interesting to compare the frequency response of the first-order FIR differentiator to that of the ideal differentiator. For the continuous-time differentiator we have $G(\omega) = j\omega$. Substituting $\omega = \Omega/T$, the system function for the ideal differentiator is obtained as

$$H_d(\Omega) = j\frac{\Omega}{T} \tag{8.145}$$

with magnitude and phases responses

$$\left|H_d(\Omega)\right| = \frac{|\Omega|}{T}, \qquad \text{and} \qquad \measuredangle H_d(\Omega) = \begin{cases} \dfrac{\pi}{2}, & \text{for } \Omega > 0 \\ -\dfrac{\pi}{2}, & \text{for } \Omega < 0 \end{cases} \tag{8.146}$$

Fig. 8.40 depicts a comparison of magnitude and phase responses of ideal and practical differentiators. Observe how the first-order FIR differentiator approximates the ideal behavior quite well in the vicinity of $\Omega = 0$, but deviates from it at higher frequencies.

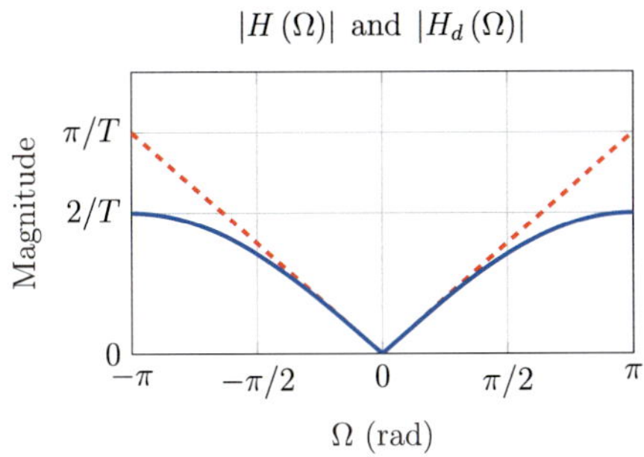

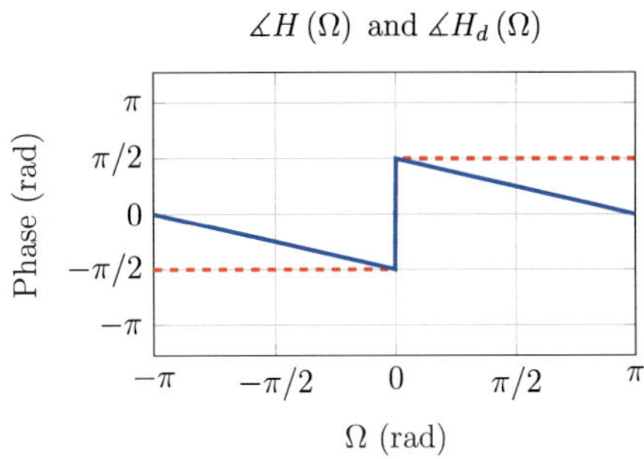

Figure 8.40 – Magnitude and phase responses of ideal and first-order differentiators.

Example 8.12: First-order backward differentiator

The first-order differentiator with impulse response in Eqn. (8.142) is used for filtering an analog sinusoidal signal as described in Fig. 8.38. The sinusoidal input signal $x_a(t)$ is

$$x_a(t) = \sin\left(2\pi f_0 t\right)$$

with $f_0 = 10$ Hz. The sampling rate used is $f_s = 200$ Hz. Determine the steady-state output signal $\hat{y}_a(t)$ of the differentiator. Sketch it along with $y_a(t)$, the exact derivative of $x_a(t)$.

Solution: The normalized frequency of the input sinusoidal signal is

$$F_0 = \frac{f_0}{f_s} = \frac{10}{200} = 0.05$$

and the sampled version of the input signal is

$$x[n] = \sin(2\pi F_0) = \sin(0.1\pi n)$$

Evaluating the system function of the differentiator at $\Omega = 0.1\pi$ yields

$$H(0.1\pi) = \frac{2}{0.005}\,\sin\left(\frac{0.1\pi}{2}\right)\,e^{-j(0.1\pi/2-\pi/2)} = 62.5738\,e^{j1.4137}$$

The steady state output of the differentiator is

$$\hat{y}[n] = 62.5738\,\sin(0.1\pi n + 1.4137)$$

which is the sampled form of the approximated derivative

$$\hat{y}_a(t) = 62.5738\,\sin(20\pi t + 1.4137)$$

The exact derivative of $x_a(t)$ is

$$y_a(t) = \frac{dx_a(t)}{dt} = 2\pi f_0\cos\left(2\pi f_0 t\right) = 62.8319\,\cos(20\pi t)$$

Fig. 8.41 provides a comparison of exact and approximate solutions for the derivative. The approximate solution $\hat{y}_a(t)$ seems to be lagging behind the exact solution $y_a(t)$. Zooming into the graph and examining the timing of zero crossings between the two signals should reveal that the lag is 2.5 ms, exactly half of the sampling interval T (see problems 8.21 and 8.22 at the end of this chapter). Remember that a length-2 linear-phase FIR filter has a delay of half a sample.

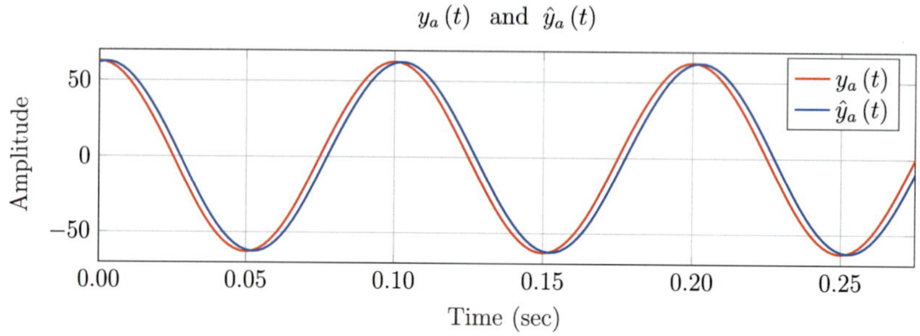

Figure 8.41 – Comparison of $y_a(t)$ and $\hat{y}_a(t)$ for Example 8.12.

Software resources: `exdt_8_12.m` , `FirstOrderDiff.mlx`

8.5.3 Fourier series design using Window functions

A simple method of designing frequency selective linear-phase FIR filters is to start with an ideal frequency response $H_d(\Omega)$, and to find the impulse response $h_d[n]$ through inverse DTFT.

$$h_d[n] = \frac{1}{2\pi}\int_{-\pi}^{\pi} H_d(\Omega)\, e^{j\Omega n}\, d\Omega\,, \qquad \text{all } n \tag{8.147}$$

In general, the resulting impulse response $h_d[n]$ is of infinite length, and must be truncated to obtain a finite-length sequence $h[n]$ for $n = 0,\ldots,N-1$.

$$h[n] = \begin{cases} h_d[n]\,, & \text{for } n = 0,\ldots,N-1 \\ 0\,, & \text{otherwise} \end{cases} \tag{8.148}$$

We know from earlier discussions that truncating an infinite-length sequence results in Gibbs oscillations in the magnitude response of the designed filter. This oscillatory behavior is especially evident right before and right after a transition band. A common solution is to use a window function in the truncation process:

$$h[n] = h_d[n]\, w[n] \tag{8.149}$$

A brief discussion of window functions was provided in Chapter 7, Section 7.2.3 in the context of spectral leakage in detecting sinusoidal components within finite-length data sequences. Truncation of $h_d[n]$ as described in Eqn. (8.148) corresponds to using a rectangular window sequence in Eqn. (8.149):

$$w_R[n] = u[n] - u[n-N] = \begin{cases} 1\,, & \text{for } n = 0,\ldots,N-1 \\ 0\,, & \text{otherwise} \end{cases}$$

The rectangular window provides the narrowest main lobe in its spectrum, however, its sidelobes are strong. They appear as ripples in the magnitude spectrum of the designed filter. Other window functions may provide better sidelobe suppression, but that is always at the expense of a widened main lobe. A wider main lobe means reduced sharpness around the transition band of the designed filter, and less precise control in the placement of transition frequencies.

Thus, the process of designing a linear-phase FIR filter with this method involves two important decisions, namely finding an appropriate $H_d(\Omega)$ as the ideal frequency response, and selecting an appropriate window function $w[n]$ for use in Eqn. (8.149) to control the tradeoff between the main lobe and the sidelobes.

We will first consider the problem of finding an appropriate $H_d(\Omega)$ for use in Eqn. (8.147). Recall that, in Section 8.5.1, four types of linear-phase FIR filters were identified. How should $H_d(\Omega)$ be chosen so that Eqn. (8.147) produces an impulse response of the desired type when it is truncated? Types I and II use impulse responses with symmetry. In contrast, types III and IV utilize antisymmetric impulse responses.

Type-I and type-II filters

For type-I and type-II filters, the impulse response exhibits positive symmetry. The frequency response is in the form (see Eqns. (8.122) and (8.131))

$$H(\Omega) = H_r(\Omega)\, e^{-j\Omega M} \tag{8.150}$$

where the amplitude response $H_r(\Omega)$ is an even function of Ω.

Lowpass filter:

For an ideal lowpass filter with a cutoff frequency of Ω_c, the amplitude response could be set as

$$H_r(\Omega) = \begin{cases} 1\,, & |\Omega| < \Omega_c \\ 0\,, & \Omega_c < |\Omega| < \pi \end{cases}$$

as shown in Fig. 8.42, and the ideal frequency response would be

$$H_d(\Omega) = H_r(\Omega)\, e^{-j\Omega M} = \begin{cases} e^{-j\Omega M}\,, & |\Omega| < \Omega_c \\ 0\,, & \Omega_c < |\Omega| < \pi \end{cases}$$

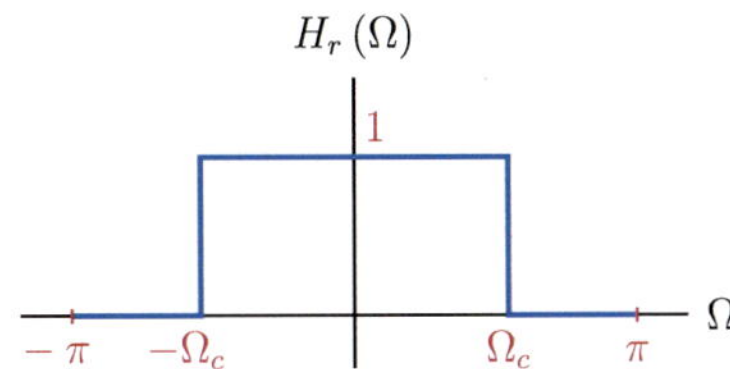

Figure 8.42 – Ideal discrete-time lowpass filter amplitude response.

$$h_d[n] = \frac{1}{2\pi} \int_{-\Omega_c}^{\Omega_c} e^{-j\Omega M}\, e^{j\Omega n}\, d\Omega = \frac{\sin\big(\Omega_c (n-M)\big)}{\pi\,(n-M)} \tag{8.151}$$

$$h_d[n] = \frac{\Omega_c}{\pi}\, \text{sinc}\left(\frac{\Omega_c}{\pi}\,(n-M)\right) \tag{8.152}$$

Truncated version of $h_d[n]$ is graphed in Fig. 8.43 for $N = 41$ and $\Omega_c = 0.3\pi$.

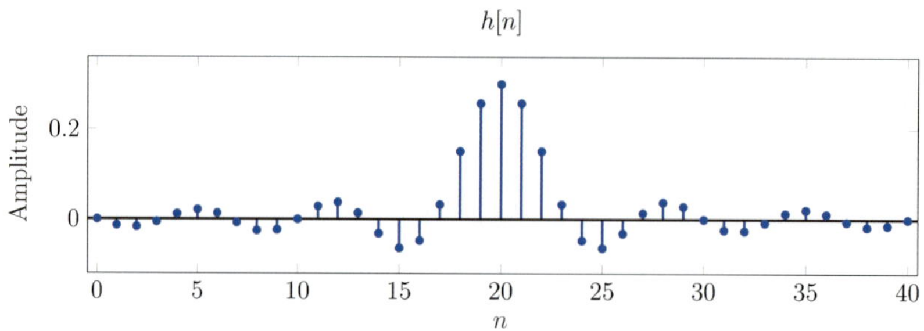

Figure 8.43 – Impulse response of the ideal lowpass filter for $\Omega_c = 0.3\pi$ truncated down to $N = 41$ samples.

Highpass filter:

$$H_r(\Omega) = \begin{cases} 0\,, & |\Omega| < \Omega_c \\ 1\,, & \Omega_c < |\Omega| < \pi \end{cases} \tag{8.153}$$

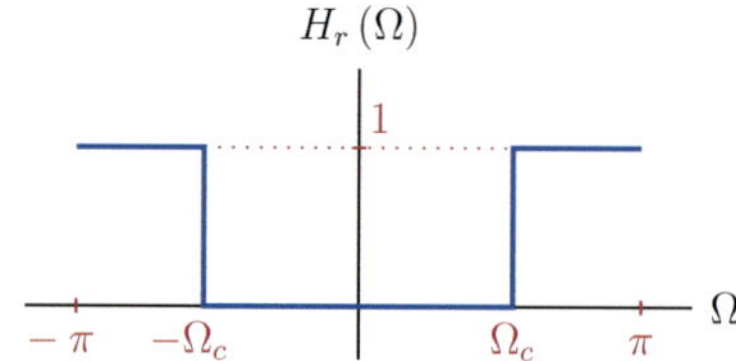

Figure 8.44 – Ideal discrete-time highpass filter amplitude response.

$$\begin{aligned} h_d[n] &= \frac{1}{2\pi} \int_{-\pi}^{-\Omega_c} e^{-j\Omega M}\, e^{j\omega n}\, d\Omega \int_{\Omega_c}^{\pi} e^{-j\Omega M}\, e^{j\omega n}\, d\Omega \\ &= \text{sinc}\,(n-M) - \frac{\Omega_c}{\pi}\, \text{sinc}\left(\frac{\Omega_c}{\pi}\,(n-M)\right) \end{aligned} \tag{8.154}$$

Eqn. (8.154) is valid for both type-I and type-II filters. If the filter is type-I (N odd), then $(n-M)$ is integer, and $\text{sinc}\,(n-M) = \delta\,(n-M)$. In this case $h_d[n]$ simplifies to

$$h_d[n] = \delta\,(n-M) - \frac{\Omega_c}{\pi}\, \text{sinc}\left(\frac{\Omega_c}{\pi}\,(n-M)\right) \tag{8.155}$$

Truncated version of $h_d[n]$ in Eqn. (8.155) is shown in Fig. 8.45 for $N = 41$ and $\Omega_c = 0.3\pi$. Type-II is generally not suitable for a highpass filter since it has a forced zero at $\Omega = \pm\pi$.

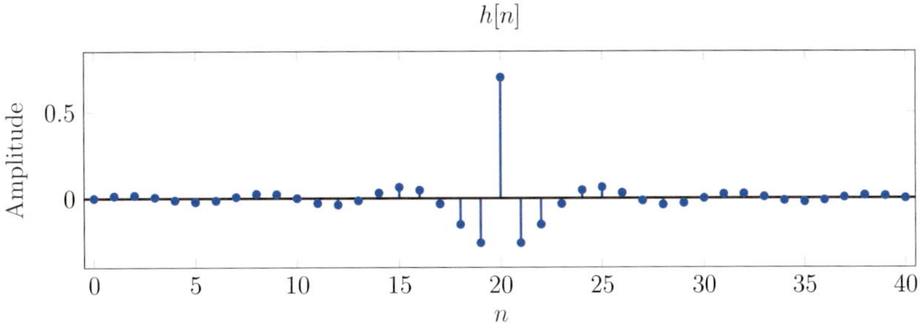

Figure 8.45 – Impulse response of the ideal highpass filter for $\Omega_c = 0.3\pi$ truncated down to $N = 41$ samples.

Bandpass filter:

Consider an ideal bandpass filter with a passband in the range $\Omega_1 < |\Omega| < \Omega_2$. The frequency spectrum of a such bandpass filter can be constructed as the difference of the frequency spectra of two lowpass filters with edge frequencies of Ω_1 and Ω_2. Using the linearity of the DTFT, the ideal impulse response for the bandpass filter can also be written as the difference of the two lowpass filter impulse responses.

$$h_d[n] = \frac{\Omega_2}{\pi}\,\text{sinc}\left(\frac{\Omega_2}{\pi}(n-M)\right) - \frac{\Omega_1}{\pi}\,\text{sinc}\left(\frac{\Omega_1}{\pi}(n-M)\right) \tag{8.156}$$

Bandstop filter:

A bandstop filter with stopband in the interval $\Omega_1 < |\Omega| < \Omega_2$ can be thought of as the sum of a lowpass filter and a highpass filter. Using Eqns. (8.152) and (8.154), the impulse response of the ideal bandstop filter is

$$h_d[n] = \frac{\Omega_c}{\pi}\,\text{sinc}\left(\frac{\Omega_c}{\pi}(n-M)\right) + \text{sinc}(n-M) - \frac{\Omega_c}{\pi}\,\text{sinc}\left(\frac{\Omega_c}{\pi}(n-M)\right) \tag{8.157}$$

Eqn. (8.157) is valid for both type-I and type-II filters. If the filter is type-I (N odd), then $(n-M)$ is integer, and $\text{sinc}(n-M) = \delta(n-M)$. In this case $h_d[n]$ simplifies to

$$h_d[n] = \frac{\Omega_c}{\pi}\,\text{sinc}\left(\frac{\Omega_c}{\pi}(n-M)\right) + \delta(n-M) - \frac{\Omega_c}{\pi}\,\text{sinc}\left(\frac{\Omega_c}{\pi}(n-M)\right) \tag{8.158}$$

Type-II is generally not suitable for bandstop filters due to its forced zero at $\Omega = 0$.

Type-III and type-IV filters

For type-III and type-IV filters, the impulse response exhibits negative symmetry. The frequency response is in the form (see Eqns. (8.133) and (8.135))

$$H(\Omega) = -j\,H_r(\Omega)\,e^{-j\Omega M} \tag{8.159}$$

where the amplitude response $H_r(\Omega)$ is an odd function of Ω. This is necessary to obtain an impulse response with negative symmetry.

As discussed in section 8.5.1, type-III and type-IV filters are not particularly good choices for frequency selective designs. Often there is no good reason to use anything other than type-I for any frequency selective filter. The main utility of type-III and type-IV impulse responses are in the design of differentiators and Hilbert transform filters. However, for the sake of completeness, we will formulate the design of a lowpass filter before moving on to the design of differentiators and Hilbert transform filters.

Lowpass filter:

For an ideal lowpass filter with a cutoff frequency of Ω_c, the amplitude response could be chosen as

$$H_r(\Omega) = \begin{cases} 1\,, & 0<\Omega<\Omega_c \\ -1\,, & -\Omega_c<\Omega<0 \\ 0\,, & \text{otherwise} \end{cases}$$

as shown in Fig. 8.46. Using $H_r(\Omega)$, the ideal frequency response can be written as

$$H_d(\Omega) = -j\,H_r(\Omega)\,e^{-j\Omega M} = \begin{cases} -j\,e^{-j\Omega M}\,, & 0<\Omega<\Omega_c \\ j\,e^{-j\Omega M}\,, & -\Omega_c<\Omega<0 \\ 0\,, & \text{otherwise} \end{cases} \tag{8.160}$$

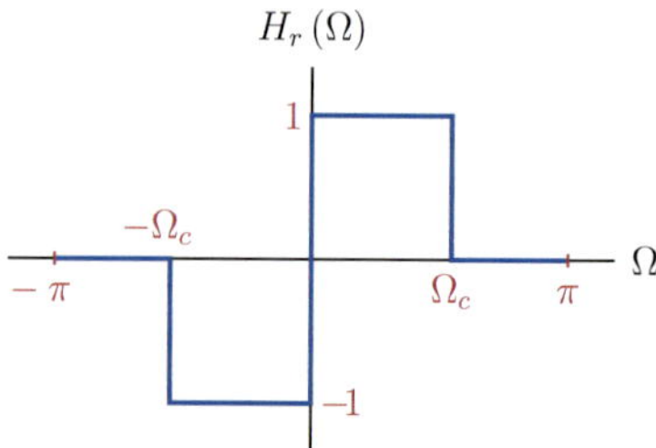

Figure 8.46 – Ideal discrete-time lowpass filter amplitude response for type-III and type-IV.

Applying the inverse DTFT integral to $H_d(\Omega)$, the ideal impulse response is found as

$$\begin{aligned} h_d[n] &= \frac{1}{2\pi}\int_{-\Omega_c}^{0} j\,e^{j\Omega(n-M)}\,d\Omega - \frac{1}{2\pi}\int_{0}^{\Omega_c} j\,e^{j\Omega(n-M)}\,d\Omega \\ &= \frac{1}{\pi(n-M)}\left[1-\cos\left(\Omega_c(n-M)\right)\right] \end{aligned} \tag{8.161}$$

It is often convenient to write $h_d[n]$ using the sinc function. Using the identities

$$1-\cos(2\alpha) = 2\sin^2(\alpha)$$

and

$$\sin(\alpha) = \alpha\,\text{sinc}\left(\alpha/\pi\right)$$

Eqn. (8.161) can be written as

$$h_d[n] = \frac{\Omega_c^2\,(n-M)}{2\pi}\,\text{sinc}^2\left(\frac{\Omega_c}{2\pi}(n-M)\right) \tag{8.162}$$

The truncated impulse response $h[n]$ obtained from Eqn. (8.162) is graphed in Fig. 8.47 for $N=41$ and $\Omega_c = 0.3\pi$.

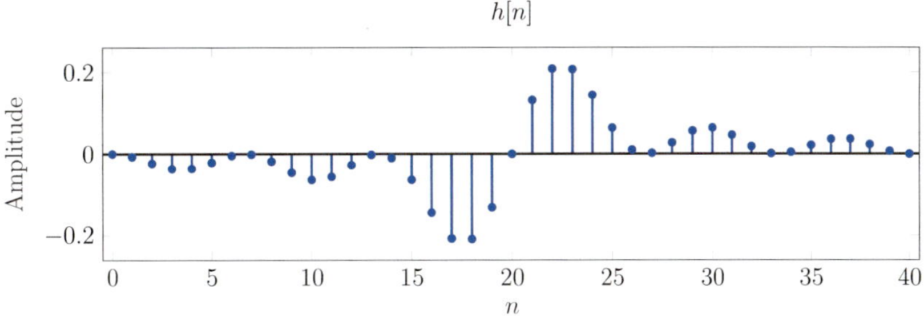

Figure 8.47 – Impulse response of the ideal type-III lowpass filter for $\Omega_c = 0.3\pi$ truncated down to $N = 41$ samples.

Differentiator:

A first-order backward differentiator for use with analog signals was designed and analyzed in Section 8.5.2. The system function $G(s) = 1/s$ for a continuous-time differentiator can be replaced with an ideal discrete-time differentiator with system function $H_d = j\Omega/T$ as shown in Fig. 8.38. The parameter T is the sampling interval used in converting the analog signal to discrete-time. The length-2 FIR differentiator provided a somewhat crude estimate of the derivative of the input signal that can be improved upon using a higher order differentiator. The first step in designing a type-III or type-IV differentiator is to select its amplitude response. Let $H_r(\Omega)$ be

$$H_r(\Omega) = \frac{\Omega}{T}$$

The ideal frequency response is

$$H_d(\Omega) = j\frac{\Omega}{T}\,e^{-j\Omega M}$$

The ideal impulse response is the inverse DTFT of $H_d(\Omega)$ which can be found through the integral

$$h_d[n] = \frac{1}{2\pi}\int_{-\pi}^{\pi} j\frac{\Omega}{T}\,e^{j\Omega(n-M)}\,d\Omega \tag{8.163}$$

Evaluation of the integral in Eqn. (8.163) using integration by parts yields (see Problem 8.23)

$$h_d[n] = \frac{\cos\big(\pi(n-M)\big)}{(n-M)\,T} - \frac{\sin\big(\pi(n-M)\big)}{\pi(n-M)^2\,T} \tag{8.164}$$

Hilbert transform filter:

The frequency response of an ideal *Hilbert transform filter* is

$$H_d(\Omega) = -j\,\mathrm{sgn}(\Omega) \tag{8.165}$$

where sgn denotes the signum function defined as

$$\mathrm{sgn}(\alpha) = \begin{cases} 1, & \alpha > 0 \\ -1, & \alpha < 0 \end{cases} \tag{8.166}$$

The corresponding amplitude response is

$$H_r(\Omega) = \mathrm{sgn}(\Omega) \tag{8.167}$$

As suggested by Eqns. (8.165) and (8.166), a Hilbert transform filter is a −90 degree phase shifter. It is essentially an all-pass filter that provides a constant magnitude spectrum along with a phase characteristic that is $-\pi/2$ radians for positive frequencies, and $\pi/2$ radians for negative frequencies. Hilbert transform filters find use especially in communication systems. The ideal impulse response can be found as

$$h_d[n] = \frac{1}{2\pi}\int_{-\pi}^{0} j\,e^{j\Omega(n-M)}\,d\Omega - \int_{0}^{\pi} j\,e^{j\Omega(n-M)}\,d\Omega$$

$$= \frac{1-\cos\left(\pi\,(n-M)\right)}{\pi\,(n-M)} \tag{8.168}$$

Using the sinc function, it can be expressed in more compact form as

$$h_d[n] = \frac{\pi\,(n-M)}{2}\,\text{sinc}^2\left(\frac{n-M}{2}\right) \tag{8.169}$$

Example 8.13: Fourier series design

Using the Fourier series method with a rectangular window, design a length-15 FIR lowpass filter to approximate an ideal lowpass filter with $\Omega_c = 0.3\pi$ rad.

Solution: We will use a type-I filter with positive symmetry in the impulse response. Center of symmetry is

$$M = \frac{15-1}{2} = 7$$

Using Eqn. (8.152), the impulse response of the ideal lowpass filter is

$$h_d[n] = \frac{0.3\pi}{\pi}\,\text{sinc}\left(\frac{0.3\pi}{\pi}\,(n-7)\right) = 0.3\,\text{sinc}\left(0.3\,(n-7)\right), \qquad \text{all } n$$

The impulse response of the FIR filter is obtained by truncating it or, equivalently, multiplying it with a rectangular window sequence.

$$h[n] = h_d[n]\,w_R[n] = 0.3\,\text{sinc}\left(0.3\,(n-7)\right), \qquad n = 0,\ldots,14$$

FIR filter coefficients are

$$h[n] = \{\underset{\underset{n=0}{\uparrow}}{0.014}, -0.031, -0.064, -0.047, 0.033, 0.151, 0.258, 0.3, 0.258, 0.151,$$

$$0.032, -0.047, -0.064, -0.031, 0.014\}$$

The magnitude response of the designed filter is shown in Fig. 8.48.

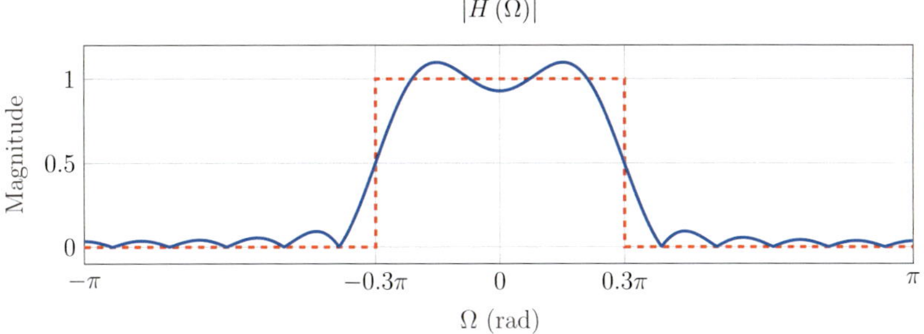

Figure 8.48 – Magnitude response of the FIR filter designed in Example 8.13.

Software resource: exdt_8_13.m

Fig. 8.48 of Example 8.13 illustrates a by-product of truncating the ideal filter impulse response, that is, the oscillatory behavior of the frequency response which is particularly evident around the cutoff frequency Ω_c. This effect is the *Gibbs phenomenon.* Recall that, in the discussion of FIR linear-phase filter types, ideal amplitude responses $H_r(\Omega)$ with sharp transitions were approximated using a finite number of sine or cosine terms (see Eqns. (8.122), (8.131), (8.133), and (8.135)). From a time-domain perspective, the FIR filter is obtained by truncating the ideal impulse response $h_d[n]$, i.e.,

$$\text{Eqn. (8.148):} \qquad h[n] = \begin{cases} h_d[n]\,, & \text{for } n = 0,\ldots,N-1 \\ 0\,, & \text{otherwise} \end{cases}$$

This act creates abrupt transitions at the two ends of $h[n]$ that lead to the oscillatory behavior of the magnitude response. An alternative way of writing the relationship in Eqn. (8.148) is to express $h[n]$ as the product of the ideal impulse response and a rectangular window sequence.

$$h[n] = h_d[n]\, w_R[n] \tag{8.170}$$

where $w_R[n]$ is defined as

$$w_R[n] = u[n] - u[n-N] = \begin{cases} 1\,, & \text{for } n = 0,\ldots,N-1 \\ 0\,, & \text{otherwise} \end{cases}$$

The spectrum of the rectangular window sequence is

$$W_R(\Omega) = \frac{\sin\left(\Omega N/2\right)}{\sin\left(\Omega/2\right)}\, e^{-j\Omega M} \tag{8.171}$$

where M is the axis of symmetry as defined earlier. Based on the multiplication property of the DTFT (see Section 3.3.5 of Chapter 3), the spectrum of $h[n]$ is the periodic convolution of the spectra of $h_d[n]$ and $w_R[n]$.

$$H(\Omega) = \frac{1}{2\pi} \int_{-\pi}^{\pi} H_d(\lambda)\, W_R(\Omega - \lambda)\, d\lambda \tag{8.172}$$

This is illustrated in Fig. 8.49.

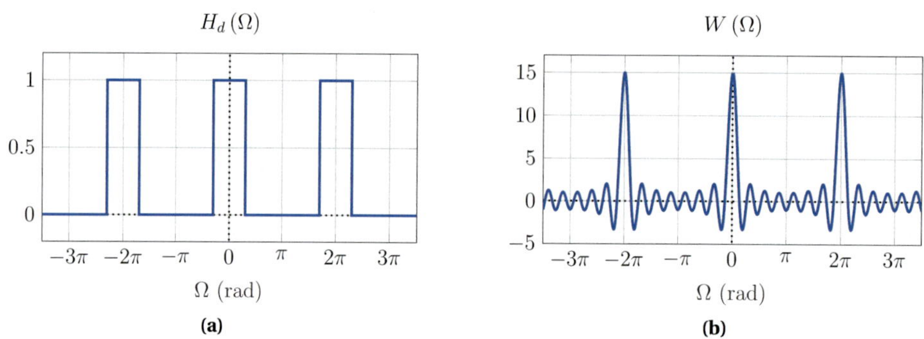

Figure 8.49 – Magnitude spectra involved in Eqn. (8.172). Parameter values $\Omega_c = 0.3\pi$ and $N = 15$ of Example 8.13 were used in the figures.

As the spectra in Fig. 8.49 reveal, the reason behind the oscillatory behavior of $|H(\Omega)|$ especially around $\Omega = \Omega_c$ is the shape of the spectrum $|W_R(\Omega)|$. The high-frequency content of $W_R(\Omega)$, on the other hand, is mainly due to the abrupt transition of the rectangular window from unit amplitude to zero amplitude at its two edges. The solution is to use an alternative window function in Eqn. (8.172), one that smoothly tapers down to zero at its edges, in place of the rectangular window. The chosen window function must also be an even function of n in order to keep the symmetry of $h_T[n]$ for linear phase. A large number of window functions exist in the literature. A few of them are listed below:

Triangular (Bartlett) window:

$$w[n] = 1 - \frac{|n-M|}{M}, \qquad n = 0,\ldots,N-1 \tag{8.173}$$

Hamming window:

$$w[n] = 0.54 - 0.46\cos\left(\frac{\pi n}{M}\right), \qquad n = 0,\ldots,N-1 \tag{8.174}$$

von Hann (Hanning) window:

$$w[n] = 0.5 - 0.5\cos\left(\frac{\pi n}{M}\right), \qquad n = 0,\ldots,N-1 \tag{8.175}$$

Blackman window:

$$w[n] = 0.42 - 0.5\cos\left(\frac{\pi n}{M}\right) + 0.08\cos\left(\frac{2\pi n}{M}\right), \qquad n = 0,\ldots,N-1 \tag{8.176}$$

The four window functions listed are shown in Fig. 8.50 for $N = 25$. Figs. 8.51 and 8.52 provide comparisons of dB magnitude spectra for several window sequences.

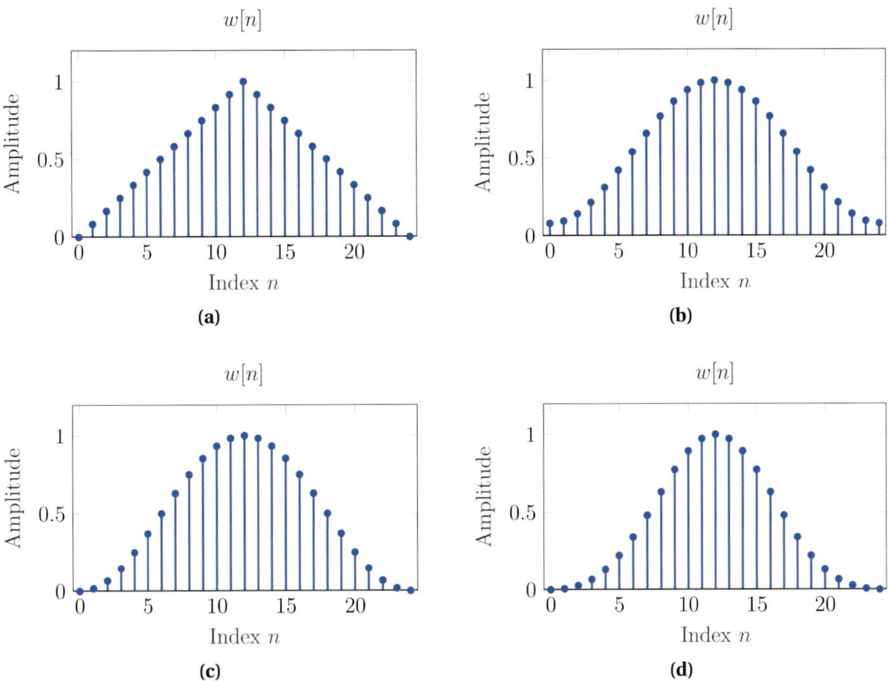

Figure 8.50 – Window functions: **(a)** Triangular (Bartlett window), **(b)** Hamming window, **(c)** von Hann (Hanning) window, and **(d)** Blackman window.

It is seen that the rectangular window has the stronger sidelobes than the other window functions. Hamming window provides side lobes that are at least 40 dB below the main lobe, and the side lobes for the Blackman window are at least 60 dB below the main lobe. The downside to side lobe suppression is the widening of the main lobe. In Hamming, von Hann and Bartlett windows, the main lobe is about twice as wide compared to the rectangular window. Blackman window main lobe is about three times as wide.

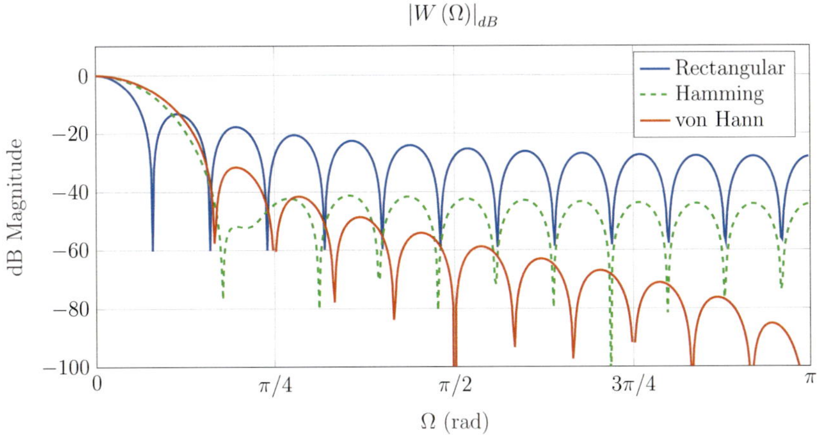

Figure 8.51 – Comparison of dB magnitude responses of rectangular, Hamming, and von Hann windows.

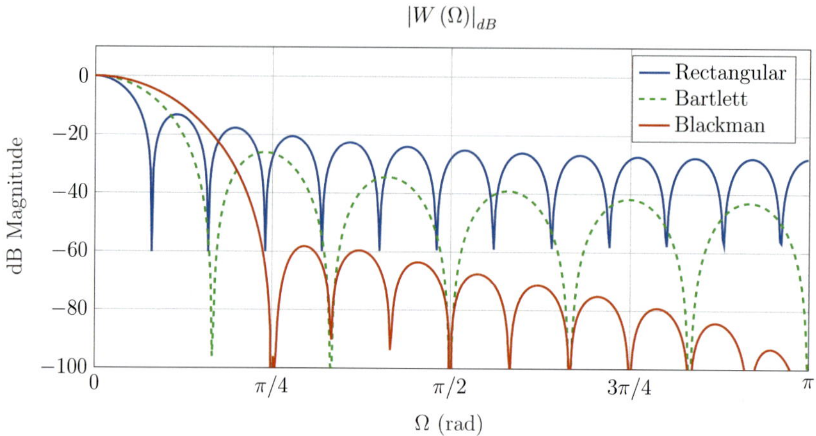

Figure 8.52 – Comparison of dB magnitude responses of rectangular, Bartlett, and Blackman windows.

Software resource: See live script `WindowFunctions.mlx`.

We are now ready to summarize the Fourier series design method for linear-phase FIR filters:

Fourier series design of FIR filters using window functions:

1. Determine the ideal impulse response $h_d[n]$ as the inverse DTFT of the ideal spectrum being approximated. Use the appropriate formula from Section 8.5.3.
2. Multiply the ideal impulse response with the selected length-N window function to obtain a finite-length impulse response $h[n] = h_d[n]\, w[n]$.

Example 8.14: Fourier series design using window functions

Redesign the filter of Example 8.13 using Hamming and Blackman windows.

Solution: Using a window function the truncated impulse response is

$$h[n] = 0.3\,\text{sinc}\big(0.3\,(n-7)\big)\;w[n]\,,\qquad n = 0,\ldots,14$$

Using the Hamming window, $h[n]$ is

$$h[n] = \{\underset{\substack{\uparrow\\ n=0}}{0.0011},\,-0.0039,\,-0.0161,\,-0.0205,\,0.0211,\,0.1251,\,0.2458,\,0.3000,\,0.2458,$$
$$0.1251,\,0.0211,\,-0.0205,\,-0.0161,\,-0.0039,\,0.0011\}$$

When the Blackman window is used, we get

$$h[n] = \{\underset{\substack{\uparrow\\ n=0}}{0.0000},\,-0.0006,\,-0.0058,\,-0.0111,\,0.0151,\,0.1081,\,0.2370,\,0.3000,\,0.2370,$$
$$0.1081,\,0.0151,\,-0.0111,\,-0.0058,\,-0.0006,\,0.0000\}$$

Magnitude responses of the two filters are shown in Fig. 8.53.

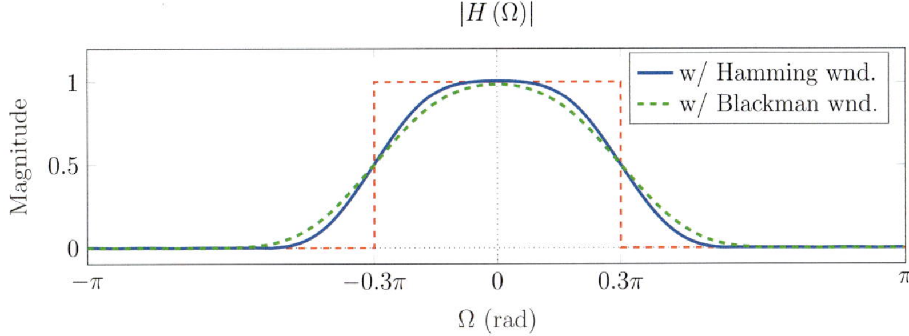

Figure 8.53 – Magnitude responses of the FIR filters designed in Example 8.13.

Software resource: exdt_8_14a.m

Example 8.15: Fourier series design of FIR differentiators

A FIR differentiator is to be designed to filter an analog signal sampled with a sampling rate of $f_s = 200$ Hz.

a. Carry out the design as a type-III filter with $N = 9$ coefficients using the Fourier series method with rectangular window. Graph the magnitude response of the resulting filter. Repeat with a Hamming window.

b. Repeat the design as a type-IV filter with $N = 8$ coefficients. Try it with rectangular and Hamming windows. Compare the results to those found in part (a).

Solution:

a. The center of symmetry is

$$M = \frac{9-1}{2} = 4$$

Coefficients of the ideal impulse response are found using Eqn. (8.164):

$$h_d[n] = \frac{200,\cos\big(\pi\,(n-4)\big)}{(n-4)} - \frac{200\,\sin\big(\pi\,(n-4)\big)}{\pi\,(n-4)^2}$$

Truncating $h_d[n]$ with a rectangular window yields

$$h_1[n] = h_d[n]\,w_R[n] = \Big\{\underset{\substack{\uparrow \\ n=0}}{-50.00},\,66.67,\,-100.0,\,200.0,\,0,\,-200.0,\,100.0,\,-66.67,\,50.0\Big\}$$

Using a Hamming window, the impulse response is

$$h_2[n] = h_d[n]\,w[n] = \Big\{\underset{\substack{\uparrow \\ n=0}}{-4.000},\,14.3154,\,-54.000,\,173.05,\,0,\,-173.05,\,54.000,$$

$$-14.3154,\,4.000\Big\}$$

Magnitude responses of the two differentiators are shown in Fig. 8.54.

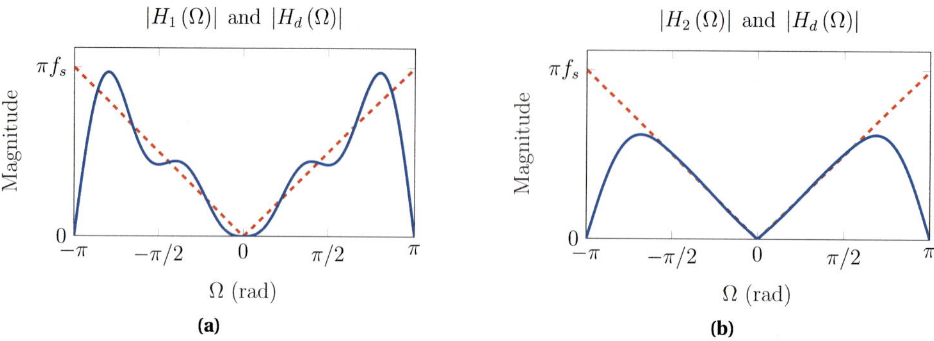

Figure 8.54 – Magnitude response of the FIR type-III differentiator **(a)** using a rectangular window, and **(b)** using a Hamming window.

b. In this case the center of symmetry is

$$M = \frac{8-1}{2} = 3.5$$

Coefficients of the ideal impulse response are found using Eqn. (8.164):

$$h_d[n] = \frac{200, \cos\big(\pi\,(n-3.5)\big)}{(n-3.5)} - \frac{200\,\sin\big(\pi\,(n-3.5)\big)}{\pi\,(n-3.5)^2}$$

Truncating $h_d[n]$ with a rectangular window yields

$$h_3[n] = h_d[n]\,w_R[n] = \Big\{ \underset{\substack{\uparrow \\ n=0}}{-5.1969}, 10.1859, -28.2942, 254.65, -254.65, 28.2942, -10.1859, 5.1969 \Big\}$$

Using a Hamming window, the impulse response is

$$h_4[n] = h_d[n]\,w[n] = \Big\{ \underset{\substack{\uparrow \\ n=0}}{-0.4158}, 2.5790, -18.1751, 243.05, -243.05, 18.1751, -2.5790, 0.4158 \Big\}$$

Magnitude responses of the two differentiators are shown in Fig. 8.55.

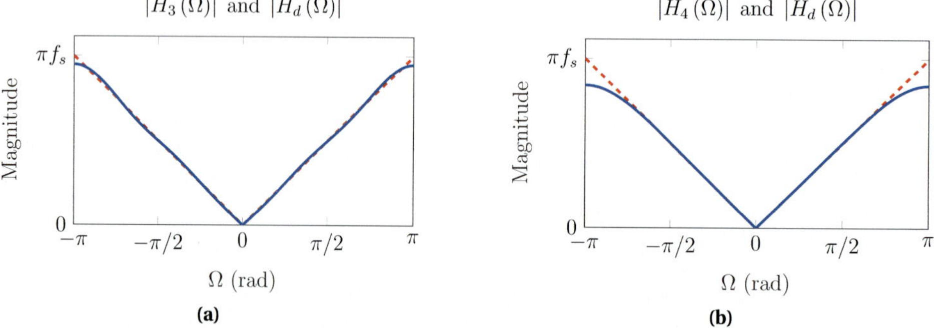

Figure 8.55 – Magnitude response of the FIR type-IV differentiator **(a)** using a rectangular window and **(b)** using a Hamming window.

Inspection of magnitude responses reveals that type-III differentiators suffer from Gibbs oscillations more significantly than type-IV differentiators. The main reason for this is the fact that type-III magnitude response has a forced zero at $\Omega = \pm\pi$ as evident in both parts of Fig. 8.54. The Hamming window reduces the Gibbs effect in $|H_2(\Omega)|$ shown in Fig. 8.54b, however, this comes at the cost of reduced accuracy for angular frequencies greater than $\pi/2$. Type-IV differentiators, on the other hand, show minimal Gibbs effect even when the rectangular window is used.

Software resources: `exdt_8_15a.m` , `exdt_8_15b.m` , `CompareFIRDiffs.mlx`

Example 8.16: Fourier series design of a Hilbert transform filter

Using the Fourier series design method with a rectangular window, design a length-18 Hilbert transform filter. Afterward repeat the design using a Hamming window. Graph magnitude responses of designed filters.

Solution: We will use type-IV linear phase in the design. The center of symmetry is

$$M = \frac{18-1}{2} = 8.5$$

Coefficients of the ideal impulse response are found using Eqn. (8.169):

$$h_d[n] = \frac{\pi\,(n-8.5)}{2}\,\text{sinc}^2\left(\frac{n-8.5}{2}\right), \qquad \text{all } n$$

Truncating $h_d[n]$ with a rectangular window yields

$$h_1[n] = \frac{\pi\,(n-8.5)}{2}\,\text{sinc}^2\left(\frac{n-8.5}{2}\right), \qquad n = 0,\ldots,17$$

Using a Hamming window, the impulse response is

$$\begin{aligned} h_2[n] &= h_d[n]\,w[n] \\ &= \frac{\pi\,(n-8.5)}{2}\,\text{sinc}^2\left(\frac{n-8.5}{2}\right)\left[0.54 - 0.46\cos\left(\frac{\pi n}{8.5}\right)\right], \quad n = 0,\ldots,17 \end{aligned}$$

Magnitude responses of the two filters are shown in Fig. 8.56.

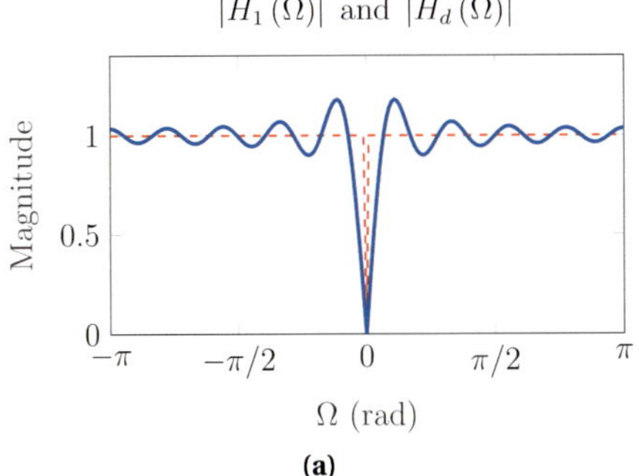

(a)

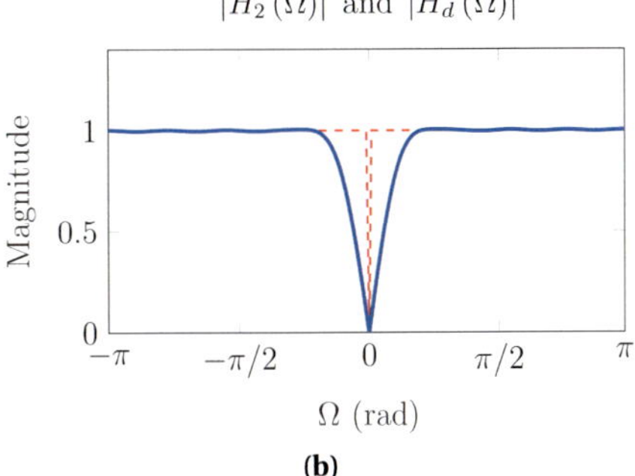

(b)

Figure 8.56 – Magnitude response of the FIR type-IV Hilbert transform filter **(a)** using a rectangular window and **(b)** using a Hamming window.

Software resources: `exdt_8_16a.m` , `exdt_8_16b.m` , `FIRHilbertTrans`

Software resource: See MATLAB Exercise 8.11.

8.5.4 Frequency sampling design

An alternative method for designing linear-phase FIR filters is the *frequency sampling* design method. It does not require the use of a window function. Instead, we relax the requirements on the ideal filter magnitude response by providing transition bands between the passbands and the stopbands of the desired filter behavior. Consider, for example, the design of a length-N type-I lowpass FIR filter with magnitude specifications

$$\left|H_d(\Omega)\right| = \begin{cases} 1\,, & 0 < \left|\Omega\right| < \Omega_1 \\ 0\,, & \Omega_2 < \left|\Omega\right| < \pi \end{cases} \tag{8.177}$$

The passband extends from $\Omega = 0$ to $\Omega = \Omega_1$. The stopband is between $\Omega = \Omega_2$ and $\Omega = \pi$. A transition band is allocated in the interval $\Omega_1 < \Omega < \Omega_2$ to avoid forcing a vertical transition in the magnitude response. Not having sharp transitions in $\left|H_d(\Omega)\right|$ significantly reduces the oscillatory behavior associated with the Gibbs effect. Fig. 8.57 illustrates the design specifications in the form of four corner points connected with line segments.

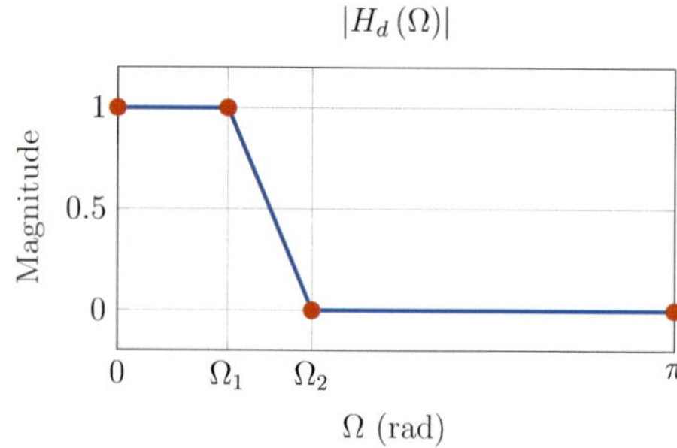

Figure 8.57 – Magnitude specifications for FIR lowpass filter.

The DTFT of any signal is 2π-periodic. We will extend the magnitude behavior in Fig. 8.57 by mirroring it to fill the frequencies in the range $\pi < \Omega < 2\pi$. The DTFT-based system function for the ideal filter is

$$H_d(\Omega) = \left|H_d(\Omega)\right| e^{-j\Omega M} \tag{8.178}$$

Now let $H_d(\Omega)$ be sampled at a set of N angular frequencies Ω_k defined as

$$\Omega_k = \frac{2\pi k}{N}\,, \qquad k = 0, \ldots, N-1 \tag{8.179}$$

to obtain the transform

$$H[k] = \left|H_d(\Omega_k)\right| e^{-j\Omega_k M}; \qquad k = 0, \ldots, N-1 \tag{8.180}$$

This is illustrated in Fig. 8.58. Recall that N is the length of the FIR filter to be designed. The set of frequencies in Eqn. (8.179) should be recognized as the DFT frequencies (see Chapter 7, Section 7.2.1). The impulse response of the FIR filter can now be computed as the inverse DFT of $H[k]$.

$$h[n] = \text{DFT}^{-1}\left\{H[k]\right\}, \qquad n = 0, \ldots, N-1 \tag{8.181}$$

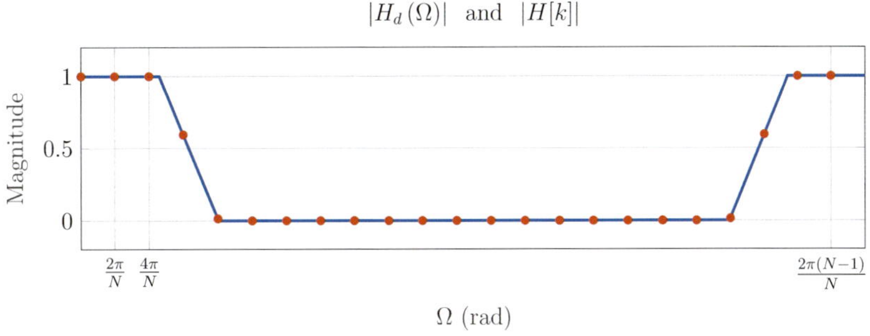

Figure 8.58 – Frequency sampling of desired magnitude response.

Contrasting the frequency sampling method with window based Fourier series design of FIR filters discussed in the previous section, the following conclusions can be made:

1. With either design method, the ideal impulse response $h_d[n]$ obtained as the inverse DTFT of $H_d(\Omega)$ is of infinite length, and is not suitable to be the impulse response of a FIR filter.
2. With the Fourier series method, $h_d[n]$ is truncated through multiplication by a window sequence so that a finite-length impulse response $h[n]$ is obtained for the final design.
3. With the frequency sampling technique, a finite-length transform $H[k]$ is obtained by sampling the ideal frequency response $H_d(\Omega)$. The impulse response $h[n]$ of the final design is obtained from $H[k]$ by inverse DFT. The actual frequency response of the designed filter can now be obtained by computing the DTFT of $h[n]$, that is

$$H(\Omega) = \sum_{n=0}^{N-1} h[n]\, e^{-j\Omega n}$$

Obviously, the response $H(\Omega)$ will be different from $H_d(\Omega)$. Instead, it will be an interpolated version of $H[k]$ defined by Eqns. (7.15) and (7.16) from Chapter 7, repeated here for convenience:

$$\text{Eqn. (7.15):} \qquad B_N(\Omega) = \frac{\sin\left(\Omega N/2\right)}{N\sin\left(\Omega/2\right)}\, e^{-j\Omega(N-1)/2}$$

$$\text{Eqn. (7.16):} \qquad H(\Omega) = \sum_{k=0}^{N-1} H[k]\, B_N\left(\Omega - \frac{2\pi k}{N}\right)$$

Example 8.17: Multiband filter design using frequency sampling method

Using the frequency sampling method, design a length-25 linear-phase multiband FIR filter to approximate the ideal magnitude response given below.

$$\left|H_d(\Omega)\right| = \begin{cases} 1\,, & 0 < |\Omega| < 0.2\pi \\ 0.5\,, & 0.3\pi < |\Omega| < 0.55\pi \\ 0\,, & 0.65\pi < |\Omega| < \pi \end{cases}$$

Solution: We will use type-I linear phase in the design. The symmetry axis is at

$$M = \frac{25-1}{2} = 12$$

For magnitude specifications, six critical frequencies and corresponding magnitude values are shown in Fig. 8.59. By mirroring Fig. 8.59 to fill the entire frequency range $0 < \Omega < 2\pi$ and then sampling the ideal magnitude spectrum at frequencies

$$\Omega_k = \frac{2\pi k}{25}\,, \qquad k = 0, \ldots, N-1$$

we obtain $\left|H[k]\right|$ as shown in Fig. 8.60.

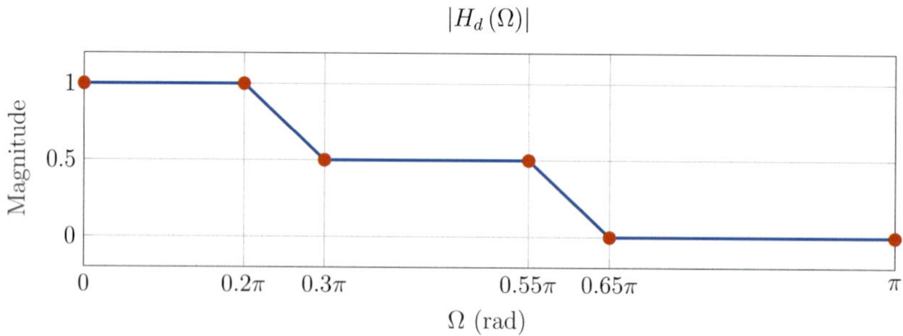

Figure 8.59 – Magnitude specifications for the multiband FIR filter of Example 8.17.

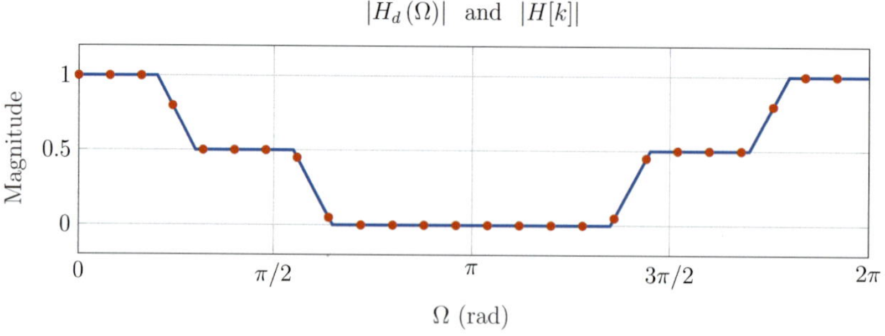

Figure 8.60 – Frequency sampling of desired magnitude response for Example 8.17.

Values of $\left|H[k]\right|$ are

$$\left|H[k]\right| = \{ \underset{\substack{\uparrow \\ n=0}}{1.0}, 1.0, 1.0, 0.8, 0.5, 0.5, 0.5, 0.45, 0.05, 0.0, 0.0, 0.0, 0.0, 0.0, 0.0, 0.0, 0.0, 0.05, 0.45, 0.5, 0.5, 0.5, 0.8, 1.0, 1.0 \}$$

The impulse response $h[n]$ of the FIR filter can now be determined using the inverse DFT:

$$h[n] = \text{DFT}^{-1}\left\{\left|H[k]\right| e^{-j12\Omega_k}\right\}$$

The impulse response and the magnitude response of the designed filter are shown in Fig. 8.61.

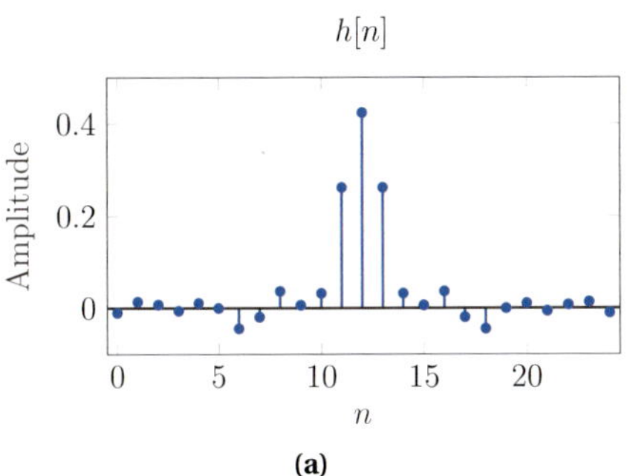

(a)

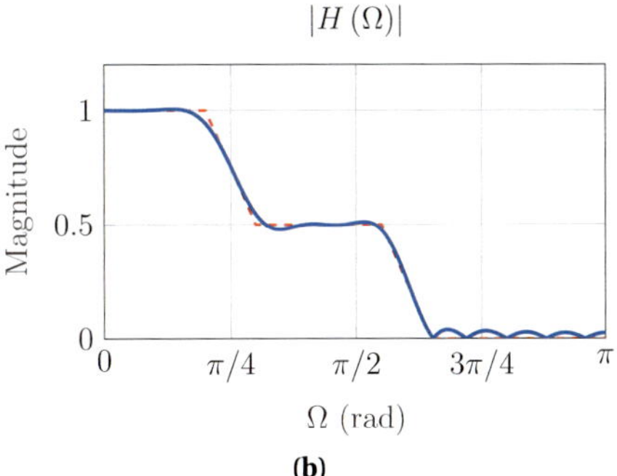

(b)

Figure 8.61 – **(a)** The impulse response and **(b)** the magnitude response of the multiband FIR filter designed in Example 8.17.

Software resources: exdt_8_17.m , FreqSampDesign.mlx

8.5.5 Least-squares design

Consider again the amplitude response of a type-I linear-phase FIR filter that was formulated in Eqn. (8.122) of Section 8.5.1 which is repeated here for convenience:

$$\text{Eqn. (8.122):} \qquad H_r(\Omega) = h[M] + 2\sum_{n=0}^{M-1} h[n]\cos\left(\Omega(n-M)\right)$$

where N is the length of the impulse response and $M = (N-1)/2$. The complete frequency response of the filter can be computed from the amplitude response as

$$\text{Eqn. (8.123):} \qquad H(\Omega) = H_r(\Omega)\, e^{-j\Omega M}$$

For a length-N FIR filter, Eqn. (8.122) has $M+1$ unknown filter coefficients. Once those coefficients $h[0], h[1], \ldots, h[M]$ are determined in some way, the remaining coefficients can be found from the symmetry condition

$$h[n] = h[2M-n], \qquad n = M+1, \ldots, 2M \tag{8.182}$$

Let's assume that, at some frequency $\Omega = \Omega_k$, we want the amplitude response of the filter to be equal to P_k. Using Eqn. (8.122) we can write the following equation:

$$2\sum_{n=0}^{M-1} h[n]\cos\big(\Omega_k(n-M)\big) + h[M] = P_k \tag{8.183}$$

Using a row vector $\mathbf{A_k}$ and a column vector $\mathbf{B}$ defined as

$$\mathbf{A_k} = \begin{bmatrix} 2\cos\big(\Omega_k(-M)\big) & 2\cos\big(\Omega_k(1-M)\big) & \ldots & 2\cos\big(\Omega_k(-1)\big) & 1 \end{bmatrix} \tag{8.184}$$

and

$$\mathbf{B} = \begin{bmatrix} h[0] & h[1] & \ldots & h[M-1] & h[M] \end{bmatrix}^T \tag{8.185}$$

Eqn. (8.183) can be expressed in the form

$$\mathbf{A_k}\,\mathbf{B} = P_k \tag{8.186}$$

using the scalar product of row vector $\mathbf{A_k}$ and column vector $\mathbf{B}$. By repeating Eqn. (8.186) for a set of K distinct frequencies Ω_k , $k = 0,\ldots,K-1$, a system of equations can be obtained. In matrix form we have

$$\mathbf{A}\mathbf{B} = \mathbf{P} \tag{8.187}$$

where the matrix $\mathbf{A}$ is constructed using row vectors $\mathbf{A_k}$, $k = 0,\ldots,K-1$.

$$\mathbf{A} = \begin{bmatrix} 2\cos\big(\Omega_0(-M)\big) & 2\cos\big(\Omega_0(1-M)\big) & \ldots & 2\cos\big(\Omega_0(-1)\big) & 1 \\ 2\cos\big(\Omega_1(-M)\big) & 2\cos\big(\Omega_1(1-M)\big) & \ldots & 2\cos\big(\Omega_1(-1)\big) & 1 \\ \vdots & \vdots & & \vdots & \vdots \\ 2\cos\big(\Omega_{K-1}(-M)\big) & 2\cos\big(\Omega_{K-1}(1-M)\big) & \ldots & 2\cos\big(\Omega_{K-1}(-1)\big) & 1 \end{bmatrix} \tag{8.188}$$

and the column vector $\mathbf{P}$ is the vector of desired amplitude response values, i.e.,

$$\mathbf{P} = \begin{bmatrix} P_0 & P_1 & \ldots & P_{K-1} \end{bmatrix}^T \tag{8.189}$$

If $K = M+1$, that is, the number of equations is equal to the number of unknown filter coefficients, then Eqn. (8.187) represents an *exact-determined* system of equations. The matrix $\mathbf{A}$ is square, and the solution should be straightforward.

$$\mathbf{B} = \mathbf{A}^{-1}\mathbf{P} \tag{8.190}$$

The frequencies Ω_k , $k = 0,\ldots,K-1$ do not have to be equally spaced; they just need to be distinct. On the other hand, if we choose the frequencies Ω_k as DFT frequencies

$$\Omega_k = \frac{2\pi k}{N}\,, \qquad k = 0,\ldots,M,$$

then the solution should match the one obtained using the frequency sampling design method, as long as the same desired amplitude response is used.

An interesting possibility is to use more equations than the number of unknown coefficients. Let $K > M+1$, and let the frequencies Ω_k be chosen as

$$\Omega_k = \frac{\pi k}{K}\,, \qquad k = 0,\ldots,K-1 \tag{8.191}$$

In this case the matrix $\mathbf{A}$ has K rows and $M+1$ columns. The column vector $\mathbf{P}$ has K elements. The resulting system of equations is referred to as *overdetermined.* There are too many equations, and it is not possible to satisfy all of them with equality. Instead, the set of equations can be written as

$$\mathbf{AB} = \mathbf{P} + \mathbf{P}_\Delta \tag{8.192}$$

so that they are satisfied in an approximate sense. The vector $\mathbf{P}_\Delta$ in Eqn. (8.192) represents the deviation of each amplitude from its desired value. The set of filter coefficients that are optimum in a *least-squares* sense are found by minimizing the squared norm of the vector $\mathbf{P}_\Delta$.

Least-squares solution:

$$\mathbf{P}_\Delta = \mathbf{AB} - \mathbf{P} \quad \longrightarrow \quad J = \|\mathbf{P}_\Delta\|^2 = \mathbf{P}_\Delta{}^T\, \mathbf{P}_\Delta$$

Determine $h[n]$, $k = 0, \ldots, M$ to minimize J

It can be shown that the least squares solution of the overdetermined system of equations in Eqn. (8.192) can be computed as

$$\mathbf{B} = \left(\mathbf{A}^T\mathbf{A}\right)^{-1}\mathbf{A}^T\,\mathbf{P} \tag{8.193}$$

The proof of Eqn. (8.193) will not be given here, but can be found in most textbooks on numerical analysis. In addition, uneven weight factors could be assigned to the equations, making some equations more important than the others in the optimization process. This will be explored in MATLAB Exercise 8.13 at the end of this chapter.

8.5.6 Parks-McClellan technique for FIR filter design

FIR filter design technique due to Parks and McClellan is also referred to as the *Remez exchange method.* It is based on the idea of finding a polynomial approximation to the desired frequency response by minimizing the largest approximation error.

The Fourier series design method discussed in the previous section, when used with a rectangular window, provides the optimum solution in the sense of mean-squared error. If the error between the desired frequency response and the actual frequency response is $E(\Omega)$, then the length-N FIR filter designed using the Fourier series method leads to the mean-squared error

$$\text{MSE} = \frac{1}{2\pi}\int_{-\pi}^{\pi} \left|E(\Omega)\right|^2 d\Omega \tag{8.194}$$

that is the smallest possible with N coefficients. On the other hand, we have seen that the magnitude response of the filter suffers from oscillatory behavior near discontinuities of the spectrum.

Better approximations can be obtained if we minimize the maximum value of the error rather than its mean-squared value. Parks-McClellan algorithm begins with an initial guess of the set of frequencies for the extrema of $E(\Omega)$. The locations of the extrema are then iteratively shifted until the maximum deviation from the ideal frequency response is made as small as possible.

Computational details of the Parks-McClellan algorithm are outside the scope of this text and may be found in references. We will provide examples of using MATLAB for designing Parks-McClellan filters in the MATLAB Exercises section.

Software resource: See MATLAB Exercise 8.15.

Summary of Key Points

- In general, the act of changing the relative amplitudes of frequency components in a signal is referred to as *filtering*, and the system that facilitates this is referred to as a filter.

- Discrete-time filters are viewed under two broad categories, namely *infinite impulse response (IIR)* filters and *finite impulse response (FIR)* filters.

- The system function of an IIR filter has both poles and zeros. The corresponding difference equation has feedback terms. The current output sample depends on one or more past samples of the output signal in addition to current and past samples of the input signal. Consequently, the impulse response of the filter is of infinite length.

- The behavior of an FIR filter is controlled only by the placement of the zeros of its system function. For causal FIR filters all of the poles are at the origin, and they do not contribute to the magnitude characteristics of the filter. The difference equation of an FIR filter has no feedback terms. The current output sample depends only on current and past samples of the input signal. The resulting impulse response is of finite length.

- For a given set of specifications, an IIR filter is generally more efficient than a comparable FIR filter. On the other hand, FIR filters are always stable. Additionally, a linear phase characteristic is possible with FIR filters whereas causal and stable IIR filters cannot have linear phase.

- A *resonant bandpass filter* is used for passing a single frequency or a very narrow band of frequencies around it, and blocking everything else. Its system function is

  Eqn. 8.6:
  $$H(z) = \frac{K\left(z^2-1\right)}{z^2-2r\cos(\Omega_0)\,z+r^2}$$

  where Ω_0 is the resonant frequency. Parameter r controls the bandwidth.

- A *notch filter* is designed for the purpose of blocking one specific frequency. The system function of a notch filter to block the angular frequency Ω_0 is

  Eqn. 8.18:
  $$H(z) = \frac{K\left(z^2-2\cos(\Omega_0)\,z+1\right)}{z^2-2r\cos(\Omega_0)\,z+r^2}$$

  Parameter r controls the width of the stopband.

- The most common method of designing an IIR filter is to find an appropriate an analog prototype filter and to convert it to a discrete-time filter by means of a transformation.

- The desired IIR filter magnitude response is specified using critical frequencies and tolerance limits. These are converted to the specifications of an appropriate analog prototype filter which is then designed and converted to a discrete-time filter.

- ☞ Design formulas and tables are widely available for designing analog prototype filters. Commonly used formulas are for Butterworth, Chebyshev type-I, Chebyshev type-II (inverse Chebyshev), and elliptic approximations.

- ☞ If a highpass, bandpass or bandstop analog filter is needed, it is obtained from a lowpass filter, referred to as a *lowpass prototype*, by means of an analog filter transformation.

- ☞ Multiple methods exist for converting an analog prototype filter to a discrete-time filter. The most important requirement in this conversion is that a stable analog prototype should yield a stable discrete-time filter. For the magnitude responses of the two filters to be similar, it is also desirable for the $j\omega$ axis of the s-plane to be mapped onto the unit circle of the z-plane. Impulse invariance and bilinear transformation are two methods that satisfy both criteria.

- ☞ Bilinear transformation is the most commonly used transformation technique for designing frequency selective IIR filters. Impulse response if limited to lowpass and bandpass filters due to the aliasing effect it causes.

- ☞ Critical frequencies for the discrete-time filter can be translated to the critical frequencies of the analog prototype through

  Eqn. 8.110:
  $$\omega_1 = \frac{\Omega_1}{T}, \qquad \text{and} \qquad \omega_2 = \frac{\Omega_2}{T}$$

  if the impulse invariance method is to be used in the conversion process. If, on the other hand, bilinear transformation is to be used, then the nonlinear warping (distortion) of the frequency axis must be taken into account. In this case the critical frequencies for the analog prototype filter are computed as

  Eqn. 8.111:
  $$\omega_1 = \frac{2}{T}\tan\left(\frac{\Omega_1}{2}\right), \qquad \text{and} \qquad \omega_2 = \frac{2}{T}\tan\left(\frac{\Omega_2}{2}\right)$$

  This is called *prewarping*.

- ☞ A length-N FIR filter has a system function that is of order $N-1$. The system function $H(z)$ has $N-1$ zeros and as many poles. All poles of the length-N FIR filter are at the origin, independent of the sample amplitudes of the impulse response. Therefore, FIR filters are inherently stable. The magnitude response of the filter is determined only by the locations of the zeros. Poles at the origin do not affect the magnitude response.

- ☞ A length-N FIR filter with impulse response $h[n]$ has a linear phase characteristic if the impulse response exhibits positive symmetry

  Eqn. 8.116:
  $$h[n] = h[N-1-n] \qquad \text{for } n = 0, \ldots, N-1$$

  or negative symmetry (may be referred to as *antisymmetry*)

  Eqn. 8.117:
  $$h[n] = -h[N-1-n] \qquad \text{for } n = 0, \ldots, N-1$$

- ☞ Depending on the length of the impulse response and the type of symmetry, linear-phase FIR filters can be categorized into four types:

  1. Type-I: N: odd, $h[n] = h[N-1-n]$ for $n = 0, \ldots, N-1$
  2. Type-II: N: even, $h[n] = h[N-1-n]$ for $n = 0, \ldots, N-1$

3. Type-III: N: odd, $h[n] = -h[N-1-n]$ for $n = 0, \ldots, N-1$
4. Type-IV: N: even, $h[n] = -h[N-1-n]$ for $n = 0, \ldots, N-1$

☞ Several methods exist for designing FIR filters. Fourier series method is based on finding the infinitely long impulse response of an ideal filter using inverse DTFT, and truncating it to have a finite-length impulse response. Truncation of the impulse response leads to Gibbs oscillations in the frequency spectrum. One method of reducing the effects of Gibbs oscillations is to multiply the impulse response with a window sequence that gradually tapers down to zero at both ends.

☞ Frequency sampling design method is based on sampling the desired magnitude response at DFT frequencies, and using the inverse DFT to obtain the impulse response. The requirements on the ideal filter magnitude response are relaxed by providing transition bands between the passbands and the stopbands of the desired filter behavior.

☞ Least-squares design method is an extension of the frequency sampling method. It is based on the idea of sampling the desired magnitude response at more frequencies than the length of the filter to be designed, thus creating an overdetermined system of equations. This overdetermined system can be solved in a least-squares sense. It is also possible to assign uneven weight factors to different areas of the magnitude response to emphasize the significance of some frequency bands over others.

☞ Parks and McClellan algorithm is also referred to as the *Remez exchange method.* It is based on the idea of finding a polynomial approximation to the desired frequency response by minimizing the largest approximation error. This results in an equiripple approximation to the desired magnitude response.

Further Reading

[1] Andreas Antoniou. *Digital Filters.* McGraw Hill New York, NY, USA: 1993.

[2] Stefan L. Hahn. *Hilbert Transforms in Signal Processing.* Boston, MA: Artech House, 1996.

[3] David A Jaffe and Julius O Smith. "Extensions of the Karplus-Strong Plucked-String Algorithm". In: *Computer Music Journal* 7.2 (1983), pp. 56–69.

[4] Kevin Karplus and Alex Strong. "Digital Synthesis of Plucked-String and Drum Timbres". In: *Computer Music Journal* 7.2 (1983), pp. 43–55.

[5] Larry D Paarmann. *Design and Analysis of Analog Filters: A Signal Processing Perspective.* Vol. 617. Springer Science & Business Media, 2005.

[6] Thomas W Parks and C Sidney Burrus. *Digital Filter Design.* Wiley-Interscience, 1987.

[7] David Ernesto Troncoso Romero and G Jovanovic. "Digital FIR Hilbert Transformers: Fundamentals And Efficient Design Methods". In: *MATLAB-A Fundam. Tool Sci. Comput. Eng. Appl* 1 (2012), pp. 445–482.

[8] R. Schaumann and M.E. Van Valkenburg. *Design of Analog Filters.* Oxford University Press, 2010.

[9] Julius O Smith. "Physical Modeling using Digital Waveguides". In: *Computer Music Journal* 16.4 (1992), pp. 74–91.

[10] Kenneth Steiglitz. *Digital Signal Processing Primer.* Courier Dover Publications, 2020.

[11] Kendall L Su. *Analog Filters.* Springer Science & Business Media, 2012.

[12] F. Taylor. *Digital Filters: Principles and Applications with MATLAB.* IEEE Series on Digital & Mobile Communication. Wiley, 2011.

[13] L.D. Thede. *Practical Analog and Digital Filter Design.* Artech House, 2005.

[14] L. Wanhammar. *Analog Filters Using MATLAB.* Springer London, Limited, 2009.

[15] Steve Winder. *Analog and Digital Filter Design.* Elsevier, 2002.

MATLAB Exercises with Solutions

MATLAB Exercise 8.1: Discrete-time processing of analog audio signals

In this exercise we will explore the idea of processing an analog signal using a discrete-time system (see Fig. 8.1). Most desktop and laptop computers have a built-in microphone available that can be used for capturing an analog audio waveform $x_a(t)$. The audio hardware within the computer converts this audio waveform to digital. Consider the script file `mexdt_8_1a.m` listed below. MATLAB function `audiorecorder()` is used for creating an audio recorder object that can interface with the audio hardware of the computer, and provide access to the discrete-time signal $x[n]$ that results from the conversion.

```
1  % Script: mexdt_8_1a.m
2  fs = 22050;                                % Sampling rate
3  nBits = 16;                                % Number of bits per sample
4  nChannels = 1;                             % Number of channels
5  sRecord = audiorecorder(fs,nBits,nChannels); % Create recorder object
6  disp('Press any key and start speaking');
7  pause;
8  disp('Recording started');
9  duration = 5;                              % Recording duration in seconds
10 recordblocking(sRecord,duration);          % Record audio
11 disp('Recording finished');
12 play(sRecord);                             % Play back the recording
13 x = getaudiodata(sRecord,'double');        % Create a vector with samples
14 plot(x);                                   % Graph the data
```

Run the script. Follow the on-screen prompts to record your voice and then listen to the recording. It seems fairly basic, yet we are actually implementing C/D and D/C conversion parts of Fig. 8.1. After specifying the conversion parameters for the sampling rate, the number of bits per sample and the number of audio channels, we call the function `audiorecorder()` on line 5 to create the object `sRecord`. The function `recordblocking()` on line 10 causes the recording process to start and continue for the specified duration of 5 seconds. Once the recording is finished, we play back the audio on line 12 using the function `play()`. In essence, we are converting an analog signal to discrete-time on line 10 so that it can be stored in the memory of a computer. Subsequently, on line 12, the discrete-time signal is converted back to analog so that it can be heard.

It's also possible to insert processing code between lines 10 and 12 of the script `mexdt_8_1a.m` to modify the signal. See the modified script `mexdt_8_1b.m` listed below:

```
1  % Script: mexdt_8_1b.m
2  fs = 22050;                                 % Sampling rate
3  nBits = 16;                                 % Number of bits per sample
4  nChannels = 1;                              % Number of channels
5  sRecord = audiorecorder(fs,nBits,nChannels); % Create recorder object
6  disp('Press any key and start speaking');
7  pause;
8  disp('Recording started');
9  duration = 5;                               % Recording duration in seconds
10 recordblocking(sRecord,duration);           % Record audio
11 disp('Recording finished');
12 x = getaudiodata(sRecord,'double');         % Create a vector with audio data
13 L = 2048;                                   % Delay amount for echo
14 r = 0.8;                                    % Gain factor for echo
15 buffer = zeros(L,1);                        % Buffer for ss_echof()
16 [y,buffer] = ss_echof(x,r,buffer);          % Add an echo
17 sound(y,fs);                                % Play back modified recording
```

In line 12, the function `getaudiodata()` is used to extract samples of the signal $x[n]$ and place them into the vector `x`. Now we can use the function `ss_echof()` that was developed in MATLAB Exercise 2.13 of Chapter 2 for adding an echo to the signal to obtain a new signal $y[n]$. Finally we play the signal in vector `y` using the function `sound()`. The buffer length chosen for the function `ss_echof()` is $L = 2048$ samples. It corresponds to a delay of $L\,T_s = L/f_s = 93$ ms.

Note that the function `sound()` is not related to the object `sRecord`, and therefore has no knowledge of the sampling rate f_s that was used in sampling $x_a(t)$. That information must be supplied through the second argument on line 17 for proper D/C conversion and playback. If you feel adventurous, change line 17 to

```
>> sound(y,1.5*fs)
```

and try again. In the two scripts listed above, we essentially used post-processing. The audio waveform was initially recorded and saved into the memory of the computer. Processing, and playback took place after the recording was completed. In practical applications the three steps of recording, processing, and playback are performed on a frame-by-frame basis. This allows the speaker to speak into the microphone while his or her voice is processed and played back simultaneously, with at most a time delay equal to the length of one frame of data (referred to as *latency*). MATLAB has a function `audioDeviceReader()` that can facilitate this type of processing. Unfortunately it is part of a toolbox that is not made available for the student edition, and will not be covered here.

Software resources: mexdt_8_1a.m , mexdt_8_1b.m

MATLAB Exercise 8.2: Writing functions for resonant bandpass filters

In this exercise, we will develop and test functions for use with resonant bandpass filters discussed in Section 8.3.1. Recall that a resonant bandpass filter is specified by means of its center frequency Ω_0 and bandpass Ω_b. Based on these specifications, the pole radius r, the gain factor K as well as the locations of zeros and poles of the system function need to be determined. We will begin by writing the function `ss_respar()` for this purpose.

```
1 function [r,gain,zrs,pls] = ss_respar(Omg0,Omgb)
2   r = 1-Omgb/2;                       % Eqn. (8.12)
3   zrs = [1;-1];                       % Eqn. (8.6)
4   pls = r*[exp(j*Omg0);exp(-j*Omg0)]; % Eqn. (8.6)
5   tmp = ss_freqz(zrs,pls,1,Omg0);     % Eqn. (8.13)
6   gain = 1/tmp;
7 end
```

Note that, in line 5, we use the function `ss_freqz()` which was developed in MATLAB Exercise 5.8 in Chapter 5. It computes the frequency response for specified values of Ω from the knowledge of pole and zero locations. Here we only need the magnitude of the system function at one frequency, $\Omega = \Omega_0$, to determine the gain factor.

Let us now solve the problem in Example 8.2 using the script `mexdt_8_2a.m` listed below. It makes use of the functions `ss_respar()` and `ss_freqz()`.

```
% Script: mexdt_8_2a.m
fs = 22050;                 % Sampling rate
f0 = 660;                   % Analog center frequency
fb = 60;                    % Analog bandwidth
Omg0 = 2*pi*f0/fs;          % Angular center frequency
Omgb = 2*pi*fb/fs;          % Angular bandwidth
[r,gain,zrs,pls] = ss_respar(Omg0,Omgb);   % Compute parameters
Omega = [-500:500]/500*pi;                 % Frequency vector from -pi to pi
[mag,phs] = ss_freqz(zrs,pls,gain,Omega); % Compute magnitude and phase
plot(Omega,mag); grid;                     % Graph magnitude
axis([-pi,pi,0,1.2]);
```

Next, we will develop a function for implementing the resonant bandpass filter that was designed. The implementation will be based on the block diagram shown in Fig. 8.6 which leads to the following equations:

$$w[n] = x[n] + 2r\,\cos(\Omega_0)\; w[n-1] - r^2\, w[n-2] \tag{8.195}$$

$$y[n] = K\left(w[n] - w[n-2]\right) \tag{8.196}$$

The function `ss_res()` is listed below. Arguments `x`, `Omg0`, `r`, and `gain` are all scalars, and are self explanatory. The last argument, `buffer` is a length-2 column vector that holds the samples $w[n-1]$ and $w[n-2]$. It is updated and returned to the calling program each time.

```
function [y,buffer] = ss_res(x,Omg0,r,gain,buffer)
  a1 = -2*r*cos(Omg0);
  a2 = r*r;
  w = x-a1*buffer(1)-a2*buffer(2);  % Eqn. (8.195)
  y = (w-buffer(2))*gain;           % Eqn. (8.196)
  buffer = [w;buffer(1)];           % Update buffer
end
```

The test script `mexdt_8_2b.m` listed below uses the resonant bandpass filter designed in Example 8.2 to extract the 660 Hz component from the guitar note in the audio file `GuitarA2.flac` and graph it.

```
% Script: mexdt_8_2b.m
[x,fs] = audioread('GuitarA2.flac');
y = zeros(size(x));    % Create an all-zero vector for the output
nSamp = length(x);     % Number of samples
buffer = zeros(2,1);   % Holds w[n-1] and w[n-2]
for n=1:nSamp
  [y(n),buffer] = ss_res(x(n),0.1881,0.9915,0.0085,buffer);
end
t = [0:nSamp-1]/fs;    % Vector of time instants
plot(t,y); grid;
```

Note that the function `ss_res()` processes one input sample at a time, and computes the corresponding output sample. It needs to be called in a loop for each sample of the input signal. Once the script completes, the extracted sinusoid can be played back by typing

```
>> sound(y,fs)
```

Software resources: `mexdt_8_2a.m` , `mexdt_8_2b.m` , `ss_respar.m` , `ss_res.m` , `GuitarA2.flac`

MATLAB Exercise 8.3: Resonators in real-time processing

The function `ss_res()` developed in MATLAB Exercise 8.2 allows implementation of a resonant bandpass filter one sample at a time. That works well in post processing single-channel audio signals. It needs to be modified for frame-based processing and/or for use with multi-channel signals. A version that is suitable for those situations is listed below in the form of the function `ss_resf()`. In this variant of the function, the argument `x` is a matrix with as many columns as the number of audio channels. Each column of `x` holds one frame of data, so the number of rows is the same as the frame size. The vector `buffer` also has as many columns as the number of channels. It has two rows that hold block diagram variables $w[n-1]$ and $w[n-2]$ for each channel.

```
function [y,buffer] = ss_resf(x,Omg0,r,gain,buffer)
  y = zeros(size(x));     % Placeholder for output frame
  frameSize = size(x,1);  % Number of samples in each frame
  a1 = -2*r*cos(Omg0);    % Compute coefficients outside loop
  a2 = r*r;
  for i=1:frameSize
    % Scalar w below is w[n] in Fig. 8.6
    w = x(i,:)-a1*buffer(1,:)-a2*buffer(2,:);  % Eqn. (8.195)
    y(i,:) = (w-buffer(2,:))*gain;             % Eqn. (8.196)
    buffer = [w;buffer(1,:)];                  % Update buffer
  end
end
```

We will use the function `ss_resf()` to extract several frequencies from the guitar note in the file `GuitarA2.flac`, and play back the result simultaneously. Before we do that, however, it is illustrative to look at the spectrum of the audio signal in question. The script `mexdt_8_3a.m` loads 1 second of audio from the file and computes the magnitude spectrum. The result is shown in Fig. 8.62. Note that, on line 7, we are graphing `mag/fs` to get an approximation to the magnitude of the continuous Fourier transform (see Chapter 7, Section 7.2.6).

```
% Script: mexdt_8_3a.m
[x,fs] = audioread('GuitarA2.flac',[1,22050]);  % Load 1 second of audio
mag = abs(fftshift(fft(x)));        % Magnitude of the spectrum
nSamp = length(x);                  % Number of samples
Fnorm = [0:nSamp-1]/nSamp-0.5;      % Normalized frequencies from -0.5 to 0.5
f = Fnorm*fs;                       % Actual frequencies from -fs/2 to fs/2
plot(f/1000,mag/fs); grid;          % Graph magnitude spectrum
axis([-11,11,0,0.015]);
xlabel('Frequency (kHz)');
ylabel('Magnitude');
title('|X_{a}(f)|')
```

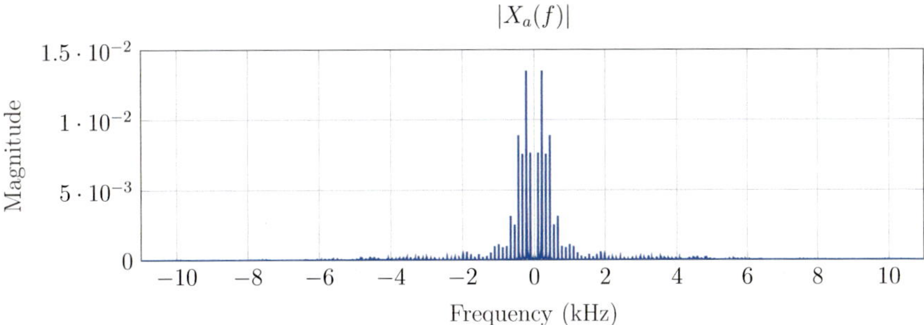

Figure 8.62 – Continuous-time Fourier transform magnitude for the guitar note A2.

The fundamental frequency is 110 Hz as expected for the note "A2". Consider a system that uses four resonant filters in parallel to extract the fundamental frequency and the harmonics 3, 5, 7 as illustrated in Fig. 8.63. The script `mexdt_8_3b.m` listed below implements this system.

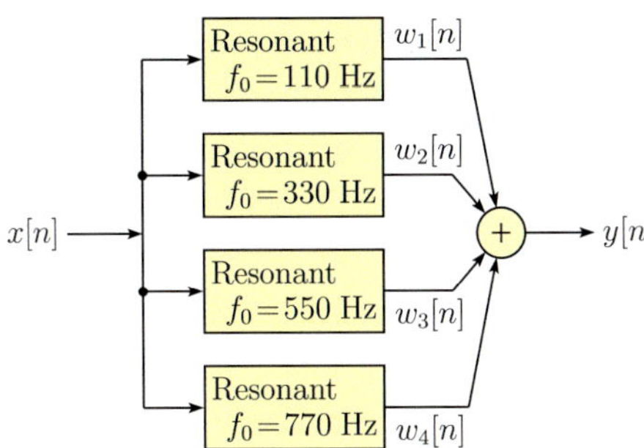

Figure 8.63 – Modifying the guitar note by adding the fundamental and the harmonics 3, 5, 7.

```
% Script: mexdt_8_3b.m
% Create an "audio file reader" object
sReader = dsp.AudioFileReader('GuitarA2.flac');
% Create an "audio player" object
sPlayer = audioDeviceWriter('SampleRate',sReader.SampleRate);
%-----------------------------------------------------------
% Design resonant bandpass filters
fs = sReader.SampleRate;
```

```
Omgb = 2*pi*60/fs;     % Bandwidth for all four filters
Omg01 = 2*pi*110/fs;  % Center frequency for filter 1
Omg02 = 2*pi*330/fs;  % Center frequency for filter 2
Omg03 = 2*pi*550/fs;  % Center frequency for filter 3
Omg04 = 2*pi*770/fs;  % Center frequency for filter 4
[r1,gain1] = ss_respar(Omg01,Omgb);  % Parameters for filter 1
[r2,gain2] = ss_respar(Omg02,Omgb);  % Parameters for filter 2
[r3,gain3] = ss_respar(Omg03,Omgb);  % Parameters for filter 3
[r4,gain4] = ss_respar(Omg04,Omgb);  % Parameters for filter 4
%------------------------------------------------------------
nChannels = info(sReader).NumChannels;  % Number of channels
buffer1 = zeros(2,nChannels);           % Buffer for filter 1
buffer2 = zeros(2,nChannels);           % Buffer for filter 2
buffer3 = zeros(2,nChannels);           % Buffer for filter 3
buffer4 = zeros(2,nChannels);           % Buffer for filter 4
while ~isDone(sReader)
  x = sReader();
  [w1,buffer1] = ss_resf(x,Omg01,r1,gain1,buffer1);
  [w2,buffer2] = ss_resf(x,Omg02,r2,gain2,buffer2);
  [w3,buffer3] = ss_resf(x,Omg03,r3,gain3,buffer3);
  [w4,buffer4] = ss_resf(x,Omg04,r4,gain4,buffer4);
  y = w1+w2+w3+w4;                      % Compute output frame
  sPlayer(y);
end
release(sReader);  % We are finished with the input audio file
release(sPlayer);  % We are finished with the audio output device
```

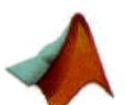

Software resources: mexdt_8_3a.m , mexdt_8_3b.m , ResonatorsRT.mlx , ss_resf.m , GuitarA2.flac

MATLAB Exercise 8.4: Writing functions for notch filters

In this exercise we will develop and test functions for designing and implementing notch filters following the design technique discussed in Section 8.3.2. First, the function `ss_notchpar()` will be developed for computing the parameter values for a notch filter. This is very similar to the function `ss_respar()` developed in MATLAB Exercise 8.2 for resonant bandpass filters.

```
function [r,gain,zrs,pls] = ss_notchpar(Omg0,Omgb)
  r = 1-Omgb/2;                          % Eqn. (8.20)
  zrs = [exp(j*Omg0);exp(-j*Omg0)];      % See Fig. 8.11
  pls = r*[exp(j*Omg0);exp(-j*Omg0)];    % See Fig. 8.11
  tmp = ss_freqz(zrs,pls,1,0);           % Eqn. (8.19)
  gain = 1/tmp;
end
```

The next step is to develop an implementation function based on the block diagram of Fig. 8.12. Functional relationships for the block diagram can be written as

$$w[n] = x[n] + 2r\,\cos(\Omega_0)\;w[n-1] - r^2\,w[n-2] \tag{8.197}$$

$$y[n] = K\left(w[n] - 2\,\cos(\Omega_0)\;w[n-1] + w[n-2]\right) \tag{8.198}$$

The code for the function `ss_notch()` is listed below. It implements a notch filter one sample at a time. The length-2 vector `buffer` holds the samples $w[n-1]$ and $w[n-2]$.

```
function [y,buffer] = ss_notch(x,Omg0,r,gain,buffer)
  a1 = -2*r*cos(Omg0);
  a2 = r*r;
  b1 = -2*cos(Omg0);
  w = x-a1*buffer(1)-a2*buffer(2);          % Eqn. (8.197)
  y = (w+b1*buffer(1)+buffer(2))*gain;  % Eqn. (8.198)
  buffer = [w;buffer(1)];                   % Update buffer
end
```

An alternative version `ss_notchf()` that can be used for frame-based processing and/or with multi-channel signals is listed below. The input argument `x` must be a matrix in which each column holds the data for one channel. Each column of the matrix `buffer` holds samples $w[n-1]$ and $w[n-2]$ for one channel.

```
function [y,buffer] = ss_notchf(x,Omg0,r,gain,buffer)
  y = zeros(size(x));       % Placeholder for output frame
  frameSize = size(x,1);  % Number of samples in each frame
  a1 = -2*r*cos(Omg0);      % Compute coefficients outside loop
  a2 = r*r;
  b1 = -2*cos(Omg0);
  for i=1:frameSize
    w = x(i,:)-a1*buffer(1,:)-a2*buffer(2,:);          % Eqn. (8.197)
    y(i,:) = (w+b1*buffer(1,:)+buffer(2,:))*gain;  % Eqn. (8.198)
    buffer = [w;buffer(1,:)];                           % Update buffer
  end
end
```

The script `mexdt_8_4.m` listed below loads the sound file `GuitarA2.flac`. It then designs a notch filter to remove the 220 Hz component from it. The magnitude spectrum of the output of the notch filter is graphed. On line 16 of the script we graph `mag/fs` to get an approximation to the magnitude of the continuous Fourier transform (see Chapter 7, Section 7.2.6).

```
1  % Script: mexdt_8_4.m
2  [x,fs] = audioread('GuitarA2.flac');
3  f0 = 220;                  % Analog center frequency
4  fb = 100;                  % Analog bandwidth
5  Omg0 = 2*pi*f0/fs;         % Angular center frequency
6  Omgb = 2*pi*fb/fs;         % Angular bandwidth
7  [r,gain,zrs,pls] = ss_notchpar(Omg0,Omgb);   % Compute parameters
8  y = zeros(size(x));        % Create an all-zero vector for the output
9  nSamp = length(x);         % Number of samples
10 buffer = zeros(2,1);       % Holds w[n-1] and w[n-2]
11 y = ss_notchf(x,Omg0,r,gain,buffer); % Apply notch filter
12 Yk = fft(y);                         % Compute FFT
13 nFFT = length(Yk);
14 mag = abs(fftshift(Yk));             % Magnitude of FFT
15 frq = ([0:nFFT-1]'/nFFT-0.5)*fs;     % Frequency vector in Hz
16 plot(frq,mag/fs);          % Graph magnitude spectrum of the output
17 xlabel('Frequency (Hz)');
18 ylabel('Magnitude');
```

Zoom into the magnitude response and confirm that the 220 Hz component is largely removed. It is also interesting to listen to various vectors with the following commands.

```
>> sound(x,fs);     % Input signal
>> sound(y,fs);     % Output signal
>> sound(x-y,fs);   % Removed by notch filter
```

Software resources: mexdt_8_4.m , ss_notchpar.m , ss_notch.m , ss_notchf.m , GuitarA2.flac

MATLAB Exercise 8.5: Case Study – ECG signal with sinusoidal interference

The electrocardiogram (ECG) is essentially a recording of the heartbeats of a patient that is sampled and digitized. Modern portable ECG systems consist of a series of sensors interfaced to a computer with the appropriate software. Typically, about a dozen sensors are placed on the patient's chest and limbs, and electrical activity due to heartbeats is recorded from various angles. The electrical voltages being recorded are quite small, usually in the μV range. They can easily be corrupted with random noise and/or sinusoidal interference. Improvement of ECG signal quality by removal or reduction of noise and interference is an active area of research in biomedical engineering. In this exercise we will focus on sinusoidal interference only, and work on an oversimplified case of removing a single tone from the signal.

Sinusoidal signals may corrupt ECG recordings through inductive and capacitive coupling effects with power lines. Every hospital room and clinic has power lines inside the walls. Even if the ECG equipment is electrically insulated, there will be some interference from 50 or 60 Hz sinusoidal signals. A MATLAB data file noisyECG.mat is provided with a recording of one of the sensors. It was sampled with a sampling rate of $f_s = 360$ Hz. It has a 60 Hz interference component that we would like to remove using the appropriate notch filter. The script mexdt_8_5a.m listed below loads the signal and graphs it.

```
% Script: mexdt_8_5a.m
load noisyECG.mat
plot(t,x); grid;
axis([0,10,0.85,1.25]);
xlabel('Time (sec)');
ylabel('Amplitude');
```

The slightly modified script mexdt_8_5b.m graphs a zoomed-in view of the data in which the 60 Hz interference is clearly visible between heartbeats. Graphs from both scripts are shown in Fig. 8.64.

```
% Script: mexdt_8_5b.m
load noisyECG.mat
plot(t(501:1000),x(501:1000)); grid;
axis([1.8,2.7,0.85,1.25]);
xlabel('Time (sec)');
ylabel('Amplitude');
```

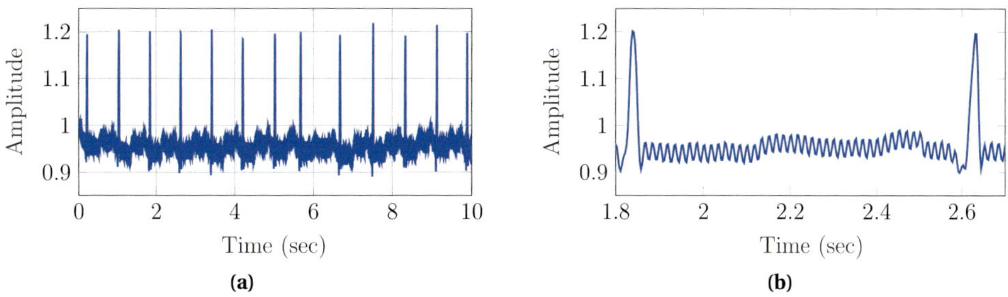

Figure 8.64 – **(a)** ECG signal for MATLAB Exercise 8.5 and **(b)** zoomed-in version to show detail.

Consider the script `mexdt_8_5c.m` listed below. Here we use the function `fft()` to compute and graph the dB magnitude spectrum of the noisy ECG signal which is shown in Fig. 8.65. The interference at 60 Hz is clearly visible in the spectrum.

```
% Script: mexdt_8_5c.m
load noisyECG.mat
fs = 360;                          % Sampling rate used in ECG data
Xk = fft(x);                       % Compute FFT
magdB = 20*log(abs(fftshift(Xk))); % dB magnitude spectrum
frq = [0:3599]/3600*fs-fs/2;       % Frequencies in Hz (from -fs/2 to fs/2)
plot(frq,magdB); grid              % Graph dB magnitude
axis([-180,180,-100,200]);
xlabel('Frequency (Hz)');
ylabel('Magnitude (dB)');
```

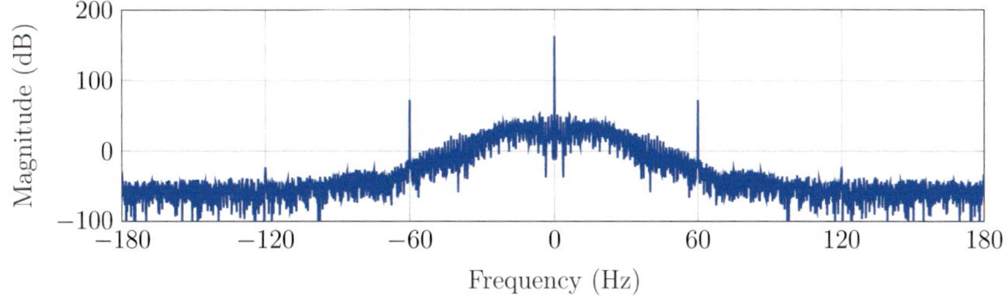

Figure 8.65 – dB Magnitude spectrum of the ECG signal with sinusoidal interference at 60 Hz.

A notch filter with a center frequency of $f_0 = 60$ Hz and a 3-dB notch bandwidth of 6 Hz will be used for removing the interference. Script `mexdt_8_5d.m` utilizes functions `ss_notchpar()` and `ss_notchf()` developed in MATLAB Exercise 8.4 for this purpose. The signal at the output of the notch filter is graphed in Fig. 8.66.

```
% Script: mexdt_8_5d.m
load noisyECG.mat
fs = 360;                          % Sampling rate used in ECG data
f0 = 60;                           % Center frequency of notch filter
fb = 6;                            % Allow 6 Hz width
Omg0 = 2*pi*f0/fs;                 % Angular center frequency
Omgb = 2*pi*fb/fs;                 % Angular bandwidth
[r,gain,zrs,pls] = ss_notchpar(Omg0,Omgb);
```

```
buffer = zeros(2,1);                    % Initialize buffer
y = ss_notchf(x,Omg0,r,gain,buffer);
plot(t,y);
```

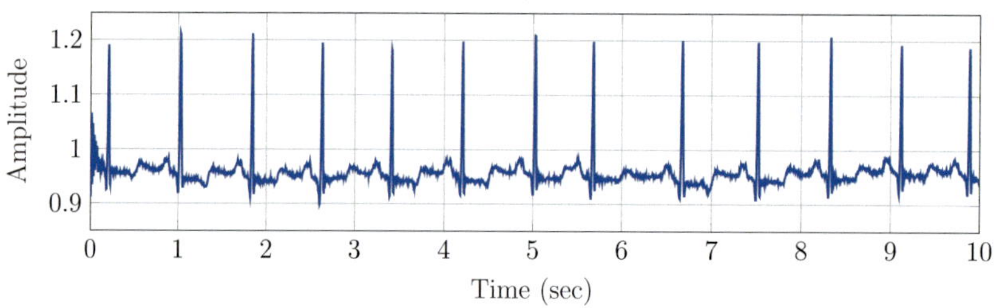

Figure 8.66 – The signal at the output of the notch filter in MATLAB Exercise 8.5.

Software resources: mexdt_8_5a.m , mexdt_8_5b.m , mexdt_8_5c.m , mexdt_8_5d.m , CleanUpECG.mlx , noisyECG.mat

MATLAB Exercise 8.6: Impulse invariant design

In Example 8.4 impulse invariance technique was used for obtaining a discrete-time filter from a Chebyshev type-I analog prototype with system function

$$G(s) = \frac{0.6250}{s^3 + 1.1542\,s^2 + 1.4161\,s + 0.6250}$$

MATLAB function `impinvar()` can be used to accomplish this task as follows:

```
>> T = 0.2;
>> num = [0.6250];
>> den = [1,1.1542,1.4161,0.6250];
>> [numz,denz] = impinvar(num,den,1/T)

numz =
   -0.0000    0.0023    0.0021

denz =
    1.0000   -2.7412    2.5395   -0.7939
```

The results obtained correspond to the z-domain system function

$$H(z) = \frac{0.0023\,z^2 + 0.0021\,z}{z^3 - 2.7412\,z^2 + 2.5395\,z - 0.7939}$$

which matches the answer found in Example 8.4. The magnitude and the phase of the system function may be graphed with the following statements:

```
>> Omg = [0:0.01:1]*pi;
>> H = freqz(numz,denz,Omg);
>> plot(Omg,abs(H));
>> plot(Omg,angle(H));
```

Software resource: mexdt_8_6.m

MATLAB Exercise 8.7: IIR filter design using bilinear transformation

Consider again the IIR lowpass filter design problem solved in Example 8.7. In this exercise we will rely on MATLAB to solve it. Recall that the specifications for the filter to be designed were given as follows:

$$\Omega_1 = 0.2\pi\,, \qquad \Omega_2 = 0.36\pi\,, \qquad R_p = 2\ \text{dB}\,, \qquad A_s = 20\ \text{dB}$$

Even though the design problem can be solved quickly with two function calls (see MATLAB Exercise 8.8 for this alternative approach), we will opt to solve it using the three steps in Section 8.4. This will allow checking the intermediate results obtained in Example 8.7. The first step is to find the specifications for the analog prototype filter. Critical frequencies ω_1 and ω_2 for the analog prototype are found using the prewarping formula. Afterward N and ω_c are determined using the function buttord().

```
>> T = 1;
>> Rp = 2;
>> As = 20;
>> omg1 = 2/T*tan(0.2*pi/2);
>> omg2 = 2/T*tan(0.36*pi/2);
>> [N,omgc] = buttord(omg1,omg2,Rp,As,'s')

N =
     4

omgc =
   0.7146
```

The last argument 's' of the function buttord() signifies that we seek the order and cutoff frequency for an *analog* Butterworth filter. Having determined N and ω_c, the analog prototype filter can be designed using the function butter().

```
>> [num,den] = butter(N,omgc,'s')

num =
         0         0         0         0    0.2608

den =
    1.0000    1.8675    1.7437    0.9537    0.2608
```

Again the last argument 's' specifies that we are designing an analog filter. The vectors num and den correspond to the analog prototype system function

$$G(s) = \frac{0.2608}{s^4 + 1.8675\,s^3 + 1.7437\,s^2 + 0.9537\,s + 0.2608}$$

If desired, magnitude and phase characteristics of the analog prototype filter may be graphed with the following lines:

```
>> omg = [0:0.01:5];
>> G = freqs(num,den,omg);
>> plot(omg,abs(G)); grid;
>> plot(omg,angle(G)); grid;
```

In step 3 the analog prototype filter is converted to a discrete-time filter using bilinear transformation. MATLAB function `bilinear()` may be used for this purpose.

```
>>  [numz,denz] = bilinear(num,den,1/T)

numz =
    0.0065    0.0260    0.0390    0.0260    0.0065

denz =
    1.0000   -2.2209    2.0861   -0.9204    0.1594
```

The system function for the discrete-time system is

$$H(z) = \frac{0.0065\,z^4 + 0.0260\,z^3 + 0.0390\,z^2 + 0.0260\,z + 0.0065}{z^4 - 2.2209\,z^3 + 2.0861\,z^2 - 0.9204\,z + 0.1594}$$

If desired, magnitude and phase characteristics of the discrete-time filter may be graphed with the following lines:

```
>> Omg = [0:0.01:1]*pi;
>> H = freqz(numz,denz,Omg);
>> plot(Omg,abs(H));
>> plot(Omg,angle(H));
```

Software resource: `mexdt_8_7.m`

MATLAB Exercise 8.8: IIR filter design using bilinear transformation – revisited

The Butterworth IIR filter of MATLAB Exercise 8.7 can be designed quickly by using the discrete-time filter design capabilities of the functions `buttord()` and `butter()` as follows:

```
>> T = 1;
>> Omg1 = 0.2*pi;
>> Omg2 = 0.36*pi;
>> Rp = 2;
>> As = 20;
>> [N,Omgc] = buttord(Omg1/pi,Omg2/pi,Rp,As);
>> [numz,denz] = butter(N,Omgc)

numz =
    0.0065    0.0260    0.0390    0.0260    0.0065

denz =
    1.0000   -2.2209    2.0861   -0.9204    0.1594
```

Note that we leave out the last argument `'s'` in order to get the discrete-time IIR filter directly. The result obtained is identical to that of MATLAB Exercise 8.7.

Important detail: In the call to `buttord()` the first two arguments we used were `Omg1/pi` and `Omg2/pi` which correspond to our normalized frequencies $2F_1$ and $2F_2$, respectively. This is due to the fact that, contrary to common practice, MATLAB normalizes frequencies with respect to the folding frequency $f_s/2$ rather than the sampling rate f_s. In our conventions the angular frequency $\Omega = \pi$ radians corresponds to the normalized frequency $F = 0.5$. In MATLAB conventions it corresponds to $\bar{F} = 1$. Therefore, in specifying a discrete-time filter to MATLAB through normalized frequencies we always use twice the conventional normalized frequency values. It is an unfortunate complication, but one we always need to be aware of when using built-in filter design functions.

Software resource: `mexdt_8_8.m`

MATLAB Exercise 8.9: A complete IIR filter design example

A Chebyshev type-I IIR bandpass filter is to be designed with the following set of specifications:

$$\Omega_1 = 0.2\pi\,, \quad \Omega_2 = 0.26\pi\,, \quad \Omega_3 = 0.48\pi\,, \quad \Omega_4 = 0.54\pi$$
$$R_p = 1\text{ dB}\,, \quad A_s = 25\text{ dB}$$

The specification diagram is shown in Fig. 8.67.

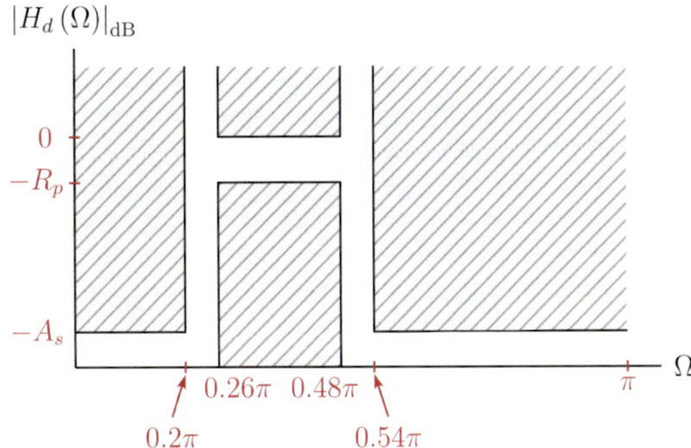

Figure 8.67 – Specification diagram for MATLAB Exercise 8.9.

Solution: The normalized edge frequencies are

$$F_1 = \frac{\Omega_1}{2\pi} = 0.1\,, \qquad F_2 = \frac{\Omega_2}{2\pi} = 0.13\,, \qquad F_3 = \frac{\Omega_3}{2\pi} = 0.24\,, \qquad F_4 = \frac{\Omega_4}{2\pi} = 0.27$$

Our first step will be to use the function `cheb1ord()` to determine the minimum filter order that allows the requirements to be satisfied. In this case the `cheb1ord()` will be used without the `'s'` parameter since we want to compute the filter order from IIR digital filter specifications and not from those of the analog prototype. The following code computes N:

```
>> Rp = 1;
>> As = 25;
>> F1 = 0.1;
>> F2 = 0.13;
>> F3 = 0.24;
>> F4 = 0.27;
>> N = cheb1ord([2*F2,2*F3],[2*F1,2*F4],Rp,As)

N =
     5
```

The result indicates that the analog prototype is of fifth order. Consequently, the bandpass IIR filter will be of order 10. At the risk of repetition, we will again emphasize the way critical frequencies are specified in built-in filter design functions of MATLAB. With the function `cheb1ord()`, the first vector in the argument list contains the two passband edge frequencies *normalized in MATLAB sense*, and the second vector in the argument list contains the two stopband edge frequencies also *normalized in MATLAB sense.*

Once N is found, the filter is designed using the `cheby1()`. Its magnitude spectrum is computed using the function `freqz()`. The magnitude response of the filter obtained is shown in Fig. 8.68.

```
>> [numz,denz] = cheby1(N,Rp,[2*F2,2*F3]);
>> Omg = [-1:0.001:1]*pi;
>> H = freqz(numz,denz,Omg);
>> plot(Omg,abs(H)); grid;
>> axis([-pi,pi,-0.1,1.1]);
>> title('|H(\Omega)|');
>> xlabel('\Omega (rad)');
>> ylabel('Magnitude');
```

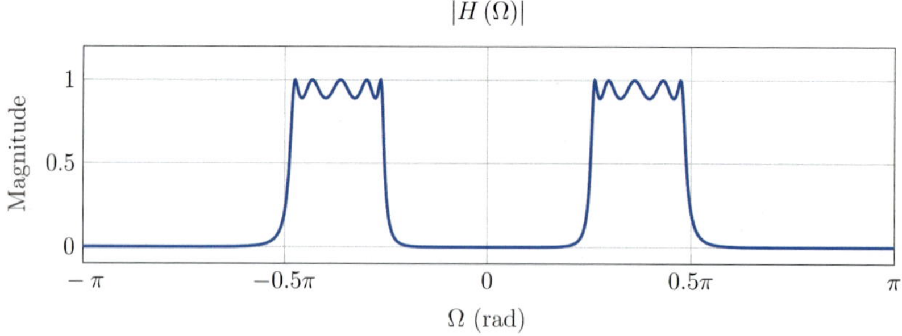

Figure 8.68 – Magnitude response of the filter designed in MATLAB Exercise 8.9.

Software resource: `mexdt_8_9.m`

MATLAB Exercise 8.10: Second-order sections for IIR design and implementation

Consider again the Chebyshev type-I bandpass filter design problem we have solved in MATLAB Exercise 8.9. The result was a system of order 10. Utilizing the built-in function `cheby()`, we have obtained numerator and denominator coefficients of the system function $H(z)$. How can this filter be implemented in either post processing or real-time? One answer is to draw a direct-form-II block diagram (see Chapter 5, Section 5.6.1) with 10 feed-forward branches and as many feedback branches. This block diagram could then be converted into MATLAB code. Obviously, this approach is impractical at best, and may even be dangerous. At a minimum, it would lead to rather complicated code that is also very specific to the filter at hand, and is not reusable. If another filter, say of order 12, is later designed, the coding exercise would have to be repeated from scratch. A more serious problem might be that the round-off errors in implementation would propagate through feedback paths connected to a long delay line. This leads to a phenomenon known as *limit-cycle oscillations.* A system that is stable on paper may become unstable in implementation.

A better solution is to represent the high-order filter in terms of second-order sections connected in cascade (see Chapter 5, Section 5.6.2). Code can be developed for a second-order section and reused as many times as needed for implementing a higher-order filter. In this exercise we will solve the design problem of MATLAB Exercise 8.9 again, this time obtaining second-order sections. Afterward we will develop a function to implement the designed filter in real time. The script `mexdt_8_10a.m` listed below is a modified version of the scripts used in the previous exercise:

```
1  % Script: mexdt_8_10a.m
2  Rp = 1;      % dB passband ripple
3  As = 25;     % dB stopband attenuation
4  F1 = 0.1;    % Corner frequencies F1 thru F4
5  F2 = 0.13;
6  F3 = 0.24;
7  F4 = 0.27;
8  N = cheb1ord([2*F2,2*F3],[2*F1,2*F4],Rp,As)  % Compute filter order
9  [zrs,pls,gain] = cheby1(N,Rp,[2*F2,2*F3])    % Alternative syntax for cheby1()
10 [sos,gain] = zp2sos(zrs,pls,gain)            % Convert to SOS
11 Omg = [-1:0.001:1]*pi;                       % Vector of angular frequencies
12 H = gain*freqz(sos,Omg);                     % Alternative syntax for freqz()
13 plot(Omg,abs(H)); grid;
14 axis([-pi,pi,-0.1,1.1]);
15 title('|H(\Omega)|');
16 xlabel('\Omega (rad)');
17 ylabel('Magnitude');
```

In line 9 an alternative syntax of the function `cheby1()` is used. It returns zeros, poles, and a gain factor as opposed to numerator and denominator coefficients. In line 10, the function `zp2sos()` is used for constructing a *SOS* matrix (SOS stands for *second-order sections*) where each row contains the coefficients of one section of the filter. The system function for the designed tenth-order filter is

$$H(z) = \prod_{i=1}^{5} H_i(z) \tag{8.199}$$

where each section is in the form

$$H_i(z) = \frac{b_{0i} + b_{1i}\, z^{-1} + b_{2i}\, z^{-2}}{1 + a_{1i}\, z^{-1} + a_{2i}\, z^{-2}}$$

The coefficients from each section of the filter form one row of the SOS matrix. For section i, the row-i of the SOS matrix is

$$\mathbf{SOS_i} = \begin{bmatrix} b_{0i} & b_{1i} & b_{2i} & 1 & a_{1i} & a_{2i} \end{bmatrix}$$

The complete SOS matrix returned by the function `zp2sos()` has 5 rows in this case, and is in the form

$$\mathbf{SOS} = \begin{bmatrix} b_{01} & b_{11} & b_{21} & 1 & a_{11} & a_{21} \\ b_{02} & b_{12} & b_{22} & 1 & a_{12} & a_{22} \\ \vdots & \vdots & \vdots & \vdots & \vdots & \vdots \\ b_{05} & b_{15} & b_{25} & 1 & a_{15} & a_{25} \end{bmatrix}$$

In order to implement this filter in real time, we will start with the function `ss_iir2()` that was developed in MATLAB Exercise 5.12 of Chapter 5, and adapt it for frame-by-frame processing. The end result will be a new function `ss_iir2f()`. The block diagram for a second-order section is shown in Fig. 8.69.

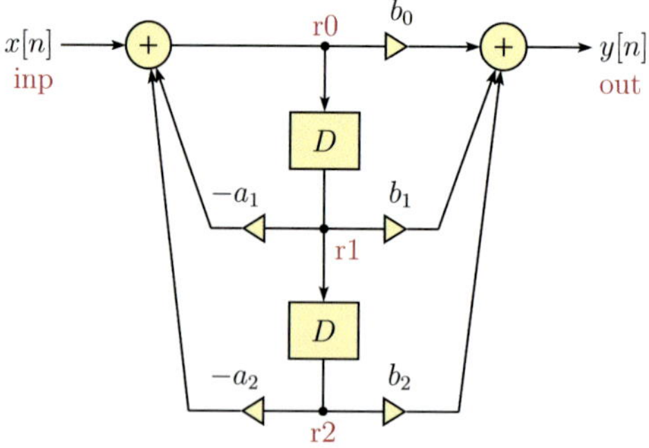

Figure 8.69 – Block diagram for second-order section.

Function `ss_iir2f()` is listed below.

```
function [out,states] = ss_iir2f(inp,coeffs,states)
  out = zeros(size(inp));      % Placeholder for output frame
  frameSize = size(inp,1);     % Number of samples in each frame
  % Extract the filter states
  r1 = states(1,:);
  r2 = states(2,:);
  % Extract the coefficients
  b0 = coeffs(1);
  b1 = coeffs(2);
  b2 = coeffs(3);
  a1 = coeffs(5);
  a2 = coeffs(6);
  for i=1:frameSize
    % Compute the output sample
    r0 = inp(i,:)-a1*r1-a2*r2;
```

```
    out(i,:) = b0*r0+b1*r1+b2*r2;
    % Update the filter states: r2 <-- r1 then r1 <-- r0
    r2 = r1;
    r1 = r0;
  end
  % Return updated states to preserve them
  states(1,:) = r1;
  states(2,:) = r2;
end
```

In script `mexdt_8_10b.m` we use the new function to filter an audio signal in real time. In the first part we quickly redesign the same bandpass filter as before. Line 5 is noteworthy: The gain factor in variable `gain` is distributed among the five filter sections by multiplying the numerator coefficients of each second-order section with the fifth root of the gain factor. In the loop between lines 17 and 25 we use the function `ss_iirf()` repeatedly, implementing the five filter sections in cascade. The output signal of each section becomes the input signal for the following section.

```
1  % Script: mexdt_8_10b.m
2  % Redesign the 10th-order Chebyshev bandpass filter to obtain SOS
3  [zrs,pls,gain] = cheby1(5,1,[0.26,0.48])  % Alternative syntax for cheby1()
4  [sos,gain] = zp2sos(zrs,pls,gain)         % Convert to SOS
5  sos(:,1:3) = sos(:,1:3)*(gain^(1/5));     % Absorb gain factor into each section
6  %-----------------------------------------------------------------
7  % Create an "audio file reader" object
8  sReader = dsp.AudioFileReader('Ballad_22050_Hz.flac','ReadRange',[1,661500]);
9  % Create an "audio player" object
10 sPlayer = audioDeviceWriter('SampleRate',sReader.SampleRate);
11 nChannels = info(sReader).NumChannels;    % Number of channels
12 states1 = zeros(2,nChannels);             % States for section 1
13 states2 = zeros(2,nChannels);             % States for section 2
14 states3 = zeros(2,nChannels);             % States for section 3
15 states4 = zeros(2,nChannels);             % States for section 4
16 states5 = zeros(2,nChannels);             % States for section 5
17 while ~isDone(sReader)
18   x = sReader();
19   [w1,states1] = ss_iir2f(x,sos(1,:),states1);
20   [w2,states2] = ss_iir2f(w1,sos(2,:),states2);
21   [w3,states3] = ss_iir2f(w2,sos(3,:),states3);
22   [w4,states4] = ss_iir2f(w3,sos(4,:),states4);
23   [y,states5]  = ss_iir2f(w4,sos(5,:),states5);
24   sPlayer(y);
25 end
26 release(sReader);  % We are finished with the input audio file
27 release(sPlayer);  % We are finished with the audio output device
```

An alternate version of the script that drops real-time playback and saves the output sound to disc instead is also provided among downloadable files.

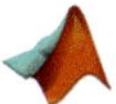

Software resources: `mexdt_8_10a.m` , `mexdt_8_10b.m` , `mexdt_8_10b_Alt.m` , `ss_iir2f.m`

MATLAB Exercise 8.11: FIR filter design using Fourier series method

Using the Fourier series design method with a triangular (Bartlett) window, design a 24th-order FIR bandpass filter with passband edge frequencies $\Omega_1 = 0.4\pi$ and $\Omega_2 = 0.7\pi$ as shown in Fig. 8.70.

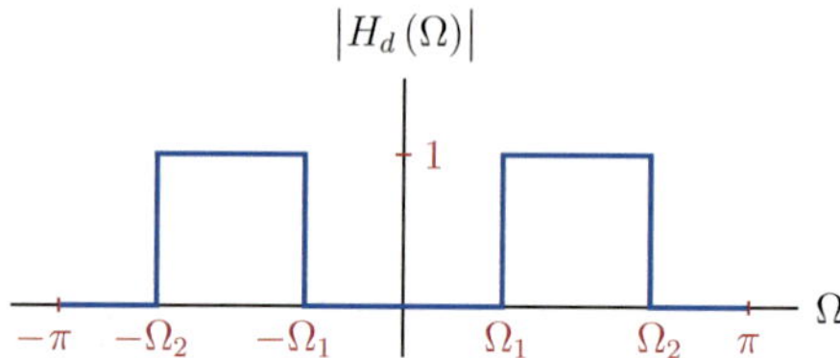

Figure 8.70 – Desired FIR magnitude response for MATLAB Exercise 8.11.

Solution: The order of the FIR filter is $N-1=24$. Normalized passband edge frequencies are

$$F_1 = \frac{\Omega_1}{2\pi} = 0.2 \qquad \text{and} \qquad F_2 = \frac{\Omega_2}{2\pi} = 0.35$$

The following two statements create a length-25 Bartlett window and then use it for designing the bandpass filter required.

```
>> wn = bartlett(25);
>> hn = fir1(24,[0.4,0.7],wn)
```

Some important details need to be highlighted. The function `bartlett()` uses the filter length N. This applies to other window generation functions such as `hamming()`, `hann()`, and `blackman()` as well. On the other hand, the design function `fir1()` uses the filter order which is $N-1$. The second argument to the function `fir1()` is a vector of two normalized edge frequencies which results in a bandpass filter being designed. (A lowpass filter results if only one edge frequency is used.) The frequencies are *normalized in MATLAB sense* which means they must be twice our normalized frequencies.

The magnitude spectrum may be computed and graphed for the bandpass filter with impulse response in vector `hn` using the following statements.

```
>> Omg = [-256:255]/256*pi;
>> H = fftshift(fft(hn,512));
>> Omgd = [-1,-0.7,-0.7,-0.4,-0.4,0.4,0.4,0.7,0.7,1]*pi;
>> Hd = [0,0,1,1,0,0,1,1,0,0];
>> plot(Omg,abs(H),Omgd,Hd); grid;
>> axis([-pi,pi,-0.1,1.1]);
>> title('|H(\Omega)|');
>> xlabel('\Omega (rad)');
>> ylabel('Magnitude');
```

The magnitude response is shown in Fig. 8.71.

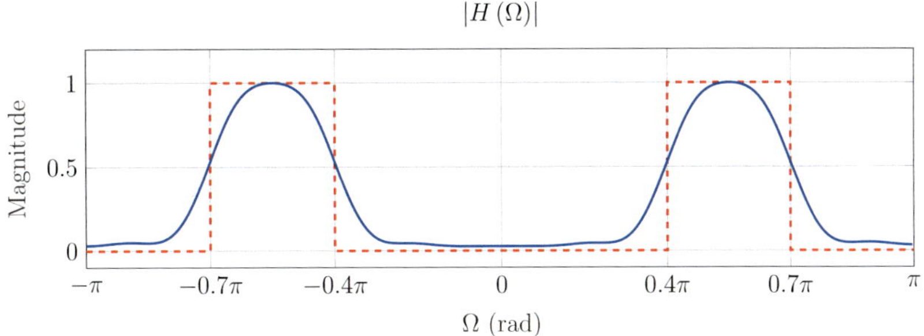

Figure 8.71 – Magnitude response of the filter designed in MATLAB Exercise 8.11.

Software resource: mexdt_8_11.m

MATLAB Exercise 8.12: FIR filter design using Fourier series method – revisited

In MATLAB Exercise 8.11 we used the built-in function `fir1()` to design a FIR bandpass filter. In this exercise we will develop our own function `ss_fir1()` for doing the same. The reason for writing our own version of the function is not because it will be better or faster than what is provided in MATLAB. Rather, our version will be based on the mathematical development of Sections 8.5.1 and 8.5.3. It will hopefully reinforce the concepts covered in those sections. The listing for the function `ss_fir1()` is given below.

```
function h = ss_fir1(N,F,fType,wType)
  Omg = 2*pi*F;                     % Angular frequencies
  n = [0:N-1]';                     % Interval for truncation
  % Find filter coefficients
  if strcmpi(fType,'highpass')      % Highpass filter
    h = Type12HP(Omg,N,n);
  elseif strcmpi(fType,'bandpass')  % Bandpass filter
    h = Type12BP(Omg(1),Omg(2),N,n);
  elseif strcmpi(fType,'diff')      % Differentiator
    h = Type34Diff(N,n);
  elseif strcmpi(fType,'hilbert')   % Hilbert transform filter
    h = Type34Hilbert(N,n);
  else                              % Lowpass filter by default
    h = Type12LP(Omg,N,n);
  end
  % Multiply with window sequence
  if strcmpi(wType,'bartlett')      % Bartlett (triangular) window
    w = bartlett(N);
  elseif strcmpi(wType,'hamming')   % Hamming window
    w = hamming(N);
  elseif strcmpi(wType,'hanning')   % Hanning (von Hann) window
    w = hann(N);
  elseif strcmpi(wType,'blackman')  % Blackman window
    w = blackman(N);
  else                              % Rectangular window by default
    w = ones(N,1);
```

```
    end
    h = h.*w;
end

function hd = Type12LP(Omgc,N,n)
  % Type-I or type-II lowpass ideal impulse response
  M = (N-1)/2;
  K = n-M;
  hd = Omgc/pi*sinc(Omgc/pi*K);  % Eqn. (8.152)
end

function hd = Type12HP(Omgc,N,n)
  % Type-I or type-II highpass ideal impulse response
  M = (N-1)/2;
  K = n-M;
  hd = sinc(K)-Type12LP(Omgc,N,n);  % Eqn. (8.154)
end

function hd = Type12BP(Omg1,Omg2,N,n)
  % Type-I or type-II bandpass ideal impulse response
  M = (N-1)/2;
  hd = Type12LP(Omg2,N,n)-Type12LP(Omg1,N,n);  % Eqn. (8.156)
end

function hd = Type12BS(Omg1,Omg2,N,n)
  % Type-I or type-II bandstop ideal impulse response
  M = (N-1)/2;
  hd = Type12LP(Omg1,N,n)+Type12HP(Omg2,N,n);  % Eqn. (8.157)
end

function hd = Type34Diff(N,n)
  % Type-III or type-IV differentiator ideal impulse response
  M = (N-1)/2;
  K = n-M+eps;  % Avoid division by zero
  hd = cos(pi*K)./K-1/pi*sin(pi*K)./(K.*K);  % Eqn. (8.164)
end

function hd = Type34Hilbert(N,n)
  % Type-III or type-IV Hilbert transformer ideal impulse response
  M = (N-1)/2;
  K = n-M;
  hd = pi/2*K.*sinc(K/2).*sinc(K/2);          % Eqn. (8.169)
end
```

Even though the listing is a bit long, it is quite simple. At the bottom of the listing there are some *local functions*, each for computing one type of ideal impulse response. For example `Type12BP()` returns the ideal filter impulse response for type-I or type-II bandpass filter, as derived in Eqn. (8.156). Other local functions compute the ideal impulse response for other filter types.

The first two arguments to the function `ss_fir1()` are the filter length N (not filter order!) and the vector `F` of normalized corner frequencies. Note that frequencies are normalized with respect to the sampling rate, and not the way MATLAB built-in functions normalize them. The first part of the function `ss_fir1()` checks the filter type parameter `ftype`, and calls the appropriate local function to get its truncated impulse response. The second part of the function checks the window type parameter `wtype`. It then generates the appropriate window sequence, and multiplies the ideal impulse response with it. As an example, the bandpass filter of MATLAB Exercise 8.11 can be obtained by typing

```
>> ss_fir1(25,[0.2,0.35],'bandpass','bartlett');
```

The filter of Example 8.13 can be designed using the following:

```
>> h = ss_fir1(15,0.15,'lowpass','rect')
```

Differentiator designs of Example 8.15 can be repeated using the following lines:

```
>> fs = 200;
>> h1 = ss_fir1(9,0,'diff','rect')*fs;
>> h2 = ss_fir1(9,0,'diff','hamming')*fs;
>> h3 = ss_fir1(8,0,'diff','rect')*fs;
>> h4 = ss_fir1(8,0,'diff','hamming')*fs;
```

No corner frequencies are needed for differentiators. We can simply enter $F = 0$ or any other value. Note that differentiators designed using function `ss_fir1()` assume a sampling rate of $f_s = 1$ Hz. In order to use them with $f_s = 200$ Hz, we simply multiply the impulse response with the sampling rate. Finally, to repeat the Hilbert transform filter designs of Example 8.16, use the following:

```
>> h1 = ss_fir1(18,0,'hilbert','rect');
>> h2 = ss_fir1(18,0,'filbert','hamming');
```

Software resources: `ss_fir1.m`

MATLAB Exercise 8.13: FIR filter design by frequency sampling

Built-in function `fir2()` can be used for designing FIR filters using the frequency sampling design method. In this exercise, we will again develop our own version of the design function, and name it `ss_fir2()`. The listing is given below.

```
function h = ss_fir2(N,F,A)
  F = [F,1-F(end-1:-1:1)];  % Mirror the set of frequencies
  A = [A,A(end-1:-1:1)];    % Mirror the amplitudes
  k = [0:N-1];              % Vector of transform indices
  Fk = k/N;                 % Frequencies used in sampling the ideal spectrum
  Ak = interp1(F,A,Fk);     % Samples of amplitude response
  M = (N-1)/2;              % Axis of symmetry
  h = ifft(Ak.*exp(-j*2*pi*Fk*M)); % Inverse FFT to get h[n]
end
```

The first argument `N` is the length of the impulse response (not filter order!). Vectors `F` and `A` contain normalized corner frequencies and corresponding amplitude response values respectively. Frequencies are normalized with respect to the sampling rate, and not the way

MATLAB built-in functions normalize them. The first element of the vector F should be 0, and its last element should be 0.5. The MATLAB statement

```
>> plot(F,A)
```

should graph the desired amplitude response. The script mexdt_8_13.m listed below designs a length-25 multiband FIR filter and graphs its magnitude response along with the desired filter characteristics specified through vectors F and A.

```
% Script: mexdt_8_13.m
N = 25;                                 % Filter length
F = [0,0.1,0.15,0.275,0.325,0.5];      % Critical frequencies (normalized)
A = [1,1,0.5,0.5,0,0];                 % Critical magnitude values
h = ss_fir2(N,F,A);                    % Design the filter
Omega = [0:0.01:1]*pi;                 % Frequencies for the graph
Hk = freqz(h,1,Omega);                 % Compute filter spectrum
% Graph the magnitude response and the specifications
plot(Omega,abs(Hk),2*pi*F,A,'r--'); grid;
axis([0,pi,0,1.2]);
title('|H(\Omega)|');
xlabel('\Omega (rad)');
ylabel('Magnitude');
```

Software resources: mexdt_8_13.m , ss_fir2.m

MATLAB Exercise 8.14: FIR filter design by least-squares method

Built-in function firls() can be used for designing FIR filters using the least-squares design method. In this exercise, we will develop our own version of the design function, and name it ss_firls(). The listing is given below.

```
function h = ss_firls(N,F,A,K,weights)
  M = (N-1)/2;                 % Axis of symmetry
  F = [F,1-F(end-1:-1:1)];     % Mirror the set of frequencies
  A = [A,A(end-1:-1:1)];       % Mirror the amplitudes
  weights = [weights,weights(end-1:-1:1)];  % Mirror the weight factors
  k = [0:K-1]';                % Vector of frequency indices
  Fk = 0.5*k/K;                % Frequencies used in setting up equations
  Pk = interp1(F,A,Fk);        % Interpolate to get samples of ideal mag. response
  Wk = interp1(F,weights,Fk);  % Interpolate to get samples of weight factors
  Omg = 2*pi*Fk;
  A = Type1CoeffMatrix(M,Omg);  % Get the coefficient matrix
  W = diag(Wk);                % Create a diagonal matrix with weight factors
  A = W*A;                     % Modify matrix A
  Pk = W*Pk;                   % Modify vector Pk
  h = A\Pk;                    % Least-squares solution
  h = [h;h(end-1:-1:1)];       % Eqn. (8.182)
end

function A = Type1CoeffMatrix(M,Omg)
```

```
% Local function to construct the coefficient matrix A
  A = [];                                     % Start with empty matrix
  for k=1:length(Omg)
    Ak = Type1CoeffVector(M,Omg(k));          % Get the row vector for Omg(k)
    A = [A;Ak];                               % Add new row to matrix A
  end;
end

function Ak = Type1CoeffVector(M,Omgk)
% Local function to build one row of the coefficient matrix
  n = [0:M-1];
  Ak = [2*cos(Omgk*(n-M)),1];                 % Eqn. (8.184)
end
```

The vectors F and A were addressed in MATLAB Exercise 8.13 in the discussion of the function ss_fir2(). The scalar K is the number of frequency points – see Eqn. (8.191). The vector weights has the weight factors indicating the relative significance of each point in the specifications; its length must be the same as that of vector A. Note that two local functions Type1CoeffVector() and Type1CoeffMatrix() are used. The former returns the vector $\mathbf{A_k}$ as defined in Eqn. (8.184). The latter repeats this for $k = 0, 1, \ldots, K-1$, and constructs the matrix **A** given by Eqn. (8.188).

The script below designs a length-25 multiband filter twice, using $K = 100$ frequency points in the least-squares optimization. In the first try, all weight factors are set to be equal to 1. In the second try, the weight factors in the stopband are increased tenfold. The dB magnitude responses of the two filters are graphed, and shown in Fig. 8.72.

```
% Script: mexdt_8_14.m
N = 25;                              % Filter length
K = 100;                             % Number of equations
F = [0,0.1,0.15,0.275,0.325,0.5];    % Critical frequencies (normalized)
A = [1,1,0.5,0.5,0,0];               % Amplitude response at critical frequencies
weights1 = [1,1,1,1,1,1];            % Weight factors for design 1
weights2 = [1,1,1,1,10,10];          % Weight factors for design 2
h1 = ss_firls(N,F,A,K,weights1);
h2 = ss_firls(N,F,A,K,weights2);
Omg = [0:255]/256*pi;
H1 = freqz(h1,1,Omg);
H2 = freqz(h2,1,Omg);
plot(Omg,20*log10(abs(H1)),Omg,20*log10(abs(H2))); grid;
```

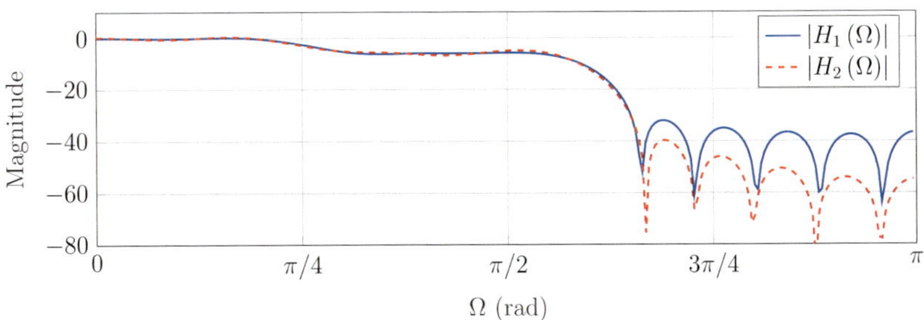

Figure 8.72 – Comparison of dB magnitude responses of the two filters designed using the least-squares method.

Software resources: mexdt_8_14.m , ss_firls.m

MATLAB Exercise 8.15: FIR filter design using Parks-McClellan algorithm

In this exercise we will use Parks-McClellan algorithm to design length-25 FIR filters with the same specifications that were used in MATLAB Exercise 8.14. The script `mexdt_8_15.m` given below is a slightly modified version of the script used in the previous exercise. We use the built-in function `firpm()` for FIR filter design with the Parks-McClellan algorithm.

There are several details that need attention. First, the argument `N` for the function `firpm()` is the filter order, not the filter length. In order to obtain a length-25 filter, we need to specify $N = 24$. Also notice how we use `2*F` for the set of corner frequencies. This is done to account for the difference in how MATLAB normalizes frequencies. Finally, there are only three weight factors, one for each band (recall that function `ss_firls()` used a weight factor for each corner frequency). The script below designs the required filter twice. In the first try, all weight factors are set to be equal to 1. In the second try, the weight factor for the stopband is set equal to 10. The dB magnitude responses of the two filters are graphed, and shown in Fig. 8.73. Notice the equiripple behavior in the stopband, and compare it to the performance of the least-squares design method.

```
% Script: mexdt_8_15.m
N = 24;                                % Filter order
F = [0,0.1,0.15,0.275,0.325,0.5];      % Critical frequencies (normalized)
A = [1,1,0.5,0.5,0,0];                 % Amplitude response at critical frequencies
weights1 = [1,1,1];                    % Weight factors for design 1
weights2 = [1,1,10];                   % Weight factors for design 2
h1 = firpm(N,2*F,A,weights1);
h2 = firpm(N,2*F,A,weights2);
Omg = [0:255]/256*pi;
H1 = freqz(h1,1,Omg);
H2 = freqz(h2,1,Omg);
plot(Omg,20*log10(abs(H1)),Omg,20*log10(abs(H2))); grid;
```

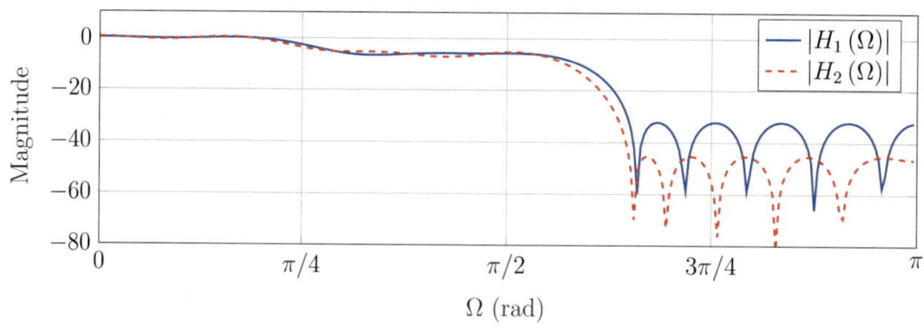

Figure 8.73 – Magnitude responses of the two filters designed using Parks-McClellan method.

Software resources: mexdt_8_15.m

MATLAB Exercise 8.16: Case Study – Plucked-string filter

Extensive research has taken place since mid-20th century on understanding the physics of how musical instruments work and how various sounds are generated, and developing mathematical models for those instruments. In more recent years, increased emphasis on the use of discrete-time systems for modeling musical instruments led to the advent of research areas known as *computer music* and *virtual instruments*. Those eventually made their way into mainstream consumer products we use today such as electronic keyboards that can simulate the sounds of a multitude of musical instruments along with a variety of sound effects.

We will begin this exercise with a very brief review of the physics of a string stretched between two points and fixed at both ends as in a guitar. Our mathematical development will be overly simplified so that we can quickly move on to the discrete-time filtering part of the exercise. More detailed derivations can be found in references [9],[10]. Once the mathematical preliminaries are in place, we will model the behavior of the string using some of the discrete-time filters we have already studied. We will also develop the MATLAB code to produce guitar sounds. Consider a piece of string stretched between points A and B as illustrated in Fig. 8.74.

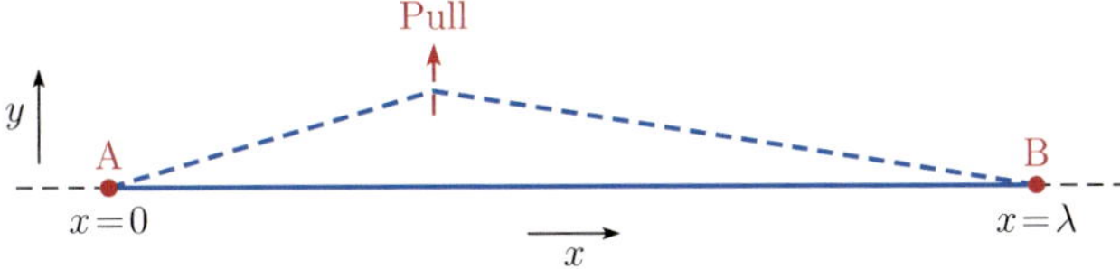

Figure 8.74 – Modeling a string stretched between two fixed points.

We will assume that, in the rest position, the string lays on the x axis with the fixed point A at $x = 0$, and the fixed point B at $x = \lambda$. If the string is plucked somewhere around the middle, a displacement occurs in the y direction, and leads to vibrations. The displacement is a function of both the lateral position x and the time t. Therefore, it can be expressed as $y(x,t)$. It is governed by the *wave equation* which is in the form

$$\frac{\partial^2 y(x,t)}{\partial t^2} = c^2 \frac{\partial^2 y(x,t)}{\partial x^2} \tag{8.200}$$

where the parameter c has the dimension of velocity. Its value depends on the physical parameters of the string such as mass density and elasticity. Let's assume that the initial displacement profile of the string at time $t = 0$ is in the shape of a function f_R as shown in Fig. 8.75. It can be shown that a solution in the form

$$y(x,t) = f_R(t - x/c) \tag{8.201}$$

satisfies the wave equation for any function f_R that is twice differentiable (see Problem 8.25 at the end of this chapter). This solution of the partial differential equation in Eqn. (8.200) represents a wave that travels to the right over time, hence the name *wave equation*.

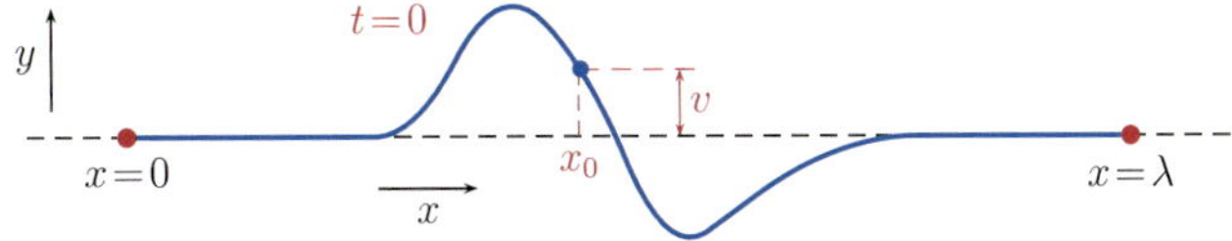

Figure 8.75 – Initial displacement profile with the function f_R.

To convince ourselves that the wave travels to the right, let's focus on one specific point with amplitude v and lateral position x_0 on the initial displacement profile shown in Fig. 8.75.

$$y(x_0, 0) = f_R(0 - x_0/c) = v \tag{8.202}$$

At a later time instant $t = t_1$, this same point appears at lateral position $x = x_1$.

$$y(x_1, t_1) = f_R(t_1 - x_1/c) = v \tag{8.203}$$

Setting the right sides of Eqns. (8.202) and (8.203) equal to each other, we have

$$f_R(0 - x_0/c) = f_R(t_1 - x_1/c) \quad \longrightarrow \quad -\frac{x_0}{c} = t_1 - \frac{x_1}{c} \quad \longrightarrow \quad x_1 = x_0 + c\,t_1 \tag{8.204}$$

confirming that the point in question moves to the right with a velocity of c. Same argument can be made about every point on the displacement profile. Similarly, a wave that moves to the left also satisfies the wave equation, so that a general solution can be written in the form

$$y(x,t) = f_R(t - x/c) + f_L(t + x/c) \tag{8.205}$$

where f_L is the term that moves to the left with a velocity of c. Eqn. (8.205) represents a solution with no boundary conditions imposed on it yet. We know that our string is fixed at two endpoints $x = 0$ and $x = \lambda$. Displacement must be equal to zero at each of these points. Imposing the condition at $x = 0$ yields

$$y(0,t) = f_R(t) + f_L(t) = 0 \quad \longrightarrow \quad f_R(t) = -f_L(t) = f(t) \tag{8.206}$$

so that the general solution simplifies to

$$y(t,x) = f(t - x/c) - f(t + x/c) \tag{8.207}$$

If we now impose the same condition at $x = \lambda$, we get

$$y(\lambda, t) = f(t - \lambda/c) - f(t + \lambda/c) = 0 \quad \longrightarrow \quad f(t - \lambda/c) = f(t + \lambda/c) \tag{8.208}$$

This is a very interesting result. It suggests that the function f must be periodic with a period of $T = 2\lambda/c$. This corresponds to a fundamental frequency ω_0 computed as

$$\omega_0 = \frac{2\pi}{T} = \frac{\pi c}{\lambda} \tag{8.209}$$

There are multiple functions that would satisfy the wave equation and the boundary conditions, some more complicated than others. For the purpose of simulating guitar sounds we will pick one that is easy to work with. Let the solution be

$$\begin{aligned} y(x,t) &= -\frac{1}{2}\sin\left(\frac{\pi c}{\lambda}(t - x/c)\right) + \frac{1}{2}\sin\left(\frac{\pi c}{\lambda}(t + x/c)\right) \\ &= -\frac{1}{2}\sin\left(\omega_0 (t - x/c)\right) + \frac{1}{2}\sin\left(\omega_0 (t + x/c)\right) \end{aligned} \tag{8.210}$$

Using the appropriate trigonometric identity, Eqn. (8.210) can be written as

$$y(x,t) = \cos(\omega_0 t)\,\sin\left(\frac{\pi x}{\lambda}\right) \tag{8.211}$$

Instead of using the fundamental frequency ω_0 in Eqn. (8.210), any harmonic $k\omega_0$ can be used, and it would still lead to a valid solution.

$$y(x,t) = \cos(k\omega_0 t)\,\sin\left(\frac{k\pi x}{\lambda}\right) \tag{8.212}$$

In fact, any linear combination of solutions of this type would also be a valid solution.

$$y(x,t) = \sum_{k=1}^{\infty} a_k \cos(k\omega_0 t)\,\sin\left(\frac{k\pi x}{\lambda}\right) \tag{8.213}$$

The wave of vibration travels a distance of 2λ before returning to its initial point and repeating the travel over and over again. The ideal solution of the wave equation rings indefinitely without getting weaker. In an actual guitar string, on the other hand, friction with air would cause the waves to get weaker as they travel back and forth, and eventually disappear. Our mathematical model does not reflect that.

Two MATLAB scripts, `WaveEqnPart1.m` and `WaveEqnPart2.m` for animated demonstrations of traveling waves are provided with the set of downloadable files. The former illustrates left and right traveling wave components $f_R(t-x/c)$ and $f_L(t+x/c)$. The latter illustrates the sum $y(x,t)$.

Now consider the feedback comb filter structure discussed in Section 5.5.7 of Chapter 5. Its block diagram is shown in Fig. 8.76. Any signal that enters through the input port reaches the output port, and is fed back to the input port after L samples of delay. Intuitively this is similar to the action of traveling waves on a string with length $\lambda = L/2$. In addition, the signal is attenuated by a factor of r each time it travels through the feedback loop, and eventually disappears (recall that $r < 1$ for stability). This makes it a bit closer to reality than the solution found in Eqn. (8.212).

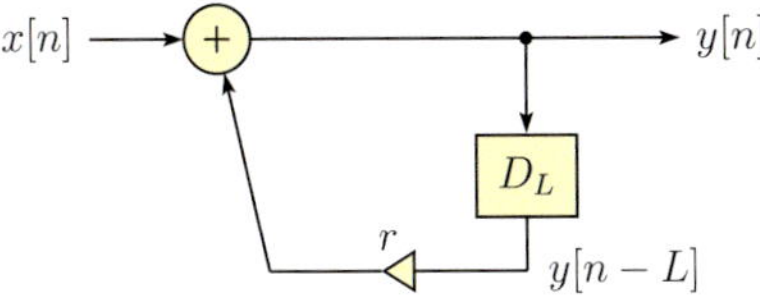

Figure 8.76 – Block diagram for feedback comb filter.

Feedback loop delay of L samples corresponds to a normalized fundamental frequency of $F_0 = 1/L$. For a sampling rate of f_s, this translates to a fundamental frequency of $f_0 = f_s/L$ in Hz. The script `mexdt_8_16a.m` listed below implements a comb filter with a length-100 delay line and a feedback gain of $r = 0.99$. It is driven by a unit impulse signal. If the output signal is played back with a sampling rate of $f_s = 22050$ Hz, the resulting fundamental frequency should be $f_0 = 220.05$ Hz. Run the script and listen to the sound. While it has some of the characteristics of the vibrating guitar string, it sounds a bit "buzzy" for the lack of a better term. It doesn't quite have the timbre we expect.

```
% Script: mexdt_8_16a.m
x = [1;zeros(22049,1)];          % Generate a unit impulse sequence
buffer = zeros(100,1);           % Buffer for 100-sample delay
y = ss_combf(x,0.995,buffer);    % Run through comb filter
sound(y,22050);                  % Play back the sound
```

In script `mexdt_8_16b.m` we compute and graph the magnitude spectrum of the sound produced in the previous script (run `mexdt_8_16b.m` without clearing the variables created by `mexdt_8_16a.m` from workspace, otherwise MATLAB will complain that it cannot find vector `y`). As expected,the fundamental frequency is at $f_0 = 220.05$ Hz, and its harmonics are also in place. So, why did we not get a realistic guitar sound? In their time-frequency analysis of instrument sounds (see Section 7.4.3 of Chapter 7 on the STFT), Karplus and Strong [4] observed that higher frequencies decay at a faster rate than lower frequencies. In our comb filter based model, however, every frequency is attenuated equally by the same gain factor r. We need to improve our model. The solution first proposed by Karplus and Strong is quite simple: Insert a lowpass filter into the loop of the comb filter as shown in Fig. 8.77a. We will refer to this new structure as the *plucked-string filter*. A computationally simple lowpass filter for use in this new setup is a length-2 moving average filter (see Section 2.3 of Chapter 2). Fig. 8.77b depicts the block diagram of the plucked-string filter with a length-2 moving average filter in its loop.

```
% Script: mexdt_8_16b.m
Yk = abs(fftshift(fft(y)));  % Magnitude response
F = [-11025:11024];           % Vector of frequencies in Hz
plot(F,Yk); grid;             % Graph the magnitude response
axis([0,1000,0,100]);
xlabel('Frequency (Hz)');
ylabel('Magnitude');
```

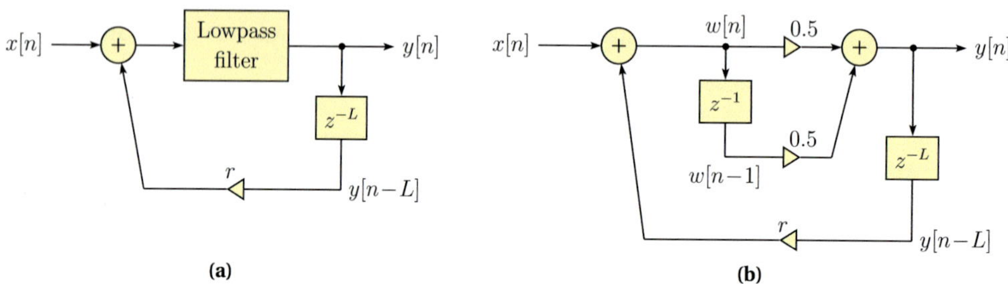

Figure 8.77 – **(a)** Comb filter with lowpass filter inserted into its loop and **(b)** using a length-2 moving average filter as a lowpass filter.

Implementation equations for the diagram in Fig. 8.77b are

$$w[n] = x[n] + r\,y[n-L] \tag{8.214}$$

and

$$y[n] = 0.5\,w[n] + 0.5\,w[n-1] \tag{8.215}$$

They lead to the function `ss_psf()` listed below. It is important to note that the delay L is not one of the arguments for `ss_psf()`. Rather, the information is passed to the function by way of the length of the vector `buffer` that must be created before the call. This is consistent with other similar functions such as ss_echo(), ss_comb(), and ss_notch() developed in previous exercises.

```
function y = ss_psf(x,r,buffer)
  y = zeros(size(x));               % Placeholder for output signal
  nSamples = size(x,1);             % Number of samples
  wnm1 = 0;                         % Variable to hold w[n-1]
  for i=1:nSamples
    w = x(i)+r*buffer(end);                 % Eqn. (8.214)
    y(i) = 0.5*(w+wnm1);                    % Eqn. (8.215)
    buffer = [y(i);buffer(1:end-1)];  % Update buffer
    wnm1 = w;                               % Update w[n-1]; w[n-1] <-- w[n]
  end
end
```

The sound test can now be repeated using the script mexdt_8_16c.m. There should be a noticeable improvement. Take note, however, that the fundamental frequency of the sound we are hearing is a bit different than that produced by the earlier script mexdt_8_16a.m. The length-2 moving average filter adds an additional half sample delay into the loop (refer to the discussion in Section 8.5.1 of this chapter). The new fundamental frequency is $f_s/(L+0.5)=22050/100.5=219.4$ Hz.

```
% Script: mexdt_8_16c.m
x = [1;zeros(44100,1)];           % Generate a unit impulse sequence
buffer = zeros(100,1);            % Buffer for 100-sample delay
y = ss_psf(x,0.995,buffer);       % Run through plucked string filter
sound(y,22050);                   % Play back the sound
```

The discussion above highlights one potential problem with the plucked-string filter, that is, the lack of a mechanism for tuning it for the exact fundamental frequency desired. The loop delay will always be an integer plus one half sample. Consider, for example, the note "*E4*" which is typically the last string on a 6-string guitar. Its fundamental frequency is $f_0=329.63$ Hz. Using $f_s=22050$ Hz it requires a loop delay of 22050/329.63 = 66.89 samples. Using the plucked-string filter of Fig. 8.77b, our choices would be either 66.5 or 67.5 sample delays corresponding to frequencies 331.58 Hz or 326.67 Hz, respectively. In order to produce the exact frequency of $f_0=329.63$ Hz we can start with $L=66$, and obtain 66.5 samples of total delay with the half sample delay coming from the moving average filter. An additional time delay of $\tau=0.39$ samples needs to be inserted into the loop. One method of doing this is to use an all-pass filter (see Section 5.5.8 of Chapter 5) as shown in Fig. 8.78.

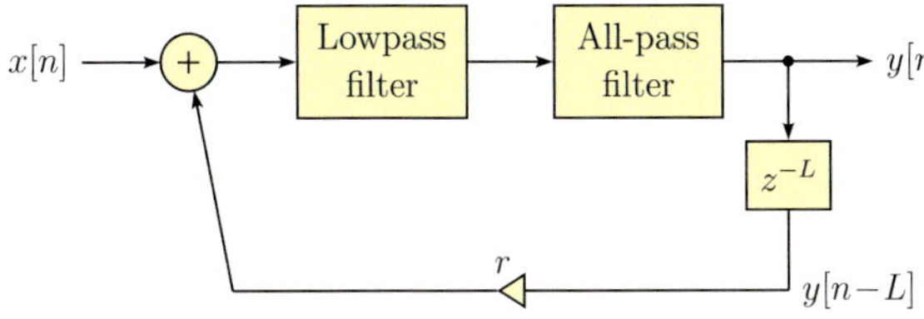

Figure 8.78 – Making the plucked-string filter tunable with the addition of an all-pass filter.

A first-order all-pass filter with a pole at $z=-a$ and a zero at $z=-1/a$ has the system function

$$H(z)=\frac{a+z^{-1}}{1+a\,z^{-1}} \tag{8.216}$$

where $|a| < 1$ for stability. A block diagram for implementing it is shown in Fig. 8.79.

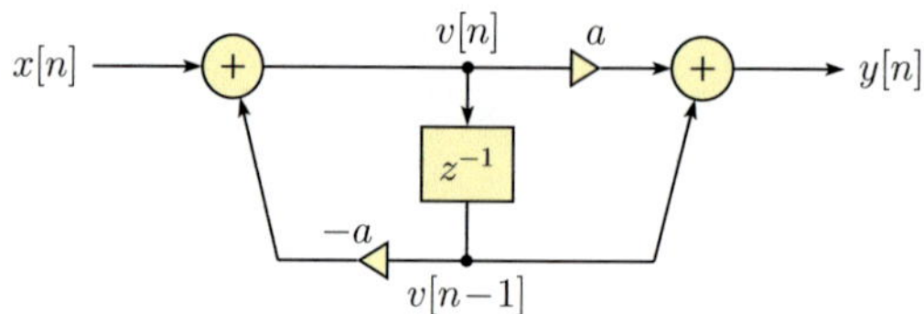

Figure 8.79 – Block diagram for first-order all-pass filter.

Evaluating the system function for $z = e^{j2\pi F}$, it can be shown that its magnitude response is $|H(F)| = 1$ for all frequencies. Its phase response is

$$\measuredangle H(F) = \frac{a + e^{-j2\pi F}}{1 + a\,e^{-j2\pi F}} \tag{8.217}$$

and the corresponding time delay is

$$\tau = -\frac{\measuredangle H(F)}{2\pi F} = -\frac{a + e^{-j2\pi F}}{2\pi F\left(1 + a\,e^{-j2\pi F}\right)} \tag{8.218}$$

In our example we need the all-pass filter time delay to be $\tau = 0.39$ at the normalized frequency $F_0 = 329.63/22050 = 0.0149$. This requires solving Eqn. (8.218) for the parameter a. There are approximate solutions in the literature for cases where $F << 1$ [3]. Instead, we will use the function `ss_apfpar()` listed below which computes the value of parameter a using linear interpolation. In line 2, we create a vector `aVec` with a set of values for the parameter a from $a = 0.01$ to $a = 0.99$. In the loop between lines 5 and 10 we compute the time delay at normalized frequency F that corresponds to each value of a in vector `aVec`. This leads to vector `tdVec` of time delay values. Finally, in line 11, we find the value of a that yields the desired time delay by linearly interpolating between the values of delay in vector `tdVec`.

```
1  function a = ss_apfpar(F,td)
2    aVec = [0.01:0.01:0.99];
3    tdVec = zeros(size(aVec));
4    Omega = 2*pi*F;
5    for i=1:length(aVec)
6      zero = -1/aVec(i);
7      pole = -aVec(i);
8      [~,phs] = ss_freqz(zero,pole,1,Omega);
9      tdVec(i) = -phs/Omega;
10   end
11   a = interp1(tdVec,aVec,td);
12 end
```

Using the function `ss_apfpar()`, the parameter a can be determined as follows:

```
>> F = 329.63/22050;
>> a = ss_apfpar(F,0.39)

a =
    0.4391
```

Fig. 8.80 depicts the tunable plucked-string filter with the all-pass tuner component added to the loop.

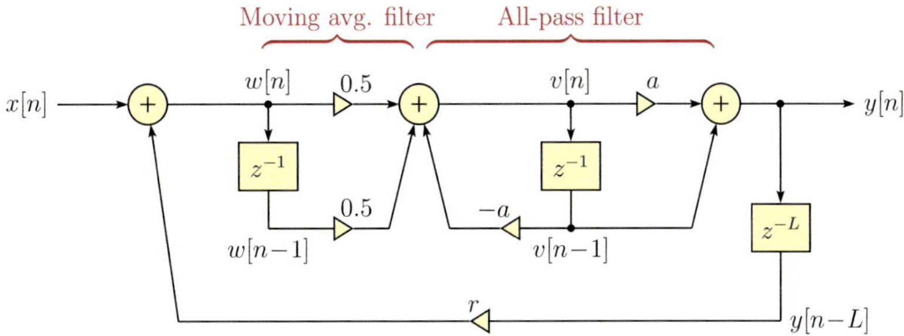

Figure 8.80 – Tunable plucked-string filter block diagram.

Implementation equations for the tunable plucked-string filter are

$$w[n] = x[n] + r\,y[n-L] \tag{8.219}$$

$$v[n] = 0.5\,w[n] + 0.5\,w[n-1] - a\,v[n-1] \tag{8.220}$$

$$y[n] = a\,v[n] + v[n-1] \tag{8.221}$$

The function `ss_tpsf()` is a slightly modified version of `ss_psf()` using Eqns. (8.219) through (8.221).

```
function y = ss_tpsf(x,a,r,buffer)
  y = zeros(size(x));              % Placeholder for output signal
  nSamples = size(x,1);            % Number of samples
  wnm1 = 0;                        % Variable to hold w[n-1]
  vnm1 = 0;                        % Variable to hold v[n-1]
  for i=1:nSamples
    w = x(i)+r*buffer(end);              % Eqn. (8.219)
    v = 0.5*w+0.5*wnm1-a*vnm1;           % Eqn. (8.220)
    y(i) = a*v+vnm1;                     % Eqn. (8.221)
    buffer = [y(i);buffer(1:end-1)];     % Update buffer
    wnm1 = w;                            % Update w[n-1]:  w[n-1] <-- w[n]
    vnm1 = v;                            % Update v[n-1]:  v[n-1] <-- v[n]
  end
end
```

The script `mexdt_8_16d.m` uses `ss_tpsf()` to generate the note “E4” with a fundamental frequency of $f_0 = 329.63$ Hz.

```
% Script: mexdt_8_16d.m
x = [1;zeros(44100,1)];                 % Generate a unit impulse sequence
buffer = zeros(66,1);                   % Buffer for 66-sample delay
y = ss_tpsf(x,0.4391,0.995,buffer);     % Run through tunable plucked string filter
sound(y,22050);                         % Play back the sound
```

Software resources: `mexdt_8_16a.m` , `mexdt_8_16b.m` , `mexdt_8_16c.m` , `mexdt_8_16d.m` , `WaveEqnPart1.m` , `WaveEqnPart2.m` , `ss_psf.m` , `ss_tpsf.m` , `ss_apfpar.m`

Problems

8.1. A resonant bandpass filter is to be designed with a normalized center frequency of $F_0 = 0.10$ and a normalized 3-dB bandwidth of $F_b = 0.01$.

- **a.** Determine the parameters Ω_0, Ω_b, r and K. Sketch the pole-zero diagram, and write the exact locations of poles and zeros of the system function.
- **b.** Write the system function $H(z)$ for the filter in simplified form.
- **c.** Roughly sketch the magnitude of the system function and indicate all critical values.

8.2.

- **a.** Using partial fraction expansion, determine the impulse response $h[n]$ of the preliminary design for a resonant bandpass filter given by Eqn. (8.5) as a function of parameters r and Ω_0. Sketch the result.
- **b.** Repeat part (a) for the system function of the completed resonant bandpass filter given by Eqn. (8.6).

8.3. A resonant bandpass filter is to be used for detecting the presence of a 600 Hz sinusoidal component in a signal that is sampled with a sampling rate of $f_s = 8$ kHz. The required 3-dB bandwidth is 30 Hz.

- **a.** Determine the parameters Ω_0, Ω_b, r, and K. Sketch the pole-zero diagram, and write the exact locations of poles and zeros of the system function.
- **b.** Write the system function $H(z)$ for the filter in simplified form.
- **c.** Roughly sketch the magnitude of the system function and indicate all critical values.

8.4. For the notch filter with system function $H(z)$ given by Eqn. (8.18), show that the 3-dB bandwidth Ω_b, that is, the frequency spacing between the two points where $|H(\Omega)| = 1/\sqrt{2}$, is approximately related to the pole radius by

$$r = 1 - \frac{\Omega_b}{2}$$

8.5. A notch filter is to be designed to remove the frequency 200 Hz in a signal sampled with a sampling rate of 1 kHz. The 3-dB width of the stopband is required to be 5 Hz.

- **a.** Determine the parameters of the notch filter. Sketch the pole-zero diagram, and write the exact locations of poles and zeros of the system function.
- **b.** Write the system function $H(z)$ for the filter in simplified form.
- **c.** Roughly sketch the magnitude of the system function and indicate all critical values.

8.6. The desired magnitude response of an IIR filter is specified as follows:

$$0.85 < \left|H(\Omega)\right| < 1 \qquad \text{for } 0 < \Omega < 0.2\pi$$

$$\left|H(\Omega)\right| < 0.1 \qquad \text{for } 0.24\pi < \Omega < \pi$$

- **a.** Sketch a specification diagram for the filter, and indicate all critical values on it.
- **b.** Determine the maximum passband ripple R_p and the minimum stopband attenuation A_s in dB. Sketch a specification diagram for the dB magnitude of the filter, and indicate all critical values.

8.7. The desired magnitude response of an IIR filter is specified as follows:

$$0.9 < \left|H(\Omega)\right| < 1 \qquad \text{for } 0 < \Omega < 0.2\pi \text{ or } 0.65\pi < \Omega < \pi$$

$$\left|H(\Omega)\right| < 0.05 \qquad \text{for } 0.25\pi < \Omega < 0.55\pi$$

a. Sketch a specification diagram for the filter, and indicate all critical values on it.

b. Determine the maximum passband ripple R_p and the minimum stopband attenuation A_s in dB. Sketch a specification diagram for the dB magnitude of the filter, and indicate all critical values.

8.8. A third-order Butterworth analog highpass filter is to be designed with a 3-dB cutoff frequency of 5 rad/s.

a. Start by designing a Butterworth analog lowpass filter with 3-dB cutoff frequency at 2 rad/s. Determine the poles of the product $H(s)\,H(-s)$. Sketch the poles on the s-plane. Select the appropriate poles to be associated with $H(s)$.

b. Construct the system function $H(s)$. Select the gain factor so that the filter has unit magnitude at $\omega = 0$.

c. Convert the filter found in part (b) to the desired highpass filter using the frequency transformation described by Eqn. (8.42) (see Fig. 8.19). Arrange it so that the behavior of the highpass filter at 5 rad/s matches that of the lowpass filter at 2 rad/s.

8.9. A third-order Chebyshev type-I analog highpass filter is to be designed. The maximum passband ripple is to be 3-dB for frequencies greater than 5 rad/s.

a. Start by designing a Chebyshev type-I analog lowpass filter with maximum passband ripple of 3-dB for frequencies less than 2 rad/s. Determine the poles of the product $H(s)\,H(-s)$. Sketch the poles on the s-plane. Select the appropriate poles to be associated with $H(s)$.

b. Construct the system function $H(s)$. Select the gain factor so that the filter has unit magnitude at $\omega = 0$.

c. Convert the filter found in part (b) to a highpass filter using the frequency transformation described by Eqn. (8.42) (see Fig. 8.19). Arrange it so that the behavior of the highpass filter at 5 rad/s matches that of the lowpass filter at 2 rad/s.

8.10. A fourth-order Chebyshev type-II analog bandpass filter is to be designed. The minimum stopband attenuation is to be 15-dB for frequencies less than 2 rad/s or greater than 3 rad/s.

a. Start by designing a second-order Chebyshev type-II analog lowpass filter with minimum stopband attenuation of 15 dB for frequencies greater than 2 rad/s. Determine the poles of the product $H(s)\,H(-s)$. Sketch poles and zeros on the s-plane. Select the appropriate poles to be associated with $H(s)$.

b. Construct the system function $H(s)$. Select the gain factor so that the filter has unit magnitude at $\omega = 0$.

c. Convert the filter found in part (b) to a bandpass filter using the frequency transformation described by Eqn. (8.46) (see Fig. 8.20). Arrange it so that the behavior of the bandpass filter at 2 rad/s and 3 rad/s matches that of the lowpass filter at 2 rad/s.

8.11. An analog filter with the system function

$$G(s) = \frac{s+3}{(s+1)\,(s+2)}$$

is to be used as a prototype for the design of a discrete-time filter through the impulse invariance technique.

a. Find the impulse response $g(t)$ of the analog prototype filter.

b. Obtain the impulse response of the discrete-time filter as a sampled and scaled version of the analog prototype impulse response, that is,

$$h[n] = T\,g(nT)$$

Use the sampling interval $T = 0.1$ s.

c. Determine the system function $H(z)$ of the discrete-time filter.

8.12. Consider again the analog filter of Problem 8.11 with the system function

$$G(s) = \frac{s+3}{(s+1)\,(s+2)}$$

This time we would like to use *unit step invariance* for converting the analog prototype to a discrete-time system. The goal is to preserve the unit step response of the filter in the conversion process.

a. Find the unit step response $g_s(t)$ of the analog prototype filter. Hint: The Laplace transform of the unit step response of the filter is

$$G_s(s) = \mathcal{L}\left\{g_s(t)\right\} = \frac{1}{s}\,G(s)$$

Expand $G_s(s)$ into partial fractions, and compute $g_s(t)$.

b. Obtain the unit step response of the discrete-time filter as a sampled and scaled version of the analog prototype unit step response, that is,

$$h_s[n] = T\,g_s(nT)$$

Use the sampling interval $T = 0.1$ s.

c. Determine the system function $H(z)$ of the discrete-time filter. Hint: Recall that the z-transform of the unit step response of the discrete-time filter is

$$H_s(z) = \mathcal{Z}\left\{h_s[n]\right\} = \frac{z}{z-1}\,H(z)$$

d. Is the system function $H(z)$ obtained here identical to the system function found in Problem 8.11? If not, why not?

8.13. Consider two analog signals $x_a(t)$ and $y_a(t)$ that have an integral relationship:

$$y_a(t) = \int_{\infty}^{t} x_a(t)\,dt$$

In the s-domain, we have

$$Y_a(s) = \frac{1}{s}\,X_a(s)$$

Therefore, the system function of an analog integrator is

$$G(s) = \frac{Y_a(s)}{X_a(s)} = \frac{1}{s}$$

a. The analog integrator is to be converted to a discrete-time system using impulse invariance. Leave the sampling interval T as a parameter. Determine the system function $H(z)$. Sketch the pole-zero diagram for $H(z)$.

b. Write the difference equation for the discrete-time system obtained. Does it look familiar? Can it be used for approximating a running integral? Comment on its expected accuracy.

8.14.

a. Convert the analog integrator with system function $G(s) = 1/s$ to a discrete-time system using unit step invariance. Leave the sampling interval T as a parameter. Determine the system function $H(z)$. Sketch the pole-zero diagram for $H(z)$.

b. Write the difference equation for the discrete-time system obtained. How does it differ from the system obtained in Problem 8.13? Can it be used for approximating a running integral? Comment on its expected accuracy.

8.15. A third-order Butterworth analog lowpass filter with a 3-dB cutoff frequency of $\omega_c = 1$ rad/s has the system function

$$G(s) = \frac{1}{(s+1)\left(s^2+s+1\right)}$$

We wish to convert this filter to a discrete-time lowpass filter with a 3-dB cutoff frequency at $\Omega_c = 0.1\pi$ using impulse invariance. Determine the appropriate sampling interval T. Afterward, follow the development in Eqns. (8.60) through (8.64) to find the system function $H(z)$. Sketch a pole-zero diagram for the filter obtained.

8.16. In Section 8.4.3 impulse invariance and bilinear transformation were discussed as viable methods for converting an analog prototype filter to a discrete-time filter. Another transformation method is the *matched z-transform* which is somewhat similar to impulse invariance.

Consider an analog prototype with the system function

$$G(s) = \frac{K \displaystyle\prod_{i=1}^{M} (s-\alpha_i)}{\displaystyle\prod_{i=1}^{N} (s-p_i)}$$

In the matched z-transform method of converting the analog prototype to a discrete-time filter, each pole at $s = p_i$ in the s-domain is mapped to a corresponding pole at

$$z = e^{p_i T}$$

in the z-domain. This part is essentially the same as what happens when impulse invariance method is used. The difference is in how the zeros of the analog prototype are treated. In matched z-transform, each zero at $s = \alpha_i$ in the s-domain is mapped to a corresponding zero at

$$z = e^{\alpha_i T}$$

in the z-domain. Furthermore, if the analog prototype has more finite poles than zeros (meaning it has some zeros at infinity), then those zeros at infinity are mapped to the point $z = -1$ which represents the highest frequency on the unit circle of the z-plane.

a. Convert the analog filter with the system function

$$G(s) = \frac{s+3}{(s+1)(s+2)}$$

to a discrete-time filter using the matched z-transform. Use $T = 0.1$ s. Determine the system function $H(z)$. Adjust the gain factor so that the resulting filter has unit magnitude at $\Omega = 0$.

b. Determine the impulse response of the discrete-time filter. How does it compare to the impulse response of the analog prototype?

8.17. An analog filter with the system function

$$G(s) = \frac{s+3}{(s+1)\,(s+2)}$$

is to be used as a prototype for the design of a discrete-time filter through bilinear transformation.

a. Using $T = 0.5$ s, convert the analog prototype system function $G(s)$ to a discrete-time filter system function $H(z)$. Determine the angular frequencies Ω_1, Ω_2 and Ω_3 that correspond to analog frequencies $\omega_1 = 1$ rad/s, $\omega_2 = 3$ rad/s, and $\omega_3 = 5$ rad/s.

b. Repeat part (a) of the problem, this time using a sampling interval of $T = 1$ s.

8.18. A third-order Butterworth analog lowpass filter with a 3-dB cutoff frequency of $\omega_c = 1$ rad/s has the system function

$$G(s) = \frac{1}{(s+1)\left(s^2+s+1\right)}$$

We would like to use bilinear transformation to convert this filter to a discrete-time filter with a 3-dB cutoff frequency at $\Omega_c = 0.1\pi$. Determine the appropriate sampling interval T. Afterward, obtain the system function $H(z)$. Sketch a pole-zero diagram for the filter obtained.

8.19. Refer to Problem 8.13 where conversion of an analog integrator to a discrete-time system through impulse invariance was considered. This problem will explore the use of bilinear transformation for the same purpose.

a. Apply bilinear transformation to $G(s)$ to find the system function $H(z)$. Leave the sampling interval T as a parameter. Sketch the pole-zero diagram for $H(z)$.

b. Write the difference equation for the discrete-time system obtained. Does it look familiar? Can it be used for approximating a running integral? Comment on its expected accuracy.

8.20. Impulse responses of several FIR filters are listed below. For each impulse response, determine if the corresponding filter has linear phase or not. If it has linear phase, identify which type it is.

a. $h[n] = \{ \underset{\underset{n=0}{\uparrow}}{5}, 4, 3, 3, 4, 5 \}$

b. $h[n] = \{ \underset{\underset{n=0}{\uparrow}}{5}, -4, 3, 2, -2, -3, 4, -5 \}$

c. $h[n] = \{ \underset{\underset{n=0}{\uparrow}}{3}, 2, 1, 4, -1, -2, -3 \}$

d. $h[n] = \{ \underset{\underset{n=0}{\uparrow}}{3}, -2, -7, 0, 7, 2 -3 \}$

e. $h[n] = \{ \underset{\underset{n=0}{\uparrow}}{2}, 4, 1, 3, 2, 1, 4, 3 \}$

f. $h[n] = \{ \underset{\underset{n=0}{\uparrow}}{6}, 4, 7, 2, 5, 2, 7, 4, 6 \}$

8.21. Refer to Exercise 8.12.

a. Using the appropriate trigonometric identity, express the approximated derivative $\hat{y}_a(t)$ as a cosine function with a time delay, so that it can be compared to the actual derivative $y_a(t)$. Express the former as a scaled and delayed version of the latter. How does the time delay between the two signals relate to the FIR differentiator used?

b. Repeat Exercise 8.12, this time using a sinusoidal signal with a frequency of $f_0 = 50$ Hz. Determine the actual derivative $y_a(t)$ and its approximation $\hat{y}_a(t)$ at the output of the FIR differentiator. Comment on the scale factor and the time delay between the two signals.

8.22.

a. Repeat Exercise 8.12 using a multitone signal

$$x_a(t) = \sin\left(2\pi f_1 t\right) + 0.8 \sin\left(2\pi f_2 t + \pi/3\right)$$

with $f_1 = 20$ Hz and $f_2 = 60$ Hz. Use the sampling rate $f = 200$ Hz. Determine the actual derivative $y_a(t)$ and its approximation $\hat{y}_a(t)$ at the output of the FIR differentiator.

b. Use MATLAB to graph the two signals and compare.

8.23.

a. Derive the expression given by Eqn. (8.164) for the impulse response of an ideal differentiator.

b. Derive the expression given by Eqn. (8.169) for the impulse response of an ideal Hilbert transform filter.

8.24. Using the Fourier series design method with a Hamming window, design a length-19 FIR filter to approximate the ideal bandpass magnitude characteristic shown in Fig. P.8.24. Carry out the calculations without the help of MATLAB, and show every step in the design.

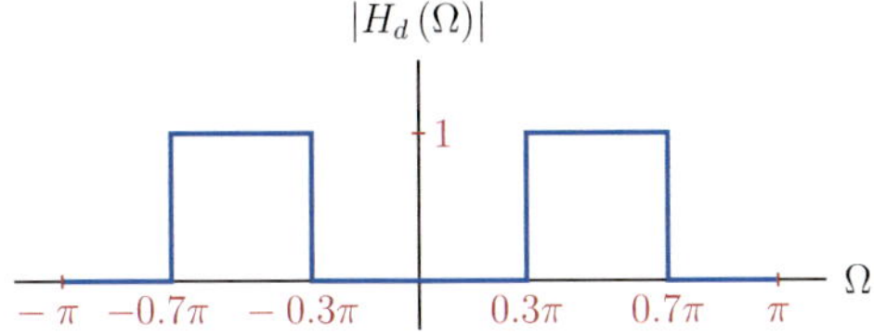

Figure P. 8.24

8.25. Show that the solution given in Eqn. (8.201) satisfies the wave equation in Eqn. (8.200) for any function f_R that is twice differentiable. Hint: Differentiate $f_R(t - x/c)$ twice with respect to x. Then differentiate it twice with respect to t. Substitute the results into the wave equation.

MATLAB Problems

8.26. The magnitude characteristic of a resonant bandpass filter can be made sharper by combining two filters in cascade. Consider the second-order resonant bandpass filter designed in Example 8.2 to extract the 660 Hz component from a signal. The system function was determined to be

$$H(z) = \frac{0.0085\left(z^2 - 1\right)}{z^2 - 1.9479\,z + 0.9830}$$

Let a fourth-order resonant bandpass filter be obtained by using two filters with system functions $H(z)$ in cascade, i.e.,

$$H_1(z) = \left[H(z)\right]^2 = \left[\frac{0.0085\left(z^2-1\right)}{z^2-1.9479\,z+0.9830}\right]^2$$

a. Sketch the pole zero diagram for the fourth-order resonant bandpass filter.

b. Write a MATLAB script to compute and graph the magnitude responses of the second- and fourth-order filters on the same coordinate system. Hint: Start with the script provided for Example 8.2, and modify it using the additional poles and zeros to obtain the fourth-order filter.

c. Zoom in to observe the behavior around the center frequency Ω_0. Estimate the 3-dB bandwidth in Hz for the fourth-order filter from the graph.

8.27. Let a continuous-time signal be defined as

$$x_a(t) = 0.8\cos\left(2\pi f_1 t\right) + \cos\left(2\pi f_2 t\right)$$

where $f_1 = 660$ Hz and $f_2 = 680$ Hz. The signal $x_a(t)$ is sampled with a sampling rate of $f_s = 22{,}050$ Hz to obtain a discrete-time signal $x[n]$. Our goal is to use a resonant bandpass filter to extract the 660 Hz component from the signal. Write a MATLAB script to perform the following steps:

a. Generate samples of $x[n]$ for $n = 0,\ldots,3999$.

b. Apply a resonant bandpass filter with center frequency $f_0 = 660$ Hz, and 3-dB bandwidth 15 Hz to the signal, and compute the output signal $y[n]$. Hint: Use functions `ss_respar()` and `ss_resf()` developed in MATLAB Exercises 8.2 and 8.3.

c. Get a fourth-order resonant bandpass filter by squaring the system function found in part (b), and apply it to the signal $x[n]$ to obtain the output $y_2[n]$. Hint: For this part, you will need to use the function `ss_resf()` for the second time. Use $y[n]$ as input to `ss_resf()` to obtain the output $y_2[n]$.

d. Graph the signals $x[n]$, $y[n]$, and $y_2[n]$ and compare. Do the signals $y[n]$ and $y_2[n]$ still contain artifacts of the 680 Hz sinusoid? Does the fourth-order filter perform better than the second-order filter?

8.28. An analog signal which is known to have sinusoidal interference at 60 Hz and its harmonics is sampled with a sampling rate of $f_s = 22{,}050$ Hz. A multiband notch filter is to be designed to suppress frequencies 60, 120, and 180 Hz.

a. Design the multiband filter as the cascade combination of three notch filters. Each one should be have a stopband width of 3 Hz. Use the function `ss_notchpar()` developed in MATLAB Exercise 8.4 to compute the parameters of each filter.

b. Compute and graph the magnitude response of the multiband filter. Use the function `ss_freqz()` developed in MATLAB Exercise 5.8 of Chapter 5 for this purpose.

c. Draw a cascade-form block diagram for the filter designed.

8.29. Refer to Problem 8.28. A signal that is corrupted by sinusoidal interference at frequencies 60, 120, and 180 Hz as described in that problem has been provided with the set of downloadable files. It is in the audio file "`Rain.flac`" which is 20 seconds long.

Listen to the file using any audio player on your computer, and pay attention to the sinusoidal interference that is present in part of the file. Write a MATLAB script to create audio reader and player objects (see MATLAB Exercise 8.3 for an example), design the three notch filters as described in Problem 8.28, and place them into the processing loop. Use functions `ss_notchpar()` and `ss_notchf()` developed in MATLAB Exercise 8.4. Process the sound, and play it back while it is being processed. Comment on the result.

8.30. Refer to Problem 8.29 where three notch filters were used in cascade to clean up an audio signal corrupted with sinusoidal interference at 60, 120, and 180 Hz. The signal with sinusoidal interference is available in the audio file "`Rain.flac`". In this case we wish to extract just the three sinusoidal components, and listen to them without the sound of rain. For this purpose, three resonant bandpass filters can be designed and connected in parallel, each extracting one frequency. The output signals of three resonators are added together to provide the total interference signal.

Write a MATLAB script to create audio reader and player objects, design the three resonant bandpass filters with appropriate center frequencies and 3-dB bandwidth of 3 Hz, and place them into the processing loop. Use functions `ss_respar()` and `ss_resf()` developed in MATLAB Exercises 8.2 and 8.3. Process the sound, and play it back while it is being processed. Comment on the result.

8.31. A lowpass filter of order 3 is to be designed by trial-and-error placement of poles and zeros on the z-plane. The passband is loosely defined as the angular frequency range $|\Omega| < 0.3\pi$. Initial poles and zeros are selected as follows:

$$\text{Poles:} \qquad p_k = 0.8,\ 0.8e^{\pm j0.2\pi}$$

$$\text{Zeros:} \qquad z_k = e^{\pm j0.4\pi},\ e^{\pm j0.8\pi}$$

a. In MATLAB, create two vectors named `pls` and `zrs` with initial poles and zeros for the system function.

b. Use the interactive pole-zero explorer in `appPoleZeroDT.m` for experimentation with positions of poles and zeros. From the checklist named "`PZ set`" select the "`Custom`" option. Enter the names of the two vectors created in part (a) to load the initial pole-zero profile from the two vectors created.

c. Observe the magnitude of the resulting system function. Make adjustments by moving poles around to get the passband as flat as possible within the range specified. The "`Select`" button selects a different pole or zero (or a conjugate pair of either) each time it is pressed. Adjustments can be made to radius or angle. Increments can be made smaller for fine tuning.

d. Once satisfied with passband behavior, make adjustments to locations of zeros in the same way.

e. When satisfied with the design, use the button "`Save PZ map`" to save final pole and zero locations to two new vectors. Use the function `ss_freqz()` to compute the magnitude and phase responses of the filter. Adjust the gain factor so that the gain at $\Omega = 0$ is unity.

8.32. Consider the design of a third-order Butterworth analog highpass filter in Problem 8.8. We wish to repeat the steps involved in the design process using MATLAB. The 3-dB cutoff frequency of the final highpass filter was specified to be 5 rad/s. Develop a MATLAB script to do the following:

a. Use function `butter()` to find numerator and denominator coefficients for a Butterworth analog lowpass filter with a 3-dB cutoff frequency of 2 rad/s. Compute the frequency response of the lowpass filter using the function `freqs()`. Graph the magnitude characteristic for $0 < \omega < 10$ rad/s. Hint: Do not forget to use the `'s'` option in function `butter()`, or MATLAB will attempt to design a discrete-time filter.

b. Transform the lowpass filter designed in part(a) to a highpass filter using the function `lp2hp()`. The frequency $\omega_{L1} = 2$ rad/s for the lowpass filter should map to the frequency $\omega_{H2} = 5$ rad/s for the highpass filter. See Fig. 8.19 and Eqn. (8.42) for details. The argument `w0` for the function `lp2hp()` is the parameter ω_0^2 in Eqn. (8.42). Compute the frequency response of the highpass filter using the function `freqs()`. Graph the magnitude characteristic for $0 < \omega < 10$ rad/s. Check to make sure design specifications are satisfied.

c. Redesign the highpass filter in one step with the function `butter()` by using `'high'` for the third argument. Compare the result to that obtained in part (b).

8.33. Consider the design of a third-order Chebyshev type-I analog highpass filter in Problem 8.9. We wish to repeat the steps involved in the design process using MATLAB. The 3-dB cutoff frequency of the final highpass filter was specified to be 5 rad/s. Develop a MATLAB script to do the following:

a. Use function `cheby1()` to find numerator and denominator coefficients for a Chebyshev type-I analog lowpass filter with a maximum of 3-dB passband ripple for frequencies less than 2 rad/s. Compute the frequency response of the lowpass filter using the function `freqs()`. Graph the magnitude characteristic for $0 < \omega < 10$ rad/s. Hint: Do not forget to use the `'s'` option in function `cheby1()`, or MATLAB will attempt to design a discrete-time filter.

b. Transform the lowpass filter designed in part(a) to a highpass filter using the function `lp2hp()`. The frequency $\omega_{L1} = 2$ rad/s for the lowpass filter should map to the frequency $\omega_{H2} = 5$ rad/s for the highpass filter. See Fig. 8.19 and Eqn. (8.42) for details. The argument `w0` for the function `lp2hp()` is the parameter ω_0^2 in Eqn. (8.42). Compute the frequency response of the highpass filter using the function `freqs()`. Graph the magnitude characteristic for $0 < \omega < 10$ rad/s. Check to make sure design specifications are satisfied.

c. Redesign the highpass filter in one step with the function `cheby1()` by using `'high'` for the fourth argument. Compare the result to that obtained in part (b).

8.34. Consider the design of a fourth-order Chebyshev type-II analog bandpass filter in Problem 8.10. We wish to repeat the steps involved in the design process using MATLAB. The specifications called for a minimum stopband attenuation of 15-dB for frequencies less than 2 rad/s or greater than 3 rad/s. Develop a MATLAB script to do the following:

a. Use function `cheby2()` to find numerator and denominator coefficients for a second-order Chebyshev type-II analog lowpass filter with a minimum stopband attenuation of 15 dB for frequencies greater than 2 rad/s. Compute the frequency response of the lowpass filter using the function `freqs()`. Graph the magnitude characteristic for $0 < \omega < 10$ rad/s.

b. Transform the lowpass filter designed in part(a) to a bandpass filter using the function `lp2bp()`. Determine transformation parameters B and ω_0 using Eqn. (8.52) with lowpass filter critical frequency 2 rad/s and bandpass filter critical frequencies of 2 rad/s and 3 rad/s. Compute the frequency response of the bandpass filter using the function `freqs ()`. Graph the magnitude characteristic for $0 < \omega < 10$ rad/s. Check to make sure design specifications are satisfied.

c. Redesign the bandpass filter in one step with the function `cheby2()` by using `'bandpass'` for the fourth argument. Compare the result to that obtained in part (b).

8.35. The analog prototype filter with the system function

$$G(s) = \frac{s+3}{(s+1)\,(s+2)}$$

was used in Problems 8.11 and 8.12 with impulse invariant and unit step invariant transformations for obtaining discrete-time filters.

a. Using the function `freqz()`, compute the frequency response of the impulse invariant design of Problem 8.11. Graph it for $-\pi < \Omega < \pi$.

b. Repeat part (a) using the unit step invariant design of Problem 8.12.

8.36. Refer to the discrete-time filter designed in Problem 8.11 using the impulse invariance method with the sampling interval $T = 0.1$ s.

a. Using the function `impinvar()`, repeat the design process of Problem 8.11. Compare the result obtained for $H(z)$ to the one found in Problem 8.11.

b. Using the function `freqs()`, compute and graph the magnitude response of the analog prototype filter in the frequency interval $-\pi/T \le \omega \le \pi/T$.

c. Using the function `freqz()`, compute and graph the magnitude response of the discrete-time filter in the angular frequency interval $-\pi \le \Omega \le \pi$.

d. Suppose that the discrete-time filter is used for processing a continuous-time signal as shown in Fig. 8.1. Determine the relationship between the frequencies ω and Ω. Based on that relationship, compute and graph the overall magnitude response of the system in Fig. 8.1 simultaneously with the magnitude response of the analog prototype filter for comparison. In other words, compute and graph $|Y_a(\omega)/X_a(\omega)|$ versus the analog prototype magnitude response $|G(\omega)|$. The graph is only meaningful in the frequency interval $-\omega_s < \omega < \omega_s$.

8.37. Repeat Problem 8.36 using a sampling interval of $T = 1$ s instead of $T = 0.1$ s. Comment on how this change affects the overall performance of the analog system in Fig. 8.1 that utilizes a discrete-time filter.

8.38. Refer to the discrete-time filter designed in Problem 8.12 using the unit step invariance method with the sampling interval $T = 0.1$ s.

a. MATLAB does not have a built-in function for converting an analog filter to a discrete-time filter using unit step invariance. We can devise a method of using the impulse invariance function `impinvar()` to facilitate step invariance design. The Laplace transform of the unit step response of our analog prototype filter is

$$G_s(s) = \frac{G(s)}{s}$$

Use the function `impinvar()` to convert $G_s(s)$ to $H_s(z)$.

b. The z-transform obtained in part (a) is the transform of the unit step response of the discrete-time filter. We have

$$H_s(z) = \frac{z}{z-1}\,H(z)$$

from which the transform $H(z)$ needs to be extracted. Do this to obtain the system function for the discrete-time filter. Compare your result with that found in Problem 8.12. Hint: One way to do this would be to use the function `tf2zpk()`. It will convert the description of the system function based on numerator and denominator polynomials to a description based on zeros, poles, and a gain factor. Afterward, delete the zero at $z = 0$ and delete the pole at $z = 1$. Use the function `zp2tf()` to get the coefficients of numerator and denominator polynomials of $H(z)$. Be careful about the fact that, unlike the function `impinvar()`, the function `zp2tf()` returns numerator and denominator coefficients in terms of positive powers of z.

8.39. Refer to the third-order Butterworth analog lowpass filter with a 3-dB cutoff frequency of $\omega_c = 1$ rad/s, used in Problem 8.15. Its system function was

$$G(s) = \frac{1}{(s+1)\left(s^2+s+1\right)}$$

a. Using impulse invariance, we wish to convert this filter to a discrete-time lowpass filter with a 3-dB cutoff frequency at $\Omega_c = 0.1\pi$. Determine the sampling rate needed. Write a script to carry out the impulse-invariance conversion using the function `impinvar()`. Afterward, use function `freqz()` to compute the frequency response of the filter obtained. Graph the magnitude of the frequency response. On the graph, mark the frequency Ω_c and check its value. Is it what was specified in the design requirements?

b. Repeat part (a), this time for a 3-dB cutoff frequency at $\Omega_c = 0.2\pi$. Determine what the new sampling rate needs to be. Graph the magnitude response again, with the frequency Ω_c marked on it. Comment on the accuracy of the 3-dB cutoff frequency.

c. Repeat for a 3-dB cutoff frequency at $\Omega_c = 0.5\pi$. Again, comment on the accuracy of the 3-dB cutoff frequency.

8.40. The analog prototype filter with the system function

$$G(s) = \frac{s+3}{(s+1)\,(s+2)}$$

was used in Problem 8.16 with matched z-transform for obtaining a discrete-time filter. Using the function `freqz()`, compute the frequency response of the filter found in Problem 8.16. Graph it for $-\pi < \Omega < \pi$.

8.41. Refer to Problem 8.16 where the matched z-transform for converting an analog prototype to a discrete-time filter was discussed.

a. Write a MATLAB function `ss_matchedz()` for the transformation. Its syntax should be

```
>> [numz,denz] = ss_matchedz(num,den,fs)
```

where `num` and `den` are vectors with numerator and denominator coefficients, and `fs` is the sampling rate in Hz. The returned vectors `numz` and `denz` hold numerator and denominator coefficients for the discrete-time filter.

b. A third-order Butterworth analog lowpass filter with a 3-dB cutoff frequency of $\omega_c = 1$ rad/s has the system function

$$G(s) = \frac{1}{(s+1)\left(s^2+s+1\right)}$$

Write a script to convert this analog prototype to a discrete-time filter using the matched z-transform. Use the function `ss_matchedz()` with a sampling rate of $f_s = 1$ Hz. Adjust the gain factor do that the filter magnitude at $\Omega = 1$ is equal to unity. Compute and graph the magnitude response of the resulting filter.

c. Use the function `impinvar()` to obtain another discrete-time filter from the same analog prototype, again using $f_s = 1$ Hz. Compute the magnitude response of this filter also. Graph the two magnitude responses on the same coordinate system to compare.

8.42. Refer to Problem 8.18. Use the function `bilinear()` to convert the analog prototype to a discrete-time filter. Use the value of T that was determined in Problem 8.18. Afterward use the function `freqz()` to compute the magnitude response. Graph it for $-\pi \le \Omega \le \pi$.

8.43. In this problem, we will test the performance of the first-order backward differentiator using a chirp signal. Refer to the magnitude response of the differentiator shown in Fig. 8.40. For low frequencies, the magnitude characteristic seems to be tangent to that of the ideal differentiator. As the frequency is increased, it deviates from the ideal.

Consider a chirp signal with linearly increasing instantaneous frequency.

$$x_a(t) = \sin\left(2\pi f_0 t + \pi c t^2\right)$$

The parameter c can be computed as

$$c = \frac{f_1 - f_0}{\tau}$$

for an instantaneous frequency that starts at f_0 and increases to f_1 over a duration of τ. The true derivative of the signal $x_a(t)$ is

$$y_a(t) = \left(2\pi f_0 + 2\pi c t\right)\cos\left(2\pi f_0 t + \pi c t^2\right)$$

which exhibits a linearly increasing amplitude.

a. Let the sampling rate be $f_s = 22{,}050$ Hz. The instantaneous frequency should sweep from 220 Hz to 6000 Hz over 2 seconds. Using MATLAB, generate sampled versions $x[n]$ and $y[n]$ of the chirp signal and its derivative for a duration of 2 seconds. Graph each signal.

b. Obtain $\hat{y}[n]$ as the output signal of the first-order differentiator when its input signal is $x[n]$. Use the function `conv()` for this. Graph the signal $\hat{y}[n]$.

c. Compare the envelopes of the signals $y[n]$ and $\hat{y}[n]$ and comment on the accuracy of the approximate derivative as a function of frequency.

8.44. Consider the specifications for an ideal bandpass filter illustrated in Fig. 8.24.

a. Using the Fourier series design method with a rectangular window, design a length-41 FIR filter to approximate the specifications. Use function `ss_fir1()` for the design. Afterward compute and graph the magnitude response for the range $0 \le \Omega \le \pi$.

b. Repeat part (a) using a Hamming window.

c. Compute dB magnitude responses of both filters, and graph them on the same coordinate system for comparison.

8.45.

a. Use the function `ss_fir1()` to design a length-24 differentiator. Use a rectangular window. Compute and graph the magnitude response of the differentiator.

b. Repeat part (a) using a Hamming window.

8.46. Refer to Problem 8.43 where the performance of the first-order backward differentiator was tested using a chirp signal and its derivative. Repeat that test using the differentiator of order 23 that was designed in part (a) of Problem 8.45. Compare the result to that of the first-order differentiator.

8.47. Refer to the discrete-time filter designed in part (a) of Problem 8.17 using the bilinear transformation method with the sampling interval $T = 0.5$ s. Write a script to perform the following steps:

a. Using the function `bilinear()`, repeat the design process of Problem 8.17. Compare the result obtained for $H(z)$ to the one found in Problem 8.17.

b. Compute and graph the magnitude response of the analog prototype filter in the frequency interval $-\pi/T \le \omega \le \pi/T$.

c. Compute and graph the magnitude response of the discrete-time filter in the angular frequency interval $-\pi \le \Omega \le \pi$.

d. Suppose that the discrete-time filter is used for processing a continuous-time signal as shown in Fig. 8.1. Compute and graph the overall magnitude response of the system in Fig. 8.1, that is, $|Y_a(\omega)/X_a(\omega)|$, simultaneously with the magnitude response of the analog prototype filter for comparison.

8.48. Refer to the length-19 FIR filter specified in Problem 8.24.

a. Write a script to compute and graph the dB magnitude response of the filter.

b. Write a script to design a length-45 filter to meet the same specifications. Compute and graph the dB magnitude characteristics of the length-15 and length-45 designs in a superimposed fashion for comparison.

c. Repeat the length-45 design using a Blackman window instead of a Hamming window. Compute and graph the dB magnitude characteristics of the two length-45 designs together and compare.

8.49. Desired amplitude response of a multiband filter is specified in Fig. 8.49. We would like to approximate this characteristic using a length-41 FIR filter designed by frequency sampling.

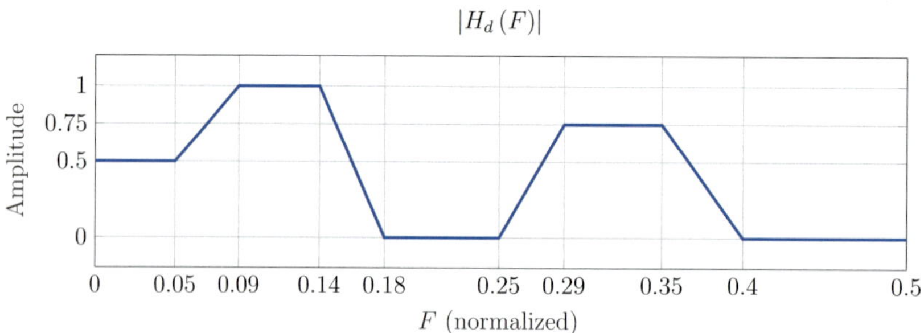

Figure P. 8.49

Write a script to design this filter through the use of the function `ss_fir2()` developed in MATLAB Exercise 8.13. Determine the appropriate vectors `F` and `A` for use with the function `ss_fir2()`. Once the impulse response is obtained, compute and graph the magnitude response of the filter overlaid with a graph of the specifications.

8.50. Refer to the desired amplitude response of a multiband filter specified in Problem 8.49, Fig. 8.49. We would like to approximate this characteristic using a length-41 FIR filter designed with the least-squares method. Write a script to design this filter through the use of the function `ss_firls()` developed in MATLAB Exercise 8.14. Determine the appropriate vectors `F` and `A` for use with the function. Use $K = 125$ frequency points for the overdetermined system of equations. Set up the weight factors to be equal to 10 between the corner frequencies of the two passbands, and equal to 1 everywhere else. Once the impulse response is obtained, compute and graph the magnitude response of the filter overlaid with a graph of the specifications.

8.51. Refer to the desired amplitude response of a multiband filter specified in Problem 8.49, Fig. 8.49. We would like to approximate this characteristic using a length-41 FIR filter designed with Parks-McClellan algorithm. Write a script to design this filter through the use of the function `firpm()`. Determine the appropriate vectors `F` and `A` for use with the function. Set up the weight factor equal to 10 for the two passbands, equal to 1 everywhere else. Once the impulse response is obtained, compute and graph the magnitude response of the filter overlaid with a graph of the specifications.

8.52.

a. Consider the script `mexdt_8_16c.m` developed in MATLAB Exercise 8.16. It generates 44,101 samples of the impulse response of the plucked string filter shown in Fig. 8.77b. The output signal is then played back with a sampling rate of $f_s = 22050$ Hz. Expand this script to compute the magnitude spectrum of the output signal in vector `y`. Graph the magnitude spectrum in terms of actual frequencies in Hz. Are the locations of the fundamental frequency and harmonics where you expect them?

b. Consider the script `mexdt_8_16d.m` which uses the tunable plucked string filter shown in Fig. 8.80. Expand this script to compute the magnitude spectrum of the output signal in vector `y`. Graph the magnitude spectrum in terms of actual frequencies in Hz. Are the locations of the fundamental frequency and harmonics where you expect them? Some of the higher-order harmonics should be slightly off on the frequency scale. Can you explain the reason for this?

MATLAB Projects

8.53. Refer to the function `ss_iir2()` developed in MATLAB Exercise 5.12 in Chapter 5 for implementing a second-order section with system function

$$H(z) = \frac{b_0 + b_1 z^{-1} + b_2 z^{-2}}{1 + a_1 z^{-1} + a_2 z^{-2}}$$

One of the input arguments of the function `ss_iir2()` is a vector of filter coefficients in a prescribed order:

$$\text{coeffs} = \left[b_{0i},\, b_{1i},\, b_{2i},\, 1,\, a_{1i},\, a_{2i}\right]$$

GUI-based MATLAB program `fdatool()` has the capability to export the coefficients of a designed filter to the workspace as a "SOS matrix" which corresponds to an implementation of the filter in the form of second-order cascade sections. Each row of the "SOS matrix" contains the coefficients of one second-order section in the same order given above.

a. Develop a MATLAB function `ss_iirsos()` for implementing an IIR filter designed using the program `fdatool()`. Its syntax should be

```
out = ss_iirsos(inp,sos,gains)
```

The matrix `sos` is the coefficient matrix exported from `fdatool()` using the menu selections *file, export.* The vector `gains` is the vector of gain factors for each second-order section. It is also exported from `fdatool()`. The vector `inp` holds samples of the input signal. The computed output signal is returned in vector `out`. The function `ss_iirsos()` should internally utilize the function `ss_iir2()` for implementing the filter in cascade.

b. Use `fdatool()` to design an elliptic bandpass filter with passband edges at $\Omega_2 = 0.4\pi$, $\Omega_3 = 0.7\pi$, and stopband edges at $\Omega_1 = 0.3\pi$, $\Omega_4 = 0.8\pi$. The passband ripple must not exceed 1 dB, and the stopband attenuation must be at least 30 dB. Export the coefficients and gain factors of the designed filter to MATLAB workspace.

c. Use the function `iirsos()` to compute the unit step response of the designed filter. Compare to the step response computed by the program `fdatool()`.

8.54. Using the tunable plucked string filter function `ss_tpsf()` developed in MATLAB Exercise 8.16, design a MATLAB-based guitar tuner. A basic 6-string guitar has the following frequencies in Hz for open strings:

E_2	A_2	D_3	G_3	B_3	E_4
82.41	110.00	146.83	196.00	246.94	329.63

Develop a script that can play proper sounds for each of the frequencies listed in the table. Your script should determine the normalized frequency of each note and the loop delay needed to generate that frequency. Adjust the parameter of the all-pass filter component for each note as needed. Develop a simple mechanism for the user to select the note to be played.

8.55. Consider the script `mexdt_8_16d.m` developed in MATLAB Exercise 8.16 for producing a synthesized guitar sound. The sound that results from running the script feels a bit flat compared to the sound of an actual guitar. We are only simulating the vibration of the string, but an actual guitar has a wooden body that acts like an echo chamber, producing reverberations.

a. Modify the script to include the simple reverberator structure that was used in MATLAB Exercise 5.14 in Chapter 5, and shown in Fig. 5.72. Use functions `ss_combf()` and `ss_allpassf()` as needed. Experiment with the parameters of comb filters and all-pass filters until you are satisfied with the sound.

b. Repeat part (a) using the Schroeder reverberator structure shown in Fig. 5.73.

APPENDIX **A**

Complex Numbers and Euler's Formula

A.1 Introduction

Complex numbers allow us to solve equations for which no real solution can be found. Consider, for example, the equation

$$x^2 + 9 = 0 \tag{A.1}$$

which cannot be satisfied for any real number. By introducing an imaginary unit[1] $j = \sqrt{-1}$ so that $j^2 = -1$, Eqn. (A.1) can be solved to yield

$$x = \mp j3$$

Similarly, the equation

$$(x+2)^2 + 9 = 0 \tag{A.2}$$

has the solutions

$$x = -2 \mp j3 \tag{A.3}$$

A general complex number is in the form

$$x = a + jb \tag{A.4}$$

where a and b are real numbers. The values a and b are referred to as the *real part* and *imaginary part* of the complex number x, respectively. Following notation is used for real and imaginary

[1] The imaginary unit was first introduced by Italian mathematician Gerolamo Cardano as he worked on the solutions of cubic and quartic equations, although its utility was not fully understood and appreciated until the works of Leonhard Euler and Carl Friedrich Gauss almost two centuries later.

DOI: 10.1201/9781003570462-A

parts of a complex number:

$$a = \mathrm{Re}\{x\}$$
$$b = \mathrm{Im}\{x\}$$

For two complex numbers to be equal, both their real parts and imaginary parts must be equal. Given two complex numbers $x_1 = a + jb$ and $x_2 = c + jd$, the equality $x_1 = x_2$ implies that $a = c$ and $b = d$.

It is often convenient to graphically represent a complex number as a point in the *complex plane* constructed using a horizontal axis corresponding to the real part and a vertical axis corresponding to the imaginary part as depicted in Fig. A.1.

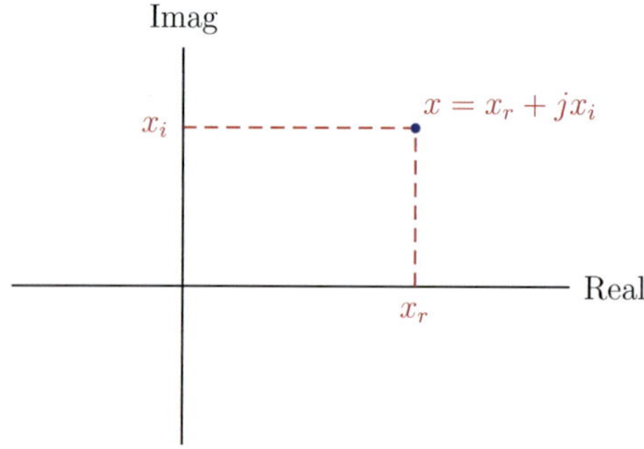

Figure A.1 – Complex number $x = x_r + jx_i$ shown as a point in the complex plane.

If the imaginary part of a complex number is equal to zero, the resulting number is said to be *purely real.* An example is $x = 3 + j0$. Similarly, a complex number the real part of which is equal to zero is said to be *purely imaginary.* An example of a purely imaginary number is $x = 0 + j5$. Purely real and purely imaginary numbers are simply special cases of the more general class of complex numbers.

Another method of graphical representation of a complex number can be found through the use of a vector as shown in Fig. A.2. Vector representation of complex numbers is quite useful since vector operations can be used for manipulating complex numbers.

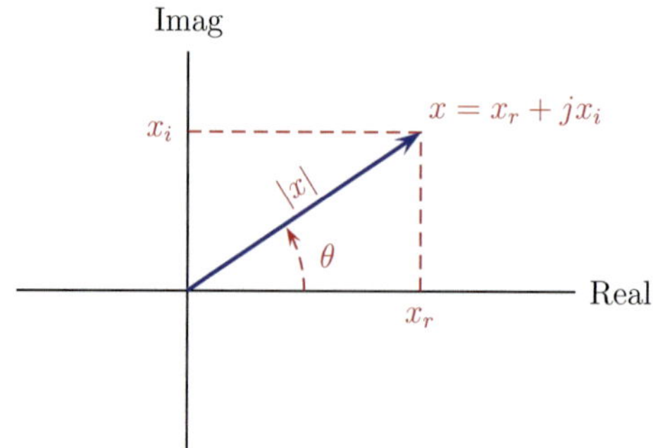

Figure A.2 – Complex number $x = x_r + jx_i$ shown as a vector.

A complex number written as the sum of its real and imaginary parts such as $x = x_r + jx_i$ is said to be in *Cartesian form*. An alternative form referred to as a *polar form* is derived from the vector representation of a complex number by using the norm $|x|$ and the angle θ of the vector instead of the real and imaginary parts x_r and x_i. The two forms of the complex number x are:

$$x = x_r + jx_i\,, \quad \text{Cartesian form}$$

$$x = |x|\, e^{j\theta}\,, \quad \text{Polar form}$$

The norm $|x|$ is the distance of the point representing the complex number from the origin. The angle θ is the angle measured counterclockwise starting with the positive real axis.

A complex number can easily be converted from Cartesian form to polar form and vice versa. Through simple geometric relationships it can be shown that

$$|x| = \sqrt{x_r^2 + x_i^2} \tag{A.5}$$

and

$$\theta = \tan^{-1}\left(\frac{x_i}{x_r}\right) \tag{A.6}$$

Alternately, real and imaginary parts can be obtained from the norm and the angle through

$$x_r = |x| \cos(\theta) \tag{A.7}$$

and

$$x_i = |x| \sin(\theta) \tag{A.8}$$

A.2 Arithmetic with Complex Numbers

Definitions of arithmetic operators are easily extended to apply to complex numbers. In Section A.2.1 addition and subtraction operators will be discussed. Multiplication and division of complex numbers will be discussed in Section A.2.2.

A.2.1 Addition and subtraction

Consider two complex numbers $x_1 = a + jb$ and $x_2 = c + jd$. Addition of these two complex numbers is carried out as follows:

$$\begin{aligned} y &= x_1 + x_2 \\ &= (a + jb) + (c + jd) \\ &= (a + c) + j(b + d) \end{aligned} \tag{A.9}$$

Real and imaginary parts of the result can be written separately as

$$\begin{aligned} \mathrm{Re}\{y\} &= \mathrm{Re}\{x_1\} + \mathrm{Re}\{x_2\} \\ \mathrm{Im}\{y\} &= \mathrm{Im}\{x_1\} + \mathrm{Im}\{x_2\} \end{aligned}$$

Subtraction of complex numbers is similar to addition. The difference $z = x_1 - x_2$ is computed as follows:

$$\begin{aligned} z &= x_1 - x_2 \\ &= (a + jb) - (c + jd) \\ &= (a - c) + j(b - d) \end{aligned} \tag{A.10}$$

Real and imaginary parts of the result can be written separately as

$$\begin{aligned}
\mathrm{Re}\{z\} &= \mathrm{Re}\{x_1\} - \mathrm{Re}\{x_2\} \\
\mathrm{Im}\{z\} &= \mathrm{Im}\{x_1\} - \mathrm{Im}\{x_2\}
\end{aligned}$$

Addition and subtraction operators for complex numbers are analogous for addition and subtraction of vectors using the parallelogram rule. This is illustrated in Fig. A.3.

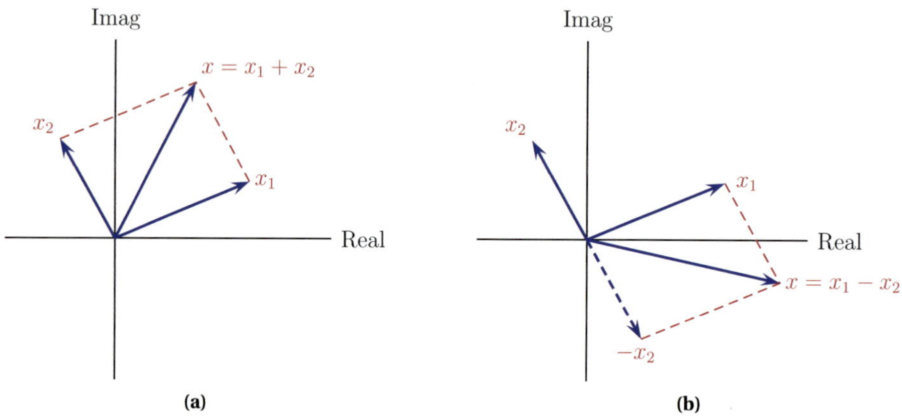

Figure A.3 – Addition and subtraction operators with complex vectors.

If the complex numbers to be added are expressed in polar form, they must first be converted to Cartesian form before they can be added. Consider two complex numbers in polar form given as

$$x_1 = |x_1|\, e^{j\theta_1}\,, \quad \text{and} \quad x_2 = |x_2|\, e^{j\theta_2} \tag{A.11}$$

The sum z of these complex numbers is found as

$$\begin{aligned}
z &= x_1 + x_2 \\
&= |x_1| \cos(\theta_1) + j\, |x_1| \sin(\theta_1) + |x_2| \cos(\theta_2) + j\, |x_2| \sin(\theta_2) \\
&= \Big[\, |x_1| \cos(\theta_1) + |x_2| \cos(\theta_2) \Big] + j \,\Big[\, |x_1| \sin(\theta_1) + |x_2| \sin(\theta_2) \Big]
\end{aligned} \tag{A.12}$$

which can be put into polar form as

$$z = |z|\, e^{j\theta_z} \tag{A.13}$$

where

$$|z| = |x_1|^2 + |x_2|^2 + 2\, |x_1|\, |x_2| \cos(\theta_1 - \theta_2) \tag{A.14}$$

and

$$\theta_z = \tan^{-1}\left(\frac{|x_1| \sin(\theta_1) + |x_2| \sin(\theta_2)}{|x_1| \cos(\theta_1) + |x_2| \cos(\theta_2)} \right) \tag{A.15}$$

A.2.2 Multiplication and division

Again consider two complex numbers $x_1 = a + jb$ and $x_2 = c + jd$. The product of these two complex numbers is computed as follows:

$$\begin{aligned} y =& x_1\, x_2 \\ =& \left(a + jb\right)\left(c + jd\right) \\ =& ac + j\,ad + j\,bc + j^2\,bd \\ =& (ac - bd) + j\,(ad + bc) \end{aligned} \tag{A.16}$$

Real and imaginary parts of the result can be written separately as

$$\begin{aligned} \text{Re}\left\{y\right\} =& \text{Re}\left\{x_1\right\}\text{Re}\left\{x_2\right\} - \text{Im}\left\{x_1\right\}\text{Im}\left\{x_2\right\} \\ \text{Im}\left\{y\right\} =& \text{Re}\left\{x_1\right\}\text{Im}\left\{x_2\right\} + \text{Im}\left\{x_1\right\}\text{Re}\left\{x_2\right\} \end{aligned}$$

Product of a complex number and its own complex conjugate is equal to the squared norm of the complex number:

$$\begin{aligned} x_1\, x_1^* =& \left(a + jb\right)\left(a + jb\right)^* \\ =& \left(a + jb\right)\left(a - jb\right) \\ =& a^2 + b^2 = \left|x_1\right|^2 \end{aligned}$$

Division of complex numbers is slightly more involved. Consider the complex number z defined as

$$z = \frac{x_1}{x_2} = \frac{\left(a + jb\right)}{\left(c + jd\right)} \tag{A.17}$$

The expression in Eqn. (A.17) can be simplified by multiplying both the numerator and the denominator by the complex conjugate of the denominator:

$$z = \frac{x_1}{x_2} = \frac{\left(a + jb\right)}{\left(c + jd\right)}\frac{\left(c - jd\right)}{\left(c - jd\right)} \tag{A.18}$$

A.3 Euler's Formula

A complex exponential function can be expressed in the form

$$e^{jx} = \cos(x) + j\,\sin(x) \tag{A.19}$$

This relationship is known as *Euler's formula* and will be used extensively in working with signals, linear systems and various transforms. If the sign of x is changes in Eqn. (A.20), we get

$$e^{-jx} = \cos(-x) + j\,\sin(-x) = \cos(x) - j\,\sin(x) \tag{A.20}$$

Using Eqns. (A.19) and (A.2) trigonometric functions $\cos(x)$ and $\sin(x)$ may be expressed in terms of complex exponential functions as

$$\cos(x) = \frac{e^{jx} + e^{-jx}}{2} \tag{A.21}$$

and

$$\sin(x) = \frac{e^{jx} - e^{-jx}}{2j} \tag{A.22}$$

Appendix B

Mathematical Relations

B.1 Trigonometric Identities

$$\cos(a \pm b) = \cos(a)\cos(b) \mp \sin(a)\sin(b) \quad \text{(B.1)}$$

$$\sin(a \pm b) = \sin(a)\cos(b) \pm \cos(a)\sin(b) \quad \text{(B.2)}$$

$$\tan(a \pm b) = \frac{\tan(a) \pm \tan(b)}{1 \mp \tan(a)\tan(b)} \quad \text{(B.3)}$$

$$\cos(a)\cos(b) = \tfrac{1}{2}\cos(a+b) + \tfrac{1}{2}\cos(a-b) \quad \text{(B.4)}$$

$$\sin(a)\sin(b) = \tfrac{1}{2}\cos(a-b) - \tfrac{1}{2}\cos(a+b) \quad \text{(B.5)}$$

$$\sin(a)\cos(b) = \tfrac{1}{2}\sin(a+b) + \tfrac{1}{2}\sin(a-b) \quad \text{(B.6)}$$

$$\cos(a) + \cos(b) = 2\cos\left(\frac{a+b}{2}\right)\cos\left(\frac{a-b}{2}\right) \quad \text{(B.7)}$$

$$\cos(a) - \cos(b) = -2\sin\left(\frac{a+b}{2}\right)\sin\left(\frac{a-b}{2}\right) \quad \text{(B.8)}$$

DOI: 10.1201/9781003570462-B

$$\sin(a) + \sin(b) = 2\sin\left(\frac{a+b}{2}\right)\cos\left(\frac{a-b}{2}\right) \tag{B.9}$$

$$\sin(a) - \sin(b) = 2\sin\left(\frac{a-b}{2}\right)\cos\left(\frac{a+b}{2}\right) \tag{B.10}$$

$$\cos(2a) = \cos^2(a) - \sin^2(a) \tag{B.11}$$

$$\sin(2a) = 2\sin(a)\cos(a) \tag{B.12}$$

$$\tan(2a) = \frac{2\tan(a)}{1-\tan^2(a)} \tag{B.13}$$

$$\cos^2(a) = \tfrac{1}{2} + \tfrac{1}{2}\cos(2a) \tag{B.14}$$

$$\sin^2(a) = \tfrac{1}{2} - \tfrac{1}{2}\cos(2a) \tag{B.15}$$

B.2 Indefinite Integrals

$$\int x e^{ax}\, dx = \frac{ax-1}{a^2}\, e^{ax} \tag{B.16}$$

$$\int \frac{1}{x^2+a^2}\, dx = \frac{1}{a}\tan^{-1}\left(\frac{x}{a}\right) \tag{B.17}$$

$$\int \frac{x}{x^2+a^2}\, dx = \frac{1}{2}\ln\left(x^2+a^2\right) \tag{B.18}$$

$$\int \frac{x^2}{x^2+a^2}\, dx = x - a\tan^{-1}\left(\frac{x}{a}\right) \tag{B.19}$$

$$\int x\cos(ax)\, dx = \frac{1}{a^2}\left[\cos(ax) + ax\sin(ax)\right] \tag{B.20}$$

$$\int x\sin(ax)\, dx = \frac{1}{a^2}\left[\sin(ax) - ax\cos(ax)\right] \tag{B.21}$$

$$\int e^{ax}\cos(bx)\, dx = \frac{e^{ax}}{a^2+b^2}\left[a\cos(bx) + b\sin(bx)\right] \tag{B.22}$$

$$\int e^{ax}\sin(bx)\, dx = \frac{e^{ax}}{a^2+b^2}\left[a\sin(bx) - b\cos(bx)\right] \tag{B.23}$$

$$\int (a+bx)^n\, dx = \frac{(a+bx)^{n+1}}{b\,(n+1)}, \qquad n > 0 \tag{B.24}$$

$$\int \frac{1}{(a+bx)^n}\, dx = \frac{-1}{b\,(n-1)\,(a+bx)^{n-1}}, \qquad n > 1 \tag{B.25}$$

B.3 Laplace Transform Pairs

Signal	Transform	ROC
$\delta(t)$	1	all s
$u(t)$	$\frac{1}{s}$	$\mathrm{Re}\{s\} > 0$
$u(-t)$	$-\frac{1}{s}$	$\mathrm{Re}\{s\} < 0$
$e^{at}\,u(t)$	$\frac{1}{s-a}$	$\mathrm{Re}\{s\} > a$
$-e^{at}\,u(-t)$	$\frac{1}{s-a}$	$\mathrm{Re}\{s\} < a$
$e^{j\omega_0 t}\,u(t)$	$\frac{1}{s-j\omega_0}$	$\mathrm{Re}\{s\} > 0$
$e^{-\lvert t\rvert}$	$\frac{-2}{s^2-1}$	$-1 < \mathrm{Re}\{s\} < 1$
$\Pi\left(\frac{t-\tau/2}{\tau}\right)$	$\frac{1-e^{-s\tau}}{s}$	$\mathrm{Re}\{s\} > -\infty$
$\cos(\omega_0 t)\,u(t)$	$\frac{s}{s^2+\omega_0^2}$	$\mathrm{Re}\{s\} > 0$
$\sin(\omega_0 t)\,u(t)$	$\frac{\omega_0}{s^2+\omega_0^2}$	$\mathrm{Re}\{s\} > 0$
$e^{at}\cos(\omega_0 t)\,u(t)$	$\frac{s-a}{(s-a)^2+\omega_0^2}$	$\mathrm{Re}\{s\} > a$
$e^{at}\sin(\omega_0 t)\,u(t)$	$\frac{\omega_0}{(s-a)^2+\omega_0^2}$	$\mathrm{Re}\{s\} > a$

B.4 z-Transform Pairs

Signal	Transform	ROC
$\delta[n]$	1	all z
$u[n]$	$\dfrac{z}{z-1}$	$\lvert z\rvert > 1$
$u[-n]$	$\dfrac{-z}{z-1}$	$\lvert z\rvert < 1$
$a^n\, u[n]$	$\dfrac{z}{z-a}$	$\lvert z\rvert > \lvert a\rvert$
$-a^n\, u[-n-1]$	$\dfrac{z}{z-a}$	$\lvert z\rvert < \lvert a\rvert$
$\cos(\Omega_0 n)\, u[n]$	$\dfrac{z\left[z-\cos(\Omega_0)\right]}{z^2-2\cos(\Omega_0)\,z+1}$	$\lvert z\rvert > 1$
$\sin(\Omega_0 n)\, u[n]$	$\dfrac{\sin(\Omega_0)\,z}{z^2-2\cos(\Omega_0)\,z+1}$	$\lvert z\rvert > 1$
$a^n\cos(\Omega_0 n)\, u[n]$	$\dfrac{z\left[z-a\cos(\Omega_0)\right]}{z^2-2a\cos(\Omega_0)\,z+a^2}$	$\lvert z\rvert > \lvert a\rvert$
$a^n\sin(\Omega_0 n)\, u[n]$	$\dfrac{a\sin(\Omega_0)\,z}{z^2-2a\cos(\Omega_0)\,z+a^2}$	$\lvert z\rvert > \lvert a\rvert$
$n\,a^n\, u[n]$	$\dfrac{az}{(z-a)^2}$	$\lvert z\rvert > \lvert a\rvert$

Closed Forms for Sums of Geometric Series

Summations of geometric series appear often in problems involving various transforms and in the analysis of linear systems. In this appendix we give derivations of closed-form formulas for infinite and finite-length geometric series.

C.1 Infinite-Length Geometric Series

Consider the sum of infinite-length geometric series in the form

$$P = \sum_{n=0}^{\infty} a^n \tag{C.1}$$

For the summation in Eqn. (C.1) to converge we must have $|a| < 1$. Assuming that is the case, let us write Eqn. (C.1) in open form:

$$P = 1 + a + a^2 + a^3 + \ldots \tag{C.2}$$

Subtracting unity from both sides of Eqn. (C.2) we obtain

$$P - 1 = a + a^2 + a^3 + \ldots \tag{C.3}$$

in which the terms on the right side of the equal sign have a common factor a. Factoring it out leads to

$$\begin{aligned} P - 1 &= a\left(1 + a + a^2 + a^3 + \ldots\right) \\ &= aP \end{aligned} \tag{C.4}$$

DOI: 10.1201/9781003570462-C

which can be solved for P to yield

$$P = \frac{1}{1-a}, \qquad |a| < 1 \tag{C.5}$$

C.2 Finite-Length Geometric Series

Consider the finite-length sum of a geometric series in the form

$$Q = \sum_{n=0}^{L} a^n \tag{C.6}$$

Unlike P of the previous section, the convergence of Q does not require $|a| < 1$. The only requirement is that $|a| < \infty$. Let us write Q in open form:

$$Q = 1 + a + \ldots + a^L \tag{C.7}$$

Subtracting unity from both sides of Eqn. (C.7) leads to

$$Q - 1 = a + \ldots + a^L \tag{C.8}$$

As before, the terms on the right side of the equal sign have a common factor a which can be factored out to yield

$$\begin{aligned} Q-1 &= a\left(1+\ldots+a^{L-1}\right) \\ &= a\left(Q - a^L\right) \end{aligned} \tag{C.9}$$

which can be solved for Q:

$$Q = \frac{1-a^{L+1}}{1-a} \tag{C.10}$$

Consistency check: If L is increased, in the limit Q approaches P provided that $|a| < 1$, that is

$$\lim_{L\to\infty}\left[Q\right] = P \quad \text{if} \quad |a| < 1 \tag{C.11}$$

C.3 Finite-Length Geometric Series (Alternative Form)

Consider an alternative form of the finite-length geometric series sum in which the lower limit is not equal to zero.

$$Q = \sum_{n=L_1}^{L_2} a^n \tag{C.12}$$

In order to obtain a closed form formula we will apply the variable change $m = n - L_1$ with which Eqn. (C.12) becomes

$$Q = \sum_{m=0}^{L_2-L_1} a^{m+L_1} \tag{C.13}$$

which can be simplified as

$$Q = a^{L_1} \sum_{m=0}^{L_2-L_1} a^m \tag{C.14}$$

The summation on the right side of Eqn. (C.14) is in the standard form of Eqn. (C.6), therefore

$$Q = a^{L_1}\left(\frac{1-a^{L_2-L_1+1}}{1-a}\right) = \frac{a^{L_1} - a^{L_2+1}}{1-a} \tag{C.15}$$

Consistency check: For $L_1 = 0$ and $L_2 = L$, Eqn. (C.15) reduces to Eqn. (C.10).

Orthogonality of Basis Functions

Orthogonality properties are used extensively in the development of Fourier series representations for the analysis of continuous-time and discrete-time signals. The use of an orthogonal set of basis functions ensures that the relationship between a periodic signal and its Fourier series representation is one-to-one, that is, each periodic signal has a unique set of Fourier series coefficients, and each set of coefficients corresponds to a unique signal. In this appendix we will summarize several forms of orthogonality properties and carry out their proofs.

D.1 Orthogonality for Trigonometric Fourier Series

Consider the two real-valued basis function sets defined as

$$\phi_k(t) = \cos(k\omega_0 t)\ ; \quad k = 1, \ldots, \infty \tag{D.1}$$

and

$$\psi_k(t) = \sin(k\omega_0 t)\ ; \quad k = 1, \ldots, \infty \tag{D.2}$$

where ω_0 is the fundamental frequency in rad/s. The period that corresponds to the fundamental frequency ω_0 is

$$T_0 = \frac{2\pi}{\omega_0}$$

It can be shown that each of the two sets in Eqns. (D.1) and (D.2) is orthogonal within itself, that is,

$$\int_0^{T_0} \phi_k(t)\,\phi_m(t)\,dt = \begin{cases} T_0/2\,, & k = m \\ 0\,, & k \neq m \end{cases} \tag{D.3}$$

and

$$\int_0^{T_0} \psi_k(t)\,\psi_m(t)\,dt = \begin{cases} T_0/2\,, & k = m \\ 0\,, & k \neq m \end{cases} \tag{D.4}$$

DOI: 10.1201/9781003570462-D

Furthermore, the two sets are orthogonal to each other:

$$\int_0^{T_0} \phi_k(t)\, \psi_m(t)\, dt = 0\,, \quad \text{for any } k, m \tag{D.5}$$

Writing Eqns. (D.3), (D.4), and (D.5) using the definitions of the basis function sets in Eqns. (D.1) and (D.2), we arrive at the orthogonality properties

$$\int_0^{T_0} \cos(k\omega_0 t)\cos(m\omega_0 t)\, dt = \begin{cases} T_0/2\,, & k=m \\ 0\,, & k\neq m \end{cases} \tag{D.6}$$

$$\int_0^{T_0} \sin(k\omega_0 t)\sin(m\omega_0 t)\, dt = \begin{cases} T_0/2\,, & k=m \\ 0\,, & k\neq m \end{cases} \tag{D.7}$$

$$\int_0^{T_0} \cos(k\omega_0 t)\sin(m\omega_0 t)\, dt = 0\,, \quad \text{for any } k, m \tag{D.8}$$

Proof of Eqn. (D.6):

Using the trigonometric identity in Eqn. (B.4), Eqn. (D.6) becomes

$$\begin{aligned}\int_0^{T_0} \cos(k\omega_0 t)\cos(m\omega_0 t)\, dt =& \frac{1}{2}\int_0^{T_0} \cos\big((k+m)\,\omega_0 t\big)\, dt \\ &+ \frac{1}{2}\int_0^{T_0} \cos\big((k-m)\,\omega_0 t\big)\, dt\end{aligned} \tag{D.9}$$

Let us first assume $k \neq m$. The integrand of the first integral on the right side is periodic with period

$$\frac{2\pi}{(k+m)\,\omega_0} = \frac{T_0}{(k+m)}$$

The integrand $\cos\big((k+m)\,\omega_0 t\big)$ has $(k+m)$ full periods in the interval from 0 to T_0. Therefore the result of the first integral is zero. Similarly the integrand of the second integral on the right side of Eqn. (D.9) is periodic with period

$$\frac{2\pi}{|k-m|\,\omega_0} = \frac{T_0}{|k-m|}$$

The integrand $\cos\big((k-m)\,\omega_0 t\big)$ has $|k-m|$ full periods in the interval from 0 to T_0. Therefore the result of the second integral is also zero. If $k=m$, then Eqn. (D.9) becomes

$$\begin{aligned}\int_0^{T_0} \cos(k\omega_0 t)\cos(m\omega_0 t)\, dt =& \int_0^{T_0} \cos^2(k\omega_0 t)\, dt \\ =& \frac{1}{2}\int_0^{T_0} \cos(2k\omega_0 t)\, dt + \frac{1}{2}\int_0^{T_0} dt \\ =& \frac{T_0}{2}\end{aligned} \tag{D.10}$$

completing the proof of Eqn. (D.6). Eqns. (D.7) and (D.8) can be proven similarly, using the trigonometric identities given by (B.5) and (B.6) and recognizing integrals that span an integer number of periods of sine and cosine functions.

D.2 Orthogonality for Exponential Fourier Series

Consider the set of complex periodic basis functions

$$\phi_k(t) = e^{jk\omega_0 t}\,; \quad k = -\infty,\ldots,\infty \tag{D.11}$$

where the parameter ω_0 is the fundamental frequency in rad/s as in the case of the trigonometric set of basis functions of the previous section, and $T_0 = 2\pi/\omega_0$ is the corresponding period. It can be shown that this basis function set in Eqn. (D.11) is orthogonal in the sense

$$\int_0^{T_0} \phi_k(t)\,\phi_m^*(t),dt = \begin{cases} T_0\,, & k = m \\ 0\,, & k \neq m \end{cases} \tag{D.12}$$

The second term in the integrand is conjugated due to the fact that we are working with a complex set of basis functions. Writing Eqn. (D.12) in open form using the definition of the basis function set in Eqns. (D.11) we arrive at the orthogonality property

$$\int_0^{T_0} e^{jk\omega_0 t}\,e^{-jm\omega_0 t}\,dt = \begin{cases} T_0\,, & k = m \\ 0\,, & k \neq m \end{cases} \tag{D.13}$$

Proof of Eqn. (D.13):

Let us combine the exponential terms in the integral of Eqn. (D.13) and then apply Euler's formula to write it as

$$\int_0^{T_0} e^{jk\omega_0 t}\,e^{-jm\omega_0 t}\,dt = \int_0^{T_0} e^{j(k-m)\omega_0 t}\,dt$$

$$= \int_0^{T_0} \cos\big((k-m)\,\omega_0 t\big)\,dt + j\int_0^{T_0} \sin\big((k-m)\,\omega_0 t\big)\,dt \tag{D.14}$$

If $k \neq m$, both integrands on the right side of Eqn. (D.14) are periodic with period

$$\frac{2\pi}{|k-m|\,\omega_0} = \frac{T_0}{|k-m|}$$

Both integrands $\cos\big((k-m)\,\omega_0 t\big)$ and $\sin\big((k-m)\,\omega_0 t\big)$ have exactly $|k-m|$ full periods in the interval from 0 to T_0. Therefore the result of each integral is zero. If $k = m$, we have

$$\int_0^{T_0} e^{jk\omega_0 t}\,e^{-jm\omega_0 t}\,dt = \int_0^{T_0} e^{j(0)\omega_0 t}\,dt = \int_0^{T_0} dt = T_0 \tag{D.15}$$

which completes the proof for Eqn. (D.13).

D.3 Orthogonality for Discrete-Time Fourier Series

Consider the set of discrete-time complex periodic basis functions

$$w_N^k = e^{-j\frac{2\pi}{N}k}\,; \quad k = 0,\ldots,N-1 \tag{D.16}$$

Parameters N and k are both integers. It can be shown that this basis function set in Eqn. (D.16) is orthogonal in the sense

$$\sum_{n=0}^{N-1} e^{j(2\pi/N)kn}\, e^{-j(2\pi/N)mn} = \begin{cases} 0\,, & k \neq m \\ N\,, & k = m \end{cases} \tag{D.17}$$

Proof of Eqn. (D.17):

The summation in Eqn. (D.17) can be put into a closed form using the finite-length geometric series formula in Eqn. (C.10) to obtain

$$\begin{aligned}\sum_{n=0}^{N-1} e^{j(2\pi/N)kn}\, e^{-j(2\pi/N)mn} &= \sum_{n=0}^{N-1} e^{j(2\pi/N)\,(k-m)n} \\ &= \frac{1 - e^{j(2\pi/N)\,(k-m)N}}{1 - e^{j(2\pi/N)\,(k-m)}}\end{aligned}$$

which can be simplified as

$$\sum_{n=0}^{N-1} e^{j(2\pi/N)kn}\, e^{-j(2\pi/N)mn} = \frac{1 - e^{j\,2\pi\,(k-m)}}{1 - e^{(j2\pi/N)\,(k-m)}} \tag{D.18}$$

Let us begin with the case $k \neq m$. The numerator of the fraction on the right side of Eqn. (D.18), is equal to zero. Since both integers k and m are in the interval $k, m = 0, \ldots, N-1$ their absolute difference $|k-m|$ is in the interval $n = 1, \ldots, N-1$. Therefore the denominator of the fraction is nonzero, and the result is zero proving the first part of Eqn. (D.17).

If $k = m$, then the denominator of the fraction in Eqn. (D.18) also becomes zero and L'Hospital's rule must be used leading to

$$\sum_{n=0}^{N-1} e^{j(2\pi/N)kn}\, e^{-j(2\pi/N)mn} = N\,, \qquad \text{for } k = m \tag{D.19}$$

Partial Fraction Expansion

The technique of partial fraction expansion is very useful in the study of signals and systems, particularly in finding the time-domain signals that correspond to inverse Laplace transforms and inverse z-transforms of rational functions, that is, a ratio of two polynomials. It is based on the idea of writing a rational function, say of s or z, as a linear combination of simpler rational functions the inverse transforms of which can be easily determined. Since both the Laplace transform and the z-transform are linear transforms, writing the rational function $X(z)$ or $X(z)$ as a linear combination of simpler functions, and finding the inverse transform of each of those simpler functions allows us to construct the inverse transform of the original function in a straightforward manner. The main problem is in determining which simpler functions should be used in the expansion, and how the weight of each should be.

In this appendix we will present the details of the partial fraction expansion technique first from the perspective of the Laplace transform for continuous-time signals, and then from the perspective of the z-transform for discrete-time signals.

E.1 Partial Fractions for Continuous-Time Signals and Systems

As an example, consider a rational function $X(s)$ with two poles, expressed in the form

$$X(s) = \frac{B(s)}{(s-a)(s-b)} = \frac{k_1}{s-a} + \frac{k_2}{s-b} \tag{E.1}$$

The rational function $X(s)$ is expressed as a weighted sum of partial fractions

$$\frac{1}{s-a}$$

and

$$\frac{1}{s-b}$$

DOI: 10.1201/9781003570462-E

with weights k_1 and k_2, respectively. The weights k_1 and k_2 are called the *residues* of the partial fraction expansion. Before we address the issue of determining the values of the residues, a question that comes to mind is: What limitations, if any, should be placed on the numerator polynomial $B(s)$ for the expansion in Eqn. (E.1) to be valid? In order to find the answer to this question we will combine the two partial fractions under a common denominator to obtain

$$\frac{k_1}{s-a}+\frac{k_2}{s-b}=\frac{(k_1+k_2)\,s-(k_1 b+k_2 a)}{(s-a)\,(s-b)} \tag{E.2}$$

It is obvious from Eqn. (E.2) that the numerator polynomial $B(s)$ may not be higher than first-order since no s^2 term or higher-order terms can be obtained in the process of combining the two partial fractions under a common denominator. Consider a more general case of a rational function with P poles in the form

$$\begin{aligned} X(s) &= \frac{B(s)}{(s-s_1)\,(s-s_2)\ldots(s-s_P)} \\ &= \frac{k_1}{s-s_1}+\frac{k_2}{s-s_2}+\ldots+\frac{k_P}{s-s_P} \end{aligned} \tag{E.3}$$

If the P terms in the partial fraction expansion of Eqn. (E.3) are combined under a common denominator, the highest order numerator term that could be obtained is s^{P-1}.

> As a general rule, for the partial fraction expansion in the form of Eqn. (E.3) to be possible, the numerator order must be less than the denominator order.

If this is not the case, some additional steps need to be taken before partial fraction expansion can be used. Suppose $X(s)$ is a rational function in which the numerator order is greater than or equal to the denominator order. Let

$$X(s)=\frac{b_Q s^Q+b_{Q-1}s^{Q-1}+\ldots+b_1 s+b_0}{a_P s^P+a_{P-1}s^{P-1}+\ldots+a_1 s+a_0} \tag{E.4}$$

with $Q \ge P$. We can use long division on numerator and denominator polynomials to write $X(s)$ in the modified form

$$X(s)=C(s)+\frac{\bar{b}_{P-1}s^{P-1}+\bar{b}_{P-2}s^{P-2}+\ldots+\bar{b}_1 s+\bar{b}_0}{a_P s^P+a_{P-1}s^{P-1}+\ldots+a_1 s+a_0} \tag{E.5}$$

where $C(s)$ is a polynomial of s. Now the new function

$$\bar{X}(s)=X(s)-C(s)=\frac{\bar{b}_{P-1}s^{P-1}+\bar{b}_{P-2}s^{P-2}+\ldots+\bar{b}_1 s+\bar{b}_0}{a_P s^P+a_{P-1}s^{P-1}+\ldots+a_1 s+a_0} \tag{E.6}$$

has a numerator that is of lower order than its denominator, and can therefore be expanded into partial fractions. Examples E.1 and E.2 will illustrate this.

Example E.1:

Consider the function

$$X(s)=\frac{(s+1)\,(s+2)}{(s+5)\,(s+6)}$$

Since both the numerator and the denominator are of second order, the function cannot be expanded into partial fractions directly. In other words, an attempt to express $X(s)$ as

$$X(s) \stackrel{?}{=} \frac{k_1}{s+5} + \frac{k_2}{s+6}$$

would fail since the s^2 term in the numerator cannot be matched by combining the two partial fractions under a common denominator. Instead, we will use long division to extract a polynomial $C(s)$ from $X(s)$ so that the difference $X(s) - C(s)$ can be expanded into partial fractions. Let us first multiply out the factors in $X(s)$ to write it as

$$X(s) = \frac{s^2+3s+2}{s^2+11s+30}$$

Now we can divide the numerator polynomial by the denominator polynomial using long division to obtain

$$\begin{array}{rrrl|rrr} & & & & 1 & & \\ \cline{5-7} s^2 & +11s & +30 & & s^2 & +3s & +2 \\ & & & & s^2 & +11s & +30 \\ \cline{5-7} & & & & & -8s & -28 \end{array} \tag{E.7}$$

Thus, the function $X(s)$ can be written in the equivalent form

$$X(s) = 1 + \frac{-8s-28}{s^2+11s+30} \tag{E.8}$$

Let

$$\bar{X}(s) = X(s) - 1 = \frac{-8s-28}{s^2+11s+30} \tag{E.9}$$

We can now expand the function $\bar{X}(s)$ into partial fractions as

$$\bar{X}(s) = \frac{k_1}{s+5} + \frac{k_2}{s+6} \tag{E.10}$$

and write $X(s)$ as

$$X(s) = 1 + \bar{X}(s) = 1 + \frac{k_1}{s+5} + \frac{k_2}{s+6} \tag{E.11}$$

Example E.2:

Consider the function

$$X(s) = \frac{(s+1)(s+2)(s+3)}{(s+5)(s+6)}$$

the factors of which can be multiplied to yield

$$X(s) = \frac{s^3+6s^2+11s+6}{s^2+11s+30}$$

Using long division on $X(s)$ we obtain

$$\begin{array}{r|rrrr} & s-5 & & & \\ \hline s^2 \quad +11s \quad +30 & s^3 & +6s^2 & +11s & +6 \\ & s^3 & +11s^2 & +30s & \\ \hline & & -5s^2 & -19s & +6 \\ & & -5s^2 & -55s & -150 \\ \hline & & & 36s & +156 \end{array} \tag{E.12}$$

Thus, the function $X(s)$ can be written in the equivalent form

$$X(s) = s-5+\bar{X}(s) = s-5+\frac{36s+156}{s^2+11s+30} \tag{E.13}$$

and can be expressed using partial fractions as

$$X(s) = s-5+\frac{k_1}{s+5}+\frac{k_2}{s+6} \tag{E.14}$$

Now that we know how to deal with a function that has a numerator of equal or greater order than its denominator, we will focus our attention on determining the residues. For the purpose of determining the residues, we will assume that the order of the numerator polynomial is less than that of the denominator polynomial. In addition, we will initially consider a denominator polynomial with only simple roots. Thus, the function $X(s)$ is in the form

$$X(s) = \frac{B(s)}{(s-s_1)(s-s_2)\dots(s-s_P)} \tag{E.15}$$

to be expanded into partial fractions in the form

$$X(s) = \frac{k_1}{s-s_1}+\frac{k_2}{s-s_2}+\dots+\frac{k_P}{s-s_P} \tag{E.16}$$

Let us multiply both sides of Eqn. (E.16) by $(s-s_1)$ to obtain

$$(s-s_1)\,X(s) = k_1+\frac{k_2\,(s-s_1)}{s-s_2}+\dots+\frac{k_P\,(s-s_1)}{s-s_P} \tag{E.17}$$

If we now set $s=s_1$ on both sides of Eqn. (E.17), we get

$$\begin{aligned} (s-s_1)\,X(s)\Big|_{s=s_1} &= k_1\Big|_{s=s_1}+\frac{k_2\,(s-s_1)}{s-s_2}\Big|_{s=s_1}+\dots+\frac{k_P\,(s-s_1)}{s-s_P}\Big|_{s=s_1} \\ &= k_1 \end{aligned} \tag{E.18}$$

Thus, the residue of the pole at $s=s_1$ can be found by multiplying $X(s)$ with the factor $(s-s_1)$ and then setting $s=s_1$. Using the form of $X(s)$ in Eqn. (E.15) the residue k_1 is found as

$$(s-s_1)\,X(s)\Big|_{s=s_1} = \frac{B(s_1)}{(s_1-s_2)\dots(s_1-s_P)} \tag{E.19}$$

which amounts to canceling the $(s-s_1)$ term in the denominator, and evaluating what is left for $s=s_1$. Generalizing this result, the residue of the n-th pole is found as

$$k_n = (s-s_n)\,X(s)\Big|_{s=s_n} \qquad n=1,\dots,P \tag{E.20}$$

Example E.3:

Find a partial fraction expansion for the rational function

$$X(s) = \frac{(s+1)\,(s+2)}{(s+3)\,(s+4)\,(s+5)}$$

Solution: The partial fraction expansion we are looking for is in the form

$$X(s) = \frac{k_1}{s+3} + \frac{k_2}{s+4} + \frac{k_3}{s+5} \tag{E.21}$$

We will use Eqn. (E.20) to determine the residues. The first residue k_1 is obtained as

$$\begin{aligned} k_1 &= (s+3)\,X(s)\Big|_{s=-3} \\ &= \frac{(s+1)\,(s+2)}{(s+4)\,(s+5)}\bigg|_{s=-3} \\ &= \frac{(-3+1)\,(-3+2)}{(-3+4)\,(-3+5)} = 1 \end{aligned}$$

The remaining two residues k_2 and k_3 are obtained similarly:

$$\begin{aligned} k_2 &= (s+4)\,X(s)\Big|_{s=-4} \\ &= \frac{(s+1)\,(s+2)}{(s+3)\,(s+5)}\bigg|_{s=-4} \\ &= \frac{(-4+1)\,(-4+2)}{(-4+3)\,(-4+5)} = -6 \end{aligned}$$

$$\begin{aligned} k_3 &= (s+5)\,X(s)\Big|_{s=-5} \\ &= \frac{(s+1)\,(s+2)}{(s+3)\,(s+4)}\bigg|_{s=-5} \\ &= \frac{(-5+1)\,(-5+2)}{(-5+3)\,(-5+4)} = 6 \end{aligned}$$

Thus, the partial fraction expansion for $X(z)$ is

$$X(s) = \frac{1}{s+3} - \frac{6}{s+4} + \frac{6}{s+5}$$

Next we will consider a rational function the denominator of which has repeated roots. Let $X(s)$ be a rational function with a pole of order r at $s = s_1$.

$$X(s) = \frac{B(s)}{(s-s_1)^r\,(s-s_2)} \tag{E.22}$$

The partial fraction expansion for $X(s)$ needs to be in the form

$$X(s) = \frac{k_{1,1}}{s-s_1} + \frac{k_{1,2}}{(s-s_1)^2} + \ldots + \frac{k_{1,r}}{(s-s_1)^r} + \frac{k_2}{s-s_2} \tag{E.23}$$

For the multiple pole at $s = s_1$ as many terms as the multiplicity of the pole are needed so that combining the terms on the right side of Eqn. (E.23) can yield a result that matches the function $X(s)$ given by Eqn. (E.22).

The residue k_2 of the single pole at $s = s_2$ can easily be determined as discussed above. We will focus our attention on determining the residues of the r-th order pole at $s = s_1$. Let us begin by multiplying both sides of Eqn. (E.23) by $(s - s_1)^r$ to obtain

$$(s-s_1)^r\,X(s) = k_{1,1}\,(s-s_1)^{r-1} + k_{1,2}\,(s-s_1)^{r-2} + \ldots + k_{1,r} + \frac{k_2\,(s-s_1)^r}{s-s_2} \tag{E.24}$$

If we now set $s = s_1$ on both sides of Eqn. (E.24), we would get

$$\begin{aligned}(s-s_1)^r\,X(s)\Big|_{s=s_1} &= k_{1,1}\,(s-s_1)^{r-1}\Big|_{s=s_1} + k_{1,2}\,(s-s_1)^{r-2}\Big|_{s=s_1} + \ldots + k_{1,r}\Big|_{s=s_1} + \left.\frac{k_2\,(s-s_1)^r}{s-s_2}\right|_{s=s_1}\\ &= k_{1,r}\end{aligned} \tag{E.25}$$

It can be shown that the other residues for the pole at $s = s_1$ can be found as follows:

$$\begin{aligned}
k_{1,r-1} &= \frac{d}{ds}\left[(s-s_1)^r\,X(s)\right]\Big|_{s=s_1}\\
k_{1,r-2} &= \frac{1}{2}\,\frac{d^2}{ds^2}\left[(s-s_1)^r\,X(s)\right]\Big|_{s=s_1}\\
&\;\;\vdots\\
k_{1,2} &= \frac{1}{(r-2)!}\,\frac{d^{r-2}}{ds^{r-2}}\left[(s-s_1)^r\,X(s)\right]\Big|_{s=s_1}\\
k_{1,1} &= \frac{1}{(r-1)!}\,\frac{d^{r-1}}{ds^{r-1}}\left[(s-s_1)^r\,X(s)\right]\Big|_{s=s_1}
\end{aligned}$$

Generalizing these results, the residues of a pole of multiplicity r are calculated using

$$k_{1,n} = \frac{1}{(r-n)!}\,\frac{d^{r-n}}{ds^{r-n}}\left[(s-s_1)^r\,X(s)\right]\Big|_{s=s_1} \qquad n = 1,\ldots,r \tag{E.26}$$

In using the general formula in Eqn. (E.26) for $n = r$ we need to remember that $0! = 1$ and

$$\frac{d^0}{ds^0}\left[(s-s_1)^r\,X(s)\right] = (s-s_1)^r\,X(s) \tag{E.27}$$

Example E.4:

Find a partial fraction expansion for the rational function

$$X(s) = \frac{s}{(s+1)^3\,(s+2)}$$

Solution: The partial fraction expansion for $X(s)$ is in the form

$$X(s) = \frac{k_{1,1}}{s+1} + \frac{k_{1,2}}{(s+1)^2} + \frac{k_{1,3}}{(s+1)^3} + \frac{k_2}{s+2} \tag{E.28}$$

The residue of the single pole at $s = -2$ is easily found using the technique developed earlier:

$$k_2 = (s+2)\,X(s)\Big|_{s=-2} = \left.\frac{s}{(s+1)^3}\right|_{s=-2} = \frac{-2}{(-2+1)^3} = 2$$

We will use Eqn. (E.26) to find the residues of the third-order pole at $s = -1$.

$$k_{1,3} = (s+1)^3\,X(s)\Big|_{s=-1} = \left.\frac{s}{(s+2)}\right|_{s=-1} = \frac{-1}{(-1+2)} = -1$$

$$k_{1,2}=\frac{d}{ds}\left[(s+1)^3\,X(s)\right]\bigg|_{s=-1}=\frac{d}{ds}\left[\frac{s}{s+2}\right]\bigg|_{s=-1}=\frac{2}{(s+2)^2}\bigg|_{s=-1}=2$$

$$k_{1,1}=\frac{1}{2!}\frac{d^2}{ds^2}\left[(s+1)^3\,X(s)\right]\bigg|_{s=-1}=\frac{1}{2!}\frac{d^2}{ds^2}\left[\frac{s}{s+2}\right]\bigg|_{s=-1}=\frac{-2}{(s+2)^3}\bigg|_{s=-1}=-2$$

Thus, the partial fraction expansion for $X(z)$ is

$$X(s)=-\frac{2}{s+1}+\frac{2}{(s+1)^2}-\frac{1}{(s+1)^3}+\frac{2}{s+2}$$

E.2 Partial Fraction Expansion for Discrete-Time Signals and Systems

The residues of the partial fraction expansion for rational z-transforms are obtained using the same techniques employed in the previous section. One subtle difference is that, when working with the inverse z-transform, we often need the partial fraction expansion in the form

$$\begin{aligned}X(z)&=\frac{B(z)}{(z-z_1)(z-z_2)\dots(z-z_P)}\\&=\frac{k_1\,z}{z-z_1}+\frac{k_2\,z}{z-z_2}+\dots+\frac{k_P\,z}{z-z_P}\end{aligned}\qquad\text{(E.29)}$$

In order to use the residue formulas established in the previous section and still obtain the expansion in the form of Eqn. (E.29) with a z factor in each term, we simply expand $X(z)/z$ into partial fractions:

$$\begin{aligned}\frac{X(z)}{z}&=\frac{B(z)}{(z-z_1)(z-z_2)\dots(z-z_P)\,z}\\&=\frac{k_1}{z-z_1}+\frac{k_2}{z-z_2}+\dots+\frac{k_P}{z-z_P}+\frac{k_{P+1}}{z}\end{aligned}\qquad\text{(E.30)}$$

The last term in Eqn. (E.30) may or may not be needed depending on whether the numerator polynomial $B(z)$ has a root at $z=0$ or not. Once the residues of the expansion in Eqn. (E.30) are found, we can revert to the form in Eqn. (E.29) by multiplying both sides with z.

Review of Matrix Algebra

In this section we present some basic definitions for matrix algebra.

Matrix

A matrix is a collection of real or complex numbers arranged to form a rectangular grid:

$$\mathbf{A} = \begin{bmatrix} a_{11} & a_{12} & \dots & a_{1M} \\ a_{21} & a_{22} & \dots & a_{2M} \\ \vdots & & & \\ a_{N1} & a_{N2} & \dots & a_{NM} \end{bmatrix} \tag{F.1}$$

A total of NM numbers arranged into N rows and M columns as shown in Eqn. (F.1) form a $N \times M$ matrix. The numbers that make up the matrix are referred to as the *elements* of the matrix. Two subscripts are used to specify the placement of an element in the matrix. The element at the in row-i and column-j is denoted by a_{ij}.

Vector

A vector is a special matrix with only one row or only one column. A matrix with only one row (a $1 \times N$ matrix) is called a *row vector* of length N. A matrix with only one column (a $M \times 1$ matrix) is called a *column vector* of length M.

Square matrix

A *square matrix* is one in which the number of rows is equal to the number of columns. An example is

$$\mathbf{A} = \begin{bmatrix} 5 & 2 & 0 & 4 \\ 3 & -1 & 0 & 0 \\ 0 & 1 & 1 & 3 \\ -7 & 4 & 3 & 5 \end{bmatrix} \tag{F.2}$$

DOI: 10.1201/9781003570462-F

Diagonal matrix

A *diagonal matrix* is a square matrix in which all elements that are not on the main diagonal are equal to zero. The main diagonal of a square matrix is the diagonal from the top left corner to the bottom right corner. Elements on the main diagonal have identical row and column indices such as $a_{11}, a_{22}, \ldots, a_{NN}$. An example is

$$\mathbf{A} = \begin{bmatrix} 5 & 0 & 0 & 0 \\ 0 & -1 & 0 & 0 \\ 0 & 0 & 1 & 0 \\ 0 & 0 & 0 & 4 \end{bmatrix} \tag{F.3}$$

Identity matrix

An *identity matrix* is a diagonal matrix in which all diagonal elements are equal to unity. For example, the identity matrix of order 3 is

$$\mathbf{I_3} = \begin{bmatrix} 1 & 0 & 0 \\ 0 & 1 & 0 \\ 0 & 0 & 1 \end{bmatrix} \tag{F.4}$$

Equality of two matrices

Two matrices **A** and **B** are equal to each other if they have the same dimensions and if all corresponding elements are equal, that is

$$\mathbf{A} = \mathbf{B} \qquad \Longrightarrow \qquad a_{ij} = b_{ij} \quad \text{for } i = 1, \ldots, M \text{ and } j = 1, \ldots, N \tag{F.5}$$

Trace of a square matrix

The *trace* of a square matrix is defined as the arithmetic sum of all of its elements on the main diagonal.

$$\text{Trace}\,(\mathbf{A}) = \sum_{i=1}^{N} a_{ii} \tag{F.6}$$

Transpose of a matrix

The *transpose* of a matrix **A** is denoted by $\mathbf{A}^T$, and is obtained by interchanging rows and columns so that row-i of **A** becomes column-i of $\mathbf{A}^T$. For example, the transpose of the matrix **A** in Eqn. (F.2) is

$$\mathbf{A}^T = \begin{bmatrix} 5 & 3 & 0 & -7 \\ 2 & -1 & 1 & 4 \\ 0 & 0 & 1 & 3 \\ 4 & 0 & 3 & 5 \end{bmatrix} \tag{F.7}$$

Determinant of a square matrix

Each square matrix has a scalar value associated with it referred to as the *determinant*. Methods for computing determinants can be found in most texts on linear algebra. As an example, the determinant of the matrix

$$\mathbf{A} = \begin{bmatrix} 5 & 2 \\ 3 & -1 \end{bmatrix} \tag{F.8}$$

is computed as

$$|\mathbf{A}| = (5)\,(-1) - (2)\,(3) = -11 \tag{F.9}$$

Minors of a square square matrix

The row-i column-j minor m_{ij} of a square matrix **A** is defined as the determinant of the sub-matrix obtained by deleting row-i and column-j from the matrix **A**. Consider, for example the square matrix

$$\mathbf{A} = \begin{bmatrix} 5 & 2 & 4 \\ 3 & -1 & 2 \\ -1 & 0 & 3 \end{bmatrix} \tag{F.10}$$

The minor m_{23} is found by deleting row-2 and column-3 and computing the determinant of the matrix left behind as

$$m_{23} = \begin{vmatrix} 5 & 2 \\ -1 & 0 \end{vmatrix} = 2 \tag{F.11}$$

Cofactors of a square square matrix

The row-i column-j cofactor Δ_{ij} of a square matrix **A** is defined as

$$\Delta_{ij} = (-1)^{i+j}\, m_{ij} \tag{F.12}$$

Using the matrix in Eqn. (F.10) as an example, the cofactor Δ_{23} is

$$\Delta_{23} = (-1)^{2+3}\, m_{23} = -2 \tag{F.13}$$

Adjoint of a square square matrix

The adjoint of a square matrix **A** is found by first transposing **A** and then replacing each element by its cofactor. For example, the adjoint of the matrix **A** in Eqn. (F.10) is

$$\text{Adj}\,(\mathbf{A}) = \begin{bmatrix} -3 & -6 & 8 \\ -11 & 19 & 2 \\ -1 & -2 & -11 \end{bmatrix} \tag{F.14}$$

Scaling a matrix

Scaling a matrix by a constant scale factor means multiplying each element of the matrix with that scale factor.

Addition of two matrices

Addition of two matrices is only defined for matrices with identical dimensions. The sum of two matrices **A** and **B** is the matrix **C** in which each element is equal to the sum of corresponding elements of **A** and **B**.

$$\mathbf{C} = \mathbf{A} + \mathbf{B} \quad \Longrightarrow \quad c_{ij} = a_{ij} + b_{ij} \quad \text{for } i = 1,\ldots,M \text{ and } j = 1,\ldots,N \tag{F.15}$$

Scalar product of two vectors

The scalar product of a $1 \times N$ row vector **x** and a $N \times 1$ column vector **y** is defined as

$$\mathbf{x} \cdot \mathbf{y} = \sum_{i=1}^{N} x_i\, y_i \tag{F.16}$$

Multiplication of two matrices

The product of an $M \times K$ matrix **A** and a $K \times N$ matrix **B** is a $M \times N$ matrix **C** the elements of which are computed as

$$c_{ij} = \sum_{k=1}^{K} a_{ik}\, b_{kj} \tag{F.17}$$

In other words, the row-i column-j element of matrix $\mathbf{C}$ is the scalar product of row-i of of matrix $\mathbf{A}$ and column-j of matrix $\mathbf{B}$.

Inversion of a matrix

The inverse of a square matrix $\mathbf{A}$ denoted by $\mathbf{A}^{-1}$ is also a square matrix. It satisfies the equation

$$\mathbf{A}^{-1}\mathbf{A} = \mathbf{A}\mathbf{A}^{-1} = \mathbf{I} \tag{F.18}$$

It is computed by scaling the adjoint matrix by the reciprocal of the determinant, that is,

$$\mathbf{A}^{-1} = \frac{\text{Adj}\,(\mathbf{A})}{|\mathbf{A}|} \tag{F.19}$$

The inverse does not exist if the determinant of $\mathbf{A}$ is equal to zero. Such a matrix is called a *singular* matrix.

Eigenvalues and eigenvectors

The scalar λ and the $N \times 1$ vector $\mathbf{v}$ are called an *eigenvalue* and an *eigenvector* of the $N \times N$ matrix $\mathbf{A}$ respectively if they satisfy the equation

$$\mathbf{A}\mathbf{v} = \lambda\mathbf{v} \tag{F.20}$$

An $N \times N$ matrix $\mathbf{A}$ has N eigenvalues and associated eigenvectors although they are not necessarily unique.

APPENDIX G

ANSWERS/PARTIAL SOLUTIONS TO SELECTED PROBLEMS

G.1 Chapter 1 Problems

1.1.

a.

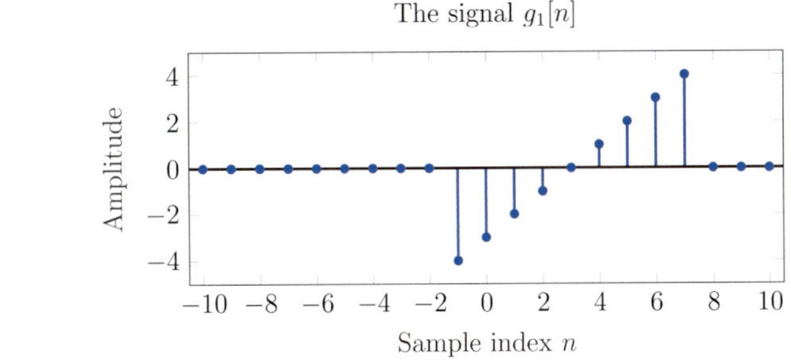

b.

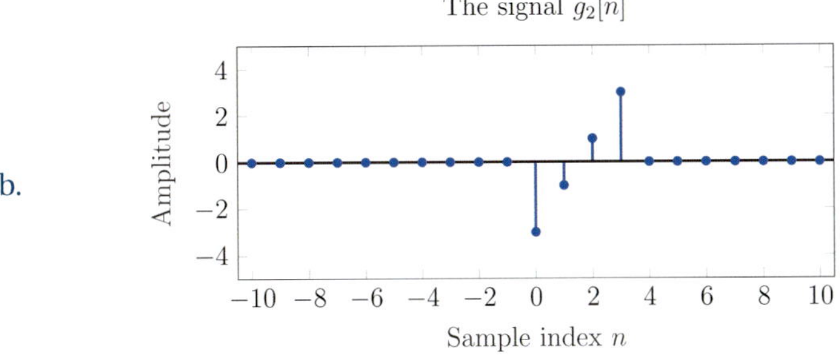

DOI: 10.1201/9781003570462-G

1.3.

a. $x_1[n] = \{ 1.2, 0.2, 0.7, 3.1, 2.7, \underset{\substack{\uparrow \\ n=0}}{2.2}, -0.7, 2.3, 4.9. 6.1, 3.3, 1.5, 0 \}$

b. $x_2[n] = \{ \underset{\substack{\uparrow \\ n=3}}{-2.76}, 0.5, 2.47, 1.98, 0.56 \}$

1.6. The period is $N = 46$ samples.

1.7.

a. $E_x = 40$

b. $E_x = 2.7708$

1.11.

a. $x[n] = 0.9220^n \left[\cos(0.2187n) + j\sin(0.2187n)\right] u[n]$

1.12.

```
x = @(n) n.*((n>=-4)&(n<=4));
n = [-15:15];
g1 = x(n-3);
stem(n,g1);
axis([-15.5,15.5,-5,5]);
title('The signal g_{1}[n]');
xlabel('Sample index n');
ylabel('Amplitude');
```

1.17.

a.

```
x = @(n) 0.8*ss_dramp(n+5)-1.2*ss_dramp(n)+0.4*ss_dramp(n-10);
n = [-15:15];
g1 = x(n).*ss_dstep(n);
stem(n,g1);
title('The signal g_1[n]');
xlabel('Sample index n');
ylabel('Amplitude');
```

b.

```
g2 = x(n).*(ss_dstep(n)-ss_dstep(n-5));
stem(n,g2);
title('The signal g_2[n]');
xlabel('Sample index n');
ylabel('Amplitude');
```

G.2 Chapter 2 Problems

2.1.

a. The system is not time-invariant.

b. The system is time-invariant.

2.2.

a. The system is time invariant.

b. The system is not time invariant.

2.5.

a. $y[n] = \{ \underset{\substack{\uparrow \\ n=0}}{2}, 3, 0, -1 \}$

b. $y[n] = \{ \underset{\substack{\uparrow \\ n=0}}{2}, -3, -1, 3, -1 \}$

2.6.

a. $y[n] = \{ \underset{\substack{\uparrow \\ n=0}}{5}, 10, 15 \}$

b. $y[n] = \{ \underset{\substack{\uparrow \\ n=0}}{3}, 12, 15, 4 \}$

2.7. $y[n] = \{ \underset{\substack{\uparrow \\ n=0}}{0.85}, 2, 2.7, 3.2, 3.5, 3.3, 2.55, 1.8, 1, 0.2, 0.1 \}$

2.12.

a. $y[n] = 0.7044\,(0.7)^n - 3.5944\,(-0.9)^n\,, \qquad n \geq 0$

b. $y[n] = 3.3333\,(-0.5)^n - 5.3333\,(-0.8)^n\,, \qquad n \geq 0$

2.13.

a.

$$y[n] = 4.5004\,(0.9220)^n \cos(0.7086n) + 2.4171\,(0.9220)^n \sin(0.7086n) \qquad n \geq 0$$

2.15.

a. $y[n] = -0.3\,(0.6)^n + 2.5\,, \qquad n \geq 0$

2.18.

a.

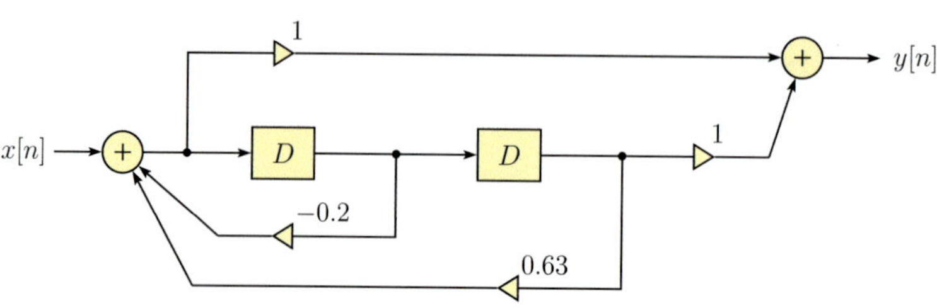

b.

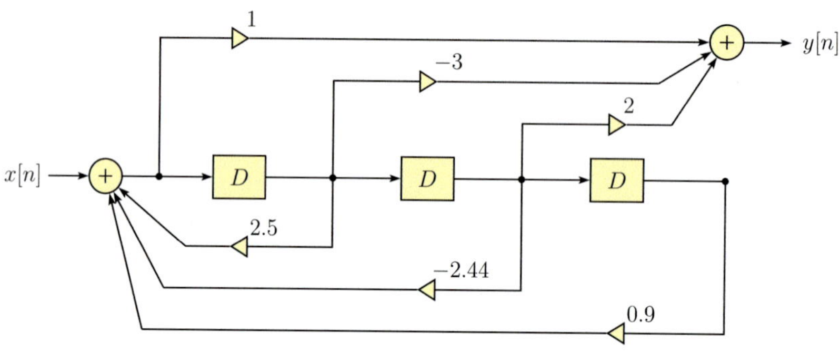

2.19

a. $y[n] = \alpha\, y[n-1] + x[n-1]$

2.23.

a. $h_{eq}[n] = h_1[n] + h_2[n] * h_3[n]$

2.24.

a. $h_{eq}[n] = h_1[n] + h_3[n] * h_2[n] + h_4[n] * h_2[n]$

2.29.

a.

```
n = [0:19];
ynm1 = 5;
ynm2 = -3;
y = [];
for i = 0:19,
  yn = -0.2*ynm1+0.63*ynm2;
  y = [y,yn];
  ynm2 = ynm1;
  ynm1 = yn;
```

```
end;
% Display the results
[n',y']
% Compute the output from analytical solution
y_anl = 0.7044*(0.7).^n-3.5944*(-0.9).^n;
[n',y_anl']
```

2.36.

a.

```
alpha = 0.7;
x = [1,zeros(1,49)];
y = zeros(1,50);
xnm1 = 0;  % The sample x[n-1]
ynm1 = 0;  % The sample y[n-1]
for n=1:50
  y(n) = alpha*ynm1+xnm1;  % y[n] = alpha*y[n-1]+x[n-1]
  ynm1 = y(n)
  xnm1 = x(n);
end
stem([0:49],y);
```

G.3 Chapter 3 Problems

3.1. $c_3 = \frac{1}{2}, \qquad c_{-3} = \frac{1}{2}, \qquad c_k = 0 \quad \text{for all other } k$

3.3.

b.

$$\begin{aligned}
\tilde{c}_0 &= 0.3750 \\
\tilde{c}_1 &= 0.2134 - j\,0.2134 \\
\tilde{c}_2 &= -j\,0.1250 \\
\tilde{c}_3 &= 0.0366 + j\,0.0366 \\
\tilde{c}_4 &= 0.1250 \\
\tilde{c}_5 &= 0.0366 - j\,0.0366 \\
\tilde{c}_6 &= j\,0.1250 \\
\tilde{c}_7 &= 0.2134 + j\,0.2134
\end{aligned}$$

3.5.

$$
\begin{aligned}
\tilde{c}_0 &= 0.1250 \\
\tilde{c}_1 &= -0.3902 - j\,0.2134 \\
\tilde{c}_2 &= 0.2500 - j\,0.1250 \\
\tilde{c}_3 &= 0.1402 + j\,0.0366 \\
\tilde{c}_4 &= -0.1250 \\
\tilde{c}_5 &= 0.1402 - j\,0.0366 \\
\tilde{c}_6 &= 0.2500 + j\,0.1250 \\
\tilde{c}_7 &= -0.3902 + j\,0.2134
\end{aligned}
$$

3.8.

$$
\begin{aligned}
\tilde{c}_0 &= 1.3333 \\
\tilde{c}_1 &= \tilde{c}_{-1} = 0.9330 \\
\tilde{c}_2 &= \tilde{c}_{-2} = 0.2500 \\
\tilde{c}_3 &= \tilde{c}_{-3} = 0.0000 \\
\tilde{c}_4 &= \tilde{c}_{-4} = 0.0833 \\
\tilde{c}_5 &= \tilde{c}_{-5} = 0.0670
\end{aligned}
$$

3.11.

a. $\tilde{y}[n] = \{\ldots,\ \underset{\substack{\uparrow \\ n=0}}{9},\ 10,\ 9,\ 6,\ 3,\ 2,\ 3,\ 6,\ \ldots\}$

3.13.

a. $X(\Omega) = 1 + \cos(\Omega) - j\,\sin(\Omega)$

c. $X(\Omega) = \left[2\,\cos(\Omega/2) + 2\,\cos(3\Omega/2)\right]\,e^{-j3\Omega/2}$

3.14.

a. $X(\Omega) = \dfrac{1}{1 - a e^{-j\Omega}}$

3.16.

a. $X(\Omega) = \dfrac{0.25\,e^{-j2\Omega}}{1 - 0.5\,e^{-j\Omega}}$

3.17.

a. $X(\Omega) = \dfrac{\frac{1}{2}\,e^{j\Omega}}{1-\frac{1}{2}\,e^{j\Omega}}$

b. $X(\Omega) = \dfrac{1}{1-0.8\,e^{j\Omega}}$

3.19.

a. $X(-\Omega)$

b. $X^*(-\Omega)$

3.21.

a. $X(\Omega) = 1+\cos(\Omega)$

3.22.

b. $X(\Omega) = j2\sin(\Omega) + j4\sin(2\Omega) + j6\sin(3\Omega) + j8\sin(4\Omega) + j10\sin(5\Omega)$

3.24.

a. $X(\Omega) = \dfrac{0.7\,e^{-j\Omega}}{\left(1-0.7\,e^{-j\Omega}\right)^2}$

3.26.

a. $X(0) = 5$

b. $X(\pi) = \dfrac{5}{9}$

3.32.

a. $\tilde{r}_{xx}[m] = \dfrac{1}{10}\cos(0.4\pi m)$

3.34.

a. $y[n] = 1.1344\,e^{j(0.2\pi n+0.1807)}$

3.35.

a. $y[n] = 0.7694\,e^{j(0.2\pi n-0.9425)}$

3.39.

a.

```
>> k = [0:7];
>> xn = [0,1,2,3,4,5,6,7];
>> ck = ss_dtfs(xn,k)
```

3.45.

a.

```
n = [0:49];
inp = ss_per(xper,n);
out = [];
ynm1 = 0;
ynm2 = 0;
xnm1 = 0;
for nn=1:50,
  xn = inp(nn);
  yn = xn+2*xnm1-ynm1-0.89*ynm2;
  xnm1 = xn;
  ynm2 = ynm1;
  ynm1 = yn;
  out = [out,yn];
end;
```

G.4 Chapter 4 Problems

4.1.

a.

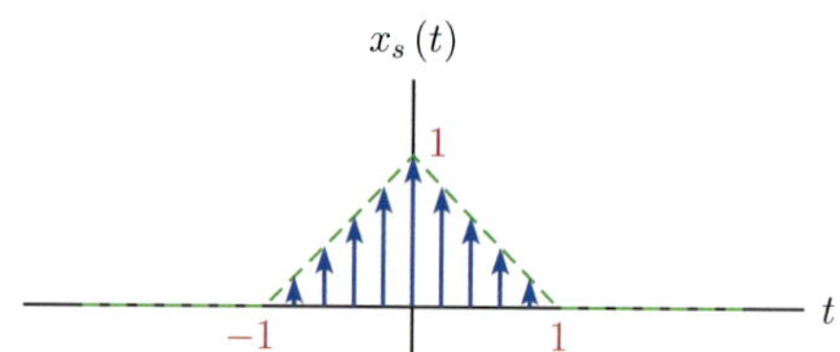

4.3.

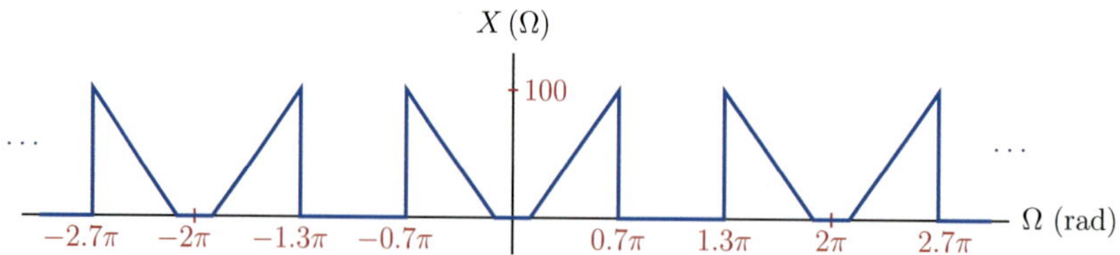

4.6. $X(\Omega) = -j\,\frac{1}{2}\,P(\Omega - \pi/15) + j\,\frac{1}{2}\,P(\Omega + \pi/15)$

4.7.

a. Not bandlimited. Cannot be sampled without loss of information.

c. Bandlimited. $f_{max} = 75$ Hz, $f_s \geq 150$ Hz.

4.8.

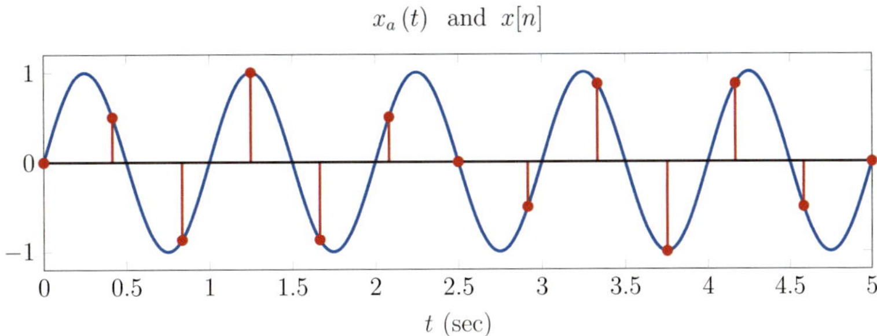

4.10.

a. $f_s = 1250$ Hz.

4.12.

a.

$$X_a(\omega) = \pi\,\delta\left(f - 100\pi\right) + \pi\,\delta\left(f + 100\pi\right) + \pi\,\delta\left(f - 120\pi\right) + \pi\,\delta\left(f + 120\pi\right)$$

4.14.

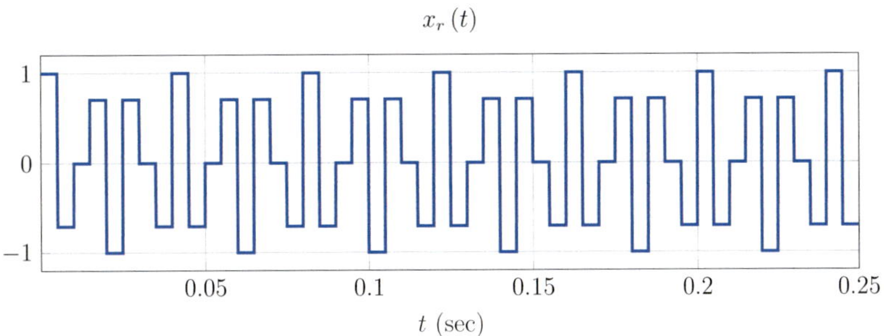

4.18.

a. Not bandlimited. Cannot be downsampled without loss of information.

c. $\Omega_{max} = \pi/3$. May be downsampled with $D = 3$.

4.21.

a.

```
Xa = @(f) 2./(1+4*pi*pi*f.*f);  % Original spectrum
f = [-7:0.01:7];
fs = 3;
Ts = 1/fs;
Xs = zeros(size(f));
```

```
for k=-5:5,
  Xs = Xs+fs*Xa(f-k*fs);
end;
plot(f,Xs);
axis([-7,7,-1,10]);
title('X_{s}(f)');
xlabel('f (Hz)');
grid;
```

4.25.

a.

```
xa = @(t) sin(2*pi*1000*t);
t = [0:5e-6:5e-3];
fs = 2400;
Ts = 1/fs;
n = [0:14];
plot(1000*t,xa(t));
axis([0,5,-1.2,1.2]);
hold on;
stem(1000*n*Ts,xa(n*Ts),'r');
hold off;
grid;
```

4.27.

a.

```
xa = @(t) exp(-abs(t));
t = [-4:0.001:4];
xzoh = ss_zohsamp(xa,0.2,0.90,t);
plot(t,xzoh);
```

G.5 Chapter 5 Problems

5.1.

a. $X(z) = \dfrac{z^2+z+1}{z^2}$ ROC: $|z| > 0$

5.2.

a. DTFT does not exist.

b. $X(\Omega) = X(z)\Big|_{z=e^{j\Omega}} = \dfrac{e^{j2\Omega}}{e^{j2\Omega}+5\,e^{j\Omega}+6}$

5.3.

a. ROC: $|z| > 0.75$

b. ROC: $\frac{1}{\sqrt{2}} < |z| < 1.25$

5.4.

a. $X(z) = \dfrac{z^{-1} - 10\,z^{-10} + 9\,z^{-11}}{\left(1 - z^{-1}\right)^2}\,, \qquad |z| > 0$

b. $X(z) = \dfrac{z^{-1} - z^{-11}}{\left(1 - z^{-1}\right)^2}\,, \qquad |z| > 1$

5.6.

a. $G(z) = \dfrac{z\left(z - 0.8598\right)}{z^2 - 1.7196\,z + 0.81}\,, \qquad |z| > 0.9$

5.7.

a. $X(z) = z^2\,, \qquad |z| < \infty$

b. $X(z) = z^{-3}\,, \qquad |z| > 0$

5.9.

a. $X(z) = \dfrac{1}{1-z}\,, \qquad |z| < 1$

b. $X(z) = \dfrac{z}{1-z}\,, \qquad |z| < 1$

5.10.

a. $X(z) = \dfrac{5\,z^3 - 6\,z^2 + 3\,z}{(z-1)^3}\,, \qquad |z| > 1$

5.11.

a. $x[0] = 1$

5.13.

a. $r_{XY}[m] = \{\,1,\ 2,\ 3,\ 3,\ \underset{\underset{m=0}{\uparrow}}{3}\ ,2,\ 1\,\}$

5.15.

a. $W(z) = \dfrac{z^2}{(z-1)\,(z-a)}\,, \qquad |z| > \max\left(1, |a|\right)$

5.17.

a. $x[n] = -(-1)^n\,u[-n-1] + (-2)^n\,u[-n-1]$

5.19.

a. $x[n] = 6.2472\,(0.922)^n \cos(0.7086n - 1.41)\; u[n]$

5.21.

a. $H(z) = \dfrac{-z}{z - (1 + c)}$

5.24.

a. $x[n] = \{ \underset{\substack{\uparrow \\ n=0}}{0}, 1, -3, 7, -15, 31, -63, \ldots \}$

5.26.

a. $h[n] = \dfrac{10}{9}\,\delta[n] - \dfrac{1}{9}\,(0.9)^n\, u[n]$

5.28.

a. $H(z) = \dfrac{z\,(z-2)}{(z+0.8)\,(z-0.7)}$

5.32.

a. $H(z) = \dfrac{5\left(z - \frac{13}{20}\right)}{z - \frac{3}{4}}, \qquad \text{ROC:} \quad |z| > \dfrac{3}{4}$

5.35.

a. $y[n] = 3.8717\,(0.8)^n\, e^{j\,(0.4\pi n + 3.0777)}$

5.38.

a. $H(z) = \dfrac{z\,(z+3)}{(z-0.6)\,(z-0.9)} \qquad |z| > 0.9$

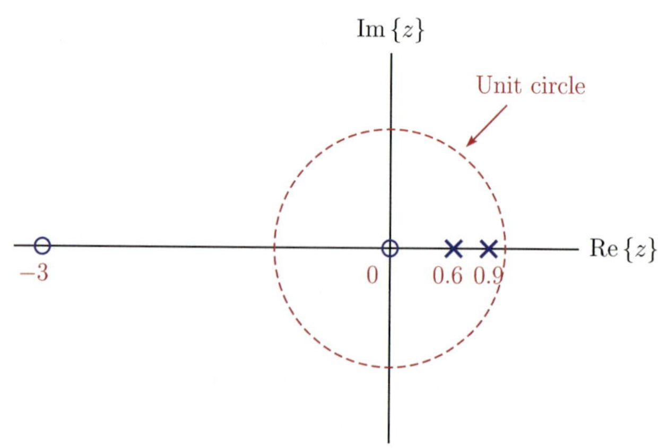

5.42.

b. $H(z) = \dfrac{z^{-1} + z^{-3}}{1 + 0.8\,z^{-1} - 2.2\,z^{-2} + 0.6\,z^{-3}}$

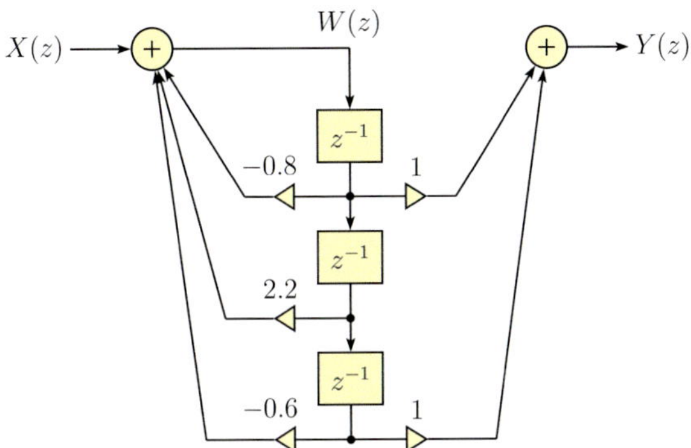

5.43.

a.

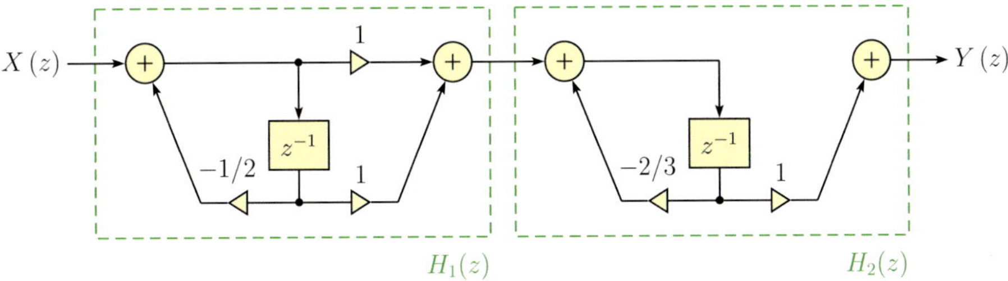

5.46.

a. $y[n] = 1.05\,(0.922)^n\cos(0.709n)\;u[n] + 5.86\,(0.922)^n\sin(0.709n)\;u[n]$

5.51.

a.

```
>> sf = zpk([0],[0.96],0.04,1)

Zero/pole/gain:
 0.04 z
--------
(z-0.96)

Sampling time: 1
```

5.52.

a.

```
% Set initial conditions to zero
ynm1 = 0;  % y[-1]
ynm2 = 0;  % y[-2]
xn = 1;    % x[0] = 1 since x[n] is a unit impulse
xnm1 = 0;  % x[-1] = 0 since x[n] is a unit impulse
out = []; % Empty vector to start
for n=0:10,
  yn = 1.2944*ynm1-0.64*ynm2+xn-0.6472*xnm1;
  out = [out,yn];
  ynm2 = ynm1;
  ynm1 = yn;
  xnm1 = xn;
  xn = 0;  % x[n] = 0 for n > 0
end;
% Display output signal
n = [0:10];
[n',out']
```

5.56.

a.

The first difference equation can be solved iteratively with the following:

```
ynm1 = 5;  % Placeholder for y[n-1], initially set to y[-1]
ynm2 = 7;  % Placeholder for y[n-2], initially set to y[-2]
y = [];    % Empty array (output stream)
for n=0:10,
  yn = 1.4*ynm1-0.85*ynm2;  % y[n] = 1.4y[n-1]-0.85y[n-2]
  y = [y,yn];               % Append to output stream
  ynm2 = ynm1;              % Update y[n-2] for next index
  ynm1 = yn;                % Update y[n-1] for next index
end;
[[0:10]',y']                % Tabulate solution
```

For the second difference equation, use

```
ynm1 = 2;   % Placeholder for y[n-1], initially set to y[-1]
ynm2 = -3;  % Placeholder for y[n-2], initially set to y[-2]
y = [];     % Empty array (output stream)
for n=0:10,
  yn = 1.6*ynm1-0.64*ynm2;  % y[n] = 1.6y[n-1]-0.64y[n-2]
  y = [y,yn];               % Append to output stream
  ynm2 = ynm1;              % Update y[n-2] for next index
  ynm1 = yn;                % Update y[n-1] for next index
end;
[[0:10]',y']                % Tabulate solution
```

G.6 Chapter 6 Problems

6.1.

a.

$$\mathbf{x}[n+1] = \begin{bmatrix} 0 & 1 \\ -0.72 & 1.7 \end{bmatrix} \mathbf{x}[n] + \begin{bmatrix} 0 \\ 3 \end{bmatrix} r[n]$$

$$y[n] = \begin{bmatrix} -0.72 & 1.7 \end{bmatrix} \mathbf{x}[n] + 3\,r[n]$$

6.2.

a.

$$\mathbf{x}[n+1] = \begin{bmatrix} -1/2 & 0 \\ 0 & -2/3 \end{bmatrix} \mathbf{x}[n] + \begin{bmatrix} 3 \\ -2 \end{bmatrix} r[n]$$

$$y[n] = \begin{bmatrix} 1 & 1 \end{bmatrix} \mathbf{x}[n]$$

6.3.

a.

$$\mathbf{x}[n+1] = \begin{bmatrix} 0 & 1 \\ -1/3 & -7/6 \end{bmatrix} \mathbf{x}[n] + \begin{bmatrix} 0 \\ 1 \end{bmatrix} r[n]$$

$$y[n] = \begin{bmatrix} 1 & 1 \end{bmatrix} \mathbf{x}[n]$$

6.4.

a.

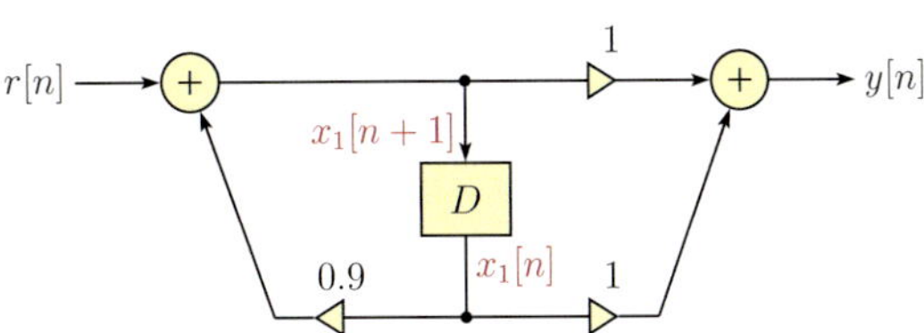

$$\mathbf{x}[n+1] = \begin{bmatrix} 0.9 \end{bmatrix} \mathbf{x}[n] + \begin{bmatrix} 1 \end{bmatrix} r[n]$$

$$y[n] = \begin{bmatrix} 1.9 \end{bmatrix} \mathbf{x}[n] + r[n]$$

6.6.

a. $\mathbf{\Phi}(z) = \dfrac{1}{(z-0.7)\,(z+0.8)} \begin{bmatrix} z^2 & -0.7\,z \\ -0.8\,z & z\,(z+0.1) \end{bmatrix}$

6.10.

a.

```
A = [-0.1,-0.7;-0.8,0];
B = [3;1];
C = [2,-1];
z = sym('z');
tmp = z*eye(2)-A;
rsm = z*inv(tmp)    % Resolvent matrix
stm = iztrans(rsm)  % State transition matrix
```

6.14.

a.

```
A = [0,1,0;0,0,1;-15,-11,-5];
B = [0;7;-32];
C = [1,0,0];
d = 0;
Ts = 0.1;
A_bar = eye(3)+A*Ts
B_bar = B*Ts
C_bar = C
d_bar = d
```

G.7 Chapter 7 Problems

7.1.

a. $X[k] = \{\underset{\substack{\uparrow \\ k=0}}{3},\ 0,\ 0\}$

7.2.

a. $X(k) = \{\underset{\substack{\uparrow \\ k=0}}{3},\ 2.4142,\ 1,\ -0.4142,\ -1,\ -0.4142,\ 1,\ 2.4142\}$

7.3.

b. $X[k] = \{\underset{\substack{\uparrow \\ k=0}}{0},\ -j\,1.1716,\ -j\,8,\ j\,6.8284,\ 0,\ -j\,6.8284,\ j\,8,\ j\,1.1716\}$

7.5.

a. $s[n] = \{\underset{\substack{\uparrow \\ n=0}}{2},\ 2,\ 1,\ 1,\ 1,\ 1,\ 1,\ 1,\ 1,\ 1\}$

7.7.

a. $\dfrac{2\pi k}{128}, \qquad k = 0, \ldots, 127$

7.9.

a. $\Omega_2 - \Omega_1 = 1.0308 - 0.9817 = 0.0491 < \Omega_r. \qquad \Rightarrow \qquad$ Only one peak will appear.

7.15.

a. $x[n-2]_{\text{mod } 4} = \{ \underset{\underset{n=0}{\uparrow}}{2}, 1, 4, 3 \}$

7.17.

a. $X[k] = \{ \underset{\underset{k=0}{\uparrow}}{3}, (0.5 - j\,1.5388), (0.5 + j\,0.3633), (0.5 - j\,0.3633), (0.5 + j\,1.5388) \}$

7.18.

a. $X[k] = \{ \underset{\underset{k=0}{\uparrow}}{15}, (2.5 - j\,3.4410), (2.5 - j\,0.8123), (2.5 + j\,0.8123), (2.5 + j\,3.4410) \}$

7.21.

a. $y[n] = \{ \underset{\underset{n=0}{\uparrow}}{9}, 5, 3, 2, 11 \}$

G.8 Chapter 8 Problems

8.1.

a.

Critical frequencies are

$\Omega_0 = 2\pi F_0 = 0.6283\,, \qquad \Omega_b = 2\pi F_b = 0.0628$

The pole radius needed to meet the specifications is

$$r = 1 - \frac{\Omega_b}{2} = 1 - \frac{0.0628}{2} = 0.9686$$

Thus, the poles of the system function are at $p_{1,2} = r\,e^{\pm j\Omega_0} = 0.7836 \pm j0.5693$, and its zeros are at $z_{1,2} = \pm 1$. The gain factor is $K = 0.0309$.

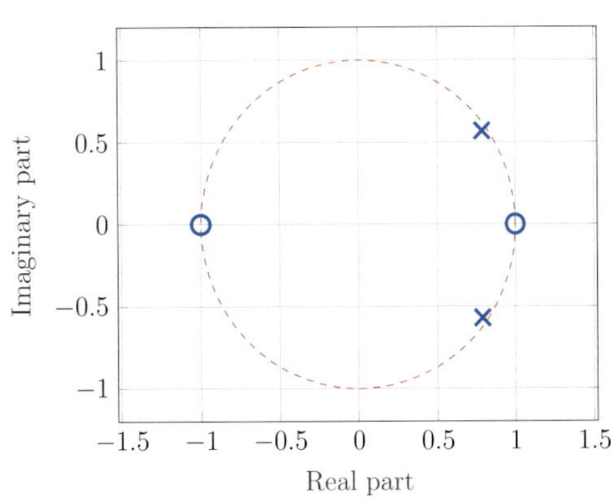

8.3.

a.

Critical frequencies are

$$\Omega_0 = 2\pi f_0 / f_s = 0.075\,, \qquad \Omega_b = 2\pi f_b / f_s = 0.0236$$

The pole radius needed to meet the specifications is

$$r = 1 - \frac{\Omega_b}{2} = 1 - \frac{0.0236}{2} = 0.9882$$

Thus, the poles of the system function are at $p_{1,2} = r\,e^{\pm j\Omega_0} = 0.8805 \pm j0.4486$, and its zeros are at $z_{1,2} = \pm 1$. The gain factor is $K = 0.0117$.

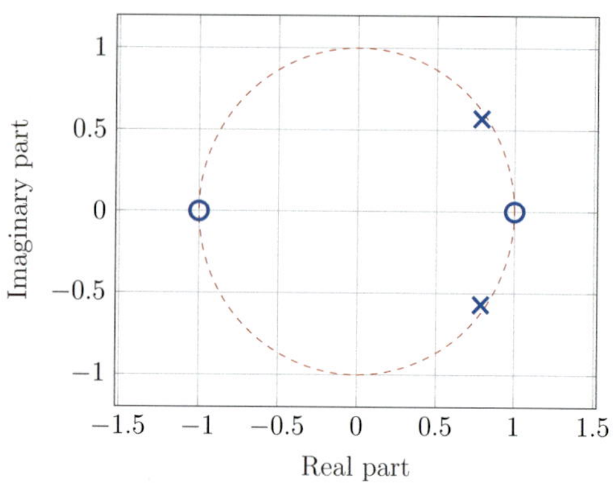

8.7.

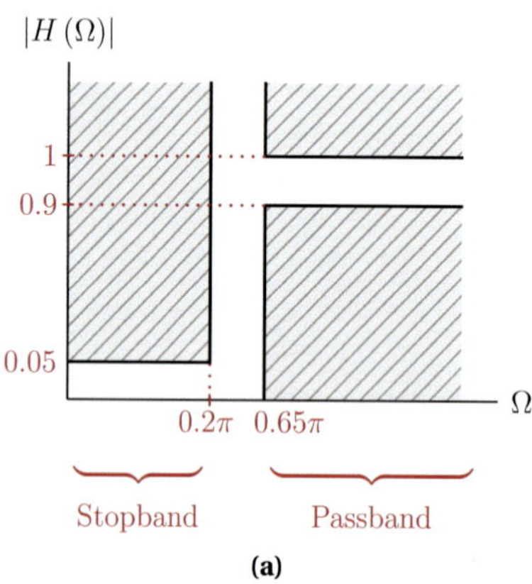

(a)

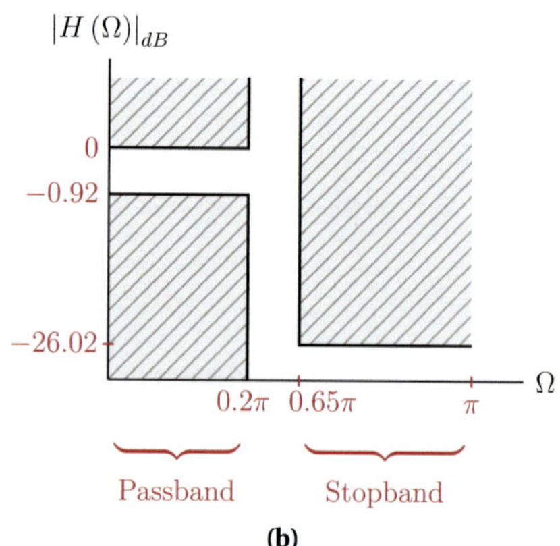

(b)

8.11.

a. $g(t) = 2\,e^{-t}\,u(t) - 2\,e^{-2t}\,u(t)$

8.15. $$H(z) = \frac{0.0125\,z^2 + 0.0101\,z}{z^3 - 2.3768\,z^2 + 1.9329\,z - 0.5335}$$

8.17.

a. $$H(z) = \frac{7\,z^2 + 6\,z - 1}{30\,z^2 - 28\,z + 6}$$

8.20.

a. Yes, type-II

b. Yes, type-IV

8.27.

a.

```
fs = 22050;      % Sampling rate
f1 = 660;        % f1 in Hz
f2 = 680;        % f2 in Hz
F1 = f1/fs;      % Normalized
F2 = f2/fs;
Omg1 = 2*pi*F1;
Omg2 = 2*pi*F2;
n = [0:7999]';
x = 0.8*cos(Omg1*n)+cos(Omg2*n);
stem(n,x);
```

8.32.

a.

```
[num,den] = butter(3,2,'s')
omg = [0:0.01:10];
G = freqs(num,den,omg);
plot(omg,abs(G)); grid;
```

8.34.

a.

```
[num3,den3] = cheby2(N,As,[omgB1,omgB4],'bandpass','s')
G2 = freqs(num3,den3,omega);
plot(omega,abs(G2)); grid;
```

8.36.

a.

```
num = [2];
den = [1,3,2];
[numz,denz] = impinvar(num,den,2)
```

8.44.

a.

```
N = 41;
h = ss_fir1(N,[0.15,0.35],'bandpass','rect');
Omg = [0:0.01:1]*pi;
Hmag = abs(freqz(h,1,Omg));
plot(Omg,Hmag); grid;
title('|H(\Omega)|');
xlabel('\Omega (rad)');
ylabel('Magnitude');
```

INDEX

absolute summable, 340, 615
addition of signals, 7
additivity rule, 75, 119, 133
aliasing, 284, 290, 637
all-pass filter, 426
    pole-zero diagram, 427, 429
amplitude, 24
amplitude response, 655, 683
analog filter transformation, 632
analog prototype filter, 627, 636
analog signal, 58
angular frequency, 24, 25, 31, 32, 43, 172, 191
anti-causal system, 133
arithmetic operations, 7
audio synthesizer, 64
autocorrelation, 234
average power, 36–38, 44

backward difference, 663
backward rectangular approximation, 642
bandlimited interpolation, 305
bandpass filter, 628, 633
bandstop filter, 628, 634
basis functions, 171, 172, 228, 243
bilateral z-transform, 344
bilinear transformation, 646
binary signal, 4
block diagram, 74, 111–113, 115, 134, 143
bounded-input bounded-output (BIBO), 130
butterfly structure, 565
Butterworth filter, 630

Cardano, Gerolamo, 735
Cartesian form, 29, 737
Cauchy residue theorem, 391
Cauchy, Augustin-Louis, 391
causal system, 129, 133, 135
causality, 74, 129, 135, 420, 627
characteristic equation, 98, 99, 102, 106
characteristic polynomial, 98, 104
    roots of, 99
Chebyshev filter, 631
    type-I, 631
    type-II, 631
Chebyshev polynomial, 631
Chebyshev rational function, 632
chirp transform algorithm (CTA), 517, 558
circular convolution, 548, 549, 572
circular convolution property
    of the DFT, 548
circular shift, 539
circular time reversal, 539
circularly conjugate antisymmetric, 541
circularly conjugate symmetric, 541
comb filter, 154, 261
    feed-forward, 154, 424
    feedback, 261
complex conjugate roots, 100
complex numbers, 735
    adding and subtracting, 737
    arithmetic with, 737
    equality of, 736
    multiplication and division of, 739
    vector representation of, 736
complex plane, 736
compression, 58
conjugate antisymmetric, 41, 42, 45, 182
    circularly, 541

conjugate symmetric, 41, 42, 45, 182
  circularly, 541
conjugation property
  of the DFT, 544
  of the DTFT, 212
constant gain factor, 7
constant offset, 7
continuous Fourier transform (CFT), 517, 535
  approximation by DFT, 535
continuous-time signal, 46
contour integral, 391
convergence, 340, 347
  region of, 340, 347, 348, 362
convolution, 7, 29, 74, 118, 120, 123, 134, 145, 148, 149, 171, 249, 404
convolution operator, 120
convolution property
  of the DTFT, 219
  of the z-transform, 380, 404
correlation property
  of the z-transform, 385
critical frequencies, 627, 639

decimation, 14
decimation-in-time, 564, 596
decimator, 310
delta modulation, 294
difference equation, 74, 79, 80, 84–87, 89, 91, 113, 116, 117, 133–135, 140, 141, 144, 239, 340, 405, 488
  constant coefficient, 90
  homogeneous, 95
  iterative solution of, 572
differential equation, 171
differentiation in frequency property
  of the DTFT, 217
differentiation in the z-domain, 377
digital signal, 4–6
direct-form realization, 434, 435
Dirichlet conditions, 348
discrete Fourier transform (DFT), 517, 518, 562
  analysis equation, 518
  circular convolution, 548
  computational complexity of, 562
  conjugation, 544
  frequency shifting, 547
  linearity property, 542
  matrix formulation of, 537
  symmetry properties, 545
  synthesis equation, 518
  time reversal, 543
  time shifting property, 542
discrete-time Fourier series (DTFS), 171, 172, 243, 517
  analysis equation, 177
  finding coefficients, 176
  linearity property, 181
  periodic convolution, 189, 550
  symmetry properties, 182
  synthesis equation, 177
  time reversal property, 182
  time shifting property, 181
discrete-time Fourier transform (DTFT), 173, 190, 192, 340, 517
  analysis equation, 193
  conjugation property, 212
  convolution property, 219
  differentiation in frequency, 217
  existence of, 197
  frequency shifting, 215
  linearity property, 206
  modulation property, 216
  multiplication property, 221
  of sampled signal, 286
  periodicity property, 180, 205
  symmetry properties, 212
  synthesis equation, 193, 228
  time reversal, 209
  time shifting, 206
discrete-time signal, 4, 6, 46, 58
discrete-time system, 73, 143, 144
discretization of continuous state-space model, 503
Dow Jones Industrial Average, 2, 3, 80, 82, 138, 140, 148
downsampling, 13, 14, 149, 307, 308
downsampling rate, 307
DTFS coefficients, 517
DTLTI system, 79, 89–91, 115, 134, 171, 243, 404, 491, 492, 615
dual-tone multi-frequency (DTMF), 484

EFS coefficients
  approximating by DFT, 533
eigenvalues, 99
electrocardiogram (ECG), 5, 696
elliptic filter, 632
encoding, 4, 6
energy signal, 38, 44, 227
energy spectral density, 228, 234

Euler's formula, 30, 35, 176, 200, 216, 244, 370, 408, 410, 739
Euler, Leonhard, 735
even component, 39–41, 45
even symmetry, 39, 42, 45
exponential Fourier series (EFS), 171, 278, 517, 533
exponential smoother, 85, 86, 90, 94, 95, 107, 110, 117, 134, 140

fast Fourier transform (FFT), 517, 519, 564, 566
feed-forward comb filter, 154, 424
feedback, 85, 90
feedback comb filter, 261, 425
filter, 615
  all-pass, 426
  bandpass, 628
  bandstop, 628
  Butterworth, 630
  Chebyshev type-I, 631
  Chebyshev type-II, 631
  elliptic, 632
  frequency selective, 615, 628
  highpass, 628
  lowpass, 628
  moving average, 79, 82–84
filter coefficients, 653
filtering, 615
finite impulse response (FIR) filter, 615, 652
first difference, 20, 43
first-order hold, 300
  reconstruction filter, 300
flat-top sampling, 297
forced response, 95, 105, 107, 110, 116, 134
forced solution, 95, 106, 107, 118
forward rectangular approximation, 642
Fourier series
  exponential, 171
Fourier transform, 340, 615
  frequency shifting property, 279
  linearity property, 279
frame-based processing, 573
frequency resolution, 517
frequency sampling, 680
frequency shifting property
  of the DFT, 547
  of the DTFT, 215
  of the Fourier transform, 279
frequency spacing, 517, 531
frequency transformation, 629
fundamental frequency, 30
fundamental period, 30–33, 44

gain amplifier, 111
Gauss, Carl Friedrich, 735
geometric mean, 634
Gibbs phenomenon, 673
Goertzel algorithm, 517, 555

Hamming window, 531
Hanning window, 270
highpass filter, 628, 633
Hilbert transform filter, 671
homogeneity rule, 75, 119, 133
homogeneous difference equation, 95, 106
homogeneous solution, 95, 99, 104, 106, 107, 118

imaginary part, 735, 737
impulse decomposition, 28, 29, 43, 118
impulse invariance, 636, 637
impulse response, 73, 74, 83, 84, 86, 115, 117, 119, 123, 128, 134, 171, 239, 404, 488
impulse sampling, 277, 286
impulse train, 299
infinite impulse response (IIR) filter, 615
information theory, 284
initial conditions, 113, 344
initial value property
  of the z-transform, 384
integrator, 111
interpolation, 15
interpolation filter, 312
interpolator, 312
inverse system, 431
  system function, 432
inverse z-transform, 353, 389, 404
inversion integral, 389
iterative solution, 94

Jacobi elliptic functions, 632

Laplace transform, 339, 391, 663
latency, 690
least-squares, 685
limit-cycle oscillations, 703
linear convolution, 572
linear phase
  type-I, 654, 683
  type-II, 657

type-III, 659
type-IV, 661, 664
linear shift, 539
linear system, 74
linear time reversal, 539
linear vs. circular convolution, 552
linearity, 74–76, 91, 133
of the DFT, 542
of the DTFS, 181
of the DTFT, 206
of the Fourier transform, 279
of the z-transform, 368
loan payments, 86
local function, 55
local variable, 136
logical indexing, 55, 57
logistic growth model, 87, 90, 94
long division, 389, 398
lowpass filter, 304, 628, 633, 634
lowpass to bandpass transformation, 633
lowpass to bandstop transformation, 634
lowpass to highpass transformation, 633

M-ary signal, 4
mathematical modeling, 7
maximum passband ripple, 628
minimum stopband attenuation, 628
minimum-phase system, 433
modes of system, 99
modulation property
of the DTFT, 216
modulo indexing, 539
moving average filter, 79, 82–84, 89, 131, 134, 135, 137, 138, 148
multiple roots, 101
multiplication by an exponential signal, 375
multiplication of signals, 7
multiplication property
of the DTFT, 221

natural response, 95, 99, 102, 104, 105, 134
natural sampling, 294
Newton-Raphson algorithm, 88, 90
non-periodic signal, 30, 36, 37
normalized average power, 227
normalized energy, 227
normalized frequency, 24, 25, 31–33, 43
notch filter, 622
numerical analysis, 7
Nyquist sampling theorem, 284, 534, 536
Nyquist, Harry, 284

object-oriented programming, 61
octave, 60
odd component, 39–41, 45
odd symmetry, 39, 42, 45
orthogonality, 176, 228
output equation, 490, 493
overlap-add method, 574, 599

Parseval's theorem, 227
Parseval, Marc-Antoine, 227
partial fraction expansion, 389, 391–393, 494
particular solution, 95, 105, 106, 118
passband, 628
passband tolerance, 628
periodic convolution property
of the DTFS, 189, 550
periodic extension, 52, 517
periodic impulse train, 277
periodic signal, 30, 36, 37, 52
periodicity
of the DTFS, 180
of the DTFT, 205
persistent variable, 136–139, 143
phase, 24
phase-shifter, 428
phase-variable canonical form, 493
picket fence effect, 521
plucked-string filter, 716
polar form, 29, 737
pole-zero placement, 417
poles, 99
of the z-transform, 353
polynomial
characteristic, 98
Chebyshev, 631
polynomial multiplication, 382, 456
polynomials
multiplication of, 382, 456
post processing, 59, 129
power series, 389
power signal, 38, 44, 227
power spectral density, 228, 229, 234
prewarping of critical frequencies, 651
processing
post, 59, 129
real-time, 59, 129
pulse train, 299
pulse-code modulation, 294

quantization, 4, 6
quantization error, 5
quantization interval, 4
quantization level, 5

radix-2, 564, 596
real part, 735, 737
real-time processing, 59, 124, 129, 573
reconstruction, 5, 299
reconstruction filter, 299
rectangular approximation method, 534, 535
rectangular window, 270, 530, 673
recursive function, 596
region of convergence, 340, 347, 348, 353, 362
Remez exchange method, 685
resampling, 306
residue formulas, 393
resolvent matrix, 501
resonant bandpass filter, 618
response
    forced, 95, 105
    natural, 95, 105
roots
    complex conjugate, 100
    multiple, 101

sample index, 1, 2
sampling, 2, 25, 275, 636
    frequency, 276
    interval, 276
    of a continuous-time signal, 275
    period, 276
    rate, 276
    the temperature, 276
sampling frequency, 2, *see* sampling
sampling interval, 2, 3, 25, 58, *see* sampling, 534, 535
sampling period, 2, *see* sampling
sampling property, 18, 43
sampling rate, 2, 25, 58, *see* sampling, 534, 536
sawtooth signal, 115
second-order system, 99, 102, 104
Shannon, Claude, 284
short time Fourier transform (STFT), 517, 559
sifting property, 19, 43
signal
    analog, 58
    average power of, 36
    binary, 4
    continuous-time, 1, 46
    digital, 4–6
    discrete-time, 1, 4, 6, 46, 58
    downsampling, 13
    energy of, 34, 44
    M-ary, 4
    non-periodic, 30, 36, 37
    periodic, 30, 36, 37, 52
    sinusoidal, 24, 43, 64, 107
    symmetry properties of, 39
    time average of, 36, 44
    time reversal, 16
    time scaling, 13
    time shifting, 11
signal adder, 111
signal classifications, 29
signals
    addition of, 7, 8
    multiplication of, 7, 10
    real vs. complex, 29
similarity transformation, 499
simulation, 111
sinusoidal signal, 26, 27, 35, 38, 64, 107
solution
    forced, 95
    homogeneous, 95
    particular, 95, 105
SOS matrix, 703
spectral leakage, 526, 527
squared norm, 37
stability, 74, 95, 130, 135, 627
stable system, 130, 135, 340
state equations, 490, 493
state matrix
    eigenvalues of, 99
state transition matrix, 502
state variable, 489
state vector, 492
state-space model, 490, 491, 493
state-space representation, 489
steady state response, 243
stem plot, 2, 46, 82
stopband, 628
stopband tolerance, 628
summation property
    of the z-transform, 387
sunspot numbers, 3, 56
superposition principle, 75, 133, 243
symmetry properties, 39
    of the DFT, 545

of the DTFS, 182
of the DTFT, 212
system
anti-causal, 133
causal, 129, 133, 135
discrete-time, 143, 144
inverse, 431
linear, 74, 75, 133
minimum-phase, 433
modes of, 99
stable, 130, 135
time invariant, 74
time-invariant, 77, 133
system function, 239, 340, 405, 420, 488, 615
obtaining from state-space model, 503
poles of, 99
system of equations
exact-determined, 684
overdetermined, 685

time average, 36, 44
time invariance, 74, 77, 78, 91, 133
time reversal, 7, 16, 17
circular, 543
time reversal property
of the DFT, 543
of the DTFS, 182
of the DTFT, 209
of the z-transform, 374
time scaling, 7, 13
time shifting, 7, 11
circular, 542
time shifting property
of the DFT, 542
of the DTFS, 181
of the DTFT, 206
of the z-transform, 372
time-invariant system, 74, 119
touch tone dialing, 484
transformation
analog filter, 632
lowpass to bandpass, 633
lowpass to bandstop, 634
lowpass to highpass, 633
transition band, 628
trapezoidal approximation to integral, 168

unilateral z-transform, 344
unit circle, 342
unit impulse funct[illegible]
unit ramp function, 22, 43, 49, 340
unit step function, 19, 43, 49, 83, 84, 86, 107, 115, 117, 143
upsampling, 15, 16, 310
upsampling rate, 311

von Hann window, 270

wave equation, 713
wide-sense stationary, 237
Wiener-Khinchin theorem, 237
Wolf number, 3
Wolf, Rudolf, 3
wrapper function, 146, 596

z-plane, 341
z-transform, 339, 615
bilateral, 344, 443
convolution property, 380, 404
correlation property, 385
differentiation in the z-domain, 377
initial value property, 384
linearity, 368
multiplication by an exponential signal, 375
properties of, 368
summation property, 387
time reversal property, 374
time shifting property, 372
unilateral, 340, 344, 443
zero padding, 517, 524, 576
zero-order hold, 299
zero-order hold reconstruction filter, 299
zero-order hold sampling, 294, 297
zeros
of the z-transform, 353
Zurich relative number, 3